ASCE Manuals and Reports on Engineering Practice No. 28

Hydrology Handbook

Second Edition

Prepared by the Task Committee on Hydrology Handbook
of Management Group D of the American Society of Civil Engineers

This is a completely re-written edition
of the Manual published in 1949.

Published by

ASCE

345 East 47th Street
New York, NY 10017-2398

Abstract:

This new edition of the *Hydrology Handbook* (Manual No. 28) incorporates the many changes and advances that have occurred in the areas of planning, development, and management of water resources since the publication of the original manual in 1949. The first six chapters, Chapters 2 through 7, relate to the natural phenomena in the hydrologic cycle, while the next three chapters describe the predictions and effects of the phenomena previously described. The final chapter reviews the applications of hydrology starting with study formulation, then reviews data management, then discusses calibration and verification of hydrologic models, and concludes with accessing accuracy and reliability of results. With this new edition, academic and practicing hydrologists have a thorough and up-to-date guide to the field of hydrologic engineering.

Library of Congress Cataloging-in-Publication Data

Hydrology handbook / prepared by the Task Committee on Hydrology Handbook of Management Group D of the American Society of Civil Engineers. -- 2nd ed.
p. cm. -- (ASCE manuals and reports on engineering practice ; no. 28)
Includes bibliographical references and index.
ISBN 0-7844-0138-1
1. Hydrology. I. American Society of Civil Engineers. Task Committee on Hydrology Handbook. II. Series.
GB661.2.H93 1996 96-41049
551.48--dc20 CIP

Library of Congress Catalog Card No: 96-41049
ISBN 0-7844-0138-1
Manufactured in the United States of America.

MANUALS AND REPORTS ON ENGINEERING PRACTICE

(As developed by the ASCE Technical Procedures Committee, July 1930, and revised March 1935, February 1962, April 1982)

A manual or report in this series consists of an orderly presentation of facts on a particular subject, supplemented by an analysis of limitations and applications of these facts. It contains information useful to the average engineer in his everyday work, rather than the findings that may be useful only occasionally or rarely. It is not in any sense a "standard," however; nor is it so elementary or so conclusive as to provide a "rule of thumb" for nonengineers.

Furthermore, material in this series, in distinction from a paper (which expresses only one person's observations or opinions), is the work of a committee or group selected to assemble and express information on a specific topic. As often as practicable the committee is under the direction of one or more of the Technical Divisions and Councils, and the product evolved has been subjected to review by the Executive Committee of the Division or Council. As a step in the process of this review, proposed manuscripts are often brought before the members of the Technical Divisions and Councils for comment, which may serve as the basis for improvement. When published, each work shows the names of the committees by which it was compiled and indicates clearly the several processes through which it has passed in review, in order that its merit may be definitely understood.

In February 1962 (and revised in April, 1982) the Board of Direction voted to establish:

> A series entitled 'Manuals and Reports on Engineering Practice,' to include the Manuals published and authorized to date, future Manuals of Professional Practice, and Reports on Engineering Practice. All such Manual or Report material of the Society would have been refereed in a manner approved by the Board Committee on Publications and would be bound, with applicable discussion, in books similar to past Manuals. Numbering would be consecutive and would be a continuation of present Manual numbers. In some cases of reports of joint committees, bypassing of Journal publications may be authorized.

MANUALS AND REPORTS OF ENGINEERING PRACTICE

#	MANUALS AND REPORTS OF ENGINEERING PRACTICE
10	Technical Procedures for City Surveys
13	Filtering Materials for Sewage Treatment Plants
14	Accommodation of Utility Plant Within the Rights-of-Way of Urban Streets and Highways
31	Design of Cylindrical Concrete Shell Roofs
33	Cost Control and Accounting for Civil Engineers
34	Definitions of Surveying and Associated Terms
35	A List of Translations of Foreign Literature on Hydraulics
36	Wastewater Treatment Plant Design
37	Design and Construction of Sanitary and Storm Sewers
40	Ground Water Management
41	Plastic Design in Steel-A Guide and Commentary
42	Design of Structures to Resist Nuclear Weapons Effects
45	Consulting Engineering-A Guide for the Engagement of Engineering Services
46	Report on Pipeline Location
47	Selected Abstracts on Structural Applications of Plastics
49	Urban Planning Guide
50	Planning and Design Guidelines for Small Craft Harbors
51	Survey of Current Structural Research
52	Guide for the Design of Steel Transmission Towers
53	Criteria for Maintenance of Multilane Highways
54	Sedimentation Engineering
55	Guide to Employment Conditions for Civil Engineers
57	Management, Operation and Maintenance of Irrigation and Drainage Systems
58	Structural Analysis and Design of Nuclear Plant Facilities
59	Computer Pricing Practices
60	Gravity Sanitary Sewer Design and Construction
62	Existing Sewer Evaluation and Rehabilitation
63	Structural Plastics Design Manual
64	Manual on Engineering Surveying
65	Construction Cost Control
66	Structural Plastics Selection Manual
67	Wind Tunnel Model Studies of Buildings and Structures
68	Aeration-A Wastewater Treatment Process
69	Sulfide in Wastewater Collection and Treatment Systems
70	Evapotranspiration and Irrigation Water Requirements
71	Agricultural Salinity Assessment and Management
72	Design of Steel Transmission Structures
73	Quality in the Constructed Project-a Guide for Owners, Designers, and Constructors
74	Guidelines for Electrical Transmission Line Structural Loading
75	Right-of-Way Surveying
76	Design of Municipal Wastewater Treatment Plants
77	Design and Construction of Urban Stormwater Management Systems
78	Structural Fire Protection
79	Steel Penstocks
80	Ship Channel Design
81	Guidelines for Cloud Seeding to Augment Precipitation
82	Odor Control in Wastewater Treatment Plants
83	Environmental Site Investigation
84	Mechanical Connections in Wood Structures
85	Quality of Ground Water
86	Operation and Maintenance of Ground Water Facilities
87	Urban Runoff Quality Manual
88	Management of Water Treatment Plant Residuals
89	Pipeline Crossings

ACKNOWLEDGMENTS

MANAGEMENT GROUP D

This Handbook of Hydrology was prepared under the direction of Management Group D of the American Society of Civil Engineers Technical Activities Committee. The Handbook was initiated in 1988 and was completed in 1995. Membership of Management Group D during this period included the following persons.

1988
George E. Hecker, *Chair*
Thomas Lee Jackson
Conrad G. Keyes, Jr.
Jerry R. Rogers
Frank Stratton
John R. Wegel

1989
Jerry R. Rogers, *Chair*
George E. Hecker
Lloyd A. Held, Jr.
Wayne B. Ingram
Conrad G. Keyes, Jr.
Frank Stratton

1990
Conrad G. Keyes, Jr., *Chair*
Catalino B. Cecilio
Wayne B. Ingram
Jerry R. Rogers
Michael L. Stevens
Frank Stratton

1991
Frank Stratton, *Chair*
Catalino B. Cecilio
Wayne B. Ingram
Conrad G. Keyes, Jr.
Michael L. Stevens
Darell D. Zimbelman

1992
Wayne B. Ingram, *Chair*
Catalino B. Cecilio
Kenneth G. Renard
Michael L. Stevens
Frank Stratton
Darell D. Zimbelman

1993
Catalino B. Cecilio, *Chair*
Mriganka M. Ghosh
Wayne B. Ingram
Kenneth G. Renard
Robert H. Wortman
Darell D. Zimbelman

1994
Darell D. Zimbelman, *Chair*
Catalino B. Cecilio
James E. Dailey
Mriganka M. Ghosh
Kenneth G. Renard
Robert H. Wortman

1995
Kenneth G. Renard, *Chair*
Philip H. Burgi
James E. Dailey
Thomas Rackford
Robert H. Wortman
Darell D. Zimbelman

Management Group D appointed a Task Committee to prepare the Hydrology Handbook.

TASK COMMITTEE MEMBERS

Thomas P. Wootton, *Chair*
Catalino B. Cecilio
Lloyd C. Fowler
Samuel L. Hui

With overall editing by

Richard J. Heggen

CHAPTER SUBCOMMITTEE MEMBERS

Each of the technical chapters in the Handbook was prepared by a subcommittee under the direction of the Task Committee. The membership of each Chapter Subcommittee included the following individuals.

Chapter 1, Introduction

Lloyd C. Fowler

Chapter 2, Precipitation

Control Group:
Clayton L. Hanson, *Chair*
Karl A. Gebhardt
Gregory L. Johnson
Marshal J. McFarland
James A. Smith

Chapter 3, Infiltration

Control Group:
Walter J. Rawls, *Chair*
David Goldman
Joseph A. Van Mullen
Timothy J. Ward

Reviewers:
Lajpat R. Ahuja
A. Osman Akan
Donald L. Brakensiek
Paul A. DeBarry
Richard J. Heggen
George V. Sabol

Chapter 4, Evaporation and Transpiration

Control Group:
Richard G. Allen, *Co-Chair*
William O. Pruitt, *Co-Chair*
Joost A. Businger
Leo J. Fritschen
Marvin E. Jensen
Frank H. Quinn

Chapter 5, Ground Water

Control Group:
Lloyd C. Fowler, *Chair*
Nazeer Ahmed
Keith E. Anderson
Clyde S. Conover
Marvin V. Damm
Donald J. Finlayson

Thomas P. Ballestero
Ronald K. Blatchley
Carl H. Carpenter
James D. Goff
A. Ivan Johnson
W. Martin Roche

Reviewers:
John F. Clerici
Calvin G. Clyde
Peter Kitandis
Porter C. Knowles
Clark C-K Liu
George E. Maddox
Stuart T. Pyle
Lou Riethmann
J. Paul Riley
Richard J. Schicht
Claire Welty

Chapter 6, Runoff, Streamflow, Reservoir Yield, and Water Quality

Control Group:
Anand Prakash, *Chair*
Richard J. Heggen
Victor M. Ponce
John A. Replogle
Henry C. Riggs

Reviewer: Rafael G. Quimpo

Chapter 7, Snow and Snowmelt

Control Group:
Douglas D. Speers, *Chair*
George D. Ashton
David M. Rockwood

Reviewer: David C. Garen

Chapter 8, Floods

Control Group:
Bi-Huei Wang, *Chair*
Michael L. Anderson
Gary R. Dyhouse
Vernon K. Hagen
Khalid Jawed
John T. Riedel
Jery R. Stedinger

Reviewers:
Arlen D. Feldman
Janet C. Herrin
Donald M. Thomas

Chapter 9, Urban Hydrology

Control Group:
David F. Kibler, *Chair*
A. Osman Akan
Gert Aron
Christopher B. Burke
Mark W. Glidden
Richard H. McCuen

Reviewers:
G. V. Loganathan
Andrew J. Reese

Chapter 10, Water Waves

Control Group:

Zeki Demirbilek, *Chair*
Robert A. Dalrymple
Robert M. Sorenson
Edward F. Thompson
J. Richard Weggel

Chapter 11, Hydrologic Study Formulation and Assessment

Control Group:

John C. Peters, *Chair*
Shawna Anderson
Patrick Atchison
John J. Buckley
David Ford
Katherine Hon
Richard H. McCuen

The following individuals participated in a final review of all or portions of this Handbook.

Raymundo Aguirre
Catalino B. Cecilio
Theodore G. Cleveland
Lloyd C. Fowler
Richard J. Heggen
Samuel L. Hui
Wayne B. Ingram
Conrad G. Keyes, Jr.
Rong-Heng Kuo
Kenneth G. Renard
J. Paul Riley
James M. Robinson, Jr.
Jerry R. Rogers
Valerie Steinmaus
Jeffry Stone
Warren Viessman, Jr.
Keh Han Wang
Thomas P. Wootton

CONTENTS

CONTENTS

LIST OF FIGURES

LIST OF TABLES

CONVERSTION TO SI UNITS

To Convert from	To	Multiply By
acre (ac)	hectare (ha)	0.4047
acres	square kilometer (km^2)	0.004047
acres	square meters (m^2)	4046.8726
acre-feet (ac-ft)	cubic meters (m^3)	1233.489
acre-feet	hectare-meter (ha-m)	0.123349
cubic foot (ft^3)	cubic meter (m^3)	0.02832
cubic feet per second (cfs)	cubic meters per second (m^3/s)	0.02832
cubic feet per second	cubic meters per minute (m^3/min)	1.69901
cubic yard	cubic meter (m^3)	0.7645
degree Fahrenheit (°F)	degree Celsius (°C) = (°F − 32) 1.8	
degree Kelven (K)	K − 273 + °C	
feet per second (ft/s)	meters per second (m/s)	0.3048
foot (ft)	meter (m)	0.3048
gallon, U.S. (gal)	cubic meters (m^3)	0.0038
gallon, U.S.	liter (l)	3.79
gallons, U.S. per minute (gpm)	cubic meters per second (m^3/s)	6.309×10^{-5}
horsepower (HP)	kilowatt (k)	0.746
inch (in)	millimeter (mm)	25.40
mile (mi)	kilometer (km)	1.609344
miles per hour (mi/hr)	kilometers per hour (km/hr)	1.609344
million gallons per day (mgd)	cubic meters per second (m^3/s)	0.0438
pound force (lbf)	newton (N)	4.45
pound force per inch (psi)	kilopascal (kpa)	6.89
pound mass (lbm)	kilogram (kg)	0.4536
short ton	kilogram (kg)	907.2
square foot (sq ft) (ft^2)	square meter (m^2)	0.0929
square inch (sq in) (in^2)	square millimeter (mm^2)	645.
square mile (sqm)	square kilometer (km^2)	2.590
square yard (sq yd) (yd^2)	square meter (m^2)	0.8361
yard (yd)	meter (m)	0.9144

Chapter 1

INTRODUCTION TO THE NEW HANDBOOK OF HYDROLOGY

I. HISTORICAL SUMMARY

In 1949, the American Society of Civil Engineers (ASCE) published a hydrology handbook titled *Manual of Engineering Practice No. 28*, which was prepared by a Committee of the Hydraulics Division. It was reprinted in 1952 and 1957. When consideration was given to reprinting *Manual 28*, the matter was referred to a Task Committee of the Hydraulics Division in 1982. After considerable discussion and review of other publications, it was recommended that *Manual 28* not be reprinted but that a new handbook on hydrology be prepared. This new publication would recognize the advances that had been made in the practice of engineering hydrology in the years since *Manual 28* was published.

With the encouragement of the Executive Committees of the Hydraulics Division and the Irrigation and Drainage Division, the Task Committee prepared a table of contents for a new handbook of hydrology and presented it along with the recommendation that the new handbook be a cooperative effort of the two Divisions. In 1986, this recommendation was approved by the Executive Committees and was transmitted to Management Group D, the Group that oversees the activities of the five ASCE technical divisions involved in water engineering. Under the leadership of Conrad G. Keyes, Jr., Management Group D accepted the report and in 1988 established a Task Committee comprising four members, two from the Hydraulics Division and two from the Irrigation and Drainage Division, to prepare a new handbook of hydrology.

II. PURPOSE OF THE NEW HANDBOOK

The 1949 hydrology handbook, *Manual 28*, has been widely distributed and used, both nationally and internationally, by educational institutions, consulting engineers, and others involved in planning, development, and management of water resources. The manual has served essentially as a baseline document; however, since its publication many changes and advances have occurred, not only in a fundamental understanding of hydrologic systems, but also in data acquisition and computation techniques and procedures. These changes and advances have been incorporated into the new *ASCE Handbook of Hydrology* so that it maintains its relevancy to academic and practicing hydrologists throughout the world. Providing a thorough and up-to-date guide for hydrologists is the purpose of this new handbook.

III. SCOPE OF THE NEW HANDBOOK

The first six chapters, Chapters 2 through 7, relate to the natural phenomena in the hydrologic cycle. Chapter 2 describes the formation and types of precipitation, variations in and measurement of precipitation, processing and interpreting precipitation records, frequency analyses of precipitation data, weather modification, and synthetic weather generation. Chapter 3 covers principles of infiltration, factors affecting infiltration and rainfall excess, infiltration and excess models, and measurement of infiltration. Chapter 4 presents the physics and theory of evaporation, interaction of surface and meteorological factors, evaporation from water surfaces, evapotranspiration from land surfaces, and regional evapotranspiration. Chapter 5 discusses ground water source and occurrence, subsurface medium, ground water movement, basin yield concepts and evaluations, recharge, ground water quality, models, and management. Chapter 6 describes the runoff process and its variability, measurement of stream flow, hydrographs, overland flow, stream flow routing, reservoir storage-yield analysis, and runoff quality. Chapter 7 covers the physical processes of snow and snowmelt; data requirements, collection, and sources; snow accumulation and distribution; snowmelt analysis and simulation; water supply forecasting; computer programs; and sample applications.

The next three chapters, Chapters 8 through 10, describe the predictions and effects of the phenomena described in the first six chapters. Chapter 8 presents flood characteristics and analysis, statistical analysis for estimating floods, estimating floods from rainfall, probable maximum flood, flood hazard, flood warning, and microcomputer software for flood analyses. Chapter 9 reviews the hydrologic impacts of urbanization, precipitation in the urban watershed, hydrologic losses in developing watersheds, urban runoff estimating methods, typical urban drainage design calculations, and computer model applications. Chapter 10 discusses wave theory, wind waves, ship-generated waves, wave-structure interaction, waves and currents, tides and tidal datums, storm surges, basin oscillations and tsunamis, water surface probability analysis, and selection of design water waves and levels.

The last chapter, Chapter 11, reviews the applications of hydrology starting with study formulation, then reviews data management, then discusses calibration and verification of hydrologic models, and ends with assessing accuracy and reliability of study results.

IV. THE HYDROLOGIC CYCLE

The constant movement of water and its change in physical state on this planet is called the water cycle, also known as natures waterwheel, or the hydrologic cycle. This cycle is depicted in Fig. 1.1. The word *cycle* implies that water derives from one source and eventually returns to that source. Water originates from the oceans and returns to the oceans. On its way, water may change its state from vapor (gas), to liquid (water), to solid (ice and snow) in any order.

A description of the hydrologic cycle can begin at any point and return to that same point. Water in the ocean evaporates and becomes atmospheric water vapor. Some of this moisture in the atmosphere falls as precipitation, which sometimes evaporates before it can reach the land surface. Of the water that reaches the land surface by precipitation, some may evaporate where it falls, some may infiltrate the soil, and some may run off overland to evaporate or infiltrate elsewhere or to enter streams. The water that infiltrates the ground may evaporate, be absorbed by plant roots and then transpired by the plants, or percolate downward to ground water reservoirs. Water that enters ground water reservoirs may either move laterally until it is close enough to the surface to be subject to evaporation or transpiration; reach the land surface and form springs, seeps or lakes; or flow directly into streams or into the ocean. Stream water can accumulate in lakes and surface reservoirs, evaporate or be transpired by riparian vegetation, seep downward into ground water reservoirs, or flow back into the ocean, where the cycle begins again.

Each phase of the hydrologic cycle provides opportunities for temporary accumulation and storage of water, such as snow and ice on the land surface; moisture in the soil and ground water reservoirs; water in ponds, lakes, and surface reservoirs, and vapor in the atmosphere. Without replenishment from precipitation, the water stored on all of the continents would gradually be dissipated by evapotranspiration processes or by movement toward the oceans.

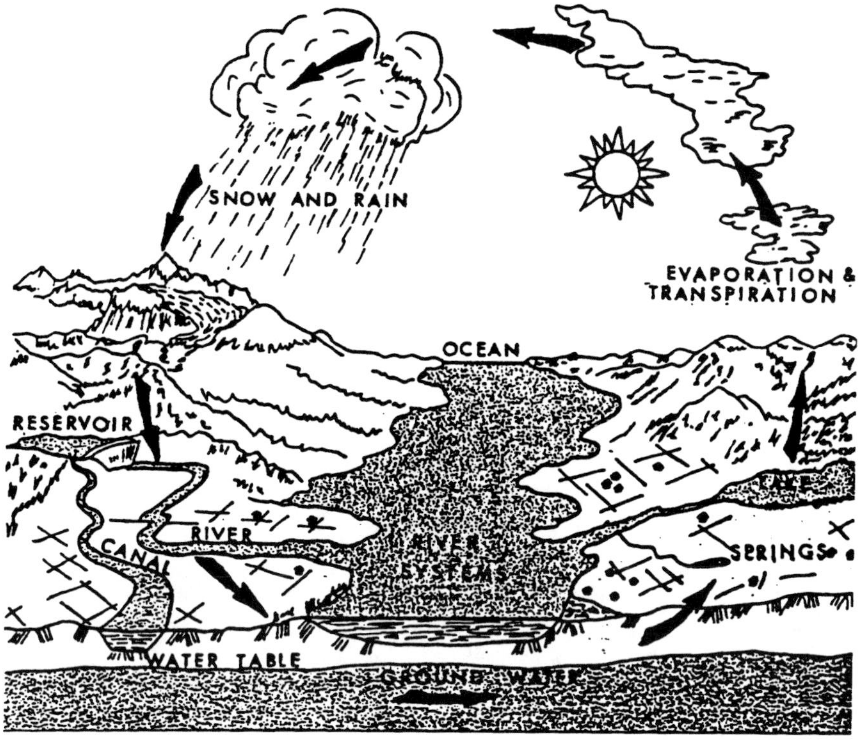

Figure 1.1—Nature's Waterwheel

The concept of water as a renewable resource stems from the hydrologic cycle. Hydrology, as the engineering science that analyzes the various components of this cycle, recognizes that the natural cycle can be altered by human and natural activities. For example, the long-term geologic forces that raise mountains can increase orographic precipitation on one side of the mountains and decrease precipitation on the other side with all of the attendant changes in stream flow, flooding, etc. In the short-term, the development and use of water modifies the natural circulatory pattern of the hydrologic cycle. For example, the use of surface water for irrigation may result in downward seepage from reservoirs, canals, ditches, and irrigated fields, adding to the ground water. Disposal of urban wastewater may recharge the ground water or change the flow in streams. All waters utilized in a non-consumptive manner may deteriorate in quality and create water quality problems in other portions of the hydrologic cycle. Diversions of stream flows impact downstream flows which, if transferred to other watersheds, impact the stream flows and ground water systems in the other watersheds. In addition, pumping from wells may reduce the flow of water from springs or seeps, increase the downward movement of water from the land surface and streams, reduce the amount of natural ground water discharge by evaporation and transpiration, induce the inflow of poorer quality water to the ground water reservoir, or have a combination of all these effects.

The natural circulation of the hydrologic cycle may be changed by actions not related to direct water use. Among these actions are weather modification activities (i.e., cloud seeding), drainage of swamps and lakes, waterproofing of the land surface by buildings and pavements, and major changes in vegetative cover (i.e., removal of forests and cultivation of additional agricultural lands).

The science of engineering hydrology attempts to account for and quantify all aspects of the hydrologic cycle. This handbook presents detailed engineering descriptions of the various phases of the hydrologic cycle.

Chapter 2
PRECIPITATION

I. INTRODUCTION

Precipitation is the primary source of the earth's water supplies. It includes all water that falls from the atmosphere to the earth's surface. Precipitation occurs in two forms that are of interest to hydrologists, liquid (rain and drizzle) and solid (snow, hail, and sleet).

Mean annual precipitation is a major climatological characteristic of a region, while other precipitation characteristics such as intensity, duration, and precipitation form (such as rain vs. snow) are attributes that are generally of most interest to practicing engineers and hydrologists. All these characteristics of precipitation are of interest because they determine the amount and timing of runoff and other hydrological concerns, such as the amount of available soil water.

This chapter deals with formation, types and distribution of precipitation, methods of measuring rainfall and snowfall, processing and interpreting precipitation records, frequency analysis, weather modification, and generation of synthetic weather records.

II. FORMATION AND TYPES OF PRECIPITATION

A. Mechanisms

Two atmospheric processes are primarily responsible for producing precipitation. These processes are generally referred to as: 1) the collision and coalescence process and 2) the ice-crystal process.

The collision and coalescence process is the means by which small water vapor droplets (created by condensation around available nuclei) increase in size in warm clouds. The term "warm" is relative, and refers to cloud conditions when temperatures are greater than 0° C. Larger cloud droplets descend more quickly toward the earth's surface under the pull of gravity than smaller droplets, and thus collisions occur as the larger droplets fall. As the water molecules in these droplets coalesce, the droplets become larger and fall more rapidly until they attain the size of raindrops (approximately 2 mm) and fall to the ground. The size of raindrops depends on several factors, including the vertical velocity (upward) supporting the droplets, the depth of the cloud, the time droplets remain suspended, and atmospheric temperature. The production of raindrops and, consequently, the precipitation that reaches the ground is a function of these factors plus the liquid water content of the cloud, the electric field in the cloud and of the droplets, and the relative size of the droplets.

Warm precipitation is usually restricted to tropical regions and some middle latitude storms in the summertime. The predominant precipitation-producing mechanism in middle and high latitudes is the ice-crystal process. In this case, clouds extend into the atmosphere above the 0° C level, and are thus known as cold clouds. Interestingly, even though ambient temperatures may be less than 0° C, liquid water droplets can still exist and are known as supercooled droplets. Such droplets can exist in a liquid phase to temperatures as low as −40° C, below which all water is in the form of ice particles. Between −40° and 0° C a combination of ice particles and liquid water droplets usually exists.

The ice-crystal process is controlled by nucleation, or the formation of an ice embryo. Homogeneous nucleation refers to the formation of an embryo from a water droplet when no foreign material is present. In this case, the droplet freezes by the simple aggregation of water molecules. This process is possible only when temperatures are near or below −40° C, which is normally the case only in high, cirrus clouds and thus is not the controlling factor in producing precipitation.

In contrast, heterogeneous nucleation occurs when foreign matter is present upon which droplets can either form and aggregate, or with which droplets can collide and freeze spontaneously. The larger the nucleating material, the faster will be the process of freezing, which can occur at higher temperatures. For instance, droplets with diameters of 10 mm will freeze at −32° C but droplets with diameters of 10,000 μm will freeze at −15° C. In general, the higher, deeper and colder the cloud, the greater the probability that ice particles are present and are the controlling mechanism for precipitation production.

B. Types of Precipitation

As ice crystals fall, they may collide and stick to one another, creating snowflakes. If air temperatures are near or below 0° C throughout the lower atmosphere, precipitation recorded on the ground will most likely be in the form of snow. If melting occurs, precipitation at the ground will be rain.

Other precipitation types are possible. If a warm layer aloft is sufficiently deep, the snow will melt and become raindrops; however, if a layer of air near the ground is near or below 0° C the rain may re-freeze. If it freezes before hitting the ground, it has a very granular, icy structure and is known as sleet. If the cold layer near the ground surface is very shallow, the rain may not re-freeze before striking the surface but may freeze on contact with surface elements whose temperature are lower than 0° C (if the water droplets in the rain are supercooled). In this case, precipitation is called freezing rain. Depending upon the location in the United States, this event is called by various names, such as an ice storm, an ice glaze, a silver thaw, or a silver frost.

Rain is typically a name reserved for drops with diameters larger than 0.5 mm. For drops with diameters smaller than 0.5 mm, the common name is drizzle, which has a much lower fall speed, or terminal velocity, than larger raindrops.

In storms, the terminal velocity of water droplets is typically countered by upward moving air. If the vertical velocity of air is positive upwards at rates near or greater than a drop's terminal velocity, the drop will remain suspended. If the upward velocity suddenly decreases, as it can in convective storms, the drops will quickly fall to earth. A sequence of these events leads to the observation of showers on the ground. Typically, raindrops have diameters between 1 and 5 mm with respective terminal velocities of between 4 and 9 m/sec (Chow et al., 1988).

Snowflakes generally have much slower terminal velocities than raindrops. The structure of snow determines its terminal velocity, while the ambient temperature where ice crystals form determines snow structure. At temperatures just slightly below freezing, snow is generally in the form of thin plates which, if they encounter temperatures at or slightly above 0° C in their descent, can melt slightly and then aggregate to form large, mushy flakes. These flakes are extremely reflective to a radar beam, as will be discussed later in this chapter.

Sleet, or ice pellets, usually have diameters around 5 mm, are transparent, and bounce as they hit the ground. In some rare cases, particularly with deep cold air trapped in a valley or on the cold side of a stationary, precipitating warm front, sleet can accumulate to depths greater than 200 mm.

Freezing rain can be among the most damaging precipitation events that occur. If it persists for more than 12 hours at sufficient intensity and with surface temperatures below −2° C, ice accumulation can be extensive. Costs of damage caused by this form of precipitation can easily run into millions of dollars through downed power lines, traffic accidents, and destruction of roofs of buildings. If the freezing droplets are smaller than 0.5 mm in diameter, the precipitation type is called freezing drizzle.

Another way in which ice particles can increase in size is by colliding with supercooled water droplets, a process known as riming. Pilots often contend with moderate to severe riming on planes when an abundance of supercooled water exists. In the atmosphere, this rimed ice will increase in size until buoyancy can no longer support it. If it is fairly small (between 2 and 5 mm) it is known as graupel, or snow pellets. These pellets are very firm and bounce like sleet on the ground surface.

In convective clouds, a graupel particle may be forced upward a number of times. If the updrafts are of sufficient strength and the quantity of supercooled water droplets is optimum, these graupel particles can grow to enormous sizes and are known as hail. Hailstones are typically no larger than 10 mm but, in severe thunderstorms with cloud tops well above the freezing level, hail has been measured with diameters exceeding 100 mm. The size of the stones is proportional to the strength and duration of the updrafts they encounter. Hail can be extremely devastating but typically falls over relatively small areas (1 to 10 km^2).

Snow grains are the remaining type of solid precipitation. These particles are smaller than graupel (less than 1 mm), are opaque and fall from stratus clouds. As such, their fall velocities are typically much less than graupel and they seldom accumulate to great depths at the land surface. In certain situations, snow grains can fall for long time periods.

C. Principal Causes of Precipitation

In middle and high latitudes, precipitation is typically the result of large scale weather systems. Large scale refers to systems with length scales usually larger than 500 km (also known as synoptic scale). Precipitation from such systems is seldom localized and amounts can be rather uniform over large areas.

Central to synoptic meteorology theory is the mid-latitude or polar-front cyclone model. First proposed by a group of Norwegian meteorologists around 1920, this model stresses the development of a low pressure center along a fairly straight, stationary line (Palmen and Newton, 1969). As the low pressure center deepens (drops in surface pressure), a wave forms with the center of the low at the apex of the fronts. Air rotates around the low cyclonically (counter-clockwise in the northern hemisphere), pulling cold air southward on the west side of the low and pushing warm air northward on the east side of the low (in the northern hemisphere). The lines of demarcation are known as fronts, which often are regions of fairly distinct changes in weather (temperature, humidity, cloudiness, wind direction, precipitation, etc.). Major precipitation areas associated with this classical cyclone model are to the north of the warm front, extending cyclonically around the low center northward and westward some distance. This area of precipitation is often rather uniform, with maximum amounts typically 50 to 250 km north of the low pressure center. Another area of precipitation in this model is along the cold front where cold air rushing in at the surface causes warmer air ahead of the front to ascend, often producing convective types of clouds. This precipitation area is typically more broken and localized, with heavy showers and thunderstorms along and immediately behind the front.

Precipitation produced by mid-latitude cyclones is a function of the quantity of water available in the atmosphere and the strength of the dynamic processes which create the clouds and vertical motions around the low. It is now known that jet streams, which are regions of high velocity air some hundreds to thousands of meters above the ground, have a controlling influence on the development and movement of mid-latitude cyclones. Rather than continuous channels of fast air, jet streams typically are cores of high velocity movement through which air parcels travel. Jet streams are the result of large thermal contrasts in the atmosphere. The polar jet usually divides very cold polar air from milder, mid-latitude air; while the subtropical jet divides very warm, moisture-laden tropical air from cooler and (usually) drier mid-latitude air. At certain times, both of these jet streams can be quite active, with their energy generating significant cyclonic storm activity from latitudes as low as 25 or 30 degrees poleward to 80 degrees. Numerical models of mid-latitude synoptic weather are very sensitive to the strength and position of these jet streams, and the transport of jet stream energy to the surface is critical in determining the fate of low pressure systems and the duration, coverage, and intensity of precipitation associated with them.

Convective precipitation can be associated with mid-latitude cold fronts but can also develop in certain situations when no synoptic scale system is active. Air mass showers and thunderstorms are examples of this type of convective activity and are most prominent in mid-latitude summer. Triggering mechanisms for this type of precipitation include upper air systems, which fail to develop a well-defined low pressure center, and very strong daytime heating in the presence of sufficient atmospheric water vapor. This latter situation can lead to extremely heavy rain and severe thunderstorms if the atmosphere is hydrostatically unstable; that is, it is basically stable in the lower boundary layer and somewhat unstable above but becomes very unstable if surface heating and other forces are sufficient to create vertical motions strong enough to penetrate through this layer. This is a potentially explosive atmosphere and is a frequent cause of severe convection and heavy precipitation developing in very short (30–60 min) time periods.

A third element contributing to precipitation development is orography, or over-mountain impact on air flow. As air flows over mountains it is forced upward. This vertical displacement of air is somewhat analogous to warm air being forced upward in the presence of a cold front. As air ascends, it expands and cools. This is a principle directly taken from the ideal gas law, sometimes referred to as the equation of state:

$$pV = mRT \tag{2.1}$$

where p is air pressure, V is volume, m is mass, R is the gas constant for 1 kg of air, and T is temperature (K) (Wallace and Hobbs, 1977). Since m/V is simply air density (ρ), this reduces to:

$$p = \rho RT \tag{2.2}$$

Both air pressure and air density decrease from the surface upward, and thus air parcels forced upward are moved into an environment of lower pressure and density. This causes the parcels to expand and cool, since temperature is indicative of molecular activity. Thus, air forced over a mountain by the wind will cool as it rises. As it cools, it may reach a level where temperature and dewpoint are equal if the absolute water vapor content of the air parcel does not change. This level is referred to as the lifting condensation level (LCL), because water vapor begins to condense into liquid water droplets and the relative humidity is around 100%. At this level, clouds form and if vertical motions are strong enough, precipitation can develop. Even on relatively clear days, a combination of daytime surface heating and wind flow leads to cloud development over mountain ridges.

If a cyclonic system is already producing clouds and precipitation, the air flow over mountains will enhance precipitation on the windward slope. Thus, it is common for precipitation to increase with elevation in mountainous areas. Because precipitation-elevation relationships change with region and with storm events, simple approaches to estimate these relationships are not necessarily reliable. Precipitation is strongly controlled by elevation in the western United States, where mountain snowpacks supply most of the water for the region.

Tropical storm systems can produce significant amounts of precipitation and cover relatively large surface areas. Such systems typically affect the United States between July and November, with peak activity in the late summer. Atlantic tropical systems originate as tropical waves off the west coast of Africa and move toward the Caribbean on the predominate easterly flow across the ocean. Some systems develop into hurricanes, which means sustained winds are in excess of 33 m/sec (74 mi/hr). The path taken by tropical storms is usually dictated by the winds aloft, which are generally weak easterly winds in this region. Strong upper level winds associated with jet streams are actually destructive to tropical storms. Larger hurricanes can sometimes have radii in excess of 500 km and their movement is much less directed by winds aloft, thus making prediction of their movement difficult.

Tropical cyclones are known as "warm core lows" because they are warmer than their surroundings from the surface upward. They are strongest at the surface, and winds and pressure gradients diminish with height. In contrast, mid-latitude or extratropical cyclones have colder temperatures in the storm center than around them, and the intensity of the low increases with increasing height in the atmosphere, thus making it a so-called "cold core low." Tropical cyclones usually have cloud heights above the freezing level, so that even in these warm systems there is typically a mixture of warm and cold precipitation-producing mechanisms in place.

The quantity of precipitation that falls from tropical systems (as well as extratropical systems) is a function of storm movement, relative location to the storm center, and storm movement relative to land masses. Even relatively weak tropical cyclones have sometimes produced extremely heavy rainfall, especially when the storm stalls over an area or moves perpendicularly into a mountain ridge. Central pressure of a hurricane is a good gage for maximum wind speeds but is a poor indicator of total precipitation. In general, the lower the pressure in the center of the storm, the more energy the storm possesses with stronger horizontal and vertical wind speeds. This condition also creates more potential for heavy precipitation; however, strong storms sometimes have relatively fast forward speeds which prevent large total accumulations, although short-duration intensities can be quite impressive.

Precipitation around a tropical cyclone, particularly one which has become fairly well-organized and concentric, usually comes from rain bands rotating cyclonically around the low pressure center. These bands of showers and thunderstorms typically increase toward the center of the storm (or "eye" in a well organized hurricane), reaching the maximum intensity in the eye wall, where a solid circle of severe thunderstorms is usually located. Often the heaviest precipitation is in these "eye wall thunderstorms" and in rain bands that are generally to the east of the cyclone center.

In the southeastern United States, tropical storms are responsible for as much as 5 to 30% of the normal precipitation in the summertime. For the remainder of the nation, the numbers are much smaller, although tropical cyclone activity in the eastern Pacific Ocean sometimes produces heavy precipitation in the Southwest. In some cases, desert locations in Arizona and southern California can receive most of their annual precipitation from the remnants of these storms. These regions have a monsoonal climate dominated by notable wet and dry seasons. Tropical storms are sometimes responsible for ending droughts, especially in the Southeast, and can actually have beneficial results.

III. VARIATIONS IN PRECIPITATION

A. Geographic Distribution

1. Latitudinal Variations Globally, average annual precipitation (Figure 2.1) is generally heaviest in tropical regions and decreases poleward, indicative of the diminishing capacity of air to hold moisture with decreasing temperature; however, there are significant deviations from this mean trend. Latitudes near 30 degrees have relatively little precipitation because of the climatic propensity for air to rise near the equator and then descend near these latitudes. Some of the world's great deserts (the Sahara, the Middle East, the Australian interior, the southwest portion of Africa, and the American Southwest) are in this region, showing that this sinking air is counter-productive to precipitation production.

Poleward moving air typically rises again in middle latitudes, reaching an average maximum at about 60 degrees latitude. Precipitation enhancement occurs in these regions, with more frequent cyclonic activity.

In addition to the cellular structure of poleward moving air (Figure 2.2), other dominant forces shaping regional precipitation are the general circulation of both the oceans and the atmosphere and their relationship to the shape and position of continents. In general, flow in the major oceans is cyclonic,

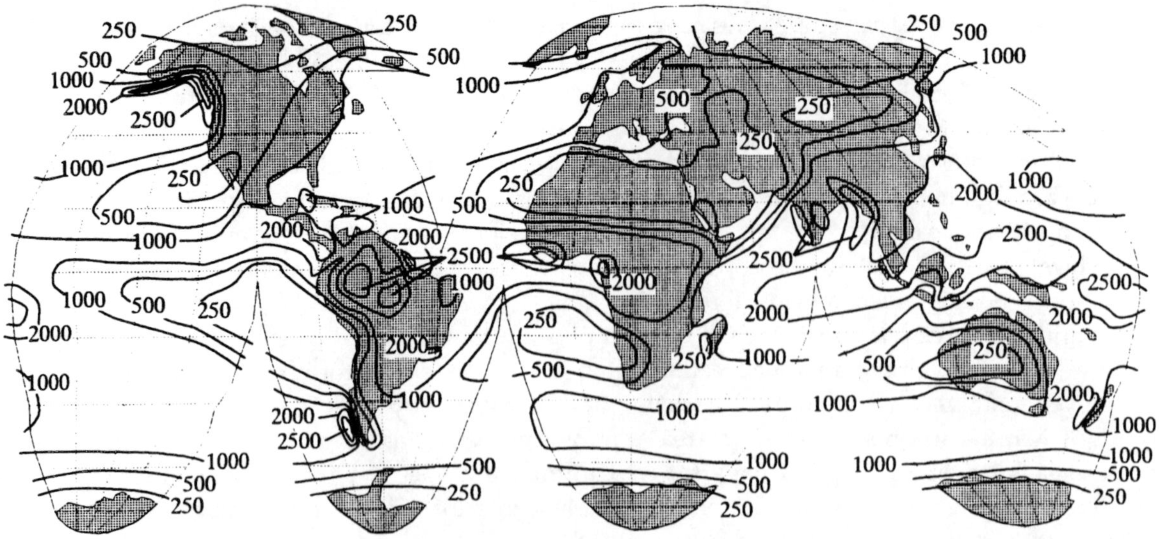

Figure 2.1.—Average Annual World Precipitation (mm)

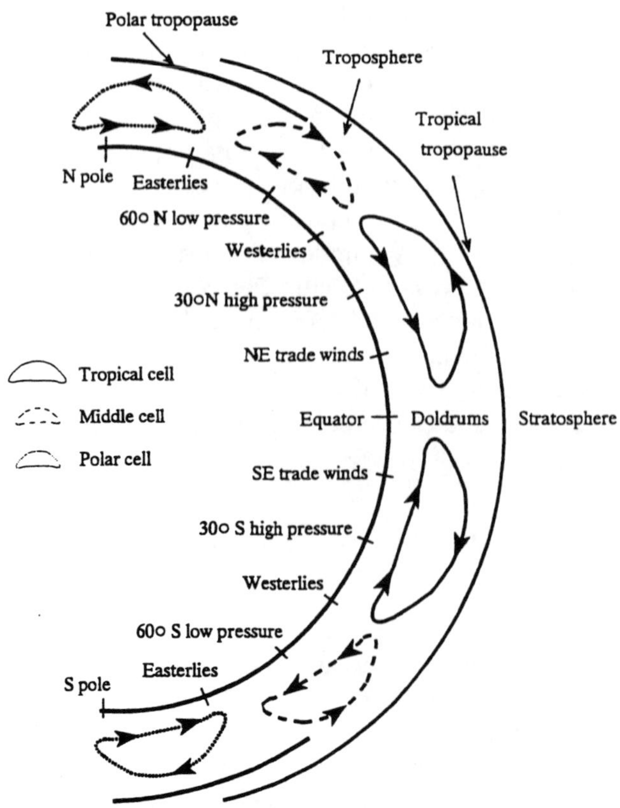

Figure 2.2.—Three Cell Atmospheric Circulation Patterns for Both Hemispheres

which means tropical water and atmospheric moisture are transported poleward along the east sides of continents. Therefore, eastern portions of North America, Australia, and the EurAsian continent have relatively more precipitation than their counterparts on the west coast, especially during their respective summers. Slight exceptions to this phenomenon do occur, such as along the north Pacific coast of the United States and Canada where a persistently strong jet stream in the winter and the circulation around the semi-permanent Aleutian low brings copious precipitation to these regions, which is enhanced by the sharp increase in elevation immediately inland from the coast.

Polar regions are typically under the influence of sinking air, plus the very cold atmosphere can hold relatively little water vapor. As a result, these areas have very little precipitation which for most if not all of the year, is locked in the solid forms of snow and ice.

2. ***Distance from a Moisture Source*** Perhaps the second most important factor in determining precipitation at a given location is its distance from a moisture source—i.e., an ocean. Continental interiors typically have less precipitation because of a minimum of precipitable water and a lack of larger, ocean derived salt particles, which are better nucleating materials than dust and other materials from land (Figure 2.3).

It is estimated that only 15% of the nearly $5.68 \times 10^{15} m^3$ of water that evaporate globally each year come from continental areas; the other 85% is evaporated from the oceans (Ahrens, 1991). The Amazon basin in South America is a notable exception to this rule. Two contributing factors to the heavy rainfall in this region are its abundant reserve of water held in the jungle and its equatorial location, where horizontal movement of storm systems is minimal. Thus, almost daily rains occur in this region and in similar interior sections of equatorial Africa.

To a lesser extent, water bodies smaller than oceans can exert influence on precipitation. The Great Lakes, for instance, are responsible for enhanced precipitation, especially in the wintertime when strong cold air advection over the relatively warm water produces significant snow squalls on windward shores. "Snow belts" are well-known in this region, but the exact location of maximum snowfall and the amount that falls is dictated by the wind direction relative to the water, the fetch over the water, the temperature contrast between the water and the advected air, and the elevation rise on the windward shore. In contrast, these same lakes can actually be responsible for precipitation reduction in the warmer months of the year because the water is relatively cooler than the land at this time, creating an air circulation pattern away from the lakes. The Great Lakes region then has a relatively higher surface pressure than the land area around it, leading to subsidence, or sinking air aloft, which is detrimental to cloud and precipitation production.

Sea breeze circulations are another example of localized wind flows created by the contrast in temperature and moisture between land and water. These diurnal circulations are dominated by an ocean to land wind flow (very regular in the tropics; pronounced in the summertime in mid-latitudes). Warm, moist air flowing inland in the Gulf Coast and Atlantic Coast regions of the United States, for instance, typically converge at the surface with warmer and drier air moving oceanward each afternoon. Surface convergence leads to upward vertical motion and, especially in the presence of daytime heating, cloud development. Precipitation is thus enhanced in regions immediately inland from coastal sections in the summertime, usually at distances from 10 to 50 km from the shoreline. This scenario is particularly apparent in the Florida peninsula where, for much of the year, these circulations create thunderstorms over the land each afternoon while satellite images reveal very little cloud development or precipitation over the nearby ocean.

3. ***Orographic Influences*** As mentioned in the section on principal causes of precipitation, precipitation is normally enhanced in the vicinity of mountains. Enhancement is dependent on several factors, including wind direction (relative to topography), wind speed, atmospheric moisture (precipitable water), elevation rise, and slope angle. For these reasons, orographic precipitation is most pronounced during the winter months in middle latitudes when atmospheric flow is strongest; however, convective precipitation in summer months also is enhanced over mountains due to diurnal winds which tend to move up slopes and through valleys during the day, and reverse their direction at night (Whiteman, 1990).

Orographic precipitation produces marked contrasts in seasonal precipitation distribution, which must be considered for any type of system design in mountainous areas. The monthly distribution of precipitation for two sites only 10 km apart in the Reynolds Creek Watershed in Idaho (Fig. 2.4) is

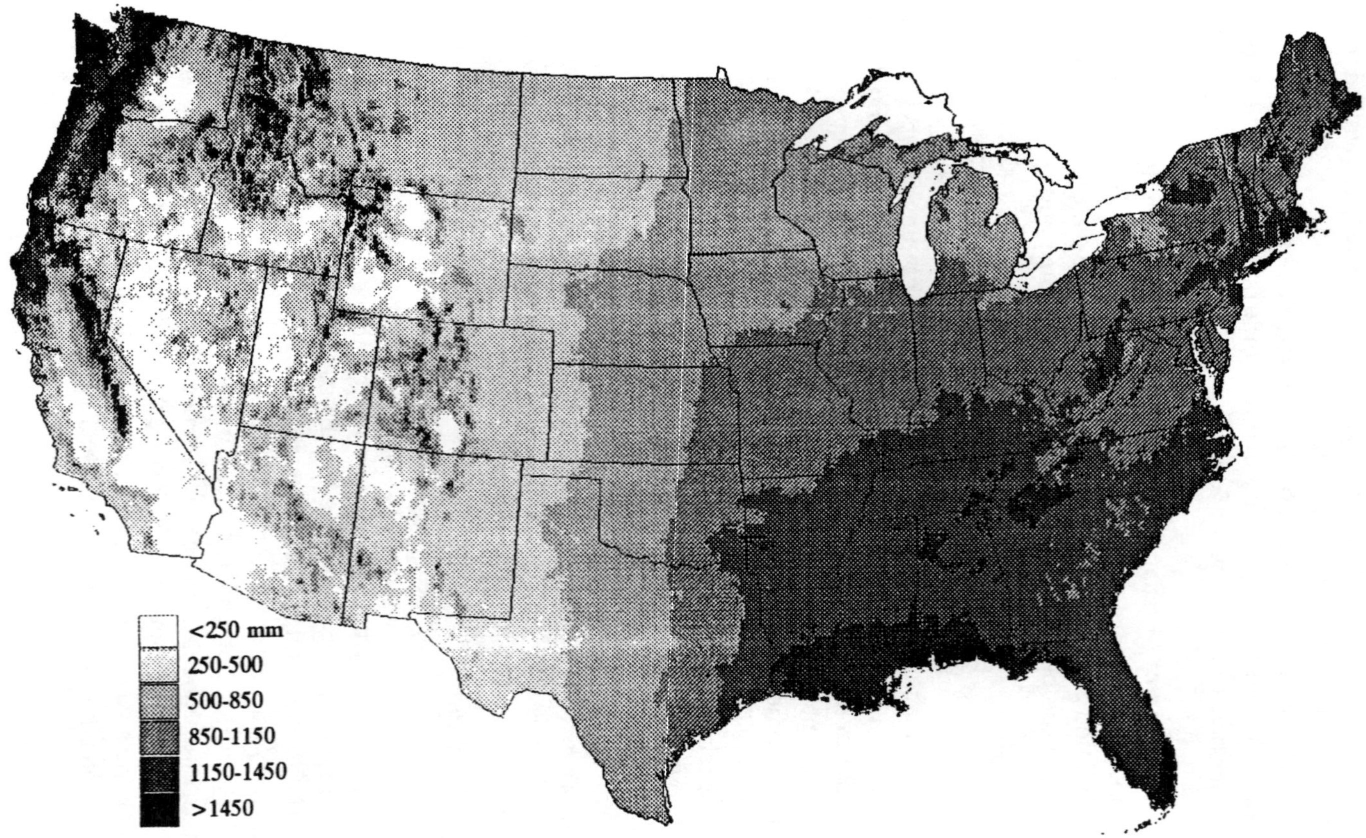

Figure 2.3.—Long-term Annual Precipitation for the Continental United States, Data from NWS/Cooperator and SCS Snotel Stations having at Least 20 yr of Valid Measurements. Map Produced using the PRISM Model, a New Statistical/Topographic Precipitation Mapping Scheme (Daly et al., 1992). For further information contact Oregon State

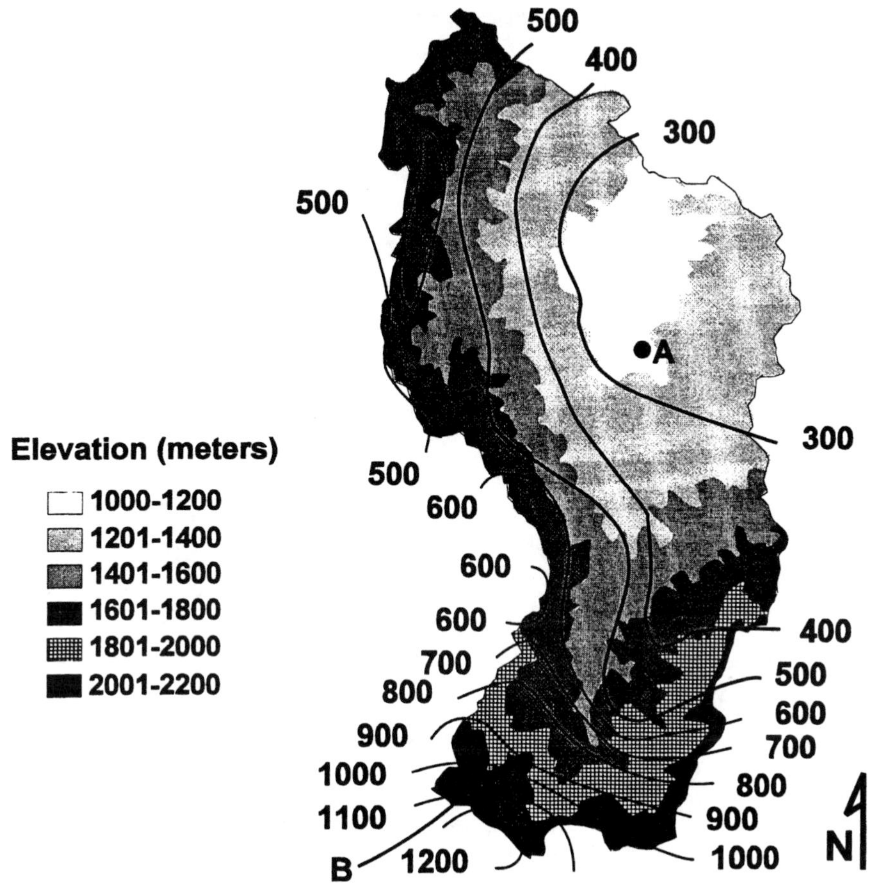

Figure 2.4.—Reynolds Creek Watershed, Idaho; Elevation Contours are Shaded, with Overlaying Isohyets of Mean Annual Precipitation (mm)

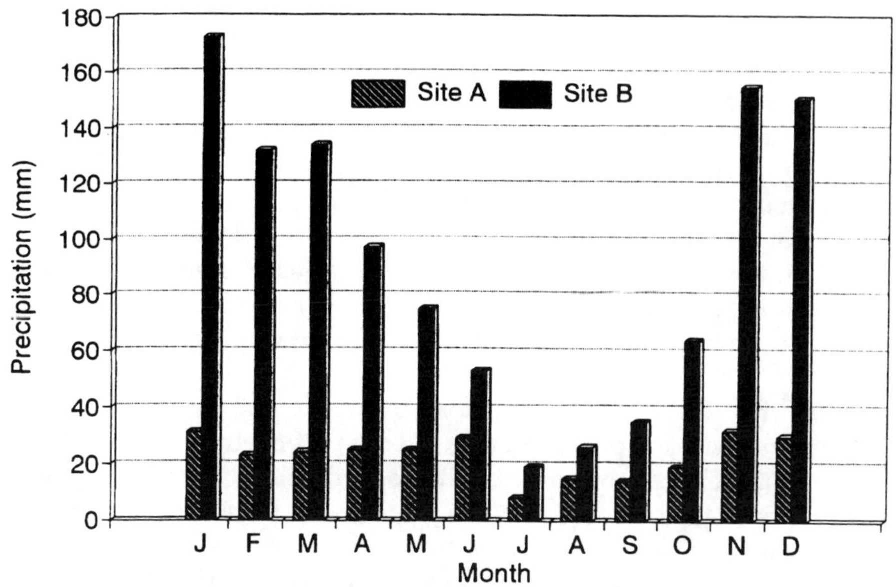

Figure 2.5.—Monthly Precipitation Distribution (mm) at sites A and B, Reynolds Creek Watershed, Idaho

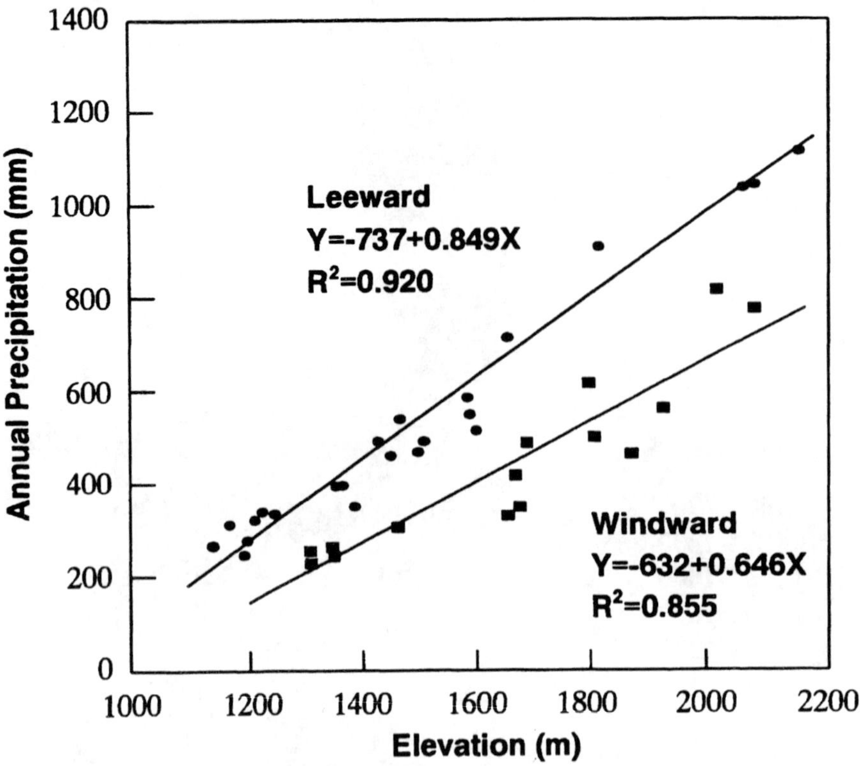

Figure 2.6.—Annual Precipitation (mm) vs. Elevation (m) for Windward and Leeward Locations, Reynolds Creek Watershed, Idaho

indicative of many mountainous areas and is shown in Figure 2.5 (Hanson and Johnson, 1993). Site A has an elevation of 1193 m with an annual precipitation of 275 mm (22% as snow) and maximum monthly precipitation in June. In contrast, site B, at an elevation of 2164 m, has an average annual precipitation of 1114 mm (76% as snow), and monthly precipitation is greatest in January. This watershed (at 43 degrees N latitude) is subject to frequent winter storms, usually accompanied by strong winds. The low elevation site (A) is downwind of site B and is in a relative precipitation "shadow." Such shadows are common on the lee-side of mountain ranges (Sumner, 1988). This is well illustrated in the map of annual precipitation for the United States (Fig. 2.3), and discussed by Daly and Neilson (1992). Where atmospheric flow is predominantly from one direction, it can lead to climatic dry zones in areas such as east of the Sierra Nevada and Cascade mountains in the western United States and even in relatively humid locations such as around Asheville, North Carolina, which is in a shadow region of the southern Appalachians.

As mentioned in the last section, precipitation-elevation relationships can sometimes be different from one side of a mountain range to the other. For example, on the Reynolds Creek Watershed, the best-fit regression line of precipitation with elevation is different for the windward and the leeward locations (Figure 2.6).

Design applications often require information about the relationship between a standard time increment for precipitation and expected precipitation in a different time interval. It is important to realize that such relationships change with elevation and other topographic factors in mountainous areas. As an example, the ratio between 1-hour precipitation and precipitation for shorter and longer durations was calculated for sites A and B at Reynolds Creek Watershed, for all return periods (Table 2.1). At the low elevation site (A) the ratio is just 1.83, while at high elevation site B the ratio is 5.01—nearly three times greater. Thus, if only 24-hour precipitation were available (as is often the case), estimated 1-hour precipitation at site A would be about half as much as the measured 24-hour amount; at site B, however,

TABLE 2.1. Ratio Between 1-hr Precipitation (mm) and Precipitation for n-min Durations at Sites A and B, Reynolds Creek Watershed, Idaho

	Site	
Duration	A	B
5 min	.45	.43
10 min	.67	.66
15 min	.77	.78
20 min	.79	.80
30 min	.87	.92
2 hr	1.09	1.20
6 hr	1.40	2.01
12 hr	1.61	3.36
24 hr	1.83	5.01

the 1-hour amount would be only one-fifth as much as the 24-hour amount. Such a difference could be crucially important in engineering design at the two sites, and such considerations should be made in estimating precipitation in mountainous regions.

B. Time Variation

Temporal precipitation variation on longer time scales is largely driven by normal fluctuations in atmospheric flow with known periodicities. For example, monsoonal climates are characterized by pronounced wet and dry seasons. Arizona and New Mexico are examples of this type of climate, with a relatively wet season usually beginning in June and continuing through the summer. Other times of the year are normally very dry, except for some precipitation increase in the winter if the subtropical jet stream is strong enough, and storm systems are actively moving east across northern Mexico and California.

Meteorologists are becoming increasingly aware of atmospheric teleconnections—that is, perturbations in the oceans or atmosphere that have an effect on the weather in locations downstream. While it may be extremely difficult to know the cause of perturbations, an understanding of the teleconnections associated with them may lead to greatly improved long-range weather forecasts.

A well-known example among these perturbations is the El Niño-Southern Oscillation (ENSO) phenomenon in the tropical Pacific Ocean. In some years, the normal east to west flow of air and water in the equatorial Pacific is reversed, increasing temperatures and precipitation in the far eastern Pacific (especially along coastal Ecuador and Peru). El Niño greatly modifies the entire energy balance in the Pacific, which is a huge energy pool for the world. As a result, changes in the weather and ocean water in the Pacific result in somewhat predictable changes in other parts of the world. Globally, some regions are positively correlated with El Niño while others are negatively correlated (Figure 2.7). In the United States, highest positive correlations are associated with winter precipitation in the Gulf Coast States and the Southwest, and with summer precipitation in the southern Rockies (Ropelewski and Halpert, 1987; Woolhiser et al., 1993). In contrast, the southern oscillation index (SOI, a measure of ENSO in which negative values indicate the presence of an El Niño event) was found to be negatively correlated with October through March precipitation in the Pacific Northwest, especially in mountain areas (Redmond and Koch, 1991). In other words, during an El Niño event, the Northwest is normally dry while the Southwest is often wet (correlations around 0.5 for both). This is but one example of atmospheric teleconnections that can directly influence temporal precipitation variability.

The seasonal distribution of precipitation in an area is an important consideration for many reasons. For example, growing season precipitation dictates the need for irrigation. In the eastern United States, precipitation is normally adequate during the growing season and, thus, irrigation usage is relatively low.

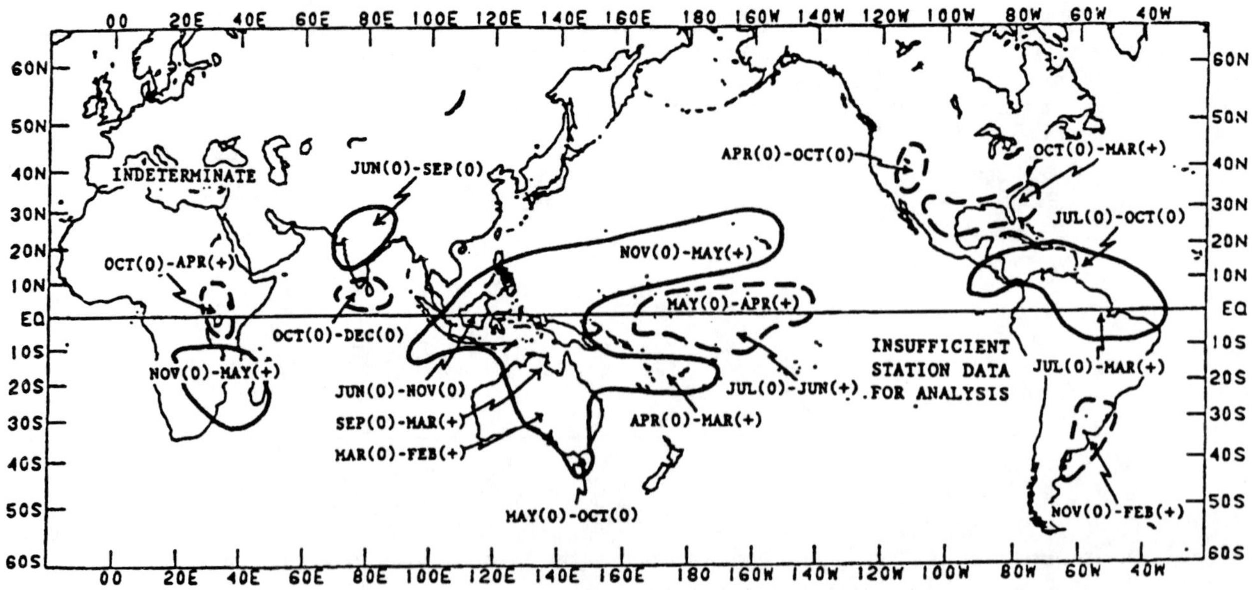

Figure 2.7.—Regional Enhancements (dashed lines) and Diminutions (solid lines) of Precipitation During El Niño Episodes. [Months of concentration are shown for each region. Year of associated anomalously high sea surface temperatures in the tropical Pacific is noted by month of concentration (0 = same year; + = subsequent year)] (from Ropelewski and Halpert, Monthly Weather Review, Vol. 115, p. 1625, American Meteorological Society, 1987)

In the western United States, irrigation is largely responsible for agricultural production. Because irrigation water supplies are normally dependent on winter snowpacks, water for streamflow and reservoir recharge are of great concern. The very dry summer seasons in the West make winter precipitation crucial to survival, while in the East, where frequent precipitation is normal, water supply systems are more vulnerable to relatively short-term droughts that occur especially during the warm season, when both water requirements and evapotranspiration are high.

Monthly precipitation distributions for selected locations in the United States are shown in Figure 2.8. Regional characteristics are quite evident. From the Northeast to the Central Gulf Coast, precipitation is nearly uniform for all months. In the Southeast, a summer to early fall precipitation maximum is noted. Very little winter precipitation with a summer maximum is characteristic of the Plains and Midwestern states. Low elevation locations in the Inter-mountain region have uniformly low precipitation year-long, while higher elevations have precipitation distributions resembling the graph in Figure 2.5. The Pacific Coast region is characterized by wet winters and extremely dry summers—exactly opposite to the Southeastern states.

On shorter time scales, precipitation variability is a result of 6-hour to 3-day synoptic events (frontal passages, cyclonic storms) and/or diurnal fluctuations, chiefly resulting from the daily cycle of insolation. Synoptic-scale events are driven by the wind flow at upper levels of the troposphere. The persistence of synoptic patterns can lead to prolonged periods of dry or wet conditions.

Diurnal precipitation variability is greatest during the warm season and is thermally driven. Thus, convective precipitation predominates during the summer and is induced by simple surface heating or in combination with thermally-forced flows, such as sea breezes or mountain/valley winds. It is also a function of available water. Therefore, the highest frequency of diurnal precipitation is found in the southeastern United States where all of these ingredients are maximized. Thunderstorms comprise a large portion of the precipitation events on these relatively short time scales, and a map of the annual number of thunderstorm days each year in the United States is shown in Figure 2.9. Obviously, thunderstorm enhancement in the eastern United States is partially due to synoptic-scale storms, but a large portion is simply due to mechanisms operating on daily time scales. Note the maximum over the interior of the Florida peninsula, a region frequently experiencing sea breeze convergence in the summer, when heat and moisture are greatest.

C. Extreme Precipitation Events

The maximum quantity of precipitation that can be expected at a location in a given period of time is related to atmospheric water vapor content (called precipitable water) and the strength of vertical motion. Precipitable water is defined to be the total mass of water in a column of atmosphere, given by:

$$m_p = \int_{z_1}^{z_2} q_v \rho_\alpha A d_z \tag{2.3}$$

where m_p is the total precipitable water, Z_1 and Z_2 are elevations, q_v is specific humidity, ρ_a is air density, and A is the cross-sectional area (Chow et al., 1988). Southeastern sections of the United States have a much greater amount of precipitation possible in any given year because precipitable water values are highest there and maximum vertical motions are typically much greater than locations further north and west. Vertical motion is a function of lifting forces related to conditional instability, tropical storms, jet stream dynamics, etc. This is not to say that vertical motions in northwestern sections of the nation for example are not sometimes quite strong; however, they typically occur in winter when precipitable water values are much lower.

The World Meteorological Organization (WMO) states the following equation as an approximation for estimating the world's extreme precipitation values:

$$P = 422T_d^{0.475} \tag{2.4}$$

where P is the precipitation depth in millimeters and T_d is the duration of the event in hours (WMO, 1983). A listing of the world's greatest recorded precipitation amounts is given in Table 2.2. These, of course, represent the current upper bounds on recorded precipitation. Most locations will never come close to receiving these extreme amounts.

Maximum expected precipitation is dependent on both the time of year and elevation. Accompanying graphs of average winter, summer, and annual precipitation (Figure 2.10), and 2-year, 6-hour precipitation versus elevation (Figure 2.11) over the Reynolds Creek Watershed show significant differences between low and high elevation sites. Other climatic regions would most likely have slightly different graphs, but seasonal and elevational dependence would be noted in most locations. These concepts are combined to calculate probable maximum precipitation (PMP). A more complete treatment of PMP is given in the section on Processing and Interpreting Precipitation Records in this chapter and in Chapter 8, Floods.

IV. THE MEASUREMENT OF PRECIPITATION

The measurement of precipitation dates back at least to the 4th century B.C., when a network of rain gages was established in India (Biswas, 1967). Rain gages were used in Palestine in the 1st century B.C., in China in the 13th century A.D., and in Korea in the 15th century (Biswas, 1970). The shape of the gages in India is not known, but the Chinese and Korean gages were cylindrical or barrel-shaped. The Korean gages were about 30 cm deep with 15 cm diameter cylinders, and had about the same characteristics (and accuracy) of many of the rain gages in widespread use today. Rain gages were first used in Europe in the

NORMAL MONTHLY TOTAL

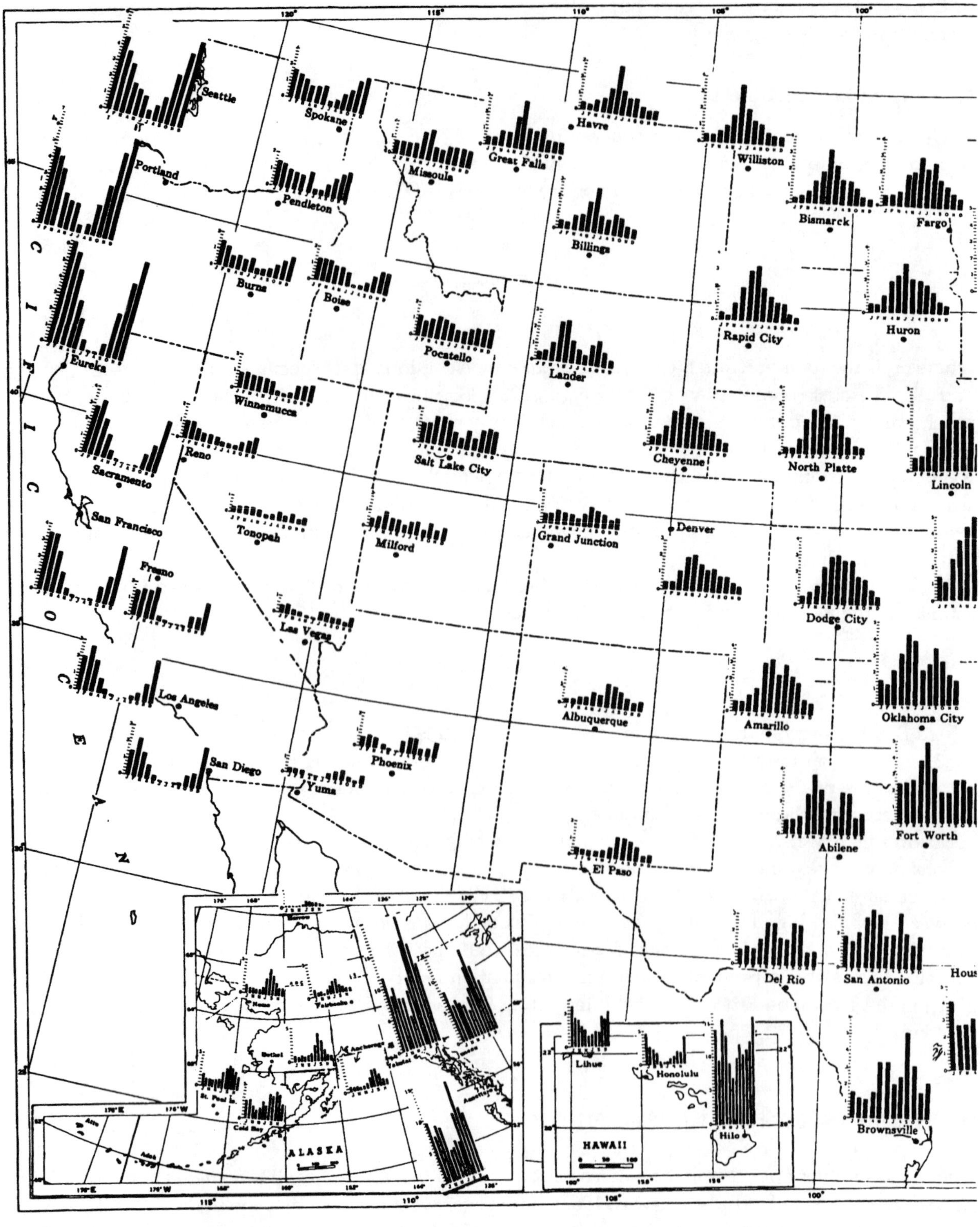

Figure 2.8.—Monthly Precipitation Distribution in the United States (1 in = 25.4 mm; from Climatic Atlas of the United States, U.S. Dept. of Commerce, 1979)

PRECIPITATION (Inches)

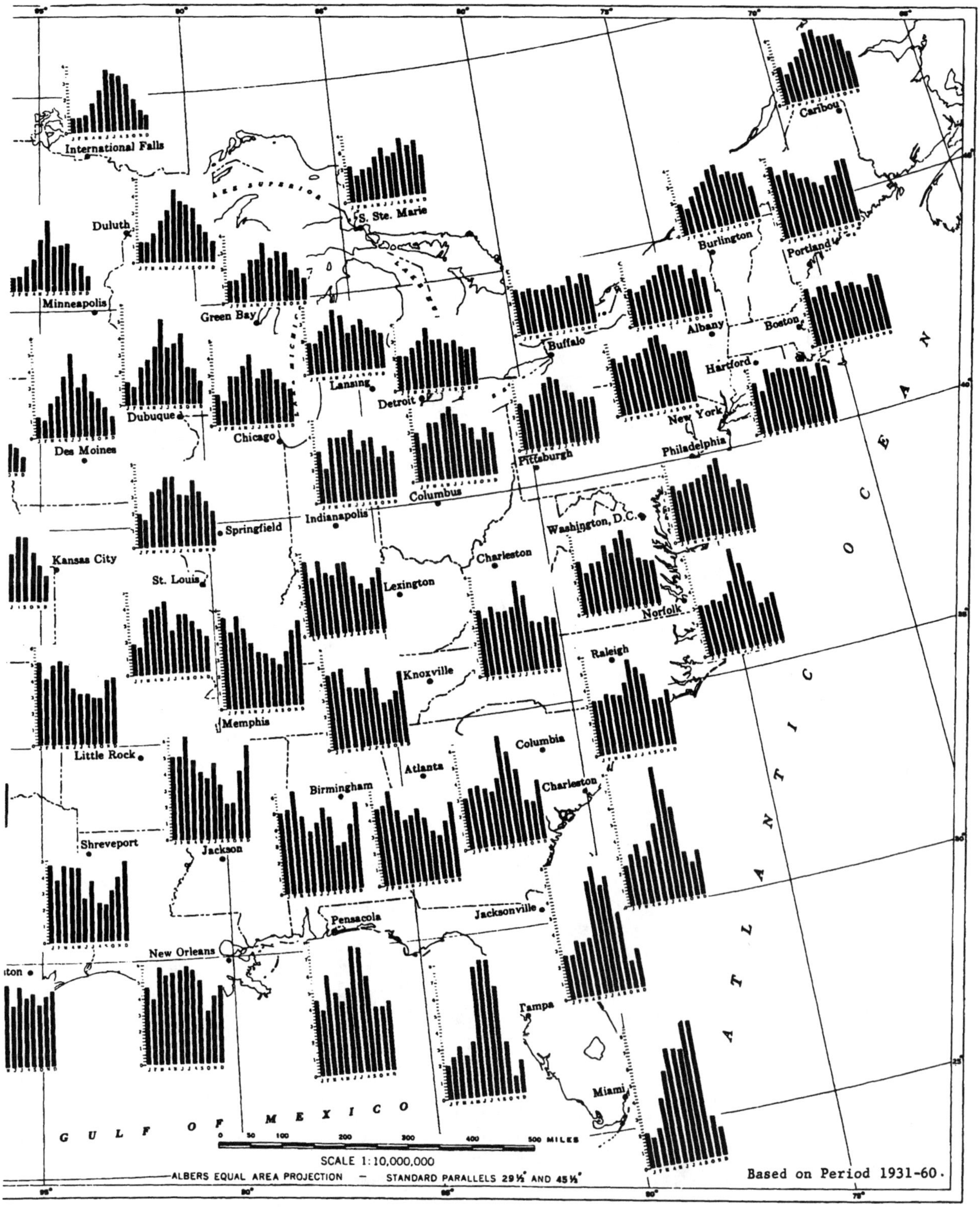

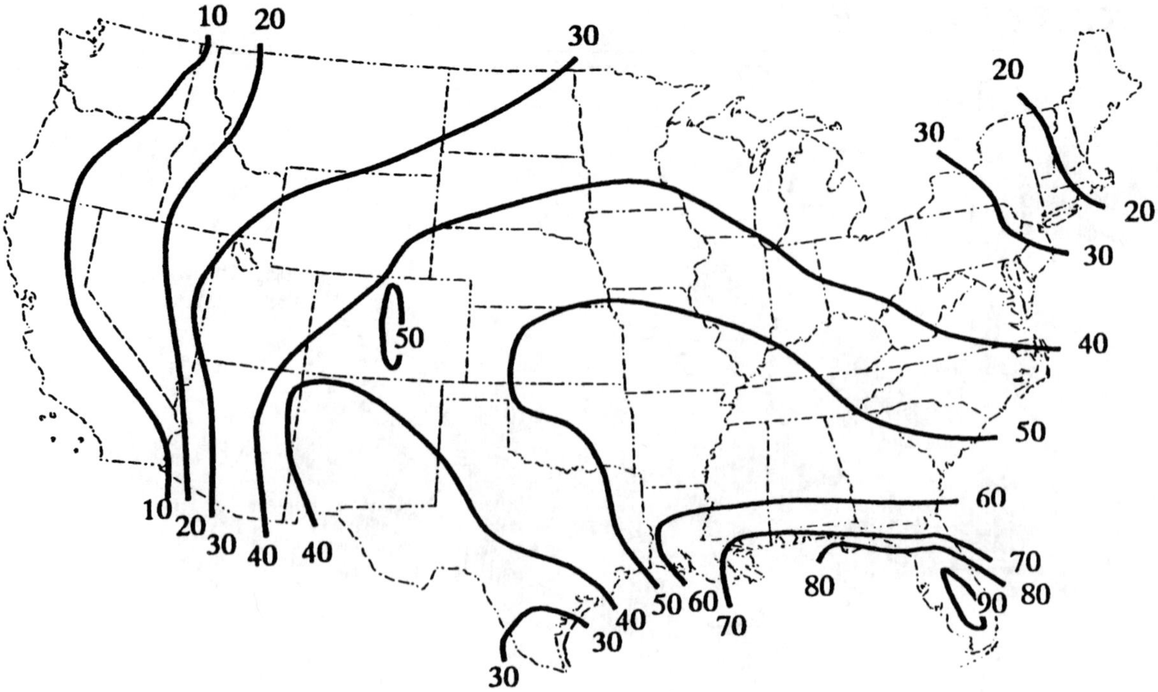

Figure 2.9.—Average Annual Thunderstorm Days for the United States

TABLE 2.2. Precipitation Extrema for the World (from Griffiths 1985, p. 125. In: David O. Houghton, 1985, Handbook of Applied Meteorology, John Wiley & Sons, Inc., New York, NY. Reprinted by permission of John Wiley & Sons, Inc.)

Parameter of Precipitation	Precipitation (in.)	Precipitation (mm)	Station
Highest annual total	1042.	26470.	Cherrapunji, India (Aug. 1 1860–July 31, 1861)
Highest annual means	460.	11680.	Mt. Waialcale, Kauai, Hawaii (32 yr)
	450.	11430.	Cherrapunji, India (74 yr)
	405.	10290.	Debundscha, Cameroons (32 yr)
Lowest annual mean	0.02	0.50	Wadi Halfa, Sudan (39 yr)
	0.03	0.75	Arica, Chile (59 yr)
Highest monthly total	366.	9300.	Cherranpunji, India (July 1861)
Highest monthly mean	106.	2700.	Cherranpunji, India (July 1861)
Highest 5-day total	150.	3810.	Cherrapunji, India (Aug. 1841)
	115.	2920.	Jamaica (Nov. 1909)
Highest 1-day total	73.62	1870.	Cilaos, La Reunion (Mar. 16, 1952)
Highest 12-hr total	52.76	1340	Belouve, La Reunion (Feb. 28–29, 1964)
Highest 4.5-hr total	31.00	787.	Smethport, Pa. (July 18, 1942)
Highest 42-min total	12.00	305.	Holt, Mo. (June 22, 1947)
Highest 20-min total	8.10	205.	Curtea-de Arges, Rumania (July 7, 1889)
Highest 1-min total	1.23	31.2	Unionville, Md. (July 4, 1956)
Highest no. of rain days in 1 yr			Cedral, Costa Rica (1968) - 355 days
			Cedral, Costa Rica (1967) - 350 days
			Behia Felix, Chile (1916) - 348 days
Highest annual total (snow)	1017.	25830	Paradise Ranger Station, Mt. Rainier, Wash. (1970–1971)
Highest annual means (snow)	582.	14780.	Mt. Rainier, Wash.
	575.	14600.	Crater Lake, Ore.
Greatest depth on ground	454.	11530.	Tamarack, Calif. (Mar. 9, 1911)
Highest monthly total (snow)	390.	9900.	Tamarack, Calif. (Feb. 1911)
Highest 12-day total (snow)	304.	7720.	Norden Summit, Calif. (Feb. 1–12, 1938)
Highest 6-day total (snow, single snowstorm)	174.	4420.	Thompson Pass, Alaska (Dec. 26–31, 1955)
Highest daily total (snow)	76.	1930.	Silver Lake, Colo. (April 14–15, 192

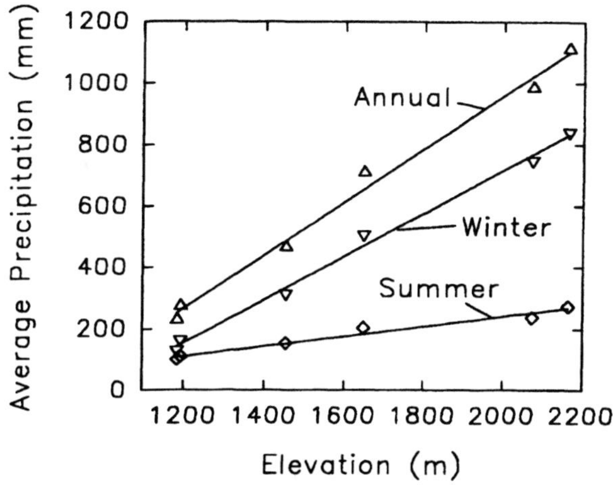

Figure 2.10.—Average Seasonal (winter and summer) and Annual Precipitation (mm) vs. Elevation (m) for Six Sites on the Reynolds Creek Watershed, Idaho

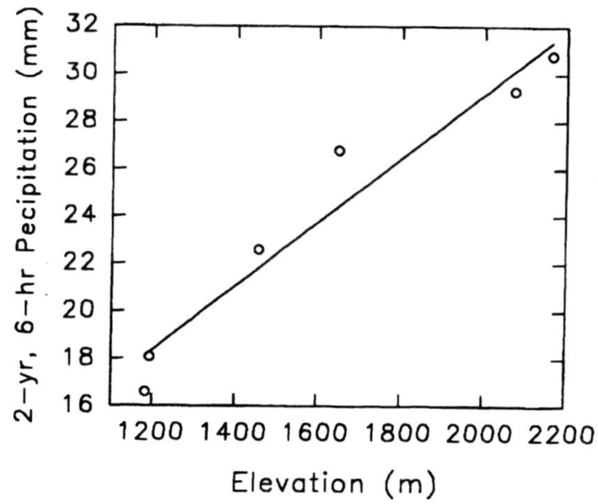

Figure 2.11.—2-yr 6-hr Precipitation (mm) vs. Elevation (m) for Six Sites on the Reynolds Creek Watershed, Idaho

17th century, including a tipping-bucket gage developed by Sir Christopher Wren and modified by Robert Hooke in 1678 (Biswas, 1970).

The 18th century was marked by the development and use of numerous designs of gages around the world. In 1802, Dalton stated (cited in Biswas, 1970):

> The rain gauge is a vessel placed to receive the falling rain, with a view to ascertain the exact quantity that falls upon a given horizontal surface at the place. A strong funnel, made of sheet iron, tinned and painted, with a perpendicular rim two or three inches high, fixed horizontally in a convenient frame with a bottle under it to receive the rain, is all the instrument required.

The measurement of the "exact quantity" of rain (or snow) that falls upon a horizontal surface has been the subject of a large number of investigations in the past two hundred years. As cited in Rodda (1967), Heberden (1769) established that elevated gages caught less rainfall than lower gages. This was shown conclusively by Jevons (1861) to be due to wind action (cited in Nipher, 1878). To eliminate the effects of wind on the collection of rainfall, Stevenson (1842) developed a gage with an orifice at ground level (cited in Rodda, 1967). This ground-level gage, also known as a pit gage, properly sited and protected against splash, is perhaps the most accurate gage for rain. A ground-level gage has been used by the WMO as a reference for the various national gages (Rodda, 1970). Ground-level gages are not suited for large networks, however, due to much the same objections listed in Nipher (1878). The pit gages fill with leaves or snow due to drifting and blowing, are prone to damage, and to float with accumulated water around the gage, and produce large measurement error with small differences in orifice height. The problems with snow measurement led Nipher to design the first shielded snow gage; a design that is still in use.

In the past 100 years, several types of gages for the measurement of precipitation have been developed. New techniques have been devised to measure precipitation with both direct sensing and remote sensing. In spite of all the latest technology, most precipitation measurements in the United States are still made with a measuring stick using a variation of the type of gage described by Dalton in 1802. For many types of rainfall or snowfall measurements, the "exact quantity that falls on a horizontal surface" must be known with a high degree of accuracy. Overall, a perfect, cost-effective system for precipitation measurement has yet to be demonstrated.

A. Uses Of Precipitation Measurements

The basic function of precipitation measurement is to sample the distribution of precipitation in space and time. For most climatological purposes, gage measurements are used as an index of the actual precipitation at a specific point. The measured amount may or may not be the "actual" precipitation that would have occurred in the absence of the gage. In hydrology, the amount of water that actually reaches the surface of the ground over an area is the desired measurement from a gage or network of gages (Rodda, 1970). For climatological purposes, the measurement period ranges from daily to monthly (e.g., with cloud index methods). Soil erosion and storm water runoff studies require rainfall intensities for durations of one hour or less. Precipitation measurement for heavy rainfall and flash flood forecasting may be needed for periods from a few minutes to several hours. For microwave communications design, the rainfall rate along a transmission path is needed with high temporal resolution, typically for 1-minute averages (Fedi, 1981; Segal, 1986). Although rapid-response rain gages have been developed, the networks are extremely limited and the data are available only for short time periods, typically only a few years. Realistically, extensive precipitation gage networks designed for one purpose (e.g., climatology) must serve other purposes (e.g., hydrology). A measurement system for one form of precipitation (e.g., rain) may be inadequate for another form (e.g., snow) or the accuracies of measurement may be substantially different. The point samples may or may not represent the spatial and temporal patterns of precipitation. In spite of these difficulties, measurements of precipitation are required. A number of measurement devices and systems, ranging from inexpensive catch cans to sophisticated radiometers on earth satellites, are in use for the sole purpose of measuring precipitation. The following section will review the devices and systems that are in operational use or have a significant potential for operational use.

B. Measurement of Precipitation with Gages

*1. **Standard Precipitation Gages in the United States*** The National Weather Service (NWS) (1989) uses two nonrecording and two recording precipitation gage designs in the climatological network. The 8 inch gage, also known as the United States Weather Bureau (USWB) standard gage, has an 8 in (20.3 cm) orifice diameter, a funnel that directs the catch into a measuring tube, and an outer container. The inner measuring container is 20 in (50.8 cm) tall and holds exactly 2 in (5.08 cm) of precipitation. Overflow collects in the outer container. The catch is read to the nearest 0.01 in (0.25 mm) with a measuring stick. The funnel is removed when snow is expected. Equivalent water in the snow is measured by melting the snow.

The 4 inch (10.2 cm) nonrecording gage is a scale design of the 8 in (20.3 cm) gage (NWS, 1989). The inner container is made of clear plastic with a calibrated scale on the side. This gage is mounted on a post or rod. The United States Forest Service has a network of nonrecording gages with a 7.56 inch (19.2 cm) orifice. The Forester gage has a funnel and measuring container that is 5 inches (12.7 cm) tall, which provides a capacity of 0.5 inches (1.27 cm). Specifications for the federal nonrecording gages are in Snow and Harley (1988). Several nonrecording gages of various designs are available from commercial sources.

Two types of weighing gages are used by the NWS (NWS, 1989). These gages record the rate and amount of precipitation. The collection portion of the gage consists of an orifice with an 8 inch (20.3 cm) diameter and a funnel that directs the catch into a collector mounted on a weighing mechanism (NWS, 1989). The Belfort (Fischer and Porter) recording gage converts the weighed precipitation to a punched tape output. The Universal recording gage converts the weighed precipitation to a strip chart, with modifications, to direct current voltage for telemetered operation. The funnels are removed when snow is expected and an antifreeze solution is used in the collector in the winter. A thin film of oil is used in the collector in the summer to prevent evaporation when the gage is not serviced every day.

Descriptions of standard gages for other nations can be found in WMO (1971) and Sevruk and Klemm (1989).

*2. **Tipping Bucket Recording Precipitation Gages*** The tipping bucket mechanism is an event counter that records the time of occurrence of each tip of a bucket, typically 0.01 inches (0.254 mm). The output may be on punch tape or transmitted as an electronic signal to a data logger for automatic weather station operation. With each tip of a bucket, a magnet activates a reed switch. By noting the time of tip, the rainfall intensity may be determined. The tipping bucket cannot resolve very low rates or determine the exact begining and ending times of the storm. The tipping bucket mechanism will result in an undercatch, especially at high rainfall rates. For a commercial tipping bucket mechanism, Williams and Erdman (1987) calibrated the gage under laboratory conditions with rainfall intensities to 594 mm/hr. Each tip represented a depth increment of 0.266 mm of rainfall. Intensities to 260 mm/hr were within 5%, but a systematic error was introduced at greater intensities. Water splashing out of the buckets was observed at intensities over 370 mm/hr, while jetting prevented the bucket from tipping at intensities greater than 594 mm/hr. The error was as high as 25% at the higher rainfall intensities. Under laboratory conditions, Costello and Williams (1991) investigated the negative bias of rainfall depth measurement at high rainfall intensities for a commercially available tipping bucket rain gage. They suggest that tipping bucket rain gages be calibrated over a wide range of intensity in order to develop a calibration curve to remove the effects of the undercatch.

3. ***Orifice Diameter for Rain Gages*** The diameter of the orifice is an important variable in the design of rain gages. This is especially true when a large number of gages are needed to create a dense network for research or monitoring purposes. The cost-effective solution for acquisition of these gages is to construct them of readily available materials. The economics of constructed materials ordinarily mandates that smaller is better, but a lower limit of orifice diameter exists for rain gages if they are to function reliably and accurately. The lower limits are determined by susceptibility to evaporation and splash losses, ability to catch hail (Kalma et al., 1969), storage capacity, and measurement sensitivity.

Small orifice gages have several different designs, as discussed in the introduction to section IV of this chapter. Small orifice gages may consist of a small cylinder of virtually any material. Some types will have funnels and measurement containers, but most do not. Some small, transparent orifice gages have a calibrated measuring scale on the side. Other gages require the observer to pour the collected rainfall into a separate container for direct measurement with a calibrated scale or for indirect measurement with a dip stick (Snow and Harley, 1988).

Huff (1955) compared four small orifice gages with a NWS standard 8-inch (20.3 cm) gage. His conclusion was that, except for light rains, the average rainfall variation in shower-type rainfall in Illinois exceeded the variations in the gage catches. The two smaller gages caught more rainfall in storms up to 5 mm than the larger two gages. Kalma et al. (1969) noted that gages with a 2.92 cm sharp-edge brass orifice caught more rainfall, especially in light rainfall events. Their gages were equipped with a funnel and a glass tube as an inner measuring container. The catch magnification ratio was 3 (compared to 10 for the NWS standard gage) and the accuracy was within 0.2 mm.

Ursic and Thames (1958) compared a network of number 10 "tin" cans (15 cm diameter, 17.8 cm high) with the NWS standard gage. They inserted a sheet metal funnel 3.81 cm below the orifice rim, but did not insert a separate measuring container. Differences in catch were within the natural variability of rainfall. Over two-thirds of the more than 500 individual measurements were within 0.25 mm of the standard deviation.

Roper (1975) examined plastic funnels 12.7 cm diameter with a sharpened rim 2.4 cm high. Rainfall catch was within 4% of the catch of the United Kingdom Meteorological Office standard gage for 95% of the rainfall occurrences. Most differences resulted from a few of the storm events, where the plastic gages caught more rain due to decreased wind effect. The overall comparison showed a 7% increase in catch for the plastic gage. Kalma et al. (1969) also noted that the small orifice gages caught more rain than the larger, standard gages. Roper (1975) noted that the plastic gages collected dew and fog, which was not recorded by the standard gage. His explanation was that both surfaces of the plastic gage were exposed to the ambient atmosphere, which resulted in a faster cooling rate than the comparable standard gage. A decreased aerodynamic effect may have been important in collection of fog droplets.

The catch of small-orifice gages with diameters as small as 2 cm is comparable with the catch of larger, standard gages for rainfall. Some measurement accuracy is lost, depending on the catch magnification ratio (Snow and Harley, 1988). Evaporation of the catch is a problem, especially in tube or wedge gages where the collected water exposed to the atmosphere (e.g., no funnel or inner measurement container is present). Wetting losses are lower for direct reading gages because the precipitation does not have to be poured into a measuring container; however, if the rainfall approaches the capacity of these gages, splash-out will occur. The decreased cross-section to the wind for the small orifice gages results in slightly more catch than in the standard gage.

4. Measurement of Precipitation on Sloping Terrain Precipitation events are normally accompanied by wind, especially with convective storms. Consequently, the precipitation particles do not fall vertically. Angles of 40 to 60 degrees from the vertical are associated with winds of only 10 m/s (Sharon, 1980). When precipitation with wind occurs over sloping terrain, the rainfall that is actually incident on the terrain deviates widely from measurements made in gages with horizontal orifices (Sharon, 1980). The measurement of precipitation may be required for water balance or chemical analysis purposes at a variety of sites that are not amenable to measurements with standard gages with ideal exposures. Rainfall-elevation distributions that are based on catches from standard gages may underestimate rainfall by 10 to 50%, depending on exposure (Sevruk, 1974b).

Two approaches to solve this problem are described in the literature. Either the gage orifice or the entire gage may be inclined such that the orifice is parallel to the ground surface. Hayes (1944) installed a pit gage (also known as a ground level gage), for rainfall measurement that was protected from splash by excelsior mats, on a mountain slope in northern Idaho. A galvanized sheet iron extension, tailored for the degree of the slope, was fitted to the gage orifice of a United States Forest Service standard gage such that the extended orifice was flush with ground level. The design, where the orifice extension is flush with or parallel to the ground surface, is frequently referred to as a stereo gage. Paired gages were established at the same elevation on both the northern and southern exposures of the ridge. Unshielded standard gages were also used for comparison. Comparison of the catch showed little difference between pit gages on either side of the ridge line and about 20% less for the standard gage on the windward side of the ridge. Sevruk (1974b) arrived at much the same conclusions from comparisons of gages at ground level with an inclined orifice, standard gages (Mougin type) with Nipher shields at 3 m above ground, and stereo gages also at 3 m above ground. Only three of the 22 horizontal shielded gages on a mountain in Switzerland yielded a rain catch comparable with ground-level gages. The elevated stereo gage improved the catch considerably over the elevated shielded gage, but still resulted in about a ten percent undercatch. In a related study, Sevruk (1974a) compared the means of gage catches for the three types of gages in three networks. The number of gages total was 56 for the elevated gages and 59 ground level gages. With the seasonal means of the elevated, shielded standard gage as the reference, the elevated stereo gages caught about 107% and the ground level gages caught about 115%. The coefficients of variation were 16 to 19% for the shielded standard gages, 5 to 8% for the elevated stereo gages, and 4% for the ground level gages. Sevruk also commented that isohyetal maps based on the ground level gages were simpler in pattern, as a result of less wind influence.

For snow measurement in northern Idaho (where a pit gage would not be practical), Helmers (1954) compared the performance of different configurations of Alter-shielded gages with a standard, unshielded gage on an exposed, windy slope. The orifice heights were at 10 ft (3 m). The gage with both the Alter shield and an inclined orifice had a catch of 239% of the unshielded standard gage. The shielded gage with an inclined orifice had a catch of 206% of the control and the standard shielded gage had a catch of 151% of the control.

For large elevated gages, the entire gage may be inclined such that the orifice is parallel to the slope. The catch of tilted gages is then converted to a horizontal standard by multiplying the secant of the angle between the plane of the orifice by the horizontal. Hayes and Kittredge (1949) compared elevated standard (NWS) gages at 0.66 m with and without shields with a tilted, standard gage and a ground level stereo gage. The tilted gages caught more rain than the horizontal gages, whether shielded (10% more) or not (20% more). The major conclusion of this work was that standard, elevated gages, whether shielded or not, are unsuitable for measurement of rainfall.

Horizontal orifice gages for rainfall measurement in mountainous terrain have a purpose, but their purpose should not be confused with the measurement of effective hydrological rainfall incident upon the ground (Sharon, 1980). Sharon states that a determination of the orographic effect on rainfall can only be separated from smaller scale effects by using uniformly installed gages with horizontal orifices. Gages with horizontal orifices measure "meteorological point rainfall" while elevated gages with inclined orifices measure "hydrological point rainfall" (Sharon, 1980). The theoretical basis for inclined gages is explained in Poreh and Mechrez (1984), who developed a two-dimensional model for the combined effect of wind and topography on rain motion and hydrological point rainfall distribution. In any event, a ground level gage with orifice extension flush with the ground most accurately measures the rainfall that would occur at the surface in the absence of a gage.

Sharon et al. (1988) used inclined small-orifice gages (2.92 cm diameter) that had orifices flush with and parallel to the ground surface to develop a model of rainfall intensity on ridges and furrows of a cultivated field. For ridges on a 1.1 to 1.2 m spacing and 0.3 to 0.4 m high, the deviations in rainfall intensity were as high as ±60% of the horizontal gage catch in the furrow. Catch ratios from either side of the furrow were as much as 4:1. In developing the model, Sharon et al. (1988) reasoned that rain falling obliquely on sloping soil surfaces is not adequately represented by a horizontal gage, even in a pit. Their model is:

$$P^* = Po\,[1 + \tan(\alpha)\tan(\beta)\cos(z\alpha - z\beta)] \qquad (2.5)$$

where P^* is the rainfall incident on the sloping soil surface, Po is the standard rainfall measurement, α is the slope angle of the ground, β is the inclination of the rainfall, $z\alpha$ is the azimuth of the slope aspect, and $z\beta$ is the azimuth of the rainfall aspect. In this expression, Po is the average of P^* in four cardinal directions.

Panicucci developed an experimental gage for the measurement of rainfall on sloping surfaces (Sevruk, 1986a). The gage has four differently sized sectors that receive the same amount of vertical rain. Each sector receives different amounts of oblique rainfall, while the sum of all sectors equals the amount measured by a conventional gage. The prevailing direction of precipitation can be estimated from a comparison of the sector amounts. Earlier "directional" gages with multiple orifices have been investigated (e.g., Lacy, 1951), but will not be reviewed here.

In summary, if rainfall measurements are needed to determine water balance for windy, sloping terrain, a ground level gage is highly recommended. It would be best to extend the gage orifice to a flush orientation with the sloping surface and to protect the installation against splash and runoff. For snow measurements on windy, sloping surfaces, a double fence (Wyoming or Tretyakov designs) surrounding a shielded gage is recommended. The double fence should be installed with tops parallel to the ground surface, as opposed to a horizontal reference. If the precipitation measurements are needed as a climatological reference, the catch from elevated horizontal gages, shielded or not, serves as a useful index. Any change in gage exposure, height, shielding or orifice inclination will invalidate the climatological index. Different precipitation networks for different purposes are recommended, and it should be readily obvious that incorporation of measurements from non-standard gages into a network of gages for hydrological (e.g., isohyet or water balance) analyses is not a good idea.

5. Measurement of Rainfall Intensity Rainfall intensities (time distributions of rainfall) are needed for the design of runoff conveyance structures and storm sewers, investigation of soil erosion processes, development and operation of hydrologic models, development and operation of military weapons systems, and design of communications systems. The most critical needs are perhaps in communication systems design. Heavy rainfall attenuates microwave propagation, especially at frequencies above 10 GHz. The physical basis of the relationship between attenuation and rain intensity is presented in Fedi

(1981). The attenuation, in dB/km, is generally expressed as a power function of the rainfall rate in mm/h (Olsen et al., 1978; Fedi, 1981; Goldhirsh, 1983). The design of microwave path links includes allowable outage time, which may be less than 60 min/yr. For critical circuits, the permitted outage may be one or two orders of magnitude smaller (Segal, 1986). Path lengths and number of repeaters can be determined accurately only if the rainfall attenuation along the paths can be predicted (Ruthroff, 1970). Instantaneous rainfall rates are not appropriate because of spatial averaging along the transmission path, so 1-minute rainfall intensities have been accepted as the most desirable (Bodtmann and Ruthroff, 1974; Segal, 1986).

Although rapid-response rain gages have been developed, the climatological data base is limited to measurements from standard recording rain gages. Rainfall intensities can be determined for periods as short as five minutes from the strip charts, punched tapes, electronic data base, or other records from the recording rain gages (e.g., Huff, 1967). The recording precipitation gages do not provide sufficient resolution to measure intensities for 1-minute periods (Hershfield, 1972). At high intensities, the tipping bucket mechanism does not respond fast enough to changes in rainfall rate even for 5-minute periods (Hershfield, 1972). Difficulties are also present in obtaining short-period accurate rainfall intensities from weighing rain gage strip charts. Manual processing of charts is a formidable task. Bodtmann and Ruthroff (1976) name two major obstacles in the derivation of accurate rainfall intensities from weighing rain gage measurements. The desired rain rate is a derivative of the recorded depth of water as a function of time but the computation of derivatives by taking differences of measured data (especially the time-of-tip from a tipping bucket) is notoriously inaccurate. Also, the distribution of rainfall rates derived from measured data is shifted toward higher rates.

The determination of 1-minute rainfall rates may be directly calculated from measurements or from statistical relationships between the rainfall rates for one minute up to several hours. One technique is to plot the rainfall intensities for periods from 5 minutes to 24 hours, and then extrapolate to 1 minute (Hershfield, 1972). Crane (1980) stated that the highest 5-minute average rain rate usually occurs from an event that also includes the highest 1-minute average rain rate. Hershfield concluded that 0.24 of the extreme 5-minute rainfall occurs in 1 minute as an average, with extremes of 0.2 to 0.6. The 1-minute rainfall intensity may also be expressed as a power function of the 5-minute intensity, with site specific regression coefficients (Segal, 1986).

Bodtmann and Ruthroff (1976) developed a procedure that was modified by Tattelman and Knight (1988) to determine 1-minute rain rates from digitized strip charts as a result of weighing rain gages. A time series was developed by interpolation for each 0.5-minute period. A three point running mean smoother was applied; then the trend was removed from the time series. The series was then fitted to a Fourier series (Fast Fourier Transform), and the high frequency terms were truncated. A cubic spline was fitted through the points in the Fourier series to produce the final smoothing. Sadler and Busscher (1989) developed an analytical procedure to derive 1-minute rainfall rates from tipping bucket time-of-tip records from an automatic weather station. The number of tips (0.5 mm/tip) in 1-minute intervals were assembled into an array of accumulated rainfall. A cubic spline was fitted by the least squares method to the curve. The resulting cubic equation was then differentiated to produce rainfall rates. This procedure produced a fairly smooth continuous curve of rainfall intensity. The distributions of rainfall intensities for several locations may be fitted to any of several models (Tattelman and Grantham, 1985).

C. Measurement Error with Precipitation Gages

For climatological purposes, the catch of a single gage may be regarded as an index of the precipitation; the true catch is not needed. However, for hydrologic assessment and modeling, the true precipitation over the watershed is necessary. Goodison (1978) stated that the conclusions in several water balance studies in Canada were that precipitation, especially snowfall, was apparently underestimated.

Point precipitation measurements have random and systematic error (Sevruk, 1986a). Random error is due to irregularities of topography and vegetation around the gage site and exposure of the gage to prevailing winds. Random error is also a result of inadequate network density to account for the natural spatial variability of rainfall. Sevruk (1986a) considers systematic error to be the most important source of error. The major source of systematic error in point precipitation measurement is due to the deformation of the wind field by the elevated gage. For snow measurement, the undercatch may exceed 50%, even with a shielded gage. Other sources of systematic error are adhesion of water to the surfaces of the funnel and measurement container, evaporation of water that adheres to the funnel, and rainsplash.

*1. **Wind*** In 1769, Heberden observed that rain gages near ground level caught more rain than nearby gages that were elevated. In 1861, Jevons established conclusively that the decreased catch was due to the effects of greater wind speed with height (cited in Weiss and Wilson, 1957; Rodda, 1967; Biswas, 1970; Sevruk, 1982). Jevons noted that any obstacle placed in the wind causes the air to sweep around the sides and over the top with increased wind speed. This increased speed deflects rain and snow particles from their original paths. Because the purpose of the gage is to record the precipitation that would fall to the surface in the absence of the gage, the result of wind is an undercatch that is more pronounced for snow than for rain (Alter, 1937). The undercatch amount is a function of the wind, the precipitation, and the gage design and placement. Because the wind speed changes rapidly near the surface, small differences in gage orifice height produce large differences (up to 10%) in precipitation catch (Sevruk, 1990). The differences are largest for snow and light rain. Folland (1988) deals extensively with the theoretical basis for airflow over and around a precipitation gage.

Two major lines of investigation have been pursued to reduce the effect of wind on the precipitation catch of a gage. The gage may be designed for improved aerodynamics or it may be shielded. In wind tunnel tests, Robinson and Rodda (1969) showed that the height of the lift of the air stream over a gage orifice appeared to increase with increasing gage diameter and with increasing sharpness of the leading edge of the gage. The increase in wind speed over the gage may be as high as 30%. An inverted funnel or trumpet has the optimum aerodynamic shape, but in heavy precipitation the flat portion either causes extensive splashing of rain or collects snow. The basic design of the gage has not changed in the past 200 years (Biswas, 1970), although Folland (1988) showed improved aerodynamics from a gage shaped like an inverted champagne glass. A steep inner wall reduced rain splash. For the fairly large cross-sections of existing gages, the wind effect is minimized only through shielding or by placing the gage in a pit so that the gage orifice is at ground level. Experiments in the mid-19th century with pit gages and shields to reduce the wind effects documented a problem that still plagues precipitation measurement today: the same gage is not ideally suited for measurement of both rain and snow.

The only gage that does not have systematic error due to wind has an orifice level with the surface (Neff, 1977). Pit gages were first used in 1842 by Stevenson, but the limitations to pit gage use were already apparent (cited in Nipher, 1878). Nipher stated that pit gages caught more rainfall than elevated gages, but had several problems. The pit surrounding the gage was filled with leaves, drifting snow or running water (which floated the gage); the pit gage system was prone to damage, and a small error in orifice height was very significant. Pit gages are suitable for reference gage purposes but are not practical for measurement networks.

Because pit gages were not practical, Nipher developed a shield to reduce the wind effects. Although earlier shield designs were developed, the shield developed by Nipher has endured. The shield in general use in the United States was developed by Alter (1937). Both the Nipher and the Alter shields deflect wind downward and around the gage orifice. Alter stated that the goal of the shield was to "leave an unaffected, undistorted movement of air over the top of the gage, comparable to the air movement over the place if the gage were not there." No shield currently in use has completely solved the wind problem (Larson, 1971).

Wind speeds normally increase upward from the earth surface as a consequence of frictional drag from surface roughness. The wind at 100 m above the surface, for example, is slowed by the effects of frictional drag of the air below 100 m. The surface 50 to 100 m below the layer of air is characterized by turbulence and eddies. For neutral atmospheric stability, the increase of wind speed with height over an open, level, smooth surface follows a logarithmic function of elevation. With increasing roughness elements, the turbulence and eddies in the wind increase and the rate of wind speed increases while height decreases. If the surface is completely rough, such as a forest canopy, the increase in wind speed with height will follow the logarithmic function of height above the zero plane displacement height, which will be above the surface but below the tops of the roughness elements. The wind speed will be very low at a height below the zero plane displacement height (Linsley et al., 1975; Rosenberg et al., 1983). Consequently, an ideal location for a precipitation gage is in a protected clearing in a forest; however, these sites are rare in reality. Gage sites based on other selection criteria and/or in the absence of ideal locations are either in open areas where the wind speeds increase dramatically above the surface or in areas where each obstruction or individual roughness element creates localized eddies, turbulence, and changes in speed in the wind flow. The gage itself is an obstruction.

Thus, two scales of eddies and turbulence in the wind affect precipitation catch in a gage. Gage turbulence is reduced through the use of shields or pits. Site turbulence up to the scale of hundreds of meters is reduced through judicious selection of gage location (Larson, 1971). A larger scale of site turbulence, such as from mountains or other topographic features, has the net effect of determining the microclimatology of the area (Weiss and Wilson, 1957). Site turbulence is reduced with the use of snow fences, turf walls, or large, elaborate site shields surrounding the gage.

In their exemplary review, Weiss and Wilson (1957) surveyed the performance of a large number of precipitation gage shields in rainfall events. In comparisons of the catch from shielded and unshielded gages, wind speed and rain intensity were major factors. With no wind, the catch was the same for both gages. With increasing wind and/or with decreasing rain intensity (i.e., smaller rain drops), the ratio of unshielded to shielded catch decreased from near 1.00 at low wind speed to values as low as 0.5 for light rain and 6 m/s wind speed. Koschmieder (1934) presented a comparison of the catches of unshielded gages at 1.1 m and pit gages. A fairly linear decrease in percentage error from 0 at no wind to 70% at 20 m/s was observed. At the other extreme, Long (1947) compared snowfall catch in shielded and unshielded snow gages in Wisconsin and observed less than 1% difference for 25 storms. The lack of commonality of controls and measurements of variables in these individual experiments precludes specific conclusions. Allerup and Madsen (1980) developed correction values for the aerodynamic effect of rainfall on unshielded Hellman gages. Their correction values show that the percentage error decreases with increasing rainfall intensity and increases with increasing wind speed. Correction values in excess of 100% occur with very light rainfall intensities and high wind. Winds of 20 m/s produce catch errors in excess of 1 mm/h at rainfall intensities of about 3 to 9 mm/h.

Snow catch in unshielded and shielded gages was also compared by Weiss and Wilson (1957). The definitive wind tunnel tests with sawdust by Warnick (1953) indicate that an unshielded gage may catch less than 10% of light to medium snow with winds in excess of 10 m/s. With winds of 4 to 6 m/s, the percentage catch improved to about 75%. Comparisons in the literature of snow catch in unshielded and shielded gages confirm that an unshielded gage is not suitable for measurement of snow. In field tests with snow catch (melted) compared to water equivalent of snow on the ground, the addition of Alter shields to the gage increased the catch to 75 % or more in the presence of wind speeds to 12 m/s. A ratio of snow catch in gages with a rigid shield (Nipher) to gages with a flexible shield (Alter) showed an exponential decrease from 1.0 at no wind to 0.5 at winds near 20 m/s. At high wind speeds, free swinging Alter shields were observed to loop, which created more turbulence than they removed (Rawls et al., 1975). Gages with a rigid Alter shield and a flexible Alter shield had essentially the same catch, so both variations are currently in use.

Goodison (1978) compared two weighing gages, Universal (Belfort) and Fischer and Porter, with Alter shields and a Canadian MSC Nipher-shielded gage, with true catch measured from a snow board at four sites in Ontario, Canada. Two sites were exposed to the wind and two sites were protected. For the Alter-shielded Fischer and Porter gage, the gage catch ratios decreased exponentially from near 1.0 at very low wind speeds to about 0.3 at wind speeds of 7 m/s at gage height. The Alter-shielded Universal gage performed somewhat better, with a gage catch ratio of about 0.4 at 7 m/s wind speed; however, the Canadian MSC Nipher-shielded gage had an entirely different pattern. It overcaught precipitation (to ratios of 1.1) at wind speeds to 4 m/s. The catch ratio decreased to about 0.7 at 7 m/s wind speeds. It should be noted that the Nipher gage was not a recording gage and had to be attended daily. For unshielded gages, the catch ratios were as low as 0.1 at 7 m/s for the Fischer and Porter gage. Goodison concluded that the undercatch of shielded gages and the lack of wind measurements at most gage locations makes properly siting gages imperative to minimize the adverse effect of wind.

Weiss and Wilson (1957) concluded that

> It has become increasingly evident that attempts to secure 'true' catch by adjoining a wind shield to the gage are vain. The importance of adequate site and environment protection characteristics is now almost universally recognized.

Shields reduce the wind effects created by the gage, but not by the site. An ideal shield minimizes snow bridging across the gage orifice and will not contribute to splash error during rain events.

Shields also are in widespread use to reduce the site effects of wind. Larson (1971) states that it is essential for all gage locations in a network to have equal protection from the wind. This is necessary if the gage catches are to be compared, used for isohyetal analyses, or used to represent the total precipitation for a watershed or other area. Where possible, natural shielding should produce uniform site protection for all gages (Larson 1971), but natural site shielding frequently is not possible, due to lack of vegetation or inaccessibility of ideal sites. Equal protection from the wind in all directions is even less possible. Brown and Peck (1962) recommended a site protection classification system, ranging from over-protected (gage too close to tall objects) through well-protected to very windy. The site protection category is important to those who use precipitation records from a network.

Reed (1915) states that low bushes and fences in the vicinity of a gage are beneficial, but the negative effect of a single bush was pointed out by Court (1960). In this study, a gage to the lee of a single small bush on a windswept ridge had a 50% increase in rain catch over the catch of an identical gage about 3 m distant. Alter (1937) found that board fences, railings or guards of almost any kind placed a few feet from the gage and level with the gage orifice were effective. Snow fences (50% density, 1.2 m high) have been used effectively to reduce the wind speed and site turbulence at gage sites (Larson 1971). Larson states that a 50% density snow fence would produce a maximum reduction in wind speed of 30 to 70% about 6 m downwind. The decrease in wind behind a snow fence results in increased snow drifting, so the gage orifice should be above the height (about 1 m) where drifting would carry snow into the orifice.

Turf walls are used around gages in the United Kingdom to reduce site effects. As described in Snow and Harley (1988), circular walls of turf centered around the gage are constructed with vertical inner walls and gently sloping outer walls. The height of the turf wall is about the same as the gage (1 ft (0.3 m) in the United Kingdom) and the diameter is three to five times the orifice height. Snow and Harley (1988) mention an alternative as a circular patch of hedge, with the gage in the central clearing. An experimental station in the Soviet Union contains a 2.5 ha area of deciduous shrubs surrounding the gage. The tops of the shrubs are maintained at the height of the gage orifice (Sevruk, 1986b). Gloyne (1955) showed that a shelterbelt of 50% density reduced the wind to 25% of its approach speed at distances to 10 times the shelterbelt height. A 30% dense shelterbelt was not effective, while the downwind effects of the 100% dense shelterbelt were limited to a few multiples of the height.

Experimental gages for snow measurement combine the shields for the site and the gage into the same installation. The Wyoming shield gage consists of two concentric shields of snow fence (50% density) around an NWS standard gage (Hanson, 1989). The distances from the gage to the inner shield top and the outer shield top are 3.05 and 6.1 m, respectively. The top of the outer fence is 3 m off the ground and the top of the inner fence is 2.4 m off the ground. The fences are canted inward, with a 30 degree angle for the outer fence and 45 degrees for the inner fence. The orifice height of the gage is at 2.4 m (Clagett, 1988). The Universal (Belfort) type gages did not have a funnel, and antifreeze and oil were used in the measuring container. With the dual-gage system as a reference for true catch, the ratios of the catches of the Wyoming gage, an Alter-shielded gage, and an unshielded gage were compared for five sites during a several year period in Idaho. For rain, the ratios were 1.02, 0.96, and 0.92, respectively. For snow, the ratios were 0.93, 0.80, and 0.60, respectively. Hanson (1989) concluded that the Wyoming gage provides reliable measurements under most climatic conditions. Clagett (1988) compared the catch of the Wyoming gage with the NWS unshielded standard gage for snowfall in Alaska. He concluded that the Wyoming gage caught twice as much snow as the NWS gage over the entire winter. When compared against *in situ* methods, the Wyoming gage caught around 80 to 90% of the total. Clagett's comparisons were similar to those of Rechard and Larson (1971), who reported that under windy conditions, an unshielded gage catches about one-third of the snowfall, a gage with an Alter shield catches about two-thirds of the total, and a gage with both site and gage shields catches about the same as the actual snowfall.

Hamon (1973) developed a method for calculating the "true" snowfall from pairs of Universal (Belfort) standard gages, one with an Alter shield and one unshielded. His dual-gage method is based on the following relationship:

$$\ln (U/A) = 1.7 \ln (U/S) \tag{2.6}$$

where U is the unshielded catch, S is the shielded catch, and A is the actual snow water equivalent. The value of 1.7 mm is empirically determined, but should be independent of wind and precipitation characteristics. Hamon concluded that the dual-gage method would provide inaccuracies of less than 10% for snow and 5% for rain.

Virtually all of the precipitation gages in the climatological network in the United States are unshielded, especially in areas where winter snows contribute less than 20% of the total precipitation. Rodda (1967) and Robinson and Rodda (1969) compared the catch from pit gages with the British Meteorological Office (BMO) standard height of 30 cm and the NWS standard height of 0.76 m. They concluded that the BMO gage had an undercatch of 3 to 7% and the NWS gage had an undercatch of 10% when compared to the pit gage. In more recent work, Rodda and Smith (1986) presented a linear relationship between mean wind and undercatch for the BMO gage at 30 cm. At 2 m/s, the percentage difference was 5%, while at 6 m/s wind speed, the percentage difference was 19%.

The effects of over-protection on gage measurement accuracy should be kept in mind. An overprotected gage may either result in overcatch or undercatch of precipitation. Undercatch occurs when precipitation that would otherwise enter the gage is intercepted by tree foliage or other site obstructions. Overcatch occurs when the intercepted precipitation, primarily snow, is blown from the interception site into the gage.

2. ***Wetting*** Rainfall that adheres to the funnel surfaces of a gage may evaporate after the end of the rainfall event. This is known as wetting loss and is unavoidable. The wetting loss varies from storm to storm and on the evaporative conditions after the rainfall event. Wetting loss also occurs due to adhesion of the water to the sides and bottom of the measuring container after it is emptied. If rain falls before this water is evaporated, the next rainfall event may be slightly overestimated (Kalma et al., 1969). Wetting loss increases as the area of gage material increases; for example, gages with a high magnification of catch have a larger wetting loss than gages that have a direct reading scale. Gages that are constructed of polished metal will have lower wetting losses than gages constructed of galvanized or painted metal (Sevruk, 1990).

Allerup and Madsen (1980) determined the wetting loss of the container and funnel for the Hellmann and Snowdon gages. Wetting of the container was about 0.1 mm and wetting of the funnel was 0.1 mm per event for both gages. The total wetting loss per precipitation day, including snow, for a Snowdon rain gage in Denmark ranged from 0.09 mm in December and January to 0.24 mm in June and July. This wetting loss amounted to about 4% of the annual precipitation. Sandsborg (1972) estimated evaporation and wetting loss to be about 1.5% for Sweden. Sevruk (1973) measured wetting loss as 2.6% of catch in ground level storage gages in Switzerland; however, wetting loss could approach 10%, depending on the frequency, type, and amount of precipitation; the drying time of the gage components; and frequency of emptying the container (Sevruk, 1982).

*3. **Evaporation*** Evaporation of the gage catch results in rainfall underestimation. Evaporation is normally insignificant but may be a serious problem for light rainfall collected in a warm gage (Snow and Harley, 1988). Sevruk (1982) states that the evaporation loss varies between 0 and 4% of the catch. Evaporation of rainfall collected in the measuring container is normally eliminated by the design of the funnel and the measuring container itself. The funnel should fit into the measuring container tightly so that container ventilation is minimal. If the funnel is removed for solid precipitation measurement, evaporation of rainfall or melted solid precipitation may be significant; however, if the measurement container is emptied daily, evaporation is not a factor except for light rainfall amounts in a very warm gage. For remote or unattended locations with recording gages, evaporation is significant without the use of an evaporation suppressing film of light oil or glycerin on the water surface (Snow and Harley, 1988). Alter (1910) described what he termed as the first general attempt to obtain seasonal precipitation in isolated regions. He used 0.20 in (5.1 mm) of olive oil to reduce evaporation in a gage without an inner tube. Comparisons of catch with the measurements from a tipping bucket gage were within 3%.

Hamilton and Andrews (1953) examined evaporation from NWS 8-inch (20.3 cm) gages without measuring tubes in southern California. Evaporation of water placed in a gage with a funnel ranged from 0.0 to 0.15 mm/day through the rainy season. Evaporation in a gage without a funnel ranged from 1.27 to 4.83 mm/day. Evaporation in a gage without a funnel, but with 0.51 mm light oil on the water surface, was equivalent to evaporation from the gage with the funnel. They noted that light oil also evaporates, so a thin film of oil would only be effective for a week or so. In order to prevent all evaporation losses under any climatic condition, 3.81 mm oil was required. Snow and Harley (1988) recommend two precautions with the use of oil or glycerin to suppress evaporation: the oil that is used should not form an emulsion with the water and it should be easily cleaned from the measuring stick.

A separate problem occurs with the use of heated tipping-bucket gages for measurement of winter precipitation (Hanson et al., 1983). The funnel and the tipping bucket mechanism are heated with either electricity or propane gas for operation when ambient temperature is below freezing. Side-by-side comparison of a heated gage with an unheated gage showed that the heated gage recorded about 30% less than the unheated gage at two sites in Idaho, and 50% less at three sites in Oregon. These differences in catch are attributed to thermal turbulence transporting snow crystals away from the gage and from evaporation of the catch, especially at low precipitation rates.

Evaporation may be reduced or eliminated in the design, maintenance, and operation of the gage. Specific recommendations are:

(a) The funnel should be as smooth as possible so that water will run freely into the measurement container (Sandsborg, 1972), and the funnel should not be painted (Snow and Harley, 1988).
(b) The exterior of the gage should be painted white to reduce the heat load (Sandsborg, 1972; Snow and Harley, 1988).
(c) The orifice edge should be sharp and the funnel should fit snugly into the measurement container.
(d) Light oil prevents evaporation of light rainfall in warm gages (Hamilton and Andrews, 1953; Snow and Harley, 1988). Light oil is routinely used in recording gages in the cooperative station network in the United States, but not in non-recording gages (NWS, 1989).
(e) With heated gages, the collector temperature should be maintained as low as possible (Hanson et al., 1983).
(f) The funnel should not be removed earlier than necessary for the collection of winter precipitation.

*4. **Condensation*** Condensation on a standard gage is a source of error only in unusually heavy dew, because dew amounts rarely exceed the amount of water required to wet the funnel (Rasmussen and Halgreen, 1978). Some gage designs are more prone to form and catch condensation than others. Small plastic gages that do not have an inner measuring container lose heat by radiation more rapidly than a standard gage, so they may collect 1 or 2 mm of condensation in a heavy dew.

*5. **Rainsplash*** Rainsplash into or from the gage may be a source of measurement error. Splash-out from the gage can be reduced, if not to zero, to a negligible minimum (1–2%, according to Sevruk, 1982) by proper design of the funnel and the rim (Ashmore, 1934). Snow and Harley (1988) report that splash-out can be minimized through the use of a funnel located "well down in the receiver," with a slope of at least 45 degrees to the vertical inner wall of the collector. Splash into the gage orifice may be significant, however. Rainsplash is produced by the impact of rain drops on the underlying surface, shields and other component parts of the gage and gage supports. Splash is greatest for large rain drops, but is not correlated with rainfall intensity.

In their investigation of airsplash of soil particles from impacting raindrops, Moss and Green (1983) described and photographically documented the behavior of water drop impact. When a large drop impacts a smooth, rigid surface, the water forms a radially expanding surface sheet of water. The velocities in the expanding sheet may be as high as six to ten times the impact velocity, which is about 9.2 m/s (Pruppacher and Beard, 1970; Epema and Riezebos, 1984; Moss and Green, 1987). An increase in the surface roughness, the kinetic energy of the raindrop, or the presence of a water film on the surface produces splashing. The structure that produces the splash droplets is a corona (shaped like a crown) on the upper surface of the expanding sheet of water. Jetting in the corona produces high-velocity splash droplets. If the water film is less than a few mm thick, soil particles can be detached and transported up to 1 m horizontally and 0.6 m vertically (Moss and Green, 1987). For a deeper film of water, splash droplets can be ejected to 0.5 m.

If the surface is not rigid, as with a dry soil, the air-filled voids of the surface reduce the impact pressures by two orders of magnitude (Nearing et al., 1987). Ashmore (1934) noted that splashing does not occur on bare soil until the surface has been moistened. In observations of rainfall splash on an inclined board above various natural surfaces, Ashmore noted vertical splash movement to 76-cm for all surfaces, with a maximum height of 1 m. Splash from a "well-kept lawn" reached 38 cm. To determine splash heights for natural surfaces, Shaw (1987, 1991) used a 5-cm-diameter central cylinder with a sheet of chromatography paper. A fluorescent fiber-reactive dye was placed in an annular ring at the base of the cylinder. Splash from the annular ring carried the dye to the chromatography paper, where it was bound. A liquid surface was used as an index for other materials, although results from only one material, glass fiber cloth, were presented. The relative amount of splash above a given height during an individual rain had a Weibull (extreme-value) probability distribution. Water drop impact with a water surface produced a maximum height of 0.8 m where one-half of the paper was covered by dye.

Splash height is correlated with the characteristics of the rainfall in addition to the surface characteristics. In his theoretical analysis of rain splash, Walklate (1989) showed that the maximum splash height is proportional to the initial velocity and diameter of the splash droplet that results from impact of a rain drop. A 1 mm diameter droplet with an initial velocity near 10 m/s can reach a height over 1 m. Walklate et al. (1989) tested splash heights from five targets subjected to large water drops near terminal velocity. Water films of two depths, plant leaves, barley straw, and barley straw with accumulated water film were used. For their experimental conditions, the maximum splash height for accumulating water films was about 0.6 m and for plant materials was about 0.4 m.

Rainsplash from the surface may be eliminated by placing the gage orifice above the level of the highest rainsplash. A tradeoff will exist between increasing the height of the orifice to avoid rainsplash and decreasing the height of the orifice to reduce wind effects. Rainsplash also may be eliminated with the use of baffles or louvers surrounding the gage and with selection of surface materials that reduce splash from an impacting rain drop. Walklate et al. (1989) prevented rainsplash in their experiments by surrounding the plant or soil target with a tray of expanded polystyrene beads. These beads allow very little splashing because the water drops pass between the individual beads. For a pit gage, Bleasdale

(1959) placed slanted louvers around the gage to deflect rain drops and rainsplash away from the gage orifice. The louvers deflected the rain into a gravel bed. The dimensions of the louver system surrounding the pit gage vary in the literature from about 0.5 m to about 1.0 m square. Rasmussen and Halgreen (1978) used a circular louver system with 0.7 m diameter, surrounded by a "tight carpet of closely cropped grass." To minimize splash-in for small, post-mounted gages, Snow and Harley (1988) recommend that the gage orifice be located well above the top of the mounting post, which should have a beveled top that slopes away from the gage. Snow and Harley also recommend that a rain gage be sited over a resilient surface, such as grass to minimize the generation of spray (splash droplets). The sharply beveled edge on the outside of the gage orifice is designed to minimize splash into the gage.

Special caution should be taken in using evaporation pans as a precipitation gage. Errors in evaporation measurement may occur on days with rainfall. The standard procedure is to subtract the gaged rainfall amount from the measured water level in the calculation of daily evaporation (NWS, 1989). It should be noted that the Class A evaporation pan, for example, is a poor rain gage for several reasons. The pan is a metal cylinder 1.21 m diameter and 25.5 cm deep, with a water level within 5 cm from the rim. The pan is placed on a wooden platform so that the base of the pan is 10 to 15 cm above the surrounding ground surface (Doorenbos, 1976; NWS, 1989). Although the pan is set fairly close to the soil surface, it presents a large cross-section to the wind. As such, it undercatches rainfall as a function of the wind speed. Splash into the pan occurs from raindrop impact on the wooden platform support of the pan and from the surface. The amount of overcatch is a function of the raindrop size distribution, rainfall intensity, and the nature of the surface. Splash out of the pan also occurs, especially in the presence of large raindrops and wind.

Nordenson and Baker (1962) screened a Class A pan with a sheet metal cylinder, "surmounted by a 15 cm ring." The cylinder, which extended 21.3 cm above the rim of the pan, was for the purpose of reducing radiation, but it also reduced splash out of the pan. For days without rain, the ratio of the pan with the splash shield to the standard exposure was 0.89. For 35 days with rainfall over 1.27 cm, the ratio decreased to 0.86. The increase in undercatch was attributed to changes in the splash characteristics, but could also have been due to increased wind influence. Nordenson and Baker did not draw any conclusions from the limited data set.

The water level in the pan also affects the quantity of splash-out from the pan. Lower water levels reduce the splash-out and also reduce the evaporation. In Nordenson and Baker (1962), the effect of a water level at 11.4 cm below the rim as opposed to the 5 cm standard (NWS, 1989) was investigated. The ratio of low water pan to standard pan evaporation was 0.91. Doorenbos (1976) notes that strong winds can blow water from the pan when the water level increases from heavy rain. Doorenbos also parenthetically states that "it is often noted that after rain the arithmetic calculation of evaporation . . . gives too high evaporation values because of splashing of rain outside the pan." Briefly summarized, the Class A pan is not a satisfactory rain gage, but it is used routinely as a rain gage in the calculation of evaporation. If rainfall is underestimated (i.e., more rain in the rain gage than in the pan), evaporation is underestimated. The magnitude of the error appears to be within 10% but could easily be much higher on days with heavy rain and high wind.

Small orifice gages are used for the measurement of "precipitation" from pressurized irrigation systems for the determination of efficiency, precipitation rate, radius of throw, and uniformity of water application (ASAE, 1990). Typically, a large number of small-orifice, non-standard gages are placed within an area to collect water during an irrigation event. Wind effects are minimized by both the small cross-section and the near-surface placement of the gages; however, splash should be anticipated as a significant source of error. Splash-in occurs if the gage orifices are placed close to a surface, such as bare ground, that produces splashing. Splash-out occurs when the gage orifice is large in relation to the depth of the gage container. An appropriate design for small-orifice gages in irrigation evaluation would be a tube gage mounted on a slender rod, with the orifice between 0.5 m and 0.9 m above the surface. If the irrigation nozzles are at or near ground level, as with turf systems, the gage orifices may be lower since the turf provides protection against splash-in.

6. Snow Plugging and Capping Wet snow occasionally sticks to the inside of the gage orifice, especially during windy conditions. The extreme action for preventing this condition is to cap the orifice, such that additional snow will not enter the gage (e.g., Schaefer and Shafer, 1983). The snow plug may slide partially into the antifreeze solution or may not fall into the bucket for measurement until a later time. If only one gage is in operation, snow plugging and capping will result in significant errors in measurements of amount and in timing. Quality control and application of a correction factor would be difficult (Goodison et al., 1989).

7. Correction of Precipitation Measurements If precipitation measurements are to be used for hydrological studies, such as watershed or regional water balance determination, the systematic errors in elevated can-type gages must be corrected (Larson and Peck, 1974; Sevruk, 1982; Allerup and Madsen, 1980; Legates and Willmott, 1990). The systematic error of 3 to 30% in precipitation measurement could lead to significant underestimation of water into the hydrologic system. In general, the systematic error is the least in tropical, heavy rainfall regimes and greatest in precipitation regimes that have a large proportion of snow or light precipitation with wind (Sevruk, 1982).

Gage measurements with different site exposures or shields should be adjusted to the same exposure and shield. Goodison (1978) interpreted his results with different snow gages as a need for unique correctional procedures for each type of gage. A record of wind speed at gage height is necessary for development and application of gage correction factors. Goodison states: "areal analysis of precipitation gage data from sites with varying exposures can provide very misleading results unless corrections for variations in gage catch are made first."

In spite of the knowledge of systematic error in precipitation measurements, correction factors have not been routinely applied to point precipitation measurements in the United States. The situation is exacerbated in the case of precipitation measurements from gages of different design, shielding, and orifice height.

Groisman et al. (1991) describe the history of corrections for bias in precipitation measurements in the Soviet Union. Sevruk (1982, 1986a) reviewed the history of the development of correction factors for precipitation gages. The first physically based correction procedures were developed in the mid-1960s in the Soviet Union as a result of a change in the national standard gage from the Nipher to the Tretyakov (also spelled Tretiyakov in Groisman et al., 1991). Initially, the use of a single reduction coefficient to adjust the old data series was attempted and rejected. In the development of correction coefficients and procedures, a standard had to be developed. An elevated gage proved unsuitable as a reference in a WMO Commission for Instruments and Observation Methods and International Association of Hydrological Sciences program in the 1950s (Sevruk, 1986a). The pit gage is now accepted as the reference gage for development of correction factors for rainfall.

The octagonal, vertical double-fence shield has been designated as the international reference for snow measurement by the WMO (Goodison et al., 1989; Sevruk, 1989). This measuring system consists of a Tretyakov gage, at a height of 3 m located at the center of two octagonal fences of 50% density. The outer fence is 12 m from the gage center and is 3.5 m high. The inner fence is 4 m from the gage center and is 3 m high. The bottom of the outer and inner fences are at 2 m and 1.5 m, respectively, above the ground.

Correction procedures are described in Sevruk (1982; 1986a), Allerup and Madsen (1986), Goodison et al. (1989), and Legates and Willmott (1990). If precipitation measurements are corrected for systematic error, Sevruk (1982) recommends that both the original and the corrected data sets be published. The correction methodology also should be published or referenced from a readily available source.

D. Direct or *in Situ* Measurements of Snow

1. Meteorological and Hydrological Snow Measurement Measurement of snow is subject to a wide variety of random and systematic errors. Simply stated, it is difficult to catch small snow particles with a gage, due to the interactions of the gage with wind. The common occurrence of blowing and drifting snow exacerbates the measurement problem. The origin of the snow that finally accumulates on the surface cannot be determined (Peck, 1972), yet it is the accumulated snow that contributes to the hydrology of that location. Consequently, the distinction between hydrological and meteorological snowfall is even more important than it is for rainfall (Sharon, 1980). Meteorological snow is the snow that falls through a horizontal plane above the boundary layer of the earth's surface. Properly exposed and shielded precipitation gages, with no wind influence, measure meteorological snow. Hydrological snow is the snow that accumulates at a specific location on the earth's surface. The snow that accumulates on the surface may be more or less than the meteorological snow depth, depending on the redistribution of the snow as it nears or reaches the surface. Meteorological snow depths or water equivalents are used extensively for index purposes. For example, the point measurements of snow accumulation in mountainous areas are correlated with stream flow from melting snowpacks for water supply forecasting. The location of the point measurement may or may not contribute to the runoff at the stream gaging location. Snow measurement for determination of the climatology of a point is also an index use. If the hydrology of the location is to be determined, then the hydrological snow should be either measured or estimated from the measured meteorological snow.

Meteorological and hydrological snow depths (or water equivalents) will be the same if the area is large enough. The scale of the area depends on the scale of the roughness elements that determine the redistribution of snow. Conceivably, the scale could range from a few meters, which include tillage ridges and furrows or differences in vegetation, to hundreds of meters in mountainous terrain. For small-scale roughness elements, random or stratified point sampling (e.g., snow courses) in the area provides an accurate estimation of the snow depth. For large scale roughness elements that typify rangeland and forest watersheds, point sampling on snow courses may be unreliable for representing the total water equivalent of the areal snowpack (Rawls et al., 1980), although the snow course could be suitable for use as an index of runoff.

Measurement systems are in widespread use for both meteorological and hydrological measurement of snow. Meteorological measurements are normally made at specific points, with catch-type gages, snow cores, snow pillows, radioactive attenuation systems, and measurement sticks. Areal measurement systems in widespread use for hydrological snow measurement include photogrammetric techniques, passive microwave from earth satellites, and airborne terrestrial gamma.

2. ***Core and Stick Snow Measurements*** Point measurements of snow are routinely made with core extractions and direct measurements of depth with a stick. The NWS instructions for observers in the cooperative station network provide a standard for these measurements (NWS, 1989). Core samples are recommended when it is obvious the gage did not catch a representative snow amount, when snow overflowed the gage, and when snow bridging or capping of the gage was observed. For these cases, the overflow can of the snow gage is forced down through the snow to the depth of the new snowfall. A piece of metal or thin wood is forced beneath the mouth of the can to prevent the snow from falling out and then the snow in the can is melted to obtain the water equivalent.

Unfortunately, this practice combines two separate measurement techniques into a single data base, without identification of the technique. The NWS gages are not shielded, so they will systematically undercatch the snowfall. The core or depth measurements are also subject to measurement error but should be more representative of both the meteorological and the hydrological snow. If the data record contains measurements from two entirely different measurement techniques, data comparability and correction of data become much more uncertain. Uncertainty is introduced by using such a record for either meteorological or hydrological purposes.

The snow depth is measured with a measuring stick at a representative location of the average depth. When new snow has fallen on old snow, the depths of both are recorded.

Snow stakes (NWS, 1989) may be used for snow depth determination in areas with deep snowfalls. The snow stake should be a sturdy rod, painted white, with inscribed calibrations. The snow stake should be placed in a properly sheltered, representative location.

3. Snow Board and Plate Snow Measurement A snow board is recommended by the NWS to facilitate the differentiation of old and new snow. A snow board is made of thin wood or other light material that is painted white (a board with higher absorptivity than snow would heat up and sink into the snow). The snow board is pushed into the snow so that the top of the board is flush with the snow surface. After each core extraction or depth measurement, the snow board should be cleaned and placed in a new location. The size of the snow board is not critical, except that it should be large enough to locate for sampling purposes and to take more than one core sample. The dimensions should be greater than a 41 cm square. Lemmela (1989) described a snow plate that he used in his point snowfall comparisons in Finland. The snow plate was a white metal disk 35 cm in diameter that was installed in the same manner as a snow board. After a snowfall, the snow plate was located by means of a long steel wire in the center. A round collector vessel was pushed through the snow to the plate and a core was extracted to obtain the water equivalent. In Lemmela's comparisons, the snow plate gave the highest weekly mean values of snow water equivalent, when compared with other point measurements.

4. Core Sampler Snow Measurement Large diameter snow samplers (e.g., 40 cm diameter of the overflow can) have a tendency to lose the snow core upon removal, so the base of the sampler must be excavated and protected for each core (Beaumont, 1966).

The federal snow sampler is a small diameter (3.8 cm) core sampler for snow depth and water equivalent measurements, typically for deep snow along a snow course. The sampler is a light-weight, graduated aluminum tube with a serrated cutter at the end for sampling deep snow packs. The water equivalent is determined by weighing the sampler with its snow core, then subtracting the weight of the sampler (SCS, 1988). Carroll and Allen (1988) state that snow tube measurements systematically underestimate true water equivalent, due to sampling difficulties associated with ice lenses, ground ice and depth of hoarfrost; however, Farnes et al. (1982) show that the federal snow sampler has a tendency to overweigh by about 10%. Beaumont (1966) states that this overestimation is a function of the sharpness of the cutter at the base of the tube. Blunt cutters are responsible for the overestimation. Powell (1987) acknowledged the published reports of oversampling, but stated

> Many times from my field experience I am convinced that the reverse is true—that the sampler does not catch all the snow in a sample, primarily because of clogging or plugging of the cutter (by layers) and the difficulty of picking up light powder snow (the sampler pushes light density snow aside).

Powell added that the sampler and the snow pillows often show very similar water equivalent values.

Based on his extensive field experience with the federal sampler, Powell (1987) identified several special problems. These included difficulties with sampling ice layers at the base of the snowpack and ice layers or lenses within a deep snowpack. Powell states that it is very difficult to obtain a sufficient core due to cutter clogging and plugging of the sampler beneath the ice layers. Snow sticking inside the sample tubes is also a major problem, especially when air temperatures are warmer than the snowpack. Ponding of free water in a ripe snowpack on snow courses is perhaps one of the most serious constraints on accurate data from the federal sampler (Powell, 1987).

5. Snow Courses for Snow Sampling Wilson (1966) and Powell (1987) addressed the location of snow courses for snow sampling. Snow course data are used as an index of water supply; therefore, a long record of snow course data is needed for correlation with historic runoff. The location of individual snow courses may have been determined by criteria that either are not known or have changed since the snow course was established. Powell argued that continual evaluation of the adequacy of each snow course is necessary, with relocation or termination as indicated by the evaluation. Presumably, snow courses that have the highest statistical correlation with streamflow records would have a high priority for continuance. Problems such as water ponding, avalanche danger, difficulty of access, ice layer formation, and undue drifting or snow erosion would target a snow course for discontinuance. Sites with a significant vegetation change or other factors that change the site exposure also would not be advisable.

Snow courses are sampled by a number of individual sample points, with a significant amount of time represented by each measurement point. The optimal number of sample points on a snow course was reviewed and summarized by Wilson (1966). Efficiency and economy of field sampling and data analysis point toward the fewest number of sample locations; however, these measurements should provide an accurate index of water equivalent on the snow course, and the historical correlations with streamflow should not be lost. Fewer core samples could also translate into greater attention to collection of each sample and thus improve the accuracy of the samples. Wilson's summary of the statistical evaluation of the sample points on a snow course includes the following:

a) Each snow course must be analyzed and treated individually.
b) Snow courses have one or more sample locations that correlate better with streamflow than other sample locations or aggregates of samples.
c) Correlation of snow water with streamflow can be improved by altering existing snow courses to include only the best samples.
d) Coefficient values between streamflow and snow water for individual samples or any grouping of samples may vary considerably depending on forecast period, historic period, gaging station, and watershed.

6. Snow Pillow Measurement of Snow A snow pillow is a hydraulic weighing platform used for measuring the snow water content that accumulates on top of the pillow (Beaumont, 1965; Cox et al., 1978; Smith and Boyne, 1981; Schaefer and Shafer, 1983). The pillows may be constructed of rubber or stainless steel and normally have a diameter or width of several meters. For the SNOTEL network, the pillow is an envelope of stainless steel, 1.2 m by 1.5 m by 2.5 cm, filled with an antifreeze solution. Three or four pillows are laid flush with the surface on a horizontal pad in a representative location. Three pillows are used when the snow water equivalent on April 1 averages less than 38 cm; otherwise four pillows are used. The pillows are hydraulically connected to a pressure transducer for snowpack weight measurement which is equal to its water equivalent (Schaefer and Shafer, 1983).

Schaefer and Shafer (1983) note several sources of random error in snow pillow measurements in addition to the usual electronic systematic errors when sensors, transducers, and transceivers are in a measurement system. The major source of random error in the measurement system is from thermal expansion and contraction of air in the pillows and plumbing lines. Thermal expansion and contraction of the fluid also will produce measurement error, but this should be small under operational conditions due to the low diurnal range of temperature at the base of a deep snowpack. A systematic error may occur from the physical nature of the snowpack. If strong ice layers form, bridging will prevent the full weight of the snowpack from registering on the hydraulic pillow (Schaefer and Shafer, 1983).

Properly calibrated snow pillows also tend to overestimate snow water equivalent. Cox et al. (1978) and Richards (1984) described measurement error due to creep with snow pillows installed on a slope. Creep is the downslope deformation of a snowpack, with a resulting additional load on the pillow. Free water moving downhill on the snow surface or on internal ice layers may accumulate on the horizontal ice layers above the pillow. Improper site drainage also would cause measurement error. Isolating the snowpack on the pillow from surrounding snowpacks with cutoff trenches and relocating the pillows were suggested as corrective actions by Richards (1984). Liquid water draining out of a snowpack on the pillow would result in an underestimate of snow water by the pillow (McGurk, 1985).

In comparisons with uncorrected federal snow sampler measurements, the three pillow configurations overweighed the water equivalent by 5%, but the four pillow configuration measurements were close to the control samples (Schaefer and Shafer, 1983). Beaumont (1965) noted that snow pillows could record snow accumulation rates as low as 0.8 mm/hr. Correlation coefficients between federal snow sampler measurements and snow pillow measurements for all the western states in the United States averaged about 0.94 (Cox et al., 1978). McGurk (1985) compared snow pillow, snow board, and Alter-shielded Belfort gage measurements for seven storms in the Sierra Nevada mountains in California. For five of the seven storms, the pillow recorded the greatest snow water equivalent. The shielded Belfort gauge recorded the least amount of snow water (by about 10%) and the snow board method was consistently in the middle.

E. Remote Sensing Measurements of Precipitation

1. Photogrammetric Measurement of Snow The area of snow coverage is relatively easy to determine from aircraft or satellite visual and near-infrared imagery. The depth of snow or the snow water equivalent cannot be readily determined from this type of analysis without knowledge of *a priori* relationships between snow-covered area and snow properties or runoff volume.

2. Terrestrial Gamma Radiation Snow Measurement The water equivalent of a snowpack can be accurately determined by measuring the attenuation of naturally occurring gamma radiation emitted from potassium, uranium, and thorium isotopes in the upper 20 cm of soil (Carroll and Allen, 1988). Water mass (not necessarily liquid) in soil and in snow attenuates the gamma radiation so that differences in the radiation emitted from bare ground and from snow-covered ground can be used to determine the snow water equivalent. The Office of Hydrology, NWS, maintains an operational Airborne Gamma Radiation Snow Survey Program based in Minneapolis, Minnesota. Each winter, over 1400 flight lines in the northern United States and southern Canada are flown by aircraft at an altitude of about 150 m above the surface. Background radiation is determined from a flight under no-snow conditions and soil moisture is determined by sampling or estimation. The soil moisture may be subjectively changed to a more representative value for the flight over snow-covered terrain. The flight lines, flown at about the time of peak snow accumulation, are typically 16 km long with a width of 300 m. The airborne snow water equivalent information is used by hydrologists to assess moisture held in the snowpack in the form of snow, ice lenses, and ground ice.

The root mean square error for mean areal snow water equivalent is less than 1.2 cm (Carroll and Allen, 1988). The error increases as the gamma count sampling rate decreases, so that typically only one estimate for the entire flight line is produced (Carroll and Carroll, 1989a). Several sources of error exist but normally are accounted for in the calculation of snow water equivalent. Variations in snow water equivalent along the flight line result in an underestimate of up to 10% (Carroll and Carroll, 1989a). These variations may result from differences in snowfall or from drifting along the flight line. Information on the distribution of the variations in snow cover can be used to improve the estimate of snow water equivalent. The potassium contained in forest biomass also causes an underestimation of snow water equivalent, particularly with deep snow. The error may reach 12%, but can be reduced if accurate estimates of the forest biomass are made (Carroll and Carroll, 1989b). Simulation studies by Carroll and Carroll (1990) that included all error sources indicate that the error in airborne snow water estimates is less than 12% for typical flight lines.

3. Estimation of Precipitation with Radar Weather radar is used extensively for qualitative assessment of precipitation and has significant potential for quantitative measurement. Radar information typically has a resolution of about 5 km on a 5- to 15-minute time scale over a fairly large area and the information is available real-time in a convenient form at a central location. Collier (1986a) regarded the maximum range of a 5.6 cm radar in northwest England to be 75 km when bright band (a highly reflective layer aloft where snow melts to form rain) situations existed. For deep convection (10 km), the effective range for precipitation measurement with radar approaches 200 km. The primary advantage of radar as a precipitation measurement system is that it provides spatial and temporal patterns, whereas gages only provide point measurements. The locations, boundaries, and intensities of the radar echoes, and their changes with time, are accurately determined either visually or digitally with radar.

These observations are the basis of mapping precipitation occurrence, as well as flash flood and heavy precipitation forecasting.

At weather radar wavelengths (normally 3 to 10 cm), large cloud droplets, rain drops, snow particles, and other solid forms of precipitation are reflective. The backscattered radiation, known as the reflectivity, Z, is highly correlated with the precipitation characteristics in the volume illuminated with the radar beam. For spherical raindrops, the reflectivity is a function of the sixth power of the raindrop diameter, so the reflectivity is primarily a function of the numbers and diameters of the larger drops. For the same air volume, the rainfall rate, R, is a function of the number of drops in each increment of diameters and the drops' diameters and their fall speeds. If an exponential function, such as Marshall-Palmer, is assumed for raindrop size distribution, the relationship between the radar reflectivity and the rainfall rate is a power function, $Z = aR^b$, where a and b are empirically fitted variables (see Krajewski and Smith, 1991, for fitting of Z-R coefficients).

With constant values of a and b (the most common are 200 and 1.6), radar estimates should be within a factor of two of true rainfall about 75% of the time (Wilson and Brandes, 1979). The value of the coefficient, a, will range from 100 in rainfall from stratiform clouds to 400 for intense convection, while the b variable is fairly constant, 1.3 to 1.6 (Austin, 1987). Fujiyoshi et al. (1990) determined coefficient values for a and b to be 427 and 1.09, respectively, for snowfall and summarized previous literature on values of the coefficients. Because radar measurement of precipitation is ordinarily conducted on an operational basis for rainfall, this discussion will focus on rainfall estimation with radar.

Changes in the relationship between reflectivity and rainfall alter the accuracy of measured rainfall. The coefficients of the Z-R relationship change with type of precipitation, precipitation intensity and ambient conditions within the precipitating cloud. For rainfall, the drop size distribution deviates from the Marshall-Palmer exponential form especially for very light or very heavy rainfall (Wilson and Brandes, 1979). The raindrop size distribution also changes rapidly in time and space within a precipitation event.

Raindrops that are not spherical also affect the basic assumptions of the Z-R relationship. Large raindrops have an oblate shape, which changes the radar cross-section and the reflectivity (Ulbrich, 1986). Pruppacher and Ritter (1971) and Pruppacher and Beard (1970) measured the shape of water drops falling at terminal velocity in a wind tunnel. Only very small raindrops were spherical. The shape of larger drops was determined by surface tension, gravity, and aerodynamic forces on the drop. Drops with diameters between 170 μm and 500 μm had the shape of an oblate spheroid (a slightly deformed sphere with a flattened base). Drops with diameters between 500 μm and 2 mm had the shape of an asymmetric oblate spheroid, with an increasingly pronounced flattened base. For drops larger than 2 mm in diameter, the base had a concave depression, which was more pronounced at larger drop sizes. The largest drop produced in the Pruppacher and Beard (1970) study had the same volume as a sphere with a diameter of about 4.5 mm. The ratio of the length of the minor (vertical) axis to the length of the major (horizontal) axis was fairly linear with drop size. For drops with an effective diameter (the spherical raindrop diameter of equivalent volume) of 1 mm, the ratio was 0.9. The ratio decreased to 0.45 for the largest drops, which had an effective diameter of 4.5 mm. The largest water drops produced in their study had a major axis of 10 mm. Under laboratory conditions, the relationship between the ratio, r, and the diameter, D, of the equivalent spherical drop was: $r = 1.03 - 0.62\,D$. Ulbrich (1986) noted that oscillations in the free atmosphere produced different relationships of r and D. He observed a relationship of: $r = 1.01 - 0.23\,D$.

The use of polarized radar to determine the raindrop size spectrum has been proposed to reduce this source of error (Ulbrich, 1986; Illingworth and Caylor, 1989; Jameson, 1989; Messaoud and Pointin, 1990). The ratio of the vertical to the horizontal polarization reflectivities contains information on the ratio of the minor to major axes of the reflectivity-weighted mean drop size. This ratio adds information on the drop size distribution, but the method is still sensitive to assumptions on the shape of the distribution.

Several error sources are present in radar estimation of precipitation. The radar beam increases in elevation with distance from the antenna. Normally, the radar scans at very low elevation angles, such as 0.5 to 1.5 degrees, to sample precipitation close to the earth's surface. If the fall speeds of the precipitation particles differ from terminal velocity as a result of the vertical motion of air (which is the same order of magnitude), the radar will either overestimate or underestimate the precipitation rate. For strong downdrafts and heavy rain, the radar may underestimate rainfall by nearly 50% (Austin, 1987). Horizontal advection of the precipitation also will cause discrepancies between gage and radar estimates. Rain may be in the radar beam volume above a measuring point, but it may not reach the surface within a kilometer or so of the measuring point.

The precipitation rate increases by coalescence and collision as snow or raindrops continue to fall through a cloud mass. If this cloud is below the radar beam, the radar underestimates the precipitation rate. Conversely, if the air is dry beneath the cloud base, a significant evaporation of rainfall occurs. The magnitude is pronounced, ranging from the commonly observed virga in semi-arid regions to 30% of heavy rain (80 mm/hr) in the first 1.6 km below the cloud base (Rosenfeld and Mintz, 1988).

With increasing range, the radar beam may intersect the 0° C isotherm, where snow melting occurs. Initially upon melting, a snowflake becomes covered with a film of water before collapsing into a smaller raindrop. This water-covered snow is much more reflective than the snow above it or the smaller accelerating raindrops below it. This produces a "bright band" about 100 m deep in the radar return. The reflectivity is typically 2–5 times that of the rain below (Smith, 1986). The high reflectivity produces a large overestimate of the precipitation rate. Smith (1986) proposed an algorithm for automatic detection and correction of the bright band effect when the precipitation is fairly widespread.

When the radar beam altitude is above the bright band, the reflectivity of the snow is lower. Smith (1984) states that the radar cross-section of a weak dielectric such as ice and snow is the same as that of a water drop of the same mass. The lower precipitation rates higher in the cloud are offset to a large extent by the slower fall rates of snow. As the snow particles fall, they increase in size and reflectivity. The lower precipitation rates of snow, in general, are responsible for the generally weaker radar return compared with rain. Consequently, reflectivity is a function of the altitude of the beam volume, but generally the radar underestimates precipitation rates when the beam is above the freezing level.

Another source of error is attenuation by intervening clouds and precipitation, especially at the shorter wavelengths. Huebner (1985) presents a method for correcting attenuation due to an intervening rain cell. Raindrops on the radome also attenuate the radar signals, especially at shorter wavelengths.

Anomalous propagation of the beam results in false echoes, especially when inversions or pronounced density gradients are present in the atmosphere. These situations must be removed using human judgment. Incomplete beam filling is a source of error that is virtually impossible to remove.

Consequently, radar measurements of precipitation must be calibrated (i.e., coefficient adjustment in the Z-R relationship) for the nature of the precipitation that is either expected or is actually occurring. Techniques for both types of calibration are currently in widespread use. Calibration of radar based on coincident precipitation rate is based on a network of automatic gages that are connected to the radar processing system through telemetry (or other means of communication).

The optimal use of radar in precipitation measurement is to combine the spatial and temporal accuracy of radar with the point accuracy of gages. Two general approaches of this combinaiton have been demonstrated (Seo et al., 1990a; 1990b). The deterministic approach is to use the information from telemetered gages to calibrate the radar measurements at specific points. The corrected radar measurements are then used to develop point precipitation estimates at ungaged locations. A common method is to calculate a ratio of the gage measurement and the radar estimate at each telemetered gage location (Barnston and Thomas, 1983). Analytical techniques are used to define an areal pattern of correction ratios for the radar estimates (Palmer et al., 1983; May, 1986). Gage measurements are also used to adjust the value of the coefficient, a, of the Z-R relationship in real time to bring the radar estimate closer to the gage measurements (Collier et al., 1983; Austin, 1987). These methods do not necessarily improve areal precipitation estimates from the gage network.

The statistical approach combines radar and gage measurements, with the effect that the radar pattern is used to extend the gage measurements throughout the measurement space. The two data sets are merged with the technique of multivariate objective analysis or cokriging (Krajewski, 1987; Creutin et al., 1988; Seo et al., 1990a; 1990b). The statistics of the radar and the gage measurement fields are used to develop a merged field that is in agreement with specified constraints.

Radar may not provide accurate measurements of precipitation amount, even with calibration of telemetered gages. The estimates may differ from gaged measurements by a factor of two or more (Wilson and Brandes, 1979). Bellon and Austin (1984) observed an inherent error of about 25% for the 600 km^2 area of Montreal, with about a 30% error for point measurements for 37 summer weather sequences over a 4-year period. For individual weather sequences, the errors ranged from less than 1% to 93% for areal rainfall estimates from calibrated radar. Collier and Knowles (1986) cautioned that even with calibration of radar estimates with precipitation gage data, it is unlikely that the resulting data will be uniformly accurate. Radar estimates of areal precipitation are superior to interpolation from a sparse telemetered gage network (May 1986). Hildebrand et al. (1979) concluded that for Illinois conditions, radar added little to areal convective rainfall estimates for gage densities greater than one gage per 100 km^2. For gage densities between one gage per 100 and 250 km^2, the accuracies of calibrated radar and interpolated gage networks were comparable, while for gage densities less than one per 250 km^2, calibrated radar may be more accurate than gages only. Wilson and Brandes (1979) state that when high accuracy rainfall measurements are needed from radar (error less than 10 to 20%), the number of gages required to provide the radar calibration would provide the necessary accuracy. Collier (1986b) determined that within 75 km of the radar site, the gage spacing would have to be less than 20 km to be more accurate than the calibrated radar.

A limit obviously exists for the number of telemetered gages that may be deployed within the operational range of the radar. Large numbers of gages are expensive and difficult to maintain and operate, even with sophisticated communications and data processing equipment (Krajewski, 1987; Creutin et al., 1988).

A network of the latest generation of weather radars, termed WSR-88D (Weather Surveillance Radar–1988 Doppler) is being deployed throughout the United States from 1991 through 1997 (Crum et al., 1993; Hudlow et al., 1991).

4. Area-Time Integral for Precipitation Estimation The Doneaud method of rainfall correlation from radar has been widely accepted and expanded since its publication in 1981 (Doneaud et al., 1981). This method is the basis for the ATI (area time integral) method and the HART (height-area rainfall threshold) method (Rosenfeld et al., 1990) of rainfall estimation. The basic premise of Doneaud's method is that a significant correlation exists between the horizontal extent of convective rainfall and the volume of the rainfall produced. The original data were developed with a 10 cm radar in North Dakota using ($Z = 155\ R^{1.88}$), with a Z threshold of 25 dBz (dBz = 10 log $\{(Z)/(1\ mm^6\ m^{-3})\}$). For 26 rain days, the correlation coefficient between the product of the area and duration of the radar echoes and the rainfall volume was 0.96.

The relationship between rainfall volume and area-time is plotted linearly on a log-log plot, with a slope slightly over 1 on the log-log scale. The relationship is: $V = k\ (\text{area-time})^b$, where b is the slope of the line. If the slope were exactly 1.0, then the average rainfall rate would be independent of the total storm volume. If the rainfall volume, V, is expressed in mm km^2, the ratio of V to area-time is the average rainfall rate in mm/hr. The average rainfall rate, R_a, is equal to $(\text{area-time})^{b-1}$. The average rainfall rate for the 26 storms was 4.6 mm/hr. Doneaud et al. (1984a) found an average rainfall rate of 4.0 mm/hr in a dry season and 4.8 mm/hr in a more humid season. The standard deviation was about 1.5 mm/hr. Correlation coefficients were 0.94 to 0.98 between rainfall volume and area-time. A total of 750 radar clusters were examined. The life cycle of a convective storm was developed. The maximum growth occurred at 56% of the total life time of the storm, with the average rainfall rate during the growth period exceeding the average rainfall rate during the decay period by 20%.

Doneaud et al. (1984b) named their method the area time integral (ATI) and defined ATI in discrete form as the product of the area and the time increment at each time step. For the lifetimes of individual storms, correlation coefficients were 0.98 between rain volume and ATI. Lopez et al. (1989) applied the

ATI technique to Florida Area Cumulus Experiment (FACE) radar data. For 12-hour accumulations over an area of 3.6×10^4 km^2, the correlations were 0.92 between rain volume deduced from radar ($Z = 300 R^{1.4}$) and ATI computed at periods ranging from 5 minutes to 3 hours. The optimum time interval was suggested as 1 hour. The authors state that, in general, larger areas of radar coverage and greater numbers of storms in the area result in less frequent radar observations without sacrificing accuracy. At time intervals greater than 3 hours, the correlation coefficients decreased rapidly. The correlations for each of the 3 years of data were within a few hundredths of each other. The average rainfall rate was 3.4 mm/hr (lower by a factor of about two from theoretical expectations). The Z threshold was lower than in the North Dakota data and the beam was wider, which could have introduced beam filling errors.

The impact of the techniques of Lopez et al. (1989) and additional work on threshold and calibration methods by Kedem et al. (1990) and Atlas et al. (1990) is that rainfall can be estimated over large areas without the need for precise rainfall rate determination from individual storms or time intervals less than the scale of 1 to 6 hours. Based on the fairly well-defined life cycles of individual storms, a storm measurement at one point in time is representative of much of the storm lifetime. For a sufficiently large number of storms in a large area, the area-integrated storm coverage at any one time is a representative sample of the area for a substantial portion of the day (Atlas et al., 1990). Doneaud et al. (1987), Kedem et al. (1990), and Atlas et al. (1990) suggested ATI methods for estimating rainfall from satellite imagery.

5. Visible and Infrared Estimation of Precipitation from Satellite Imagery Earth satellites with sensors operating in the visible to near infrared, thermal infrared, and microwave windows of the atmosphere have numerous capabilities for precipitation measurement. Extensive treatments of satellite remote sensing for precipitation measurement and other aspects of hydrology are in Barrett (1981), Atlas and Thiele (1981), Deutsch et al. (1981), and Engman and Gurney (1991). Barrett (1981) summarized the advantages and special attributes of satellite systems as follows.

a) Satellite systems provide complete high-resolution global coverage typically on the scale of a few kilometers for visible and infrared and tens of kilometers for passive microwave.
b) Satellite systems yield spatially continuous data, which contrasts strongly with irregular networks having different sensors and exposures.
c) Satellites can investigate the distributions of selected elements much more homogeneously than *in situ* observing networks with large numbers of instrument packages.
d) A high temporal frequency on the scale of minutes or hours is provided for equatorial to temperate areas with geostationary satellites. The temporal frequency on the scale of hours is provided for high latitude areas with polar orbiting satellites. Systems of polar orbiters provide daily coverage of equatorial areas.
e) Satellites provide a view of the atmosphere and earth surface from above rather than from within. They do not modify the environment of the measured precipitation elements.
f) Satellite data have different physical bases than surface-based data; therefore, they add new sources of information.
g) Satellite data are available for large areas, in convenient digital form, at central locations for processing, analysis, and archiving.
h) Remote sensing from satellites for data-sparse areas is cost-effective.

Consequently, a number of techniques for precipitation measurement from satellite sensor systems are in operational use by national agencies. Satellite-based techniques provide precipitation estimates for areas, typically 1500 km^2 or more, where other sources of data are limited or when data comparability is a concern.

The following review of precipitation measurement will focus on the techniques in operational use or suited for operational use, as described in technical journals. Passive microwave sensing of snow and rain are addressed separately.

Rainfall rates and amounts can be estimated from the visible reflection and thermal infrared emission from cloud tops. The basic premise is that the intensity of convective rainfall is correlated with the visible brightness and the radiative temperature of the cloud tops. Colder and brighter tops represent both deeper and more intense convection that in turn is correlated with rainfall intensity and volume. A strong correlation exists between either the brightness or the temperature of the cloud top and the vertical extent of the cloud, which in turn is highly correlated with precipitation rate. Significant variations in these relationships occur among climatic regimes (Adler and Mack, 1984). Precipitation efficiency is higher in areas with less vertical wind shear, increased ambient moisture content, and higher stability than in areas with vertical wind shear, drier ambient conditions, and decreased stability. Adler and Mack (1984) contrasted storms in Florida with those in Oklahoma, representing these two climatic regimes respectively, to develop a technique for adjustment. This technique relies on the incorporation of environmental variables, such as vertical wind shear, moisture, and temperature, into the rainfall rate retrievals.

Validation and calibration of satellite methods should be based on rain gage network information, radar estimates of rainfall, or aviation weather reports (O'Sullivan et al., 1990). The major point is that satellite measurement techniques are verified against areal estimates of rainfall for areas around 10^5 km^2 (Bellon and Austin, 1986). When satellite techniques are used to determine convective rainfall over land, fairly accurate estimates of the areal precipitation are needed. Otherwise, the inherent natural variability of rainfall and the imprecision of estimating accurate areal rainfall from point estimates creates a problem. Flitcroft et al. (1989) addressed these problems and recommended that the point rainfall totals be transformed to the grid areas or pixels of the satellite coverage. Careful attention to the development of areal precipitation fields from point measurements is indicated, which is beyond the scope of this chapter.

Cloud top temperatures are measured in the 10.5 to 12.5 μm thermal infrared (IR) window of the atmosphere. The cloud top has the temperature of the ambient environment, so very cold tops of convective systems represent very deep and intense convection. The primary source of error is that rainfall is not directly sensed by the thermal radiometers (Griffith, 1987). Consequently, thermal infrared measurements are best suited for convective rainfall where the basic assumptions are valid. Winter and transitional season cirrus associated with the subtropical jet would produce erroneous precipitation indications if the technique were based only on cloud top temperatures (Turpeinen et al., 1987). In the visible channel (usually between 0.55 and 0.75 μm), the cirrus is not as bright as the tops of convective cells, so bispectral techniques confer certain advantages but suffer a daylight-only limitation. The expanding cirrus shield in the decreasing stage of convection also increases the cloud area considerably after the convection has ended and the precipitation has decreased (Richards and Arkin, 1981; Negri and Adler, 1987).

Griffith developed two diagnostic, automatic techniques for use with the visible and infrared spin scan radiometer (VISSR) on the Geostationary Operational Environmental Satellites (GOES) (Griffith et al., 1978; Griffith et al., 1981; Griffith, 1987). The life-history technique incorporates information about the life cycle of the convective cloud in the rainfall estimation. The streamlined technique produces rainfall estimates from single images. Both techniques are suited for precipitation estimation for scales from the individual convective cloud to large areas.

In the life-history approach of Griffith, the assumptions are: 1) raining clouds have cloud top temperatures colder than $-19°$ C, 2) rainfall rate is directly proportional to the cloud area on a given image, 3) rainfall intensity at a point is inversely proportional to the temperature of the cloud top, and 4) rainfall distribution in time is a function of the stage of the life cycle of the cloud (Griffith, 1987). These assumptions presume that convective cloud development follows a more-or-less established pattern. In order to estimate rainfall from a given cloud, the maximum areal extent of the cloud has to be determined. In the streamlined approach, the maximum areal extent of the cloud is specified beforehand; otherwise the techniques are the same. The differences between the two techniques are small when compared with rain gage network estimates of areal rainfall.

The distribution of rainfall is determined empirically with the Griffith approach. One-half of the rain volume is assigned to the area with the coldest 10% of the cloud area, and the remaining one-half of the rain volume is assigned to the next warmer 40% (Griffith, 1987). Modifications of the life history or streamlined approaches could incorporate actual rainfall areas and intensities from calibrated radar (Barrett, 1981). The method's accuracy depends upon the nature of the convective rainfall. Small, short-lived convection is usually underestimated or missed. In a comparison with gaged rainfall estimates for August 1979 in the central third of the United States, Griffith reported an accuracy equivalent to a gage network with density of about one gage per 2300 km^2. Residual mean square errors (RMSE) were 1 mm and 14 mm for daily area and point rainfall estimates, using gaged amounts as a reference.

The life-history method of Stout et al. (1979) incorporated a term for rate of cloud area change, with an implied assumption that cloud area and cloud area rate of change are correlated with precipitation. This method used cloud brightness in the visible and cloud top temperature only to identify clouds that were likely to have precipitation. From there, rainfall on an hourly basis was determined from the cloud area and the rate of change of the cloud area from either visible or infrared imagery. Precipitation was determined from shipboard radar in the tropical north Atlantic with a Z-R relationship determined from previous research. Their results compared favorably with those determined by the cloud top temperature techniques. Correlations were around 0.85 for hourly rainfall amounts over a 10^5 km^2 area and with the other life cycle methods.

The Scofield-Oliver technique is directed toward identifying heavy rainfall areas, with application to operational forecasting. This technique is based on decision-tree logic from visual image analysis in the visible and MB-enhanced GOES infrared (IR) channels (Scofield and Oliver, 1979; Scofield, 1983), supplemented by information from conventional sources. First, the IR images are examined visually to determine areas of active convection. Secondly, estimates are made of the rainfall intensities for these areas at half-hour intervals. The estimates are made from an equation with factors for cloud top temperature and cloud growth or divergence aloft, overshooting top, cluster or line merger, saturated environment, and precipitable water in the storm's environment. The information for these factors was developed from an analysis of the visible and IR images. Scofield (1983) and Spayd (1983) discussed modifications that would account for heavy rain from relatively warm cloud top convective systems.

Cloud indexing methods in general relate the frequency of highly reflective clouds (HRC) or cloud top temperatures colder than a threshold, to the total precipitation within a specified area for a specified period. The area is generally large, cells of up to several degrees latitude and longitude, and the time scale is typically five days or more. Indexing methods are strictly empirical; the precipitation in the grid as determined from gages or radar is related to the satellite data through regression. The satellite data may include subjectively assigned cloud types and empirically assigned indices of precipitation potential. Although the methodology is fairly simple, it has widespread applications for climate definition, especially in data sparse areas such as oceans and deserts. The development of cloud indexing methods is summarized in Barrett (1981), and a recent review of index methods for climate-scale precipitation determination is provided by Arkin and Ardanuy (1989). They assign the niche for index methods covering areas over 10^4 km^2 for five-day or longer periods

*6. **Passive Microwave Snow Measurement*** Snowpack properties are also measured operationally with passive microwave radiometers on earth-orbiting satellites. Passive microwave radiation is independent of solar radiation, so day and night data sets can be combined. The atmospheric influence on emitted radiation is negligible except when very large cloud water droplets or raindrops are present. Ice clouds and most layered clouds are essentially transparent. The antenna diameter is limited for earth satellites, so the typical footprint diameter for a passive microwave radiometer is in the range of 10 to 50 km. The radiation received at the antenna is integrated over the footprint, so small-scale variations are averaged. The maximum utility for snowpack determination with passive microwave radiometry is for large, relatively flat areas such as prairies and steppes.

At microwave wavelengths, water has a very high dielectric constant that is wavelength- and temperature-dependent. The value of the real component (the permittivity) of the complex dielectric constant is 19 at 0° C and 36 at 20° C at 1.55 cm (19 GHz). At 0.81 cm (37 GHz), the permittivity value is 9 at 0° C and 17 at 20° C (Ulaby et al., 1986). In contrast, the permittivity of air and dry soils at these temperatures is typically in the low single digits. As water is added to dry soil with air spaces, the permittivity increases rapidly. An increase in permittivity decreases the thermal emissivity at 1.55 cm wavelength from about 0.95 for dry soil to less than 0.6 for a saturated soil with water in surface detention storage (Schmugge, 1983; Wilke and McFarland, 1986). Ice and snow have a permittivity near that of dry soil and air.

In the microwave region of the electromagnetic spectrum, the radiation emitted from a surface layer is proportional to the first power of the actual temperature, according to the Rayleigh-Jeans approximation of Planck's Law (Ulaby et al., 1981). If the atmospheric attenuation (normally less than 10%) and upwelling and reflection from downwelling radiation are neglected, the energy received at an airborne or spaceborne radiometer is equal to the product of the emissivity and the temperature (in degrees K). This energy is expressed as the brightness temperature. The emissivity is inversely proportional to the permittivity, so adding moisture to soil decreases the brightness temperature.

The index of refraction of a layer and the reflection coefficient between a surface layer and the air are functions of the permittivity, so the energy emitted from the layer will be a function of the viewing angle. Additionally, the energy emitted will be polarized, so the orientation of the radiometer antenna will determine the energy received (Burke et al., 1984). Incidence angles at the Brewster angle, near 50 degrees, produce the maximum separation between vertically and horizontally polarized brightness temperatures.

The loss factor is the complex component of the dielectric constant. The loss factor is high for water but very low for ice and snow. For a low dielectric loss factor, the radiation propagating through the medium is attenuated primarily by scattering (Kunzi et al., 1982; Burke et al., 1984). Consequently, a dry snowpack is a volume scatterer of the radiation emitted from the surface layer. This Mie scattering by snow particles is wavelength dependent, with a maximum near 0.81 cm (37 GHz) (Stiles et al., 1981). The volume scattering at this wavelength is in turn dependent on the crystalline structure of the snowpack. Temperature and wind conditions at time of snowfall, thaw and freeze cycles of the surface layer, snow grain metamorphosis such as the formation of hoar crystals at the base of a deep snowpack, and density changes in a snowpack change the crystalline structure and the microwave response of the snowpack. The addition of only a few percent content of liquid water increases the dielectric loss such that the wet snow surface becomes a simple emitter, as opposed to a volume scatterer (Kunzi et al., 1982; Matzler et al., 1984).

These various microwave physics of a snowpack are the bases of methods to retrieve snowpack properties from earth satellites such as the NIMBUS-7 and the newer satellites in the Defense Meteorological Satellite Program (DMSP). The Scanning Multichannel Microwave Radiometer (SMMR) on NIMBUS-7 and the Special Sensor Microwave Imager (SSM/I) have been used to determine snow cover extent, snow depth, snow water equivalent, and ripeness of the snowpack (Stiles et al., 1981; Kunzi et al., 1982; Burke et al., 1984; Hallikainen, 1984; McFarland et al., 1987; Foster et al., 1987; McFarland and Neale, 1991).

7. ***Passive Microwave Measurement of Rainfall*** Information available from passive microwave measurements is directly related to precipitation, without significant influence from non-precipitating clouds such as jet stream or anvil cirrus. Stratiform and frontal precipitation can be measured with passive microwave data, but not easily (if at all) from visible and infrared observations. Low resolution, on the order of 10 to 50 km, is perhaps the most significant of the difficulties that exist. Reviews of passive microwave remote rainfall sensing are provided by Barrett (1981), Wilheit (1986), and Engman and Gurney (1991).

Snow particles and water droplets are both attenuators and emitters of radiation in microwave wavelengths. The attenuation is from absorption and scattering. Rainfall estimation with passive microwave radiometers may be based either on the attenuation or the emission characteristics of raindrops (Wilheit, 1986; Spencer et al., 1989). Passive microwave emission over oceans is characterized by very cold, polarized, and fairly uniform brightness temperatures. In contrast, the brightness temperatures from raindrops are relatively unpolarized and much higher, due to the higher emissivity of the more-or-less spherical raindrops. Precipitating clouds over oceans appear warm against a cold background. The brightness temperatures increase in a non-linear fashion with increasing rainfall rates (Wilheit, 1986). Passive microwave response behavior to rain is most pronounced at longer wavelengths, such as the 1.55 cm (19.35 GHz) of the DMSP Special Sensor Microwave Imager (see Hollinger et al. (1990) for a description of the DMSP SSM/I). At shorter wavelengths, such as 0.35 cm (85.5 GHz), the ice particles are volume scatterers of the emitted radiation (Spencer et al., 1989)

F. Data Comparability

Although sensor technology for the direct or indirect measurement of precipitation continues to develop, the major information sources on precipitation are historical data bases developed from gage networks. These data bases are used often to investigate new hypotheses, which may lead to the development of new sensors and networks. Information from a new technology may have to be combined with information from a historical technology. A gage network designed to support one primary objective, such as the documentation of the climate, may have inadequacies for other objectives, such as documentation of the water balance. A gage network designed to measure rain (e.g., ground level gages) is not satisfactory to measure snow. Data may be available from networks with different gage designs and exposures (e.g., NWS and United States Forest Service networks).

With all of these potential problems, data comparability is a stated or implied goal of a number of networks, and is very important if different data sources are available (Jansen, 1983). If data comparability must be established between two data sets, two steps are necessary: 1) to judge the adequacy of the data sets individually and combined to evaluate the new hypothesis and 2) to determine the uncertainty of each data set and of the combined data set (Jansen, 1983). The uncertainty includes the totality of systematic error and bias that determine the confidence with which the hypothesis is to be evaluated.

Jansen states that the process of assuring that the data base is adequate to test a hypothesis is basic to scientific analysis. This process requires an understanding of what the measurements actually represent, the accuracy and reliability of the data, the data processing methods, the quality assurance/quality control procedures and the characteristics of the sensors, sites, and exposures. Most of this information is either poorly documented or not documented at all for historical data sets (Jansen, 1983). Consequently, the comparability between historical data sets cannot be determined.

Jansen (1983) specifies the minimum responsibilities for the establishment of sensors and networks for the collection of data to support new investigations. He separates the roles of data gatherers and data analysts. The responsibilities of the data gatherer are: 1) to design the experiment to meet the objectives, 2) to document all objectives, assumptions, procedures, changes, and results completely in an accessible form, and 3) to design, maintain, and report on a comprehensive quality control program for the experiment. The responsibilities of the data analyst are: 1) to pose a specific hypothesis or objective, with consideration given to selection of data requirements, 2) to determine the acceptable level of uncertainty for a reliable solution or evaluation, 3) to use all of the documentation and quality control results to determine if the total uncertainty of the combined data bases is within acceptable limits, and 4) to report the results of these evaluations along with the conclusions.

The individual sensors and the network of sensors should include the following elements (Gill, 1964), as cited in Mazzarella (1985): 1) a clear understanding of the operation principles of the basic sensor and a knowledge of its dynamic response by the gatherer and the analyst, 2) a general understanding of the operation principles of the indicating or recording system by the gatherer and the analyst, 3) static and dynamic calibration of the sensor system, 4) proper installation, including site selection, and proper use of the sensors, 5) routine servicing, 6) periodic maintenance, 7) periodic calibration checks, and 8) alertness for small clues that may indicate errors developing in the system.

For precipitation measurement with gages, static and dynamic calibrations are the most difficult. Laboratory calibrations are insufficient because they cannot test for all atmospheric couplings, such as effects of wind on catch, nor can they reveal a future drift in response (Brock, 1985). Intercomparisons conducted at one location for all sensors or gages is recommended, but can normally only be accomplished at the start and end of an experiment. The alternative is to make intercomparisons at each station, in rotation, with reference sensors (Brock, 1985). This procedure is very difficult with precipitation measurements, because the best reference gages are ground level gages for rain and double fence shielded gages for snow. Efforts may be warranted to develop and agree upon portable reference standards to calibrate precipitation gages and to quantify site effects. Calibrations against a reference sensor will verify the measurement accuracy of the individual gages, but the question of static and dynamic biases between stations in a network is even more important, according to Brock. This is especially true for hydrologic investigations, where knowing the true precipitation over a watershed is required.

V. PROCESSING AND INTERPRETING PRECIPITATION RECORDS

A. Processing Precipitation Data

Precipitation record processing normally refers to the data reduction methods employed for obtaining reliable data from the "raw" instrumental record. In the past, data processing was very time-consuming compared with today's computerized procedures; however, it is important to review some of the basic types of precipitation data processing techniques to gain a better understanding of how each technique may influence data records. Hanson and Wight (1986) give examples where incorrect data were entered into the record. They suggest some suspicion of a record until it is closely examined.

1. *Personal Observation* Personal observation remains one of the more common techniqes for precipitation measurement and involves volunteers who, because of their location and consistency, remain dedicated to making measurements on a routine agenda. The NWS has supported procedures for observation since 1892 and has periodically updated them to keep observations as consistent as possible. Observations are made daily at a consistent time, generally between 5 p.m. and 8 p.m. or between 6 a.m. and 8 a.m., using a non-recording gage or a weighing recording gage (United States Dept. of Commerce, 1970). Time and depth of precipitation are recorded along with other observations that might be important. Some simple rules for recording personal observation are followed such as emptying gages only once every 24 hours, and crossing out mistakes rather than over-writing or erasing. Records are sent monthly to the NWS, National Climatic Data Center (NCDC) in Asheville, North Carolina for processing.

All of the precautions and standards for consistent data collection and processing minimize problems; however, complications do occur, usually when an observer deviates from the standards or fails to notice irregularities. Some examples include:

a) Shortened measuring stick—the measuring stick being used is slightly shorter than it should be, resulting in consistently high readings.
b) Consistently incorrect readings from the measuring stick—the operator uses the wrong end of the stick or consistently misreads the hundredths marking.
c) The operator uses the previous year's data due to being absent at observation time.
d) The operator relocates the precipitation gage without notifying anyone.
e) Incorrect dates and values are recorded.
f) A different, untrained individual replaces the operator without noting it on the data form.

These types of errors are difficult to detect without careful examination of the field data forms. Generally, the NWS does an outstanding job maintaining representative records and assuring quality data.

2. *Chart Recorders* Chart recorders (using pen traces) have been the standard recording device in the United States for many years, even when mounted in a universal tipping bucket gage (Brakensiek et al.,

1979). Chart processing requires meticulous manual reduction or can be automated using computer-aided drafting (CAD) programs and digitizing tablets. Generally, a technician examines the overall quality of the trace to see if any major quality control problems may have occurred (such as insects, birds, small animals, wind, snow capping, etc.). The technician locates an event, marks the lowest pen trace to represent 0-depth and marks successive points on the trace at the approriate time intervals. These marks, showing depth and time, are then transferred to the record which can then be used to compute storm amount, storm intensity, etc. A detailed discussion on precipitation gage operation, maintenance, and data reduction can be found in the field manual by Brakensiek et al. (1979).

Weighing gages offer very good resolution of high-intensity storms, depending on the rate at which the chart is driven. In addition, the pen traces from weighing gage charts can offer many clues about how a gage is operating. Tipping bucket charts offer some advantages for data reduction, because they are quickly processed in most cases; however, high-intensity precipitation may result in an unreadable chart if the tips are too close together. The tipping bucket chart does not reflect some of the quality control indicators given above for the weighing chart.

3. Digital Recorders The computer has provided an entirely different form of data recording and processing called digitizing. Digital recorders use magnetic storage devices or computer chips (Random Access Memory, RAM) to hold information generated by the analog weighing or tipping instrument. The analog information is transformed into digital data with transducers so the information can be stored as a binary signal. The digital recorder can be set at a variety of time-increment recording rates depending on the desired resolution. The advantages of digital recorders include ease of use, reliability and speed of data processing. The disadvantage of digital recorders over analog recorders can be an inadequate recording-time interval and the inability to see the instrument trace while conducting field maintenance; however, most newer recorders are programmed to increase the reading rate at the initiation of an event. Such recorders decrease data processing time significantly and increase the amount of storage available over conventional fixed-time digital recorders. Portable computers can be used in the field to examine digital information graphically. Therefore, it is difficult to identify significant disadvantages of digital recording equipment. As with analog recorders, precipitation recording systems using digital recorders require a very good quality control program to be successful. Frequent inspection of the equipment is required to maintain consistent data and minimize loss of records.

B. Station Relocation Considerations

Stations may be relocated over their life of operation. Record adjustment may be necessary to accommodate reasonable analyses and comparisons. These techniques are used for mean annual precipitation. These simple techniques are useful when studying long-term change and should be the first analysis performed when investigating climate change. Kincer's (1938) recommendation (cited from Linsley et al., 1949) holds true:

> In using precipitation in the solution of hydrologic problems, it is necessary to ascertain that time trends in the data are due to meteorological causes

and are not due to changes in location, tree growth, instrumentation, or technique. Station historical descriptions should be carefully examined for information that might suggest some change in these and other factors.

1. Double Mass Curve The double mass curve technique is a reliable procedure for checking the consistency of a precipitation record, e.g., when a station is moved. This technique compares long-term annual or seasonal precipitation of a group of comparison stations to the station being evaluated. Key assumptions are that the stations have regional consistency over long time periods, the comparison stations do not have inconsistencies, and the application is for a yearly or seasonal precipitation. Some seasons of the year may have more inconsistencies than others. Therefore, seasonal analysis may provide better results than using total annual values. The accumulated annual or seasonal values for the comparison stations are plotted against the accumulated annual or seasonal value for the evaluation station. Abrupt deviations in the slope of the double mass plot suggest some change not related to climatic variables. For more information about this very useful method of quality checking data see Chow (1964), Linsley et al. (1975), Searcy and Hardison (1960), and Weiss and Wilson (1953).

C. Estimation of Missing Records

The NWS recommends two arithmetic methods to estimate missing records as developed by Paulhus and Kohler (1952). Both of the methods require data from three adjacent stations. The first requires that the three adjacent stations have mean annual precipitation within 10% of the station with missing data, and the arithmetic mean of the three stations is used as the missing record. The second method, called the normal-ratio method, is used where the mean annual precipitation of one or more of the adjacent stations exceeds the station with missing data by more than 10% and is calculated as

$$P_x = \frac{1}{3}\left[\frac{N_x P_a}{N_a} + \frac{N_x P_b}{N_b} + \frac{N_x P_c}{N_c}\right] \tag{2.7}$$

where P_x is the precipitation for the station with missing records and $P_{a,b,c}$ are the adjacent stations' precipitation values. $N_{a,b,c,x}$ are the long-term mean annual precipitation values at the respective stations.

Another method, the reciprocal-distance (inverse distance) method is also used to estimate missing records. As given by Wei and McGuinness (1973), Simanton and Osborn (1980), Tung (1983; 1984), and Hanson (1984) it uses the equation:

$$P_x = \frac{\left[\frac{P_1}{D_1^k} + \frac{P_2}{D_2^k} + \frac{P_3}{D_3^k} + \frac{P_h}{D_n^k}\right]}{\left[\frac{1}{D_1^k} + \frac{1}{D_2^k} + \frac{1}{D_3^k} + \frac{1}{D_n^k}\right]} \tag{2.8}$$

where $P_{1,2,3,..h}$ are the amounts at adjacent stations 1 to h, $D_{1,2,3,..h}$ are the distances between the station with missing data and the adjacent stations 1 to h, h is the number of stations used, and k is a constant used in weighing the distance.

Regression analysis and time series analysis provide additional alternatives to estimating missing records (Salas 1993).

D. Temporal and Spatial Extrapolation of Precipitation Data

One of the challenges for engineers involved in hydrologic analysis and design is obtaining representative precipitation information. Gaging stations are not typically within close proximity to a project site; nor do they usually contain a sufficient period of record to allow direct use of the station record in project design. An engineer must rely on surrounding stations, often with incomplete or non-overlapping record periods to construct a suitable surrogate record from which to complete analysis and design.

The spatial techniques for estimating missing records described above are useful in extrapolating precipitation records to ungaged areas provided the adjacent stations used are representative of the ungaged area. Simple weighting and statistical methods include arithmetic mean, inverse distance (shown above), the Thiessen method (1911), and isohyetal methods. More complex statistical methods include polynomial, multiquadratic, a variety of optimal estimating techniques such as objective analysis and kriging, and approaches that consider elevation and orography. Reviews of some of the methods are provided in Hall and Barclay (1975), Creutin and Obled (1982), Tabios and Salas (1985), Singh and

Chowdhury (1986), and Salas (1993). Other approaches use physical modeling and statistical means to estimate unknown locations.

Significant errors can be expected where complex terrain is encountered. Other factors that should be considered are seasonal and diurnal periodicity (Riley et al., 1987). Finally, where hydrologic analysis is critical, nothing is superior to on-site precipitation data. Design engineers should stress the importance of on-site meteorology and the need to build this requirement into project planning.

The following paragraphs present brief descriptions of methodologies used in the temporal and spatial extrapolation of precipitation data. They are intended to supply the engineer with basic information on available methodology, and experienced hydrologists and/or statisticians should be appropriately consulted. The treatment of frequency analysis is discussed later in this chapter.

1. ***Thiessen Method*** The Thiessen (1911) method is a technique for approximating the distribution area around precipitation gages for the purpose of distributing average precipitation depths over an area. A Thiessen polygon network is manually constructed using perpendicular bisectors to lines between gages and carefully removing overlapping bisectors until an even spatial distribution is obtained. The area of each polygon is multiplied by the representative gage's depth and summed over the total area of interest. This sum is divided by the total area to obtain the average depth.

A number of investigators have developed automated approaches for constructing Thiessen polygons (Croley and Hartmann, 1985; Diskin, 1969; Shih and Hamrick, 1975). As with the arithmetic average method, the Theissen method does not account for anything other than the distribution of area and must be applied with this constraint in mind.

2. ***Isohyetal Method*** This method described in Linsley et al. (1949) is still a useful and fairly accurate technique for determining the spatial distribution of precipitation (France, 1985). The isohyetal map is constructed by plotting the stations and their appropriate precipitation values on a topographic base map (Linsley et al., 1949). Human judgment is used as to the placement of isohyetal contours affected by topography. Average estimated precipitation for an area is computed by the weighted average for the area in question.

The number of gages and the skill of the analyst obviously make a difference in the resolution of isohyetal maps. This point was demonstrated by Hamlin (1983).

3. ***Regression*** Linear and multiple regression offers a straightforward and useful tool in extrapolating precipitation records to areas having incomplete records. Regression relationships between two or more stations are developed with the existing data. Their relationships are then used to fill in missing records or to provide an estimate of what a station's record might have been based on the independent stations. An excellent discussion with criteria for improving estimators of parameters is included in Salas (1993).

4. ***Polynomial Interpolation*** This method utilizes a polynomial function, fitted to the stations. A least squares statistical treatment resolves the weighting of a location based on its coordinates. Unwin (1969) presents an example of this method compared with regression analysis. Chidley and Keys (1970) present an example of the procedure and compare it to arithmetic, Thiessen, and isohyetal methods. Their method can include the effect of other parameters such as altitude and exposure; however, Tabios and Salas (1985) found that polynomial interpolation gave the poorest result when compared to Thiessen and inverse distance.

5. ***Objective Analysis and Kriging*** Objective analysis was developed by Gandin (1965) and was recommended by the WMO (1970). Gandin summarizes the objective analysis procedure as:

> interpolation between values of analyzed elements at the nodes [grid points] of a regular predetermined network; elimination and at least partial correction of errors which show up when data

from various stations are compared; matching of the fields of meteorological elements and possibly the smoothing of these fields. The results of the objective analysis are then recorded in the computer storage, in the form of values of the elements at the grid points.

Similar to objective analysis, kriging is a spatial interpolation technique that arrives at interpolated surfaces with minimum error. The technique is used regularly in geotechnical analysis and has been applied to hydrologic analysis by a number of authors (Montmollin et al., 1980; Chua and Bras, 1982; Creutin and Obled, 1982; Tabios and Salas, 1985; Dingman et al., 1988; Yates, 1986; Yates and Warrick, 1986a, 1986b; Seo et al., 1990). The advantage of kriging over ordinary least-squares techniques is that the estimated values at observation sites are equal to the actual measurement (Gebhardt et al., 1988). In using kriging, precipitation at a site is considered a function of a predictable trend and a random component. Kriging may treat the trend as a constant (simple kriging), or may describe the trend as a polynomial (universal kriging). The variance of the difference between rain gage measurements is related to the distance between rain gages, called a semivariogram. The direction (of the distance) is assumed not to affect rain gage values. Gebhardt et al. (1988) suggests that the data should be categorized into similar sets, possibly separating ridge top data from valley data, where practical. Montmollin et al. (1980) and Chua and Bras (1982) present methods for analyzing precipitation in mountainous areas using kriging. Grundy and Miesch (1988) point out that "there is no cookbook method of analyzing spatial statistics." Therefore, kriging should be undertaken only with help from experienced statisticians. Nevertheless, it does provide a contour map of the accuracy of the interpolated surface.

Creutin and Obled (1982) and Tabios and Salas (1985) provide valuable comparisons of the statistical techniques for distributing station information over space. Tabios and Salas (1985) performed a comparison of spatial interpolation techniques in Nebraska and northern Kansas that included Thiessen polygon, polynomial interpolation, inverse distance, multiquadratic interpolation, optimal interpolation, and kriging. The area involved about 52,000 km^2, with 30 years of precipitation records from 29 stations. Optimal interpolation and kriging techniques were the best of all the procedures. Polynomial interpolation gave the poorest results. In mean estimation, both the Thiessen and inverse distance techniques were comparable, with the latter giving smaller errors of interpolation. Creutin and Obled (1982) found that the more sophisticated techniques (such as objective analysis and kriging) provide a better estimation where there were intense and strongly varying rainfall events when compared to more common techniques. They recommended that, as in optimal interpolation, an approach that leads to stable calibration is preferable if the interpolation is to be conducted repeatedly.

Kriging has been shown to be of limited usefulness, as are all conventional extrapolation methods discussed above, where precipitation relationships with other spatial variables (notably elevation) are inconsistent. In this case other approaches are required, such as the approach outlined by Daly et al. (1994). Singh and Chowdhury (1986) reviewed thirteen methods, some discussed above, on two sites in New Mexico, and found little differences in the ability of most of the methods to produce comparable results.

6. Probable Maximum Precipitation (PMP) Probable maximum precipitation (PMP) represents the best judgment of the upper limit of precipitation that might occur (Gilman 1964) and is directed towards very costly projects where hydrologic events have the potential for serious damage. Large dams are good examples of such projects where spillways are sized to accommodate an approximate worst-case situation. Huschke (1959) defines PMP as "the theoretically greatest depth of precipitation for a given duration that is physically possible over a particular drainage basin at a particular time of year." The National Weather Service changed this definition by replacing "over a particular drainage" with "over a given size storm area at a particular geographical location." Wang (1984) noted that the "modification recognizes the fact that there is a difference in the storm depths over a storm area and those over a specific basin of the same size because storm patterns do not coincide exactly with the shape of the basin." The World Meteorological Society (WMO, 1986) defines PMP as "theoretically the greatest depth of precipitation for a given duration that is physically possible over a given size storm area at a particular geographical location at a certain time of year." The World Meteorological Society (WMO, 1986) pointed out that the PMP was once termed "maximum possible precipitation" (MPP) and was changed because the latter implicated a more stable upper limit with no regard to uncertainty. The WMO (1973) stated that:

> Procedures for estimating PMP, whether meteorological or statistical, are admittedly inexact, and the results are approximations. Different, but equally valid, approaches may yield different estimates of PMP.

Procedures for estimating PMP are grouped into non-orographic, orographic, and statistical methods, with a fourth method termed "generalized" for use in compiling large PMP maps from smaller subparts. Wang (1984) reports that estimates on a site-specific basis have involved hydrometeorological analysis, application of models, and statistical methods.

The selection of PMP analysis is a major concern. Wang (1984) points out some serious problems with PMP analysis and was later supported by Flook (1987):

> There are several approaches for estimating PMP, each of which has its merits and limitations. There is often considerable question as to what approach should be used and to what extent analyses should be made for a project. To be considered are the nature and scope of the project, stage of the study, topographical and meteorological characteristics of the basin, and meteorological and climatic data available for analysis. It is the responsibility of the practicing engineer or hydrologist, or both, to carefully evaluate the basin characteristics, available data and type and scope of the project for the selection of most appropriate procedure. In any case, it should be understood that current knowledge of storm mechanisms and their precipitation producing efficiencies are inadequate to permit precise evaluation of extreme precipitation. Therefore, any PMP estimate must be considered as an approximation which can vary over a range of values depending on the input data used, and assumptions made in deriving the estimate. Some conservatism should be considered in the analysis depending on potential consequence of under-estimating the PMP.

Riedel and Schreiner (1980), when comparing PMP and maximum observed rainfall, found that east of the 105th meridian, approximately one-fourth of the storms studied (177 of 675) were greater than or equal to the 50% PMP. Their study also included maps showing ratios of 25.9 km^2 PMP to 100-year occurrence. They suggest that the PMP to 100-year ratios give general guidance to approximate PMP magnitudes, but it is not possible to assign recurrence intervals to the PMP. An excellent discussion on PMP can be found in Smith (1993).

A map of PMP for a 6-hour period covering 25.9 km^2 (10 mi^2) is shown in Fig. 2.12. This is taken from the United States National Weather Service (NWS) Hydrometeorological Report No. 51 (Schreiner and Riedel, 1978). For regions west of 105 W. longitude, consult United States Dept. of Commerce, Hydrometeorological Report No. 49 (1977); United States Weather Bureau, Hydrometeorological Report No. 43 (1966), technical papers no. 38 (1960) and 49 (1964), and Frederick et al. (1977).

In some areas of the United States, such as the Northwest, extreme rainfall events are not necessarily the cause of major flooding. In these areas, snowmelt which may or may not occur with rainfall is a cause of major runoff events; therefore, the NWS developed procedures to estimate the frequency of "water available for runoff" (Richards et al., 1983). Their studies resulted in a set of depth-duration-frequency maps for the northwest United States that represent water available for runoff rather than probabilities of extreme rainfall.

Another type of design storm is the Standard Project Storm (SPS) utilized by the United States Army Corps of Engineers (1965). Gilman (1964) stated that the project storm was developed as a guide to project design not requiring the large margin of safety that is built into the PMP. Gilman (1964) also stated that

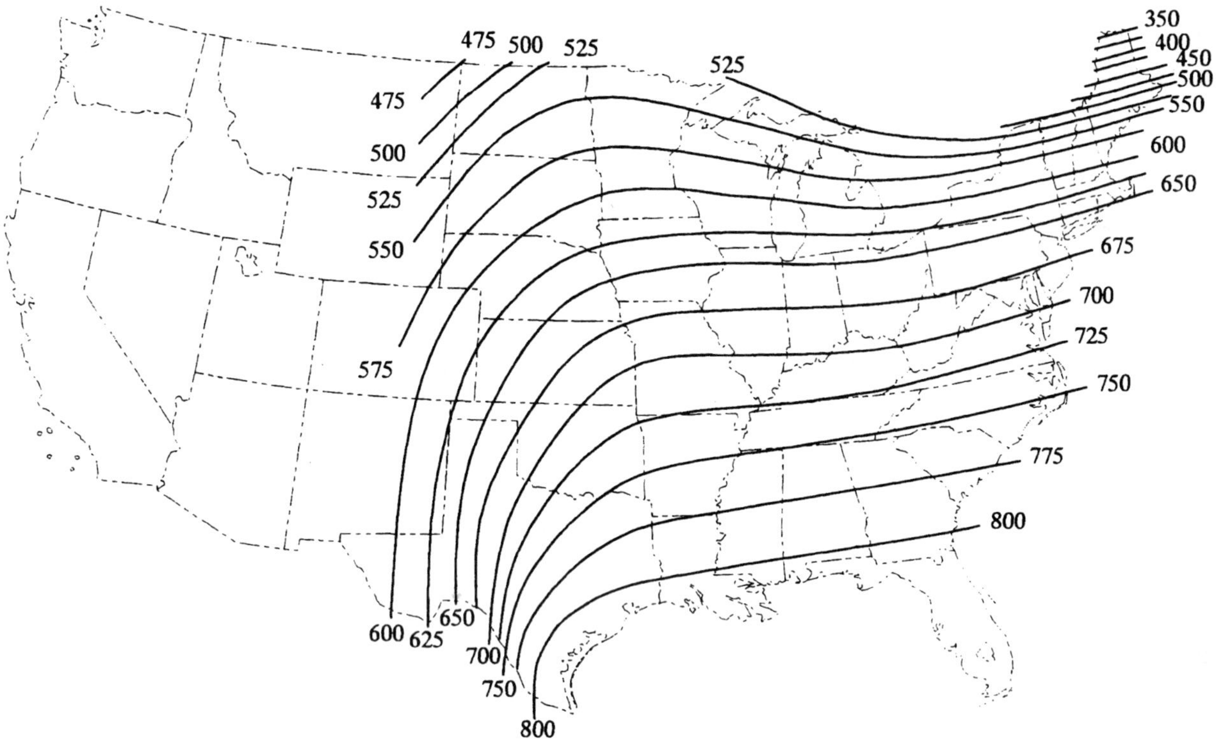

Figure 2.12.—All-season Probable Maximum Precipitation (PMP, mm) for 6 hr, 25.9 km² (10 mi.²), United States East of 105W Longitude (from Schreiner and Riedel, 1978)

"the SPS estimate for a particular drainage area and season of year in which snowmelt is not a major consideration should represent the most severe flood-producing rainfall depth-area-duration relationship and isohyetal pattern of any that is considered reasonably characteristic of the region in which the drainage basin is located, giving consideration to the runoff characteristics and existence of water regulation structures in the basin." The SPS index isohyets used in developing the SPS were originally developed by approximating 40 to 60% of the PMP.

VI. PRECIPITATION FREQUENCY ANALYSIS

Chapter 8, Sec. III, Statistical analysis for estimating floods, provides additional discussion of frequency analysis. The discussion that follows summarizes conventional analyses of precipitation records.

Precipitation frequency analyses are used extensively for design of engineering works that control storm runoff. These include municipal storm sewer systems, highway and railway culverts, and agricultural drainage systems. A common application of precipitation frequency analysis is development of a "design storm," that is, a precipitation pattern used in water resource systems design.

The objective of precipitation frequency analysis is to develop useful relationships between four variables associated with precipitation events: 1) the size of the area of interest (km²; often specified by a catchment area), 2) the duration of interest (in hours), 3) the frequency of occurrence, or "return interval," of the event (in years), and 4) the precipitation accumulation (mm). The most common way of presenting precipitation frequency analysis results is to represent rainfall accumulation in terms of the other three variables. For example, the product of a precipitation frequency analysis could be the 100-year, 24-hour rainfall accumulation for a 259 km² area.

The language and methodology of precipitation frequency analysis come from probability and statistics. The connection between probability of an event and return interval of an event is central to precipitation frequency analysis and arises as follows. If an event has probability of occurrence p for a year, then the return interval T is the average time, in years, between events and is given by p^{-1}. The 100-year storm, for example, is the storm magnitude that is exceeded with probability 0.01 in any given year.

A. Rain Gage Data for Frequency Analysis

The types of data typically available for frequency analyses include the following:

(a) Continuous records of daily rain gage accumulations (published monthly in "Climatological Data" and available through the National Climatic Data Center (NCDC) in Asheville, North Carolina).
(b) Continuous records of clock-hour rain gage accumulations (published in "Hourly Precipitation Data" and available through NCDC).
(c) Annual maximum N-min rain gage accumulations (time increments N are typically 5, 15, and 60 min; data are available through NCDC).
(d) Rainfall accumulation data from dense networks of rain gages (e.g., data from experimental watersheds of the Agricultural Research Service of USDA, which are described in Thurman et al., 1984).

The spatial density of rain gage networks is varied; the list given above generally represents decreasing spatial density for numbers 1 through 3. Data from dense experimental networks, which are needed for computing depth-area correction factors, are often far from the area for which precipitation frequency computations are needed.

Rain gage data that are used for precipitation frequency analysis are typically available in the form of annual maxima or converted to this form in the case of continuous records of clock-hour or daily data. An alternative data format for frequency studies is "partial duration series" (also referred to as peaks over threshold data). Analyses using partial duration series typically give higher design storm values than annual peak analyses for short return intervals (2–5 years). There is little or no difference between the two types of analyses for long return intervals (10 years or longer) (Stedinger et al., 1993). Analyses for short return periods that are based on annual peak observations often contain adjustments to reflect results based on partial duration series analyses (Frederick et al., 1977).

B. Frequency Analysis Techniques

The "Gumbel distribution" has been consistently used by the NWS (Gumbel, 1958; Myers and Zehr, 1980) for precipitation frequency analysis. The Gumbel distribution is specified by the following formula:

$$F(x) = e^{-e^{-\frac{(x-a)}{b}}} \tag{2.9}$$

where $F(x)$ is the cumulative distribution function and a and b are unknown parameters. If $F(x)$ represents the distribution of maximum annual rainfall (for a specific duration), then the rainfall accumulation [Q(T)] with return period T is given by

$$Q(T) = (a - b)\ln\left[\ln\left(\frac{T}{T-1}\right)\right]. \tag{2.10}$$

The parameters a and b can be estimated by the "method of moments," given a sample $X_1, \ldots, X_n$ of annual maximum rainfall values for n years. The parameter estimators are given by

$$\hat{b} = \frac{\sqrt{6}\hat{s}}{\pi} \tag{2.11}$$

$$\hat{a} = \hat{\mu} - 0.577\hat{b} \tag{2.12}$$

where

$$\hat{\mu} = n^{-1}\sum_{i=1}^{n} X_i \tag{2.13}$$

is the sample mean and

$$\hat{s} = \left[(n-1)^{-} \sum_{i-1}^{n} (X_i - \hat{\mu})^2 \right]^{1/2} \tag{2.14}$$

is the sample standard deviation.

Table 2.3 contains annual maximum hourly rainfall accumulations from Oklahoma City, Oklahoma. The sample mean of the observations is 38 mm. The sample standard deviation of the observations is

TABLE 2.3. Annual Maximum Values of Hourly Rainfall at Oklahoma City, OK, for the 41-yr Period 1948–1988

Year	Month	Day	Hour	Accumulation (mm)
1948	6	21	18	37
1949	5	17	20	31
1950	8	15	21	47
1951	6	10	9	37
1952	5	23	4	26
1953	12	2	20	31
1954	5	1	18	31
1955	6	16	1	83
1956	10	14	16	32
1957	5	24	20	42
1958	6	20	5	33
1959	9	3	8	40
1960	7	29	22	38
1961	7	13	22	29
1962	8	31	12	22
1963	7	13	16	42
1964	5	10	19	38
1965	8	6	15	27
1966	8	18	22	36
1967	4	12	2	40
1968	5	7	1	28
1969	6	14	1	49
1970	5	29	7	53
1971	8	14	4	19
1972	7	2	15	33
1973	11	19	20	27
1974	6	8	6	44
1975	7	29	13	33
1976	5	30	2	20
1977	5	19	17	32
1978	6	21	13	47
1979	5	20	17	38
1980	6	22	3	23
1981	7	27	22	70
1982	9	14	21	26
1983	5	13	23	27
1984	10	27	1	51
1985	9	13	16	35
1986	10	2	9	48
1987	5	27	18	40
1988	6	29	19	62

13 mm. From Equations 2.11 and 2.12, the parameter estimates are $\hat{a} = 32.2$, $\hat{b} = 10.1$. From equation 2.10, the 2-year, hourly rainfall accumulation is computed to be 36 mm; the 100-year, hourly accumulation is 79 mm. These values are based on clock hour.

Alternative frequency analysis techniques have been integrated into standard engineering practice. The generalized extreme value (GEV) distribution, which contains the Gumbel distribution as a special case, has been used in the United Kingdom (National Environment Research Council (NERC), 1975) and the Netherlands (Buishand, 1989). Alternative methods for estimating parameters, which are superior for small sample sizes, are described in Hosking (1990). Wallis (1982), Buishand (1984), and Schaefer (1990) develop point precipitation frequency products using a "regional frequency analysis" procedure.

C. Point Precipitation Frequency Analysis

The standard source of precipitation frequency procedures and results in the United States for many years was the work by Hershfield (1961), which is commonly referred to as TP-40. Recent updates to TP-40 include the works by Frederick et al. (1977) and Miller et al. (1973). The work of Frederick et al. (1977) is typically referred to as HYDRO-35. These reports provide point precipitation frequency maps for durations ranging from 5 minutes to 10 days and return intervals ranging from 2 years to 100 years. The standard steps in producing these maps are the following:

a) Estimate return intervals for N-min intensities using the Gumbel distribution.
b) Apply corrections for conversion to partial duration series results, as needed (Frederick et al., 1977).
c) Develop isopluvial maps from frequency results for gage sites, using interpolation procedures and subjective judgement.

Figure 2.13 shows 100-year, 24-hour precipitation in the United States from TP-40. Figures 2.14a and 2.14b show 2-year, 60-minute precipitation and 100-year, 60-minute precipitation from HYDRO-35. TP-40

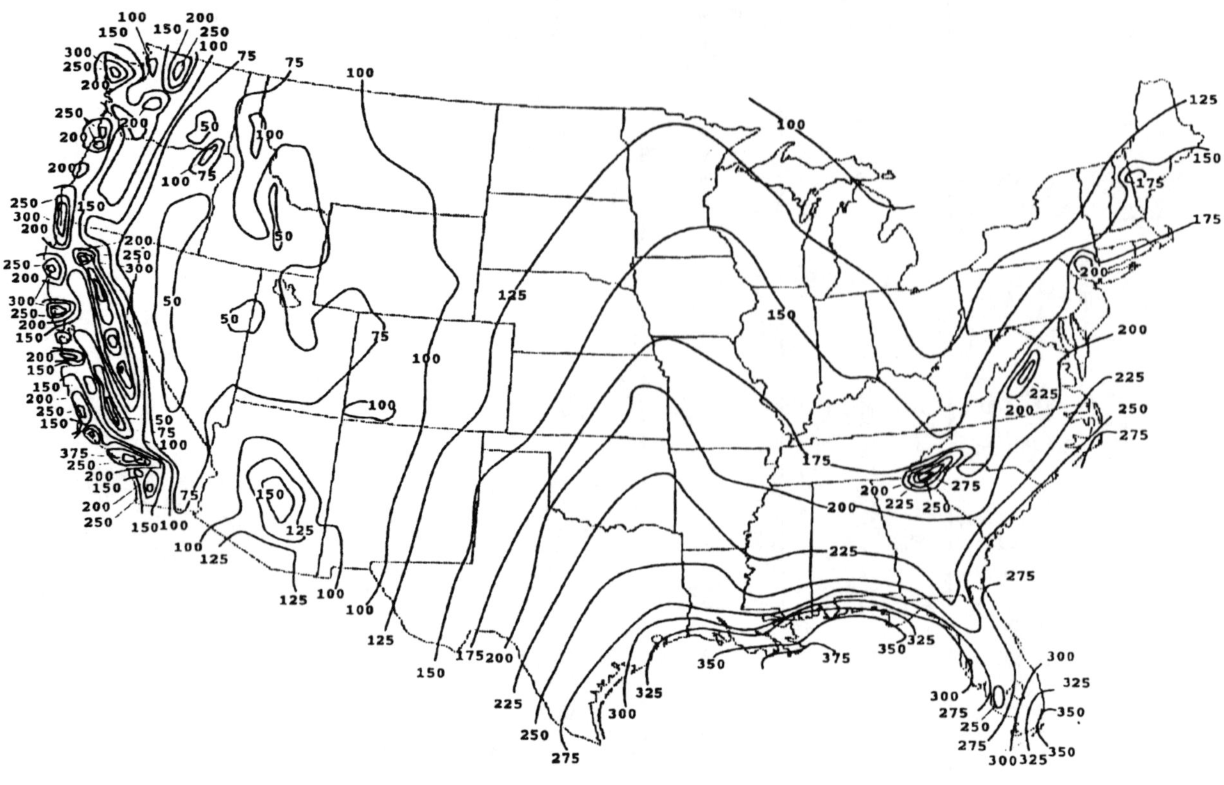

Figure 2.13.—100-yr 24-hr Precipitation (from Hershfield, 1961)

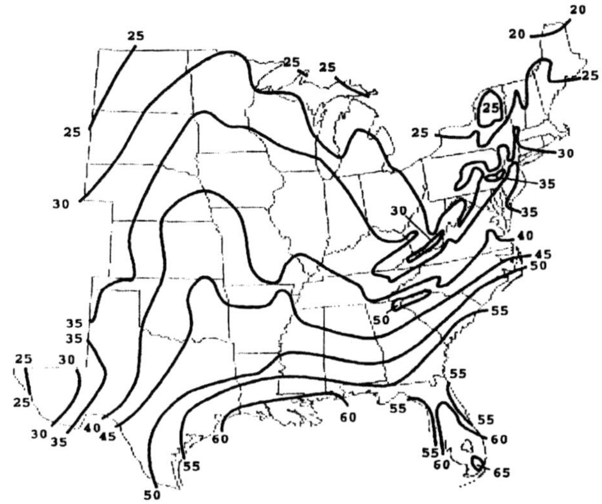

Figure 2.14a.—2-yr 60-min Precipitation for the Eastern United States (from Frederick et al., 1977)

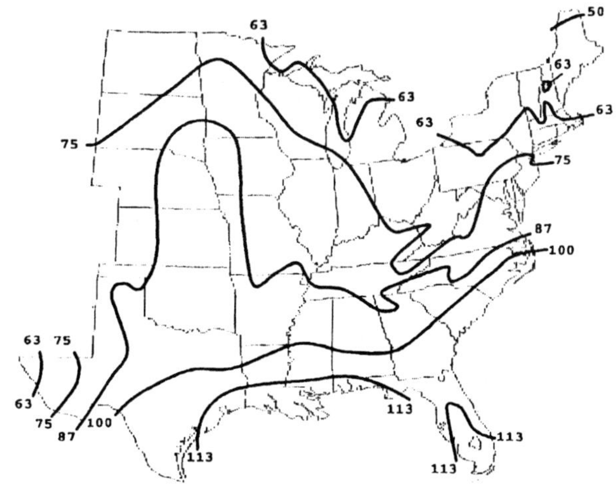

Figure 2.14b.—100-yr 60-min Precipitation for the Eastern United States (from Frederick et al., 1977)

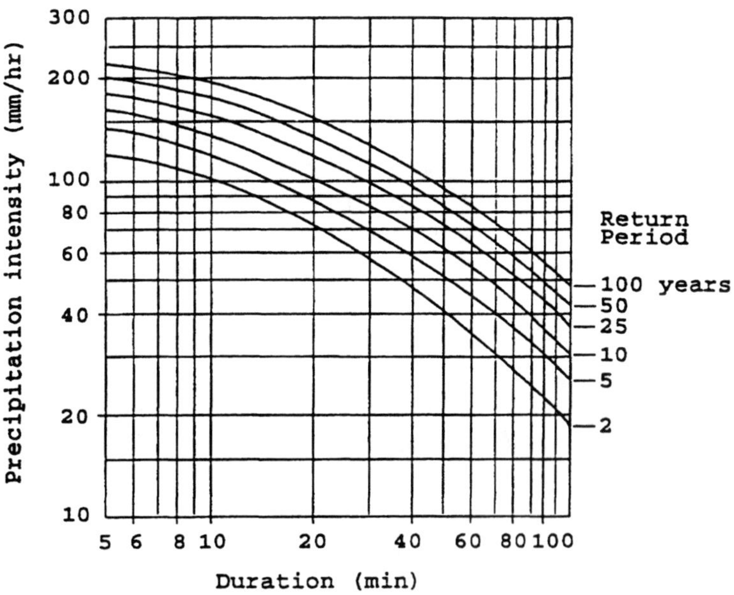

Figure 2.15.—Intensity-Duration-Frequency Curves for Chicago, IL

and HYDRO-35 provide results at time intervals ranging from 5 minutes to 2 days and for return intervals of 2 years and 100 years. Results from TP-40 were extended in the United States Weather Bureau (1964) to cover time intervals from 2–10 days. Empirical equations have been developed for relating these results to precipitation frequency values for return intervals and time intervals for which results are not explicity presented (Chow et al., 1988).

Precipitation frequency results are often summarized by "intensity-duration-frequency relationships" or IDF curves. Figure 2.15 illustrates the form of IDF curves for Chicago, Illinois. IDF curves can either be developed from precipitation frequency analysis using local data or derived from standard publications, such as HYDRO-35.

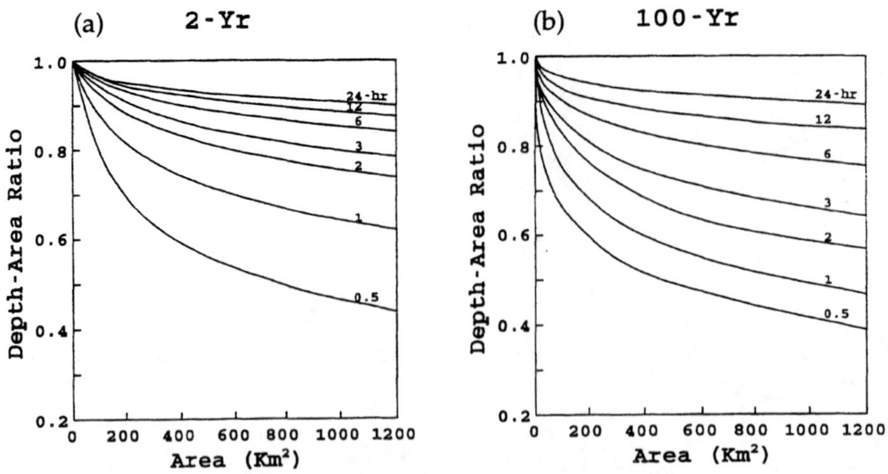

Figure 2.16.—Depth-area Correction Factors for 2-yr and 100-yr Storms and Specified Durations (from Myers and Zehr, 1980)

D. Frequency Analysis for Area-Averaged Precipitation

TP-40 and its successors provide point precipitation frequency values. For design purposes, it is necessary to convert point values to areal values. This conversion is generally accomplished by applying an "area correction factor" to point precipitation values.

Dense networks of rain gages have been used to develop depth-area-duration correction factors, as described in Zehr and Myers (1984) and Osborn et al. (1980). Figures 2.16a and 2.16b illustrate the correction factors presented in Myers and Zehr (1980). For the 2-year, 1-hour storm, the area correction factor at 1000 km^2 is 64% of the point value. For the 100-year, 1-hour storm, the area correction factor drops to 49% of the point value for an area of approximately 1000 km^2.

Area correction factors can vary significantly from location to location. Validation is required if area correction factors are to be used far from the location in which they were developed.

Rodriguez-Iturbe and Mejia (1974) have developed procedures for developing area correction factors that utilize information on the spatial correlation of precipitation (Bras and Rodriquez-Iturbe, 1985). The information required for development of correction factors is more easily regionalized than are standard procedures.

E. Storm Hyetographs

For design applications, it is often necessary to specify a temporal pattern, or "hyetograph," associated with rainfall depths of a given duration and frequency. For example, 5-year, 60-minute rainfall depth may be used in the form of 5-minute increments in hydrologic models.

Approaches to specify temporal patterns of design rainfall depths are given in Wenzel (1982), Pilgrim and Cordery (1975), Huff and Changnon (1964), and Huff (1967). Pilgrim and Cordery (1975) warn that approaches that overly smooth the temporal rainfall patterns are unsuited for design applications, because the relative variability of intensities is the factor which often has the greatest effect on the design hydrograph. Procedures for developing patterns that incorporate the average variability of storm rainfall are described in Pilgrim and Cordery (1975).

Two important points noted by Pilgrim and Cordery (1975) and Huff and Changnon (1964) are that the variability of intensities decreased with decreasing exceedance probability and that the majority of extreme storms have multiple peaks of high rainfall intensity. Multiple peaks in rainfall intensity are typical of the flash-flood producing storms described by Chappell (1989).

F. New Technologies for Precipitation Frequency Analysis

The density of rain gages has been a significant limitation in development and implementation of precipitation frequency analysis procedures. Radar provides a potentially important source of precipitation data for frequency analyses. A network of weather radars, termed WSR-88D (Weather Surveillance Radar—1988 Doppler; the development title for the system was Next Generation Weather Radar, or NEXRAD), will be deployed throughout the United States from 1991 to 1997 (Crum et al., 1993). A description of the hydrometeorological capabilities of the system is given in Hudlow et al. (1991). Use of radar data for precipitation frequency analysis has been proposed by Frederick et al. (1977).

Procedures developed for converting point precipitation frequency values to areal values are, at best, ad hoc. Models of the spatial (and temporal) distribution of rainfall have been developed for directly carrying out areal precipitation frequency analyses (Bras and Rodriguez-Iturbe, 1985; Gupta and Waymire, 1979; Smith and Karr, 1990).

Increasingly, design problems require information on very rare hydrologic events, i.e., events with return intervals much longer than 100 years. PMP (Hansen, 1987) has been the traditional tool for addressing these design problems. New frequency analysis procedures, which exploit some of the tools of PMP, have been developed for assessing rainfall magnitudes with very long return intervals. Especially noteworthy are stochastic storm transposition techniques (Fontaine and Potter, 1989; Foufoula-Georgiou, 1989).

VII. WEATHER MODIFICATION

Weather modification, known as cloud seeding in many localities, is generally thought of as a means of augmenting rainfall and snowfall, decreasing the amount and/or size of hail, and as a method of dissipating fog on airport runways. Weather modification, as we understand it today, started in the 1940's with experiments using dry ice by Irving Langmuir and Vincent Schaefer (Dennis, 1980; Elliott, 1983; Havens et al., 1981). In 1946, Schaefer dropped dry ice pellets into stratocumulus clouds over Massachusetts and the seeded path turned to snow, which left a hole in the clouds. The dry ice pellets cooled the air to −40° C or lower which caused ice crystals to form. These ice crystals grew by deposition and, when they became large enough, fell as precipitation. In 1947, Vonnegut demonstrated that silver iodide smoke particles could be used as a seeding agent because silver iodide particles have a similar crystalline structure to ice crystals. Silver iodide crystals act as effective ice nuclei at temperatures of −4° C and lower. Other substances, such as common salt, lead iodide, cupric sulfide, some organics and dry ice are used occasionally; however, silver iodide is the most commonly used cloud seeding agent. Silver iodide is generally delivered to the clouds by air currents in the form of smoke from either ground burners or burners that are mounted on aircraft (Committee on Weather Modification, 1983). Since the first cloud seeding experiments in the 1940's, numerous seeding experiments and field operations in some 30 countries around the world have taken place (Todd and Howell, 1985a). Todd and Howell also state that between 3 and 10% of the area of the contiguous United States has been under a cloud seeding program during the 25-year prior to their report. In a summary report of seeding projects from around the world, Todd and Howell (1985b) found that precipitation increased by about 20% for hygroscopic and ice-phase nucleants. They also found a median increase of 11% to 17% for all other studies. In general, the scientific consensus is that dissipation of supercooled fog and stratus is an operational technology that can be used to open airport runways, etc. (Dennis, 1980; Todd and Howell, 1985a).

Summary reports other than the one by Todd and Howell (1985b) suggest that precipitation can be increased 10-15% from specific orographic cold cloud systems in the western United States but decreasing orographic precipitation is also possible (Dennis, 1980; Todd and Howell, 1985a). Neyman et al. (1972) analyses of the 7-year Arizona experiment in the Santa Catalina mountains showed a 30% decrease of rain. Chappell (1972) found that snowfall augmentation in mountainous areas of the western United States resulted in increases of 10% to 30% in wintertime precipitation; however, later analysis by Rango and Hobbs (1987) of seeding done in the Colorado Rockies indicated that increased precipitation by winter seeding could not be confirmed. In a western Montana study, Super (1986) found that seeding with silver iodide only increased winter precipitation in just a small percent of the cases, but never decreased precipitation.

Results of seeding convective clouds are very inconclusive with some studies showing increased precipitation, some showing decreases and most not being able to show any results because of inadequate statistical design, poor seeding procedures, or other problems with the cloud seeding projects (Dennis, 1980). Aircraft seeding of silver iodide resulted in rainfall increases of up to 26% in some areas of Israel (Rosenfeld and Farbstein, 1992); however, in other areas of Israel precipitation was not affected when seeding was performed on days when dust from the Sahara-Arabian deserts was observed. This indicates that there are many environmental influences that can determine the outcome of seeding operations. In a cloud seeding experiment in west Texas, Rosenfeld and Woodley (1989) found a positive response to silver iodide seeding but the sample was small, which prevented rigorous statistical analysis. Flueck et al. (1986) found a positive effect to seeding for enhancing convective rainfall in two studies in Florida, but as in the Rosenfeld and Woodley (1989) study these results could not be statistically substantiated.

A review of the literature on weather modification indicates that cloud seeding projects to suppress hail have been conducted in numerous locations around the world (Dennis, 1980; Todd and Howell, 1985b). Atlas (1977) and Dennis (1980) found that in general data from most hail suppression experiments could not be used to show whether hail increased, decreased or had no effect; however, Dennis (1980) does say that updraft seeding resulted in a modest 20 to 30% suppression. This is supported by results presented by Todd and Howell (1985b) who found that there was a decrease in hail of up to 50% from their review of numerous seeding programs around the world. Johnson et al. (1994) found a decrease of about 49% in crop-hail insurance loss payments in six western North Dakota counties due to glaciogenic seeding material that was delivered to summer convective clouds during the 1976 through 1988 period. An evaluation of the effectiveness of silver iodide ground generators for reducing hail in southwestern France by Dessens (1987) showed that hail damage had been reduced 41% during the course of the 30-year project. These results were also based on hail insurance reports. Mezeix (1987) questioned Dessens' results because of a lack of statistical significance at a high enough probability level, but Dessens (1987) suggests that Mezeix's comments only strengthened the case for reduced hail damage due to cloud seeding. In a recent study of hail suppression by silver iodide in Yugoslavia, Mesinger and Mesinger (1992) show that the frequency of hail was reduced in the order of 15-20%, which supports the conclusions of Dennis (1980) and Todd and Howell (1985b).

Another form of weather modification is "inadvertent weather modification," which results from a wide range of human activities such as urban-produced air pollution and heating, dust from overgrazing in the drier areas of the world, burning agricultural and forest lands and even contrails that are caused by high altitude aircraft (Changnon, 1980). The temperature in urban areas is generally warmer than in surrounding rural areas and urban areas are often associated with local increases in precipitation and severe weather and decreases in visibility and solar radiation. The effect of urban areas on local weather is directly related to the size of the city and the amount of industry located in and around the city. The METROMEX study of St. Louis, Missouri showed that total annual rainfall has increased by 30% (Changnon, 1980). A more recent evaluation of precipitation data from St. Louis showed that environmental effects of the city increased both the amount and frequency of precipitation mostly during fall months (Changnon et al., 1991). Precipitation increases of up to 10% have been observed in the vicinity of cooling towers and increases in the order of 25% have occurred in large irrigated areas such as central Washington and Nebraska.

VIII. SYNTHETIC WEATHER GENERATION

Many hydrologic processes such as runoff and erosion depend on local weather conditions. Therefore, precipitation and other weather data are needed to assess the effects of climate on engineering projects and as inputs to hydrologic models. At many sites, climate records are not sufficient in length or are not available for making the desired analyses. This is especially true in mountainous areas, where weather-measuring stations may not represent conditions in nearby watersheds. Therefore, it is often desirable to have the capability to generate climatic data series that have the appropriate statistical characteristics for specific locations or for watersheds.

Several different types of weather simulation models have been developed, most of which have been developed for crop growth models; however, some have been developed for water quality and hydrologic studies. These models range from those that were developed to generate monthly and annual precipitation (Srikanthan and McMahon, 1982), to some that can be used to generate daily precipitation (Adamowski and Smith, 1972; Carey and Haan, 1978; and Richardson, 1978) and those that have the ability to generate several weather variables on a daily time interval (Bruhn et al., 1980; Hanson et al., 1994; Hungerford et al., 1989; Larsen and Pense, 1982; Nicks and Gander, 1993, 1994; Pickering et al., 1994; Woolhiser et al., 1988; and Young, 1994). A detailed discussion on modeling daily precipitation is presented by Woolhiser (1992).

One method of simulation is to extrapolate weather data from a climatic measuring site to another site that has no data. Hungerford et al. (1989) developed a model to provide inputs to ecological models "that is designed to extrapolate routine National Weather Service (NWS) data to adjacent mountainous terrain." Their model simulates daily solar radiation, air temperature, humidity and precipitation. The two stochastic simulation models USCLIMATE (Hanson et al., 1994) and CLIGEN (Nicks and Gander, 1993; 1994) were developed to simulate daily climatic data. USCLIMATE can be used to simulate precipitation probabilities and data for daily precipitation, maximum and minimum air temperature, and solar radiation at any given location within the contiguous United States. Daily temperature and solar radiation data are conditioned on the precipitation process which is described by a Markov chain-mixed exponential model. Parameters for specific sites within a region can be accessed directly or they can be estimated for points between sites. CLIGEN was developed to generate weather data for more than 1100 sites in the United States, Puerto Rico, and the Pacific Islands. The procedure that determines whether a day is wet or dry is based on a two-state Markov chain, and the amount of precipitation per wet day is based on a skewed normal distribution. Generated air temperature and solar radiation are conditioned on whether the day is wet or dry. CLIGEN can be used to generate dew point temperature, and wind speed and direction. CLIGEN also has procedures for computing storm duration, peak storm intensity, and storm disaggregation for rainfall intensity patterns. Econopouly et al. (1990) and Hershenhorn and Woolhiser (1987) also present procedures for disaggregation of daily rainfall, i.e., generating short-term intensities from daily rainfall.

These models are used to simulate weather variables at a certain point, and no provisions exist in them to account for spatial dependence between sites. This is a shortcoming of the present models when their output is used in watershed or larger area hydrologic studies.

IX. REFERENCES

Adamowski, K,. and Smith, A. F. (1972). "Stochastic generation of rainfall." *J. Hydr. Div.*, ASCE, 98(HY11), 1935–1945.

Adler, R. F. and Mack, R. A. (1984). "Thunderstorm cloud height-rainfall rate relationships for use with satellite estimation techniques." *J. Clim. Appl. Meteor.*, 23(2), 280–296.

Ahrens, C. D. (1991). *Meteorology today, an introduction to weather, climate, and the environment.* West Publishing Company, St. Paul, MN.

Allerup, P. and Madsen, H. (1980). "Accuracy of point precipitation measurements." *Nordic Hydrol.*, 11(2), 57–70.

———. (1986). "On the correction of liquid precipitation." *Nordic Hydrol.*, 17(4-5), 237–250.

Alter, J. C. (1910). "Seasonal precipitation measurements." *Monthly Wea. Rev.*, 38(2), 1885–1886.

———. (1937). "Shielded storage precipitation gages." *Monthly Wea. Rev.*, 65(7), 262–265.

American Society of Agricultural Engineers Standard. (1990). "Procedure for sprinkler testing and performance reporting." ASAE S398.1., ASAE, St. Joseph, MO.

Arkin, P. A. and Ardanuy, P. E. (1989). "Estimating climate-scale precipitation from space. A review." *J. Climate,* 2(11), 1229–1238.

Ashmore, S. E. (1934). "The splashing of rain." *Quart. J. Royal Meteor. Soc.,* 60, 517–522.

Atlas, D. (1977.) "The paradox of hail suppression." *Science,* 195(4274), 139–145.

Atlas, D., Rosenfeld, D., and Short, D. A. (1990). "The estimation of convective rainfall by area integrals. 1. The theoretical and empirical basis." *J. Geophys. Res.,* 95(D3), 2153–2160.

Atlas, D. and Thiele, O. W., eds. (1981). *Precipitation measurements from space.* NASA, Goddard Space Flight Center, Greenbelt, MD.

Austin, P. M. (1987). "Relation between measured radar reflectivity and surface rainfall." *Monthly Wea. Rev.,* 115(5), 1053–1070.

Barnston, A. G. and Thomas, J. L. (1983). "Rainfall measurement accuracy in FACE: a comparison of gage and radar rainfalls." *J. Clim. Appl. Meteor.,* 22(12), 2038–2052.

Barrett, E. C. (1981). *The use of satellite data in rainfall monitoring.* Academic Press, New York, NY.

Beaumont, R. T. (1965). "Mt. Hood pressure pillow snow gage." *Proc. Western Snow Conference,* 33rd Annual Meeting, Colorado Springs CO, 29–36.

Beaumont, R. T. (1966). "Snow accumulation." *Proc. Western Snow Conference,* 34th Annual Meeting, Seattle WA, 3–6.

Bellon, A. and Austin, G. L. (1984). "The accuracy of short-term radar rainfall forecasts." *J. Hydrol.,* 70(1), 35–49.

Bellon, A. and Austin, G. L. (1986). "On the relative accuracy of satellite and raingage rainfall measurements over middle latitudes during daylight hours." *J. Clim. Appl. Meteor.,* 25(11), 1712–1724.

Biswas, A. K. (1967). "Development of rain gages." *J. Irrig. and Drain. Div.,* ASCE, 93(IR3), 99–124.

Biswas, A. K. (1970). *History of hydrology,* American Elsevier Pub. Co. Inc., New York, NY.

Bleasdale, A. (1959). "The measurement of rainfall." *Weather,* 14, 12–18.

Bodtmann, W. F. and Ruthroff, C. L. (1974). "Rain attenuation on short radio paths: theory, experiment, and design." *Bell System Tech. J.,* 53, 1329–1349.

Bodtmann, W. F. and Ruthroff, C. L. (1976). "The measurement of 1 min rain rates from weighing raingage recordings." *J. Appl. Meteor.,* 15(11), 1160–1166.

Brakensiek, D. L., Osborn, H. B., and Rawls, W. J. (1979). "Field manual for research in agricultural hydrology." *Agriculture Handbook 224,* U.S. Dept. of Agriculture, Washington, D.C.

Bras, R. and Rodriguez-Iturbe, I. (1985). *Random functions and hydrology.* Addison-Wesley, Reading, MA.

Brock, F. V. (1985). "Ground-based observing systems." *Handbook of applied meteorology,* D. D. Houghton, ed., John Wiley & Sons, NY, 341–351.

Brown, M. J. and Peck, E. L. (1962). "Reliability of precipitation measurements as related to exposure." *J. Appl. Meteor.,* 1(2), 203–207.

Bruhn, J. A., Fry, W. E., and Fick, G. W. (1980). "Simulation of daily weather using theoretical probability distributions." *J. Appl. Meteor.,* 19(9), 1029–1036.

Buishand, T. A. (1984). "Bivariate extreme-value data and the station year method." *J. Hydrol.,* 69(4), 77–95.

Buishand, T. A. (1989). "Statistics of extremes in climatology." *Statistica Neerlandica,* 43(1), 1–29.

Burke, H-H. K., Bowley, C. J., and Barnes, J. C. (1984). "Determination of snowpack properties from satellite passive microwave measurements." *Remote Sens. of Environ.,* 15, 1–20.

Carey, D. I. and Haan, C. T. (1978). "Markov process for simulating daily point rainfall." *J. Irrig. and Drain. Div.,* ASCE, 104(IR1), 111–125.

Carroll, T. R. and Allen, M. (1988). *Airborne gamma radiation snow water equivalent and soil moisture measurements and satellite areal extent of snow cover measurements. A user's guide, Version 3.0.* National Remote Sensing Hydrology Program, Office of Hydrology, National Weather Service, Minneapolis, MN.

Carroll, S. S. and Carroll, T. R. (1989a). "Effect of uneven snow cover on airborne snow water equivalent estimates obtained by measuring terrestrial gamma radiation." *Water Resour. Res.*, 25(7), 1505–1510.

Carroll, S. S. and Carroll, T. R. (1989b). "Effect of forest biomass on airborne snow water equivalent estimates obtained by measuring terrestrial gamma radiation." *Remote Sens. of Environ.*, 27(13), 313–319.

Carroll, S. S. and Carroll, T. R. (1990). "Estimating the variance of airborne snow water equivalent estimates using computer simulation techniques." *Nordic Hydrol.*, 21(1), 35–46.

Changnon, S. A. (1980). "Is your weather being modified?" *J. Irrig. and Drain. Div.*, ASCE, 106(IR1), 37–48.

Changnon, S. A., Shealy, R. T., and Scott, R. W. (1991). "Precipitation changes in fall, winter, and spring caused by St Louis." *J. Appl. Meteor.*, 30(1), 126–134.

Chappell, C. F. (1972). "Orographic cloud seeding as a water resource." *Proc. Irrigation and Drainage Division Specialty Conference*, ASCE, New York, NY, 389–397.

Chappell, C. F. (1989). "Quasistationary convective events." Mesocale Meteorology, P.S. Ray, ed., American Meteorology Society, Boston, MA.

Chidley, T. R. E. and Keys, K. M. (1970). "A rapid method of computing areal rainfall." *J. Hydrol.*, 12(1), 15–24.

Chow, V. T., ed. (1964). *Handbook of applied hydrology*. McGraw-Hill Book Co., New York, NY.

Chow, V. T., Maidment, D. R., and Mays, L. W. (1988). *Applied hydrology*. McGraw-Hill Book Co., New York, NY.

Chua, S. and Bras, R. L. (1982). "Optimal estimators of mean areal precipitation in regions of orographic influence." *J. Hydrol.*, 57(1), 23–48.

Clagett, G. P. (1988). "The Wyoming windshield—an evaluation after 12 years of use in Alaska." *Proc. Western Snow Conference*, 56th Annual Meeting., Kalispell, MT, 113–123.

Collier, C. G. (1986a). "Accuracy of rainfall estimates by radar, part I: Calibration by telemetering raingauges." *J. Hydrol.*, 83(3), 207–223.

Collier, C. G. (1986b). "Accuracy of rainfall estimates by radar, Part II: Comparison with raingauge network." *J. Hydrol.*, 83(3), 225–235.

Collier, C. G. and Knowles, J. M. (1986). "Accuracy of rainfall estimates by radar, Part III: Application for short-term flood forecasting." *J. Hydrol.*, 83(3), 237–249.

Collier, C. G., Larke, P. R., and May, B. R. (1983). "A weather radar correction procedure for real time estimation of surface rainfall." *Quart. J. Royal Meteor. Soc.*, 109, 589–608.

Committee on Weather Modification. (1983). "Guidelines for cloud seeding to augment precipitation." *J. Irrig. and Drain. Engrg.*, ASCE, 109(1), 111–182.

Costello, T. A. and Williams, H. A. Jr., (1991). "Short duration rainfall intensity measured using calibrated time-of-tip data from a tipping bucket raingage." *Ag. and Forest. Meteor.*, 57(1-3), 147–155.

Court, A. (1960). "Reliability of hourly precipitation data." *J. Geophys. Res.*, 65(12), 4017–4024.

Cox, L. M., Bartee. L. D., Crook, A. G., Farnes, P. E., and Smith, J. L. (1978). "The care and feeding of snow pillows." *Proc. Western Snow Conference*, 46th Annual Meeting, Otter Rock, OR, 40–47.

Crane, R. K. (1980). "Prediction of attenuation by rain." *IEEE Trans. Comm.*, COM-28(9), 1717–1733.

Creutin, J. D., Delrieu, G., and Lebel, T. (1988). "Rain measurement by raingage-radar combination: A geostatistical approach." *J. Atmos. Ocean. Tech.*, 5(1), 102–115.

Creutin, J. D. and Obled, C. (1982). "Objective analyses and mapping techniques for rainfall fields: An objective comparison." *Water Resour. Res.*, 18(2), 413–431.

Croley, T. E. II and Hartmann, H. C. (1985). "Resolving Thiessen polygons." *J. Hydrol.*, 76(1), 363–3797.

Crum, T. D., Alberty, R. L., and Burgess, D. W. (1993). "Recording, archiving, and using WSR-88D data." *Bull. Amer. Meteor. Soc.*, 74(4), 645–653.

Daly, C. and Neilson, R. P. (1992). "A digital topographic approach to modeling the distribution of precipitation in mountainous terrain." *Interdisciplinary Approaches in Hydrology and Hydrogeology*, M. E. Jones and A. Laenen, eds., American Institute of Hydrology, Minneapolis, MN, 437–454.

Daly, C., Neilson, R. P., and Phillips, D. L. (1994). "A statistical-topographic model for mapping climatological precipitation over mountainous terrain." *J. Appl. Meteor.*, 33(2), 140–158.

Dennis, A. S. (1980). *Weather Modification by Cloud Seeding*. Academic Press, New York, NY.

Dessens, J. (1987). "Reply Hail in Southwestern France. Part II: Results of a 30-year hail prevention project with silver iodide seeding from the ground." *J. Clim. Appl. Meteor.*, 26(12), 1777–1778.

Deutsch, M., Wiesnet, D. R., and Rango, A., eds. (1981). *Satellite hydrology*, AWRA, Minneapolis MN.

Dingman, S. L., Seely-Reynolds, D. M., and Reynolds R. C., III. (1988). "Application of kriging to estimating mean annual precipitation in a region of orographic influence." *Water Resour. Bull.*, 24(2), 329–339.

Diskin, M. H. (1969). "Thiessen coefficients by a Monte Carlo procedure." *J. Hydrol.*, 8(3), 323–335.

Doneaud, A. A., Ionescu-Niscov, S., and Miller, J. R. (1984a). "Convective rain rates and their evolution during storms in a semiarid climate." *Monthly Wea. Rev.*, 112(8), 1602–1612.

Doneaud, A. A., Ionescu-Niscov, S., Priegnitz, D. L., and Smith, P. L. (1984b). "The area-time integral as an indicator for convective rain volumes." *J. Clim. Appl. Meteor.*, 23(4), 555–561.

Doneaud, A. A., Miller, J. R., Johnson, L. R., Vonder Haar, T. H., and Laybe, P. (1987). "The area-time integral technique to estimate convective rain volumes over areas applied to satellite data—a preliminary investigation." *J. Clim. Appl. Meteor.*, 26(1), 156–169.

Doneaud, A. A., Smith, P. L., Dennis, A. S., and Sengupta, S. (1981). "A simple method for estimating convective rain volume over an area." *Water Resour. Res.*, 17(6), 1676–1682.

Doorenbos, J. (1976). "Agro-meteorological field stations." *Irrigation and Drainage Paper 27*. Food and Agriculture Organization of the United Nations, Rome.

Econopouly, T. W., Davis, D. R., and Woolhiser, D. A. (1990). "Parameter transferability for a daily rainfall disaggregation model." *J. Hydrol.*, 118(1-4), 209–228.

Elliott, R. D. (1983). "Section 1—summary. Committee on Weather Modification of the Irrigation and Drainage Division. Guidelines for cloud seeding to augment precipitation." *J. Irrig. and Drain. Engrg.*, ASCE, 109(1), 112–116.

Engman, E. T. and Gurney, R. J. (1991). *Remote sensing in hydrology.* Chapman and Hall, New York, NY.

Epema, G. F. and Riezebos, H. (1984). "Drop shape and erosivity. Part I: Experimental set up, theory, and measurements of drop shape." *Earth Surface Processes and Landforms*, 9(6), 567–572.

Farnes, P. E., Peterson, N. R., Goodison, B. E., and Richards, R. P. (1982). "Metrication of manual snow sampling equipment." *Proc. Western Snow Conference*, 50th Annual Meeting, Reno, NV, 120–132.

Fedi, F. (1981). "Prediction of attenuation due to rainfall on terrestrial links." *Radio Science*, 16(5), 731–743.

Flitcroft, I. D., Milford, J. R., and Dugdale, G. (1989). "Relating point to area average rainfall in semiarid West Africa and the implications for rainfall estimates derived from satellite data." *J. Appl. Meteor.*, 28(4), 252–265.

Flook, L. R. (1987). "Discussion on: Estimation of probable maximum precipitation: Case Studies." *J. Hydr. Engrg.*, ASCE, 113(3), 409–411.

Flueck, J. A., Woodley, W. L., Barnston, A. G., and Brown, T. J. (1986). "A further assessment of treatment effects in the Florida area cumulus experiment through guided linear modeling." *J. Clim. Appl. Meteor.*, 25(4), 546–564.

Folland, C. K. (1988). "Numerical models of the raingauge exposure problem, field experiments and an improved collection design." *Quart. J. Royal Meteor. Soc.*, 114, 1485–1516.

Fontaine, T. A. and Potter, K. W. (1989). "Estimating probabilities of extreme rainfalls." *J. Hydr. Engrg.*, ASCE, 115(11), 1562–1575.

Foster, J. L., Hall, D. K., and Chang, A. T. C. (1987). "Remote sensing of snow." *EOS*, 68(32), 682–684.

Foufoula-Georgiou, E. (1989). "A probabilistic storm transposition approach for estimating exceedance probabilities of extreme precipitation depths." *Water Resour. Res.*, 25(5), 799–815.

France, P. W. (1985). "A comparison of various techniques for computation of areal rainfall." *IE (I) Journ.*-CI, 66, 74–78.

Frederick, R. H., Meyers, V. A., and Auciello, E. P. (1977). "Five- to 60-minute precipitation frequency for the eastern and central United States." *NOAA Tech. Mem. NWS HYDRO-35,* National Weather Service, NOAA, U.S. Dept. of Commerce, Washington, D.C.

Fujiyoshi, Y,. Endoh, T., Yamada, T., Tsuboki K., Tachibana, Y., and Wakahama, G. (1990). "Determination of a Z-R relationship for snowfall using a radar and high sensitivity snow gages." *J. Appl. Meteor.,* 29(2), 147–152.

Gandin, L. S. (1965). *Objective analysis of meteorological fields,* translated from Russian by R. Hardin, Israel Program for Scientific Translation, Jerusalem, Israel.

Garen, D. C., Johnson, C. L., and Hanson, C. L. (1994). "Mean areal precipitation for daily hydrologic modeling in mountainous regions." *Water Resour. Bull.,* 30(3), 481–491.

Gebhardt, K., Bohn, C., and Gordon, N. (1988). "Estimating daily precipitation in the northern Great Basin." *Saval Ranch Research and Evaluation Project Final Report,* Bureau of Land Management, Reno, NV, 4–25.

Gill, G. C. (1964). "Data validation." *Symposium on environmental measurements.* Public Health Service Pub. No. 999-AP-15. U.S. Dept of Health, Education, and Welfare. Washington, D.C.

Gilman, C. S. (1964). "Rainfall." *Handbook of Applied Hydrology,* V. T. Chow, ed., McGraw-Hill Book Co., New York, NY, 9.1–9.68.

Gloyne, R. W. (1955). "Some effects of shelter-belts and wind-breaks." *Meteor. Mag.,* 84, 272–281.

Goldhirsh, J. (1983). "Yearly variations of rainfall-rate statistics at Wallops Island and their impact on modeled slant path attenuation distributions." *Trans. Antenn. and Propag.,* IEEE, AP-31, 918–921.

Goodison, B. E. (1978). "Accuracy of Canadian snow gage measurements." *J. Appl. Meteor.,* 17(10), 1542–1548.

Goodison, B. E., Sevruk, B., and Klemm, S. (1989). "WMO solid precipitation measurement intercomparison: objectives, methodology, analysis." *Atmospheric Deposition.* IAHS. Pub. No. 179, 58–64.

Griffith, C. G. (1987). "Comparison of gauge and satellite rain estimates for the central United States during August 1987." *J. Geophys. Res.,* 92(8), 9551–9566.

Griffith, C. G., Augustine, J. A., and Woodley, W. L. (1981). "Satellite rain estimation in the U.S. high plains." *J. Appl. Meteor.,* 20(1), 53–66.

Griffith, C. G., Woodley, W. L., Gruge, P. G., Martin, D. W., Stout, J., and Sikdar, D. N. (1978). "Rain estimation from geosynchronous satellite imagery—visible and near infrared studies." *Monthly Wea. Rev.,* 106(8), 1153–1171.

Griffiths, J. F. (1985). "Climatology." *Handbook of Applied Meteorology,* D. D. Houghton, ed., John Wiley & Sons, New York, NY, 125.

Groisman, P. Ya., Koknaeva, V. V., Belokrylova, T. A., and Karl, T. R. (1991). "Overcoming biases of precipitation measurement: A history of the USSR experience." *Bull. Amer. Meteor. Soc.,* 72(11), 1725–1733.

Grundy, W. D. and Miesch, A. T. (1988). "Brief Descriptions of STATPAC and related statistical programs for the IBM personal computer." *USGS Open File Report* 87-411-A, U.S. Dept. Interior, Washington, D.C.

Gumbel, E. J. (1958). *Statistics of extremes.* Columbia University Press, New York, NY.

Gupta, V. and Waymire, E. C. (1979). "A stochastic kinematic study of subsynoptic space-time rainfall." *Water Resour. Res.,* 15(3), 637–644.

Hall, A. J. and Barclay, P. A. (1975). "Methods of determining areal rainfall from observed data." *Prediction in catchment hydrology; A national symposium on hydrology,* Chapman, T. G. and F. X. Dunin, eds., sponsored by the Australian Academy of Science, Nov. 25–27.

Hallikainen, M. T. (1984). "Retrieval of snow water equivalent from Nimbus-7 SMMR data: Effect of land-cover categories and weather conditions." *J. Ocean Engineer.,* IEEE, OE-9(5), 372–76.

Hamilton, E. L. and Andrews, L. A. (1953). "Control of evaporation from rain gages by oil." *Bull. Amer. Meteor. Soc.,* 34(5), 202–204.

Hamlin, M. J. (1983). "The significance of rainfall in the study of hydrological processes at basin scale." *J. Hydrol.,* 65(1-3), 73–94.

Hamon, W. R. (1973). "Computing actual precipitation." *Proc. Distribution of precipitation in mountainous areas,* Vol I. WMO/OMM No. 326., World Meteorological Organization, Geneva, Switzerland, 159–174.

Hansen, E. M. (1987). "Probable maximum precipitation for design floods in the United States." *J. Hydrol.,* 96(1), 267–278.

Hanson, C. L. (1984). "Discussion of 'Point rainfall estimated for a mountainous region' by Yeou-Koong Tung." *J. Hydr. Div., ASCE,* 110(12), 1874–1877.

Hanson, C. L. (1989). "Precipitation catch measured by the Wyoming shield and the dual-gage system." *Water Resour. Bull.,* 25(1), 159–164.

Hanson, C. L., Cumming, K. A., Woolhiser, D. A., and Richardson, C. W. (1994). "Microcomputer program for daily weather simulations in the contiguous United States." *ARS-114,* Agricultural Research Service, U.S. Dept. of Agriculture, Washington, D.C.

Hanson, C. L. and Johnson, G. L. (1993). "Spatial and temporal precipitation characteristics in Southwest Idaho." Management of Irrigation and Drainage Systems, R. G. Allen and C. M. Neale, eds., ASCE, New York, NY, 394–401.

Hanson, C. L. and Wight, J. R. (1986). "Acquiring and analyzing climatic and weather data for monitoring and trend analysis." *Use of cover, soils and weather data in rangeland monitoring symposium proceedings,* Smith E. L., S. S. Coleman, C. E. Lewis, and G. W. Tanner, eds., Society for Range Management, 39th Annual Meeting, Kissimmee, Florida, 3–13.

Hanson, C. L., Zuzel, J. F., and Morris, R. P. (1983). "Winter precipitation catch by heated tipping bucket gages." *Trans. ASAE,* 26(5), 1479–1480.

Havens, B. S., Jiusto, J. E., and Vonnegut, B. (1981). "Early history of cloud seeding." *J. Weather Modification,* 13(1), 14–88.

Hayes, G. L. (1944). "A method of measuring rainfall on windy slopes." *Monthly Wea. Rev.,* 72(5), 111–114.

Hayes, G. L., and Kittredge, J. (1949). "Comparative rain measurements and rain-gage performance on a steep slope adjacent to a pine stand." *Trans. AGU,* 30(2), 295–301.

Heberden, W. (1769). "Of the quantities of rain which appear to fall, at different heights, over the same spot of ground." *Phil. Trans. Royal Society,* 59, 359–362.

Helmers, A. E. (1954). "Precipitation measurements on wind-swept slopes." *Trans. AGU,* 35:471–474.

Hershenhorn, J. and Woolhiser, D. A. (1987). "Disaggregation of daily rainfall." *J. Hydrol.,* 95(1), 299–322.

Hershfield, D. M. (1961). "Rainfall frequency atlas of the United States for durations from 30 minutes to 24 hours and return periods from 1 to 100 years." *U.S. Weather Bureau Tech. Paper No. 40,* U.S. Dept. Commerce, Washington, D.C.

Hershfield, D. M. (1972). "Estimating the extreme-value 1-minute rainfall." *J. Appl. Meteor.,* 11(6), 936–940.

Hildebrand, P. H., Towery, N., and Snell, M. R. (1979). "Measurement of convective mean rainfall over small areas using high density raingages and radar." *J. Appl. Meteor.,* 18(10), 1316–1326.

Hollinger, J. P, Pierce, J. L., and Poe, G. A. (1990). "SSM/I instrument evaluation." *Trans. Geo. and Remote Sensing,* IEEE, GE-28(5), 781–790.

Hosking, J. R. M. (1990). "L-Moments: Analysis and estimation of distributions using linear combinations of order statistics." *J. Royal Statistical Soc. B,* 52(1), 105–124.

Hudlow, M. D., Smith, J. A., Walton, M. L., and Shedd, R. C. (1991). "NEXRAD–New era in hydrometeorology in the United States." *Hydrological Applications of Weather Radar,* I. D. Chuckie and C. G. Collier, eds. Ellis Horwood, Ltd., 602–612.

Huebner, G. L. (1985). "Use of radar for precipitation measurements." *Hydrological forecasting,* M. G. Anderson and T. P. Burt, eds., John Wiley and Sons, New York, NY, 77–100.

Huff, F. A. (1955). "Comparison between standard and small orifice raingages." *Trans. AGU,* 36(4), 689–694.

Huff, F. A. (1967). "Time distribution of rainfall in heavy storms." *Water Resour. Res.,* 3(4), 1007–1019.

Huff, F. A. and Changnon, S. A., Jr. (1964). "A model 10-inch rainstorm." *J. Appl. Meteor.,* 3(5), 587–599.

Hungerford, R. D., Nemani, R. R., Running, S. W., and Coughlan, J. C. (1989). "MTCLIM: A mountain microclimate simulation model." *Forest Service Research Paper INT-414,* U.S. Dept. of Agriculture, Washington, D.C.

Huschke, R. E. (1959). *Glossary of Meteorology.* American meteorological Society, Boston, MA, 446.

Illingworth, A. J. and Caylor, I. J. (1989). "Polarization radar estimates of raindrop size spectra and rainfall rates." *J. Atmos. Ocean. Tech.,* 6(6), 939–949.

Jameson, A. R. (1989). "Theoretical analysis and meteorological interpretation of the role of raindrop shape on microwave attenuation and propagation phase shifts: Implications for the radar measurement of rain." *J. Atmos. Ocean. Tech.*, 6(1), 76–88.

Jansen, J. J. (1983). "Data comparability: an objective for precipitation chemistry monitoring." *Sampling and analysis of rain.* ASTM STP 823. Amer. Soc. of Testing and Materials, Philadelphia, PA, 61–71.

Jevons, W. S. (1861). "On the deficiency of rain in an elevated rain gauge as caused by wind." *London, Edinburgh, and Dublin Phil. Mag.*, 22, 421. (cited in Rodda 1967).

Johnson, L. R., Smith, P. L., and Mielke, P. W. Jr. (1994). "Some evaluations of the North Dakota cloud modification project." Paper No. 942017, ASAE, St. Joseph, MO.

Kalma, J. D., Lomas, J., Thaller, M., and Shashoua, Y. (1969). "An accurate small orifice rain gage." *Water Resour. Res.*, 5(1), 300–305.

Kedem, B., Chiu, L. S., and Karni, Z. (1990). "An analysis of the threshold method for measuring area-average integrals." *J. Appl. Meteor.*, 29(1), 3–20.

Kincer, J. B. (1938). "Determination of dependability of rainfall records by comparison with nearby records." *Trans. AGU*, 19, 529–532.

Koschmieder, H. (1934). "Methods and results of definite rain measurements." *Monthly Wea. Rev.*, 62(1), 5–7.

Krajewski, W. F. (1987). "Cokriging radar-rainfall and rain gage data." *J. Geophys. Res.*, 92(D8), 9571–9580.

Krajewski, W. F. and Smith, J. A. (1991). "On the estimation of climatological Z-R relationships." *J. Appl. Meteor.*, 30(10), 1436–1445.

Kunzi, K. F., Patil, S., and Rott, H. (1982). "Snow-cover parameters retrieved from Nimbus-7 Scanning Multichannel Microwave Radiometer (SMMR) data." *Trans. Geoscience and Remote Sensing*, IEEE, GE-20, 452–67.

Lacy, R. E. (1951). "Observations with a directional raingauge." *Quart. J. Royal Meteor. Soc.*, 77, 283–292.

Larsen, G. A., and Pense, R. B. (1982). "Stochastic simulation of daily climatic data for agronomic models." *Agron. J.*, 74(3), 510–514.

Larson, L. W. (1971). *Shielding precipitation gages from adverse wind effects with snow fences.* Water Resources Series No. 24, Water Resources Research Institute, University of Wyoming, Laramie, WY.

Larson, L. W. and Peck, E. L. (1974). "Accuracy of precipitation measurements for hydrologic modeling." *Water Resour. Res.* 10(4), 857–863.

Legates, D. R. and Willmott, C. J. (1990). "Mean seasonal and spatial variability in gauge-corrected, global precipitation." *Inter. J. Clim.*, 10(2), 111–127.

Lemmela, R. (1989). "Point snowfall measurements with different methods." *Precipitation measurement,* B. Sevruk, ed., WMO/IAHS/ETH Workshop on Precipitation Measurement, St. Moritz, Swiss Fed. Inst. of Tech., Zurich, Switzerland, 97–100.

Linsley, R. K., Jr., Kohler, M. A., and Paulhus, J. L. H. (1949). *Applied hydrology.* McGraw-Hill Book Co., New York, NY.

Linsley, R. K., Kohler, M. A., and Paulhus, J. L. H. (1975). *Hydrology for engineers,* 2nd Ed. McGraw-Hill Book Co., New York, NY.

Long, T. L. (1947). "A comparison of snowfall catch in shielded and unshielded precipitation gages." *Bull. Amer. Meteor. Soc.*, 28(3), 151–153.

Lopez, R. E., Atlas, D., Rosenfeld, D., Thomas, J. L., Blanchard, D. O., and Holle, R. L. (1989). "Estimation of areal rainfall using the radar echo area time integral." *J. Appl. Meteor.*, 28(11), 1162–1175.

Matzler, C., Aebischer, H., and Schanda, E. (1984). "Microwave dielectric properties of surface snow." *J. Ocean. Engineer.*, IEEE, OE-9(5), 366–371.

May, B. R. (1986). "Discrimination in the use of radar data adjusted by sparse gauge observations for determining surface rainfall." *Meteor. Mag.*, 115(1365), 101–115.

Mazzeralla, D. A. (1985). "Measurements today." *Handbook of Applied Meteorology,* D. D. Houghton, ed., John Wiley Interscience, New York, NY, 283–328.

McFarland, M. J. and Neale, C. M. U. (1991). "Land parameter algorithm validation and calibration." *DMSP Special Sensor Microwave/Imager calibration/validation. Final Report Vol. II,* Naval Research Laboratory, Washington DC., 9–1 to 9–108.

McFarland, M. J., Wilke, G. D., and Harder, P. H. (1987). "Nimbus 7 SMMR investigation of snowpack properties in the northern Great Plains for the winter of 1978-79." *Trans. Geo. and Remote Sensing,* IEEE, GE-25(1), 35–46.

McGurk, B. J. (1985). "Precipitation and snow water equivalent sensors: an evaluation." *Proc. Western Snow Conference,* 54th Annual Meeting, Phoenix AZ, 71–80.

Mesinger, F. and Mesinger, N. (1992). "Has hail suppression in eastern Yugoslavia led to a reduction in the frequency of hail?" *J. Appl. Meteor.,* 31(1), 104–111.

Messaoud, M. and Pointin, Y. B. (1990). "Small time and space measurements of the mean rainfall rate made by a gage network and a dual-polarization radar." *J. Appl. Meteor.,* 29(9), 830–841.

Mezeix, J. F. (1987). "Comments on 'Hail in southwestern France. Part II: Results of a 30-year hail prevention project with silver iodide seeding from the ground.' " *J. Clim. Appl. Meteor.,* 26(12), 1774–1776.

Miller, J. R., Frederick, R. H., and Tracey, R. J. (1973). "Precipitation frequency atlas of the western United States, Vol. I: Montana; Vol. II: Wyoming; Vol. III: Colorado; Vol. IV: New Mexico; Vol. V: Idaho, Vol. VI: Utah; Vol. VII: Nevada; Vol. VIII: Arizona; Vol. IX: Washington; Vol. X: Oregon; Vol. XI: California." *NOAA Atlas 2,* National Weather Service, NOAA, U.S. Dept. of Commerce, Silver Spring, MD.

Montmollin, F. A., Olivier, R. J., Simard, R. G., and Zwahlen, F. (1980). "Evaluation of a precipitation map using a smoothed elevation-precipitation relationship and optimal estimates (kriging)." *Nordic Hydrol.,* 11(3-4), 113–120.

Moss, A. J. and Green, P. (1983). "Movement of solids in air and water by raindrop impact. Effects of drop-size and water-depth variation." *Aust. J. Soil Res.,* 21, 257–269.

Moss, A. J. and Green, P. (1987). "Erosive effects of the large water drops (gravity drops) that fall from plants." *Aust. J. Soil Res.,* 25, 9–20.

Myers, V. A. and Zehr, R. M. (1980). "A methodology for point-to-area rainfall frequency ratios." *NOAA Technical Rpt. NWS 24,* National Weather Service, NOAA, U.S. Dept. of Commerce, Washington, D.C.

National Weather Service (NWS) (1989). "Cooperative station observations." National Weather Service Observing Handbook No. 2. Office of Systems Operations, Observing Systems Branch, NOAA, U.S. Dept. of Commerce, Silver Spring, MD.

Natural Environment Research Council (NERC) (1975). "Flood studies report (5 volumes)." NERC, London.

Nearing, M. A., Bradford, J. M., and Holtz, R. D. (1987). "Measurements of waterdrop impact pressures on soil surfaces." *Soil Sci. Soc. Amer. J.,* 51(5), 1302–1306.

Neff, E. L. (1977). "How much rain does a rain gage gage?" *J. Hydrol.,* 35(3-4), 213–220.

Negri, A. J. and Adler, R. F. (1987). "Infrared and visible satellite rain estimation. Part I. A grid cell approach." *J. Clim. Appl. Meteor.,* 26(11), 1551–1564.

Neyman, J., Osborn, H. B., Scott, E. L., and Wells, M. (1972). "Re-evaluation of the Arizona cloud-seeding experiment." *Proc. National Academy of Science,* 69(6), 1348–1352.

Nicks, A. D. and Gander, G. A. (1993). "Using CLIGEN to stochastically generate climate data inputs to WEPP and other water resources models." *Proc. Federal Interagency Workshop on Hydrologic Modeling Demands for the 90's,* J. S. Burton ed., U.S. Geological Survey Water Resources Investigations Rpt. 93-4018, U.S. Govt. Printing Office, Washington, D.C., 7–14 to 7–21.

Nicks, A. D. and Gander, G. A. (1994). "CLIGEN: A weather generator for climate inputs to water resource and other models." *Proc. 5th International Conference on Computers in Agriculture,* D. G. Watson, F. S. Zazueta, and T. V. Harrison, eds. ASAE, St. Joseph, MO, 7-14–7-21.

Nipher, F. E. (1878). "On the determination of true rainfall in elevated gages." *Proc. Amer. Assoc. Advance,* Science, St. Louis, MO, 103–108.

Nordenson, T. J. and Baker, D. R. (1962). "Comparative evaluation of evaporation instruments." *J. Geophys. Res.,* 67(2), 671–679.

Olsen, R. L., Rogers, D. V., and Hodge, D. B. (1978). "The aR^b relation in the calculation of rain attenuation." *Trans. Antennas Propag.,* IEEE, AP-26, 318–329.

Osborn, H. B., Lane, L. J., and Myers, V. A. (1980). "Watershed relationships for southwestern thunderstorms." *Trans. ASAE,* 23(1), 82–87 and 91.

O'Sullivan, F., Wash, C. H., Stewart, M., and Motell, C. E. (1990). "Rain estimation from infrared and visible GOES satellite data." *J. Appl. Meteor.*, 29(3), 209–223.

Palmen, E. and Newton, C. W. (1969). *Atmospheric circulation systems.* Academic Press, New York, NY.

Palmer, S. G., Nicholass, C. A., Lee, M. J., and Bader, M. J. (1983). "The use of rainfall data from radar for hydrometeorological services." *Meteor. Mag.*, 112, 333–346.

Paulhus, J. L. H. and Kohler, M. A. (1952). "Interpretation of missing precipitation records." *Monthly Wea. Rev.*, 80(8), 129–133.

Peck, E. L. (1972). "Snow measurement predicament." *Water Resour. Res.*, 8(1), 244–248.

Pickering, N. B., Hansen, J. W., Jones, J. W., Wells, C. M., Chan, V. K., and Godwin, D. C. (1994). "WeatherMan: A utility for managing and generating daily weather data." *Agron. J.*, 86(2), 332–337.

Pilgrim, D. and Cordery, I. (1975). Rainfall temporal patterns for design floods. *J. Hydr. Div.*, ASCE, 101 (HY1), 81–95.

Poreh, M. and Mechrez, E. (1984). "The combined effect of wind and topography on rainfall distribution." *J. Hydrol.*, 72(1-2), 1–23.

Powell, D. R. (1987). "Observations on consistency and reliability of field data in snow survey measurements." *Proc. Western Snow Conference*, 55th Annual Meeting, Vancouver, BC, 69–77.

Pruppacher, H. R. and Beard, K. V. (1970). "A wind tunnel investigation of the internal circulation and shape of water drops falling at terminal velocity in air." *Quart. J. Royal Meteor. Soc.*, 96, 247–256.

Pruppacher, H. R. and Ritter, R. L. (1971). "A semi-empirical determination of the shape of cloud and rain drops." *J. Atmo. Sci.*, 28, 86–96.

Rango, A. L. and Hobbs, P. V. (1987). "A reevaluation of the Climax cloud seeding experiments using NOAA published data." *J. Clim. Appl. Meteor.*, 26(7), 757–762.

Rasmussen, K. R. and Halgreen, C. (1978). "Some errors in precipitation measurement." *Nordic Hydrol.*, 9(3-4), 145–160.

Rawls, W. J., Jackson, T. J., and Zuzel, J. F. (1980). "Comparison of areal snow storage sampling procedures for rangeland watersheds." *Nordic Hydrol.*, 11(2), 71–82.

Rawls, W. J., Robertson, D. C., and Zuzel, J. F. (1975). "Comparison of precipitation gage catches with a modified Alter and a rigid Alter type windshield." *Water Resour. Res.*, 11(3), 415–417.

Rechard, P. A. and Larson, L. W. (1971). "The use of snow fences for shielding precipitation gages." *Proc. Western Snow Conference*, 139th Annual Meeting, Billings, MT, 56–61.

Redmond, K. R. and Koch, R. W. (1991). "Surface climate and streamflow variability in the western United States and their relationship to large-scale circulation indices." *Water Resour. Res.*, 27(9), 2381–2399.

Reed, W. G. (1915). "Note on the effects of raingage exposure." *Monthly Wea. Rev.*, 43(7), 318–322.

Richards, F. and Arkin, P. (1981). "On the relationship between satellite-observed cloud cover and precipitation." *Monthly Wea. Rev.*, 109(5), 1081–1093.

Richards, F., Miller, J. F., Zurndorfer, E. A., and Foat, N. S. (1983). "Water available for runoff for 1 to 15 days duration and return periods of 2 to 100 years for selected agricultural regions in the northwest United States." *NOAA Technical Report NWS 36*, National Weather Service, U.S. Dept. of Commerce, Washington, D.C.

Richards, R. P. (1984). "Snow pillow overloading by slope induced creep." *Proc. Western Snow Conference*, 52nd Annual Meeting, Sun Valley, ID, 184–187.

Richardson, C. (1978). "Generation of daily precipitation over an area." *Water Resour. Bull.*, 14(5), 1035–1047.

Riedel, J. T. and Schreiner, L. C. (1980). "Comparison of generalized estimates of probable maximum precipitation with greatest observed rainfalls." *NOAA Technical Report NWS 25*, National Weather Service, Office of Hydrology, U.S. Dept. of Commerce, Washington, D.C.

Riley, G. T., Landin, M. G., and Bosart, L. F. (1987). "The diurnal variability of precipitation across the central Rockies and adjacent Great Plains." *Monthly Wea. Rev.*, 115(6), 1161–1172.

Robinson, A. C. and Rodda, J. C. (1969). "Rain, wind and the aerodynamic characteristics of rain gages." *Meteor. Mag.*, 98(1161), 113–120.

Rodda, J. C. (1967). "The systematic error in rainfall measurement." *J. Instit. of Water Engrg.*, 21, 173–177.

Rodda, J. C. (1970). "The precipitation component of the water balance—its problems and prospects." *WMO Bull.*, 19(2), 102–105.

Rodda, J. C. and Smith, S. W. (1986). "The significance of the systematic error in rainfall measurement for assessing wet deposition." *Atmos. Environ.*, 20(5), 1059–1064.

Rodriguez-Iturbe, I. and Mejia, J. M. (1974). "On the transformation of point rainfall to areal rainfall." *Water Resour. Res.*, 10(4), 729–735.

Ropelewski, C. F. and Halpert, M. S. (1987). "Global and regional scale precipitation associated with EL Nino/Southern Oscillation." *Monthly Wea. Rev.*, 115(8), 1606–1626.

Roper, I. J. (1975). "A simple rain-gauge for dense network design." *Weather*, 30(10), 329–336.

Rosenberg, N. J., Blad, B. L., and Verma, S. B. (1983). *Microclimate: The biological environment*. John Wiley & Sons, New York, NY.

Rosenfeld, D. Atlas, D., and Short, D. A. (1990). "The estimation of convective rainfall by area integrals, 2. The height-area rainfall threshold (HART) method." *J. Geophys. Res.*, 95(D3), 2161–2176.

Rosenfeld, D. and Farbstein, H. (1992). "Possible influence of desert dust on seedability of clouds in Israel." *J. Appl. Meteor.*, 31(7), 722–731.

Rosenfeld, D. and Mintz, Y. (1988). "Evaporation of rain falling from convective clouds as derived from radar measurements." *J. Appl. Meteor.*, 27(3), 209–215.

Rosenfeld, D. and Woodley, W. L. (1989). "Effects of cloud seeding in west Texas." *J. Appl. Meteor.*, 28(10), 1050–1080.

Ruthroff, C. L. (1970). "Rain attenuation and radio path design." *Bell System Tech. J.*, 49, 121–135.

Sadler, E. J. and Busscher, W. J. (1989). "High-intensity rainfall rate determination from tipping-bucket rain gauge data." *Agron. J.*, 81(6), 930–934.

Salas, J. D. (1993). "Analysis and modeling of hydrologic time series." *Handbook of Hydrology*, D. R. Maidment, ed., McGraw-Hill, New York, NY, 19.1–19.72.

Sandsborg, J. (1972). "Precipitation measurements with various precipitation gauge installations." *Nordic Hydrol.*, 3(3), 80–106.

Schaefer, G. L. and Shafer, B. A. (1983). "A critical analysis of five years of SNOTEL performance in the Rocky Mountains." *Pro. International symposium in hydrometeorology*, A. I. Johnson and R. A. Clark, eds., AWRA, Bethesda, MD., 31–37.

Schaefer, M. G. (1990). "Regional analysis of precipitation annual maxima in Washington State." *Water Resour. Res.*, 26(1), 119–131.

Schmugge, T. J. (1983). "Remote sensing of soil moisture: Recent advances." *Trans. Geosci. Remote Sensing*, IEEE, GE-21, 336–341.

Schreiner, L. C. and Riedel, J. T. (1978). "Probable maximum precipitation estimates, United States east of the 105th meridian." *NOAA Hydrometeorological Rpt. No. 51*, National Weather Service, U.S. Depart. of Commerce, Washington, D.C.

Scofield, R. A. (1983). "A satellite technique for estimating rainfall from flash flood producing thunderstorms." *Proc. International symposium in hydrometeorology*, A. I. Johnson and R. A. Clark, eds., AWRA, Bethesda MD, 121–128.

Scofield, R. A. and Oliver, V. J. (1979). "A satellite technique for estimating rainfall from thunderstorms and hurricanes." *Satellite hydrology*, M. Deutsch, D. R. Wiesnet and A. Rango, eds., American Water Resources Association, Sioux Falls, SD, 70–76.

Searcy, J. K. and Hardison, C. H. (1960). "Double-mass curves, manual of hydrology: Part 1. General surface-water techniques." *Geological Survey Water-Supply Paper 1541-B*, U.S. Dept. of Interior, Washington, D.C.

Segal, F. (1986). "The influence of raingage integration time on measured rainfall-intensity distribution functions." *J. Atmos. Ocean. Tech.*, 3(4), 662–671.

Seo, D., Krajewski, W. F., and Bowles, D. S. (1990a). "Stochastic interpolation of rainfall data from rain gages and radar using cokriging, 1, Design of experiments." *Water Resour. Res.*, 26(3), 469–477.

Seo, D., Krajewski, W. F., Azimi-Zonooz, A. and Bowles, D. S. (1990b). "Stochastic interpolation of rainfall data from rain gages and radar using cokriging, 2, Results." *Water Resour. Res.*, 26(5), 915–924.

Sevruk, B. (1973). "Initial wetting losses incurred by ground level storage gages." *Water Resour. Res.*, 9(3), 759–763.

Sevruk, B. (1974a). "Comparison of mean rain catch of various gauge networks." *Nordic Hydrol.*, 5(2), 50–63.

Sevruk, B. (1974b). "The use of stereo, horizontal, and ground level orifice gages to determine a rainfall-elevation relationship." *Water Resour. Res.*, 10(6), 1138–1141.

Sevruk, B. (1982). "Methods of correction for systematic error in point precipitation measurement for operational use." *WMO, Operational Hydrology Report No. 21*, Geneva, Switzerland.

Sevruk, B. (1986a). "Correction of precipitation measurements—a summary report." *Correction of Precipitation Measurements*, B. Sevruk, ed., Zurcher Geographische Schriften No. 23, Geneva, Switzerland, 13–23.

Sevruk, B. (1986b). "Effect of wind and intensity of rain on the rain catch." *Correction of Precipitation Measurements*, B. Sevruk, ed., Zurcher Geographische Schriften No. 23, Geneva, Switzerland, 251–256.

Sevruk, B. (1989). "Precipitation measurement." WMO-IAHS-ETH Workshop on Precipitation Measurement, Swiss Federal Institute of Technology, Zurich, Switzerland.

Sevruk, B. (1990). "Accuracy of precipitation data." *Observed Climate Variations and Change: Contributions in Support of Section 7 of the 1990 IPCC Scientific Assessment*, D. E. Parker, ed., Intergovernmental Panel on Climate Change, Geneva, Switzerland, XXIII.1–XXIII.12.

Sevruk, B. and Klemm, S. (1989). "Types of standard precipitation gauges." *Precipitation Measurement*, B. Sevruk, ed., WMO-IAHS-ETH Workshop on Precipitation Measurement, Swiss Federal Institute of Technology, Zurich, Switzerland, 227–232.

Sharon, D. (1980). "The distribution of hydrologically significant rainfall incident on sloping ground." *J. Hydrol.*, 46(1), 165–188.

Sharon, D., Morin. J., and Moshe, Y. (1988). "Micro-topographical variations of rainfall incident on ridges of a cultivated field." *Trans. ASAE*, 31(6), 1715–1722.

Shaw, M. W. (1987). "Assessment of upward movement of rainsplash using a fluorescent tracer method and its application to the epidemiology of cereal pathogens." *Plant Path.*, 36, 201–213.

Shaw, M. W. (1991). "Variation in the height to which tracer is moved by splash during natural summer rain in the UK." *Ag. and Forest Meteor.*, 55(1-2), 1–14.

Shih, S. F. and Hamrick, R. L. (1975). "Modified Monte Carlo technique to compute Thiessen coefficients." *J. Hydrol.*, 27(3-4), 339–356.

Simanton, J. R. and Osborn, H. B. (1980). "Reciprocal-distance estimate of point rainfall." *J. Hydr. Div.*, ASCE, 106(HY7), 1242–1246.

Singh, V. P. and Chowdhury, P. K. (1986). "Comparing some methods of estimating mean aerial rainfall." *Water Resour. Bull.*, 22(2), 275–282.

Smith, C. J. (1986). "Reduction of errors caused by bright bands in quantitative rainfall measurements made using radar." *J. Atmos. Ocean. Tech.*, 3(1), 129–141.

Smith, F. W. and Boyne, H. S. (1981). "Snow pillow behavior under controlled laboratory conditions." *Proc. Western Snow Conference*, 49th Annual Meeting, St. George, UT, 13–22.

Smith, J. A. (1993). "Precipitation." *Handbook of Hydrology*, D. R. Maidment, ed., McGraw-Hill, New York, NY, 3.1–3.47.

Smith, J. A. and Karr, A. F. (1990). "A statistical model of extreme storm rainfall." *J. Geophys. Res.*, 95(D3), 2083–2092.

Smith, P. L. (1984). "Equivalent radar reflectivity factors for snow andice particles." *J. Clim. Appl. Meteor,.* 23(8), 1258–1260.

Snow, J. T. and Harley, S. B. (1988). "Basic meteorological observations for schools: rainfall." *Bull. AMS*, 69, 497–507.

Soil Conservation Service (SCS). (1988). "Snow surveys and water supply forecasting." *Agric. Inform. Bull. 536*, Soil Conservation Service, U.S. Dept. of Agriculture, Washington, D.C.

Spayd, L. E. (1983). "Estimating rainfall using satellite imagery for warm-top thunderstorms embedded in a synoptic scale cyclonic circulation." *Proc. International Symposium in Hydrometeorology,* A. I. Johnson and R. A. Clark, eds., AWRA, Bethesda, MD, 139–146.

Spencer, R. W., Goodman, H. M., and Hood, R. E. (1989). "Precipitation retrieval over land and ocean with the SSM/I: Identification and characteristics of the signal." *J. Atmos. Ocean. Tech.,* 6(2), 254–273.

Srikanthan, R. and McMahon, T. A. (1982). "Simulation of annual and monthly rainfalls—A preliminary study at five Australian stations." *J. Appl. Meteor.,* 21(10), 1472–1479.

Stedinger, J. R., Vogel, R. M., and Foufoula-Georgiou, E. (1993.) "Frequency analysis of extreme events." *Handbook of Hydrology,* D. R. Maidment, ed., McGraw-Hill, New York, NY, 18.1–18.66.

Stevenson, T. (1842). "On the defects of rain gages with description of an improved form." *Edinburgh New Phil. J.,* 33, 12.

Stiles, W. H., Ulaby, F. T., and Rango, A. (1981). "Microwave measurements of snowpack properties." *Nordic Hydrol.,* 12(3), 143–166.

Stout, J. E., Martin, D. W., and Sikdar, D. N. (1979). "Estimating GATE rainfall with geosynchronous satellite images." *Monthly Wea. Rev.,* 107(5), 585–598.

Sumner, G. (1988). *Precipitation process and analysis.* John Wiley & Sons, New York, NY.

Super, A. B. (1986). "Further exploratory analysis of the Bridger Range winter cloud seeding experiment." *J. Climate Appl. Meteor.,* 25(12), 1926–1933.

Tabios, G. Q. and Salas, J. D. (1985). "A comparative analysis of techniques for spatial interpolation of precipitation." *Water Resour. Bull.,* 21(3), 365–380.

Tattelman, P. and Grantham, D. G. (1985). "A review of models for estimating 1 min rainfall rates for microwave attenuation calculations." *Trans. COM.,* IEEE, 33(4), 361–372.

Tattelman, P. and Knight, R. W. (1988). "Analyses of 1-min rain rates extracted from weighing raingage recordings." *J. Appl. Meteor.,* 27(8), 928–938.

Thiessen, A. H. (1911). "Precipitation for large areas." *Monthly Wea. Rev.,* 39(7), 1082–1084.

Thurman, J. L., Roberts, R. T., and Burford, J. B. (1984). "Retrieval procedures for hydrologic data from ARS experimental watersheds in the United States." *USDA Agricultural Research Service ARS-2,* Beltsville, MD.

Todd, C. J. and Howell, W. E. (1985a). "Weather modification." *Handbook of Applied Meteorology,* D. D. Houghton, ed., John Wiley & Sons, New York, NY, 1065–1139.

Todd, C. J., and Howell, W. E. (1985b). *World atlas and catalog of reported results of precipitation management by cloud seeding.* C. J. Todd and W. B. Howell, Golden, CO.

Tung, Y. (1983). "Point rainfall estimation for a mountainous region." *J. Hydr. Engrg.,* ASCE, 109(10), 1386–1393.

Tung, Y. (1984). "Closure of 'Point rainfall estimation for a mountainous region.'" *J. Hydr. Engrg.,* ASCE, 110(12), 1877.

Turpeinen, O. H., Abidi, A., and Belhouane, W. (1987). "Determination of rainfall with the ESOC index." *Monthly Wea. Rev.,* 115(11), 2699–2706.

Ulaby, F. T., Moore, R. K., and Fung, A. K. (1981). *Microwave remote sensing. Vol I. Microwave remote sensing fundamentals and radiometry.* Addison-Wesley Pub. Co., Reading, MA.

Ulaby, F. T., Moore, R. K., and Fung, A. K. (1986). *Microwave remote sensing. Vol III. From theory to application.* Artech House, Inc., Norwood, MA.

Ulbrich, C. W. (1986). "A review of the differential reflectivity technique of measuring rainfall." *Trans. Geo. and Remote Sensing,* IEEE, GE-24(6), 955–965.

U.S. Army Corps of Engineers (1965). "Standard project flood determinations." Washington, D.C.

U.S. Dept. of Commerce (1970). "Weather Bureau observing handbook No. 2. Substation Observations." Weather Bureau, Silver Spring, MD.

———. (1977). "Probable maximum precipitation estimates, Colorado River and Great Basin drainages." *Hydrometeorological Report No. 49,* NOAA, National Weather Service, Silver Spring, MD.

———. (1979). "Climatic atlas of the United States." NOAA, National Climatic Center, Asheville, NC.

U.S. Weather Bureau (1960). "Generalized estimates of probable maximum precipitation west of the 105th meridian." *Tech. Paper No. 38,* U.S. Dept. of Commerce, Washington, D.C.

———. (1964). "Two- to ten-day precipitation for return periods of 2 to 100 years in the contiguous United States." *Tech. Paper No. 49,* U.S. Dept. of Commerce, Washington, D.C.

———. (1966). "Probable maximum precipitation, northwest states." *Hydrometeorological Report No. 43,* U.S. Dept. of Commerce, Washington, D.C.

Unwin, D. J. (1969). "The areal extension of rainfall records: an alternative model." *J. Hydrol.,* 7, 404–414.

Ursic, S. J. and Thames, J. L. (1958). "An inexpensive rain gage." *J. Soil and Water Conserv.,* 13(5), 231–232.

Walklate, P. J. (1989). "Vertical dispersal of plant pathogens by splashing. Part I. The theoretical relationship between rainfall and upward rain splash." *Plant Path.,* 38, 56–63.

Walklate, P. J., McCartney, H. A., and Fitt, B. D. L. (1989). "Vertical dispersal of plant pathogens by splashing. Part II. Experimental study of the relationship between raindrop size and the maximum splash height." *Plant Path.,* 38, 64–70.

Wallace, J. M. and Hobbs, P. V. (1977). *Atmospheric science: An introductory survey.* Academic Press, New York, NY.

Wallis, J. (1982). "Probable and improbable rainfall in California." *IBM Res. Rep. RC 9350,* Yorktown Heights, NY.

Wang, B. (1984). "Estimation of probable maximum precipitation: Case studies." *J. of Hydr. Engrg.,* ASCE, 110(10), 1457–1472.

Warnick, C. C. (1953). "Experiments with windshields for precipitation gages." *Trans. AGU,* 34(3), 379–388.

Wei, T. C. and McGuinness, J. L. (1973). "Reciprocal distance squared methods: A computer technique for estimating areal precipitation." *Agricultural Research Service NC-8,* U.S. Dept. of Agriculture, Washington, D.C.

Weiss, L. L. and Wilson, W. T. (1953). "Evaluation of the significance of slope changes in double mass curves." *Trans., American Geophysical Union,* 34(6), 893–896.

Weiss, L. L. and Wilson, W. T. (1957). "Precipitation gage shields." *IASH Pub.,* 43(1), 462–484.

Wenzel, H. G. (1982). "Rainfall for urban stormwater design." Urban Storm Water, D. F. Kibler, ed., *Water Resourc. Mono. 7,* AGU, Washington, D.C.

Whiteman, C. D. (1990). "Observations of thermally developed wind systems in mountainous terrain." Atmospheric processes over complex terrain, W. Blument, ed., *Meteor. Mono. 23, No. 45.,* AMS, Boston, MA.

Wilheit, T. T. (1986). "Some comments on passive microwave measurement of rain." *Bull. AMS,* 67(10), 1226–1232.

Wilke, G. D. and McFarland, M. J. (1986). "Correlations between Nimbus-7 Scanning Multichannel Microwave Radiometer data and an antecedent precipitation index." *J. Clim. Appl. Meteor.,* 25(2), 227–238.

Williams, R. G. and Erdman, M. D. (1987). "Low-cost computer interfaced rain gauge." *Computers and Electronics in Agric.,* 2, 67–73.

Wilson, J. A. (1966). "Determination and uses of best individual sampling points on individual snow courses." *Proc. Western Snow Conference,* 34th Annual Meeting, Seattle WA, 82–86.

Wilson, J. W. and Brandes, E. A. (1979). "Radar measurement of rainfall—a summary." *Bull. AMS,* 60, 1048–1058.

Woolhiser, D. A. (1992). "Modelling daily precipitation—Progress and problems." *Statistics in the Environmental and Earth Science,* A. Walden and P. Guttorp, eds., Edward Arnold, London, UK, 71–89

Woolhiser, D. A., Hanson, C. L., and Richardson, C. W. (1988). "Microcomputer program for daily weather simulation." *Agricultural Research Service ARS-75,* U.S. Dept. of Agriculture, Washington, D.C.

Woolhiser, D. A., Keefer, T. O., and Redmond, K. T. (1993). "Southern oscillation effects on daily precipitation in the southwestern United States." *Water Resour. Res.,* 29(4), 1287–1295.

World Meteorological Organization (WMO) (1970). "The planning of meteorological station networks." *Tech. Note 111,* Geneva, Switzerland.

———. (1971). "Guide to meteorological instruments and observing practices." *4th ed. WMO No. 8.* TP 3. World Meteorological Organization, Geneva, Switzerland.

———. (1973). "Manual for estimation of probable maximum precipitation." *Operational Hydrology Report,* WMO No. 332, 1st ed., Secretariat of the World Meteorological Organization, Geneva, Switzerland.

———. (1983). "Guide to hydrological practices." *Vol. II, WMO No. 168,* 4th ed., Geneva, Switzerland.

———. (1986). "Manual for estimation of probable maximum precipitation." *Operational Hydrology Report,* WMO No. 332, 2nd ed., Secretariat of the World Meteorological Organization, Geneva, Switzerland.

Yates, S. R. (1986). "Disjunctive kriging. 3. Cokriging." *Water Resour. Res.,* 22(10), 1371–1376.

Yates, S. R. and Warrick, A. W. (1986a). "Disjunctive kriging. 1. Overview of estimation and conditional probability." *Water Resour. Res.,* 22(5), 615–621.

———. (1986b). "Disjunctive kriging. 2. Examples." *Water Resour. Res.,* 22(5), 623–630.

Young, K. C. (1994). "A multivariate chain model for simulating climatic parameters from daily data." *J. Appl. Meteor.,* 33(6), 661–671.

Zehr, R. M. and Myers, V. A. (1984). "Depth-area ratios in the semi-arid southwest United States." *NOAA Tech. Mem. NWS HYDRO-35,* National Weather Service, NOAA, U.S. Dept. of Commerce, Silver Spring, MD.

X. GLOSSARY

Cyclone—An area of closed cyclonic circulation with a low-pressure center.

El Niño—A warm southward-moving ocean current along the Pacific Coast of Ecuador and Peru.

Graupel—Precipitation consisting of ice particles with a snow-like structure that are 2 to 5 mm in diameter.

Hail—Precipitation in the form of large ice stones formed in cumulonimbus clouds.

Jet Stream—Relatively strong winds concentrated within a narrow band approximately at the tropopause in middle latitude.

Lifting Condensation Level—The level at which a parcel of dry air lifted adiabatically becomes saturated.

Partial Duration Series—A series composed of all events during the period of record which exceeds a set criterion.

Precipitable Water—The amount of water vapor in a vertical column of air with unit area extending between any two levels.

Probable Maximum Precipitation—The theoretically greatest depth of precipitation for a given duration that is physically possible over a particular drainage area at a particular time of year.

Return Period—The average time interval between the occurrence of an event of a given quantity and the occurrence of an equal or larger event.

Snow Sampler—A set of light, jointed metal tubes used for obtaining snow samples and a spring scale graduated to directly read the depth of water contained in the sample.

Standard Project Storm—The largest storm to occur for a particular drainage area and season of year in which snowmelt is not a major consideration representing the most severe flood-producing rainfall depth-area-duration relationship and isohyetal pattern of any that is considered reasonably typical of that region.

Terminal Velocity—The constant speed that a free falling body obtains when the drag force of the medium on the body equals the net gravitational force on the body.

Virga—Streaks of water or ice particles that fall out of a cloud but evaporate before reaching the earth's surface as precipitation.

Chapter 3

INFILTRATION

I. INTRODUCTION

Infiltration is commonly defined as the process of water entry at the land surface into a soil from a source such as rainfall, irrigation, or snowmelt. The rate of infiltration is generally controlled by the rate of soil water movement below the surface. Rainfall excess is the portion of applied water that leaves the surface site not as infiltration but as runoff. As can be seen in Figure 3.1, the classic components of the hydrologic cycle for an event are 1) evaporation, 2) interception and depression storage, 3) infiltration, and 4) rainfall excess. The difference between rainfall excess and infiltration models is that rainfall excess models lump the losses (infiltration, evaporation, interception, and depression storage) together while infiltration models only describe the infiltration portion. Since evaporation, interception, and depression storage are normally minor compared to the infiltration portion during an event, rainfall excess models can be considered synonymous with infiltration models. This chapter of the handbook presents current knowledge and practice of modeling infiltration and rainfall excess.

II. PRINCIPLES OF INFILTRATION

The time-dependent rate of infiltration into a soil is governed by the Richards Equation (Richards, 1931), subject to given antecedent soil moisture conditions in the soil profile, the rate of water application at the soil surface, and the conditions at the bottom of the soil profile. In general, the initial soil water potential varies with soil depth, z. The initial conditions) at time t equal 0 and can be expressed as follows

$$h(z, t) = f(z); \quad t = 0 \tag{3.1}$$

where the profile of matric potential head (h) varies with depth (z).

The boundary condition at the soil surface depends upon the rate of water application. For a rainfall event with intensities less than or equal to the saturated hydraulic conductivity of the soil profile, all the rain infiltrates into the soil without generating any runoff. For higher rainfall intensities, all the rain infiltrates into the soil during early stages until the soil surface becomes saturated. After this point, the infiltration is less than the rain intensity and runoff begins. These conditions may be expressed as:

$$-K(h)\frac{\partial h}{\partial z} + 1 = R; \quad \theta(0, t) \leq \theta_s, \ t \leq t_p \tag{3.2}$$

$$h = h_o; \quad \Theta(0, t) = \Theta_s, \ t > t_p, \tag{3.3}$$

where R is the rainfall intensity, h_o is a small positive ponding depth on the soil surface, and t_p is the ponding time. These conditions also accommodate time varying rainfall intensities, as well as when

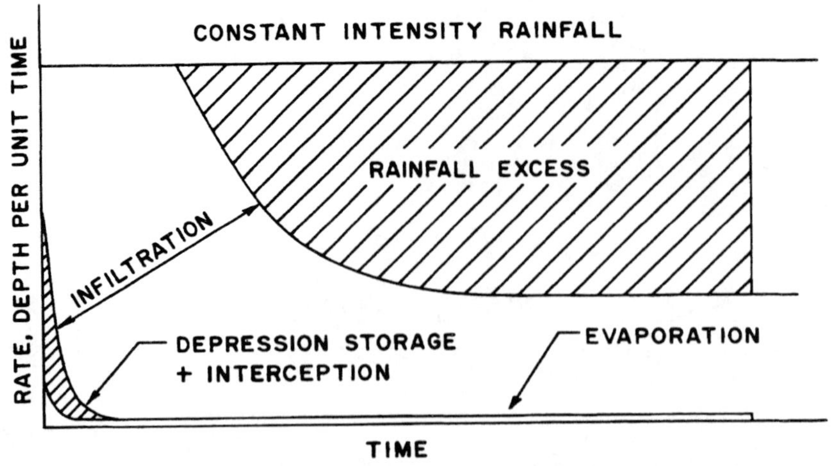

Figure 3.1.—Schematic of Rainfall Excess Components (Sabol et al., 1992).

rainfall intensity is smaller than the saturated hydraulic conductivity (K_s) of the soil throughout the storm. For a surface ponded-water irrigation, condition Equation 3.3 will apply from time zero on.

The surface boundary conditions equations 3.2 and 3.3 apply at any point in the field during rainfall. In a long sloping field, some infiltration may continue to occur in lower parts of the field even after the rainfall stops, due to continued overland flow from upper parts. During this phase, conditions described by equations 3.2 and 3.3 still apply after rainfall is replaced by the overland flow per unit area at the point of interest. To obtain the overland flow rates, hydrodynamic equations of overland flow need to be solved interactively with infiltration (Woolhiser, et al., 1990).

The lower boundary condition depends upon the depth of the unsaturated profile. For a deep profile, a unit-gradient flux condition is commonly applied at a depth, L, below the infiltration-wetted zone:

$$q(L, t) = K(\Theta, L); \qquad t > 0. \tag{3.4}$$

For a shallow profile, a constant pressure head is assumed at the water table depth L:

$$h(L, t) = 0; \qquad t > 0. \tag{3.5}$$

The Richards Equation 3.1 subject to the general conditions described in Equations 3.2 to 3.5 in a layered soil profile does not have any known analytical or closed-form solutions for infiltration. However, the solutions can be obtained by using finite-difference or finite-element numerical methods (Rubin and Steinhart, 1963; Mein and Larson, 1973).

For non-layered soils, uniform initial soil moisture distribution, and limited surface boundary conditions, some closed-form solutions are available. The pioneering work of Philip (1957) provided a series solution for vertical infiltration into a semi-infinite homogeneous soil, with a constant initial moisture content Θ_i and a constant matric potential h_o maintained at the soil surface. Recently, Swartzendruber (1987) presented a solution that holds for both small to intermediate and large times.

III. FACTORS AFFECTING INFILTRATION/RAINFALL EXCESS

Factors which affect infiltration have been divided into the following categories: 1) Soil; 2) Surface; 3) Management; and 4) Natural (Brakensiek and Rawls, 1988). Depending on their importance for the specific application, these categories should be accounted for when applying infiltration models. Published research results are used to illustrate the relative importance of the various factors. Methods for incorporating the effects of the factors into infiltration models will be discussed in the Infiltration/Rainfall Excess Models for Practical Applications section.

A. Soil

Soil factors encompass both soil physical properties including particle size, morphological, and chemical properties and soil water properties including soil water content, water retention characteristic, and hydraulic conductivity. Soil water properties are closely related to soil physical properties.

1. *Soil Physical Properties*

a. Soil Texture. Soil texture is determined from the size distribution of individual particles in a soil sample. Soil particles smaller than 2 mm are divided into three soil texture groups: sand, silt, and clay. Fig. 3.2 shows the particle size, sieve dimension, the defined size class, and the limits for the basic soil texture classes for the U.S. Department of Agriculture (USDA) (Soil Conservation Service, 1982b). The soil texture groups which have the greatest effect on infiltration (Rawls et al., 1991) are the percentages of sand, silt, clay, fine sand, coarse sand, very coarse sand, and coarse fragments (0.2 cm). Fig. 3.3a and 3.3b illustrate that coarse textured soils (gravelly sandy loam) normally have a higher infiltration rate than fine textured soils (clay loam). Coarse fragments in the soil normally increase the infiltration rate of the soil.

b. Morphological Properties. The morphological properties having the greatest effect on infiltration are bulk density, organic matter, and clay type. These properties are closely related to soil structure and soil surface area. As bulk density increases, soil porosity and infiltration decrease. Soil organic matter (1.74

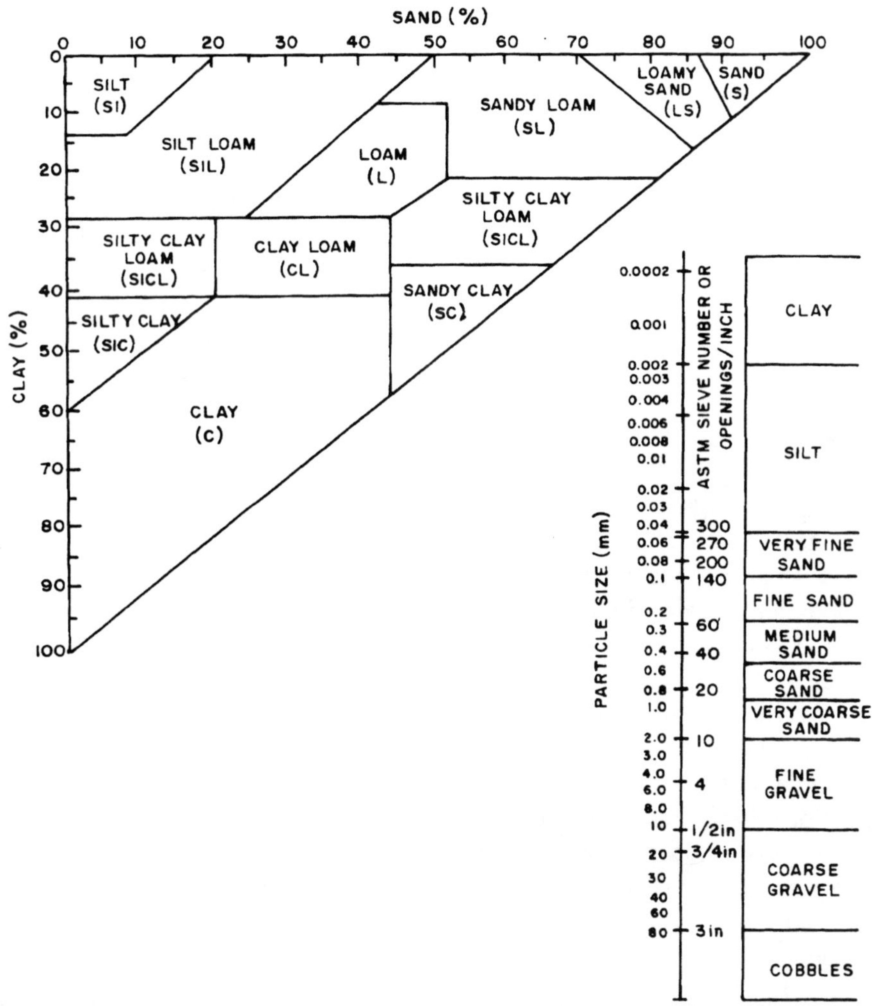

Figure 3.2.—USDA Soil Textural Triangle and Particle Size Limits (SCS, 1986).

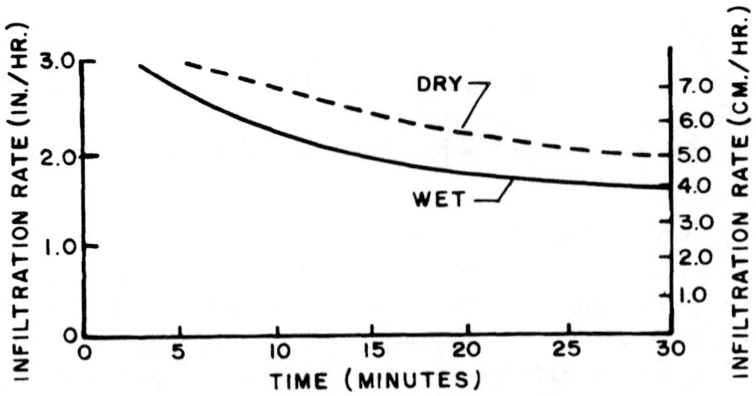

Figure 3.3a.—Infiltration Curves for the Artemisa arbuscula/Poa secunda (low) Community, Coils Creek Watershed, Clay Loam (Blackburn, 1973).

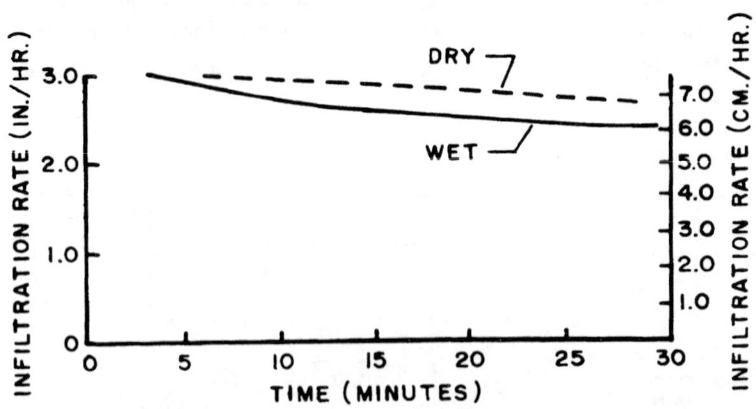

Figure 3.3b.—Infiltration Curves for Symphoricarpos longiflorus/Artemisia tridentata/Agropyron spicatum/Wyethia mollius Community, Coils Creek Watershed, Gravelly Sandy Loam (Blackburn, 1973).

times the percent of organic carbon) is inversely related to bulk density; thus as organic matter increases, infiltration increases.

The clay mineralogy, or clay type, has a significant effect on infiltration for soils containing a large percentage of clay (10%). For example, expandable clays such as montmorillionitic have a significantly lower infiltration rate than nonexpendable clays such as kaolinitic.

Soil layers due to natural profile development, crusts, or compaction limit or modify infiltration rates. Fig. 3.4 illustrates that a soil with a porous layer overlying a less porous layer results in a final infiltration rate that approaches the final rate of the lower layer. Also, the crusted surface, which is synonymous to a less porous layer overlying a more porous layer, produces a significantly lower final infiltration rate.

Macroporosity is natural or man-induced through "channels" connected from the soil surface to some depth in the soil profile. They may be cracks, worm holes, tillage marks, or intentional soil slots. Fig. 3.5 shows that compaction severely reduces the volume of macroscopic porosity and reduces infiltration rates. Natural cracks in a montmorillonitic clay can increase the infiltration rate from 0.05 cm/hr to over 5 cm/hr.

Other morphological properties such as the thickness of the soil horizon and soil structure are derived from soil survey descriptions and a quantitative description of their effects on soil water movement have not been determined.

c. Chemical Properties. Chemical properties of the soil are important because they affect the integrity of the soil aggregates (group of soil particles bound together). Chemical soil properties should be

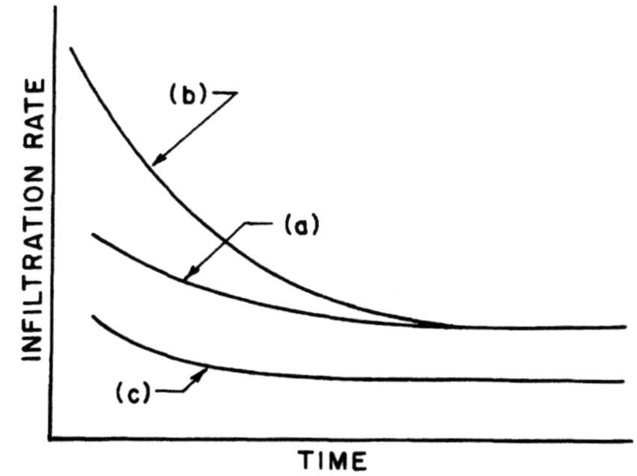

Figure 3.4.—Infiltration Rate for: a) Uniform Soil, b) With a More Porous Upper Layer, and c) Soil Covered With a Surface Crust (Hillel, 1971).

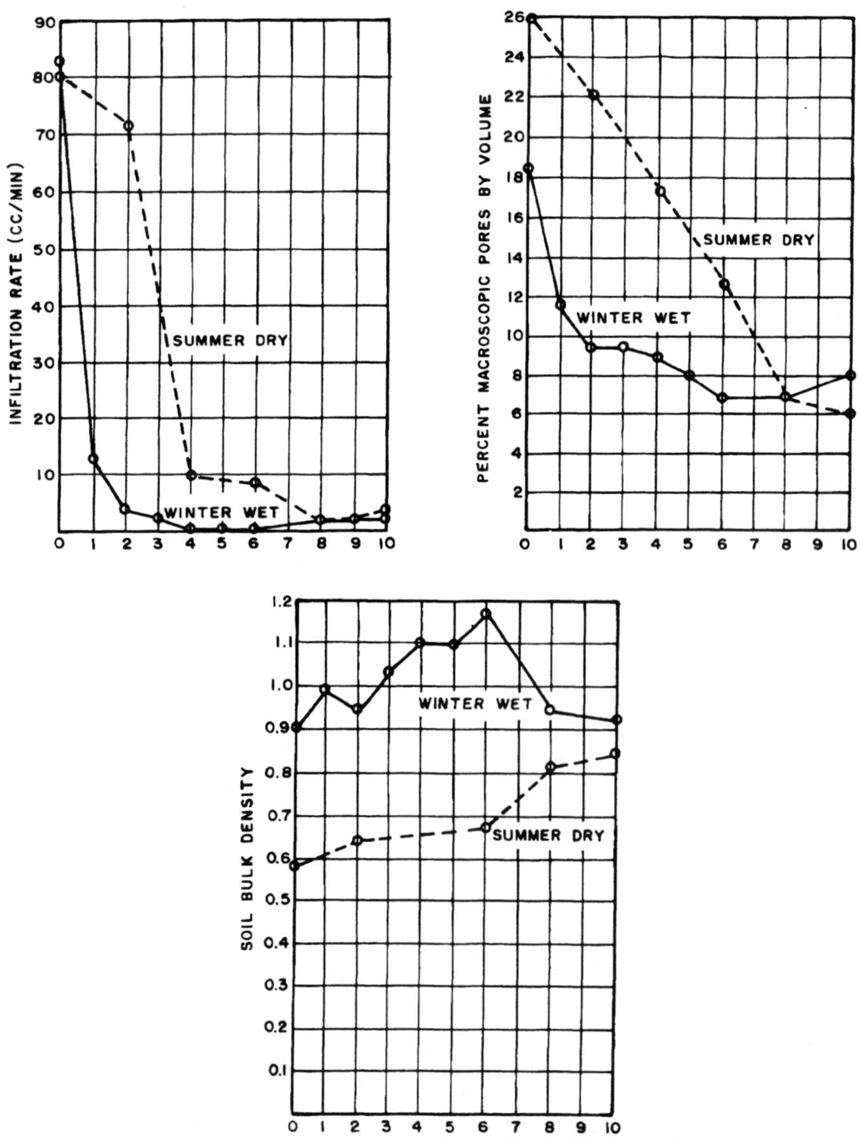

Figure 3.5.—a) Infiltration Rates, b) Volume of Macroscopic Pores, and c) Soil Bulk Density Influenced by Soil; Compaction (Lull, 1959).

considered when they are outside normal ranges (Rawls, et al., 1991). Also, chemistry of the infiltrating water can have an effect on infiltration.

2. *Soil Water Properties*

a. Soil Water Content. Fig. 3.3a and 3.3b illustrate that the higher the water content the smaller the infiltration rate.

b. Water Retention Characteristic. The water retention characteristic of the soil describes the soil's ability to store and release water and is defined as the relationship between the soil water content and the soil suction or matric potential. Fig. 3.6 illustrates the water retention relationship for two contrasting soil textures. Note that the sandy loam soil retains less water than the clay soil.

c. Hydraulic Conductivity. The hydraulic conductivity is the ability of the soil to transmit water and depends upon both the properties of the soil and the fluid (Klute and Dirkson, 1986). Total porosity, pore size distribution, and pore continuity are the important soil characteristics affecting hydraulic conductivity. The hydraulic conductivity at or above the saturation point is referred to as "saturated hydraulic conductivity" and is directly related to infiltration.

The hydraulic conductivity of the sandy loam soil decreases more rapidly with decrease in matrix potential than the hydraulic conductivity of clay soil. Thus, at lower matrix potential (or higher suctions) the hydraulic conductivity of the clay soil is higher. The rate of change of matrix potential in the sandy soil also decreases much more rapidly than that of the clay soil.

Hysteresis is caused by entrapment of air in the soil during wetting, and can cause the hydraulic conductivity to decrease; however, its effect is normally small and for practical applications has mostly been neglected.

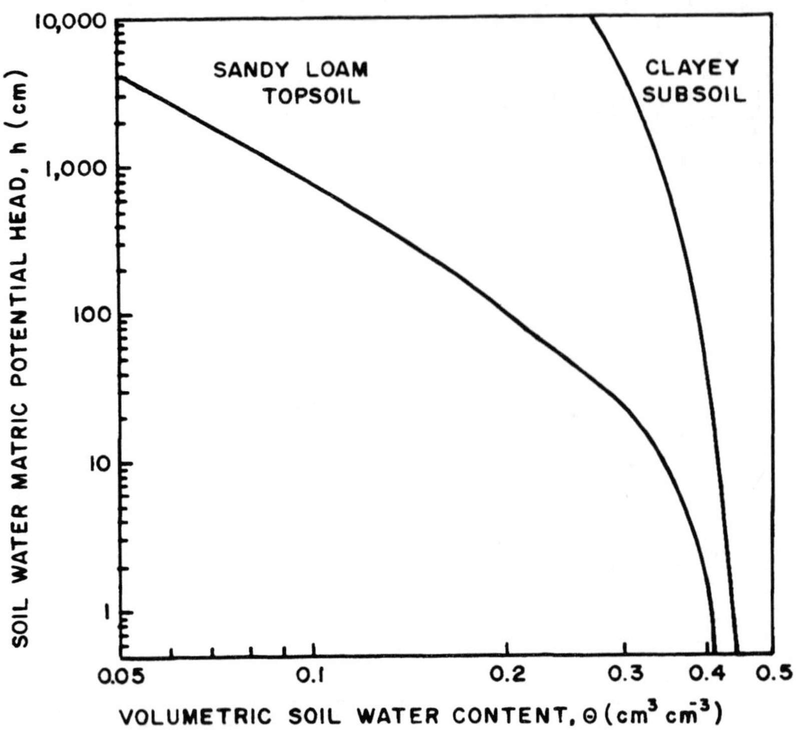

Figure 3.6.—The h(θ) Relationship of Sandy loam and Clayey Horizons of Cecil Soil (Ahuja et al., 1985).

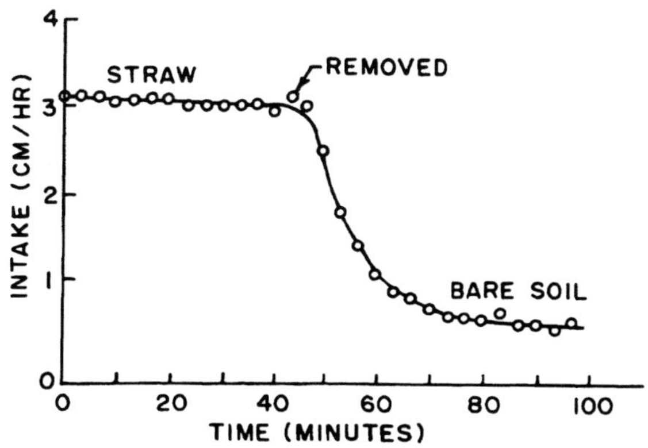

Figure 3.7.—Effect of Straw Cover and Bare Soil on Infiltration Rates (Gifford, 1977).

B. Surface

The surface factors are the factors that affect the movement of water through the air–soil interface. Factors at the soil surface are associated with materials that cover the surface, modify the shape of the surface and remove water by surface storage from its cycle of movement into the soil.

1. Cover Cover factors are materials that protect the soil surface. Lack of cover, or bare soil, is associated with the formation of a surface crust. A surface seal (crust) can develop under the action of raindrops (Summer and Stewart, 1992). The impact of raindrops can break down surface soil structure and cause the movement of soil fines into surface or near surface porosity. Once formed, a crust can impede infiltration. Fig. 3.7 shows the relative effect of surface cover removal on the infiltration rate. Fig. 3.8 indicates a strong relationship between total organic cover and infiltration rates. Fig. 3.9 exhibits the drastic effect of a surface crust on infiltration.

The effect that bare rangeland soil has on infiltration is shown in Fig. 3.10. Note the interaction with soil moisture. Fig. 3.11 indicates the effect of shrub canopy to increase infiltration. Fig. 3.12 points out the effect of surface cover, gravel, bare, litter, or crown cover on total infiltration. Bare and gravel cover reduce infiltration. Figs. 3.13 and 3.14 show that a grass or rock cover decreases the effect of trampling to reduce infiltration. The effect of cover, clipped cover, and bare ground on the final infiltration rate is shown in Fig. 3.15. Surface cover protects the ground surface from crusting and enhances infiltration rates.

2. Configuration Surface configurations can be natural or man-made. Fig. 3.16 is an example of the effect of natural soil mounds under shrubs on rangeland in Nevada. Compared to interspace areas, the dune areas have much higher infiltration rates. The soil of the dune areas is higher in organic matter and has a lower bulk density (higher porosity).

Man-made configurations result from various kinds of tillage. In tillage, this is referred to as surface roughness and is measured as random roughness (Zobeck and Onstad, 1987). Table 3.1 (Freebairn, 1989) exhibits the random roughness change due to several kinds of tillage on Minnesota silt loam soil. Note that tillage also reduces bulk density. Fig. 3.17 shows the tendency for an increasing random roughness to be associated with higher infiltration rates. The effect of roughness due to tillage is enhanced when protected from rainfall impact. Fig. 3.18 shows the surface roughness effect due to clods on infiltration and how this influence declines with exposure to rainfall. The dashed line, referring to surface cover with a furnace filter, is evidence of the efficiency of cover to sustain infiltration rates.

3. Storages Surface storage results from interception and/or depressions. Interception is caused by vegetation or other types of surface cover. Tables and equations for agricultural areas are found in Horton (1919) and Onstad (1984). Both interception and depressions reduce rainfall excess and may be very significant for small storm events.

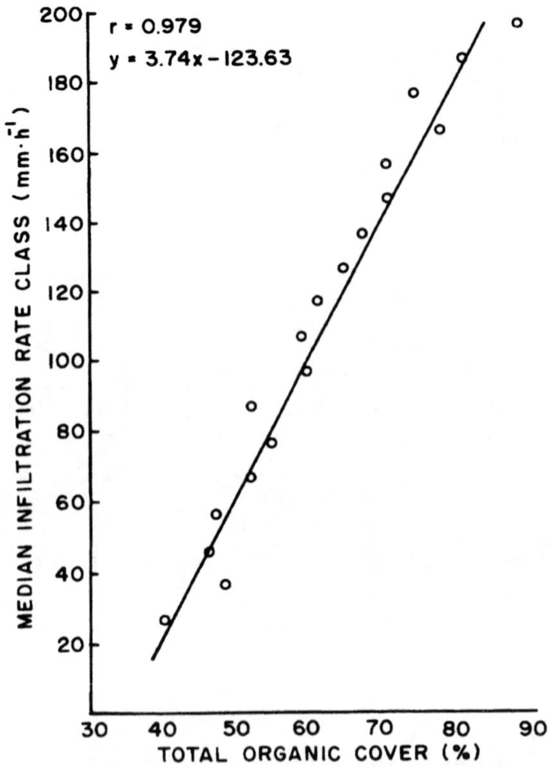

Figure 3.8.—Relationship of Median Infiltration Rate Classes with Total Organic Cover (%), Edwards Plateau, TX (Thurow, 1985).

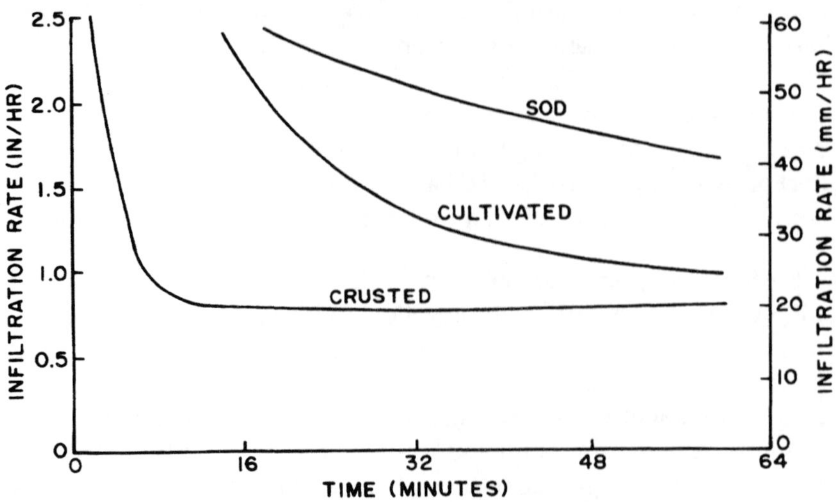

Figure 3.9.—Effect of Surface Sealing and Crusting on Infiltration Rate for a Zanesville Silt Loam (Skaggs and Khaleel, 1982).

C. Management

Management, as a factor affecting infiltration, modifies the soil properties or the soil surface condition. The effects of these factors have already been discussed in previous sections. In this section, the effect of management practices on infiltration will be presented.

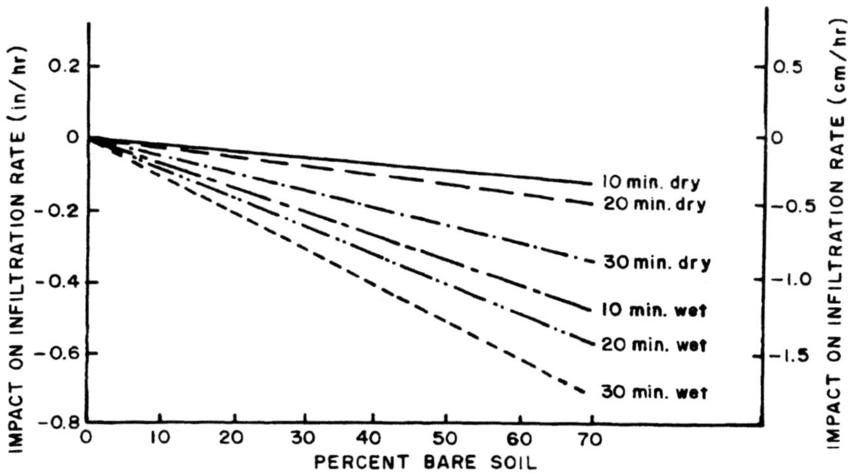

Figure 3.10.—Impact of Percent Bare Soil on Infiltration Rates After Various Time Intervals (Gifford, 1977).

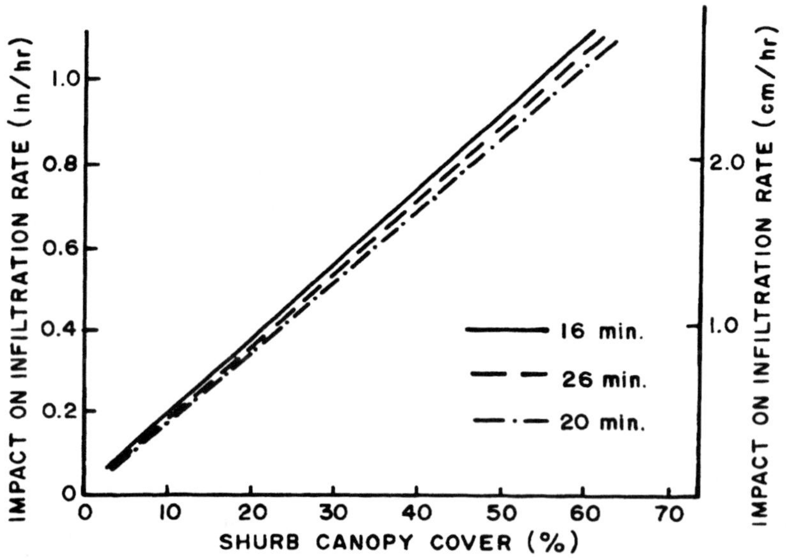

Figure 3.11.—Impact on Shrub Canopy on Infiltration Rates After Various Time Intervals (Gifford, 1977).

1. ***Agriculture*** Agricultural management systems involve different types of tillage, vegetation, and surface cover. Fig. 3.19 illustrates tillage practices (moldboard plow, chisel plow, and no till) influence on infiltration. Brakensiek and Rawls (1988) reported that moldboard plowing would increase soil porosity from 10 to 20% depending on soil texture and hence, increase infiltration rates over non-tilled soils. Rawls (1983) reported that increasing the organic matter of the soil lowers the bulk density, increases porosity and hence, increases the infiltration. Fig. 3.20 gives the steady-state infiltration rate of bare ground, soybean vegetation, and residue cover interactions for planting, midseason, and harvest. The bare soil steady-state rate decreases between planting and midseason and remains stable primarily as a result of crusting. Residue maintains a high steady-state rate until harvest, while the canopy and canopy residue combination increase the steady-state rate.

Table 3.2 (North Central Regional Committee 40, 1979) presents a grouping of major Midwest agricultural soils under tillage or grass management. There is a tendency for the same soil under grass

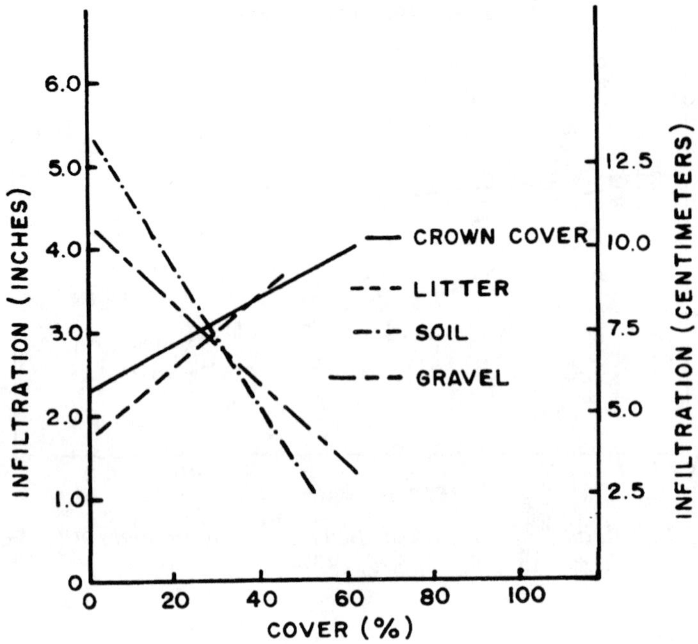

Figure 3.12.—Relationship of Various Covers to Infiltration Rates (Gifford, 1977).

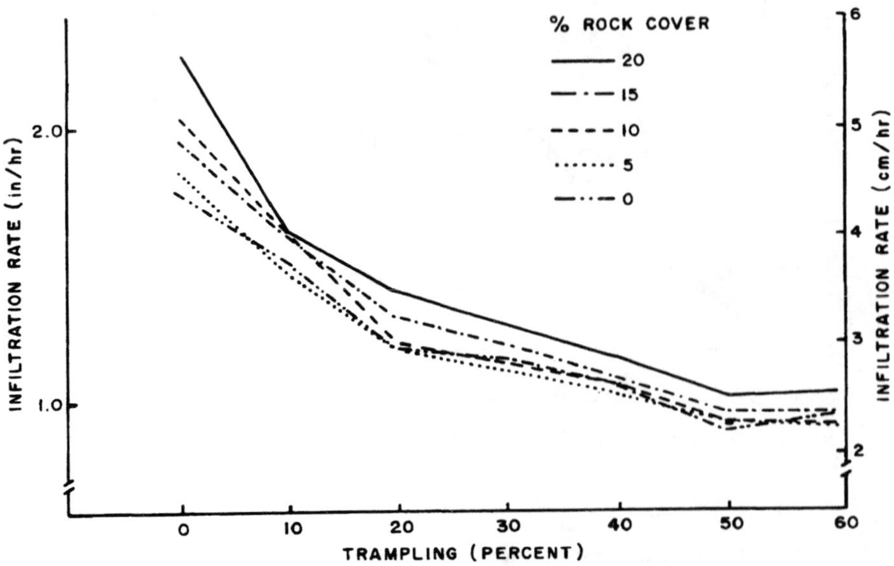

Figure 3.13.—Infiltration Rate at 30 Minutes as a Function of Rock Cover and Trampling (Dadkah and Gifford, 1980).

management to have higher infiltration rates than when under tilled management. Fig. 3.19 clearly depicts the influence of tillage management on infiltration rates. Fig. 3.21 indicates that the influence of tillage management decreases as the surface is exposed to rainfall. A surface cover reduces the degradation of the tillage affect. Rawls et al. (1983) developed a diagram from tillage data for estimating the effect of plowing to increase soil porosity, hence, increasing infiltration rates.

Additions to the soil that increase organic matter would increase infiltration. As shown in Fig. 3.11, the soil area under shrubs where litter accumulates and soil organic matter increased supported higher

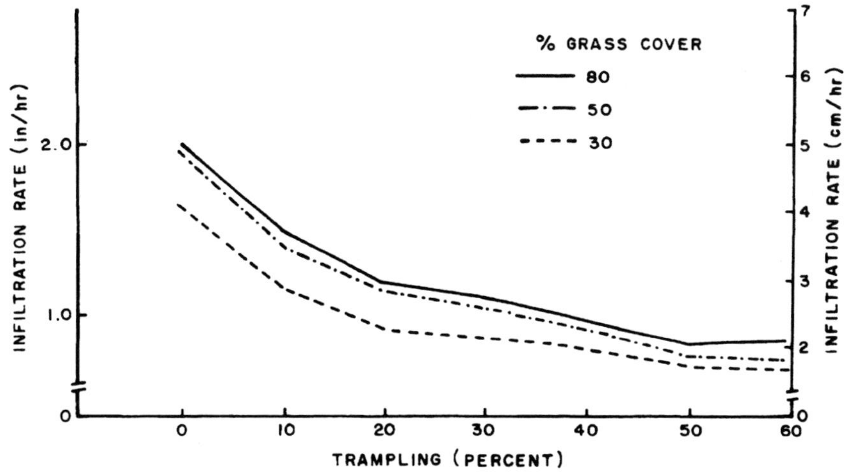

Figure 3.14.—Infiltration Rate at 30 Minutes as a Function of Grass Cover and Trampling (Dadkhah and Gifford, 1980).

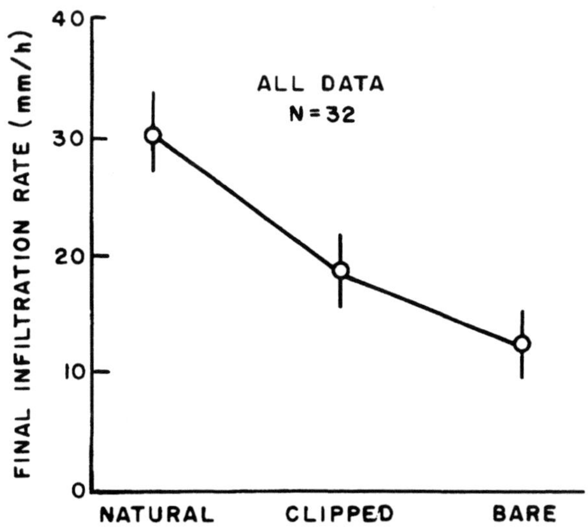

Figure 3.15.—Final Infiltration Rates Shown as a Mean From Five Sites in Arizona and Nevada for Very Wet Runs with Natural, Clipped, and Bare Cover (Lane et al., 1987).

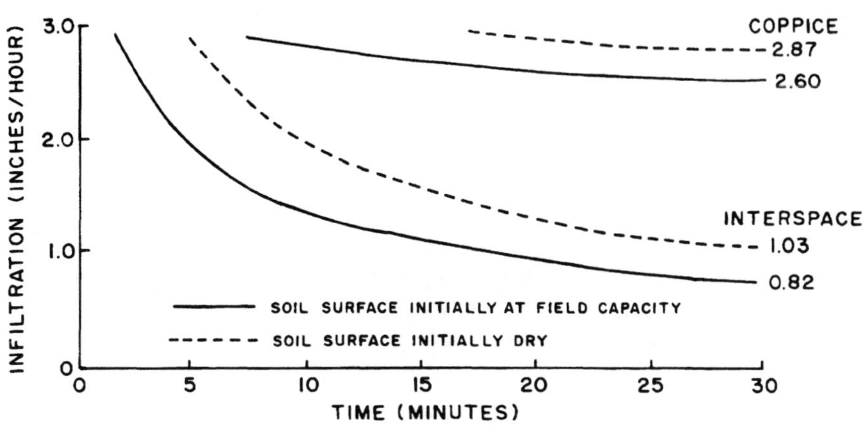

Figure 3.16.—Infiltration Curves for the Artemisia tridentata Community, Dickwater Watershed, Coppice Dune and Interspace (Blackburn, 1973).

TABLE 3.1. Influence of Tillage on Random Roughness (Freebairn, 1989).

Port Byron silt loam-Lawler farm. Rochester, MN					
Tillage meth	Plot	RR before tillage mm	RR after tillage mm	Bulk den before gm/cm^3	Bulk den after gm/cm^3
Chisel plow	E	9	17	1.16	1.07
	F	9	18	1.16	1.07
	G	9	16	1.16	1.07
	H	9	21	1.16	1.07
Moldboard plow	I	9	22	1.16	0.97
	J	9	19	1.16	0.97
	K	9	17	1.16	0.97
	L	9	25	1.16	0.97
Fall chisel	AA	7	24	1.16	1.15
	BB	7	18	1.16	1.15
Fall plow	EE	7	42	1.16	0.99
	FF	7	27	1.16	0.99

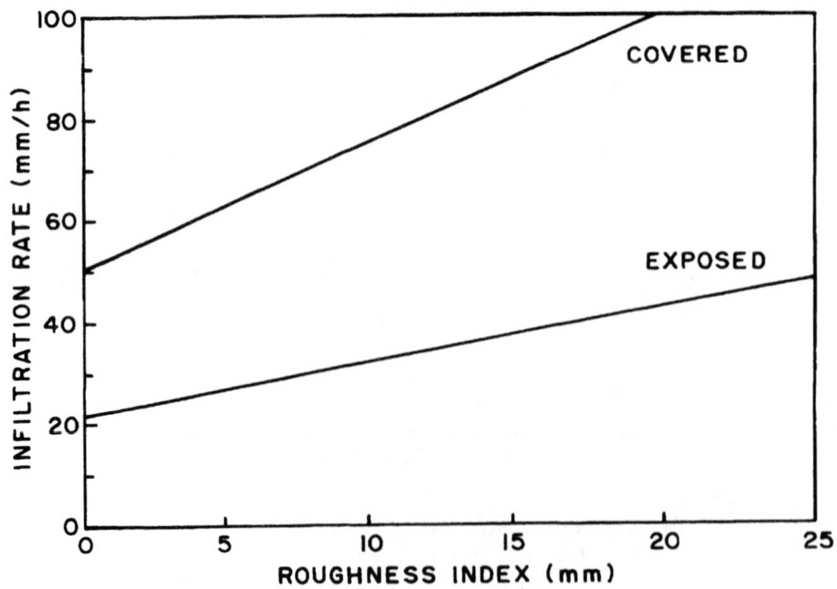

Figure 3.17.—Infiltration Rate vs. Roughness Index for Covered and Exposed Agricultural Plots (Freebairn, 1989).

infiltration rates. Rawls (1983) reports organic matter to be a major factor that lowers the bulk density (increased porosity); hence, increasing organic matter increases the hydraulic conductivity.

2. ***Irrigation*** Many of the factors in previous sections apply to infiltration under irrigation practices. Sprinkler systems are similar to rainfall infiltration processes. Border irrigation is similar to ponded infiltration processes. Furrow irrigation factors have been investigated by Trout and Kemper (1983) and by Kemper et al. (1988). Sub-irrigation systems involve factors common to soil moisture movement.

3. ***Rangeland*** The type of vegetation on rangelands is an important factor determining infiltration rates. Fig. 3.22 reveals a major difference of mean infiltration rates for three vegetation types. In Fig. 3.23, a significant effect on total infiltration is seen between successional stages of rangeland vegetation. Fig.

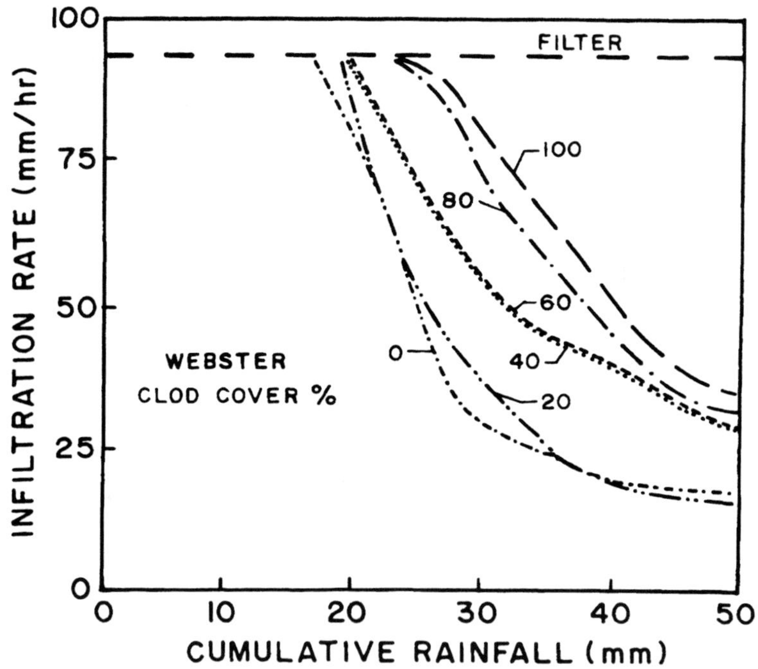

Figure 3.18.—Infiltration Rates Webster Clay Loam with the Surface Covered with Various Percentages of 25-50 mm Clods or Furnace Filter (Freebairn, 1989).

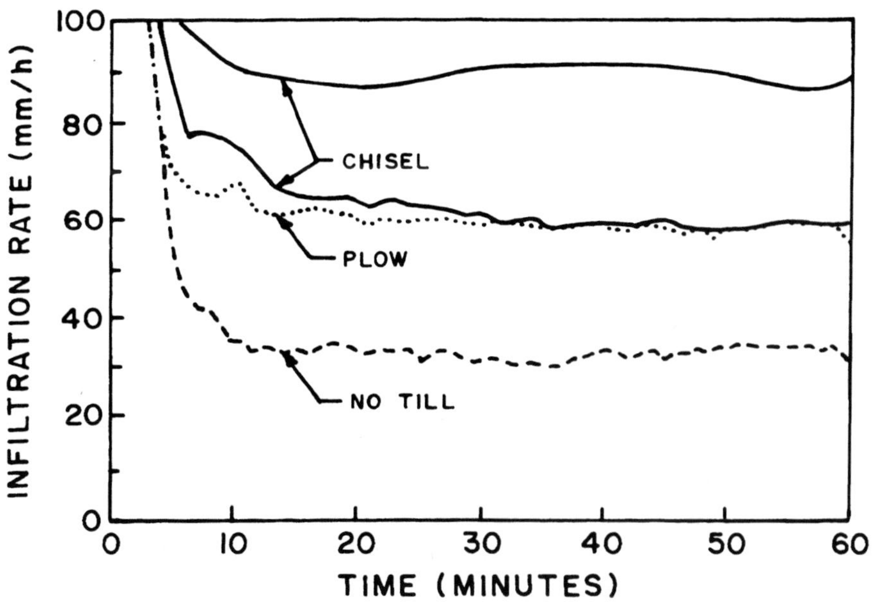

Figure 3.19.—Infiltration Rates for Port Byron Silt Loam, Two Months After Chisel and Moldboard Plow Tillage and Four Months After Planting Under No-till (Freebairn, 1989).

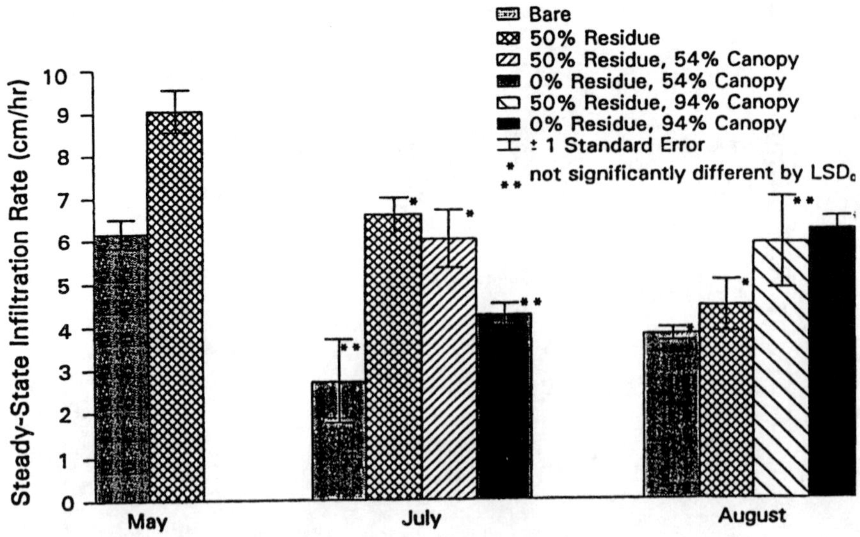

Figure 3.20.—Seasonal Effects of Agricultural Practices on Steady-State Infiltration Rates (Rawls et al., 1993a).

3.24 shows an improvement of infiltration rates going from bare ground to grass to shrub vegetation types.

Grazing practices also influence infiltration. Fig. 3.25 provides evidence that grazing reduces the final infiltration rate. Heavy grazing is associated with a greater reduction of the final infiltration rate. From studies in New Mexico, presented in Fig. 3.26, a difference is reported between several grazing treatments. Fig. 3.27 clarifies the effect of stocking density on infiltration rates. The effect of grazing systems on infiltration is shown in Fig. 3.28. The three grazing systems are MCG-moderate continuous, HCG-heavy continuous, and SDG-short duration grazing. Effects of rangeland improvements on runoff are summarized in Rangeland Hydrology (Branson et al., 1981). Fig. 3.29 shows an effect of burning on infiltration rates.

D. Natural

Natural factors include natural processes such as precipitation, freezing, change of seasons, temperature, and moisture which vary with time and space and interact with other factors in their effect on infiltration. The temporal and spatial variability effects will be discussed.

*1. **Temporal*** The effect of cumulative antecedent rainfall on exposed and 50% residue-covered agricultural soil is shown in Fig. 3.20, indicating a decrease in the steady-state infiltration rate with continued exposure to the action of rainfall. For bare soil, it seems that a stable steady-state infiltration rate is achieved between planting and midseason, indicating that a stable crust is achieved early in the growing season and maintained thereafter. Fig. 3.20 demonstrates that the steady-state infiltration rate increases as canopy cover increases over the growing season. Also, Fig. 3.20 indicates that canopy cover and residue cover do not cause additive increases in the steady-state infiltration rate.

Increases in rainfall intensity expand surface disturbance caused by the rain drops and the buildup of a ponding head. This usually increases the bare soil infiltration. For bare soil with canopy cover, this intensity effect is dissipated by the growing crop canopy.

Soil temperature influences infiltration through its effect on the viscosity of water. Lee (1983) found that freezing the soil with a high moisture content decreases infiltration to almost zero, while freezing the soil at a low moisture content increases infiltration by twice its normal rate.

TABLE 3.2. Alphabetical Listing of Some Major Soils of the North Central Region of the United States and Alaska and Their Mean Equilibrium Infiltration Rates, As Measured with a Sprinkling Infiltrometer (North Central Regional Committee 40, 1979).

Soil	Rate (inches per hour)		Test location
	Tilled surface	Grass surface	
Bearden silty clay loam	.2	.9	N. Dakota
Blount silty clay loam	.3	.3	Ohio
Bodenburg silt loam	.4	.4	Alaska
Canfield silt loam	.4	1.1	Ohio
Cincinnati silt loam	1.1	1.1	Indiana
Cisne silt loam	.6	.6	Illinois
Clermont silt loam	.8	.8	Indiana
Dudley clay loam	.4	1.0	S. Dakota
Elliott silt loam	1.5	1.5	Illinois
Emmet loamy sand	1.5	1.5	Michigan
Fayette silt loam	.2	.6	Wisconsin
Flanagan silt loam	1.9	3.6	Illinois
Grundy silt loam	1.1	1.1	Iowa
Holdrege silt loam	1.2	1.2	Nebraska
Hoytville clay	.3	1.0	Ohio
Ida silt loam	.7	1.5	Iowa
Keith very fine sandy loam	.7	1.5	Nebraska
Kenyon silt loam	1.3	1.3	Minnesota
Knik silt loam	.4	.4	Alaska
Kranzburg silty clay loam	.8	1.0	S. Dakota
Miami sandy loam	1.2	1.2	Michigan
Minto silt loam	.4	.4	Alaska
Moody silt loam	.4	1.4	Iowa
Morton loam	.4	1.0	N. Dakota
Plainfield loamy sand	3.7	4.4	Wisconsin
Port Byron silt loam	.5	2.0	Minnesota
Russell silt loam	.5	.5	Indiana
Sharpsburg silty clay loam	1.3	1.3	Nebraska
Sims loam	1.1	2.5	Michigan
Sinai silty clay loam	.6	1.1	S. Dakota
Webster clay loam	1.0	.6	Minnesota
Withee silt loam	.2	1.1	Wisconsin

Fig. 3.30 shows a seasonal influence of native rangeland infiltration rates. Fig. 3.20 indicates a seasonal effect for agricultural lands on the final infiltration rates.

2. ***Spatial*** The effect of spatial variability on infiltration is a measure of the difference between "point" infiltration and apparent infiltration rates associated with composite areas or watersheds. The variability of soil, surface, and management factors over an area is a result of the same practices or factors that have been discussed in previous sections. Smith (1983) states "there is no set of parameters which, when used with point infiltration models will produce the same response as the net (ensemble) from the area."

Smith and Hebbert (1979) studied the effect of a linear variation of hydraulic conductivity along a plane "watershed" on runoff for two rates of uniform rainfall. Their results clearly indicate the effect of different spatial conductivity patterns on runoff hydrographs, i.e., rainfall excess. It is also apparent that a strong interrelationship is at work between the spatial infiltration patterns and rates of rainfall. A random infiltration pattern tends to occur on the plane which produces a runoff hydrograph similar to a uniform pattern of infiltration for at least higher rainfall rates. However, for lower rainfall rates, the impact of random infiltration variability on runoff (rainfall excess) is still a major factor.

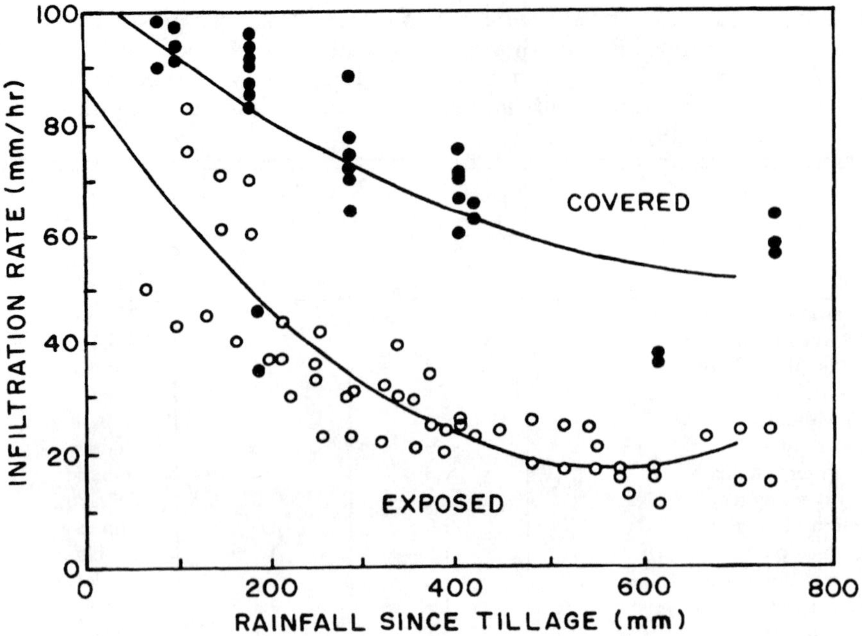

Figure 3.21.—Infiltration Rates vs. Antecedent Rainfall Since Tillage for Covered and Exposed Plots, Webster Clay Loam (Freebairn, 1989).

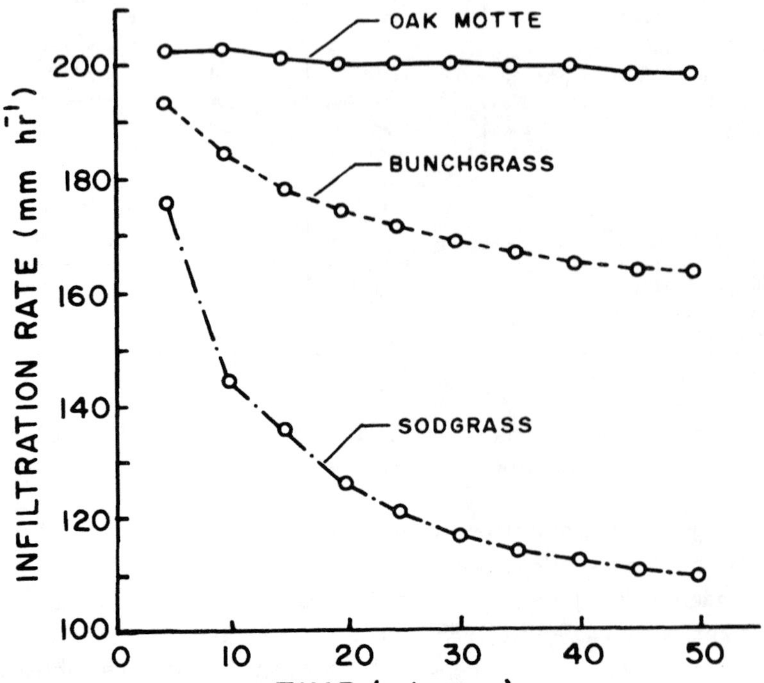

Figure 3.22.—Mean Infiltration Rates for Three Vegetation Types, Edwards Plateau, TX (Thurow et al., 1986).

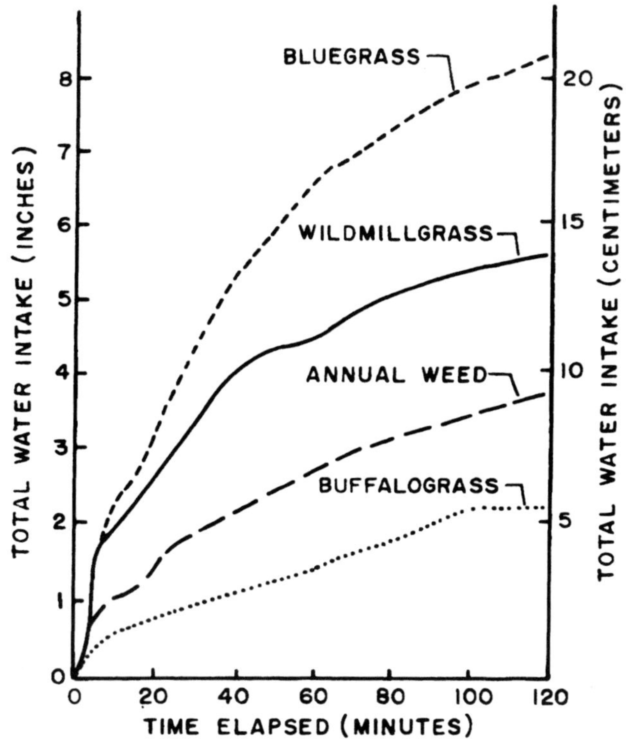

Figure 3.23.—Infiltration Amounts for Four Successional Stages of Rangeland Vegetation (Gifford, 1977).

III. INFILTRATION/RAINFALL EXCESS MODELS FOR PRACTICAL APPLICATIONS

Infiltration models for field applications usually employ simplified concepts which predict the infiltration rate or cumulative infiltration volume. This assumes that surface ponding begins when the surface application rate exceeds the soil surface infiltration rate. The rainfall excess models that lump all losses (infiltration, depression storage, interception) are strictly empirical models.

A. Rainfall Excess Models

Rainfall excess is the part of rainfall that is not lost to infiltration, depression storage, and interception. While a number of models have been proposed for estimating rainfall excess, the most commonly-used models are the index models and the SCS curve number model.

Index models are relatively simple methods that can be useful when performing gaged analysis (e.g., when rainfall and runoff data are available) or when a simple method is commensurate with the data available for estimating loss values. The most commonly used index models are 1) phi index, 2) initial and constant loss rate and 3) constant proportion loss rate (Pilgrim and Cordery, 1992).

1. Phi Index The phi index, ϕ, is probably the most widely used index model. The loss rate defined by the phi index is shown in Fig. 3.31 and is described mathematically as

$$f(t), = I(t), \quad \text{for } I(t) < \phi, \tag{3.6}$$

where f(t) is the loss rate, I(t) is the rainfall intensity, t is time, and ϕ is a constant called the ϕ index. The phi index can be estimated from storm data by separating baseflow from the total runoff and then determining the ϕ which causes the rainfall excess to equal the total runoff. The advantage of this method

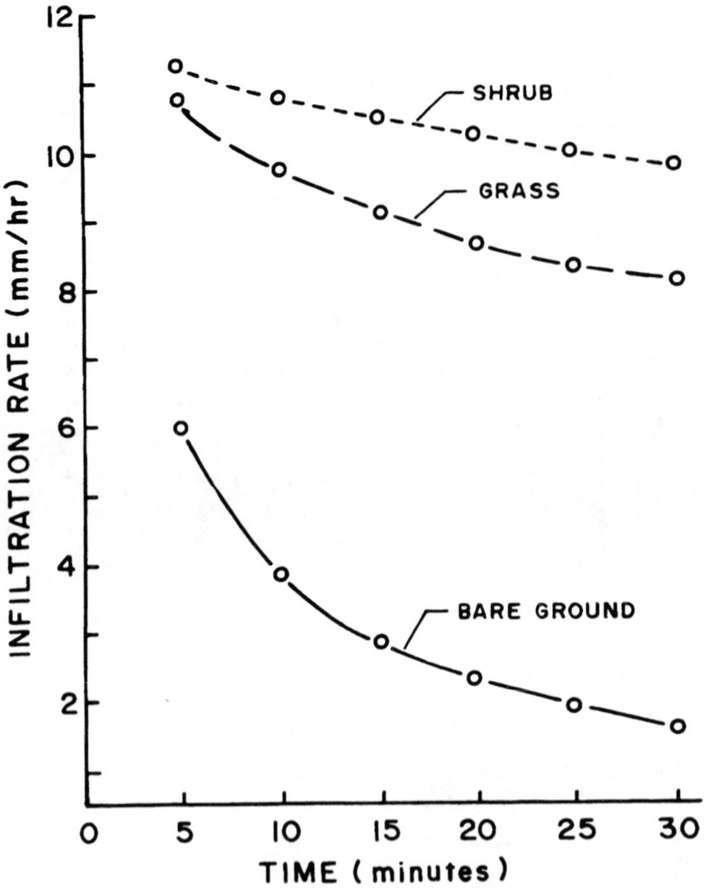

Figure 3.24.—Mean Infiltration Rates for Shrub (GRBI), Grass (CHRO), and Bare Ground (BAGR) (Mbakaya, 1985).

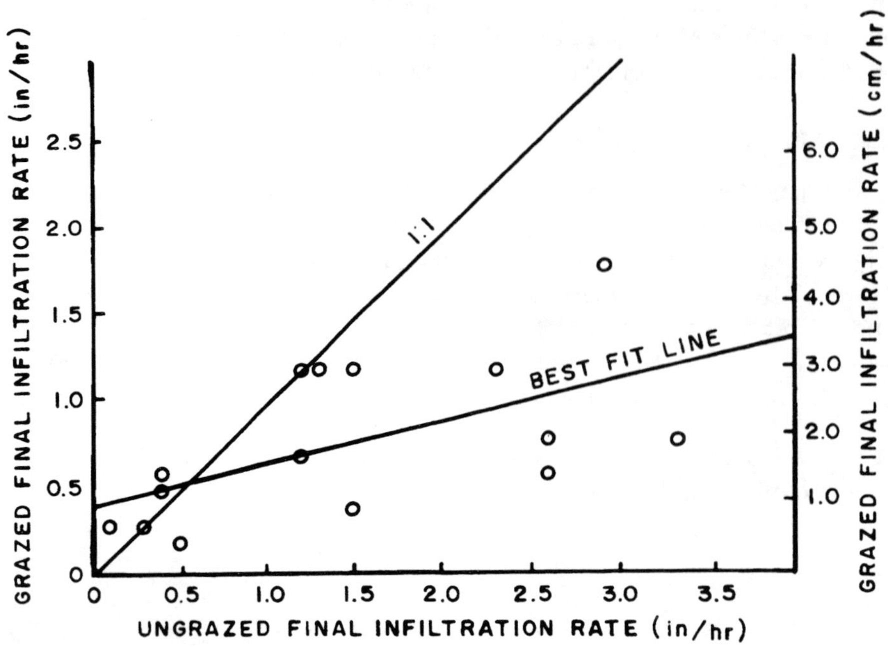

Figure 3.25.—Relationships Between Final Infiltration Rates on Heavily Grazed Areas (Gifford and Hawkins, 1978).

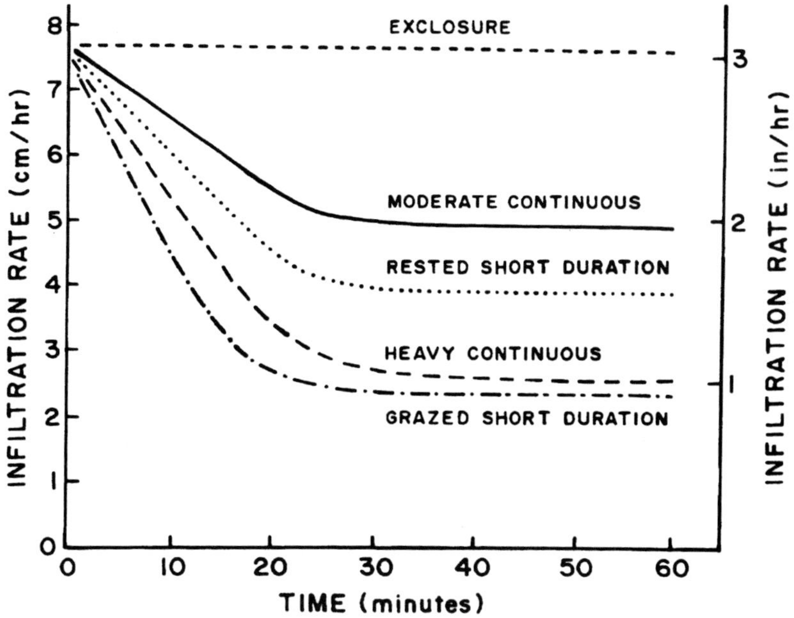

Figure 3.26.—Mean Infiltration Rates for Various Grazing Treatments at Fort Stanton, NM (Weltz and Wood, 1986).

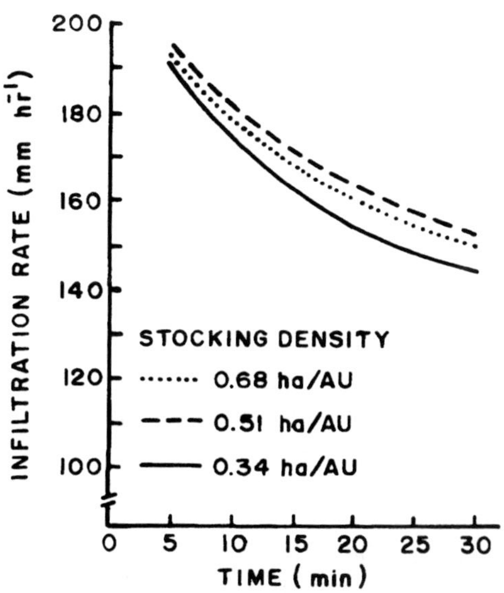

Figure 3.27.—Mean Infiltration Rates for Pastures Grazed at Three Stocking Densities (Warren et al., 1986).

is that it requires only a single parameter. The disadvantages are that it requires rainfall runoff records and the ϕ is dependent on the watershed and storm conditions from which it was determined.

2. ***Initial and Constant Loss Rate*** The loss rate function associated with this procedure is shown in Fig. 3.32 and is described mathematically as

$$
\begin{aligned}
f(t) &= I(t) \quad \text{for } P(t) < IA \\
f(t) &= I(t) - C \quad \text{for } I(t) > C,\ P(t) \geq IA \\
f(t) &= I(t) \quad \textit{for } I(t) \leq C,
\end{aligned}
\tag{3.7}
$$

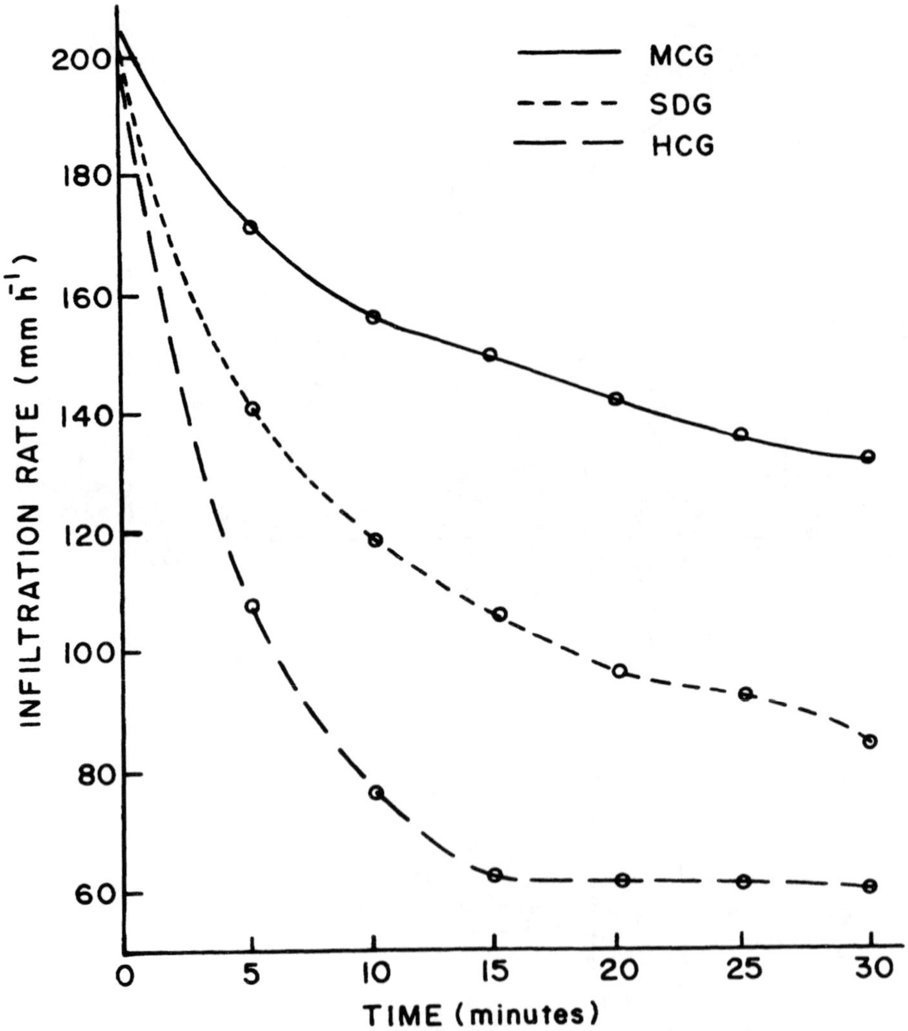

Figure 3.28.—Infiltration Rates for Three Grazing Systems on the Edwards Plateau: MCG-Moderate Continuous, HCG-Heavy Continuous, and SDG-Short Duration (Thurow, 1985).

where P(t) is the cumulative rainfall volume at time t from the beginning of rainfall, I(t) is the rainfall intensity, IA is the initial loss, and C is the constant loss rate. This method is a crude approximation to a typical infiltration curve that decays from some initial high rate to a final constant infiltration rate. The initial loss might be considered to represent the total loss due to surface factors and volume infiltrated prior to attaining the soils long-term infiltration rate. This method is similar to the phi index and has similar advantages and disadvantages. Table 3.3 presents some currently used loss rates (Sabol et al., 1992).

3. ***Constant Proportion Loss Rate*** The loss rate function associated with this procedure is shown in Fig. 3.33 and is described mathematically as

$$f(t) = CP * I(t), \tag{3.8}$$

where f(t) is the loss rate, I(t) is rainfall intensity, and CP is a constant ranging from 0 to 1. The advantages and disadvantages are the same as the phi index; however, it does not realistically represent the infiltration process. Calculation of CP requires rainfall and runoff records.

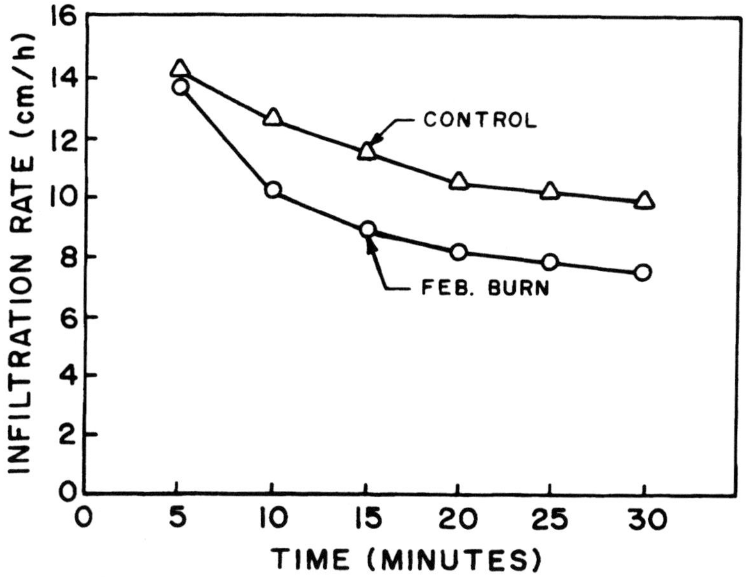

Figure 3.29.—Mean Infiltration Rates of February Burn and Control Areas, Kinoko, Kenya (Cheruiyot, 1984).

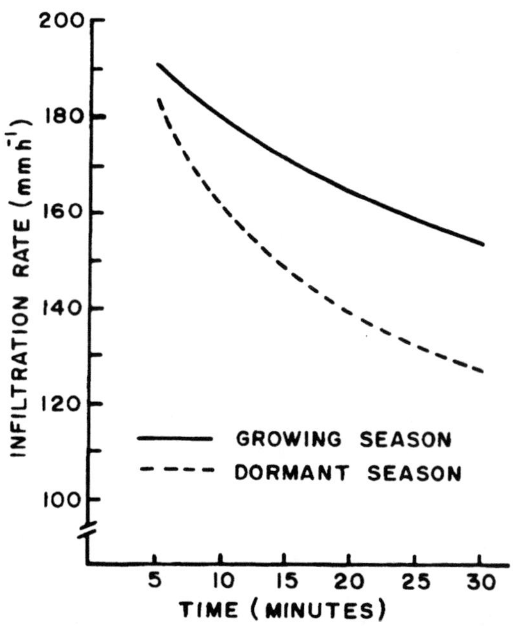

Figure 3.30.—Mean Infiltration Rates During the Growing and Dormant Seasons Near Sonora, TX (Warren et al., 1986).

4. SCS Runoff Curve Number Model The SCS Runoff Curve Number (CN) method is described in detail in Chapter 4 of the Soil Conservation Service National Engineering Handbook (1972). The SCS runoff equation is

$$Q = \frac{(P - I_a)^2}{(P - I_a) + S}, \tag{3.9}$$

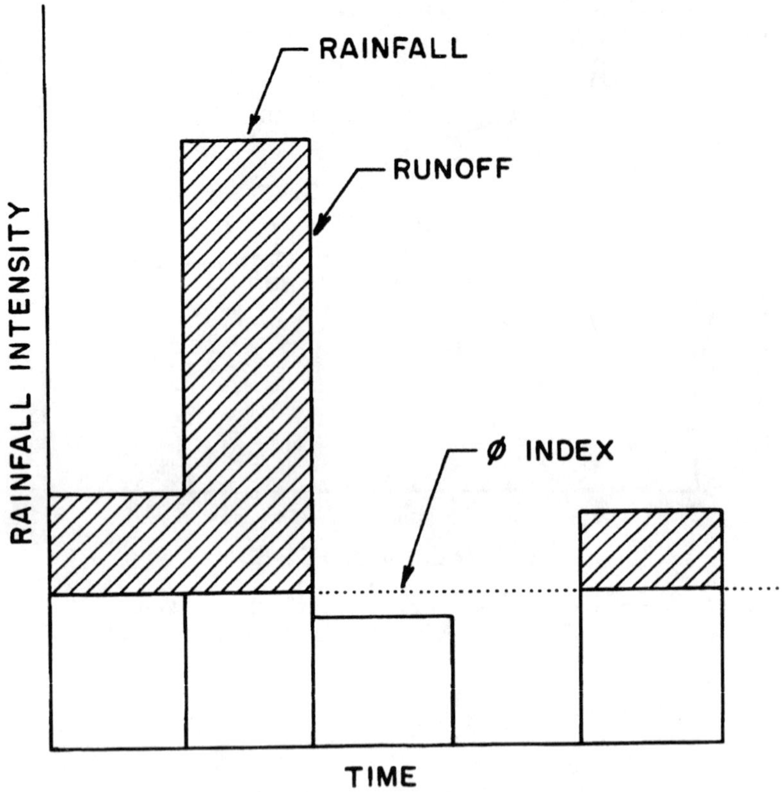

Figure 3.31.—ϕ Index Loss Rate.

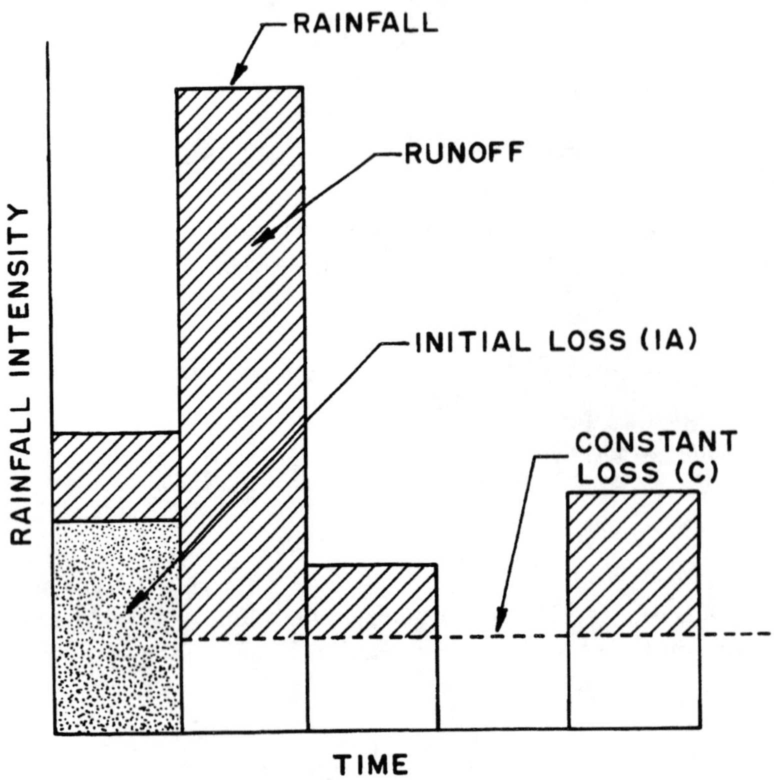

Figure 3.32.—Initial and Constant Loss Rate.

TABLE 3.3. Constant Loss Rates (Sabol et al., 1992).

Hydrologic soil group[1]	Uniform Loss Rate, Inches/hour	
	Musgrave (1955)	U.S. Bureau of Reclamation (1987)
A	0.30–0.45	0.30–0.50
B	0.15–0.30	0.15–0.30
C	0.05–0.15	0.05–0.15
D	0.00–0.05	.0–0.05

1. U.S. Soil Conservation Service

where Q is the runoff (in), P is the rainfall (in), S is the potential maximum retention after runoff begins (in), and I_a is the initial abstraction (in).

Initial abstraction is all losses before runoff begins. It includes water retained in surface depressions and water intercepted by vegetation, evaporation, and infiltration. I_a is highly variable but according to data from many small agricultural watersheds, I_a was approximated by the following empirical equation:

$$I_a = 0.2\,S. \tag{3.10}$$

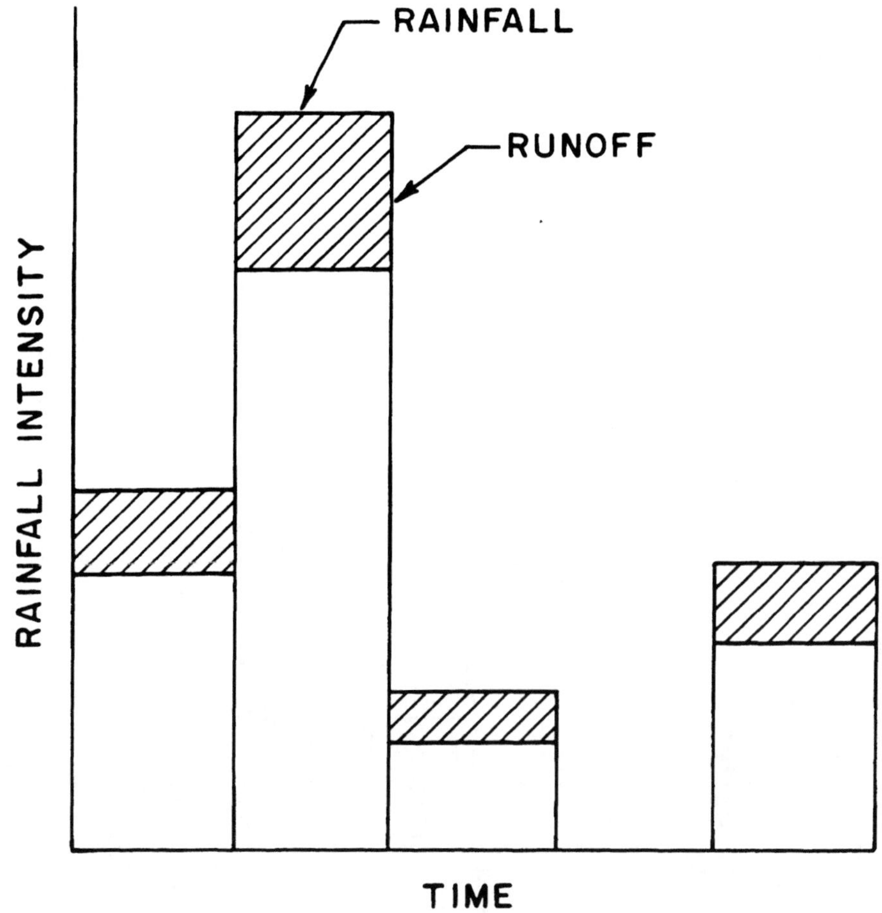

Figure 3.33.—Constant Proportion Loss Ratio.

By eliminating I_a as an independent parameter, this approximation allows the use of a combination of S and P to produce a unique runoff amount. Substituting Equation 3.10 into Equation 3.9 gives

$$Q = \frac{(P - 0.2\,S)^2}{(P + 0.8\,S)} \tag{3.11}$$

where the parameter S is related to the soil and cover conditions of the watershed through the curve number, CN. A graphical solution of Equation 3.11 is given in Fig. 3.34. CN has a range of 40 to 100, and S is related to CN by

$$S = \frac{1000}{CN} - 10. \tag{3.12}$$

Equation 3.12 calculates S in units of inches. The major factors that determine CN are the hydrologic soil group, cover type, treatment, hydrologic condition, and antecedent runoff condition. The values of CN's in Table 3.4 (a to d) represent average antecedent runoff conditions for urban, cultivated agricultural, other agricultural, and arid and semiarid rangeland uses (Soil Conservation Service, 1986). The following sections explain how to determine factors affecting the CN.

a. Hydrologic Soil Groups The SCS has classified all soils into four hydrologic soil groups (A, B, C, and D) according to their infiltration rate, which is obtained for bare soil after prolonged wetting. The four groups are defined as follows.

Group A soils have low runoff potential and high infiltration rates even when thoroughly wetted. They consist chiefly of deep, well- to excessively-drained sands or gravels. The USDA soil textures normally included in this group are sand, loamy sand, and sandy loam. These soils have a transmission rate greater than 0.3 in/hr.

Group B soils have moderate infiltration rates when thoroughly wetted and consist chiefly of moderately deep to deep, moderately well- to well-drained soils with moderately fine to moderately coarse

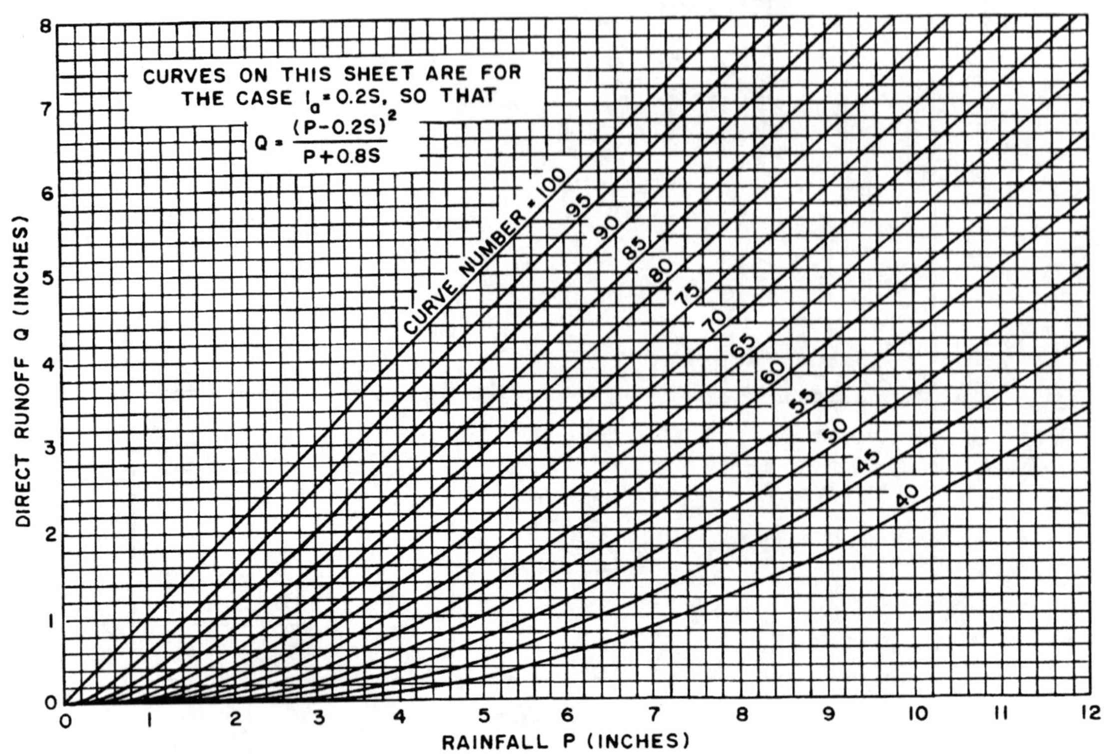

Figure 3.34.—Solution of Runoff Equation (SCS, 1972).

TABLE 3.4a. Runoff Curve Numbers for Urban Areas[1] (Soil Conservation Service, 1986).

Cover description		Curve numbers for hydrologic soil group—			
Cover type and hydrologic condition	Average percent impervious area[2]	A	B	C	D
Fully developed urban areas (vegetation established)					
Open space (lawns, parks, golf courses, cemeteries, etc.)[3]:					
Poor condition (grass cover < 50%)		68	79	86	89
Fair condition (grass cover 50% to 75%)		49	69	79	84
Good condition (grass cover > 75%)		39	61	74	80
Impervious areas:					
Paved parking lots, roofs, driveways, etc. (excluding right-of-way)		98	98	98	98
Streets and roads:					
Paved; curbs and storm sewers (excluding right-of-way)		98	98	98	98
Paved; open ditches (including right-of-way)		83	89	92	93
Gravel (including right-of-way)		76	85	89	91
Dirt (including right-of-way)		72	82	87	89
Western desert urban areas:					
Natural desert landscaping (pervious areas only)[4]		63	77	85	88
Artificial desert landscaping (impervious weed barrier, desert shrub with 1- to 2-inch sand or gravel mulch and basin borders)		96	96	96	96
Urban districts:					
Commercial and business	85	89	92	94	95
Industrial	72	81	88	91	93
Residential districts by average lot size:					
1/8 acre or less (town houses)	65	77	85	90	92
1/4 acre	38	61	75	83	87
1/3 acre	30	57	72	81	86
1/2 acre	25	54	70	80	85
1 acre	20	51	68	79	84
2 acres	12	46	65	77	82
Developing urban areas					
Newly graded areas (pervious areas only, no vegetation)[5]		77	86	91	94
Idle lands (CN's are determined using cover types similar to those in table 2-2c).					

[1]Average runoff condition, and $I_a = 0.2S$.
[2]The average percent impervious area shown was used to develop the composite CN's. Other assumptions are as follows: impervious areas are directly connected to the drainage system, impervious areas have a CN of 98, and pervious areas are considered equivalent to open space in good hydrologic condition. CN's for other combinations of conditions may be computed using Fig. 3.35 or 3.36.
[3]CN's shown are equivalent to those of pasture. Composite CN's may be computed for other combinations of open space cover type.
[4]Composite CN's for natural desert landscaping should be computed using figures 3.35/3.36 based on the impervious area percentage (CN = 98) and the pervious area CN. The pervious area CN's are assumed equivalent to desert shrub in poor hydrologic condition.
[5]Composite CN's to use for the design of temporary measures during grading and construction should be computed using Fig. 3.35 or 3.36 based on the degree of development (impervious area percentage) and the CN's for the newly graded pervious areas.

textures. The USDA soil textures normally included in this group are silt loam and loam. These soils have a transmission rate between 0.15 to 0.3 in/hr.

Group C soils have low infiltration rates when thoroughly wetted and consist chiefly of soils with a layer that impedes downward movement of water and soils with moderately fine to fine texture. The USDA soil texture normally included in this group is sandy clay loam. This soil has a transmission rate between 0.05 to 0.15 in/hr.

Group D soils have high runoff potential. They have very low infiltration rates when thoroughly wetted and consist mainly of clay soils with a high swelling potential, soils with a permanent high water

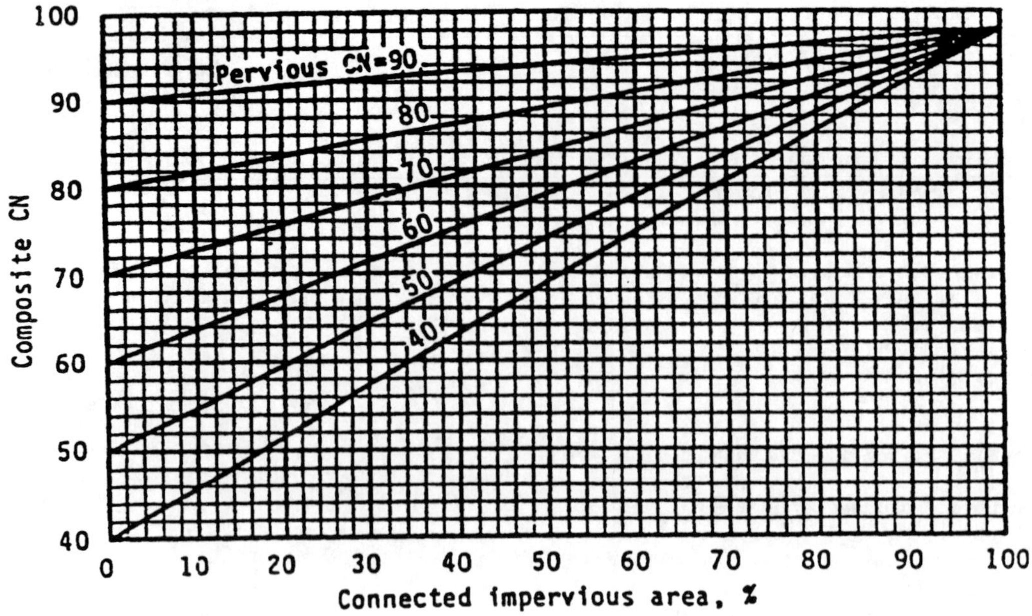

Figure 3.35.—Composite CN with Inconnected Impervious Areas and Total Impervious Areas Less Than 30%. (SCS 1986)

table, soils with a claypan or clay layer at or near the surface and shallow soils over a nearly impervious material. The USDA soil textures normally included in this group are clay loam, silty clay loam, sandy clay, silty clay, and clay. These soils have a very low rate of water transmission (0.0 to 0.05 in/hr). Some soils are classified in group D because of a high water table that creates a drainage problem; however, once these soils are effectively drained, the soils are placed into another group.

A list of most of the soils in the United States and their respective hydrologic soil group classification is given in Soil Conservation Service (1982a). Maps and soil reports are available on a county basis for most of the country and can be obtained from the library or SCS offices.

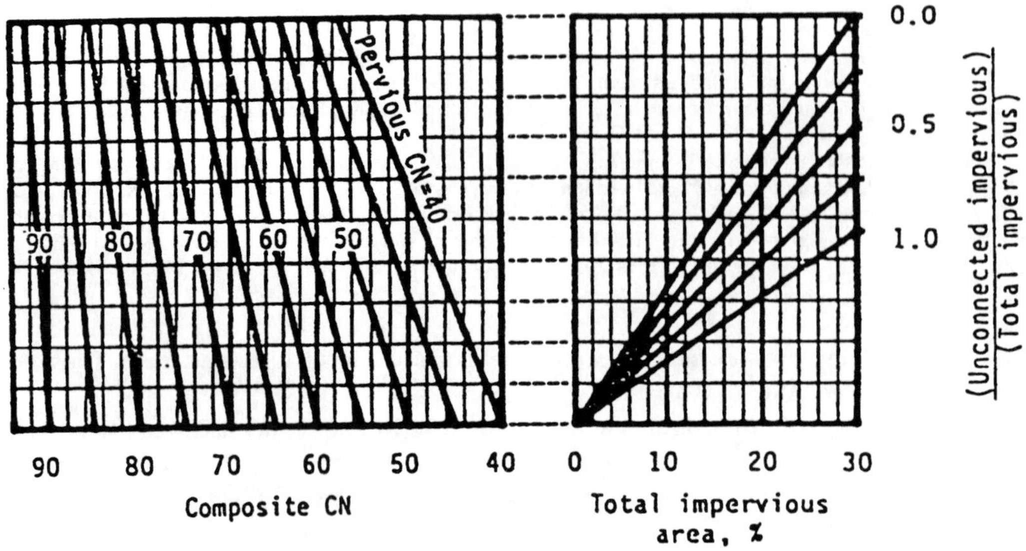

Figure 3.36.—Composite CN with Inconnected Impervious Areas and Total Impervious Areas Less Than 30%. (SCS 1986)

TABLE 3.4b. Runoff Curve Numbers for Cultivated Agricultural Lands[1] (Soil Conservation Service, 1986).

Cover description			Curve numbers for hydrologic soil group—			
Cover type	Treatment[2]	Hydrologic condition[3]	A	B	C	D
Fallow	Bare soil	—	77	86	91	94
	Crop residue cover (CR)	Poor	76	85	90	93
		Good	74	83	88	90
Row crops	Straight row (SR)	Poor	72	81	88	91
		Good	67	78	85	89
	SR + CR	Poor	71	80	87	90
		Good	64	75	82	85
	Contoured (C)	Poor	70	79	84	88
		Good	65	75	82	86
	C + CR	Poor	69	78	83	87
		Good	64	74	81	85
	Contoured & terraced (C&T)	Poor	66	74	80	82
		Good	62	71	78	81
	C&T + CR	Poor	65	73	79	81
		Good	61	70	77	80
Small grain	SR	Poor	65	76	84	88
		Good	63	75	83	87
	SR + CR	Poor	64	75	83	86
		Good	60	72	80	84
	C	Poor	63	74	82	85
		Good	61	73	81	84
	C + CR	Poor	62	73	81	84
		Good	60	72	80	83
	C&T	Poor	61	72	79	82
		Good	59	70	78	81
	C&T + CR	Poor	60	71	78	81
		Good	58	69	77	80
Close-seeded or broadcast legumes or rotation meadow	SR	Poor	66	77	85	89
		Good	58	72	81	85
	C	Poor	64	75	83	85
		Good	55	69	78	83
	C&T	Poor	63	73	80	83
		Good	51	67	76	80

[1]Average runoff condition, and $I_a = 0.2S$.
[2]*Crop residue cover* applies only if residue is on at least 5% of the surface throughout the year.
[3]Hydrologic condition is based on combination of factors that affect infiltration and runoff, including (a) density and canopy of vegetative areas, (b) amount of year-round cover, (c) amount of grass or close-seeded legumes in rotations, (d) percent of residue cover on the land surface (good ≥ 20%), and (e) degree of surface roughness.
Poor: Factors impair infiltration and tend to increase runoff.
Good: Factors encourage average and better than average infiltration and tend to decrease runoff.

b. Treatment Treatment is a cover-type modifier used in Table 3.4 to describe the management of cultivated agricultural lands. It includes mechanical practices, such as contouring and terracing, and management practices, such as crop rotations and reduced or no tillage.

c. Hydrologic Condition Hydrologic condition indicates the effects of cover type and treatment on infiltration and runoff and is generally estimated from density of plant and residue cover on sample areas. A good hydrologic condition indicates that the soil usually has a low runoff potential for the given hydrologic soil group, cover type, and treatment. Some factors to consider when estimating the effect of cover on infiltration and runoff are: 1) canopy or density of lawns, crops, or other vegetative areas; 2) amount of year-round cover; 3) amount of grass or close-seeded legumes in rotations; 4) percent of residue cover; and 5) degree of surface roughness. Several factors, such as the percentage of impervious

TABLE 3.4c. Runoff Curve Numbers for Other Agricultural Lands[1] (Soil Conservation Service 1986).

Cover description		Curve numbers for hydrologic soil group—			
Cover type	Hydrologic condition	A	B	C	D
Pasture, grassland, or range—continuous forage for grazing.[2]	Poor	68	79	86	89
	Fair	49	69	79	84
	Good	39	61	74	80
Meadow—continuous grass, protected from grazing and generally mowed for hay.	—	30	58	71	78
Brush—brush-weed-grass mixture with brush the major element.[3]	Poor	48	67	77	83
	Fair	35	56	70	77
	Good	[4]30	48	65	73
Woods—grass combination (orchard or tree farm).[5]	Poor	57	73	82	86
	Fair	43	65	76	82
	Good	32	58	72	79
Woods.[6]	Poor	45	66	77	83
	Fair	36	60	73	79
	Good	[4]30	55	70	77
Farmsteads—buildings, lanes, driveways, and surrounding lots.	—	59	74	82	86

[1]Average runoff condition, and $I_a = 0.2S$.

[2]*Poor*: < 50% ground cover or heavily grazed with no mulch.
Fair: 50 to 75% ground cover and not heavily grazed.
Good: > 75% ground cover and lightly or only occasionally grazed.

[3]*Poor*: < 50% ground cover.
Fair: 50 to 75% ground cover.
Good: > 75% ground cover.

[4]Actual curve number is less than 30; use CN = 30 for runoff computations.

[5]CN's shown were computed for areas with 50% woods and 50% grass (pasture) cover. Other combinations of conditions may be computed from the CN's for woods and pasture.

[6]*Poor*: Forest litter, small trees, and brush are destroyed by heavy grazing or regular burning.
Fair: Woods are grazed but not burned, and some forest litter covers the soil.
Good: Woods are protected from grazing, and litter and brush adequately cover the soil.

area and the means of conveying runoff from impervious areas to the drainage system, should be considered in computing CN for urban areas.

d. Antecedent Runoff Condition Antecedent runoff condition (ARC) is an index of runoff potential for a storm event. The ARC is an attempt to account for the variation in CN at a site from storm to storm. CN for the average ARC at a site is the median value as taken from sample rainfall and runoff data. The CN's in Table 3.4 are for the average ARC, which is used primarily for design applications. Soil Conservation Service (1982a) and Rallison (1980) give a detailed discussion of storm-to-storm variation and a demonstration of upper and lower enveloping curves.

e. Curve Number Limitations Curve numbers describe average conditions that are useful for design purposes. If the rainfall event used is a historical storm that departs from average conditions, the modeling accuracy decreases. The runoff curve number equation should be applied with caution when recreating specific features of an actual storm. The equation does not contain an expression for time. Therefore, it does not account for rainfall duration or intensity although Equation 3.11 can be applied to the cumulative rainfall at a number of points within the cumulative rainfall hyetograph. Thus, an excess rainfall hyetograph can be generated for the storm.

The user should understand the assumptions reflected in the initial abstraction term I_a and should ascertain that these assumptions apply to the situation at hand. I_a, which consists of interception, initial infiltration, surface depression storage, evapotranspiration, and other factors, is generalized as 0.2S based

TABLE 3.4d. Runoff Curve Numbers for Arid and Semiarid Rangelands[1] (Soil Conservation Service, 1986).

Cover description		Curve numbers for hydrologic soil group—			
Cover type	Hydrologic condition[2]	A[3]	B	C	D
Herbaceous—mixture of grass, weeds, and low-growing brush, with brush the minor element.	Poor		80	87	93
	Fair		71	81	89
	Good		62	74	85
Oak-aspen—mountain brush mixture of oak brush, aspen, mountain mahogany, bitter brush, maple, and other brush.	Poor		66	74	79
	Fair		48	57	63
	Good		30	41	48
Pinyon-juniper—pinyon, juniper, or both; grass understory.	Poor		75	85	89
	Fair		58	73	80
	Good		41	61	71
Sagebrush with grass understory.	Poor		67	80	85
	Fair		51	63	70
	Good		35	47	55
Desert shrub—major plants include saltbush, greasewood, creosotebush, blackbrush, bursage, palo verde, mesquite, and cactus.	Poor	63	77	85	88
	Fair	55	72	81	86
	Good	49	68	79	84

[1]Average runoff condition, and $I_a = 0.2S$. For range in humid regions, use table 3.4c.
[2]*Poor*: < 30% ground cover (litter, grass, and brush overstory).
Fair: 30 to 70% ground cover.
Good: > 70% ground cover.
[3]Curve numbers for group A have been developed only for desert shrub.

on data from agricultural watersheds. This approximation can be especially important in an urban application because the combination of impervious areas with pervious areas can imply a significant initial loss that may not take place. The opposite effect, a greater initial loss, can occur if the impervious areas have surface depressions that store some runoff. To use a relationship other than $I_a = 0.2S$, one must use rainfall-runoff data to establish new S or CN relationships for each cover and hydrologic soil group. Runoff from snowmelt or rain on frozen ground cannot be estimated using these procedures.

The CN procedure is less accurate when runoff is less than 0.5 in. As a check, another procedure should be used to determine runoff. The SCS runoff procedures apply only to direct surface runoff and do not include subsurface flow or high ground water levels that contribute to runoff. These conditions are often related to Hydrologic Soil Group.

Finally, Type A soils and forest areas that have been assigned relatively low CN's are shown in Table 3.4. Good judgment and experience based on stream gage records are needed to adjust CN's as conditions warrant. When the weighted CN is less than 40, another procedure should be used to determine runoff.

B. Infiltration Models

The evolution of infiltration modeling has taken three directions: the empirical, the approximate (meaning approximation to the physically-based models), and the physical approach. Most of the empirical and approximate models treat soil as a semi-infinite medium with the soil saturating from the surface down. Physically-based models specify appropriate boundary conditions and normally require detailed data input. The Richards equation is the physically-based infiltration equation used for describing water flow in soils. Solving this equation mathematically is extremely difficult for many flow problems. Until recent advances in numerical methods and personal computing power, use of the Richards equation was not feasible for practical applications. Ross (1990) developed a software package using the Richards Equation for simulating water infiltration and water movement in soils where water is added as precipitation and removed by runoff, drainage, evaporation from the soil surface, and transpiration by vegetation. The practical use of the program is enhanced by the development of procedures for predicting soil hydraulic properties previously discussed.

The following sections summarize the commonly used empirical models developed by Kostiakov (1932), Horton (1919), and Holtan (1961) and approximate models developed by Green and Ampt (1911), Philip (1957), Morel-Seytoux and Kanji (1974), and Smith and Parlange (1978), including methods to estimate their parameters.

1. *Kostiakov Model* Kostiakov (1932) proposed a simple infiltration model relating the infiltration rate, f_p, to time, t, which was presented by Skaggs and Khaleel (1982) as

$$f_p = K_k t^{-\alpha}, \tag{3.13}$$

where K_k and α are constants which depend on the soil and initial conditions and may be evaluated using the observed infiltration rate–time relationship.

Use of Kostiakov's model is limited by its need for a set of observed infiltration data for parameter evaluation; thus, it cannot be applied to other soils and conditions which differ from the conditions for which parameters K_k and α are determined. The Kostiakov model has primarily been used for irrigation applications.

2. *Horton Model* A three-parameter empirical infiltration model was presented by Horton (1940) and has been widely used in hydrologic modeling. Horton found that the infiltration capacity (F_p) to time (t) relationship may be expressed as

$$F_p = f_c + (f_o - f_c)e^{-\beta t} \tag{3.14}$$

where f_o is the maximum infiltration rate at the beginning of a storm event, and it reduces to a low and approximately constant rate of f_c as the infiltration process continues and the soil becomes saturated. The parameter β controls the rate of decrease in the infiltration capacity (Skaggs and Khaleel, 1982). Horton's equation is applicable only when effective rainfall intensity (i_e) is greater than f_c. Parameters f_o, f_c, and β must be evaluated using observed infiltration data. Generalized parameter estimates are given in Table 3.5.

Wide-scale application of this model is limited because of the parameters' dependence on specific soil and moisture conditions. These parameters can be related to the physically based parameters of the Green-Ampt equation (Morel-Seytoux, 1988; 1989).

3. *Holtan Model* Holtan (1961) developed an empirical equation on the premise that soil moisture storage, surface connected porosity, and the effect of root paths are the dominant factors influencing the infiltration capacity. Holtan et al. (1975) modified the equation to be

$$f = (GI)(A)S_\alpha^{1.4} + f_c, \tag{3.15}$$

TABLE 3.5. Parameter Estimates for Horton Infiltration Model (Horton, 1940).

Soil and cover complex	f_o, mm h^{-1}	f_c, mm h^{-1}	β, min^{-1}
Standard agricultural (bare)	280	6-220	1.6
Standard agricultural (turfed)	900	20-290	0.8
Peat	325	2-29	1.8
Fine sandy clay (bare)	210	2-25	2.0
Fine sandy clay (turfed)	670	10-30	1.4

where f is the infiltration capacity (in/hr), GI is the growth index of crop in percent maturity varying from 0.1 to 1.0 during the season, A is the infiltration capacity (in hr^{-1}) per $(in)^{1.4}$ of available storage and is an index representing surface-connected porosity and the density of plant roots which affect infiltration (Table 3.6), S_a is the available storage in the surface layer (a horizon) in inches, and f_c is the constant infiltration rate when the infiltration rate curve reaches asymptote (steady infiltration rate). Musgrave (1955) relates f_c to different hydrologic soil groups (Table 3.7).

Holtan's equation computes the infiltration capacity based on the actual available storage (S_a) of the surface layer (a horizon). This equation is easy to use for predicting rainfall infiltration and the values for the input parameters can be obtained from tables for known soil type and land use. A major difficulty with the use of Holtan's equation is the evaluation of the depth of the top layer (control depth). Holtan et al. (1975) suggest using plow layer or depth to the impeding layer as the control depth.

The following steps may be used in applying Holtan's equation:

(a) Find f_c for a given hydrologic soil group using Table 3.7 as used by Skaggs and Khaleel (1982).
(b) Find A from Table 3.6 as used by Skaggs and Khaleel (1982).
(c) Estimate GI based on crop maturity stage using known data or observing it in the field, following Holtan et al. (1975).
(d) Computation of the initial available storage (S_a) requires measured or predicted initial water content (Θ_i), saturated water content (Θ_s) and depth of the surface layer (d). The value of S_a has to be recalculated for each time step in which infiltration is being computed. The initial available storage (S_{ao}) and available storage for the other time steps (S_{ai}) may be determined as:

$$S_{ao} = (\theta_s - \theta_i)\,d \tag{3.16}$$

TABLE 3.6. Estimates of Vegetative Parameter "A" in Holtan Infiltration Model (Frere et al., 1975).

Land use or cover	Basal area rating*	
	Poor condition	Good condition
Fallow†	0.10	0.30
Row crops	0.10	0.20
Small grains	0.20	0.30
Hay (legumes)	0.20	0.40
Hay (sod)	0.40	0.60
Pasture (bunch grass)	0.20	0.40
Temporary pasture (sod)	0.20	0.60
Permanent pasture (sod)	0.80	1.00
Woods and forests	0.80	1.00

*Adjustments needed for "weeds" and "grazing."
†For fallow land only, poor condition means "after row crop" and good condition means "after sod."

TABLE 3.7. Estimates of Final Infiltration Rate for Holtan Infiltration Model (Musgrave, 1955).

Hydrologic soil group	f_c, cm/h
A	0.76
B	0.38-0.76
C	0.13-0.38
D	0.0-0.13

and

$$S_{ai} = S_{ai-1} - F_{i-1} + (f_c \Delta t)_{i-1} + (ET \Delta t)_{i-1}, \tag{3.17}$$

where F is the amount of water infiltrated during Δt, $(f_c\Delta t)$ is the drainage at a rate of f_c up to the limit of S_{ao}, ET is the evapotranspiration during Δt, and i is a subscript indicating the time step. For the second time step $S_{ai-1} = S_{ao}$, considering S_{ao} is available storage for the first time step.

4. *Green-Ampt Model* The Green-Ampt (1911) model is an approximate model utilizing Darcy's law. The original model was developed for ponded infiltration into a deep homogeneous soil with a uniform initial water content. Water is assumed to infiltrate into the soil as piston flow resulting in a sharply-defined wetting front which separates the wetted and unwetted zones as shown in Fig. 3.37. Neglecting the depth of ponding at the surface, the Green-Ampt rate equation is

$$f = K(1 + (\phi - \theta_i)S_f / F) \tag{3.18}$$

and its integrated form is

$$K(t) = F - S_f (\phi - \theta_i) \ln (1 + F)/[(\phi - \theta_i)S_f], \tag{3.19}$$

where K is the effective hydraulic conductivity (cm/hr), S_f is the effective suction at the wetting front (cm), ϕ is the soil porosity (cm³/cm³), θ_i is the initial water content (cm³/cm³), F is accumulated infiltration, [cm], and f is the infiltration rate (cm/hr). Equation 3.18 assumes a ponded surface so that infiltration rate equals infiltration capacity.

Mein and Larson (1973) developed the following system for applying the Green-Ampt model to rainfall conditions. Just prior to surface ponding, the rainfall rate, R [L/T], equals the infiltration rate, f, and the cumulative infiltration at time to ponding, F_p, equals the rainfall rate times the time to surface ponding t_p. Thus, the infiltration rate for steady rainfall is

$$f = R \quad \text{for } t < t_p \tag{3.20}$$

$$f = K + K S_f (\phi - \theta_i)/F_p \quad \text{for } t > t_p, \tag{3.21}$$

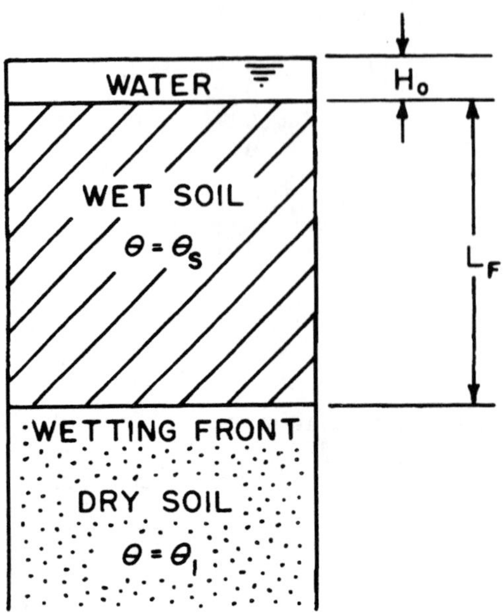

Figure 3.37.—Green-Ampt Model.

where $t_p = F_p/R$ and $F_p = (S_f(\phi) - \theta)/(R/K - 1)$. The integrated form that is analogous to Equation 3.19

$$K(t - t_p + t_p') = F - [(\phi S - \theta_i)S_f \ln (1 + F/(\phi S - \theta_i)S_f)], \tag{3.22}$$

where t_p' is the equivalent time to infiltrate volume F_p under initially surface ponded conditions. It can be calculated from Equation 3.19. Generally, the Green-Ampt models can be applied by incrementing F and solving for f in Equation 3.21 and then using Equation 3.20 for f.

The Green-Ampt Equations 3.18 and 3.19 for homogeneous soils can be extended to describe infiltration into layered soils, when the hydraulic conductivity of the successive layers decreases with depth (Childs and Byborbi, 1969; Hachum and Alfaro, 1980). As long as the wetting front is in the top layer, the equations remain the same. After the wetting front enters the second layer, the effective hydraulic conductivity, K, is set equal to harmonic mean K_h for wetted depths of layers 1 and 2 $(K_1K_2)^{1/2}$, and the capillary head, S_f, is set equal to S_f of the second layer. This principle is then carried through to third and succeeding layers.

For a layered soil, in which the saturated hydraulic conductivity of a subsoil layer is greater than that of a layer above (typically a crusted soil), the above Green-Ampt Equations cannot be used after the wetting front enters the higher-K layer. For such cases, it may be assumed that infiltration through the higher-K layer continues to be governed by harmonic mean K of the upper layers (Moore and Eigel, 1981).

One of the most common forms of soil layering is the formation of a crust on the soil surface caused by raindrop impact. The thickness of the crust is generally very small, e.g., 1.5 to 3.0 mm (McIntyre, 1958) and usually develops during the first 10 cm of rainfall. Ahuja (1983) developed a Green-Ampt approach based on physical principles to handle a developing crust; however, for practical applications it can be assumed that a stable crust exists on a bare soil.

For infiltration under ponded conditions, the soil in the wetted zone is nearly saturated. The wetted zone then develops a viscous resistance to air flow, which reduces infiltration rate. To account for this effect, Morel-Seytoux and Khanji (1974) introduced a correction factor to the Green-Ampt equation for a homogeneous soil. The correction factor varies with the soil type and ponding depth ranging from 1.1 to 1.7 with an average of 1.4. The correction factor can be used to reduce the Green-Ampt infiltration rate (dividing the rate by the correction factor) when trapped air in the soil is a problem.

To apply the Green-Ampt model, the effective hydraulic conductivity, K, the wetting front suction, S_f, the porosity, ϕ, and the initial moisture content, θ_i, must be measured or estimated. These parameters can be determined by fitting to experimental infiltration data; however, for specific application purposes, it is easier to determine the parameters from readily available data such as soils and land use data.

Average values for the Green-Ampt wetting front suction, S_f, saturated hydraulic conductivity, K_s, and porosity, ϕ, are given in Table 3.8 for the eleven USDA soil textures. These values can be used as a first estimate; however, if more detailed soil properties are available, more refined estimates can be made using the following prediction equations.

The porosity, ϕ, can be determined from measured bulk density or estimated bulk density determined from Fig. 3.38 using percent sand, percent clay, and percent organic matter available from most soil analyses. Also, for first estimates of porosity, Fig. 3.39 can be used. If the cation exchange capacity of the clay, which is an indicator of the shrink-swell capacity of the clay, is available, the bulk density at the water content for 33 kPa tension can be estimated by the following:

$$\begin{aligned} BD = {} & 1.51 + 0.0025\,(S) - 0.0013\,(S)\,(OM) \\ & - 0.0006\,(C)\,(OM) - 0.0048\,(C)\,(CEC), \end{aligned} \tag{3.23}$$

where BD is the bulk density of < 2 mm material (g/cm^3), S is the percentage of sand, C is the percentage of clay, OM is the percentage of organic matter (1.7) (% organic carbon), and CEC is (cation exchange capacity of clay)/(% clay) (ranges 0.1–0.9).

Coarse fragments (> 2 mm in size) in the soil affect the porosity, and adjustments should be made using the equation (Brakensiek et al., 1986a):

$$\phi_c = \phi\, CFC, \tag{3.24}$$

TABLE 3.8. Green-Ampt Parameters (Rawls et al., 1993a).

Soil texture class	Porosity ϕ	Wetting front soil suction head S_f, cm	Saturated hydraulic conductivity* K_s, cm/h
Sand	0.437 (0.374-0.500)	4.95 (0.97-25.36)	23.56
Loamy sand	0.437 (0.363-0.506)	6.13 (1.35-27.94)	5.98
Sandy loam	0.453 (0.351-0.555)	11.01 (2.67-45.47)	2.18
Loam	0.463 (0.375-0.551)	8.89 (1.33-59.38)	1.32
Silt loam	0.501 (0.420-0.582)	16.68 (2.92-95.39)	0.68
Sandy clay loam	0.398 (0.332-0.464)	21.85 (4.42-108.0)	0.30
Clay loam	0.464 (0.409-0.519)	20.88 (4.79-91.10)	0.20
Silty clay loam	0.471 (0.418-0.524)	27.30 (5.67-131.50)	0.20
Sandy clay	0.430 (0.370-0.490)	23.90 (4.08-140.2)	0.12
Silty clay	0.479 (0.425-0.533)	29.22 (6.13-139.4)	0.10
Clay	0.475 (0.427-0.523)	31.63 (6.39-156.5)	0.06

*K_s can be modified to obtain the Green-Ampt K. For bare ground conditions K can be taken as $k_s/2$.

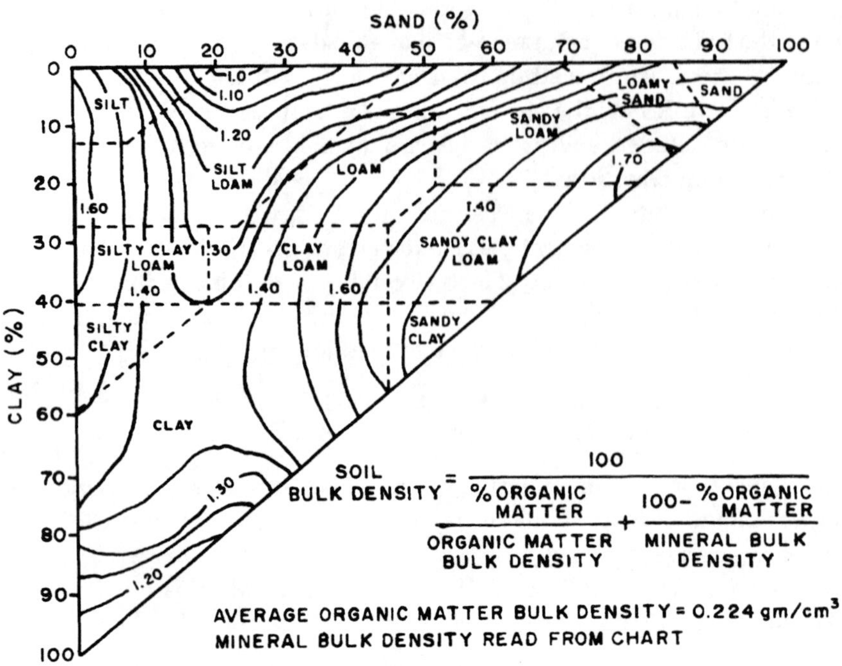

Figure 3.38.—Soil Bulk Density (Rawls, 1983).

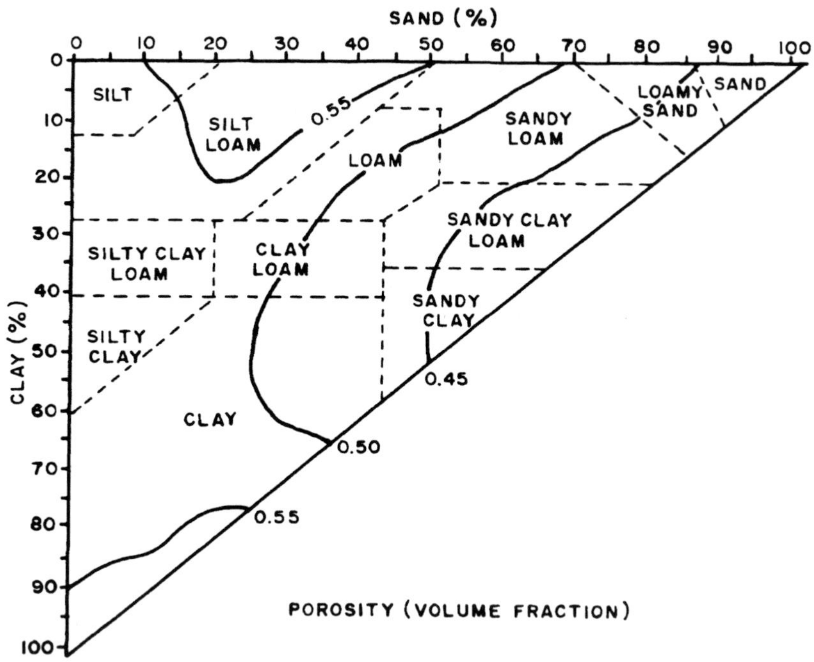

Figure 3.39.—Porosity Classified According to Soil Texture (Rawls et al., 1990).

where ϕ_c is the porosity of the soil with coarse fragments (% vol), ϕ is the porosity of the soil without coarse fragments (% vol), CFC = 1 − VCF/100, VCF is the volume of coarse fragments (> 2 mm) computed from (WCF/2.65) (100)/((100 − WCF)BD) + WCF/2.65, WCF is the percent weight of coarse fragments, and BD is the bulk density of the soil fraction less than 2 mm (g/cm³).

The initial water content (θ_i) should be measured, or it can be estimated from moisture retention relationships (Rawls and Brakensiek, 1983). A good estimate of wet, average, and dry initial water contents is the water content held at −10 kPa, −33 kPa and −1500 kPa, respectively; however, this is dependent on location, e.g., 23 in the western rangeland average is −1500 kPa water content while in the eastern part of the United States it is closer to the −33 kPa water content.

The Green-Ampt wetting front suction parameter (S_f) can be estimated from the Brooks Corey parameter water retention parameters (Rawls and Brakensiek, 1983) as:

$$S_f = \frac{2 + 3\lambda}{1 + 3\lambda} \frac{(h_b)}{2}, \tag{3.25}$$

where S_f is the Green-Ampt wetting front suction (cm), λ is the Brooks-Corey pore size distribution index, and h_b is the Brooks-Corey bubbling pressure.

Rawls and Brakensiek further simplify Equation 3.25 by relating the Green-Ampt wetting front suction parameter to soil properties in the following equation.

$$\begin{aligned} S_f = \text{Exp}\ [&6.53 - 7.326\ (\phi) + 0.00158\ (C^2) + 3.809\ (\phi^2) \\ &+ 0.000344\ (S)\ (C) - 0.04989\ (S)\ (\phi) + 0.0016\ (S^2)\ (\phi^2) \\ &+ 0.0016\ (C^2)\ (\phi^2) - 0.0000136\ (S^2)\ (C) - 0.00348\ (C^2)(\phi) \\ &- 0.000799)\ (S^2)\ (\phi)], \end{aligned} \tag{3.26}$$

where S is the percentage of sand, C is the percentage of clay, and ϕ is the porosity (% vol).

A graphical representation of the Green-Ampt wetting front suction parameter is shown in Fig. 3.40.

The wetted front suction parameter S_f was related to the hydraulic conductivity by Van Mullem (1989):

$$S_f = 4.903\ (K_s + 0.02)^{-.4932}. \tag{3.27}$$

This equation permits the estimation of S_f from K_s for various cover types and conditions.

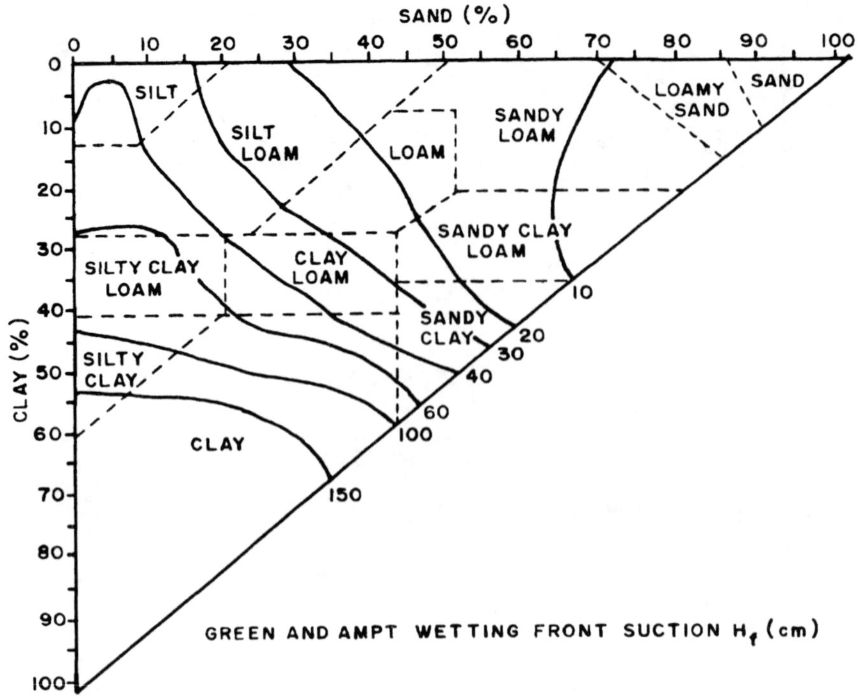

Figure 3.40.—Green-Ampt Wetting Front Suction Classified According to Soil Texture (Rawls et al., 1990).

The saturated hydraulic conductivity of the soil matric can be obtained in several ways other than measuring it. Selection of saturated hydraulic conductivity prediction techniques depends upon the availability and the level of information on physical and hydraulic soil properties (Mualem, 1986). If only soil texture classes are available, the saturated hydraulic conductivities and corresponding unsaturated hydraulic conductivity curves can be obtained from Fig. 3.41 (Saxton et al., 1986). If specific soil texture information is available, the saturated hydraulic conductivity for undisturbed conditions can be obtained from Fig. 3.42. If characteristics of the water retention curve are available, then the technique developed by Ahuja et al. (1985, 1988, 1989), where the saturated hydraulic conductivity is related to an effective porosity [(ϕ_e), total porosity obtained from soil bulk density minus the soil water content at −33 kPa matric potential], is represented by the following generalized Kozeny–Carman equation:

$$K_s = B\,\phi_e^n, \tag{3.28}$$

where ϕ_e can be set equal to 4 and B equals 1058 when K_s has units of cm h^{-1}. For soils with sand greater than 65 percent and/or clays greater than 40 percent, the saturated hydraulic conductivities may vary by an order of magnitude or greater.

The Marshall (1958) saturated hydraulic conductivity equation using the equivalent pore radius for the Sierpinski carpet at the first recursive level from Tyler and Wheatcraft (1990) becomes

$$K_s = 4.41\,(10^7)(\phi^x/n^2)(R_1^{\,2}), \tag{3.29}$$

where K_S is the saturated hydraulic conductivity, cm h^{-1}, R_1 is the largest pore radius, ϕ is the total porosity, x is the pore interaction exponent, and n is the total pore size classes.

In Equation 3.29, the exponent x was set equal to 4/3 as proposed by Millington and Quirk (1961). The maximum micropore radius, R_1, is calculated by

$$R_1 = 0.148/h_b, \tag{3.30}$$

where h_b is the geometric mean soil bubbling pressure (cm) (Rawls and Brakensiek, 1986).

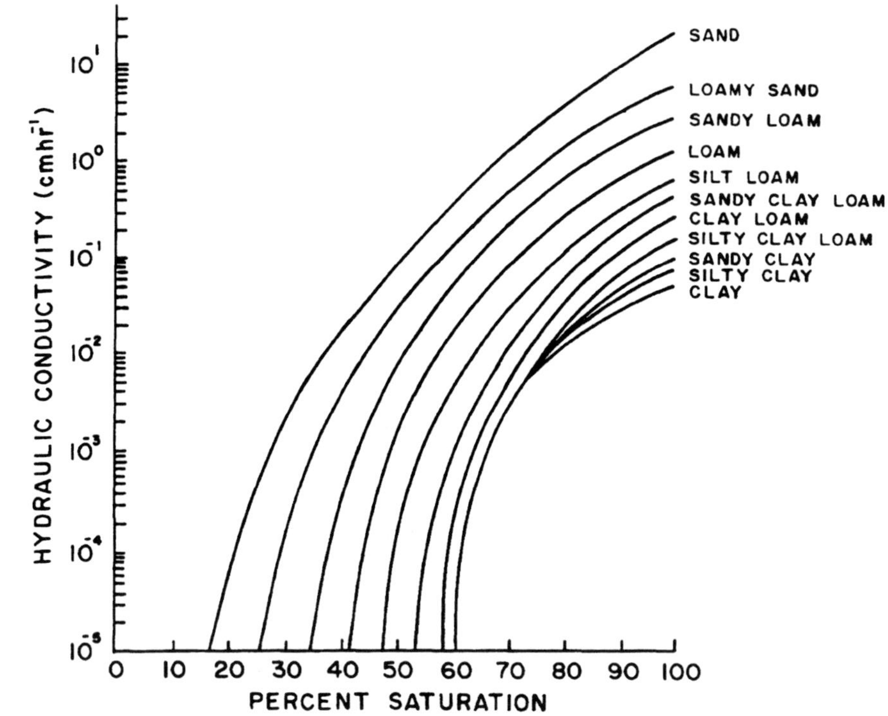

Figure 3.41.—Hydraulic Conductivity Curves Classified by Soil Texture (Saxton et al., 1986).

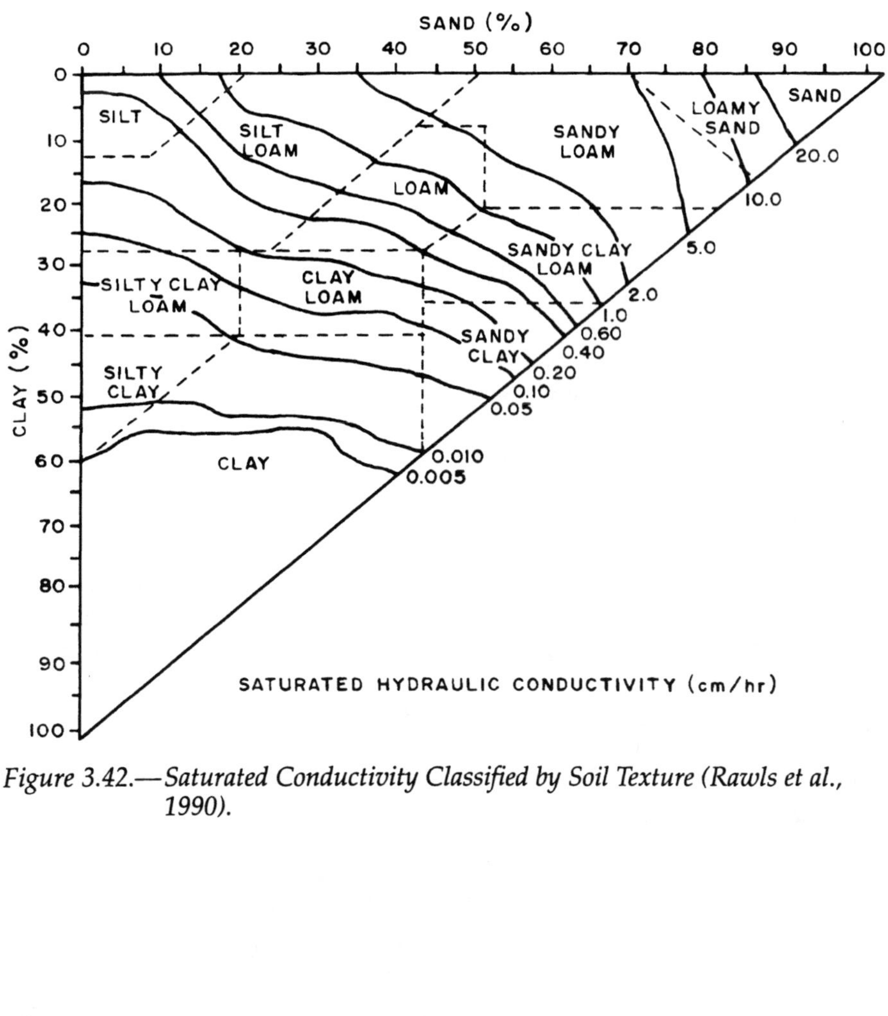

Figure 3.42.—Saturated Conductivity Classified by Soil Texture (Rawls et al., 1990).

Green and Corey (1971) proposed the following equation for estimating the n-value;

$$n = m\,[\Theta_s/(\Theta_s - \Theta_l)], \tag{3.31}$$

where Θ_s is the total porosity, Θ_l is the porosity at a lower water content, and m = 12 (Marshall, 1958). Ahuja et al. (1985) assumed that the porosity at a lower water content is equal to the water held at a tension of −33 kPa or what is commonly known as field capacity.

The parameters in Equation 3.29 can be estimated from soil properties, for example, matrix porosity by methods given by Rawls (1983), the −33 kPa water content, the Brooks and Corey bubbling pressure, and pore size index by methods presented by Rawls and Brakensiek (1986).

Macropores are defined as large voids in the soil, such as decayed root channels, worm holes and structural cleavages or cracks. Under surface ponded conditions, much water can be conducted through a single or interconnected macropores, thus bypassing the soil matrix resulting in an overestimation of the saturated hydraulic conductivity of the soil matrix during infiltration. The hydraulics of the macropore system are not modeled by the classic soil physics models used to model the soil matrix. The measurement of saturated hydraulic conductivity is normally composed of both macropore and matrix flow. Knowledge of the matrix and macropore saturated hydraulic conductivity is critical to realistically describing the flow system (Ahuja et al., 1988).

Equation 3.29 can also be used for calculating the saturated hydraulic conductivity of macropores. The value of ϕ that is the areal porosity of the macropores, ϕ_m, that have a radius greater than 0.2 mm. The exponent x is assumed to have the same value as the matrix (x = 1.333). R_1 is the radius of the maximum macropore. The calculation of n by Equation 3.31 is not appropriate for macropores because all the porosity at surface ponding contributes to K_s. Rawls et al. (1993b) assume that n = m and calibrate n-values to soil properties through multiple linear regression producing the following equations:

$$n = 87.37 + 74\,(R_1 - 52)\,D; \quad R^2 = 0.88; \quad SDE = 3.82, \tag{3.32}$$

where R_1 is the maximum macropore radius (cm) and D is the fractal dimension of the soil texture (Brakensiek and Rawls, 1992).

A practical approach to handling macroporosity is to set the effective hydraulic conductivity equal to the saturated hydraulic conductivity (K_s) times a macroporosity factor. Rawls et al. (1989) and Brakensiek and Rawls (1988) developed two macroporosity factors for areas that do not undergo mechanical disturbance on a regular basis (for example, rangeland) and one for areas that do undergo mechanical disturbance on a regular basis, (for example, agricultural areas). The prediction equations for the macroporosity factor (A) for undisturbed rangeland areas R is

$$A = \mathrm{Exp}\,[2.82 - 0.099\,(S) + 1.94\,(BD)]. \tag{3.33}$$

For undisturbed agricultural areas the equation is

$$A = \mathrm{Exp}\,[0.96 - 0.032\,(S) + 0.04\,(C) - 0.032\,(BD)], \tag{3.34}$$

where S is the percentage of sand, C is the percentage of clay, and BD is the bulk density of the soil (< 2 mm), g/cm^3. The macroporosity factor in Equations 3.33 and 3.34 is constrained to be at least 1.0.

For the bare area outside a plant canopy, the soil is assumed to be crusted and the effective hydraulic conductivity is equal to the saturated hydraulic conductivity (K_s) times a crust factor. Rawls et al. (1990) developed the following relationship for the crust factor:

$$CRC = SC/(1 + \psi_i/L), \tag{3.35}$$

where CRC is the crust factor, SC is the correction factor for partial saturation of the soil subcrust. From Table 3.9, this factor is equal to 0.736 + 0.0019 (% sand); ψ_i is the matric potential drop at the

TABLE 3.9. Mean Steady–State Matric Potential Drop ψ_i across Surface Seals by Soil Texture (Rawls et al. 1990).

Soil texture	Matric potential drop ψ_i, cm	Reduction factor for subcrust conductivity (SC)
Sand	2	0.91
Loamy sand	3	0.89
Sandy loam	6	0.86
Loam	7	0.82
Silt loam	10	0.81
Sandy clay loam	5	0.85
Clay loam	8	0.82
Silty clay loam	10	0.76
Sandy clay	6	0.80
Silty clay	11	0.73
Clay	9	0.75

crust/subcrust interface (cm). From Table 3.9, the value is 45.19 − 46.68 (SC), and L is the wetting front depth (cm).

The crust factor in equation 3.35 incorporates the effects of a crust into a one layer Green-Ampt model. Also, it does not predict a transient crust conductivity as it assumes that a steady-state crust conductivity can be predicted by

$$K_c = K_s \, SC \, Z_c(\psi_i + Z_c), \tag{3.36}$$

where Z_c is the crust thickness (cm) (can be assumed equal to 0.5 cm), and K_s is the saturated hydraulic conductivity of the soil (cm/hr).

Coarse fragments (> 2 mm in size) in the soil, in addition to their effect in reducing porosity, also affect the saturated hydraulic conductivity of the soil. The saturated hydraulic conductivity of the soil matrix should be multiplied by the following correction for coarse fragments (Brakensiek et al., 1986b):

$$\text{Coarse Fragment Correction} = \left(\frac{1 - \text{percent by weight of coarse fragments}}{100}\right). \tag{3.37}$$

Freezing the soil also affects the saturated hydraulic conductivity. Rawls and Brakensiek (1983) generalized the work of Lee (1983) and developed a saturated hydraulic conductivity correction factor for frozen soil which is

$$FSC = 2.0 - 1.9\,(\theta_f/\theta_{33}), \tag{3.38}$$

where FSC is the frozen soil hydraulic conductivity correction, θ_f is the percent volume of soil water at time of freezing, and θ_{33} is the percent volume of soil water held at −33 kPa matric potential. When θ_f is greater than θ_{33}, FSG equals 0.001.

Van Mullen (1989) adjusted K_s for management and cover conditions by relating it to the SCS curve number. The change in curve number from an initial calibrated value was transformed to an adjusted value of K_s by varying the porosity. For areas where soil data are lacking, K_s may be estimated from CN by

$$K_s = 21462\, e^{-.16061\, CN} \tag{3.39}$$

for values of CN between 55 and 94.

To incorporate the surface effects it is recommended that the unit area be divided into the following three categories: 1) the area that is bare and outside of canopy cover, 2) the area that has ground cover, and 3) the bare area under canopy. An effective hydraulic conductivity should then be developed for each area. Compute the infiltration separately for each area and then sum the three infiltration amounts weighted according to their areal cover to obtain the area infiltration. This method of determining the infiltration assumes that the three areas do not interconnect. If the areas do interconnect, this method overpredicts infiltration.

For the area that is bare under canopy, the effective hydraulic conductivity can be assumed to be equal to the soil saturated hydraulic conductivity (K_s). For the bare area outside canopy, the effective hydraulic conductivity is assumed to be equal to the final crusted conductivity. The area that has ground cover has an effective conductivity equal to the macropore hydraulic conductivity.

A conceptual approach is to consider the soil porosity as two domains, the macropores and the soil matrix, with an interaction between the two. The conceptual basis of the model is similar to that of Hoogmoed and Bouma (1980).

5. Philip Model Philip (1957) proposed that the first two terms of his series solution could be used as an infiltration model. The equation is:

$$f = st^{1/2} + A, \tag{3.40}$$

where f is the infiltration rate (cm/hr), t is time from ponding, s is the sorptivity, and A is a parameter with dimension of conductivity. Parameters in Equation 3.40 can be evaluated from experimental infiltration data using regression analysis; however, using the following approximations, the parameters can be estimated from soils data. The parameter "s" can be approximated using the following equation developed by Youngs (1964):

$$s = (2\,(\phi - \theta_i)\,K_s\ Sf)^{1/2}, \tag{3.41}$$

where S_f is the effective wetting front suction (cm) which can be estimated from soils data using Equation 3.26; ϕ is the total porosity that can be estimated from the soil bulk density using Equations 3.23 and 3.24; θ_i is the initial water content that can be measured or estimated from water retention data according to the degree of wetness, and K_s (cm/hr) is the effective conductivity that is estimated using procedures given for estimating the effective conductivity in the Green-Ampt model. The A parameter in Equation 3.41 was found by Youngs (1964) to range from 0.33 *K to K. The generally recommended value is K.

6. Morel-Seytoux and Khanji Model Morel-Seytoux and Khanji (1974) developed a model that incorporates the viscous resistance. The model is

$$f = \frac{\frac{K}{B}\,(\phi - \theta_i)\,(H_o + H_c) + F}{F}, \tag{3.42}$$

where f is the infiltration rate (cm/hr), F is the cumulated infiltration (cm), K is the effective hydraulic conductivity which can be estimated from the procedures used for estimating the Green-Ampt effective conductivity, B is the viscous resistance correction factor which ranges between 1 and 1.7, H_c (cm) is the effective capillary drive which can be estimated using procedures used for estimating the Green-Ampt wetting front suction, H_o is normally considered to be zero, ϕ is the volume fraction, which is estimated from the soil bulk density using Equation 3.26, and θ_i is the initial water content (volume fraction) which is measured or estimated from water retention data depending on the degree of wetness. The parameters can also be determined from infiltration data using regression methods.

*7. **Smith-Parlange Model*** Smith and Parlange (1978) developed two models, one for slowly varying hydraulic conductivity with water content which is analogous to the Green-Ampt model and one for rapidly varying hydraulic conductivity with water content near saturation. The model for slowly varying hydraulic conductivity is

$$f = K\,[C_o / (KF) + 1]. \tag{3.43}$$

The model for rapidly varying hydraulic conductivity is

$$f = K\frac{e^{F/C_0}}{e^{F/C_0} - 1}, \tag{3.44}$$

where f is the infiltration rate (cm/hr), F is the cumulated infiltration volume (cm), K is the effective hydraulic conductivity which can be estimated using the procedures given for estimating the Green-Ampt effective hydraulic conductivity, and C_0 is the sorptivity which can be estimated by the following equation:

$$C_o = [2\,S_f\,K\,(\phi - \theta_i)]^{1/2}, \tag{3.45}$$

where ϕ (volume fraction) is the total porosity estimated from the soil bulk density, θ_i (volume fraction) is the initial water content (which can be measured or estimated from moisture retention data depending on the degree of wetness), S_f is the wetting front suction which can be estimated using procedures given for estimating the Green-Ampt wetting front suction, and K (cm/hr) is the effective hydraulic conductivity which can be estimated using procedures given for estimating the Green-Ampt effective hydraulic conductivity. The parameters can also be determined from infiltration data using regression methods.

C. Applications of Infiltration/Rainfall Excess Models

Many rainfall/runoff simulation models have been developed that utilize infiltration/rainfall excess models (Renard et al., 1983).

For decision makers, designers, and operators, a comprehensive mathematical rainfall/runoff computer simulation model can be an invaluable tool. Most models are designed so that if all input parameters are carefully chosen, the simulated processes should secure satisfactory results; however, to be able to utilize the models with confidence and reliability, they should be calibrated against observed data. Once calibrated, the model can be utilized in the design of culverts, storm sewers, bridges, channels, dams, detention basins, flood control projects, and for other hydrologic applications.

One of the more useful applications of hydrologic modeling is evaluating the changes in hydrologic response which may be expected when surface conditions are changed by agricultural and urban development. Infiltration models may be used to estimate changes not only in peak flow but in groundwater flow, crop production, and pesticide and nutrient transport. To predict changes, the initial calibration may not be as important as the variations predicted by varying the parameters of the model.

Until now, the emphasis has been placed on developing computer models that simulate the peak rate of runoff. Runoff control has been geared towards managing or controlling the peak rate of runoff through some sort of surface structure such as detention basins; however, with increasing development and environmental pressure along with efficient use of valuable real estate, engineers and hydrologists are beginning to utilize subsurface storage and relying on infiltration capacity of the soils to manage storm water runoff. Controls that promote the infiltration capacity of the soils for storm water management include infiltration basins, infiltration trenches, dry wells, swales, filter strips, and porous pavement.

In response to the increasing use of systems that dispose of a portion of the runoff into the ground, states such as Maryland (Maryland Dept. of Natural Resources, 1984) have developed standards and specifications for infiltration practices.

TABLE 3.10. Summary of Commonly Used Engineering Design Models and Their Rainfall Excess or Infiltration Component.

Design condition	Model name	Infiltration/ rainfall excess methodology	Reference
Drainage	Drainmod	Green-Ampt	(Skaggs, 1980)
Erosion	WEPP	Green-Ampt	(Lane and Nearing, 1989)
Water Quantity	TR-20	SCS Curve No	(Soil Conservation Service, 1982a)
Water Quantity	TR-55	SCS Curve No	(Soil Conservation Service, 1986)
Water Quantity	PSRM	SCS Curve No	(Aron, 1987)
Water Quantity	HEC-1	Initial & Uniform SCS Curve No Holtan Green-Ampt	(U.S. Army Corps Engineers, 1981a)
Water Quantity	SSARR	Index	(Burnash et al., 1973)
Water Quality	STORM	SCS Curve No	(U.S. Army Corps Engineers, 1977)
Water Quality	SWMM	Horton Green-Ampt	(U.S. Environmental Protection Agency, 1985)
Water Quantity	NWSRFS	Index	(National Weather Service, 1979)
Water Quantity	DR3M	Green-Ampt	(Alley and Smith, 1985)
Water Quantity	PRMS	Green-Ampt	(Leavesly et al., 1983)
Water Quantity	CREAMS	SCS Curve No	(Knisel, 1980)
Water Quantity	SWRB	SCS Curve No	(Arnold & Williams, 1985)
Water Quantity	EPIC	SCS Curve No	(Williams, et al., 1984)
Water Quantity	SWIM	Richards	(Ross 1990)
Water Quantity	KINEROS	Smith & Parlange	(Woolhiser et al., 1990)
Water Quantity	HYMO	SCS Curve No Green-Ampt	(Williams and Hahn, 1973)

A summary of the infiltration methodologies utilized in the most commonly used engineering design models is summarized in Table 3.10. As can be seen in Table 3.10, the SCS curve number method and the Green-Ampt infiltration model are the most popular methods in use.

IV. MEASUREMENT OF INFILTRATION

Infiltration and soil water properties can vary temporally and spatially. Selection of measurement techniques and data analysis techniques should consider these effects.

A. Areal Measurement

Areal infiltration estimation is normally determined from hydrograph analysis of rainfall-runoff data from a watershed. The advantage of this method is that it gives an integrated measure of infiltration over a large area under natural rainfall and climate conditions. Methods of hydrograph analysis that can be used to determine areal infiltration are described in Chow (1964). This type of analysis requires measurement of rainfall and runoff rates. Details on this procedure are given by Brakensiek et al. (1979). The main disadvantage is that the results are specific and cannot be transferred to other watersheds.

B. Point Measurement

Point infiltration measurements are normally made by applying water at a specific site to a finite area and measuring or calculating the soil intake rate. There are primarily four types of instrumentation used for measuring point infiltration. The four types of infiltrometers are ring or cylinder, sprinkler, tension, and furrow. An infiltrometer must be selected that duplicates the system being investigated. For example, ring infiltrometers should be used to determine infiltration rates for inundated soils such as flood irrigation or pond seepage. Sprinkler infiltrometers should be used where the effect of rainfall on surface conditions influences the infiltration rate. Tension infiltrometers are used to determine the infiltration rates of the macropores. Furrow infiltrometers should be used when the effect of flowing water is important.

Infiltrometers are primarily used for measuring the infiltration rate; however, other auxiliary data should also be collected to meet the goal of the project. The types of auxiliary data which should be considered are: 1) physical plot characteristics such as length, width, area, aspect, landscape position, slope, elevation, soil series, soil profile description, and physical and chemical properties of the soil profile; 2) tillage information such as date, type, implement, depth, speed, direction, soil moisture condition, random roughness, bulk density, amount and energy of rain between tillages, and general tillage history for the previous three years; 3) crop conditions such as crop type, description, grain yield, and residue yield for the previous three years; 4) cover information, such data as type, canopy cover, canopy height, leaf area index, surface residue in percent cover, and weight and rock cover; and 5) surface soil conditions such as date, crust thickness, amount of surface cracks, amount of pores greater than 1 mm, and depth of frozen ground.

1. Ring or Cylinder Infiltrometers These infiltrometers are usually metal rings with a diameter typically of 30 cm and a height of about 20 cm. The ring is driven into the ground about 5 cm and water is applied inside the ring with a constant head device. Intake measurements are recorded until a steady infiltration rate is observed (Bouwer, 1986). To help eliminate the effect of lateral spreading, many researchers use what is called a double ring infiltrometer. A double ring infiltrometer is essentially a ring infiltrometer with a larger ring around it. Both rings have water applied to them; however, the infiltration measurement is taken with the inner ring. The outer ring serves as a buffer to help minimize the lateral spreading effect. ASTM standards for constructing ring infiltrometers are given in Lukens (1981). The advantages of this method are that only a small area is needed for measurements, it is inexpensive to construct, simple to run, and does not have a high water requirement. This method usually produces higher steady-state infiltration rates than rainfall simulators.

2. Sprinkler Infiltrometers Sprinkler infiltrometers have been designed to emulate some aspects of natural rainfall such as drop size distribution, drop velocity at impact, range of intensities, angle of impact, nearly continuous and uniform raindrop application, and the capability to reproduce rainstorm intensity distribution and duration (Meyer, 1979). Neff (1979) listed the advantages and disadvantages of rainfall simulators as follows.

Advantages

(a) Rainfall simulators are cost-efficient. Because of the degree of control that can be exercised over simulator operation, the cost per unit of data collected is quite low when compared to the unit cost of long-term field installed experiments depending on natural rainfall. Long-term experiments require not only the cost of initial instrumentation, but also a great deal of personnel time for plot and instrument maintenance and servicing during periods in which little or no data is being collected.

(b) Rainfall simulators provide a maximum of control over when and where data are to be collected; plot conditions at test time; and, within design limitations, rates and amounts of rain to be applied. If an investigator must depend on natural rainfall, it may take many years to collect data with the required combinations of rainfall amounts and intensities, land management sequences, and crop growth stages for valid analysis and interpretation. The degree of control afforded by rainfall simulators provides a technique for collecting a great deal of data in a relatively short time.

Disadvantages

(a) Rainfall simulators are expensive to construct and use because of the cost of components and assembly, and the number of people required to operate them.

(b) The areas treated are small, ranging from a fraction of a square meter to several hundred square meters, depending on the simulator design. These small areas may or may not be representative of the general area of design. For example, such factors as rodent holes and large bushes and plants on the plots can have a disproportionate effect on the results.

(c) Most simulators do not produce drop-size distributions that are representative of natural rainfall. Simulators with tube-type drop formers produce drops within a narrow range of sizes, and drop size can be adjusted only by changing the size of the tubes. Simulators with the nozzle-type drop formers produce drops over a wide range of sizes, but they are often smaller than natural rainfall drops.

(d) Most simulators do not produce rainfall intensities with the temporal variations representative of natural rainfall. Some simulators can produce different intensities, but they are usually varied between runs, and not within runs.

(e) Some simulators do not produce drops that approach the terminal velocity of corresponding size drops of natural rainfall. The lower velocities, in combination with smaller drop-size distributions, result in kinetic energy smaller than that produced by natural rainfall. Kinetic energy of simulators with nozzle-type drop formers and free falling drops may be only 40–50% of natural rain; however, some simulators are designed with nozzles pointed downward and the drops are applied under pressure and approach the energy of natural rain.

Neff (1979) further states that even though there are more disadvantages than advantages, the key factor is whether the advantages are more important. In many instances simulated rain is the only way to obtain results in a reasonable time period and under controlled conditions. The size of the plot to represent the conditions to be evaluated, the water requirement, portability, operation unaffected by wind, operation in sloping terrain, and the purpose of the research influence the selection of the appropriate rainfall simulator.

Agassi et al. (1988) report that the water quality should be the same as rain since the water applied can have a significant effect on the surface seal formation. Peterson and Bubenzer (1986) inventoried the different rainfall simulators reported in the literature and categorized them according to whether the simulated rain was produced by a nozzle or by a drop former. The inventory includes the appropriate reference, type of nozzle, nozzle pressure, nozzle movement and spray pattern, drop size, intensity range, and plot size. Peterson and Bubenzer (1986) also give details for constructing a rainfall simulator. Renard (1985) summarizes, from the erosion view, the history, perspective, and future of rainfall simulators indicating that they are an integral part of erosion research. Currently, the most widely used infiltrometer is the rotating boom developed by Swanson (1965).

3. ***Tension Infiltrometer*** The tension infiltrometer is made up of three main components as illustrated in Fig. 3.43. The first component is a tension control tube that contains three air entry ports which independently determine the amount of tension in the system by varying the distance between the air entry point and the water level. The second component is a large tube with a scale used for measuring the change in water level. The third component is a contact plate at the base of the device with a wire mesh, outer nylon mesh, and a micromesh pulled taut around the base. The diameter of the base plate is the same size as the inner ring of the double ring infiltrometer. The large tube is filled with water by submerging the base plate in a container of water and drawing water up to the tube by means of a vacuum pump fastened to a port at the top of the tube. (Watson and Luxmore, 1986; Clothier and White, 1981; Perroux and White, 1988; Ankeny et al., 1988).

4. ***Furrow Infiltrometer*** Infiltration measurements for furrow irrigation systems can be made with a blocked furrow infiltrometer, recirculating furrow infiltrometer, or inflow–outflow measurement. Details on how to run furrow infiltration tests are presented in American Society of Agricultural Engineers (1989) and Kincaid (1986). The following is a brief description of the three furrow infiltration methods.

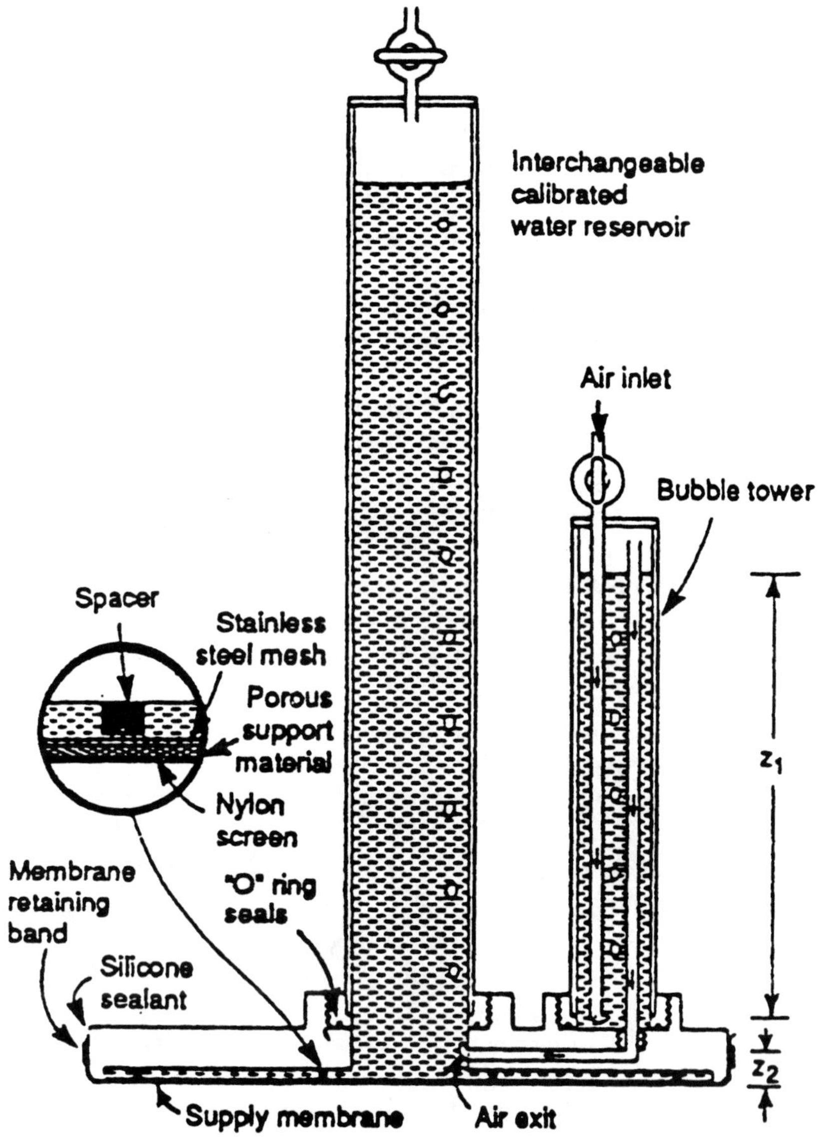

Figure 3.43.—Tension Infiltrometer (Clothier and White, 1981).

The blocked furrow infiltrometer test consists of: 1) blocking off a section of furrow, usually up to five meters in length, and then 2) ponding water in the furrow section to a depth approximating the observed depth during irrigation. The water level in the furrow is kept relatively constant by continually adding water until a steady infiltration rate is achieved or the time exceeds the expected duration of the irrigation.

The recirculating furrow infiltrometer is a modification of the blocked furrow infiltrometer with water being continuously recycled over the furrow segment in an attempt to better simulate the flow conditions of an actual irrigation.

The inflow-outflow test determines the infiltration rates along a furrow by taking the difference in measured flow rates at the beginning and end of a segment. Flumes or other suitable flow measuring devices are used for these measurements. The length of segment depends on the infiltration characteristics of the furrow. The volume of water going into storage must also be considered in the analysis.

V. REFERENCES

Agassi, M., Shainberg, I., and Morin, J. (1988). "Effects on seal properties of changes on drop energy and water salinity during a continuous rainstorm." *Aust. J. Soil Res.* 26:1–10.

Ahuja, L.R. (1983). "Modeling infiltration into crusted soils by the Green-Ampt approach." *Soil Sci. Soc. Am. J.* 47:412–418.

Ahuja, L.R., Naney, J.W., and Williams, R.D. (1985). "Estimating soil water characteristics from simpler properties or limited data." *Soil Sci. Soc. Am. J.* 49:1100–1105.

Ahuja, L.R., Barnes, B.B., Cassel, D.K., Bruce, R.R., and Nofziger, D.L. (1988). "Effect of assumed unit gradient during drainage on the determination of unsaturated hydraulic conductivity and infiltration parameters." *Soil Sci.* 145:235–243.

Ahuja, L.R., Cassel, D.K., Bruce, R.R., and Barnes, B.B. (1989). "Evaluation of spatial distribution of hydraulic conductivity using effective porosity data." *Soil Sci.* 148:404–411.

Alley, W.M. and Smith, P.E. (1985). "U.S. Geological Survey, distribution routing rainfall-runoff-model-version II User's Manual," Reston, Virginia. American Society of Agricultural Engineers. (1989). "Evaluation of Furrow Irrigation Systems." *ASAE Engineering Practice EP419.*

Ankeny, M.D., Kaspar, T.C., and Horton, R. (1988). "Design for an automated tension infiltrometer." *Soil Sci. Soc. Am. J.* 52:893–896.

Arnold, J.G. and Williams, J.R. (1985). "Validation of SWRRB—Simulator of water resources in rural basins." *In Proc. Symp. on Water Management, Am. Soc. of Civil Eng.*, 107–114.

Aron, G. (1987). "Penn. State runoff model for IBM-PC." *Pennsylvania State University and Institute for Research on Land and Water Resources*. State College, PA.

Blackburn, W.H. (1973). "Infiltration rate and sediments production of selected plant communities and soils in five rangelands in Nevada." *Agr. Exp. Sta., College of Agri., Univ. of Nevada at Reno, NV.*, 86.

Bouwer, H. (1986). "Intake rate: methods of soil analysis Part I physical and mineralogical methods." In *A. Klute (ed) Am. Soc. Agronomy Second Edition, Monograph 9, Chapt. 32.*

Brakensiek, D.L. and Rawls, W.J. (1988). "Effects of agricultural and rangeland management systems on infiltration," in *Modeling Agricultural, Forest, and Rangeland Hydrology-Proc. of the 1988 International Symposium*, Am. Soc. Agr. Engrs. St. Joseph, MI, 247.

Brakensiek, D.L. and Rawls, W.J. (1992). "Commentary on fractal processes in soil water retention." *Water Resour. Res.* 28:601–602.

Brakensiek, D.L., Osborn, H.B., and Rawls, W.J. (1979). "Field manual for research in agricultural hydrology." *U.S. Dept. of Agric. Handbook 224*, 550.

Brakensiek, D.L., Rawls, W.J., and Stephenson, G.R. (1986a). "Determining the saturated hydraulic conductivity of soil containing rock fragments." *Soil Sci. Soc. Amer.* 50(3):834–835.

Brakensiek, D.L., Rawls, W.J., and Stephenson, G.R. (1986b). "A note on determining soil properties for soils containing rock fragments." *J. of Range Mang.* 39(5):408–409.

Branson, F.A., Gifford, G.F., Renard, K.G., and Hadley, R.F. (1981). *Rangeland hydrology*, Kendall/Hunt, Dubuque, IA 1981, Chp. 4, 47–72.

Brutsaert, W. (1967). "Some methods of calculating unsaturated permeability." *Trans. Am. Soc. Agr. Engrs.* 10(3):400–404.

Burnash, J.C., Ferral, R.L., and McGuire, R.A. (1973). "A generalized streamflow simulation system, conceptual modeling for digital computers." *California Department of Water Resources*. Sacramento, CA.

Cheruiyot, S.K. (1984). "Infiltration rates and sediment production of a Kenya grassland as influenced by vegetation and prescribed burning." M.S. Thesis Range Science Dept., Texas A. & M., College Station, TX, 30.

Childs, E.C. and Bybordi, M. (1969). "The vertical movement of water in stratified porous material. 1. Infiltration." *Water Resour. Res.* 5:2–.

Clothier, B.E. and White, J. (1981). "Measurement of sorptivity and soil water diffusivity in the field." *Soil Sci. Soc. Am. J.* 45:242–245.

Chow, V.T. (1964). *Handbook of applied hydrology. McGraw-Hill*, New York, NY.

Dadkhah, M. and Gifford, G.F. (1980). "Influence of vegetation, rock cover and trampling on infiltration rate and sediment production." *Water Resources Bulletin*. 16:979–987.

Freebairn, D.M. (1989). "Rainfall and tillage effects on infiltration of water into soil." Ph.D. Thesis, Univ. of Minn., 63.

Frere, M.H., Onstad, C.A., and Holtan, H.N. (1975). "ACTMO - An agricultural chemical transport model." *U.S. Dept. of Agriculture*, ARS-11-3, Washington, DC.

Gifford, G.F. (1977). "Vegetation allocation for meeting site requirements," in *Development of Strategies for Rangeland Management*, Westnew Press. 56.

Gifford, G.F. and Hawkins, R.H. (1978). "Hydrologic impact of grazing on infiltration: A critical review." *Water Resour. Res*. 14:305–313.

Green, R.E. and Corey, J.C. (1971). "Calculation of hydraulic conductivity: A further evaluation of some predictive methods." *Soil Sci. Soc. Am. J.*, 35:3–8.

Green, W.H., and Ampt, G.A. (1911). "Studies on soil physics: 1. Flow of air and water through soils." *J. Agr. Sci.*, 4. 1–24.

Hachum, A.Y. and Alfaro, J.F. (1980). "Rain infiltration into layered soils: Prediction." *J. Irrig. and Drain. Div., Am. Soc. Civil Engr*. 106:311–321.

Hillel, D. (1971). Soil and water physical principles and processes. Academic Press, New York, NY. 288 p.

Holtan, H.N. (1961). "A concept for infiltration estimates in watershed engineering." *U.S. Dept. of Agriculture, Bull. 41-51*. 25 p.

Holtan, H.N., Stilner, G.J., Henson, W.H., and Lopez, N.C. (1975). "USDA-HL-74 Model of watershed hydrology." *U.S. Dept. of Agriculture., Tech. Bull. No. 1518*.

Hoogmoed, W.B., and Bouma, J. (1980). "A simulation model for predicting infiltration into cracked clay soil." *Soil Sci. Soc. Am. J*. 44:458–461.

Horton, R.E. (1919). "Rainfall interception." *Monthly Weather Review, 49*, 603–623.

Horton, R.E. (1940). "An approach toward a physical interpretation of infiltration-capacity." *Soil Sci. Soc. Am. Proc*. 5:399–417.

Kemper, W.D., Trout, T.J., Humpherys, A.S., and Bullock, M.S. (1988). "Mechanisms by which surge irrigation reduces furrow infiltration rates in a silty loam soil." *Trans. of the ASAE*. 31(SW):821–829.

Kincaid, D.C. (1986). "Intake Rate: Border and furrow. Methods of soil analysis Part I Physical and mineralogical methods." In: A. Klute (ed) *2nd edition. Chapter 34*.

Klute, A. and Dirksen, C. (1986). "Hydraulic conductivity and diffusivity—Laboratory methods." In: A. Klute (ed) *Methods of Soil Analysis, Part I Physical and Mineralogical Methods*. Am. Soc. of Agron., Monog. No. 9, 2nd edition. 687–734.

Knisel, W.G. (ed), (1980). "CREAMS: A field scale model for chemicals, runoff, and erosion from Agricultural Management Systems," *Soil Conservation Service, Report 26*, U.S. Dept. Agr., Washington, DC, 643.

Kostiakov, A.N. (1932). "On the dynamics of the coefficient of water-percolation in soils and on the necessity for studying it from a dynamic point of view for purposes of amelioration." *Trans. Sixth Comm. Intl. Soil Sci. Soc.*, Russian Part A:17–21.

Lane, L.J., Simanton, J.R., Hakonson, T.E., and Rommey, E.M. (1987). "Large-plot infiltration studies in desert and semiarid of the Southwestern U.S.A." In: *Proc. of the Intern. Conf. on Infiltration Development and Application. Water Resources Research Ct.*, University of Hawaii at Manoa, Honolulu, HI, 365–376.

Lane, L.J. and Nearing, M. (1989). "USDA Water erosion prediction project, Hillslope Profile Version," Rept. #2. Nat'l Soil Erosion Research Lab, West Lafayette, IN.

Leavesly, G.H., Lichty, R.W., Troutman, B.M., and Saindon, L.G. (1983). "Precipitation runoff modeling system: User's Manual." *USGS Water Resources Investigations Report 83-4238*, Denver, CO.

Lee, H.W. (1983). "Determination of infiltration characteristics of a frozen palouse silt loam soil under simulated rainfall." Ph.D. Dissertation, Univ. of Idaho Graduate school.

Lukens, R.P. (ed) (1981). "Annual book of ASTM standards Part 19 Soil and rock, Building Stones." 509–514.

Lull, H.W. (1959). "Soil Compaction on forest and range lands." *U.S. Forest Service*, Washington DC. 7.

Maryland Dept of Natural Resources. (1984). "Maryland standards and specifications for stormwater management infiltration practices." Annapolis, MD.

Marshall, T.J. (1958). "A relationship between permeability and size distribution of pores." *J. Soil Science*, 9:1–9.

Mbakaya, D.S. (1985). "Grazing systems effects on infiltration rates and sediment production of abused grassland Buchuma, Kenya." *M. S. Thesis*, Range Science Dept., Texas A&M, College Station, TX.

McIntyre, D.S. (1958). "Permeability measurements of soil crusts formed by raindrop impact." *Soil Sci.*, 85:185–189.

Mein, R.G. and Larson, C.L. (1973). "Modeling infiltration during a steady rain." *Water Resour. Res.*, 9:384–394.

Meyer, L.D. (1979). "Methods for attaining desired rainfall characteristics in rainfall simulations." Proceedings of the Rainfall Simulator Workshop. *USDA-ARS-ARM*. 35–48.

Millington, R.J. and Quirk, J.P. (1961). "Permeability of porous solids." *Trans. Faraday Soc.* 57:1200–1206.

Moore, I.D. and Eigel, J.D. (1981). "Infiltration into two-layered soil profiles." *Trans. Am. Soc. Agr. Eng.* 24:1496–1503.

Morel-Seytoux, H.J. (1988). "Recipe for a simple but physically based approach to infiltration under variable rainfall conditions." (1988) *Hydrology Days*, No. 1005, Ft Collins, CO. 226–247.

Morel-Seytoux, H.J. (1989). *Unsaturated flow in hydrologic modeling: Theory and practice.* Kluwer Academic, Boston, MA. 531.

Morel-Seytoux, H.J. and Khanji, J. (1974). "Derivation of an equation of infiltration." *Water Resour. Res.* 10(4):795–800.

Mualem, Y. (1986). "Hydraulic conductivity of unsaturated soils—prediction and formulas." In A. Klute (ed) *Methods of Soil Analysis, Part 1—Physical and Mineralogical Methods. Am. Soc. Agron.*, No. 9, 2nd ed. 799–823.

Musgrave, G.W. (1955). "How much rain enters the soil." USDA Yearbook of Agriculture, Washington, DC. 151–159.

National Weather Service, Office of Hydrology. (1979). *National Weather Service River Forecast System User's Manual.* Silver Spring, Maryland.

Neff, E.L. (1979). "Why rainfall simulation?" Proceedings of the Rainfall Simulator Workshop. *USDA-ARS-ARM-10.* 3–8.

North Central Regional Committee 40. (1979). "Water infiltration into representative soils of the North Central Region." Agr. Exp. Sta., Univ. of Ill. Bull. 760 (NC Reg. Res. Pub. No. 259). 23.

Onstad, C.A. (1984). "Depressional storage on tilled soils in Trans. of the ASAE." 27(SW):729–731.

Perroux, K.M. and White, I. (1988). "Designs of disc permeameters." *Soil Sci. Soc. Am. J.* 52:1205–1215.

Peterson, A. and Bubenzer, G. (1986). "Intake rate sprinkler infiltrometer." In: A. Klute (ed) *Methods of Soil Analysis Part I—Physical and Mineralogical Methods.* 2nd edition, Chapter 33.

Philip, J.R. (1957). "The theory of infiltration. 4. Sorptivity and algebraic infiltration equations." *Soil Sci.* 84:257–264.

Pilgrim, D.H. and Cordery, I. (1993). "Flood runoff," Chapter 9 *Handbook of Hydrology*, Ed. D.R. Maidment. *McGraw-Hill, Inc.*

Rallison, R.E. (1980). "Origin and evolution of the SCS runoff equation." Proceeding, Symposium of Watershed Management. ASCE, Boise, ID. 912–914.

Rawls, W.J. (1983). "Estimating soil bulk density from particle size analysis and organic matter content." *Soil Sci.* 135(2):123–125.

Rawls, W.J. and Brakensiek, D.L. (1983). "A procedure to predict Green-Ampt infiltration parameters." *Advances in Infiltration*, Am. Soc of Agr. Eng. 102–112.

Rawls, W.J., Brakensiek, D.L., and Soni, B. (1983). "Agricultural management effects on soil water processes: Part I Soil Water Retention and Green-Ampt Parameters." Trans. Am. Soc. Agr. Engrs. 26(6):1747–1752.

Rawls, W.J. and Brakensiek, D.L. (1986). "Prediction of soil water properties for hydrologic modeling." *Watershed Management in the Eighties*, ASCE. 293–299.

Rawls, W.J., Brakensiek, D.L., and Savabi, R. (1989). "Infiltration parameters for rangeland soils." *J. of Range Management* 42:139–142.

Rawls, W.J., Brakensiek, D.L., Simanton, J.R., and Kohl, K.D. (1990). "Development of a crust factor for the Green-ampt model." *Trans Am. Soc. of Agri. Eng.* 33:1224–1228.

Rawls, W.J., Gish, T.J., and Brakensiek, D.L. (1991). "Estimating soil water retention from soil physical properties and characteristics." *Advances in Soil Sci.* 16:213–234.

Rawls, W.J., Ahuja, L.R., Brakensiek, D.L., and Shirmohammadi, A. (1993a). "Chapter 5: Infiltration and Soil Water Movement." In *Handbook of Hydrology* ed. D.R. Maidment. McGraw Hill, New York, NY.

Rawls, W.J., Brakensiek, D.L., and Logsdon, S.D. (1993). "Predicting saturated hydraulic conductivity using fractal principles." *J. Soil Sci. Soc. of America.* 57:1193–1197.

Renard, K.G. (1985). "Rainfall simulators and USDA erosion research. History perspective and future." *Proc. of the Rainfall Simulator Workshop Society of Range Management.* Denver, CO. 3–7.

Renard, K.G. (1993). USGS Water Resources Investigations Report, 93-4018.

Renard, K.G., Rawls, W.J., and Fogel, M.M. (1982). "Currently available models." In *Hydrologic Modeling of Small Watersheds.* Eds. C.T. Haan, H.P. Johnson and D.L. Brakensiek. ASAE Monographs No. 5, St. Joseph, MI. P. 507–522.

Richards, L.A. (1931). "Capillary conduction of liquids in porous mediums." *Physics*, 1. 318–333.

Ross, P.J. (1990). *"SWIM—A simulation model for soil water infiltration and movement* (Reference Manual)." CSIRO Division of Soils, Davies Laboratory, Townsville, Qld 4814, Australia.

Rubin, J. and Steinhardt, R. (1963). "Soil water relations during rain infiltration: 1. Theory." *Soil Sci. Soc. Am. Proc.*, 27:246–251.

Sabol, G.V., Rumann, J.M., Khalili, D., and Waters, S.D. (1992.) *Drainage design manual for Maricopa County, Arizona, Volume 1, Hydrology: Flood Control District of Maricopa County*, Phoenix, AR.

Saxton, K.E., Rawls, W.J., Romberger, J.S., and Papendick, R.I. (1986). "Estimating generalized soil water characteristics from texture." *Trans. of Am. Soc. of Agri. Engrs.* 50(4):1031–1035.

Skaggs, R.W. (1980). *Rainmod-Reference Report: Methods for design and evaluation of drainage-water management systems for soil with high water tables*, USDA-SCS, South National Tech Ct., Forth Worth, TX, 330.

Skaggs, R.W., and Khaleel, R. (1982). "Infiltration. In: C.T. Haan (ed-in-chief) Hydrologic modeling of small watersheds," ASAE Monog. No. 5. Am. Soc. Agr. Engrs., St. Joseph, MI. 4–166.

Smith, R.E. (1983). "Flux infiltration theory for use in watershed hydrology." In: *Proc. of the Nat. Conf. on Advances in Infiltration.* Am. Soc. Agr. Engrs., St. Joseph, MI. 313–323.

Smith, R.E. and Parlange, J.Y. (1978). "A parameter-efficient hydrologic infiltration model." *Water Resour. Res.* 14(3)533–538.

Smith, R.E. and Hebbert, R.H.B. (1979). "A Monte Carlo analysis of the hydrologic effects of spatial variability of infiltration." *Water Resour. Res.* 15(2)419–429.

Soil Conservation Service (SCS) (1972). "National Engineering Handbook, Section 4 Hydrology. Chapts. 7, 8, 9, and -10." Washington, DC.

Soil Conservation Service (SCS) (1982a). "Technical Release 20." Washington, DC.

Soil Conservation Service (SCS) (1982b). "Procedures for collecting soil samples and methods of analysis for soil survey." *Soil Survey Investigations Report No. 1.* Washington DC.

Soil Conservation Service (SCS) (1986). "Urban hydrology for small watersheds." Technical Release 55. Washington DC.

Summer, M.E. and Stewart, B.A. (1992). "Soil crusting: Chemical and physical properties." *Advances in Soil Science.*

Swanson, N.P. (1965). "Rotating boom rainfall simulator." *Transaction American Society of Agricultural Engineers.* 8(1):71–72.

Swartzendruber, D. (1987). "A quasi solution of Richards' equation for downward infiltration of water into soil." *Water Resour. Res.* 5:809–817.

Thurow, T.E. (1985). "Hydrologic relationship with vegetation and soil as affected by selected livestock grazing systems and climate on the Edwards Plateau." Ph.D. Thesis Range Science Dept., Texas A & M, College Station, TX.

Thurow, T.E., Blackburn, W.H. and Taylor, C.A. (1986). "Hydrologic characteristics of vegetation types as affected by livestock grazing systems, Edwards Plateau, Texas." *J. Range Manag.* 39:506–510.

Trout, T.J. and Kemper, W.D. (1983). "Factors which affect furrow intake rates." In: Proc. of the Nat. Conf. on Advances in Infiltration. *Am. Soc. Agr. Engrs.*, St. Joseph, MI. 302–312.

Tyler, S.W. and Wheatcraft, S.W. (1990). "Fractal processes in soil water retention." *Water Res. 26:1047:1054.*

U.S. Army Corps of Engineers. (1977). *Storage treatment overflow runoff model, STORM—User's Manual*. Washington, DC.

U.S. Army Corps of Engineers. (1981). *Stream flow synthesis and reservoir regulation—User's Manual*, Portland, Oregon.

U.S. Environmental Protection Agency. (1985). *Storm water management model—User's Manual*. Washington, DC.

Van Mullem, J.A. (1989). "Applications of the Green-Ampt Infiltration Model to Watersheds in Montana and Wyoming." MS Thesis, Montana State University, Bozemann, MT.

Warren, S.D., Blackburn, W.H., and Taylor, Jr C.A. (1986). "Soil hydrologic response to number of pastures and stocking density under intensive rotation grazing." *J. Range Management* 39:501.

Watson, K.W. and Luxmoore, R.J. (1986). "Estimating macroporosity in a forest watershed by use of a tension infiltrometer." *Soil Sci. Soc. Am. J.* 50:578–587.

Weltz, M., and Wood, M.K. (1986). "Short duration grazing in Central New Mexico: effects on infiltration rates." *J. Range Management* 39:366.

Williams, J.R. and Hann, Jr R.W. (1973). *HYMO: Problem oriented language for hydrologic modeling—User's Manual*, U.S. Dept. Agri., Agriculture Research Service Bulletin S-9.

Williams, J.R., Jones, C.A., and Dyke, P.T. (1984). "A modeling approach to determining the relationship between erosion and soil productivity," Trans. *Am. Soc. Agric. Eng.*, 27(1):129–144.

Woolhiser, D.A., Smith, R.E., and Goodrich, D.C. (1990). *Kineros, A kinematic runoff and erosion model: Documentation and user Manual*. U.S. Dept Agri., Agricultural Research Service Bulletin ARS-77, 130.

Youngs, E.G. (1964). "An infiltration method measuring the hydraulic conductivity of unsaturated porous materials." *Soil Science*. 97:307–311.

Zobeck, T.M. and Onstad, C.A. (1987). "Tillage and rainfall effects on random roughness: A review." *Soil and Tillage Research* 9:1–20.

VI. GLOSSARY

Depression storage—Amount of water that can be stored on the soil surface.

Evapotranspiration—Process by which water moves from the soil to the atmosphere.

Hydraulic Conductivity—The ability of the soil to transmit water.

Infiltration—Movement of water into the soil; can be expressed as a rate or accumulated amount.

Infiltrometers—Equipment to measure infiltration rates.

Interception—Amount of water caught and stored on vegetation.

Macropores—Large voids in the soil, such as decayed root channels, worm holes, and structural cleavages or cracks.

Porosity—Fraction of the total volume of soil occupied by pores.

Rainfall excess—The part of rainfall which is not lost to infiltration, depression storage, or interception.

Soil texture—The size distribution of individual particles smaller than 2mm in a soil sample.

Soil Water Content—The ratio of the mass of water to the mass of dry soil.

Surface ponding—The point at which the total soil surface is ponded with water; begins when the surface application rate exceeds the soil surface infiltration rate.

Water Retention Characteristic—Soil's ability to store and release water; defined as the relationship between the soil water content and the soil suction or matric potential.

Chapter 4

EVAPORATION AND TRANSPIRATION

I. INTRODUCTION

Evaporation from soil and water surfaces and transpiration from growing plants are major processes in the primary component of the hydrologic cycle that returns precipitated water to the earth's atmosphere as vapor. The primary force driving evaporation and transpiration is energy input from the sun. Transpiration is a special case of evaporation. Evaporation occurs at the surface of moist cells within plant tissues, and the water vapor then diffuses into intercellular leaf spaces and diffuses through stomates to the atmosphere. As a result, plants have some direct control of the process in contrast to evaporation from water surfaces. Other factors that effect evaporation from free water surfaces and transpiration from vegetated surfaces involve differences in reflectance of radiation, which affects radiant energy input, differences in heat storage capacities, and the aerodynamic roughness of water and vegetation which affects the transfer of sensible and latent heat.

The purpose of this chapter is to present current technology for estimating evaporation (E) from water bodies and evapotranspiration (ET) from soil and plants and to summarize principles and methodology for use by engineers involved in solving E and ET problems and in conducting E and ET studies. Evaporation and transpiration have been studied for centuries. For reviews of the history on the subject, see Brutsaert (1982) and Jensen et al. (1990).

Major progress in linking the processes with energy exchange came about in the twentieth century. The classic work of Penman (1948) laid the foundation for relating evapotranspiration to meteorological variables. Penman combined the energy balance component required to sustain evaporation with a mechanism to remove water vapor (sink strength). Many investigators, including Penman, continued to expand the theory of the combination equation since 1950 with special emphasis on the aerodynamic aspects. A detailed summary of the work in the 1950's and early 1960's was presented by Rijtema (1965) and Monteith (1965). Today, the most basic and practical equation for estimating evapotranspiration is commonly known as the "Penman-Monteith" method.

II. PHYSICS AND THEORY OF EVAPORATION

A. Surface–Air Energy Exchanges

1. Physical Properties of Water and Air Physical properties of liquid water and water vapor along with basic terminology are summarized in this section. Some important water properties are shown in Table 4.1.

Saturation vapor pressure, e^o, is an important parameter in hydrology. At the water–air interface there is a continuous flow of molecules from the water surface to the air, and a return flow to the liquid surface.

TABLE 4.1. Physical Properties of Liquid Water (Van Wijk and de Vries, 1963).

Temperature (°C)	Density (kg m^{-3})	Surface Tension (N m^{-1})	Dynamic Viscosity (10^{-3} Pa s)	Heat of vapori-zation (MJ kg^{-1})	Specific heat (kJ kg^{-1}°C^{-1})	Thermal conductivity (J m^{-1} s^{-1}°C^{-1})
−10	997.94	—	—	2.525	4.271	—
−5	999.18	0.0764	—	—	—	—
0	999.87	.0756	1.792	2.501	4.218	0.561
4	1000.00	.0750	—	2.492	4.205	.596
5	999.99	.0748	1.519	2.489	4.202	.574
10	999.73	.0742	1.308	2.477	4.192	.586
15	999.13	.0734	1.140	2.466	4.186	.595
20	998.23	.0727	1.005	2.453	4.182	.603
25	997.08	.0719	.894	2.442	4.180	.611
30	995.68	.0711	.801	2.430	4.178	.620
35	994.06	.0703	.723	2.418	4.178	.628
40	992.25	.0695	.656	2.406	4.178	.632
45	990.24	.0687	.599	2.394	4.179	.641
50	988.07	.0679	.549	2.382	4.181	.645

When equilibrium with pure water exists, the two flows are equal and the air is saturated with water vapor. The partial pressure exerted by the vapor at this time is called the saturation vapor pressure. The vapor pressure at equilibrium depends on the liquid water pressure, temperature, and its chemical content (solutes). Typical values of saturation vapor pressure and density of water vapor over a plane surface of pure water at the same temperature and pressure are summarized in Table 4.2. Saturation vapor pressure in kPa can be calculated following Tetens (1930) and Murray (1967) for T in °C as:

$$e^{o} = \exp\left(\frac{16.78\ T - 116.9}{T + 237.3}\right). \tag{4.1}$$

This expression calculates within 0.1% of values in the Smithsonian Meteorological Tables for temperatures in the range of 0 to 50°C (Allen et al., 1989). The slope of the saturation vapor pressure curve, Δ, kPa °C^{-1}, is obtained by differentiating Equation 4.1 (Allen et al., 1989):

$$\Delta = \frac{4098\ e^{o}}{(T + 237.3)^{2}}. \tag{4.2}$$

Absolute humidity, ρ_v, is the water vapor density, i.e., the mass of water vapor per unit volume of moist air. It can be calculated from the ideal gas law:

$$\rho_v = \frac{P_v}{R_v T}, \tag{4.3}$$

where ρ_v is the absolute humidity, kg m^{-3}, P_v is the vapor pressure, N m^{-2} or Pa, R_v is the gas constant for water vapor (461.5 J kg^{-1} K^{-1}), and T is the absolute temperature in K (K = 273.2 + °C). For ρ_v in kPa the density in kg m^{-3} is:

$$\rho_v = \frac{2.167\ P_v}{T}. \tag{4.4}$$

Specific humidity, q, is the relative density of water vapor, i.e., the mass of water vapor per unit mass of moist air:

TABLE 4.2. Physical Properties of Water Vapor (List, 1984; ASTM, 1976).

Temperature (°C)	Saturation vapor pressure over water (kPa)	Saturation vapor density over water (10^{-3} kg m^{-3})	Diffusion coefficient at P = 100 kPa (10^{-5} m^2 s^{-1})
0	0.611	4.85	2.26
5	.872	6.80	—
10	1.227	9.40	2.41
15	1.704	12.83	—
20	2.337	17.30	2.57
25	3.167	23.05	—
30	4.243	30.38	2.73
35	5.624	39.63	—
40	7.378	51.19	2.89
45	9.586	65.50	—
50	12.340	83.06	—

1 kPa = 10 mb = 7.501 mm Hg
10^{-3} kg m^{-3} = 10^{-6} g cm^{-3}

$$q = \frac{\rho_v}{\rho_a} = \frac{m_v}{m_v + m_a} = \frac{0.622\, e}{P - 0.378\, e}, \tag{4.5}$$

where q is the specific humidity, kg kg^{-1}, ρ_a is the density of moist air, m_v is the mass of water vapor, m_a is the mass of dry air, P is the total atmospheric pressure in kPa, e is the vapor pressure in kPa, and 0.622 is the ratio of the molecular mass of water to the apparent molecular mass of dry air.

Mixing ratio, r, is the dimensionless ratio of the mass of water vapor to a unit dry air mass. At the same temperature, and when e is small relative to the total pressure P, the mixing ratio is:

$$r = \frac{m_v}{m_a} \simeq 0.622 \frac{e}{P}. \tag{4.6}$$

A more precise equation (than Equation 4.1) for calculating saturation vapor pressure, e^o, of air over water at pressure P and temperature is:

$$e^o = \frac{r^o}{0.622 + r^o} P, \tag{4.7}$$

where r^o is the saturation mixing ratio.

Saturation deficit, d_a, or vapor pressure deficit of the air is:

$$d_a = e^o - e. \tag{4.8}$$

Relative humidity, RH, is the dimensionless ratio of actual vapor pressure to saturation vapor pressure, usually expressed in percent. RH and d_a are defined as:

$$RH = 100 \frac{e}{e^o} \tag{4.9}$$

and

$$d_a = e^o \left[\frac{100 - RH}{100}\right]. \tag{4.10}$$

Relative humidity by itself has limited utility in E and ET calculations unless air temperature or one of the vapor terms corresponding to the same time scale and interval as RH is given.

Dew point temperature is the temperature to which a parcel of air must be cooled at constant pressure and constant vapor content until saturation occurs, or the temperature at which saturation vapor pressure is equal to the actual vapor pressure of the contained vapor, and r is the mixing ratio defined in Equation 4.6. At dew point temperature, condensation normally occurs:

$$e_d^o = \frac{r}{0.622 + r} P. \tag{4.11}$$

Wet bulb temperature, T_w, is the temperature that a moist evaporating surface may approach when the radiation energy balance is zero. In practice, the wet bulb temperature represents the equilibrium temperature of a thermometer covered with a cloth that has been wetted with pure water in air moving at least 4.6 m/s. The wet bulb thermometer is cooled until heat drawn from the air equals the gain of latent heat due to evaporation. The vapor pressure of the ambient air is:

$$e = e_w^o - \frac{Pc_p}{0.622\ \lambda}(T - T_w) = e_w^o - \gamma(T - T_w), \tag{4.12}$$

where $e_w{}^o$ is saturation vapor pressure at T_w, P is atmospheric pressure, c_p is the specific heat of air at constant pressure, λ is the latent heat of vaporization, T is ambient air temperature, and γ is referred to as the psychrometric constant for fully aspirated psychrometers. At sea level, normal values are: P (101.3 kPa at one atm.), c_p (1.003×10^{-3} MJ kg^{-1} °C^{-1}), and λ (2.453 MJ kg^{-1} at 20 °C). The pyschrometric constant, γ, is computed as:

$$\gamma = \frac{c_p P}{\lambda \epsilon}, \tag{4.13}$$

where, for $c_p = 1.013$ kJ kg^{-1} K^{-1} for moist air (Brutsaert, 1982), P is in kPa, λ is in kJ kg^{-1}, and $\epsilon = 0.622$. γ has units of kPa °C^{-1}. Mean atmospheric air pressure, P, can be computed using the ideal gas law (Burman et al., 1987) as:

$$P = P_o \left(\frac{T_o - \alpha\,(z - z_o)}{T_o} \right)^{\frac{g}{\alpha_a R}}, \tag{4.14}$$

where P_o and T_o are known atmospheric pressure in kPa and absolute temperature in K at elevation z_o in m, and z is the elevation of the location or instrument in m above mean sea level. The assumed constant adiabatic lapse rate, α_a, normally is taken as 0.0065 K m^{-1} for saturated air or 0.01 K m^{-1} for non-saturated air; g is gravitational acceleration, 9.81 m/s and R is the specific gas constant for dry air, 287.0 J kg^{-1} K^{-1}. Values for P_o, T_o, and z_o are commonly those for the standard atmosphere at sea level, which are 101.3 kPa, 288 K, and 0 m, respectively (List, 1984). Equation 4.14 is relatively insensitive to the value of α_a for elevations up to 3,000 m. Air density, ρ_a, can be computed as

$$\rho_a = \frac{1000\ P}{T_v R} \tag{4.15}$$

where P is in kPa, R is 287.0 J kg^{-1} K^{-1}, and T_v is virtual temperature in K. T_v can be computed as follows (Jensen et al., 1990):

$$T_v = \frac{T}{1 - 0.378\frac{e}{P}} \tag{4.16}$$

where T is air temperature, K, and e is mean vapor pressure of the air, kPa.

Latent heat of vaporization, λ, in MJ kg^{-1} can be computed following Harrison (1963) as:

$$\lambda = 2.501 - 2.361 \times 10^{-3}\, T, \tag{4.17}$$

where T is mean air temperature in °C. If available, mean surface temperature or wet bulb temperature can be used to compute the value of λ, which better represents conditions at the evaporating surface.

2. ***Standard Atmosphere (Lower Atmosphere)*** Characteristics of a standard atmosphere (average conditions for the United States at latitude 40°N) are given in Table 4.3. Characteristics of standard atmospheres at other latitudes are summarized by Burman et al. (1987).

B. Radiation Balance

Historical and recent studies of evaporation (E) and evapotranspiration (ET) clearly show, where soil water is not limiting, that the primary variable controlling the rate of E and ET is solar radiation impinging on the evaporating surfaces. Part of the solar radiation, R_s, is reflected back to the atmosphere and part of that which is absorbed by the surface is reradiated back as long-wave radiation, although this loss of energy is compensated in part by the downcoming long-wave from the sky. The radiation balance determines the net radiant energy, R_n, available at the evaporating surfaces. The most accurate methods of estimating E and ET require determining the radiation balance, or net radiation for the surface. Many studies have shown that daily net radiation is closely related to the daily rate of E from shallow water bodies and ET when soil water is not limiting, especially in warm to hot subhumid and humid climates. During daytime, the responses of E or of ET to changes in solar and net radiation are closely linked even over periods as short as five minutes. Examples in the next section reveal that the actual fraction of R_n going into E or ET can vary markedly depending upon air mass conditions above the surface.

1. ***Solar Radiation*** The principal source of heat energy for E and ET is solar radiation, R_s. When measured at the earth's surface, R_s includes both direct and diffuse short-wave radiation and may be called "global radiation." The most recent measurement of the solar constant, or the rate at which solar radiation is received on a surface normal to the incident radiation outside the earth's atmosphere is 1.367 kJ m^{-2} s^{-1} (London and Frohlich, 1982; Lean, 1989).

On cloudless days, the atmosphere is relatively transparent to solar or short wave radiation. About 70 to 80 percent of extraterrestrial radiation reaches the earth's surface in semiarid areas. The balance is reflected from dust particles or is absorbed by various gases in the atmosphere. Techniques for estimating atmospheric transmittance are presented by Idso (1969; 1970; 1981) and Davies and Idso (1979). Solar

TABLE 4.3. Standard Lower Atmosphere (Adapted from List, 1984).

Altitude (m)	Temperature (°C)	Pressure (kPa)	Pressure (mb)	Density (kg m^{-3})
0	15.0	101.3	1013	1.226
500	11.75	95.5	955	1.168
1,000	8.50	89.9	899	1.112
1,500	5.25	84.6	846	1.058
2,000	2.00	79.5	795	1.007
3,000	−4.50	70.1	701	.909

radiation can be measured directly with instruments ranging from a spring-wound chart and bi-metallic sensors to complex sensors and automated data-logging systems.

Estimates of daily cloudless day solar radiation received at the earth's surface, R_{so}, can be made using extraterrestrial radiation, R_A, and atmospheric transmittance. Clear sky short-wave radiation, R_{so}, can be approximated under most conditions as

$$R_{so} = k_A \; R_A, \tag{4.18}$$

where k_A is a clearness index which is largely a function of station elevation (atmospheric thickness) and atmospheric turbidity. For conditions of low turbidity, k_A can be computed as a function of station elevation for locations less than 6,000 m using a relationship developed by Allen et al. (1994b):

$$k_A = 0.75 + 0.00002\,E, \tag{4.19}$$

where E is elevation in m. This relationship was obtained by integrating Equation 4.15 (air density) with respect to elevation and assuming constant turbidity with elevation, absorption, and scattering of solar radiation in proportion to air density. Atmospheric pressure in Equation 4.15 was computed using Equation 4.14. Equation 4.19 predicts $k_A = 0.75$ at sea level, which agrees with Equation 4.27 for $n/N = 1$. Equation 4.19 also agrees with long-term measurements by Wright (1978) at Kimberly, Idaho, elevation 1,200 m, where $R_{so}/R_A = 0.78$. For areas of high turbidity due to pollution or airborne dust, k_A may need to be reduced by up to 10% depending on the relative turbidity.

Estimates of R_{so} for a given area can also be developed from an envelope curve plotted through measured radiation data on cloudless days. The relationship $R_{so} = k_A R_A$ is useful for verifying correct operation of pyranometers. Ordinarily, R_{so} should plot as an upper envelope of measured R_s. Reflection of solar radiation by adjacent clouds can occasionally increase measured R_s to levels which are above R_{so}.

Extraterrestrial radiation, R_A, can be computed for a location as a function of latitude and day of the year using the following mathematical equations from Duffie and Beckman (1991) (Jensen et al., 1990):

$$R_A = \frac{24(60)}{\pi} G_{sc}\, d_r [\omega_s \sin(\phi)\sin(\delta) + \cos(\phi)\cos(\delta)\sin(\omega_s)] \tag{4.20}$$

where R_A is daily extraterrestrial radiation in MJ m^{-2} d^{-1} and ϕ is latitude of the station in radians (negative for southern latitudes). The declination, δ, in radians can be estimated as:

$$\delta = 0.4093 \sin\left(2\,\pi\, \frac{(284 + J)}{365}\right), \tag{4.21}$$

where J is the day of the year (January 1st = 1). The term d_r is the relative distance of the earth from the sun, where:

$$d_r = 1 + 0.033 \cos\left(\frac{2\pi}{365} J\right). \tag{4.22}$$

The sunset hour angle, ω_s, in radians can be calculated using Equation 4.23 or 4.24:

$$\omega_s = \arccos[-\tan(\phi)\tan(\delta)] \tag{4.23}$$

$$\omega_s = \frac{\pi}{2} - \arctan\left[\frac{-\tan(\phi)\tan(\delta)}{[1 - \tan^2(\phi)\tan^2(\delta)]^{1/2}}\right]. \tag{4.24}$$

G_{sc} in the R_A equation is the solar constant of 0.0820 MJ m^{-2} min^{-1} (1367 W m^{-2}) as recommended by the International Association of Meteorology and Atmospheric Physics (London and Frohlich, 1982). The $-\tan(\phi)\tan(\delta)$ expression in Equation 4.23 must be limited to less than or equal to 2.0 in extreme latitudes (> 55°) during winter months. If the $-\tan(\phi)\tan(\delta)$ expression is less than -1.0 in extreme latitudes (> 55°) during summer months, then the $-\tan(\phi)\tan(\delta)$ expression should be set equal to $[\tan(\phi)\tan(\delta) - 2.0]$. Equations for estimating hourly extraterrestrial and clear-sky radiation can be found in Allen (1996) and Hottel (1976) as summarized in Appendix B of ASCE Manual 70 (Jensen et al., 1990). Heermann et al. (1985) developed an empirical equation for estimating daily R_{so} in the United States; however, the Heermann equation is applicable only to northern latitudes between 25 and 65 degrees during growing season periods and is therefore not presented here. Equation 4.18 is recommended instead.

Solar radiation, R_s, on a given day is affected mainly by cloud cover. Therefore, R_s can be estimated using R_{so} and either degree of cloud cover or percent of sunshine. More reliable R_s estimates are obtained using recorded percent sunshine as compared to observed cloud cover data because observed cloud cover data are qualitative. Constants developed for a linear equation by Fritz and MacDonald (1949) are similar to those obtained in Canada and Australia:

$$R_s = (0.35 + 0.61S)R_{so}, \tag{4.25}$$

where S represents the ratio of actual to possible sunshine. Equation 4.25 allows for the fact that no recording instrument gives a full record of sunshine even on cloudless days. Where cloud cover is not strongly influenced by local orographic features, measurement of daily R_s at a single station may be used for estimates over large areas (100–200 km in diameter) without significant error over 5- or 10-day periods.

Solar radiation also can be estimated from extraterrestrial radiation, R_A. Black et al. (1954) correlated extraterrestrial solar radiation and duration of sunshine as recorded by Marvin and Campbell-Stokes sunshine recorders. The resulting equation based on 32 stations is:

$$R_s = (0.23 + 0.48S)R_A. \tag{4.26}$$

Doorenbos and Pruitt (1977) recommended a generalized form of the Penman R_s equation:

$$R_s = \left(0.25 + 0.5\frac{n}{N}\right)R_A. \tag{4.27}$$

The quantity n/N was defined by Doorenbos and Pruitt (1977) as "the ratio between actual measured bright sunshine hours and maximum possible sunshine hours." In practice, n/N and S are usually assumed to be the same quantity. Doorenbos and Pruitt presented a table to convert cloudiness expressed in eighths (octas) or in tenths to n/N values. Experimentally-derived constants for Equation 4.27 are presented in an appendix table for a number of specific locations by Doorenbos and Pruitt (1977). Variable N in Equation 4.27 can be calculated from the sunset hour angle as:

$$N = \omega_s\left(\frac{24}{\pi}\right) \tag{4.28}$$

where ω_s is sunset hour angle in radians, calculated using Equation 4.23 or 4.24.

Hargreaves and Samani (1982) proposed estimating R_s from the range in daily air temperature:

$$R_s = K_{RS}\,(T_x - T_n)^{0.5}\,R_A \tag{4.29}$$

where T_x and T_n are maximum and minimum daily air temperature, °C. Variable K_{RS} is an empirical coefficient equal to about 0.16 for interior regions (Hargreaves and Samani, 1982) and about 0.19 for coastal regions (Hargreaves, 1994). Equation 4.29 performs best using mean monthly data. When applied to daily data, Equation 4.29 tends to overestimate for cloudy days. R_s predicted by Equation 4.29 should be limited to $\leq R_{so}$. Bristow and Campbell (1984) developed an equation similar to Equation 4.29; however, their formulation is more complex to apply and has similar accuracy.

2. *Solar Radiation Data Base, United States* The National Renewable Energy Laboratory (NREL) has compiled a mean monthly solar radiation data base and associated 30-year (1961-1990) record for the United States (NREL, 1992). These data include measured and modeled solar radiation and meteorological data for 239 locations in the United States and its possessions. They are available on magnetic media (disks and tape) from the National Climate Data Center in Asheville, North Carolina. Future releases will be produced on CD-ROM.

3. *Net Radiation* Net radiation requires the measurement or estimates of both incoming and reflected short wave radiation (< 4 μm) and net long-wave radiation. The atmosphere is much less transparent to long-wave radiation as compared to short-wave radiation. Water vapor, carbon dioxide, and ozone are good absorbers and emitters at some infrared wave lengths. Radiant energy absorbed by these various gases is emitted in all directions according to the radiation law.

$$R = \epsilon\sigma T^4, \tag{4.30}$$

where ϵ is emissivity, σ is the Stefan-Boltzmann constant, and T is the absolute temperature, K. Values of the Stefan-Boltzmann constant are $\sigma = 5.675 \times 10^{-8}\,\text{J m}^{-2}\,\text{K}^{-4}\,\text{s}^{-1}$, or $\sigma = 4.903 \times 10^{-9}\,\text{MJ m}^{-2}\,\text{K}^{-4}\,\text{d}^{-1}$.

Because of the high transparency of the atmosphere to the 8 to 13 μm wavelengths, there is a net loss of radiant energy to the atmosphere within this range during cloudless and partly cloudy 24-hour periods. For practical purposes, only the long-wave radiation components are important between sunset and sunrise. A small amount of diffuse short-wave sky radiation is involved just after sunset and prior to sunrise, respectively.

Net radiation flux can be measured directly using hemispherical net radiometers or estimated from net short-wave and net long-wave components.

$$R_n = R_s \downarrow - \alpha R_s \uparrow + R_L \downarrow - R_L \uparrow = (1 - \alpha)R_s \downarrow - R_b \uparrow, \tag{4.31}$$

where R_s is short-wave radiation or solar radiation, and R_L is long-wave radiation. The arrows indicate direction; $(1 - \alpha)R_s$ represents the net short-wave radiation received by a water, soil, or vegetated surface; α is the short-wave reflectance or albedo and R_b is the net back, or net outgoing, long-wave radiation.

Rather than estimating net radiation for each vegetated surface, or for each crop and growth stage, it is most common to measure or estimate the net radiation for a well-watered reference crop like grass or alfalfa. The resulting estimated rate of *ET* for other crops is then related to the reference crop. Mean daily reflectance coefficients (albedo) for most green field crops with a full cover ranges from 0.20 to 0.25. A value of 0.23 for α is commonly used for vegetated surfaces with a full cover. Full cover exists when the leaf area index, which is the leaf area per unit land area, is greater than about 3.

Wright (1982) presented an equation for estimating net radiation received by alfalfa in Kimberly, Idaho in which the albedo varies during the growing season (Jensen et al., 1990, Eq. 6.67, p. 137). The albedo values for inland water bodies range from 0.04 to 0.09. Bolsenga and Vanderploeg (1992) suggested a range from 0.07–0.15 depending upon water-surface conditions. Brutsaert (1982) summarized albedo values as reported earlier by a number of researchers (Table 4.4).

Long-wave radiation emitted by a surface can be estimated using Equation 4.30 if the surface temperature is known. The emissivity of a mixture of vegetation and soil surfaces, ϵ_{vs}, is about 0.98. Table 4.5 lists reported ϵ_{vs} values for a number of surfaces. The emissivity of water surfaces is about 0.97. The incoming long-wave radiation is more difficult to estimate since it must be an integrated value based on variations

TABLE 4.4. Approximate Mean Albedo Values For Various Natural Surfaces (Brutsaert, 1982).

Nature of surface	Albedo
Deep Water	0.04-0.08
Moist dark soils; plowed fields	0.05-0.15
Gray soils; bare fields	0.15-0.25
Dry soils; desert	0.20-0.35
White sand; lime	0.30-0.40
Green grass and other short vegetation (e.g., alfalfa, potatoes, beets)	0.15-0.25
Dry grass; stubble	0.15-0.20
Dry prairie and savannah	0.20-0.30
Coniferous forest	0.10-0.15
Deciduous forest	0.15-0.25
Forest with melting snow	0.20-0.30
Old and dirty snow cover	0.35-0.65
Clean, stable snow cover	0.60-0.75
Fresh dry snow	0.80-0.90

TABLE 4.5. Values of the Emissivities, ϵ, of Some Natural Surfaces.

Nature of surface	Emissivity
Bare soil (mineral)	0.95-0.97
Bare soil (organic)	0.97-0.98
Grassy vegetation	0.97-0.98
Tree vegetation	0.96-0.97
Snow (old)	0.97
Snow (fresh)	0.99

in water vapor and temperature, with elevation above the ground surface, degree of cloud cover and temperature of the clouds. It can be directly measured during dark hours with a total hemispheric radiometer (upward facing), or indirectly computed during daylight hours by subtracting R_s from the radiometer readings. Dense clouds act as a black body and essentially block the long-wave window (8 to 13 μm) to outer space and they emit long-wave radiation to the ground. The temperature at the cloud base is required to calculate this component of incoming thermal radiation.

Calculation of net thermal radiation becomes complex for partly cloudy conditions. A detailed analysis of the factors affecting net radiation involves both the atmospheric and surface emittances and cloud cover. Calculation of R_n on sloping terrain and related references is discussed in the section "Evapotranspiration from Land Surfaces—Direct Penman-Monteith."

Kimball et al. (1982) derived a complex model involving the areas of cloud cover and cloud temperature and transmittance in the 8 to 13 μm wavelength window. For a more complete discussion, see Davies and Idso (1979) and Dong et al. (1992).

Since measured incoming long-wave radiation data generally are not available, and temperature and vapor pressure as a function of elevation above the ground surface likewise are generally not available, most estimates of incoming long-wave radiation are based on air temperature and vapor pressure at instrument shelter height. The effects of cloud cover on long-wave atmospheric radiation and simple empirical equations to estimate net long-wave radiation are presented below.

The early work by Brunt (1932) relates clear-sky atmospheric emittance, ϵ_a, to vapor pressure at screen height, e_d, which takes the general form:

$$\epsilon_a = a_b + b_b\sqrt{e_d}, \tag{4.32}$$

where a_b and b_b are empirical coefficients.

For cloudy conditions, the net long-wave or thermal radiation, R_b, is influenced by the degree of cloud cover, its elevation and temperature.

$$R_b = R_L \uparrow - R_L \downarrow = R_{vs} \uparrow - (R_a + R_c)\downarrow, \tag{4.33}$$

where R_L is long-wave radiation, R_{vs} is long-wave radiation emitted by the vegetated or soil surface, R_a is full-spectrum long-wave radiation emitted by a cloudless sky and R_c is additional long-wave radiation emitted by clouds and transmitted to the earth's surface through the 8–13 μm wave band. In practice, the two long-wave radiation terms in Equation 4.33 are usually calculated using only instrument height data. Since air temperature at instrument height is used to estimate both incoming and outgoing long-wave radiation, then:

$$R_b = f\,(\epsilon_{vs} - \epsilon_a)\,\sigma T^4, \tag{4.34}$$

where ϵ_a is atmospheric emittance, ϵ_{vs} is vegetative and soil emittance, T is air temperature in K, and f is a factor to adjust for cloud cover. Equation 4.34 assumes that the vegetated surface temperature and air temperature at instrument height are equal.

The two emittances may be combined into a single term known as the net emittance, ϵ', which represents the difference between atmospheric and the combined vegetative and soil emittance when a single temperature is used. Values of ϵ_{vs} for common, fully developed crops range from 0.94 to 0.995. A canopy value of 0.98 is recommended for most radiation balance calculations over vegetated surfaces. A value of 0.97 is recommended for water.

Wright and Jensen (1972) developed the following expression for f in Equation 4.34 which represents an average of day-time and night-time conditions:

$$f = a\frac{R_s}{R_{so}} + b. \tag{4.35}$$

The Brunt form of the net emittance expression ϵ' is:

$$\epsilon' = \epsilon_{vs} - \epsilon_a = a_1 + b_1\sqrt{e_d}, \tag{4.36}$$

where e_d is the saturation vapor pressure at dewpoint temperature.

Combining Equations 4.34, 4.35, and 4.36 results in a general equation for net outgoing long-wave radiation.

$$R_b = \left(a\frac{R_s}{R_{so}} + b\right)\left(a_1 + b_1\sqrt{e_d}\right)\sigma T^4. \tag{4.37}$$

For 24-hour or longer time steps, T^4 in Equation 4.37 should be replaced by $(T_x^4 + T_n^4)/2$ for improved accuracy, where T_x and T_n are mean maximum and minimum daily air temperatures in K at screen or shelter height for the period. Experimental values of the coefficients for use in Equation 4.37 are presented in Table 4.6.

In practice, the net long wave radiation is first estimated for clear skies using air temperature and humidity data measured at screen height as follows:

$$R_{bo} = \epsilon'\sigma T^4 = \left(a_1 + b_1\sqrt{e_d}\right)(4.903 \times 10^{-9})\frac{(T_x^4 + T_n^4)}{2}, \tag{4.38}$$

where R_{bo} is net long-wave radiation for clear skies, ϵ' is the net emissivity when using only instrument height temperature, e_d is the saturation vapor pressure at mean dewpoint temperature in kPa, 4.903×10^{-9} is the Stefan-Boltzmann constant in MJ m^{-2} d^{-1} K^{-4} for units of R_{bo} in MJ m^{-2} d^{-1}, and T_x and T_n

TABLE 4.6. Experimental Coefficients for Net Long-Wave Radiation Equations 4.35 to 4.38 (Jensen et al., 1990).

Region	(a,	b)	(a_1,	b_1)
Davis, California	(1.35,	−0.35)[1]	(0.35,	−0.145)[2]
Southern Idaho	(1.22,	−0.18)[3]	(0.325,	−0.139)[3]
England	(not available)		(0.47,	−0.206)[4]
England	(not available)		(0.44,	−0.253)[5]
Australia	(not available)		(0.35,	−0.133)[6]
General	(1.3,	−0.3)[7]	(0.39,	−0.158)[8]

For e in mb, divide b_1 values by $\sqrt{10}$.
[1]Pruitt (1976).
[2]Goss and Brooks (1956).
[3]Wright and Jensen (1972).
[4]Monteith and Szeicz (1962).
[5]Penman (1948).
[6]Fitzpatrick and Stern (1965).
[7]Suggested for arid areas.
[8]Budyko (1956).

are in K. If Equation 4.38 is applied on an hourly or shorter time step, then T_x and T_n can be set equal to the mean temperature for the time step.

Since a single level temperature measurement is used, $(a_1 + b_1 \sqrt{e_d})$ represents ϵ' or the difference between the emittance for the reference crop and the effective emittance for the atmosphere. The dewpoint temperature does not change greatly during the day, and a single level dewpoint observation during the day is adequate for most estimates of R_{bo} for eventual calculation of E or ET.

When solar radiation data are available, they can be used to adjust the net long-wave radiation for clear skies or for partly cloudy conditions.

$$R_b = \left(a \frac{R_s}{R_{so}} + b\right) R_{bo}. \tag{4.39}$$

The coefficients a and b are derived empirically and $a + b = 1.0$. Smith et al. (1991) and Allen et al. (1994b) have recommended using $a = 1.35$, $b = -0.35$, $a_1 = 0.34$, and $b_1 = -0.14$ at most locations for consistency in standardized reference ET calculations.

Where humidity measurements are not available, minimum air temperatures may be taken as the dewpoint temperature in semihumid and humid areas. In arid areas, the dewpoint temperature may be several degrees less than the minimum air temperature. Simple calibration relationships can be developed to enable using minimum air temperature as a substitute for dewpoint temperature.

C. Energy Balance

In cold humid climates, or in cold winter months of temperate to semiarid zones, only 50 to 60 percent of the net radiation, R_n, may be converted to latent heat. In hot, arid climates, latent heat may exceed net radiation by 10 to 50 percent with sensible heat derived from the air and converted to latent heat. In spite of these relationships, the heat energy balance approach to determining or estimating E and ET is recognized as a reliable and conservative method. A thorough understanding of the factors controlling the energy balance of a water body or land area enables accurate estimates or predictions of E and ET.

Whereas the radiation balance considers only radiation terms in developing the net exchange of radiation (Equation 4.33), the energy balance relates to the various ways in which R_n is balanced by the inputs or outputs of energy from non-radiative parameters. The vertical energy balance at the soil or water surface or at the "effective surface" of a crop is the sum of sensible heat flux to or from the air and soil (or of water), latent heat, net radiation, and other miscellaneous fluxes. The major components of the vertical energy balance can be expressed as follows:

$$R_n = \lambda E + H + G, \qquad (4.40)$$

where R_n is the net radiation flux at the surface, λE is the latent heat flux (positive during evaporation), H is the sensible heat exchange from the surface to air (positive if the air is warming), and G is the sensible heat exchange from the surface to the soil or water (positive if the soil or water is warming). The miscellaneous terms such as heat storage within the foliage, photosynthesis, and respiration are generally insignificant relative to magnitudes of R_n, λE, H, and G, and are neglected in Equation 4.40. For daily or longer periods, G can be ignored, although with significant variation of R_n and/or air mass conditions from one day to the next, this may not be advisable. For shorter periods, e.g., hourly, G must be taken into account. Some of the minor terms such as changes in heat storage in a canopy may also need attention, especially in forest canopies. If R_n and sensible heat flux densities H and G can be measured or estimated reliably, then the latent heat flux density, λE, can be computed from Equation 4.40.

Calculation of R_n has been discussed earlier in this section. Description and equations for calculating H and λE are presented later in the sections on Energy Balance and Mass Transfer Methods. Calculation of G is discussed in the following section.

D. Sensible Heat Flux Density—Soil

The magnitude of soil heat storage or release can be significant over a few hours but is usually small from day to day because heat stored early in the day as the soil warms is lost late in the day or at night when the soil cools. The rate of heat storage or release at any depth z can be expressed by:

$$G = K_{Ts} \frac{\partial T_s}{\partial z}, \qquad (4.41)$$

where T_s is the temperature of the soil and K_{Ts} is the thermal conductivity of the soil in J m^{-1} s^{-1} °C^{-1}. Therefore, G in Equation 4.41 has units of J m^{-2} s^{-1} (W m^{-2}) for T_s in °C and z in m. Typical values of K_{Ts} are presented in Table 4.7. K_{Ts} for soils varies widely with soil moisture content. The effect of soil moisture is to provide thermal continuity between soil particles, thereby increasing bulk K_{Ts}. Brutsaert (1982,

TABLE 4.7. Thermal Properties of Soil Constituents at 20°C and Standard Atmospheric Pressure (van Wijk and de Vries, 1963).

Soil material	Density heat ρ [Mg m^{-3}]	Specific heat [kJ kg^{-1}C^{-1}]	Volumetric heat capacity C_s [MJ m^{-3} C^{-1}]	Thermal conductivity λ [J m^{-1} s^{-1} C^{-1}]	Thermal diffusivity α [10^{-6} m^2 s^{-1}]
Quartz	2.65	0.733	1.93	8.37	4.3
Minerals*	2.65	0.733	1.93	2.93	1.5
Organic matter*	1.3	1.926	2.51	0.25	0.1
Water	1.00	4.187	4.19	0.59	0.142
Air	0.0012	1.005	0.0012	0.026	0.021

*Approximate average values
1000 kg m^{-3} = 1 g cm^{-3}
4186.8 J kg^{-1} C^{-1} = 1 cal g^{-1} C^{-1}
4.1868 MJ m^{-3} C^{-1} = 1 cal cm^{-3} C^{-1}
0.41 868 J m^{-1} s^{-1} C^{-1} = 0.41 868 W m^{-1} C^{-1} = 10^{-3} cal cm^{-1} s^{-1} C^{-1}
10^{-6} m^2 s^{-1} = 10^{-2} cm^2 s^{-1}

Figure 6.9) presented curves of thermal conductivity as a function of volumetric water content for four widely different soils. These values ranged from 0.1 to 2.5 J m^{-1} s^{-1} K^{-1}.

Soil or water heat flux density can also be estimated by monitoring the change in temperature of a soil or water profile over time:

$$G = C_s \int_0^{z_s} \frac{\partial T_s}{\partial t} dz, \tag{4.42}$$

where C_s is the heat capacity per unit volume of soil, MJ m^{-3} °C^{-1}, t is time, and z_s is the depth of soil which responds to increased or decreased temperature change. C_s can be estimated from the following equation by de Vries (1963):

$$C_s = 1.93\ V_m + 2.51\ V_{om} + 4.19\ \theta, \tag{4.43}$$

where V_m, V_{om}, and θ represent the volume of minerals, organic matter, and water per unit volume of soil, respectively. For example, if a soil has 50% solids and negligible organic matter and a volumetric water content of 0.27, its heat capacity will be 2.1 MJ m^{-3} °C^{-1}. Approximate values for soils and soil components are given in Table 4.7. Utilization of Equation 4.42 in combination with soil heat flux plates is common.

Since the magnitude of daily soil heat flux under a crop canopy over 10- to 30-day periods or longer is relatively small, the daily value of G normally can be neglected for most practical estimates involving energy balance; however, the total value of G over the entire period may be significant, especially for 30 days or longer.

An approximation of the soil heat flux involved over long periods of time (≥30 days) can be obtained by assuming that the soil temperature to a depth of 2 m changes approximately with average air temperature and that the average volumetric heat capacity for the soil is 2.1 MJ m^{-3} °C^{-1},

$$G = -\ 4.2\ \frac{T_{i-1} - T_{i+1}}{\Delta t} = 4.2\ \frac{T_{i+1} - T_{i-1}}{\Delta t}, \tag{4.44 a}$$

where G is the average daily soil heat flux in MJ m^{-2} d^{-1}, T equals the mean air temperature in °C for time period i, and t equals the time in days between the midpoints of the two periods. In real time predictions, where T_{i+1} is unknown, G for monthly or longer time periods can be predicted as:

$$G = 4.2\ \frac{T_i - T_{i-1}}{\Delta t}. \tag{4.44b}$$

Additional information can be found in articles by van Wijk and de Vries (1963), de Vries (1963), and Brutsaert (1982).

For application to short-periods, e.g., when hourly data for G are required, other approaches must be used. One method commonly used in research is the use of heat flux plates installed near the soil surface (e.g., at 0.01 m). For full cover vegetation situations where temperature gradients even near the surface are minor, heat flux plates can give reasonably accurate data for G; however, for many situations, strong temperature gradients exist near the soil surface and differences in thermal conductivities of the plate and of the soil can produce large errors in estimating G. Also, with the strong thermal gradients during day and night, condensation and evaporation near the upper or lower surface of the flux plate can be a significant problem. Therefore, soil heat flux plates should be placed below depths of soil moisture evaporation (generally 0.08 to 0.15 m) to avoid errors associated with H to λE conversion between the surface and the plate.

In measurements of G in water, care must be taken to avoid effects of absorption of short-wave radiation by the plate. Therefore, the plate should generally be buried in the lake substrate. In applications where soil is saturated or is very dry, the user should check with the manufacturer to ascertain whether the plate calibration holds.

Measurements by soil heat flux plates, G_{zs}, at some depth z_s are usually adjusted to represent G at the soil or water surface by incorporating Equation 4.42:

$$G = G_{z_s} + C_s \int_0^{z_s} \frac{\partial T_s}{\partial t}\, dz, \tag{4.45}$$

where $\partial T_s/\partial t$ is the change in temperature within the z_s thick layer above the heat flux plate. Temperature is usually measured with thermocouples or buriable thermistors. Tanner (1960) developed resistance thermometry systems for obtaining the average 0 to 0.1-m soil temperature using a string of ten or more units to provide average temperature changes with time in the upper 0.1 m of soil.

Another method more widely applicable involves calorimetry to determine changes in soil heat storage within a zone from the surface to some depth z, where the flux is negligible or can be estimated reliably, e.g., with Equation 4.42. Both profiles of soil temperature and soil moisture content must be known making this method a difficult one at best albeit one of the most reliable, presuming reliable profile data are available.

Soil heat flux density for dense, short vegetation such as for grass, wheat, or alfalfa can often be approximated during daylight periods as $G = 0.1\ R_n$ (Clothier et al., 1986; Choudhury et al., 1987; Allen et al., 1994b). A universal relationship proposed by Choudhury et al. (1987) and Choudhury (1989) for estimating G under daytime conditions is:

$$G = 0.4\ e^{-0.5\, LAI}\ R_n \tag{4.46}$$

where LAI is the leaf area index and e is the natural number, with G having the same units as R_n. Equation 4.46 estimates $G = 0.1\ R_n$ for an $LAI = 2.8$, which is typical for clipped grass. Pruitt (1994), in an analysis of late spring and summertime data collected over clipped grass in Davis, California, found Equation 4.46 to provide close estimates of G during daytime for a range of LAI from around 2.8 up to 4.5. During moderate- to strong-advection days at Davis, G was found to be almost negligible or even reversed in direction during daylight hours. A very low fraction of G/R_n (0.036), was used by Verma et al. (1986) during daylight for a deciduous forest.

For nighttime under grass forage, Allen et al. (1994b) found $G = 0.5\ R_n$ to be valid. In Davis, a $G = 0.3\ R_n$ was found at night. In all of these reported ratios, it should be noted that the effects of seasonal variations are not addressed. The seasonal lag in soil or water bodies behind solar radiation, can be expected to produce higher ratios of G/R_n in spring than in fall months with equivalent solar or net radiation amounts.

In computerized evapotranspiration models where soil surface temperatures are calculated from surface heat exchange equations, G can be estimated using the "force-restore" method (Lin 1980). A thin surface layer is used to simulate surface temperature for use in estimating long-wave radiation and in "forcing" turbulent heat transfer to the air. A deeper second layer is used to "restore" the surface layer temperatures.

III. INTERACTION OF SURFACES AND METEOROLOGICAL FACTORS

A. Energy Balance—Air-Mass Interactions

1. Introduction Penman (1956) defined Potential Transpiration (evapotranspiration) as "the amount of water transpired in unit time by a short green crop, completely shading the ground, of uniform height and never short of water;" however, others (Monteith, 1981) have defined potential transpiration or potential *ET* as evapotranspiration from vegetation having a saturated surface (surface resistance, r_s, = 0). The latter definition, although correct, has proven to be difficult to measure and to apply in the field, especially in the absence of weather data that reflect a potential *ET* environment (measured over a wet surface). Therefore, use of the term potential *ET* has been discouraged as a climatic index and for use in *ET* estimation.

Wright and Jensen (1972) indicated that maximum *ET* for nonsaturated conditions may be approximated by *ET* from a well watered crop of alfalfa with a height of at least 0.2 m. They suggested the use of the term reference crop *ET*. Doorenbos and Pruitt (1975; 1977) described a reference crop evapotranspiration (ET_o) as "the rate of evapotranspiration from an extensive surface of 8–15 cm tall, green grass cover of uniform height, actively growing, completely shading the ground and not short of water." Subsequently, this definition specified the grass cover as being that of one of the cool-season grasses with roughness, density, leaf area, and canopy resistance characteristics similar to perennial ryegrass (*Lolium perenne*) or alta fescue (*Festuca arundinacea* Schreb. 'Alta'), since warm-season grass varieties, such as Bermuda (*Cynodon dactylon*), exercise considerable control over transpiration.

Grass has served as a standard reference crop largely due to the preponderance of grass in lysimeter studies around the world. Also, in much of the world, several species of cool-season grasses remain green throughout the year. With grass, it is possible to maintain a near-constant canopy with a Leaf Area Index (*LAI*) of ~3, and with weekly or more frequent mowing, growth differences within and outside of a lysimeter, whether due to fertility or soil moisture differences, can be minimized.

Despite its widespread use, there are disadvantages in using grass as a reference. Because grass is short and aerodynamically quite smooth, the *ET* losses can be considerably lower than those for most other types of vegetation with rougher canopies, particularly in hot, dry, and windy conditions. Hence, well-irrigated alfalfa, some 0.3 to 0.5 m in height, has been the other major crop used as a reference with the symbol ET_r (Jensen, 1968; Jensen and Haise, 1963; Jensen et al., 1990). There are few vegetative canopies that may have higher *ET* rates than 0.3 to 0.5-m tall alfalfa. Thus, in a two-step procedure (using ET_r and a crop coefficient K_c), there should be, for full-cover conditions, a narrower range in K_c values than when ET_o (grass) is used as a reference. This would especially be the case when a wide range of climates and crops is involved.

There are also some disadvantages in the use of alfalfa as a reference. Cuttings destroy the entire canopy and actual evapotranspiration is drastically reduced for about a week following each cutting. Trampling of the crop can present much more of a problem than with a grass turf. Alfalfa can be difficult to grow in some tropical climates and at high altitudes if validation is required. In addition, various alfalfa varieties may exhibit different stem erectness and stomatal control. These variations may create differences in ET_r as large as 10% among alfalfa varieties (Wright, 1988).

The Working Group on Water Requirements of the International Commission on Irrigation and Drainage (ICID, 1985) provides an updated terminology listing for evapotranspiration and water requirements. They define reference crop evapotranspiration as: "evapotranspiration from a given well-adapted crop selected for comparative purposes under given weather conditions and with adequate fetch (sufficient to make edge effects relatively unimportant) and for a standardized watering regime appropriate for this crop and the region concerned."

The Working Group's definition of potential evaporation is: "the evaporation from a given surface when all surface-atmospheric interfaces are wet (saturated) so that there is no restriction due to either biological control or soil water content on the water vapor loss from the surface area." (ICID, 1985). They pointed out that its magnitude depends primarily on atmospheric conditions and surface albedo but also varies with geometric characteristics of the surface. In terms of hydrology, this term would apply only following precipitation, condensation (dew), fog interception, or a crop irrigated by sprinkling. Within Chapter 4, potential evaporation is identified as E_o and relates to the above definition.

In the following section, data are presented and equations are used to illustrate, in general, some of the basic interactions relative to *ET* that take place between plants and the atmosphere.

2. *Latitude, Temperature, Humidity and Wind*

a. R_n vs R_s Relationships. Net radiation of cropped surfaces during daylight is primarily a function of incident solar radiation. A number of the important *ET* equations involve the input of total 24-hour net radiation, which is related not just to R_s but to the length of the night and to nighttime surface and atmospheric conditions. The examples in Fig. 4.1 for grass or grass-clover crops show that there can be a wide range in the fraction of solar radiation involved in 24-hour net radiation, although it is evident from Fig. 4.1(a) that in spite of the latitude differences and obvious differences in climate that must exist between Copenhagen, Davis, California, and Zaire, the data could be represented by a single regression relationship. The offset indicated relates to the need for some 3 to 5 MJ m^{-2} day^{-1} of R_s at Copenhagen and Davis, respectively, to make up for the negative R_n of the long nighttime periods of winter. Even though r^2 values exceed 0.92, the results presented in Fig. 4.1(b) show a wide range of R_n/R_s fractions during the year. The Copenhagen results are unique in that the long winter nights can result in negative R_n exceeding in magnitude the daytime positive R_n, thus producing a negative fraction of R_n/R_s. The 10-year results of Aslyng and Jensen (1965) reveal an average negative 24-hour R_n for the four months of November through February.

The only location where R_n/R_s remains fairly steady throughout the year is at the tropical site (Zaire). The slightly lower midsummer fractions in Zaire as compared to the other locations probably relates to the 12-hour days in Zaire with much longer days at the higher latitudes.

Because of the wide variation of R_n/R_s with seasons except in the tropics, a requirement of considerable empirical adjustment might be expected in the use of solar radiation-based evaporation equations; however, in interpreting the R_n/R_s patterns of Fig. 4.1, it should be noted that except for moderate to strong wind conditions, *ET* is largely a daytime phenomena, and hence can be expected to be more strongly related to daytime net radiation than to 24-hour total R_n. The R_n^{+}/R_s curve for Davis in Fig. 4.1 (b) reveals rather minor seasonal variability, especially in comparison to R_n/R_s for Davis. Thus, a closer correlation between *ET* and R_s can be expected than that which might be inferred from the R_n/R_s relationships.

b. Effect of Weather on Fraction of R_n used in ET. A-priori reasoning might lead one to conclude that for an effectively wet surface (i.e., a very thin film of water on the surface or a complete canopy of vegetation with insignificant stomatal resistance), most of the net radiation minus soil heat flux ($R_n - G$) would be converted to latent heat λE with only minor amounts of sensible heat, *H*. Surface temperature should remain at or near air temperature, and near-neutral stability conditions within the turbulent boundary layer might be expected during much of the daytime period. Just such a concept concerning the partition of net radiation for vegetative surfaces was prevalent in earlier literature. Researchers suggested potential evapotranspiration [as defined by Penman (1948) or by Thornthwaite (1948)] as a condition under which essentially all of the ($R_n - G$) would be used in the evaporative process; however, in reality this is rarely the case, although the effect of advection in producing ratios of latent heat, λE to R_n well over 1.0, has long been recognized. In the early literature, ratios less than 1.0 seemed to be expected only where significant stomatal resistances were likely, or where plant cover was only partially shading the ground or just upwind from cold ocean or lake surfaces.

The following examples of seasonal patterns of *ET* and of $\lambda ET/R_n$ reveal that the above assumptions were far too simplistic. The results shown in Fig. 4.2 are for studies where, for the most part, full cover, vegetative surfaces with low stomatal resistances were involved. Data for September and October in Phoenix, Arizona are missing due to lack of full cover and/or low soil moisture conditions. For Davis, 100 out of over 1200 days on record were excluded from the analysis because of strong advection. The Copenhagen curve is based on 1970 daily data (Jensen, 1972) as supplied to FAO for development of Irrigation and Drainage Paper 24 (Doorenbos and Pruitt, 1975; 1977).

The climates of Phoenix, Davis, and Copenhagen, due to latitude and other factors, represent a wide range of weather conditions. Although advection (both heating and cooling) was no doubt a factor at times, the normal seasonal lag of temperature and saturation vapor pressure deficit behind radiation

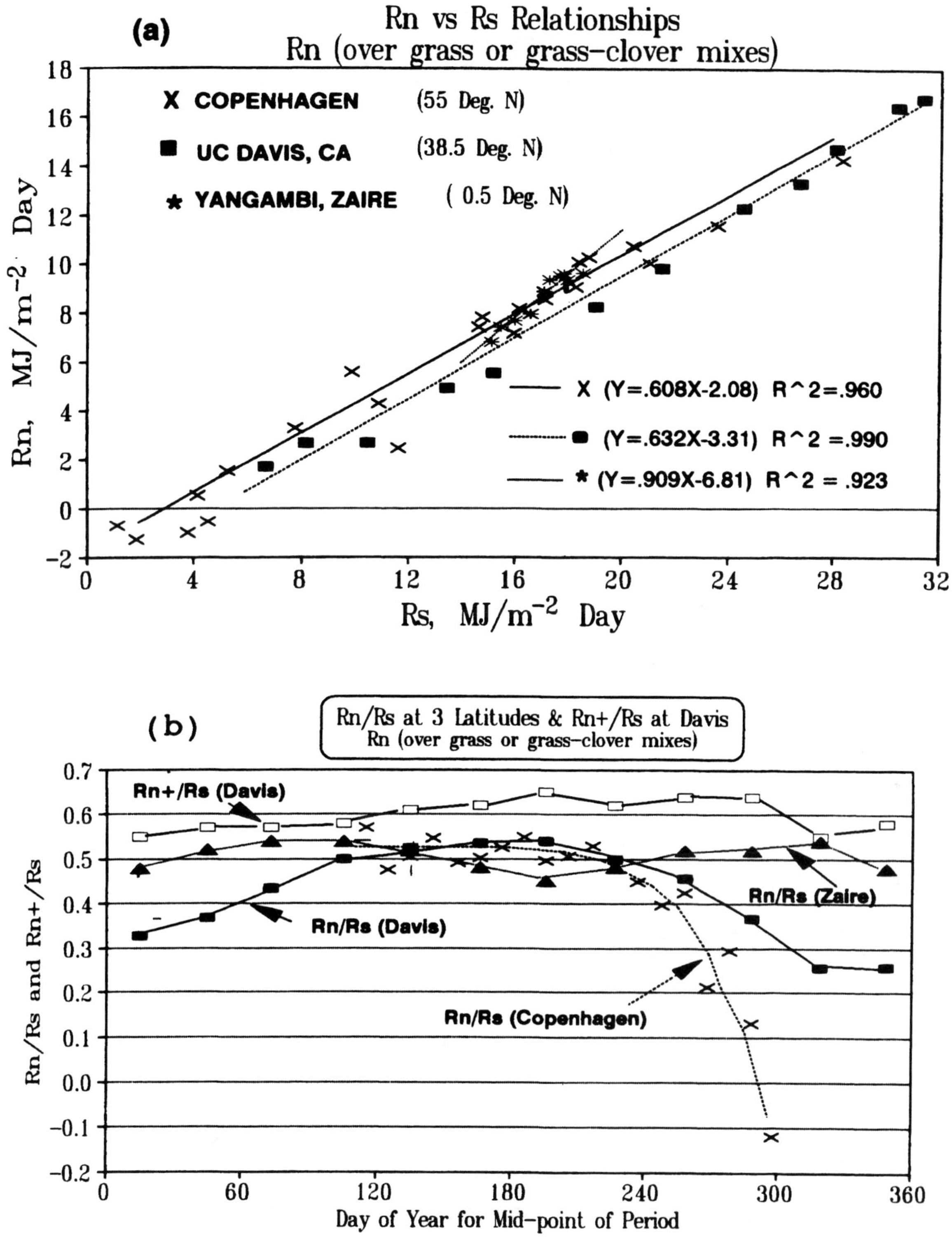

Figure 4.1—(a) Regression Relationship of R_n vs R_s for Three Locations; (b) R_n/R_s and R_n^+/R_s Fractions as a Function of Day of Year for Copenhagen, Denmark, Davis, CA, and Yangambi, Zaire (Jensen, 1972; Pruitt, 1971; Bultot and Griffiths, 1972).

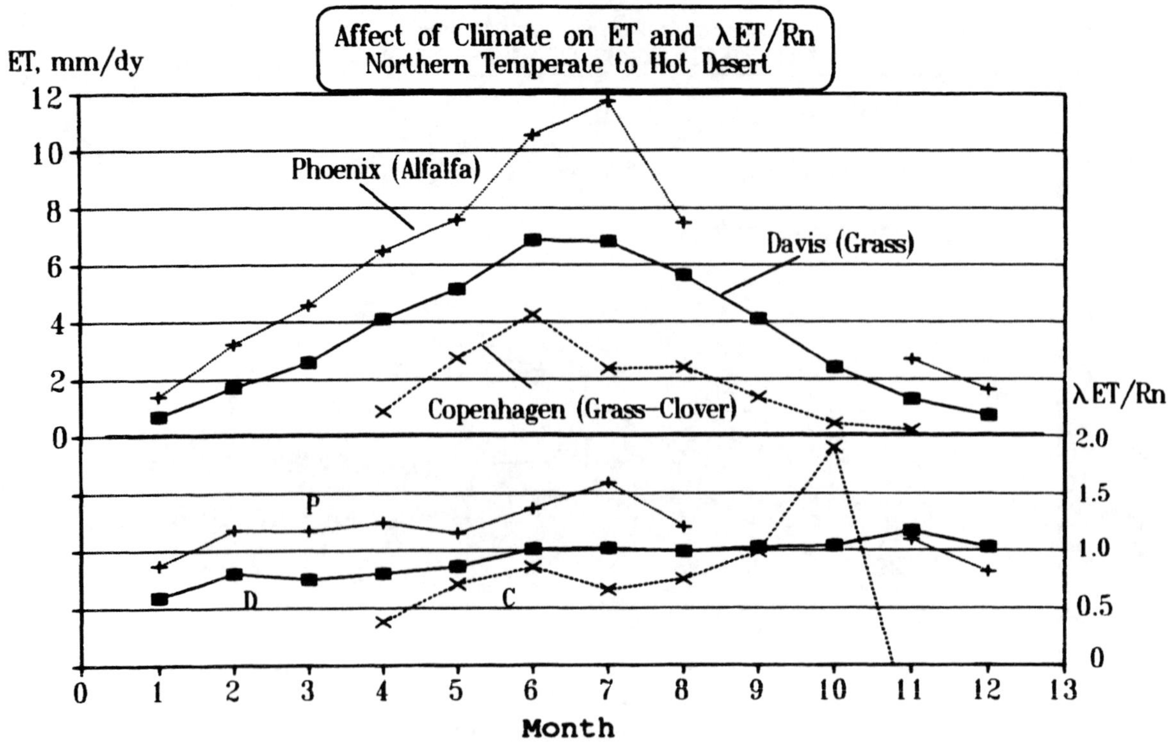

Figure 4.2—Evapotranspiration and the Fraction of 24-hour R_n Involved in λET for Cropped Surfaces at Phoenix, AZ, Copenhagen, and Davis, CA (Pruitt, 1971; Van Bavel, 1966).

produced a strong trend for lower $|\lambda E/R_n|$ in late winter and spring months than for summer and fall months.

In Fig. 4.2, curves for $\lambda ET/R_n$ were based on 24-hour values of R_n. In Fig. 4.3(a), Davis, California data for January 1960 to June 1963 show not only the fractions $\lambda ET/R_s$ and $\lambda ET/R_n$, but the fraction of daytime positive net radiation used in ET, $\lambda ET/R_n$. During the last half of the year, the lower curves are distinctly different in shape than the upper curve based on total 24-hour net radiation. Since days of strong advection were not included in this analysis, the data are primarily for days when some 95-100% of the total ET occurred during daylight hours. The $\lambda ET/R_n$ curve thus reveals that well-irrigated, frequently-mowed grass in the climate at Davis does not, in the daytime, use up all the net radiation as latent heat, and that significant sensible heat is lost to the air and soil during daytime periods, especially during non-summer months. This has important implications, not only for the energy balance, but also in relation to stability conditions that prevail during daytime within the turbulent boundary layer. This in turn relates to the question of whether or not stability-related correction expressions are required for accuracy in estimating E or ET.

The results in Fig. 4.3(b–d) show a clear seasonal cyclic or lag effect in ET-radiation relationships. One of the early papers revealing such trends (Pruitt, 1964) was based on only 1960 out of the 3-½ years of data presented in Fig. 4.3.

From the preceding figures it is obvious that the net radiation partitioned into latent and sensible heat fluxes is highly dependent upon prevailing air-mass conditions. Presenting a basic understanding of the effect of various climatic factors on ET is difficult. We would like simple answers to questions such as "what is the effect of wind on evapotranspiration, or of radiation, or of humidity?" However, it is obvious that these and other factors, including surface conditions and the level of radiation itself, are all interactive.

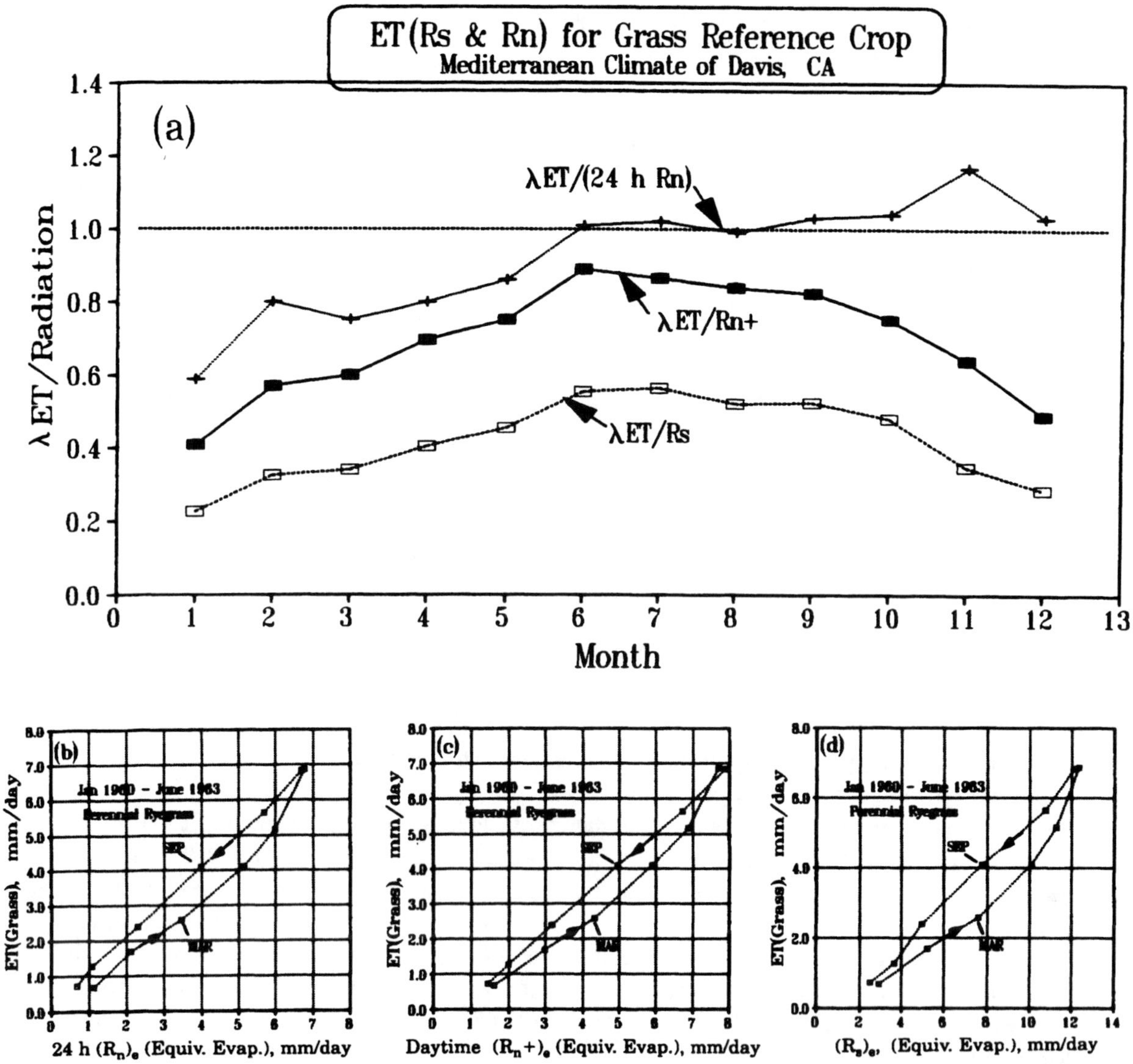

Figure 4.3—(a) Mean Monthly Fractions of Solar and Net Radiation Used as λET by Perennial Ryegrass and (b − d), the Cyclic relationships of ET Versus $(R_n)_e$, $(R_n+)_e$, and $(R_s)_e$, (equiv. mm. evap.), Jan 1960 - June 1960-63. Davis, CA (Pruitt and Angus, 1960; Pruitt, 1971).

3. ***Equations Illustrating Weather Effects on Potential Evaporation*** The combination energy balance-aerodynamic equations, such as those proposed by Penman (1948) or Slatyer and McIlroy (1961), offer excellent models for illustrating how the various weather parameters interact to affect energy partitioning at evaporating surfaces. These models are based on empirically-determined aerodynamic wind functions while the more theoretically based Penman-Monteith (Monteith, 1965) method has a wind function term based on the log-law wind profile, and includes a leaf or canopy resistance term.

The Penman-Monteith equation will be used later to present examples which help in evaluating not only the effects of air-mass conditions, but surface resistance as well, on net radiation partitioning at the surface into the various energy balance terms. At this point, the simpler McIlroy equation is used to cover some aspects of the above and to introduce a now widely-used method for estimating *ET* for non-advection conditions (Priestley and Taylor, 1972).

Since the McIlroy equation is used merely to illustrate the weather effects, units for the terms are not needed. It can be expressed as:

$$\lambda E = \left[\frac{s}{s+\gamma}\right](R_n - G) + h\,(D_z - D_s), \tag{4.47}$$

where λE is the latent heat flux density, γ is the psychrometric constant, and s is the slope of the saturation vapor pressure versus temperature taken at the average of surface temperature and the wet-bulb temperature of the air at height z. D_z is the wet-bulb depression at height z, D_s is the same for air at the surface, and h is a wind-speed dependent heat-transfer coefficient which might be expected to vary with stomatal resistance, crop roughness, and stability conditions of the air mass above the surface. For a hypothetical effectively moist surface, D_s would drop out and, except for the empiricism of the wind-function term normally determined under non-wet surface conditions, the equation would be a predictor of potential evaporation, E_0, as defined earlier (ICID, 1985).

Considering the effects of temperature alone on the energy balance, the term $s/(s+\gamma)$ in Equation 4.47 is temperature sensitive, e.g., at a barometric pressure of 101.3 kPa, varying from 0.4 at 0°C to 0.85 at 40°C. Hence, under very humid conditions (or climates) where the second half of the equation would have little significance, the equation predicts that the fraction of R_n used as λE could be expected to be greater in the warm fall months as compared to cooler spring months, or for cold or cool morning periods as compared to afternoon periods. Also in less humid climates the warmer fall temperatures (or afternoon periods) would result in greater wet-bulb depressions, D_z, producing even higher fall or afternoon ratios of $\lambda E/R_n$. In locations well away from the equator, well-defined cyclic effects can be expected in the relationships of λE vs. R_n (or R_s) on an annual basis (Fig. 4.3) as illustrated for both annual and diurnal periods more than three decades ago by Pruitt (1964).

An interesting case to consider, and one which can provide an excellent estimate of the lowest $\lambda E/R_n$ to be expected for potential evaporation cases, is the use of the first half of Equation 4.47. This portion of the equation offers an estimate of so-called "Equilibrium Evaporation," a condition described by Slatyer and McIlroy (1961) for cases where air in the lower meter or so is saturated or nearly saturated. Predicted effects of wind become zero, (except as it affects temperature) and the ratio of $\lambda E/R_n$ is defined by $[s/(s+\gamma)]$, a function of temperature alone for any given barometric pressure. Such conditions would seldom exist for a daily or longer period; however, on a diurnal basis in many climates, RH is nearly 100 percent for the first hour or so of sunlight.

The well-known Priestley-Taylor (1972) equation essentially provides a compromise between "equilibrium evaporation" and the requirement of including an aerodynamic term such as that in Equation 4.47. The Priestley-Taylor equation can be expressed as:

$$\lambda E = \alpha\left(\frac{\Delta}{\Delta+\gamma}\right)(R_n - G), \tag{4.48}$$

where Δ is the slope of the saturation vapor pressure vs air temperature relationship and α is a multiplier which essentially compensates for the lack of an aerodynamic term typical of combination aerodynamic-energy balance equations. Note that the $s/(s+\gamma)$ of McIlroy's equation is equivalent to the more commonly used $\Delta/(\Delta+\gamma)$, although the temperature base used for the latter is the dry-bulb air temperature.

Priestley and Taylor analyzed data obtained in earlier studies conducted over both land and ocean surfaces and found that a value for α of 1.26 in Equation 4.48 provided estimates of ET for the land studies in good agreement with measured ET; however, this was the case only when data from land-based studies were restricted to include the first few days immediately following rain over an extensive upwind area. See Jensen et al. (1990) for a comprehensive analysis of this and other methods using world-wide data.

4. Summary Although net radiation at an evaporating surface is no doubt the most important single factor affecting *ET* over daily or longer periods, studies in the early 1960s clearly showed a strong out-of-phase relationship between *ET* and R_n, both diurnally and seasonally. It is clear from the results considered herein that the partitioning of net radiation into various components of the energy balance is very much a function of other climatic factors such as wind, temperature, and humidity as well as plant and soil factors. Actual *ET* by crops can be significantly affected by stomatal resistance as illustrated by the Penman-Monteith equation, where under some conditions increased wind may even result in decreased *ET*. The degree of plant cover, stage of maturity, and moisture status of the soil are also important factors.

IV. EVAPORATION FROM WATER SURFACES

A. Introduction

Evaporation from water surfaces is a major component of the hydrologic cycle and must be considered in a wide variety of hydrologic and water resource studies. Because of its nature, evaporation from water surfaces is rarely measured directly, except over relatively small spatial and temporal scales (Jones, 1992). Evaporation from a water surface is most commonly computed indirectly by one or more techniques. The techniques described here are applicable to most engineering and hydrologic studies and include measurements from evaporation pans, water balance, energy balance, mass transfer procedures, and a combined technique. The selection of the best technique to use for a particular computation is largely a function of the data availability and the required accuracy of the computed evaporation.

B. Methods of Estimating Water Surface Evaporation

1. Pan Evaporation Evaporation from water bodies can be estimated from a network of evaporation pans sited throughout the country. The standard evaporation pan in the United States is the Class A Pan which is 1.21 m in diameter and 0.254 m deep placed 0.15 m above ground level on an open-timber framework. Two other commonly used pans are the Soviet GGI-3000 (0.3 m^2) pan and the GGI-20 m^2 tank, both placed in the soil with only 0.075 to 0.1 m of rim above the soil surface. Details can be found in the "Guide to hydrometeorological practices" (World Meteorological Organization 1970). Daily evaporation is computed from the water balance of the pan, which is the difference between the change in storage in the pan and inputs of precipitation and water added to maintain an adequate supply in the pan for evaporation. Due to differing thermal characteristics between the pan and large water bodies, the pans tend to overestimate the total amount of evaporation and distort the seasonal distribution. Thus, on a seasonal basis, pan evaporation usually peaks several months before the peak evaporation of deep lakes. Pan estimates are also suspect during late fall and early winter periods when ice may form in the pan but not on large water bodies. Annual lake evaporation estimates are obtained by multiplying the annual pan data by an appropriate coefficient. These coefficients have been computed for a number of water bodies and, for a United States National Weather Service (NWS) Class A pan, tend to range in value from 0.65–0.85 (World Meteorological Organization, 1973; U.S. Department of Commerce, 1968). The coefficient is higher under humid conditions and lower under arid or dry conditions. A coefficient of 0.70 is applicable when water and air temperatures are approximately equal; however, research over the last several decades points to the need for caution in the use of general pan evaporation data to estimate lake evaporation or crop evapotranspiration (*ET*).

Early studies in the United States in semi-arid to arid climates revealed a very significant response in evaporation to pan size, but now it is clear that this result was primarily due to a decreasing effect of local advection with increasing size of the water surface area. Pruitt and Doorenbos (1977b) reported almost no difference (4 percent) between evaporation from a 0.62-m diameter Soviet pan and a 5-m diameter Soviet pan when both were located in a 5-ha irrigated grass field. On the other hand, at a dry site in

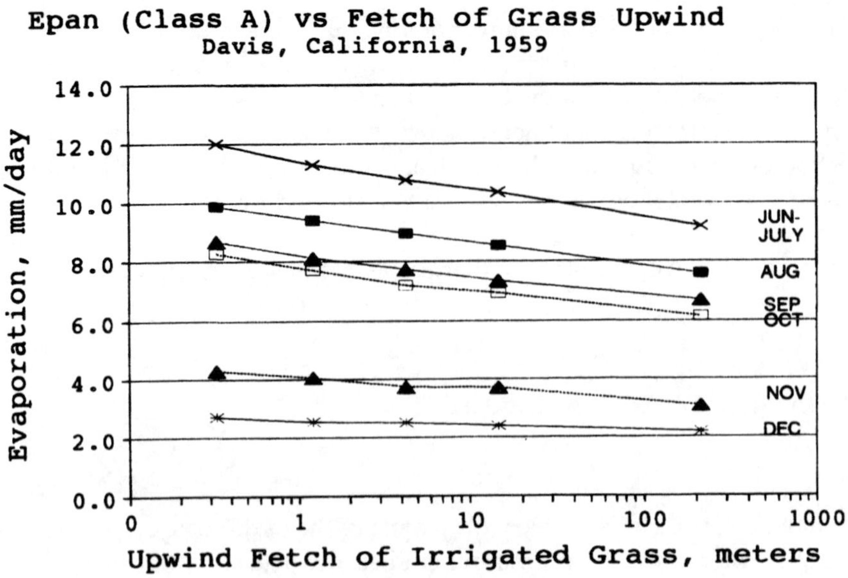

Figure 4.4—Evaporation from NWS Class A Pans Plotted Against the Upwind Fetch of Irrigated Land Cropped to 0.07 to 0.15-m tall Grass, Davis, CA (Pruitt, 1966).

Nevada, evaporation from the smaller Soviet pan averaged some 1.6 times that of the larger pan, although when corrected for the net heat transfer from soil to the pans, a factor of 1.45 resulted (Hounam 1973).

Young (1947) presented a good discussion of the problem of local pan environment in relation to estimating lake evaporation. Soon thereafter, studies in India by Ramdas (1957) and studies at Prosser, Washington and in California (Pruitt, 1960; 1966; State of California, 1975; 1979) provided clear evidence that unless local environment of pans was taken into account, the estimation of crop ET (or of lake evaporation) was subject to errors of up to 35%. Fig. 4.4 presents some of the early results that illustrate the problem associated with the environment just upwind. These results came from a study involving four Class A pans located within a large fallow field, with three of them placed within various-sized circular areas, flood irrigated frequently and planted to grass. The fifth pan located within a 5-ha irrigated grass field had a minimum upwind fetch of grass or irrigated pasture land of some 200 m. Data from the one pan in the fallow field that had no surrounding grass were plotted as if having a 0.3-m fetch of grass in order to use a log-linear plot. From this and similar studies (e.g., Ramdas, 1957; Pruitt, 1960; Stanhill, 1962), Pruitt developed recommended pan K_p values for estimating ET_o and in turn evaporation from shallow water bodies. Taken into account were the effects on pan coefficients of upwind fetch (both dry and moist), mean relative humidity, and total daily wind on K_p (Jensen, 1974; Doorenbos and Pruitt, 1975; 1977; Jensen et al., 1990) (see Table 4.10 later in this chapter).

Mean monthly, seasonal, and annual Class A pan evaporation for individual stations throughout the United States have been tabulated by Farnsworth and Thompson (1982) and Farnsworth et al. (1982). The second reference, "Evaporation Atlas for the Contiguous 48 States," contains maps with isolines of pan coefficients recommended for use with the evaporation maps in estimating so-called "free water surface evaporation" (FWS) from shallow water bodies. Fig. 4.5 presents a portion of the Annual FWS Evaporation map for the 48 states from that publication. Seasonal maps for the periods May–October and November–April are included in this National Weather Service report.

2. ***Water Budget Procedure*** The water budget procedure is a simple method for computing evaporation from a water surface for monthly, seasonal, or annual time periods. All of the terms in the water

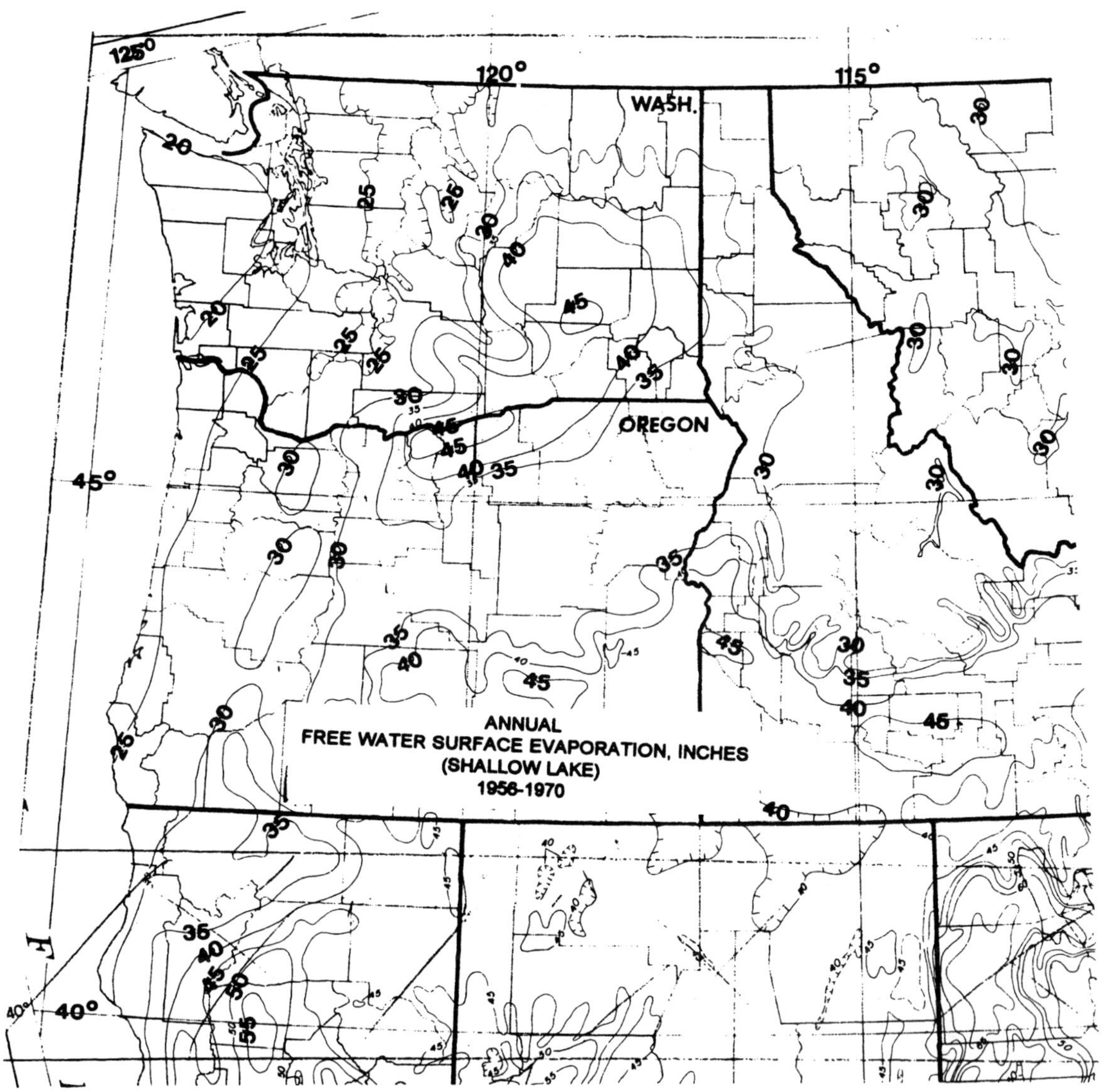

Figure 4.5—A portion (NW USA) of a Map of Estimated Free Water Surface (FWS) Annual Evaporation for the 48 States (Farnsworth et al., 1982).

balance, except for evaporation, are either measured or estimated with evaporation computed as a residual. The water balance procedure is based upon the hydrologic equation

$$Inflow = Outflow + \Delta Storage. \tag{4.49}$$

Inflow terms consist of precipitation on the water surface, runoff from the land basin, major channel inflow from outside the immediate drainage basin, ground water inflow, and diversions into a free water body (lake, reservoir, pond, etc.) from outside its basin. Outflow terms consist of evaporation from the water surface, major channel flow out of the water body, diversions out of the water body, and ground water flow from the water body. Thus, the water balance equation can be expressed in terms of water depth evaporated from the surface, E, as:

$$E = P_r + R + Q_i + G_w + D_i - Q_o - D_o - \Delta S, \tag{4.50}$$

where P_r is the precipitation on the water surface, R is the runoff inflow from the drainage basin, Q_i and Q_o are major channel flows into and out of the water body, respectively, G_w is the net ground water flow into the water body, D_i and D_o are diversions into and out of the water body, respectively, and ΔS is the change in water body storage.

For most calculations, the terms are expressed in units of depth in mm or cm over the water body surface for a specified time interval. The computations are most accurate when the evaporation is of the same order of magnitude or larger than the other terms of the water balance. The ground water component is usually the most difficult term to estimate.

Water balance data sources are the National Climatic Data Center, NOAA regional data centers and state climatologists for precipitation data, the U.S. Geological Survey, state and provincial water agencies, and flood and conservation districts for runoff, major channel flows, and diversion data.

3. *Aerodynamic Method* Aerodynamic methods are among the most widely applied to determine evaporation from lakes and large reservoirs. The basic equations were tested on Lakes Hefner and Mead in the 1950's (U.S. Geological Survey, 1954; 1958; Harbeck, 1962). The basic Lake Hefner form is

$$E = M\,(e_s - e_z)\,u_z, \tag{4.51}$$

where E is the evaporation in mm per day, M is the mass transfer coefficient, e_s is the saturation vapor pressure at the surface water temperature T_s (kPa), e_z is the vapor pressure of the air at level z (kPa), and u_z is the wind velocity at level z in m s^{-1}. The vapor pressure of the air e_z is equivalent to the saturation vapor pressure of the air at the dew point temperature and is usually computed by

$$e_z = 0.01\,RH\,(e_z^o), \tag{4.52}$$

where $e_z{}^o$ is saturation vapor pressure at air temperature (T_z) at level z (kPa), and RH is the relative humidity (%). Values of $e_z{}^o$ can be obtained from standard meteorological tables (List, 1963; Berry et al., 1973), or estimated from Equation 4.1 using water surface temperature for T_s.

The mass transfer coefficient, M (Equation 4.51), is a function of the measurement height of the meteorological measurements, the atmospheric stability, barometric pressure, roughness of the water surface, size of lake, and the density and viscosity of the air. For the Lake Hefner study (U.S. Geological Survey, 1954), M, an empirical coefficient, is 0.097 for wind, temperature, and humidity measurements at the 8 meter level and where vapor pressure is expressed in kPa. The basic mass transfer equation requires measurements of water temperature, air temperature, humidity or vapor pressure, and wind speed. The wind speed and vapor pressure gradients can be adjusted from one level to another level over the same surface by the logarithmic law (Oke, 1978; Allen et al., 1989; Jensen et al., 1990), provided that the effect of stability is negligible:

$$u_2 = u_1 \frac{\ln\left(\dfrac{z_2 - d}{z_{om}}\right)}{\ln\left(\dfrac{z_1 - d}{z_{om}}\right)}, \tag{4.53}$$

where z_1 and z_2 are the measurement heights for level 1 and level 2 (m), respectively, z_{om} is the roughness height for momentum transfer (m), and d is the zero-plane displacement (m). For many water surface applications, z_{om} can be taken as 0.0001 m and d is generally 0 for water. The wind level adjustments are relatively insensitive to the value of z_{om}. The difference between using 0.0001 m and 0.001 m is approximately 5 percent which is well within the application accuracy.

Most mass transfer research over the past 25 years has been directed towards the bulk aerodynamic equation. The bulk equations can consider the impact of atmospheric stability as well as air mass characteristics on the evaporation rate. This form of equation can be written as

$$E = 86400\ C_E\ \rho_a\ (q_s - q_z)\ u_z, \tag{4.54}$$

where E is the evaporation (mm/day), C_E is the bulk evaporation coefficient for level z, ρ_a is the density of air (kg m^{-3}), q_s is the saturation specific humidity at the water surface temperature (kg$_w$ kg$_a{}^{-1}$), q_z is the specific humidity of the air at level z (kg$_w$ kg$_a{}^{-1}$), u_z is the wind velocity at level z (m s^{-1}), 86,400 (the number of seconds in a day). The specific humidity is defined in Equation 4.5 where units of e and P are the same.

Equations 4.51 and 4.54 are nearly equivalent as shown by expressing Equation 4.54 in the form:

$$E = 86400\ C_E\ \rho_a\ 0.662\ \frac{(e_s - Ke_z)}{P - 0.378\ e_s}\ u_z, \tag{4.55}$$

where K is $(P-0.378\ e_s)/(P-0.378\ e_z)$. For typical values of P, e_a, and e_s, K approaches 1.0. The pressure correction term 0.378 e_s can also be neglected. The combined error of the two assumptions, on the order of 1 percent or less, is well within data limits. Thus, the mass transfer coefficient M can be expressed in terms of the bulk evaporation coefficient C_E as

$$M = 0.622\ (86400)\ \rho_a\ \frac{C_E}{P} = 53740\ \frac{\rho_a}{P}\ C_E, \tag{4.56}$$

which shows a dependence of M on the air density, ρ_a, and the pressure, P, as well as on the bulk evaporation coefficient C_E.

The solution of Equations 4.51 and 4.54 is dependent upon the selection of the bulk evaporation coefficient, C_E, or the mass transfer coefficient M. As mentioned earlier, the Lake Hefner study recommended a value for M of 0.097 for measurements at the 8 m level (P in kPa). Extensive research conducted during the International Field Year for the Great Lakes (IFYGL) (Aubert and Richards, 1981) indicate an average value for C_E of 0.00125 at the 8 m level. The IFYGL Atmospheric Boundary Layer Panel (Aubert and Richards, 1981) recommended $C_E = (1.5 \pm 0.3) \times 10^{-3}$ for the 3-m level, with all values in this range being equally possible. Donelan et al. (1974), using a wind set-up procedure, determined the following values of C_E at the 4 m level based on atmospheric stability: $C_E = 0.00131$ for neutral conditions, $C_E = 0.00159$ for unstable conditions, and $C_E = 0.00135$ as the best overall estimate. Converting to the 8-m level yields values for C_E of 0.00115, 0.00140, and 0.00119 for neutral, unstable, and overall conditions, respectively. Parameters to which evaporation is most sensitive are the water surface and dewpoint temperatures or humidity, but the largest potential error is introduced by the wind speed, followed by the water surface and dewpoint temperatures or humidity.

For detailed studies on large lakes it may also be desirable to use ratios of over-lake to over-land wind speed and humidity and to consider atmospheric stability (Phillips and Irbe, 1978). Quinn (1979) and Liu and Schwab (1987) discussed the estimation of C_E using wind speed and air-sea temperature differences.

*4. **Energy Balance Method*** The energy balance procedure as applied to lakes is based upon conservation of energy principles for a water body. As with the water balance procedure, the evaporation is computed as a residual. This procedure is the most data-intensive of the standard evaporation procedures, but it has wide applicability to many differing water bodies. The energy balance for a water body may be expressed as

$$\Delta Q_t = R_n - \lambda \rho_w\ E - H + Q_V - Q_w, \tag{4.57}$$

where R_n is net radiative energy (MJ m^{-2}d^{-1}), ΔQ_t is the change in lake energy storage (MJ m^{-2} d^{-1}), $\lambda \rho_w E$ is the energy utilized by evaporation (MJ m^{-2} d^{-1}), ρ_w is density of liquid water (kg m^{-3}), λ is the latent heat (MJ kg^{-1}), E is the evaporation rate (m d^{-1}), H is the energy convected from the water body as sensible heat (MJ m^{-2} d^{-1}), Q_v is the net energy advected into the water body (MJ m^{-2} d^{-1}), and Q_w is the energy advected by the evaporated water (MJ m^{-2}d^{-1}).

The net radiative energy flux at a surface was covered earlier (Equation 4.31). Also presented were methods for estimating the various components not measured. A common method for calculating R_n, with revisions to relate to earlier terminology, can be expressed as:

$$R_n = (1 - \alpha)\, R_s + (1 - \alpha_t)\, R_L \downarrow - \epsilon\, (4.90 \times 10^{-3})\, (T_s + 273.16)^4 \tag{4.58}$$

for R_n, R_s, and R_L in MJ m^{-2} d^{-1} where α is the solar albedo (0.07–0.15), dependent upon surface conditions (Bolsenga and Vanderploeg, 1992), α_t is the long wave albedo [equivalent to $(1 - \epsilon)$ or 0.03 (WMO, 1970)], 4.90×10^{-9} is the Stefan-Boltzmann constant (MJ m^{-2} K^{-4} d^{-1}), ϵ is the long wave emissivity [0.92-0.97, with $\epsilon = 0.97$ typically used (Oke, 1978)], and T_s is the water surface temperature (°C).

In addition to procedures covered earlier, the short-wave and long-wave radiation may be measured or can be estimated by a number of empirical equations (Croley, 1989; Derecki, 1976; Gray, 1970; Sellers, 1965). As the sensible heat (H) and the latent heat (λE) in Equation 4.57 are not measured, they are usually expressed in terms of the Bowen ratio β, which is the ratio of sensible to latent heat. The Bowen ratio (see also Equation 4.78) is expressed here as

$$\beta = \frac{H}{\lambda E} = \gamma \frac{(T_s - T_z)}{(e_s^o - e_z)}, \tag{4.59}$$

where β is the Bowen ratio, T_s is the water surface temperature, °C, T_z is the air temperature at level z, °C, $e_s{}^o$ is the saturation vapor pressure at T_s, kPa, e_z is the vapor pressure at level z, kPa, and γ is the psychrometric constant (Equation 4.13), kPa °C^{-1}. Parameter β is dimensionless.

The energy advected by the evaporating water, Q_w, (Equation 4.57) i.e., by the removal of water from the system, although less than 1% of λE, is accounted for by:

$$Q_w = \rho_w\, C_w\, E\, (T_s - T_b), \tag{4.60}$$

where ρ_w is the density of liquid water (kg m^{-3}), C_w is the specific heat of water (MJ kg^{-1} °C^{-1}), T_s is the temperature of the evaporated water (°C), T_b is a base temperature (assumed to be 0 °C), and E is the evaporation rate (m day^{-1}). Equation 4.60 accounts for the warming of water from the base temperature to the water surface evaporating temperature.

The net energy advected into a water body, Q_v, is composed of the energy transported into the water body by runoff, precipitation, and out-of-basin inflows. The energy advected out of a water body is composed of any outflows from the lake and the melting of solid precipitation at the lake surface. This is expressed on an energy-per-unit- area-per-time step basis for a base operating temperature equal to 0°C as:

$$Q_v = (V_r\, T_r + V_p\, T_p + V_i\, T_i - V_o\, T_o)\, \frac{\rho_w\, C_w}{A} - \frac{V_{sp}\, \rho_w\, \lambda_f}{A}, \tag{4.61}$$

where V_r is the volume of runoff (m^3), T_r is the temperature of the runoff (°C), V_p is the volume of precipitation (m^3), T_p is the temperature of the precipitation (°C), V_i is the volume of out-of-basin inflow (m^3), T_i is the temperature of the inflow (°C), V_o is the volume of outflow (m^3), T_o is the temperature of the outflow (°C), A is the surface area of the water body (m^2), V_{sp} is the volume of solid precipitation (m^3), ρ_w is the water density (kg m^{-3}), and λ_f is the latent heat of fusion (MJ kg^{-1}). The energy advected into a lake is usually relatively small compared with other energy terms and may be neglected in many studies.

The evaporation is then computed from the relationship

$$E = \frac{R_n - \Delta Q_t + Q_v}{\rho_w\, [(1 + \beta)\, \lambda + C_w\, (T_s - T_b)]}, \tag{4.62}$$

where E is the evaporation rate (m day^{-1}), R_n is net radiation (MJ m^{-2} day^{-1}), ρ_w is the density of water (kg m^{-3}), λ is the latent heat of vaporization (MJ kg^{-1}), and β is the Bowen ratio. The latent heat of vaporization can be found in standard meteorological tables (List, 1963) or computed by Equation 4.17. It should be noted that the computation of E becomes unstable as the Bowen Ratio approaches -1.

Data to use the energy balance method can be obtained from the National Climatic Data Center for radiation components from satellite analysis, municipal water intakes, or in situ measurements for water temperatures and from thermistor strings or water intakes for the changes in energy storage. A detailed energy budget for Lake Ontario is given by Aubert and Richards (1981). This includes an evaluation of all energy terms.

5. Combination Methods For many applications, a combination of the aerodynamic and energy balance procedures is highly desirable. The most common of the combined equations was derived by Penman (1948; 1956) which in its general form is written as:

$$\lambda E = \frac{\Delta (R_n - G) + \gamma E_a}{\Delta + \gamma}, \tag{4.63}$$

where λE is the latent heat of evaporation (MJ m^{-2} d^{-1}), γ is the psychrometric constant (kPa °C^{-1}), (Equation 4.13), Δ is the slope of the saturation vapor pressure curve at temperature T_a (kPa °C^{-1}), (Equation 4.2), $(R_n - G)$ is the net radiation minus soil or water heat flux density (MJ m^{-2} d^{-1}), and E_a is a bulk aerodynamic expression containing an empirical wind function.

$$E_a = 6.43\,(a_w + b_w\,u_z)\,(e_e^o - e_z), \tag{4.64}$$

where a_w and b_w are empirical wind function coefficients and u_z is wind speed at the z height (m s^{-1}). $e_z{}^o$ and e_z are defined in Equation 4.52. Various values for the empirical coefficients a_w and b_w have been proposed (Penman, 1948; Cuenca and Nicholson, 1982). Penman (1963) proposed $a_w = 0.5$ and $b_w = 0.54$ for open water for $z = 2$ m and for $e_z{}^o$ computed using mean daily air temperature.

An important point to consider in use of combination equations (and others as well) is that the use of weather data collected over a typical weather station on land cannot be expected to give as good an accuracy as computations using data collected over the water surface. First, the wind at a height z over aerodynamically smooth water will be significantly higher than wind at height z over a typical weather station site. Temperature and humidity may also be quite different over the two sites. Further discussion of effects of surfaces on weather data is included later in the section on Evapotranspiration from Land Surfaces. This section also includes suggestions for adjusting non-characteristic humidity, temperature, and wind data according to the evaporating surface.

6. Simulation Studies For many complex studies, such as determining the impacts of potential climate change or variability on lake evaporation, it is necessary to combine the energy balance and mass transfer procedures (Croley, 1989; Hostetler and Bartlein, 1990). In this type of approach, the mass or bulk transfer equations are substituted for λE and H in the energy balance and solved for the water temperature T_s and the resulting energy stored in the lake ΔQ. The evaporation is determined using the computed T_s in the bulk transfer equation. Depending upon the accuracy desired and the overlying assumptions, many simplifications may be used to simplify the procedure. This type of technique is currently widely applied in many water resource and climate-change studies.

V. EVAPOTRANSPIRATION FROM LAND SURFACES

A. Introduction

Perhaps the greatest challenge in estimating evapotranspiration from land surfaces is in predicting ET in situations where it is much less than the potential rate due to low soil water availability. Generally, when estimating ET in areas where precipitation is greater than the potential ET rate for vegetative cover

(ET_c), or where full irrigation is practiced, ET can be readily estimated using the reference ET concept along with crop coefficients. It can also be estimated by direct application of the Penman-Monteith or some other resistance-based method using "well-watered" surface resistances (r_s); however, when precipitation is less than ET_c for extended time periods, actual evapotranspiration, ET_a, usually falls below the potential ET for the particular surface (ET_c) and becomes more difficult to estimate.

Usually, a soil water balance is required to estimate ET for water-short areas (where precipitation is less than ET_c). Two methods are discussed in this section which can be used to reduce the evapotranspiration estimate according to available soil moisture. One of these methods applies to $K_c\ ET_o$ computations (crop coefficient approach). The other method applies to direct application of the Penman-Monteith method.

Other approaches to estimating ET from land surfaces during water shortages include making direct measurements of ET. The challenge here is in extrapolating point or localized measurements to large spatial scales and in extrapolating measurements to other time periods. ET measurement can be difficult and time-consuming and is therefore generally used to develop and calibrate weather-based ET estimation methods which are then applied to larger scales and to different time periods.

This section of the chapter includes methods developed for well-watered agricultural surfaces summarized from FAO-24 (Doorenbos and Pruitt, 1977) and from ASCE Manual 70 (Jensen et al., 1990) with extensions to natural vegetation which may be short of water.

B. Volumetric Measurements for Estimating Land Surface ET

1. Introduction Because it is most desirable to know what the actual ET is from a field, watershed, or region, all ET methods must be considered as providing only estimates of the average expected ET for the area under consideration. As might be expected, the larger and more diverse the area, the more uncertain the estimate.

Fortunately, since the process of evaporation is to a large extent controlled by solar radiation (or more directly, net radiation), the variation of ET within areas with a uniformly effectively-moist surface can be expected to be rather minor, barring the presence of advective edge effects. This is normally the case for some time following extensive precipitation in an area or following irrigations in a farm field. As soil moisture becomes depleted in areas receiving and/or storing less than the average water received in a region, variability will increase. The uniformity of precipitation or water applied by irrigation would also be a factor. The problem becomes difficult when the areas under consideration, e.g., watersheds or regions, involve a wide variety of vegetation species, vegetation heights, aerodynamic roughness, leaf development, rooting depth, soil type, and differing responses to soil moisture levels and weather conditions.

2. The Soil Water System A unit volume of soil contains solids (V_s), water (V_w), and air and water vapor (V_{av}), and can be expressed as:

$$V = V_s + V_w + V_{av}. \tag{4.65}$$

The soil water content is commonly expressed on a volume basis, θ, or on a mass basis, w, usually expressed as $m^3\ m^{-3}$ or $kg\ kg^{-3}$, respectively, or in percent:

$$\theta = \frac{V_w}{V} \tag{4.66}$$

$$w = \frac{m_w}{m_s}, \tag{4.67}$$

where m_w and m_s are the masses of water and soil, respectively. In practice, the soil water content is determined by drying at 105°C.

Conversion of soil water content on a mass to a volumetric basis requires knowing the bulk density of the dry soil matrix, ρ_b, where $\rho_b = m_s/V$, from which

$$\theta = \frac{\rho_b}{\rho_w} w, \tag{4.68}$$

where ρ_w is the density of liquid water.

The volumetric water content represents the equivalent depth of water per unit soil depth. Prior to the use of the neutron moisture probe, which is calibrated to indicate θ, most field studies involved gravimetric determination of w. To determine the volumes of water involved, Equation 4.68 was used, and in most cases ρ_b, once determined during a study, was considered to remain constant at a given depth throughout the season or study. This assumption is reasonable below the plow depth, but can result in substantial measurement error in the total soil water in the plowed layer. The error is not great for determining changes in soil-water content over 5- to 15-day periods; however, ρ_b should be adjusted for water content changes in swelling and cracking soils.

3. Soil Water Balance As applied to watersheds, the hydrologic water balance usually involves estimating water yields, where one of the components being estimated independently is the ET. On the other hand, for small watersheds where inter-basin ground water flow is highly restricted and runoff can be measured, ET can be determined from the hydrologic balance if average watershed precipitation is known. In recent years, this approach has been vital in development of water and salt balance data for large basins, especially in relation to irrigation projects. An example is the water balance evaluation of the large Imperial Valley in California (Oster et al., 1984).

The basic soil water balance equation for a specific time period, Δt, can be written as:

$$\Delta\theta \cdot z_s = P - Q_r - ET_a - DP + GW, \tag{4.69}$$

where $\Delta\theta$ is the change in mean moisture content of the soil profile (m^3 m^{-3}), z_s is the depth of the soil profile (mm), P is total precipitation + irrigation depth (mm), Q_r is surface runoff (mm), ET_a is actual evapotranspiration (mm), DP is deep percolation flux density at the lower boundary of the soil profile (mm), and GW is upward movement of ground water at the lower boundary of the soil profile (mm). Generally, the soil profile of interest is the plant root zone which, by definition, is the upper portion of the soil where all moisture extraction by plant roots occurs. In this case, DP and GW in Equation 4.69 are flux densities into and out of the bottom of the root zone.

$\Delta\theta$ can be defined as:

$$\Delta\theta = -\frac{\int_0^{z_r} \theta_{1(z)}\, dz - \int_0^{z_r} \theta_{2(z)}\, dz}{z_r}, \tag{4.70}$$

where $\theta_{1(z)}$ is volumetric moisture content at depth z (m^3 m^{-3}) at $t = 1$, $\theta_{2(z)}$ is volumetric moisture content at depth z (m^3 m^{-3}) at $t = 2$, and z_r is the effective depth of the root zone (mm). Equation 4.69 can be rearranged in terms of Equation 4.70 to predict total soil moisture in the root zone or to greater depths, if necessary, as:

$$\int_0^{z_r} \theta_{2(z)}\, dz = \int_0^{z_r} \theta_{1(z)}\, dz + \Delta\theta \cdot z_r = \int_0^{z_r} \theta_{1(z)}\, dz + P - Q_r - ET_a - DP + GW. \tag{4.71}$$

Reduction of ET_a below potential levels due to soil water deficiency is discussed in a later section.

Generally, deep percolation flux density is estimated according to the drained upper limit, θ_{DU}, of the soil. θ_{DU} is equivalent to the older term of field capacity and represents the moisture content to which the soil will drain under a unit hydraulic gradient (gravity) 2 to 3 days after wetting. θ_{DU} represents the upper limit of the soil's long-term ability to retain water. Average values and ranges of θ_{DU} are given in Table 4.8 for general soil texture classes along with values for the lower limit of extractable water (wilting point), θ_{LL}, and total retained water available for plant extraction (available water (AW)). Available water is calculated as $AW = \theta_{DU} - \theta_{LL}$. One item to note from the data in Table 4.8 is that even though values for θ_{DU} and θ_{LL} vary widely with soil texture, values for available water vary relatively little, especially for moderately coarse to fine textured soils. Therefore, if θ_{DU} can be determined from field observations of soil moisture one to two days following significant rain or irrigation, then θ_{LL} can be approximated as $\theta_{DU} - AW$, where AW is taken from Table 4.8. When possible, specific values of θ_{DU} and θ_{LL} should be determined from field studies (Ratliff et al., 1983).

When the calculation time step for Equation 4.71 is one day, DP is often assumed to occur if total θ of the root zone is greater than θ_{DU} on any day. In this case:

$$\begin{aligned} &\text{If } \int_0^{z_r} \theta_{2(z)}\, dz > \theta_{DU} \cdot z_r, \\ &\quad \text{then } DP = \int_0^{z_r} \theta_{2(z)}\, dz - \theta_{DU} \cdot z_r \\ &\quad \text{and } \theta_{2(z)} = \theta_{DU} \text{ for all } z_r, \\ &\quad \text{otherwise } DP = 0. \end{aligned} \tag{4.72}$$

One may wish to delay application of Equation 4.72 until 2 days after a major precipitation or irrigation event to account for some ET extraction from moisture above θ_{DU} during the first two days while the soil profile is draining to the drained upper limit.

Generally, when solving the soil water balance equation for ET, one should sample θ to depths well below the root zone and attempt to control the influx of P and irrigation to reduce the occurrence and rate of DP. Often, periods with little or no precipitation are selected for evaluation. Occasionally, changes in soil texture below the root zone or the presence of bedrock can hydraulically reduce the rate of deep percolation and improve measurement accuracy.

TABLE 4.8. General Averages and Ranges of Soil Water Parameters for Agricultural Soils (Jensen et al., 1990).

	WATER CONTENT, VOLUME BASIS, $m^3\ m^{-3}$					
	Drained upper limit		Lower limit of extractable water		Available water	
Texture Class	Average	Range	Average	Range	Average	Range
Coarse						
Sand	0.12	0.07-0.17	0.04	0.02-0.07	0.08	0.05-0.11
Loamy Sand	0.14	0.11-0.19	0.06	0.03-0.10	0.08	0.06-0.12
Moderately Coarse						
Sandy Loam	0.23	0.18-0.28	0.10	0.06-0.16	0.13	0.11-0.15
Medium						
Loam	0.26	0.20-0.30	0.12	0.07-0.16	0.15	0.11-0.18
Silt Loam	0.30	0.22-0.36	0.15	0.09-0.21	0.15	0.11-0.19
Silt	0.32	0.29-0.35	0.15	0.12-0.18	0.17	0.12-0.20
Moderately Fine						
Silty Clay Loam	0.34	0.30-0.37	0.19	0.17-0.24	0.15	0.12-0.18
Fine						
Silty Clay	0.36	0.29-0.42	0.21	0.14-0.29	0.15	0.11-0.19
Clay	0.36	0.32-0.39	0.21	0.19-0.24	0.15	0.10-0.20

Situations where saturated conditions (ground water) exist near the root zone should be avoided. In the absence of a shallow water table ($GW = 0$) and with adequate sampling depth, a record of soil moisture change with time can yield reliable measurements of *ET* at the measurement site.

The careful use of a neutron probe in most cases can provide relatively accurate data, although gravimetric sampling can often provide data of adequate accuracy, e.g., in *ET* studies or for irrigation scheduling programs. The difficulty with either of these procedures (neutron probe and gravimetric sampling) is in estimating the moisture flux density (*DP* or *GW*) at the bottom of the measured profile. Often this difficulty limits the usefulness of applying the soil water balance equation to determine *ET*.

Usually soil moisture measurements are made at discrete depth increments, for example, each 0.15 m. Total soil moisture in the profile can be estimated from discrete measurements as:

$$\int_0^{z_r} \theta_z \, dz = \sum_{i=1}^{n} d_i \, \theta_i, \tag{4.73}$$

where θ_i is the volumetric soil water content representing layer i ($m^3\ m^{-3}$), d_i is the depth (thickness) of layer i (mm), and n is the total number of layers measured between 0 and z_r.

Table 4.9 gives an example of results for a 6½-day period when *ET* for a dense crop of ladino clover was estimated from a well-replicated (18 sites within a 0.2-ha plot) gravimetric soil sampling study. Even with 18 sampling sites and nearly ideal soil and underlying conditions (sampling was from the surface to bedrock), the standard deviation of estimated *ET* was some 13% of the indicated mean depletion for the period, although the standard error of the mean *ET* was only 3.1% of the mean. One might expect a lower standard deviation had the measurements been made with a neutron probe at the same number of neutron access sites. At this particular site, however, the variation in the depth of this relatively shallow

TABLE 4.9. An Example of a Gravimetric Soil Sampling Program to Determine *ET*. Washington State University Irrigation Experiment Station, Prosser, WA. (Pruitt, 1955).

Sampling Site	Ave. Depth z (m)	June 4 w (%)	June 4 θ·z (mm)	June 10 w (%)	June 10 θ·z (mm)	Difference 6/4-6/10 Δθ·z (mm)
1	1.16	12.9	205	8.5	135	70
2	1.18	15.8	256	12.2	198	58
3	1.31	15.2	274	12.1	218	56
4	1.18	15.3	248	11.0	178	70
5	1.19	14.6	237	11.0	179	58
6	1.20	15.5	255	11.6	191	64
7	1.15	14.8	233	11.1	175	58
8	1.19	16.7	272	12.5	203	68
9	1.22	16.2	270	12.3	205	65
10	1.08	17.0	250	12.3	181	69
11	1.09	17.9	268	11.8	177	91
12	1.11	17.0	259	12.7	194	66
13	1.03	17.3	243	12.6	177	66
14	1.02	18.7	262	12.8	179	83
15	1.02	19.8	276	15.2	212	64
16	0.99	14.1	191	9.2	124	67
17	1.00	15.1	208	9.4	129	78
18	1.01	16.3	226	11.9	165	61
Ave.			246		179	67
					Std. Dev.	8.75
					100*(8.75)/67 =	13.0%

soil (0.99-1.31 m) presents a problem in interpreting both the uppermost and lowest-layer readings. This would not be a problem for deeper soils, although evaluation of the upper 0.2–0.3 m probe readings is problematic.

Tensiometers, soil moisture electrical resistance blocks, or heat dissipation electronic sensors can be valuable in computing a soil moisture balance. They give a record of matrix potential (or soil moisture tension) and, when soil moisture characteristic curves are available, can provide a means of following volumetric moisture content changes. For an in-depth discussion of some of these methods, a paper by Campbell and Campbell (1982) is recommended. Bertuzzi et al. (1994) present a recent detailed analysis of various sampling strategies for estimating *ET*.

a. Some Typical Problems. Estimating *ET* by gravimetric, neutron scattering, or water potential measurement of soil moisture can involve inaccuracy due to such factors as:

(1) Variability of bulk density and moisture holding characteristics of the soil;
(2) Possible deep percolation losses (or gains by capillary rise);
(3) The need for neutron probe calibration between soil types;
(4) (For some vegetation types) The difficulty of obtaining the samples or taking the readings without significantly altering the plant cover at the sampling site. *ET* from trampled plants can be drastically reduced as compared to *ET* by plants in a pristine condition. Fig. 4.6 illustrates the considerable effort one agency took to avoid this problem when taking neutron moisture readings.
(5) A related serious problem, the altering of density, aeration, and infiltration characteristics of the surface soil. Water intake and/or aeration at and near the sampling site can be quite different from general field areas, and plant growth is very likely to be affected. This may reduce soil moisture extraction, leading to an *ET* estimation unrepresentative of actual field *ET*.
(6) Minimum time steps for estimating *ET* with the soil water balance equation should be greater than or equal to 7 days so that the magnitude of estimated ET is significantly greater than the random errors associated with measurements of moisture content.

Figure 4.6—A Technician Taking a Neutron-Probe Reading Using a Light, Above-crop Platform to diminish Trampling of a Crop of Alfalfa and Compaction of Soil at the Access Tube Site, San Joaquin Valley, CA (courtesy of N. MacGillivray, California Department of Water Resources, Fresno).

4. *Tanks and Lysimeters*

a. Some Positive Aspects. Highly-sensitive weighing or floating lysimeters offer the only absolute method of precisely measuring water loss from soil and crop canopy surfaces. Because of this, lysimeters have had very important input in the development and testing of the more theoretical micrometeorological methods for estimating *ET*. Several figures presented earlier offer evidence of the important role lysimeters have played in providing accurate half-hourly or hourly *ET* data. The water lost even from very simple lysimeters may be precisely determined over 24-hour or longer periods. For detailed coverage of lysimeter types and designs, the following references are recommended reading: WMO (1966), Aboukhaled et al. (1982), and Allen, et al. (1991a). The latter reference is the proceedings of an international symposium on lysimeters and contains not only the descriptions of many lysimeter types in use around the world, but the results of research involving many of them.

b. Some Problems. It should be realized that any lysimeter, no matter how precise a measuring system, provides a measure of *ET* from a very limited plant sample size. Therefore, one can only expect to obtain reasonable estimates of actual field *ET* from lysimeters when plant density, height, and leaf area of vegetation inside the lysimeter are close to that of the surrounding field. When plant density or leaf area differs, then even though the results may be valuable in plant physiology or other studies, the actual *ET* obtained will generally not represent actual field *ET*. Wind disturbance of readings can also produce significant errors.

Differences between inside lysimeter and outside field conditions such as soil characteristics, moisture availability, fertility, soil temperature, and rooting characteristics can cause differences in crop growth, plant cover, leaf temperature and maturation and thus produce differences in < I > between field and lysimeter. Differences in plant height in particular can produce erroneous data with the errors related significantly to the lysimeter area; the smaller the lysimeter size, the larger the possible error. Differences in maturation are typical in fairly shallow lysimeters unless irrigation is continued close to the season's end.

Some of the seriously flawed lysimeter results have been produced by a so-called "bloom effect" on small lysimeters where the area of exposed plant canopy has exceeded the assumed effective area of the lysimeter. Tanner (1967) provided an excellent discussion on lysimeters, and was probably one of the earliest to recognize how serious this problem can be. More recent coverage of this and other aspects of lysimetry was detailed by Pruitt and Lourence (1985) and Allen et al. (1991b). Examples are frequent in the literature of so-called crop or pan coefficients ranging from 1.3 to 1.8 or higher. H.L. Penman's early comment may be a bit unrealistic (crop evaporation should never exceed evaporation from an open water surface); however, any pan coefficient greater than 1.1 for weekly or longer periods for a well-exposed, unscreened Class A pan should lead to a strong suspicion that the *ET* determinations (whether by lysimetry or meteorological techniques) are in error, or that they represent conditions other than continuous expanses of vegetative surfaces. The same cautionary statement could be applied in the case of finding crop coefficients greater than 1.1 $ET_{alfalfa}$ or 1.3 ET_{grass}. This point is discussed further in the section on cover or crop coefficients.

In recent years, a number of sensitive lysimeters capable of hourly *ET* determination have been installed in forests (Fritschen et al., 1973; Reyenga et al., 1988; and Dunin et al., 1985). In spite of the problems noted above, at least in some cases, the results with lysimeters of around 10 m^2 have been shown to provide *ET* measurements for forest trees in good agreement with Bowen ratio ET determination (Dunin et al., 1985; 1991).

Another serious problem with lysimeters relates to foot and even farm machinery traffic in and around lysimeters as they are serviced. This can be especially serious with grain or other similar, closely-spaced tall crops.

Crop height differences inside and outside a lysimeter, as produced by differing degrees of lodging, can be a problem. Again, the smaller the lysimeter, the greater the possible error in estimating actual field losses. Pruitt and Lourence (1985) provide an example that illustrates this problem, although in this case, with 29 m^2 lysimeters, errors of only 5 to 10 percent were noted for a barley crop in spite of extreme differences between lodging inside and outside the lysimeter.

In general, weighing lysimeters are necessary to accurately determine *ET*. Nonweighing lysimeter types, where *ET* is determined by observing changes in the water table level, are difficult to interpret due to changes in soil moisture above the water table caused by water extraction by plant roots and changes in the capillary fringe above the water table with changing temperature (Hill and Allen, 1991). Hydraulic weighing lysimeters are often plagued by hysteresis of the hydraulic measurement systems and thermal problems (Kruse and Neale, 1988), although such problems are likely insignificant for weekly or longer determinations.

With recent improvements in electronic load cells, low-cost electronic weighing lysimeters, which are suitable for daily measurements of *ET*, have been developed for remote operation (McFarland et al., 1983; Allen and Fisher, 1990). Some non-weighing lysimeters have been developed where the lysimeter acts merely as a hydraulic barrier to water movement into or out of the lysimeter and the change in soil moisture with time is measured by keeping records of all water added to the lysimeter (irrigation and precipitation), as well as any drainage or runoff. In more recent times, neutron probes have been used to obtain better running records of losses by *ET*. Just a few examples of the literally hundreds of such lysimeters around the world are those described by Harrold and Dreibelbis, 1958; Van Hylckama, 1968; Niel and Zunino, 1971; Sarraf and Bovée, 1973; Boman, 1991; and Evans et al., 1991.

Another difficulty with lysimeters is in applications with non-well-watered vegetation (e.g., in natural rainfall areas where precipitation is less than *ET*). In these applications, differences in soil textures, soil profile depth, fertility, salinity, and thermal conditions inside lysimeters can make it difficult to control moisture conditions and, consequently, *ET* inside lysimeters to match conditions directly outside the lysimeter. In most cases, undisturbed, natural monolith soil profiles are required in lysimeters to accurately measure natural *ET* in semiarid and arid regions (Armijo et al., 1972; Schneider and Howell, 1991).

5. *Evaporimeters*

a. Introduction. The use of evaporation pans to estimate evaporation from water bodies was introduced earlier along with some of the factors that can produce a wide range of results as related to the local environment, placement, size of pan, etc. Pans and atmometers have been used more widely in estimating crop water use than in lake evaporation estimation. Pruitt and Doorenbos (1977a) listed fifteen references that involved some eight different countries where the pan evaporation method has been used significantly in irrigation scheduling schemes. In terms of the percentage of irrigable land within a country where evaporation pans have played an important part, the country of Israel probably leads the list. Stanhill (1962) provided an early reference for Israel, and Shalevet et al. (1981) later summarized ET/E_p relationships for many crops.

b. ET_o From Pan Evaporation. Pan evaporation data can be used to estimate ET_o from the simple proportional relationship of:

$$ET_o = k_p \; E_p, \tag{4.74}$$

where k_p is dependent on the pan type involved and a number of other factors as discussed earlier. Pruitt and Doorenbos (1977a) gave the background of the development of the FAO-24 Pan Evaporation method which was first published in Jensen (1974) and subsequently by Doorenbos and Pruitt (1975; 1977), Burman et al. (1980), Jensen et al. (1990), and Shuttleworth (1992). The method was based on data from many studies, primarily studies concerned with the effect of pan type, local environment, and general weather conditions on k_p (Ramdas, 1957; Pruitt, 1960; 1966; Stanhill, 1962).

The FAO Pan method provided tables of recommended k_p's as a function of pan type (Class A and Colorado sunken), upwind fetch of both dry- and wet-surface environments, mean relative humidity, and daily wind run. Table 4.10 reproduces the original tables for the Class A pan as developed by Pruitt in

TABLE 4.10. Suggested Values of k_p for Relating Evaporation from a USA Class A Pan to *ET* from 0.08-0.15 m Tall, Well-watered Grass Turf (ET_o). Based on Original FAO Analysis Leading to FAO-24 Pan Method (Doorenbos and Pruitt, 1975; 1977).

	CASE A: (Pan surrounded by Short, Green, Growing Crop, preferably grass or a wet soil surface)				**Case B:** (Pan surrounded by Dry, Bare Soil or Dry, Dormant Grass)			
		(Rhmax + Rhmin)/2				(RHmax + RHmin)/2		
2-m Wind km/day	Wet upwind fetch,m	Low < 40%	Med 40-70%	High > 70%	Dry upwind fetch,m	Low < 40%	Med 40-70%	High > 70%
Light	0	0.55	0.64	0.73	0	0.7	0.8	0.85
< 175 km/d	10	0.66	0.75	0.82	10	0.6	0.7	0.8
(< 2 m/s)	100	0.74	0.81	0.85	100	0.55	0.65	0.75
	1000	0.77	0.83	0.87	1000	0.5	0.6	0.7
Moderate	0	0.50	0.58	0.65	0	0.65	0.75	0.8
175-425 km/d	10	0.60	0.68	0.75	10	0.55	0.65	0.7
(2-5 m/s)	100	0.66	0.73	0.78	100	0.5	0.6	0.65
	1000	0.70	0.77	0.81	1000	0.45	0.55	0.6
Strong	0	0.45	0.52	0.59	0	0.6	0.65	0.7
425-700 km/d	10	0.52	0.60	0.67	10	0.5	0.55	0.65
(5-8 m/s)	100	0.58	0.66	0.71	100	0.45	0.5	0.6
	1000	0.62	0.70	0.75	1000	0.4	0.45	0.55
Very Strong	0	0.40	0.46	0.52	0	0.5	0.6	0.65
> 700 km/d	10	0.45	0.53	0.61	10	0.45	0.5	0.55
(> 8 m/s)	100	0.50	0.59	0.65	100	0.4	0.45	0.5
	1000	0.55	0.63	0.68	1000	0.35	0.4	0.45

the FAO analysis, rounded off to the nearest 0.01 rather than to 0.05. Fig. 4.7 presents sketches explaining an idealized situation for both wet (CASE A) and dry (CASE B) upwind fetches.

Many weather stations do not easily fit into the simplified classifications suggested in Fig. 4.7. One exception is illustrated by a weather station planted to Bermuda grass that is dormant during several months of the year (CASE B) but presents a green, growing crop during other months (CASE A). Also, for weather stations located in bare soil environments, the CASE B classification might apply very well during dry seasons of the year, but CASE A is more appropriate during consistently rainy periods. Another complicated situation was envisioned by Doorenbos and Pruitt (1977), who suggest that when pans are placed in a small enclosure surrounded by tall crops, the coefficients for dry, windy climates should be increased by up to 30%, but for calm, humid conditions, a 5 to 10% increase should be sufficient to compensate for wind shielding of the pan by the tall crop.

In addition to the coefficient variation with wind and humidity, there is also an interaction with radiation intensity. Since smaller coefficients under drier and windier conditions are largely the result of greater response of pans to sensible heat advection as compared to crops (short, smooth ones), the relative

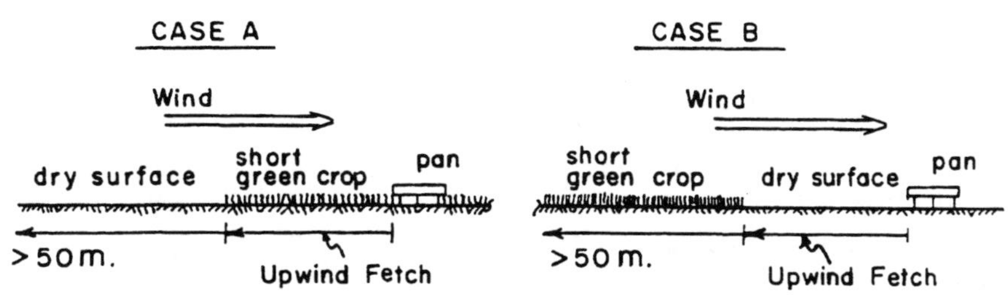

Figure 4.7—Example Environments of Surroundings of Evaporation Pans for FAO-24 Method (Doorenbos and Pruitt, 1975; 1977).

effect (hence, coefficients) is greater under the lower-radiation conditions of fall, winter, and spring than under the higher radiation of midsummer (Jensen et al., 1990).

The coefficients of Table 4.10 apply to National Weather Service (NWS) Class A pans mounted on a standard wooden platform such that the top of the pan is about 0.4 m above the surrounding ground level. For the CASE A siting, the grass should be irrigated during dry spells, frequently enough to insure low leaf resistances at all times. Also presumed is the maintenance of a water level within a zone ranging from 0.05 to 0.075 m below the top of the pan. If the pan is protected from birds, e.g., with 12.5-mm (0.5-in) mesh screen, coefficients should be increased 5 to 10% (Stanhill, 1962). Pans made of monel metal, or older galvanized metal pans which have lost their original reflectance characteristics, may need a reduction of k_p of up to 5%. In general, the water should be kept clean although turbidity differences appear to produce little difference in evaporation from Class A pans. The need to avoid contamination by oil-related products should be evident.

For those using computer programs in estimating ET_o from the FAO Pan method, polynomial equations based on the data in Table 4.10 were developed by Allen and Pruitt (1991) and first published in Jensen et al. (1990, App. E). The equation for CASE A estimation of k_p is:

$$k_p = 0.108 - 0.000331\, u_2 + 0.0422 \ln (Fetch) + 0.1434 \ln (RH_{mean}) - 0.000631\, [\ln (Fetch)]^2\, [\ln (RH_{mean})] \tag{4.75}$$

where u_2 is wind run at 2-m height in km d^{-1}, *Fetch* is the upwind fetch in meters, and RH_{mean} is the average of maximum and minimum relative humidity in percent. For CASE B situations (dry fetch upwind), the suggested equation is:

$$\begin{aligned} k_p = {} & 0.61 + 0.00341\, RH_{mean} - 0.00000187\, u_2\, RH_{mean} - 0.11 \times 10^{-6} u_2\, Fetch \\ & + 0.0000378\, u_2 \ln (Fetch) - 0.0000332\, u_2 \ln (u_2) \\ & - 0.0106\, [\ln (u_2) \ln (Fetch)] + 0.00063\, [\ln (Fetch)]^2\, [\ln (u_2)]. \end{aligned} \tag{4.76}$$

Limits should be imposed on the above equations to avoid unreasonable estimates. They are:

$$30 <= RH_{mean} <= 84\%\ ;\ 84 <= u_2 <= 700\ km\ d^{-1}\ and\ 1 <= Fetch <= 1000\ m.$$

Thom et al. (1981) developed a "reasonably physically-based" approach using Class A pan evaporation which would avoid the obvious high degree of empiricism in the FAO-Pan method. It may well become a valuable method, although the paper did not deal specifically with upwind fetch conditions and distances, factors of importance in semi-arid to arid conditions.

C. Energy Balance and Mass Transfer Methods

1. *Bowen Ratio Method* The Bowen Ratio Energy Balance (BREB) method rearranges the energy balance equation (Equation 4.40) in order to cancel aerodynamic transport terms. This permits determination of λE by measuring air temperature and vapor pressure at two elevations above the surface in addition to R_n and G. The BREB equation for application to ET from vegetation is:

$$\lambda E = \frac{R_n - G}{1 + \beta}, \tag{4.77}$$

where the Bowen ratio β (Bowen, 1926) is the ratio of H to λE. The aerodynamic transport equations for H and λE are expressed later in this section as Equations 4.91 and 4.90. Under most measurement conditions, the transport coefficients (inverse of aerodynamic resistances) can be considered to be equal for H and λE (Sellers, 1965; Tanner, 1968; Dyer, 1974). Therefore, β can be expressed in a finite difference form as:

$$\beta = \frac{H}{\lambda E} = \frac{c_p P}{\lambda \epsilon} \frac{(\Delta T + \Gamma \Delta z)}{\Delta e} = \gamma \frac{[T_2 - T_1 + \Gamma (z_2 - z_1)]}{e_2 - e_1}, \tag{4.78}$$

where T_2 and e_2 are air temperature and vapor pressure at height z_2, and T_1 and e_1 are air temperature and vapor pressure at height z_1. γ is the pyschrometric constant (Equation 4.13) and Γ is the adiabatic lapse rate, generally taken as 0.01°C m^{-1}. Γ can generally be neglected if the distance between the Bowen ratio measurement heights is less than 2 m. Units of c_p, P, and λ in Equation 4.78 should provide for units of γ in kPa °C^{-1} so that, for T in °C and e in kPa, β is dimensionless.

Generally, the z_1 height should be at least 0.3 m above the crop canopy for a smooth, dense canopy and should be placed further above the canopy for tall, sparse crops where microscale turbulence among individual plants can disturb exponentially shaped temperature and vapor profiles. Generally, the z_2 height is 1 to 2 m above z_1.

Bowen ratio systems are available commercially and are in widespread use for measuring λE above both agricultural and natural vegetation. Generally $T_2 - T_1$ is measured using differentially wired fine-wire thermocouples, which may be naturally or artificially aspirated, or using precision resistance thermometers that are shielded from the sun and require artificial aspiration. Vapor pressure is generally measured at each height using aspirated wet- and dry-bulb psychrometers or electronic hygrometers, or by suctioning air samples from each height through a common chilled mirror hygrometer. On some systems, the z_2 and z_1 sensors are automatically exchanged with one another every 10 minutes or so in order to cancel instrumentation biases. Generally, λE is computed or summarized every 20 to 30 min and is summed over a daylight or 24-hour period to provide daily estimates of ET. Problems with naturally aspirated (exposed) thermocouples include contamination by dust and spider webs that increase radiation loading and thermal bias between T_1 and T_2 (Allen et al., 1994c).

The net radiation measurement in the BREB calculation should be made from an elevation high enough to measure an average representative surface condition similar to the condition upwind of the measurements. Soil heat flux density is generally measured at 0.08 to 0.15 m depths below the surface using soil heat flux plates. The 0.08 to 0.15 m depth is recommended to insure that the soil flux density is measured below the zone of soil moisture vaporization and to reduce the influence of vertical conduction of heat near the soil heat flux plate. Sensible heat absorption and release above soil heat flux plates is estimated by measuring soil temperature change above the plate. The corrected soil heat flux density, G, is computed according to Equation 4.46.

Important advantages of the BREB method are the ability to measure λE even from non-potential surfaces and the elimination of wind or turbulent transfer coefficients. The disadvantages are sophistication and fragility of sensors and datalogging equipment and the numerical instability of Equations 4.77 and 4.78 during periods of β near -1. The requirement for adequate upwind fetch also places limits on the method.

The question of whether or not correction for air stability conditions is needed in the use of the Bowen ratio-energy balance method is open to question. Most of the evidence suggests that correction for stability is not needed and that the usual assumption of the near-equality of transfer coefficients for heat and water vapor throughout a wide range of stability conditions is realistic (Cellier and Brunet, 1992).

The Bowen ratio method is recognized as one of the most accurate ET equations if R_n, G, and the gradients of temperature and humidity can be accurately measured. For tree crops and forest canopies, accurate measurement of the gradients at a height far enough above the canopy to avoid effects of individual trees is very difficult due to the very small gradients involved. Nevertheless, the Bowen ratio method has served and continues to serve as one of the major methods employed for forest research (McNaughton and Black, 1973; Thom et al., 1975; McIlroy and Dunin, 1982; Denmead and Bradley, 1985; Fritschen and Simpson, 1985; Dunin et al., 1985; 1991). Accuracies of well-designed and operated BREB systems have been estimated to be approximately 10 percent (Sinclair et al., 1975; Seguin et al., 1982).

a. Computations. The numerical instability of the BREB approach that develops as β approaches -1 (sensible heat transport toward the surface in near equilibrium with latent heat transport away from the surface) should be addressed by employing another estimation method during such periods. Fortunately, this generally occurs during periods of low R_n (dawn and dusk) when actual ET is normally low. Regardless, if estimates from Equation 4.77 are not rejected outright during such periods, very serious errors, even for daily totals, can result. For example, as $\beta \rightarrow 0$, the predicted $\lambda E \rightarrow \infty$ unless $(Rn - G) = 0$. Angus and Watts (1984), in a comprehensive error analysis relating to sensor-produced errors, showed the potential for error in calculation of λE increases very rapidly as β drops below -0.2, with accuracy requirements for all sensors becoming very demanding.

The severe problem with β very close to -1 is illustrated in Fig. 4.8. Section (a) shows the excellent capabilities of the BREB method in estimating actual ET during most hours of the day; however, the problem of $\beta \rightarrow -1$ is also clearly shown. Although in magnitude, errors are minimal for the 1700–1800 hours, it is obvious that 50 to 100 percent errors have developed even prior to β reaching -0.5, a confirmation of the Angus and Watts (1984) results. The data from eight days of an early micrometeorological study at Davis, California (March to October) are shown in Fig. 4.8(b), with twelve half-hour periods when $(-1.5 < \beta < -0.5)$ separated from the other 142 periods. The results also show the excellent response of the BREB method with these tests conducted under a wide range of conditions from cool, humid, to hot, dry, to a cold, dry, strong-wind day in March. Again, there were periods of rather drastic errors in estimating ET. Pruitt et al. (1987) utilized an hourly Penman equation to derive ET_o for periods affected by the problem when $\beta \rightarrow -1$. To obtain crop ET estimates for those problem periods, the ET_o was then multiplied by a crop coefficient, K_c, determined during daylight periods prior to or following the periods of unstable BREB estimates. Much earlier, Tanner (1967) suggested that simple combination equations should be substituted for BREB estimates during problem periods. This does mean that additional instrumentation for measuring wind speed for at least one height is required.

b. Applications of the Bowen Ratio method. Early applications of the BREB system are discussed in Tanner (1960), Pruitt (1963) and Fritschen (1965; 1966). McNaughton and Black (1973), Thom et al. (1975) and Dunin et al. (1985; 1991) have used the BREB to measure λE from forests. Malek et al. (1987) used BREB systems to measure long-term λE losses from desert playas in western Utah, and Duell (1988) reported applications to Owens Valley, California. Ashktorab et al. (1989; 1994) utilized a micro BREB system to measure evaporation from bare, moist soil and from soil underneath a canopy of tomato plants, where lower and upper thermocouples and vapor pressure intakes were positioned 1 and 6 cm above the soil surface.

Dugas et al. (1991) compared four Bowen Ratio system designs over irrigated spring wheat in Arizona. Lower temperature and vapor pressure arm positioning ranged from 0.1 to 0.6 m above the canopy surface. Upper arm positioning ranged from 0.8 to 1 m above the lower arm. The range of daytime λE among the four systems over a two day period was less than 12 percent and was within 1 percent during one of the test days. Measured net radiation and soil heat flux densities were identical for the four systems. Garratt (1978a) discussed minimum height settings of the lower Bowen Ratio arm above vegetative surfaces to avoid influences caused by individual roughness elements. Jacobs and van Boxel (1988) discussed minimum heights above row crops to avoid influences of the row structure. Cellier and Brunet (1992) showed BREB accuracy to be unaffected by location of the lower T and e measurements to within 0.5 m of a corn canopy.

The BREB approach is somewhat more difficult to apply to evaporation measurements from water bodies due to the difficulty in measuring the $R_n - G$ term. Estimation of G in the case of water requires measurement of the change in water temperature profile with time. This can be complicated in deep water bodies due to convective currents. In shallow water bodies, heat flux density into substrate may require measurement. Applications of the BREB approach to water have been discussed by Andersen (1954), Keijman (1974), and Hoy and Stephens (1979).

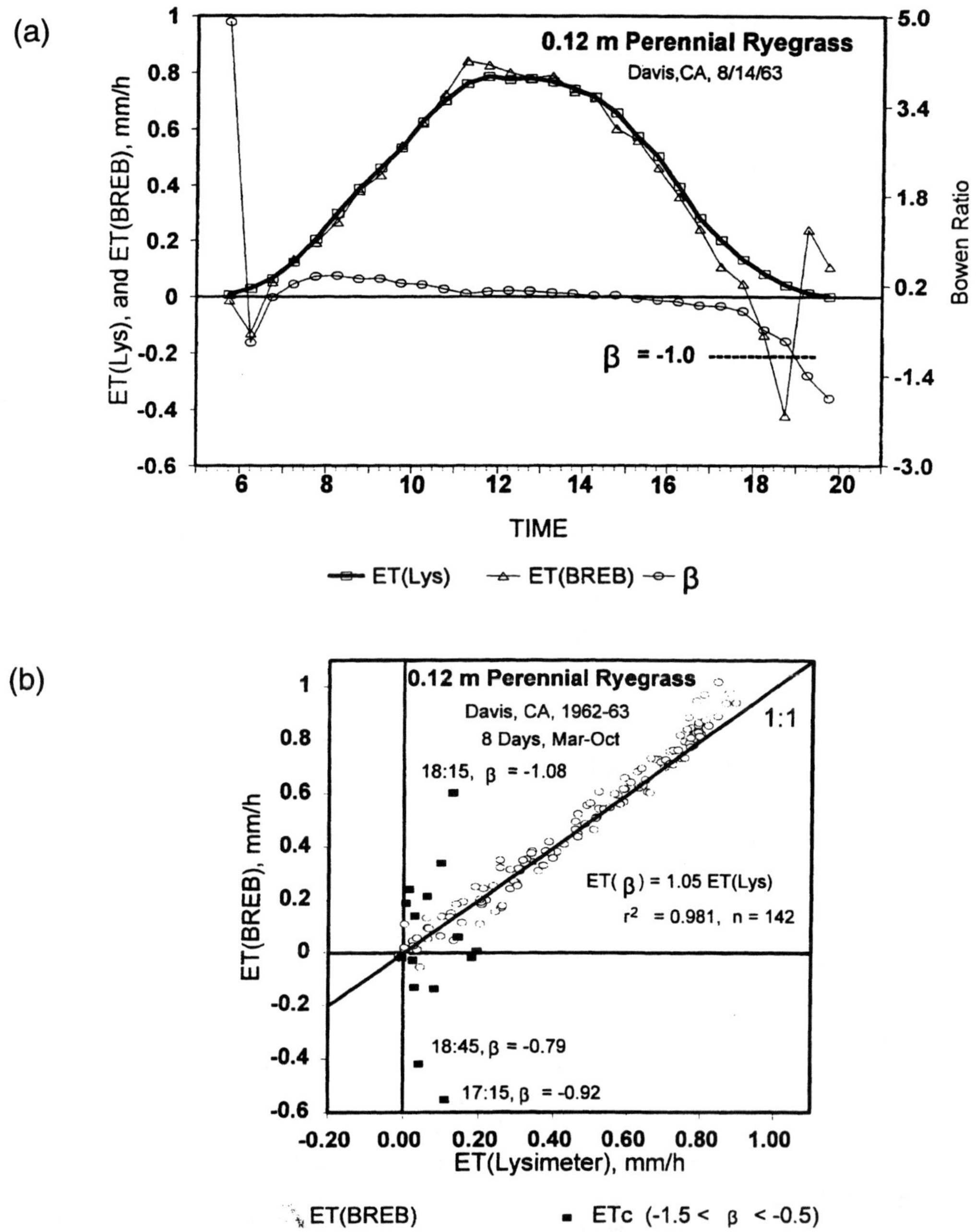

Figure 4.8—Examples of Early Tests of the BREB Method: (a) Comparison of ET of Perennial Ryegrass with BREB Calculations for 14 August, 1962 at Davis, CA, (b) Regression of ET(BREB) on ET(Lysimeter) for 142 Half-hr Periods for 8 days during 1962-63 (Pruitt, 1963).

2. ***Eddy Correlation*** *ET* can be measured directly using eddy correlation sampling of the boundary layer (Swinbank, 1951) according to the relationship:

$$\lambda E = \rho\, \overline{w'q'} = \frac{0.622}{P} \rho\, \overline{w'e'}, \tag{4.79}$$

where q' is the instantaneous deviation of specific humidity from mean specific humidity (q), e' is the instantaneous deviation of vapor pressure from mean vapor pressure (e), and w' is the instantaneous deviation of vertical wind velocity from mean vertical wind velocity (w). The overbar indicates means of the products of the instantaneous deviations over 1- to 5-min periods.

Residual latent heat λE can also be computed from the energy balance equation as $\lambda E = R_n - G - H$ (Equation 4.40), where sensible heat flux density is measured by eddy correlation as:

$$H = \rho c_p \overline{w'T'}, \tag{4.80}$$

where T' is the instantaneous deviation of air temperature from mean temperature (T).

The concept of eddy correlation draws on the statistical correlation between vertical fluxes of vapor or sensible heat within upward and downward legs of turbulent eddies. This requires high speed measurement of T, w, and e or q, usually at frequencies of 10 Hz (10 times per second) using quick response sensors. The vertical component of wind, w, is generally measured using a sonic anemometer, and T is measured using ultra fine wire thermocouples (on the order of 13 micron diameter). Specific humidity is measured using quick response hygrometers such as Lyman-alpha or Krypton hygrometers (Buck, 1976; Tanner, 1988; Campbell and Tanner, 1985). All measurements must be made at the same point in order to measure characteristics of the same eddy. Eddy correlation systems are commercially available.

Computing λE as a residual of the energy balance by measuring H, as in Equation 4.80, has the advantage of eliminating the requirement for the quick response hygrometer, which can be expensive and can require frequent maintenance. The disadvantage is the need to measure R_n and G which can be problematic under some conditions such as with sparse or heterogeneous vegetation and over water surfaces. Equation 4.80 can serve as an energy balance closure check on Equation 4.79.

Advantages of the eddy correlation method are the direct sampling of the turbulent boundary layer and measurement of ET over potential or non-potential surfaces. The eddy correlation has disadvantages similar to other boundary layer sampling techniques, including complex instrumentation and fetch requirements. The sonic anemometer must be plumbed perfectly so the component of horizontal wind speed does not bias measurements. In general, the eddy correlation method requires personnel who are well-trained in both electronics and biophysics. Instrumentation is relatively fragile and expensive. Additional details on the eddy correlation method can be found in articles by Dyer (1961), Businger et al. (1967), McBean (1972), Brutsaert (1982), and Weaver et al. (1986).

3. *Fetch Requirement* It has generally been suggested that upwind fetch for boundary layer instrumentation be on the order of 100 m for each m of z_2 above the top of the canopy (Brutsaert, 1982) to establish an internal equilibrium boundary layer that is representative of the surface energy exchange being measured. Brutsaert (1982) has provided theoretical considerations of boundary layer development that can be used to estimate minimum fetch requirements as a function of surface roughness. If one assumes that the lower 10 percent of an internal boundary layer downwind of a surface discontinuity has reached full adjustment to a new equilibrium with the surface (Peterson, 1969; Brutsaert, 1982), then Brutsaert's equation 7.39 can be expressed for minimum required fetch as:

$$x_f = \left(\frac{30\,(z - d)}{(z_{om})^{0.125}} \right)^{1.14}, \tag{4.81}$$

where x_f is the minimum fetch distance required for complete boundary layer development (m), z is the maximum sensor height above the ground (m), d is zero plane displacement (m), and z_{om} is momentum roughness height of the surface (m). Equation 4.81 estimates x_f for near-neutral conditions. The exponent (1.14) should be increased for situations of increasing stability and can be decreased for conditions of increasing instability (Brutsaert, 1982). Equation 4.81 indicates that as surface roughness increases, the fetch requirement (distance required for the new equilibrium to occur) decreases, as shown in Table 4.11, where z_{om} and d for vegetation are estimated as 0.123 and 0.67 times canopy height. Equation 4.81

TABLE 4.11. Minimum Recommended Upwind Fetch Distances, m, for Various Types of Surface Cover (Equation 4.81).

Height and type of surface cover	z = 1 m	z = 2 m	z = 3 m	z = 12 m
Water (d = 0, z_{om} = 0.0001 m)	180	400	630	3,000
0.12 m Grass	80	190	300	1,500
0.5 m Alfalfa	45	130	220	1,200
1.5 m Cattails	n/a	60	140	950
10 m Dense Trees	n/a	n/a	n/a	320

estimates x_f for discontinuities between surfaces of similar roughness and may not predict well for large changes in vegetation height, such as transversing from forest to grass or vice versa.

The values for minimum fetch in Table 4.11 computed using Equation 4.81 indicate that Equation 4.81 follows the 100:1 rule for a relatively wide range of vegetation and equipment heights. According to the equation, the 100:1 rule represents most closely 0.12 m tall vegetation. Equation 4.81 becomes very sensitive to roughness as $z_{om} \to 0$ so that fetch requirements become large (e.g., for open water).

As an example of application, for a Bowen ratio installation with the z_2 height at 1 m above the ground surface having 0.1 m tall vegetation, the minimum fetch requirement under neutral conditions according to Equation 4.81 would be 80 m. The 100:1 ratio (relative to equipment height above zero plane displacement) would suggest a fetch of 100 m. A fetch distance of 80 m would require a minimum vegetation stand size of 2 ha (5 acres) with the BREB system placed at the center. This requirement (which increases as vegetation height increases) makes it difficult to use the BREB or any other boundary layer sampling system to measure λE from small, isolated vegetative stands.

As a confirmation on the above estimates, Pruitt (1963) found no difference in BREB calculations of ET for 0.1 to 0.15 m-tall grass using z_1 and z_2 heights of 0.25 and 0.5 m, respectively, and z_1 and z_2 heights of 0.5 and 1.0 m, when upwind fetch of grass was 180-200 m. On the other hand, Dyer and Pruitt (1962), in tests on the same field with sensors of an eddy correlation system located at 4.0 m, reported the fetch of grass to be highly inadequate for sensors at that height. Equation 4.81 suggests a fetch length of 430 m for z_2 = 4 m over a 0.1-m crop.

An expression for predicting the fraction of λE sensed at a specific instrument height that was generated from a specific distance of upwind fetch is presented by Gash (1986), Schuepp et al. (1990), and Shuttleworth (1992):

$$F = \exp\left(\frac{(z-d)\left(1 - \ln\left(\frac{(z-d)}{z_{om}}\right)\right) - z_{om}}{k^2 x_f \left(1 - \frac{z_{om}}{(z-d)}\right)} \right), \tag{4.82}$$

where F is the fraction of vapor and sensible heat flux densities at the z height that was contributed by fetch having an upwind length of x_f m. Variable k in Equation 4.82 is the von Kármán constant (0.41). Equation 4.82 represents F for conditions having neutral stability and overpredicts F for stable conditions. It underpredicts F for unstable (buoyant) conditions. Equation 4.82 is useful for assessing quality of measurements by BREB and eddy correlation sensors, where F represents the fraction of λE or H measured that was generated by the fetch of vegetation or surface over which the sensors were located. Clearly, F should be close to 1.0 in order for measurements to be completely representative of the measurement surface. Equation 4.82 predicts $F \approx 0.8$ for values of x_f calculated using Equation 4.81, indicating that it is more conservative. When applied to BREB systems with $z = z_2$, Equation 4.82 is conservative, since the lower BREB measurement will have a higher F (for $z = z_1$). Equation 4.82 predicts $F = 0.91$ and $F = 0.54$ for $z = 1$ m and $z = 4$ m, respectively, for 200 m fetch over 0.1-m grass. These calculations agree with observations by Pruitt (1963) and Dyer and Pruitt (1962) described in the previous paragraph concerning adequacy of fetch.

4. *Mass Transfer With Direct Use of Aerodynamic Expressions*

a. Introduction. The Bowen Ratio (BREB) method is considered to be superior to the direct use of turbulent transfer equations for estimating transfers of water vapor above a surface. Its superiority is due primarily to constraints provided in the BREB method by inclusion of the radiation balance in the method. Errors in estimating gradients of temperature or humidity are not directly related to the estimates of λE or H as is the case in strictly aerodynamic methods.

However, there are situations where net radiation or heat transfer below the surface are not readily measurable. For forests, brushlands, or orchards, accounting for changes in heat storage of the canopy can also be of importance for periods shorter than 24 hr. Hence, consideration should be given to the use of aerodynamic methods that circumvent the above problems.

The classic flux-gradient relationships based on turbulent boundary-layer similarity theory are well-documented and easily applied (at least for short crops) within what has been described in recent years as the inertial sublayer. However, in a layer below the inertial sublayer, described as the roughness sublayer, it has become evident that the commonly accepted flux-gradient relationships may underestimate fluxes by up to a factor of two or more. This is evident from studies over forests by Thom et al. (1975), Raupach (1979), and Denmead and Bradley (1985); over bushland by Garratt (1978b); and even over tall agricultural crops, e.g., corn (*Zea mays*) by Mukammal et al. (1966) and Cellier and Brunet (1992). The underestimation of fluxes is caused by more efficient momentum, heat and vapor transfer within the roughness sublayer than is predicted by logarithmic similarity (*k*-theory) with general stability corrections. The increased efficiency is due to increased turbulence and turbulent wakes near individual roughness elements and thermal "seeding" within the canopy (Thom et al., 1975; Cellier and Brunet, 1992). Hence, the aerodynamic method has limitations over tall crops and especially over forests. For more information, the reader is referred to the latter two articles.

The commonly accepted flux-gradient relationships were obtained in studies conducted over grass or other short, relatively smooth-surface vegetation. They can be expected to provide relatively good accuracy in estimating *ET* for such surfaces where wind, temperature, and humidity sensors can be conveniently located above the roughness sub-layer (two or more times the canopy height (Cellier and Brunet, 1992; Jacobs et al., 1989)). For taller tree crops, bushland, and especially forests, the use of classical relationships requires positioning of the instruments so far above the surface that three major problems may occur. Not only is the instrument height and servicing a problem, but extreme accuracy of sensors is required due to the small gradients. Additionally, the upwind fetch requirements may become extreme. Various suggestions have been made to make use of the aerodynamic method even for sensor locations within the surface sublayer; however, the empiricism involved is considerable. Cellier and Brunet (1992), in a study over corn involving measurements throughout the entire sublayer and above, report the derivation of simple generalized flux-profile relationships that can be used to calculate the fluxes of heat and water vapor. They indicate that their new relationships are shown to be compatible with most of the previously published results obtained over a wide range of canopies. One encouraging aspect of their paper relates to the BREB method (Equations 4.77 and 4.78), where they reported no problem with the method, even down to the lowest two heights and well within the roughness sublayer.

The reader is encouraged to review the Cellier and Brunet (1992) paper and references contained therein as well as the book by Brutsaert (1982) before proceeding to estimate ET from measurements within the roughness sublayer. For one, the reader should be aware of the need for integrated forms of the stability functions when the difference in measurement heights is several meters or more, as would be required in forestry studies. This would apply to the aerodynamic method in the inertial sublayer as well.

The following presentation of aerodynamic methods relates to cases where measurements are made in the inertial layer. Various forms of stability functions are presented for use with hourly or shorter calculation timesteps.

b. Background and Correction for Air Mass Stability. In calculating *ET* for short time periods (hourly or less) using aerodynamic approaches, the inclusion of stability-correction expressions in the log-law or other wind profile equations becomes vital as evidenced by Rossby and Montgomery (1935), Monin and Obukhov (1954), Swinbank (1955), Panofsky (1963), Businger (1966), Dyer (1967), Dyer and Hicks (1970), Webb (1970), Pruitt et al. (1973) and Oke (1978). Basically, the aerodynamic estimation of water vapor or sensible heat typically involves use of the Monin-Obukhov universal wind profile expression involving the modification of the log-law wind profile to provide applicability under non-adiabatic conditions. It can be expressed as:

$$\frac{\partial u}{\partial z} = \frac{u_*}{k\,(z-d)}\,\phi_M, \tag{4.83}$$

where $\partial u/\partial z$ is the gradient of the horizontal component of wind speed at height z, u_* is the friction velocity in m s^{-1} (defined by Eq. 4.102 and 4.106), k is the von Kármán constant (taken as 0.41 herein), and d is the zero plane displacement of the wind profile by the canopy. The Monin-Obukhov ϕ_M function is equal to 1.0 for neutral buoyancy conditions in the equilibrium boundary layer, less than 1.0 for unstable conditions, and greater than 1.0 for stable conditions. Monin and Obukhov related ϕ_M to their dimensionless height ratio, z/L, in a Taylor series expression. Subsequently, only the first two terms have found wide usage, that is

$$\phi_M = 1 + \alpha_{mo}\,(z/L), \tag{4.84}$$

where z/L is a stability related function (Obukhov, 1946) and α_{mo} is a constant determined experimentally by Monin and Obukhov (1954). The scale length, L, in meters, as redefined by Businger and Yaglom (1971) and Brutsaert (1982) is:

$$L = \frac{-\,u_*^3\,\rho}{k\,g\left(\dfrac{H}{T_a c_p} + 0.61\,ET\right)}, \tag{4.85}$$

where u_* is in m s^{-1}, ρ is air density in kg m^{-3} (Equation 4.15), g is acceleration of gravity, 9.81 m s^{-2}, H is the sensible heat flux density, J m^{-2} s^{-1} (negative if to the surface), T_a is mean air temperature at the z height (K), and c_p is specific heat of moist air (J kg^{-1} K^{-1}). *ET* is evaporation flux density in kg m^{-2} s^{-1} (or mm s^{-1}).

An alternate stability parameter that is generally more convenient to employ in practice is the Richardson number, *Ri* (Thom et al., 1975). Whereas z/L requires direct input or estimates of u_* and H, *Ri* requires only temperature and wind gradients although, as with z/L, some correction for humidity gradients is desirable. *Ri* can be expressed as:

$$Ri = \frac{g\,(T_2 - T_1 + \Gamma\,(z_2 - z_1))\,(z_2 - z_1)}{0.5\,(T_2 + T_1)\,(u_2 - u_1)^2}, \tag{4.86}$$

where g is in m s^{-2}, T_1 and T_2 are absolute air temperatures, K, at heights z_1 and z_2 above the surface, and Γ is the dry adiabatic lapse rate (0.01 K m^{-1}). Values of T and u should be based on mean profile data, preferably involving periods of 30 minutes or less. Γ is usually ignored for investigations near the surface where $(z_2 - z_1)$ involves only a distance of a meter or two.

Equations involving water vapor (*E*) or heat transfer (*H*) can be written respectively as:

$$\frac{\partial q}{\partial z} = \frac{q_*}{k\,(z-d)}\,\phi_v \tag{4.87}$$

and

$$\frac{\partial \theta_T}{\partial z} = \frac{\theta_T^*}{k\,(z-d)}\,\phi_H, \tag{4.88}$$

where q_* is the friction specific humidity defined by $\lambda E/(\rho\lambda u_*)$, θ_T^* is the friction potential temperature defined by $H/(\rho c_p u_*)$, and ϕ_V and ϕ_H are stability functions analogous to ϕ_M in Equation 4.83.

Potential temperature, θ_T, for micrometeorological studies is defined as:

$$\theta_T = T + \Gamma\, z, \tag{4.89}$$

where θ_T and T are in K.

The potential temperature is the temperature that would result if air were brought adiabatically to the surface pressure level (Brutsaert, 1982). It provides for inclusion of the effect of lapse rate stability and H calculations. The use of θ_T in place of T has little effect on the calculation of sensible heat where sensors are within a few meters of one another. If virtual temperature, T_v (see Equation 4.16) is used in Equation 4.89 in place of T, then a virtual potential temperature, θ_{T_v}, is calculated. Use of T_v in the equation for Ri (Equation 4.86) may be important when relatively large Δq's occur between measurement heights and where errors in measurements of q are much less than the value of Δq, so that correction for Δq does not introduce additional error.

Discrete heights can be used in measurement of u, q, and θ_T resulting in a form of evaporation equation first derived by Thornthwaite and Holzman (1939). When stability functions are added the flux of water vapor can be expressed as:

$$E = -\rho\, k^2 \frac{(u_2 - u_1)\,(q_2 - q_1)}{[\ln\,[(z_2 - d)\,/\,(z_1 - d)]]^2}\,\frac{1}{(\phi_M\ \phi_V)}, \tag{4.90}$$

with E in mm s^{-1} (over a sq. m area = kg m^{-2} s^{-1}), air density, ρ, is in kg m^{-3}, mean horizontal wind speeds are in m s^{-1} at heights z_2 and z_1 in m, and q is the mean specific humidity in kg kg^{-1} at the same two heights above the surface (soil or water). Absolute humidity, ρq, is more convenient to measure, e.g., with infrared hygrometers; hence, the direct use of $(\rho q_2 - \rho q_1)$ in kg m^{-3} is common, eliminating the ρ in the first part of Equation 4.90. If vapor pressure data are directly available, the specific humidity term in Equation 4.90 should be replaced by $(e_2 - e_1)$, and the ρ should be multiplied by $0.622/P$. Both vapor pressure, e, and barometric pressure, P, should be in kPa (10 mb = 1 kPa).

Again, for applications over low canopy heights (or soil or water surfaces), the transfer of sensible heat H can be obtained from:

$$H = -\rho\, c_p\, k^2 \frac{(u_2 - u_1)\,(\theta_{T_2} - \theta_{T_1})}{[\ln\,[(z_2 - d)\,/\,(z_1 - d)]]^2}\,\frac{1}{(\phi_M\ \phi_H)}, \tag{4.91}$$

where θ_{T_2} and θ_{T_1} are potential temperatures at the z_2 and z_1 heights. With H in J m^{-2} s^{-1} and T in °C or K, c_p should be expressed in J kg^{-1} K^{-1} (1013).

The commonly accepted forms of stability correction for the unstable case are those described as the Businger-Dyer formulations (e.g., Dyer and Hicks, 1970; Dyer, 1974; Businger, 1988). With the assumption of equality of ϕ_H and ϕ_V, they are:

$$\phi_M = (1 - 16\ z/L)^{-1/4} \quad ; \quad \phi_H, \phi_V = (1 - 16\ z/L)^{-1/2}. \tag{4.92}$$

Although agreement is not universal in the literature, the equivalence of Ri and z/L for unstable conditions is assumed by many when Ri is calculated with z_2 and z_1 above the surface. When z_1 is taken at the surface ($z_1 = d + z_{om}$), then the calculated Ri (using Equation 4.86) represents an integrated or bulk Ri between the surface and the z_2 height, and is substantially lowered by the generally large values of u^2 in the denominator. In this case, the equality between Ri and z/L is no longer valid. Therefore, the recommendation in this handbook is that z/L can be replaced by Ri, only when the z_1 measurement height is above the surface. When z_1 is at the surface, then the z/L parameter should be used along with the integrated forms for stability correction (ψ_{sh} and ψ_{sm}) discussed in the next section.

For the case where both z_1 and z_2 are above the surface and where $z_2 - z_1$ is less than about 3 m, a stability correction factor, F_{sc}, for unstable conditions ($Ri < 0$) can be calculated as:

$$F_{sc} = \frac{1}{(\phi_M \ \phi_H)} = \frac{1}{(\phi_M \ \phi_V)} = (1 - 16\ Ri)^{3/4}. \tag{4.93}$$

Stewart and Thom (1973) identified $(\phi_M\phi_H)^{-1}$ and $(\phi_M\phi_V)^{-1}$ as stability factors, F. Raupach (1979) later suggested the term "influence factor" for F, since the terms should account for surface as well as stability effects. To avoid mixing of identities, these terms are labeled as F_{sc}.

For the stable case ($Ri > 0$, $\phi_M > 1$), the preferred expression appears to be that of Webb (1970). With an assumed equivalence of ϕ_H, ϕ_V, and ϕ_M, it can be shown that, for z_1 and z_2 both above the surface and $z_2 - z_1$ is less than about 3 m:

$$\phi_H, \phi_V = \phi_M = (1 - 5.2\ Ri)^{-1}. \tag{4.94}$$

Then the stability factor (or influence function), F_{sc}, for calculating both H and E under stable conditions is:

$$F_{sc} = \frac{1}{\phi_M \ \phi_H} = \frac{1}{\phi_M \ \phi_V} = (1 - 5.2\ Ri)^2. \tag{4.95}$$

Clearly, as 5.2 $Ri \rightarrow 1.0$, the value of $F_{sc} \rightarrow 0$. A limit for Ri of 0.1 is suggested here, leading to a lower limit for F_{sc} of 0.23. Placing this limit on F_{sc} appears reasonable in light of results shown by Thom et al. (1975, Fig. 1).

c. Calculating ET Using ($R_n - G - H$). Especially for cases where λE exceeds H, the use of Equation 4.91 along with $R_n - G$ to estimate E has great appeal. The equation for ET in mm s^{-1} for this method can be expressed as:

$$ET = \frac{R_n - G - H}{\lambda}, \tag{4.96}$$

where ET is in mm s^{-1}, R_n, G, and H are in J m^{-2} s^{-1}, and λ (Equation 4.17) is in J kg^{-1}.

If data for R_n and G are available (measured or calculated) this method avoids the measurement of humidity, still one of the more uncertain micrometeorological measurements. Although humidity data used to calculate virtual air temperature data (T_v) would improve the accuracy of Ri, the improved accuracy of H estimates would, for most conditions, be insignificant.

Fig. 4.9 provides an example comparison of measured ET for 0.12 m fescue grass with ET as calculated from Equations 4.90, 4.91, and 4.96. Sensor heights were 0.5 and 2.0 m as indicated by the curve for Ri. As compared to the Bowen ratio estimates of ET (Fig. 4.8), the estimates by the direct aerodynamic method (Equation 4.90) are obviously inferior, with considerable unsteadiness. This perhaps reflects mostly on the degree of accuracy involved in measuring wind and humidity differences, yet both the wind and absolute humidity data used for the two heights were based on smoothed profiles drawn through data points for 8 to 10 measurement heights above the surface. The actual data for each separate sensor were reported by Stenmark and Drury (1970).

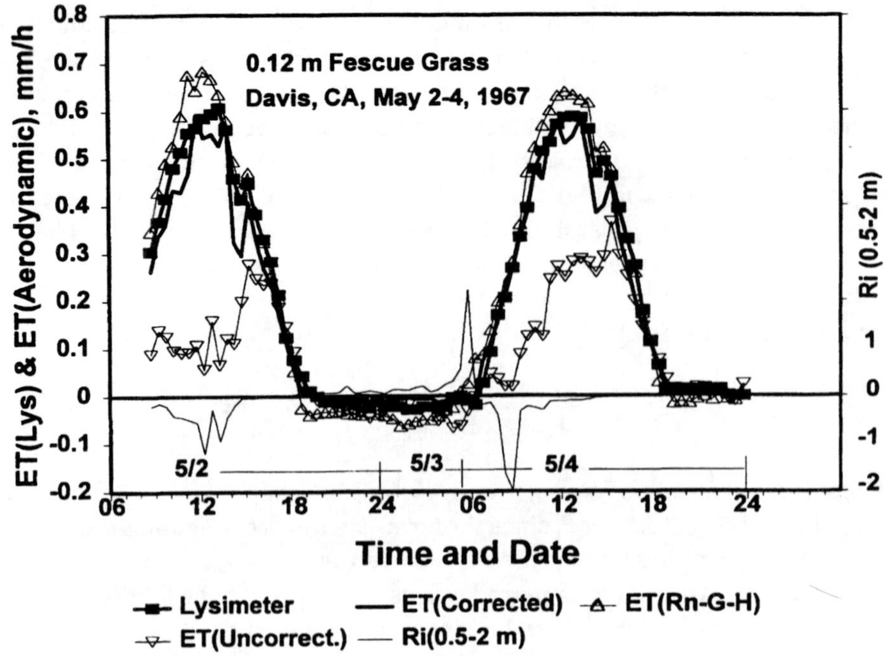

Figure 4.9—A Comparison of Measured ET(grass) with Aerodynamically Calculated ET, Corrected and Uncorrected for Stability. ET(Corrected) is based on Equation 4.90 with stability correction from Eq. 4.93. Curve ET (uncorrected) is based on Eq. 4.90 but without stability correction. ET(R_n-G-H) data are based on Equations 4.91 and 4.96 (Pruitt, 1994; Morgan et al., 1971).

The most striking feature of Fig. 4.9 is the proof of the need for stability correction of the aerodynamic approach on cool, relatively humid spring days. On May 4, between 0700 and 0930 h, air temperature ran from 11 to 14 °C and *RH* from 90 to 72%. This resulted in $H/(R_n - G)$ ratios of 0.48 to 0.40, and ratios of $\lambda E/(R_n - G)$ of 0.52 to 0.60 (Pruitt et al., 1968). The resulting strongly unstable conditions produced, in the analysis leading to Fig. 4.9, F_{sc} values from Equation 4.93 of 10 or greater between 0800 and 0900 h. These results can hardly be related significantly to stomatal control since the grass was still wet from a very heavy dew at least up to 0800 h.

The results in Fig. 4.9 indicate that, at least for the reported study, the correction for unstable conditions as in Equation 4.93 may fall short by some small amount. The over-estimation of *ET* by Equation 4.96 is likely the result of underestimation of *H* by Equation 4.91, although errors in R_n and *G* and even in the lysimeter measurements of *ET* (grass) may be involved. Note, however, that the inclusion of the energy balance in Equation 4.96 results in a much smoother curve for the *ET* estimates by this combined aerodynamic method.

*5. **Sensible Heat From Temperature Differences*** As indicated with the eddy correlation and aerodynamic methods, λE can be determined indirectly by solving the energy balance for λE, having computed *H* by other means. One promising method for determining *H* is the use of air temperature at some height *z* along with surface temperature using infrared thermometers. Hatfield et al. (1984) and Katul and Parlange (1992) are examples for reference. The former provides results of testing the method at a number of United States locations and with several crops.

Sensible heat is commonly estimated by expressing the sensible heat equation in a difference form and by using an aerodynamic resistance term, r_a (s m^{-1}), leading to:

$$H = -\frac{\rho c_p}{r_a}\left(\theta_{T_z} - \theta_{T_o}\right), \tag{4.97}$$

where θ_{T_z} and θ_{T_o} are mean values for potential temperature at the z height and at the surface (K), calculated from T_z (air temperature at height z) and T_o (surface temperature) using Equation 4.89. The aerodynamic resistance, r_a, between the surface and height z can be estimated by:

$$r_a = \frac{\left(\ln\left(\frac{z-d}{z_{om}}\right) - \psi_{sm_{(z/L)}}\right)\left(\ln\left(\frac{z-d}{z_{oh}}\right) - \psi_{sh_{(z/L)}}\right)}{k^2\ u_z}, \tag{4.98}$$

where d is the zero-plane displacement, z_{om} is the surface roughness length for momentum transfer (see later sections for their estimation), z_{oh} is the surface roughness length for sensible heat transfer (approximated for most vegetative surfaces as 1/10 of z_{om}), *ln* is the natural logarithmic function, and $\psi_{sm_{z/L}}$ and $\psi_{sh_{z/L}}$ are integrated stability functions for describing effects of buoyancy or stability between the surface and height z on sensible heat and momentum transfer. A similar equation for r_a, but where wind, temperature, and humidity measurements are at different heights, is presented as Equation 4.110 in the following section on Combination Methods. Equation 4.98 was developed by equating Equation 4.97 and Equation 4.99 (introduced in the next section) and solving for r_a.

Integrated forms of stability correction, as presented in the next section, should be used in air-to-surface determinations of heat, vapor, and momentum fluxes. As with the purely aerodynamic estimates of H or E, measurements of u and T should be based on 30-min mean data or less. They should not be applied with 24-hr means due to errors in averaging T_z and T_o over long periods.

a. Integrated Forms of Stability Correction. The integrated stability-correction expressions $\psi_{sm_{z/L}}$ and $\psi_{sh_{z/L}}$ are recommended for all surface-to-air calculations of H and λE, in place of the gradient ϕ functions. They are certain to be required for study over tall crops such as tree crops and forests. Another factor requiring attention in applying surface-to-air models for such applications is that the adiabatic lapse rate, Γ (approx. 0.01 °C m^{-1}), must be taken into account, e.g., in the calculation of H, and use of virtual air temperature, T_v, and potential virtual temperature, θ_{T_v}, may be important.

Brutsaert (1982) provides a comprehensive coverage of the subject, where his discussion and formulas relate to the use of the Monin and Obukhov (1954) z/L stability parameter. The use of z/L requires an iterative solution to solve for several unknown parameters with closure on the energy balance. Since most of the commonly used stability correction expressions were developed in relation to z/L, its use is recommended. A recent study by Katul and Parlange (1992) illustrates the use of iterative solutions for z/L.

Briefly, and with some simplifying assumptions, the following presents the calculation of sensible heat flux H using integrated forms of stability correction for the aerodynamic method involving T_z (or θ_{T_z}) and T_o (or θ_{T_o}) or involving T_2 (or θ_{T_2}) and T_1 (or θ_{T_1}) along with the use of z/L. For applications above the surface ($z_2, z_1 > d + z_{om}$), the assumption that $Ri = z/L$ is generally valid for unstable conditions and Ri can be used in place of z/L to avoid an iterative solution; however, when z_1 is taken at the surface ($d + z_{om}$) or when stable conditions exist, then the integrated stability correction forms should be applied using z/L rather than Ri for reasons discussed earlier relating to a surface-to-air bulk Richardson number.

b. Surface-to-Air Temperature Differences. When surface temperature is used, the profile expression involving sensible heat flux density, including correction for stability, is:

$$H = a_h\ k\ u_*\ \rho\ c_p \frac{(\theta_{T_o} - \theta_{T_z})}{\left[\ln\left[\frac{z-d}{z_{oh}}\right] - \psi_{sh_{(z/L)}}\right]}, \tag{4.99}$$

where a_h, the ratio of k for heat to k for momentum, is generally assumed to be equal to or close to 1.0

and θ_{T_o} and θ_{T_z} are mean values for potential temperature at the surface and z heights, computed using Equation 4.89. Units for the z, d, and z_{oh} terms in Equation 4.99 must cancel and other terms must combine to provide H with units normally of J m^{-2} s^{-1}. The term $\psi_{sh_{(z/L)}}$ (dimensionless) is the integrated stability function for heat transfer. With the assumptions that roughness lengths are normally much smaller than L, and that the Businger-Dyer expressions are valid, $\psi_{sh_{(z/L)}}$ for unstable conditions ($z/L < 0$) can be expressed as (Paulson 1970):

$$\psi_{sh_{(z/L)}} = 2 \ln \left[\frac{(1 + x^2)}{2}\right] \tag{4.100}$$

with

$$x = (1 - 16\ z/L)^{1/4} \tag{4.101}$$

Normally, z/L in Equation 4.101 is calculated following Brutsaert (1982) as $(z - d)/L$, where d, the zero-plane displacement length, is included and L is the Monin-Obukhov stability length (Equation 4.85). In the case of surface-to-air differences, z in Equation 4.101 is the height of the air temperature and wind measurements.

The other unknown in Equation 4.99 is u_* which can be obtained for surface-to-air applications from:

$$u_* = \frac{k\, u_z}{\ln\left[\dfrac{z - d}{z_{om}}\right] - \psi_{sm_{(z/L)}}}, \tag{4.102}$$

where $\psi_{sm_{(z/L)}}$ (dimensionless) is the integrated stability correction function for momentum transfer, again, assuming that the roughness lengths are normally much smaller than L (Paulson, 1970). The integral form of the expression for $\psi_{sm_{(z/L)}}$ for unstable conditions for use with $T_o - T_z$ is:

$$\psi_{sm_{(z/L)}} = 2 \ln [(1 + x)/2] + \ln [(1 + x^2)/2] - 2 \arctan (x) + \pi / 2 \tag{4.103}$$

with x obtained from Equation 4.101 and where arctan (x) is in radians.

For stable conditions, agreement as to appropriate stability functions is less certain. Fortunately, the method being considered, i.e., the estimation first of H using temperature differences and then solving for ET, allows for some uncertainty. Also, except for hot, arid, or desert climates, stable conditions primarily relate to late afternoon to early morning periods when ET is minimal.

Assuming $\phi_M = \phi_H = \phi_V$ for stable conditions ($z/L > 0$) and acceptance of the functions of Webb (1970), the recommended relationships for integrated stability forms are given as:

$$\psi_{sm_{(z/L)}} = \psi_{sh_{(z/L)}} = -5.2\ (z/L) \tag{4.104}$$

for $0 < z/L < 1.0$ and where, again, z/L should be computed as $(z - d)/L$. For $z/L > 1$, Brutsaert (1982) recommends setting $\psi_{sm_{(z/L)}}$ and $\psi_{sh_{(z/L)}}$ to about -5 or -6.

c. Calculation Steps. To calculate H and subsequently ET from the energy balance using the integrated stability functions and z/L, one would follow an iterative procedure, where one first uses Equations 4.102 and 4.99 to solve for H and Equation 4.96 to solve for ET, assuming neutral stability conditions ($\psi_{sh_{(z/L)}}$ and $\psi_{sm_{(z/L)}}$ are initially set equal to 0). Variable L is then calculated using Equation 4.85 with the initial estimates of H and ET. $\psi_{sh_{(z/L)}}$ and $\psi_{sm_{(z/L)}}$ are then solved using Equations 4.100, 4.103, or 4.104, depending on the sign of z/L, and Equations 4.102 and 4.99 are then resolved for u_* and H. ET for use in Equation 4.85 is recalculated using the energy balance equation, Equation 4.96, and the process is repeated until the value of z/L (and u_* and H) becomes numerically stable. Generally only three to four iterations are required.

d. Air-to-Air Temperature Differences (When $\Delta z > 3$ m). When air temperature and wind are measured at two heights in the profile (rather than measuring T_o), and the z_1 and z_2 elevations are separated by large distances ($z_2 - z_1 >$ about 3 m), then the integrated forms of stability correction should be employed, where the following variations on Equations 4.99 and 4.102 are used. The profile expression involved for sensible heat flux density, including correction for stability, is:

$$H = -a_h\, k\, u_*\, \rho\, c_p \frac{(\theta_{T_2} - \theta_{T_1})}{\left[\ln\left(\frac{z_2 - d}{z_1 - d}\right) - \psi_{sh_{(z_2/L)}} + \psi_{sh_{(z_1/L)}}\right]}, \tag{4.105}$$

where θ_{T_2} and θ_{T_1} are potential temperatures at z_2 and z_1 heights, computed from T_2 and T_1 using Equation 4.89. Units for the z, d, and z_{oh} terms in Equation 4.105 must cancel and other terms must combine to provide H with units normally of J m^{-2} s^{-1}. The terms $\psi_{sh_{(z_2/L)}}$ and $\psi_{sh_{(z1/L)}}$ (dimensionless) are integrated stability functions for sensible heat transfer calculated using Equation 4.100 or 4.104, where z/L is equal to $(z_2 - d)/L$ and $(z_1 - d)/L$ for $\psi_{sh_{(z_2/L)}}$ and $\psi_{sh_{(z1/L)}}$, respectively.

Friction velocity, u_*, can be obtained from wind speed measurements at two heights as:

$$u_* = \frac{k\,(u_2 - u_1)}{\left[\ln\left(\frac{z_2 - d}{z_1 - d}\right) - \psi_{sm_{(z_2/L)}} + \psi_{sm_{(z_1/L)}}\right]}, \tag{4.106}$$

where terms $\psi_{sm_{(z_2/L)}}$ and $\psi_{sm_{(z1/L)}}$ (dimensionless) are integrated stability functions for momentum transfer and are calculated using Equation 4.103 or 4.104 with z/L equal to $(z_2 - d)/L$ and $(z_1 - d)/L$, respectively. Variables u_2 and u_1 are mean horizontal wind speed measurements at the z_2 and z_1 heights. Units for u_*, u_2, and u_1 are equivalent. In situations where T_1 and T_2 are measured at the z_1 and z_2 heights, but where only one measurement for u_z is made, then u_*, since it is constant with height, can be calculated using Equation 4.102, provided that an estimation of surface roughness, z_{om}, can be made.

The same iterative procedure for calculating H, u_* and z/L is followed as for surface to air measurements described previously.

e. Estimating ET from Air-to-Air Humidity Differences. As in the estimation of H from Equation 4.105 for cases where $(z_2 - z_1) >$ about 3, ET can be estimated from the following:

$$E = -a_v\, k\, u_*\, \rho \frac{(q_2 - q_1)}{\left[\ln\left(\frac{z_2 - d}{z_1 - d}\right) - \psi_{sv_{(z_2/L)}} + \psi_{sv_{(z_1/L)}}\right]}, \tag{4.107}$$

where a_v is the ratio of k for water vapor to k for momentum, commonly assumed to be 1.0. and steps used in solving for H in Equation 4.105 are used in solving for ET, assuming equality of the ψ_{sh} and ψ_{sv} functions. Refer to Equation 4.90 for units involved.

f. Advantages, Disadvantages, and Precautions. The energy balance equation with H estimated using Equations 4.97, 4.99, or 4.105 has the advantage of ET estimation in areas where precipitation is less than ET and therefore where plant moisture stress occurs, since no knowledge of surface resistance is required. The estimation of r_a or use of surface temperature does require measurement of wind speed and perhaps surface roughness. Estimation of equivalent surface roughness, z_{om} and z_{oh}, along with measurement of a representative, average θ_{T_o} in Equations 4.99 through 4.102 can be problematic in tall, sparse canopies such as those found in forest and natural settings. Even in short agricultural crops, sensing the effective surface temperature can be a problem, where at night, θ_{T_o} may be influenced by leaf layers closest to the warm soil surface. For crops with sparse canopies, proper integration of soil and vegetation temperatures is especially difficult.

An additional disadvantage of the energy balance approach is that T_1 and T_2 or T_o and T_z measurements must be specific to the surface in question. Therefore, historic measurements of air temperature over standard weather surfaces are not useful with this method.

As with the Bowen ratio and eddy correlation methods, and all boundary layer sampling techniques for that matter, it is critical that a large fetch of like-vegetation occurs upwind of all sensors. This is necessary since the surface temperature-energy balance method, as with all boundary layer methods, assumes that equilibrium, one-dimensional vapor, wind, and temperature profiles have developed. Divergence from equilibrium energy exchange conditions may cause substantial estimation errors due to decoupling of the energy exchange surface from the boundary layer profile and resulting divergence of flux densities within the boundary layer. Fetch requirements for each sensor should be adhered to as discussed previously.

Estimating ET as $R_n - G - H$ generally requires the use of stability correction factors discussed previously, especially over dry surfaces. An example illustrating the seriousness of ignoring air profile stability conditions when using $ET = R_n - G - H$ is that reported by Pruitt and Lourence (1966) showing under-estimation of actual grass ET during calm, mid-morning periods and an over-estimation of ET by a factor of 2 to 3 during late-afternoon stable periods with colder, heavier air near the surface. During periods with winds greater than 4 to 5 m s^{-1}, stability correction was no longer needed over the transpiring crop.

g. Examples. Examples showing effects of stability correction on estimation of r_a using surface and air temperature measurements are given in Figure 4.10 for 0.23-m and 0.12-m high fescue grass at Logan, Utah and at Davis, California, respectively. Surface temperatures at both sites were made using IR thermometers; however, at Davis, T_o was based on averages of 5-cm and 10-cm leaf thermocouples for $R_n < 0$ since the warm nighttime soil surface appeared to dominate the IR readings. At both locations, surface temperature was lower than air temperature during nighttime and late-afternoon periods, causing moderately strong stability of the boundary layer. Correction for this stability using Equation 4.104 with r_a calculated using Equation 4.98 (using an iterative solution for z/L) increased calculations of r_a dramatically, especially at Logan during nighttime periods where z/L became quite large due to low wind speeds ($u_2 < 1$ m s^{-1}). With the almost calm, cooler and more humid morning conditions between 0800 and 1400 h at both sites, which normally produce ratios of $\lambda E/(Rn - G) < 1.0$, T_o exceeded T_z and unstable conditions resulted. These unstable conditions lasted until mid-afternoon when z/L became positive. The corrected values of r_a during some morning periods (made using Eqs. 4.98 and 4.100 to 4.103) ranged from 50 to 80 s m^{-1} lower than the uncorrected values for both locations.

The effects of stability and instability on estimation of r_a shown in Fig. 4.10 appear to be quite significant; however, it is evident from Fig. 4.11, where ET was estimated using the energy balance, Equation 4.96, with H calculated using Equation 4.99, that only for the Davis study was there a large need for stability correction due to 15–20% over-estimation of ET, e.g., at mid-day, by the uncorrected log-law values. The corrections applied to calculate H for stability agreed very well for the Davis study with independent estimates of H made by subtracting lysimeter measurements of ET from $R_n - G$. Use of the calculated H in the energy balance resulted in under-estimation of ET (more negative values) at Davis during nighttime hours, although dewfall was occurring in this case. This underestimation may have been caused by errors in the measurements for R_n or G.

For the Logan results, only at night were there differences of more than a few percent in corrected and non-corrected estimates of ET since the grass forage was transpiring during the daytime at near-maximum rates, so that little instability occurred and H was small. The magnitude of H at Davis was more

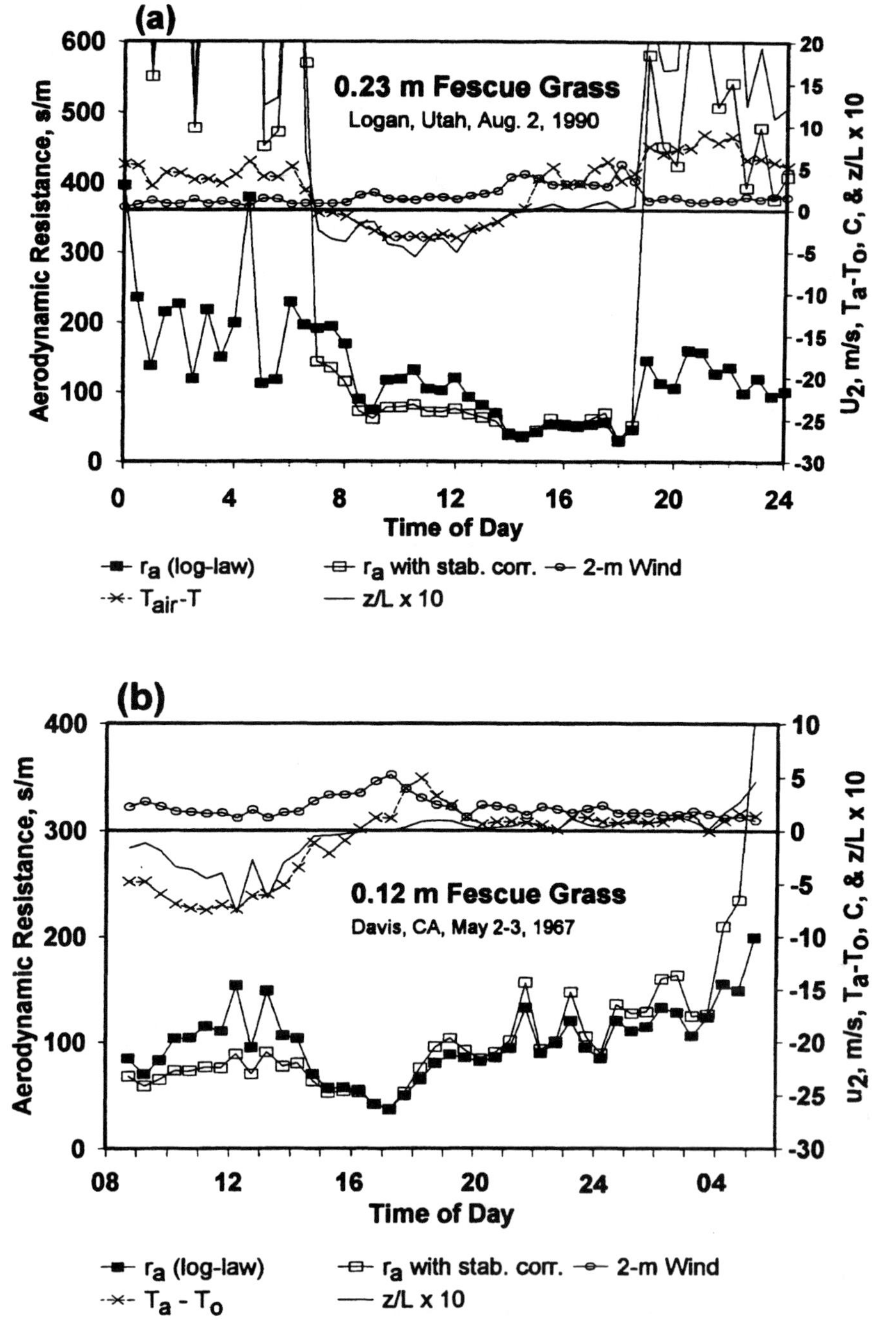

Figure 4.10—Aerodynamic Resistance (r_a) over Grass Computed with and without Stability Correction for Studies at a) Logan, UT, and b) Davis, CA (Allen, 1993; Pruit, 1994; Morgan et al., 1971).

than double that at Logan, no doubt due to cooler and more humid conditions of early May leading to greater instability at Davis. In applications over moisture-stressed natural vegetation, H may become large relative to λE and $T_o - T_z$ will increase. Such effects will increase the importance of applying stability corrections. The primary benefit of applying stability corrections at Logan was during the nighttime hours (for example, 20 to 24 hours) when the uncorrected H resulted in estimation of ET of about 0.05 mm h^{-1}, whereas ET estimated using corrected H and ET measured by lysimeter indicated

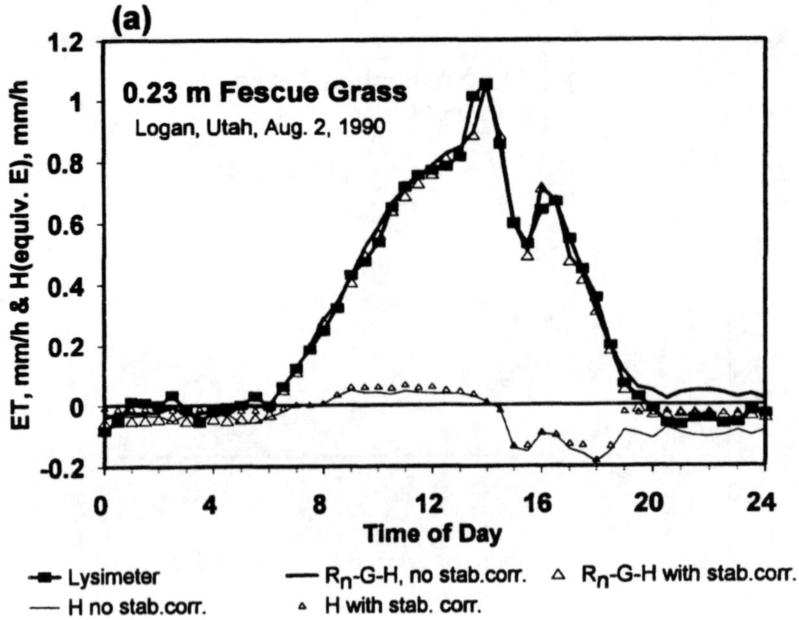

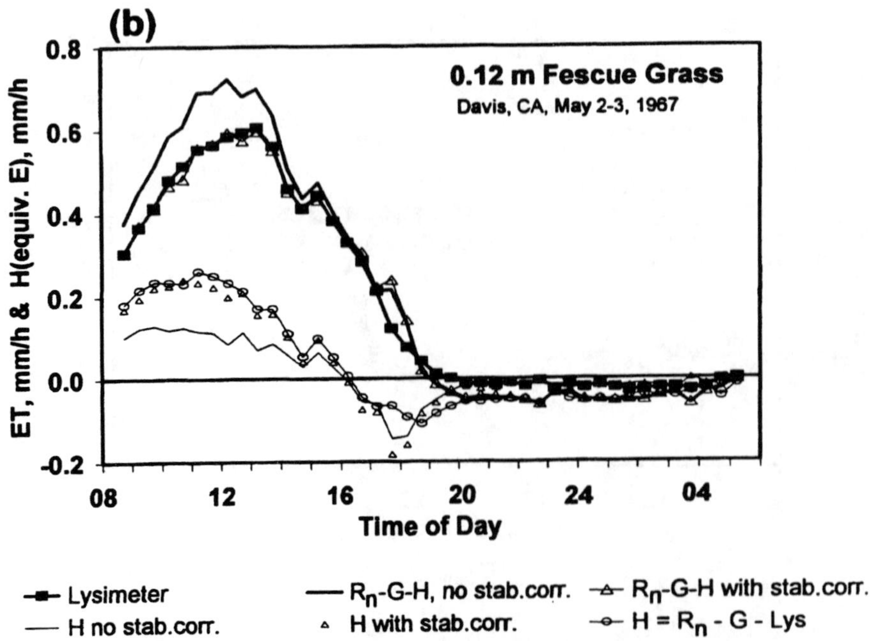

Figure 4.11—Comparison of ET(Lys) with ET Calculated from $R_n - G - H$ with H Calculated Using the Log-law and Corrected Log-law with Integrated Stability Expressions at (a) Logan, UT and (b) Davis, CA (Allen and Fisher, 1990; Pruitt, 1994; Morgan et al., 1971).

negative *ET* during this period. These errors are small, but would cause an error in the 24-hr sum for *ET* of almost 1 mm d^{-1} if the errors occurred over an 8-hr night.

D. Reference Crop ET Methods

The combination of the sensible and latent heat transfer equations (Equations 4.90 and 4.91) with the energy balance equation (Equation 4.96) following simplifying assumptions by Penman (1948) results in the well known Penman equation:

$$ET_o = \frac{\Delta}{\Delta + \gamma}(R_n - G) + \frac{\gamma}{\Delta + \gamma}(a_w + b_w u_2)(e_z^o - e_z), \tag{4.108}$$

where R_n is net radiation and G is soil heat flux density, Δ is the slope of the saturation vapor pressure curve, γ is the psychrometric constant, $e_z{}^o$ is mean saturation vapor pressure at air temperature, e_z is actual vapor pressure of the air, u_2 is wind speed at the 2 m height, and a_w and b_w are empirical wind function coefficients.

1. *The Penman-Monteith Equation* Monteith (1965) and Rijtema (1965) introduced a surface resistance parameter into the Penman equation and replaced the linear wind function term with an aerodynamic resistance parameter. The result is what is termed the Penman-Monteith equation:

$$\lambda ET = \frac{\Delta (R_n - G) + \rho_a c_p (e_z^o - e_z)/r_a}{\Delta + \gamma\left(1 + \frac{r_s}{r_a}\right)}, \tag{4.109}$$

where units of λET, R_n, and G are in J m^{-2} s^{-1} (or W m^{-2}), ρ_a is air density, kg m^{-3}, computed using Equation 4.15, c_p is specific heat of dry air, J kg^{-1} °C^{-1}, generally taken as 1010 J kg^{-1} °C^{-1}, and variables $e_z{}^o$ and e_z are in kPa. Equations for calculating $e_z{}^o$ and e_z for use with 24-hour calculations are given later in this section. Units for Δ and γ in Equation 4.109 are kPa °C^{-1}. For convenience, Δ is generally computed using Equation 4.2 at mean air temperature. Aerodynamic resistance to vapor and heat transfer, r_a, in Equation 4.109 is computed using Equation 4.98 or 4.110. Bulk surface resistance, r_s, is defined in the following paragraphs. Both r_a and r_s have units of s m^{-1}.

Aerodynamic resistance between the surface and height z for use in Equation 4.109 that uses the integrated stability correction forms is calculated as:

$$r_a = \frac{\left(\ln\left(\frac{z_u - d}{z_{om}}\right) - \psi_{sm(z_u/L)}\right)\left(\ln\left(\frac{z_{T,e} - d}{z_{oh}}\right) - \psi_{sh(zT,e/L)}\right)}{k^2\, u_z}, \tag{4.110}$$

where measurement heights for u_z (at height z_u) and temperature and vapor pressure (at height $z_{T,e}$) differ. The integrated stability functions in Equation 4.110 must be solved using $z/L = z_u/L$ for $\psi_{sm_{(z_u/L)}}$ and $z/L = z_{T,e}/L$ for $\psi_{sh_{(zT,e/L)}}$. These stability parameters can be set equal to zero for calculation of reference crop evapotranspiration ET_o or ET_r and for time steps greater than or equal to 24 hours.

Evapotranspiration expressed as a depth per time can be computed from λET as $ET = \lambda ET/\lambda$. If λ is computed using Equation 4.17, with units expressed in J kg^{-1}, then $ET = \lambda ET/\lambda$ will have units of mm s^{-1}. For equivalent ET in mm d^{-1} (for 24-hr time periods), multiply mm s^{-1} by 86,400 s d^{-1}. For mean air temperature T = 20 °C, λ has the value of 2.45 × 10^6 J kg^{-1}, which is a commonly used value for λ. When applied to 24-hr periods, R_n, G, r_a, and r_s in Equation 4.109 should represent 24-hr averages for these parameters (Jensen et al., 1990).

The combination equations that adhere to the form of Equation 4.108 (those by Penman (1948, 1963), Doorenbos and Pruitt (1977), Wright and Jensen (1972), and Wright (1982)) produce straight-line relationships for ET as a function of wind speed; however, some findings have indicated that a curvilinear relationship might be more appropriate, e.g., Rijtema (1965) and Pruitt and Doorenbos (1977a, pp. 112–114)). With the inclusion of the surface resistance term, the Penman-Monteith equation projects a curvilinear wind function, where the rate of increase in *ET* decreases with increasing wind speed. The traditional Penman form with linear wind function may overpredict *ET* under moderate to strong winds if the winds continue throughout the night and RH_{max} remains low; however, in the Penman-Monteith method, aerodynamic resistance is decreased with higher wind, resulting in an increase in $\gamma\ (1 + r_s/r_a)$ and, in turn, a decrease in $\Delta/(\Delta + \gamma(1 + r_s/r_a))\ (R_n - G)$. This has the effect of reducing the influence of the aerodynamic term of the Penman-Monteith equation on *ET* estimates for a given surface resistance, r_s. The method is thus less susceptible to problems under the conditions mentioned earlier than are the traditional Penman models. The Penman-Monteith has the further advantage of flexibility in assigning roughness and surface resistance values to fit various types and heights of vegetation. This extends its direct application to a wide variety of surfaces as well as to soil moisture conditions.

The Penman-Monteith equation has been widely used to represent reference ET_o for use in irrigation management (Allen et al., 1989; Jensen et al., 1990; Martin et al., 1993; Allen et al., 1994a; b). In addition, when used with algorithms to increase surface resistance r_s (bulk stomatal resistance) under conditions of low soil moisture, the Penman-Monteith equation can be used to estimate actual evapotranspiration, ET_a, for a wide variety of natural vegetation under nonpotential moisture conditions.

For daily or longer calculation timesteps, $e_z{}^o$ in the Penman-Monteith equation is computed as (Allen et al., 1989; Jensen et al., 1990):

$$e_z^o = \frac{e^o(T_{max}) + e^o(T_{min})}{2}, \tag{4.111}$$

where T_{max} and T_{min} are maximum and minimum daily air temperature for the period at the *z* height and e^o is the saturation vapor pressure function. Actual vapor pressure of the air, e_z, is generally computed as:

$$e_z = e^o(T_d) \tag{4.112}$$

where T_d is mean daily or early morning dewpoint temperature.

2. ***Aerodynamic and Surface Parameters*** When used to predict *ET* from a dense grass or alfalfa reference, aerodynamic resistance in the Penman-Monteith equation can be calculated using the following parameter estimations (Allen et al., 1989; Jensen et al., 1990):

$$z_{om} = 0.12\ h \tag{4.113}$$

$$z_{oh} = 0.1\ z_{om} \tag{4.114}$$

$$d = 0.67\ h, \tag{4.115}$$

where z_{om}, z_{oh}, *d*, and mean plant height, *h*, are in m. Aerodynamic resistance, r_a, is computed using Equation 4.98 or 4.110. Additional equations for predicting z_{om} and *d* and general values for z_{om} for various types of vegetation are listed in the section on Evapotranspiration from Land Surfaces—Direct Penman-Monteith.

Surface resistance, r_s, for densely growing vegetation is generally computed as:

$$r_s = \frac{r_l}{LAI_{eff}}, \tag{4.116}$$

where r_l is the bulk stomatal (or surface) resistance of the vegetation per unit *LAI* (s m^{-1}) and LAI_{eff} is the effective leaf area index contributing to *ET*. Parameter r_l is the inverse of g_l, the stomatal conductance per unit leaf area, which is discussed in the section on Evapotranspiration from Land Surfaces—Direct

Penman-Monteith. The value of r_l is generally taken as 100 s m^{-1} for many well-watered agricultural crops (Monteith, 1965; Szeicz and Long, 1969; Allen et al., 1989; Jensen et al., 1990) when calculations are made on a 24-hr basis. When *ET* calculations are made on an hourly or shorter basis, r_l should be reduced to about 70–80 s m^{-1}, assuming that stability corrections are made or that buoyancy conditions are nearly neutral. When stability corrections are not made, then artificially decreasing r_l to less than 70 s m^{-1} for unstable conditions and artificially increasing r_l to greater than 80 s m^{-1} for stable conditions may be required for hourly calculations. Similarly, if 24-hour calculations are made under conditions of strong aerodynamic instability (z/L, $Ri < -0.05$), then the 24-hour value used for r_l may need to be artificially reduced to account for the effects of instability in reducing the magnitude of aerodynamic resistance, r_a. When deficient soil moisture reduces *ET* to less than maximum levels, r_l must be increased following procedures outlined in the section on Evapotranspiration from Land Surfaces—Direct Penman-Monteith.

A "minimum" or "unconstrained" r_s, $r_{s_{min}}$ is defined and listed in Table 4.20 in the section on Evapotranspiration from Land Surfaces—Direct Penman-Monteith as the surface resistance when all environmental parameters (notably R_s, T, VPD, and soil moisture) are at optimum levels. Under these conditions, values for a "minimum" r_l, $r_{l_{min}}$, may be about 40 s m^{-1} for many agricultural crops; however, values for $r_{l_{min}}$ should only be used in Equation 4.116 in combination with stomatal conductance reducing functions (g()) (see Evapotranspiration from Land Surfaces—Direct Penman-Monteith), since the environmental parameters are rarely all at optimum levels necessary to sustain the 40 s m^{-1} value.

a. Effective Leaf Area Index. A standardized LAI_{eff} has been established for grass and alfalfa reference crops (Allen et al., 1989; Jensen et al., 1990; Allen et al., 1994b) as:

$$LAI_{eff} = 0.5\ LAI. \tag{4.117}$$

The 0.5 multiplier suggests that generally only the upper half of a dense canopy such as dense stands of grass and alfalfa is active in heat and vapor transport and is the zone of major net radiation absorption (Szeicz and Long, 1969). The 0.5 multiplier has been supported by field research in wheat by Choudhury and Idso (1985).

Ben-Mehrez et al. (1992) presented an expression for LAI_{eff} that accounts for larger ratios of LAI_{eff}/LAI at small LAI and smaller ratios of LAI_{eff}/LAI when LAI is large:

$$LAI_{eff} = \frac{LAI}{0.3\ LAI + 1.2}. \tag{4.118}$$

Equation 4.118 was based on data from Shuttleworth (1991) and Rochette et al. (1991) for semidense agricultural crops and predicts $LAI_{eff} = 0.67\ LAI$ at $LAI = 1$ and $LAI_{eff} = 0.24\ LAI$ at $LAI = 10$. The equation predicts LAI_{eff} = 0.4 to 0.5 LAI at LAI's between 3 and 4, which are common for grass and alfalfa references, so that it is in general agreement with Equation 4.117 when used with standard reference vegetation heights. Equation 4.118 is intuitively attractive for practical use, since it automatically reduces effective LAI as LAI becomes large and as shading of interior leaves increases and air exchange inside canopies decreases. The equation may not apply to sparse vegetation such as in forests.

The LAI for dense, clipped grass was computed following Allen et al. (1989) as

$$LAI = 24\ h, \tag{4.119}$$

where LAI is defined as the total leaf area (one side) per unit land area, $m^2\ m^{-2}$ and h is mean crop height, m, limited in Equation 4.116 to ≤ 0.4 m. The LAI for an alfalfa reference crop can be estimated using an equation from Allen et al. (1989):

$$LAI = 5.5 + 1.5 \ln (h), \tag{4.120}$$

where h is in m and $h > 0.03$ m. Additional relationships for predicting LAI are given in the section on Evapotranspiration from Land Surfaces—Direct Penman-Monteith.

b. Stability Correction. The combination energy balance-aerodynamic methods (combination equations), which are used to estimate surface-to-air transfers, are far less affected by stability conditions than are the purely aerodynamic equations (Equations 4.90 and 4.91), especially for near-potential evaporation conditions and when estimating reference crop *ET*; however, for watersheds and other non-irrigated environments where a combination of soil types and depths, infrequent precipitation, etc., reduces evaporation or transpiration to levels much lower than reference *ET*, it is recommended that hourly calculations with combination equations include correction expressions. Under non-well-watered conditions, reduction in surface conductance functions (increases in surface resistance) may be an even more important consideration.

Examples of applying the Penman-Monteith equation to grass cover are included in the section on Evapotranspiration from Land Surfaces—Direct Penman-Monteith. The first of these examples (shown later in Fig. 4.23), showing half-hourly calculations with and without stability correction at Logan, Utah and Davis, California for the same days as in Fig. 4.10 and 4.11, indicates that correction for stability over the well-watered grass surfaces was not nearly as important and critical for the Penman-Monteith as it was in estimating ET as $R_n - G - H$ (shown in Fig. 4.11). The presence of $1/r_a$ in both the numerator and denominator of the Penman-Monteith equation dampens the effect of aerodynamic stability or instability (and wind speed, for that matter) on the *ET* estimate.

3. Reference K_c ET_o Approach The most common application to date of the Penman or Penman-Monteith-type equation for estimating evapotranspiration has been a two-step process. The two steps are 1) the determination of a reference ET_o, which has generally been for either clipped grass (Doorenbos and Pruitt, 1977) or alfalfa (Wright and Jensen, 1972; Wright, 1982) and 2) multiplying ET_o by a crop coefficient, K_c, to obtain actual *ET* for a particular crop or surface condition, where:

$$ET_c = K_c \, ET_o. \tag{4.121}$$

The K_c used in Equation 4.121 must be the K_c associated with the specific reference type used (grass or alfalfa).

The two-step K_c ET_o approach has been successfully and widely applied for several reasons. One reason is the ready "visualization" of the crop coefficient, K_c, that generally has fixed limits given by potential energy availability. Another reason for the widespread use of the K_c ET_o approach has been the standardization of the reference estimate (ET_o) and the general transferability of K_c curves. Relating ET_o to a specific reference and utilizing K_c's in a two-step process has the additional major advantage of giving the individual making calculations a mental representation of the process. In addition, it is relatively straightforward to select consistent crop coefficients and to validate reference equations in new areas.

The alternative to using the K_c ET_o approach is the direct application of the Penman-Monteith or other multi-layer combination equation to estimate actual *ET* directly. This approach, while having the promise of more accurate estimates through more direct and graphic description of surface characteristics, is generally more difficult to apply and usually includes uncertainty in the exact nature of surface parameters. Both the K_c ET_o approach and direct Penman-Monteith approach are discussed in the following sections.

4. Reference ET_o Calculations A range in height in the reference crop definitions (both for grass and for alfalfa) has been necessary to allow for the use of field *ET* measurements in calibrating and validating ET_o equations; however, it is necessary to standardize reference crop heights when applying ET_o and ET_r equations such as the Penman-Monteith equation. This is necessary to specifically characterize the surface parameters (roughness and bulk stomatal resistance) and to produce standardized climatic references. Allen et al. (1989), Jensen et al. (1990), Smith et al. (1991), and Allen et al. (1994a; b) have suggested adapting a fixed height of 0.12 m for the grass reference and 0.5 m for the alfalfa reference in the Penman-Monteith equation.

5. Penman-Monteith as a Grass Reference When a suggested height of $h = 0.12$ m is used to predict z_{om} and *LAI* for use in the aerodynamic and surface resistance equations, the following consolidated form of the Penman-Monteith can be derived for estimating grass reference ET_o assuming that stability effects are small (Allen et al., 1989; Smith et al., 1991; Allen et al., 1994b):

$$ET_o = \frac{0.408\Delta(R_n - G) + \gamma \dfrac{900}{T + 273} u_2(e_z^o - e_z)}{\Delta + \gamma(1 + 0.34\, u_2)}, \tag{4.122}$$

where ET_o is grass reference *ET* expressed in mm d^{-1}, and R_n and G are expressed in MJ m^{-2} d^{-1} for use with 24-hr calculation time steps. T is mean daily air temperature (°C), Δ and γ have units of kPa °C^{-1}, u_2 is mean 24-hr wind speed at 2 m above the ground (m s^{-1}), and e_z^o and e_z are in kPa and are based on measurements at 1.5 to 2 m. Wind measured at heights other than 2 m above the ground surface can be adjusted to 2 m for use in Equation 4.122 using Equation 4.53 or 4.182. When applied to hourly calculation time steps, the 900 coefficient in Equation 4.122 becomes 37 for ET_o in units of mm h^{-1} and R_n and G in units of MJ m^{-2} h^{-1}. All other coefficients and parameter units remain the same.

6. Penman-Monteith as an Alfalfa Reference The Penman-Monteith equation can be applied as an alfalfa reference using Equations 4.113–4.117 and 4.120. These equations can be consolidated into Equation 4.109 to form a single equation for a 0.5-m-tall reference alfalfa crop (Allen et al., 1989):

$$ET_r = \frac{0.408\Delta(R_n - G) + \gamma \dfrac{1700}{T + 273} u_2(e_z^o - e_z)}{\Delta + \gamma(1 + 0.40\, u_2)}, \tag{4.123}$$

where all units and definitions are the same as in Equation 4.122 (ET_r in mm d^{-1} and R_n and G in MJ m^2 d^{-1}), and ET_r denotes alfalfa reference *ET*. For hourly calculations where ET_r is in mm h^{-1}, R_n and G are in MJ m^2 h^{-1}, and aerodynamic stability effects are assumed to be small, the 1700 coefficient changes to 70. Units for u_2 and e remain the same.

7. 1985 Hargreaves Equation The Hargreaves equation (Hargreaves and Samani, 1982; 1985) is suggested as a means for estimating ET_o in situations where data banks are limited and only maximum and minimum air temperature data are available. The form of the 1985 Hargreaves equation is:

$$ET_o = 0.0023\,(T_{max} - T_{min})^{0.5}\,(T_{mean} + 17.8)\,(R_A), \tag{4.124}$$

where T_{max} and T_{min} are maximum and minimum daily air temperature in °C, T_{mean} is mean daily air temperature $((T_{max} + T_{min})/2)$, and R_A is average daily extraterrestrial radiation (Equation 4.20). ET_o in equation 4.124 has the same units as R_A.

The 1985 Hargreaves equation was the highest ranked temperature related method for calculating ET_o reported in the ASCE Manual 70 analysis (Jensen et al., 1990). Allen (1992) found Equation 4.124 to estimate well in a wide range of latitudes and climates for periods of 5 days or longer without significant error.

The principal feature of the Hargreaves equation is the inclusion of the difference between maximum and minimum daily air temperature. The temperature difference provides an indication of general humidity and cloudiness. This occurs due to the fact that when humidity levels are high, differences between T_{max} and T_{min} are relatively low and *ET* rates are relatively low. The lower temperature differences occur because higher dewpoint temperatures under conditions of high humidity hold T_{min} to relatively higher levels. When cloud cover occurs, maximum temperatures are less due to decreased solar radiation, and minimum temperatures are usually higher due to increased long-wave emittance and reflection by nighttime cloud cover.

One often overlooked advantage of an equation such as the 1985 Hargreaves equation relative to more complex equations is the reduced data requirement. For the 1985 Hargreaves equation, only maximum and minimum air temperature measurement is required. This is advantageous in regions where solar radiation, humidity, and wind data are lacking or are of low or questionable quality. Generally, air temperature can be measured with less error and by less trained individuals than the other three parameters required by combination equations. Equation 4.124 can be calibrated against the Penman-Monteith equation (Equation 4.122) where data are available to produce a "regionally" calibrated temperature equation.

An alternative to using a temperature-based equation such as Equation 4.124 when data are lacking is to employ the Penman-Monteith equation using estimates for missing variables. Generally, applying Equation 4.122 on a daily or monthly basis with R_s estimated using Equation 4.29, $T_d = T_{min}$, and $u = 2$ m s^{-1} gives estimates similar to Equation 4.124 for agricultural weather sites. Examples comparing ET_o estimates made using estimated R_s, T_d, and T_{min} are included near the end of this chapter.

8. ET_o Software Programs Software programs for computing ET_o by various common and popular ET_o equations have been made available by Allen (1991) and Snyder and Pruitt (1994). These personal computer-based programs are useful for calculating ET_o, comparing estimates among various reference equations and validating ET_o computations.

VI. EVAPOTRANSPIRATION FROM LAND SURFACES—GENERAL APPLICATIONS

A. The "Crop" Coefficient

Estimating ET in hydrologic studies goes beyond agricultural crops and extends to various natural vegetation forms or types under rainfed conditions. For these reasons, the term K_c might be more appropriately referred to as a "cover" coefficient rather than as a "crop" coefficient. Usage and application would be the same; however, to be consistent with past literature and applications, this handbook adheres to the use of the term "crop" coefficient. The reader should understand that this term applies to nonagricultural vegetation as well as to bare soil.

By expressing ET_c and ET_o in terms of the Penman-Monteith equation as shown in Equation 4.125, where the additional c subscripts represent characteristic values for the actual vegetation and the additional o subscripts represent the same for the reference crop, one can visualize that the value of K_c depends on the relative roughness, leaf area, and albedo (in net radiation calculation) of the actual vegetative surface in relation to the same characteristics of the grass or alfalfa reference surface.

$$K_c = \frac{\lambda ET_c}{\lambda ET_o} = \frac{\dfrac{\Delta(R_{nc} - G_c) + \rho c_p (e_z^o - e_z)/r_{ac}}{\Delta + \gamma\left(1 + \dfrac{r_{sc}}{r_{ac}}\right)}}{\dfrac{\Delta(R_{no} - G_o) + \rho c_p (e_z^o - e_z)/r_{ao}}{\Delta + \gamma\left(1 + \dfrac{r_{so}}{r_{ao}}\right)}}. \tag{4.125}$$

In addition, the relative proportions of net radiation, wind, temperature, and vapor pressure deficit affect the value of K_c to some degree. Clearly, the more similar a vegetative cover is to the reference condition, the closer the value of K_c will be to 1.0 and the less variation in the value of K_c with changing weather conditions.

1. Field Scale Applications When applying Equation 4.125 under well-watered conditions, where a majority of energy for the ET process is from net radiation, the K_c for large expanses of similar vegetation does not exceed 1.0 by more than 0.1 relative to an alfalfa reference and by more than 0.3 relative to a grass reference. This is due to the fact that the reference crops provide a nearly maximum sink for short-wave radiation and have large LAI and roughness to promote energy and vapor transfer. Therefore, an expanse of reference crop (especially alfalfa) will approach the maximum conversion of available energy into λET, so that the ratio of any other tall crop to alfalfa will be near 1.0. This observation arises from viewing the maximum values for K_c reported by Wright (1982), where none of Wright's K_c's, based on an alfalfa reference, exceeded 1.0. In the case of grass, where the cover is shorter and LAI may be less, values of K_c's may approach 1.25 for tall, dense crops under arid and semiarid conditions.

Limiting K_c to approximately 1.0 for an alfalfa reference-base or to 1.3 for a grass reference-base applies to large expanses of vegetation (>50 m diameter) and is significant when evaluating field measurements of evapotranspiration. Higher values of K_c indicate problems with field measurements. Measurement problems include 1) improper computation of vegetation area in lysimeter studies (Allen et al., 1991b), 2) violation of necessary fetch requirements in boundary layer (energy balance) measurements, or 3) weather data collection difficulties and errors. The first two problems have been discussed in previous sections.

2. Small Expanses of Vegetation When ET is measured from small expanses of vegetation, the internal boundary layer above the vegetation may not be in equilibrium with the new surface and may not have developed up to the height of instrumentation. In addition, small expanses of tall vegetation surrounded by shorter cover can exhibit a "clothesline effect." In these cases, ET from the isolated stands, on a per unit area basis, may be significantly greater than the corresponding ET_r or ET_o computed for an alfalfa or grass reference, assuming an infinitely large fetch of similar reference vegetation. An example of these situations would be ET from a single row of trees surrounded by short vegetation, or ET from a narrow strip of cattails along a stream channel. Allen et al. (1992) reported K_c values for small (6 m-wide) stands of cattails and bulrushes surrounded by grass pasture equal to 1.6 to 1.8 during midseason, relative to an alfalfa reference. These measurements exhibited a strong clothesline effect. Coefficients were only 1.15 for a cattail wetland which was 200 m in diameter (Allen et al., 1994c).

Pruitt (1976) reported K_c values for a 4.2 m-tall Monterey pine tree (*Pinus radiata*) varying from 1.4 in February to March to 2.0 during spring and summer and approaching 3.0 during late fall and dry, early winter months. The tree was growing in a 1.83 m by 2.44 m hydraulic-pillow lysimeter located within a 1-ha dry, non-cropped field. It was near the middle of a 10-tree wind row oriented normal to prevailing winds.

The preceding results indicate the importance of knowing the type of setting for which ET estimates are needed. If ET estimates are needed for small, isolated stands of vegetation, then the K_c may exceed 1.0 for an alfalfa reference and 1.25 for a grass reference by up to 50 to 80 percent. If ET estimates are to represent large expanses of vegetation or small stands of vegetation surrounded by mixtures of other vegetation having similar roughness and moisture conditions, then K_c's will generally be less than or equal to 1.0 for alfalfa and 1.25 for grass references.

3. Crop (Cover) Coefficients Because of the two reference crop definitions (grass and alfalfa), two families of K_c curves for agricultural crops have been developed. These are the alfalfa-based K_c curves by Wright (1981; 1982) and grass-based curves by Pruitt (Doorenbos and Pruitt, 1977; Jensen et al., 1990). The user must exercise caution in not mixing grass-based K_c's with an alfalfa reference and vice-versa. Usually K_c's based on alfalfa reference can be "converted" for use with a grass reference by multiplying by a factor ranging from 1.10 to 1.25, depending on climate (1.10 for humid, calm conditions and 1.25 for arid, moderately windy conditions). Grass-based K_c's are presented in this handbook due to the large number of K_c's that have been presented in FAO-24 and elsewhere for agricultural crops. Alfalfa based K_c's for eight irrigated crops in southern Idaho have been presented by Wright (1981; 1982).

Generalized crop coefficient curves for estimating crop ET_c for crops or other vegetation are shown in Fig. 4.12. The K_{cb} curve represents basal crop coefficients for conditions where the soil surface is visually dry, so that soil evaporation is minimal but where the availability of soil water does not limit plant growth or transpiration. This curve represents a minimum ET_c situation for adequate soil moisture. The spikes in Fig. 4.12 indicate occurrences of precipitation or irrigation which wet the soil surface and temporarily increase total ET_c for a one- to five-day period. These spikes reduce to the K_{cb} curve when the soil surface dries. The spikes generally approach a maximum value of 0.8 to 1.0 for an alfalfa ET_r base and 1.0 to 1.2 for a grass ET_o base. The K_{cm} curve in Fig. 4.12 represents a mean crop coefficient, which

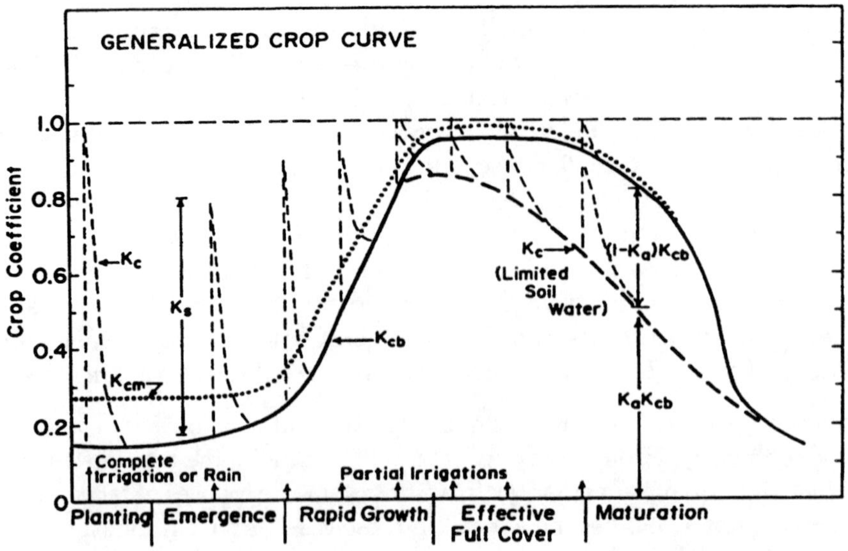

Figure 4.12—Generalized Cover Coefficient Curves Showing the Effects of Growth Stage, Wet Surface Soil and Limited Available Soil Water (Wright, 1982; Jensen et al., 1990).

includes effects of wet soil (spikes). The final, limited soil water curve in the figure represents the decrease in ET_c when plant water uptake is limited by low soil moisture availability.

a. Basal Crop Coefficients. Basal crop coefficients primarily represent the transpiration component of *ET*. Their use provides for adjustment for wet soil effects after rain or irrigation. This results in more accurate estimates of ET_c on a daily basis for use in soil moisture modelling. The total crop coefficient, K_c, is computed from K_{cb} as:

$$K_c = K_{cb}K_a + K_s, \tag{4.126}$$

where K_a is a dimensionless coefficient dependent on available soil water and K_s is a coefficient to adjust for increased evaporation from wet soil immediately after rain or irrigation. The value of K_a is 1 unless available soil moisture limits transpiration, in which case it has a value less than 1. The values for K_s represent the spikes shown in Fig. 4.12.

b. Surface Evaporation. Wright (1982) suggested the following expression for K_s:

$$K_s = (K_1 - K_a K_{cb})\left(1 - \left(\frac{t}{t_d}\right)^{1/2}\right) f_w, \tag{4.127}$$

where K_1 is the maximum value for K_c normally occurring after rain or irrigation. K_1 is generally 0.8 to 1.0 for an alfalfa-based K_c and 1.0 to 1.2 for a grass-based K_c. Variable t in Equation 4.127 represents the days after major rain or irrigation and t_d is the usual number of days required for the soil surface to visually appear dry. Variable f_w is the relative fraction of the soil surface wetted by rainfall or irrigation. Normally, for precipitation, sprinkler, and surface irrigations that wet the entire soil surface, $f_w = 1.0$. Equation 4.127 simulates the time-based reduction in evaporation as the soil dries. General values for t_d are 3 days for sandy soils, 5 days for silty soils, and 7 days for clayey soils (Wright, 1982). Variable t must be limited to between $0 <= t <= t_d$. If soil wetting occurs in the morning, then t is generally set equal to 0 for that day. Most of the evaporation occurs in the first few days after the soil is wetted. Daily precipitation totals that are less than about 0.3 ET_o can generally be assumed evaporated and can be ignored.

Generally, application of the $K_c = K_{cb} K_a + K_s$ procedure requires a daily water balance computation similar to that described in Equation 4.69, with K_s, K_{cb}, and ET_o computed for each day. In order to limit total evaporation and transpiration from the surface layer of soil during a drying cycle (ΣET_s) to the water available, a separate water balance should be conducted for the upper 10 to 15 cm soil layer from which evaporation occurs. In this layer, water can be depleted by both surface evaporation and root extraction (transpiration). The ratio of root extraction from the evaporation layer can be estimated as proportional to the ratio of $(z_e/z_r)^{0.6}$, where z_e is the depth of the evaporation layer (normally 0.10 to 0.15 m) and z_r is the total depth of the root zone ($z_e \leq z_r$). This relationship is based on the commonly used 40-30-20-10% root extraction pattern for quartile rooting depths (top to bottom) in moist soils. Total daily moisture depletion from the evaporation layer can then be estimated as:

$$ET_s = \left(\left(\frac{z_e}{z_r}\right)^{0.6}(K_{cb}K_a) + K_s\right)ET_o \tag{4.128}$$

In situations with few roots in the top layer between plant rows during development or where roots are periodically destroyed by tillage, root extraction from the surface layer is slight and $ET_s = K_s\, ET_o$. The daily soil water balance of the evaporation layer is equal to:

$$\theta_{e_2} z_e = \theta_{e_1} z_e + P - Q_r - ET_s, \tag{4.129}$$

where θ_{e_2} and θ_{e_1} are volumetric moisture contents of the evaporation layer at the beginning of days 2 and 1, respectively, and P, Q_r, and ET_s represent precipitation, runoff, and ET_s on day 1. Parameter θ_{e_2} is subject to the following limits:

$$\mathit{If}\ \theta_{e_2} > \theta_{DU}\ \mathit{then}\ \theta_{e_2} = \theta_{DU} \tag{4.130}$$

$$\mathit{If}\ \theta_{e_2} < \theta_{LE}\ \mathit{then}\ K_s = 0\ \mathit{and}\ ET_s = 0, \tag{4.131}$$

where θ_{LE} is the lower limit of moisture in the evaporation layer (m^3 m^{-3}). Generally, the evaporating layer will dry to below wilting point, approaching air dry. The value of θ_{LE} will vary with soil texture, crust formation and vegetation or mulch cover. A reasonable approximation for the evaporation layer is $\theta_{LE} \approx 0.5\ \theta_{LL}$, where θ_{LL} is the lower limit of plant extractable water listed in Table 4.8. The value for θ_{DU} used in Equation 4.130 may be set a few percentage points above normal values listed in Table 4.8 to compensate for extra moisture retained in the evaporation layer above θ_{DU} for the first few days after wetting. A complete soil water balance of the full z_r root zone (including z_e) is needed to compute K_a and ET_c.

c. Moisture Stress Reduction. K_a can be computed following FAO 33 (Doorenbos and Kassam, 1979) as:

$$K_a = \frac{\theta - \theta_{LL}}{\theta_t - \theta_{LL}} \tag{4.132}$$

for $\theta < \theta_t$, where θ is mean volumetric soil moisture in the root zone (m^3 m^{-3}), and θ_t is the threshold θ below which transpiration is decreased due to moisture stress. $K_a = 1.0$ for $\theta \geq \theta_t$. Variable θ_{LL} is the soil moisture at the lower limit of moisture extraction by plant roots (m^3 m^{-3}). This variable is synonymous with the term "wilting point." Typical values for various soil texture classes are listed in Table 4.8. The threshold soil moisture, θ_t, can be estimated from the relationship:

$$\theta_t = (1 - F_{ns})(\theta_{DU} - \theta_{LL}) + \theta_{LL}, \tag{4.133}$$

where F_{ns} is the average fraction of available soil moisture that can be depleted before moisture stress and ET reduction occur (F_{ns} = fraction for no stress). Variable θ_{DU} is the moisture content at the drained limit of the soil (field capacity), (m^3 m^{-3}). Values for all θ parameters should represent averages for the effective root zone. Values for θ_{DU} and θ_{LL} are listed in Table 4.8 and values for F_{ns} for agricultural crops are listed in Table 4.12.

F_{ns} normally varies from 0.30 depletion of available moisture ($\theta_{DU} - \theta_{LL}$) for shallow rooted plants at high rates of ET_o (> 8 mm d^{-1}) to 0.70 for deep rooted plants at low rates of ET_o (< 3 mm d^{-1}). A value of 0.50 for F_{ns} is common. After computation of K_c, ET_c is computed using Equation 4.121.

TABLE 4.12. Mean Crop Coefficients, K_c, for Subhumid Climates[1], Mean Maximum Height, Ranges of Maximum Effective Rooting Depth, and Soil Moisture Depletion Fractions for No Stress, F_{ns} (primary source: Doorenbos and Pruitt 1977). For use with ET_o from Eq. 4.122.

Crop	K_{c1}[2]	K_{c2}	K_{c3}	Max. Ht. (m)	Max. Root (m)	F_{ns}
Alfalfa Hay[3]	0.40	1.20	1.15	0.7	1.0-2.0	0.55
Artichokes	0.90	1.00	0.95	1.5	0.6-0.9	0.45
Asparagus	0.30	0.95	0.30	0.4-2	1.2-1.8	0.45
Avocado					0.6-0.9	0.40
Banana, 1st year	0.50	1.10	1.10	2	0.5-0.8	0.35
2nd year	0.70	1.15	1.10	4	0.5-0.8	0.35
Barley, Wheat, Oats	0.30	1.15	0.25	1	1.0-1.5	0.55
Beans, green	0.40	1.00	0.90	0.4	0.5-0.7	0.45
Beans, dry and Pulses	0.40	1.15	0.35	0.4	0.6-0.9	0.45
Beets, table	0.40	1.05	0.95	0.4	0.6-1.0	0.50
Berries	0.30	1.05	0.50	1.5	0.6-1.2	0.50
Carrots	0.50	1.05	0.80	0.3	0.5-1.0	0.35
Castorbeans	0.40	1.15	0.55	0.5		
Celery	0.35	1.05	1.00	0.6	0.3-0.5	0.20
Citrus, no ground cover						
70% canopy	0.65	0.80	0.75	4	1.2-1.5	0.50
50% canopy	0.60	0.75	0.70	3	1.1-1.2	0.50
20% canopy	0.45	0.55	0.55	2	0.8-1.1	0.50
Citrus, with cover or weeds						
70% canopy	0.80	0.90	0.85	4	1.2-1.5	0.50
50% canopy	0.85	0.90	0.85	3	1.1-1.2	0.50
20% canopy	0.95	0.95	0.95	2	0.8-1.1	0.50
Clover hay[3]	0.40	1.20	1.10	0.6	0.6-0.9	0.50
Coffee, bare	0.90	0.95	0.95	2-4	0.9-1.5	
Coffee, weeds	1.05	1.10	1.10	2-4	0.9-1.5	
Conifer Trees	1.00	1.05	1.05	10		
Corn, Field	0.40	1.15	0.55,0.35[4]	2.2	1.0-1.7	0.60
Corn, Sweet	0.40	1.15	1.05	1.5	0.8-1.2	0.50
Cotton	0.40	1.20	0.70	1.5	0.8-1.7	0.60
Cucumber, Fresh Market	0.35	1.00	0.75	0.3	0.7-1.2	0.50
Machine harvest	0.35	1.00	0.90	0.3	0.7-1.2	0.50
Crucifers[5]	0.40	1.05	0.90	0.3	0.4-0.6	0.40
Dates	0.90	0.95	0.95	8	1.5-2.5	0.50
Deciduous Orchard, killing frost						
Apples, Cherries, no cover	0.45	0.95	0.70[6]	4	1.0-2.0	0.50
Peaches, Stone Fruit, no cov.	0.45	0.85	0.65[6]	3	1.0-2.0	0.50
Apples, Cherries, grnd cover	0.50	1.20	0.95[6]	4	1.0-2.0	0.50
Peaches, Stone Fruit, cover	0.50	1.15	0.85	3	1.0-2.0	0.50
Deciduous Orchard, no killing frost						
Apples, Cherries, no cover	0.60	0.95	0.75[6]	4	1.0-2.0	0.50
Peaches, Stone Fruit, no cov.	0.55	0.85	0.65[6]	3	1.0-2.0	0.50
Apples, Cherries, grnd cover	0.80	1.20	0.85[6]	4	1.0-2.0	0.50
Peaches, Stone Fruit, cover	0.80	1.15	0.85[6]	3	1.0-2.0	0.50
EggPlant	0.40	1.05	0.90	1	0.8	0.45
Flax	0.30	1.10	0.25	0.8	1.0-1.5	0.50
Grapes, Concord	0.50	0.85	0.45	2	1.0-2.0	0.45
Grapes, Wine, Table	0.30	0.70	0.45	2	1.0-2.0	0.45
Grass Pasture, Rotation	0.40	0.85	0.85	0.15	0.5-1.5	0.60
Grass, Pasture, Poor Manage.	0.30	0.75	0.75	0.10	0.5-1.5	0.60
Grass[7], mowed, cool season	0.90	0.95	0.95	0.07	0.5-1.0	0.40
mowed, warm season	0.80	0.85	0.85	0.07	0.5-1.0	0.50
Groundnuts	0.40	1.05	0.60	0.6	0.5-1.0	0.50
Hops	0.30	1.05	0.85	7	1.0-1.2	0.50
Kiwi	0.30	1.05	1.05	3-10		
Lentil	0.30	1.15	0.30	0.4	0.6-0.8	
Lettuce	0.30	1.05	0.95	0.4	0.3-0.5	0.30
Melons	0.40	1.05	0.75	0.4	0.8-1.5	0.40
Millet	0.30	1.10	0.30	1.5		
Mint	0.60	1.15	1.10	0.8	0.4-0.8	0.40
Olives	0.65	0.70	0.70	6-10	1.2-1.7	0.65
Onion, dry	0.50	1.05	0.85	0.5	0.3-0.5	0.25
Onion, green	0.50	1.00	1.00	0.5	0.3-0.4	0.25

TABLE 4.12. *(Continued)*

Crop	K_{c1}[2]	K_{c2}	K_{c3}	Max. Ht. (m)	Max. Root (m)	F_{ns}
Open Water, <1 m or in tropics	1.05	1.05	1.05	0.005		
Open Water, >1 m depth	0.75	0.75	1.25[8]	0.005		
Palm Trees	0.95	1.00	1.00	8	0.7-1.0	0.65
Peas, Fresh	0.40	1.15	1.10	0.4	0.6-1.0	0.40
Dry/Seed	0.40	1.15	0.30	0.4	0.6-1.0	0.40
Peppers, fresh	0.35	1.05	0.90	0.7	0.5-1.0	0.25
Pineapple	0.50	0.30	0.30	1.2	0.3-0.6	0.50
Pistachios	0.30	1.15	0.45	3-6	0.3-0.5	0.40
Potato	0.40	1.15	0.75[9]	0.8	0.4-0.6	0.35
Pumpkin	0.40	1.00	0.80	0.4	0.9-1.2	0.40
Radishes	0.30	0.90	0.85	0.3	0.3	0.30
Rice	1.10	1.20	0.90	1	0.5	0.20[10]
Safflower	0.35	1.15	0.25	1	1.0-2.0	0.60
Sisal	0.40	0.50	0.50	2	1.0-2.0	0.70
Sorghum, grain	0.30	1.00	0.55	1	1.0-2.0	0.55
Sorghum, sweet	0.30	1.20	1.05	2-4	1.0-2.0	0.55
Soybeans	0.35	1.15	0.50	0.9	0.6-1.3	0.55
Spinach	0.30	1.00	0.95	0.3	0.3-0.5	0.20
Squash	0.30	0.95	0.75	0.3	0.6-0.9	0.50
Strawberries	0.40	0.85	0.75	0.2	0.2-0.3	0.20
SugarBeet	0.35	1.20	0.95[11]	0.6	0.7-1.2	0.55
SugarCane	0.40	1.25	0.75	3-4	0.8-1.8	0.65
Sunflower	0.30	1.15	0.35	2	0.8-1.5	0.45
Sweet Potatoes	0.40	1.15	0.65	0.8	1.0-1.5	0.65
Tea, nonshaded	0.95	1.00	1.00	1		
Tea, shaded	1.10	1.15	1.15	1		
Tomato	0.40	1.20	0.65	1	0.4-0.8	0.40
Turnip	0.50	1.10	0.95	0.4	0.5-0.8	0.40
Walnut Orchard	0.50	1.05	0.65	4-5	1.7-2.4	0.50
Wetlands - temperate climate						
Cattails, Bulrushes, killing frost	0.30	1.20	0.30	2		
Cattails, Bulrushes, no frost	0.60	1.20	0.60	2		
Short Veg., no frost	1.05	1.10	1.10	0.3		
Reed Swamp, standing water	0.80	0.90	0.90	1		
Reed Swamp, moist soil	0.60	0.70	0.70	1		
Winter Wheat	0.30	1.15	0.25	1	1.5-1.8	0.55

[1]Values for K_{c2} and K_{c3} represent those for a subhumid climate ($RH_{min} \sim 45\%$) with moderate wind speed (averaging 2 m s^{-1}). For more humid or arid conditions or for more or less windy conditions, K_{c2} and K_{c3} should be modified as follows:

$$K_{c2} = K_{c2\,table} + (0.1 + 0.04\,u_2 - 0.004\,RH_{min})(h/3)^{0.3}$$

$$K_{c3} = K_{c3\,table} + (0.1 + 0.04\,u_2 - 0.004\,RH_{min})(h/3)^{0.3} \qquad \text{when } K_{c3} \geq 0.4$$

$$K_{c3} = K_{c3\,table} + 0.001\,(RH_{min} - 45) \qquad \text{when } K_{c3} < 0.4$$

where h is the maximum plant height, m, u_2 is mean wind speed at 2 m over grass, m s^{-1} and RH_{min} is minimum daytime relative humidity, %.

[2]General values for K_{c1} under infrequent soil wetting (~ each 10 days). K_{c1} is a function of wetting interval and potential evaporation rate during the initial and development periods and is more accurately estimated using Eq. 4.139 and 4.140 or by employing "basal" calculations with calculated K_s.

[3]The coefficients for hay crops represent immediately following cutting; full cover; and immediately before cutting, respectively. An overall mean coefficient for alfalfa hay which averages cutting and noncutting periods is 1.05 and for clover is 1.00. K_{c3} for alfalfa seed, which is never cut for forage or hay is about 0.5.

[4]The first K_{c3} value is for harvest at high grain moisture. The second K_{c3} value is for harvest after complete field drying of grain (to about 18% moisture, wet mass basis).

[5]Crucifers include cabbage, cauliflower, broccoli, and Brussel sprouts.

[6]The K_{c3} value represents the K_c prior to leaf drop.

[7]Cool season grass varieties include dense stands of bluegrass, ryegrass, and fescue. Warm season varieties include bermuda grass and St. Augustine grass. The 0.95 values for cool season grass represent a 0.06 to 0.08 m height under general turf conditions. Where careful water management is practiced and rapid growth is not required, K_c's for turf can be reduced by 0.10.

[8]For deep water in temperate latitudes where large temperature changes in the water body occur during the year, initial and peak period evaporation is low as radiation energy is absorbed into the deep water body as sensible heat. During the fall and winter periods (K_{c3}), heat is released from the water body which increases the evaporation lake above that for grass. K_c's listed for evaporation from shallow and deep water bodies should be used with caution. The user is referred to the section on Evaporation from Water Surfaces.

[9]The K_{c3} value for potatoes is about 0.40 for long season potatoes with vine kill.

[10]F_{ns} for rice is 0.20 of saturation (Doorenbos and Kassam 1979).

[11]The K_{c3} value for sugar beets is 0.60 if no irrigation occurs during the last month.

The mean soil moisture balance for the root zone should be computed on a daily basis using Equations 4.71 and 4.72. Computations are made for the complete z_r root depth, including the evaporation layer. ET_a in Equation 4.71 is calculated as $K_c ET_o$ where K_c is computed using Equation 4.126.

d. Mean Crop Coefficients. In basin-wide water balance studies or irrigation system planning, use of mean crop coefficients may be more useful and convenient than computing a daily K_c based on K_{cb} and K_s. The mean crop curve, K_{cm}, shown in Fig. 4.12 lies above the basal curve by an amount that depends on the frequency of soil wetting. When a mean coefficient is used, usually no additional adjustment is made for the effects of surface soil wetness. Adjustments can be made for the effects of limited soil moisture as:

$$K_c = K_a K_{cm}, \tag{4.134}$$

where K_a is defined in Equation 4.132. Values of K_{cm} during partial crop cover are dependent on precipitation frequency and irrigation practices that wet all or part of the soil surface. K_{cm} curves can be generated from K_{cb} curves for known or simulated precipitation or irrigation frequencies following the K_{cb} approach previously outlined. Additional information on usage is presented in the following section, Using the FAO Cover Coefficient in Basal Calculations.

4. *FAO Grass-Based Crop Coefficients* The FAO-24 report (Doorenbos and Pruitt, 1977) presents a relatively simple procedure for constructing crop (cover) coefficients. In the FAO procedure, a K_c curve such as that shown in Fig. 4.13 is constructed by:

(a) Dividing the growing season into four parts describing crop growth stages:

(1) initial period,
(2) Crop Development period,
(3) Midseason period,
(4) Late Season period.

(b) Placing straight line segments through each of these periods with the lines through the Initial and Midseason periods placed horizontally.

Only three tabularized values of K_c are required to describe and construct an FAO K_c curve. These are K_c of the initial period (K_{c1}), K_c of the midseason (K_{c2}), and K_c at the time of harvest (K_{c3}). K_{c2} represents a mean K_c expected for the midseason period, not the absolute peak K_c reached by the crop.

The four crop growth stages are generally characterized in terms of benchmark crop growth stages or cultivation practices. Definitions of these stages are given in Table 4.13.

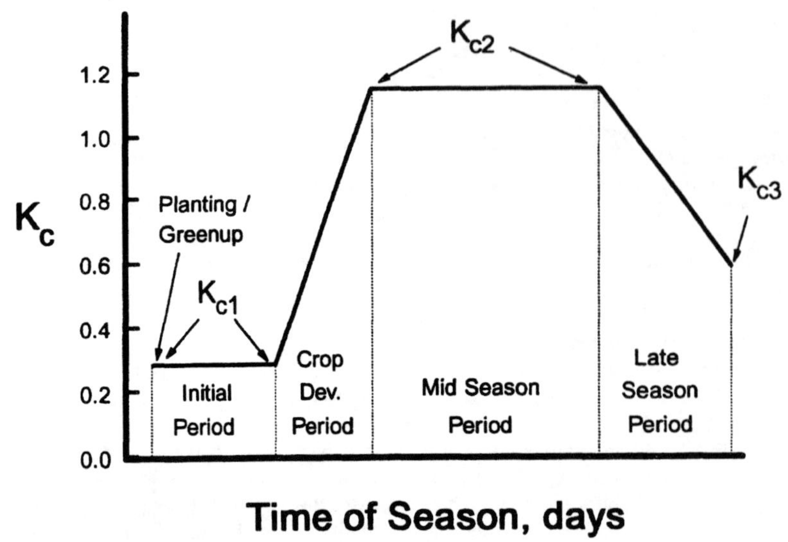

Figure 4.13—FAO Crop Coefficient Curve and Stage Definitions (Doorenbos and Pruitt, 1977).

TABLE 4.13. General Benchmark Growth Stages for Defining FAO Crop Stages.

Initial:	*planting* to *10 percent ground cover* (Highly dependent on crop and time of year)
Crop Development:	*10 percent ground cover* to *effective cover* (effective cover = initiation of flowering for many crops)
Mid Season:	*effective cover* to *start of maturity* (start of maturity is often indicated by yellowing or senescence of leaves, browning of fruit, etc.)
Late Season:	*start of maturity* to *harvest*

Values for K_{c1}, K_{c2}, and K_{c3} are listed in Table 4.12 for various agricultural crops. Most information in Table 4.12 was taken from FAO-24 (Doorenbos and Pruitt, 1977) and an upcoming revision of FAO-24 with additional information from Wright (1982), Snyder et al. (1989a; b) and Jensen et al. (1990).

The K_c values in Table 4.12 represent K_c values for condition of moderate humidity ($RH_{min} \approx 45\%$) and calm to moderate wind speed (0 to 5 m s^{-1} at the 2 m height over grass averaging 2 m s^{-1}). Adjustments should be made to values from Table 4.12 for situations having more or less humidity or wind. These adjustments are described in the next section. In addition, the K_c values presented in Table 4.12 represent potential water use by healthy, well-watered, disease free, dense stands of vegetation. The effect of any reduction in stand density, height, or leaf area of vegetation relative to that attained under pristine conditions caused by disease, water shortage, etc., should be accounted for by reducing the K_{cb}'s listed in the tables. The amount of decrease should be made according to the amount of effective (green) leaf cover (or leaf area) relative to that for healthy vegetation.

a. Adjustment of K_c for Climate. The ratio of ET_c to grass ET_o for many crops increases as wind speed increases and as minimum daily relative humidity decreases. This is due primarily to differences in roughness between taller agricultural crops and clipped grass. Increased roughness and perhaps leaf area makes the aerodynamic aspects of vapor transport more important and significant, thereby increasing K_c. Therefore, for climates with RH_{min} < > 45 percent or with mean wind speeds at 2 meter over grass (u_2) < > 2.0 m s^{-1}, K_{c2} from Table 4.12 should be adjusted as follows:

$$K_{c2} = K_{c2_{table}} + [0.1 + 0.04\, u_2 - 0.004\, RH_{\min}]\left[\frac{h}{3}\right]^{0.3}, \tag{4.135}$$

where $K_{c2_{table}}$ is the value for the K_{c2} taken from Table 4.12 and h is the mean maximum plant height, m, during the midseason period ("full" cover). Equation 4.135 is valid for h up to 20 m. The mean values of h are given in Table 4.12 for the crops and vegetation listed. Where posssible, h should be obtained from field observations.

The K_{c3} at the time of termination of the growing period is adjusted for climate with RH_{min} < > 45%, or with mean wind speed at 2 m (u_2) < > 2.0 s^{-1}, as

$$K_{c3} = K_{c3_{table}} + [0.1 + 0.04\, u_2 - 0.004\, RH_{\min}]\left[\frac{h}{3}\right]^{0.3} \quad (\text{for } K_{c3_{table}} \geq 0.4) \tag{4.136a}$$

$$K_{c3} = K_{c3_{table}} + 0.001\,(RH_{\min} - 45) \quad (\text{for } K_{c3_{table}} < 0.4), \tag{4.136b}$$

where $K_{c3_{table}}$ is the value for K_{c3} taken from Table 4.12.

Minimum daily relative humidity, RH_{min}, is defined as the average minimum daily relative humidity during a growth stage. It is calculated as:

$$RH_{min} = \frac{e^o(T_d)}{e^o(T_{max})}\,100. \tag{4.137}$$

When dewpoint temperature or other hygrometric data are not available, then RH_{min} can be estimated by substituting T_{min} for T_d. Then:

$$RH_{min} \cong \frac{e^o(T_{min})}{e^o(T_{max})} 100. \tag{4.138}$$

This substitution of T_{min} for T_d is reasonable under humid, semihumid, and irrigated conditions.

b. Estimating Initial K_{c1}. Values for K_{c1} in Table 4.12 represent mean soil wetting conditions and are only approximate. These K_{c1}'s should only be used for making approximate estimates of *ET* during planning studies. Accurate estimates of K_{c1} must consider the frequency with which the soil surface is wetted, as this significantly affects the *ET* rate during the initial and development periods which may be predominately evaporation. K_{c2} and K_{c3} are less affected since vegetation during these periods are generally near full ground cover so that effects of surface evaporation are small.

FAO-24 (Doorenbos and Pruitt, 1977) presented a figure for estimating the K_{c1} value as a function of the frequency of irrigation or precipitation and the average value of ET_o during the initial period and first half of the development period. Regression equations which represent the FAO figure have been presented by Cuenca (1987). For soil wetting intervals (I_w) (from irrigation or precipitation) less than 4 days:

$$K_{c1} = [1.286 - 0.27 \ln(I_w)] \exp[(-0.01 - 0.042 \ln(I_w))\, ET_{oi}]. \tag{4.139}$$

For $I_w \geq 4$ days:

$$K_{c1} = 2(I_w)^{-0.49} \exp[(-0.02 - 0.04 \ln(I_w))\, ET_{oi}], \tag{4.140}$$

where ET_{oi} is average grass reference ET_o during the initial period and first half of the development period (mm d^{-1}). As the occurrence of wetted soil due to irrigation or precipitation increases (I_w becomes smaller), the value of the K_c coefficient increases during both the initial and crop development periods.

A second and more accurate procedure for estimating mean K_{c1} is to apply the basal K_c approach to the FAO K_c curves, assuming that the basal K_{c1} is 0.15 when the initial condition is nearly bare soil. This procedure is described in a following section.

c. Lengths of Growth Stages. The FAO-24 publication provided general lengths of growth (development) stages for various types of climates and locations. This information has been condensed into Table 4.14 and supplemented from other sources (primarily Wright (1982), Snyder et al. (1989a; b), and Jensen et al. (1990)).

The rate of vegetative development and attainment of effective full cover is affected by weather conditions, especially by mean daily air temperature (Ritchie and NeSmith, 1991). Therefore, the length in time between planting and effective full cover for various crops or other vegetation varies with climate, latitude, elevation, planting date (if cultivated), and variety. Generally, once effective full cover for a plant canopy has been reached, the rate of phenological development (flowering, seed development, ripening, and senescence (dying of leaf tissue)) often proceeds at a rate that is dependent on plant genotype and is less dependent on weather (Wright, 1982). In some situations, the emergence of vegetation, greenup, and attainment of effective full cover can be predicted using cumulative degree-based regression equations or more sophisticated plant growth models (Sinclair, 1984; Flesch and Dale, 1987; Ritchie and NeSmith, 1991; Ritchie, 1991).

Local observations of plant stage development should be used when possible, with values in Table 4.14 used as a guide and for comparison. Local information can be obtained from farmers, ranchers, agricultural extension agents, and local researchers. When determining stage dates from local observations, the following guidelines may be helpful.

Effective full cover for row crops (e.g., beans, sugar beets, potatoes, and corn) is generally considered to occur when leaves of plants in adjacent rows intermingle so that soil shading becomes nearly complete (or when plants reach nearly full size, if no intermingling occurs). If, for some reason, shading of the soil does not become complete for crops which usually almost completely shade the soil, then the value for

K_{c2} should be scaled down accordingly (perhaps 0.5% decrease in K_{c2} for each 1% of unshaded soil). This may occur due to reduction in plant growth due to disease, grazing, pests, moisture stress or with cultural practices calling for vegetation free strips between crop rows.

Because it is difficult to visually determine when densely sown vegetation such as winter and spring cereals and grasses reach effective full cover, the more easily detectable stage of heading is used (Wright, 1982). For dense grasses, effective full cover occurs at about 0.10–0.15 m height. For thin stands of grass (dry rangeland), grass height may approach 0.3 to 0.5 m before effective full cover is reached, depending on what value is used for K_{c2}. Densely planted forage such as alfalfa and clover reach effective full cover at about 0.3 to 0.4 m.

For many agricultural plants, effective full cover is considered to occur when the leaf area index, *LAI*, approaches 3.0 (Ritchie, 1972; Wright, 1982; Ritchie and NeSmith, 1991). *LAI* is defined as the average total area of leaves (one side) per unit area of ground surface.

The length of the initial and development periods may be relatively short for deciduous trees and shrubs which develop leaves in spring at relatively fast rates. The K_{c1} selected for trees and shrubs should reflect the ground condition prior to leaf bloom (K_{c1} is affected by the amount of grass or weed cover, wetness, mulch density, etc.).

The end of the midseason and beginning of the late season is usually marked by senescence (browning or dying) of leaves, often beginning with the lower leaves of plants. Senescence, along with less efficient stomates in leaf surfaces, causes a reduction in K_c. The length of the late season may be relatively short (on the order of 5 to 10 days) for vegetation killed by frost or for agricultural crops which are harvested fresh. The value of K_{c3} used after the termination of plant growth or following harvest should reflect the condition of soil (surface moisture, mulch cover) and condition of the vegetation following plant death or harvest.

d. FAO K_c Example. An example application for using the FAO K_c procedure under mean soil wetness conditions is presented in Fig. 4.14 for spring barley planted at Logan, Utah (latitude of about 42°). The initial, development, midseason, and late season stages were taken from Table 4.14 to have lengths equal to 20, 25, 60, and 30 days, respectively, for an early spring planting in a high latitude. Values for K_{c1}, K_{c2}, and K_{c3} were selected from Table 4.12 as 0.3, 1.15, and 0.25. Mean wind speed at Logan during the mid and late seasons is about 2 m s^{-1} and average RH_{min} is about 30 %. Therefore, K_{c2} is adjusted using Equation 4.135 as $K_{c2} = 1.15 + (0.1 + 0.04\,(2) - 0.004\,(30))(1/3)^{0.3} = 1.19$. K_{c3} is adjusted using Equation 4.136b as $K_{c3} = 0.25 + 0.001\,(30 - 45) = 0.23$. K_{c3} is slightly lower than the value in Table 4.12 (which is for $RH_{min} = 45\%$), since the adjustment assumes that the soil surface and crop during the late season would be slightly drier under the more arid conditions.

ET_o during the initial and development periods (April) averages 4 mm d^{-1}, and the irrigation and precipitation wetting interval during April is 8 days. Therefore, from Equation 4.140, K_{c1} equals approximately 0.5. This is an improved estimate@tr = of K_{c1} for Logan over that listed in Table 4.13. An FAO K_c curve can then be constructed based on the three K_c values of 0.5, 1.19, and 0.23 and the four lengths of growth stages (20, 25, 60, and 30 days) as shown in Figure 4.14.

5. Using the FAO Cover Coefficients in Basal Calculations Estimates of K_c and ET_c made using the basal K_{cb} approach with calculations of soil evaporation made on daily time steps can be up to 10 to 20% more accurate than mean K_c estimates made using values in Table 4.12 only, especially for the first few days following soil wetting. This is especially true for the initial and development periods.

The FAO crop coefficients can be used in basal calculations by selecting K_{c2} and K_{c3} from Table 4.12 and by setting $K_{cb1} = K_{cb_{min}}$, $K_{cb2} = K_{c2}$ and $K_{cb3} = K_{c3}$ to represent conditions having a nearly bare soil surface. Appropriate adjustment to K_{c2} and K_{c3} should be made using Equations 4.135 to 4.136b to reflect general humidity and wind speed conditions. The basal curve is then drawn using K_{cb1}, K_{cb2}, and K_{cb3} following the same procedures described previously for the general FAO K_c curve. $K_{cb_{min}}$ is used to represent residual evaporation from bare soil. A general value of $K_{cb_{min}}$ for bare soil is 0.15. For grasses, brush and trees, the $K_{cb_{min}}$ at "green up" or leaf development may be on the order of 0.3.

TABLE 4.14. Lengths of Crop Development Stages for Various Planting Periods[1] and Climatic Regions[2] (primary source: Doorenbos and Pruitt, 1977).

Crop	A[1] B[2]	C[3]	A[1] B[2]	C[3]	A[1] B[2]	C[3]
Alfalfa, 1st cutting cycle	ES:M:	8/20/30/0	ES:HL:	10/30/35/0		
other cutting cycles	S :M:	4/11/15/0	S :HL:	5/18/22/0		
Artichokes	P :M :	40/40/250/30	LS:M :	20/40/220/30		
Asparagus	LC:WW:	50/30/100/50	LC: M:	90/30/200/45		
Banana, 1st year	S :M :	120/90/120/60				
2nd year	S :M :	90/90/90/90				
Barley, Wheat, Oats	Su:WW:	15/30/65/40	LF:LL:	15/25/50/30	ES:HL:	20/25/60/30
Beans, green	LW:M:	20/30/30/10	LSu:M:	15/25/25/10		
Beans, dry and Pulses	Su:LL:	15/25/35/20	LS:HL:	20/30/35/20		
Beets, table	EC:WW:	25/30/25/10	ES:M :	25/30/25/10	S :M :	15/25/20/10
Carrots	EC:WW:	20/30/80/20	ES:M :	25/35/40/20	S :HL:	20/30/30/20
Castorbeans	S :HL:	25/40/65/50				
Celery	EC:WW:	25/40/95/20	LC:WW:	30/55/105/20	Su:M :	25/40/45/15
Citrus	Su:M :	120/90/90/60				
Corn, Field	LC:WW:	25/40/45/30	Su:LL:	20/35/40/30	ES:HL:	30/40/50/50
Corn, Sweet	EC:WW:	20/30/50/10	LS:M :	20/25/25/10	ES:HL:	30/40/30/10
Cotton	S :LL:	30/50/60/55	S :HL:	30/50/55/45		
Crucifers[4]	LW:M:	25/35/25/10	S :M:	20/30/20/10	LSu:M:	30/35/90/40
Cucumber	EC:LL:	25/35/50/20	LSu:LL:	20/30/40/15	S :HL:	25/35/50/20
Deciduous orchard	S :LL	20/70/120/60	S :HL:	20/70/90/30		
Eggplant	EC:WW:	30/40/40/20	LS:M :	30/45/40/25		
Flax	EC:LL:	30/40/100/50	S :HL:	25/35/50/40		
Grapes (wine)	S :LL:	20/40/120/60	S :HL:	20/50/90/20	S :HL:	30/60/40/80
Grass Pasture	S :LL:	10/20/—/—	S :HL:	10/20/—/—		
Groundnuts	LC:T:	25/35/45/25	LS:M:	35/45/35/25	LS:HL:	35/35/35/35
Lentil	EC:WW:	25/35/70/40	S :HL:	20/30/60/40		
Lettuce	EC:WW:	30/40/25/10	LC:WW:	35/50/45/10	S :M:	20/30/15/10
Melons	LC:WW:	30/45/65/20	LS:M:	25/35/40/20		
Millet	Su:LL:	15/25/40/25	S :HL:	20/30/55/35		
Onion, dry	EC:WW:	20/35/110/45	S :M:	15/25/70/40		
Onion, green	EC:WW:	20/45/20/10	S :M:	25/30/10/5		
Peas, fresh	ES:WW:	20/25/35/15	LW:M:	25/30/30/15	LS:M:	15/25/35/15
Peppers, fresh	EC:WW:	30/40/110/30	S:M,HL:	30/35/40/20		
Pistachios	LW:M :	20/60/30/40				
Potato	EC:WW:	25/30/30/20	ES:HL:	30/35/50/30	LS:HL:	25/30/45/30
Radishes	LW,M:	10/10/15/5	S :M:	5/10/15/5	Su:HL:	5/10/15/5

The first step in applying the FAO K_c in basal computations is to construct the K_{cb} curve using K_{cb1}, K_{cb2}, and K_{cb3} as described above. Equations 4.126, 4.127 and 4.132 for computing K_c, K_s, and K_a are then applied on a daily time step. K_s should be reduced if evaporable soil moisture in the upper 5 to 15 cm of soil is depleted before $t = t_d$, using Equation 4.131 in conjunction with Equations 4.128-130. K_1 should be set equal to 1.2 to 1.25 for use with the grass-based FAO coefficients and to 1.0 when used with alfalfa-based basal coefficients.

An example of applying the FAO K_c procedure in a basal fashion is shown in Fig. 4.15 for a snap bean crop which was harvested as dry seed. The data in the figure were measured using a precision lysimeter system at Kimberly, Idaho (data from J.L. Wright 1990). The agreement between the estimated daily K_c's from Equation 4.126 (thin, continuous line) and 24-hour measurements (symbols) is relatively good. Measured K_c's were higher following wetting by rainfall or irrigation, as expected. The second and third wet soil evaporation spikes in Fig. 4.15 were less than 1.25 since these were due to wetting by furrow irrigation where f_w, the fraction of soil surface wetted, was 0.5. Some of the day-to-day variation in measured K_c's shown in Fig. 4.15 was due to measurement variation in the lysimeter (about 0 to 5%); however, the majority of variation is due to random errors in ET_o estimation caused by time-scale effects in the daily ET_o equation and by errors in weather measurements. Other errors are due to the assumption

TABLE 4.14. *(Continued)*

Crop	A[1] B[2]	C[3]	A[1] B[2]	C[3]	A[1] B[2]	C[3]
Rice	W :T:	30/30/60/30	Su:T:	30/30/80/40	S :M :	30/30/60/30
Safflower	EC:WW:	35/55/60/40	LW:HL:	25/35/55/30	S :HL:	20/35/45/25
Sorghum	ES:LL:	20/35/45/30	LS:LL:	20/35/40/30	S :HL:	20/30/40/30
Soybeans	W :T:	15/15/40/15	S :HL:	20/35/60/25	LS:HL:	20/25/75/30
Spinach	EC:WW:	20/30/40/10	LW:M:	20/20/25/5	S :M:	20/20/15/5
Squash (W.Pumpkin)	LW:M	20/30/30/15	LS:HL:	25/35/35/25	Su:HL:	20/35/30/25
Squash (Zucc., C.Neck)	EC:WW:	25/35/25/15	S :M	25/35/25/15	Su:M:	20/30/25/15
Sugar Beet	EC:WW:	45/75/80/30	ES:HL:	30/45/60/45	LS:HL:	25/35/50/50
SugarCane, virgin	:T:	50/70/220/140	:LL:	35/60/190/120	:HL:	75/105/330/210
ratoon	:T:	30/50/180/60	:LL:	25/70/135/50	:HL:	35/105/210/70
Sunflower	S: M:	25/35/45/25	LS:HL:	20/35/45/25		
Tomato	EC:WW	35/45/70/30	LC:WW:	30/40/40/25	S: M:	30/40/45/30
Wetlands—Temperate Climate						
Cattails, Bulrush, killing frost	S :HL	10/30/80/20				
Cattails, Bulrush, no frost	S :LL	180/60/90/30				
Short Veg., no frost	S :HL	180/60/90/30				
Winter wheat	EW:HL:	20[5]/70[5]/40/25				

[1]P - Perennial ES - Early Spring S - Spring
LS - Late Spring Su - Summer LSu - Late Summer
LF - Late Fall EW - Early Winter W - Winter
LW - Late Winter EC - Early Cool Season LC - Late Cool Season
[2]T - Tropics WW - Warm winter M - Mediterranean LL - Low latitudes HL - High latitudes, temperate
"Warm Winters" represent regions such as tropics, subtropics, or low Middle East desert
"Low Latitudes" represent interior regions with generally full-year growing seasons.
"High Latitudes" represent regions having temperate or continental types of climates with occurrence of killing frosts during winter periods.

[3]The four values separated by "/" are the predicted lengths of the four FAO growth stages, namely, 1) initial, 2) development, 3) midseason, and 4) late season.

[4]Crucifers include cabbage, cauliflower, broccoli, and Brussel sprouts. Wide range in length of season is due to varietal and species differences.

[5]These periods for winter wheat lengthen in frozen climates according to days having zero growth potential.

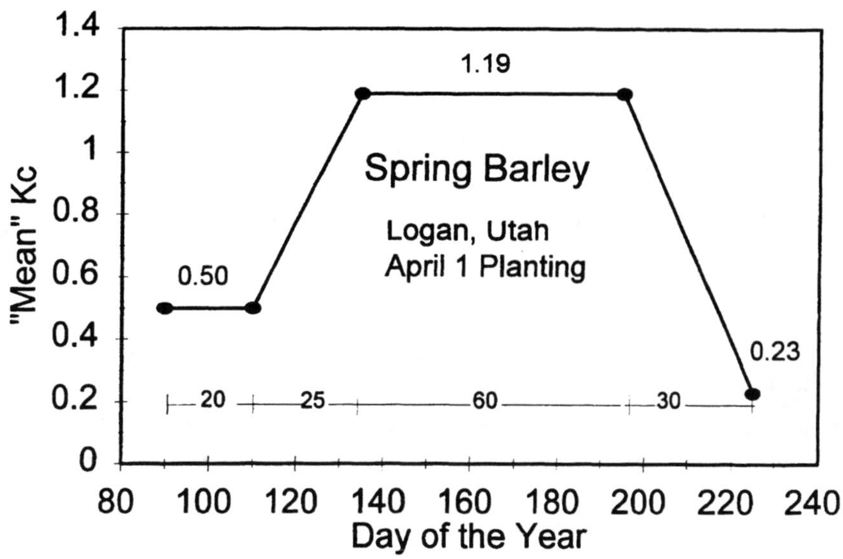

Figure 4.14—Example Construction of an FAO-24 Crop or Cover Coefficient Curve.

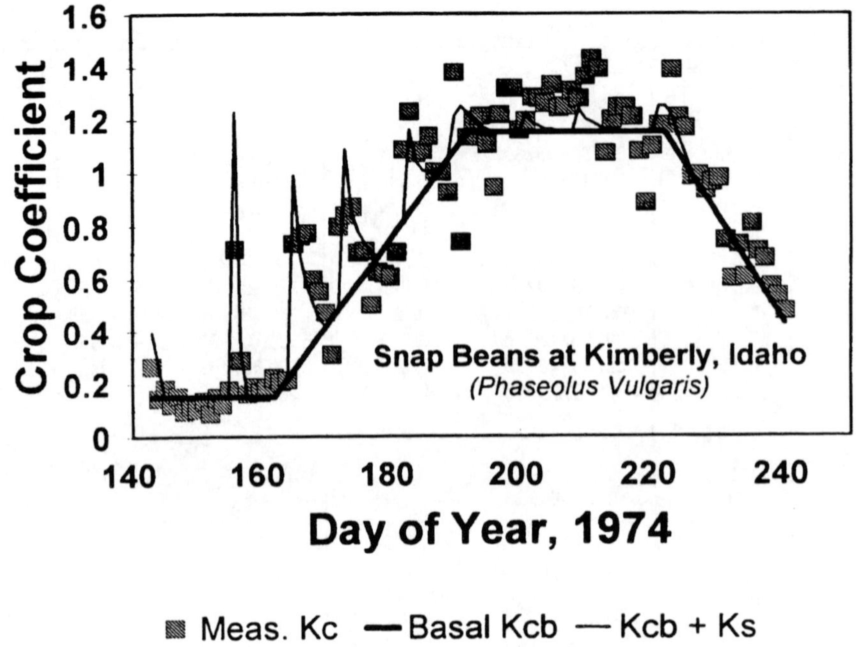

Figure 4.15—Measured and Predicted Daily Crop Coefficients for a Dry, Edible Bean Crop at Kimberly, ID. The Basal Crop Curve (K_{cb}) Was Derived from Table 4.12 and setting $K_{cl} = K_{cb_{min}} = 0.15$. (Data from Wright, 1990).

of a relatively constant ratio of ET_c to ET_o (constant K_{cb}) from day to day, whereas in reality, K_{cb} will vary with changes in wind speed and humidity from day to day.

6. Estimating K_c Curves for Natural Vegetation The two-step K_cET_o approach provides a simple, convenient way to characterize the estimated cover ET_c from natural vegetation. This is done by developing cover (crop) coefficient curves which represent the ratios of ET_c to ET_o during various growth stages. Estimation of lengths of growth stages was described in the section on Lengths of Growth Stages. Estimation of the K_c during the peak growth period (K_{c2}) should be made according to the amount of ground shaded by vegetation, the density, and height of plants, and the amount of stomatal regulation under moist soil conditions. The value of K_c for conditions of low soil moisture availability is generally determined using Equation 4.132.

The K_c development process should adhere to upper limits for K_c of 1.1 for an alfalfa reference and 1.25 for a grass reference for stands of vegetation larger than 500 to 2000 m^2. *ET* from these stands is governed by energy exchange principles and by the principle of conservation of energy as discussed in the previous sections on Field Scale Applications and Small Expanses of Vegetation. K_c's for small stands (< 500 m^2) should also adhere to these limits when the vegetation height and leaf area are less than or equal to those of surrounding vegetation and when soil moisture availability is similar. Only under conditions of "clothesline effects" (where vegetation height is greater than surroundings) or "oasis effects" (where vegetation has higher soil moisture availability than surroundings) will peak K_c's exceed the limits stated above. The user should exercise caution when extrapolating *ET* measurements from small vegetation stands or plots to large stands or regions, as overestimation of regional *ET* may occur.

As discussed in the section on Lengths of Growth Stages, the value for K_{c2} for natural vegetation should be reduced when plant density or leaf area are low (LAI < 3). Where LAI can be measured or approximated, a peak basal K_{cb2} for vegetation can be approximated under normal conditions as:

$$\hat{K}_{cb2} = K_{cb_{min}} + (K_{cb_h} - K_{cb_{min}})\left(\frac{LAI}{3}\right)^{0.5} \quad \text{for } LAI < 3.0, \tag{4.141}$$

where $\hat{K}_{cb2}$ is an approximation for K_{cb2}, K_{cb_h} is the estimated basal K_c during peak plant growth (nearly maximum plant height) and $K_{cb_{min}}$ is the minimum basal K_c for bare soil ($K_{cbmin} \approx 0.15$). For large stand size (> about 500 m^2), K_{cb_h} (for use with grass reference ETo) can be approximated as a function of climate and mean plant height as:

$$K_{cb_h} = \min(1.0 + 0.1h, 1.20) + (0.1 + 0.04\, u_2 - 0.004\, RH_{min}) \left(\frac{h}{3}\right)^{0.3}, \tag{4.142}$$

where u_2 is wind speed at 2 m height, m s^{-1}, RH_{min} is mean daily minimum relative humidity, and h is mean maximum plant height at "full cover," m. The "min" function selects the minimum expression in the parentheses. Equation 4.142 is a generalized approximation of K_{cb_h}, where the mean stomatal control by vegetation under well watered conditions is similar to that of most agricultural vegetation (i.e., $r_l \approx$ 100 sm^{-1}). When the mean stomatal control by vegetation is greater than for agricultural vegetation then Equation 4.142 should be reduced by about 10 or 20 percent for each doubling of r_l above 100 sm^{-1}. The form of Equation 4.142 adheres to trends followed in the K_c tables contained in FAO-24 (Doorenbos and Pruitt, 1977), where the K_c of tall vegetation increases as the wind speed increases and as the relative humidity decreases, due to aerodynamic factors. For small, isolated stand sizes, K_{cb_h} may need to be increased beyond the value estimated using Equation 4.142 as discussed in previous sections.

The relationship in Equation 4.141 is similar to one used by Ritchie (1974). The 0.5 exponent represents the effect of microscale advection (transfer) of sensible heat from dry soil surfaces between plants to plant leaves, thereby increasing ET per unit leaf area. For $LAI > 3$, K_{cb2} can be estimated as $K_{cb2} = K_{cb_h}$. The LAI used in Equation 4.141 should be the "green" LAI representing only healthy leaves which can be active in vapor transfer.

Where only estimates of the fraction of ground surface covered by vegetation are available, the following approximation for K_{cb2} ($\hat{K}_{cb2}$) can be made:

$$\hat{K}_{cb2} = K_{cb_{min}} + (K_{cb_h} - K_{cb_{min}}) (F_{gc})^{\frac{1}{(1+h)}}, \tag{4.143}$$

where F_{gc} is the fraction of ground cover and h is the mean maximum vegetation height, m. K_{cb_h} is estimated using Equation 4.142. Equation 4.143 should be used with caution, as it is only a rough estimate of the maximum K_{cb} expected during peak plant growth and water use periods having a dry soil surface. The K_{cb2} estimated by Equation 4.143 should be applied as a basal coefficient, since the actual K_c may increase to beyond 1.0 following precipitation, even if K_{cb2} is small, due to surface evaporation among the sparse vegetation. In addition, K_c should be reduced using K_a from Equation 4.132 when soil moisture is low.

Equation 4.143 suggests that as h increases, total leaf area and resulting net radiation capture increases, thereby increasing K_c. In addition, as h increases, more opportunity for microadvection of heat from soil to canopy occurs and turbulent exchange within the canopy increases for the same amount of ground coverage. Both of these increases heighten the relative magnitude of K_{cb2}. Values for K_{cb3} can be scaled from K_{cb2} in proportion to the health and leaf condition of the vegetation at termination and the length of the late season period (i.e., whether leaves senesce slowly or are killed by frost). The F_{gc} parameter and h are probably the simplest indices to estimate in the field. Again, Equations 4.141 and 4.143 should be used only as general or preliminary estimates of K_{cb}.

In summary, the convenience of the K_c approach is in its adherence to energy conservation principles and the implicit incorporation of environmental, boundary layer, and physiological factors affecting the ET_c from the particular vegetation. Many of these factors must be explicitly estimated when using a direct estimation equation such as the Penman-Monteith equation discussed in the next section.

VII. EVAPOTRANSPIRATION FROM LAND SURFACES—DIRECT PENMAN-MONTEITH

Direct application of the Penman-Monteith or other resistance-based combination methods has the advantage of allowing the user to alter the physically-based parameters of surface roughness and bulk surface resistance to represent characteristics of the surface or vegetation type in question. This allows the user to objectively characterize a surface via visual observation, remote sensing, or photography and then to estimate the corresponding roughness and bulk surface resistance parameters using approaches discussed in subsequent sections. Advantages of the Penman-Monteith equation are the requirement of weather measurements at one reference height only, rather than requiring the measurement of gradients which is more difficult. In addition, the Penman-Monteith equation appears to be less sensitive to omission of stability correction adjustments as compared to gradient methods. The Penman-Monteith equation can be used to estimate *ET* on hourly, daily, and even monthly calculation time steps (Allen et al., 1994b).

Disadvantages of the Penman-Monteith equation for application in hydrology include problems in estimating *ET* from sparse or multi-layered vegetation such as for forests in semi-arid regions. In these situations, mean heights and locations of sinks for momentum and radiation are different from mean heights and locations of sources of λE and H. These variations create differences in eddy diffusivities (aerodynamic resistances) within and directly above canopies and also create problems in characterizing average relationships between leaf stomatal resistance and solar radiation (Stewart, 1983; Denmead, 1984).

Other disadvantages of the Penman-Monteith and other combination equations include difficulties in quantifying bulk surface resistance for complex canopies, especially when soil moisture is limiting. In addition, significant errors may be introduced when weather data used in the *ET* equation are measured above surfaces which are significantly different from those being quantified, for example, using weather measurements taken over clipped grass to estimate *ET* from forest vegetation.

A. Types of Applications

The direct application of the Penman-Monteith equation utilizes the basic Penman-Monteith combination equation (Equation 4.109) with aerodynamic resistance, r_a, computed using Equation 4.110. Bulk stomatal resistance is generally estimated as a function of leaf area index (*LAI*) and an average stomatal resistance. Alternatively, a minimum stomatal resistance ($r_{l_{min}}$) for well-ventilated, sunlit leaves with no soil moisture limitation can be divided by multiplicative stomatal conductance reducing functions, $g()$, which are based on environmental variables (Stewart, 1989) as discussed in a following section.

1. *Roughness Length and Zero-Plane Displacement* Roughness length for momentum transfer, z_{om}, and zero plane displacement, d, required in Equation 4.110 can be estimated for relatively dense vegetation canopies ($LAI > 2$) using Equations 4.144 and 4.145 (Brutsaert, 1975; Monteith, 1981; Allen et al., 1989; Jensen et al., 1990).

$$z_{om} = 0.12\ h \tag{4.144}$$

$$d = 0.67\ h, \tag{4.145}$$

where h is the mean maximum height of the plant canopy. Units for z_{om}, d, and h are the same. For taller vegetation or sparse vegetation, more complex and elaborate equations may be required (Tanner, 1968; Lettau, 1969; Plate, 1971; Brutsaert, 1975; Monteith, 1981). For example, Equation 4.146 by Lettau (1969):

$$z_{om} = 0.5\ h^{*}\ \frac{s}{S} \tag{4.146}$$

where h is the vegetation height in m, h^* is the effective obstacle height in m, s is the silhouette area (area "seen" by wind), m^2, and S is the horizontal plane area (total ground area divided by the number of roughness elements), m^2.

Thom (1971) and Garratt and Hicks (1973) proposed the following equation for estimating z_{om} for forest canopies:

$$z_{om} = \sigma\,(h - d) \tag{4.147}$$

or alternatively

$$\frac{z_{om}}{h} = \sigma\,(1 - \frac{d}{h}), \tag{4.148}$$

where σ, an empirical factor, was suggested to be equal to 0.36, which is equivalent to the relationship suggested by Brutsaert (1982) for dense, uniform plant stands:

$$z_{om} = \frac{1}{e}\,(h - d), \tag{4.149}$$

where e is the natural number (2.718). When d is estimated as 0.67 h, then Equation 4.147 with $\sigma = 0.36$ and Equation 4.149 both reduce to $z_{om} = 0.12\,h$, which is equivalent to Equation 4.144.

Measurements of z_{om}/h and d/h by Leuning and Attiwill (1978) for 27-m eucalyptus trees having projected areas of 80% of the ground surface, measurements by Jarvis et al. (1976) for thirteen coniferous forest regions having mean tree heights of 10 to 28 m, and measurements by de Bruin and Moore (1985) for 18-m tall Scots pine (*Pinus sylvestris* L.) indicate a value for $\sigma = 0.22$. Canopy densities in these studies were not specified. Moore (1974) determined a value of $\sigma = 0.27 \pm 0.07$ for 105 published estimates of d, z_{om}, and h for vegetation types ranging from smooth grass to forest.

The utility of Equation 4.147 is that it is applicable to canopies where zero-plane displacement varies with plant density and height. Seginer (1974) discussed the variation in z_{om}/h and d/h with increasing canopy density from a theoretical point of view. It is logical that z_{om}/h varies with stand density, since as stand density increases, more roughness elements are present to effect momentum transfer to the surface. However, as density increases beyond a particular limit, the logarithmic wind profile begins to become displaced upward in the canopy (d increases) so that less of the lengths of vegetation are effective in momentum transfer and z_{om}/h begins to decrease.

Following the logic presented above, Perrier (1982) proposed relationships which predict z_{om}/h and d/h to vary with LAI for well-defined canopies:

$$\frac{z_{om}}{h} = (1 - e^{\frac{-a\,LAI}{2}})\,e^{\frac{-a\,LAI}{2}} \tag{4.150}$$

$$\frac{d}{h} = 1 - \left(1 - \left[e^{\frac{-a\,LAI}{2}}\right]\right)\left(\frac{2}{a\,LAI}\right), \tag{4.151}$$

where LAI is total leaf area index ($m^2\ m^{-2}$) and a is an adjustment factor for LAI distribution within the canopy. Parameter $a = (2\,f)$ for $f \geq 0.5$ and $a = (2\,(1-f))^{-1}$ for $f < 0.5$, where f is the proportion of LAI lying above $h/2$. Estimations by Equations 4.150 and 4.151 are shown in Fig. 4.16 for $f = 0.5$ (uniform canopy structure), $f = 0.7$ (top heavy canopy), and $f = 0.3$ (sparsely topped canopy). Also included in Fig. 4.16 are z_{om}/h and d/h reported by Jarvis (1976) for coniferous forests and z_{om}/h estimated by Equation 4.148 using d estimated by Equation 4.151 with f set equal to 0.7 to resemble structures of pine trees. The

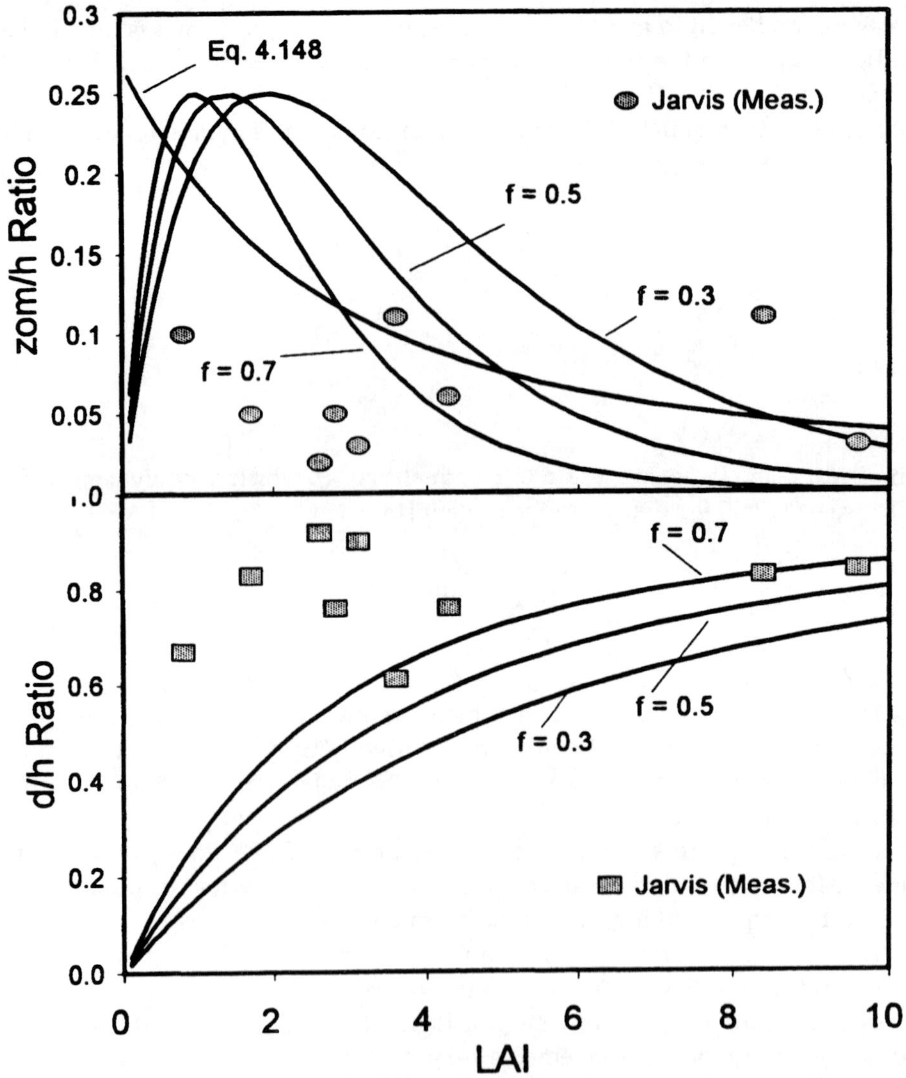

Figure 4.16—Ratios of z_{om}/h and d/h Computed Using Equations 4.150 and 4.151 by Perrier (1982), and z_{om}/h Estimated Using Equation 4.148 with σ = 0.27 and with d Estimated Using Equation 4.151 with f = 0.7 Measurements are from (Jarvis, 1976).

agreement in Fig. 4.16 between Equations 4.150, 4.151 and measured values by Jarvis for coniferous forests is poor for $LAI < 3$, where d was underestimated and z_{om} was overestimated.

It is difficult to recommend which z_{om}/h relations are best to estimate z_{om} for sparse or tall canopies. Equations 4.150 and 4.151 may adequately estimate z_{om}/h and d/h for dense forests and for sparse vegetation where LAI and h are large. The Perrier functions are intuitively reasonable, as they estimate small d for low LAI and a peak in z_{om} as stand density thickens with $d/h < 0.5$. The ratio z_{om}/h estimated for $LAI > 5$ may become too low for $f > 0.3$. This can be corrected by establishing a lower limit on z_{om}/h in Equation 4.150 (e.g., 0.10) for $LAI > 5$. The use of Equation 4.147 in conjunction with d estimated using Equation 4.151 may be a useful approach, with σ in the range of 0.20 to 0.40; however, this equation combination estimates large z_{om}/h for small values of LAI (< 0.5), which may be unreasonable. At values of $LAI < 0.5$, r_s estimated for use in the Penman-Monteith equation should be relatively high, thereby keeping ET estimates low.

As a sample application of Equation 4.146, Lettau (1969) estimated s and S for 0.1 m clipped fescue grass cover as 4 cm^2 and 16 cm^2, so that Equation 4.146 estimated z_{om} = 0.013 m. Equation 4.144 estimates z_{om} = 0.012 m for 0.1 m grass cover, which is equivalent. For full cover alfalfa, however, where h = 0.5 m, s = 60 cm^2, and S = 90 cm^2, z_{om} is estimated by Equation 4.146 to be 0.17 m, whereas Equation 4.144 estimates z_{om} = 0.06 m which is in agreement with findings of Allen et al. (1989) and Jensen et al. (1990). This indicates that Equation 4.146 may apply best to sparse vegetation (Lettau, 1969).

Examples of values for z_{om} and ratios of z_{om}/h and d/h reported in the literature are listed in Table 4.15 from various sources. These are general values and can be used to gain an idea of typical values for various types of vegetation.

Research has shown that for many crops, d and z_{om} can vary considerably with wind speed. Szeicz et al. (1973) found z_{om} to range from under 0.01 m at 1.5 m s^{-1} to 0.07 m at 5 m s^{-1} for a crop of sorghum. The zero-plane displacement, d, of flexible crops is particularly susceptible to varying wind; however, for the Penman-Monteith method the estimate of r_a is less sensitive to d when wind, T, and e measurements are taken at elevations which are greater than twice the crop height.

a. Roughness Lengths of Scalars. Generally, roughness lengths for sensible heat and vapor transfer, z_{oh} and z_{ov}, are less than z_{om} due to effects of bluff body transfer (pressure effects) on momentum transfer which impact the value of z_{om} but do not significantly impact values for z_{oh} and z_{ov}. As pointed out by

TABLE 4.15. Examples of Roughness Lengths for Various Surfaces and Associated Ratios of z_{om}/h (primary source: Brutsaert, 1982).

Surface description	z_{om}, m	z_{om}/h	d/h
Mud flats, ice	0.00001		
Smooth airport runway	0.00002		
Large water surface	0.0001-0.0006		
Grass lawn (h ≤ 0.1 m)	0.01	0.1	
Grass at airport	0.0045		
Prairie grass	0.00665		
0.075 m artificial grass	0.01	0.13	
≤0.1 m, thick grass	0.023	0.23	
≤0.5 m, thin grass	0.05	0.1	
0.18 m wheat stubble	0.024	0.14	
grass, scattered bushes	0.04		
1-2 m vegetation, Cape Canaveral	0.2	0.13	
10-15 m trees, Cape Canaveral	0.4-0.7	0.04	
Deciduous Trees[1]	0.015, 0.2, 1.0		
Orchards[1]	0.015, 0.2, 0.3		
Scots pine *Pinus sylvestris* (h = 18 m)[2]		0.069	0.69
Coniferous forests, general (h = 10 − 28 m)[3]		0.02-0.16 (mean = 0.07)	0.67-0.8
27 m eucalyptus[4]	1.9-4.1	0.07-0.15	
Savannah scrub (25% trees, 65% dry grass, 10% burnt grass, sand)	0.4		
Cotton (0.4-1.2 m)[5]	0.03-0.13	0.10	
Snap Beans (0.1-0.6 m)[6]	0.05-0.2	0.08-0.5 (mean = 0.13)	
Maize (0.2-2 m)[7]	0.01-0.14	0.065	
Impervious Urban[8]	1.0		
Bare Rock, Bare Soil[8]	0.005		

[1]MORECS values by Thompson et al. (1981) for defoliated trees, trees at leaf emergence, and trees at full leaf, respectively.
[2]de Bruin and Moore (1985).
[3]Jarvis et al. (1976).
[4]Leuning and Attiwill (1978).
[5]Pieri and Fuchs (1990).
[6]Allen et al. (1995).
[7]Jacobs et al. (1989).
[8]MORECS value by Thompson et al. (1981).
All others are from Brutsaert (1982).

Brutsaert (1982), z_{oh} and z_{ov} can range from values slightly greater than z_{om} for aerodynamically smooth surfaces, e.g., smooth soil or water, to z_{oh}/z_{om} and z_{ov}/z_{om} ratios of 1/7 to 1/12 for permeable-rough surfaces, a category into which most vegetated surfaces fall. In this handbook, we suggest using:

$$z_{oh} = z_{ov} = 0.1\ z_{om} \tag{4.152}$$

for most agricultural surfaces. The ratios z_{oh}/z_{om} and z_{ov}/z_{om} are affected to some degree by the shape of the surface roughness elements and Reynold's number (Garratt and Hicks, 1973). In-depth discussions of the relationship between z_{oh}, z_{ov}, and z_{om} are given by Thom (1972), Garratt and Hicks (1973), Brutsaert (1982), and Verma (1989), where the sublayer Stanton number B^{-1} is used to represent the differences among the roughness lengths. Basically, the ratio z_{oh}/z_{om} decreases to very small values with increasing wind speed and roughness for nonpermeable rough surfaces such as plowed soil and rigid vegetation with large leaves such as beets, cabbage, or vineyards (Brutsaert, 1982; Verma, 1989), whereas z_{oh}/z_{om} is relatively constant for surfaces or vegetation having more porous or fibrous roughness elements, such as grass, corn, and forests (Brutsaert, 1982).

According to data summarized by Garratt and Hicks (1973), the following mean values can be used for general estimates of z_{oh}/z_{om} under conditions of neutral stability: short grass and agricultural crops: $z_{oh}/z_{om} = 0.08$ to 0.14 (mean = 0.10); pine forests: $z_{oh}/z_{om} = 0.35$; lake evaporation: $z_{oh}/z_{om} = 1$. Other observations have included heterogeneous forests with 1 m tall grass understory where $z_{oh}/z_{om} = 0.08$ (Garratt, 1978b). Ratios of z_{ov}/z_{om} can be expected to be similar.

b. Determination of Roughness from Wind Profile Measurements. Roughness length for momentum transfer, z_{om}, can generally be determined for unfamiliar surfaces from wind profile measurements. For most measurements, a recommended minimum of four anemometers are placed above the vegetation canopy with increased spacing between instruments of height in order to better sample the wind profile. Although two anemometers are sufficient to determine z_{om} using the following equation, measurements with four anemometers are necessary to confirm adherence of the wind profile to a logarithmic relationship during neutral periods and to verify integrity of the anemometers used to solve for z_{om}. An equation that works well for solving for z_{om} was recommended by Monteith and Unsworth (1990):

$$z_{om} = \exp\left(\frac{u_2 \ln(z_1 - d) - u_1 \ln(z_2 - d)}{u_2 - u_1}\right), \tag{4.153}$$

where u_2 and u_1 are wind speeds at z_2 and z_1 elevations above the ground surface and d is the zero plane displacement height. z_{om} in Equation 4.153 has the same units as z and d, which is normally m.

When more than two anemometers are present, then Equation 4.153 can be solved for various combinations of two anemometers. If the wind speed profile follows a logarithmic relationship, then the value for z_{om} solved for each combination should be in agreement. If four anemometers are present in the measurement profile, there will be six independent solutions for z_{om}.

Generally, z_{om} is determined from wind speed profile data for only neutral conditions, as stability or instability of the boundary layer profile will influence the wind profile shape. Neutral conditions can be verified by computing a discrete Richardson number, Ri, between the two sensor elevations using Equation 4.86. Generally, a range of $-0.01 < \mathrm{Ri} < 0.01$ is used to indicate neutral buoyancy (Monteith and Unsworth, 1990). Measurements of u and T for profile analyses are normally averaged over periods of 30 minutes or less.

Cellier and Brunet (1992) have presented fairly conclusive evidence that suggests that wind profile measurements must be made at elevations which are at least 2 to 3 times the canopy height or 3 times individual plant or tree spacings, whichever is greater, in order to obtain valid results from Equation 4.153. Jacobs et al. (1989) recommended placing instruments above the $d + 10\ z_{om}$ elevation. At elevations below those recommended, wind, temperature, and vapor density profiles deviate from the classic

Monin-Obukhov profile shapes, and application of Equation 4.153 may underestimate the value for z_{om}. The minimum elevation applies to the lowest anemometer setting, so that sensor masts placed over tall vegetation must be quite high. The thickness of the "roughness sublayer" in which this problem occurs is greater for sparse canopies such as rangeland, arid land forests, and sparsely planted orchards so that the potential for having wind profile measurements within the roughness sublayer increases with decreased vegetation density. For more detailed information concerning this precaution, the reader can consult Cellier and Brunet (1992), Thom et al. (1975), Raupach (1979), Garratt (1980; 1992), Cellier (1986), van de Griend and van Boxel (1989), and Raupach et al. (1991).

c. Determination of Zero-Plane Displacement from Wind Profile Measurements. When three or more measurements of wind speed are available, zero-plane displacement, d, can be obtained under neutral stability conditions from the following equation by Monteith and Unsworth (1990):

$$\frac{(u_1 - u_2)}{(u_1 - u_3)} = \frac{\ln(z_1 - d) - \ln(z_2 - d)}{\ln(z_1 - d) - \ln(z_3 - d)}, \tag{4.154}$$

where u_1, u_2, and u_3 are wind speeds at z_1, z_2, and z_3 heights. z and d have the same units. Equation 4.154 is solved implicitly using a Newton-secant or other solution method. Again, the same precautions in minimum height of anemometer placement must be exercised for determination of d as they are exercised for z_{om}.

2. *Bulk Surface (Stomatal) Resistance* Stomatal resistance of leaves is a dynamic parameter which changes with sunlight illumination, vapor pressure deficit, leaf temperature, and soil water potential (Jarvis, 1976; Stewart, 1989; Price and Black, 1989; Stewart and Verma, 1992). For a single-layer model as the Penman-Monteith equation, the bulk surface resistance, r_s, represents the integration of resistance to vapor movement through leaf stomates, vapor diffusion from soil, and vapor transport within a dense canopy. For sparse crops, an integrated value for r_s is computed by considering resistances for exposed soil and vegetation in parallel, as discussed in a later section. The single-layer Penman-Monteith equation should be considered as only an approximate estimate for ET from sparse vegetation or from crops during early development where the leaf area index is less than 1.0 due to complexities in energy exchange between the soil and canopy. In these situations, the user should consider using multi-layered equations or even the more simplified, but potentially more robust, $K_c\ ET_o$ approach.

For vegetation where $LAI > 1$, the effect of soil evaporation is generally low relative to transpiration and r_s is dominated by the vegetation. Under these conditions, r_s can usually be estimated using Equation 4.116 with effective leaf area, LAI_{eff}, estimated from Equation 4.117 or Equation 4.118. For dense canopies, LAI_{eff} is the upper sunlit portion of the canopy where most heat and vapor transport activity occurs. Equation 4.117 has been used by Szeicz and Long (1969) and Allen et al. (1989) to predict LAI_{eff} for dense pine forest canopies and for wheat and grass and alfalfa reference surfaces. Equation 4.118 (Ben-Mehrez et al., 1992) may serve as a more universal estimate of LAI_{eff} for LAI ranging from sparse to very dense.

Parameter r_l in Equation 4.116 is average stomatal resistance for a single LAI layer ($LAI = 1$) characteristic to ambient soil moisture and other environmental factors such as R_s, VPD, and T. Monteith (1965; 1981) and Sharma (1985) suggested a value of $r_l = 100\ \mathrm{s\ m^{-1}}$ for many agricultural crops under mean daytime conditions. Allen et al. (1989) and Jensen et al. (1990) found that $r_l = 100\ \mathrm{s\ m^{-1}}$ approximated well for grass and alfalfa for 24-hr calculation time steps. Valle et al. (1985) measured average values of $r_l = 120\ \mathrm{s\ m^{-1}}$ for soybeans under ambient conditions. Vanderkimpen et al. (1995) found that $r_l = 100$ s $\mathrm{m^{-1}}$ estimated well for 24-hr calculations of ET from snap beans and wheat.

When stomatal conductance reducing functions, g(env.) (introduced in a later section) are employed, then r_l can be estimated as $r_l = r_{l_{min}}/\mathrm{g(env.)}$, where $r_{l_{min}}$ is a minimum or unconstrained r_l when all environmental parameters are at optimum levels (g(env.) = 1) and where g(env.) is the composite conductance reducing factor for various environmental parameters, introduced in a later section. For most vegetation, r_l is about 1.5 to 3 times the value for $r_{l_{min}}$ since R_s, VPD, and T are usually at lower or higher levels than optimum for full stomatal opening.

Leaf resistance is generally greater for trees than for agricultural crops. In conifers, stomatal pores are usually located at the bottom of an indentation in the leaf epidermis which is filled with wax tubes, resulting in higher r_l relative to nonconifers. Jarvis et al. (1976) summarized reported measurements of r_l for various types of conifers. Minimum values of r_l ranged from 120 to 300 s m^{-1} for new needles of most conifer types to 1100 to 1400 s m^{-1} for 3- and 4-year-old needles (Table 4.16). The larger r_l for conifers enables trees to avoid rapid dehydration due to generally large LAI and small aerodynamic resistance r_a. Jarvis et al. (1976) calculated general ratios of r_s/r_a between 20 to 40 for Sitka spruce and 20 to 70 for Scots pine. Stewart (1988) reported similar values. These ratios contrast with ratios of r_s/r_a for field crops which are much lower, often between 1 and 2 (Monteith, 1965).

a. Estimating Leaf Area. Leaf area indices can often be estimated for various vegetation types as a function of mean plant height, shape, and stand density. Due to the wide diversity in canopy structures and leaf design, *LAI* vs. *h* relationships are unique to specific vegetation types. Allen et al. (1989) suggested Equation 4.119 for dense stands of clipped grass and Equation 4.120 for alfalfa. The equation for clipped grass indicates a proportional increase in leaf area with increasing height of grass due to uniform vertical leaf extension with height. This would probably not be the case with nonclipped range or pasture grass where much of the plant height is due to stem and seed head extension rather than leaf extension. Equation 4.120 for alfalfa suggests an exponentially decreasing rate of increase in LAI with increase in h as stem extension with less leaf development occurs in maturing plants. Alfalfa is generally grown in densely planted stands so that for h greater than approximately 0.2 m, leaf and stem development is essentially one-dimensional (vertical).

Allen et al. (1995) found $LAI = 10\ h^2$, where *h* is plant height in m, to predict well for snap beans (*Phaseolus vulgaris*) planted in 0.6 m rows for $LAI \leq 5.5$ and $h \leq 0.75$ m. The quadratic increase in *LAI* with *h* for snap beans planted in rows was due to the roundness of the plant and extension of leaves in two directions (width and height) with increasing height. This type of relationship may apply to other bush type vegetation planted in rows.

The following relationships were determined for maize, cotton, and grass prairie (Allen, 1993) based on data presented by Jacobs et al. (1989), Pieri and Fuchs (1990), and Verma et al. (1992), respectively: for tall maize, $LAI = 1.6\ h$; for cotton, $LAI = 2.5\ h$ for $h \leq 0.4$ m and $LAI = 3.4 + 2.4 \ln(h)$ for $0.4 < h < 1.2$ m ($r^2 = 0.91$); for rainfed grass prairie in Kansas, LAI = 5 h before leaf senescence. These relationships suggest that simple height-based functions are possible for most agricultural types of crops. The utility of basing *LAI* on *h* is that *h* is a relatively easy parameter to observe or to recollect, and is required for estimating z_{om}. In rangeland and other situations having sparse plant density, plant density may need to be added to *LAI–h* relationships as discussed in the following paragraph.

Leaf area indices for rangelands, brushlands, and forests vary widely depending on grass, shrub, or tree density, maturity, and type. Rauner (1976) reports total *LAI* for deciduous forests ranging from 4 to about 7. Jarvis et al. (1976) list *LAI* for ten conifer species based on projected canopy area at midday. The *LAI* per projected tree area (LAI_{pa}) ranged from 2.5 to 4.5 for many pine species and ranged from 8 to 10 for *Picea abies* and *Picea sitchensis* species of spruce. An areal average LAI was calculated by Jarvis et al. (1976) as:

$$LAI = LAI_{pa}\ A_s\ S_d, \tag{4.155}$$

where LAI_{pa} is the projected area *LAI* (*LAI* based on the shaded ground area during midday), A_s is the midday shaded area per plant (m^2 $plant^{-1}$), and S_d is stand density (plants m^{-2}).

For annual types of vegetation, the increase in LAI during the growing season can often be expressed as a function of time or as a function of thermal units (Jagtap and Jones 1989a), similar to the approach used for crop coefficients. Allen et al. (1995) developed simple, empirical development functions for *LAI*

TABLE 4.16. Typical Values of the Resistance per Unit Leaf Area, r_l, and Bulk Stomatal Resistance, r_s, for Various Canopy Types. Parameters $r_{l_{min}}$ and $r_{s_{min}}$ are minimum daytime values[a] with g(env.) = 1. Adapted from Garratt (1992) and other sources.

Canopy type	r_l s m^{-1}	$r_{l_{min}}$ s m^{-1}	r_s s m^{-1}	$r_{s_{min}}$ s m^{-1}	Reference
Forest					
Tropical			125-150		Shuttleworth (1984)
			80		Shuttleworth (1989)
				50	Dorman and Sellers (1989)
Ficus Benjamin		45			Bailey et al. (1993)
Deciduous			100-160	60 (LAI = 5)	Verma et al. (1986)
			70-100		Shuttleworth (1989)
				50	Dorman and Sellers (1989)
			100-150		Perrier (1982)
		150			Choudhury and Monteith (1986)
Aspen	400				Sikka (1993)
Eucalyptus	200-400				Körner et al. (1979)
Maple	400-700	250			Körner et al. (1979)
Coniferous					
Pinus resinosa (Pine)					Waggoner and Turner (1971), as cited by Jarvis et al. (1976)
current year needles[b]		310			
1 year old		430			
2 year old		860			
3 year old		890			
4 year old		1,100			
Picea sitchansis (Sitka Spruce)					Ludlow and Jarvis (1971) and Neilson et al. (1972) as cited by Jarvis et al. (1976)
Current year needles[b]		120			
1 year old		480			
2 year old		650			
3 year old		1,400			
Mean		300	80	40 (LAI=12)	Milne (1979)
Pseudotsuga menziesii (Douglas Fir)		120-300			Jarvis et al. (1976)
General conifers			90-150		Jarvis et al. (1976)
			70-100	56	Shuttleworth (1989)
				43	Noihan and Planton (1989)
				60	Dorman and Sellers (1989)
				30-60	Kelliher et al. (1993)
			200-300		Perrier (1982)
			100 ± 30		Shuttleworth (1989)
Other					
Savannah			100		Garratt (1978b)
Shrubs		240			Sikka (1993)
Sagebrush (*Artemisia Tridentata*)	1000				Körner et al. (1979)
Mesophytes[c]	100-200				Gates (1980)
		150-500			Cowan and Milthorpe (1968)
Xerophytes[d]		800-2000			Cowan and Milthorpe (1968)
Grassland				40	Dorman and Sellers (1989)
		30-50		40-50 (LAI = 1.3)	Kelliher et al. (1993)
Kansas Prairie (big bluestem) (*Andropogon geradii*)				40 (LAI = 3)	Stewart and Verma (1992)
and Indian grass (*Sorghastrum nutrans*)				65 (LAI = 1.3)	Stewart and Gay (1989)
temperate			100		Perrier (1982)
subtropical			200		
Crops, General	50-320				Slayter (1967)
			40-130		Perrier (1982)
				20-75	Sellers and Dorman (1987)
				20-150	Noilhan and Planton (1989)
			30-35	30	Dorman and Sellers (1989)
	100				Monteith (1965), Sharma (1985)
Grain Sorghum	200	150	100-140 (LAI=2)		Szeicz et al. (1973)
		150			Choudhury and Monteith (1986)

TABLE 4.16. *(Continued)*

Canopy type	r_l s m^{-1}	$r_{l_{min}}$ s m^{-1}	r_s s m^{-1}	$r_{s_{min}}$ s m^{-1}	Reference
Snapbeans (*Phaseolus Vulgaris*)		130			Choudhury and Monteith (1986)
Soybeans	120				Valle et al. (1985)
				50 (LAI = 3.4)	Grant and Baldocchi (1992)
Maize		70	80	40	McGinn and King (1990)
		70		25 (LAI = 3.5)	Jacobs et al. (1989)
		70			Rochette et al. (1991)
	160				Körner et al. (1979)
Barley	150-250 (young to old)		45	40	Szeicz and Long (1969)
			70		Perrier (1982)
Wheat			50	30	Hatfield (1985)
			60		Perrier (1982)
		50			Choudhury and Idso (1985)
Alfalfa	80	50			Cowan and Milthorpe (1968)
			40		Perrier (1982)
			30 (day-time)	20	McGinn and King (1990)
Cotton		20			Stanhill (1976)
Sugar Beets		50			Cowan and Milthorpe (1968)
	100				Körner et al. (1979)
Citrus			250		Perrier (1982)
0.15 m Clipped grass	100-150		80-120	70	Szeicz and Long (1969)
0.10 to 0.12 Irrig. & Clipped Grass	75	40	40-60	25-30	Pruitt (1960)

[a] $r_{l_{min}}$ and $r_{s_{min}}$ represent minimum daytime values of r_l and r_s which occur when all environmental variables are optimum. For example, full levels of irradiance, low vapor pressure deficit, optimum leaf temperature, etc. These levels will occur in dense canopies only for sunlit leaves and only when all other variables are optimum. $r_{l_{min}}$ and $r_{s_{min}}$ should only be used in conjunction with the g( ) functions.
[b] Current year, 1, 2, 3, and 4 year refer to the age of needles measured.
[c] Mesophytes are tropical and temperate plants needing plentiful water for survival.
[d] Xerophytes are plants structurally adapted for life and growth with a limited water supply.

and h vs. cumulative growing degree days for snap beans and wheat, where the dependent variables in the equations were LAI/LAI_{max} and h/h_{max}, respectively. LAI_{max} was maximum LAI of green leaves during the season and h_{max} was maximum obtained plant height. Use of LAI_{max} and h_{max} allowed for application of the functions to other cultivar types. Development of temperature-based functions facilitated the application of the Penman-Monteith equation in a continuous manner with time.

b. General Values of Bulk Surface Resistance r_s. Transpiration increases with increasing LAI until nearly complete closure of the canopy occurs. For agricultural crops, maximum transpiration is often attained after an $LAI = 3$ is reached (Ritchie, 1972; Tanner and Jury, 1976; Wright, 1982; Sharma, 1985). Increases in LAI above 3 generally effect little increase in ET, as most available energy (radiation and sensible heat) has been converted into λE within the canopy when $LAI = 3$. This provides the rationale behind using Equation 4.117 or Equation 4.118 for predicting LAI_{eff}. In general, the Penman-Monteith equation predicts little increase in ET for $LAI > 3$.

Typical values for r_s and r_l reported in the literature are listed in Table 4.16. Additional values for r_s used by the British Meteorological Office rainfall and evaporation calculation system (MORECS) are listed in Table 4.17 and mean "green" LAI values are listed in Table 4.18 (Thompson et al., 1981). Values for r_l in column 1 and r_s in column 3 of Table 4.16 represent constrained surface resistance for single LAI and complete canopies and are intended for use in daily, monthly, or hourly ET calculations where g functions are not employed. Values for $r_{l_{min}}$ and $r_{s_{min}}$ in Table 4.16 represent minimum r_l and r_s where all

TABLE 4.17. Generalized Daytime Values of Bulk Surface Resistance for Dense Green Cover in Great Britain Having Adequate Soil Moisture (Thompson et al., 1981).

Type of cover	r_s, s m^{-1}
Grass and Riparian vegetation	80, 80, 60, 50, 40, 60, 60, 70, 70, 70, 80, 80 (Jan-Dec)
Cereals	40
Potatoes and Sugar Beets	40
Deciduous trees	80
Conifers	70[1]
Upland vegetation	120 Jan-Mar, 100 Apr-Sept, 120 Oct-Dec
Bare soil	100
Water	0

[1]At zero vapor pressure deficit and 20°C and assumed independent of soil surface resistance.

TABLE 4.18. Maximum Leaf Area Indices for Dense Green Cover in Great Britain as used in MORECS (Thompson et al., 1981).

Type of cover	Green leaf area index
Grass and Riparian vegetation	2.0, 2.0, 3.0, 4.0, 5.0, 5.0, 5.0, 5.0, 4.0, 3.0, 2.5, 2.0 (Jan-Dec)
Cereals	5.0
Potatoes	4.0
Sugar beets	4.0
Deciduous trees	6.0
Conifers	6.0

environmental variables (namely R_s, T, VPD, and θ) are at optimum levels. Therefore these values for $r_{l_{min}}$ and $r_{s_{min}}$ should only be used in conjunction with the g functions.

The MORECS r_s values in Table 4.17 were developed to support direct application of the Penman-Monteith equation across Great Britain. Parallel resistance calculations were used to integrate effects of soil evaporation and plant transpiration as discussed in a following section. Calculations in MORECS were made on daily time steps for daylight periods, so that values of r_s in Table 4.17 represent general daytime values of r_s. Mean monthly values of r_s listed for grass and riparian vegetation decrease during spring months with leaf development and increase late in the growing season as leaves senesce (die off) or go dormant. Additional values for r_l, which were summarized from a wide range of literature sources, can be found in Körner et al. (1979). All values in Table 4.16 and 4.17 may be biased by the procedure used to determine them.

For hourly or shorter calculation time steps, it may be useful to adjust r_l for levels of R_s, VPD, T, and θ as described in the following section. For hydrologic moisture balance studies, where water balances and ET are estimated on daily or longer calculation time steps, mean 24-hr or daytime values for r_s can be used where r_s is estimated as the quotient of mean values for r_l and effective leaf area, with additional adjustment for limited soil moisture.

c. Estimation of Surface Resistance for Hourly or Shorter Periods. For most vegetation, leaf resistance decreases with increasing irradiance (solar radiation) and may increase with increasing vapor pressure deficit. In addition, leaf resistance increases with decreasing soil moisture due to partial stomatal closure. Jarvis (1976), Stewart (1988; 1989), Price and Black (1989), Stewart and Verma (1992), and others have proposed computing bulk surface resistance for short time periods in terms of meteorological variables and soil moisture deficit. Generally, computations are expressed in terms of bulk surface conductance, g_s, which is the reciprocal of r_s ($g_2 = 1/r_s$).

Turner (1991) has reviewed environmental influences on stomatal conductance and the various challenges in measuring and determining specific relationships. Collatz et al. (1991) present an excellent discussion on general physiological interactions related to stomatal conductance and describe a unified stomatal conductance function which integrates physiological interactions among photosynthesis rate, carbon dioxide concentration, temperature and relative humidity at the leaf surface, and solar radiation. A single, integrated process is used to associate levels of R_s and T with photosynthesis rate which, in turn, is used to predict the effect of relative humidity on stomatal opening using an iterative solution process; however, this procedure is beyond the scope of this handbook, and the following simplified stomatal conductance functions are presented for general hydrologic use. The following functions should be considered to reflect current state-of-the-art applications for simulating hydrologic processes at the time of this writing. Substantial advancements are being made in this area, and the reader should consult current literature in order to update what is presented in this chapter.

A widely used surface conductance model proposed by Jarvis (1976) uses a multiplicative means for incorporating the effects of specific environmental factors (Stewart, 1989):

$$g_s = g_{\max}\, g(env.) = g_{\max}\left[g(LAI)\; g(R_s)\; g(VPD)\; g(T)\; g(\theta)\right], \tag{4.156}$$

where g_{max} is the maximum bulk surface conductance when all other functions are not limiting. Functions $g(LAI)$, $g(R_s)$, $g(VPD)$, $g(T)$, and $g(\theta)$ represent reducing functions for leaf area, solar radiation, vapor pressure deficit, air temperature, and soil moisture, respectively. Variable g(env.) represents the conjugate effect of the individual functions. All reducing functions in Equation 4.156 are limited by $0 \le g(\,) \le 1$, including the accumulative g(env.). The form of Equation 4.156 assumes that each environmental parameter will independently reduce g_s (increase r_s), with the reduction accelerated by changes in other parameters.

Price and Black (1989) present an alternative equation to Equation 4.156, where they find better agreement to measured data when the reduction in g_s is based on only the most limiting environmental parameter. The equation suggested by Price and Black is:

$$g_{\max}\, g(env.) = g_{\max}\left[minimum(g(LAI),\; g(R_s),\; g(VPD),\; g(T),\; g\,(\theta))\right] \tag{4.157}$$

where the $g(\,)$ functions are the same as those defined for Equation 4.156 and are also limited to $0 \le g(\,) \le 1$. In the examples provided near the end of this chapter, the form of Equation 4.156 is employed.

When g_{max} is not measured, g_{max} can be estimated as $g_l \cdot LAI_{eff}$, where LAI_{eff} is effective leaf area, estimated using Equation 4.117 or 4.118 and g_l is the maximum stomatal conductance per unit leaf area in the same units as g_{max}. Estimates of g_l can be made from Table 4.16, where $g_l = 1/r_{l_{min}}$. In addition, g_{max} can be estimated from Tables 4.16 and 4.17, where $g_{max} = 1/r_{s_{min}}$. Values for g_l and g_{max} decrease with leaf age (senescence), as noted in Table 4.16 for conifer needles over a multi-year period. Conductances of leaves of annual plants may decrease with time through the growing season (Burrows and Milthorpe, 1976), especially if damage from insects, environmental stress (moisture and temperature), disease, or air pollution occurs.

Forms of reducing functions that have been suggested for use in Equations 4.156 and 4.157 include the following series of equations. These functions are plotted in Fig. 4.17 for $g(R_s)$, $g(VPD)$, $g(T)$, and $g(\theta)$.

For g(LAI) Stewart (1989) proposes:

$$g(LAI) = \frac{LAI}{LAI_{\max}}, \tag{4.158}$$

where LAI_{max} is the maximum value of LAI during the year. The values for LAI and LAI_{max} in Equation 4.158 should be set equal to "effective" LAI, as defined in Equation 4.117 or 4.118, to account for effects of shading on r_a and r_l within tall, dense canopies and the potential for nonlinearity between LAI_{max} relative to total LAI (Equation 4.118).

For $g(R_s)$ Stewart (1988, 1989) proposes:

$$g(R_s) = \frac{R_s(1000 + K_R)}{1000\ (R_s + K_R)}, \tag{4.159}$$

where R_s is global short wave radiation (W m^{-2}) and K_R is an empirical factor ranging from 40 to 100 W m^{-2} for Scots pine (*Pinus Sylvestris*) (Stewart, 1988). Price and Black (1989) proposed:

$$g(R_s) = 1 - e^{-0.003\ R_s}, \tag{4.160}$$

for a 22-year-old stand of Douglas fir (*Pseudotsuga Menziesii*).

Stewart (1988; 1989) proposes a function for predicting effects of specific humidity deficit on g(VPD), which in terms of vapor pressure deficit is expressed as:

$$g(\text{VPD}) = 1 - K_{\text{VPD}}\text{VPD} \quad \text{for } 0 < \text{VPD} < \text{VPD}_c \tag{4.161}$$

$$g(\text{VPD}) = 1 - K_{\text{VPD}}\text{VPD}_c \quad \text{for VPD} \geq \text{VPD}_c \tag{4.162}$$

with K_{VPD} = 0.50 kPa^{-1} and VPD_c, a threshold vapor pressure deficit, = 1.5 kPa for Scots pine. Stewart and Gay (1989) found K_{VPD} = 0.12 to 0.19 kPa^{-1} for rainfed grass prairie in Kansas. Price and Black (1989) proposed

$$g(VPD) = 2.24\ e^{(-1.6\ VPD^{0.45})} \tag{4.163}$$

for Douglas fir, where g(VPD) ≤ 1. Kelliher et al. (1993) propose a means for computing *VPD* at the leaf surface, which eliminates effects of r_a on the decoupling of the $g(VPD)$ vs. *VPD* relationships that are based on *VPD* computed above the canopy. These adjustments may improve the consistency and transferability of the above *VPD* relationships.

For air temperature Stewart (1988; 1989) proposes a relationship after Jarvis (1976):

$$g(T) = \frac{(T - T_L)\ (T_H - T)^a}{(K_T - T_L)\ (T_H - K_T)^a} \tag{4.164}$$

where

$$a = \frac{(T_H - K_T)}{(K_T - T_L)} \tag{4.165}$$

and T is air temperature °C; T_L and T_H represent lower and upper limits of stomatal activity, where T_L was taken as 0°C and T_H as 40°C by Stewart (1988) for a Scots pine forest; and K_T is an empirical "optimum" conductance temperature fitted from field data. K_T ranged from 17 to 19°C for Scots pine. Articles by Ritchie and NeSmith (1991) and Norman and Arkebauer (1991) suggest values of T_L, K_T, and T_H equal to about 0, 25, and 40°C for wheat, 12, 35, and 48°C for corn and 5, 28, and 46°C for soybeans. Information in McArthur et al. (1975) suggests values of 16, 27, and 35°C for cotton.

Conductance functions based on soil moisture, θ, are generally used in hydrologic studies as surrogates for more direct and perhaps accurate functions based on leaf water potential or soil water potential. The surrogate $g(\theta)$ functions are used since estimates of θ are easily made available from soil water balance calculations, whereas leaf water and soil water potentials must be measured or inferred from soil characteristic curves and plant models. For operational *ET* estimates, $g(\theta)$ are adequate for predicting the reduction in *ET* with reduction in soil moisture availability.

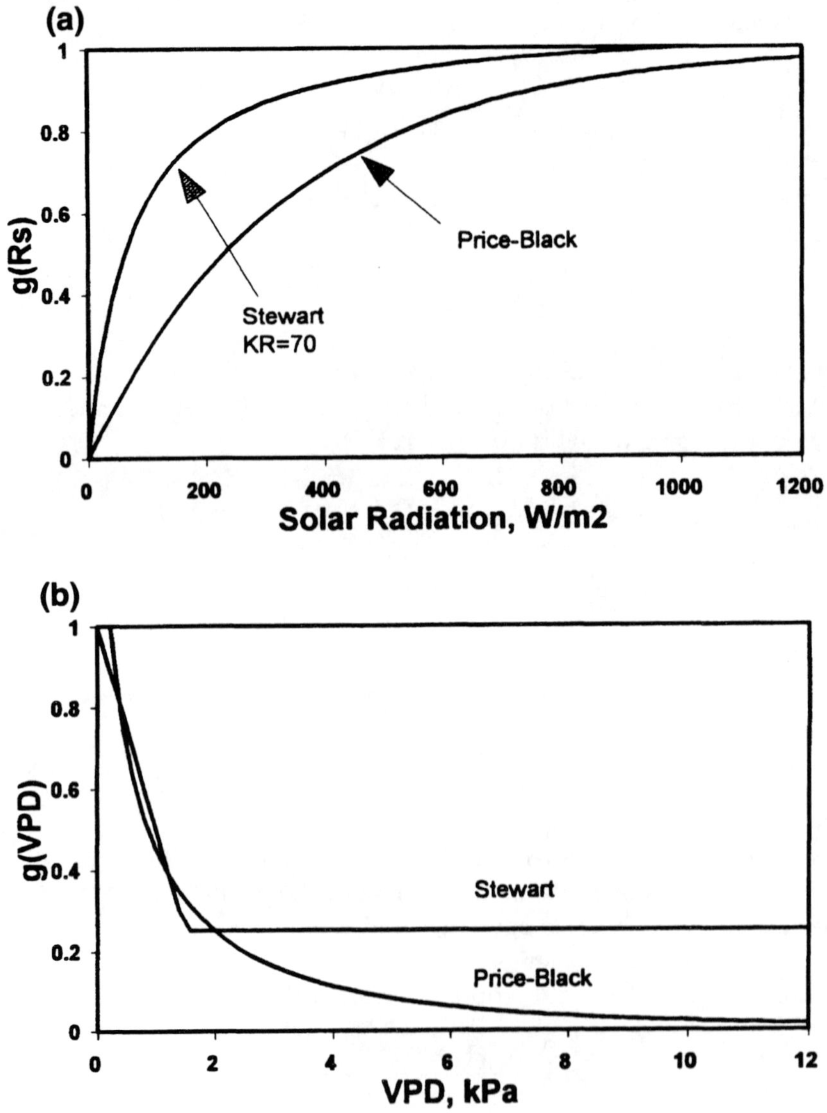

Figure 4.17—Surface Conductance Functions for a) Solar Radiation, b) Vapor Pressure Deficit, c) Air Temperature, and d) Soil Moisture For Various Surfaces (Scots Pine, Douglas Fir, Grass Prairie, and Fescue). (Continued on next page)

For the effect of decreasing soil moisture, three primary functions are recommended. The first is a function similar to that used by Stewart (1988) which is of the form:

$$g(\theta) = 1 - e^{-K_\theta \theta_e}, \tag{4.166}$$

where K_θ is an empirical factor equal to about 6.7 for Scots pine. A second form of the function is by Stewart and Verma (1992):

$$g(\theta) = \frac{1 - e^{-K_\theta \theta_e}}{1 - e^{-K_\theta}}, \tag{4.167}$$

where K_θ ranges from 0.25 to 0.12 for grass prairie. Variable θ_e in Equations 4.166 and 4.167 represents the effective fraction of available soil moisture, calculated as:

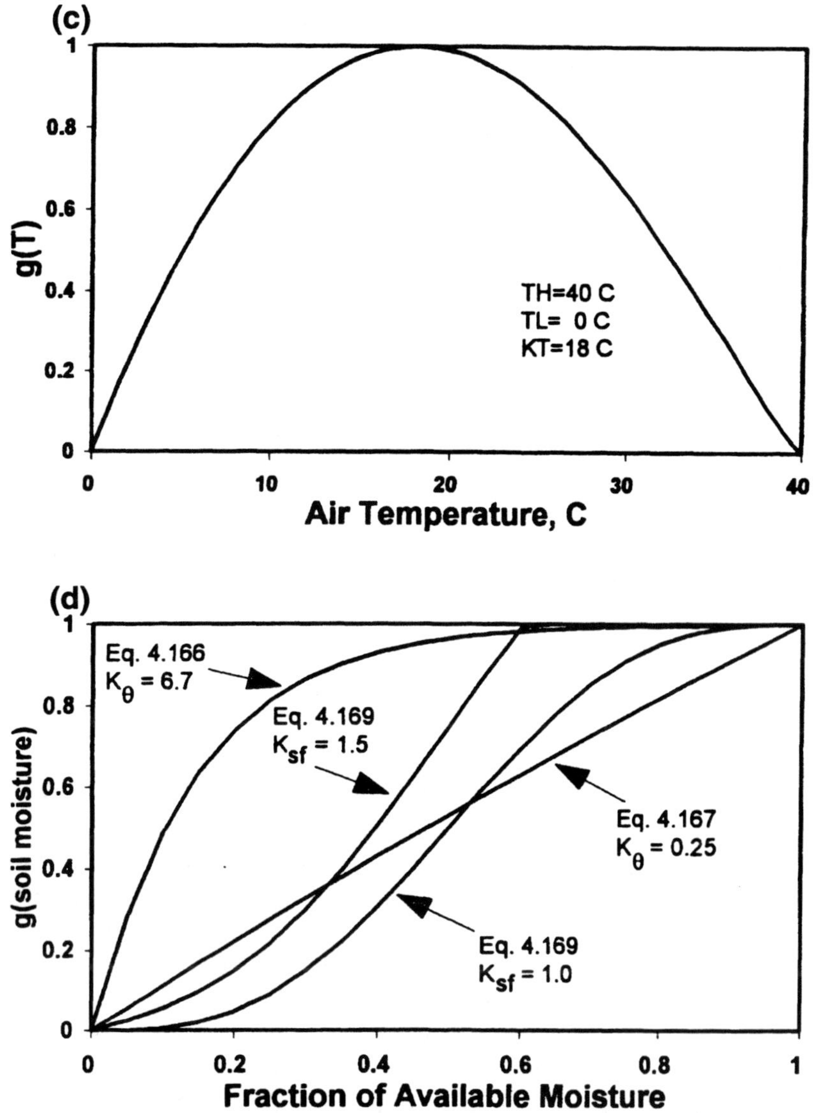

Figure 4.17—Continued from previous page

$$\theta_e = \frac{\theta - \theta_{LL}}{\theta_{DU} - \theta_{LL}}, \tag{4.168}$$

where θ is the mean volumetric soil moisture in the root zone (m^3 m^{-3}), θ_{LL} is the mean volumetric soil moisture at the lower limit of moisture extraction by plant roots (m^3 m^{-3} (Table 4.8)), and θ_{DU} is the mean volumetric soil moisture at the drained upper limit (m^3 m^{-3} (Table 4.8)).

The third $g(\theta)$ function is a sinusoidal relationship:

$$g(\theta) = K_{sf}\theta_e - \frac{\sin(2\pi\theta_e)}{2\pi}, \tag{4.169}$$

where $g(\theta) \leq 1$, and θ_e is defined as in Equation 4.168. K_{sf} is an empirical "tenacity" factor ($K_{sf} \geq 1$) which describes the ability of plants to extract soil moisture before any reduction in $g(\theta)$. K_{sf} ranges from 1 for sensitive plants to 1.5 for moderately sensitive plants to 3 for insensitive (tenacious) plants. The value for $g(\theta)$ from Equation 4.169 must be limited to less than or equal to 1 during computations. Equation 4.169 was proposed by Allen (1993) for fescue grass.

Equation 4.169 predicts values for $g(\theta)$ that reduce slowly from 1 at high levels of θ_e, reaching maximum rates of reduction near $\theta_e = 0.5\ K_{sf}$, and approach 0 when θ_e is approximately 0.2 (Fig. 4.17d). This type of response is realistic, as most plants leave stomates nearly wide open at high levels of θ_e (all other environmental factors at optimum) to maximize transpiration and photosynthesis, but close stomates almost completely before θ_e approaches θ_{LL}. They then "hold on" for sometimes long periods of low θ_e, transpiring at low subsistence rates controlled by diffusion through nearly closed stomates and through the leaf epiderm (Turner, 1979; Jones, 1979; Burrows and Milthorpe, 1976). In some situations, the user may wish to reduce the value for K_{sf} towards 1.0 during periods having high *ET* rates to reflect difficulty of low soil hydraulic conductivity in supplying high *ET* rates.

The natures of Equations 4.166 to 4.169 do not conform to that for the K_a function used in the crop coefficient approach (Equation 4.132 and 4.133), since g_s can reduce (and r_s can increase) by nearly 40 percent in many agricultural situations before significant reductions in λE, as predicted by the Penman-Monteith equation, occur. Effects of applying Equations 4.166 to 4.169 in a daily soil water balance are demonstrated in a following section. Substantial differences exist between the three $g(\theta)$ function forms as shown in Fig. 4.17d. However, as shown in the later example, all three functions provide reasonable results.

Applications of both approaches for reducing ET(g_s vs. θ and K_a vs. θ) are conservative in that they are both based on a soil moisture balance. Therefore, errors in over or underestimating g_s or K_a generally cause smaller relative errors in predicted ET_a over long periods in low rainfall areas, where most precipitation is evaporated during the growing season. Therefore, in an arid or semiarid situation, errors in g_s or K_a may affect the estimated timing in *ET* reduction, but may not significantly affect the total estimate of cumulative *ET*. This may not be the case in semihumid regions, where soil moisture deficits occur over short terms, so that errors in the nature of the $g(\theta)$ and $K_r(\theta)$ functions are more critical.

d. Minimum Stomatal Conductance. Utilization of some of the g( ) functions described in this section can result in g( ) = 0, notably for Equation 4.158 (for LAI = 0), Equations 4.159 and 4.160 (for $R_s = 0$), and Equation 4.164 (for $T \le T_L$ or $T \ge T_H$); however, g( ) = 0 implies that total surface conductance, g_s, equals 0 so that r_s is infinity. This never occurs in nature (except perhaps in the desert), since all vegetation will have some cuticular conductance (about 0.0003 m s^{-1} per unit LAI (Rochette et al., 1991). In addition, a minimum residual g_s (or maximum residual r_s) must be retained to account for evaporation from soil beneath vegetation, for example, by making use of Equation 4.170. As indicated in a later section, r_s is about 50 s m^{-1} for moist bare soil and exceeds 2000 s m^{-1} for dry soil. Therefore, depending on the soil surface moisture conditions and the amount of transfer of sensible heat and radiation through the vegetation cover to the soil, some residual g_s should be employed even for dense vegetation. In analyzing ET data at Davis from grass irrigated six days earlier, Pruitt (1994) found night time values of r_s averaging almost 200 s m^{-1}. This was during a 2-hr period around midnight, (R_s and $g(R_s) = 0$), yet with a 5-m s^{-1} wind at 2 m, 20°C air temperature, a RH of 60%, and a VPD of 0.95 kPa, ET averaged 0.08 mm h^{-1}. A value of r_s = 200 s m^{-1} is equivalent to a minimum g_s of 0.005 m s^{-1} and was used during nighttime periods whenever the value of g_{max} g(env.) fell below this value. When soil surface moisture is measured or can be predicted by a simulation model (using a soil surface layer balance or time since wetting), then the minimum value for g_s can be varied to fit the actual moisture conditions.

e. Application. Other surface conductance or surface resistance functions have been proposed which can be substituted into Equation 4.156 or 4.157 (Szeicz and Long, 1969; Szeicz et al., 1973; Hatfield, 1985; Choudhury, 1983; 1989). In general, unique K coefficients should be determined for g() functions for specific vegetation types. Stewart (1988) demonstrated difficulty in applying some g( ) functions among different years for the same type of vegetation.

Improvements in application of the g(T) and g(VPD) functions can be made when surface temperature and VPD at the leaf surface are used rather than T and VPD at some reference height (Collatz et al., 1991), since the leaf responds most directly to environmental conditions at the leaf surface rather than to conditions in the boundary layer above. However, this increases the difficulty and computational complexity involved in applying the Penman-Monteith equation, since one must conjunctively and iteratively utilize the energy balance equation (Equation 4.40) along with the Penman-Monteith equation to solve for H and then T_o and $g(T_o)$. Dickinson et al. (1984) describe three current regional evapotranspiration models that employ various forms of the g( ) functions described in this section. These models have been used as components of general circulation models (GCM) to assess effects of climate change. Excellent reviews of mechanics and processes of stomatal conductance and environmental interactions have been given by Allaway and Milthorpe (1976) and Burrows and Milthorpe (1976).

It is recommended that all g functions which impact the value of g_s be utilized in estimating g(env.), since the absence of any one of the parameters may bias estimates of g_s upward and may result in overestimation of ET. Note that in Table 4.16, values for r_l (which represent average g(env.) conditions) are generally 2 to 3 times the values given for $r_{s_{min}}$ for the same type of vegetation. Therefore, one should expect, under normal conditions, that g_s calculated from Equations 4.156 and 4.157, when averaged over all daylight periods (preferably weighted according to ET flux density), to be only about one-half the value of g_{max}, especially in semiarid regions. This should occur when all relevant g functions are included in the g_s calculation.

An example of employment of g functions is shown in Fig. 4.18a, where g functions for R_s, T, and VPD are plotted during a typical summer day in Logan, Utah and a spring day in Davis, California (Fig. 4.18b), where ET was near maximum levels and soil moisture was high. These are the same days and locations used in ET calculations shown in Fig. 4.11. Only data past 0800 hours were available for Davis on the May 2, 1967.

The $g(R_s)$ function (Equation 4.159) with $K_R = 100$ W m^{-2} estimated $g(R_s)$ near 1 during most daylight periods at both locations, indicating sufficient sunlight to open stomates. Equation 4.159 was relatively insensitive to the value used for K_R. Air temperature was also at near optimum levels (g(T) = 1) during daylight hours except at Logan during midafternoon when temperatures exceeded optimum levels and g(T) reduced to about 0.7 (Fig. 4.18a). Values for T_L, K_T, and T_H in Equation 4.164 were set at 0, 20, and 40°C for the grass cover at both sites. Low values for g(T) during early morning hours at Logan occurred when T was less than K_T.

The primary reducing function at both Logan and Davis during the days shown in Fig. 4.18 was g(VPD) (using Equation 4.161), which reached a minimum level (about 0.55) during afternoon periods at Logan and approached this minimum level during afternoon periods at Davis. Coefficients used in the g(VPD) equation were $K_{VPD} = 0.3$ kPa^{-1} and $VPD_c = 1.5$ kPa. The $K_{VPD} = 0.3$ kPa^{-1} best fit g_s values calculated as residuals from the lysimeter data at both locations and is halfway between the $K_{VPD} = 0.5$ kPa^{-1} used by Stewart (1988) for Scots pine and the $K_{VPD} = 0.12$ to 0.19 kPa^{-1} used by Stewart and Gay (1989) for natural prairie. The VPD_c value was the same as that used by Stewart (1988) for Scots pine. The net effect of the conductance reducing functions was to predict g(env.), using Equation 4.156 with g(LAI) and $g(\theta) = 1$, equal to about 0.4 to 0.7 during daylight periods at Logan (Fig. 4.18a) and g(env.) equal to about 0.55 to 0.7 during daylight periods at Davis (Fig. 4.18b). The g_{max} at both locations ($g_{max} = 0.07$ m s^{-1} for 0.23 m grass at Logan and $g_{max} = 0.036$ m s^{-1} for 0.12 m grass at Davis) was calculated as $g_l \cdot LAI_{eff}$ where $LAI_{eff} = 0.5$ LAI (Equation 4.117). The value for g_l was set equal to the maximum g_l computed as $1/r_{l_{min}}$, where $r_{l_{min}}$ equaled 40 s m^{-1} (Table 4.16, last row). This value for $r_{l_{min}}$ compares to a value of $r_l = 75$ s m^{-1} which best fit lysimeter measurements of ET when no g functions were employed. A minimum value for g_s was set equal to 0.005 m s^{-1} to account for evaporation from moist soil. Calculations of ET using the Penman-Monteith equation with r_s computed with and without g functions are presented for these same two days and locations near the end of this chapter.

Values of g_{max} fitted from ET measurements have ranged from 0.018 to 0.024 m s^{-1} for Scots pine (Stewart, 1988), equivalent to $r_{s_{min}} = 40$ to 55 s m^{-1}. Price and Black (1989) estimated g_{max} for the Douglas fir forest as 0.022 m s^{-1}, which is equivalent to $r_{s_{min}} = 45$ s m^{-1}. The LAI of the Douglas fir was about 3, while the understory had an LAI of about 2. Equation 4.116 with $r_{l_{min}} = 120$ s m^{-1} (Table 4.16) would predict $r_{s_{min}} = 48$ s m^{-1} for $LAI_{eff} = 0.5$ LAI (Equation 4.117), which is in agreement with the measured values.

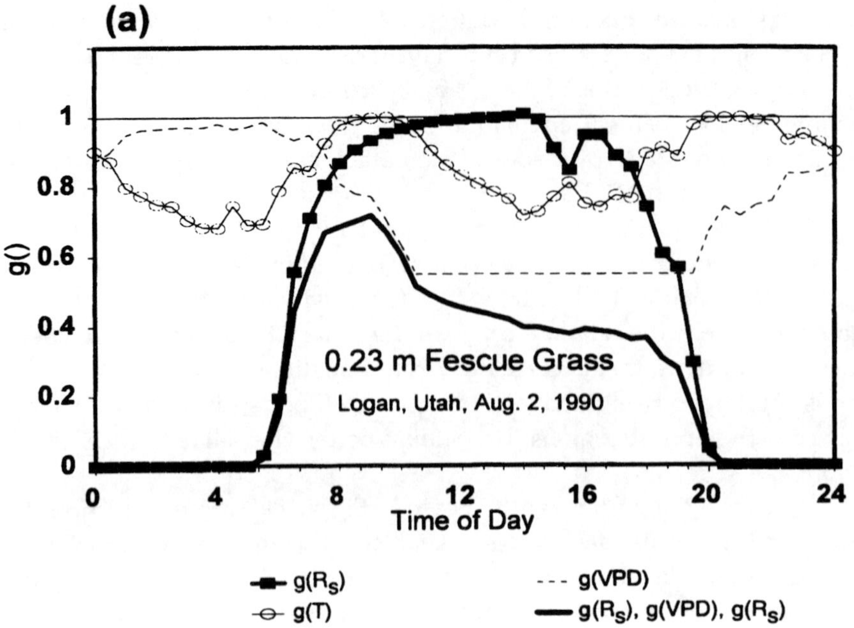

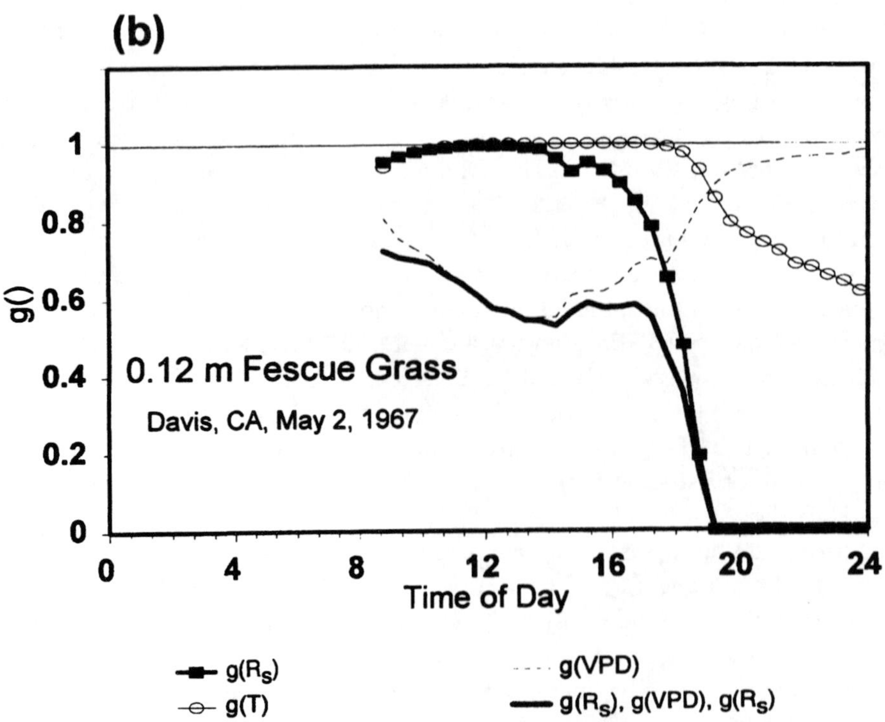

Figure 4.18—Stomatal Conductance Functions for (a) August 2, 1990 at Logan, UT and (b) for May 2, 1967 at Davis, CA Equations 4.159 - 4.164.

In application to big bluestem (*Andropogon geradii*) and Indian grass (*Sorghastrum nutrans*) prairie in Kansas, Stewart and Gay (1989) used g_{max} = 0.016 m s^{-1}, which is equivalent to $r_{s_{min}}$ = 63 s m^{-1}. Coefficients used by Stewart and Gay in Equation 4.159 ranged from K_R = 190 for LAI = 0.5 to K_R = 500 for LAI = 1.3.

f. Constant vs. Variable Estimates of r_s. For 24-hour or longer calculation time steps, use of a constant value for r_s in the Penman-Monteith equation is generally adequate, with adjustments made only for low soil moisture. In some situations, increases in daily values for r_s can be made to account for stomatal closure during periods of low or very high air temperature during early spring, late fall, and midsummer (using Equation 4.164 or a similar equation). The value used for r_s for 24-hour calculations should represent mean effects of $g(R_s)$ and g(VPD) during the day. Perrier (1982) concluded that high values of r_s, which occurred during the beginning and end of a day did not substantially affect the cumulative ET estimates for the whole day due to low energy availability (R_n and VPD) during early and late day periods and due to the occurrence of dew during morning hours in some locations.

Use of a constant value for r_s in the Penman-Monteith equation for calculation timesteps of less than 24 hours is often acceptable if the calculations are to be summed over daily or longer periods so that errors in specific hourly estimates are not relevant; however, where coefficients for Equations 4.158 to 4.169 are available or can be fitted to field measurements, then the stomatal conductance functions described in Equations 4.158 to 4.169 should be employed along with $r_{s_{min}}$ for hourly (or shorter) calculation timesteps. The impact of using constant r_s without g( ) functions vs. r_s computed using g( ) for half-hourly data is demonstrated in the application examples in a following section.

3. Soil Evaporation and r_s Evaporation from soil generally dominates the computation of ET and r_s for low amounts of leaf area and should therefore be considered in ET estimates when LAI is less than 2. Complex models for simulating the flow of heat and evaporation of water from bare and partially vegetated soils have been introduced and are included in Horton and Wierenga (1983), Novak (1989), and Horton and Chung (1991); however, these models are generally complex for routine application. Generally, soil moisture evaporation can be determined for most hydrologic studies by varying surface resistance with cumulative evaporation.

Grant (1975) proposes calculating an integrated surface resistance, r_s, for uniform vegetation cover which considers vegetative surface resistance and soil resistance in parallel. His equation is of the form:

$$\frac{1}{r_s} = \frac{1 - K_r^{LAI}}{r_{sn}} + \frac{K_r^{LAI}}{r_{ss}}, \tag{4.170}$$

where r_{sn} is the bulk stomatal (surface) resistance for a fully developed vegetative canopy (s m^{-1}) and r_{ss} is the soil resistance to evaporation, s m^{-1}. K_r is a radiation extinction coefficient which helps to characterize the net exposure of vegetation and soil. Values for K_r equal to 0.85 and 0.70 are suggested by Monteith and Unsworth (1990) for beans and wheat, respectively.

Perrier (1982) suggests a relationship for nonuniform vegetation cover of the form:

$$\frac{1}{r_s} = F_{gc}\frac{LAI_{eff}}{r_l} + (1 - F_{gc})\frac{1}{r_{ss}}, \tag{4.171}$$

where F_{gc} is the fraction of soil covered and/or shaded by vegetation, and r_l is stomatal resistance, s m^{-1}, per unit *LAI*.

Soil resistance to evaporation, r_{ss}, varies significantly with soil moisture. Minimum values occur when the soil surface is nearly saturated. Values increase nearly exponentially as the top portion of the soil profile dries due to evaporation. The thickness of the soil layer dried by evaporation is a function of soil hydraulic and thermal properties.

Early in a drying cycle, when soil near the surface is moist, water is transported toward the surface by liquid transport according to hydraulic properties of the soil. During this period, evaporation occurs at rates close to potential. This stage of evaporation is often referred to as the first stage of evaporation. After the soil surface has dried to the extent that hydraulic transport to the surface lags evaporative demand, then second stage drying begins. As the hydraulic conductivity near the soil surface decreases, evaporation occurs at lower rates and is limited by conduction of sensible heat (soil heat flux density) downward into the soil and by diffusion of vapor from below the surface to the surface. Because the hydraulic conductivity and thermal conductivity of soil decrease with decreasing moisture, the rate of evaporation decreases with drying until it reaches a very low rate. At this point, r_{ss} assumes a maximum value.

For most soils, r_{ss} can be expressed as a function of the amount of water evaporated from the soil relative to the maximum amount of water which can be evaporated:

$$r_{ss} = r_{ssn} + (r_{ssx} - r_{ssn})\left(\frac{\Sigma E_s}{(\Sigma E_s)_x}\right)^{K_{rs}}, \tag{4.172}$$

where r_{ssn} is minimum soil resistance (s m^{-1}) when the soil surface is wet from rainfall or irrigation, and r_{ssx} is maximum soil resistance (s m^{-1}) when the upper soil layer has been depleted of all evaporable water. ΣE_s is the cumulative depth of water evaporated from the soil surface (mm) and $(\Sigma E_s)_x$ is the maximum cumulative depth of evaporable water (mm). The exponent K_{rs} in Equation 4.172 ($K_{rs} > 1$) simulates the length of first stage drying where r_{ss} remains close to r_{sn} for 1 to 3 days while the soil supplies water for evaporation at nearly the potential rate and then begins a rapid increase in r_{ss} as the moisture in the soil surface nears depletion. Values for r_{ssn}, r_{ssx}, and K_{rs} were 50 s m^{-1}, 2000 s m^{-1}, and 10 for a bare Portneuf silt loam soil in Idaho (data from Wright, 1991).

Generally, evaporation dries the upper soil profile to below the lower limit of extractable water (θ_{LL}). An approximate estimate of total $(\Sigma E_s)_x$ is to multiply the depth of soil dried by evaporation (generally about 0.1 to 0.15 m) by the difference between volumetric moisture content at θ_{DU} (drained upper limit) and volumetric moisture content halfway between θ_{LL} and oven-dry θ_{OD}:

$$(\Sigma E_s)_x = \left(\theta_{DU} - \frac{\theta_{LL} + \theta_{OD}}{2}\right) Z_e \tag{4.173}$$

where θ_{DU} is the volumetric moisture content of the soil after drying in an oven, and θ_{DU}, θ_{LL}, and θ_{OD} all have units of mm m^{-1}. Usually $\theta_{OD} = 0$. Parameter Z_e is the depth of the evaporating layer of the soil. For example, if a silt loam soil with $\theta_{DU} = 310$ mm m^{-1}, $\theta_{LL} = 140$ mm m^{-1}, and $\theta_{OD} = 0.00$ were dried by evaporation to a depth of 0.12 m, the total surface evaporable water $(\Sigma E_s)_x$, would be approximately:

$$(\Sigma E_s)_x = \left(310 - \frac{140 + 0}{2}\right) 0.12 = 29 \text{ mm}$$

The value of $(\Sigma E_s)_x$ varies with soil texture, structure and organic cover. General values for $(\Sigma E_s)_x$ are approximately 20 mm for a bare, sandy textured soil, 30 mm for a bare, silt loam, and 40 mm for a bare, clay-textured soil. $(\Sigma E_s)_x = 30$ mm for a bare Portneuf silt loam soil, based on data from Wright (1991). Denmead (1984) suggested that evaporation from a forest floor covered with litter may be only one-half of what is expected from a bare exposed soil.

ΣE_s is generally determined on a daily calculation time step basis using a moisture balance of the upper 0.10 to 0.15 m soil layer. A general equation for ΣE_s is:

$$\Sigma E_{s_i} = \Sigma E_{s_{i-1}} - P_i - I_i + \frac{ET_a - ET_b}{F_e}, \tag{4.174}$$

where ΣE_{s_i} is cumulative soil evaporation on day i, $\Sigma E_{s_{s_{i-1}}}$ is cumulative soil evaporation on day i−1, P is precipitation on day i, I is mean irrigation depth on day i, and ET_a is actual ET on day i computed using

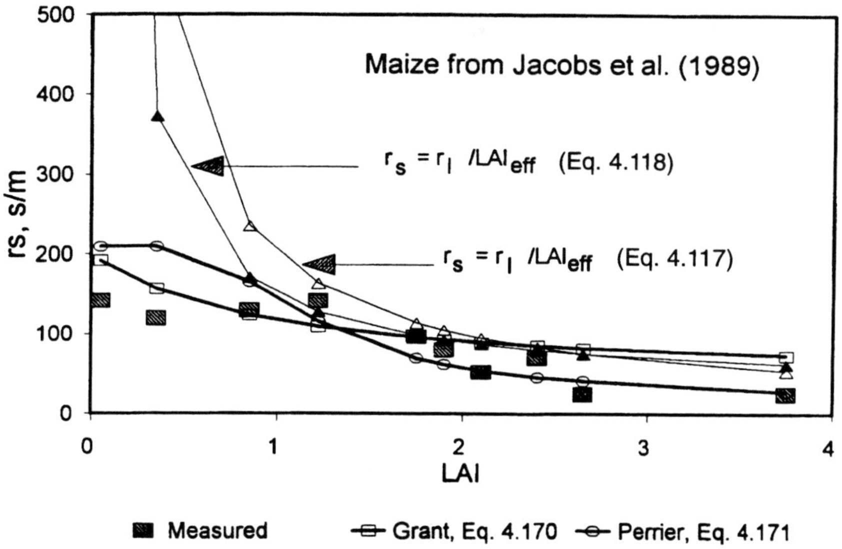

Figure 4.19—Examples of Calculations for Total Surface Resistance, Equations 4.116, 4.170, and 4.171 (Measurements from Jacobs et al., 1989).

the Penman-Monteith equation where r_s includes the effects of r_{ss}. ET_b is a basal ET rate that reflects a dry soil surface and is computed using the Penman-Monteith equation where r_s is calculated using $r_{ss} = r_{sx}$ for a dry soil surface. The difference $ET_a - ET_b$ represents the soil evaporation component. F_e is the fraction of the total soil surface contributing to evaporation. F_e may be less than 1 when part of the soil is shaded by dense vegetation. $\Sigma E_{s\,i}$ in Equation 4.174 has limits of 0 and $(\Sigma E_s)_x$.

Fig. 4.19 shows an example of estimating surface resistance using resistance data for maize reported by Jacobs et al. (1989). K_r in Equation 4.170 was estimated as 0.7. Surface resistance of the canopy at maximum LAI, r_{sn}, was estimated as 60 s m^{-1} based on $r_l = 100$ s m^{-1} and LAI_{eff} from Equation 4.118 for LAI = 3.8. Calculations were relatively insensitive to the value of K_r between 0.6 and 0.8. Equation 4.171 was applied assuming $r_l = 100$ s m^{-1} and $r_{ss} = 200$ s m^{-1}, and F_{gc} was assumed to range from 0.05 at LAI = 0.5 to 0.95 when LAI = 4. No information was presented by Jacobs et al. concerning occurrences of soil wetting. Therefore, $r_{ss} = 200$ s m^{-1} was used as an approximation for the entire period.

Equation 4.170 and Equation 4.171 both replicated measured r_s relatively well, based on estimates for r_{ss} and LAI_{eff}. Equation 4.170 agreed more closely with observed measurements of r_s during periods of significant soil moisture evaporation (LAI > 1) than did Equation 4.171. If the occurrences of soil wetting had been known, then Equation 4.172 could have been applied, where ΣE_s would have been computed on daily time steps using Equation 4.174.

The effect of ignoring soil evaporation in computing r_s (e.g., using Equation 4.116 with Equation 4.117 or 4.118, only) is significant at LAI < 1 as shown in Fig. 4.19. The effect of ignoring soil evaporation was less when LAI > 1.5.

4. Evaporation of Intercepted Rainfall When leaves are wet from rain, r_l effectively becomes zero and evapotranspiration occurs at substantially higher rates than from dry leaves, all other micrometeorological factors being equal. Evaporation of intercepted rainfall can be a significant portion of the water balance, especially in forests with large total leaf area and where single leaf stomatal resistances are large.

Interception is usually measured as a difference between gross precipitation and precipitation reaching the soil. In forests, the annual interception may be 10-40% of total precipitation (Zinke, 1967; Sharma, 1985), depending on canopy storage and rainfall amount, intensity, and frequency. Rutter (1975) and Rutter et al. (1975) suggested that canopy storage during an individual rainfall event may vary from 0.8

to 1.5 mm for established forests, with no distinct difference between storage capacity of forests and herbaceous species. Dickinson (1984) approximated maximum canopy storage, S as

$$S = 0.2\ \text{LAI} \tag{4.175}$$

where S is in mm and LAI is total leaf area index.

The potential increase in total ET due to evaporation of intercepted rainfall can be inferred from the Penman-Monteith equation (Equation 4.109) as:

$$\frac{E_i}{ET_c} = \frac{\Delta + \gamma\left(1 + \frac{r_s}{r_a}\right)}{\Delta + \gamma}, \tag{4.176}$$

where E_i is potential evaporation (or ET) during evaporation of intercepted rainfall and ET_c is total ET during a similar period having a dry canopy, but with the same T, RH, and wind measurements. Because r_a of forests is generally small and r_s of forests is generally large relative to agricultural crops, the ratio of r_s/r_a is large for forests, characteristically ranging between 10 and 70 or more (Jarvis et al., 1976). This is in contrast to agricultural crops which commonly have ratios close to 1. Therefore, the ratio E_i/ET_c can be as high as 5/1 for forests as compared to ratios of perhaps 1.6/1 for agricultural crops. These ratios assume that the canopies are exposed to the same weather conditions under wet and dry scenarios. However, this is seldom the case where the large evaporative flux density from a wet forest canopy will invariably humidify and cool the equilibrium boundary layer above the canopy, thereby reducing the vapor deficit gradient and consequently the ratio of E_i/ET_c. The modification of the weather of the boundary layer is discussed later in the following section on weather measurement. It must also be understood that the period of E_i may be quite short, depending on the storage capacity of the canopy.

Rutter et al. (1975) propose a generalized method for predicting evaporation of rainfall interception in forests. A variation of their proposed method is included here. If variable S is defined as the surface storage capacity of a canopy (maximum retained interception), expressed as a mm depth over the land surface, then when the amount of water on a canopy, C, equals or exceeds S, then evaporation of intercepted precipitation, E_c, is set equal to E_i as computed using the Penman-Monteith equation (Equation 4.109) with $r_s = 0$. When $C < S$, then E_c is reduced in proportion to the ratio of C to S as

$$E_c = E_i \frac{C}{S}. \tag{4.177}$$

When $C \geq S$, it is assumed that transpiration during interception, T_i, is zero, and that all evaporative energy is consumed by evaporation of intercepted water. When $C < S$, T_i is estimated as:

$$T_i = \left(1 - \frac{C}{S}\right) ET_c \tag{4.178}$$

where ET_c is the ET rate from dry foliage, computed using Equation 4.109 with $r_s > 0$. Total evapotranspiration from the canopy, ET_i, at any time is computed as

$$ET_i = E_c + T_i. \tag{4.179}$$

When $C \geq S$, $ET_i = E_c = E_i$, and $T_i = 0$. When $C < S$, $ET_i < E_i$, and $T_i \leq ET_c$. When $C = 0$, $ET_i = T_i = ET_c$, and $E_c = 0$. This approach essentially assumes that evaporation from the forest floor during the interception event is zero and that all evaporation occurs within the wetted canopy. If this assumption is not true, for example in sparse canopies, then the soil evaporation component of the evapotranspiration flux should be included using procedures detailed in the preceding section on soil evaporation.

In the Rutter model, it is assumed that (1 − p) P rainfall is added to the canopy (C term) during each computation time step, where p is the fraction of rain that falls between gaps in the canopy and P is rainfall depth during the time step. The drainage rate from the canopy, D, during a time step is computed as

$$D = e^{(a + bC)}, \qquad (4.180)$$

where D has units of P and a and b are empirical coefficients. Rutter et al. (1975) suggested a value for b = 3.7 for most canopies for C in mm. Coefficient a is computed by noting canopy drainage rate when C = S. Rutter et al. (1975) determined D for a Corsican pine stand equal to 0.002 mm min^{-1} when C = S and suggested estimating D for other canopies as 0.0019 S for S in mm. Table 4.19 lists typical values of p and S for forest stands evaluated by Rutter et al. (1975).

Due to the dynamic nature of the Rutter model in accounting for storage of water on the vegetative canopy with new additions of P and losses of D with time, it is important to use short measurement time steps for P and short computation time steps in the model. Rutter et al. (1975) suggested using hourly measurement and computation time steps, with time steps longer than 3 hours providing poor results.

Generally, when computing ET on daily calculation time steps, the increase in total ET due to evaporation of intercepted precipitation, E_{it}, can be approximated by the following:

$$E_{it} = (ET_i - ET_c)\, t_p + S, \qquad (4.181)$$

where ET_i is the potential rate of ET when leaves are wet, estimated using Equation 4.109 with $r_s = 0$, and ET_c is the ET rate when leaves are dry ($r_s > 0$). ET_i and ET_c should be computed using wind, humidity, and air temperature measured during or surrounding the precipitation event. Parameter t_p is the duration of the precipitation event and S is canopy storage capacity. If E_{it} and S have units of mm and ET_i and ET_c have units of mm h^{-1}, then t_p must have units of h. If the time of day of the precipitation event is unknown, then 24-hr averages of ET_i and ET_c can be used as a general approximation. The total evaporation of intercepted precipitation, E_{it}, converted to mm h^{-1} units, can be added to ET_c for the period encasing the precipitation event to obtain total ET for the period. It is important to note that both ET_i and ET_c are almost always less during a rainfall event as compared to periods with no rainfall due to reduced net radiation, R_n, and generally increased humidity and decreased temperature during rainfall. This is discussed in more detail in the next section.

TABLE 4.19. Values of Fraction of Rain Falling between Gaps in the Canopy, *p*, and Canopy Storage Capacity, *S*, per m² of Land Surface (after Rutter et al., 1975).

Stand type	Mean tree height m	p	*S* mm
Corsican pine	20	0.25	1.05
Douglas fir	24	0.09	1.2
Norway spruce	10	0.25	1.5
Hornbeam leafy	17	0.35	1.0
Hornbeam leafless	17	0.55	0.65
Oak leafy	15	0.45	0.88
Oak leafless	15	0.80	0.28
Oak defoliated	15	0.85	0.18

*5. **Humidity, Air Temperature, Wind, and Solar Radiation Measurements*** When applying the Penman-Monteith or other boundary layer equations, it is important to measure wind, humidity, and temperature over the surface in question. This is important, as the rates of λE and H from a vegetative surface affect the shapes and magnitudes of vapor and air temperature profiles. In the same manner, the roughness of an evaporating surface affects the shape and magnitude of wind velocities above the surface. Vapor and air temperature profiles established above a surface function as feedback mechanisms to λE and H and play important roles in dampening differences in r_s and r_a among vegetation types.

As the value of r_s decreases (higher bulk stomatal conductance), λE increases and consequently H (if positive) decreases. The result is relatively higher vapor concentration in the equilibrium boundary layer above the surface and cooler temperatures. These two changes work together in reducing the vapor pressure deficit of the boundary layer, thereby dampening the increase in λE demand (negative feedback). An increase in the value of r_s works in the opposite direction, where the VPD of the boundary layer is increased due to reduced vapor concentration and increased air temperature, resulting in higher λE demand on the vegetation (positive feedback). The higher λE demand of the boundary layer partially dampens decreases in λE caused by increases in surface resistance.

The functioning of boundary layer feedback processes is important and may present problems when RH and T data are measured over surfaces which have aerodynamic and surface characteristics and soil moisture availability which are different from the surface in question. Errors introduced can be major, especially in the case of predicting evaporation of intercepted precipitation from tall, aerodynamically rough forest canopies using weather data from adjacent grassed weather sites. In this type of situation, the smoother aerodynamics and lower λE of the grassed site result in higher wind speeds, higher temperatures, and lower vapor pressures. When transposed over a tall vegetated surface with low r_s, overestimation of ET occurs.

In situations where ET is estimated for natural rainfed vegetation using a direct equation application such as with the Penman-Monteith equation, then humidity, air temperature, and wind measurements should be made over the same type of vegetation. In situations where the K_c ET_o approach is utilized, then humidity, temperature, and wind measurements should be made over a well-watered, agricultural surface, preferably clipped grass. This is necessary, as most ET_o equations have been calibrated against weather data gathered over well-watered clipped grass (Doorenbos and Pruitt, 1977; Wright, 1982; Jensen et al., 1990), and K_c's are usually computed using this standardized ET_o.

It is generally difficult to obtain humidity, temperature, and wind measurements over unique types of vegetative canopies, especially when estimating ET for historical periods. Often, such data are available only from grassed or barren weather sites. Weather measurement sites are often located in nonagricultural, non-grassland, or non-forested sites such as in cities or near airport runways and buildings. Under these circumstances the user should make allowances for the uncharacteristic nature of the humidity, temperature, and wind data and perhaps assign some type of uncertainty level to the ET estimates.

All weather station equipment used in estimating ET should conform to accepted standards and specifications regarding equipment type, placement, maintenance, calibration, and reporting frequency. Ley and Elliot (1993) have reported on proposed standards for agricultural weather stations and discuss recommended standards and specifications.

Measurements of meteorological data should be screened for integrity and accuracy before use. Calibration and maintenance of sensors should be confirmed and documented. Air temperature and humidity data should be evaluated using procedures discussed in the next section. Daily and hourly solar radiation data should be plotted against clear sky solar radiation (R_{so}) to assess whether maximum measurements of R_s follow the R_{so} "envelope." Occasionally, R_s can exceed R_{so} when clouds are present outside of the view angle of the sun, and reflect additional radiation toward the sensor. R_{so} can be estimated from extraterrestrial radiation (R_A) using Equation 4.18. Jensen et al. (1990), Allen et al. (1994b) and Allen (1996) contain methods for computing R_A and R_{so} for hourly periods. Monthly values for R_s can be compared against R_s estimated using Equations 4.27 and 4.29 to assess general data integrity.

6. Adjustment of NonCharacteristic Humidity, Air Temperature, and Wind Data

a. Air Temperature and Humidity. When air temperature and humidity data are required for use in an ET_o calculation, they may need to be adjusted to account for dryness of the weather station environment. Allen et al. (1983) and Allen and Pruitt (1986) developed empirical procedures for adjusting air temperature measurements from weather stations located in environments where ET was less than maximum rates due to limited soil moisture or green vegetation. The objective of the adjustments was to create air temperature data sets reflective of well-watered environments over clipped grass for use in an ET_o equation. An aridity rating procedure was used to characterize effects of dryness of the local station environment as well as the region on required air temperature adjustments. In the application to weather stations in Idaho, daily maximum and minimum air temperatures were adjusted downward by as much as 4.5°C for stations having nonirrigated, nontranspiring surfaces surrounded by similar conditions. The magnitude of adjustments by Allen and Pruitt were based on paired observations of air temperatures by Allen et al. (1983).

Ley and Allen (1994) applied more complex methods for adjusting air temperature and humidity data from nonreference stations using energy balance and water balance relationships. These methods required measurement of R_s, T, humidity, wind, and precipitation on a daily basis.

Humidity measurements over arid, nonirrigated surfaces are in general lower than those made over well-watered grassed surfaces or over other well-watered vegetation. If humidity data are to be used to estimate ET_o, which by definition is ET from a well-watered surface, then humidity measurements should be made in a well-watered environment.

In an environment having healthy vegetation and adequate soil moisture, minimum daily air temperature, T_{min}, usually approaches dewpoint temperature, T_d, especially if the wind dies down by early morning. Air temperatures decrease during nighttime hours due to surface cooling caused by long-wave emission and evaporation caused by a positive VPD. Under well-watered conditions where evaporable water is present, air temperature usually decreases at night until the dewpoint is reached, provided that wind speed is relatively calm so that large amounts of warm, dry air (in an arid region) are not transported to the surface. When near-surface air temperature approaches T_d, T is prevented from decreasing below T_d by condensation of vapor from the air and the corresponding heating effect of released latent heat. Under these conditions, the relationship $T_{min} = T_d$ generally holds and can be used to assess the relative impact of station environment on humidity and air temperature representativeness for ET_o calculations.

When soil moisture or vegetation cover limits ET to less than ET_o, T_{min} may remain above T_d. This occurs due to two factors. The first factor is the large reservoir of sensible heat in the atmosphere which is transferred toward the surface during the night, reducing the effect of cooling by long wave emission. The large reservoir of heat is created during the daytime due to large flux densities of sensible heat, H, resulting from low ET. The second factor is the effect of lower soil moisture availability on evaporative cooling during nighttime hours.

Under most evaporative conditions, T_d is higher during daytime hours than during nighttime hours due to humidification of the boundary layer by ET. Consequently, daytime measurements of absolute humidity may exceed nighttime values by 5 to 20 percent. Therefore, the relationship between T_{min} and T_d may not be strictly one-to-one when T_d is based on 24-hr or daylight averages. In these cases, T_{min} measurements can be expected to go a few degrees below 24-hr average T_d measurements.

An example of the effect of dryness of weather station environment on proximity of T_{min} to T_d is shown in Fig. 4.20 where mean monthly differences between T_{min} and T_d are plotted against ratios of mean monthly precipitation and ET_o for three data sets (62 weather stations in Sudan, Africa (a); 26 Weather Service locations across the United States (b); and two stations in southern Idaho (c)). The ratios of P to ET_o provide an indication of the relative wetness of the station during a particular month and the likelihood of healthy, transpiring vegetation.

Figs. 4.20a, b, and c indicate that $T_d \approx T_{min}$ where vegetation of the weather sites are in a reference (well-watered) condition (monthly $P/ET_o > 1$), which would include conditions within most large

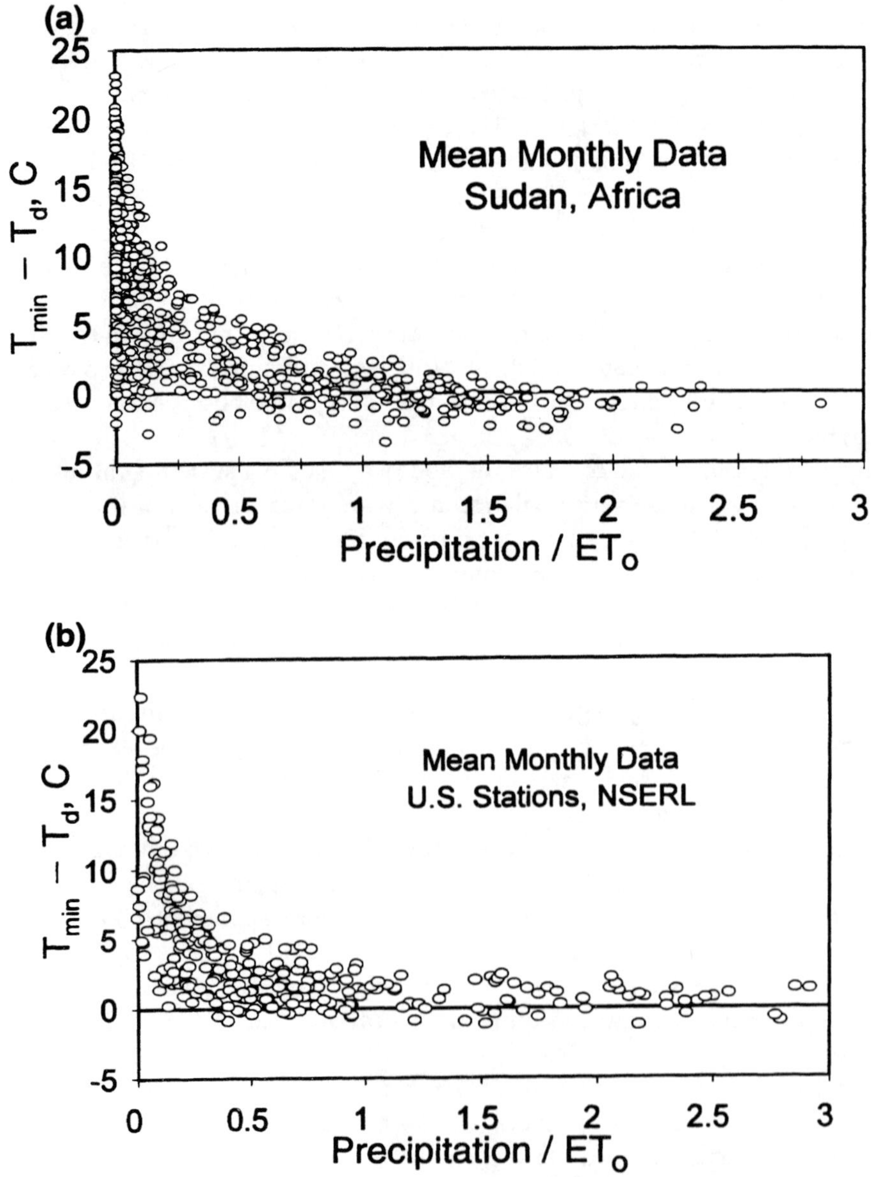

Figure 4.20—Monthly Mean T_{min} − Monthly Mean T_d for a) Weather Stations in Sudan, Africa, b) 26 US Weather Service Stations across the United States, and c) Two Weather Stations in Southern Idaho. (Continued on next page)

irrigation projects. The average absolute deviation between T_{min} and T_d when $P/ET_o > 1$ is about 2°C. When measured in a dry environment (monthly P/ET_o near zero), T_d can be as much as 20°C less than T_{min}. Daily differences between T_{min} and T_d are similar to the monthly averages shown.

The effect of irrigation on $T_{min} - T_d$ is shown in Fig. 4.20c, where data from Kimberly and Boise, Idaho are measurements from a semiarid climate (annual rainfall is less than 250 mm per year), but where the Kimberly location is over irrigated grass surrounded by over 150,000 ha of irrigated land. The Boise location is the Boise airport which is surrounded by predominately nonirrigated rangeland for a 10 km distance and then irrigated land beyond that in the predominate upwind direction. The effect of surrounding the weather station with well-watered vegetation is obvious. Values for $T_{min} - T_d$ would increase for both of these locations if they had been surrounded by large expanses of rangeland only.

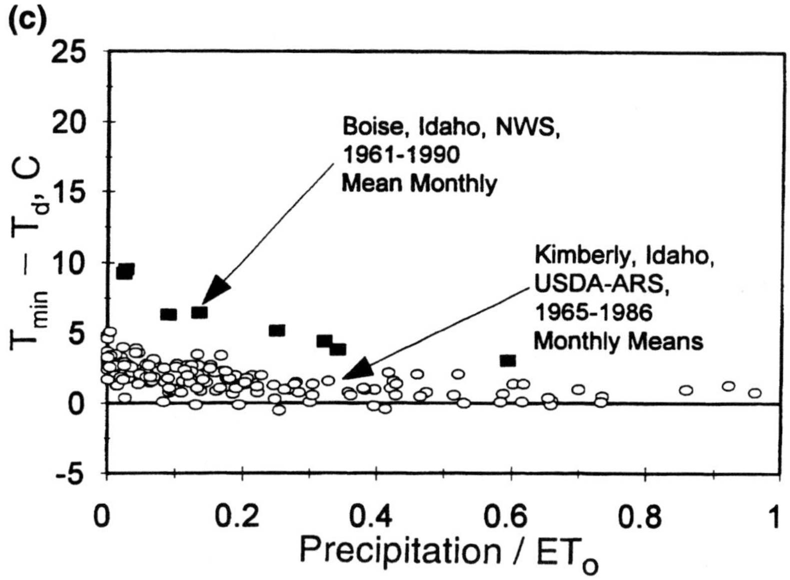

Figure 4.20—Continued from previous page

Graphical comparisons of T_{min} vs. T_d can be used as a check on data integrity and representativeness. When data are to represent a well-watered reference setting, then the relationship $T_d \approx T_{min}$ can be employed. When measured T_{min} exceeds T_d due to station dryness, and data representing a reference setting are desired, then adjusted T_{max}, T_{min}, and T_d can be approximated under conditions where night time wind speeds are calm ($< 2\ ms^{-1}$) by subtracting 0.5 ΔT from T_{max} and T_{min} and by adding 0.5 ΔT to T_d, where $\Delta T = T_{min} - T_d$ (Allen, 1994, 1996). However, when applying the Penman-Monteith equation directly over dry vegetation, no adjustment of temperature and humidity data should be made.

b. Wind Speed. Aerodynamic resistance computed by Equation 4.110 assumes that the wind speed at some reference height is measured above the surface in question and is therefore governed by exponentially-shaped profiles characteristic of the surface roughness and measurement height. When wind measurements are made over 0.12 m grass, for instance, wind speed measured at 2 m above the grass will generally be significantly higher than that measured at 2 m above an adjacent forest of 10-m trees.

Rutter et al. (1975) found wind speeds measured at 2 m above a 20-m tall pine forest to be only 0.45 of wind speeds measured at 2 m above a nearby grassed area. Allen and Wright (1996) found wind speeds at 2 m height over 0.7 m tall snap beans to be 0.7 of speeds measured at 2 m height over 0.12 m grass 0.8 km away. Wind speeds at 2 m height over 1 m tall winter and spring wheat were 0.6 times speeds measured at 2 m height over 0.12 m grass. These differences were found to significantly affect estimates of ET made with aerodynamic methods (Vanderkimpen et al., 1996).

Rutter et al. (1975) and Allen et al. (1989) proposed adjusting wind measurements for differences in surface roughness and zero plane displacement by extrapolating upward from the measurement surface to some height of a regional internal boundary layer, Z_{IBL_R}, where the mean horizontal wind velocity of the equilibrium boundary layer is an integration of regional effects and is presumed to be generally independent of underlying surface roughness of a specific location. The wind velocity at the Z_{IBL_R} height was then extrapolated downward from the Z_{IBL_R} height to some elevation above the surface in question using aerodynamic parameters characteristic of the new surface.

Allen and Wright (1996) improved this procedure by combining equations for adjusting wind velocities for both roughness and wind sensor elevation differences into one equation which considers the relative heights of internal boundary layers developed over both the weather station surface and the surface in question. The equation is of the form:

$$u_{z_v} = u_{z_w} \frac{\ln\left(\frac{z_{IBL_W} - d_W}{z_{om_W}}\right) \ln\left(\frac{z_{IBL_V} - d_R}{z_{om_R}}\right) \ln\left(\frac{z_V - d_V}{z_{om_V}}\right)}{\ln\left(\frac{z_W - d_W}{z_{om_W}}\right) \ln\left(\frac{z_{IBL_W} - d_R}{z_{om_R}}\right) \ln\left(\frac{z_{IBL_V} - d_V}{z_{om_V}}\right)}, \tag{4.182}$$

where u_{z_v} is the wind speed adjusted to represent conditions at the z_v height over vegetation or water surface v, and u_{z_w} is the original wind speed measurement at the z_w height over the weather station surface w. Variable Z_{IBL_w} is the height of the internal boundary layer developed over surface w which is in equilibrium with surface w. Variable Z_{IBL_v} is the height of the internal boundary layer developed over surface v which is in equilibrium with surface v. These parameters are estimated using Equation 4.183. Variables Z_{om_w}, Z_{om_v}, and Z_{om_R} are roughness lengths for momentum transfer characterizing the weather surface, the vegetation v, and the integrated region. Variables d_w, d_v and d_R are zero-plane displacement heights for the weather surface, vegetation (or water surface) v, and the region. Units of all z and d variables are the same (m) and units for u are in m s^{-1}. The z_{om} and d parameters can be estimated using Equations 4.144 to 4.151.

Values for Z_{IBL_w} and Z_{IBL_v} may range from only 1 to 2 m for small fields of 1 ha to more than 100 m for very large stands of homogeneous vegetation. On a regional scale, the height of a regional internal boundary layer, Z_{IBL_R}, may vary from tens of meters to perhaps 2000 m and may average about 100 to 400 m (Brutsaert, 1982). Brutsaert (1982) presents an equation to predict boundary layer growth downwind of a discontinuity between two surfaces of similar but unequal roughness:

$$z_{IBL} = d + 0.33\, z_{om}^{\;\;0.125}\, x_f^{0.875}, \tag{4.183}$$

where z_{IBL} is the height of the perturbed IBL above a surface of new roughness (z_{om}) and zero plane displacement (d). Variable x_f in Equation 4.181 is the horizontal distance downwind of the surface discontinuity (horizontal fetch). Variables z_{IBL}, d, z_{om} and x_f in Equation 4.183 have units of m. Equation 4.183 can be applied to estimate both Z_{IBL_w} and Z_{IBL_v}. Equation 4.183 is a rearrangement of Equation 4.81, except that the difference (z−d) in Equation 4.183 is first multiplied by a factor of 10 to approximate the height of the full z_{IBL} which is generally 10 times that of the fully mixed internal sublayer, where flux densities of H and λE are in equilibrium with the new surface (Brutsaert, 1982).

Limitations in the application of Equation 4.182 are that $z_w < Z_{IBL_w}$ and $z_v < Z_{IBL_v}$. If these conditions do not exist, then one should use alternate equations presented in Allen and Wright (1996). Other limitations in Equation 4.182 are that all d's must be less than the corresponding elevations (z) to prevent negative values and extrapolation of wind speed to below zero plane displacement heights.

In summary, in application of Equations 4.182 and 4.183, u_{z_v} is wind speed predicted to have occurred at the z_v elevation if the ground cover was vegetation of type v rather than type w. The postulation of Equation 4.182 is that wind speeds measured over surfaces can be extrapolated upward from z_w using a logarithmic relationship (assuming near neutral conditions) to some Z_{IBL_w} height using aerodynamic roughness properties characteristic of the weather measurement surface, and then extrapolated further upward or downward using aerodynamic properties of the region, and then extrapolated back downward close to the surface using a logarithmic shape characteristic of roughness properties of the vegetation in question. This approach suggests that wind speeds above Z_{IBL_w} and Z_{IBL_v} are driven by the general speed of an air mass, but that logarithmic profiles in equilibrium with the underlying surfaces tie the regional or local air mass and momentum sink to the ground. This is a simplification of actuality, especially during periods of boundary layer thermal instability and due to the fact that wind profiles above intermixed vegetation stands do not remain completely separate of each other. However, the approach does provide an approximate adjustment mechanism which has some application value.

7. Estimating Weather Data for Reference Estimates When reference estimates of ET_o are required for areas where a compliment of R_s, T_d, and wind data (u) are not available, then Equation 4.124 by Hargreaves and Samani (1985) can be employed to calculate ET_o. Equation 4.124 can be regionally calibrated against the Penman-Monteith equation (Equation 4.122) at locations having sufficient data with which to calculate the Penman-Monteith equation. An alternative to using Equation 4.124 when data are not available is to estimate values for R_s, T_d, and u and to then apply the Penman-Monteith equation (Equation 4.122) with these estimates.

When only T_{max} and T_{min} measurements are available, estimates of daily or monthly R_s can be made using Equation 4.29 and daily or monthly T_d can be estimated as $T_d = T_{min}$ as discussed in a previous section. Often, for purposes of applying Equation 4.122, mean monthly or daily wind speed can be approximated as $u_2 = 2$ m s^{-1}. This approximation can be increased or decreased for regions based on known wind characteristics.

Examples showing the nature of errors introduced by estimating R_s, T_d, and u as described above are presented in Fig. 4.21 where daily and 5-day estimates of ET_o at Eaton, Colorado during 1994 were made using Equation 4.122 with estimated R_s, T_d and u ($u_2 = 2$ m s^{-1}), and are plotted against estimates made using Equation 4.122 with measured R_s, T_d, and u (Figs. 4.21a and c). As a comparison, daily and 5-day estimates of ET_o calculated using Equation 4.124 (Hargreaves) are plotted against those using Equation 4.122 (using measured R_s, T_d, and u) in Figs. 4.21b and d. As expected, substantial deviation occurred between daily estimates of ET_o calculated using only T_{max} and T_{min} and ET_o estimates calculated using measured R_s, T_d, u, and T. This was true for both Equation 4.122 (Fig. 4.21a) and Equation 4.124 (Fig. 4.21b). However, deviations were reduced when five-day averages of ET_o were calculated. In both cases, estimates by the methods using temperature only (Equation 4.122 with estimated R_s, T_d, and u and Equation 4.123) followed the 1:1 line against Equation 4.122 with measured R_s, T_d, and u. The deviation shown in Fig. 4.21 may be acceptable when calculating daily soil water balances where the soil moisture reservoir provides for substantial buffering against effects of errors in specific daily ET estimates, provided accuracy in ET over a 5- to 10-day period is good. The Eaton weather station was an electronic agricultural weather station located in an irrigated environment.

Fig. 4.22 shows the effect of making estimates of monthly ET_o over a long time frame (8 year period) at Davis, California, using Equation 4.122 (Penman-Monteith grass reference) with estimated R_s, T_d, and u ($u_2 = 2$ m s^{-1}). These estimates are plotted against calculations made using Equation 4.122 with measured R_s, T_d, and u for comparison. Fig. 4.22b shows estimates made using Equation 4.124 (Hargreaves) vs. estimates by Equation 4.122 for comparison. In general, the two procedures provided similar results with deviations from estimates made by Equation 4.122 with measured data averaging about 15 percent for Equation 4.122 with estimated data and about 10 percent for Equation 4.124. The Davis site was located at a University of California research center and was surrounded for the most part by irrigated crops.

In general, Equations 4.122 with estimated R_s, T_d, and u and 4.124 provide relatively good estimates of grass ET_o on 5-day to 30-day calculation steps and are useful for estimating ET_o in areas where only maximum and minimum daily air temperature data are available. However, the reader must bear in mind that the $T_d = T_{min}$ relationship does not hold for nonreference conditions (conditions where limited soil moisture reduces local ET so that air temperature is increased and dewpoint temperature is decreased).

8. Penman-Monteith Calculation Examples The following section includes example applications of the Penman-Monteith method for estimating ET directly, without the use of crop coefficients. The first example is for fescue grass at Logan, Utah and Davis, California with calculations made on a half-hourly basis. The second example is for grass forage at Logan, Utah under moisture-stress conditions with calculations made on a daily (24-hour) basis.

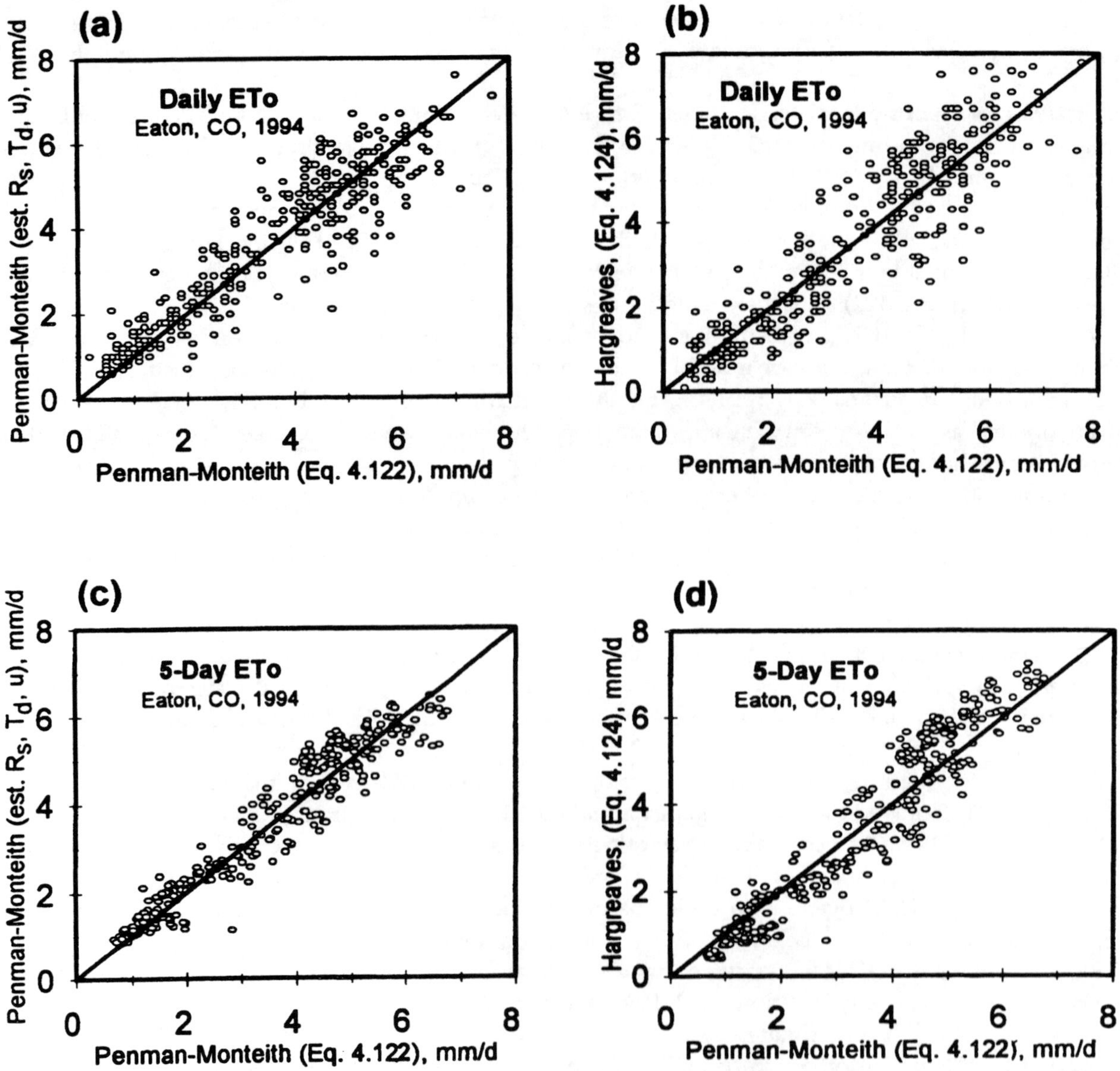

Figure 4.21—Grass Reference ET_o Calculated using Equation 4.122 with R_s Estimated From Equation 4.29, $T_d = T_{min}$ and $u_2 = 2\ m\ s^{-1}$ and ET_o Calculated using Equation 4.124 vs. Equation 4.122 using measured R_s, T_d, and u_2 for Daily Calculations (a and b) Five-Day Calculation Time Steps (c and d) at Eaton, CO in 1994. (Weather data from Crookston, 1994)

a. Half-Hourly Calculations with the Penman-Monteith Equation. The Penman-Monteith method with resistance algorithms for grass (Equations 4.110, 4.113 to 4.117 and 4.119) was applied to half-hourly measurements of micrometeorological data at both Logan, Utah and Davis, California and calculations were compared to lysimeter measurements. The details of the data sets are the same as those described for the energy-balance method applications shown in Figs. 4.10 and 4.11, with aerodynamic resistances computed using the same integrated stability corrections (Equations 4.100 to 4.104) based on infrared measurements of surface temperature. The micrometeorological data in this example were collected over the grass crops. Results of the application are shown in Fig. 4.23a for the day of August 2, 1990 at Logan, Utah and in Fig. 4.23b for the days of May 2–3, 1967 at Davis, California.

The grass at Logan was primarily fescue and was grown as forage. The grass height on August 2 was

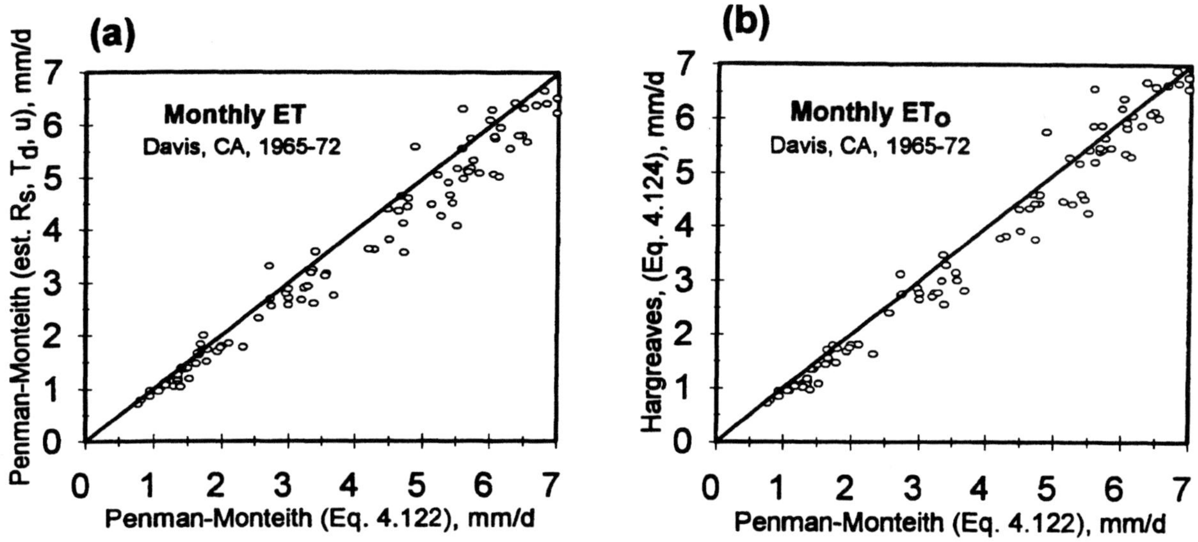

Figure 4.22—Grass Reference ET_o Calculated using a) Equation 4.122 with R_s Estimated From Equation 4.29, $T_d = T_{min}$ and $u_2 = 2$ m s^{-1} and b) Equation 4.124 vs. Equation 4.122 using measured R_s, T_d, and u_2 for Monthly Calculation Time Steps at Davis, CA between 1965 and 1972.

0.23 m. The LAI at Logan was estimated from Equation 4.119 as 5.5 m^2 m^{-2}. The grass at Davis was alta fescue and was clipped as turf. The grass height on May 2–3 was 0.12 m and the LAI measured on May 2 was 2.94 m^2 m^{-2}. Bulk surface resistance, r_s, at both locations was calculated using Equation 4.116 with LAI_{eff} calculated using Equation 4.117. Net radiation was calculated at Logan from measured R_s using the standard procedures described at the beginning of the chapter due to faulty measurement of long-wave radiation during nighttime periods by the net radiometer used. Net radiation at Davis was an average of measurements by three net radiometers.

Two procedures were used to approximate r_l. In the first procedure, a constant value of $r_l = 75$ s m^{-1} was used during all time periods to represent a mean daily value for r_l for use with hourly calculations with integrated stability correction (see last entry of Table 4.16). Calculations of r_a were made with and without stability correction. Resulting calculations by the Penman-Monteith equation (Equation 4.109) with these resistances are labeled "P-M, no stab.corr." and "P-M, Int.stab.corr." In the second procedure, a value of $r_{l_{min}} = 40$ s m^{-1} was used in conjunction with the stomatal conductance reducing functions (g( )) to provide a variable r_l during the calculation periods. Calculations for r_a were made using integrated stability correction functions. Resulting calculations by the Penman-Monteith equation are labeled "P-M, Int.stab.corr.,g( )" in Fig. 4.23.

Values for the g( ) functions and composite g(env.) were presented earlier in Fig. 4.18. g(θ) was set equal to 1.0 to reflect high soil moisture conditions at each site. As indicated in Fig. 4.18, the products of $g(R_s)$ g(VPD) g(T) calculated for August 2, 1990 at Logan and May 2, 1967 at Davis reached highs of about 0.75 during midmorning hours and then decreased to about 0.4 to 0.6 during afternoon hours when both VPD and T increased. In the first procedure with constant r_l, the resulting values for r_s were 27 s m^{-1} at Logan and 50 s m^{-1} at Davis, the differences being due to differences in LAI. In the second procedure, $r_{s_{min}}$ equaled 14 s m^{-1} ($g_{max} = 0.07$ m s^{-1}) and 27 s m^{-1} ($g_{max} = 0.037$ m s^{-1}) at Logan and Davis, respectively. The inclusion of the g( ) functions increased the values for r_s in the second procedure from 14 s m^{-1} to 35 s m^{-1} during afternoon hours at Logan and from 27 s m^{-1} to about 50 s m^{-1} at Davis. Surface resistance during nighttime hour ($R_n < 0$) was set equal to 200 s m^{-1} to approximate r_s for damp soil beneath the grass canopies (see discussions in the section on Soil Evaporation and r_s).

In essence, the Penman-Monteith calculations fit lysimeter measured ET well using both procedures for estimating r_s (constant $r_l = 75$ s m^{-1} throughout the calculation periods and $r_l = r_{l_{min}}$ of 40 s m^{-1}

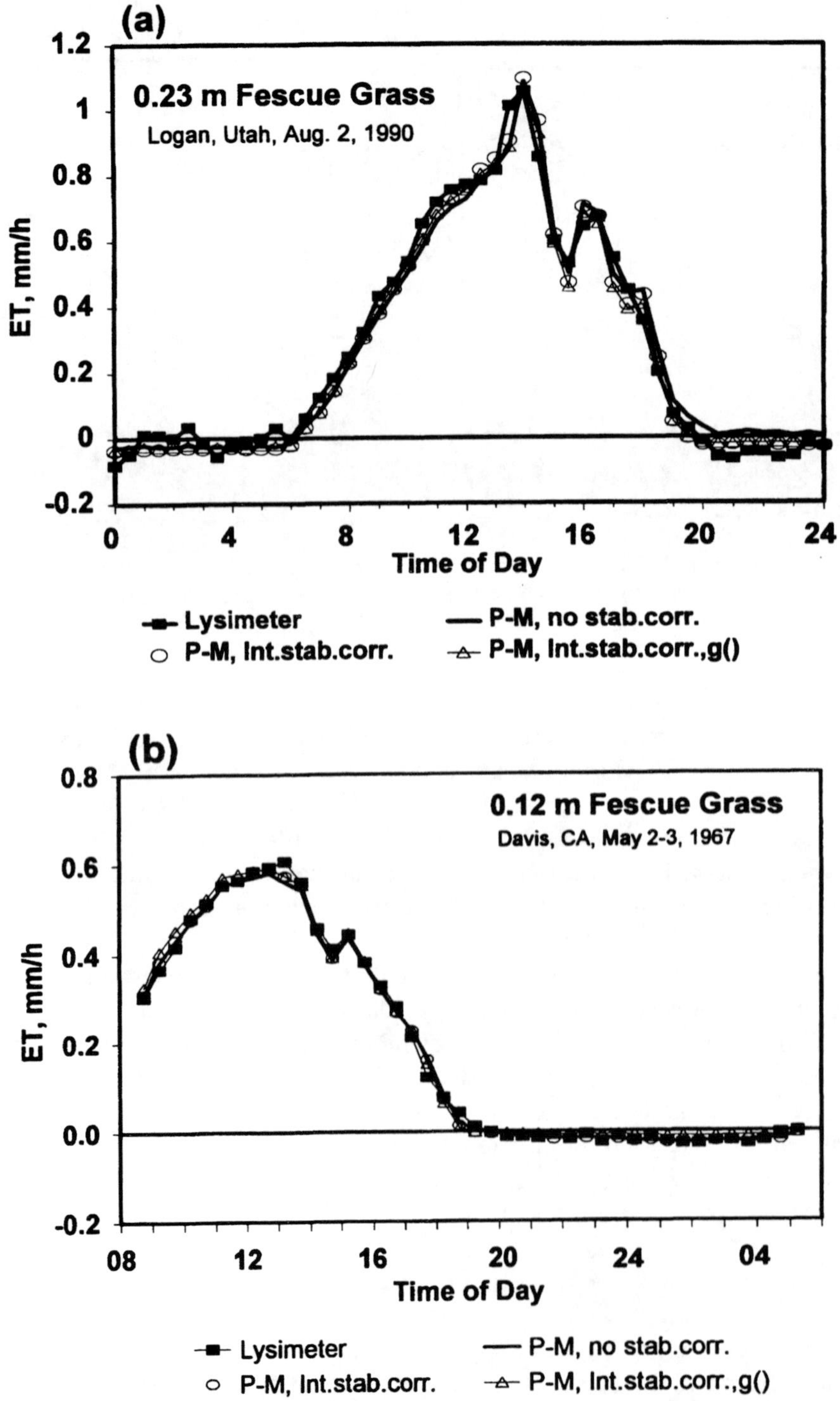

Figure 4.23—Comparisons between Measured ET and ET Estimated using the Penman-Monteith (P-M) Equation (Equation 4.109) with and without Integrated Stability Correction with $r_l = 75$ s m^{-1}, and with Integrated Stability Correction and $r_l = r_{l_{min}}$ of 40 s m^{-1} Divided by $g(R_s)g(T)\,g(VPD)$ for a) 0.23 m Fescue Grass at Logan, UT and b) 0.12 m Fescue Grass at Davis, CA (Allen, 1993; Pruitt, 1994; Morgan et al., 1971).

divided by $g(R_s)g(T)g(VPD)$). Calculations were influenced little by inclusion of stability corrections at either location, even though values for the Monin-Obukhov z/L numbers ranged from −0.5 to over 2 at Logan and −0.7 to 0.1 at Davis. This is in contrast to the $ET = R_n - G - H$ calculations presented in Fig. 4.11 which were significantly influenced by stability corrections, especially at Davis. The seeming lack of influence of stability correction on the Penman-Monteith estimates stems from the presence of $1/r_a$ in both the numerator and in the denominator of the Penman-Monteith equation, so that a major amount of change in $1/r_a$ due to stability correction is self-cancelling, depending on the relative magnitudes of $R_n - G$, $e_z{}^o - e_z$ and r_s. This was not the case with the energy balance calculations which were shown in Fig. 4.11. Under strong conditions of instability brought on by moisture stress, effects of stability correction on Penman-Monteith estimates would become noticeably stronger.

There was slight underestimation during the highest half-hour period at Davis at about 1300 hours and there was a slight lag in estimates relative to lysimeter measurements during morning hours at Logan. The Penman-Monteith equation predicted slight, net condensation of moisture (dew) during portions of nighttime periods at both locations, especially when corrected for stability. The gain in weight due to condensation was also indicated by most of the nighttime lysimeter records, especially at Davis. Under these conditions, r_s should technically be set to 0. However, without the use of a highly sensitive lysimeter or a leaf wetness sensor, it is often difficult to predict when condensation occurs. In the examples shown in Figs. 4.23a and 4.23b, use of $r_s = 0$ rather than $r_s = 200$ s m^{-1} made very little difference in the ET predictions at night since aerodynamic resistances were high due to very light wind and stable conditions, along with very small VPD's.

The dips in calculated and measured ET during midafternoon periods at both locations were due to cloud fronts which passed overhead, reducing net radiation and air temperatures.

b. Daily Calculations with the Penman-Monteith Equation under Conditions of Soil Moisture Stress. The Penman-Monteith equation was applied to a three-month data set of micrometeorological and lysimeter data from Logan, Utah which were collected during a March 20–June 30, 1992 study. The grass cover was tall fescue grown for forage. All irrigation of the lysimeter and surrounding 40-ha field was withheld during 1992 so that the grass crop experienced progressively increasing moisture stress caused by progressively decreasing soil moisture. Fig. 4.24 exhibits daily (24-hour) calculations of ET made using the Penman-Monteith equation (Equation 4.109), where the P-M (no stress) calculations were made using r_s calculated using Equation 4.116 with $r_l = 100$ s m^{-1}. This curve represents the upper limit of ET under conditions of limited moisture stress. The other curves in the figure represent calculations by the Penman-Monteith equation where r_s was reduced by dividing r_s by the product $g(T)g(\theta)$. $g(\theta)$ was calculated using the three methods described in the section on stomatal conductance functions (Equations 4.166, 4.167, and 4.169). Because the computation time steps were 24-hour, the $g(R_s)$ and $g(VPD)$ functions, which apply to instantaneous values of R_s and VPD, were set equal to 1.0. However, because the first part of the season (late March and April) was cool, the g(T) function was applied to the daily time steps to adjust for low stomatal conductance caused by low daytime temperatures. Parameters used in Equation 4.164, were $T_L = 0°C$, $K_T = 20°C$ and $T_H = 40°C$. Because calculations were made on a 24-hour basis, a mean value of daytime air temperature for use in Equation 4.164 was approximated for each 24-hour period as $T = (2\ T_{max} + T_{min})/3$. In addition, no corrections for boundary layer instability were made since calculations were made on 24-hour timesteps.

The grass height for the Logan data set ranged from 0.01 m on March 20 to 0.60 m in June. Leaf area index for the tall fescue was predicted as LAI = 0.20 h, where h is mean plant height in m, based on a set of LAI measurements made at h = 0.4 m. LAI was linearly reduced after June 1 to account for leaf senescence and firing of leaves near the bottoms of plants caused by effects of the drought conditions. Total LAI was reduced by 1.5 after April 25 due to a hard frost (−5°C) on that date. Because of the tall height of the forage and large LAI, LAI_{eff} in Equation 4.116 was computed using Equation 4.118 by Ben-Mehrez et al. (1992).

Parameters applied in the $g(\theta)$ functions were $K_\theta = 1.2$ in Equation 4.166, $K_\theta = 0.25$ in Equation 4.167, and $K_{sf} = 1.2$ in Equation 4.169. A value of $K_\theta = 1.2$ was required in Equation 4.166 rather than the $K_\theta =$

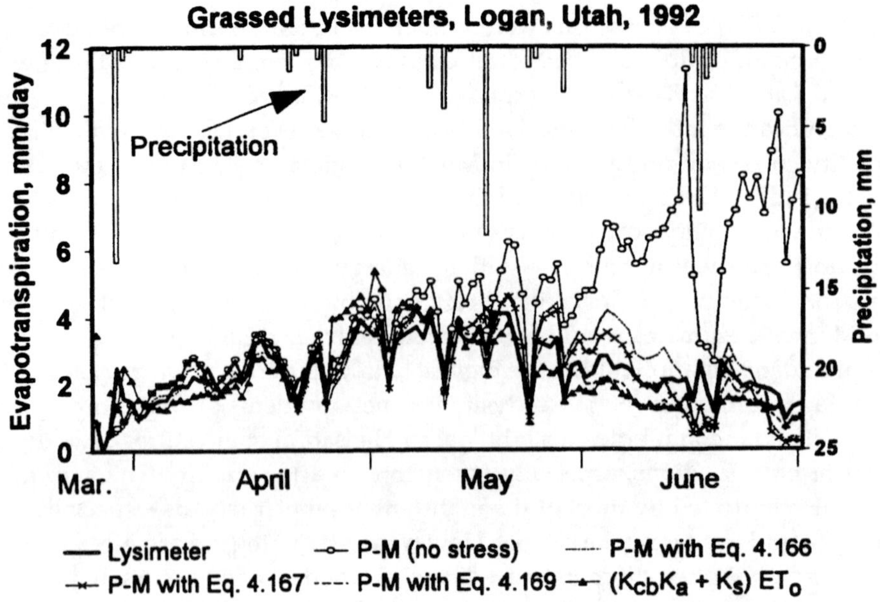

Figure 4.24—Comparison between Measured ET and ET Estimated using the Penman-Monteith (P-M) Equation (Equation 4.109) for Fescue Grass Forage at Logan, UT, March - June, 1992 under Conditions of Low Soil Moisture and with 24-hour Calculation Time Steps (Allen, 1993).

6.7 suggested by Stewart (1988) for Scots pine, in order to better follow measured data. The value $K_\theta = 0.25$ in Equation 4.167 which was suggested by Stewart and Verma (1992) for Kansas prairie worked well for the fescue at Logan. The values for θ_{DU} and θ_{LL} for the clay loam soil were 320 and 140 mm m^{-1}. The rooting depth for the fescue grass was taken as 1 m.

The ET = K_c ET_o approach was also applied to this data set, with K_c estimated using Equation 4.126 ($K_c = K_{cb}K_a + K_s$). The K_{cb} curve was constructed using $K_{c1} = 0.4$ and $K_{c2} = K_{c3} = 1.1$. The K_s values were calculated using Equation 4.127 and K_a was calculated using Equations 4.132 and 4.133 where θ_{DU} and θ_{LL} had the same values as for the g(θ) functions and F_{ns} was set equal to 0.4 following Table 4.12. ET_o was calculated using Equation 4.122.

All three g(θ) functions performed relatively well in reducing the estimated ET according to soil moisture availability. Equation 4.167 followed lysimeter ET most closely, with Equation 4.166 overestimating lysimeter measurements during June when soil moisture was low and with Equation 4.169 estimating well during April and May, but underestimating measurements during June by about 30 percent. The K_cET_o procedure overestimated lysimeter ET by about 20% during late April and early May since no reduction in K_c occurred until 40 percent of available soil moisture was depleted. The procedure underestimated lysimeter measurements during June by about 30%. In general, all four moisture–stress procedures or functions evaluated show usefulness in estimating effects of dry soil moisture.

*9. **Measurement and Estimation of ET on Sloping Lands*** Many studies of hydrologic processes under natural conditions involve sloping land surfaces. On sloping surfaces, the potential amount of solar energy converted into ET increases for southward facing slopes (in the northern hemisphere) and decreases for northward facing slopes. The increase or decrease is related to the change in the angle between the slope and the angle of the sun (Swift, 1976; Perez et al., 1986). Under most conditions of small slope (less than about 15°), ET calculations can be made on a horizontal basis with little error. However, the user must bear in mind that the ET flux density calculated horizontally (for example, expressed in mm or $m^3\ m^{-2}$) is based on a per unit horizontal land area. To convert to a unit area basis parallel to the land slope, a trigonometric adjustment must be made.

When calculations are made on a with-slope basis, then net radiation measurements or calculations must be made parallel to the slope, or calculations of R_n must be adjusted for the effect of slope. With slopes greater than about 15 percent, horizontally measured R_n adjusted for slope angle can deviate from R_n measured parallel to the slope, since the upper surface of a horizontally-placed radiometer may "see" some long-wave and reflected short-wave radiation from the ground or vegetation surface uphill (Nie and Kanemasu, 1989, Fritschen and Qian, 1990). In addition, wind speeds used in ET equations which are applied on a with-slope basis must represent wind movement parallel to the slope, and z heights used in resistance calculations must be expressed as perpendicular to the slope. Examples of applications of ET equations for sloped areas are included in Dozier and Frew (1990) and Neale et al. (1995).

Other problems in calculating or measuring ET in areas having variable land slope include convergence or divergence of wind streamlines, especially downwind of abrupt changes in slope. This can cause severe errors in eddy correlation and Bowen ratio measurements. In addition, density-induced drainage of air in down-slope directions during periods of low solar radiation and in up-slope directions during periods of high solar radiation, can substantially affect local wind velocities, air temperatures, and vapor gradients, especially during calm periods. In application of ET equations over areas having large changes in elevation, air temperatures can be adjusted during extrapolation over areas with changing elevation using lapse correction (about $0.01^0\ m^{-1}$ for nonsaturated air).

10. Multi-Level Resistance Equations One of the drawbacks of the Penman-Monteith equation, especially for use in sparse canopies, is the implicit assumption that the mean vertical plane of momentum sink within the canopy is at the same height ($d + z_{om}$) as that for sensible heat (source or sink) and latent heat (source). This is not generally true for sparse canopies or for dense canopies having intermittent gaps with exposed lower story vegetation or bare soil. Under these conditions, a significant amount of solar radiation can penetrate to levels below $d + z_{om}$ resulting in the generation of sensible heat at lower levels with the possible conversion to latent heat. The combination of bulk stomatal resistance with soil resistance into a single surface resistance, r_s, as is done for the Penman-Monteith equation implies that these two evaporating surfaces are at the same vertical level. This is never the case and consequently ignores the aerodynamic resistance between the soil surface and mean height of canopy moisture evaporation.

In most situations, the errors induced by the single exchange plane assumption of the Penman-Monteith are partially accounted for during analysis of research data by varying the apparent z_{om}/h ratio, effective leaf area, single leaf resistance, or radiation extinction coefficient, K_r, to improve the fit. However, because the transport resistances within the canopy are dynamically linked to wind speed, canopy structure, radiation levels, and soil moisture, the fitted parameters listed above may not transfer well to other growth stages or locations. Wind speed and movement within canopies was discussed and estimated by Ben-Mehrez et al. (1992).

Various multi-layer models have been developed which attempt to take the spatial and vertical distribution of sources and sinks within and below a plant canopy into account. This is done by dividing the canopy and sometimes the underlying soil into a number of layers, each with its own surface resistance, aerodynamic resistance, leaf area, and radiation extinction and exposure characteristics. In multi-layer models, two processes are evaluated simultaneously: exchanges between a leaf or soil and the surrounding air and the transfer from one layer into the next, through the canopy air space. Recent multi-layer models that deal with incomplete canopies and below-canopy fluxes include the Jagtap and Jones (1989a; b) model developed for soybeans, and models by Shuttleworth (1979) and Lhomme (1988a; b).

Two major impediments to application of multilayer models have generally been the large quantity of detailed data required to describe various interactions within and among layers and the iterative solutions of large, implicit systems of equations. Chen (1984) and Lhomme (1988a) have introduced expressions of flux equations which can be solved with direct, explicit computations. In addition, a number of simplified multi-layer models have been proposed by Katerji and Perrier (1985), Shuttleworth and Wallace (1985), Lascano et al. (1987), and Choudhury and Monteith (1988). Most of these models are of a two-layer nature, in that only two different source/sink elevations are recognized, namely the plant canopy and the soil surface. The Shuttleworth and Wallace (1985) model has received application and testing by Lafleur and Rouse (1990) in subarctic wetlands and by Wallace et al. (1990) with dryland millet.

Specific equations and procedures for applying multi-layer models are beyond the scope of this handbook. The reader is referred to literature sources mentioned in this section for details of each model and simulation approach. Generally, the single layer Penman-Monteith equation provides acceptable results in most general hydrologic and water balance studies.

VIII. REGIONAL EVAPOTRANSPIRATION

A. Introduction

Regional evapotranspiration is a necessary input to world climate models, regional water use and supply, calculation of food supply, etc. Therefore, relatively accurate evaluation is necessary. The methods for determining evapotranspiration at specific locations cited in this chapter and listed in this section are valid for regional evapotranspiration. The problem here is the extension of information from a point source to a larger area. This problem is somewhat reduced by using data from watershed studies. In addition, remote sensing may be used to assess information from large areas.

Vegetation can be divided into cultivated and natural. Cultivated vegetation has been selected over the years for maximum production. Consequently, the stomatal resistance is generally low (unless developed for drought tolerance) and plants transpire at near potential rates when water is not limiting. Natural vegetation has been selected over the years for survival. Thus its stomatal (or surface) resistance may be an order of magnitude larger than for cultivated vegetation. Therefore, for ET computation purposes, regions should be divided into cultivated and natural vegetation. Regions also can be divided into humid and arid areas. In the United States, the Mississippi River is an approximate division between humid and arid regime (humid toward the east and arid toward the west).

For cultivated vegetation, evapotranspiration generally is energy-dependent in semihumid to humid areas and water- and energy-dependent in semiarid and arid areas. Therefore, a simple measurement of net radiation in humid areas and precipitation in arid areas can yield first order approximations of evapotranspiration from cultivated but nonirrigated vegetation. Evapotranspiration may be somewhat less than R_n for natural vegetation in humid areas, ranging from only 0.4 of R_n during periods of the year with daytime average temperature of 0°C but approaching 1.0 during hot summer and fall months (see discussion of Figs. 4.2 and 4.3 as well as Equations 4.47 and 4.48). Equation 4.47 can be useful to indicate the minimum possible ratio of $\lambda ET/R_n$ under potential loss conditions, although for the higher latitudes (N or S), the use of daytime R_n only would be appropriate to avoid estimating negative ratios (see Fig. 4.2.). In zones well away from the equator, a well defined cyclic relationship between λE and R_n can be expected with higher ratios in fall than in spring.

B. Theory

Of the methods presented in previous sections of this chapter, the most promising for regional application are the water balance method, lysimeters, energy balance methods, and mass transfer or modeling methods. Application of some of these methods to regional studies is discussed here.

One of the energy balance methods is the Bowen Ratio Energy Balance method (BREB), a combination of the transport and the energy balance equations. This method was presented in Equations 4.77 and 4.78 and operational procedures were discussed in the section on Evapotranspiration from Land Surfaces.

The assumptions of the Bowen ratio method are: 1) transport of H and λE is one-dimensional (vertical), 2) the surface is homogeneous, dense and shallow (i.e., heat, water, and momentum sources and sinks are indistinguishable), and 3) steady-state conditions exist. These assumptions are less demanding for BREB than for other meteorological techniques because the gradient measurements for the BREB are usually made close to the surface and wind measurements are not required. Recommendations on minimum fetch requirements are discussed in the section on Evapotranspiration from Land Surfaces.

For wet surfaces of large extent having minimal advection, evaporation may approach an equilibrium condition. Priestley and Taylor (1972) developed an empirical relationship for equilibrium conditions which was presented previously as Equation 4.48:

$$\lambda E_{eq} = \alpha \frac{\Delta(R_n - G)}{\Delta + \gamma}, \tag{4.184}$$

where λE_{eq} is equilibrium evaporation or evapotranspiration for a large wet area. As has been discussed previously, Priestley and Taylor concluded that the value of $\alpha = 1.26$ in semihumid to humid areas. The value of $\Delta/(\Delta + \gamma)$ at 25°C is approximately 0.74. Therefore, under these conditions, from Equation 4.48, $\lambda E = 0.93(R_n - G)$ or equilibrium latent heat flux is approximately equal to net radiation.

Caution is needed in use of Equation 4.48 during late-fall to late-winter months in zones well away from the equator where the long nights of negative R_n can negate to some extent the response of λET to R_s and the daytime positive R_n (see Nakayama et al., 1983 and Pruitt and Doorenbos, 1977b). The use of 24-hour R_n resulted in experimentally determined values of α close to 2.0 in winter months even under low or non-advection conditions at Davis, California. For higher latitude locations, e.g., Copenhagen, Denmark, the equation breaks down completely during winter periods when 24-hr R_n becomes negative even though some ET is produced by daytime positive net and solar radiation. Under arid or semiarid conditions with advection of heat in air masses, Equation 4.48 can significantly underestimate potential ET (Jury and Tanner, 1975; Jensen et al., 1990; Jones and Kiniry, 1986; Steiner et al., 1991).

When the water supply is limited, evapotranspiration is less than potential. The Penman-Monteith type of resistance-based combination equation (Equation 4.109) can be applied in these types of situations. A number of r_s values for different canopies have been included in Table 4.16. Typical values are 100 s m^{-1} for forests, 40 s m^{-1} for grassland and 30 to 50 s m^{-1} for crops. These are minimum values assuming that sufficient water is available in the soil. The value of r_s will generally increase with decreasing soil moisture, increasing humidity (vapor) deficit, and decreasing R_n or R_s. Application of the Penman-Monteith equation and procedures for increasing r_s to reflect effects of decreasing soil moisture are discussed in detail in the section on Evapotranspiration from Land Surfaces—Direct Penman-Monteith.

The value of r_a is relatively large for smooth surfaces and relatively small for rough surfaces such as forests. Consequently, the effect of r_s is felt strongest over forests. For $r_s = 40$ s m^{-1}, $ET_c/ET_i \approx 0.9$ for grassland and ≈ 0.5 forest, where ET_c is the Penman-Monteith estimate for the particular cover under well-watered conditions ($r_s = 40$ s m^{-1}), and ET_i is the Penman-Monteith estimate with the same r_a value but with $r_s = 0$ (representing evaporation immediately following rain). The increased sensitivity of evaporation from forests to r_s has been discussed in the section on Evaporation of Intercepted Rainfall.

C. Applications

1. Complimentary Approach to Estimating ET Whenever ET falls below the maximum rate for the particular surface due to low availability of soil moisture, the component of sensible heat flux density to the air (H) increases, following Equation 4.40. This results in increased heating of the boundary layer of air above the surface which is manifested in two ways: 1) air temperatures increase and 2) vapor pressure deficits increase. These two increases in turn increase the potential demand placed by the environment on the vegetation and form a type of feedback process which further increases moisture stress of the vegetation. The total increased evaporative demand can be estimated by inserting the elevated temperatures and reduced vapor pressures measured over moisture-stressed vegetation into a Penman-type equation. If this estimate, termed ET_{ps}, for ET potential over stressed vegetation is compared to some type of ET estimate which represents ET under conditions of full soil moisture availability, especially under regionally potential conditions such as represented by the Priestley-Taylor equation (E_{eq}), then the difference, $ET_{ps} - E_{eq}$, can be used to represent the shortage of evapotranspiration due to deficit soil moisture. In this case, actual ET from the water short vegetation, ET_a can be estimated as:

$$ET_a = ET_{eq} - (ET_{ps} - E_{eq}) = 2\,E_{eq} - ET_{ps}. \tag{4.185}$$

Equation 4.185 is known as the complementary approach and was first introduced by Bouchet (1963) and further developed by Brutsaert and Stricker (1979) and Morton (1983). A difficulty with the complementary method is in estimating the rate of E_{eq}, especially in semiarid areas where E_{eq} as estimated with the Priestley-Taylor equation is less than the equilibrium, well-watered ET rate approached in semiarid regions even under moist soil conditions over long periods of time. For further information on the complementary method, the reader is referred to the articles by Brutsaert and Stricker (1979) and Morton (1983).

2. *Application of Energy Balance Models to Forests and Grasslands* Several models have been developed for estimating evapotranspiration from large catchment areas. These models generally integrate various resistances affecting vapor and heat transfer within complex canopies and may consider the development of multistoried canopies over time. The models include a forest ET–CO_2 uptake and water balance model by Running and Coughlan (1988), a forest transpiration–photosynthesis–CO_2 uptake model by Price and Black (1989), and a rangeland–snow melt–ET–heat and water flow model by Flerchinger and Pierson (1991). Some components of the Price and Black model were discussed in the section on surface conductance functions. Energy–balance–crop growth models have been developed for agricultural crops and include the CERES-Maize model (Jones and Kiniry, 1986), CERES-Wheat (Ritchie and Otter, 1985), ENWATBAL (Lascano et al., 1987) and SOYGRO (Jones et al., 1987).

Sikka (1993) and Sikka et al. (1993) modified and applied the Running-Coughlan model to a 210 km^2 catchment area in northern Utah and utilized a fuzzy clustering algorithm along with LandSat thematic mapper (TM) data and digital elevation data to predict the variation in LAI and vegetation types for hydrologic units within the watershed. Vegetation was comprised of a mixture of conifer, aspen, shrubs, and sagebrush, with average LAI's for hydrologic units ranging from 0.6 to 2.5. Sikka concluded that the modified Running-Coughlan model provided acceptable results as long as computerized calibrations were made on the specific watershed using historical weather and streamflow data.

In general, evapotranspiration–vegetation development–energy balance models require rather extensive data describing the phenological development of vegetative canopy cover, photosynthetic–stomate interactions, and penetration of light within canopies. Therefore, these models are usually applied in specialized studies where accurate fluxes are needed during short time periods or in studies examining sensitivity of ET and other fluxes to changes in vegetation, climate, or hydrology.

3. *Approximating Monthly Stream Flow for Ungaged Watersheds* Oftentimes, ET information is needed to estimate mean streamflow from ungaged watersheds. In locations where streamflows are not gaged, approximate estimates of monthly stream discharges can sometimes be made by employing a very general and basic hydrologic balance equation

$$Q = P - ET - GW - \Delta\ Storage, \tag{4.186}$$

where Q is the monthly stream discharge from a watershed, having units equal to P, ET, and GW, GW is ground-water discharge from the basin, and ΔStorage is the net change in soil moisture or ground-water storage in the watershed, with the same units as Q. In simple applications of Equation 4.186 to watersheds where ground-water flow discharge from the watershed is small, the last two terms of Equation 4.186 can be ignored, so that Q can be estimated as the difference between P and ET. Hargreaves (1993; 1995) found good agreement between monthly Q estimated as $Q = P - ET_o$ and measured Q for several watersheds in Central America and in the United States, where ET_o was calculated grass reference ET. ET_o provided a general approximation of ET throughout the watershed areas under the assumption that ET approaches ET_o when $P > ET_o$ during any month.

4. Data Sources in Cultivated Areas An accurate and online data base is required to fully utilize ET prediction methods. Since the 1970's, there has been a rapid expansion of nonfederal automated weather stations (AWS) across the United States and Canada. A total of 20 states have or are acquiring a network of AWS. They include: Alabama, California, Colorado, Florida, Georgia, Idaho, Louisiana, Minnesota, Nebraska, Nevada, New Mexico, New York, North Carolina, Ohio, Oklahoma, Oregon, South Dakota, Utah, Washington, and Wisconsin (Meyer and Hubbard, 1992). Most of the AWS are part of a national network of Regional Climate Centers. A total of seven Regional Climate Centers exist and are located in Nevada, Nebraska, Louisiana, Ohio, South Carolina, and New York (Changnon et al., 1990).

According to Meyer and Hubbard (1992), data from these stations are used for crop water-use (Meyer et al., 1989), irrigation scheduling (Ley and Evans, 1990; Martin et al., 1990), livestock management (Hahn, 1981), integrated pest management (Jones et al., 1987), crop canopy temperature estimates (Sagar et al., 1988), forestry management (Running, 1981), Robinson and Hubbard, 1990), frost and freeze warnings and forecasts (Martsolf, 1981; Ley and Kroeger, 1988), crop growth monitoring (Arkin and Dugas, 1981), crop consulting and determination of crop insurance rates (Snyder, 1991), and drainage design and management (Curry et al., 1988).

Data and calculations from most of the AWS are available in real time. Therefore, for the area covered, either measured or modeled evapotranspiration could be available. Models yield reasonable estimates of evapotranspiration and soil moisture for several crops (Robinson and Hubbard, 1990). These results must be applied to the area of the various crop surfaces represented to obtain regional estimates. Satellite imagery could be used to supply areal extent of the various crops in an area around the AWS.

5. Data Sources in Natural Vegetation In addition to the AWS, programs like NASA's FIFE program have provided data and model development of evaporative fluxes over natural grasslands such as in central Kansas in relation to remote sensing. The Department of Energy Atmospheric Radiation Measurement program contains the Cloud and Radiation Testbed (CART) which proposes the measurement of the evaporative flux by the energy balance Bowen ratio (BREB) technique at ten locations in Oklahoma (Anonymous, 1990). The measurements are automated and the results are available in real time. This data set contains information necessary for further model development. Several other programs [Long Term Ecological Research Program, ParkNet, National Atmospheric Deposition Program/National Trends Network, Multistate Atmospheric Power Production Pollution Study, etc. (Anonymous, 1990)] are collecting long-term data over various ecosystems.

IX. SELECTING THE APPROPRIATE EVAPOTRANSPIRATION METHOD

The type of method used to calculate or estimate evapotranspiration depends on the type of application and the quantity and accuracy of data available. For large scale applications including basin-scale water balances, application of a crop coefficient–reference ET (K_c ET_o) approach with monthly calculation time steps may be appropriate. ET_o for monthly timesteps can generally be calculated using the Hargreaves air temperature method (Equation 4.124) or using the Penman-Monteith equation with estimated R_s, T_d, and wind (see section on Estimating Weather Data for Reference Estimates) when these variables are not measured. For applications to smaller areas, especially those having relatively homogeneous surface cover, the K_c ET_o and direct Penman-Monteith approaches may be equally appropriate. Calculations are usually made on an hourly or 24-hour basis.

In studies of effects of environmental stresses or climate change, energy balance methods using surface temperature measurements and/or the Penman-Monteith method with hourly or shorter time steps are the most appropriate. In studies where effects of changes in vegetation type, density, height, or water availability are evaluated, the Penman-Monteith equation or more complex multi-layered energy balance equations should be used, since changes in vegetation characteristics can be more directly and readily incorporated into the calculations for aerodynamic and surface resistances. Where specific documentation of ET is required for small areas, direct measurements must usually be made using micrometeorological methods including the Bowen ratio and eddy correlation methods or soil water balance methods including lysimeters and direct soil profile measurement. Various precautions that must be exercised when measuring ET have been discussed in the relevant sections of this chapter.

Generally, as one progresses from the K_c ET_o approach to the direct Penman-Monteith approach, the quantity of data required increases, the required calculation timestep decreases, and the requirement for representativeness and accuracy of data increases. It is difficult to apply the direct Penman-Monteith equation using historical data from standard weather stations, since the historical weather data reflect ET flux densities of the weather measurement surface. When applied to some other vegetation type, the direct Penman-Monteith may over or under predict, since the effect of feedback mechanisms between ET and vapor content and temperature of the boundary layer are not reflected in the data.

Energy balance methods, including the combination equation (Penman-Monteith equation) and energy balance equation (Equation 4.96), are generally more dependable and consistent in estimating ET as compared to direct aerodynamic methods (for example Equation 4.90), since the inclusion of energy availability in the former equations places an upper limit on ET estimates. In addition, instrumentation requirements are often less rigorous and demanding.

The K_c ET_o procedure can sometimes be the best overall approach to estimating ET since the use of ET_o provides a measure of general energy available for evapotranspiration, and the K_c coefficient can incorporate a wide assortment of physiological and environmental factors affecting ET rate, including plant density, soil evaporation, and effects of small stand size. This implicit incorporation makes the K_c ET_o approach robust in application and fairly transferrable to other climates, regions, and time periods. Physiological and environmental factors need to be explicitly identified and incorporated during direct application of the Penman-Monteith and other directly applied boundary layer equations, making these approaches usually more data intensive. However, application of the direct Penman-Monteith equation can often serve as a valuable check on validity of K_c values and can in some instances be used to develop K_c's for new types or variations of vegetation.

Usually, calculations made using the Penman-Monteith equation on a 24-hr basis are sufficient for hydrologic studies. Under most conditions, 24-hr estimates of ET_o calculated using the Penman-Monteith equation for 24-hr timesteps provide estimates of total 24-hr ET_o which are within 5 to 10 percent of 24-hr sums made using hourly timesteps for any given day, with differences varying randomly, depending on the general proportions of daytime to nighttime wind speeds and daylengths (Allen et al., 1994b). Results are similar for calculations made using the direct Penman-Monteith equation. The exception is for situations where the stomatal conductance functions (g()) or stability corrections are to be applied. In these situations, calculations should be made on an hourly or shorter basis. Calculations of ET_o on a monthly basis are generally equivalent to those made on a 24-hr basis and summed over the month (Jensen et al., 1990; Allen et al., 1994b).

Energy balance methods must generally use hourly or shorter time steps since measurements of surface temperature, which are highly correlated with solar radiation, are involved, and since corrections for stability are generally employed, especially under conditions of moisture stress.

X. REFERENCES

Aboukhaled, A., Alfaro, A. and Smith, M. (1982). "Lysimeters." Irrigation and Drainage Paper 39, FAO, Rome, 68 p.

Allaway, W.G. and Milthorpe, F.L. (1976). "Structure and functioning of stomata." 57-102. In Kozlowski, T.T. (ed.) *Water Deficits and Plant Growth*, Academic Press, NY, Vol IV:57-102.

Allen, R.G. (1991). "REF-ET. Reference evapotranspiration computation software." Dept. Biological and Irrigation Engineering, Utah State University, Logan, UT, 30.

Allen, R.G. (1992). "Evaluation of a temperature difference method for computing grass reference evapotranspiration." Report submitted to Water Resources Development and Management Service Land and Water Development Division, UN-FAO, Rome, Italy. Revised January 8, 1993.

Allen, R.G. (1993). Personal communication.

Allen, R.G. (1994). "Adjusting air temperature and humidity measurements from arid environments to reflect irrigated conditions." Preliminary Consultant's Report submitted to Water Resources Development and Management Service Land and Water Development Division, UN-FAO, Rome, Italy.

Allen, R.G. (1996). "Assessing integrity of weather data for reference evapotranspiration estimation." *J. Irrig. and Drain. Engrg.*, ASCE, 122(2): 97–106.

Allen, R.G., Brockway, C.E., and Wright, J.L. (1983). "Weather station siting and consumptive use estimates." *J. Water Resour. Plng. and Mgmt.*, ASCE 109(2): 134-146.

Allen, R.G. and Fisher, D.K. (1990). "Low-cost electronic weighing lysimeters." *Trans. Am. Soc. Agric. Engrs.*, 33(6):1823-1833.

Allen, R.G., Hill, R.W., and Vemulapali, S. (1994c). "Evapotranspiration parameters for variably-sized wetlands" Paper presented at the 1994 Summer Meeting of ASAE. No. 942132, 24 p.

Allen, R.G., Howell, T.A., Pruitt, W.O., Walter, I.A., and Jensen, M.E. (eds.). (1991a). *Lysimeters for evapotranspiration and environmental measurements.* ASCE, New York, NY. 444 p.

Allen, R.G., Jensen, M.E., Wright, J.L., and Burman, R.D. (1989). "Operation estimates of evapotranspiration." *Agron. J.*, 81:650-662.

Allen, R.G., Prueger, J.H., and Hill, R.W. (1992). "Evapotranspiration from isolated stands of hydrophytes: cattail and bulrush." *Trans. Am. Soc. Agric. Engrs.*, 35(4):1191-1198.

Allen, R.G. and Pruitt, W.O. (1986). "Rational use of the FAO Blaney-Criddle formula." *J. Irrig. and Drain. Engrg.*, ASCE, 112(2):139-155.

Allen, R.G. and Pruitt, W.O. (1991). "FAO-24 reference evapotranspiration factors." *J. Irrig. and Drain. Engrg.*, ASCE, 117(5):758-773.

Allen, R.G., Pruitt, W.O. and Jensen, M.E. (1991b). "Environmental requirements of lysimeters." 170-181. In Allen, R.G., Howell, T.A., Pruitt, W.O., Walter, I.A., and Jensen M.E. (eds.) *Lysimeters for Evapotranspiration and Environmental Measurements.* ASCE, New York, NY. 170-181.

Allen, R.G., Smith, M., Perrier, A., and Pereira, L.S. (1994a). "An update for the definition of reference evapotranspiration." *ICID Bulletin.* In press. 34 p.

Allen, R.G., Smith, M., Pereira, L.S., and Perrier, A. (1994b). "An update for the calculation of reference evapotranspiration." *ICID Bulletin.* In press. 58 p.

Allen, R.G., Vanderkimpen, P.J., and Wright, J.L. (1995). "Development of resistance parameters for operational application of the Penman-Monteith equation." *Agr. and For. Meteorol.* Accepted for publication.

Allen, R.G. and Wright, J.L. (1996). "Transferring wind measurements from weather stations to agricultural crops." *J. Hydrologic Engineering*, ASCE, Accepted for publication.

Am. Soc. Testing Mat's. (1976). "Standard for metric practice." Am. Soc. for Testing Mat's., ANSI/ASTM E #380-76m, 37 p.

Andersen, E.R. (1954). "Energy-budget studies." Water loss investigations: Lake Hefner Studies, Tech. Report. Prof. Paper 269. Geol. survey, US Dept. Interior. pp. 71-119.

Angus, D.E. and Watts, P.J. (1984). "Evapotranspiration—How good is the Bowen ratio method?" *Agric. Water Mgmt.*, 8, 133-150.

Anonymous. (1990). "Atmospheric radiation measurement program plan." DOE/ER-0441, U.S. Department of Energy, Washington, D.C.

Arkin, G.F. and Dugas, W.A. (1981). "Making weather and climate dependent crop management decisions." Computer Techniques and Meteorological Data Applied to Problems of Agriculture and Forestry: A Workshop, A.

Weiss, Ed., Department of Agricultural Meteorology, University of Nebraska-Lincoln, 223-237. [National Technical Information Service, Accession No. PB82192436, Springfield, VA.].

Armijo, J.D., Twitchell, G.A., Burman, R.D., and Nunn, J.R. (1972). "A large undisturbed weighing lysimeter for grassland studies." *Trans. Amer. Soc. Agric. Engrs.*, 5(5):827-830.

Ashktorab, H., Pruitt, W.O., Paw U, K.T., and George, W.V. (1989). "Energy balance determinations close to the soil surface using a micro-Bowen ratio system." *Agr. and For. Meteorol.* 46, 259-274.

Ashktorab, Hossein, Pruitt, W.O. and Paw U, K.T. (1994) "Partitioning of Evapotranspiration using lysimeter and micro-Bowen-ratio system." *J. Irrig. and Drain. Engrg.*, ASCE, 120(2):450-464.

Aslyng, H.C. and Jensen, S.V.E., (1965). "Radiation and energy balance at Copenhagen, 1955-1964." Hydrotechnical Laboratory, Royal Veterinary and Agricultural College, Copenhagen. 40 p.

Aubert, E.J., and Richards, T.L. (eds.) (1981). "The international field year for the great lakes." Great Lakes Environmental Research Laboratory, National Oceanic and Atmospheric Administration, Ann Arbor, MI.

Bailey, B.J., Montero, J.I., Biel, C., Wilkinson, D.J., Anton, A., and Jolliet, O. (1993). "Transpiration of *Ficus benjamina*: comparison of measurements with predictions of the Penman-Monteith model and a simplified version." *Agr. and For. Meteorol.* 65:229-243.

Ben-Mehrez, M., Taconet, O., Vidal-Madjar, D., and Valencogne, C. (1992). "Estimation of stomatal resistance and canopy evaporation during the HAPEX-MOBILHY experiment." *Agr. and For. Meteorol.* 58:285-313.

Berry, F.A., Bollay, E., and Beers, N.R. (eds.) (1973). *Handbook of meteorology*. McGraw-Hill Inc., New York, NY.

Bertuzzi, P., Bruckler, L., Bay, D., and Chanzy, A. (1994). "Sampling strategies for soil water content to estimate evapotranspiration." *Irrig. Sci.*, 14:105-115.

Black, J.N., Bonython, C.W., and Prescott, J.A. (1954). "Solar radiation and duration of sunshine." *Proc. Roy. Meteorol. Soc.*, 80:231-235.

Bolsenga, S.J. and Vanderploeg, H.A. (1992). "Estimating photosynthetically available radiation into open and ice-covered freshwater lakes from surface characteristics; a high transmittance case study." *Hydrobiologia*. In press.

Boman, B.J. (1991). "Alfalfa ET measurements with drainage lysimeters." pp. 264-271 in *Lysimeters for Evapotranspiration and Environmental Measurements*, Allen, R.G. et al., (ed). ASCE, New York, NY.

Bouchet, R.J. (1963). "Evapotranspiration reele et potentielle, signification climatique," Publ. 62, p. 134-142, General Assembly Berkeley, Int. Ass. Sci. Hydrol., Grentbrugge, Belgium (in French).

Bowen, I.S. (1926). "The ratio of heat losses by conduction and by evaporation from any water surface." *Phys. Rev.* 27:779-787.

Bristow, K.L. and Campbell, G.S. (1984). "On the relationship between incoming solar radiation and daily maximum and minimum temperature." *Agr. and For. Meteorol.* 31:159-166.

Brunt, D. (1932). "Notes on radiation in the atmosphere: I" *Quart. J. Roy. Meteorol. Soc.*, 58:389-420.

Brutsaert, W. (1975). "The roughness length for water vapor, sensible heat and other scalars." *J. Atm. Sci.*, 32:2028-2031.

Brutsaert, W. (1982). *Evaporation into the atmosphere.* D. Reidel Publishing Co. Dordrecht, Holland. 300 p.

Brutsaert, W. and Stricker, H. (1979) " An advection-aridity approach to estimate actual regional evapotranspiration." *Water Resources Res.*, 15(2):443-450.

Buck, A. (1976). "The variable path Lyman-alpha hygrometer and its operating characteristics." *Bull. Am. Meteorol. Soc.* 51:1113-1118.

Budyko, M. I. (1956). *The heat balance of the earth's surface.* (Translated by N. A. Stepanova, 1958). U.S. Dep. Com. Weather Bur. PB 131692. 259 p.

Bultot, F. and Griffiths, J.F., (1972). "The equatorial wet zone." Chapt. 8 In: Climates of Africa, *World Survey of Climatology*. Elsevier Publishing Co., Amsterdam. 10:259-311.

Burman, R.D., Nixon, P.R., Wright, J.L., and Pruitt, W.O. (1980). "Water requirements." 189-232. In Jensen, M.E. (ed.) *Design and operation of farm irrigation systems*. Amer. Soc. Agric. Engr., St. Joseph, Mich.

Burman, R.D., Jensen, M.E., and Allen, R.G. (1987). "Thermodynamic factors in evapotranspiration." p. 28-30. In James, L. G. and English, M. J. (ed.) Proc. Irrig. and Drain. Spec. Conf., ASCE, Portland, Oregon, July.

Burrows, F.J. and Milthorpe, F.L. (1976). "Stomatal conductance in the control of gas exchange." 103-152. In Kozlowski, T.T. (ed.) *Water Deficits and Plant Growth*, Academic Press, New York, NY, Vol IV.

Businger, J.A., (1966). "Transfer of heat and momentum in the atmospheric boundary layer." Proc. Arctic Heat Budget and Atmospheric Circulation, Santa Monica, CA. RAND Corp. p. 305-332.

Businger, J.A. (1988). "A note on the Businger-Dyer profiles." *Boundary-Layer Meteorol.* 42:145-151.

Businger, J.A., Miyake, M., Dyer, A.J., and Bradley, E.F. (1967). "On direct determination of turbulent heat flux near the ground." *J. Appl. Meteorol.* 6(6):

Businger, J.A. and Yaglom, A.M. (1971). "Introduction to Obukhov's paper on 'Turbulence in an atmosphere with a non-uniform temperature'," *Boundary-Layer Meteorol* 2:3-6.

California, State of. (1975) *Vegetative water use in California*. Dept. of Water Resources Bull. No. 113-3, 104 p.

California, State of. (1979) *Evaporation from water surfaces in California*. Dept. of Water Resources Bull. No. 73-79. 163 p.

Campbell, G.S. and Campbell, Melvin D. (1982). "Irrigation scheduling using soil moisture." 25-42. In Hillel, Daniel (ed.) *Advances in Irrigation*. Academic Press, New York, NY.

Campbell, G.S. and Tanner, B.D. (1985). "A krypton hygrometer for measurement of atmospheric water vapor concentration." *Moisture and humidity* ISA, Research Triangle Park, NC. pp 609-612.

Cellier, P. (1986). "On the validity of flux-gradient relationships above very rough surfaces." *Boundary-Layer Meteorol.* 36:417-419.

Cellier, P. and Brunet, Y. (1992). "Flux-gradient relationships above tall plant canopies." *Agr. and For. Meteorol.* 58:93-117.

Changnon, S. A., Lamb, P.J., and Hubbard, K.G. (1990). "Regional climate centers: New institutions for climate services and climate-impact research." *Bul. Am. Meteorol. Soc.* 71(4):527-537.

Chen, J. (1984). "Uncoupled multi-layer model for the transfer of sensible and latent heat flux densities from vegetation." *Boundary-Layer Meteorol.* 28:213-225.

Choudhury, B.J. (1983). "Simulating the effects of weather variables and soil water potential on a corn canopy temperature." *Agric. Meteorol.* 29:169-182.

Choudhury, B.J. (1989). "Estimating evaporation and carbon assimilation using infrared temperature data: vistas in modeling. In *Theory and Applications of Optical Remote Sensing*, Ghassem, A. (ed)., John Wiley & Sons. New York. p. 628-690.

Choudhury, B.J. and Idso, S.B. (1985). "Evaluating plant and canopy resistances of field-grown wheat from concurrent diurnal observations of leaf water potential, stomatal resistance, canopy temperature, and evapotranspiration flux." *Agr. and For. Meteorol.*, 34:67-76.

Choudhury, B.J. and Monteith, J.L. (1986). "Implications of stomatal response to saturation deficit for the heat balance of vegetation." *Agr. and For. Meteor.*, 36:215-225.

Choudhury, B.J. and Monteith, J.L. (1988). "A four-layer model for the heat budget of homogenous land surfaces." *Quart. J. Roy. Meteorol. Soc.*, 114:373-398.

Choudhury, B.J., Idso, S.B., and Reginato, R.J. (1987). "Analysis of an empirical model for soil heat flux under a growing wheat crop for estimating evaporation by an infrared temperature based energy balance equation." *Agr. and For. Meteor.*, 39:283-297.

Clothier, B.E., Clawson, K.L., Pinter, P.J., Moran, M.S., Reginato, R.J., and Jackson, R.D. (1986). "Estimates of soil heat flux from net radiation during the growth of alfalfa." *Agr. and For. Meteorol.* 37:319-329.

Collatz, C.J. Ball, J.T., Grivet, C., and Berry, J.A. (1991). "Physiological and environmental regulation of stomatal conductance, photosynthesis and transpiration: a model that includes a laminar boundary layer." *Agr. and For. Meteorol.*, 54:107-136.

Cowan, I.R. and Milthorpe, F.L. (1968). "Plant factors influencing the water status of plant tissues." 137-193. In Kozlowski, T.T. (ed.) *Water Deficits and Plant Growth*, Academic Press, New York, Vol I.

Croley, T.E., II. (1989). "Verifiable evaporation modeling on the Laurentian Great Lakes." *Water Resources Res.*, 25(5):781-792.

Crookston, M. (1994). Personal communication.

Cuenca, R.H. (1987). *Irrigation system design: An engineering approach.* Prentice Hall, Englewood Cliffs, NJ. 552 p.

Cuenca, R.H. and Nicholson, M.T. (1982). "Application of Penman equation wind function." *J. Irrig. and Drain. Engrg.*, ASCE 108(IR1):13-24.

Curry, R.B., Klink, J.C., Holman, J.R., Elwell, D.L., and Sciarini, M.J. (1988). "Current Ohio experience with an automated weather station network." *Appl. Eng. Agric.*, 4:150-155.

Davies, J.A. and Idso, S.B. (1979). "Estimating surface radiation balance and its components." 183-210. In Barfield, B.J. and Gerbes, J.F. (eds.) *Modification of the Aerial Environment of Plants.* Am. Soc. Agric. Engr. Mono. No. 2. 544 p.

Denmead, O.T. (1984). "Plant physiological methods for studying evapotranspiration: problems of telling the forest from the trees." *Agric. Water Manag.* 8:167-189.

Denmead, O.T. and Bradley, E.F. (1985). "Flux-gradient relationships in a forest canopy." 421-442. In Hutchison, B. A., and Hicks, B. B., (eds.) The Forest Atmosphere Interaction. Dordrocht, Holland, D. Reidel Publishing Co.

Derecki, J.A. (1976). "Lake Erie terrestrial radiation." *Water Resources Res.*, 12(5):979-984.

de Bruin, H.A.R. and Moore, C.J. (1985). "Zero-plane displacement and roughness length for tall vegetation, derived from a simple mass conservation hypothesis." *Boundary-Layer Meteorol.* 31:39-49.

de Vries, D.A. (1963). "Thermal properties of soils." 210-235. In van Wijk, W.R. (ed.). *Physics of Plant Environment*, North Holland Pub. Co.,

Dickinson, R.E. (1984). "Modeling evapotranspiration for three-dimensional global climate models." In *Climate Processes and Climate Sensitivity*, Hansen, J.E. and Takahashi, T. (ed.), AGU Geophysical Monograph 29. p. 58-72.

Donelan, M.A., Elder, F.C., and Hamblin, P.F. (1974). "Determination of the aerodynamic drag coefficient from wind set-up (IFYGL)." Proc. Conf. Great Lakes Res., 17th, 778-788.

Dong, A., Grattan, S.R., Carroll, J.J., and Prashar, C.R.K. (1992). "Estimation of daytime radiation over well-watered grass." *J. Irrig. and Drain. Engrg.*, ASCE, 118 (3):466-479.

Doorenbos, J. and Kassam, A.H. (1979). "Yield response to water." FAO Irrig. and Drain. Paper No. 33, FAO, Rome, Italy. 193 pp.

Doorenbos, J. and Pruitt, W.O., (1975). "Crop water requirements." *Irrigation and Drainage Paper No. 24*, FAO Rome, Italy. 179 p.

Doorenbos, J. and Pruitt, W.O., (1977). "Crop water requirements." *Irrigation and Drainage Paper No. 24*, (rev.) FAO, Rome, Italy. 144 p.

Dorman, J.L. and Sellers, P.J. (1989). "A global climatology of albedo, roughness length and stomatal resistance for atmospheric general circulation models as represented by the simple biosphere model (SIB)." *J. Appl. Meteor.*, 28:833-855

Dozier, J. and Frew, J. (1990). "Rapid calculation of terrain parameters for radiation modeling from digital elevation data." *IEEE Trans. Geoscience and Remote Sens.* 28:963-969.

Duell, L.F.W. (1988). "Estimates of evapotranspiration in alkaline scrub and meadow communities of Owens Valley, California using the Bowen-ratio, eddy-correlation, and Penman combination methods." U.S.G.S. Open-File report 88-92. 78 p.

Duffie, J.A. and Beckman, W.A. (1991). *Solar engineering of thermal processes.* 2nd ed., John Wiley and Sons, NY. p. 1-146.

Dugas, W.A., Fritschen, L.J., Gay, L.W., Held, A.A., Matthias, A.D., Reicosky, D.C., Steduto, P., and Steiner, J.L. (1991). "Bowen ratio, eddy correlation, and portable chamber measurements of sensible and latent heat flux over irrigated spring wheat." *Agr. and For. Meteorl.*, 56:1-20.

Dunin, F.X., McIlroy, I.C., and O'Loughlin, E.M. (1985). "A lysimeter characterization of evaporation by eucalypt forest and its representativeness for the local environment." 271-291. In Hutchison, B.A., and Hicks, B.B. (eds.). *The Forest Interaction.* Proc. Forest Environmental Measurements Conf., Oak Ridge Tennesee, 1983.

Dunin, F.X., Reyenga, W., and McIlroy, I.C. (1991). "Australian lysimeter studies of field evaporation." 237-245. In: Allen, R.G., Howell, T.A., Pruitt, W.O., Walter, I.A., and Jensen, M.E. (eds.). *Lysimeters for Evapotranspiration and Environmental Measurements.* ASCE, New York, NY.

Dyer, A.J. (1961). "Measurements of evaporation and heat transfer in the lower atmosphere by an automatic eddy-convection technique." *Quart. J. Roy. Meteorol. Soc.*, 87:401-412.

Dyer, A.J. (1967). "The turbulent transport of heat and water vapour in an unstable atmosphere." *Quart. J. Roy. Meteorol. Soc.*, 93:501-508.

Dyer, A.J. (1974). "A review of flux-profile relationships." *Boundary Layer Meterol.* 7, 363-372.

Dyer, A.J. and Hicks, B.B. (1970). "Flux-gradient relationships in the constant flux layer." *Quart. J. Roy. Meteorol., Soc.* 96:715-721.

Dyer, A.J. and Pruitt, W.O. (1962). "Eddy flux measurements over a small irrigated area." *J. Applied Meteorol.* 1(4):471-473.

Evans, R.G., Kroeger, M.W., and Mahan, M.O. (1991). "Installation and operation of large drainage lysimeters on grapes." pp. 388-396 in *Lysimeters for Evapotranspiration and Environmental Measurements*, Allen, R.G. et al., (ed). ASCE, New York, NY.

Farnsworth, R.K. and Thompson, E.S. (1982). "Mean monthly, seasonal, and annual pan evaporation for the United States." NOAA Technical Report NWS 34.

Farnsworth, R.K., Thompson, E.S., and Peck, E.L. (1982). "Evaporation atlas for the contiguous 48 United States." NOAA Technical Report NWS 33.

Fitzpatrick, E.A. and Stern, W.R. (1965). "Components of the radiative balance of irrigated plots in a dry monsoonal environment." *J. Applied Meteorol.*, 4:649-660.

Flerchinger, G.N. and Pierson, F.B. (1991). "Modeling plant canopy effects on variability of soil temperature and water." *Agr. and For., Meteorol.* 56:227-246.

Flesch, T.K. and Dale, R.F. (1987). "A leaf area index model for corn with moisture stress reductions." *Agron. J.*, 19:1008-1014.

Fritschen, L.J. (1965). "Accuracy of evapotranspiration determinations by the Bowen ratio method." *Bull. Int. Soc. Sci. Hydrol.*, 2:38-48.

Fritschen, L.J. (1966). "Evapotranspiration rates of field crops determined by the Bowen ratio method." *Agron. J.*, 58:339-342.

Fritschen, L.J. and Qian, P. (1990). "Net radiation, sensible and latent heat flux densities on slopes computed by the energy balance method." *Boundary-Layer Meteorol.*, 53:163-171.

Fritschen, L.J., and Simpson, J.R. (1985). "Evapotranspiration from forests: Measurement and modeling." 393-404. In *Advances in Evapotranspiration.*, Public. No. 74-85, Am. Soc. Agric. Engrs., St. Joseph, MI.

Fritschen, L.J., Cox, L., and Kinerson, R. (1973). "A 28-meter Douglas fir in a weighing lysimeter." *Forest Science* 4:256-261.

Fritz, S. and MacDonald, J.H. (1949). "Average solar radiation in the United States." *Heating and Ventilating*, 46:61-64.

Garratt, J.R. (1978a). "Flux profile relations above tall vegetation." *Quart. J. Roy. Meteorol. Soc.*, 104:199-211.

Garratt, J.R. (1978b). "Transfer characteristics for a heterogeneous surface of large aerodynamic roughness." *Quart. J. Roy. Meteorol. Soc.*, 104:491-502.

Garratt, J.R. (1980). "Surface influence upon vertical profiles in the atmospheric near-surface layer." *Quart. J.Roy. Meteorol. Soc.*, 106:803-819.

Garratt, J.R. (1992). *The atmospheric boundary layer*. Cambridge Univ. Press, 316 p.

Garratt, J.R. and Hicks, B.B. (1973). "Momentum, heat and vapor transfer to and from natural and artificial surfaces." *Quart. J. Roy. Meteorol. Soc.*, 99:680-687.

Gash, J.H.C. (1986). "A note on estimating the effect of a limited fetch on micrometeorological evaporation measurements." *Boundary-Layer Meteorol.* 35:409-413.

Gates, D.M. (1980). *Biophysical ecology*. Springer-Verlag, New York. 611 pp.

Goss, J.R., and Brooks, F.A. (1956). "Constants for empirical expressions for downcoming atmospheric radiation under cloudless skies." *J. Meteorol.*, 13: 482-488.

Grant, D.R. (1975). "Comparison of evaporation from barley with Penman estimates." *Agr. Meteorol.*, 15:49-60.

Grant, R.F. and Baldocchi, D.D. (1992). "Energy transfer over crop canopies: simulation and experimental verification." *Agr. and For. Meteorol.*, 61:129-149.

Gray, D.M. (1970). *Handbook on the principles of hydrology*. Secretariat, Canadian National Committee for the International Hydrological Decade. Ottawa, Canada.

Hahn, G.L. (1981). "Use of weather data in the rational selection of livestock management practices." Computer Techniques and Meteorological Data Applied to Problems of Agriculture and Forestry: A Workshop, A. Weiss, Ed., Department of Agricultural Meteorology, University of Nebraska-Lincoln, 362-382.

Harbeck, G.E., Jr. (1962). "A practical field technique for measuring reservoir evaporation utilizing mass-transfer theory." U.S. Geological Survey Prof. Paper 272-E:100-105.

Hargreaves, G.H. (1993). "Estimating runoff from precipitation and temperature." 402-408. In Allen, R.G. and Neale, C.N. (eds) *Management of Irrigation and Drainage Systems*, Proceedings of the 1993 National Conference on Irrigation and Drainage Engineering, ASCE.

Hargreaves, G.H. (1994). Personal communication.

Hargreaves, G.H. (1995). "Estimating monthly runoff from an ungauged watershed." *J. Irrig. and Drain, Engrg.*, ASCE, submitted.

Hargreaves, G.H. and Samani, Z.A. (1982). "Estimating potential evapotranspiration." Tech. Note, *J. Irrig. and Drain. Engrg.*, ASCE, 108(3):225-230.

Hargreaves, G.H. and Samani, Z.A. (1985). "Reference crop evapotranspiration from temperature." *Appl. Eng. in Agric.*, 1(2):96-99.

Harrison, L.P. (1963). "Fundamental concepts and definitions relating to humidity." In Wexler, A. (ed.) *Humidity and Moisture*. Vol. 3. Reinhold Publishing Company, New York, NY.

Harrold, L.L. and Dreibelbis, F.R. (1958). "Evaluation of agricultural hydrology by monolith lysimeters." Tech. Bull. No. 1179, USDA., Washington, D.C. 166 p.

Hatfield, J.L. (1985). "Wheat canopy resistance determined by energy balance techniques." *Agron. J.*, 77:279-283.

Hatfield, J. L., Reginato, R. J., and Idso, S. B. (1984). "Evaluation of canopy temperature-evapotranspiration models over various crops." *Agric. Meteorol.*, 32: 41-53.

Heermann, D. F., Harrington, G. J., and Stahl, K. M. (1985). "Empirical estimation of clear sky solar radiation." *J. Climate and Applied Meteorol.*, 24(3):206-214.

Hill, R.W. and Allen, L.N. (1991). " Water table lysimeter data interpretation." pp. 254-263, In Allen, R.G. et al., (eds) *Lysimeters for Evapotranspiration and Environmental Measurements*, ASCE, New York, NY.

Horton, R. and Wierenga, P.J. (1983). "Estimating the soil heat flux from observations of soil temperature near the surface." *Soil Sci. Soc. Am. J.* 47, 14-20.

Horton, R. and Chung, S.O. (1991). "Soil heat flow." 397-438 In Hanks, R.J. and Ritchie, J.T. (eds.) *Modeling Plant and Soil Systems*, Agronomy Series No. 31, Am. Soc. Agron., Madison, WI, Chapter 17.

Hostetler, S.W. and P.J. Bartlein. (1990). "Simulation of lake evaporation with application to modeling lake level variations of Harney-Malheur Lake, Oregon." *Water Resources Res.*, 26(10):2603-2612.

Hottel, H.C. (1976). "A simple model for estimating the transmittance of direct solar radiation through clear atmospheres." *Solar Energy* 18:129.

Hounam, C. E. (1973) "Comparison between pan and lake evaporation." World Meteorological Organization, 354, Technical Note 126.

Hoy, R.D. and S.K. Stephens. (1979). "Field study of lake evaporation: Analysis of data from phase 2 storages and summary of phase 1 and phase 2." Austria. Water Resources Council, Dept. of National Devel., Tech. Paper No. 41. 177 p.

Idso, S.B. (1969). "Atmospheric attenuation of solar radiation." *J. Atm. Sci.*, 26, 1088-1095.

Idso, S.B. (1970). "The transmittance of the atmosphere for solar radiation on individual clear days." *J. Appl. Meteorol.*, 9, 239-241.

Idso, S.B. (1981). "A set of equations for full spectrum and 8-14 μm and 10.5-12.5 μm thermal radiation from cloudless skies." *Water Resources Res.*, 17:295-304.

Int'l Comm. on Irrig. and Drain (ICID) (1985). "On updated evapotranspiration and crop water requirement terms for the 1967 ICID Multilingual Technical Dictionary on Irrigation and Drainage," *ICID Bulletin*, 34:1.

Jacobs, A.F.G., Halbersma, J., and Przybyla, C. (1989). "Behaviour of the crop resistance of maize during a growing season." 165-173. In *Estimation of Areal Evapotranspiration*, Black, T.A., Spittlehouse, D.L., Novak, M.D., and Price, D.T. (ed)., IAHS Pub. No. 177.

Jacobs, A.F.G. and van Boxel, J.H. (1988). "Computational parameter estimation for a maize crop." *Boundary-Layer Meteorol.* 42:265-279.

Jagtap, S.S. and Jones, J.W. (1989a). "Evapotranspiration model for developing crops." *Trans. Am. Soc. Agric, Engrs.*, 32:1342-1350.

Jagtap, S.S. and Jones, J.W. (1989b). "Stability of crop coefficients under different climate and irrigation management practices." *Irrig. Science*, 10:231-244.

Jarvis, P.G. (1976). "The interpretation of the variations in leaf water potential and stomatal conductance found in canopies in the field." *Phil. Trans. Roy. Soc.*, London., B 273: 593-610.

Jarvis, P.G., James, G.B., and Landsberg, J.J. (1976). "Coniferous forest." 171-240, In Monteith, J.L. (ed.) *Vegetation and the Atmosphere*, 2. Academic Press, London.

Jensen, M.E. (1968). "Water consumption by agricultural plants." 1-22. In Kozlowski, T.T. (ed.) *Water Deficits and Plant Growth*, Vol. II, Academic Press, NY.

Jensen, M.E. (1972). Personal communication.

Jensen, M.E. (ed.) (1974). "Consumptive use of water and irrigation water requirements." *Irrig. and Drain. Div. Report*, ASCE, 215 p.

Jensen, M.E. and Haise, H.R. (1963). "Estimating evapotranspiration from solar radiation." *J. Irrig. and Drain. Div.*, ASCE, 89:15-41.

Jensen, M.E., Burman, R.D., and Allen, R.G. (1990). *Evapotranspiration and irrigation water requirements*. ASCE Manuals and Reports on Engineering Practice No 70. 350 p.

Jones, A.L., Fisher, P.D., and Croft, B.A. (1981). "Implementing pest management models using real time weather data." 223-237. In Weiss, A. (ed) *Computer Techniques and Meteorological Data Applied to Problems of Agriculture and Forestry: A Workshop*, Department of Agricultural Meteorology, University of Nebraska-Lincoln.

Jones, C.A. and Kiniry, J.R. (1986). (ed). *CERES-Maize: A simulation model of maize growth and development*. Texas A&M Univ. Press, College Station, TX.

Jones, F.E. (1992). *Evaporation of water: with emphasis on applications and measurements*. Lewis Publishers, Inc. Chelsea, MI.

Jones, H.G. (1979). "Stomatal behaviour and breeding for drought resistance." 406-428. In Mussell, H. and Staples, R.C. (eds.) *Stress Physiology in Crop Plants*, John Wiley & Sons, New York, NY.

Jones, J.W., Boote, K.J., Jagtap, S.S., Wilkerson, G.G., Hoogenboom, G., and Mishoe, J.W. (1987). "SOYGRO V5.4 Technical Documentation." Agric. Eng. Dep. Res. Rep., Univ. of Florida, Gainesville, FL.

Jury, W.A. and Tanner, C.B. (1975). "Advection modification of the Priestley and Taylor evapotranspiration formula." *Agron. J.*, 67:569-572.

Katerji, N. and Perrier, A. (1985). "Determination de la resistance globale d'un couvert vegetal a la diffusion de vapeur d'eau et de ses differentes composantes." *Agr. and For. Meteorol.*, 34:105-120.

Katul, G.G. and Parlange, M.B. (1992). "Estimation of bare soil evaporation using skin temperature measurements." *J. Hydrol.*, 132:91-106.

Keijman, J.Q. (1974). "The estimation of the energy balance of a lake from simple weather data." *Boundary-layer Meteorol.* 7:399-407.

Kelliher, F.M., Leuning, R., and Schulze, E.D. (1993). "Evaporation and canopy characteristics of coniferous forests and grasslands." *Oecologia* 95:153-163.

Kimball, B.A., Idso, S.B., and Aase, J.K. (1982). "A model of thermal radiation from partly cloudy and overcast skies." *Water Resources Res.*, 18(4):931-936.

Körner, C., Scheel, J.A., and Bauer, H. (1979). "Maximum leaf conductance in vascular plants." *Photosynthetica*, 13(1):45-82.

Kruse, E.G. and Neale, C.M.U. (1988). "Sources of error in hydraulic weighing lysimeter measurements." *Trans. Am. Soc. Agric. Engrs.*, 32(1):81-88,96.

Lafleur, P.M. and Rouse, W.R. (1990). "Application of an energy combination model for evaporation from sparse canopies." *Agric. and Forest Meteorol.* 49:135-153.

Lascano, R.J., Van Bavel, C.H.M., Hatfield, J.L., and Upchurch, D.R. (1987). "Energy and water balance of a sparse crop: simulated and measured soil and crop evaporation." *Soil Sci. Soc. Am. J.* 51, 1113-1121.

Lean, J. (1989). "Contribution of ultraviolet irradiance variations to changes in the sun's total irradiance." *Science*, 244:197-200.

Lettau, H.H. (1969). "Note on aerodynamic roughness-parameter estimation on the basis of roughness-element description." *J. Appl. Meteorol.*, 8:828-832.

Leuning, R. and Attiwill, P.M. (1978). "Mass, heat, and momentum exchange between a mature Eucalyptus forest and the atmosphere." *Agric., Meteorol.* 19:215-241.

Ley, T.W. and Allen, R.G. (1994). "Energy and soil water balance analyses of arid weather sites." 1994 Summer Meeting of the Am. Soc. Agric. Engr., Kansas City, Mo., ASAE Paper No. 94-2078. 26 p.

Ley, T.W. and Elliott, R.L. (1993). "Standards for automated agricultural weather stations." 977-984, In: R.G. Allen and C.M. Neale (ed.) *Management of Irrigation and Drainage Systems: An Integrated Approach*, Proceedings, National Conference on Irrigation and Drainage Engineering, ASCE, New York, NY.

Ley, T.W. and Evans, R.G. (1990). "Washington public agriculture weather system." In *Visions of the Future*, Proc. Third National Irrig. Symp., ASAE, St. Joseph, MI. p. 261-269.

Ley, T.W. and Kroeger, M.W. (1988). "Washington public agricultural weather system." ASAE Paper #PNR88-106, 12 pp. [Available from the American Society of Agricultural Engineers, St. Joseph, MI.]

Lhomme, J.P. (1988a). "Extension of Penman's formulae to multi-layer models." *Boundary-Layer Meteorol.* 42:281-291.

Lhomme, J.P. (1988b). "A generalized combination equation derived from a multi-layer micrometeorological model." *Boundary-Layer Meteorol.* 45, 103-115.

Lin, J.D. (1980). "On the force-restore method for prediction of ground surface temperature." *J. Geophysical Res.* 85:3251-3254.

List, R.J. (1963). *Smithsonian meteorological tables.* Smithsonian Miscellaneous Collections Volume 114. Smithsonian Institution, Washington, DC.

List, R.J. (1984). *Smithsonian meteorological tables*, 6th rev. ed., Smithsonian Institution, Washington, DC. 539 p.

Liu, P.C. and Schwab, D.J. (1987). "A comparison of methods for estimating u^* from given u^z and air-sea temperature differences." *J. Geophys. Res.*, 92(C6):6388-6494.

London, J. and Fröhlich, C. (eds.) (1982). "Extended abstracts presented at the symposium on the solar constant and the spectral distribution of solar irradiance." IAMAP Third Scientific Assembly, August 17-28, 1981, Hamburg, Federal Republic of Germany. 41 p.

Ludlow, M.M. and Jarvis, P.G. (1971). *J. Appl. Ecol* 8:925-953.

Malek, E., Bingham, G.E., and Hipps, L.E. (1987). "Daily and seasonal evapotranspiration losses from the margin and playa of a west desert valley." 18th Conf. on Agr. and For. Meteorol., Sept., 14-18, 1987. W. Lafayette, IN, Am. Meteorol. Soc. Boston, Mass. p. 121-124.

Martin, E.C., Shayya, W.H., Bralts, V.F., Loudon, T.L., and Johnson, J.A. (1990). "Irrigation scheduling: a statewide program in Michigan." 700-706, In *Visions of the Future*, Proc. Third National Irrig. Symp., ASAE, St. Joseph, MI.

Martin, D.L., Gilley, J.R., Carmack, W.J., and Hardy, L.A. (1993). "SCS methods for determining irrigation water requirements." pp 1013-1038. In R.G. Allen and Neale, C.M. (eds.) Proceedings of the ASCE National Conference on Irrigation and Drainage Engineering.

Martsolf, J.D. (1981). "Satellite frost forecasting system for Florida." Computer Techniques and Meteorological Data Applied to Problems of Agriculture and Forestry: A Workshop, A. Weiss, (ed)., Department of Agricultural Meteorology, University of Nebraska-Lincoln, 223-237.

McArthur, J.A., Hesketh, J.D., and Baker, D.N. (1975). "Cotton." 297-325. In Evans, L.T. (ed.) *Crop Physiology*, Cambridge Univ. Press.

McBean, G.A. (1972). "Instrument requirements for eddy correlation measurements." *J. Appl. Meteorol.* 11:1078-1084.

McFarland, M.J., Worthington, J.W., and Newman, J.S. (1983). "Design, installation and operation of a twin weighing lysimeter for fruit trees." *Trans. Am. Soc. Agric. Engrs., 26(6)*:1717-1721.

McGinn, S.M. and King, K.M. (1990). "Simultaneous measurements of heat, water vapor and CO_2 fluxes above alfalfa and maize." *Agr. and For. Meteorol.*, 49: 331-349.

McIlroy, I.C. and Dunin, F.X. (1982). "A forest evaporation technique comparison experiment." The First National Symposium on Forest Hydrology, Melbourne. The Institute of Engineers, Australia, May.

McNaughton, K.G. and Black, T.A. (1973). "A study of evapotranspiration from a Douglas fir forest using the energy balance approach." *Water Resources Res.*, 9:1579-1590.

Meyer, S.J. and Hubbard, K.G. (1992). "Nonfederal automated weather stations and networks in the United States and Canada: A Preliminary Survey." *Bul. Am. Meteorol. Soc.* 73(4):449-457.

Meyer, S.J., Hubbard, K.G., and White, D.A. (1989). "Estimating potential evapotranspiration: The effect of random and systematic error." *Agric. Forest Meteorol.* 46:285-296.

Milne, R. (1979). "Water loss and canopy resistance of a young sitka spruce plantation." *Bound. Layer. Met.* 16:67-81.

Monin, A.S. and Obukhov, A.M. (1954). "The basic laws of turbulent mixing in the surface layer of the atmosphere." *Akad. Nauk. SSSR Trud. Geofiz. Inst.*, No. 24 (151):163-187.

Monteith, J.L. (1965). "Evaporation and the environment." 205-234. In *The state and movement of water in living organisms*, XIXth Symposium. Soc. for Exp. Biol., Swansea, Cambridge Univer. Press.

Monteith, J.L. (1981). "Evaporation and surface temperature." *Quart. J. Roy. Meteorol. Soc.*, 107:1-27.

Monteith, J.L. and Szeicz, G. (1962). "Radiative temperature in the heat balance of natural surfaces." *Quart. J. Roy Meteorol. Soc.*, 88:496-507.

Monteith, J.L. and Unsworth, M.H. (1990). *Principles of environmental physics*, 2nd Ed. Edward Arnold, London, New York. 291 p.

Moore, C.J. (1974). "A comparative study of forest and grassland micrometeorology." Ph.D. Thesis, The Flinders, Univ. of S. Australia, 237 p.

Morgan, D.L., Pruitt, W.O., and Lourence, F.J. (1971). "Analyses of energy, momentum and mass transfers above vegetative surfaces." *Research and Development Tech. Rept*, ECOM 68-G10-F, Univ. of California, Davis, 142 p.

Morton, F.I. 1983. Operational estimates of areal evapotranspiration and their significance to the science and practice of hydrology. *J. Hydrol.*, 66:1-76.

Mukammal, E.I., King, K.M., and Cork, H.F. (1966). "Comparison of aerodynamic and energy budget techniques in estimating evaporation from a cornfield." *Archiv. Met. Geophys. u. Bioklim.*, Ser. B, 14:384-395.

Murray, F.W. (1967). "On the computation of saturation vapor pressure." *J. Appl. Meteor.*, 6:203-204.

Nakayama, K., Pruitt, W.O., and Chandio, B. (1983). "Some aspects of the Priestley and Taylor model to estimate evapotranspiration." *The Tech. Bull. of the Faculty of Horticulture*. Chiba Univ. No. 32:25-30.

National Renewable Energy Laboratory (NREL). (1992) "Users manual national solar radiation data base - 1961 - 1990." NSRDB - Vol. 1, 105p. + App.

Neale, C.M.U. Artan, G.A., Tarboton, D.G., and Hanson, C.H. (1995). "A spatially distributed energy balance model: development, validation and application." Proc. 25th Conf. Agric. and For. Meteorol., Am. Meteorol. Soc., Jan. 1995, Dallas, TX.

Neilson, R.E., Ludlow, M.M., and Jarvis, P.G. (1972). "Photosynthesis in Sitka spruce (*Picea sitchensis* (Bong.) Carr.)," *J. Appl. Ecol.* 9:721-745.

Nie, D. and Kanemasu, E.T. (1989). "A comparison of net radiation on slopes." 142-143. In Proc. 19th Conf. Agric. and For. Meteorol., Am. Meteorol. Soc., March 1989, Charleston, S. Carolina.

Niel, P. and Zunino, J.P. (1971). "Essais sur les besoins en eau des cultures au cours desannées 1969 et 1970." L'Irrigant. Revue de la Soc. du Canal de Provence No. 55 (Sept-Dec.).

Noilhan, J. and Planton, S. (1989). "A simple parameterization of land surface processes for meteorological models," *Monthly Weather Rev.* 117:536-549.

Norman, J.M. and Arkebauer, T.J. (1991). "Predicting canopy light-use efficiency from leaf characteristics." 7:125-143. In Hanks, R.J. and Ritchie, J.T. (eds.) *Modeling Plant and Soil Systems*, Agronomy Series No. 31, Am. Soc. Agron., Madison, WI.

Novak, M.D. (1989). "The use of analytical solutions to determine the evapotranspiration from bare and vegetated surfaces." 133-152. In Black, T.A., Spittlehouse, D.L., Novak M.D., and Price, D.T. (eds.). *Estimation of Areal Evapotranspiration*, IAHS Pub. No. 177.

Obukhov, A.M. (1946). "Turbulence in an atmosphere with non-uniform temperature," *Trudy Instit. Teoret. Geofiz:* AN-S.S.S.R., No. 1 (English Trans.: (1971)), *Bound. Layer Meteorol.* 2:7-29.

Oke, T.R. (1978). *Boundary layer climates.* John Wiley & Sons, New York, NY. A Halsted Press Book.

Oster, J.D., Meyer, J.L., Hermsmeier, L., and Kaddah, M. (1984). "Imperial Valley: irrigation, drainage and runoff." *Soil and Water* No. 60, Fall 1984. Cooperative Extension, Univ. of California., 6701 San Pablo Ave., Oakland, Ca. 94608.

Panofsky, H.A. (1963). "Determination of stress from wind and temperature measurements." *Quart. J. Roy. Meteorol. Soc.*, 89:85-94.

Paulson, C.A. (1970). "The mathematical representation of wind speed and temperature profiles in the unstable atmospheric surface layer." J. Appl. Meteorol. 9:857-861.

Penman, H.L. (1948). "Natural evaporation from open water, bare soil and grass." *Proc. Roy. Soc. London*, A193:120-146.

Penman, H.L. (1956). "Estimating evaporation." *Trans. Am. Geophys. Union*, 37: 43-50.

Penman, H.L. (1963). *Vegetation and hydrology*. Tech. Comm. No. 53, Commonwealth Bureau of Soils, Harpenden, England. 125 pp.

Perez, R., Stewart, R., Arbogast, C., Seal, R., and Scott, J. (1986). "An anisotropic hourly diffuse radiation model for sloping surfaces: description, performance, validation, site dependency evaluation." *Solar Energy* 36:481-497.

Perrier, A. (1982). "Land surface processes: vegetation." pp 395-448 in Eagleson, P. (ed.) *Land Surface Processes in Atmospheric General Circulation Models* Cambridge Univ. Press.

Petersen, E.W. (1969). "On the relation between the shear stress and the velocity profile after a change in surface roughness," *J. Atmos. Sci.* 26:773-774.

Phillips, W.D. and Irbe, J.G. (1978). "Land-to-lake comparison of wind temperature and humidity on Lake Ontario during the International Field Year for the Great Lakes (IFYGL)." Rep. CLI-2-77, Atmos. Environ. Serv., Environ. Canada, Downsview, Ontario.

Pieri, P. and Fuchs, M. (1990). "Comparison of Bowen ratio and aerodynamic estimates of evapotranspiration." *Ag. and For. Meteorol.*, 49:243-356.

Plate, E.J. (1971). "Aerodynamic characteristics of atmospheric boundary layers." *U.S. Atomic Energy Comm., Critical Review Series*, TID-25465.

Price, D.T. and Black, T.A. (1989). "Estimation of forest transpiration and CO_2 uptake using the Penman-Monteith equation and a physiological photosynthesis model." 213-227 In Black, T.A., Spittlehouse, D.A., Novak, M.D., and Price, D.T. (eds.) *Estimation of Areal Evapotranspiration*, IAHS Pub. No. 177.

Priestley, C.H.B. and Taylor, R.J. (1972). "On the assessment of surface heat flux and evaporation using large-scale parameters." *Monthly Weather Rev.* 100:81-92.

Pruitt, W.O. (1955). "1955 Annual report." Washington Agricultural Experiment Station, Prosser, Washington. Washington State Univ., Pullman, WA. p. 50-66.

Pruitt, W.O. (1960). "Relation of consumptive use of water to climate." *Trans. Am. Soc. Agric. Engrs.*, 3(1):9-13, 17.

Pruitt, W.O. (1963). "Application of several energy balance and aerodynamic evaporation equations under a wide range of stability." Final Report to USAEPG on Contract DA-36-039-SC-80334, Univ. Calif., Davis, pp. 107-124.

Pruitt, W.O. (1964). "Cyclic relations between evapotranspiration and radiation." *Trans. Am. Soc. Agric. Engrs.*, 7(3):271-275, 280.

Pruitt, W.O. (1966). "Empirical method of estimating evapotranspiration using primarily evaporation pans." *Proc. Conf. on Evapotranspiration and its Role in Water Resources Management*. Chicago. Dec. ASAE, pp. 57-61.

Pruitt, W.O. (1971). "Factors affecting potential evapotranspiration." 82-102 In Monke, E.J. (ed). *Biological Effects in the Hydrological Cycle*. Proceedings of the Third International Seminar for Hydrology Professors. Purdue University, West Lafayette, IN. July 18-30.

Pruitt, W.O. (1976). Personal communication.

Pruitt, W.O. (1994). Personal communication.

Pruitt W.O. and Angus, D.E. (1960). "Large weighing lysimeter for measuring evapotranspiration." *Trans. Am. Soc. Agric. Engrs.*, 3(2):13-15,18.

Pruitt, W.O. and Lourence, F.J. (1966). "Tests of aerodynamic, energy balance and other evaporation equations over a grass surface." In *Investigations of Energy, Momentum and Mass Transfers near the Ground.* Final Report, USAEPG Contract. University of California, Davis, CA. p. 37-63.

Pruitt, W.O. and Doorenbos, J. (1977a). "Background and development of methods to predict reference crop evapotranspiration (ETo)." In Appendix II, Crop Water Requirements, *Irrigation and Drainage Paper No. 24.* (rev.) FAO, Rome, Italy. pp. 108-119.

Pruitt, W.O. and Doorenbos, J. (1977b). "Empirical calibration, a requisite for evapotranspiration formulae based on daily or longer mean climatic data?" International Round Table Conference on "Evapotranspiration," Budapest, Hungary. ICID, 20 p.

Pruitt, W.O. and Lourence, F.J. (1985). "Experiences in lysimetry for ET and surface drag measurements." 51-69. In *Advances in Evapotranspiration.*, Public. No. 74-85, Am. Soc. Agric. Engrs., St. Joseph, MI.

Pruitt, W.O., Morgan, D.L., and Lourence, F.J. (1968). "Energy momentum and mass transfers above vegetative surfaces." Technical Report ECOM-0447(E)-F. University of California, Davis, CA. 59 p.

Pruitt, W.O., Morgan, D.L., and Lourence, F.J. (1973). "Momentum and mass transfers in the surface boundary layer." *Quart. J. Roy. Meteorol. Soc.*, 99:370-386.

Pruitt, W.O., Swann, B.D., Held, A., Sutton, B., Matista, A., and Hsiao, T.C. (1987). "Bowen ratio and Penman: Australian-California tests." 149-158 In Larry G. James and Marshall J. English, (eds) *Irrigation Systems for the 21st Century*, Proc. of the ASCE National Conference on Irrigation and Drainage Engineering, Portland, OR.

Quinn, F.H. (1979). "An improved aerodynamic evaporation technique for large lakes with application to the International Field Year for the Great Lakes." *Water Resources Res.*, 15(4):935-940.

Ramdas, L.A. (1957). "Evaporation and potential evapotranspiration over the Indian Sub-Continent." *Indian J. Agr. Sci.*, 27(2):137-149.

Ratliff, L.F., Ritchie, J.T., and Cassel, D.K. (1983). "Field-measured limits of soil water availability as related to laboratory-measured properties." *Soil Sci. Soc. Am. J.* 47:770-775.

Rauner, J.L. (1976). "Deciduous forests." 241-264. In Monteith, J.L. (ed.) *Vegetation and the Atmosphere*, 2. Academic Press, London.

Raupach, M.R. (1979). "Anomalies in flux-gradient relationships over forest." *Boundary Layer Meteorol.* 16:467-486.

Raupach, M.R., Antonia, R.A., and Rajagopalan, S., (1991). "Rough-wall turbulent boundary layers." *Appl. Mech. Rev.* 44:1-25.

Reyenga, W., Dunin, F.X., Bautovich, B.C., Rath, C.R., and Hulse, L.B. (1988). "A weighing lysimeter in a regenerating eucalypt forest: design, construction, and performance." *Hydrol. Processes*, 2:301-314.

Rijtema, P.E., (1965). "Analysis of actual evapotranspiration." Agric. Res. Rep. No. 69, Centre for Agric. Publ. and Doc., Wageningen, 111 p.

Ritchie, J.T. (1972). "Model for predicting evaporation from a row crop with incomplete cover." *Water Resour. Res.*, 8:1204-1213.

Ritchie, J.T. (1974). "Evaluating irrigation needs for southeastern U.S.A." Proc. 1974 Irrig. and Drain. Specialty Conf., ASCE, p. 262-273.

Ritchie, J.T. (1991). "Wheat phasic development." 31-54. In Hanks, R.J. and Ritchie, J.T. (eds), *Modeling Plant and Soil Systems*, Agronomy Series No. 31, Am. Soc. Agron., Madison, WI.

Ritchie, J.T. and NeSmith, D.S. (1991). "Temperature and crop development." 5-29. In Hanks R.J. and Ritchie, J.T. (eds.). *Modeling Plant and Soil Systems*, Agronomy Series No. 31, Am. Soc. Agron., Madison, WI.

Ritchie, J.T. and Otter, S. (1985). "Description and performance of CERES-Wheat: a user-oriented wheat yield model." p. 159-175 in ARS Wheat Yield Project. ARS-38. Natl. Tech. Info. Serv., Springfield, VA.

Robinson, J.M. and Hubbard, K.G. (1990). "Soil water assessment model for several crops in the High Plains." *Agron. J.*, 82:1141-1148.

Rochette, P., Pattey, E., Desjardins, R.LO., Dwyer, L.W., and Stewart, D.W. (1991). "Estimation of maize canopy conductance by scaling up leaf stomatal conductance." *Agric. For. Meteorol.*, 54:241-261.

Rossby, C.G. and Montgomery, R.B. (1935). "The layer of frictional influence in wind and ocean currents," *Papers Phys. Oceanogr. Meteor.*, Woods Hole Oceanographic Institution, M.I.T., 3:3.

Running, S.W. (1981). "The influence of microclimate on forest productivity: A system to predict the biophysical site quality of forest land." 223-237 In Weis, A. (ed.) *Computer Techniques and Meteorological Data Applied to Problems of Agriculture and Forestry: A workshop*, Department of Agricultural Meteorology, University of Nebraska-Lincoln.

Running, S.W. and Coughlan, J.C. (1988). "A general model of forest ecosystem processes for regional applications: I. Hydrologic balance, canopy gas exchange and primary production processes." *Ecological Modeling* 42:125-154.

Rutter, A.J. (1975). "The hydrological cycle in vegetation." 11-154 In Monteith, J.L. (ed.) *Vegetation and the Atmosphere*, Vol.1. Academic Press, NY.

Rutter, A.J., Morton, A.J., and Robins, P.C. (1975). "A predictive model of rainfall interception in forests: II. Generalization of the model and comparison with observations in some coniferous and hardwood stands." *J. Appl. Ecol.* 12:367-380.

Sagar, R.M., Hubbard, K.G., Norman, J.M., and Holtzer, T.O. (1988). "Estimation of corn canopy temperature and water budget using automated weather station data." *CAMAC Progress Report* 88-4, Department of Agricultural Meteorology, University of Nebraska-Lincoln, 143 p.

Sarraf, S. and Bovée, A.C.J. (1973). "Evapotranspiration of bananas." Magon Série Tech. No. 20, IRA, Lebanon (in French). 18 p.

Schneider, A.D. and Howell, T.A. (1991). "Large, monolithic, weighing lysimeters." 37-45 In Allen, et al. (ed.) *Lysimeters for Evapotranspiration and Environmental Measurements*, ASCE, New York, NY.

Schuepp, P.H., Leclerc, M.Y., MacPherson, J.I., and Desjardins, R.L. (1990). "Footprint prediction of scalar fluxes from analytical solutions of the diffusion equation." *Bound. Layer Meteorol.* 50:355-373.

Seginer, I. (1974). "Aerodynamic roughness of vegetated surfaces." *Boundary-Layer Meteorol.* 5, 383-393.

Seguin, B., Brunet, Y., and Perrier, A. (1982). "Estimation of evaporation: a review of existing methods and recent developments." In European Geologic Society Symposium on Evaporation. Leeds, U.K., August, 1982, 21 p.

Sellers, W.D. (1965). *Physical Climatology*. Univ. of Chicago Press, Chicago. 272 p.

Sellers, P.J. and Dorman, J.L. (1987). "Testing the simple biosphere model (SIB) using point micrometeorological and biophysical data." *J. Clim. Appl. Meteor.* 26, 622-651.

Shalhevet, J., Mantell, A., Bielorai, H., and Shimshi, D. (eds.) (1981). *Irrigation of field and orchard crops under semi-arid conditions*. International Irrigation Information Center, Bet Dagan, Israel. 132 p.

Sharma, M.L. (1985). "Estimating Evapotranspiration." 213-281. In Hillel, D (ed.) *Advances in Irrigation, Volume 3*. Academic Press, Inc.

Shuttleworth, W.J. (1979). "Below-canopy fluxes in a simplified one-dimensional theoretical description of the vegetation-atmosphere interaction." *Boundary-Layer Meteorol.* 17, 315-331.

Shuttleworth, W.J. (1984). "Eddy correlation of energy partition for Amazonian forest." *Quart. J. Roy. Meteorol. Soc.*, 110:1143-1162.

Shuttleworth, W.J. (1989). "Micrometeorology of temperate and tropical forest." Phil. *Trans. Roy. Soc. London* B 324:299-334.

Shuttleworth, W.J. (1991). "Evaporation models in hydrology." 93-120. In Schmugge, T.J. and André, J.C. (eds.) *Land Surface Evaporation, Measurement and Parameterization*. Springer-Verlag, New York.

Shuttleworth, W.J. (1992). "Evaporation." Chapt. 4. In Maidment (ed) *Handbook of Hydrology*. McGraw-Hill Inc., New York, NY.

Shuttleworth, W.J. and Wallace, J.S. (1985). "Evaporation from sparse crops - an energy combination theory." *Quart. J. Roy. Meteorol. Soc.*, 111:839-853.

Sikka, A.K. (1993). "Modeling the influence of climate change on evapotranspiration and water yield." Unpublished Ph.D. dissertation, Dept. Civil Engineering, Utah State Univ. 330 p.

Sikka, A.K., Gunderson, R.W., Limaye, A.S., and Riley, J.P. (1993). "Application of distributed parameter watershed model for determining the effects of climate change on water resources." 464-471. In Allen, R.G. and Neale, C.M. (ed.) *Management of Irrigation and Drainage Systems: An Integrated Approach*, Proceedings, National Conference on Irrigation and Drainage Engineering, ASCE, New York, NY.

Sinclair, T.R. (1984). "Leaf area development in field-grown soybeans." *Agron. J.*, 76:141-146.

Sinclair, T.R., Allen, L.H., and Lemon, E.R. (1975). "An analysis of errors in the calculation of energy flux densities above vegetation by a Bowen ratio profile method." *Boundary-Layer Meteorol.* 8:129-139.

Slayter, R.O. (1967). *Plant-water relationships*. Academic Press. London. 366 p.

Slatyer, R.O. and McIlroy, I.C. (1961). "Practical microclimatology." C.S.I.R.O., Melbourne, UNESCO, Paris. 340 p.

Smith, M., Allen, R., Monteith, J., Perrier, A., Pereira, L., and Segeren, A. (1991). "Report of the expert consultation on procedures for revision of FAO guidelines for prediction of crop water requirements." UN-FAO, Rome, Italy, 54 p.

Snyder, R.L. (1991). Personal communication.

Snyder, R.L. and Pruitt, W.O. (1994). "ET_O—Computer software and manual." Dept. Land, Air and Water Resources, Univ. California, Davis, CA.

Snyder, R.L., Lanini, B.J., Shaw, D.A., and Pruitt, W.O. (1989a). "Using reference evapotranspiration (ET_O) and crop coefficients to estimate crop evapotranspiration (ET_C) for agronomic crops, grasses, and vegetable crops." Cooperative Extension, Univ. California, Berkeley, CA, Leaflet No. 21427, 12 p.

Snyder, R.L., Lanini, B.J., Shaw, D.A., and Pruitt, W.O. (1989b). "Using reference evapotranspiration (ET_O) and crop coefficients to estimate crop evapotranspiration (ET_C) for trees and vines." Cooperative Extension, Univ. California, Berkeley, CA, Leaflet No. 21428, 8 p.

Stanhill, G., (1962). "The control of field irrigation practice from measurements of evaporation." *Israel J. Agr. Res.*, 12:51-62.

Stanhill, G. (1976). "Cotton." 121-150. In Monteith, J.L. (ed.) *Vegetation and the atmosphere*, 2. Academic Press, London.

Steiner, J.L., Howell, T.A., and Schneider, A.D. (1991). "Lysimetric evaluation of daily potential evaporation models for grain sorghum." *Agron J.*, 83:240-247.

Stenmark, Ernest B. and Drury, Lorne D. (1970). "Micrometeorological field data from Davis, California; 1967 cooperative field experiment runs." Res. and Dev. Tech. Report, ECOM-6053, Fort Huachuca, AZ. AD724613. 576 p.

Stewart, J.B. (1983). "A discussion of the relationships between the principal forms of the combination equation for estimating crop evaporation." *Agric. Meteorol.* 30:111-127.

Stewart, J.B. (1988). "Modelling surface conductance of pine forest." *Agric. For. Meteorol.* 43:19-35.

Stewart, J.B. (1989). "On the use of the Penman-Monteith equation for determining areal evapotranspiration." 3-12. In: Black, T.A., Spittlehouse, D.L., Novak, M.D., and Price, D.T. (ed)., *Estimation of Areal Evapotranspiration*, IAHS Pub. No. 177.

Stewart, J.B. and Gay, L.W. (1989). "Preliminary modelling of transpiration from the FIFE site in Kansas." *Agric. and Forest Meteorol.* 48:305-315.

Stewart, J.B. and Thom, A.S. (1973). "Energy budgets in pine forest." *Quart. J. Roy. Meteorol. Soc.*, 99:154-170.

Stewart, J.B. and Verma, S.B. (1992). "Comparison of surface fluxes and conductances at two contrasting sites within the FIFE area." *J. Geophys. Res.* 97(D17):18,623-18,628.

Swift, L.W. (1976). "Algorithm for solar radiation on mountain slopes." *Water Resources Res.* 12:108-112.

Swinbank, W.C. (1951). "The measurement of vertical transfer of heat and water vapor by eddies in the lower atmosphere." *J. Meteorol.* 8:135-145.

Swinbank, W.C. (1955). "An experimental study of eddy transports in the lower atmosphere." Tech. Paper No. 2, C.S.I.R.O. Division of Meteorological Physics. Melbourne, Australia.

Szeicz, G. and Long, I.F. (1969). "Surface resistance of crop canopies." *Water Resources. Res.*, 5:622-633.

Szeicz, G., Van Bavel, C., and Takami, S. (1973). "Stomatal factor in the water use and dry matter production by sorghum." *Agric. Meteorol.* 12:361-389.

Tanner, B.D. (1988). "Use requirements for Bowen ratio and eddy correlation determination of evapotranspiration." 605-616. In Allen, R.G. and Neale, C.M. (ed.) *Management of Irrigation and Drainage Systems: An Integrated Approach*, Proceedings, National Conference on Irrigation and Drainage Engineering, ASCE, New York, NY.

Tanner, C.B. (1960). "Energy balance approach to evapotranspiration from crops." *Soil Sci. Soc. Am. Proc.* 24:1-9.

Tanner, C.B. (1967). "Measurement of evapotranspiration." 534-574. In Hagan, R.M., Haise, H.R., and Edminster, T.W. (eds.). *Irrigation of Agricultural Lands.* ASA Mono. No. ll, Am. Soc. of Agron., Madison, WI.

Tanner C.B. (1968). "Evaporation of water from plants and soils." 73-106. In Kozloski, T.T. (ed.) *Water deficits and plant growth.* Vol. I, Academic Press, New York, NY.

Tanner, C.B. and Jury, W.A. (1976). "Estimating evaporation and transpiration from a crop during incomplete cover." *Agron. J.*, 68:239-242.

Tetens, O. (1930). "Uber einige meteorologische Begriffe." *Z. Geophys.*, 6:297-309.

Thom, A.S. (1971). "Momentum absorption by vegetation." *Quart. J. Roy. Meteorol. Soc.*, 97:414-428.

Thom, A.S. (1972). "Momentum, mass and heat exchange of vegetation." *Quart. J. Roy. Meteorol. Soc.*, 98:124-134.

Thom, A.S., Stewart, J.B., Oliver, H.R., and Gash, J.H.C. (1975). "Comparison of aerodynamic and energy budget estimates of fluxes over a pine forest." *Quart. J. Roy. Meteorol. Soc.*, 101:93-105.

Thom, A.S., Thony, J.L., and Vauclin, M. (1981). "On the proper employment of evaporation pans and atmometers in estimating potential transpiration." *Quart. J. Roy. Meteorol. Soc.*, 107:711-736.

Thompson, N., Barrie, I.A., and Ayles, M. (1981). "The Meteorological Office rainfall and evaporation calculation system: MORECS." Great Britain Hydrological Memorandum No. 45. 67 p.

Thornthwaite, C.W. (1948). "An approach toward a rational classification of climate." *Geograph. Rev.*, 38,55.

Thornthwaite, C.W. and Holzman, B. (1939). "The determination of evaporation from land and water surfaces." *Monthly Weather Review* 67:4-11.

Turner, N.C. (1979). "Drought resistance and adaptation to water deficits in crop." 343-372. In Mussell, H. and Staples, R.C. (eds.) *Stress Physiology in Crop Plants*, John Wiley & Sons, New York, NY.

Turner, N.C. (1991). "Measurement and influence of environmental and plant factors on stomatal conductance in the field." *Agr. and For. Meteor.*, 54:137-154.

U.S. Department of Commerce. (1968). "Climatic Atlas of the United States." National Climatic Data and Information Center, Ashville, N.C.

U.S. Geological Survey. (1954). "Water-loss investigations: Vol. 1, Lake Hefner studies." Technical Report, U.S. Geological Survey Circ. 229 (also Prof. Paper 269).

U.S. Geological Survey. (1958). "Water-loss investigations: Lake Mead studies." U.S. Geological Survey Prof. Paper 298.

Valle, R., Mishoe, J.W., Jones, J.W., and Allen, L.H. (1985). "Transpiration rate and water use efficiency of soybean leaves adapted to different CO_2 environments." *Crop Science* 25, 477-481.

Van Bavel, C.H.M. (1966). "Evaporation and energy balance of alfalfa." Part I, Annual Rpt., USAERDA, Research Report No. 381, U.S. Water Conservation Laboratory, Phoenix, AZ.

Van Hylckama, T.E.A. (1968). "Water level fluctuation in evapotranspirometers." *Water Resources Res.*, 4:761-768.

van de Griend, A.A. and van Boxel, J.H. (1989). "Water and surface energy balance model with a multilayer canopy representation for remote sensing purposes." *Water Resources Res.*, 25:949-971.

Vanderkimpen, P.J., Allen, R.G., and Wright, J.L. (1996). "Evaluation of a semi-empirical form of the Penman-Monteith equation for the prediction of crop evapotranspiration." *Agric. and For. Meteorol.* Accepted for publication.

van Wijk, W.R., and de Vries, D.A. (1963). "Periodic temperature variations in a homogeneous soil." 102-143. In van Wijk, W.R. (ed.) *Physics of the Plant Environment*. North-Holland Publishing Co., Amsterdam.

Verma, S.B. (1989). "Aerodynamic resistances to transfers of heat, mass and momentum." 13-20. In Black, T.A., Spittlehouse, D.L., Novak, M.D., and Price, D.T. (eds.) *Estimation of Areal Evapotranspiration*, IAHS Pub. No. 177.

Verma, S.B., Baldocchi, D.D., Anderson, D.E., Matt, D.R., and Clement, R.J. (1986). "Eddy fluxes of CO_2, water vapour, and sensible heat over a deciduous forest." *Boundary Layer Meteorol.* 36:71-91.

Verma, S.B., Kim, J., and Clement, R.J. (1992). "Momentum, water vapor, and carbon dioxide exchange at a centrally located prairie site during FIFE." *J. Geophys. Res.* 97(D17):18,629-18,639.

Waggoner, P.E. and Turner, N.C. (1971). "Transpiration and its control by stomata in a pine forest." Bull. Conn. Agric. Exp. Station 726.

Wallace, J.S., Roberts, J.M., and Sivakuma, M.V.K. 1990. "The estimation of transpiration from sparse dryland millet using stomatal conductance and vegetation area indices." *Agric. and Forest Meteorol.* 51:35-49.

Weaver, H.L., Weeks, E.P., Campbell, G.S., Stannard, D.I., and B.D. Tanner. (1986). "Phreatophyte water use estimated by eddy-correlation methods." Proc. Water Forum '86, ASCE, New York, NY. 847-854.

Webb, E.K. (1970). "Profile relationships: the log-linear range, and extension to strong stability." *Quart.J. Roy. Meteorol. Soc.* 96:67-90.

World Meteorological Organization. (1966). "Measurement and estimation of evaporation and evapotranspiration." Tech. Note No. 83, Working Group on Evaporation Measurement, WMO, Geneva, Switzerland. 121 p.

World Meteorological Organization. (1970). "Guide to Hydrometeorological Practices." WMO No. 168.TP.82. Geneva, Switzerland.

World Meteorological Organization. (1973). "Comparison Between Pan and Lake Evaporation." Technical Note No. 126, WMO Publication No. 354. Geneva, Switzerland.

Wright, J.L. (1978). Personal communication.

Wright, J.L. (1981). "Crop coefficients for estimates of daily crop evapotranspiration." *Irrig. Scheduling for Water and Energy Conserv. in the 80's*, Am. Soc. of Agric. Engrs. Dec. p. 18-26.

Wright, J.L. (1982). "New evapotranspiration crop coefficients." J. Irrig. and Drain. Div., ASCE, 108(2):57-74.

Wright, J.L. (1988). "Daily and seasonal evapotranspiration and yield of irrigated alfalfa in southern Idaho." *Agron. J.*, 80:662-669.

Wright, J.L. (1990) Personal communication.

Wright, J.L. (1991). Personal communication.

Wright, J.L. and Jensen, M.E. (1972). "Peak water requirements of crops in Southern Idaho." *J. Irrig. and Drain. Div.*, ASCE, 96(1):193 -201.

Young, A.A. (1947). "Evaporation from water surfaces in California." California Dep. of Public Works, Div. of Water Resources and the U.S. Dept. of Agric., SCS. Bull. N. 54, 68 p.

Zinke P.J. (1967). "Forest interception studies in the United States." 137-161. In Sopper, W.E. and Lull, H.W. (eds.) *Forest Hydrology*. Pergamon, Oxford.

XI. GLOSSARY

Adiabatic equilibrium—a vertical distribution of pressure and temperature such that an air parcel displaced adiabatically will continue to possess the same pressure and temperature as its surroundings. In an adiabatic atmosphere, the potential temperature is constant with height.

Advection—horizontal transport of heat energy by large-scale motions of the atmosphere. Evidenced by a horizontal gradient of temperature or humidity in the direction of the mean wind.

Aerodynamic—refers to forces acting upon the water, soil, or vegetation surface by moving air.

Aerodynamically rough surface—a surface whose roughness elements are sufficiently large that the turbulent boundary layer reaches the surface.

Aerodynamically smooth surface—a surface whose roughness elements are sufficiently small to be entirely embedded in the laminar sublayer.

Aerodynamic methods—methods of estimating *E* or *ET* based on measurements of wind speed at one or more heights along with humidity. Varies from empirical bulk transfer expressions (surface to air) to gradients measured in the turbulent boundary layer, used along with assumed mathematical expressions for wind profile shapes with correction for non-adiabatic conditions.

Albedo (α)—the ratio of electromagnetic radiation reflected from a water, soil, or vegetation surface to the amount incident upon it. In practice, the value is applied primarily to solar radiation. Can be expressed as a percentage.

Bar—a unit of pressure equal to 10^6 dynes cm^{-2} or 100 kilopascals (1000 mb).

Bowen ratio (β)—the ratio of sensible heat to latent heat flux. The fluxes are considered positive when upward.

Bowen Coefficient—as used in Bowen ratio water-body evaporation equation; a function of the specific heat and latent heat of water.

Class A pan—the United States National Weather Service official evaporation pan, 1.21 m (47.5″) in diameter, 0.254m (10″) deep, mounted on an open-framed, wooden platform such that the top of the pan is 0.41 m (16″) above the soil surface. Water level in the pan is maintained between depths of 0.203 m and 0.178 m (8″ and 7″, or refilled daily to a point gage with point set at a 0.203 m depth of water.

Combination methods—combination of aerodynamic transfer expressions along with the energy balance, e.g., the Penman (1948) method with empirically derived wind functions. Example of a more theoretically-based method is the Penman-Monteith with wind function based on the log-law wind profile, and with the inclusion of a canopy resistance term.

Crop coefficient or cover coefficient (K_c)—the ratio of evapotranspiration occurring with a specific crop at a specific stage of growth to reference crop evapotranspiration at that time. Note: careful consideration of the basis of published K_c values is critical, i.e., based on ET_o (grass) or ET_r (alfalfa).

Crop coefficient (basal), K_{cb}—represents a condition when the soil surface is visually dry, so that soil evaporation is minimal, but the availability of soil water does not limit plant growth or transpiration.

Day length—the length of day from sunrise to sunset expressed in hours.

Deep percolation—the drainage of soil water by gravity below the maximum effective depth of the root zone.

Dew point—the temperature to which a given parcel of air must be cooled at constant pressure and at constant water vapor content until saturation occurs, or the temperature at which saturation vapor pressure of the parcel is equal to the actual vapor pressure of the contained water vapor.

Effective precipitation—the portion of precipitation that remains on the foliage or in the soil that is available for evapotranspiration and reduces the withdrawal of soil-water by a like amount.

Emissivity (ϵ)—the ratio of the emittance of a given surface at a specified wavelength and temperature to the emittance of an ideal black body at the same wavelength and temperature. For energy balance computations under natural conditions, the emissivity is commonly taken as the ratio of the total emittance of a surface to the total emittance of an ideal black body.

Energy balance method—as applied to water bodies, a procedure based upon energy conservation determinations for the body of water. For land surfaces, air temperature and humidity gradients are used to calculate the Bowen ratio, an estimate of the ratio of sensible to latent heat, with the energy balance equation then used to solve for λE or λET.

Evaporation (E)—the physical process by which a liquid or solid is transformed to the gaseous state.

Evapotranspiration (ET)—the combined processes by which water is transferred from the earth's surface to the atmosphere; evaporation of liquid or solid water plus transpiration from plants (normally expressed as a depth of water in mm, but can be expressed as a latent heat flux as W m^{-2}, MJ m^{-2} d^{-1}, or KJ m^{-2} h^{-1}.)

Extraterrestrial radiation—solar radiation received on top of the earth's atmosphere, or that which would be received on a horizontal plane at the earth's surface if there were no atmosphere.

Fetch—the upwind distance of a surface similar in nature to an area of interest, over which meteorological measurements are taken.

Flux density—the rate of flow of any quantity, usually a form of energy, through a unit area of specified surface, commonly identified only as a flux.

Free water surface (FWS) evaporation—measured or estimated evaporation from free water surfaces, largely related to expected evaporation from very shallow bodies of water.

GGI-20 m^2 tank—large standard evaporation pan of the former Soviet Union, 20 m^2 in area, 2 m deep, and sunken in the ground as in the GGI-3000 pan.

GGI-3000 pan—a small standard evaporation pan of the former Soviet Union, 0.3 m^2 in area, 0.6 to 0.8 m deep, and sunken in the soil with only 0.075 to 0.1 m of the rim above the soil surface.

Global radiation—the total of direct solar radiation and diffuse sky radiation received by a unit horizontal surface essentially for wave lengths of 0.3 to 4 microns.

Humidity (absolute)—mass of water vapor per cubic meter of air, also called water vapor density.

Humidity (relative)—the dimensionless ratio of actual vapor pressure of the air to saturation vapor pressure, commonly expressed in percentage.

Humidity (specific)—mass of water vapor per mass of moist air.

Joule—the unit of energy or work done when the point of application of 1 newton is displaced a distance of 1 meter in the direction of the force; 1 joule = 1 watt second.

Laminar flow—a flow in which fluid (or air) moves smoothly in streamlines in parallel layers of sheets (non-turbulent flow).

Latent heat (λE)—the heat released or absorbed per unit mass of water in a reversible, isobaric-isothermal change of phase.

Leaf area index (LAI)—the area of leaves (one side only) per unit area of soil surface.

Long-wave radiation (R_L)—electromagnetic radiation with a wavelength greater than 4 microns.

Mass transfer coefficient (M)—experimentally derived as a function of measurement height of the meterological measurements, the atmospheric stability, barometric pressure, surface roughness, size of water body, and the density and viscosity of the air.

Mixing ratio—the ratio of the mass m_v of water vapor to the m_a of dry air with which the water is associated.

Net radiation—the difference of the downward (+) and upward (−) solar and long-wave radiation flux passing through a horizontal plane just above the ground surface (or the vegetation surface).

Neutral stability—the state of an unsaturated or saturated column of air in the atmosphere when its environmental lapse rate of temperature is equal to the dry-adiabatic lapse rate or the saturation adiabatic lapse rate, respectively. Under such conditions, a parcel of air displaced vertically will experience no buoyant acceleration.

Potential evaporation (Penman, 1956)—the amount of water transpired in unit time by a short green crop, completely shading the ground, of uniform height and never short of water.

Potential evaporation (E_o) (ICID, 1985)—the evaporation from a given surface when all surface-atmospheric interfaces are wet (saturated) so that there is no restriction due to either biological control or soil water content on the water vapor loss from the surface area.

Pyranometer—a general name for actinometers that measure the combined intensity of incoming direct solar radiation and diffuse sky radiation.

Reference crop evapotranspiration, ET_o (Doorenbos and Pruitt, 1975; 1977)—the rate of evapotranspiration from an extensive surface of 8 to 15 cm tall, green grass cover of uniform height, actively growing, completely shading the ground and not short of water (grass should be a cool-season variety).

Reference crop evapotranspiration, ET_r (Jensen, et al., 1990)—the rate at which water, if readily available, would be removed from the soil and plant surfaces expressed as the rate of latent heat transfer per unit area, λET_r, expressed as a depth of water evaporated and transpired from a reference crop, ET_r. The leaf surfaces of the reference crop are typically not wet.

Richardson number—a dimensionless number associated with the shearing flows of a stratified fluid and commonly used in aerodynamic equations to correct for nonadiabatic conditions, both stable and unstable.

Roughness length for heat and water vapor, z_{oh} and z_{ov}—integration constants relating to the profiles of air temperature and specific humidity, respectively.

Roughness length for momentum, z_{om}—an integration constant of the log-law wind profile. Graphically, it may be visualized as the zero velocity intercept of the straight line resulting from a semi-logarithmic plot of mean velocity data versus elevation in the dynamic (or inertial) sublayer.

Saturation deficit (also called vapor pressure deficit, VPD)—the difference between the actual vapor pressure and saturation vapor pressure at the existing temperature.

Sensible heat—same as enthalpy; used in meteorology in contrast to latent heat.

Short-wave radiation (R_s)—electromagnetic radiation with a wavelength of 0.3 to 4 microns. Also a term used loosely to distinguish solar and diffuse sky radiation from long-wave radiation.

Solar constant—the rate at which solar radiation is received outside the earth's atmosphere on a surface normal to the incident radiation. (1367 W m^{-2}, 0.0820 MJ m^{-2} min^{-1} or 1.959 cal cm^{-2} min^{-1}; London and Frohlich, 1982).

Specific heat—the heat capacity of a system per unit mass.

Stefan-Boltzmann law—one of the radiation laws that states that the amount of energy radiated per unit time from a unit surface area of an ideal black body is proportional to the 4th power of the absolute temperature T of the black body $R = \sigma T4$ where R is the emittance and σ is the Stefan-Boltzman constant (5.67×10^{-8} W m^{-2} K^{-4}).

Stomatal resistance—the resistance to flow of water vapor from inside to outside of leaf, a highly complex function of sunlight illumination, vapor pressure deficit along with wind speed, leaf temperature, and soil water potential. Stage of plant growth, leaf age, and senescence are other factors.

Transpiration—the process by which water in plants is transferred as water vapor to the atmosphere.

Vapor pressure—the partial pressure of water vapor in the atmosphere.

Water budget procedure—method for computing evaporation from water bodies for monthly, seasonal, or annual periods, based on the hydrologic equation.

Wet bulb depression—the difference in degrees Celsius between the dry bulb temperature and the wet bulb temperature of the air.

Zero-plane displacement—an empirically determined constant introduced into the logarithmic wind speed profile to extend its applicability to very rough surfaces or to take into account the displacement of a profile. It can be considered to indicate the mean level at which momentum is absorbed by the individual elements of the plant community, e.g., the level of action of the bulk aerodynamic drag on the community (Thom, 1971).

Chapter 5
GROUND WATER

I. INTRODUCTION

Ground water comprises the subsurface water that occurs within the saturated zone of the earth's surface, or within the part of the earth's crust where all openings are filled with water. It is distinguished from water on and above the earth's crust, and from water occurring below the ground surface but above the saturated zone (or occurring within the unsaturated zone). Nevertheless, ground water is only one phase of the hydrologic cycle and cannot be considered separately or in isolation from all the other phases, especially when managing water resources of an area.

The amount of water stored in the earth's crust is immense. The volume of water stored in the rocks of the world's land areas may be on the order of 8 billion km^3, of which half is at depths less than 800 m. This volume is about 35 times the combined storage of all the world's rivers, fresh water lakes, reservoirs, and inland seas, and is about one-third the volume of water stored in arctic and antarctic ice fields, in the glaciers of Greenland and great mountain systems of the world.

Despite its magnitude, most ground water is not visible because it is stored below the earth's surface; however, a small percentage of this underground supply is visible as naturally-discharging springs, or as streams and rivers which are sustained by ground water when direct runoff from precipitation ceases.

This lack of visibility tends to promote some mysticism about ground water and a disregard for the need to protect it from pollution. Ground water is not mystic, and it does need protection. While ground water is a resource of immense magnitude, it is not evenly available and is exhaustible. Excessive continued exploitation may result in damage to ground water resources through lower ground water levels, aquifer consolidation and resulting land surface subsidence, water quality degradation, and plain exhaustion of the supply. Spilling or dumping of wastes on the earth's surface invites pollution of ground water. It is easier and much less costly to prevent pollution of ground water than to restore its quality after pollution.

This chapter presents the basic hydrologic facts about ground water and some of the management techniques that are used to maintain the viability of ground water resources. Most of this material has been taken directly from ASCE Manual 40 "Ground Water Management," 1987 edition. ASCE Manual 40 contains further discussion of ground water aspects related to design and operation of artificial recharge projects, surface and subsurface exploration and testing, development and use, legal and organizational considerations, use of aquifers for purposes other than water supply, land-surface subsidence, economics and implementation of management plans, and monitoring.

II. SOURCE AND OCCURRENCE OF GROUND WATER

On the basis of the sources from which they are derived, subsurface waters may be classified as meteoric, magmatic, connate, and metamorphic (White, 1957). Meteoric water is the water in the circulatory system of the hydrologic cycle, which includes evaporation from the water bodies and soil of the earth's surface, transport as vapor in the atmosphere, precipitation, overland runoff, stream flow, infiltration into the soil mantle, and downward percolation, as well as subsurface movement. Practically all the fresh water available comes from this circulating system, and meteoric waters are therefore of dominant interest to hydrologists.

For fresh water supplies, the remaining classes are of little practical value, although they are of considerable concern when encountered in wells as they generally are too saline for potability. Magmatic waters derive from molten rock (magma), and are found in some volcanic areas and some thermal springs. Connate waters were trapped in interstices of sedimentary rocks at the time of deposition, or at a later period that is still, in a geological sense, in the distant past. These waters may be encountered in deep wells (notably oil wells, where they commonly are found as brines) and in some mineralized springs. Metamorphic waters are those waters brought into existence by the heat, pressure, and recrystallization that created metamorphic rocks. They also may be encountered in some springs. These last three classes of water are generically related, respectively, to the three major classes of rocks (igneous, sedimentary, and metamorphic), and may similarly grade one into another. Some classifiers of subsurface water also list so-called juvenile waters, which have existed in the deep interior of the earth and are believed to be approaching the surface for the first time, but such waters have no characteristics that would distinguish them from the classes mentioned previously.

Subsurface space may be divided vertically into zones based upon the relative proportion of pore space that is occupied by water: a "saturated zone" in which all pores are filled with water, and an overlying "unsaturated zone" in which the pores contain gases (chiefly air and water vapor) in addition to any water that may be held by molecular attraction. During recharge periods, water moves under the force of gravity downward through the unsaturated zone to the saturated zone below (Remson et al., 1959). The water table has been defined as the upper limit of the saturated zone and is that locus of points in which the ground water is at atmospheric pressure. So far as water use is concerned, practically all vegetation, except for phreatophytes, depends on water in the unsaturated zone because the roots require aeration as well as water. In contrast, wells, springs, and streams (and therefore, people) obtain water from the saturated zone.

The upper part of the unsaturated zone, immediately beneath the land surface, first receives the infiltration from precipitation or some other source of surface water supply. This zone of soil water, extending downward from the land surface as far as plant roots penetrate, is of prime interest to agriculturists because of its ability to retain water for plant use. In places where the surficial materials are gravel, coarse sand, or very permeable rock, water retention may be negligible, and hence there is likely to be little or no vegetative cover present.

Beneath the zone of soil water is an intermediate zone, where water is retained solely by molecular forces, and where downward percolation presumably occurs, at least intermittently, toward the saturated zone. This zone comprises an indefinite zone in which there is no practical means of extraction. In some places, this zone is hundreds of meters thick; in others it is completely absent. In the latter case, the water table is found within the soil water zone.

At the base of the unsaturated zone, water may rise by capillary forces, the amount of rise ranging from practically zero in coarse material to as much as 2.5 m or more in clay materials. All pores may be saturated near the base of this capillary fringe; the number of pore spaces that are filled with water decreases in an upward direction. Water in the capillary fringe exists at pressures less than atmospheric pressure. In areas of shallow water table, the capillary fringe may extend upward to the root zone of plants and even to the land surface, thus permitting discharge of water by evapotranspiration. Because of the variation in capillarity among pores of different sizes, the upper limit of actual saturation must be an irregular surface, difficult to observe or to delineate. Therefore, so far as practical use is concerned, it is desirable to refine the definition of a water table so that it marks the upper limit of saturation at atmospheric pressure. Thus, its position can be observed in supercapillary openings such as wells, sumps, shafts, and caves penetrating below the water table.

The saturated zone extends downward from the water table to depths where interconnected pores still exist. Such depths vary with compression of the rocks and appear to exceed depths penetrated by oil wells (more than 12,000 m). As a rule, the depth reached by the fresh circulating water included in the concept of the hydrologic cycle is much less, although some wells obtain fresh water at depths exceeding 1,000 m below the land surface.

Numerous complications are introduced into this simple picture by the great variations in the size of pores and in the rate of water movement through them. Because the permeability of rock materials varies so much, water may accumulate in some parts of the unsaturated zone sufficiently to cause saturation, at least temporarily, in a perched zone, which would then form a perched water table. Within the saturated zone, some rocks are of such a low permeability that gravity water movement is negligible and no determinable water table exists.

Geologic materials that are sufficiently permeable to transmit water readily under a pressure differential and which contain water are called aquifers. By contrast, aquitards are the materials that are sufficiently impermeable to preclude, or severely restrict, such movement for all practical purposes. An important result of the great variability in permeability of material is the occurrence of water in confined and unconfined states.

Among the saturated materials capable of yielding water to wells, springs or streams, a classic distinction is made between unconfined (or water table) aquifers and confined (or artesian) aquifers.

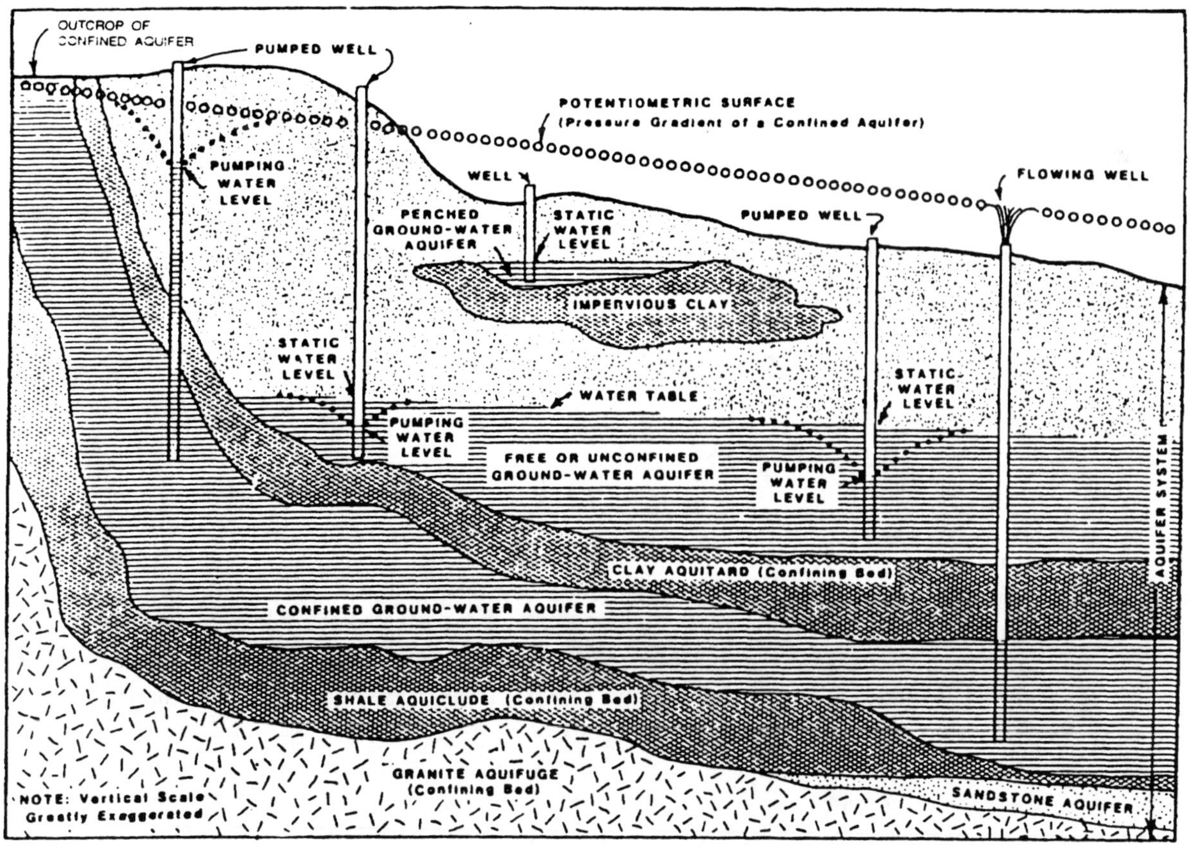

Figure 5.1—Diagrammatic Cross Section Showing Free, Confined, and Perched Ground Water Conditions

Unconfined ground water is found in the saturated zone whenever the upper surface of the zone forms a water table under atmospheric pressure, free to rise and fall with changes in volume of stored water. Confined water is found in aquifers which are separated from the unsaturated zone or another saturated zone by materials of markedly less permeability (see Fig. 5.1). The water pressure throughout the saturated thickness of a confined aquifer is everywhere in excess of atmospheric pressure.

Obviously, the distinction between confined and unconfined water results chiefly from the variations in permeability of geologic materials. Variations in permeability are commonplace. Although there are many exceptions, unconfined ground water is most likely to be found only in the shallowest wells, while water in deep aquifers is likely to have some degree of confinement. Because permeability is a relative term, confinement is also relative. Materials with identical characteristics may form an aquifer in one place and a confining layer above a much more permeable aquifer in another place.

An artesian well is one in which the water level in the well rises above the top of the aquifer in which it is contained. If the water rises above the land surface, a flowing artesian well is formed. Pumping of confined water causes rapid pressure reduction in the aquifer for considerable distances from the discharging well.

For hydrologists concerned with quantitative determination of ground water storage and movement, an important reason for distinguishing between confined and unconfined water is that the hydraulics of flow involved are different and require different testing techniques and formulas, as outlined by D.K. Todd (1980). The difference is comparable, in many respects, to that between flow in pipes and open-channel flow. It should be noted that the conditions of confinement can be changed. Pumping from wells in confined aquifers may eventually dewater the upper part of the aquifer, first in the immediate vicinity of the wells and then in progressively broader areas, until the entire aquifer becomes unconfined. The converse may also occur in recharge operations where water is injected into aquifers overlain by less permeable rocks.

For confined aquifers, the material is not dewatered or drained unless the hydraulic head drops below the top of the aquifer. Consequently, the term "specific yield" does not apply to confined conditions. A correlative term, the storage coefficient or storativity, is applied to confined aquifers and is defined as the volume of water an aquifer releases from or takes into storage per unit of surface area of the aquifer per unit change in head. Storage coefficients are usually on the order of 5×10^{-2} to 1×10^{-5}.

The water obtained from a confined aquifer results from a slight expansion of the water and a very small compression of the porous medium. Artesian aquifers are apparently all more or less compressible and elastic, though they differ widely in the degree and relative importance of these properties. In general, the properties of compressibility and elasticity are of most consequence in aquifers that have low permeability, slow recharge, and high head. If it were not for the elasticity, the supplies of water that could be recovered through artesian wells for intermittent or fluctuating needs would be much smaller.

III. GROUND WATER RESERVOIRS

The term "ground water reservoir" is widely and loosely used to denote places where ground water is accumulated under conditions that make it suitable for development and use. As there is no standard or generally accepted definition of the term, it is necessary in each use to describe the particular characteristics and limiting conditions of the reservoir. "Ground water basin," "aquifer," "water-bearing zone" (or formation), "ground water district," and "ground water area" are other terms of similarly broad and diverse usage, applied to geologic formations ranging from a few meters to thousands of meters in thickness, distributed over areas ranging from a few hectares to thousands of square kilometers, and capable of yields to wells ranging from a few liters per hour (l/h) to several cubic meters per second (m^3/s). Although it is vague, "ground water reservoir" is an excellent term for general use and is parallel in concept to "surface water reservoir" as it conveys the concept of water in storage.

A. Essential Hydrologic Characteristics

A ground water reservoir must contain formations that are not only porous, but are sufficiently permeable to yield water by gravity to springs and wells. Traditionally, ground water reservoirs, where first encountered, have been filled with water by nature, sometimes to overflowing. In contrast, surface reservoirs characteristically start out empty, and acquire stored water only after a considerable investment in study, planning, design, and construction. The addition of water to permeable formations within the unsaturated zone illustrates the use of a ground water reservoir in a similar fashion. Conversely, the reservoir would still exist even if the water were pumped out of it. Thus, the potential for storage, rather than the existence of storage, is an essential characteristic of ground water reservoirs.

Another essential characteristic of ground water reservoirs is movement of the water, both in the direction of decreasing head and generally from areas of recharge toward areas of discharge. In contrast to surface reservoirs, no reduction of velocity generally occurs in the subsurface areas designated as reservoirs; in fact, the best ground water reservoirs may have a greater velocity of movement because of the excellent permeability of the rocks; however, in comparison with stream velocities, the movement of ground water (even in the best ground water reservoirs) is so slow that practically all of it can be regarded as held in storage. The chief exception to this generalization is the flow in subterranean channels of limestone, especially in areas of karst topography or volcanic regions. Such underground streams may accept and carry the turbulent and even turbid flow of surface streams and often have equivalent velocities.

Many, if not most, ground water reservoirs contain a complicated array of permeable rock materials, which may occur in layers, conduits, compartments or lenses, each bounded by less permeable or impermeable materials. Because of the existence of many relatively permeable barriers to ground water movement, the water in some permeable rock formations is likely to be confined under artesian pressure. Unconfined water may flow laterally into rock formations where it is confined. On the other hand, confined water may leak through semi-permeable barriers and into rock formations to which it is not confined. The permeable and impermeable rocks, conduits, barriers, faults, and other geologic structural features as well as erosional and weathering effects are all parts of the geologic framework of a ground water reservoir.

Source and disposal of water vary greatly among ground water reservoirs. Where the principal inflow is from precipitation, the outflow constitutes most of the water found in streams during low rainfall periods, and the ground water reservoirs might logically be delimited by the confines of a river basin, as in the Middle Loup River basin in Nebraska. As a contrasting example, the Southern High Plains in Texas and New Mexico divide their limited natural outflow among tributaries of the Mississippi, Rio Grande, and some intrastate rivers of Texas. Some ground water reservoirs are recharged principally by a single source, but their discharge is distributed over a wide area, as in the Colorado River Delta downstream from Yuma, Arizona. Others are recharged from many streams, but have a single central area of natural disposal, such as the numerous closed basins found in the southwestern United States. Many ground water reservoirs extend across the divide separating major stream basins, and some, such as Playas Valley in New Mexico, extend across the Continental Divide.

B. Principal Types of Aquifers

The aquifers comprising the principal ground water reservoirs in the United States include both unconsolidated or partially consolidated gravel and sand and also include consolidated rocks, of which limestone, basalt, and sandstone are most noteworthy. Other rock types, such as silt and clay, shale, volcanic rocks, crystalline and metamorphic rocks, are relatively impermeable, although each of these can be sufficiently permeable to constitute an aquifer of small yield in some localities. As a general rule, the delineation of aquifers is a delineation of rock types, for the classification of rocks as sedimentary, metamorphic, or igneous, as loose or consolidated, or as soluble or insoluble quickly reveals some of their characteristics with respect to the occurrence of ground water.

1. Gravel and Sand Gravel and sand are usually of alluvial, glacial, lacustrine, marine, or aeolian deposition. They may underlie large plains and valleys or be local in character and of small areal extent.

Alluvial deposits by perennial streams may be fairly well-sorted (more uniform size of particles) and therefore permeable. The stream beds may be underlain by well-sorted sand and gravel that have been carried hundreds of kilometers from their sources. Shifting channels result in lateral distribution of these coarser stream-bed deposits, and as a result of a long period of aggradation, the entire valley floor may be underlain at various depths with sand and gravel deposits of different thicknesses. When floods overflow the stream banks, coarse material is deposited adjacent to the main stream channel, and silt and clay particles settle out in the more distant overflow areas. Mountain streams deposit the coarsest material closest to the point where the stream debouches from the canyon mouth. The stream deposits become finer with increasing distance from the point of initial deposition.

Ephemeral streams from mountains in arid areas deposit gravel and sand with much less sorting. Stream gradients and valley slopes are steep, often tens of meters per kilometer, with resulting velocities sufficient to transport large boulders. The boulders are, of course, deposited at the canyon mouth, and owing to the ephemeral nature of the flow, the remaining detrital material is deposited at the same time. This may result in sediment containing large boulders with some sand and gravel but with a clay content high enough to result in relatively low permeability.

The central and lower portions of these alluvium-filled valleys may consist of fine-grained materials deposited as stratified lake beds or playa clays. Often these clay beds are interbedded with layers of fine sand or even with stream deposits. The lower valley fill may also contain considerable soluble mineral matter in the form of caliche or other calcium carbonate salt and gypsum.

Where a stream flows into a lake or sea, the current is abruptly reduced. The coarser sediment settles rapidly, while finer materials are transported further into the body of relatively still water. Additional incursions of sand and gravel are carried to the outer edge of the previous accumulation and form rather steeply inclined beds, which are the front or face of a "delta" accumulation. Meanwhile, the silt and clay are carried still farther into the still water body and settle to form layers over the bottom of the lake or sea. As deposition continues, the delta front deposits are built progressively farther over the bottom beds. The gradient of the stream is modified by the delta formation, and sediment is deposited upstream from the original mouth of the stream. Thus, the delta grows outward by deposition of underwater beds and upwards by deposition of top beds. Deposition also occurs upstream in the original channel of the stream that provides the source material.

The deltas of larger streams are formed in the manner described in the preceding paragraph. The original underwater deposits are augmented until their surface rises above the water level. The Colorado River Delta, extending from Yuma, Arizona, across the Gulf of California, is an example of such deposition. The sand and gravel of this delta are the aquifers from which wells produce irrigation water for thousands of hectares in Mexico.

Many large ground water reservoirs are found in sand and gravel that have been deposited in extensive plains formed by parallel, coalescing, or distributary streams. Examples of these reservoirs include the Coastal Plain, formed by streams emptying into the Gulf of Mexico and the Atlantic Ocean, and the Great Plains, formed by streams draining the Rocky Mountains.

Some exceedingly permeable sands and gravel are glaciofluvial (deposited by streams flowing from glaciers). Ancient valleys that have been filled with these glacial outwash materials contain excellent aquifers, and such valleys, with the stratified deposits in glacial lakes, provide important water storage in the glaciated regions of the northern United States.

Although gravel and sand deposits are preeminent as aquifers, they constitute a very small proportion of all unconsolidated sediments. Many of these sediments are composed of a heterogeneous mixture of sands and gravels or even boulders, with predominating silt and clay that reduces the permeability so as to place these sediments in the class of aquicludes rather than aquifers. This becomes particularly noticeable in formations in which lime deposition or chemical disintegration of rock particles into clay-like materials has occurred with partial or complete closures of the pores. Both deposition and chemical disintegration processes greatly reduce the permeability of an aquifer, and their effects are particularly noticeable with increased age and depth of formations.

2. Sandstone and Conglomerate Sandstone and conglomerate are the consolidated equivalents of sand and gravel. In these rocks, the individual particles have been cemented together, thus reducing the porosity to such an extent that they may yield little water. The best known sandstone aquifers are those which are only partially cemented. These yield most of their water from the pores between the grains, although fractures and joints may serve as collecting channels. Good sandstone wells may yield hundreds of liters of water per minute, but as a rule they yield much less than those in formations of loose sand and gravel. Extensive beds of sandstone underlie portions of the United States, and in some places comprise the only aquifer. Dakota sandstone, which derives its name from the two states where it constitutes an important source of ground water, is an aquifer of this type.

3. Limestone The limestones that are most important as ground water reservoirs are those in which a sizable proportion of the original rock has been dissolved and removed. Many very large springs issue from limestone. Of the 78 or more springs in the United States having a maximum discharge of more than 3,000 l/s, about half occur in limestone, including 27 in Florida alone. Fractures and joints in limestone also yield water in small to moderate amounts.

Most limestones are of marine origin, and many are composed largely of the fossil remains of marine organisms. If fresh water now occurs in any of these beds—that is, if the beds constitute a usable ground water reservoir—sea water must have been flushed out, and water from rain and snow must have circulated through the rock.

Solution cavities and channels in limestone may be large enough so that wells reaching them are very productive. On the other hand, rapid flow through limestone may be a health hazard if the water is polluted before it enters the limestone or receives pollution thereafter. The solution channels may be separated by considerable amounts of solid rock, so that some wells may penetrate deep into the limestone without yielding water, even though productive wells may be close by.

4. Lavas and Other Volcanic Rocks Basalt, a dark-colored volcanic rock, is an important aquifer in some sections of the United States, especially in the west, where enormous flows of lava have spread out over large areas in successive sheets of varying thickness. The hydraulic conductivity of a basalt aquifer depends on the presence of fractures, cracks, and tubes or caverns. The hydraulic conductivity of some basalt is so great that under natural conditions there is practically no slope to the water table or to the pressure surface under artesian conditions.

Other volcanic rocks, such as rhyolites and other more silicious rocks, do not usually yield water in quantities comparable to those secured from basalt. Most shallow intrusive rocks in the form of sills, dikes, or plugs are low in permeability, and many of them are impervious enough to act as barriers to ground water movement.

Crystalline and metamorphic rocks, including granite, basic igneous rocks, gneiss, schist, quartzite, and slate, originated generally at great depths below the earth's surface and are relatively impermeable. In areas where rocks of these types supply water for domestic or stock wells, the water comes chiefly from the zone of weathering or from jointed or faulted rocks. Whether the development of permeability results from weathering or fracturing, it is likely to be greatest near the surface and progressively less with increasing depth.

C. Principal Types of Basins

Closed basins in the southwestern United States are generally rimmed by mountain ranges and have centripetal drainage, which forms a lake in the lowest part of the basin whenever runoff from precipitation is sufficient; however, many of these basins are in arid regions where there is no stream flow except after occasional storms, and where water principally moves from the bordering mountains toward the lowlands as ground water. In such basins, the mountains may constitute ground water divides as well as topographic divides. The low part of the basin constitutes a sink from which ground water is discharged by evapotranspiration. Thus, a ground water basin exists which is coextensive with a drainage basin; however, this may be the exception rather the rule. Many ground water basins do not conform with surface drainage basins, either in area or in pattern of flow.

If ground water basins are to be subdivided, their sub-basin boundaries would logically be the relatively impermeable barriers that make such subdivision necessary. The Santa Ana River valley in southern California contains a ground water basin that is an excellent example of natural compartmentalization by faults and other barriers. Pumping draws more heavily upon some compartments than others, revealing the skeleton of less permeable barriers and the varying rates at which water can flow through, around, or over them. For management purposes, a ground water basin may have to be subdivided using barriers to flow as well as political boundaries. Such approaches are not often desirable but may be required for practical reasons.

The term "basin" may also express contrasting degrees of confinement as, for example, in an artesian basin. Particularly in the valley fill of arid basins, where there are likely to be large variations in the permeability of sediments tapped by a single well, some water may be confined, and some unconfined in one season and confined in another, while in many other areas water can be described as "semiconfined." In broad perspective, a ground water basin containing water under confined conditions is not a complete and self-supporting hydrologic unit, for it generally does not receive water by downward movement through the overlying confining bed. Even if the confining bed were moderately permeable, the artesian pressures would oppose such movement; however, if the artesian head is below ground level or below the free water table, as it is in many areas, the artesian aquifer may receive recharge through the confining bed. This phenomenon happens in many regions. An artesian basin, therefore, is replenished in areas where the water table is above the artesian head, primarily in permeable recharge areas where the confining bed is absent or somewhat permeable. Although it is important to differentiate between confined and unconfined waters (because of the differences in ground water hydraulics) it should be recognized that an artesian basin is actually only a part of the complete underground system necessary for yielding perennial supplies to wells or springs.

In a free or unconfined ground water basin, the water table represents the height to which the basin has been filled. Changes in the water level in wells that tap the basin are direct indications of the change in the volume of water in storage. A summation of the products of the changes in water level, the average specific yield of the aquifers in the zone of water level change, and the area dewatered or recharged (as the case may be) gives the change in the quantity of stored water.

The water table slopes in the direction of ground water flow, from areas of recharge at higher elevations to areas of discharge at lower elevations. If the water moves into confined aquifers, the boundary between the unconfined and the confined ground water basins is logically the edge of the confining bed. In many ground water reservoirs, the terms "forebay" and "outcrop" have been used to designate a free or unconfined ground water area that serves as recharge area to an artesian or confined area (see Fig. 5.1).

In a confined ground water basin, the aquifer is completely saturated and confined by overlying impervious or relatively impervious materials. Changes in the potentiometric surface, determined by water level or pressure measurements in wells that tap the confined aquifer, represent changes of pressure in the aquifer. Such changes in pressure may reflect changes in head of the recharge, outcrop, or discharge area. They may also represent changes in atmospheric pressure, but the largest recorded changes have resulted from pumping in the vicinity of the observation wells.

IV. THE SUBSURFACE MEDIUM

The medium within which ground water occurs is porous, contains secondary openings such as fractures or bedding planes, or both. The availability, movement, and quality of ground water frequently depend directly upon the type of medium within which it is found.

A. Porous Medium

Some openings in geologic materials are "primary" in the sense that they were formed contemporaneously with the materials, e.g., the spaces between individual particles of gravel, sand, clay, volcanic tuff, and breccia, or the gas cavities formed during cooling of volcanic rocks. In sedimentary rocks, porosity depends upon the shape, arrangement, and degree of sorting of the particles. For example, spheres of uniform size have a porosity ranging from 26% to 47.6%, depending upon their arrangement. In sediments, except for some limestones, the porosity is generally greatest at the time of deposition and is then reduced by compaction (closer packing of particles) and cementation (deposition of mineral matter around the particles). Even though the process of consolidation may also include the development of secondary openings by fracturing, the net product of consolidation is generally a reduction in porosity. A very large proportion of the water yielded by wells comes from primary openings in unconsolidated or semiconsolidated sediments, chiefly gravel and sand. In several localities, water withdrawal has resulted in further compaction of fine-grained materials, as shown by subsidence of the land surface, and has thus reduced the overall porosity in subsided zones.

B. Medium with Secondary Openings

Fissures, fractures, bedding planes, and solution cavities are examples of "secondary" openings in geologic materials through which ground water can move. Such openings are created after the particular geologic material has been emplaced and are caused by cooling, folding, faulting, or chemical action. Fractures usually are larger and more numerous near the surface, because increasing pressure with depth tends to minimize fracture aperture size. Solution cavities are common in limestone and other soluble rocks. Cavities may result from ground water solution of the rock minerals, but fractures may have provided the original path enabling the water to make contact with the minerals. Weathering is an important agent in formation of secondary openings in crystalline and cemented sedimentary rocks in near-surface exposures. Plant roots and animal and insect borings may also create secondary openings in rocks and soils.

1. Porosity Porosity is the ratio of the total volume of pores in rock or soil material to the total volume of the material. The porosity of most rock materials ranges from 1% to 50% by volume. Typical value ranges of porosities for some rock materials are given in Table 5.1. In aquifer material, porosities less than 5% are considered small, those between 5% and 20% are considered medium, and those greater than 20% are large. Pores may range in size from large caverns to the microscopic spaces between crystals. Pores are classified as "subcapillary" (so small that water is held primarily by molecular force), "capillary" (small enough to hold water by surface tension), and "supercapillary" (large enough to allow gravity drainage or to yield water to wells).

Porosity is an inherent property of the rock material independent of the presence or absence of water, which can be determined by laboratory measurements (Davis, 1969); however, for ground water studies, total porosity is less important than effective porosity, or the interconnected pore space that can be drained by gravity (Dagan, 1967). If a unit volume of material is saturated (all pore space filled with water) the volume of water that is yielded by gravity drainage is the specific yield, and the volume retained by molecular forces is termed the specific retention. Both values are expressed as percentages of the total volume of rock and together constitute the total porosity of the given rock material.

The porosity of unconsolidated rock materials *in situ* is frequently changed by human activities. For example, in several localities, extraction of ground water has resulted in consolidation of sediments, primarily of clay deposits, as demonstrated by subsidence of the land surface (Poland, 1969, 1981; Poland and Davis, 1969). Also, collection of samples by drilling or some types of core barrels may disturb soils from their natural *in situ* structure. Accordingly, laboratory tests on such disturbed samples may show porosity and specific yield values that incorporate some error (Johnson, 1967).

TABLE 5.1. Typical Value Ranges for Porosities and Specific Yield for Various Aquifer Materials.

Aquifer materials	Porosity (in %)	Specific yield
Clay	45-55	1-10
Fractured Crystalline Rock	0-10	
Glacial Till	10-25	
Gravel	25-40	15-30
Limestone	1-20	0.5-5
Sand	25-40	10-30
Sand and Gravel	10-35	15-25
Sandstone	5-30	5-15
Shale	0-10	0.5-5
Silt	35-50	
Vesicular Basalt	10-50	

2. ***Specific Retention*** The specific retention of a rock or soil is the ratio of the volume of water which the rock or soil, after being saturated, retains against the pull of gravity to the volume of the rock or soil. Ideally, the definition implies that gravity drainage is complete; however, the quantity of water held in pores above the water table during gravity drainage depends upon particle size, distance above the water table, time of drainage, and other variables. Lowering of the water table and infiltration occur over such short periods of time that gravity drainage is rarely or never complete. Nevertheless, specific retention is a useful though approximate measure of the residual moisture-holding capacity of the unsaturated zone in the region above the capillary fringe.

3. ***Specific Yield*** The specific yield of a rock or soil is the ratio of the volume of water that the saturated rock or soil yields by gravity to the volume of the same rock or soil. The definition implies that gravity drainage is complete. In the natural environment, specific yield is generally observed as the change that occurs in the amount of water in storage per unit area of unconfined aquifer as the result of a unit change in head. Such a change in storage is produced by the draining or filling of pore space and also depends upon particle size, rate of change of the water table, time, and other variables. Hence, specific yield is only an approximate measure of the relation between storage and head in unconfined aquifers. It is equal to porosity minus specific retention. Typical value ranges for specific yield of some rock materials are given in Table 5.1.

Specific yield is an important concept in ground water hydrology (Lohman, 1972; Youngs, 1969). The average specific yield of saturated sediments, multiplied by the total volume of those sediments, gives the volume of water contained in those sediments that can be recovered by gravity drainage (Johnson, 1967). No time factor is included in the definition of specific yield, but many experiments have shown that draining is initially most rapid, decreases with time, and may continue for months or even years (Prill et al., 1965). Fine-grained materials have a lower specific yield than coarser materials, even though their porosity may be greater, and the drainage process is likely to continue for a much longer period of time than for coarse materials. The difficulty in individual determination of specific yield and the problem of extrapolating the results to the volume and variety of rocks in a ground water reservoir lead to the conclusion that excellent opportunities exist for refining existing techniques and developing new ones for determining usable storage of a ground water reservoir. Some techniques for determining specific yield (Johnson, 1967) and their limitations are:

(a) Draining long columns of saturated materials in the laboratory (Johnson et al., 1963). For accuracy, it is essential to reproduce the natural conditions as specific yield is related not only to the size and uniformity of the material, but also to the arrangement of the grains. The limitation to this method is that it is very difficult to obtain undisturbed samples that truly represent natural conditions.

(b) Saturating, in the field, a considerable body of material situated above the water table and allowing it to drain downward naturally. After drainage, the moisture content of the material is determined by collecting core samples, using a nuclear moisture meter or other techniques.
(c) Measuring moisture content retained immediately above the capillary fringe after the water table has receded an appreciable distance, as is common in summer and autumn. This technique is essentially the same as the field saturation and draining method.
(d) Ascertaining the volume of sediment drained by heavy pumping and keeping a record of the quantity of water that is pumped. This method requires accurate records of water pumped and the magnitude and extent of the decline in water table. It is accurate only if there is no recharge to, and no other discharge from, the aquifer during the period of pumping.
(e) Ascertaining the volume of sediments saturated by a measured amount of precipitation or seepage from one or more streams. This method requires accurate records of water recharge and the magnitude and extent of the rise in water table. It is accurate only if there is no discharge from, and no other recharge to, the aquifer.
(f) Making indirect determinations with small samples in the laboratory by the application of centrifugal force (Johnson et al., 1963). This technique is also used to determine moisture equivalent.
(g) Making particle-size analyses and estimating from there the specific retention and the specific yield. The accuracy of the estimate depends upon sufficient statistical data showing the relation of specific yield to particle-size distribution based on actual test data. In certain areas the sediments may be composed of materials that have enough similarity to permit the development of reliable relationships. In a number of instances, this method has been done fairly successfully.

4. Permeability, Hydraulic Conductivity, and Transmissivity The permeability of a porous medium describes the ease with which a fluid will pass through it, and indicates its capacity to transmit water under a head differential. No rock material is absolutely impermeable in the strictest sense of the word; permeability is a relative term. Rock materials of low porosity characteristically have low permeability as well, but high porosity does not necessarily produce high permeability. Rock materials with isolated interstices, such as some volcanic rocks, may be quite porous, but necessarily have very little permeability. It is common to find rocks that have very little permeability in one direction, such as the vertical axis, and a much higher permeability in another direction, such as along the horizontal axis; i.e., they are anisotropic with respect to permeabilities.

The permeability, k, is dependent only on the properties of porous media and is independent of the fluid properties that govern the flow. Earlier researchers have defined k as the specific permeability proportional to the square of representative grain diameter of the medium, or the constant of proportionality depending upon porosity, packing, size distribution, and shape of grains (Reeve, 1957). Several formulas have been developed that include variables designed to express the properties of the porous media, but they have encountered difficulty due to the complexity of variations in porous media (Dagan, 1967). The specific permeability is commonly expressed in "Darcy" units, first introduced in the petroleum industry. These units depend upon the medium but not upon the fluid that flows through that medium. The U.S. Geological Survey has adopted the term "intrinsic permeability" for k for use in ground water studies.

Hydraulic conductivity, K, in consistent units has been adapted to replace permeability in the inconsistent units gpd per ft^2 (gallons per day per square foot). Hydraulic conductivity is the flow of a unit volume of water per day under a unit hydraulic gradient through a unit cross-sectional area at prevailing temperatures. In common units, hydraulic conductivity is expressed as centimeters per second or meters per day, under a unit hydraulic gradient. Table 5.2 gives the relationship of the various units used for hydraulic conductivity and permeability. The relationship between hydraulic conductivity and permeability is:

$$K = \frac{k\rho g}{\mu} \tag{5.1}$$

TABLE 5.2. Hydraulic Conductivity–Permeability Conversion Factors.[a,b]

Unit to be converted (1)	Centimeters per second (2)	Feet per day (3)	Feet per year (4)	Darcy (5)	Meinzer (gpd per sq ft) (6)
Centimeters per second	1	2.835×10^{3}	$1{,}0348 \times 10^{6}$	1.033×10^{3}	2.12×10^{4}
Feet per day	3.53×10^{-4}	1	365	3.64×10^{-1}	7.48
Feet per year	9.67×10^{-7}	2.74×10^{-3}	1	9.99×10^{-4}	2.05×10^{-2}
Darcy	9.68×10^{-4}	2.75	1.001×10^{3}	1	20.50
Meinzer (gpd per sq ft)	4.72×10^{-5}	1.34×10^{-1}	48.8	4.88×10^{-2}	1

[a] Multiply unit at left by number in column to get unit at top of column. All units based on temperature of 60°F or 15.6°C.
[b] From Johnson, A.I., "Application of Laboratory Permeability Data," U.S. Geological Survey open-file report, 1963.

where K is the hydraulic conductivity, k is the permeability, ρ is the density, g is gravity, and μ is viscosity. K has the units of length divided by time (L/T) and k has the units of length squared (L^2).

Typical value ranges of hydraulic conductivity, K, for various rock materials are given in Table 5.3.

Rock materials within the unsaturated zone have less hydraulic conductivity than do the same rock materials if they are saturated. This relationship is logical because the movement of water is impeded by the air that occupies some of the available void space. Generally, the lower the amount of water in the void space, the lower the hydraulic conductivity down to a threshold point (approximately representative of the moisture content at specific retention) where there is essentially no liquid flow.

Transmissivity, as defined and used by the U.S. Geological Survey, is the product of the hydraulic conductivity multiplied by the thickness of the saturated portion of the aquifer. It represents the quantity of water that flows through a unit width of the saturated portion of the aquifer under a unit hydraulic gradient and at prevailing water temperature.

V. MOVEMENT OF WATER

Movement in the unsaturated zone is usually vertical but may have a significant horizontal component because of the effect of permeability on the rate and direction of movement. Thus, the unsaturated zone may include layers of materials having contrasting permeability or some layers in which horizontal permeability greatly exceeds vertical permeability.

TABLE 5.3. Typical Value Ranges for Hydraulic Conductivity, K, for Various Aquifer Materials.

Aquifer materials	Hydraulic conductivity (m/day)
Clay, Unweathered Marine	10^{-4}—10^{-7}
Fractured Igneous and Metamorphic Rocks	10—10^{-3}
Glacial Till	10—10^{-7}
Gravel, Fine to Coarse	10^{5}—10^{2}
Karst Limestone	10^{3}—10^{-1}
Limestone, Unjointed Crystalline	10^{-1}—10^{-5}
Sand, Fine to Coarse	10^{3}—10^{-2}
Sandstone, Cemented and Unjointed	10^{-2}—10^{-5}
Sandstone, Friable	1—10^{-3}
Shale	10^{-4}—10^{-8}
Silt or Loess	1—10^{-4}
Tuff	3—3×10^{-5}
Unfractured Igneous and Metamorphic Rocks	10^{-5}—10^{-9}
Vesicular Basalt	10^{3}—10^{-3}

The movement of water in the saturated zone ordinarily has a large horizontal component in the direction of decreasing head; however, movement in the saturated zone, even in unconsolidated granular materials is seldom uniform in direction or in velocity. Water may move around deposits of impervious material or follow the stratified courses of the more permeable material. Water follows the path of least resistance and moves with less restriction through sections which have the higher permeability.

A. Velocity

When water moves through a uniformly permeable material at very low velocities, all molecules move in more or less parallel lines. As the velocity increases, a value termed "higher critical velocity" is reached, at which time the streamline or laminar flow ceases and movement becomes turbulent. When the velocity of turbulent flow is gradually reduced, a velocity known as "lower critical velocity" is reached at which the flow again becomes laminar. At velocities intermediate between the higher and lower critical velocities, the flow is unstable. Although turbulent flow may occur in large joints, fractures, or in solution openings in limestone, ground water in granular materials ordinarily moves at laminar velocities. Turbulent flow in sand and gravel aquifers is virtually nonexistent, except possibly at the point of water entry into some pumped wells.

The velocity of laminar flow follows Darcy's law, which states that velocity varies directly with the hydraulic gradient, or

$$v = K\frac{\Delta h}{l} \tag{5.2}$$

in which v is velocity, Δh is the difference in head between two points in the path of movement, l is distance along the path of movement, and K is hydraulic conductivity. For water moving in the zone of saturation, the factor $\Delta h/l$ is the hydraulic gradient.

The velocity of flow is influenced by the viscosity of the water, which varies inversely with the temperature. For example, an increase in water temperature from 4°C to 32°C approximately doubles the velocity under the same hydraulic gradient. Although the temperature of ground water is fairly constant throughout the year, its temperature increases with depth. In areas where shallow ground water is derived by rapid infiltration from a recharge source, the ground water temperature may vary in response to the changes in surface water temperature. In some localities, this factor is of practical importance.

Hydraulic gradients and, by inference, the directions of ground water movement are commonly portrayed on two-dimensional maps. The position of the water table is indicated by the static water level in wells tapping unconfined water. The form of the water table is commonly shown on maps by contour lines connecting points of equal elevation of the water table. As with the contours on topographic maps, the direction of maximum gradient (and therefore the horizontal component of ground water movement) is typically perpendicular to the contours. As the direction of water movement is down-gradient, the areas of recharge constitute highs on the maps. Influent streams, for example, commonly form ground water ridges. Conversely, areas of discharge are lows on the map, including valleys under effluent streams (where the contours bend upstream). Closed depression contours are common in the vicinity of discharging wells. In contrast to inferences that may be drawn from topographic maps concerning stream velocities, the spacing of ground water contours may be related inversely to the rates of water movement. Commonly, a steepened water table slope indicates decreased hydraulic conductivity or thinning of the aquifer and a flattened slope indicates increased permeability or thickening of the aquifer.

Water levels in wells tapping a confined aquifer are the basis for constructing contour maps or profiles of a potentiometric, or pressure, surface for the aquifer. Such a surface indicates the potential energy head of the water in the aquifer. A map of the potentiometric surface can be the basis for inferences as to hydraulic gradients and recharge and discharge areas, similar in concept to those drawn from water table maps; however, an isolated closed high, which might be interpreted as a point source of water, is more likely to indicate a well that taps a different aquifer. In addition, particularly in confined aquifers, a closed depression may be an indication of upward movement through confining deposits rather than horizontal movement into that location. Because the map indicates only the horizontal component of movement, it cannot portray actual movement through an aquifer that has the structure of a syncline (inverted siphon), for instance.

In any confined aquifer capable of perennial supply to wells, the confinement must end somewhere to permit replenishment of the supply from precipitation, stream flow or other surface source, or the confining layer must leak sufficiently to provide a source of recharge water to the wells. A map that shows the hydraulic gradients throughout a complete hydrologic unit might portray one or more potentiometric surfaces in some areas and a water table in other areas.

Dyes, chemical ions, and radioactive tracers have been used to measure the direction and velocity of ground water movement through porous media with varying degrees of success. Base exchange of cations, if the porous media have this capability, may seriously impair the usefulness of certain chemicals. As summarized by Koufman and Orlob (1957), the ideal tracer would be one that is capable of detection in very low concentrations, does not react with injected or displaced waters, is not removed from solution or acted upon by the porous medium, is readily obtainable and economical, and is safe for use with potable water sources.

B. Quantity of Flow

If the velocity of movement has been measured, it may be used in computing the quantity of flow. This quantity of water discharged in a unit time, Q, is equal to the product of the average pore water velocity, V_{eff}, the average effective porosity, η_e, of the media through which the water flows, and the gross cross-sectional area, A, of the aquifer:

$$Q = V_{eff}\,\eta_e\,A. \tag{5.3a}$$

Effective porosity is the portion of interconnected pore spaces through which water can move. The effective porosity and cross-sectional area may be estimated roughly from the analysis of material encountered in well logs and geological studies.

Alternatively, the quantity of flow may be determined by the product of the hydraulic conductivity, K, the hydraulic gradient, I, along the flow direction, and the gross cross-sectional area, A, through which the water flows:

$$Q = KIA. \tag{5.3b}$$

This is essentially a restatement of Darcy's law.

C. Flow of Ground Water in Three Dimensions

While Darcy's law is generally applied to the determination of average horizontal velocity of flow resulting from head differential over an appreciable distance, it is equally applicable to water flow in any direction occurring in any part of the aquifer water mass.

In general form, Darcy's law may be written as

$$\nu = -K\frac{\partial h}{\partial s} \tag{5.4}$$

in which ν is Darcy velocity (flux), K is hydraulic conductivity in the s direction (∂s = distance changes), and h is the total energy or head at the point under consideration (∂h = head changes). Flow along the three principal coordinate axes (anisotropic) may consequently be described as (Kashef, 1976; Reeve, 1957)

$$\nu_x = -K_x \frac{\partial h}{\partial x}, \quad \nu_y = -K_y \frac{\partial h}{\partial y}, \quad \nu_z = -K_z \frac{\partial h}{\partial z}. \tag{5.5}$$

In isotropic aquifers, $K_x = K_y = K_z$. Continuity under conditions of steady flow without change in time and density is expressed by

$$\frac{\partial \nu_x}{\partial x} + \frac{\partial \nu_y}{\partial y} + \frac{\partial \nu_2}{\partial z} = 0 \tag{5.6}$$

which, in view of the hydraulic conductivity K, becomes

$$\frac{\partial^2 h}{\partial x^2} + \frac{\partial^2 h}{\partial y^2} + \frac{\partial^2 h}{\partial z^2} = 0 \tag{5.7}$$

which is a form of the well-known Laplace equation for ground water flow through an isotropic homogeneous aquifer under steady-state conditions.

D. Flow Nets

Flow nets (Cederegren, 1967; Todd, 1980) are graphical representations of the pattern of ground water flow in both vertical and horizontal projections. The net is composed of two families of curves, one of which represents the flow lines, while the other represents the variation in head in the aquifer or flow cross section (potential lines). The mathematical basis for the flow net can be derived from the Laplace equation, the third term eliminated with the assumption of two-dimensional flow.

Flow nets are two sets of curves or lines for isotropic media that form an orthogonal pattern of squares or rectangles. As most ground water situations involve curvilinear paths of flow, the form of net tends to assume a curvilinear pattern; however, the points of flow and head intersection or equipotential lines for isotropic media should be right angles, and the mean distance between opposite sides of the resulting rectangles should be equal for homogeneous media. Repeated divisions of such orthogonal patterns should produce subdivisions that approach the shape of true squares.

The use of flow net principles facilitates the solution of many problems involving water movement through either confined or unconfined aquifers. Flow lines drawn perpendicular to water table or potentiometric contours indicate the general direction of ground water flow in a specific area (see Fig. 5.2). Knowing the direction of ground water movement has become increasingly important in considering the danger of contaminating ground water supplies. A flow net can indicate the direction of contaminant travel. Areas of recharge and discharge can also be defined with a flow net map. Flow lines diverge in areas of recharge and converge in areas of discharge. In areas of uniform ground water flow, wide contour spacing (flat gradients) indicate areas of higher hydraulic conductivity than those areas with narrow contour spacing (steep gradients). Flow nets can be helpful in locating water wells by indicating areas with a good source of supply (recharge) and favorable hydraulic conductivity.

VI. GROUND WATER BASIN YIELD CONCEPTS

Water is universally classified as a renewable resource. Renewal of water supplies is inherent in the hydrologic cycle, and it is obvious that precipitation replenishes soil moisture and streamflow. In concept, ground water is also recognized as a generally renewable resource because virtually all usable supplies

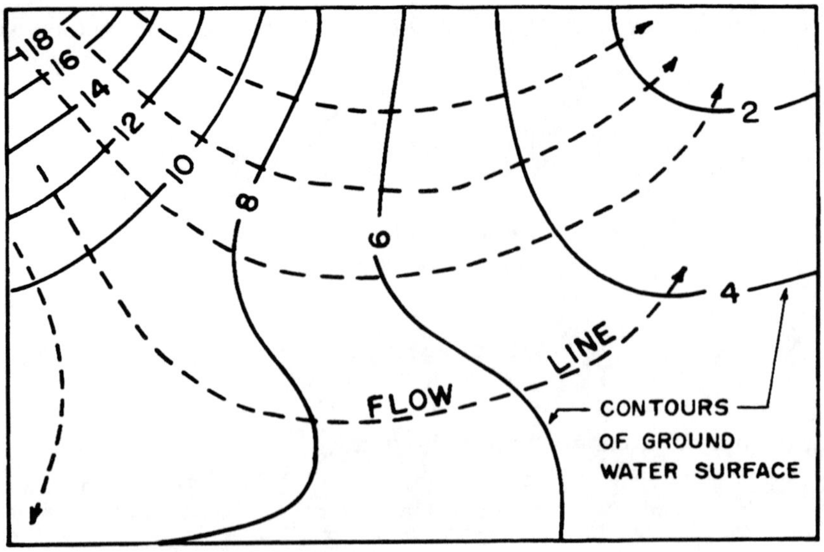

Figure 5.2—Typical Flow Net

are atmospheric in origin. Detailed studies of many ground water areas have shown conclusively that ground water supplies are recharged by precipitation. There are no known examples of fresh ground water resource areas that are isolated from such possible replenishment, even though the replenishment may have been ancient.

Nevertheless, quantity of ground water replenishment is limited. The average annual precipitation on the land area of the United States is about 6×10^{12} m^3. Of this amount, a portion evaporates, a portion is used in transpiration, and a portion runs off to the oceans. Only a small fraction of the 6×10^{12} m^3 is available for ground water replenishment. Nearly 2.5×10^{14} m^3 of fresh ground water are estimated to reside within 1,000 m of the land's surface. Apparently, the quantity of ground water that can be considered available as a renewable resource is limited by the quantity of precipitation replenishment. The replenishment is a small fraction of the total precipitation per year plus the reduction of nonbeneficial losses from the ground water reservoir.

A perennial supply of water can be assured by not withdrawing more water than is available over a long term. This concept holds true for either ground water or surface water projects. Determining the perennial yield from ground water sources is more complicated than determining yield from surface water projects because it is difficult to differentiate replenishable ground water from accumulated stored ground water.

A. Perennial Yield

The perennial yield of a ground water reservoir is defined as "the maximum quantity of water which can be withdrawn annually from a ground water supply under a given set of conditions without causing an undesirable result." The words "undesirable result" are key to the definition. Adverse or deleterious side effects of ground water development, such as sea water intrusion, land surface subsidence, salt water coning, and others are included. As the definition states, perennial yield is determined for a specified set of operating conditions. Any change in conditions, such as changes in economics, land use, or importation of new water supplies requires calculation of a new yield. The term "safe yield" is often used instead of the preferred "perennial yield."

The concept of perennial yield of a ground water reservoir stems from the principle of renewable resources, as shown in the hydrologic cycle, and is based on an analogy to surface reservoir operation. The perennial yield from a designated ground water reservoir can be achieved if the artificial discharge by wells are patterned to reduce discharge from the reservoir and to induce recharge in an equal amount, and if storage is utilized only to provide regulation of the fluctuating inflow to meet the demand on the wells. In practice, three complicating factors must be taken into account.

First, if the water table is close to the ground surface, either initially or during the operating period, some recharge to the water-bearing formation may be rejected. As water cannot be added to storage when a reservoir is full, the excess water must be evaporated or wasted. Development of a previously unused or underused aquifer usually results in increased recharge from surface streams and precipitation or a reduction in ground water discharge to a surface stream. Therefore, use of an aquifer replenished by meteoric water may diminish surface runoff so that the yield from an aquifer reduces the water available from surface sources. Jenkins (1968) discusses the theory of stream depletion by the operation of wells and provides sample calculations. Conover (1954, 1961a, 1961b) portrays the effects of ground water use on surface water and the benefits of coordinated surface water–ground water use.

Second, an aquifer contains a supply of water that can be mined, regardless of whether the aquifer is replenished. The perennially available water supply therefore may be substantially augmented for a very long time indeed—50 years of mining is not an unusual circumstance.

Third, the patterns of distribution, rates, and volumes of pumpage from wells in a developed ground water reservoir commonly differ from the ideal pattern for obtaining the maximum perennial replenishment. The maximum yield then may depend on the practicability of locating or relocating wells rather than on any characteristic of the natural resource. Pumping from a ground water reservoir in quantities greater than the perennial yield may modify one or more of the items in the hydrologic equation, thereby reducing or increasing the yield over time.

The perennial yield of an aquifer, in some instances, can be substantially augmented by engineering controls. For example, more water can be made available through artificial recharge by spreading basins or injection wells or lowering ground water levels to reduce evapotranspiration to capture rejected recharge or surface water from streams. The concept of changing the perennial yield of an area requires applied water management, including the services of specialists in hydrology, geology, and engineering for evaluation. The amount of ground water production quantified by the phrase "perennial yield" is fixed at any point in time only in the sense that no more money may be available for engineering construction, or that legally no more water may be obtained from any source. Should these constraints be changed, for example, by the importation of water and the utilization of underground storage, the perennial yield could be increased. Under such circumstances, the concept of a specific quantity for perennial yield is not a fixed amount but a changing quantity depending on the set of conditions applicable to the aquifer area. It involves consequences of pumping, such as economics, environmental effects, change in quality, and deformation of aquifers, which are deemed acceptable or safe. In contrast, overdraft involves any factor that may be termed unsafe; it involves more than just a continuing decline in storage or ground water mining.

B. Mining Yield

Water can be mined from an aquifer as long as the water table or potentiometric surface is consistently lowered in elevation over a period of years as a result of a withdrawal of ground water in excess of the induced recharge plus the reduced ground water discharge. This assumes that proper allowances are made for climatic variations.

In many areas—the High Plains areas in Colorado, Nebraska, Kansas, Texas, and New Mexico, for instance—the replenishment of ground water is not only negligible but additional recharge cannot be induced by lowering the water table. Further, the natural discharge is so diffuse, small, and distant from the centers of pumpage that, in practical terms, the discharge is minimally affected by the pumpage. In this sense, the water essentially is being mined. Stored water in this area was estimated by the Texas Water Development Board (Texas Board of Water Engineers, 1951) as about 3.54×10^{11} m^3, of which possibly 2.22×10^{11} m^3 to 2.47×10^{11} m^3 was considered recoverable. The quantity available is presently a 40-year or 50-year supply at the 1969 rate of mining, with no side effects such as land surface subsidence or salt water encroachment to be considered. Water management in this instance involves keeping track of ground water pumpage and the quantity remaining in the aquifer, leading an educational campaign to reduce use and waste of water and conducting a political effort to import a new, replacement water supply. In addition, more economic uses of water, such as industrial rather than agricultural, and arbitrary reduction in the rate of pumpage may be of value. The establishment of well-spacing regulations also could be useful.

Aquifers are usually variable in thickness and depth. Water-mining operations are, consequently, variable in rate and duration, depending on the location of wells in the aquifer. Property owners whose wells tap the lowest elevations of the aquifer normally enjoy wells possessing the largest yields and longest lives. Those whose properties overlie thin or shallow aquifers find that their wells lose capacity more quickly.

Under identical conditions of hydrogeology, different results would be obtained if the same quantity of water were to be used for industrial processes, cooling tower makeup, or drinking water rather than for agriculture. The water deposit would be minable for a very long time in such cases. Therefore, even under the classic conditions existing in the High Plains area of the Central United States, the life of a water mining operation cannot be predicted from engineering data alone. A management plan including regulation of well-drilling, well-spacing, and pumpage rates would provide an estimate of the useful life of the deposit and would assist in the development of the resource by providing a definite life for investment purposes. Under the circumstances existing in mined areas, public management of ground water resources might be concentrated on education, determination of the rate of ground water decline, and the depletion of the ground water reserve for tax purposes.

C. Sustained Yield

The term "sustained yield" is most useful when applied to well fields in situations that Thomas (1957) has termed "water-course aquifers." Such aquifers generally underlie the floodplain of a major river system and are in hydraulic contact with the water in the river. The sustained yield is the minimum rate of pumpage sustainable under all conditions of river discharge (and temperature) by a specified well field that taps the alluvial aquifers.

Sustained yield is not a fixed quantity. It can be changed by water management efforts even if the results of constructing and operating injection wells, spreading grounds, or recharge pits to improve recharge conditions are excluded. Additional wells can be constructed and operated while evapotranspiration is reduced, and the yield of the system (assuming the permeability and area of the river bed is not restricting the natural recharge) can be increased, albeit in progressively smaller, more expensive, steps (Conover, 1954).

D. Deferred Perennial Yield

Another ground water basin yield concept is deferred perennial yield. In a ground water basin or district, the quantity of water available for all uses might be limited by legal restrictions or the price of water; however, it is possible that a significant quantity of water can be obtained from aquifer storage. The withdrawal of this water might be cheaper than importing an equal quantity from outside the basin. Lowering the potentiometric surface often produces no undesirable side effects.

Therefore, extractions at a rate greater than the ultimate steady-state supply of ground water may be acceptable. For an initial period of years, the pumpage may be far greater than the anticipated steady-state supply of water. The economic use of water withdrawn from storage frequently results in a strengthened economy capable of importing water from a greater distance and at a higher cost than was originally anticipated. The danger in using this approach is that the proposed new source of water may not be available in the future due to changing social and economic values. Also, managers must be aware of possible damage to natural habitats that may be caused by extensive lowering of ground water levels;

however, the strengthened economy may sustain conversion of uses requiring large quantities of water to uses with low water requirements, such as light manufacturing or high technology activities, thereby reducing future water uses.

E. Maximum Perennial Yield

Maximum perennial yield means the maximum quantity of water perennially available if all available methods and sources are utilized for recharging the aquifer and if the extractions are located so that the maximum reduction in natural aquifer discharge is achieved. In the final analysis, the maximum perennial yield depends on the quantity of water economically, legally, and politically available for capture by the user. This quantity is not fixed because, given adequate aquifer storage and the economic incentive to build and operate recharge and extraction facilities, the problem then becomes one of ground water basin management.

To achieve the maximum perennial yield, it is necessary to "unitize" the aquifer for the most efficient, economical production of water. Instead of individually operated well fields, each well field owner receives a share in the common pool. The manager specifies the pattern of well development, negotiates for imported water on behalf of the users, and operates the entire system in a coordinated fashion. The only limit to the operation is the ability to import and distribute more water. Admittedly, such a unit operation is complex and highly technical, and it becomes increasingly complicated as water quality becomes more critical and disposal or reuse of the water receives more attention.

VII. EVALUATION OF GROUND WATER BASIN YIELD

A. The Hydrologic Balance

The hydrologic balance, or hydrologic equation, is a statement of the conservation of matter applied to a ground water basin. All water entering an area during any given period of time must either go into storage within its boundaries, be consumed, exported or flow out, either on the surface or underground during that period. The hydrologic balance is expressed in Table 5.4.

The hydrologic balance for a virgin ground water basin, unmodified by human activities, would include all items of inflow and outflow in Table 5.4 except for the import and export of water. In a short-term budget (a month, a season, a year, or even several years) substantial differences may occur between natural inflows and outflows, items 1 through 3 in Table 5.4. The differences are accounted for in the balance by changes in storage, items 5 through 7. Budgets for successive short periods of time would show an unsteady state resulting from short-term climatic variations. The longer the time period chosen for the budget, the more nearly the budget will indicate the steady-state conditions of the average climate. Over a long enough period of time, changes in storage become insignificant, and items 1 through 3 express the overall natural balance between inflow and outflow.

If an area is in hydrologic balance, how is it possible to increase the quantity of water taken from the area to meet human needs without destroying the long-term hydrologic balance of the area? If ground

TABLE 5.4. The Hydrologic Balance of a Ground Water Basin.

Inflow	Outflow
1. Surface	1. Surface
2. Subsurface	2. Subsurface
3. Precipitation	3. Consumptive uses
4. Imported water	4. Exported water
5. Decreased surface storage	5. Increased surface storage
6. Decreased soil moisture storage	6. Increased soil moisture storage
7. Decreased ground water storage	7. Increased ground water storage
TOTAL INFLOW = TOTAL OUTFLOW	

water is to be developed in an area to meet an increased local human need, the consumptive use in the area, outflow item number 3, will increase. Such an increase can be expected to upset the hydrologic balance of the area unless the increase is compensated elsewhere. Either the inflow to the area must be increased, other outflows must be decreased, or there must be a combination of the two. Changes in storage items, numbers 5 through 7, cannot be considered in the long-term balance. Surface, subsurface, and precipitation inflows, items 1 through 3, cannot normally be increased, although cloud seeding has increased precipitation in some areas. Water can be imported into an area, number 4, but would be more likely utilized directly to meet increased human needs rather than to sustain ground water levels.

Long-term hydrologic balance in an area of increased human needs can be achieved by reducing surface and subsurface outflows and nonbeneficial consumptive uses in the area, outflow numbers 1 through 3. As ground water is pumped out for new or increased human needs, the ground water table is usually lowered. Lowering the ground water table lowers lake and swamp levels and reduces their areas, thereby reducing surface and subsurface outflows from the area. It also reduces the availability of water to nonbeneficial consumptive uses, outflow numbers 1 through 3. Reductions in surface and subsurface outflows from an area may adversely affect adjacent areas and may have to be compensated in the form of increased exports from the area, outflow item 4. In this situation, the only source of water to meet increased human needs comes from reduction in nonbeneficial consumptive uses. Nonbeneficial uses vary from place to place for monetary and social reasons. In making a hydrologic balance, the present and probable future meaning of nonbeneficial use should be examined.

Annual or seasonal hydrologic budgets are commonly used by reservoir operators, irrigation districts, cities, and other water distributors. These budgets are usually kept in the form of balance sheets. A balance sheet for a ground water project might include the quantities of water pumped, quantities delivered to users, losses, and changes in ground water levels. Successive balance sheets usually show marked changes in individual quantities depending largely on changes in precipitation. Above-normal precipitation usually results in lower human needs, together with higher ground water levels. Below-normal precipitation usually has the opposite effect. These conditions are indicated in Fig. 5.3 which compares ground water level, precipitation, streamflow, and ground water withdrawals from the basin at Milford, Utah (Burden et al., 1989).

A hydrologic budget defining water supply in terms of annual or seasonal precipitation is needed to evaluate existing and potential development of dependable water supplies in a given area. Precipitation records are more readily available for longer periods of time than are ground water records. In cold regions snow melt presents problems needing complex solutions if less than annual periods are used. If the relation between ground water conditions and precipitation can be reasonably well-established, the long-term ability of a ground water source to supply human needs can be better evaluated by means of the precipitation records.

Unfortunately, such a relation is difficult to establish. Available basic data, such as precipitation, pumping, and ground water levels, may not be concurrently available for an extended period of time. Determinations made using such basic data may be quickly invalidated as additional basic data are acquired. Available basic data may not be representative of the overall area. Ground water level data, for example, may be adversely affected by operation of nearby wells. This complication is indicated in Fig. 5.4 which shows seasonal fluctuations in artesian pressure due to the opening and closing of a cluster of flowing artesian wells in the Central Sevier Valley, Utah. Pressures drop abruptly in March and April when the wells are opened to flow. When they are capped in August and September, the pressure essentially recovers to its previous level (Young and Carpenter, 1965). Some basic data items may be accurately measured, while other items may be poorly measured or may not exist. When relatively small items in a given area have to be derived from differences in large items, gross errors may occur because of inaccuracies in the large-item data. Such errors lead to poor relationships between precipitation and water supply.

Natural drainage basins are usually selected as areas in which to develop the hydrologic balance relating precipitation to yield. The natural basin limits frequently represent physical boundaries of both surface and ground water flow directions. Where the ground water flow direction differs from the

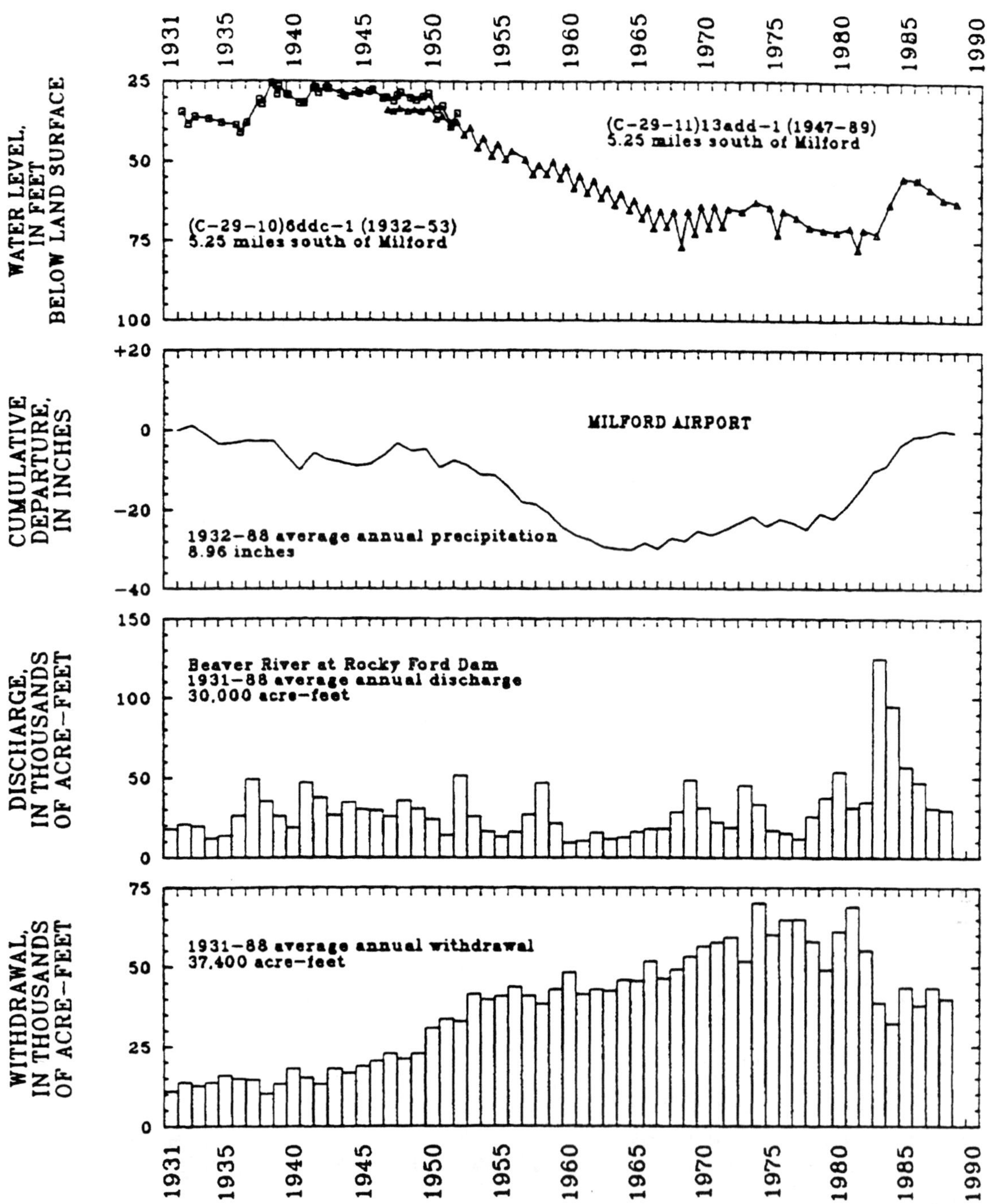

Figure 5.3—Relation of Water Levels in Wells to Cumulative Departure from the Average Annual Precipitation, to Discharge of the Beaver River and to Annual withdrawal from Wells.

direction of surface water flow, it may be more useful to use the ground water basin limits rather than the surface water basin limits, depending on the individual characteristics of the area being studied.

Once an area has been selected, it is necessary to assemble all available hydrologic records to evaluate the inflow and outflow items and balance them to obtain a useful hydrologic balance. The result of accumulating all available information is frequently referred to as a hydrologic inventory.

An independent evaluation of each of the inflow and outflow items of the hydrologic balance is desirable in a comprehensive hydrologic investigation. If the determined inflows and outflows are not

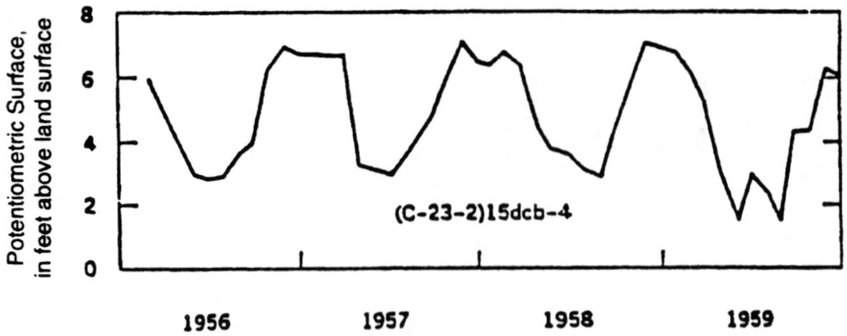

Figure 5.4—Fluctuations in Artesian Pressure due to Opening and Closing Artesian Wells in Central Sevier Valley, Utah.

in balance when compared (within the limits of accuracy of the data involved), adjustment must be made. The value or values inherently subject to the largest probable error are usually adjusted to effect a balance. If the lack of balance exceeds reasonable limits of error in the basic data, further investigation and special studies may be required. It is frequently impossible to independently evaluate all of the inflow and outflow items of the hydrologic inventory. As a result, engineering judgment is required in preparing a hydrologic balance and evaluating the dependable water supply from a specific ground water area.

Time lag must be recognized and adjusted in using basic hydrologic data. If, for example, water levels in a number of wells are used in determining changes in ground water storage, the levels taken during periods of minimum pumping and minimum or stable natural recharge are usually the most reliable because these can be expected to be periods of least change in ground water storage. While precipitation may fluctuate rapidly with time, the effect of an unusual period of precipitation may not be reflected in a change in the ground water level for months or even years.

In most cases, the dependable water supply of a specific ground water area is needed for long-time average climatic conditions as well as for severe drought and excessive precipitation periods. It is normally impossible to make a direct estimate of long-term conditions because of the relatively short periods for which basic data are available. The climatic and other conditions prevailing during the period for which basic data are available must be compared with long-term climatic and other conditions that are expected to prevail, and study period results must be adjusted accordingly.

B. Perennial Yield Estimates

Perennial yield of an aquifer is the withdrawal of water from the aquifer at a new equilibrium rate established by pumping. Perennial yield exceeds recharge or discharge of an aquifer under natural equilibrium by the quantity of induced recharge captured from evapotranspiration, precipitation, and surface water. Ground water studies provide perennial yield estimates of a ground water area. The perennial yield is used to plan the development of an unused ground water reservoir or to modify the management of a used ground water reservoir.

Ground water reservoirs undergo constant change, even in their natural state without human development. Water levels in unconfined aquifers and potentiometric levels in confined aquifers rise and fall with changing recharge and discharge rates. Both long-term and short-term changes in recharge and discharge rates are caused by corresponding variations in precipitation and streamflow. Removal of water from a ground water reservoir to meet human needs usually accentuates natural changes.

Lowering the water level in an unconfined aquifer reduces the amount of water in storage, but may increase the recharge from precipitation and influent streams, increase subsurface inflow from adjacent areas, reduce discharge from effluent streams, reduce subsurface outflow, and reduce areas of moist land and the accompanying evapotranspiration losses. A rise in water level may have the opposite effects. A change in the potentiometric level of a confined aquifer that increases the hydraulic gradient between areas of recharge and pumping may increase the rate of recharge. A change in the potentiometric level that reduces the hydraulic gradient between the areas of pumping and natural discharge reduces the rate of natural discharge. Lower pressures in a confined aquifer may also cause consolidation of the confining strata and increase the yield of water from pore storage; at the same time such pressure loss may cause subsidence of the overlying lands.

Perennial yield of a ground water area may be dictated by considerations other than the physical and climatic conditions of the area. Preexisting development and established water withdrawal patterns may determine perennial yield. The cost of well construction and pumping may limit the degree of water level lowering that can be expected. Pumping may have to be limited to preclude damage to overlying developed areas from possible land surface subsidence. Excessive lowering of the water level may cause salt water or other unsuitable water intrusion, or may damage natural or protected habitats. Also, legal restrictions may exist that prohibit interference with established water rights.

Perennial yield may change with time and with changing economic or social conditions. A change of use from irrigation to domestic, municipal, or industrial use changes the value of the water and may increase economical pumping lifts. Greater pumping lifts may increase perennial yield by increasing storage capacity, increasing recharge, and reducing outflow. Increased economic pumping lifts may also result from increases in the value of irrigated crops.

It is preferable to reserve the concept of perennial yield for use with an entire ground water reservoir or a natural subdivision thereof. Perennial yields have been estimated for single wells or for relatively small well fields in a large ground water reservoir. The accuracy of such estimates is obviously limited by the assumptions that must be made concerning the effects of development of the reservoir that may be made by others. If the total yield of the reservoir is estimated, the share of the total that should be reasonably produced by a single well, or a small number of wells, could be more reasonably estimated.

Perennial yield is difficult to estimate even for a fully developed ground water reservoir because of inadequate data. It is almost always necessary to estimate some of the necessary data by indirect methods. As a rule, the older, more developed ground water areas have more and better hydrologic and geologic data available than the undeveloped areas.

1. Calculation by Hydrologic Balance Calculation of perennial yield from a ground water reservoir is similar to determining the yield of a surface reservoir. The inflow into a reservoir minus the outflow from a reservoir equals the change in storage in the reservoir over the time period of interest. Estimated inflows into a ground water reservoir for a series of discrete time periods are compared with a series of trial yield patterns of withdrawal from the reservoir to obtain fluctuations in the ground water reservoir levels. Surface and subsurface outflows must be estimated for each reservoir level. The maximum trial yield that can be withdrawn without producing undesirable results is the perennial yield.

Subsurface inflow and outflow from a ground water reservoir can seldom be estimated accurately. An average subsurface inflow or outflow for the average ground water reservoir level is usually sufficient for estimating a given range of water level fluctuation.

If, from previous experience, the ground water reservoir is known to have sufficient capacity within a desirable water level operating range to allow the established long-time mean withdrawal to continue without undesirable consequences, then it is necessary to calculate only the long-time mean withdrawal rate. No reservoir operation studies are needed. In this case, the perennial yield is the surplus (increase) of water in storage plus net recharge. Pumpage may be used in lieu of net recharge.

In the Livermore Valley of California, a ground water inventory (California Department of Water Resources, 1963) was made for a 9-year period of record with the results (in millions of cubic meters per year) shown in Table 5.5.

This table shows that the 9-year average yield was about 215.8×10^6 m^3 and could have been 265.7×10^6 m^3 without a decrease of water in storage. (The change in storage, $+49.9 \times 10^6$ m^3, could be pumped increasing the pumpage, -215.8×10^6 m^3, to -265.7×10^6 m^3. The first four items in Table 5.5 are inflow

TABLE 5.5. Results from a Ground Water Inventory in Livermore Valley, CA (California Department of Water Resources, 1963).

Item	Million cubic meters
Recharge from rain and applied water	97.7
Stream recharge	130.3
Artificial recharge	6.5
Subsurface inflow	31.2
Change in storage	49.9
Pumpage	215.8

to the basin and total 265.7 × 10^6 m^3.) The perennial yield for this basin to include future years would require adjustment of each of the recharge amounts to land–use conditions of the years being considered. The perennial yield would be 263.7 × 10^6 m^3 plus or minus the adjustment and experimental error.

2. Calculation from Limited Data Sometimes it is necessary to make reconnaissance estimates of ground water availability in sparsely populated regions where little or no ground water data are available. Exploratory drilling information, operating wells, and detailed geologic studies may be unavailable, and even the existence of a ground water aquifer may be uncertain. Reconnaissance perennial yield estimates can be derived only indirectly from the other hydrologic cycle constituents (precipitation, runoff) and in an indirect estimate of natural ground water discharge based on observation of springs, seepage areas, marshes, ponds, and phreatophytic vegetation sustained by ground water. If such reconnaissance estimates are favorable for an area where water use is growing, additional investigations of the extent and characteristics of a potential aquifer should be undertaken.

In more heavily populated areas, where limited ground water development has taken place, available ground water records may be limited to water levels in a few wells. The records may or may not indicate the depth and extent of an aquifer, and specific yields are usually unavailable. Such records can only be used to derive mean or weighted mean water levels during an inventory period. Arithmetic mean water levels would be appropriate for evenly-spaced wells and a weighted mean for unevenly distributed wells.

Average water level changes in an area may be correlated with discharge from the ground water reservoir or the difference between discharge and recharge. In the simplest case, where it is found that no net change occurs in average water levels from the beginning to the end of an inventory period, the average annual net draft during the period is the perennial yield. This is true provided that 1) the average water supply to the ground water reservoir during the inventory period is equal to the long-term water supply under conditions of future development, including streamflow modifications, 2) if a confined aquifer exists, land surface subsidence during the inventory and the reduction in aquitard water storage are determined, and 3) there has been no rejection of potential recharge during the inventory period because ground water levels were too high.

In an alternative technique, the average annual draft is plotted against average annual change in ground water level for several periods of time. An example is shown in Fig. 5.5. Each point is the average annual amount of water pumped for five consecutive years plotted against the average annual change in water levels during the same five years. Progressive 5-year periods were used to smooth out annual variations in water supply. The plotted points in Fig. 5.5 fall very nearly on a straight line, with zero average annual change in ground water level intersecting an average annual draft of 12,000 acre-ft. (14.8 × 10^6 m^3). If the ground water recharge during the study period approximates the long-term mean recharge, the perennial yield can be expected to be 12,000 acre-ft. (14.8 × 10^6 m^3) per year as indicated by this analysis.

In an area of steadily falling ground water levels, all points may fall below zero change in water level. The yield that would produce no significant change in water level would have to be inferred by

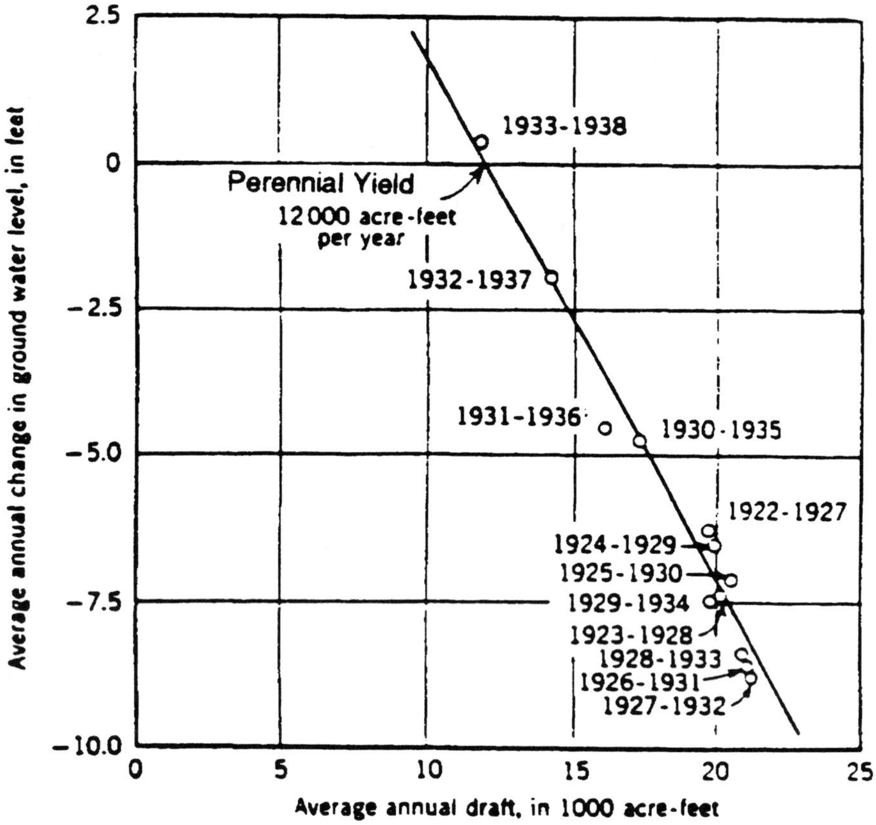

Figure 5.5—Determination of Perennial Yield, Pasadena Basin, California.

extrapolation of the line fitting the plotted points. Care should be exercised in applying such extrapolation, because the apparent relationship may not be linear over an extended water-level range.

Perennial yield was determined for an area near Cedar City Valley in southwestern Utah (Thomas, 1952) by comparing average annual inflow with average annual draft. The inflow, or recharge, to the ground water reservoir was found to come chiefly from Coal Creek, and draft consisted primarily of water pumped by irrigation wells. It was found that the ground water levels remained unchanged if the annual runoff from Coal Creek was 12.3×10^6 m^3 greater than the irrigation pumping draft for the period 1934 through 1951. The 12.3×10^6 m^3 appears to be a nearly constant natural ground water loss. The perennial ground water yield, therefore, fluctuates with the runoff of Coal Creek. In the 18-year period 1934 through 1951, the average annual runoff of Coal Creek was 29.6×10^6 m^3. Incomplete records indicate an average runoff of 49.2×10^6 m^3 for Coal Creek during the 11-year period 1916 through 1926. The runoff was only 19.7×10^6 m^3 per year during the 1953 through 1957 drought period. Irrigation well pumping during the drought period averaged about 18.5×10^6 m^3 annually, a quantity far less than the perennial yield that would be indicated during the 1916 through 1926 period, but far greater than the perennial yield indicated for the 1953 through 1957 period. A perennial yield equal to the long-term average annual runoff of Coal Creek minus 12.3×10^6 m^3 per year can be realized if the ground water aquifer has enough storage to equate the inflow fluctuations without excessive ground water level fluctuations and if all other hydrologic relationships remain substantially constant.

*3. **Calculation from Pressure Trough*** A direct method of determining perennial yield of a confined aquifer was developed in an investigation of salt water intrusion into the Salinas River Valley in California (California Department of Public Works, 1935). This method is based on Darcy's law but does not require a separate determination of the aquifer's hydraulic conductivity and cross-sectional area. Fig. 5.6 shows a simplified cross section through the Salinas Valley confined aquifer area.

The 180-ft aquifer was heavily pumped for irrigation by about 660 wells. Water is also pumped for municipal and industrial purposes. Only 4% of the total pumping was from the 400-ft aquifer during the 1943 through 1945 study period. Excessive salinity caused rapid abandonment of many of the 180-ft aquifer wells near the coast in 1943 and 1944. The Salinas Valley investigation was undertaken to ascertain the reasons for and extent of possible salt water intrusion into the aquifer.

Sea water intrusion was confirmed by discovery of a trough in the potentiometric surface, as shown in Fig. 5.6, and by well water quality analysis. When such a trough exists, all water pumped from the aquifer on the inland side of the bottom of the trough is supplied by ground water flowing toward the ocean. Water pumped between the bottom of the trough and the ocean shoreline is supplied by ground water moving inland from the ocean. As the 180-ft aquifer is not sealed off from the ocean, sea water intrusion may approach, but cannot extend beyond, the farthest inland position of the bottom of the trough, except for the toe of the sea water wedge as shown in Fig. 5.6.

When pumping exceeds the perennial yield of the aquifer, a pressure surface trough begins to form a short distance inland from the shoreline. The trough deepens and the bottom of the trough moves inland with increased summer season pumping. When pumping decreases, the trough becomes more shallow and the bottom of the trough moves toward the shoreline. The perennial yield of the aquifer is the yield that does not cause the formation of a trough and that does not cause excessive loss of fresh water into the ocean. It corresponds to the hydraulic gradient that reaches the shoreline just above the ocean levels and that has no inland pressure trough.

The total pumping at the time the pressure trough was detected was about 6.9 m^3/sec. Total pumping from the aquifer Q reduced by seaward pumping between the bottom of the trough and the shore Q_0 equaled the pumping from the inland fresh water source Q_i:

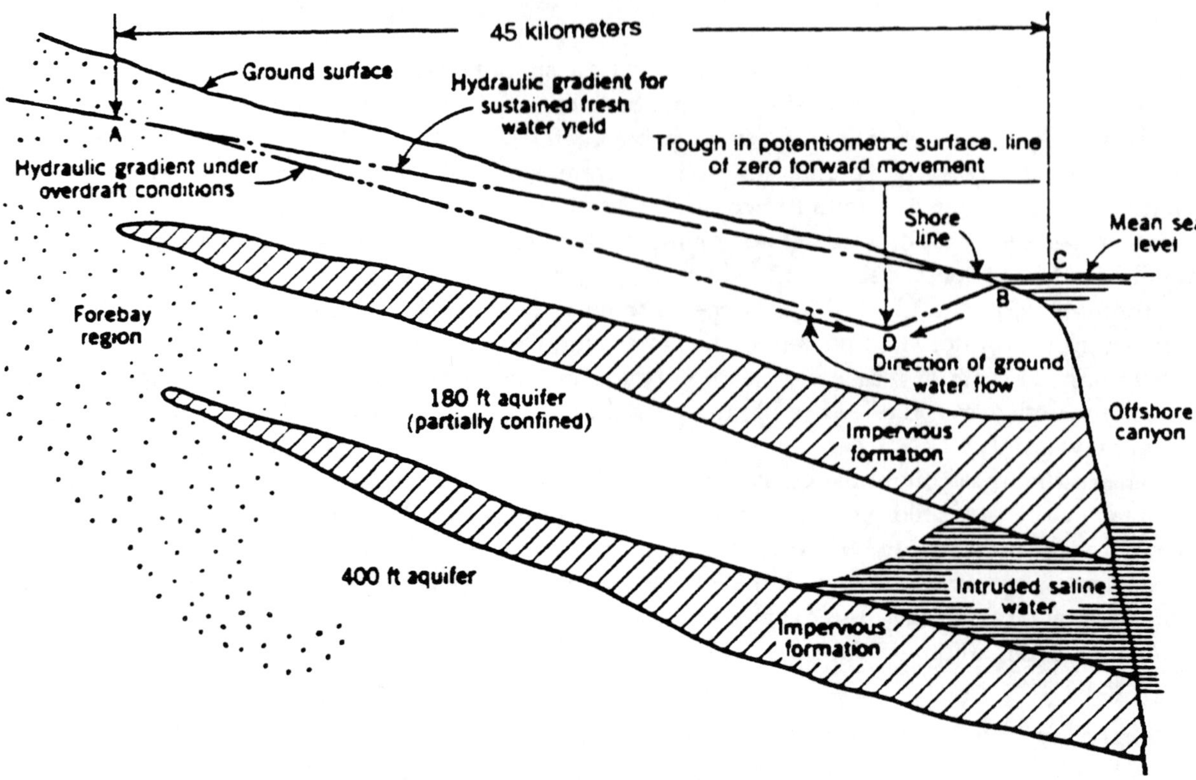

Figure 5.6—Hydraulic Gradients and Potentiometric Surface of 180-ft Aquifer.

$$Q_i = Q - Q_0. \tag{5.8}$$

Pressure trough elevations were taken, and its average location was determined to obtain distance L and head differential Δh from the forebay region of the aquifer to the bottom of the trough. Using

$$Q_i = KA\frac{\Delta h}{L} \tag{5.9}$$

in which K is the average hydraulic conductivity, A is the average net cross-section area of the aquifer, Δh is the average difference in elevation between the forebay and trough levels, and L is the average distance from the forebay to the bottom of the trough. Solving for KA, Equation 5.9 becomes

$$KA = \frac{Q_i L}{\Delta h}. \tag{5.10}$$

Values of KA for the inland side of the aquifer were computed for three trough locations as shown in Table 5.6.

The average KA was taken to be 8,787. The perennial yield of the fresh water aquifer using this value of KA, the distance L from the forebay to the shore, and the Δh corresponding to the gradient for sustained fresh water yield, was computed as follows:

$$\text{Perennial yield } Q = KA\frac{\Delta h}{L} \tag{5.11}$$

$$8787 \times \frac{100}{135{,}000} = 6.5 \text{ m}^3/\text{sec}.$$

The maximum draft on the aquifer was 9.3 m^3/sec, a rate 2.8 m^3/sec more than the perennial yield. The annual overdraft rate was estimated to be 24.7×10^6 m^3 for the prevailing pumping periods.

The foregoing calculated overdraft was greater than the seaward overdraft, the draft between the bottom of the trough and the shore, Q_0, which was 1.6 m^3/sec. The calculated perennial yield is based on reestablishing a fresh water hydraulic gradient that would reduce the fresh water flow from the forebay while at the same time preventing sea water encroachment.

The perennial yield calculated by this method is qualified by the existing pattern of pumping draft. In this example, there was evidence that the yield could be increased by changing the pattern of withdrawal.

4. *Calculation by Hydraulic Formula* Perennial yield may be determined by Darcy's law where the hydraulic conductivity, hydraulic gradient, and cross-sectional area of the water-bearing formation are known. This method was used to determine that the perennial yield of a well field near Hamilton, Ohio, adjacent to the Miami River could be substantially increased by pumping more water and lowering the ground water level, thereby inducing more recharge from the Ohio River.

Three interconnected aquifers were separately analyzed in a quantitative study of the Roswell, New Mexico, ground water reservoir (Hantush, 1955). These aquifers included a limestone artesian aquifer

TABLE 5.6. Values of KA.

Date (1)	Continuous pumping, Q_i in cu m/sec (2)	Average pressure surface slope, Δh/L (3)	Hydraulic conductivity times area, KA (4)
4/29/45	6.7	101.0/132,000	8,787
5/20/45	6.8	101.5/131,700	8,787
6/30/45	7.1	104.5/130,000	8,787

overlain by a leaky formation, an unconfined limestone aquifer to the west of the artesian aquifer, and an unconfined alluvial aquifer overlying the artesian aquifer. The unconfined limestone aquifer constitutes the 18,000 km^2 recharge area for the artesian aquifer. The overlying alluvial aquifer is recharged by upward leakage from the artesian aquifer as well as by infiltration from the Pecos River, irrigation, and precipitation. The study shows that the perennial yield from the Roswell ground water reservoir is limited by the distribution of wells in both the alluvial and artesian aquifers. Hantush (1955) also presents the need to allow the alluvial aquifer to continue discharging water to the Pecos River to satisfy existing surface water rights and the need for salinity control near the eastern edge of the artesian aquifer.

5. ***Calculation of Mining Yield*** Mining a ground water reservoir occurs when water is removed from a reservoir without the expectation of its being replenished. Total water available by mining is the quantity of water in the reservoir; however, the quantity mined may be limited to the quantity which can be economically extracted from the reservoir or the quantity that can be legally extracted. Annual mining yield is the annual draft on the reservoir that is expected to exhaust the reservoir in a specified number of years.

Mining occurs whenever the long-term average draft on a ground water reservoir exceeds perennial yield. Limited ground water mining also occurs during the initial development of a ground water reservoir. The ground water level is initially lowered a limited depth to reduce or cut off flows from points of natural discharge in order to direct more water for beneficial use. Such initial ground water level lowering may dry up springs, ponds, and flowing wells, and may thereby significantly modify the local environment. Initial mining may be limited by legal restrictions.

Ground water mining can also occur intermittently in a reservoir whose withdrawals are regulated to the perennial yield, which is conservatively determined to minimize shortages during dry periods. Higher than normal withdrawals may be allowed for such a reservoir during infrequent, protracted dry periods.

The Ogallala Aquifer is an outstanding example of a mined ground water reservoir (Banks, 1953). The aquifer area underlies most of the state of Nebraska, western Kansas, eastern Colorado, western Oklahoma, eastern New Mexico, and northern Texas. The region underlain wholly or in part by the Ogallala Aquifer totals about 57,000,000 ha.

Development of the aquifer started in the late 1930s. Irrigated agriculture expanded rapidly in the area following World War II, from about 1,417,000 ha mostly in Texas and Nebraska in 1950 to over 6,075,000 ha in 1980. Less than 8.6×10^9 m^3 of water were withdrawn from the aquifer in 1950 compared to more than 26.9×10^9 m^3 withdrawn in 1980. Irrigation efficiency improvements reduced the application of irrigation water by about 30% from 0.61 m to 0.43 m during the same period.

Annual natural aquifer recharge is estimated to vary from 10 to 24 mm over much of the area to about 150 mm in some sandy surface areas in Nebraska. The aquifer was determined to have slightly over 3.7×10^{12} m^3 of water remaining in storage in 1977. This storage is expected to fall to nearly 2.8×10^{12} m^3 by the year 2020 if present trends continue. This 2×10^{10} m^3 per year average mining overdraft on the aquifer is causing and will cause undesirable effects over much of the region, and a number of regulatory and water supply enhancement plans are under consideration to minimize the adverse effects of depleting the aquifer.

In 1986, the U.S. Congress passed PL 99-662, the Ogallala Aquifer Research and Demonstration Program. The program has led to proposals for aquifer management. Obviously, due to the interstate nature of the aquifer and the differing objectives of individual states, an aquifer management plan for the Ogallala in addition to overall aquifer management must be sensitive to individual needs. Unfortunately, as the system is currently overtaxed, something must give or else the aquifer will slowly dry up. A recent management plan from Nebraska includes optimal irrigation management, development of new irrigation technologies, cultivation of crop varieties that use less water, water policy alternatives, drought management measures, and increased data collection (Larson, 1989).

The Edwards Aquifer in south central Texas has developed ground water management plans which include: registration and permitting of new wells, determination of the validity of water (surface and ground) rights claims, transfer/retirement of pumping rights as new supplies are developed, and issuance and withdrawal of water rights permits. Hydrologically, the plan attempts to limit the pumping of the Edwards Aquifer to 75% of the average annual recharge (approximately 5.5×10^8 m^3/y). This regional water plan allows the Edwards Underground Water District to implement specific measures to best manage the ground water resources of the Edwards Aquifer. Specific measures include estimating the total projected water demands, estimating ground water pumping, conservation, wastewater reuse, surface water projects, artificial recharge, drought management, and financing and implementation (Jensen, 1988).

6. Accuracy of Estimates The general status and accuracy of water-supply estimates made using the principles of ground water hydrology are summarized well by Baker (1951):

> Ground water hydrology is not an exact science, not so much because of a present lack of scientific or technical knowledge concerning the principles in regard to the behavior of ground water as because of lack, in any estimates made, of basic data of a quality and a quantity which will allow results to be achieved of a precision attained in other engineering and scientific computations. If ground water had a value—per unit of quantity—approaching that of petroleum, it would be entirely feasible to make such field investigations and study as would be necessary to allow the determination of safe yield or other phenomena in a fairly precise manner. Under present conditions, however, the ground water hydrologist usually must accept the basic data which he finds available—much of it of poor or uncertain quality—and must endeavor to supplement this with such additional data as it is feasibly and financially possible for him to secure. To such data as he ultimately secures, he must apply sound and mature judgment; and through utilization of all of these, arrive at his preliminary estimate. To this estimate must be applied a factor of safety which should vary inversely with the quality and quantity of basic data that has been used. With time, more data of better quality usually becomes available, and subsequent estimates of greater accuracy can be made.

It should be recognized that any estimate of ground water yield will probably be superseded in the future as more data and better methods of analysis become available. Ground water yield estimates based on numerous long-term records and sound estimates of future development can be expected to be more dependable and less subject to future change. Human activities, which create the need for water in the first place, will probably continue to modify the hydrologic balance of an area. Changes that may affect the hydrologic balance include the extent and timing of the ground water draft, the transfer of water from agricultural use to urban use, increases in impermeable areas, modifications of surface runoff patterns, modifications in subsurface inflow and outflow, the import or export of water, and alteration of ground water gradients. These and other possible changes in hydrologic conditions must be evaluated in planning methods of ground water development and in estimating ground water yield. Agencies managing ground water basins must continuously study the hydrologic and physical conditions that affect the performance of the basin.

VIII. RECHARGE

Recharge to the ground water body is naturally accomplished by the processes of infiltration and percolation, described subsequently in this chapter. Infiltration is the process whereby precipitation and surface water move downward into the soil mantle or rock materials. Percolation is the vertical and lateral movement of water through various openings in the geologic materials (i.e., through pores in the soil column or in sedimentary materials, or through fissures and fractures in more consolidated rock materials) in response to gravity or differences in pressure that create a hydraulic (energy) gradient. Depending upon the size and degree of interconnection of the openings in the geologic materials and the energy gradient, the velocity of ground water movement may range from a few centimeters per year to a thousand or so meters per year. In fact, some highly pervious volcanic rocks and carbonate rocks containing solution channels transmit moving waters comparable in size and velocity to some rivers.

Infiltration and downward percolation to the underlying ground water body may result from either natural or artificial availability of water on the surface of the ground. Natural sources of recharge include: 1) rain, directly or from overland flow en route to a watercourse, 2) melting snow or ice, or 3) water in stream channels, lakes, or other natural water bodies. Artificial sources of recharge may include: 1) leakage from reservoirs, conduits, septic tanks, and similar water-related structures, 2) irrigation, or other water applications including deliberate flooding of a naturally porous area, 3) effluents discharged to evaporation or percolation ponds, and 4) injection, as through wells or similar structures. Natural sources furnish the chief supply to ground water, but in the western United States, where irrigation and artificial recharge are practiced in many localities, contributions of water from such sources are substantial.

The optimum use of available surface and ground water resources is an essential element of any comprehensive plan for water-supply development. In many instances, underground storage capacity can be used for cyclic storage of conserved surface water supplies, both local and imported. Aritificial recharge of ground water basins, denoting the planned introduction of surface waters into permeable underground formations, may be accomplished independently of, or in conjunction with, naturally occurring infiltration and percolation to permeable aquifers. Artificial recharge offers a potential solution to many of the complex problems associated with restoring and operating ground water reservoirs, and of preventing or minimizing overdraft and its attendant adverse consequences.

A. Natural Recharge

Natural recharge to the saturated zone in a ground water reservoir may come from a number of sources, including: 1) stream bed percolation, 2) deep percolation of rainfall, 3) leakage from ponds, lakes, and reservoirs, and 4) subsurface inflow.

Stream bed percolation and deep penetration of rainfall are generally the two most important sources of natural recharge. In areas where stream channels traverse areas of coarse surface and subsurface deposits, such as the upper and central portions of alluvial fans, appreciable quantities of water usually percolate. When the water table is a considerable distance below the stream bed, permeability and width of the stream channel deposits under the stream are the prime factors in determining percolation rates. After the ground water mound builds up to the stream channel, the hydraulic gradient of the mound and the thickness of the deposits become other controlling factors.

Deep percolation of rainfall is influenced by many factors, including: 1) texture and gradation of surface and near-surface deposits and their vertical permeability, 2) nature and consumptive use of the vegetative cover, 3) frequency, intensity, and volume of rainfall, 4) topography, and 5) temperature. In some ground water reservoirs, stream bed and rainfall percolation may play a minor role in natural recharge. Other sources, such as subsurface inflow or incidental recharge stemming from uses of water on the surface, may become of major importance. Incidental recharge resulting from the use of water includes deep percolation of applied irrigation water, water from cesspools and septic tank leach fields, leakage from water supply and sewage conduits, and waste water discharged to the ground surface. Some of the conditions affecting natural ground water recharge are illustrated in Fig. 5.7.

B. Artifical Recharge

Artificial recharge is the deliberate process of replenishment of stored ground water through works provided primarily for that purpose. It should be noted that intent is an essential element in this definition. Many projects designed for disposal of return irrigation waters, sewage, cooling water, or other wastes also augment ground water supplies; however, such ground water replenishment is generally incidental to the primary function of these works, and they are not considered true artificial recharge projects.

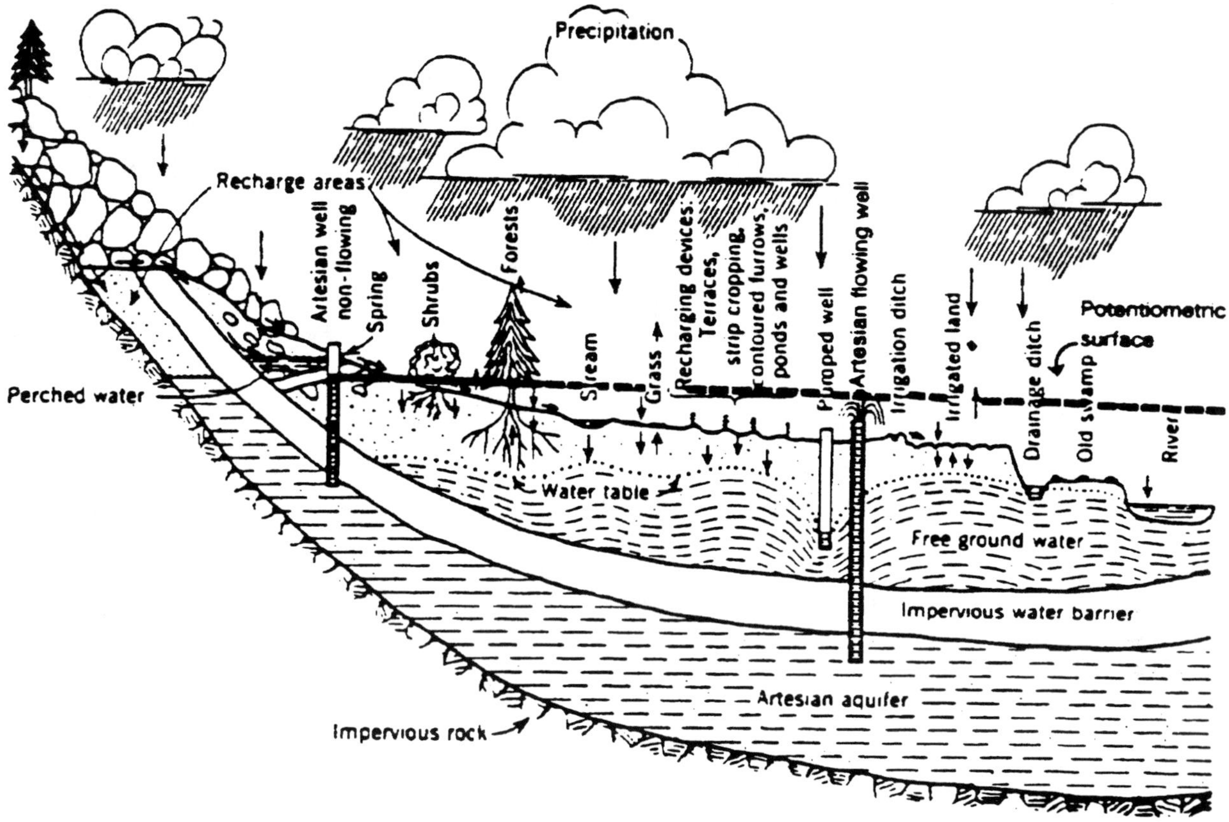

Figure 5.7—Ground Water Replenishment and Withdrawal.

The primary purposes of most artificial recharge activities are to 1) combat adverse conditions resulting from overdevelopment of ground water, 2) increase conservation of local and imported waters, 3) increase the quantity, or yield, of ground water available for use, and 4) allow the ground water basin storage to be used for peak use periods. Specific objectives are outlined as follows.

1. Augmentation of natural recharge to the ground water reservoir to compensate for human activities that tend to reduce natural recharge. Activities that reduce natural recharge include:
   a. Lining of stream channels for flood protection or erosion control.
   b. Collecting sewage and industrial wastes of suitable mineral quality and discharging them to saline waters or out of the ground water basin.
   c. Sealing of natural recharge areas with impervious surfaces, such as sidewalks, streets, airports, parking lots, and buildings.
   d. Diverting surface water supplies which might otherwise infiltrate naturally in stream channels.
2. Coordination of surface and ground water reservoir operation.
3. Reduction of ground water overdraft.
4. Amelioration of problems of adverse salt balance, control of sea water intrusion and local bodies of poor quality water, and eliminating problems of ground water pollution.
5. Maintenance or raising of water levels to avoid deepening wells, increased pumping costs, and obsolescence of equipment.
6. Reduction or cessation of critical land surface subsidence problems in areas where ground water withdrawal results in consolidation of sediments.
7. Store off-season water supplies in the ground water basin for use during peak water demand periods.

Artificial replenishment of ground water is accomplished primarily through works designed to maintain high infiltration capacities, increase the wetted area, and lengthen the period of infiltration beyond that which exists under natural conditions. Although no two projects are identical, most utilize variations or combinations of the basic operating works: basins, modified stream beds, pits, ditches and furrows, flooding, and injection wells. Upstream surface reservoirs are often provided to regulate flood flows, temporarily store floodwaters, limit flows to the absorptive capacity of the spreading works, and remove a large portion of the detritus.

Another method of artificial recharge utilizes the operational flexibilities of water-supply systems that have available surface water supplies. This is the "in lieu" method. The "in lieu" method leaves water underground, which would have been extracted by supplying surface water, directly to the agency or individuals who would have extracted the ground water. Leaving water underground accomplishes in a very efficient and effective manner the same effect as putting it underground.

1. ***Preliminary Planning Considerations*** To develop, operate, and maintain an economical and efficient recharge project, proper planning is in order. The cultural, hydrologic, and geologic conditions at potential sites must be carefully evaluated. These conditions may limit the methods of artificial recharge to be considered and will affect the design, operation, and maintenance of the project. An important factor governing project operation is the infiltration capacity at potential sites. Although this rate primarily depends upon soil and geologic conditions at the selected site, cultural and hydrologic conditions also have a strong influence on the rate.

2. ***Cultural Considerations*** The cultural development in the vicinity of the proposed site has a decided effect upon the capital cost of a given project. This, in turn, is directly related to the economic feasibility of the project and the selection of the method of recharge. Special consideration must be given to water rights availability in project development. In general, where water spreading is contemplated in urban areas, techniques must be employed that result in a high degree of control over the supply to be spread. These techniques include construction and operation of injection wells, pits, or shafts, and basin-type projects. Levees, when used, must be designed to prevent leakage; fencing to protect property and prevent unauthorized entry (as by children, with consequent danger of drowning) must be provided; and continuous surveillance and maintenance for elimination of odor and insect nuisance must be accomplished. In urban areas, consideration should be given to joint use of recharge ponds for ground water recharge and compatible recreational uses.

In agricultural areas, flooding techniques and applications of surplus water supplies during winter periods constitute an additional method of artificial recharge. Other water spreading methods not utilizing control facilities should be limited to remote areas where possible damage from uncontrolled flow is minor in effect. The purchase and maintenance of natural recharge areas, to forestall urbanization or other development that may reduce recharge capability, is particularly effective.

3. ***Surface Hydrologic Considerations*** The quantity, quality, and character of the water to be infiltrated should be carefully evaluated. In addition to the selection of a site and method of artificial recharge, the quantity of water available for recharge has direct bearing on the need for, and size of, a project.

The water supply available may vary widely within a short period of time or may remain almost constant for a long period. The availability usually can be ascertained from a review of recorded streamflow for the selected supply source. Generally, fluctuating storm runoffs cause problems related to inefficient project operation, as large volumes of water containing high concentrations of silt and debris must be handled within a relatively short period of time. Operational criteria may also require bypassing flows of unacceptable quality if infiltration rates can be maintained only by using this method.

Any of the spreading methods can be used for replenishment when the source is a fluctuating flow. Excessive fluctuations can be handled best by increasing the storage capacity of the project. The basin and pit techniques have the advantage of providing temporary storage capacity, dependent on the area covered and depth of water. At the other extreme, injection wells have very little storage capacity and require a rather uniform supply of water with high chemical, biological, and physical quality.

Water quality, primarily with respect to silt and clay contents, has an important bearing on site selection and recharge method. This is particularly true with regard to injection well methods, where almost all silt and clay must be removed for successful operation. Quality of water used in connection with deep pits or shafts, with no facilities for desilting, should be nearly as good. Relatively good quality water is desired for basin projects. On the other hand, properly designed ditch and furrow projects have been used advantageously with waters of high silt and clay content.

4. Geologic and Subsurface Hydrologic Considerations Where surface recharge methods are to be used, the selection of a proposed artificial recharge site or the evaluation of the effectiveness of an existing site should be based on a detailed knowledge of the geologic conditions at the site. Physical properties of surface and subsurface deposits determine, to a large extent, the sustained infiltration capacities that can be maintained and the volumes of water that can be recharged to the ground water body. In addition, location of geologic structures and subsurface deposits which might form a complete or partial barrier to ground water movement, and the position and hydraulic gradient of the existing water table or potentiometric surface, affect the efficiency and suitability of an artificial recharge project.

Infiltrating water first must pass through the surface deposits prior to reaching the water table. Therefore, the properties of the various soil horizons are critical in the establishment of initial and sustained infiltration capacities. The more important chemical, physical, and biological properties of the soil include: 1) texture, 2) permeability, 3) clay-pan and iron or lime-hardpan development in the subsoil horizon, 4) depth of soil profile, 5) organic matter, 6) degree of compaction due to use of construction equipment, and 7) the chemistry of the interaction between soil and recharge water. In general, the coarser the texture of the surface deposits, the higher are the initial and sustained infiltration capacities. In many cases, development of clay-pan and iron-hardpan layers in the subsoil horizon is the major limitation to maintenance of high sustained capacities.

Aside from the influence of soil horizons, the determination of sustained infiltration capacities, maximum storage capacity and effective transmission rates of ground water from beneath the recharge site to other parts of the ground water basin is based on various physical and structural features of the underlying subsurface deposits. The important features of the subsurface deposits that must be evaluated include:

(a) The physical character and permeability of subsurface deposits above the water table. This determines the allowable sustained infiltration capacity.
(b) The permeability, specific yield, thickness of the deposits, and position and allowable fluctuation of the water table. These factors establish the subsurface water storage capacity in the vicinity of the project.
(c) The transmissivity of the aquifer and the hydraulic gradients of the water table. These factors affect the rate of movement of ground water from spreading sites to areas of ground water draft.
(d) The underground structural and lithologic barriers existing in the area. Such barriers will form an impediment to both vertical and lateral movement of ground water.
(e) The chemistry of the native ground water in relationship to that of the recharge water must be considered to minimize problems associated with possible chemical reactions or ion exchange phenomena.

5. Infiltration Capacity One of the major problems associated with the evaluation of an artificial recharge site is the determination of infiltration capacity (see Chapter 3, Infiltration). Such capacities are critical, for they dictate the method and size of the recharge site and the operation and maintenance techniques to be utilized.

Many factors affect infiltration capacities at artificial recharge projects. Natural factors include the composition of surface soils and the geologic and subsurface hydrologic conditions discussed previously.

The quality of the water and the procedures used in the construction, operation, and maintenance of projects also affect infiltration capacities, although these factors can ordinarily be managed to maintain favorable capacities.

Of the various factors mentioned in the preceding paragraph, the most significant effect results from soils and geologic conditions. For example, in the upper coarse portion of alluvial fans, where surface deposits consist mainly of sand, gravel, cobbles, and boulders, high to moderate infiltration capacities are attained; however, the highest units of infiltration frequently occur at a location on the lower slope of the alluvial fan, where sands of uniform grain size are predominant, and the ratio of voids to total mass of surface deposits is at a maximum. Surface deposits on the central portion of alluvial fans are characterized by sandy soils, with some silt and clay, and display moderate to low infiltration capacities. Fine-grained silt or clay soils, or soils of variable surface texture underlain by extensive clay-pan or iron-hardpan layers, occur on the distal portion of alluvial fans or on older elevated and dissected alluvial deposits and are generally characterized by low infiltration capacities.

Infiltrometers and small experimental plots have been used successfully to determine the infiltration characteristics of the surface soils at a site. For a given quality of water and type of material, the infiltration capacities are affected by the length of recharge period, size of test plot, condition of soil, and length of drying period for the spreading area. Thus, care should be taken in extrapolation of preliminary test measurements to estimates of potential project performance. In general, it has been found that the initial infiltration capacity decreases rapidly, followed by an increase to nearly the initial value, after which a gradual continuous reduction occurs until basins are taken out of service for drying and maintenance.

The soil type may be used as a guide in obtaining an estimate of the infiltration capacities to be expected for a project site. Soil types may be identified from available soil maps or by field reconnaissance supplemented by hand auger borings to determine the soil profile. Individual studies at potential recharge sites are recommended as long-time infiltration capacities are site-specific.

After the applied recharge water has passed through the soil zone, geologic and subsurface hydrologic conditions control the sustained infiltration capacity because layers of less permeability become the controlling hydraulic resistance in the profile and override the initial surface restriction state. Subsurface conditions should be thoroughly investigated to determine the total useable storage capacity and the rate of water movement from the spreading grounds to the areas of ground water draft.

IX. GROUND WATER QUALITY

The generally low velocity of ground water movement, tending to inhibit mixing and diffusion in a ground water basin, is one of the principal factors influencing ground water quality. Impervious materials form localized barriers to ground water movement and prevent uniform dispersion of water throughout the aquifer. As a result, considerable variation in water quality under natural conditions can be expected to occur in the ground water reservoir.

Probably the most significant characteristic of ground water quality deterioration is the fact that it is usually a cumulative, insidious process that is easier and much less costly to prevent than to correct once pollutant concentrations have reached problem proportions. It is particularly important, therefore, that ground water quality not be allowed to deteriorate to the point where use begins to be adversely affected. In many cases, maintenance of water quality becomes a limiting consideration in ground water basin operation. History has shown that many ground water basins suffer quality deterioration over a period of time. Therefore, the prudent manager should have standby plans on how to deal with degradation from various natural and human sources.

A. Water Quality Requirements

Major factors in considering the suitability of a water supply are water-quality requirements and limitations associated with its contemplated uses (Durfor and Becker, 1964; Hem, 1959; Keith, 1976).

Various criteria have been developed covering all categories of water quality, including bacterial content (Allen and Geldreich, 1975), physical characteristics, and chemical constituents. Normally, water-quality problems of bacterial content or physical characteristics can be alleviated. On the other hand, the removal or neutralization of undesirable chemical characteristics is often both difficult and expensive (Cheh et al., 1980; Costle, 1979). In many instances, the chemical content constitutes a major limitation on the utilization of ground water (Craun et al., 1976; Crawford, 1972).

B. Waste Disposal Considerations

Many ground water quality problems stem from the fact that people live and conduct most of their activities on lands overlying ground water basins. Inevitably, much of the waste that results from these activities finds its way to the ground water below, commingling with and impairing the quality of ground waters (U.S. Environmental Protection Agency, 1977). Because of the potential harm due to the various contaminants entering ground water, and because of the great difficulty in correcting a polluted ground water condition, there is great national interest in the topic of ground water contamination and its control (Nielsen, 1983; U.S. Public Health Service, 1961). Evaluation of the effects of a waste discharge on ground water quality is unique to each case. Factors include the local hydrogeological conditions, type and volume (mass) of contaminant, its reactions with bacteria and with the soil media, and the hydraulic conditions governing the velocity, plume shape, and dispersion of the ground water.

Ground water pollution may be defined as the artificially induced degradation of natural ground water quality. The term "contamination" may be generally synonymous with the term "pollution," though it may have a more specific connotation tied to the intended use of the water.

The travel of various pollutants depends upon the physical and chemical characteristics of the pollutants and of the aquifer systems. The polluting substances may be particulates, e.g., bacteria and viruses, or chemical, e.g., various dissolved substances and organic molecules. Particulates usually are removed within a short distance of travel by the kinds of factors that are active in the filtration process, e.g., straining and adsorption. In gravel material, these particulate substances may travel far; however, removal may be virtually complete in sandy loam or clay soils due to filtering and adsorption. Dissolved chemical substances, on the other hand, are not filtered. Some may interact with bacteria (Alexander, 1981; McKenna, 1977), and others may interact chemically with clay fractions in the soil. Nitrates may be reduced by anaerobic bacteria. Organic molecules are subject to adsorption, particularly on clay particles, as well as biodegradation. Cations are subject to ion change by the clays. Most anions, on the other hand, such as chlorides and nitrates, are not subject to any change whatsoever. There is no interaction with any soil fraction.

The travel of pollution through ground water aquifers is a function of both the hydraulic characteristics of the flow, e.g., convection patterns and dispersion, and the chemical interactions with the soil systems and aquifer materials. For some kinds of interaction the aquifer has a limited capacity. For example, when all of the adsorption sites are occupied, or when the cationic change capacity is exhausted and the aquifer material is essentially at equilibrium with the chemicals within the ground water solution, no further interaction will take place. The pollution will then move down-gradient past these occupied sites to areas where the equilibrium condition does not exist. If the pollution is emitted as a slug instead of continuously, desorption may occur as uncontaminated water reaches the sites previously occupied. The resulting desorption will cause a contaminant wave to travel downstream, but its velocity will be less than that of the ground water.

C. Other Water Quality Factors

Ground water quality is subject to deterioration from many sources in addition to the effects from waste disposal. Soluble salts in the watershed surface and underground formations as well as faulty well construction are among the many widely divergent potential causes of degradation. Consideration must be given to all appropriate factors in evaluating problems involved in the maintenance of ground water quality.

1. Watershed and Aquifer Influences Water is the universal solvent. Consequently, it is continually subject to degradation as it moves across or through the soil mantle and aquifer materials. Soluble minerals are acquired in ever-increasing concentrations from the formations with which water comes in contact. The quantity and characteristics of these dissolved minerals are influenced largely by the composition of formations encountered, the duration of contact, and the magnitude of flow.

The waters of the San Joaquin Valley, California, illustrate the influence of these factors. The Sierra Nevada Range, to the east of the valley, is composed of relatively insoluble crystalline granitic rock overlain by equally resistant volcanics and metamorphics. Runoff from the abundant precipitation on mountain slopes is calcium bicarbonate in type, with total dissolved mineral content generally less than 100 mg/l. The underground basins on the east side of the valley, replenished by these surface flows, contain correspondingly high quality ground waters. Normal mineral content of ground water ranges between 150 mg/l and 350 mg/l, with calcium bicarbonate predominating. The west side of the San Joaquin Valley, on the other hand, is bordered by the Coast Range, composed largely of marine sedimentary formations. Runoff is meager, and drainage from these easily eroded sediments is poor in quality. The mineral content of surface water ranges from 200 mg/l to over 6,000 mg/l, largely composed of sulfates and chlorides. Ground waters on the west side of the valley are likewise highly mineralized, of poor quality, and unsuitable for many beneficial uses.

Depending upon the specific composition of water chemistry and rock mineral type, there is a great degree of variability in the possible final chemical equilibrium state of the rock/water system. Important factors in the development of ground water chemistry include: total dissolved solids, microorganisms, water residence time, recharge, evaporation, temperature, pressure, pH, and oxidation reduction potential.

Another condition that may impact water quality is the radioactive element composition of the rock. Much recent attention has been focused on rock uranium endowment owing to the related problems with uranium progeny (radon, for example). The decay of radioactive elements can yield progeny which are more easily dissolved into the water phase than the parent material. This may then result in attendant water quality concerns.

2. Connate Waters Many water-bearing formations that were deposited in marine environments contain entrapped saline waters (connates) underlying or adjoining fresh ground water bodies. Under equilibrium conditions, fresh water floats on these connates with little or no intermingling. Pumping of ground waters, with resultant lowering of ground water levels, tends to disturb this balance and may induce movement of saline water into the fresh water–bearing formation; however, the fresh waters can usually be pumped without appreciable degradation of the remaining supply if precautions are taken to avoid overly deep wells and to avoid heavy concentrations of wells that tend to create excessive drawdowns or "pumping holes." Connates in formations contiguous to developed basins can be kept from intruding the normally good quality water by maintaining, where possible, a greater hydraulic head in the fresh water body than that which exists in the connate aquifer at the point of contact. Where this is not possible, restrictions in use of the ground water basin must be accepted, or physical solutions similar to those used to counteract sea water intrusion must be used.

Saline waters that issue from faults or fissures as highly mineralized springs and seeps can percolate to fresh ground waters and thus may be a source of degradation. These flows are usually of small magnitude, and their effects are generally localized. From an aquifer management standpoint, such sources should be sealed or the waters conveyed across infiltration areas in impervious drainage channels.

3. Water Wells Every water well, active or abandoned, is a potential channel for direct introduction of pollutants into a water-bearing formation. The use of wells for disposal of sewage or industrial waste obviously invites water quality problems. To a lesser degree, the entry of surface drainage through improperly sealed or abandoned wells can contribute to deterioration of ground water quality.

Water wells can be instrumental in inducing widespread degradation in a ground water reservoir where the water-bearing formation consists of a series of aquifers, separated by impermeable strata, containing waters of differing quality. Interconnection of aquifers can result where improperly constructed, defective, or abandoned wells puncture these impermeable seals. If hydrostatic pressure differentials exist between the aquifers, ground waters under the higher head will tend to move into the formation containing waters under lower head wherever the separating stratum is punctured and hydraulic continuity is established. In this manner, ground waters of inferior quality can intrude and degrade waters in another aquifer. Each well at which this interchange occurs becomes a focal point from which degradation may spread in all directions, although the greatest effect will usually occur in the direction of ground water movement.

The mode of well construction has a decided effect on the possibility of quality deterioration. Gravel-packed wells completed in multiple aquifers by their very nature become passageways between aquifers unless special precautions are taken to provide impermeable seals at appropriate points in the gravel envelope.

Abandoned water, oil, or natural gas wells, if not properly sealed, can create even more serious problems. Merely filling an abandoned hole indiscriminately with trash, drill cuttings, rocks, or rubble only serves to conceal a potential source of degradation. The continuity of impervious strata between aquifers must be restored by using a positive sealant, such as bentonite clay or cement grout, to assure that there can be no interconnection of aquifers through the bore. This is a necessary precaution, even for wells perforated in only one aquifer but penetrating other aquifers, because casings will eventually deteriorate and thus permit interconnection of ground water bodies.

4. Sea Water Intrusion Of all the water on earth, less than 4% is usable fresh water, including the 0.1% that exists as water vapor in the atmosphere. In the quest for fresh ground water, the vastly more abundant saline water is encountered in many places, both along the ocean shore and in regions several kilometers inland. Saline water is also found in the vicinity of inland saline lakes and in rocks that are saturated with unusable water, whether briny or merely brackish. Due to the higher density, saline waters are generally found at greater depth than fresh waters, but in some places saline water is encountered in rocks overlying fresh aquifers.

The principles of occurrence and movement of water in porous media apply to saline as well as fresh water. Strictly speaking, only the salt water that contaminates fresh water reservoirs as a result of human activity is saline intrusion or encroachment. (The same terms are also used loosely to apply to saline water found to exist where it is not expected, even though the human factor is not proved.)

Wherever a fresh water aquifer is in hydraulic connection with the ocean or other large body of saline water, the possibility exists that such water may invade the fresh water body (California Department of Water Resources, 1958; Cooper, 1964; Henry, 1959; Lee and Cheng, 1974; Newport, 1975). Sea water, because of its greater density, tends to move inland under the fresh water body and a fresh water–salt water interface is created. This interface occurs at a depth below sea level generally about forty times the height of the fresh ground water table above sea level, based on sea water specific gravity of 1.025. This is the Ghyben–Herzberg principle. The resulting hydrostatic equilibrium between two fluids of different densities is expressed as follows:

$$h_s = \frac{P_f}{P_s - P_f}(h_f) \tag{5.12}$$

where h_s is the depth of fresh water below sea level, P_f is the density of the fresh water, P_s is the density of sea water, and h_f is the height of the fresh water table above sea level. Thus, in an island comprised of uniformly permeable rocks, the fresh ground water reservoir would resemble an iceberg as it floats on the salt water with most of its bulk below sea level; but the proportion below sea level, on the basis of the average densities of ocean water and fresh water, would be about 40:1. This type of ground water occurrence is often called a fresh water lens.

M.K. Hubbert (1940) pointed out that the Ghyben–Herzberg relation constitutes a static equilibrium existing only in the natural state, and that the flow of ground water toward the sea creates a dynamic equilibrium, under which the depth of the fresh water–salt water interface near the ocean is greater than that indicated by the Ghyben–Herzberg relation. The prevalence of anisotropy in most sedimentary rocks allows fresh water wedges to extend far beneath seashore boundaries.

Field data from two areas where an exceedingly permeable aquifer is saturated by fresh water overlying salt water indicate that the interface is not sharp. In the Pearl Harbor area near Honolulu, Hawaii, there was a zone of diffusion with a vertical thickness of 300 m or more, and in the vicinity of Miami, Florida, a zone of diffusion had a horizontal breadth of about 600 m at the base of the aquifer. In both areas, springs discharge salty waters from the zone of diffusion. H.H. Cooper (1959) states that where a zone of diffusion exists, "the salt water is not static but flows perpetually in a cycle from the floor of the sea into the zone of diffusion and back to the sea, and this flow tends to lessen the extent to which the salt water occupies the aquifer."

In several recent research efforts (Dagan and Bear, 1968; Liu et al., 1983), numerical modeling techniques were used to investigate the flow and salinity transport in coastal aquifers. It was concluded that the evolution of the zone of diffusion may be more important in the management of coastal aquifers than the traditional sustainable yield approach.

The successful development of ground water in a Ghyben–Herzberg lens system hinges largely on meticulous control of pumping to avoid creating a condition of persisting deep drawdown of the water table (Chandler and McWorter, 1975; Dagan and Bear, 1968; Das Gupta, 1983; Sahni, 1972; Schmorak and Mercado, 1969). The encroachment of sea water can be largely avoided by utilization of judiciously located shallow wells, infiltration galleries, or tunnels, where drawdowns are controlled.

A modified Ghyben–Herzberg relationship may exist where an artesian aquifer is in hydraulic continuity with the ocean (Collins and Gelhar, 1971; Rumer and Shian, 1968). Under natural conditions, the ground water may be under considerable hydrostatic pressure generated from a remote inland forebay area, with potentiometric levels well above sea level. The fresh water head will usually be adequate to limit the intrusion of sea water to a relatively short distance inland from the point of contact. When the ground water basin is developed and fresh water levels drop, the sea water has a tendency to move landward, displacing the fresh water in the aquifer. A landward sloping fresh water–salt water interface is formed, with a depth governed by the Ghyben–Herzberg relationship. As water levels continue to decline, the prism of sea water, which has been designated the sea water wedge, continues to advance until conditions of equilibrium are reestablished (Baumann, 1957; Bear and Dagan, 1962a; Bear and Dagan, 1962b; Bear and Dagan, 1964; California Department of Water Resources, 1960b; Mercer et al., 1980; Pinder and Cooper, 1970; Segol et al., 1975; Shamir and Dagan, 1971; University of California, 1955).

Extensive research, coordinated with the actual operation of a system of works to minimize or repel saline intrusion, has been conducted in southern California (California Department of Water Resources, 1957, 1958, 1960a). The results of this work indicate the feasibility of maintaining a barrier to saline intrusion by the injection of fresh water through wells located along, or landward from, a fresh water–salt water interface. Similar work has been performed in Florida (Kashef, 1976) and Hawaii (Larson et al., 1977). Also in Florida, salinity control structures, dams with vertically moving gates, have been constructed on most sea level canals on the southeast coast to successfully control salt water encroachment from the canals into the surficial, highly permeable Biscayne Aquifer (Kohout, 1960). At the same time, the canals transport fresh water from inland areas to replenish the coastal part of the aquifer.

Another approach to control sea water intrusion involves the physical severance of the hydraulic connection with the ocean. In effect, a subsurface dam might be formed by reducing the permeability in a section of the aquifer, thereby blocking ground water movement. This dam could be accomplished by actually constructing a cutoff wall using sheet piling, cement grout, bentonite clay, and so forth. An impervious zone might also be created by injecting emulsified asphalt, silica gel, calcium acrylate, various plastics, or similar materials into the dam.

In an artesian aquifer where a sea water wedge has developed, sea water can move inland generally as far as a landward hydraulic gradient exists in the ground water basin. No matter how low the ground water levels, normally no progressive landward ground water movement occurs where the water table or potentiometric surface slopes seaward. Thus, a ridge or trough in the ground water table or potentiometric surface near the coast will effectively limit the inland movement of sea water. Such a protective trough can be developed by means of a line of pumping wells, properly spaced along the coast (California Department of Water Resources, 1969).

Salt water intrusion can be reversed by concentrated recharge of the aquifer that creates a ground water mound or ridge near the coast. A series of spreading grounds or injection wells, or a combination of both, can be utilized as dictated by the geologic conditions encountered. Little waste of fresh water would result if the height of the ridge were carefully controlled and kept as nearly as possible at the minimum elevation required to protect the aquifer. Most of the fresh water utilized to maintain the ridge would also tend to augment recharge to the ground water basin and be available for reuse. The feasibility of establishing and maintaining a ground water ridge in a pressure aquifer by injecting water through wells has been demonstrated in Los Angeles and Orange Counties in California, and in other locations as well (California Department of Public Works, 1951; California Department of Water Resources, 1966b; Kashef, 1976; Larson et al., 1977).

Each of the aforementioned methods for control of sea water intrusion has certain characteristics that affect ground water basin operation. Maintenance of water levels above sea level would reduce pumping costs; however, such maintenance would severely restrict the utilization of ground water storage in coastal aquifers, many of which are hundreds of meters thick and extend far below sea level. This could tend to limit natural ground water recharge, both from subsurface inflow and percolation of surface runoff, as available storage might not be adequate for full conservation of local water resources. The pumping trough overcomes these disadvantages to some extent, as water levels on the landward side can be maintained below sea level; however, the depth of the trough limits the elevation to which the water table can be lowered, and thus limits the usable ground water storage. Both the ground water ridge and subsurface barrier would permit full utilization of underground storage, inasmuch as inland ground water levels could be maintained lower than sea level.

Many other factors are involved in evaluating the feasibility of protecting a ground water basin from sea water intrusion. These include the capital and annual costs of physical works required (such as wells, supply and distribution lines, subsurface barriers, and so on), the availability and cost of supplemental water supplies, water rights and other legal considerations, drainage and salt balance requirements, potentialities for conservation of flood runoff and wastewaters, and the value of the storage capacity involved. All of these factors must be considered in determining the most advantageous method of sea water intrusion control.

5. Salt Balance Considerations Large quantities of chemicals, in the form of mineral salts, can be carried into an area by tributary inflow, both surface and subsurface, and as a result of human activities. Simultaneously, salts are usually removed from an area in outflowing waters. The relation of salt outflow to salt input is termed "salt balance." An adverse salt balance condition, or one in which inflow exceeds outflow, is manifestly undesirable as it indicates a possible progressive accumulation of salts in the area.

Salts percolated to the saturated zone are not solely attributable to the salinity of water supply to a basin (California Department of Water Resources, 1967). Other sources include: 1) salts leached from fertilizers and manure, 2) agricultural wastewater whose salinity has been raised by evaporation and transpiration, 3) municipal waste whose salinity has been increased by a few hundred parts per million due to domestic and commercial uses and recharge of ion-exchange water softeners, and 4) streamflow, precipitation, and imported recharge water. Further, phreatophytes may increase ground water salinity by extracting and transpiring water without effecting a proportionate removal of salts from the ground water.

In irrigated areas, where soils are tight and water supplies limited, most of these salts are retained in the soil horizon. Some salts may be absorbed on soil particles, and some may form insoluble compounds that precipitate from the soil water; however, large quantities usually remain in the soil water solution, which, due to evapotranspiration stress and dissolution of soil minerals, can become concentrated to levels harmful to vegetative growth (American Society of Civil Engineers, 1990). In pervious soils, where sufficient water is available for leaching purposes, accumulation of salts in the soil horizon can be avoided; however, the salts leached from the soil may be carried down to ground water and may accumulate there, ultimately rendering the ground waters unusable. Under such conditions, maintaining of a favorable salt balance can become a necessary prerequisite to successfully continue the operation of a ground water reservoir.

The quantity of salt directly removed from an area by consumptive use is negligible. In the transpiration process through which water is consumed by vegetation, only an insignificant quantity of the dissolved minerals in the applied water is utilized. The principal medium for disposal of salts is outflowing water in the form of surface runoff, drainage, exported water, subsurface outflow, or exported sewage. The portion of the effluent flow consisting of rising water or subsurface outflow can be profoundly influenced by prevailing water levels in a ground water reservoir. Where levels are depressed for long periods of time, this outflow may be reduced below the amount necessary for adequate salt disposal. A major consideration in operation of a ground water reservoir is the level at which the water table must be kept to induce sufficient outflow for maintenance of favorable salt balance conditions. Where it is undesirable to maintain such ground water levels, it may be necessary to resort to pumping of poor quality ground water for export from the area to assure adequate outflow.

The salt concentration in water tends to increase with each cycle of use to which the water is subjected. In a system of interconnected ground water reservoirs, where waters from upgradient ground water basins are utilized and become sources of recharge to downgradient ground water basins, the extent to which ground waters can be reused depends to a large degree on the quality of the original water source and the rate at which dissolved minerals are added with each reuse. This reuse factor may be a major consideration in evaluating the relative economics of water supplies that differ in mineral content, particularly where high-cost import supplies are involved.

Concurrent problems involved in the study of salt balance are: 1) the salts involved in both inflow and outflow should be chemically similar and occur in the same relative proportions, otherwise there is a possibility of excessive concentration of a single salt in the basin, and 2) determining and identifying the degradation of ground water stemming from adverse salt balance in overdrawn basins is difficult. Long-term records of mineral analyses of water from a network of observation wells, combined with similar data for surface water inflow and outflow, are vital to the solution of these and similar problems.

D. Water Quality Monitoring

In general, a ground water monitoring program should be used to establish the variability of background water quality and then to indicate potential problems so that timely corrective actions can be taken. The monitoring program would then be needed to evaluate the effectiveness of any remediation measures.

The collection of basic data, though important, should not be an end to itself, but should be an integral part of a comprehensive, balanced program of continuing examination and interpretation of all factors involved in protection and maintenance of ground water quality. Constant and homogenous data over an extended period of time are necessary to determine the source of the contaminant and to aid in repairing any damage to the ground water basin.

In formulating an effective plan for control and maintenance of ground water quality, full consideration must be given to geology, hydrology, and cultural development within the area of the ground water reservoir. Water quality characteristics must be established and trends observed. A comprehensive investigative program for evaluation of these factors is essential. Ground waters should be sampled and analyzed regularly. Sampling points should include shallow and deep wells selected with due consideration to areal distribution, geologic and hydrologic influences, pattern of pumpage, and waste disposal practices. Samples should be obtained when water levels are at their lowest and also, if possible, during the period of replenishment when water levels are high.

In areas vulnerable to sea water intrusion, particular attention should be given to changes in chloride concentration. Total dissolved solids content is significant where salt balance problems exist. Volatile and dissolved organics, phenols, boron, and heavy metals are usually critical where large volumes of industrial waste are involved. Domestic sewage disposal particularly influences nitrate, phosphate, and detergent content. Waste disposal is not the only source of contaminants. Storage and handling of various products inevitably include spills and leaks, some of which may go unnoticed for an extended period. Thus, ground water quality observations, in order to be properly evaluated and interpreted, must be augmented with adequate information on geology, hydrology, rate and pattern of ground water extractions, and waste-disposal practices.

X. GROUND WATER MODELS

Historically, many types of models have been used for simulating ground water distribution and flow conditions in water-bearing materials. The first simulation attempts used analog models of physical systems such as sand boxes, electrical conductivity sheets, and resistance–capacitance network models. More recently, digital computers have become highly advanced and numerical analysis techniques have been developed that take full advantage of their capabilities. With these newly available tools, mathematical ground water basin models and their application in ground water management have become commonplace. Analytical and numerical modeling techniques, such as finite difference and finite element techniques, are routinely applied to localized and comprehensive ground water basin management problems.

The easy accessibility of current computerized ground water models led to their early application in the overall ground water system investigation. In the first stages of their application, ground water models were used only in basins where the geologic structure was well-defined and the hydrology was well understood. Now, mathematical ground water models are often applied in basins where the aquifer structure and hydrology are only approximately known. With these applications, as in all ground water modeling situations, the model itself is used as a tool to guide ground water analysts in their investigations. Once a rudimentary ground water model has been developed, it quickly reveals needs for additional field analysis, data collection, and geologic and hydrologic studies to build an adequate data base. The accuracy of model predictions is a function of the size and quality of the data base of aquifer characteristics. One of the benefits derived from constructing a mathematical model is an improved appreciation of the relative importance of the various elements of hydrologic and geologic data. Hydrogeologic studies provide the geometry and hydraulic characteristics of the ground water basin. A hydrologic study provides detailed water balance information for use in the verification of a model.

A. Model Formulation

The basis for all ground water flow models is a generalized equation that relates storativity, hydraulic conductivity, and water inflow–outflow characteristics. Combining Darcy's law and the law of conservation of mass gives

$$\frac{\partial}{\partial x}(K_x \frac{\partial h}{\partial x}) + \frac{\partial}{\partial y}(K_y \frac{\partial h}{\partial y}) + \frac{\partial}{\partial z}(K_z \frac{\partial h}{\partial z}) = S_s \frac{\partial h}{\partial t} + Q_n \qquad (5.13)$$

in which h is the hydraulic head, K_x, K_y, K_z are hydraulic conductivity in the x, y, and z directions, respectively, S_s is the specific storage, and Q_n is the net volume flux.

Equation 5.13 is applicable to ground water flow in three dimensions and is the basis for many simplified equations that portray specific ground water systems. Besides Equation 5.13, any ground water model application involves boundary conditions that specify flow conditions at the boundary of the study area. These boundary conditions are either: 1) constant head boundaries, 2) constant flow boundaries, or 3) boundaries where flow is a function of the difference between some external head and the head in the ground water basin.

One of the most common of the simplifications that can be made to Equation 5.13 is for an idealized, confined single aquifer system in which the vertical (z) flow terms are dropped from the equation. This simplification requires that several assumptions apply to the aquifer, the most important of them being that vertical variations in aquifer properties and head are negligible, and that the density of water is relatively constant in space.

B. Solution Techniques

The first ground water models were analogs of simple ground water systems. For instance, sand tanks were used to portray ground water movement in unconfined conditions. More complex ground water situations could be simulated by analog electrical resistor networks or analog computers (Walton and Prickett, 1963). While these analog models were useful, they were physically and economically restricted to small-scale, idealized ground water problems. They also lacked flexibility of use and were difficult to adapt from one aquifer to another.

With the widespread availability of digital computers, large-scale ground water problems are now easily solvable (Mercer and Fause, 1981). Early use of digital computers required extensive computer programming to make a ground water model functional. As ground water modeling techniques have advanced, many different solution strategies have been developed to solve the systems of mathematical equations, and libraries of successfully applied computer programs are available. An inventory of the computer programs available to ground water modelers is presented in the American Geophysical Union Water Resources Monograph, "Ground Water Management: The Use of Numerical Models" (Bachmat et al., 1980). The International Ground Water Modeling Center, located at the Colorado School of Mines, Golden, Colorado, maintains a current directory of ground water models.

Of the numerical modeling techniques that are available, the two most commonly used are the finite difference (Gray, 1982; Trescott et al., 1976; Anderson and Wang, 1982) and finite element methods (Lewis, 1982; Zienkiewiez, 1977). Both these techniques will be generally described in the following paragraphs, along with the analog technique for model solution.

1. *Finite Difference Method* The finite difference method derives its name from the manner in which the method solves a system of equations. The ground water basin is divided into a mesh of polygons, often rectangular. Nodes are then defined either in the center of the blocks formed by the grid lines (see Fig. 5.8(a), block centered) or at the intersections of the lines (see Fig. 5.8(b), mesh centered). Within any block or at any node, the ground water head and all material properties are assumed to be the same. Flows between nodes can then be computed using Darcy's law and the gradient defined as the difference in heads between neighboring nodes. With these assumptions, a system of equations can be constructed, with one equation for each node. Once the equations are set up, they may be solved using any one of the many available matrix inversion subroutines.

There are many variations of the general finite difference description just given. As stated previously, nodes in the network may be specified either at the center of blocks or at the intersection of grid lines. Block-centered nodes are most suitable when boundary conditions are mainly specified head values.

Besides these two versions of the finite difference method, there is a third major variation of the technique. In this variation, known as the integrated finite difference method, nodes are centered in the middle of polygons and flows are assumed to occur only perpendicular to the boundaries of the polygons. The integrated finite difference method was one of the first developed for digital computers and has been applied extensively.

Finite difference models are probably the most popular method of ground water simulation. This is because they are relatively easy to understand and to code into usable computer programs. A very simple

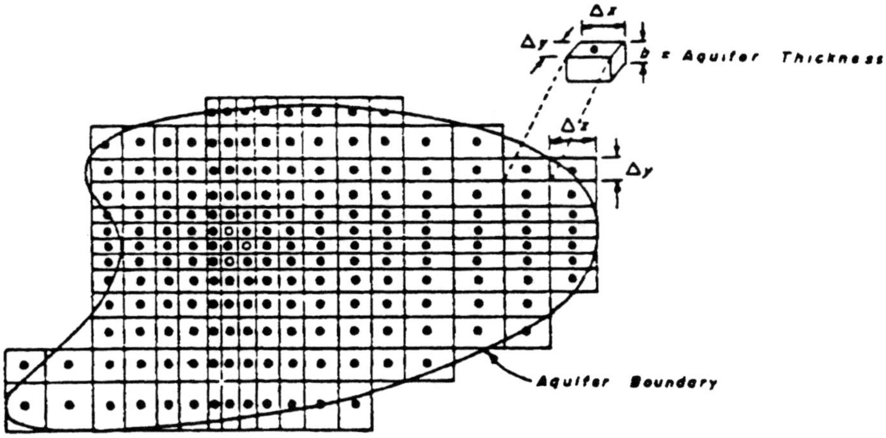

(a.) FINITE DIFFERENCE GRID WITH BLOCK CENTERED NODES.

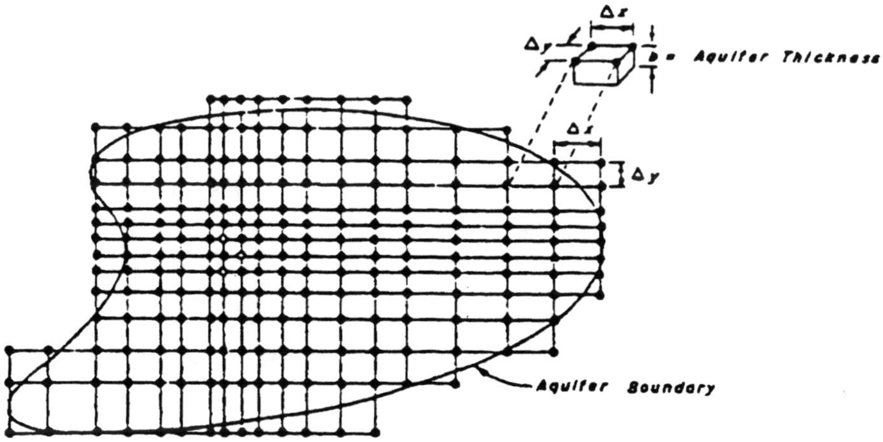

(b.) FINITE DIFFERENCE GRID WITH MESH CENTERED NODES.

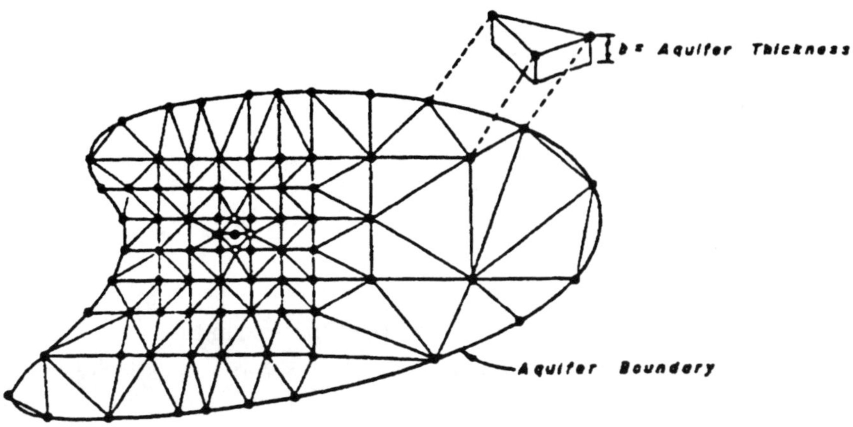

(c.) FINITE ELEMENT MESH WITH TRIANGULAR ELEMENTS.

Figure 5.8—Two-Dimensional Finite Difference and Finite Element Configurations for an Aquifer Study.

finite difference model may require only two pages of computer coding and can be applied to large-scale simulation. Because of this relative simplicity, the finite difference method is also well-adapted to use on microcomputers with limited memory.

2. Finite Element Method The finite element method is a generalized procedure for solving differential equations describing physical systems. The method proceeds by dividing a ground water system into a mesh of triangular, quadrilateral, or other shapes (see Fig. 5.8(c)), each of which is called an element. The ground water surface (head) in each of these elements is determined by the values of the heads at the nodes (corners) of the elements.

Mathematically, the finite element technique uses polynomial functions, called basis functions. Several basis functions are used at each node, one for each element that touches the node. These basic functions may be linear or quadratic, and by combining the basic functions for all the nodes bounding an element, a continuous head surface is formed. The finite element procedure uses the basis functions to integrate the effects of areal net recharge and storage change and develops a solvable system of equations with the unknowns being ground water heads at the nodes. Once this set of equations is formed, standard matrix solution techniques are employed to derive the solution.

In general, the finite element method requires more complicated computer programming and more computer run time than the finite difference method. Its advantage lies in its triangular and quadrilateral elements which can be used to accurately emulate irregular geographic and geologic features, such as recharging rivers, faults, aquifer boundaries, and water agency boundaries. The finite element model can be very detailed in areas of interest and relatively general in other areas where ground water hydrology and geology is more uniform. Because of this variation, a well-designed finite element model may have significantly fewer elements than a comparable finite difference model. This advantage makes the finite element method potentially more efficient in basins where geographic patterns of ground water hydrology and geology are irregular.

3. Analog Techniques Long before digital computers were available, many physical systems were simulated using physical analogs. The first attempt to simulate ground water storage and movement is credited to Pavlovskii in 1918. Early use of electrical analog simulation techniques resulted from similarity between Darcy's law and Ohm's law. Darcy's law can be written as

$$V = -K\frac{dh}{ds} \tag{5.14}$$

in which V is the velocity of flow for a specific discharge, K is the hydraulic conductivity, h is the total head, and s is the distance along the direction of flow.

Ohm's law can be expressed as

$$i = y\frac{dv}{dx} \tag{5.15}$$

in which i is the electric current, y is the conductivity per unit length of conductor in the direction of flow, v is the electric potential or voltage, and x is the length of conductor in the direction of current flow. Thus, the flow of ground water resulting from gravitational forces can be expressed by the same differential equation that describes current flow in electrodynamic systems. This similarity has led to several types of electric analog models. These models include conductive liquid analogs, conductive solid sheet analogs, resistance–capacitance network analogs, and complex analog computer models.

The advantage of analog techniques is that they force the modeler to physically create the model. This construction process may ultimately provide the modeler with a better sense of the physical system being simulated. The disadvantages are that a real-life analog ground water model is bulky, expensive to build, difficult to modify, and incapable of simulating common situations such as transient multilayer leaky aquifer systems. With the ready availability of digital computers and already developed programs, analog techniques are not widely used at this time.

C. Model Development

A basin-wide model will normally involve the application of a general-purpose ground water program that solves the fundamental mathematical equations relating the physical parameters of the basin. The following steps are typical of those followed by ground water analysts in creating a specific basin ground water model:

(a) Subdivide the ground water basin into a mesh of nodes or elements. (Nodes are used for finite difference models and elements are used in finite element models; they both serve the same general purpose.) The type of mesh network is dictated by the physical characteristics of the basin, the ultimate use of the model, and the solution technique employed. A rectilinear pattern is commonly used for the finite difference solution technique, while triangular- or quadrilateral-shaped elements are used for the finite element approach (see Figs. 5.8(a), (b), and (c)).

(b) Analyze geologic and well pumping data and determine transmissivity and storativity variations in the aquifer. Usually this information is needed for each node, grid point, or element. Also, the data are used to identify and characterize the boundary conditions.

(c) Analyze historical surface hydrologic data and determine the net deep percolation for a given time increment for each nodal area for the study period, preferably 10 years or more. This step may be a computer subprogram to facilitate changes in the data base and recomputation of input to the model.

(d) Tabulate historical potentiometric heads for each nodal area during the study period. These historical water level records, such as hydrographs or maps, are based on recorded measurements of ground water level elevations.

1. *Determination of Number and Locale of Nodes or Nodal Areas* The locations of the nodes and size of the nodal areas should portray variations in replenishment, transmission, storage, and need for more detailed definition of ground water levels in a local area. To provide better overall accuracy and stability in computation, the nodes should be more dense and nodal areas smaller where the gradient (rate of change) in water level elevations changes rapidly in time or space.

2. *Determination of Transmissivity and Storativity Factors* Results from hydrogeologic investigation of the basin are used to estimate transmission and storage factors. Drillers' logs of water wells and electric logs of oil wells located throughout the area are analyzed to delineate the water-producing zones. Estimates of the hydraulic properties of these zones may be obtained by means of aquifer pumping tests, specific capacity tests, and geologic logs of wells (Chow, 1952; Walton, 1962).

Storativity and transmissivity values are assigned to all water-bearing sediments. These values are the first approximation of the storativity and transmissivity used in the basin model. Both the probable range of values and the best approximation should be determined.

One procedure for obtaining the transmissivity factor that affects the subsurface flow through the aquifer is outlined as follows:

(a) Prepare maps showing lines of equal transmissivity for the saturated water-bearing sections of the aquifer.

(b) From these maps, obtain appropriate values of transmissivity in the water-bearing material in the region of each of the model's nodes.

(c) In a highly faulted basin an alternative procedure would be to compute transmissivities separately for each node or element.

There may be a negligible effect on the water level responses of using a uniform storativity factor for the basin set within reasonable limits rather than a factor that varies from node to node. Tests on two models in California, in the Los Angeles Coastal Plain and San Gabriel Valley, have shown that the difference in water level responses between uniform versus nonuniform storativity factors was on the order of 50 mm or so (California Department of Water Resources, 1966a). These small differences probably occurred because the time frame was large; where seasonal or daily fluctuations in water level are of importance in the analysis, the storativity factor becomes more important because net flow through the system is more a function of changes in aquifer water storage. Even for the long term, a 50 mm difference may be important. Because the number of ground water equations employed in this process is less than the number of unknowns, several sets of model parameters can produce the same aquifer heads. When the resultant model is used to project into the future, the variations in storativity may become important. Therefore, the use of the most realistic variations in storativity from node to node may become important.

3. Preparation of Historical Net Deep Percolation and Water Level Elevation Data In developing the model, annual quantities of net deep percolation for each node or nodal area are needed for the entire study period. The following approach is one that has been used successfully.

(a) Estimate a factor expressing the quantity of deep percolation of precipitation as a percentage of the total precipitation in the basin. Multiply this factor by the precipitation estimated at each node or element using the Thiessen polygon method.
(b) Estimate a factor expressing the quantity of deep percolation of delivered water as a percentage of water delivered to the basin for the various types of land use. Then determine the amount of delivered water for each node or element on the basis of land use. Multiply these delivered water quantities by the percolation factor to obtain percolation of delivered water for each basin subarea. In a more detailed investigation, a soil-moisture inventory may be made of the root zone that takes into account precipitation, applied water, crop or urban development, and soil parameters.
(c) Obtain annual quantities of water recharged in each node or element from artificial recharge project records.
(d) Determine total quantities of stream recharge for each reach between existing gaging stations, estimate by extrapolation from measured flows on adjacent streams, or estimate by averaging numerous values of point measurements of infiltration over an area. Apportion this recharge to each node or element in direct proportion to the length of the river channel traversing each nodal area.
(e) Add the quantities derived in steps a through d to obtain total deep percolation to the zone of saturation for the study period.
(f) Determine ground water extractions from each nodal area.
(g) Estimate net subsurface inflow into or outflow out of the modeled ground water basin for nodes or elements on the periphery of the model.

Historical water well hydrographs and contour maps of equal ground water elevation for the test period are used to develop representative ground water level elevations for each node or element in the model. These representative nodal water level elevations are used both to begin the simulation and for comparison during the verification of the model.

D. Family of Models Technique

In large basins, a family of models approach can be used. The first model would be a somewhat generalized approach, with the basin broken into a few areas and with parameters and boundaries being first approximations. Using this model provides the investigator a feeling for the system, giving more insight into which parameters are important and where to put resources into improving the model inputs. Successive models would subdivide the basins into smaller areas, define the inputs and parameters, and lead to a final, more sophisticated model.

E. Hydraulic Model Calibration and Verification

In order to use a ground water model for future predictions, evidence must be developed to demonstrate its ability to reasonably simulate the historic past.

The purpose of the historic hydrologic study is to provide the net deep percolation information for calibration and verification of the ground water model as well as to evaluate the aquifer hydraulic characteristics and boundary conditions which were used. Water levels within a ground water system respond to outside stimuli such as recharge and pumping. The response of this system to the stimuli is shown in the water level histories of the system. If the ground water model is a good simulation of the prototype, then the model should be able to reproduce historic water level conditions.

The general procedure in calibration is to estimate the range of ground water flow parameters (and solute transport parameters, if it is a water quality model) and then test the model using first the "best guess" values of the basin parameters to obtain system response to external stresses over time. The result is computed values of the system state variables. System response may be expressed in terms of changes in water levels, gradients, fluxes, or water quality concentrations. Then a comparison is made between the computed values and the known histories of the basin. If a good job has been done in developing the data, the initial run may result in a fairly close match; however, some adjustment of the parameters of storativity, transmissivity, or net deep percolation almost always is required. During the course of the data development, limits on the reasonable variation of the properties of the system are established. Based on hydrologic judgment, selective changes are made within these limitations and additional computer runs are made. The results are again compared to the historic water levels. This diagnostic process is repeated until a sufficiently good match exists between the computer and the historic water-level elevations. The process of adjusting the model parameters until a good fit with historical measurements is achieved is called calibration. There are some systematic calibration techniques that are being developed to reduce the trial and error process to a minimum. Continued work on these techniques will result in improved models and model use in the future. During this period of calibration and verification, it is advantageous to have the investigators who develop the data assist with the verification of the system. A tremendous insight into the dynamics of the system is gained during the verification process. Once the model has been verified to the modelers' satisfaction and within the significance of the data and the ultimate use of the model, it is ready to be used to predict the behavior of the basin under alternative plans for basin management.

F. Model Application

A calibrated and verified ground water model can be used for future predictions. For example, the drawdown of ground water levels at various distances and pumping durations for a well pumping a certain quantity of water can be presented in a set of mathematically or computer generated curves. The model also allows a comprehensive and rational evaluation of alternative management plans by studying responses of the ground water to various inputs. To make a complete economic evaluation of these plans, a system submodel using optimization techniques such as linear and dynamic programming can be built into the model (California Department of Water Resources, 1959). As a result, a management plan selected by modeling analysis would be most economic and also meet system constraints.

G. Ground Water Quality Models

In the past, water quality did not receive the attention it now does in connection with water planning. Because of pollution problems and shortages of water supplies, water quality has become a great concern and can exert a significant effect on water resources management.

A management plan in which quantity alone is considered may not be adequate, because existing and future water quality may limit the use of some waters. Furthermore, water quality problems may affect the economics of the alternative management plans to such a degree that what is supposedly the most economical plan might prove to be the most expensive because of special water treatment required. To complete economic evaluations of alternative management plans, an assessment of the magnitude of economic consequences due to changes in water quality is essential. This assessment requires determination of the magnitude of water quality changes under various alternative water management plans.

Surface water systems are comparatively simple and accessible, and many reliable investigations have been conducted for the purpose of predicting future surface water quality. On the other hand, the development of ground water quality prediction methods has progressed more slowly.

Conceptually, the prediction of water quality in ground water basins may be accomplished by developing a mathematical model that integrates all physical characteristics, surface and subsurface, of the ground water basins to simulate physical conditions. Such a model requires sufficient knowledge of the various influences on water quality, comprehensive data concerning these influences, and a computational tool.

1. Basin Salt Balance Procedure A mass balance approach may be used as a preliminary type of general water quality model. The basin "salt balance" is a form of quality model in which entire ground water basins are treated as a unit. It takes into account all salts percolated from the ground surface to the saturated zone and extractions therefrom.

The basin salt balance approach is of a general nature and should be used only for general short-term projections (less than 20 to 30 years) and small areas (less than 800 to 1,200 ha). In the case of large basins, the quality picture presented by a general salt balance evaluation is too broad for meaningful interpretation.

A basic assumption usually made in the salt balance procedure is that mineral content of a given ground water basin is equally distributed, and that any addition of salts is immediately dispersed in equal density throughout the ground water body. As this does not happen, the basin salt balance procedure does not give a precise solution for each unit of the ground water basin. It is a generalized informative procedure that indicates whether basin salinity is increasing or decreasing and whether existing problems will likely increase, or if salinity is not a problem now if it will become a problem in the future.

2. Detailed Modeling Endeavors The transport and fate of ground water pollutants are now receiving much attention in the scientific and political arenas. Presented herein is the commonly accepted ground water quality formulation; however, there is some debate as to the accuracy of the theoretical models compared to the actual movement of contaminants (National Research Council, 1984, 1990; Rai and Zachara, 1986; Kincaid and Morrey, 1984). It must be understood that when one enters the realm of ground water quality modeling, much more information is necessary over and above that required for ground water hydraulic modeling. To start, the exact water quality parameter(s) must be known and well-characterized. It is common to view a suite of chemicals as simply total dissolved solids, as in evaluation of the movement of landfill leachate. Other situations may warrant the evaluation of specific chemical species, for example, nitrate or a dissolved organic chemical. There are also instances where particles (microorganisms for example), dissolved gasses (radon, oxygen, etc.) or immiscible contaminants (gasoline, etc.) may be of interest. In each case, information regarding the characteristics of the constituent in water are necessary: 1) whether the constituent is conservative or non-conservative and 2) whether the constituent is miscible or immiscible with water. Once these questions are answered, the next important element is knowledge of how the constituent was (is) introduced into the ground water environment: 1) instantaneous release, 2) continuous release, 3) intermittent release, and 4) point source or distributed source. These represent some of the boundary conditions for the ultimate modeling endeavor.

For most ground water quality modeling efforts, the hydraulics can be performed independent of the water quality model. That is, the water quality model uses as part of its data input the potential surface and fluxes of the aquifer system for which water quality is to be modeled. These hydraulics can either be assumed, measured, or obtained from modeling endeavors themselves.

3. Model Formulation The general formulation of contaminant transport in aquifers is with the advection–dispersion equation. The equation accounts for contaminant mass continuity by incorporating all the pertinent transport, source, and sink terms for a constituent. These terms can broadly be subdivided into two classes: physical mass transport and chemical/biological mass transport. The physical mass transport terms include mass loading/pumping, advection, diffusion, and dispersion (de Marsily, 1986; Javandel et al., 1984; Kincaid et al., 1984; National Research Council, 1990; Thomann and Mueller, 1987; Walton, 1985). Mass loading/pumping is the source or sink of the contaminant of interest. Advection is the movement of the contaminant due to the ground water flow itself. Diffusion is the movement of constituents in response to spatial gradients in constituent concentration (contaminant will flow from locations of high concentration to low concentration). Dispersion is fluid mixing due to the nonhomogeneities of the aquifer structure on a micro- and macro-scale.

Chemical/biological mass transport processes are (usually) changes in character of a constituent in a closed system. Normally, this occurs with an increase in some other type of constituent (i.e., aerobic biodegradation of gasoline normally yields carbon dioxide and water). In some instances it may be just as important to model the progeny of chemical/biological processes as it is to model the original constituents of interest. Chemical/biological transport processes include: biodegradation, volatilization, radioactive decay, sorption/desorption, dissolution/precipitation, acid/base reactions, complexation, hydrolysis/substitution, and redox reactions.

Biodegradation reactions are those that involve microorganisms that change the chemical nature of the original constituents. Microorganisms may perform these reactions in a spectrum of pH and redox conditions and in both aerobic and anaerobic environments.

Volatilization is the mass exchange of a chemical across the air/water interface. The chemical flow may be to either phase, controlled by the Henry's constant, vapor concentration of the chemical, water concentration of the chemical, and chemical fraction dissolved in the water.

Radioactive decay is the decline of radioactive activity of a radionuclide through nuclear reactions. Sorption/desorption reactions are solid/water chemical interactions whereby chemicals move from one phase to the other. The chemical moves from water to solid (sorption) or solid to water (desorption) in response to the concentration of the chemical in water and solid, solids concentration, and the sorption and desorption rate constants.

Dissolution/precipitation reactions are processes which add (dissolution) or remove (precipitation) contaminants from ground water. Acid/base reactions are chemical reactions involving the transfer of hydrogen ions (protons). This type of reaction directly relates to ground water pH conditions.

The complexation process involves the combination of charged chemical species (cations and anions). This process can increase or decrease the dissolved chemical concentration of a particular species.

Hydrolysis/substitution processes involve reactions of halogenated organic compounds with either water, hydrogen ion, hydroxyl ion, or with another anion (substitution). In these cases, a chemical bond is cleaved and then a new one formed with one of the listed groups. Lastly, redox processes are reactions that involve the transfer of electrons.

In all of these chemical/biological processes there is a theoretical mathematical equation that exists to describe each process (Kincaid et al., 1984; National Research Council, 1990; Thomann and Mueller, 1987; Walton, 1985). Often these equations include rate coefficients that describe the process; however, extreme caution should be taken when selecting published values of rate coefficients. Often, the rate coefficients are specific to the setting in which they were measured and may have little or no relation to the situation that is to be modeled. This is especially true of some laboratory-evaluated rate coefficients, which were developed under tightly controlled situations. The laboratory setting may not have the specified and important characteristics of chemical speciation and complexation as the field situation (American Water Works Association, 1981; Hern and Melancon, 1986; Howard, 1989a, 1989b; National Research Council, 1984; Rai et al., 1984; Rai and Zachara, 1986).

The generic form of the two dimensional advection–dispersion equation is:

$$\frac{\partial C}{\partial t} = D_x \frac{\partial^2 C}{\partial x^2} + D_y \frac{\partial^2 C}{\partial y^2} - V_x \frac{\partial C}{\partial x} - V_y \frac{\partial C}{\partial y} \pm C^{*} \Sigma \pm L \qquad (5.16)$$

where ∂, represents the partial differential operator, x, y represent horizontal spatial coordinates, C represents concentration, t represents time, D represents the aquifer dispersion coefficient, C^* represents the sink or source of solute, V represents the actual fluid velocity, and L represents all the pertinent transport (chemical/biological reaction) terms. Equation 5.16 was developed for a constant porosity (effective specific yield) system and a dissolved (i.e., chloride) or solid (i.e., clay particles) constituent with constant dispersive coefficients and velocities. The equation is subject to boundary and initial conditions regarding the constituent of interest. In very simple cases, this (or the one-dimensional versions) is solved to yield analytical equations for predicting the constituent concentration in space and time.

The dispersion coefficients incorporate diffusion, mechanical dispersion, and macrodispersion. Because of the latter, dispersion coefficients are typically scale-dependent with significant changes in the coefficients occurring in the near field (within 10 m) of the source. Diffusion seems to dominate in the far field (greater than 100 m) (Anderson and Wang, 1982; de Marsily, 1986; Javandel et al., 1984; National Research Council, 1990).

It is important to recognize that in any site-specific application, all, some, or none of the terms in Equation 5.16 may be operating. The exact applicability of terms must be established early on in a contaminant transport study.

The transport, source, or sink terms of Equation 5.16 represent either boundary conditions for loading or chemical/biological processes. In order to develop either analytical or numerical solutions to the advection–dispersion equation, these terms must be described mathematically. For example, a first-order decay process would be:

$$L = -KC \qquad (5.17)$$

where K is a first-order decay rate and the other terms are previously described. This very simple model can be used to describe biodegradation, hydrolysis, and radioactive decay. Other mathematical models have been developed for the remaining chemical/biological transport phenomena. Again, obtaining accuracy between modeled and field concentrations may be difficult due to the problems with the accuracy/dependability of rate coefficients in natural environments (Anderson and Wang, 1982; de Marsily, 1986; Javandel et al., 1984; National Research Council, 1990; Thomann and Mueller, 1987).

For contaminants which are immiscible in water, separate hydraulic equations must be written for the immiscible phase(s) as well as for the water phase. These equations are linked by that fraction of the porosity that each fluid occupies (the pore fluid content). That is, the entire effective pore volume is completely filled; however, there may be two or more immiscible fluids in a given pore space. In turn, fluid pressure and fluid permeability are related to the fluid content. Owing to the relative density of the immiscible contaminant compared to that of water, given a large spill volume, the contaminant will either sink (dense nonaqueous phase liquid, DNAPL) through the water phase down to the lower aquifer boundary or float (light nonaqueous phase liquid, LNAPL) on top of the ground water. In LNAPL spills, the floating contaminant can actually reside on top of the capillary fringe (Fiedler, 1989). For small volumes of immiscible fluid spills, these fluids will spread until they reach relative immobility (Abriola, 1984; Corey, 1977; Fiedler, 1989; Hern and Melancon, 1986) in trying to satisfy the specific retention of the porous medium for that fluid. Details of multi-phase flow dynamics may be found in either Corey (1977) or Abriola (1984).

4. Solution Techniques As with the hydraulic equations that describe ground water flow, there are two basic classes of solution techniques for ground water quality models: analytical and numerical. The analytical methods make use of simplifying assumptions in order to allow integration of the advection–dispersion equation. Numerous solutions exist for one-, two-, and three-dimensional ground water flow and contaminant transport, as well as for various source/sink and chemical/biological reactions (Anderson and Wang, 1982; de Marsily, 1986; National Research Council, 1990; Walton, 1985).

As with ground water hydraulics, the more complex the system, the more appropriate are the numerical techniques for analysis of ground water quality situations. Common methods for numerical computations are through the use of finite element and finite difference techniques. There are other techniques which also are of utility, including: boundary elements, method of characteristics, stochastic functions, integrated finite difference, boundary integral, and flow path network, to name a few (Anderson and Wang, 1982; Bachmat et al., 1980; Hern and Melancon, 1986; Javandel et al., 1984; Kincaid and Morrey, 1984; Morrey et al., 1986; National Research Council, 1990; Walton, 1985).

Caution should be used when selecting a model (Anderson and Wang, 1982; Walton, 1985). It is very important to have strong familiarity with a model in order to know the model's strengths and weaknesses. Experience with modeling in general is paramount. There are many decisions that must be made prior to the actual model conception. These decisions include: what field, laboratory, or published data to collect; how to collect field data; where to collect field data; what is the model grid design; what computers are available and which one is to be used; what codes are available for the computer to be used; how much verification data is necessary; what is the calibration data; how does model sensitivity affect all modeled scenarios; and how accurate are the model input parameters. Experience with various models allows for early and accurate decisions to be made. Once some of these decisions are made, certain models will be precluded from use.

Quite often the solution technique for ground water quality modeling endeavors is to follow a multi-level approach in which various levels of effort and modeling are utilized. At the first level, existing information is used for very simple contaminant transport considering only advection and dilution. The results of this preliminary effort can highlight the approximate zone involved in transport and fate of a constituent as well as the weaknesses and limitations of the existing data base. This level requires basic information on ground water flow and the mass of constituent in the aquifer. Level one results may indicate the need for further field investigation. At the second level, analytical or semi-analytical models are used which include more of the chemical/biological as well as dispersion terms in the advection–dispersion equation. These models cannot account for aquifer nonhomogeneity or anisotropy, but they can reveal a more accurate representation of the affected area with a relatively minor degree of additional input data. In some cases, due to the small extent of the affected aquifer, the analytical models may represent the actual system very well.

At the third level, a numerical model is selected that can accommodate the geometry, hydraulics, and chemical characteristics which are to be simulated. Although constant hydraulic and geometric characteristics can be used in the numerical models, their real advantage is their utility in nonhomogeneous and anisotropic conditions. This means that the more spatial information that is gathered, the more accurate the model will be. Numerical models need more information than analytical models. If this information is not available, use of the analytical models may be the more effective approach. If simple, uniform ground water flow gradients are found for homogeneous, isotropic aquifers, analytical models are more effective. If variable, rapidly changing ground water gradients are found, even with homogeneous and isotropic conditions, numerical models are superior to analytical models.

One last note on model selection. With the large selection of models now available, one should consider the ultimate utility of the modeling results. There is an advantage to using well-known (benchmark) models that have been extensively used. There are also advantages to using models that the nontechnical community can understand. Both of these concepts deal with the issue of familiarity. The more familiar a concept or model, the more it is likely to be understood, used, and accepted. Thus, when selecting a model, if there is only one model that can be used for a particular problem, the choice is obvious;

however, if there is a group of models that can perform the necessary simulation, the more familiar model is the better choice.

5. Determination of Model Parameters Model requirements generally include: system geometry, physical boundary conditions, water quality boundary conditions, water quality initial conditions, aquifer system hydraulics (potentials and velocities), dispersion coefficients, and chemical/biological rate coefficients (i.e., retardation factors, decay rates, distribution coefficients, etc.). System geometry and physical boundary conditions are best evaluated by drilling logs and remote sensing or geophysical methods. Relying solely on published/map data reduces the effectiveness of the modeling endeavor. Water quality initial and boundary conditions are either assumed, measured, recreated (from historical records) or, in some cases, already known. The more accurate the mass continuity of initial and boundary conditions are, the more accurate will be the modeling results.

As discussed previously, aquifer hydraulics usually limit water quality models, and thus this information may be developed independent of the water quality model. Aquifer hydraulics may be either measured, assumed, or developed through aquifer hydraulic modeling.

The dispersion coefficients are best measured in the field at the site to be modeled. Unfortunately, in some instances, this may be impractical. Published methodologies may be used to estimate the dispersion coefficients (Anderson and Wang, 1982; de Marsily, 1986; National Research Council, 1990). Again, this is a scale-dependent parameter, much more so than the hydraulic parameters, and thus more accuracy is obtained from site-specific, field-measured values. Normally field methodologies use single or multiple wells and tracers in order to develop the dispersion coefficients. As with other aquifer field methodologies for evaluation of aquifer parameters, the field exercise should closely simulate the system for which the parameters are to be utilized. For example, if hydraulic and dispersion aquifer coefficients are to be evaluated in a field exercise, and the application/model were for a landfill closure, more passive field techniques (borehole dilution, natural gradient tracer test, etc.) will generate parameters more relevant to system performance (landfill closure) than more dynamic field techniques (pumping test, two well pumping dispersion test, etc.). In this example, if the landfill is to be closed by capping, and no leachate collection or interceptor well/drain system were to exist, the system will perform under a passive natural gradient condition with climatologic and vegetative stresses dominating, not under artificial pumping/forced gradient conditions. Thus, more reliable field parameters would be developed from passive field exercises.

Chemical/biological rate coefficients may be evaluated in laboratory or field situations or selected from the literature (American Water Works Association, 1981; Howard, 1989a; 1989b; Rai and Zachara, 1986; Rai et al., 1984).

6. Determination of Grid System Detail As with the modeling of aquifer hydraulics, the number of nodes and size of each nodal area must accurately portray the system to be modeled in sufficient detail to accommodate all relevant features but not in excess so as to become cumbersome. Normally, the most detailed definition is required where gradients (hydraulic and chemical) are the steepest in time and space. Model stability and solution convergence depends, in large part, on the grid definition. Therefore, it may be necessary to evaluate the model performance by using a few grid systems which vary in detail until the user is convinced that one grid is able to yield satisfactory results over the spectrum of situations that are to be modeled.

7. Model Calibration and Verification Model calibration is a very important step in the overall modeling endeavor. Calibration is the step whereby model parameters are adjusted in order to predict observed values of aquifer concentrations. Thus, this step requires the availability of measurements or other known values throughout the aquifer system given some initial and boundary conditions. The model is first run using the initial estimate for dispersion and chemical/biological rate coefficients, and initial and boundary conditions, then comparisons between model results and known data are made. If the two compare well, then the model coefficients and boundary conditions are appropriate. If the two do not compare well, then the model coefficients and/or boundary conditions must be modified. At this point, there should be no doubts about the accuracy of the known data. That is, if the modeled information does not compare well with the known data, this is not the point in the analysis where one should explain away known data. The known data that is used for calibration efforts must have already passed reliability tests; otherwise, the data should not be used for calibration.

The water quality coefficients are diagnostically modified and then the model rerun until the parameters of comparison (gradients, concentrations, and/or fluxes) nearly match those of the known data set(s). In most cases, first a conservative constituent is modeled and checked against known data. This allows calibration of dispersion, advection and source/sink coefficients. Next, a nonconservative constituent (the constituent of interest or one very similar to it) is simulated and checked against known data, allowing calibration of the chemical/biological rate coefficients.

Verification is a step that is often deleted because whatever data exists is generally used for the calibration task. The verification task requires that a set of known data, data not used in the calibration step, be compared to model results. The model results in this case result from the use of the final calibrated parameters. If the two compare well, then the model can be used for predictive purposes; however, if the comparisons are unfavorable, then there are still problems with the calibrated parameters. At the point of successful calibration and verification, the model is then used for prediction.

XI. GROUND WATER MANAGEMENT

History shows that the usual practice in ground water development and utilization has been to treat problems and initiate action programs on an individual, piecemeal basis. The actors involved often fail to recognize the benefits of complete management of the ground water resource and that ground water is integrated with the total water resource system. Each user has individual goals and these may be inconsistent, or even in conflict, with the goals of other users in the area. For example, an irrigation district may wish to maintain high ground water levels to reduce pumping costs, while another water agency may wish to lower levels in the dry season to permit storage of percolated storm runoff. This practice of lowering levels may be in conflict with an agency which is operating an injection well-type sea water intrusion barrier and wishes to maintain high ground water levels to reduce recharge water requirements and operating costs. One agency may wish to export all waste water to satisfy a certain water quality objective, while another may wish to treat and recharge waste water to augment the ground water supply. Some users may be overpumping ground water and taking a smaller quantity of imported water in an effort to strengthen their ground water water rights in conflict with the goals of the seller of imported water and those striving to keep ground water storage levels high.

Conflicting goals of water agencies and users can usually be reconciled through interaction in the development of a ground water management plan by all interested parties or their elected and appointed representatives (American Society of Civil Engineers, 1987). Often it is desirable to create an agency with boundaries at least contiguous with the borders of the area to be managed, vested with the authority necessary to implement a plan of orderly development of the ground water resources. Not only must technical problems of hydrology and geology be solved, but attention must also be given to their legal, organizational, social, and political aspects.

Worldwide studies of ground water and detailed studies in the United States have identified several significant facts that all water managers should know, which are listed below.

1. Ground water is a resource of immense magnitude, but it is unevenly available and exhaustible.
2. Ground water and surface water resources are interrelated and should always be considered together.
3. Natural replenishment of the vast supply of underground water occurs slowly; therefore, excessive continued exploitation may require decades to centuries for natural replenishment. Artificial

replenishment can be used to replace ground water extracted for use more quickly than nature alone may replace it.

4. In terms of biologic or organic quality, ground water is generally better than surface water, but in some areas surface disposal of wastes has resulted in contamination of ground water. In terms of chemical quality (dissolved solids), surface water is generally better than ground water. In arid climates, surface water is generally better than ground water in headwater areas, but poorer downstream. In humid climates, surface water is generally better chemically, but not biologically.
5. The cost of developing ground water is generally lower than the cost of developing surface water. Ground water may be developed on a "pay-as-you-go" or "pay-as-you-grow" basis. Surface water development generally involves a large initial capital investment.
6. Ground water becomes more valuable as new surface water projects become more difficult to develop. Value is followed by more regulation, particularly for urban uses. This requires extensive monitoring and management techniques that increase the cost of supplying ground water.
7. The United States and many other nations have not made the maximum effective use of this vast resource of generally high quality water (McGuinness, 1963; Solley et al., 1983); however, the use of ground water is increasing worldwide.
8. In coastal areas, offshore extensions of deep confined aquifers may contain considerable quantities of fresh water (Johnston et al., 1982). In the past few years, there has been increased interest in this unused source of water; however, care must be exercised in the utilization of this supply to avoid saline contamination.
9. For storage purposes, underground reservoirs have many advantages over surface reservoirs:
   a. They cost nothing to construct; however, as with the dead storage in a surface reservoir, there is a cost for installing wells and pumping the water out for use, and a cost for maintenance of wells and pumps.
   b. They do not accumulate silt, but surface siltation of replenishment areas may significantly curtail infiltration rates.
   c. They have minimal or no evaporation losses, except in areas of shallow water table where vegetative use of water may be high and saline residues may accumulate.
   d. In general, they have relatively uniform temperature and mineral quality; however, there may be large variations from place to place within the reservoir.
   e. They are free of turbidity, except in some limestone areas and volcanic areas with high secondary permeability and where large underground conduits occur.
   f. They do not preempt the land surface that is useful for other purposes.
   g. They are not subject to eutrophication.
   h. They have unlimited life if properly managed; however, indiscriminate overdevelopment or poor practices can dewater aquifers and cause consolidation of sediments, resulting in land surface subsidence and degradation of quality.

Often in the past, individuals and agencies responsible for developing water supplies either ignored these basic ground water facts in their planning, or they proceeded as though they were ignorant of them. This approach to water development ignores the requirement that successful management of the world's water resources must be three-dimensional, adding to the two-dimensional development of surface water the third dimension of depth to include ground water. It is no longer sufficient to repeat the same procedures on an ever-grander scale. Maximum utilization of the world's water resources cannot be achieved merely by building more dams, reservoirs, and canals, or merely by pumping more ground water. Successful management demands full consideration of all implications of aquifer development, water quality, hydrogeochemical aspects and, where appropriate, the coordinated operation of surface and ground water supplies and storage capacities.

The tools, techniques, and qualified technical specialists to explore and evaluate ground water reservoirs are now available to water managers. All that is needed is effort on the part of water managers to use these tools and personnel so that society can benefit from the accessibility, availability, dependability, and generally uniform quality of this large supply of water.

After identifying the problems that threaten the ground water supply of an area and deciding that they should be solved, the next step for water managers is to develop the objectives for an action program to avoid, solve, or mitigate the difficulties. Advice from many specialists—engineering (including hydrologic), legal, fiscal, geologic, environmental, and economic—is required in this difficult task.

Questions that may arise during the deliberations are:

(a) Should the problems be approached on the basis of political or hydrologic boundaries or a combination thereof?
(b) How much of the cost of water service should be paid by the user through water bills, and how much through ad valorem or pumping taxes?
(c) Are the occurrence and use of ground and surface waters sufficiently unrelated so that they can be managed separately, or are they sufficiently related to require coordinated management?
(d) When should legal, environmental, organizational, and political influences be considered—before, during, or after formulation of alternative management plans?

Noneconomic forces inevitably influence the implementation of a plan and may even override economic factors. Many planners elect to disregard noneconomic aspects until the alternatives are compared on the basis of costs and benefits. In this manner, the decision makers can evaluate the extra cost of revisions required by legal, institutional, social, and political forces.

Formulation of management objectives is often more complex than the technical task of developing alternative plans to meet them. Among the major obstacles are the misconceptions mentioned previously. Many people oppose implementation of a management plan because it breaks with traditional practices. Others will object to any plan because it connotes "control" and an infringement on what they consider to be their prerogatives or rights. Management, if it truly performs its function, considers the effect of various objectives on voter and water user reaction, economic growth, capital investment, water rights, and relations with other public bodies.

A. Alternative Plans for Ground Water Management

Alternative plans for water resources management should be designed to provide for the existing uses of water and to provide future water needs, whenever and wherever they may occur, with suitable water and within the framework of optimum utilization of the available water resources. Successful water management plans must be physically feasible and economically viable, have social acceptability, meet with legal conformance, and engage organizational effectiveness. In the implementation of a management scheme, the successful manager develops programs of operation that come closest to satisfying these requirements. Ground water must be managed in conformance with these water resources management concepts.

1. Management Elements The elements to be considered in a ground water management program depend on local conditions. The process of developing the best alternative plan requires a consideration of the physical elements of a management plan: recharge, extractions, and surface water (local and imported). These should be considered from the viewpoint of both water quantity and quality.

Any undesirable condition that could result from planned basin operation should also be identified and its impact evaluated. Waterlogging, excessive pumping lifts, water quality degradation, habitat and other environmental damage, and land surface subsidence are examples.

Two time frameworks should be adopted. First, within the constraints imposed by current legal, political, and financial conditions, a short-range management program can be evolved. Second, a long-range program can be best investigated by removing from consideration those restrictions that can be changed with time. Thus, in developing the alternative plans for ground water management, consideration will be given to the full spectrum of variation available for each.

a. Patterns and Schedules of Recharge The need for recharge in a ground water management plan is influenced by two factors. The first is the potential demand for ground water, and this, in turn, is influenced by economic and legal considerations, both of which can be altered by management processes. The second factor is that of the physical limitations of the ground water basin, and the availability of additional water for recharge.

Once the future need for water is estimated, portions of the total demand to be met from the ground water basin are determined, together with available or potential surface supplies, for various alternative management plans. Having accounted for natural return to and losses from the basin, the pattern and schedule for recharge can be defined to meet this need for ground water. Cost of ground water and cost and availability of supplemental replenishment water (either imported, reclaimed waste water, or converted sea water), costs of recharge and of recovery must also be estimated. Definition of a suitable recharge pattern and schedule often is not possible without a trial and error operation or ground water modeling techniques.

The physical limitations of the basin must be considered simultaneously with the study of a plan for water supply operation. These limitations include storage capacity, rechargeability, transmissivity, and land area appropriate for replenishment activity. Modeling techniques allow systematic consideration of physical limitations of the basin.

b. Patterns and Schedules of Pumping from Wells The rate of withdrawal, schedule of extractions, and location of points of ground water withdrawal are essential elements in any ground water management plan.

Where ground water is the only water supply, the pumping schedule must simply meet the water needs, unless surface (system) storage is provided, in which case a more uniform extraction rate is possible. The schedule of extraction can be more complicated where ground water is produced as an integrated unit of a total water supply system which includes surface distribution facilities and imported water supply. It is important to note that dual capacity—that is, the ability to take water either from aquifers or surface supply at the discretion of the operator—gives a management program much more flexibility.

Generally, the water manager, after assessing the total water need within a given basin, determines ratios of ground water use versus the surface supply use (local or imported). This determination is based on capability of existing wells, availability of ground water, availability of a surface supply, quantity of surface (system) storage available, overall water conditions, quality considerations, and costs of production. Generally, ground water is the most attractive source, particularly in areas where the alternatives are either costly imported water or expensive treatment to make local supply potable.

Because ground water is usually so attractive from both a cost and a quality viewpoint, water mining with its attendant problems may occur without some regulation or accommodation to extractions. The manager should structure the program so that the maximum freedom in utilizing the basin is available to all pumpers on an equitable basis, either quantitatively or financially.

In a water management program intended to satisfy future water needs and solve specific problems, such as sea water intrusion, the investigator would determine the need to modify the existing extraction and distribution pattern. In coastal areas where water mining might induce sea water intrusion, regulated pumping can be used to control the extent of intrusion. If a seaward hydraulic gradient is maintained, the incursion can be eliminated. This usually reduces the potential for optimum ground water basin operation. As an alternative, a series of wells can be constructed parallel to the coast and operated by the management agency. Extraction of water at this location will form a trough in the ground water level, and the landward migration of sea water will be halted.

Another application of extraction regulation as a water management tool is modifying a basin pumping pattern to implement a recharge program. In those regions where waterlogging or constricted geological conditions can impair a recharge operation, systematic redistribution of basinwide extraction patterns can create favorable gradients that would increase the capability for recharge.

Ideally, the pattern of extraction should be arranged to develop the maximum sustained yield of the basin without causing any adverse condition, while still taking advantage of the storage capacity and the ability of the aquifers to distribute water from the point of recharge to the points of extraction. Obviously, the economics of such an extraction pattern in relation to the required distribution system must be considered.

The ground water supply is frequently reduced by subsurface outflow and rising water which leaves the basin as surface flow. If the extraction pattern is not set by other considerations, such as surface water rights and wetlands preservation, additional water may be salvaged by locating production to intercept basin outflow.

c. Use of Imported Water As the needs of the water system grow and exceed the water naturally available to the ground water basin, it is frequently necessary to supplement the "perennial yield" of the basin through use of imported water. The extent of this supplementation depends, once again, on the demands placed upon the basin and the economics and availability of imported water. Usually during the initial years of operation of large import projects, excess capacity is available at only a nominal additional cost. If adequate ground water storage is available, sizable saving can be obtained through storing or "banking" the imported water underground for future use. This type of operation is frequently overlooked, or may be difficult to develop, for political and financial reasons.

d. Maintaining Storage Capacity for Infiltration of Storm Flows Alternative plans of operation must include both long-term and short-term water supply conditions. If the basin is maintained at full capacity year round, the benefits of storage capacity are lost. Thus, the degree of variation in underground storage should depend upon the potential peak water supplies available for recharge and the recharge capacity of the basin. Maintaining a full basin with decreased well-pumping costs must be compared with the cost of imported water to keep it full. In addition, the cost of the potential loss of natural yield through waste of runoff to the ocean or inland sinks when the basin is full should be considered. Each area, with its particular financial and political problems, should be analyzed for optimum operation. The decision to be made about ground water level versus storage level is probably more sensitive to contemporary constraints imposed by local attitudes than any other factor in ground water management.

e. Control of Ground Water Levels The maintenance of ground water elevations in a managed area may become a choice between waterlogging in the upper reaches or overdraft in the lower reaches of the system. In order to maintain a hydraulic gradient across the basin, the upper reaches at the point of recharge are waterlogged, with the attendant problems of damage to structures and loss of water through evapotranspiration, whereas the lower reaches are purposely overdrafted to induce effective basin-wide recharge. Sometimes waterlogging can occur in both upper and lower reaches, with the pumping depression located between them.

Financial or physical remedial action can be taken in areas of raised or lowered ground water levels. For example, selective pumping in areas where waterlogging can occur will lower high water levels; or restricting land use near recharge areas reduces the need for and costs of controlling rising ground water adjacent to these areas. In areas of pumping depression created by a management scheme, well owners could be given financial support to offset the associated costs of additional pump lift or supplied emergency water to offset loss of access to water during extended dry periods. In general, plans resulting in waterlogging should be avoided. If this is not practicable, costs of wells or other water level control measures to reduce waterlogging should be included in the cost of the management program, just as the expense of surface distribution facilities to supply areas denied water because of excessive drawdown is included.

f. Water Quality The main thrust of water resources planning in the past has been development of additional supplies to meet water needs. In other words, investigations have been quantity-oriented, often at the expense of water quality considerations. Today, quality and quantity are considered inseparable. Water must be of suitable quality for the intended beneficial uses. In many cases, the important factor becomes the quality rather than the quantity of water. Use and reuse of ground water, if natural recharge is not sufficient, causes degradation of quality with time. Quality objectives must be set, and quality of water must be maintained within the limitations imposed by those objectives.

g. Land Surface Subsidence In the course of developing a water resources management plan that involves the lowering of ground water levels below their historical elevations, the investigator should attempt to determine the nature and extent of consolidation of deposits which might occur and its economic (Fowler, 1981) impact because of land surface subsidence.

Subsidence of the land surface has been observed at a number of locations throughout the world (Poland, 1981). It is related, in large part, to several human activities, including:

(a) Extraction of ground water.
(b) Withdrawal of oil, gas, and associated brackish and saline waters.
(c) Application of water to unconsolidated moisture-deficient deposits.
(d) Dewatering of peat soils and deeper peat deposits.
(e) Loads imposed by engineering structures causing foundation settlement.
(f) Subsurface mining.

In addition to subsidence due to human activities, the following natural causes can result in a decline of the ground surface:

(a) Solution and weathering.
(b) Tectonic activity.
(c) Volcanic activity.

One or more of these human and natural causes that result in subsidence may take place in a specific area of ground water development during any period of time.

Agencies and individuals involved in planning the management of ground water reservoirs, or in day-to-day mechanics of operating a ground water reservoir, should familiarize themselves with the types of subsidence, and with the cause, areal extent, magnitude, and chronology of damage associated with each type. Subsidence may be the critical factor in managing some ground water reservoirs. For such reservoirs, any subsidence monitoring program must be designed so that cause and effect are associated; there is no use from a subsidence standpoint in monitoring the water levels in a deep aquifer if the subsidence is due to the application of water on unconsolidated moisture-deficient deposits.

2. Integration into Total Water Resources System Integration of ground water management plans into the total water resources system and the total environment should be considered during formulation of plans for a specific basin. The first step toward integration is to interlock or rationally combine the ground water and surface water facilities to permit their coordinated use.

The coordinated operation of a ground water basin involves the use of a basin for transmitting and storing varying portions of the area's water supply by coordinating the aquifer functions with surface facilities, such as reservoirs and pipelines, to meet the water requirements of an area.

The existing level of economic development in any area can be attributed in some measure to the availability and utilization of local water resources. In the arid southwest, much of the economy has been created through overdrafting the ground water supplies until imported surface water projects could be planned, financed, and constructed. In other areas of the United States with abundant surface water supplies, the potential of the ground water resources has not been fully examined.

Significant attributes of ground water basins are the quantity of fresh ground water in storage and the basin's storage capacity and transmission characteristics. Alternative plans of conjunctive operation should be studied to assure the optimum use of these attributes to satisfy the physical needs for water. At the extreme ends of the operational spectrum, needs may be met with water delivered through surface facilities only, or solely through the aquifers.

Before defining the coordinated operation of the ground water basin, an analogy between physical characteristics of the ground water basin and surface water system should be described. The rate of deep percolation and subsurface flow into the ground water reservoir is roughly analogous to the rate of inflow into a surface reservoir. Storage capacity of the ground water basin is comparable to the storage capacity of the surface reservoir. The transmission characteristics of the aquifers of the ground water basin may be compared to the characteristics of the distribution system in the area of use. Finally, the potentiometric head and ground water table in the ground water basin are analogous to the hydraulic grade line in the surface distribution system. By use of equations that numerically describe the flow characteristics of the ground water basin and surface distribution network, it is possible to calculate and integrate capabilities of these water delivery media to meet the physical objectives.

Facilities must be provided so that the largest flow rate requirement is satisfied at all locations and at required pressures. In brief, the coordinated operation of a ground water basin must include determining the changing requirements for surface and ground water supplies and facilities to meet the largest flow rate requirements of each year. In instances where the water use of an area is met by both imported and locally pumped ground water, changes in the quantity of water used from one source affects the quantities required from the other. At the same time, capacities of facilities to deliver imported water through the surface system also affect the capacities of booster pumps and storage facilities of the local distribution system required to regulate the water supplies to meet the fluctuating water use. The cost of operation depends on the number and type of facilities provided, the manner of operating these facilities, and the amounts of water used from both imported and ground water sources.

In developing a water system for municipal and industrial purposes, it is considered good practice to utilize more than one source of supply. In rapidly developing areas where there are space limits and use of water is high, operation of a ground water system in coordination with a surface water supply is mandatory. The proper balance between use of a ground water system and supplementary surface features for storage and transmission for any given region is unique to each area and a function of the amounts of water and space available. A coordinated system of supply using ground water and surface water is desirable, because it provides the means of capitalizing on the inherent advantages of each which are hydrologically complementary (Conover, 1961a).

Generally, for practical and economic reasons, ground water supplies are used in the early stages of the development of an area, for either agriculture or some limited urban use. As the transition from rural to urban use and industrialization occurs, and the requirements for water supply increase and become more complex, a supplemental imported supply may be needed. By establishing traditions of coordinated use of all available supplies as they occur or as they become available, the manager greatly enhances the opportunity for optimization of the total water system.

The manager must evaluate the effect that variations in operational plans will have on existing and future import and distribution facilities; how implementation of the plan will affect the staging, or itself be influenced by the timing, of future import facilities; and the degree of influence each additional water supply source, such as reclaimed waste water and saline water conversion, will have on the overall plan.

Sufficient emergency storage should be provided within the ground water basin so that any prolonged failure of importation facilities will not seriously jeopardize the water supply to the local area. Prior to developing any plan, it is advantageous to project conservative estimates of future water supply conditions and not to depend on chances that above-normal water supplies will be available. Plans of operation should not cause undue hardship to individuals or inhabitants of local areas within the water supply service district without providing some economic relief. During severe drought conditions, some or all of the service area must share the burden of any deficit.

In addition to utilizing the underground reservoir in conjunction with an imported water supply, the basin can be coordinated with other sources of water to control both quantity and quality. Development of surface reservoirs can be undertaken to store normal and flood flows for later artificial recharge of the ground water basin. Reservoirs located in the mountainous area upstream of the basin will require downstream spreading facilities. In California, several projects including Santa Felicia Dam in Ventura County and Twitchell Dam in San Luis Obispo County were constructed to store water for later recharge of aquifers.

Reservoirs constructed over ground water basins and built without linings or cutoffs provide additional direct recharge to the ground water basin. Small dams have been constructed on southern Texas water courses to enhance the recharge of the Edwards Plateau limestone. Wadi Hanifah and Wadi Nimar dams in Saudi Arabia provide similar recharge capability.

The aquifer system can also be used as a treatment facility. Imported and local surface water usually must be treated prior to delivery to the customer. In lieu of passage through a treatment plant and direct delivery, these waters can be spread and later repumped, thus utilizing the ground water basin as a treatment and transmission facility. This is not to say that waste water containing organic or toxic chemicals can be adequately treated in this manner. The ground water supply can be contaminated by infiltration of surface waters containing hazardous materials.

Numerous facilities have been constructed for flood control with the additional benefits of flood flow conservation through artificial recharge. Projects constructed by the United States Army Corps of Engineers, USDA Soil Conservation Service, and local flood control districts fall into this category.

Ground water management has a role in enhancing the local environment. Although some additional costs may result from environmental enhancement, these expenses help gain public acceptance of management planning. Multiple use of water resources should be a goal of water managers, and the citizens of an area generally accept that it is not only justifiable but highly desirable to utilize the water supply improvements for recreation.

There are many methods of improving the environment through development of a water resources management system. Among these are the development of recreational lakes associated with artificial recharge projects. These replenishment facilities can provide swimming, boating, and fishing, and, with proper beautification by planting trees and appropriate landscaping, become attractive regional parks. In every case care must be taken to avoid potential water quality pollution problems that may be associated with recreation. Replenishment water, in certain areas, can be released in intermittent streambeds to rejuvenate ecosystems by providing running water year round, and thereby enhance a naturally depressed environment.

Use of the ground water system reduces the need for surface storage tanks and surface distribution systems which can be unsightly. Unlined streams and flood control channels should be encouraged wherever possible instead of completely lined channels. They not only provide increased natural infiltration, but also offer a more natural setting. Landscaping, hiking, and riding trails could be provided adjacent to channels. To reduce noise pollution, submersible pumps and underground facilities can be substituted for the use of surface pumping facilities.

In determining the amount of surface and subsurface storage required for the system, it is necessary to take into account the cyclic variations in precipitation on lands over and tributary to the ground water basin and in the import source area. These cyclic variations should then be compared to the variability and nature of the water needs.

Past practice has been to design conservatively so that there would be sufficient water during prolonged droughts. In water short areas, this practice may not be possible or practical. The acceptance of some deficiencies during prolonged droughts should be explored and, if found acceptable, a drought management plan should be developed.

3. ***Economic Comparison of Management Plans*** Three questions must be answered before any water project is undertaken: "Should it be done?" "Can it be done?" and "Is it affordable?"

The first question refers to the social desirability of the project. What improvement or degradation in the environment will result? Will it satisfy water needs in an effective and economic manner? What are the human and ecological benefits and detriments? Among the benefits might be a more secure source of water, an improvement in water quality or quantity, new areas for recreation, improved fire protection, and a possible reduction in population density. Land surface subsidence from substantial lowering of ground water levels could be a detriment.

The second question refers to the physical, or engineering, feasibility of the project. Can the work be accomplished with proven technology? (A water project is not an experimental laboratory; if it does not work properly, it is a permanent drain on the resources of the community.) Obtaining the answer to the second question means answering questions like: Is water available at the present time and in the future for the project? Is the hydrogeology favorable? Is the water quality satisfactory for the end use of the water, or must it be upgraded prior to use? Are there any legal or environmental restrictions that would affect the engineering justification of any plan? Will the project produce beneficial or deleterious side effects, and if so, what are they?

The third question requires an evaluation of the costs and benefits of the project and a comparison of the two. Do the benefits derived from the project exceed its cost and is the community capable and willing to pay the costs?

Once several alternative projects have been proposed, an estimate of the costs involved with each must be made. The capital costs for construction of physical works are some of the major costs to be encountered. After a project is working, the operating and maintenance costs become the important economic concerns. As the project is used, parts of it wear out and must be replaced. This cost is a very real future cost that must be included in the evaluation of costs.

A ground water management program is usually justified by its ability to prevent damage that inevitably will occur if the existing practice continues (for instance, salt water intrusion, land surface subsidence, water quantity depletion, increased cost of ground water extraction). It may also be justified by potential improvement in the utilization of the aquifer as a storage reservoir and local water distributary. The value of the benefits to be derived from or by the project must be estimated. As these benefits are likely to occur over a period of time, the present worth of the benefit stream must be determined.

When the determination of the costs and benefits has been completed for all alternatives, an objective comparison can be made. The comparison evaluation should be made with community input to assure acceptance for the eventual selection of the most desirable alternative projects. The final comparison of the costs and benefits should include evaluations of:

(a) The long-term water requirements to satisfy water demands at several levels of water cost, and to prevent or alleviate such spillover effects as salt water encroachment in aquifers, land-surface subsidence, and increase in the total dissolved solids in ground water.
(b) A drought water supply plan, along with its savings over meeting full water needs in all years.
(c) The differences between cost to the nation and cost to the local agency (owner).
(d) The feasible alternatives both structural and nonstructural (for example, prohibition or reduction of ground water extractions or increase in pumpage charge) to accomplish the water–management objectives.
(e) The qualitative descriptions of the diffuse benefits and costs accruing to each proposed procedure. (Note: The difficulty involved in this evaluation is recognized—it falls in the category of, "it can't be done, but it has to be done anyway.")

4. Financial Feasibility Financial feasibility should not be confused with economic justification. A project may be economically attractive and financially infeasible. Such a situation results in diffuse benefits, and the cost can be neither allocated nor paid. Financial feasibility can be defined as the willingness and ability of the beneficiaries to pay for the services and products of the project. Moreover, a project may be economically attractive, yet the borrowing power of the sponsoring organization may be so limited by statute that the project cannot be financed. Various financing alternatives must be explored and matched to the benefit–cost analysis as well as the buyers' capability to meet debt service.

B. Implementation of Management Plans

Little good comes from a desirable plan that is not implemented. Once implemented, it must be monitored so that modifications can be made to be certain that the plan attains its original goals and objectives. A ground water management plan that cannot be changed should never be constructed. The hydrologic future is too uncertain, and economic and legal futures and social desires are all subject to major change. Flexibility in the plan, adjustment as a result of monitoring, or changes in water needs are the keys to continuing success of any management plan.

XII. REFERENCES

Abriola, L.M. (1984). *Multiphase migration of organic compounds in a porous medium, A mathematical model*, Springer-Verlag Berlin.

Alexander, M. (1981). "Biodegradation of chemicals of environmental concern." *Science* Vol. 211, 132-138.

Allen, M.J. and Geldreich, E.E. (1975) "Bacteriological criteria for groundwater quality." *Ground Water*, Vol. 13, No. 1, 45-51.

American Society of Civil Engineers. (1987). *Ground water management*, ASCE Manuals and Reports on Engineering Practice No. 40.

American Society of Civil Engineers. (1990). "Agricultural salinity assessment and management." ASCE Manuals and Reports on Engineering Practice No. 71.

American Water Works Association. (1981). *Proceedings AWWA Seminar on organic chemical contaminants in groundwater: Transport and removal*, Conference in St. Louis, MO, June 7, AWWA, Denver, CO.

Anderson, M.P. and Wang, H.F. (1982). *Introduction to groundwater modeling, finite difference and finite element methods*, W.H. Freeman and Company, New York, NY.

Bachmat, Y., et al. (1980). *Groundwater management: the use of numerical models*, Water Resources Monographs **5**, American Geophysical Union, Washington, DC.

Baker, D.M. "Safe yield of ground water reservoirs." *Assemblee Generale de Bruxelles, Assoc. Intl. D'Hydrologie Scientifique*, Vol. 2, 160-164.

Banks, H.O. (1953). "Utilization of underground storage reservoirs." *Transactions*, ASCE, Vol. 118, 220.

Baumann, P. (1957). "Basin recharge." *Transactions*, ASCE, Vol. 122, 458-502.

Bear, J. and Dagan, G. (1962a). "The transition zone between fresh and salt waters in coastal aquifers," *Progress Report No. 1*, Technical Research and Development Foundation, Haifa, Israel.

Bear, J. and Dagan, G. (1962b). "The transition zone between fresh and salt waters in coastal aquifers." *Progress Report No. 2*, Technical Research and Development Foundation, Haifa, Israel.

Bear, J. and Dagan, G. (1964). "Moving interface in coastal aquifers." *Journal of the Hydraulics Division*, ASCE, Vol. 90, No. HY4, Proc. Paper 3988, 193-216.

Burden, C.B. et al. (1989). "Ground water conditions in Utah." Utah Division of Water Resources, Cooperative Investigations Report No. 29, 64.

California Department of Public Works, Division of Water Resources. (1935). "Salinas Basin investigation." *Bulletin No. 52*, February.

California Department of Public Works, Division of Water Resources. (1951). "Proposed investigational work for control and prevention of sea water intrusion into ground water basins."

California Department of Water Resources. (1957). "Report by Los Angeles County Flood Control District on investigational work for preventional and control of sea water intrusion, West Coast Basin experimental project, Los Angeles County." *Appendix B, Bulletin No. 63.*

California Department of Water Resources. (1958). "Sea-water intrusion in California." *Bulletin No. 63.*

California Department of Water Resources. (1959). "Investigation of alternative aqueduct systems to serve Southern California." *Bulletin No. 78.*

California Department of Water Resources. (1960a). "Laboratory and model studies of sea-water intrusion." *Appendix C, Bulletin No. 63.*

California Department of Water Resources. (1960b). "Intrusion of salt water into the ground water basins of Southern Alameda County." *Bulletin 81.*

California Department of Water Resources. (1963). "Alameda County investigation." *Bulletin No. 13.*

California Department of Water Resources. (1966a). "Planned utilization of ground water basins, San Gabriel Valley—Appendix A: Geohydrology." *Bulletin No. 104-2*, 122-123.

California Department of Water Resources. (1966b). "Santa Ana gap salinity barrier, Orange County." *Bulletin 147-1.*

California Department of Water Resources. (1967). "Progress report on water quality prediction procedure for the coastal plain of Orange County." *Office Report.*

California Department of Water Resources. (1969). "Sanitary landfill studies." *Bulletin 147-5.*

Cederegren, H.A. (1967). *Seepage, drainage, and flow nets*, John Wiley & Sons, New York, NY, 27-29.

Chandler, R.L. and McWorter, D.B. (1975). "Upconing of the salt-water-fresh-water interface beneath a pumping well." *Ground Water*, Vol. 13, No. 4, 354-359.

Cheh, A.L., Skochdopole, J., Koski, P., and Cole, L. (1980). "Nonvolatile mutagens in drinking water: Production by chlorination and destruction by sulfite," *Science*, 90-93.

Chow, V.T. (1952). "On the determination of transmissivity and storage coefficients from pumping test data." *Transactions*, American Geophysical Union, Vol. 33, 397-404.

Collins, M.A. and Gelhar, L.W. (1971). "Seawater intrusion in layered aquifers." *Water Resources Research*, Vol. 7, No. 4, 971-979.

Conover, C.S. (1954). "Ground water conditions in the Rincon and Mesilla Valleys and adjacent areas in New Mexico." *USGS Water-Supply Paper 1230.*

Conover, C.S. (1961a). "Ground water resources—development and management," *USGS Circular 442.*

Conover, C.S. (1961b). "Groundwater in the arid and semi arid zone; its source, development, and management." *Publication 57, Groundwater in Arid Zones*, International Association of Hydrologic Sciences, 513-529.

Cooper, H.H. (1959). "A hypothesis concerning the dynamic balance of fresh water and salt water in a coastal aquifer." *Journal of Geophysical Research*, Vol. 65, No. 4, 461-467.

Cooper, H.H. et al. (1964). "Sea water in coastal aquifers." *USGS Water-Supply Paper 1613-C*, 84.

Corey, Arthur T. (1977). *Mechanics of heterogeneous fluids in porous media*, Water Resources Publications, Fort Collins, CO.

Costle, D.M. (1979). "Dibromochloropropane (DBCP) suspension order and notice of intent to cancel." *Federal Register*, **44** (219): 65135-6117.

Craun, G.F., McCabe, L.J., and Hughes, J.M. (1976). "Water disease outbreaks in the U.S." *Journal of the American Water Works Association*, Vol. 68, 420-424.

Crawford, M.D. (1972). "Hardness of drinking water and cardiovascular disease." *Proceedings of the Nutrition Society*, Vol. 31, 347-353.

Dagan, G. (1967). "A method of determining the permeability and effective porosity of unconfined anisotropic aquifers." *Water Resources Research*, Vol. 3, 1059-1071.

Dagan, G. and Bear, J. (1968). "Solving the problem of local interface upconing in a coastal aquifer by the method of small perturbations." *Journal of Hydraulics Research*, Vol. 6, No. 1, 15-44.

Das Gupta, A. (1983). "Steady interface upconing beneath a coastal infiltration gallery." *Ground Water*, Vol. 21, No. 4, 465-474.

Davis, S.N. (1969). "Porosity and permeability of natural materials." in *Flow Through Porous Media*, R.J.M. DeWiest, ed., Academic Press, New York & London, 53-89.

de Marsily, G. (1986). *Quantitative hydrogeology—groundwater hydrology for engineers*, Academic Press, San Diego, CA.

Durfor, C.N. and Becker, E. (1964). "Public water supplies of the 100 largest cities in the United States; 1962." *USGS Water-Supply Paper 1812.*

Fiedler, F. (1989). *An Investigation of the relationship between actual and apparent gasoline thickness in a uniform sand aquifer*, MS Thesis, University of New Hampshire, Durham, NH.

Fowler, L.C. (1981). "Economic consequences of land surface subsidence." *Journal of the Irrigation and Drainage Division*, ASCE, Vol. 107, No. IR 2, Proc. Paper 16335, June.

Gray, W.G. (1982). "Comparison of finite difference and finite element methods." *NATO/ASI Mechanics of Fluids in Porous Media—New Approach in Research*, University of Delaware.

Hantush, M.S. (1955). "Preliminary quantitative study of the Roswell ground water reservoir, New Mexico." New Mexico Inst. of Mining and Tech., Socorro, NM.

Hem, J.D. (1959). "Study and interpretation of the chemical characteristics of natural waters." *USGS Water-Supply Paper 1473.*

Henry, G.R. (1959). "Salt intrusion into fresh water aquifers." *Journal of Geophysical Research*, Vol. 64, No. 11, 1911-1919.

Hern, S.C. and Melancon, S.M., eds. (1986). *Vadose zone modeling of organic pollutants*, Lewis Publishers, Inc., Chelsea, MI.

Howard, Philip H., ed. (1989a). *Handbook of environmental fate and exposure data for organic chemicals, Volume I, Large production and priority pollutants*, Lewis Publishers, Inc., Chelsea, MI.

Howard, Philip H., ed. (1989b) *Handbook of environmental fate and exposure data for organic chemicals, Volume II, Solvents*, Lewis Publishers, Inc., Chelsea, MI.

Hubbert, M.K. (1940). "Theory of ground-water motion." *Journal of Geology*, Vol. 48, No. 8, 785-944.

Javandel, I., Doughty, C., and Tsang, C.F. (1984). *Groundwater transport: Handbook of mathematical models*, AGU Water Resources Monograph Series 10, American Geophysical Union, Washington, DC.

Jenkins, C.T. (1968). "Techniques for computing rate and volume of stream depletion by wells." *Ground Water*, Vol. 6, No. 2, March-April, 37-46.

Jensen, R. (1988). "A new approach to regional water management," In *Texas Water Resources*, Texas Water Resources Institute, College Station, TX, Vol. 14, No. 3.

Johnson, A.I. (1963). "Application of laboratory permeability data." USGS open file report.

Johnson, A.I. (1967). "Specific yield—compilation of specific yields for various materials." *USGS Water-Supply Paper 1662-D.*

Johnson, A.I., Prill, R.C. and Morris, D.A. (1963). "Specific yield—column drainage and centrifuge moisture content." *USGS Water-Supply Paper 1662-A.*

Johnson, A.I., Moston, R.P., and Morris, D.A. (1967). "Physical and hydrologic properties of water-bearing deposits in subsiding areas of Central California." *USGS Professional Paper 497-A.*

Johnston, R.H., Bush, D.W., Krause, R.E., Miller, J.A., and Sprinkle, C.L. (1982). "Summary of hydrologic testing in tertiary limestone aquifer, Tenneco offshore exploratory well—Atlanta OCS, Lease-Block 427 (Jacksonville NH 17-5)," *USGS Water-Supply Paper 2180.*

Kashef, A.I. (1976). "Control of salt-water intrusion by recharge wells." *Journal of the Irrigation and Drainage Division*, ASCE, Vol. No. 102, IR4, Proc. Paper 12617, 445-447.

Keith, L.H., ed. (1976). *GC/MS determination of volatiles for the national organics reconnaissance survey of drinking water: An identification and analysis of organic pollutants in water*, Ann Arbor Science.

Kincaid, C.T. and Morrey, J.R. (1984). *Geohydrochemical models for solute migration. Volume 2: Preliminary evaluation of selected computer codes.* Electric Power Research Institute, EA-3417, Vol. 2, Research Projects 2485-2, 1619-1, Palo Alto, CA.

Kincaid, C.T., Morrey, J.R., and Rogers, J.E. (1984). *Geohydrochemical models for solute migration. Volume 1: Process description and computer code selection*, Electric Power Research Institute, EA-3417 Vol. 1, Research Project 1619, Palo Alto, CA.

Kohout, F.A. (1960). "Cyclic flow of salt water in Biscayne Aquifer of Southeastern Florida." *Journal of Geophysical Research*, Vol. 65, No. 7, 2133-2141.

Koufman, W.J. and Orlob, G.T. (1957). "Use of radio active and chemical tracers in research on ground water." *Conference Proceedings on New Research Methods in Hydrology*, Scripps Inst. of Oceanography, La Jolla, CA, February.

Larson, Pat, ed. (1989). "Aquifer preservation goal of council." in *Water Current*, University of Nebraska Water Center, Lincoln, NE.

Larson, S.P., Papadopoulos, S.S., Cooper, H.H., Jr., and Burnham, W.L. (1977). "Simulation of wastewater injection into a coastal aquifer system near Kahului, Maui, Hawaii." *Hydraulics in the Coastal Zone, Proceedings*, 25th Annual Hydraulics Division Specialty Conference, ASCE, pp. 107-116.

Lee, C. and Cheng, R.T. (1974). "On seawater encroachment in coastal aquifers." *Water Resources Research*, Vol. 10, No. 5, 1039-1043.

Lewis, R.W. (1982). "Finite element models for flow through porous media." *NATO/ASI Mechanics of Fluids in Porous Media—New Approach in Research*, University of Delaware.

Liu, C.C.K., Lau, L.S., and Mink, J.F. (1983). "Ground water model for a thick fresh-water lens." *Groundwater*, Vol. 21, No. 21, 293-300.

Lohman, S.W. (1972). "Ground-water hydraulics." *USGS Professional Paper 708*.

McGuinness, C.L. (1963). "The role of ground water in the national water situation." *USGS Water-Supply Paper 1800*, 1963.

McKenna, E.J. (1977). "Biodegradation of polynuclear aromatic hydrocarbon pollutants by soil and microorganism." Amherst, Chemical Engineers, 70th Annual Meeting, November 13-17, New York.

Mercer, J.W. and Fause, C.R. (1981). "Ground water modeling." National Water Well Association.

Mercer, J.W., Larson, S.P., and Faust, C.R. (1980) "Simulation of saltwater interface motion." *Ground Water*, Vol. 18, No. 4, 374-385.

Morrey, J.R., Kincaid, C.T., and Hostetler, C.J. (1986). *Geohydrochemical models for solute migration. Volume 3: Evaluation of selected computer codes*, Electric Power Research Institute, EA-3417 Vol. 3, Research Project 2485-2, Palo Alto, CA.

National Research Council, Committee on Ground Water Modeling Assessment, Frank W. Schwartz, Chairman. (1990). *Ground water models scientific and regulatory applications*. National Academy Press, Washington, DC.

National Research Council, Geophysics Study Committee, Charles L. Drake, Chairman. (1984). *Groundwater contamination*. National Academy Press, Washington, DC.

Newport, R.D. (1975). "Salt water intrusion in the United States." R.S. Kerr Environmental Research Laboratory, Ada, OK.

Nielsen, D.M., ed. (1983). "Aquifer restoration and ground water monitoring." *Proceedings of Third National Symposium*, National Water Well Association.

Pinder, G.F. and Cooper, H.H. (1970). "A numerical technique for calculating the transient position of the saltwater front." *Water Resources Research*, Vol. 6, No. 3, 875-882.

Poland, J.F. (1969). "Status of present knowledge and need for additional research on compaction of aquifer systems." *Land Subsidence*, Vol. 1, International Association of Scientific Hydrology, Publication 88, 11-21.

Poland, J.F. (1981). "Subsidence in United States due to ground water withdrawal." *Journal of the Irrigation and Drainage Division*, ASCE, Vol. 107, No. IR1, Proc. Paper 16329, 115-135.

Poland, J.F. and Davis, G.H. (1969). "Land subsidence due to withdrawal of fluids." *Reviews in Engineering Geology*, Vol. 2, Geological Society of America.

Prill, R.C., Johnson, A.I., and Morris, D.A. (1965). "Specific yield—laboratory experiments showing the effect of time on column drainage." *USGS Water-Supply Paper No. 1662-B*.

Rai, D. and Zachara, J.M. (1986). *Chemical attenuation rates, coefficients, and constants in leachate migration. Volume 1: A critical review*, Electric Power Research Institute, EA-3356, Vol. 1, Research Project 2198-1, Palo Alto, CA.

Rai, D., Zachara, J.M., Schmidt, R.A., and Schwab, A.P. (1984). *Chemical attenuation rates, coefficients, and constants in leachate migration. Volume 2: An annotated bibliography*, Electric Power Research Institute, EA-3356, Vol. 2, Research Project 2198-1, Palo Alto, CA.

Reeve, R.C. (1957). "The measurement of permeability in the laboratory in drainage of agricultural lands." J.N. Lathim, ed., *Agronomy Monograph No. 7*, American Society of Agronomy, Madison, WI, 414-441.

Remson, I., Randolph, J.R., and Barksdale, H.C. (1959). "Zone of aeration and its relationship to ground water recharge." *Journal*, American Water Works Association, Vol. 51, No. 3.

Rumer, R.R., Jr. and Shian, J.C. (1968). "Salt water interface in a layered coastal aquifer." *Water Resources Research*, Vol. 4, No. 5, 1235-1247.

Sahni, B.M. (1972). "Salt water coning beneath fresh water wells." *Water Management Tech. Report No. 18*, Colorado State University, Fort Collins, CO.

Schmorak, S. and Mercado, A. (1969). "Upconing of fresh water—sea water interface below pumping wells, field study." *Water Resources Research*, Vol. 5, No. 6, 1290-1311.

Segol, G., Pinder, G.F., and Gray, W.G. (1975). "A Galerkin finite element technique for calculating the transient position of the saltwater front." *Water Resources Research*, Vol. 11, No. 2, 343-347.

Shamir, V. and Dagan, G. (1971). "Motion of the sea water interface in coastal aquifers: A numerical solution." *Water Resources Research*, Vol. 7, No. 3, 644-657.

Solley, W.B., Chase, E.B., and Mann, W.B. (1983). "Estimated use of water in the United States in 1980." *USGS Circular 1001*, 1983.

Texas Board of Water Engineers. (1951). "Water supply of the Houston Gulf Coast Region." *Bulletin No. 5101*, 1951; also see *Bulletin 5201*, 1952.

Thomann, R.V. and Mueller, J.A. (1987). *Principles of surface water quality modeling and control*, Harper & Row Publishers, Inc., New York, NY.

Thomas, H.E. (1952). "Cedar City Valley." In *Status of development of selected ground water basins in Utah*. Utah State Engineer's Office, *Tech. Publication 7*, 22-34.

Thomas, H.E. (1957). *The conservation of ground water*, McGraw-Hill Book Co., Inc., New York, NY.

Todd, D.K. (1980). *Ground water hydrology, 2nd ed.*, John Wiley & Sons, New York, NY.

Trescott, P.C., Pinder, G.F., and Larson, S.P. (1976). "Finite difference model for aquifer simulation in two dimensions with results of numerical experiments." *USGS Automated Data Processing and Computations Book 7*.

U.S. Environmental Protection Agency. (1977). "The report to Congress—waste disposal practices and their effects on ground water." EPA-570/9-77-001.

U.S. Public Health Service. (1961). "Proceedings of the 1961 symposium—ground water contamination." Robert A. Taft Sanitary Engineering Center, *Technical Report W61-5*.

University of California. (1955). "Report on laboratory and model studies of sea water intrusion." *Technical Bulletin 11*, May.

Walton, W.C. (1962). "Selected analytical methods for well and aquifer evaluation." Illinois State Water Survey, *Bulletin No. 49*.

Walton, W.C. (1985). *Practical aspects of groundwater modeling flow, mass and heat transport, and subsidence analytical and computer models, 2nd ed.*, National Water Well Association, Worthington, OH.

Walton, W.C. and Prickett, T.A. (1963). "Hydrogeologic electric analog computers." *Journal of the Hydraulics Division*, ASCE, Vol. 89, No. HY6, Proc. Paper 3695, November, 67-91.

Wang, H.F. and Anderson, M.P. (1982). *Introduction to ground water modeling—Finite difference and finite element models*, W.H. Freeman & Co., San Francisco, CA.

White, D.E. (1957). "Magmatic, connate, and meteoric waters." *Geological Society of America Bulletin*, Vol. 68, 1659-1682.

Young, R.A. and Carpenter, C.H. (1965). "Ground water conditions and storage in the Central Sevier Valley, Utah." *USGS Water-Supply Paper 1787*, 43.

Youngs, E.G. (1969) "Unconfined aquifers and the concept of the specific yield." *Bulletin*, International Association for Scientific Hydrology, Vol. 14, No. 2, 191-196.LLs

Zinenkiewiez, O.C. (1977). *The finite element method, 3rd ed.*, McGraw-Hill, London.

XIII. NOTATIONS

A = area
C = concentration
°C = degrees Celsius
C* = sink or source of solute
D = aquifer dispersion coefficient
g = acceleration of gravity
h = total energy or head at a point
ha = hectare
h_f = height of fresh ground water table above sea level
h_s = depth of fresh water below sea level
I = hydraulic gradient
i = electric current
K = first-order decay rate
K = hydraulic conductivity
k = permeability
km = kilometer
km^2 = square kilometer, length or distance
L = transport terms (chemical/biological reaction)
l = distance
l = liter
l/s = liters per second
m = meter
mg/l = milligrams per liter
m/d = meters per day
mm = millimeters
m/s = meters per second
m^3 = cubic meter
m^3/d = cubic meters per day
m^3/s = cubic meters per second
P_f = density of fresh water
P_s = density of sea water
p = density
Q = quantity of water discharged in unit time
Q_n = net volume flux
S_s = specific storage
s = distance along direction of flow
t = time
V = velocity
V_{eff} = average pore water velocity

v = electric potential or voltage

x = horizontal spatial coordinate

x = length of conductor

y = conductivity per unit length of conductor

y = horizontal spatial coordinate

z = vertical spatial coordinate

Δh = difference in head between two points

$\Delta h/1$ = hydraulic gradient

∂ = partial differential operator

η_e = average effective porosity

μ = viscosity

ν = velocity

XIV. GLOSSARY

Abandoned Well—A well whose use has been permanently discontinued or which is in such a state of disrepair that it cannot reasonably be used.

Absorption—The process by which substances in gaseous, liquid, or solid form dissolve or mix with other substances.

Acre-foot—The volume of water required to cover one acre to a depth of one foot: 43,560 cubic feet, 325,851 U.S. gallons, or 1,233 cubic meters.

Adsorption—Adherence of gas molecules, ions, or molecules to the surface of the solids.

Alluvial Fan Deposit—A cone-shaped deposit of alluvium made by a stream where it runs out onto a level plain or meets a slower stream. The fans generally form where streams issue from mountains upon the lowland.

Alluvium—A geological term describing beds of sand, gravel, silt, and clay deposited by flowing water.

Alluvium (older)—Sand, gravel, silt, and clay deposits with an age range of hundreds of thousands to more than 1,000,000 years.

Alluvium (younger)—Sand, gravel, silt, and clay deposits of recent geologic age.

Anisotropic Mass—A mass having different properties in different directions at any given point.

Anticline—A convex upward fold, the core of which contains stratigraphically older rocks.

Applied Water—Water delivered onto the land in the case of irrigation use, or to an individual's meter (or its equivalent) in the case of urban use; does not include direct precipitation.

Appropriative Right—A concept in water law under which users who demonstrate earlier use of water from a particular source are said to have rights over all later users of water from the same source.

Aquiclude—A formation which, although porous and capable of absorbing water, does not transmit it at rates sufficient to furnish an appreciable supply for a well or spring.

Aquifer—A water-bearing formation that provides a ground water reservoir; a formation, group of formations, or part of a formation that contains sufficient permeable material capable of yielding significant quantities of water to wells and springs.

Aquifer Continuity—The degree to which ground water stored in one aquifer or portion of an aquifer is able to move into another aquifer or portion of an aquifer.

Aquifer Outcrop Area—See Outcrop Area.

Aquifer System—A heterogenous body of interbedded permeable and poorly permeable material that functions regionally as a water-yielding unit; it comprises two or more permeable beds separated at least locally by confining beds that impede ground water movement but do not greatly affect the regional hydraulic continuity of the system; includes both saturated and unsaturated parts of permeable material.

Aquifer Test—A test involving the withdrawal of measured quantities of water from or addition of water to a well and the measurement of resulting changes in head in the aquifer both during and after the period of discharge or addition; made to determine hydrologic properties of the aquifer.

Aquifuge—A rock that contains no interconnected openings or interstices and therefore neither absorbs nor transmits water; an impermeable rock.

Aquitard—A confining bed that retards but does not prevent the flow of water to or from an adjacent aquifer; a leaky confining bed. It does not readily yield water to wells or springs, but may serve as a storage unit for ground water.

Area of Influence of a Well—The area surrounding a well within which the potentiometric surface has been lowered; may be a transient or steady-state condition, depending upon volume and duration of pumping.

Artesian—The condition in which the water level stands above the top of the aquifer, synonymous with confined.

Artesian Aquifer—See Confined Aquifer.

Artesian Well—A well deriving its water from an artesian or confined water body. The water level in an artesian well stands above the top of the artesian water body it taps. If the water level in an artesian well stands above the land surface the well is a flowing artesian well.

Artificial Recharge—The deliberate replenishment of the ground water supply by means of spreading basins, recharge wells, irrigation, or other means to induce infiltration of surface water. May include the substitution of surface water deliveries to water users who would otherwise extract ground water, thereby leaving water underground (this is the "in lieu" method).

Bank Storage—The storage of water in an aquifer adjacent to and interconnecting with a surface water body so that a change in a stage of the adjacent surface water body causes a change in storage of water in the aquifer.

Base Runoff—Sustained or fair weather runoff. In most streams, base runoff is composed largely of ground water effluent. The term base flow is often used in the same sense as base runoff.

Bedrock—A general term for the consolidated (solid) rock that typically underlies soils or other unconsolidated surficial materials.

Brackish Water—Water that is intermediate in salt content between fresh water and saline water; neither fresh nor saline, but in between; contains between 1,000 and 10,000 milligrams per liter of dissolved solids.

Capillarity—The rise or movement of water in the interstices of a soil or rock due to capillary forces associated with the fluid property of surface tension.

Capillary Fringe—The lower subdivision of the unsaturated zone immediately above the water table in which the interstices are filled with water under pressure less than that of the atmosphere; being continuous with the water below the water table but held above it by surface tension, its upper boundary with the intermediate belt of the unsaturated zone is indistinct.

Capture—Water withdrawn artificially from an aquifer is derived from a decrease in storage in the aquifer, a reduction in the previous discharge from the aquifer, an increase in the recharge, or a combination of these changes. The decrease in discharge plus the increase in recharge is termed capture.

Casing—A tubular retaining structure that is installed in the well bore to maintain the well opening.

Clay—1) A fine-grained geologic material (grain size less than 0.004 mm in diameter) which has very low permeability; or 2) sedimentary deposits composed primarily of these particles.

Coefficient of Permeability—See Permeability Coefficient.

Concrete Casing—Large-diameter casing made of concrete and installed in dug wells or in wells drilled with a bucket auger.

Conductivity, Hydraulic—See Hydraulic Conductivity.

Conductor Casing—A tubular retaining structure installed in the upper portion of a well between the wall of the drilled hole and the inner well casing.

Cone of Depression—A depression in the water table or potentiometric surface of a ground water body that is in the shape of an inverted cone and develops around a well which is being pumped. It defines the area of influence of the pumping well.

Confined—Refers to ground water under sufficient pressure to rise above the aquifer containing it when the aquifer is penetrated by a well. The difference between the water level in a well and the top of the aquifer is the hydrostatic head. Confined ground water is also known as artesian water.

Confined Aquifer—An aquifer bounded above and below by beds of distinctly lower permeability than that of the aquifer itself; an aquifer containing confined ground water.

Confined Ground Water—Ground water contained in a confined aquifer. Confined ground water moves under the control of the difference in head between the intake and discharge areas of the water body.

Confining Bed—A body of impermeable or distinctly less permeable material stratigraphically adjacent to one or more aquifers.

Conjunctive Operation or Use—A term used to describe operation of a ground water basin in coordination with a surface water system. The purpose is to artificially recharge the basin during years of above-average precipitation so that the water can be withdrawn during years of below-average precipitation, when surface supplies are below normal.

Connate Water—Water entrapped in the interstices of a sedimentary rock at the time it was deposited. These waters may be fresh, brackish, or saline in character. Usually applies only to water found in geologically older formations.

Consumptive Use—The total water loss from a water supply or system by evaporation or transpiration from a vegetated or nonvegetated surface, commercial or industrial process, and all domestic and municipal uses. The difference between the quantity of water withdrawn from a source and the quantity returned to the source or another source of usable water.

Consumptive Use of Irrigation Water—The amount of irrigation water, excluding precipitation or other naturally supplied water, that is lost by the evapotranspiration process. See Evapotranspiration.

Contact—The surface between two rocks of different type or age.

Crop Water Requirement—The total water required, from planting to harvest of a given crop in a specific climate regime, needed for evapotranspiration and provided by rainfall and irrigation when adequate soil water is maintained so that it does not limit plant growth or crop yield.

Darcy's Law—An empirical law based on experimental evidence for the flow of fluids with the assumption that the flow is laminar and that inertia can be neglected. It states that the velocity of the flow through a formation is directly proportional to the hydraulic gradient.

Deep Percolation—The drainage of soil water downward by gravity below the maximum effective depth of the root zone toward storage in subsurface strata.

Depletion—Loss of water from surface or ground water reservoirs at a rate greater than that of recharge. See Streamflow Depletion.

Destroyed Well—A well that is no longer in existence. It should have been properly filled so that it can neither produce water nor act as a vertical conduit for the movement of water.

Discharge—The volume of water that passes a given location within a given period of time; outflow; the flow of water from a well, a pump, a pipe, a drainage basin, an aquifer.

Discharge Area—An area in which water is lost from the saturated zone.

Drain Tile—Pipes of various materials, in short lengths, laid in covered trenches underground, in most cases quite loosely and with open joints, to collect and carry off excess ground water or to dispose of wastewater in the ground.

Drain Well—A shaft or well drilled through an impermeable stratum to allow water to drain through to a permeable stratum. Also called absorbing well, negative well, dead well.

Drainage Area—The area of a watershed within which all surface runoff drains by gravity into a stream channel or lake upstream of a given location.

Drainage Basin—A part of the earth's surface occupied by a drainage area or drainage areas.

Drainage Divide—The boundary of a drainage area. A topographic ridge dividing one drainage area from another.

Drainage Well—A well installed to drain surface water, storm water, or treated waste water into underground strata. Also a water well constructed to remove excess water from the formation and result in lower ground water levels.

Drawdown—The vertical distance the free water elevation is lowered, or the reduction of the pressure head due to the removal of free water.

Drilled Well—A well for which the hole is excavated by mechanical means such as the rotary or cable tool methods.

Dug Well—A shallow, large-diameter well usually constructed by excavating with hand tools instead of by drilling or driving.

Effective Hydraulic Conductivity—The rate of flow of water through a porous medium that contains more than one fluid, such as water and air in the unsaturated zone. It should be specified in terms of both the fluid type and content and the existing pressure.

Effective Porosity—The amount of interconnected pore space available for fluid transmission. It is expressed as a percentage of the total volume occupied by the interconnecting interstices.

Effluent Stream—A stream or reach of a stream whose flow is being increased by inflow of ground water. A gaining stream.

Electrical Conductivity (EC)—The measure of the ability of water to conduct an electrical current the magnitude of which depends on the concentration of minerals in the water. Related to total dissolved solids.

Equipotential Line—The trace of a line connecting points of equal potential on a potentiometric surface or a two-dimensional cross section.

Evaporation—The physical process by which a liquid, such as water, in a stream, lake, or moist soil is transformed into a gaseous state. It may be expressed as the total or the mean rate in units of mass or volume per unit area, or as an equivalent depth of water, for the period concerned.

Evapotranspiration—The total water vapor loss from an area by evaporation and transpiration from plants over a given time period. It includes the evaporation of water from soil, from dew, and from intercepted precipitation, as well as transpiration from plants.

Fault—A fracture in the earth's crust, with displacement of material on one side of the fracture with respect to the other. Frequently acts as a barrier to movement of ground water, but in some situations may act as a conduit.

Fault Line—The trace of a fault plane with the ground surface or with a horizontal plane.

Fault Plane—The more or less planar surface of a fault along which dislocation has taken place.

Fault Surface—In a fault, the surface along which displacement has occurred.

Fault Trace—The surface expression of a fault.

Fault Zone—A fault that is expressed as a zone of numerous small fractures or of breccia or fault gouge. A fault zone may be hundreds of meters in width.

Fine-grained—1) Said of a crystalline rock in which the individual minerals have an average diameter or less than 1 mm; 2) said of a soil in which silt or clay predominate.

Flowline—The path a water particle follows during seepage under laminar flow conditions.

Flownet—A graphical representation of flowlines and equipotential lines used in the study of seepage phenomena.

Fold—A bend in rock strata. An anticline is an upward fold; it influences ground water by inducing flow away from the fold axis. A syncline is a downward fold; it influences ground water by inducing flow toward the fold axis.

Forebay, Ground Water—See Outcrop Area.

Formation—A geologic term that designates a specific group of beds of strata which have been deposited in sequence one above the other and during a specific period of geologic time.

Free Water Elevation—Also termed ground water surface, ground water elevation, free water surface, phreatic surface, and water table. The elevation at which pressure in the water is zero with respect to atmospheric pressure.

Fresh Water—Water that generally contains 0 to 1,000 milligrams per liter of dissolved solids.

Gaining Stream—See Effluent Stream.

Geohydrology—A term, often used interchangeably with hydrogeology, referring to the hydrologic or flow characteristics of subsurface waters. See Hydrogeology.

Ghyben–Herzberg Principle—A principle that accounts for the existence of a body of fresh water floating on sea water within an aquifer because of the different densities. Generally speaking, fresh water extends to a depth about forty times the height that the fresh water table is found above sea level. Conversely, a lowering of the fresh water table by 30.5 cm (1 ft) causes sea water to rise 13 m (40 ft) within the aquifer.

Gouge—Finely abraded material occurring between the walls of a fault, the result of grinding movement.

Gravel—1) Geologic material, regardless of mineral composition, with diameter greater than 2 mm; or 2) unconsolidated sediment composed primarily of these particles. Gravel frequently is found as lens-shaped units within sandy deposits.

Gravel-Packed Well—A well in which filter material (sand, gravel, etc.) is placed in the annular space between the casing and the borehole to increase the effective diameter of the well, and to prevent fine-grained material from entering the well during pumping.

Gravel-Packed with Slotted Casing—Includes all wells having a filter envelope and factory-milled or torch-cut slots.

Gravel-Packed with Well Screen—Includes all wells having a filter envelope and either wire-wrapped well screen or louvered perforations.

Gravel-Wall Well—A type of well used in a water-bearing formation containing a large proportion of fine-grained material which permits the passage of water at low velocity. Gravel is introduced around the screen or intake section of the well to increase the specific capacity and to prevent extremely fine material from flowing into the well. Depending on details of construction, such wells are called gravel well, gravel-envelope well, gravel-filled well, or gravel-packed well.

Ground Water—1) That part of the subsurface water that is in the saturated zone; or 2) loosely, all subsurface water as distinct from surface water.

Ground Water Barrier—A body of material that is impermeable, or has a relatively low permeability, and which occurs below the land surface in such a position that it impedes the horizontal movement of ground water and consequently causes a pronounced difference in the level of the water table on opposite sides of it.

Ground Water Basin—A ground water basin consists of an area underlain by permeable materials which are capable of storing or furnishing a significant water supply; the basin includes both the surface and the permeable materials beneath it.

Ground Water Basin Management—The planned use of a ground water basin as to yield, storage space, transmission capability, and ground water in storage. It includes: 1) protection of natural recharge and use of artificial recharge, 2) planned variations in amount and location of pumping over time, 3) coordinated use of ground water storage with surface water supplies, and 4) protection and planned maintenance of ground water quality.

Ground Water Body—All ground water, whether unconfined or confined, contained within a ground water basin.

Ground Water Budget—A numerical account of the recharge, discharge, and changes in storage of an aquifer, part of an aquifer, or system of aquifers. The ground water equation.

Ground Water Discharge—1) Release of water from the saturated zone by any means; 2) the water or the quantity of water released.

Ground Water Divide—A ridge in the water table or other potentiometric surface from which ground water moves away in both directions.

Ground Water Equation—The equation that balances the ground water budget.

Ground Water Geology—The science of subsurface water, with emphasis on the geologic aspects (see Hydrogeology).

Ground Water Hydrology—The branch of hydrology that deals with ground water—occurrence, movement, replenishment, depletion, and character.

Ground Water Mound—A rounded, mound-shaped surface in a water table or other potentiometric surface that builds up as a result of the downward percolation of water.

Ground Water Movement—The movement, or flow, of water in the saturated zone, whether naturally or artificially induced.

Ground Water Outflow—The discharge from a drainage basin, or from any area, occurring as ground water.

Ground Water Pumpage—The quantity of ground water pumped.

Ground Water Ridge—The ridge or rise in the water table that develops beneath an influent stream.

Ground Water Runoff—That part of the runoff that has passed into the ground, has become ground water, and has been discharged into a stream channel as spring or seepage water.

Ground Water Storage—1) The quantity of water in the saturated zone; or 2) water available only from storage as opposed to capture.

Ground Water Subbasin—A discrete unit of a ground water basin.

Ground Water Trough—A depression in the water table or other potentiometric surface caused by a flow of ground water into a stream, drainage ditch, or a line of pumping wells.

Ground Water Withdrawal—The process of withdrawing ground water from a source; also, the quantity of water withdrawn.

Head—1) The pressure of a fluid on a given area, at a given point represented by the height of the fluid above the point; 2) the water-level elevation in a well, or elevation to which water of a flowing well rises in a pipe extended high enough to stop the flow. See Hydraulic Head.

Homogeneity—Synonymous with uniformity. A material is homogenous if its hydrologic properties are identical everywhere.

Horizontal Well—Wells drilled horizontally into hillsides and Ranney-type collectors and infiltration galleries. Wagon-wheel wells are in this category.

Hydraulic Conductivity—For an isotropic porous medium and homogenous fluid, the volume of water at the existing kinematic viscosity that moves in unit time under a unit hydraulic gradient through a unit area measured at right angles to the direction of flow. Replaces the term coefficient of permeability.

Hydraulic Gradient—Slope of the water or potentiometric surface. The change in static head per unit of distance in a given direction. If not specified, the direction generally is understood to be that of the maximum rate of decrease in head.

Hydraulic Head—The height above a datum plane (such as sea level) of the column of water that can be supported by the hydraulic pressure at a given point in a ground water system. For a well, the hydraulic head is equal to the distance between the water level in the well and the datum plane. See Head.

Hydrogeology—The science that deals with subsurface waters and related geologic aspects of surface waters. Also used in the more restricted sense of ground water geology only. See Geohydrology.

Hydrograph—A graph showing stage, flow, velocity, or other characteristics of water with respect to time. A stream hydrograph commonly shows rate of flow; a ground water hydrograph shows water level or head.

Hydrologic Balance—See Hydrologic Budget.

Hydrologic Budget—An accounting of the inflow to, outflow from, and storage in a hydrologic unit such as a drainage basin, aquifer, soil zone, lake, or reservoir; the relationship between evaporation, precipitation, runoff, and the change in water storage, expressed by the hydrologic equation.

Hydrologic Equation—The equation that balances the hydrologic budget.

Hydrology—The science that relates to the distribution and circulation of naturally occurring water on and under the earth's surface.

Imbition—The absorption of water by a granular rock or any other porous medium under the force of capillary attraction and in the absence of any pressure.

Impermeability—The condition of a rock, sediment, or soil that renders it incapable of transmitting fluids under pressure.

Impermeable—A condition of geologic material that renders it incapable of transmitting significant quantities of water under the pressure differences ordinarily found in subsurface water.

Impervious Strata—A geologic unit that does not transmit water in sufficient quantity to furnish an appreciable supply to wells or springs.

Imported Water—Water transported into a watershed from a different watershed. Native water is water naturally within a watershed.

Inactive Well—A well not routinely operated but capable of being made an operating well with a minimum of effort.

Infiltration—The flow or movement of water through the soil surface into the ground.

Infiltration Capacity—The maximum or limiting infiltration rate.

Infiltration Rate—The rate at which a soil under specified conditions absorbs falling rain, melting snow, or surface water; expressed in depth of water per unit time (centimeters per second, inches per hour).

Influent Stream—Stream or reach of a stream that loses water into the ground.

Injection Well—Well used for introducing water into an aquifer. Technique used to stop sea water intrusion, replenish an aquifer, or dispose of cooling water.

Interstice—A pore or open space in rock or granular material, not occupied by solid matter. It may be occupied by air, water, or other gaseous or liquid material. Also called void, void space.

Interstitial Water—Water contained in the interstices of rock materials.

Intrinsic Permeability—A measure of the relative ease with which a porous medium can transmit a liquid under a potential gradient. It is a property of the medium alone and independent of the nature of the liquid and of the force field causing movement. It is a property of the medium that depends upon the shape and size of the pores.

Isotropic—Said of a medium whose properties are the same in all directions.

Isotropy—That condition in which all significant properties are independent of direction.

Laminar Flow—Flow in which head loss is proportional to the first power of the velocity.

Lava Tube—A hollow space beneath the surface of a solidified lava flow, formed by the withdrawal of molten lava after the formation of the surficial crust.

Leaching Requirement—The fraction of water entering the soil that must pass through the root zone in order to prevent soil salinity from exceeding a specific value which would adversely affect plant growth or crop yield.

Leaky Aquifer—See Semiconfined Aquifer.

Lens—A geologic deposit bounded by converging surfaces (at least one of which is curved), thick in the middle and thinning out toward the edges, resembling a convex lens.

Losing Stream—See Influent Stream.

Magmatic Water—Water that exists in or is derived from a magma or molten rock.

Mathematical Model—Mathematical equations that simulate responses of a ground water basin to changes in recharge and discharge patterns and timing.

Meteoric Water—Water that is in or derived from the atmosphere.

Mineral Spring—A spring whose water contains large quantities of mineral salts, either those commonly occurring in the locality or of a rare or uncommon character.

Mining—The process, deliberate or inadvertent, of extracting ground water from a source at a rate so that the ground water level declines persistently, threatening actual exhaustion of the supply.

Net Water Use—That portion of applied water which is consumptively utilized for beneficial purposes or otherwise irrecoverably lost; does not include that portion of applied water subject to possible reuse.

Observation Well—A well used to observe water levels or to obtain water quality samples.

Outcrop Area—The area of a ground water basin where a confined aquifer is exposed at the surface and where recharge to the confined aquifer takes place. Also known as the "Forebay Area" in California.

Outcrop, Water Table—A place or location where the water table intersects the surface of the ground.

Overdraft—The condition of a ground water basin where the amount of water withdrawn exceeds the amount of water captured over the basin over a period of time. The use of water in excess of the perennial yield.

Perched Ground Water—Unconfined ground water separated from an underlying body of ground water by an unsaturated zone. Its water table is a perched water table.

Percolation—The movement, under hydrostatic pressure, of water through the interstices of rock or soil, except the movement through large openings such as caves.

Percolation Rate—The rate, expressed as either velocity or volume per unit of time, at which water percolates through a porous medium.

Perennial Yield—The maximum quantity of water that can be withdrawn annually from a ground water supply under a given set of conditions without causing an undesirable result.

Perforations—Openings in a well casing to allow the entrance of ground water into the well. Perforations may be made either before or after installation of the casing.

Permeability—The capacity of a geologic material for transmitting a fluid. The degree of permeability depends upon the size and shape of the openings and the extent of the interconnections.

Permeability Coefficient—The rate of flow of water through a unit cross-sectional area under a unit hydraulic gradient at the prevailing temperature (field permeability coefficient) or adjusted to a temperature of 15°C (60°F).

Permeable—A condition of a geologic material that renders it capable of transmitting a significant quantity of water without impairment of its structure.

pH—Hydrogen ion activity. A number used to express the acidity of solutions including water. A pH value lower than 7 indicates an acidic solution, a value of 7 is neutral and a value higher than 7 indicates an alkaline solution.

Phreatic Water—A term that originally was applied only to water occuring in the upper part of the saturated zone under water–table conditions (synonym of unconfined ground water, or well water), but has come to be applied to all water in the saturated zone.

Phreatophyte—A plant that habitually obtains its water supply from the saturated zone, either directly or through the capillary fringe.

Piezometer—A facility emplaced to measure and record ground water levels.

Piezometric Surface—An obsolete term replaced by potentiometric surface.

Porosity—1) An index of the void characteristics of a soil or rock material; degree of perviousness. 2) The ratio usually expressed as a percentage of the volume of the interstices, whether isolated or connected, in a given quantity of material to the total volume of the material.

Potable Water—Water that is safe and palatable for human use. Fresh water in which any concentrations of pathogenic organisms and dissolved toxic constituents have been reduced to safe levels, and which is, or has been treated to be, tolerably low in objectionable taste, odor, color, or turbidity and of a temperature suitable for the intended use.

Potentiometric Surface—An imaginary surface representing the elevation and pressure head of ground water and defined by the level to which water rises in a well or piezometer. The water table is a particular potentiometric surface.

Pressure Head—Hydrostatic pressure expressed as the height of a column of water that the pressure can support, expressed with reference to a specific level such as land surface.

Pump Tax, Ground Water Charge, Production Assessment, Replenishment Assessment—Assessments levied on the amount of ground water pumped.

Pumpage—The quantity of water (or other liquid) pumped over a period of time.

Pumping—The extraction of water (or other liquid).

Pumping Lift—The distance water must be lifted in a well from the pumping level to ground surface.

Recharge—Replenishment of ground water by downward infiltration of water from rainfall, streams, and other sources. Natural Recharge is that recharge which occurs without assistance or enhancement by humans. Artificial Recharge is that recharge which occurs when people deliberately modify the natural recharge pattern to increase recharge.

Recharge Area—The area that contributes water to an aquifer. Normally considered to be the natural area of recharge, as contrasted with a constructed recharge basin.

Recharge Basin—A basin constructed on the ground surface to receive the discharge from streams, storm drains, or other sources for the purpose of replenishing ground water supply.

Return Flow—The part of the water diverted from a stream which is not consumptively used, but eventually returns to the surface streamflow either by overland flow (tailwater) or by deep percolation to the water table and subsequent movement into the stream as baseflow.

Riparian Rights—A concept of water law under which authorization to use water in a stream is based on ownership of the land adjacent to the stream.

Safe Yield—See Perennial Yield.

Salaquifer—An aquifer that contains salt water or water too salty for normal uses.

Saline—Consisting of or containing salts (minerals), the most common of which are potassium, sodium, or magnesium in combination with chloride, nitrate, or carbonate.

Saline Water—Water containing more than 10,000 milligrams per liter of dissolved solids.

Salt Water Encroachment—Displacement of fresh surface or ground water by the advance of salt water due to its greater density, usually in coastal and estuarine areas, but also by movement of brine from beneath a playa lake toward wells discharging fresh water. Encroachment occurs when the total head of the salt water exceeds that of adjacent fresh water.

Salt Water Intrusion—The movement of salt water into fresh water aquifers. See Salt Water Encroachment.

Sand—1) Geologic material with grain sizes ranging from 0.0625 mm to 2 mm; or 2) a sediment composed primarily of these particles.

Saturated—A condition in which the interstices of a material are filled with a liquid, usually water. It applies whether the liquid is under greater than or less than atmospheric pressure, as long as all connected interstices are full.

Saturated Rock—A rock that has all its interstices or void spaces filled with water.

Saturated Soil—Soil that has its interstices or void spaces filled with water to the point at which runoff occurs.

Saturated Zone—That part of the water-bearing material in which all voids, large and small, are ideally filled with water under pressure greater than atmospheric.

Saturation—1) A condition reached by a material, whether it be in solid, gaseous, or liquid state, that holds another material within itself in a given state in an amount such that no more of the material can be held within it in the same state. The material is then said to be saturated or in a condition of saturation. 2) The condition of a liquid when it has taken into solution the maximum possible quantity of a given substance at a given temperature and pressure.

Sedimentary—Said of rocks formed from sediments. Includes such rock types as sandstone, conglomerate, shale, etc.

Seep—An area, generally small, where water, or another liquid such as oil, percolates slowly to the land surface. For water, it may be considered as a synonym of a seepage spring, but it is used by some for flows too small to be considered a spring.

Seepage—1) The act or process involving the slow movement of water or other fluid through a porous material such as soil; or 2) the amount of fluid that has been involved in seepage.

Seepage Line—The upper free water surface of the zone of seepage.

Semiconfined Aquifer—An aquifer that is bounded above by a semipervious layer and below by a layer that is either impervious or semipervious.

Semiperched—With respect to ground water or a water table, having a greater pressure head than an underlying body of water not separated from it by an unsaturated rock. The semiperched ground water belongs to the same saturation zone as the underlying body of ground water, and where it occurs there is only one water table, which is called a semiperched water table.

Semiperched Ground Water—Unconfined water separated by a low-permeability, but saturated, bed from a body of confined water whose hydrostatic level is below the water table.

Semipervious Layer—A geologic formation or layer which has a low, though measurable, hydraulic conductivity.

Silt—1) Geologic material with grain sizes ranging from 0.004 mm to 0.0625 mm; or 2) a sediment composed primarily of these particles.

Soil Moisture—Water, or moisture, contained in the soil mantle.

Soil Porosity—The percentage of the soil (or rock) volume that is not occupied by solid particles, including all pore space filled with air and water.

Soil Water—Water in the soil mantle, available for use by plants.

Specific Capacity—The rate of discharge of water from the well divided by the drawdown of water level within the well.

Specific Conductance—A measure of the ability of a water to conduct an electrical current, expressed in micromhos per centimeter at 25°C. Specific conductance is related to the type and concentration of ions in solution and can be used for approximating the dissolved-solids content of the water. Commonly, the concentration of dissolved solids (in milligrams per liter) is about 65% of the specific conductance (in micromhos). This relation is not constant from supply to supply, and it may even vary in the same source with changes in the composition of the water.

Specific Conductivity—With reference to the movement of water in soil, a factor expressing the volume of transported water per unit of time in a given area.

Specific Discharge or Specific Flux—For ground water, the rate of discharge of ground water per unit area measured at right angles to the direction of flow.

Specific Permeability—A factor expressing the permeability of a material; it equals a constant times the square of representative pore diameter.

Specific Retention—The ratio of the volume of water which the rock or soil, after being saturated, retains against the pull of gravity to the volume of the rock or soil.

Specific Storage—The volume of water that a unit volume of aquifer releases from storage because of expansion of the water and compression of the void spaces and grains under a unit decline in average head within the unit volume.

Specific Yield—The ratio of volume of water which the rock or soil, after being saturated, yields by gravity to the volume of the rock or soil.

Spreading Water—Discharging native or imported water to a permeable area for the purpose of encouraging it to percolate to the saturated zone. Spreading, artificial recharge, and replenishment all refer to operations used to place water in a ground water basin.

Spring—A place where ground water flows naturally from a rock or the soil onto the land surface or into a body of surface water.

Static Head—The height above a standard datum of the surface of a column of water or other liquid that can be supported by the static pressure at a given point. It is the sum of the elevation head and the pressure head, the velocity head being negligible under conditions to which Darcy's law can be applied.

Storage Capacity—The volume of space below the land surface that can be used to store ground water. Total Storage Capacity is the total volume of space that could be used to store ground water. Available Storage Capacity is that volume of the total storage capacity that does not presently contain ground water and is therefore available to store recharged water.

Storage Coefficient—The volume of water an aquifer releases from or takes into storage, per unit surface area of the aquifer per unit change in head.

Storativity—See Storage Coefficient.

Stratum—In geology, a single bed or layer of rock that is more or less homogeneous.

Streamflow Depletion—The net decrease (or increase) in streamflow caused by the diversion of water from the stream, which at any time is equal to the diversion rate minus the return flow rate to the same stream. Over a long period of time (several years) the streamflow depletion is equal to the consumptive use of irrigation water, but, because of ground water storage and lag time associated with return flows, these parameters may be quite different for periods of several months or even a year or more. At times when no diversion is being made but return flows from previous irrigation are still occurring, the effect may become a streamflow accretion rather than depletion.

Subsurface Inflow—Ground water movement through the subsurface into a ground water basin.

Subsurface Outflow—Ground water movement through the subsurface out of a ground water basin.

Sustained Yield—The volume of ground water that can be extracted annually from a ground water basin without causing adverse effects.

Syncline—A fold, the core of which contains the stratigraphically younger rocks; it is concave upward.

Thermal Water—Water, generally of a spring or geyser, whose temperature is appreciably above the local mean annual air temperature.

Total Dissolved Solids (TDS)—The quantity of mineral (salts) in solution in water, usually expressed in milligrams per liter.

Total Head—The sum of the elevation head, pressure head, and velocity head of a liquid. For ground water, however, the velocity–head component is generally negligible.

Transmissivity—The rate at which water of the prevailing kinematic viscosity is transmitted through a unit width of the aquifer under a unit hydraulic gradient; equal to the hydraulic conductivity times the aquifer thickness.

Transpiration—Water absorbed and transpired and used directly in the building of plant tissue.

Unconfined (Free) Ground Water—Unconfined water is found in the saturated zone whenever the upper surface of the zone forms a water table under atmospheric pressure, free to rise and fall with changes in volume of stored water.

Unconsolidated Deposits—A sediment that is loosely arranged or unstratified, or whose particles are not cemented together.

Unconsolidated Material—See Unconsolidated Deposits.

Underground Water—See Ground Water.

Unsaturated Zone—The zone between the land surface and the water table. It includes the capillary fringe and may contain water under pressure less than that of the atmosphere.

Upconing—The process by which saline water underlying fresh water in an aquifer rises upward into the fresh water zone as a result of pumping water from the fresh water zone.

Usable Storage Capacity—The quantity of ground water of acceptable quality that can be economically withdrawn from storage.

Vadose Zone—See Unsaturated Zone.

Valley Floor—The central portion of a ground water basin; an area of low-to-negligible relief suitable for agricultural or urban development.

Water-Bearing—An adjective used to describe bodies of rock that contain or yield water.

Water Budget—See Hydrologic Budget.

Water Quality—1) The fitness of water for use; and 2) the physical, chemical, and biological characteristics of water.

Watershed—A region or area bounded peripherally by a topographic divide that drains to a stream or lake.

Water Table—1) The upper surface of a saturated zone except where that surface is formed by an impermeable body; 2) locus of points in soil water at which the pressure is equal to atmospheric pressure; or 3) the surface where ground water is encountered in a well in an unconfined aquifer. The water table is a particular potentiometric surface.

Water Well—1) A well that extracts water from the saturated zone or that yields useful supplies of water; or 2) a well that yields ground water information (see Piezometer) or that replenishes ground water.

Well—An artificial excavation (pit, bore, hole, tunnel), generally cylindrical in form and often walled in, sunk (drilled, dug, driven, bored, or jetted) into the ground to such a depth as to penetrate water-yielding rock or soil and to allow the water to flow to or be pumped to the surface or recharged into the subsurface; a water well.

Well Casing—The casing or tubular lining of a bored or drilled well.

Well Construction—The physical characteristics of a water well; e.g., method of drilling well, depth and diameter of casing, depth and type of perforations, size and extent of filter envelope, length of well seal, etc.

Well Log—1) A record made by the driller, geologist, or engineer of a well which lists geologic materials encountered during drilling and information on the construction of the well (such as casing perforations and sanitary seal); or 2) a log obtained by special devices lowered in a well, showing such information as resistivity, radioactivity, spontaneous potential, and acoustic velocity as a function of depth; especially a lithologic record of the rocks penetrated.

Well Screen—A wire-wound tube or factory perforated casing installed in a well that maximizes the entry of water from the producing zone and minimizes the entrance of sand and other aquifer materials.

Zone of Aeration—See Unsaturated Zone.

Zone of Saturation—See Saturated Zone.

CHAPTER 6

RUNOFF, STREAM FLOW, RESERVOIR YIELD, AND WATER QUALITY

I. INTRODUCTION

Runoff, or surface runoff, refers to all the waters flowing on the surface of the earth, either by overland sheet flow or by channel flow in rills, gullies, streams, or rivers. Stream flow refers to the flow in natural streams. Runoff and stream flow are continuous processes by which water is constantly flowing from higher to lower elevations by the action of gravitational forces. In this way, overland flow concentrates into small streams, which in turn combine to form larger streams and rivers. Eventually, rivers flow into oceans completing the hydrologic cycle.

Runoff is usually expressed in terms of either volume or flow rate. The usual units of runoff volume are cubic meters, cubic feet, or acre-feet. Flow rate or discharge is usually expressed in cubic meters per second (m^3/s) or cubic feet per second (cfs). Discharge at a cross-section or gaging station usually varies in time; therefore, its value at any time is the instantaneous or local discharge. Instantaneous values of discharge can be integrated over a period of time to give the runoff volume for the entire period.

Runoff is also expressed in terms of depth units, by dividing the runoff volume by the catchment/watershed/basin area to obtain an equivalent spatially averaged runoff depth. In certain applications of flood hydrology runoff can also be expressed as: a) peak discharge per unit drainage area ($m^3/s/km^2$; cfs/mi^2), b) peak discharge per unit runoff depth ($m^3/s/cm$; cfs/in), and c) peak discharge per unit drainage area per unit runoff depth ($m^3/s/km^2/cm$; $cfs/mi^2/in$).

This introductory section sets the stage for the remainder of the chapter. It contains two subsections: 1) Description of Physical Processes and 2) Variability of Runoff. Description of Physical Processes seeks an understanding of the runoff processes as a framework for the detailed calculations that will follow. Variability of Runoff describes the aspects inherent to the study of runoff, namely its local, spatial, temporal, seasonal, regional, and geographical variability.

A. Description of Runoff Process

Surface runoff comprises all the waters flowing on the surface of the earth in response to precipitation. Surface runoff eventually goes on to constitute stream flow; however, stream flow comprises both surface and subsurface flow. The subsurface flow component of stream flow consists of exfiltration from interflow and ground water flow.

Surface runoff occurs when water originating in precipitation (rainfall and snow) flows freely on the surface of the earth, driven by gravitational forces. Typically, the first manifestation of surface runoff is overland sheet flow. In overland flow, water flows under laminar or mixed laminar/turbulent flow conditions as sheet flow on the surface, usually at very small depths and velocities.

Given the topographical irregularity of the earth's surface, overland flow soon concentrates into rill flow; in turn, rill flow concentrates into gully and stream flow. Rill flow is the flow in small rivulets. These feature greater depths and velocities than the adjacent or contributing overland flow. The distinction between rills and gullies is primarily one of scale; rills can be obliterated by common agricultural practices. Gullies, on the other hand, are a more permanent feature of the landscape. The distinction between gullies and streams is largely one of age. Gullies are young in a geologic time frame while streams are older.

1. ***Overland Flow Generation*** Three distinct physical processes and models have been developed to explain the process of overland flow generation. These are: a) Hortonian overland flow, b) saturation overland flow, and c) overland flow due to surface crusting. In Hortonian overland flow (named after Robert E. Horton, the renowned American hydrologist), the excess rainfall intensity (i.e., the rainfall intensity minus the rate of infiltration) flows on the ground surface first as overland sheet flow, later as rill and/or gully flow, eventually concentrating into streams and rivers. This type of overland flow is typical of highly intensive rainstorms on arid and semiarid catchments, which often lack enough vegetation cover (and associated litter and shallow root environment) to retain moisture for extended periods. As soon as the rainfall intensity exceeds the capacity of the soil to absorb water, the flow runs on the ground surface toward lower elevation.

Saturation overland flow occurs when the soil profile is saturated. In this case, overland flow occurs—regardless of the difference between rainfall intensity and infiltration rate—due to the soil's inability to absorb any more water. It follows that saturation overland flow occurs when long-duration general storms cover large basins having relatively shallow soil mantles. Typically, saturation overland flow occurs as a result of a storm couplet (i.e., a sequence of two storm events). The first storm saturates the soil mantle; the second storm—following on the wake of the first—goes on almost entirely to constitute overland flow and surface runoff.

The third process of overland flow and surface runoff generation is caused by the phenomenon of surface crusting. This mode of surface runoff generation occurs when highly intensive storms fall on arid and semiarid catchments devoid of protective vegetation cover. The excess rainfall intensity leads to Hortonian flow and, depending on the upper soil layer's texture and structure, to splash erosion and associated entrainment of silt-sized particles in the flow. The entrained sediments are transported downwards, segregated by size, and eventually deposited on the surface to form an impermeable crust. Once the surface crust is developed, infiltration is reduced to negligible amounts, and essentially all subsequent rainfall is converted into surface runoff. The formation of a surface crust is, by nature, highly random and dependent on a suitable combination of climate, storm type, vegetation cover, soils, and topographic relief.

2. ***Stream Flow Generation*** Stream flow generation differs from overland flow generation in that it includes both surface and subsurface runoff. Classical hydrology defines stream flow in terms of three components: a) surface runoff, b) interflow, and c) ground water flow. Notwithstanding classical hydrology, recent theories of stream flow generation have emphasized the timing, rather than the path, of the components of stream flow. Two components are recognized under this framework: a) quickflow, consisting of surface runoff (overland, rill, and gully), fast interflow, and rain falling directly on the channel network and (b) baseflow, consisting of slow interflow and ground water flow. This permits the separation of stream flow into two, rather than three, components, simplifying the process of modeling and computation.

3. ***Hillslope Hydrology*** Hillslope hydrology refers to the hydrologic process taking place on hillslopes. These processes are intrinsically related to stream flow generation. The question to be answered in this

regard is whether the referred path of runoff is on the surface, as overland flow, or through the subsurface, by subsurface storm flow or fast interflow.

Theories of stream flow generation from hillslopes range from Hortonian overland flow (surface flow) to throughflow (subsurface interflow). Hortonian flow is applicable to arid and semiarid regions with poorly vegetated slopes and relatively thin soil covers. Throughflow is applicable to humid and subhumid regions with heavily vegetated areas and thick permeable soil layers. Between these two extremes lies a gamut of theories that consider a mix of Hortonian overland flow and throughflow.

The partial area flow concept (Ponce, 1989) is a combination of throughflow in the upper hillslopes and overland flow in the lower hillslopes. Interest in the partial area concept arose from the recognition that stream flow estimates were improved by assuming that only a small portion of the watershed is able to contribute to stream flow during the hydrograph peak. The size of the partial areas is a function of storm depth, rainfall intensity, and antecedent moisture. Generally, partial areas may be expected to encompass between 5 and 20% of the total watershed.

Refinements of the partial area concept have led to the variable source area model of hillslope hydrology. These variable source areas are envisioned as comprising low-lying lands adjacent to streams and rivers and concentrated near watershed outlets. Variations in the extent of source areas are dictated by antecedent moisture conditions, soil-moisture storage capacity, and rainfall intensity. When the upper soil horizon is saturated, both throughflow and overland flow occur, with eventual exfiltration to stream banks.

The variable source area model is a dynamic version of the partial area concept. This dynamism is manifested in soil moisture and runoff changes occurring annually, seasonally, between storms, and during storms. The variable source areas are largely responsible for overland flow, whereas the remainder of the watershed acts primarily as a reservoir to provide baseflow and to maintain saturation of the source areas.

4. Runoff Concentration An important characteristic of surface runoff is its concentration property. To describe this property, assume that a storm falling on a given watershed produces a temporally and spatially uniform effective rainfall intensity. In such a case, surface runoff eventually concentrates at the watershed outlet, provided the effective rainfall duration is longer than or equal to the time of concentration. Runoff concentration implies that the flow rate at the outlet gradually increases until rainfall from the entire watershed has had time to travel to the outlet and is contributing to the flow at that point. At that time, the maximum, or equilibrium flow rate is reached, implying that the surface runoff has concentrated at the outlet. The time that it takes a parcel of water to travel from the farthest point in the divide to the watershed outlet is referred to as the time of concentration.

The equilibrium flow rate is calculated by multiplying the effective rainfall intensity by the watershed area:

$$Q = 2.78\, I_e A \tag{6.1}$$

in which Q is the equilibrium flow rate in l/s, I_e is the effective rainfall intensity in mm/hr, and A is the watershed area, in ha.

The process of runoff concentration can lead to three distinct types of watershed response assuming constant rainfall excess and insignificant infiltration. The first type occurs when the effective rainfall duration is equal to the time of concentration. In this case, runoff concentrates at the outlet, reaching its maximum rate after an elapsed time equal to the time of concentration. Since rainfall stops at this time, subsequent flows at the outlet are no longer concentrated because not all the watershed is contributing. Therefore, the flow gradually starts to recede back to zero, as shown in Fig. 6.1(A), where I_e is effective rainfall intensity (mm/hr), Q is the flow rate (l/s), Q_e is the equilibrium flow rate (l/s), $Qmax$ is the maximum flow rate (l/s), t is time (hr), t_c is concentration time (hr), and t_r is the duration of effective rainfall (hr).

The second type of watershed response occurs when the effective rainfall duration is longer than the time of concentration. In this case, the runoff concentrates at the outlet, reaching its maximum rate after an elapsed time equal to the time of concentration. Since rainfall continues, the whole watershed continues to contribute to flow at the outlet, and subsequent flows remain concentrated and equal to the equilibrium value. After rainfall stops, the flow gradually recedes back to zero, as shown in Fig. 6.1(B).

The third type of response occurs when the effective rainfall duration is shorter than the time of concentration. In this case, the flow at the outlet does not reach the equilibrium value. After rainfall stops, the flow recedes back to zero. The requirement that volume be conserved and recession time be equal to the time of concentration leads to the idealized flat-top response shown in Fig. 6.1(C).

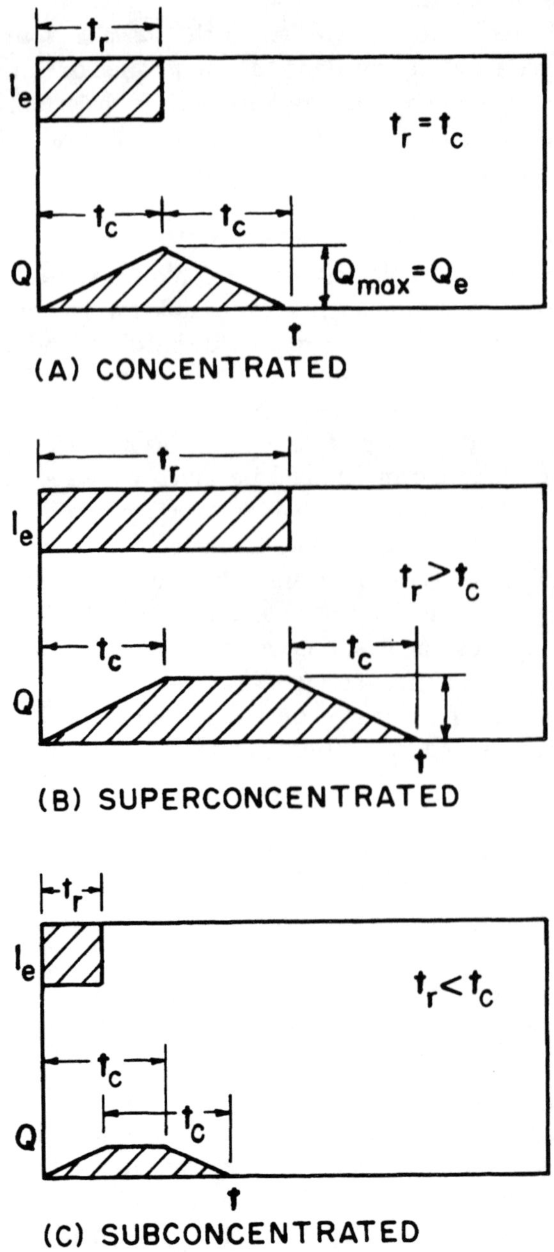

Figure 6.1—Effect of Rainfall Duration on Runoff.

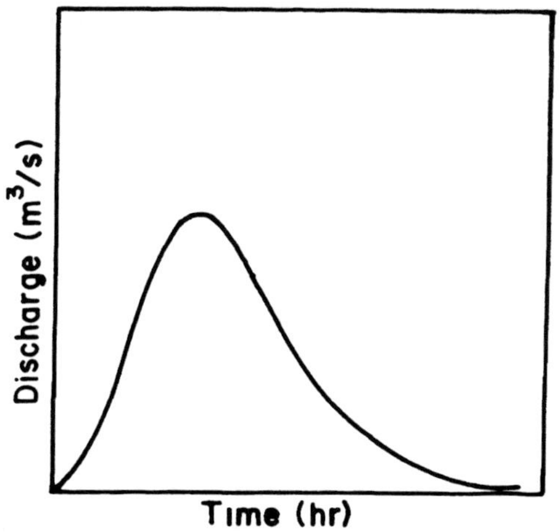

Figure 6.2—Typical Single-Storm Hydrograph.

5. ***Runoff Diffusion*** In nature, watershed response shows a more complex behavior than that which may be attributed solely to runoff concentration. Runoff rates from a watershed are governed by the natural processes of convection and diffusion. Convection refers to simple translation or wave travel. Diffusion is the mechanism acting to spread the flow rates in time and space.

The net effect of runoff diffusion is to reduce the flow rates to levels below those that could be attained by convection only. In practice, diffusion acts to smooth out watershed response. The resulting response function is usually continuous, and it is referred to as the stream flow hydrograph, runoff hydrograph, or simply "the hydrograph." Typical single-storm hydrographs have a shape similar to that shown in Fig. 6.2.

6. ***Rating Curves*** It is known that river stage varies with discharge, but the exact nature of this relationship is not readily apparent. Given a long and essentially prismatic channel reach, a single-valued relationship between stage and discharge at a cross-section defines the equilibrium rating curve. For steady flow, there is a single value of stage for each value of discharge. In this case, the equilibrium rating curve can be calculated with either the Manning or Chezy formulas. This property of uniqueness of the rating qualifies the channel reach as a channel control.

Rating curves are a useful and practical tool in hydrologic analysis, allowing the direct conversion of stage to discharge and vice versa. Discharge can be obtained from the rating curve by the simple procedure of measuring the stage. Conversely, if discharge is known, for instance, at a watershed outlet, stage at the outlet can be readily determined from a suitable rating.

There are several ways to determine an equation for the rating. Invariably, they are based on curve-fitting stage-discharge data. A widely used equation is the following:

$$Q = a(h - h_o)^b \tag{6.2}$$

in which Q is discharge (cms), h is the gage height (m), h_o is the reference height (m), and a and b are constants. Several values of reference height may be attempted. The proper value is that which makes the stage-discharge data plot as close as possible to a straight line on logarithmic paper, with the values of a and b determined by regression.

Other flow conditions, specifically nonuniform and unsteady, can cause deviations from the equilibrium rating. In particular, flood wave theory justifies the presence of a loop in the rating curve. Intuitively, the rising limb of the flood wave hydrograph has a steeper water surface slope than that of equilibrium flow, leading to greater flows at lower stages. Conversely, the receding limb has a milder water surface slope, resulting in smaller flows at higher stages (Fig. 6.3); however, the loop effect is likely to be small.

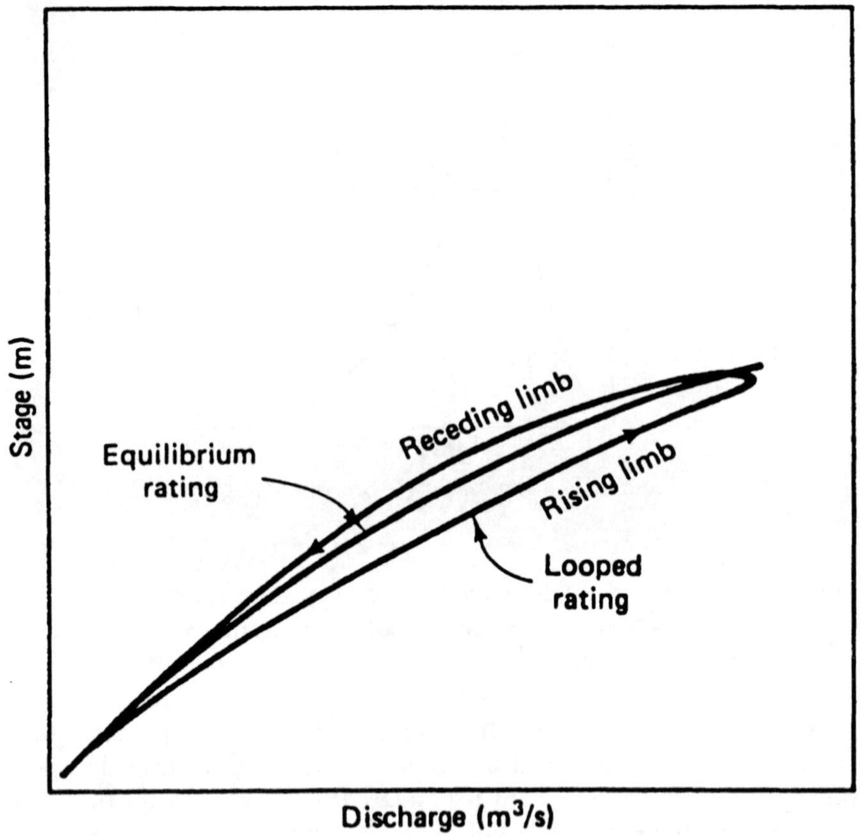

Figure 6.3—Equillibrium and Looped Rating Curves.

Two other mechanisms have a bearing in the evaluation of stage-discharge relations: the short-term and the long-term sedimentation effects. The short-term effects are due to the fact that the amount of boundary friction varies with flow rate. Rivers flowing on loose boundaries composed of gravel, sand, and silt constantly try to minimize their changes in stage. This is accomplished through the following mechanism: during low flow, the bottom friction consists not only of grain friction but also of form friction, the latter due to bed features such as ripples and dunes during high flows, the swiftness of the current acts to obliterate the bed features, reducing the form friction to a minimum, with only the grain friction remaining. The reduced friction during high flows gives rivers the capability to carry a greater discharge for a given stage. This explains the shift from low-flow rating to high-flow rating in natural river channels.

The long-term sedimentation effect is due to the fact that rivers continuously subject their channels to endless cycles of erosion and deposition, depending on the sediment load they transport. Some very active rivers may be eroding; others may be aggrading. Some geomorphologically active rivers may substantially change their cross sections during major floods. Invariably, shifts in rating are the net result of these natural geomorphic processes.

B. Variability of Runoff

Runoff and stream flow vary not only seasonally but also annually and with geographic location. Stream flow data for individual basins show that variability in runoff volume is significantly longer than that in the rainfall volume for the same period. On a global basis, total annual runoff volume is about 30% of the total precipitation volume. The difference is due to the abstractive processes, i.e., interception, infiltration, surface storage, evaporation, and evapotranspiration. The value of 30% does not account for seasonal or geographical variability.

1. Seasonal Variability The typical basin in a temperate region shows runoff volumes varying during the year, low during the dry season and high during the wet season; however, basins in more extreme climates may show different behavior. For instance, in the ephemeral streams of typical arid regions, runoff is nonexistent during periods of no precipitation. For these streams, runoff and stream flow occur only in direct response to precipitation. Conversely, in humid and tropical climates, rivers show substantial quantities of runoff throughout the year, with comparatively little variability between the seasons.

The seasonal variability of stream flow can be traced to the relative contributions of direct and indirect runoff. During the wet season in temperate regions, indirect runoff is a small but nevertheless measureable fraction of direct runoff. In arid regions, particularly for ephemeral streams, indirect runoff is either negligible or does not exist. For humid regions, indirect runoff is substantial throughout the year and may be of the same order of magnitude as direct runoff.

The seasonal variability of runoff and stream flow can be further explained in the following way: ground water reservoirs serve as the mechanism for the storage of large quantities of moisture, which are slowly transported to lower elevations (or zones of lower potential). Some of the moisture is eventually released back to the surface waters. The process is slow and is therefore subject to substantial diffusion. The net effect is a permanent contribution from ground water to surface water in the form of dry-weather flow. Therefore, to evaluate the seasonal variability of runoff and stream flow, it is necessary to examine the relation between surface water and ground water.

2. Annual Variability Year-to-year stream flow variability shows some of the same features as those of seasonal stream flow variability. For instance, large basins show runoff variability from one year to the next as a function of the state of moisture at the end of the first year and of the precipitation quantities added during the second year. During dry years, rainfall replenishes the basin's stored moisture, with little of it showing as direct runoff. This results in the low levels of runoff that characterize dry years. Conversely, during wet years, the basin's storage capacity fills up quickly, and any additional precipitation is almost entirely converted into runoff, producing high stream flow levels that are characteristic of wet years.

Runoff and stream flow variability are intrinsically linked to the relative contribution of direct and indirect runoff involving the mechanisms of surface and ground water flow, and their temporal and spatial variability. The dearth of data for all relevant phases of the hydrologic cycle makes the evaluation of stream flow variability on mechanistic principles a rather complex process. A practical alternative is the reliance on statistical tools to develop estimates of runoff and stream flow variability based on records. This alternative has led to the concept of flood frequency, expressed as the average period of time (the return period) that it takes a certain flood magnitude to recur at a given location. An annual flood series is developed from daily discharge measurements at a given station. This is accomplished either by selecting the maximum daily flow for each of n years of record (the annual maxima series), selecting the greatest flow values in the entire n-year record (the annual maxima series), or selecting the greatest flow values in the entire n-year record, regardless of when they occurred (the annual exceedence series). The statistical analysis of the flood series allows the calculation of the flow rate associated with one or more selected frequencies.

3. Daily-Flow Analysis The variability of stream flow can also be expressed in terms of the day–to–day fluctuation of flow rates at a given station. Some streams show great variability from day to day, with high peaks and low valleys succeeding one another endlessly. Others show very little day–to–day variability, with high flows not very different from low flows. The reason for this difference in behavior can be attributed to differences in the nature of watershed response. Small basins are likely to have steep gradients and to concentrate flows with negligible runoff diffusion, producing hydrographs that show a large number of high peaks and correspondingly low valleys. Conversely, large basins are likely to have milder gradients and to concentrate flows with substantial runoff diffusion, resulting in a smooth hydrograph with low peaks and comparatively high valleys. A practical way to evaluate day-to-day stream flow variability is the flow–duration curve (see section on flow duration curves).

4. Geographical Variability of Stream Flow Stream flow varies from one basin to another and from one geographic region of a certain climate to another of a given climate. Two variables help describe the geographical variability of stream flow: a) basin area and b) mean annual precipitation.

Intuitively, the volume available for runoff is directly proportional to the basin area; however, this concept is limited by the available precipitation. The mean annual precipitation determines the type of climate and therefore has a direct bearing on the variability of stream flow (see Chapter 2). The basin area is important, not only because of the potential runoff volume but also because larger basins have milder overall gradients. This causes increased runoff diffusion, increasing the chances for infiltration and loss of surface water to ground water. The net effect is that the peak discharge per unit basin area is inversely related to basin area, as in the following equation:

$$q_p = cA^{-m} \tag{6.3}$$

in which q_p is peak discharge per unit basin area, A is the basin area in km^2, and c and m are constants to be determined by regression. An example of this trend is represented by the Creager curves (Creager et al., 1945). The equation for Creager's enveloping equation is as follows:

$$q_p = C_1(0.386\mathrm{A})^m \tag{6.4}$$

$$m = 0.9358\, A^{-0.048} \tag{6.5}$$

in which q_p is peak discharge per unit basin area, ((m^3/s)/km^2), A is the basin area (km^2), $C_1 = \mathrm{C/A}$, and C is a constant varying in the range 38–130.

II. MEASUREMENT OF STREAM FLOW

The measurement of stream flow usually involves the determination of average velocity in the flow section and the application of this average velocity to an area of flow that is separately determined. This requirement for two types of data collection introduces several sources of possible error in the final determination of discharge rate. Many of the flow measurement techniques described herein deal primarily with the determination of average flow velocity. The errors quoted most often relate to velocity determination. The exceptions are flumes and weirs where velocity and area are combined to compute the discharge rate. When the area is defined by an irregular natural channel that is difficult to measure, significant errors in estimated stream flows may result due to errors or inaccuracies in the measurement or computation of flow areas.

A. Direct Measurement

This method involves measuring the width, depth, and velocity of flow in each of about 20 to 30 subsections of a stream cross section, computing the discharge in each subsection as the product of width, depth, and mean velocity, and summing the discharges in the subsections.

1. Current Meters The hydrographer selects the locations of the verticals at which velocity measurements are to be taken. The hatched area in Fig. 6.4 is a subsection whose vertical boundaries are halfway between the adjacent verticals. The verticals are located so as to obtain subsections which have about the same discharge, or ones for which the velocity in a subsection does not vary appreciably laterally from

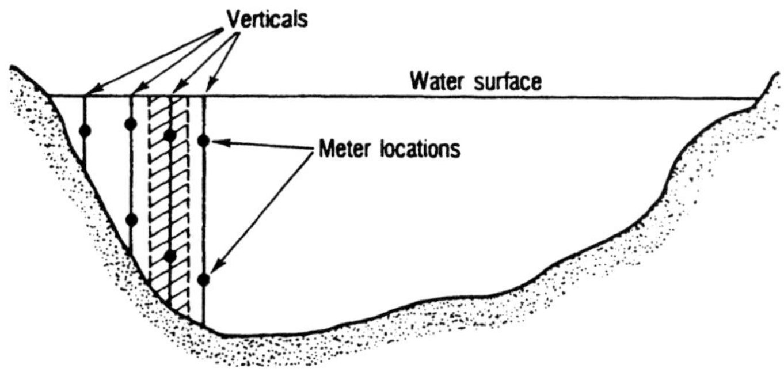

Figure 6.4—Stream Cross Section Showing Meter Locations for a Discharge Measurement (Riggs, 1985).

that measured. Fig. 6.4 also shows meter locations at which velocity measurements are made. These locations are at 0.2 and 0.8 of the depth except for shallow depths for which the meter is set at 0.6 depth from the water surface. Studies have shown the mean of velocities at 0.2 and 0.8 depth to be virtually the mean velocity in the vertical; similarly, the velocity at 0.6 depth is very close to the mean in the vertical. In special situations, the flow profile may warrant checking to verify that it is indeed "standard." This is done by collecting data at 10 or 20 depth points in one or more verticals and comparing this average velocity determination with the previously mentioned determinations (Riggs, 1985; Herschy, 1985).

Current meters commonly used in the United States are vertical-axis Price meters. The standard meter (Fig. 6.5) is for general use, and the Pygmy meter (two fifths the size of the standard meter) is used for shallow depths. Meters are rated by towing them at several constant velocities in a long tank containing still water.

When measuring velocity, the number of revolutions of the meter rotor is obtained by counting clicks (one click per one or five revolutions) in an earphone. Clicks are produced by an electrical circuit between the meter and the earphone. Elapsed time is measured with a stopwatch. The number of revolutions and the elapsed time can also be determined electronically with a current-meter digitizer, which also displays the results.

For measurements of shallow streams, the hydrographer usually wades directly into the stream. In the wading method, a tag line is stretched across the stream at the selected cross section. This line has marks

Figure 6.5—Standard Price Current Meter Mounted Above a Sounding Weight.

at selected intervals for horizontal control (usually 1 ft, 5 ft, 1 m, etc.). The current meter is mounted on a wading rod that is used to measure depth and to hold the current meter at the desired location or locations in the vertical.

Deeper streams are usually measured from bridges, cableways, or occasionally from boats. Bridge railings and cableways are marked at intervals for horizontal control. A heavy tag line is used for locating and holding the position of a boat on a cross section. The meter is suspended on a light cable above a sounding weight (Fig. 6.5). A reel mounted on the bridge, cableway or boat provides vertical location control. Flags at intervals on the cable and a counter on the reel are used to indicate the depth sounded and to set the meter to the desired positions in the verticals. The weight also helps maintain the line in a near vertical position.

When a stream is ice covered, measurements of depth and velocity may be made through holes cut in the ice along a selected cross section. The type of meter suspension depends on the water depth. Velocity distribution in the vertical under ice cover is similar to the velocity distribution in a pipe. Velocity is measured at 0.6 depth for depths under 0.75 m, and at 0.2 and 0.8 depth for greater depths. For greater precision, it may be desirable to thoroughly define the vertical velocity curves at one or more verticals in the section.

Reliability of discharge measurements ranges widely. Best results are obtainable in a smooth channel where the velocity is moderate, changing only slightly between adjacent verticals; the velocity is perpendicular to the cross section; and the discharge of the stream does not change during the measurement. Under these conditions, the potential accuracy of the measurement is rated excellent. Carter and Anderson (1963) discuss the accuracy of discharge measurements in more detail.

Portable electromagnetic velocity sensors are commercially available and are used in many of the flow metering situations that commonly use the mechanical meters discussed above. The sensor is a probe, 30 to 44 mm in diameter, which is pointed upstream. The basic procedure to determine channel discharge from these point velocity readings is the same as described for the mechanical meters. Water flowing past the probe interacts with an electromagnetic field generated inside the probe to produce a voltage in the water near the probe. This voltage is a function of water velocity. Auxiliary equipment facilitates data gathering and computation of the final discharge rate for a channel section. Its size allows it to be handled and used much like the Pygmy current meter except that it has selectable flow ranges that can generally allow it to also substitute for the larger mechanical meter for flows exceeding about 1 m/s. The necessary shielded cable that is used with it may limit its application from bridges.

These probes operate on electromagnetic principles. The probe consists of an electromagnet inside a molded plastic housing with two electrodes spaced 180° apart and about one probe radius from the rounded end of the probe. Water flowing around the probe end and along the probe interacts with the electromagnetic field to produce a small voltage in the water near the probe. This voltage is detected by the electrodes and is a function of the flow velocity. In one version, the result is processed and a point velocity is displayed on a panel meter.

With mechanical meters, the velocity errors are somewhat controlled by operator choices of count length and propeller selection. In the electromagnetic versions, these choices are averaging–time, corresponding to count length, and flow-range, which corresponds to propeller selection. Stray electrical currents, particularly when operating in urban environments, have been reported by field users.

Errors in area determination and velocity distribution frequently dominate the final discharge results when using the mechanical meters, particularly on small channels, meaning that the point velocities can be relatively accurate when compared to these other sources of error. The electromagnetic probes, on the other hand, may have problems of stable calibration and electrical interference that may introduce uncertainties in individual velocity determinations, adding another degree of uncertainty to the final results.

Because the reliability of velocity measurements obtained from the electromagnetic sensor is not comparable with the reliability of the mechanical meter, the International Standards Organization (ISO) has not yet approved this type of meter for use in gaging streams. A further limitation is the need for a shielded cable between the probe and the recorder or indicator. This requirement appears to limit use of the electromagnetic velocity sensor to the gaging of small streams. The reader is referred to the section titled Electromagnetic Gaging Stations for an application of this technique.

2. ***Moving-Boat Method*** The employment of the moving-boat method permits the discharge of a large stream to be measured within a few minutes. The conventional velocity–area approach for determining discharge is used, but the data are collected at 30–40 verticals observed from a boat while it traverses the cross section.

A rate indicator unit is used in conjunction with a current meter rating table to obtain the vector combined velocity of the stream and the boat. An angle reading, representing the angle a vane makes with the cross-sectional path, defines the direction of the vector (Fig. 6.6).

Prior to measuring, two range markers on each side of the cross section are established, anchored floats are located on the cross section 15 to 20 m from each shore, and the width of the cross section is measured.

Also required are two or more vertical-velocity characterizations made in the cross section. Whereas the standard gage points may have one or two samples in each vertical, the vertical-velocity characterizations can be made with enough samplings at vertical intervals to reliably define the velocity profile in the vertical. This may require 10 to 20 point velocity determinations throughout the vertical. These are used to adjust velocities measured at a constant distance below water surface to mean velocities in the vertical.

Equipment on the boat includes an angle indicator, a propeller-type current meter and a sonic sounder. Also required is a rate indicator which shows the current-meter velocity and provides a method of automatically selecting measuring points at intervals of distance traveled.

The traverse of the cross section is made without stopping and data are collected at regular intervals. The boat operator maintains a course along the cross section by "crabbing" into the direction of flow sufficiently to keep on line. The force exerted on the current meter is the combination of two forces acting simultaneously, one force due to the movement of the boat, and the other from the stream flow normal to the cross section. Data collected at each sampling point include the sensed vector velocity of the boat and stream and the angle of the propeller and vane to the line across the stream, which permits computation of the stream velocity.

Details of the procedures and the computation of discharge are given by Smoot and Novak (1969) and Herschy (1985). Advantages of the moving-boat method over conventional methods include the short time required and the consequent reduction of interference with marine traffic. Reliability of results is comparable to that for conventional measurements, if the results of several traverses are averaged. In normal practice, a series of at least 6 runs, each with 30 or 40 verticals, are averaged.

Instruments currently used for the moving boat are more automated than those described above. Furthermore, an acoustic doppler current profiler that provides a vertical profile of horizontal velocities

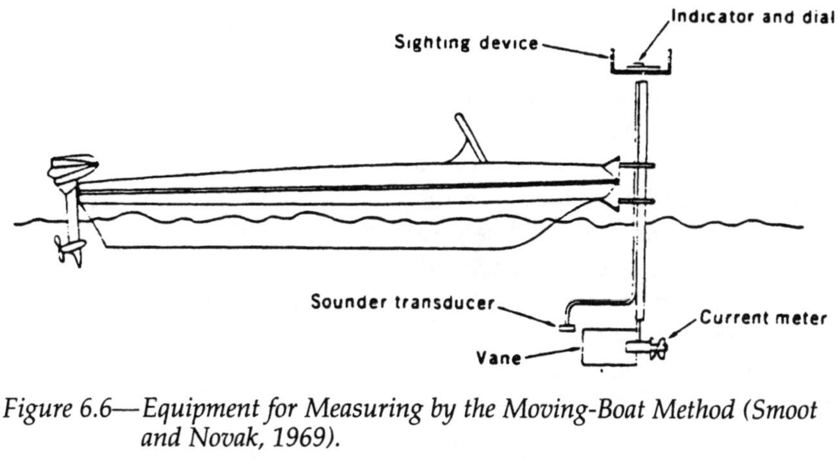

Figure 6.6—Equipment for Measuring by the Moving-Boat Method (Smoot and Novak, 1969).

and the direction of water movement is being developed. This system will likely develop as the new moving-boat system.

3. *Other Methods* Tracer-dilution methods are useful, primarily for small streams where there is considerable turbulence, making it difficult to use a current meter. These methods are also particularly useful when the cross-sectional area is difficult to define, such as boulder-filled mountain streams, ice-covered streams, or sand-channel streams. See Kilpatrick and Cobb (1985) for details.

Volumetric methods and portable weirs or flumes may be used to measure small flows of 500 l/s or less, but they are usually impractical in very rough channels or channels with very flat slopes. Permanently installed weirs or flumes usually produce reliable discharges once properly installed.

Unusual flood discharges have been estimated by timing drift in several parts of a cross section and by surveying the cross section after a flood. Results tend to be unreliable.

B. Indirect Measurements

It is frequently impossible to measure extreme flood peaks by the current-meter method because of inability to reach the site, the occurrence of floating or submerged debris, or the submergence or destruction of the measuring structure. Consequently, many peak discharges must be determined after the passage of a flood by indirect methods that are based on hydraulic equations relating the discharge to the water-surface profile and to the geometry and roughness of the channel. The water-surface profile is defined by high-water marks on both banks of the stream reach. See Benson and Dalrymple (1967) for the various types of high-water marks, and for general field and office procedures for making indirect measurements. The common types of indirect measurements are slope-area, contracted-opening, and flow-over-dams, weirs, and embankments.

1. *Slope-Area* This is the most commonly used form of indirect measurement. The basic equation in the United States is the Manning formula

$$Q = (1/n)\,A\,R^{2/3}\,S^{1/2} \tag{6.6}$$

in which Q is the discharge, (m^3/s), A is the cross sectional area (m^2), and R is the hydraulic radius (m). In United States customary units,

$$Q = (1.486/n)\,A\,R^{2/3}\,S^{1/2} \tag{6.7}$$

where Q is the discharge (cfs), A is the cross sectional area (sq ft), R is the hydraulic radius (ft), S is the friction slope (dimensionless), and n is a roughness coefficient.

A field survey of the selected reach involves locating and surveying high-water marks for defining water surface profiles on both banks, surveying three or more cross sections, and estimating the roughness coefficients for the reaches between adjacent cross sections. From these data, the discharge can be computed as described by Dalrymple and Benson (1967) and Herschy (1985).

The reach selected should be reasonably straight with uniform channel geometry, uniform channel roughness, a smooth water-surface profile, and no wide overflow sections. These conditions should also exist for some distance upstream of the reach. Length of the reach should be at least 75 times the mean depth in the channel. Fall in the reach should be greater than the velocity head and greater than 15 cm. A minimum of three cross sections are recommended.

Selection of the appropriate roughness coefficients requires some experience. Guidelines for selecting roughness coefficients in natural channels are given in many texts, for example, Chow (1959) and also Barnes (1967), who compiled the results of many verifications of the Manning roughness coefficient in natural channels. The roughness coefficient of a natural channel is a function of bed roughness, bank irregularity, effect of vegetation (if any), depth of water, channel slope, and other factors. Consequently, the value of n computed from known water-surface profiles, channel geometry, and discharge may not be transferred with confidence to another site unless the dominant factors affecting n are common to the two sites.

Reliability of a discharge computed by the slope-area method is reduced if there are poor or questionable flood marks, if the water-surface profiles are irregular, and if the areas of adjacent cross sections are much different. Furthermore, the areas at cross sections, as measured after a flood, are assumed to be as they existed at the time of the peak discharge, but this is unlikely to be correct for channels with erodible beds. All of these factors, as well as the uncertainty in the estimates of n, need to be considered in evaluating the reliability of a particular slope-area measurement. Some indication of the reliability of peak discharges computed by the slope-area method is given by Riggs (1976) in which he uses estimates of the roughness coefficients recorded on the field notes of some surveys of channel reaches for which the coefficients were later verified from known peak discharges. These estimates of n (or averages of several) were available at 27 sites and were used to compute discharges from the data provided by the field surveys. The discharges, based on these estimates of n, were found to be within 30% of the known discharges. The estimates of n for these 27 sites were made by hydrologists with extensive experience in this type of work. Personnel with less experience might select less desirable reaches and make less suitable estimates of n.

2. ***Control in Channels*** Recording of stream flow usually depends on exploiting some type of channel control and recording a stream flow stage at a gaging station. This stage must be related to channel velocity and channel flow area. While area is a matter of channel geometry relative to stage, the flow velocity depends on a velocity control mechanism that can be mathematically characterized. For open channels, controls are frequently accomplished by channel control, which means that the flow velocity is controlled by channel friction. This rate of energy conversion to heat energy results in a decrease in total energy in the direction of flow (Chow, 1959). This slope of the energy line is used in equations such as the Manning formula to characterize flow in open channels. The errors of uncertainty in this control characterization can be considerable, 10 to 20%, or higher, because it usually requires an estimation of the friction value or current meter ratings to calibrate the channel section, which may not apply after storm flow passage. Because section calibrations usually change seasonally and during storm events, channel control is not as reliable as critical flow control.

Critical flow control, sometimes called section control, depends on an overfall such as a permanent riffle or rock ledge, a steep channel section, that causes supercritical flow downstream of a subcritical flowing reach, or a sufficient contracted opening, such as a bridge, that causes critical velocity. In this case, the contraction causes a rise in the upstream water surface that is just enough to pass the discharge through the opening at critical flow (Chow, 1959). Weirs and flumes are critical flow devices. The errors of uncertainty can be less than ±5% at well-characterized contractions or overfalls.

When suitable natural controls are not available, an artificial control may sometimes be attempted; but for large streams, artificial controls are expensive, likely to be damaged by floods and may cause changes in the approach channel.

3. ***Contracted–Opening Method*** Stream width contractions commonly occur at bridges. The contraction creates an abrupt drop in water-surface elevation between an approach section and the contracted section. This drop, the geometries of the approach and the contracted sections, and an estimate of the discharge coefficient are used to compute discharge. The equation is

$$Q = CA_3 \sqrt{2g\ (\Delta h + \alpha_1 \frac{V^2}{2g} - hf)} \tag{6.8}$$

in which Q is the discharge (m^3/s), C is the coefficient of discharge, A_3 is the gross area of contracted section (m^2), Δh is the difference in elevation between approach section and contracted section (m), α_1

$V^2/2g$ is the weighted average velocity head in the approach channel section, where V is the average velocity, Q/A, α_1 is a velocity distribution coefficient (Chow, 1959) that takes into account the variation in velocity in that section, and h_f is the head loss due to friction between the approach and the contracted sections (m).

The characteristics of a contracted section site should be carefully considered. A contraction should have a stable channel; the fall in water surface, as defined by high water marks, should be more than 15 cm; the fall should be more than four times the computed friction loss between the two sections; and the bridge geometry should be reasonably close to one of the types for which discharge coefficients have been defined. If these requirements cannot be met by any available contracted-opening site, some other method such as slope-area should be considered.

See Matthai (1967) for field and computation procedures including procedures for computing the discharge coefficient for selected bridge abutment geometries. Matthai also suggests methods for computing discharge through multiple-opening contractions and discharges that include flow over a road as well as through a contracted opening. The methods presented in Matthai (1967) are based on laboratory studies and field verifications.

4. Flow Through Culverts The peak discharge through culverts can be determined from high-water marks that define the headwater and tailwater elevations (Brater and King, 1976). For convenience in computation, culvert flow has been classified into six types based on the location of the control section and the relative heights of the headwater and tailwater elevations (Chow, 1959). Discharge equations have been developed for each type of flow by application of the continuity and energy equation between the approach section and the terminal section. Bodhaine (1968) describes the site information needed, how to determine the type of flow that occurred, and how to compute the discharge.

The peak discharge through a culvert is equivalent to the peak discharge of the stream a short distance above the culvert only if the flood caused minimal poundage upstream from the culvert. Therefore, poundage corrections are usually needed.

5. Flow Over Dams and Embankments A weir, dam, or embankment forms a control section at which discharge is related to the upstream water-surface elevation. The peak discharge at the control section can usually be determined from a field survey of high-water marks and the geometry of the particular structure.

The two most important elements of the flow computation are the head and the discharge coefficient of the structure; however, if a high velocity of approach makes the velocity head a substantial part of the total head, the computed discharge may be unreliable. Discharge coefficients have been derived for some types of dams and weirs. Thus, if the structure is one for which the discharge coefficient can be selected with confidence, and if the head on the structure can be accurately determined, then a reliable discharge can be computed.

Procedures for measuring peak discharge using dams, weirs, and embankments are given by Hulsing (1967). Also included in that report are discharge coefficients and formulas for three general classes of weirs—sharp-crested, broad-crested, and round-crested—and for highway embankments and weirs of unusual shape.

6. Weirs and Flumes More recently, a more complete treatment of fixed-section broad-crested weirs and long-throated flumes has become available (Bos et al., 1991). These configurations can be reliably rated by computer model (Clemmens et al., 1987) to within ±2% plus readout method error. Primarily developed for concrete-lined irrigation canals, the devices have been successfully used to measure discharges less than 25 m^3/s. Features are a) accuracy, b) economy of construction, c) flexibility of cross-sectional shape, d) fair to good sediment carrying capacity, and e) low required head loss, usually less than 10% of the stream flow depth. A typical configuration for the broad-crested weir style is shown in Fig. 6.7.

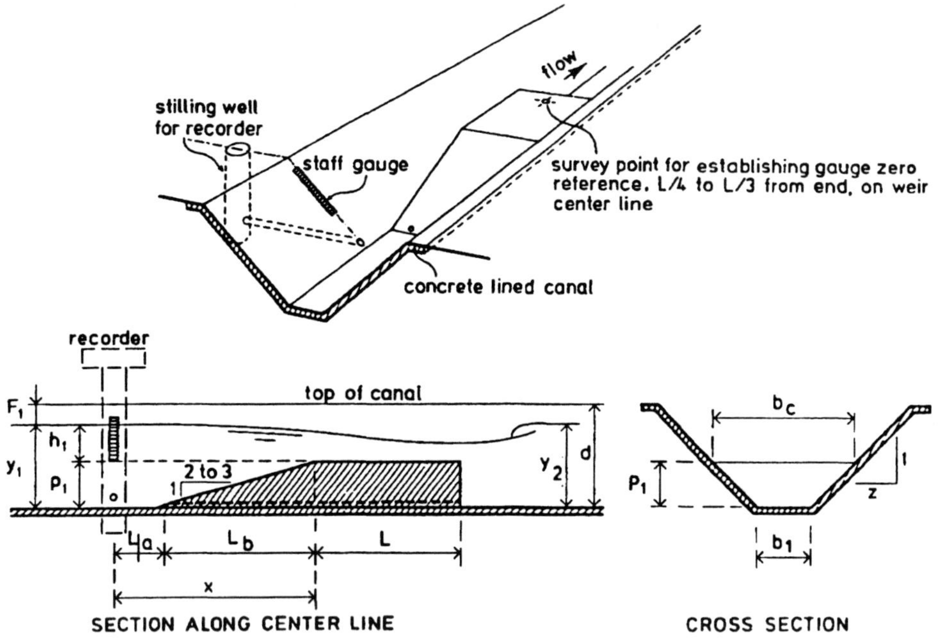

Figure 6.7—Typical Layout for Trapezoidal Long-Throated Flume (broad-crested weir style) Commonly Used in Lined Channels.

The broad-crested sill in a section of circular culvert or round pipe flowing partly full is shown in Fig. 6.8. The format shown is for a portable version that is practical for up to about 100 l/s.

Larger sizes have been designed for use in corrugated metal and concrete pipe culverts. Usually a sill height of one-quarter of the pipe diameter is used unless high downstream water levels force the use of a higher sill. The calibration for specific dimensions is derived by the modeling processes in Clemmens et al. (1987).

Sediments tend to clear these broad-crested weirs during the rising phase of a storm hydrograph, and appear to continue clearing during steady flow; however, on the falling leg of the hydrograph, the approach channel upstream from the sill may tend to fill with sediments. This is usually a problem with all weirs and flumes. Absolute velocities can be reliably computed for points throughout these devices, which can sometimes be used by the design engineer to evaluate sediment moving capabilities of known size or geomorphic source.

The computer calibration method is also applicable to side contractions. Complex sections for wide ranges in flow rate that approach 1:1000 (low to high flow rate ratio) also have been installed. In general,

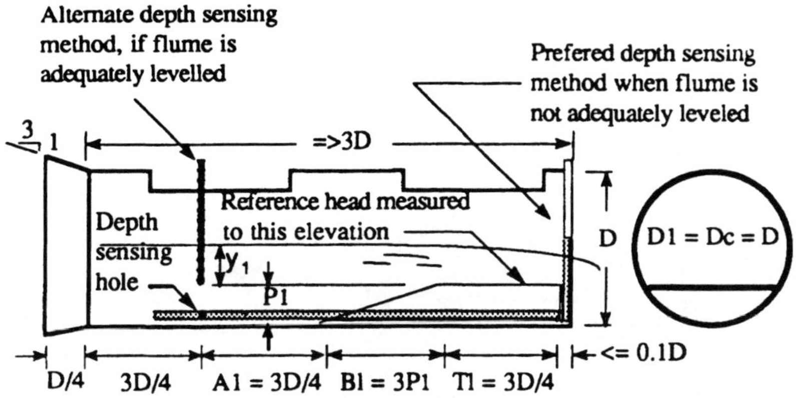

Figure 6.8—Broad-Crested Weir Style of Flume in a Round Pipe Flowing Partly Full.

the sediment carrying capacity of flumes does not appear to improve by using side contractions rather than bottom sills. A detailed treatment is given in Bos et al. (1991). Pre-computed tables are given for various rectangular and trapezoidal channels with sills in the previously mentioned references and in Clemmens et al. (1988) and Replogle et al. (1991). The latter also contains some precomputed sizes for circular pipes.

C. Continuous Records of Stream Flow

1. Conventional Gaging Stations Gaging stations typically obtain a continuous record of stream water stage. For channel control situations, this involves a continuous record of stream stage and the subsequent conversion to discharge through a rating curve. An example of a gaging station is shown in Fig. 6.9.

The general locations of gaging sites are usually dictated by water management considerations or by the hydrologic network. For channel control situations, an ideal location for a gaging station is a reach of stable channel that is isolated from downstream backwater effects by a steep section of channel, a permanent riffle, or a rock ledge a short distance downstream from the gaging station. At such a location, both reliable stage and calibration discharge measurements can usually be obtained. See Rantz (1982) for more detailed considerations in specific site selection.

2. Sensing and Recording Stage Water-surface elevation (stage) is usually referred to some arbitrary gage datum and is also called gage height. Gage height at a gaging station may be sensed by a float in a stilling well or by a gas-purge system that transmits the pressure head of water in the stream to a manometer or pressure transducer. Staff gages, mounted in the stream, are used to make periodic checks on the sensing equipment.

These sensings are recorded as a continuous trace of the gage height (on an analog recorder), as punches on a tape at intervals of a few minutes (on the digital recorders), or as computer bits on data acquisition systems. The gage-height record may also be recorded and stored temporarily on an electronic data collection platform and transmitted by satellite to computer data files. This and other applications of computer processing of hydraulic and hydrologic data is discussed in Herschy (1985).

3. Stage-Discharge Relations A stage-discharge relation is also called a rating or a rating curve. The relation is usually determined empirically by means of measurements of discharge and the corresponding stages. At a new station, many measurements are needed to define the relation throughout the entire range of stages. Periodic measurements are needed thereafter to verify the defined rating and identify changes, if needed.

When the discharge record at a new gaging station is needed before enough periodic data have been obtained to define the rating, the rating can be estimated using a survey of the downstream channel and

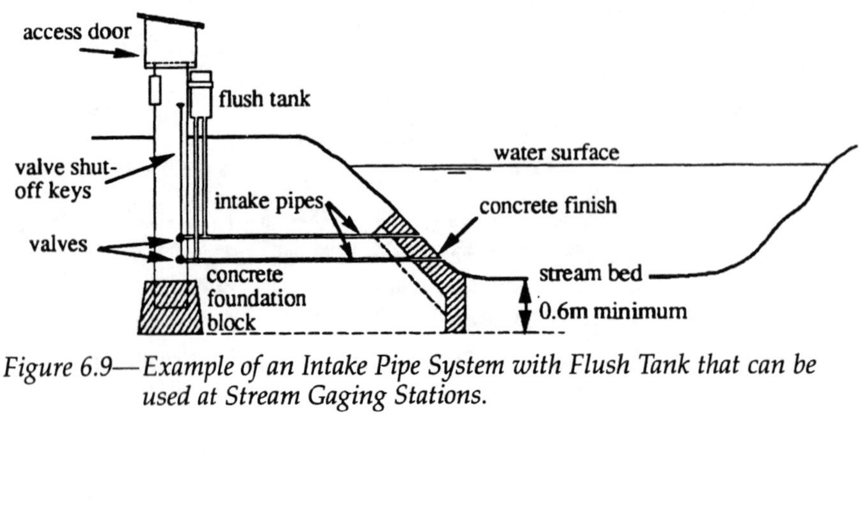

Figure 6.9—Example of an Intake Pipe System with Flush Tank that can be used at Stream Gaging Stations.

evaluations of channel roughness. The technique, known as the step-backwater method, is described in many hydraulics texts, e.g., Chow (1959), Davidian (1984), and verified by Bailey and Ray (1966).

Stage-discharge relations may change temporarily or permanently with changes in the physical characteristics of the channel. Consequently, as mentioned previously, periodic independent discharge measurements should be made to verify the established rating or to indicate changes. A temporary change in a rating may be caused by growth of vegetation in the channel, formation of ice or lodging of drift. Permanent changes usually result from scour or fill produced by a flood, or by some construction activity.

4. Discharge Relations Using Slope As Parameter No simple relation exists between stage and discharge in channels where water surface slope is variable. Variable slopes may be caused by a) backwater from a confluent stream or from tides, or by manipulation of gates at a downstream dam or b) by rapidly changing discharge in a channel having a relatively flat slope. The effect of the latter is to produce a loop curve on which the discharge for a given stage is greater when the stream is rising than it is when the stage is falling (Fig. 6.3).

Discharge ratings using stage and slope are of two types: constant-fall ratings and variable-fall ratings. See Rantz (1982), Kennedy (1984), and Herschy (1985) for procedures to develop these ratings. Constant-fall ratings are preferred where backwater is present at all stages and at all times (Chow, 1964; Scott, 1982; Herschy, 1985). Variable-fall ratings are used where backwater is intermittent.

To obtain the continuous record of slope needed for these ratings, an additional record of stream stage is collected at an auxiliary gage downstream from the base gage. Channel slope need not be measured where variable slope is caused by rapidly changing discharge. Methods are available to adjust the simple stage-discharge relation for the effect of changing discharge. Some of these methods are briefly discussed below. A detailed discussion is contained in Herschy (1985).

Many sites can be evaluated by stage-fall discharge ratings. In simple form, these consist of a reference gage at which the stage is observed continuously and current meter measurements are made occasionally, and a second gage downstream where stage is also observed continuously. Based on the same datum and the synchronized times of the two, the water surface fall in the distance between the two gages provides a measure of water surface slope.

Under backwater situations, the fall as measured between the two gages is used as a third parameter. It is variously referred to as the stage-fall-discharge method, or stage-slope-discharge method.

Stage-fall-discharge methods, as mentioned above, are broadly divided into a) constant fall ratings (with the unit-fall method as a special case), and b) variable-fall ratings.

The unit-fall-discharge method assumes that the ratio of Q/Q_r is proportional to the square root of the so-called fall ratio, F/f_c, where Q_r is the discharge corresponding to the stage at the first gage for a constant fall f_c, F is the measured water surface fall between the two gages, and Q is the current-meter discharge determination for that stage.

This square root assumption is most likely when the channel is approximately the same shape at both gage locations, if not the same size. The procedure is modified in the more general constant-fall rating procedure in that two curves are used. The first curve is the relation between stage and discharge for a constant fall of some specific value. The second is the relation between measured fall, F, and the discharge ratio, Q/Q_r. A unique feature of the constant fall method is that the two reference gages need not be at the same datum (Herschy, 1985).

Among the variable-fall ratings, the limited-fall rating is particularly applicable for situations when backwater is not always present. The general procedure is to deduce nonbackwater ratings based on measurements when backwater is observed. Later, the reverse procedure is used to obtain discharge rates that include corrections for the observed backwater. Because only stages are observed, a separate means is needed to detect backwater. This experience is usually obtained during the development of the initial ratings. When the backwater is variable, local experience is probably best in determining when to apply the backwater corrections.

These methods are subject to high levels of uncertainty and in complex cases can yield only approximate results.

D. Gaging Stations Using Index Velocity

1. Ultrasonic Meter Stations An ultrasonic velocity meter (UVM) measures the velocity of flowing water by means of an ultrasonic signal that moves faster downstream than upstream.

The major components of most UVM's are two transducers, one on each side of the channel, installed at the same elevation with the line between the two transducers diagonal to the stream, and a console on the streambank. The console provides the pulses to the system, records the responses to the pulses, and computes and records the index velocities with their corresponding times of occurrence.

These index velocities and the corresponding stages are calibrated with current meter measurements where possible.

UVM stations are used on large streams where variable backwater occurs and where stream slopes are too flat to provide discharge measurements accurate enough for defining an acceptable stage-fall-discharge relation. Theory and application are described by Laenen (1985).

2. Vane Meter Stations When flow in a small tidal stream may be in either direction and velocities are low, a vane (or deflection) meter may be used as an index of velocity. At such a site, the stage and angle of the vane are recorded continuously. Discharge measurements are used for calibration.

Vanes have been proposed for indicating flow rate per unit width based on measuring the horizontal force exerted on a triangular blade that is held rigidly, thus preventing deflection, which complicates the hydraulics. Restoring the blade to a vertical position with gravity weights by rotating the blade assembly about a horizontal axis is effective, but the blade shape is no longer triangular and is costly to fashion if the device is to indicate flow per unit width independent of flow depth. However, if a force cell of some sort is used in conjunction with a mounting about a vertical axis (door-hinge style) then only a simple triangular blade is required to indicate flow rate per unit width, independent of flow depth (Replogle, 1968).

3. Electromagnetic Gaging Stations This method permits the computation of a continuous discharge record on streams with highly unsteady flows and variable water-surface slopes. Continuous records of velocity at one point in the cross-section are obtained by an electromagnetic meter. A stage record at the cross-section is also obtained. Discharge measurements are used to relate discharge to stage and point velocity, stage being an index of cross-section area and point velocity an index of mean velocity in the cross-section. This relation is used to compute the continuous discharge record. Rantz (1982) describes the method in more detail.

One style uses an electromagnetic probe mounted on the floor of the channel that senses the near-wall velocity. Another sensor, usually in the form of a pressure transducer, senses flow depth. Previously developed data regarding the channel cross-sectional shape and the probable velocity profile allow a microprocessor unit to translate the two pieces of information into average section velocity and area for an instantaneous indication of discharge rate.

4. Discharge Records Downstream From Dams With Movable Gates Where other methods are impractical, the discharge through movable gates can be obtained by calibration with discharge measurements. Records of gate positions can then be used to compute a continuous record of discharge (Collins, 1977).

E. Partial Record Stations

1. Crest Stage Stations The purpose is to obtain, at little cost, a record of instantaneous flood peak discharges that are used to define a flood-frequency curve for the site. A crest stage gage is installed upstream from a high-water control. During a flood, the water rises in the gage pipe and carries at its surface some powdered cork that is deposited on a stick in the pipe at the peak stage. After the flood, the hydrographer measures and records the height, erases the mark, and recharges the gage with cork. Discharges, corresponding to some of the peak stages, are defined by indirect discharge measurements. These discharge measurements are used to define a stage-discharge relation for the range of peak stages recorded. From this relation and the record of peak stage, a record of peak discharges is obtained.

2. Low-Flow Partial Record Stations Occasional measurements of low flow are made at a designated site over a period of two or more years. These measurements are correlated with the concurrent low flows of a nearby gaged stream for which a low-flow frequency curve has been defined. The object of operating the partial-record station, to obtain an estimate of the 7-day 10-year low flow at the site, is accomplished by transferring the 7-day 10-year low flow of the gaged stream through that correlation (Riggs, 1972).

An example of the procedure and an empirical test of the reliability of the estimates of the 7-day 10-year estimate at a partial-record station are reported by Riggs (1985). Estimates are quite reliable if the concurrent base flows of the two streams are closely related, if the needed downward extrapolation of the relation is small and if the estimate is larger than 30 l/s.

3. Mean-Flow Sites Mean flow can be estimated from monthly measurements made at a site for a year. The measurements should be made about the middle of each month and should be used with concurrent discharges and computed monthly mean flow at a nearby gaged stream. See Riggs (1985) for further elaboration and an example of the application.

III. HYDROGRAPHS

A hydrograph is a graphical representation of the variation of a water flow quantity (discharge or stage) with respect to time. Depending on the represented flow quantity, the hydrograph is designated as surface runoff, base flow, stream flow, low flow, flood flow, or stage hydrograph. The runoff or discharge hydrograph at a point represents the response of the contributing watershed to rainfall and/or snow melt, antecedent precipitation, infiltration, evaporation, evapotranspiration, and topography. Depending on the time scale used, one can develop a continuous stream flow hydrograph for several successive years, an annual stream flow hydrograph or an event-based flood or storm flow hydrograph. Continuous stream flow hydrographs are used for reservoir operation studies and power generation studies for run-of-the-river hydroelectric power plants. The event-based hydrographs result from relatively short-duration events rather than long-term (e.g., a month or a year long) weather patterns. Event-based hydrographs are used to develop design flood hydrographs for spillways and floodways, e.g., probable maximum flood, standard project flood, and 100-year flood hydrographs. These hydrographs are defined by the storm that produces them, e.g., 6-, 24-, 48-, or 96-hour probable maximum precipitation, standard project storm, and 100-year precipitation. This section describes methods to develop event-based or storm flow hydrographs and their components.

A. Hydrograph Components

A typical storm runoff hydrograph is shown in Fig. 6.10. It consists of three distinct components: the approach limb AB, the rising limb BD, and the receding limb DH. Addition of a fourth component, i.e., baseflow, to the storm runoff hydrograph results in the complete stream flow hydrograph. The shape of the approach limb depends on the antecedent conditions, distribution of rainfall intensity, hydraulic length of basin, ground slope, and hydrologic conditions of soils in the lower portions of the drainage basin. The rising limb depends on the same characteristics along with the duration and temporal and spatial distribution of rainfall in the entire watershed. The receding limb begins with the decrease in the amount of or the cessation of surface flow to the channel and represents discharge due to drainage of water stored in the channels and floodplains. The receding limb, particularly in the lower part, also

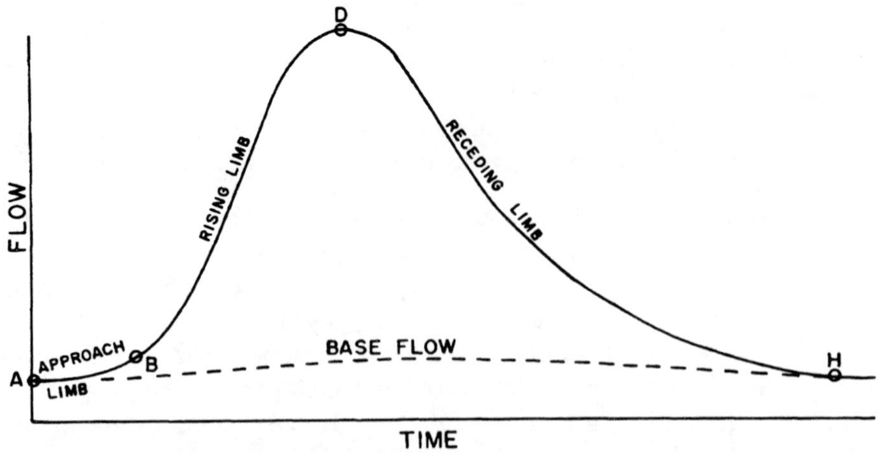

Figure 6.10—Typical Runoff Hydrograph.

includes withdrawals from bank storage and interflow (lateral flow) through surface soils. The lower portion of the receding limb exponentially decreases to the baseflow of the stream that is contributed only by the discharge of ground water.

The approach and rising limbs and upper portion of the receding limb of a surface runoff hydrograph are generally estimated by the unit hydrograph procedure. The recession limb is estimated by empirical and heuristic methods or analytical and numerical ground water flow models (see Fig. 6.11).

B. Drainage Basin Effects

The hydraulic characteristics of a watershed have two basic effects on the shape of the hydrograph at the outlet of the watershed. The first effect is due to the size, shape, hydraulic length (length measured along the average flow path of surface runoff), slope, and drainage pattern of the watershed. This effect is reflected in the unit hydrograph of the basin and includes the travel times of overland and channel flows and detention and storage of runoff in surface features, e.g., lakes, swamps, and drainage channels. The second effect is due to infiltration and detention in surface soils and other superficial features, interception and evapotranspiration by plants, and evaporation from soil surfaces. This effect is reflected in precipitation losses and transforms total precipitation to precipitation excess or effective precipitation.

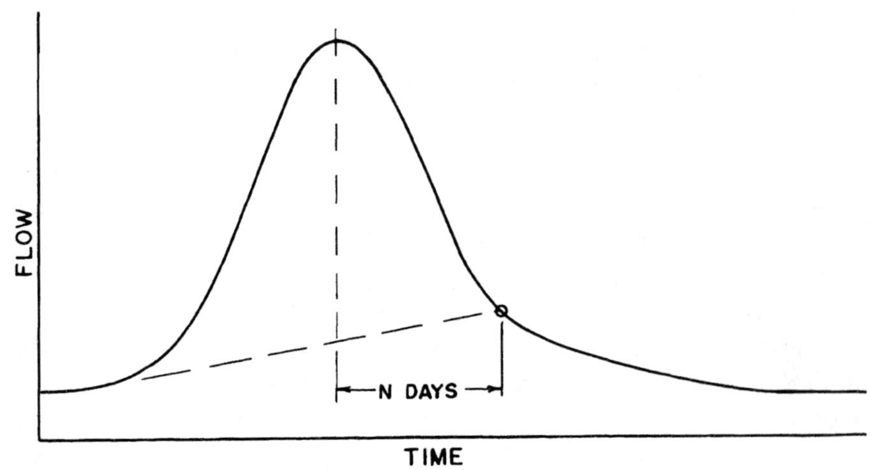

Figure 6.11—Empirical Truncation of Recession Limb.

C. Estimation of Precipitation Losses

Precipitation losses include initial abstraction, infiltration, interception, evapotranspiration, and depression storage. Methods to estimate infiltration loss include field measurements using infiltrometers, computation of average rate of infiltration (W - index or Φ index) using time-intensity graphs of rainfall and observed volumes of the corresponding storm runoff, and use of calibrated theoretical models (e.g., Philip, Horton, and Green and Ampt models; see Chapter 3). Methods to estimate evapotranspiration loss include field measurements, use of relevant evapotranspiration data and use of theoretical evapotranspiration models (Chow, 1964; Ponce, 1989; Eagleson, 1970).

There are four commonly used methods to account for precipitation losses in a rainfall-runoff analysis (Barfield et al., 1981; USACE, 1987; Ponce, 1989).

1) Constant initial loss followed by uniform loss rate. In this method, the entire precipitation is lost until the prescribed initial loss is satisfied. Thereafter, precipitation loss occurs at the prescribed uniform rate. This method is suitable for watersheds where calibrated or experimentally verified data for initial and uniform loss rates are available for various antecedent soil moisture and climatic conditions.
2) Empirical Loss Rate Function. In this method, the loss rate, L_R, in mm or inches per hour

$$L_R = (A + D)\, P^e \tag{6.9}$$

in which P is precipitation depth in mm or inches during the selected time interval, e is the exponent relating precipitation to loss rate, usually varying from 0 to 1.0, and A and D are variable loss rate coefficients, such that,

$$A = S/R^{0.1C} \tag{6.10}$$

and

$$D = 0.2\, D_R \left(1 - \frac{C}{D_R}\right)^2. \tag{6.11}$$

S is the starting or maximum value of the exponentially varying component of the loss rate coefficient (A), usually a function of the infiltration capacity or soil type, land use, and vegetal cover; R is the ratio of loss rate coefficient on the exponential loss rate curve to that corresponding to 25.4 cm (10 inches) more of accumulated loss, usually dependent on the capacity of the surface soils of the basin to absorb precipitation; C is accumulated rain loss up to the current time interval; D_R is the parameter defining the component D of the variable loss rate coefficient and its variation with cumulative rain loss; and D_R is primarily a function of the antecedent soil moisture deficiency and is usually storm-dependent.

Because of dimensional incongruity, there is no direct conversion for the coefficients of this method between the English and metric units. In each case, the coefficients have to be estimated by separate calibrations. Physical visualization of the values of the coefficients of this method is rather difficult. Therefore, its application is limited to watersheds for which calibrated values are known *a priori* or reliable coefficients can be obtained by parameter optimization.
3) Soil Conservation Service (SCS) Curve Number Method (USDA, 1985). In this method, a curve number (CN) is assigned to each watershed or portion of watershed based on soil type, land use and treatment, and antecedent moisture condition. The SCS has classified more than 4,000 soils into four hydrologic soil groups. Local SCS offices can provide pertinent information on soils in their respective jurisdictions. The definitions of the four hydrologic soil groups are reproduced below:

a) Soils having high infiltration rates even when thoroughly wetted and consisting chiefly of deep, well- to excessively-drained sands or gravels. These soils have a high rate of water transmission and a low runoff potential.
b) Soils having moderate infiltration rates when thoroughly wetted and consisting chiefly of moderately deep to deep, moderately well- to well-drained soils with moderately fine to moderately coarse textures. These soils have a moderate rate of water transmission.
c) Soils having slow infiltration rates when thoroughly wetted and consisting chiefly of soils with a layer that impedes downward movement of water, or soils with moderately fine to fine texture. These soils have a slow rate of water transmission.
d) Soils having very slow infiltration rates when thoroughly wetted and consisting chiefly of clay soils with a high swelling potential, soils with a permanent high water table, soils with a claypan or clay layer at or near the surface, and shallow soils over nearly impervious material. These soils have a very slow rate of water transmission and a high runoff potential.

The moisture condition resulting from precipitation 5 to 30 days preceding the storm under consideration is referred to as the antecedent moisture condition (AMC). Antecedent moisture conditions are divided into the following three groups;

- AMC-I—A condition of watershed soils where the soils are dry but not to the wilting point, and when satisfactory plowing or cultivation takes place. This condition is not considered applicable to design basis flood computations.
- AMC-II—The average case for annual floods, that is, an average of the conditions which have preceded the occurrence of the maximum annual flood on numerous watersheds.
- AMC-III—When heavy rainfall or light rainfall and low temperatures have occurred during the 5 days previous to the given storm, and the soil is nearly saturated.

Curve numbers (CNs) for various soils and land uses for the average watershed condition, AMC II, are given in Chapter 3. Curve numbers for one antecedent moisture condition may be converted to a different one by the use of Table 6.1. Alternatively, curve numbers for AMC I (CN_I) and AMC III (CN_{III}) can be computed from those for AMC II (CN_{II}) using the following equations (Hawkins et al., 1985):

$$CN_I = \frac{CN_{II}}{2.3 - 0.013\, CN_{II}} \tag{6.12}$$

$$CN_{III} = \frac{CN_{II}}{0.43 + 0.0057\, CN_{II}}. \tag{6.13}$$

It should be noted that the curve number method is an empirical method to estimate surface runoff. Precipitation loss is estimated as the difference between precipitation and computed surface runoff. The equations used to estimate surface runoff or rainfall excess are as follows:

$$Q = \frac{(P - 0.2S)^2}{P + 0.8S}, \; P > 0.2S \tag{6.14}$$

$$S = \frac{2540}{CN} - 25.4 \tag{6.15}$$

in which Q is surface runoff (*cm*), S is the potential soil retention (*cm*), and P is precipitation (*cm*). The curve numbers have been found to decrease with increasing rainfall. The precipitation depth, P, and curve number, CN, used in Equations 6.14 and 6.15 are independent of storm duration, and

TABLE 6.1. Corresponding Runoff Curve Numbers for Three AMC Conditions.

AMCII	AMCI	AMCIII	AMCII	AMC1	AMCIII
100	100	100	60	40	78
99	97	100	59	39	77
98	94	99	58	38	76
97	91	99	57	37	75
96	89	99	56	36	75
95	87	98	55	35	74
94	85	98	54	34	73
93	83	98	53	33	72
92	81	97	52	32	71
91	80	97	51	31	70
90	78	96	50	31	70
89	76	96	49	30	69
88	75	95	48	29	68
87	73	95	47	28	67
86	72	94	46	27	66
85	70	94	45	26	65
84	68	93	44	25	64
83	67	93	43	25	63
82	66	92	42	24	62
81	64	92	41	23	61
80	63	91	40	22	60
79	62	91	39	21	59
78	60	90	38	21	58
77	59	89	37	20	57
76	58	89	36	19	56
75	57	88	35	18	55
74	55	88	34	18	54
73	54	87	33	17	53
72	53	86	32	16	52
71	52	86	31	16	51
70	51	85	30	15	50
69	50	84			
68	48	84	25	12	43
67	47	83	20	9	37
66	46	82	15	6	30
65	45	82	10	4	22
64	44	81	5	2	13
63	43	80	0	0	0
62	42	79			
61	41	78			

Source: USDA 1985.

the resulting precipitation loss includes initial abstraction, infiltration, interception, evaporation, and surface detention.

4) Holton Loss Rate Method. This method uses the following equation for the rate of infiltration in mm or inches per hour, f,

$$f = G \cdot A \cdot F^b + f_c \tag{6.16}$$

in which G is a growth index representing the relative maturity of the ground cover varying from near 0.0 when the crop is planted to 1.0 when the crop is full-grown; A is the infiltration capacity in mm or inches per hour that is related to the cover crop and soil texture and is interpreted as an index of the pore volume directly connected to the soil surface; F is the equivalent depth of unfilled pore space in mm or inches in the surface soil layer that is available for storage of infiltrated water, ranging from maximum available soil moisture capacity to 0.0; f_c is the constant rate of percolation below the surface layer in mm or inches per hour; and b is an exponent with a typical value of 1.4. Holton et al. (1975) have given the values of A for several vegetation types. The values of f_c range from 11.4 to 7.6 mm/hr (0.45 to 0.30 in/hr) for hydrologic soil group A, 7.6 to 3.8 mm/hr (0.30 to 0.15 in/hr) for B, 3.8 to 1.27 mm/hr (0.15 to 0.05 in/hr) for C, and 1.27 mm/hr (0.05 in/hr) or less for D. The quantity of infiltrated water during a time interval is taken to be the quantity of available water (i.e., rain or snowmelt), or the estimated loss in mm (inches), F, during the time interval, whichever is smaller. The quantity, F, is computed by an iterative process as follows:

$$F = \frac{f_1 + f_2}{2} \Delta t \tag{6.17}$$

f_1 is the infiltration rate in mm/hr (inches/hour) at the beginning of the time interval computed from Equation 6.16 with known depth of unfilled pore space at that time (i.e., S_1), and f_2 is the infiltration rate in mm/hr (inches/hour) at the end of the time interval computed with an assumed value of unfilled pore space at the end of the time interval (i.e., S_2). The value S_2 is successively refined using the following relationship,

$$S_2 = S_1 - F + f_c \Delta t. \tag{6.18}$$

This method has a tractable physical basis; however, a number of soil and cover parameters must be known from previous studies, experiments, or calibration to use the method for a practical situation. In addition to the ranges indicated previously, default values of the parameters of the method are given in the HEC-1 Users Manual (USACE, 1987). Using these values as initial estimates, the unit graph and loss rate optimization option of the HEC-1 model can be used to obtain values that will best reconstitute the observed runoff hydrograph for a sub-basin.

D. Hydrograph Recession and Baseflow Separation

The upper portion of the recession limb of a hydrograph is contributed by 1) the portion of infiltrated water that passes through the shallower zones of the soil to reach the stream within a relatively short time after the storm without reaching the ground water table (i.e., interflow) and 2) exfiltration from bank storage of water that infiltrates into the river banks during high river stages. The portion of the recession limb from the inflection point to the point of return to baseflow usually plots as a straight line or a series of straight lines on semi-logarithmic paper, with flow plotted on the log scale. This leads to the following forms of recession equations:

$$Q(t) = Q_o e^{-t/ts} \tag{6.19}$$

or

$$Q(t) = Q_o K^t \tag{6.20}$$

in which $Q(t)$ is the flow at any time t after the inflection point on the recession limb, Q_0 is the flow at the inflection point, t_s is a recession constant known as the time of storage, and K is a recession constant. Values of the recession constants, t_s and K, are obtained by analysis of observed storm runoff hydrographs.

Baseflow is the dry weather flow of perennial streams contributed by subsurface flow from the water table due to infiltration antecedent to the runoff period under study, supplemented by any recharge resulting from deep percolation of infiltrated water during the period under study. In most cases, baseflow constitutes a small portion of the total storm runoff. Therefore, minor errors in its estimation do not significantly affect design-basis flood or storm runoff hydrographs; however, it is necessary to know the baseflow hydrograph to determine surface runoff and unit hydrographs from a given stream flow hydrograph and to develop a complete flood hydrograph using the unit hydrograph for a basin. Some commonly used methods for baseflow separation are described in the following paragraphs.

1) Empirical Truncation of Recession Limb (Linsley et al., 1958). In this method, the recession limb of the storm runoff hydrograph is truncated N days after the time of peak flow, where,

$$N = 1.21\, A^{0.2} \quad \text{where } A \text{ is in } km^2$$

$$N = A^{0.2} \quad \text{where} A \text{ is in } mi^2 \tag{6.21}$$

   in which A is the drainage area. The values of N given by Equation 6.21 should be slightly reduced for mountainous watersheds and increased for long narrow or relatively flat basins. This point on the recession limb is connected backwards to the beginning of the approach limb by a straight line (Fig. 6.12).
2) Extension of Curve Preceding the Approach Limb. In this case, the portion of the hydrograph preceding the approach limb is smoothly extended to a point in time under the peak of the storm runoff hydrograph. From this point, a straight line is drawn to a point N days after the peak on the recession limb. This implies that there is subsurface flow from the stream to the banks as the water surface in the stream rises resulting in gradual decrease in baseflow during the rising time of the hydrograph (Fig. 6.12).
3) Backward Extension of Lower Portion of Recession Limb. In this method, the lowest portion of the recession limb is smoothly extended backwards to the time of inflection on the recession limb. This point is joined to the beginning of the approach limb by an arbitrarily drawn rising limb. This approach represents situations where ground water contribution is relatively large and reaches the stream fairly rapidly (Fig. 6.13).
4) Use of Semilogarithmic Plots of Hydrographs (Dooge, 1973). When a storm flow hydrograph is plotted on semi logarithmic paper with flows on the log scale, the approach limb and upper and lower portions of the recession limb may be approximated by separate straight lines. The lower

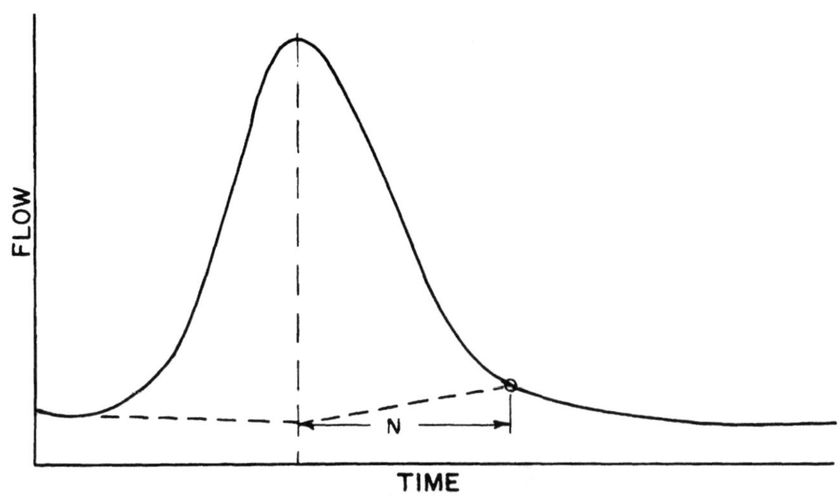

Figure 6.12—Extension of Curve Preceding Approach Limb.

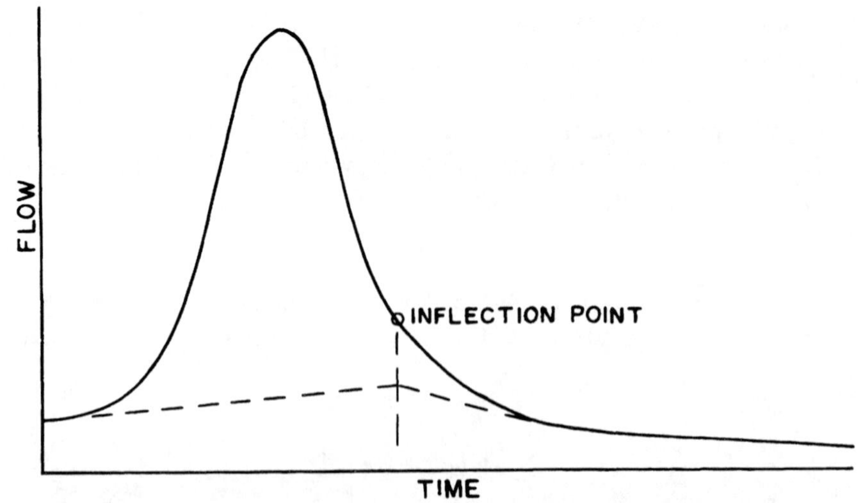

Figure 6.13—Backward Extension of Lower Portion of Recession Limb.

straight line of the recession limb is extended backwards up to the time of the hydrograph peak. This point is then connected back to the beginning of the approach limb or its arbitrarily drawn straight line extension to the right. The resulting obtuse-angle triangle on this semilogarithmic plot approximates baseflow (Fig. 6.14).

Depending upon the availability of data and judgment, any one of these methods can be used to estimate baseflows.

E. Time Parameters

Time parameters commonly used for hydrograph analyses include the time of concentration, lag time, time to peak, time to equilibrium, and the time–area curve. The time of concentration, t_c, is defined as the time required for a particle of water to travel from the most distant point in the watershed to the outlet or location under consideration. The lag time, t_L, is a weighted time of concentration so that

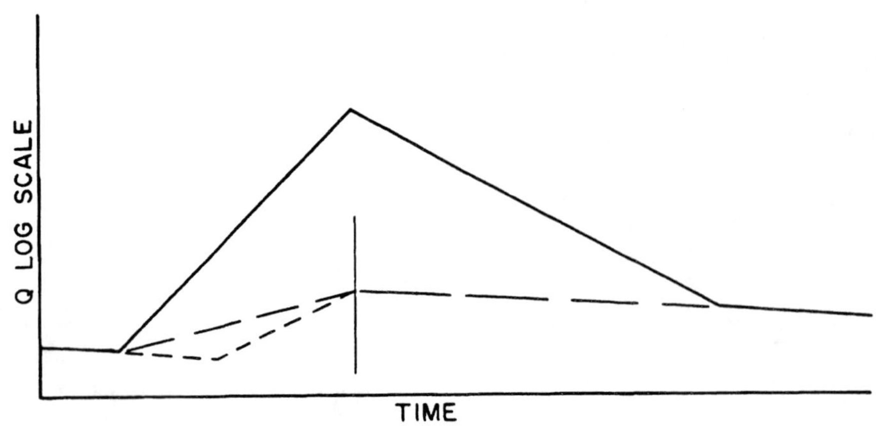

Figure 6.14—Semilogarithmic Plot of Hydrograph.

$$t_L = \frac{\sum_{i=1}^{N} A_i \, Q_i \, t_i}{\sum_{i=1}^{N} A_i \, Q_i} \tag{6.22}$$

in which A_i is the *ith* increment of watershed area, Q_i is rainfall excess from area A_i, t_i is travel time from the center of A_i to the location under consideration, and N is the number of incremental areas into which the entire watershed is divided to estimate lag time. Different investigators have defined lag time in different ways. In hydrograph analysis, lag time is commonly defined as the time from the center of mass of rainfall excess to the peak of the surface runoff hydrograph. For average natural watershed conditions and approximately uniform distribution of runoff in the watershed (USDA, 1985),

$$t_L = 0.6\, t_c. \tag{6.23}$$

When runoff is not uniformly distributed, the watershed can be divided into smaller segments within which runoff is nearly uniform so that the empirical relationship of Equation 6.23 can be applied. The time to peak, t_p, is the time from the beginning of surface runoff to the peak of the surface runoff hydrograph. The use of the parameters time to peak and time–area curve are described later in the section on Unit Hydrographs. The time to equilibrium is related to the S-hydrograph concept described subsequently. It is the time when the ordinate of the S-hydrograph attains a constant value equal to the rate of rainfall excess, i.e., the runoff rate attains a constant maximum value due to continuous effective rainfall at a constant rate. The time to equilibrium, t_e, is defined as the time since the beginning of surface runoff to the time when the rate of surface runoff equals the rate of rainfall excess. The time–area curve is a plot of the cumulative area of the watershed contributing surface runoff to the basin outlet as a function of time.

There are a number of methods for estimating the time of concentration and lag time (USDA, 1985; Prakash, 1987; USBR, 1987). Since the location of the hydrograph peak is dependent on the time of concentration, the hydrologist should make a judicious choice from values of t_c estimated by several different methods applicable to the watershed being analyzed.

1) Stream Hydraulics Method (USDA, 1972). This method is applicable to watersheds or portions thereof where well-defined water courses can be identified and surface runoff rather than baseflow is the dominant component of the hydrograph. The steps of computations are as follows:
   a) Identify existing water course in the basin, measure the hydraulic lengths of different reaches, and estimate the 2-year peak or bankfull discharge for each stream reach.
   b) Use the Manning formula to estimate average velocity for each reach, and compute the corresponding travel times.
   c) Add the travel times for successive stream reaches. The total travel time for the 2-year peak or bankfull discharges is assumed to represent t_c for average floods. For larger floods, higher velocities should be used to estimate travel times.
2) Upland Method (USDA, 1972). This method is applicable to overland flows and flows through grassed waterways, over paved areas and through small upland gullies in small watersheds [800 ha (2000 ac) or less]. The travel time for each type of flow is estimated using the velocities shown in Fig. 6.15. The velocity of surface runoff along terrace channels can be assumed to be 0.46 m/s (1.5 ft/sec). The method can be applied to portions of large watersheds and used in conjunction with other methods.
3) Kirpich Method (USBR, 1977). This method was originally developed for small watersheds [0.4047 to 45.3 ha (1 to 112 ac)] in Tennessee, but has also been applied to larger basins. The Kirpich equation for time of concentration, t_c, in hours is as follows:

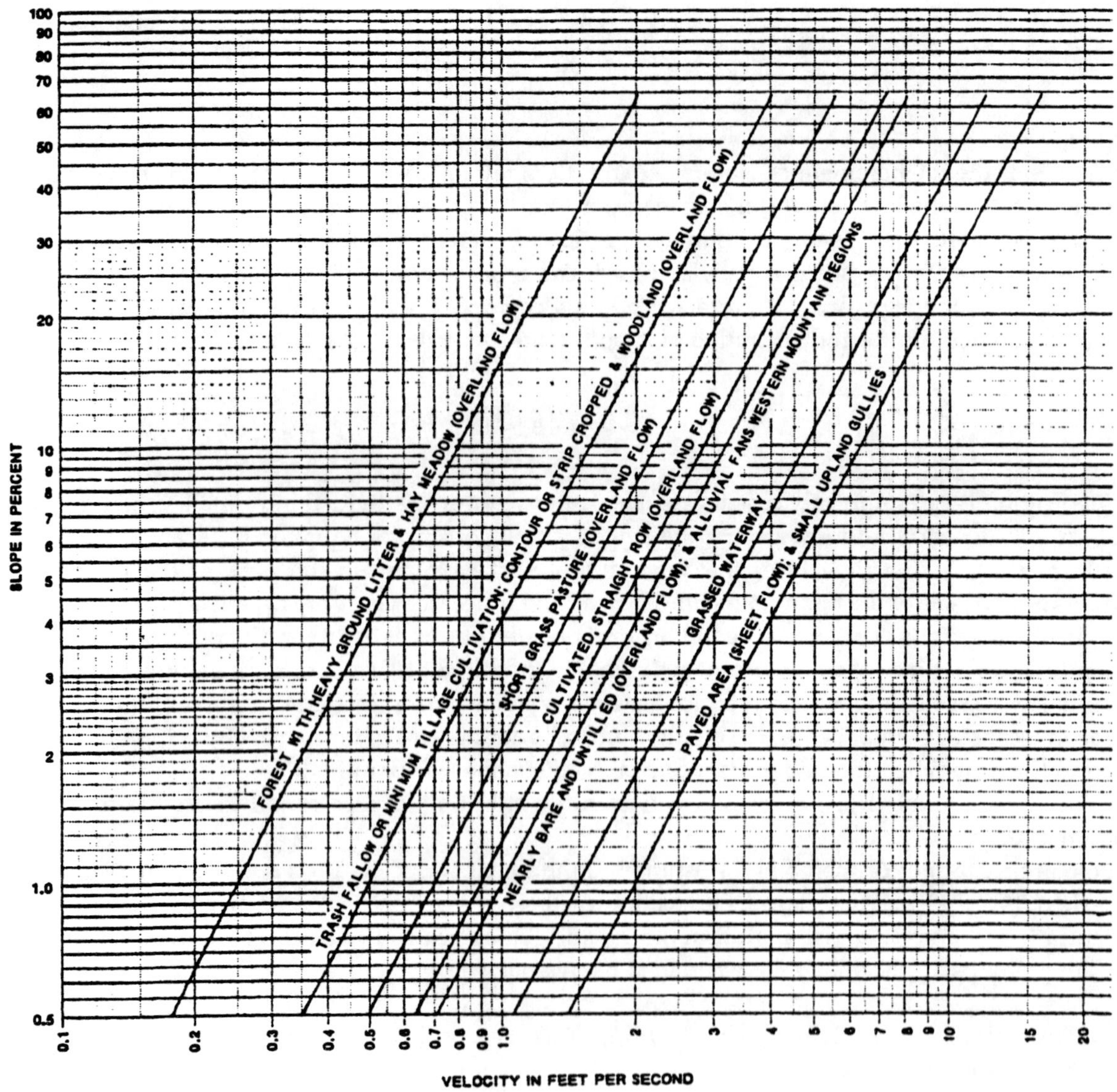

Figure 6.15—Velocities for Upland Method of Estimating T_c.

$$t_c = \left(\frac{0.87\ L^3}{H}\right)^{0.385} \quad (S.\,I.\ Units) \tag{6.24a}$$

$$t_c = \left(\frac{11.9\ L^3}{H}\right)^{0.385} \quad (U.\,S.\ Customary\ Units) \tag{6.24b}$$

in which L is the length of the longest watercourse in km or mi, and H is the difference in elevation between the upper end of the watershed and the location under consideration in m or ft.

4) Kerby-Hathaway Formula (McCuen et al., 1984). This formula was calibrated for predominantly overland flows on small watersheds (less than 4.047 ha or 10 ac). The equation for the time of concentration, t_c, in hours is as follows:

$$t_c = 0.02407\ L_f^{0.47}\ n^{0.47}\ S_f^{-0.235} \quad (S.\,I.\ Units) \tag{6.25a}$$

$$t_c = 0.01377\ L_f^{0.47}\ n^{0.47}\ S_f^{-0.235} \quad (U.\,S.\ Customary\ Units) \tag{6.25b}$$

in which L_f is the straight line distance in m or ft from the most distant point on the watershed to the point under consideration measured parallel to the slope, S_f is the mean slope of the basin (m/m or ft/ft), and n is the Manning roughness coefficient or retardance coefficient. Typical values of retardance coefficient n are shown in Table 6.2.

5) Curve Number Method (USDA, 1972). This method was developed for areas less than 800 ha (2000 ac) and is applicable to heavily forested watersheds with steep channels, watersheds with relatively large subsurface or interflow component, meadows, smooth land surfaces, and large paved parking lots. In practice, the method can be applied to larger watersheds with a judicious choice of the curve number. In this method, the lag time, t_L, in hours is computed by the following equation:

$$t_L = \frac{1.347\ L_f^{0.8}\ (S + 2.54)^{0.7}}{1900\ S_p^{0.5}} \quad (S.I.\ Units) \tag{6.26a}$$

$$t_L = \frac{L_f^{0.8}\ (S + 1)^{0.7}}{1900\ S_p^{0.5}} \quad (U.S.\ Customary\ Units) \tag{6.26b}$$

in which L_f is the hydraulic length of watershed in m or ft, and S_p is the average watershed land slope in percent.

6) Synthetic Unit Hydrograph Lag Time (USBR, 1987). Based on unit hydrograph data for a number of watersheds of different sizes and shapes, the U.S. Bureau of Reclamation has developed the following equation for lag time, t_L, in hours:

$$t_L = 14.4328\ K_n \left[\frac{L\ L_{ca}}{S^{0.5}}\right]^{0.33} \quad (S.I.\ Units) \tag{6.27a}$$

$$t_L = 26\ K_n \left[\frac{L\ L_{ca}}{S^{0.5}}\right]^{0.33} \quad (U.S.\ Customary\ Units) \tag{6.27b}$$

in which L is the length of the longest watercourse from the point under consideration to the boundary of the drainage basin in km (mi), L_{ca} is the length along the longest watercourse from the point under consideration to a point opposite the centroid of the drainage basin in km (mi), S is the overall slope of the longest watercourse along L in m/km (ft/mi), and typical values of K_n are given in Table 6.3 (1 mile = 1.609 km and ft/mi = 0.189 m/km).

7) Snyder's Method (Chow, 1964). Snyder developed the following empirical equation to estimate the basin lag, t_L, in hours,

$$t_L = 0.7517\ C_t\ (L\ L_{ca})^{0.3} \quad (S.I.\ Units) \tag{6.28a}$$

$$t_L = C_t\ (L\ L_{ca})^{0.3} \quad (U.S.\ Customary\ Units) \tag{6.28b}$$

TABLE 6.2. Typical Values of Retardance Coefficient.

Type of Surface	Retardance Coefficient, n
Smooth impervious surface	0.02
Smooth bare packed soil	0.10
Poor grass, cultivated row crops, or moderately rough bare surface	0.20
Pastured or average grass	0.40
Deciduous timberland	0.60
Coniferous timberland, deciduous timberland with deep forest litter or dense grass	0.80

TABLE 6.3. K_n Values for Watersheds in Different Regions.

Region	Watershed Size km² (mi²)	K_n*
Great Plains in Colorado, Kansas, Oklahoma, Nebraska, New Mexico, Wyoming, and North Dakota.	5.18 to 10,282.3 (2.0 to 3,970)	0.070 for basins with considerable overland flow. 0.030 for basins with well-defined drainage network
Rocky Mountain Watersheds in Colorado, Wyoming, Utah, Oregon, Montana, Idaho, and New Mexico.	3.37 to 6,503.5 (1.3 to 2,511)	0.260 for 100-year floods 0.130 for general storm PMF 0.050 for thunderstorm events
Southwest Desert, Great Basin, and Colorado Plateau in Arizona, California, and parts of Colorado.	5.96 to 12,250.7 (2.3 to 4,730)	0.070 for basins with coniferous forests at higher elevation. 0.042 for desert terrains.
Sierra Nevada, California	53.35 to 5,374.2 (20.6 to 2,075)	0.150 for basins with substantial coniferous growth. 0.064 for basins with well-developed drainage networks.
Coast and Cascade Ranges in California, Oregon, and Washington.	18.65 to 1,978.8 (7.2 to 764)	0.150 for basins with very heavy coniferous growth extending into the overbank flood plain. 0.080 for low lying basins with considerably sparser vegetation.
Urban basins in California, Texas, Kentucky, Virginia, and Maryland.	0.52 to 238.3 (0.2 to 92.0)	0.033 for basins with low density or partial development with only minor flood water collection facilities. 0.013 for basins with high density development combined with a good collection system.

*Values between indicated upper and lower limits may be used for basins with intermediate characteristics.

Source: USBR 1987.

in which L and L_{ca} are as defined for Equation 6.27. The basin lag is defined as the time from the midpoint of rainfall excess to the peak of the unit hydrograph. The coefficient C_t has been found to be 2.0 for watersheds in the fairly mountainous Appalachian Highlands, 0.4 in southern California, 0.7 to 1.0 in the Sierra Nevada, and 8.0 in sections of states bordering the eastern Gulf of Mexico. An analysis of 20 basins in the North and Middle Atlantic States led to the following relationship for C_t,

$$C_t = \frac{0.6}{\sqrt{S}} \qquad (6.29)$$

in which S is the basin slope.

F. Unit Hydrograph

1. Definition The unit hydrograph of a drainage basin is defined as a hydrograph of direct runoff resulting from a unit depth of effective rainfall uniformly distributed over the basin area during a specified period of time known as the unit duration. The unit hydrograph is the most common method to develop surface runoff hydrograph for a basin from known depths and duration of rainfall excess. It represents the hydrologic response of the basin to a unit input of rainfall excess. The unit hydrograph theory is based on the following assumptions: a) The effective rainfall is uniformly distributed both in time and space and b) the time base (duration) of a unit hydrograph for a specified unit duration is constant. Hydrographs for effective rainfall depths other than unity can be obtained by linear multiplication of the unit hydrograph ordinates (UHOs); and the hydrograph for a storm consisting of a series of effective rainfall depths lagged by the unit duration can be obtained by linear superposition or convolution.

2. *Derivation by Direct Method* This method involves development of the unit hydrograph for a basin from known rainfall hyetographs and flow hydrographs at its outlet. The steps of computations for this method are as follows:

a) Identify relatively isolated and reasonably intense storms in the basin, prepare a weighted average mass curve of precipitation for each storm from all gages in the basin, and use these mass curves to prepare average hyetographs for all selected storms with the same time increment for each hyetograph [plot of rainfall rate, mm/h (in/hr), versus time (hr)].
b) Perform baseflow separation for each storm flow hydrograph, planimeter the area representing direct runoff between the storm flow and baseflow hydrographs, and convert it to depth units (effective rainfall depth) dividing by the watershed area in appropriate units.
c) Divide each hyetograph into effective rainfall and abstractions by a horizontal line such that the area of effective rainfall above this line is equal to the previously estimated depth of effective rainfall. Measure the time duration of effective rainfall on the abscissa of the hyetograph. This gives the unit duration of the unit hydrograph.
d) Divide the ordinates of the direct runoff hydrograph by the estimated effective rainfall to obtain the corresponding unit hydrograph ordinates (UHOs).
e) Plot all unit hydrographs with the same unit duration on the same graph and obtain an average unit hydrograph by judgment.
f) Check that the area under the average unit hydrograph represents unit rainfall excess times the watershed area converted to proper units.

3. *Derivation by Indirect (Synthetic) Methods* There are a number of methods to develop synthetic unit hydrographs. Some of these methods are presented in the following paragraphs. The choice of one or the other method depends on the availability of calibrated coefficients for these methods for the watershed under investigation or other similar watersheds in the region.

a) Triangular Unit Hydrograph (USDA, 1972; USBR, 1977). This method is based on the assumption that the unit hydrograph for a basin can be approximated by a triangle (Fig. 6.16).

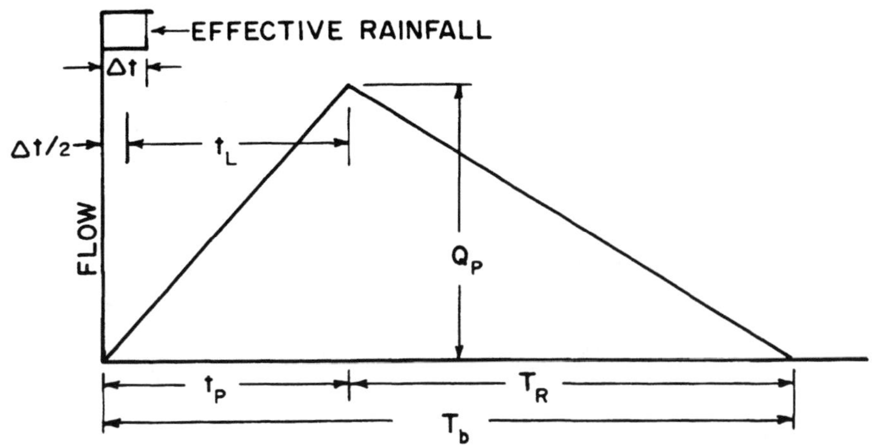

Figure 6.16—Triangular Unit Hydrograph.

Analysis of a number of hydrographs by the Soil Conservation Service (USDA, 1972) resulted in the following average relationships:

$$T_r = 1.67\ t_p \tag{6.30}$$

$$T_b = 2.67\ t_p \tag{6.31}$$

$$t_p = t_L + \Delta t/2 = 0.6t_c + \Delta t/2 \tag{6.32}$$

Defining Q as effective rainfall,

$$Q_p = 2\ AQ/T_b = \frac{2AQ}{2.67\ t_p}. \tag{6.33}$$

When A is the drainage area in km² (mi²), Q_p is the peak flow in m³/s (cfs), $Q = 1$ cm (1 inch), and T_b is in hours,

$$Q_p = \frac{2.08\ A}{t_p} \quad (S.I.\ Units) \tag{6.34a}$$

$$Q_p = \frac{484\ A}{t_p} \quad (U.S.\ Customary\ Units). \tag{6.34b}$$

b) Clark's Unit Hydrograph. This method involves the preparation of a time–area curve for the basin. The basin area is divided in several zones by isochrones which are loci of points of equal travel times to the basin outlet. The time interval between adjacent isochrones preferably should be an integer multiple or fraction of the unit duration of the unit hydrograph so as to yield whole numbers in minutes or hours; otherwise, area values for different multiples or fractions of unit periods can be obtained by interpolation. If the time–area curve for a specific watershed is not available, then the following generalized equations may be used as an approximation (USACE, 1987):

$$\overline{A}(i) = 1.414\ \overline{T}^{1.5} \qquad 0 \le \overline{T} < 0.5 \tag{6.35}$$

$$1 - \overline{A}(i) = 1.414\ (1 - \overline{T})^{1.5} \qquad 0.5 \le \overline{T} \le 1.0 \tag{6.36}$$

in which $\overline{A}\ (i)$ is the cumulative area up to isochrone for time t as a fraction of total basin area and $\overline{T}$ is time t expressed as a fraction of the time of concentration, t_c. Both $\overline{A}\ (i)$ and $\overline{T}$ are dimensionless.

The method of unit hydrograph development involves two computational steps. The first step computes the contribution of unit depth of effective rainfall during the unit duration, Δt, from areas between successive isochrones. These contributions commence as the time since the beginning of effective rainfall equals or exceeds the travel time for a particular isochrone. This component, designated as the translation hydrograph is computed as follows:

$$P(t) = \sum_{i=1}^{N} A\left[t - (i-1)\frac{\Delta t}{N}\right] I(i) \tag{6.37}$$

in which Δt is the unit duration of the unit hydrograph, $\Delta t/N$ is the time step of computations for isochrones, which is the interval between successive computed UHOs, $P(t)$ is the contribution of translation hydrograph at time, t, $A(t)$ is the area between isochrones at t and $t - \Delta t/N$, t is time chosen as an integer multiple of $\Delta t/N$, and N is the number of time steps used to divide Δt for computational purposes (usually equal to the number of isochrones within Δt). Herein, $I(i) = 1/\Delta t \times \Delta t/N = 1/N$ is a fraction of unit effective rain in cm (in) that occurs in any time step, i, within Δt.

The second step accounts for storage effects of the watershed by routing $P(t)$ through a linear reservoir (see the section on Stream Flow Routing). This is done by the following recursive relationship.

$$U(t) = C_o\ P(t) + C_1\ P\left(t - \frac{\Delta t}{N}\right) + C_2\ U\left(t - \frac{\Delta t}{N}\right) \tag{6.38}$$

in which $U(t)$ = UHO at time t and C_0, C_1, and C_2 are linear reservoir routing coefficients computed as follows:

$$C_0 = C_1 = \frac{\frac{\Delta t}{N}}{2K + \frac{\Delta t}{N}} \tag{6.39}$$

$$C_2 = 1 - 2\,C_o. \tag{6.40}$$

K is the storage coefficient in units of time so that

$$S = KO \tag{6.41}$$

in which S is watershed storage and O is outflow from the watershed. With A in km^2 (mi^2), Δt in hrs, and a unit rain of 1 cm (1 in), $P(t)$ is in km^{2}cm$/(\Delta t/N)$ hr [mi^2 in$/(\Delta t/N)$ hr], and the conversion factor to express $P(t)$ in m^3/s (cfs) is $2.78/(\Delta t/N)$ [$645.33/(\Delta t/N)$].

The HEC-1 model (USACE, 1987) uses a slightly different computational algorithm for Clark's unit hydrograph. In this model, the time step of computations is the same as the unit duration of the unit hydrograph, Δt. Then,

$$U(t) = 0.5\,[Q(t - \Delta t) + Q(t)] \tag{6.42}$$

$$Q(t) = 2C_0B(t) + C_2\,Q(t - \Delta t) \tag{6.43}$$

$$B(t) = A\,[\overline{A}(t) - \overline{A}(t - \Delta t)]. \tag{6.44}$$

Areas between isochrones, Δt apart from one another, are obtained by interpolation of the time area curve (Equation 6.44).

With A in km^2 (mi^2), Δt in hours, and a unit rain of 1 cm (1 in), $U(t)$ is in km^2·cm$/\Delta t$·hr, and the conversion factor to express $U(t)$ in m^3/s (cfs) is $2.7778/\Delta t$ ($645.3333/\Delta t$).

The process of unit hydrograph computations using both methods is illustrated in the following example.

4. *Example of Unit Hydrograph Computations* The drainage area of a watershed is 160 km^2 (61.78 mi^2). The time of concentration and storage coefficient (K) are estimated to be 3.0 and 2.0 hours, respectively. The time–area diagram with 30-minute isochrones is given below.

Time (hr)	0.5	1.0	1.5	2.0	2.5	3.0
Cumulative Area (km^2)	16	48	96	128	144	160

Develop

a) a 1-hour unit hydrograph for the watershed using Equations 6.37 to 6.41
b) a 30-minute unit hydrograph using the HEC-1 procedure.

Solution:

a) Computation of 1-hour unit hydrograph

$$\Delta t = 1 \text{ hr},\ N = 2,\ \Delta t/N = 0.5 \text{ hr},\ I(i) = 0.5 \text{ cm}$$

$$C_0 = C_1 = \frac{0.5}{2 \times 2 + 0.5 =} 0.1111$$

$$C_2 = 1 - 0.2222 = 0.7778$$

Computations of $P(t)$ using Equation 6.37 are illustrated in Table 6.4.
Computations of $U(t)$ using Equation 6.38 are illustrated in Table 6.5.

b) Computation of 30-minute unit hydrograph
In this case, $N = 1$ and $\Delta t = 0.5$ h.

$$C_0 = C_1 = \frac{0.5}{2 \times 2 + 0.5} = \frac{1}{9}$$

$$C_2 = \frac{7}{9}$$

The computations of UHOs using Equations 6.42, 6.43, and 6.44 are illustrated in Table 6.6.
The UHOs of Table 6.5 can also be obtained from those of Table 6.6 using Equation 6.68.

c) SCS Dimensionless Unit Hydrograph (USDA, 1985; USACE, 1987). After analyzing a large number of unit hydrographs for watersheds of different sizes and in different locations, the SCS developed a generalized dimensionless unit hydrograph in terms of t/t_p and Q/Q_p in which t_p is the time from the beginning to the peak of unit hydrograph in hours, Q_p is the peak of UHOs in m³/s, and Q is UHO in m³/s at time, t, in hours. The coordinates of the dimensionless unit hydrograph are reproduced in Table 6.7.
The equations defining t_p and Q_p are as follows:

$$t_p = 5t_r \tag{6.45}$$

$$t_p = \frac{t_r}{2} + t_L \tag{6.46}$$

TABLE 6.4. Computation of P(t).

Ordinate No. (1)	t (hr) (2)	A (km²) (3)	$A(t)\cdot I(i)$ (4)	$A(t-\Delta t/N)\cdot I(i)$ (5)	$P(t)$ (km² cm/0.5 hr) (6)
1	0	0	0	0	0
2	0.5	16	8	0	8
3	1.0	32	16	8	24
4	1.5	48	24	16	40
5	2.0	32	16	24	40
6	2.5	16	8	16	24
7	3.0	16	8	8	16
8	3.5	0	0	8	8
9	4.0	0	0	0	0

TABLE 6.5. Computation of 1-hr UHO/s, Using Clarks Method

UHO No. (1)	t (hr) (2)	$C_0 \cdot P(t)$ (3)	$C_1 \cdot P(t-\Delta t/N)$ (4)	$C_2 \cdot U(t-\Delta t/N)$ (5)	$U(t)$ km²cm/0.5 hr (6)	$U(t)$ m³/sec (7)
1	0	0	0		0	0
2	0.5	0.8889	0	0	0.8889	4.938
3	1.0	2.6667	0.8889	0.6914	4.2470	23.594
4	1.5	4.4444	2.6667	3.3032	10.4143	57.857
5	2.0	4.4444	4.4444	8.1000	16.9888	94.382
6	2.5	2.6667	4.4444	13.2135	20.3246	112.914
7	3.0	1.7778	2.6667	15.8080	20.2525	112.514
8	3.5	0.8889	1.7778	15.7519	18.4186	102.326
9	4.0	0	0.8889	14.3256	15.2145	84.525
10	4.5	0	0	11.8335	11.8335	65.742
11	5.0	0	0	9.2038	9.2038	51.132
12	5.5	0	0	7.1585	7.1585	39.769
13	6.0	0	0	5.5678	5.5678	30.932
14	6.5	0	0	4.3305	4.3305	24.058
15	7.0	0	0	3.3682	3.3682	18.712
16	7.5	0	0	2.6197	2.6197	14.554
17	8.0	0	0	2.0376	2.0376	11.320
18	8.5	0	0	1.5848	1.5848	8.804
19	9.0	0	0	1.2326	1.2326	6.848
20	9.5	0	0	0.9587	0.9587	5.326
21	10.0	0	0	0.7457	0.7457	4.143
22	10.5	0	0	0.5800	0.5800	3.222
23	11.0	0	0	0.4511	0.4511	2.506
24	11.5	0	0	0.3509	0.3509	1.949
25	12.0	0	0	0.2729	0.2729	1.516
26	12.5	0	0	0.2123	0.2123	1.179
27	13.0	0	0	0.1651	0.1651	0.917
28	13.5	0	0	0.1284	0.1284	0.713
29	14.0	0	0	0.0999	0.0999	0.555
30	14.5	0	0	0.0777	0.0777	0.432
31	15.0	0	0	0.0604	0.0604	0.336
32	15.5	0	0	0.0470	0.0470	0.261
33	16.0	0	0	0.0366	0.0366	0.203
34	16.5	0	0	0.0284	0.0284	0.158
35	17.0	0	0	0.0221	0.0221	0.123
			TOTAL		159.9234	

By definition, the area under the unit hydrograph should be 160 km²·cm. The computation of UHOs should be continued until the summation of the values in column 6 equals 160 km²·cm within an acceptable truncation error.

$$T_b = 5\ t_p \tag{6.47}$$

$$Q_p = \frac{2.08\ A}{t_p} \quad (S.I.\ Units) \tag{6.48a}$$

$$Q_p = \frac{484\ A}{t_p} \quad (U.S.\ Customary\ Units) \tag{6.48b}$$

in which t_r is the unit duration of the unit hydrograph in hours. Equation 6.48 assumes a triangular unit hydrograph with a base of 2.67 t_p (see subsection (a)). Equations 6.45 and 6.46 give $t_r/t_L = 2{:}9$ compared to a ratio of 2:11 for Snyder's method (Equation 6.51). Since there is some empiricism

TABLE 6.6. Computation of 30-min Unit Hydrograph Using HEC-1 Procedure.

UHO no. (1)	t (hr) (2)	$B(t) = A[\bar{A}(t)-\bar{A}(t-\Delta t)]$ (km²) (3)	$2\ C_0B(t)$ (4)	$C_2Q(t-\Delta t)$ (5)	$Q(t)$ (6)	$U(t)$ km²cm/Δt·hr (7)	$U(t)$ m³/sec (8)
1	0	0	0	0	0	0	0
2	0.5	16	3.5556	0	3.5556	1.7778	9.877
3	1.0	32	7.1111	2.7655	9.8766	6.7161	37.312
4	1.5	48	10.6667	7.6818	18.3485	14.1125	78.403
5	2.0	32	7.1111	14.2711	21.3822	19.8654	110.363
6	2.5	16	3.5556	16.6306	20.1862	20.7842	115.468
7	3.0	16	3.5556	15.7004	19.2560	19.7211	109.562
8	3.5	0	0	14.9769	14.9769	17.1165	95.091
9	4.0	0	0	11.6487	11.6487	13.3128	73.960
10	4.5	0	0	9.0601	9.0601	10.3544	57.524
11	5.0	0	0	7.0467	7.0467	8.0534	44.741
12	5.5	0	0	5.4808	5.4808	6.2638	34.799
13	6.0	0	0	4.2628	4.2628	4.8718	27.066
14	6.5	0	0	3.3155	3.3155	3.7892	21.051
15	7.0	0	0	2.5787	2.5787	2.9471	16.373
16	7.5	0	0	2.0057	2.0057	2.2922	12.734
17	8.0	0	0	1.5600	1.5600	1.7829	9.905
18	8.5	0	0	1.2133	1.2133	1.3867	7.704
19	9.0	0	0	0.9437	0.9437	1.0785	5.992
20	9.5	0	0	0.7340	0.7340	0.8389	4.660
21	10.0	0	0	0.5709	0.5709	0.6525	3.625
22	10.5	0	0	0:4440	0.4440	0.5075	2.819
23	11.0	0	0	0.3453	0.3453	0.3947	2.193
24	11.5	0	0	0.2686	0.2686	0.3070	1.705
25	12.0	0	0	0.2089	0.2089	0.2388	1.326
26	12.5	0	0	0.1625	0.1625	0.1857	1.032
27	13.0	0	0	0.1264	0.1264	0.1445	0.803
28	13.5	0	0	0.0983	0.0983	0.1124	0.624
29	14.0	0	0	0.0765	0.0765	0.0874	0.486
30	14.5	0	0	0.0595	0.0595	0.0680	0.378
31	15.0	0	0	0.0463	0.0463	0.0529	0.294
32				0.0360	0.0360	0.0412	0.229
33				0.0280	0.0280	0.0320	0.178
34				0.0218	0.0218	0.0250	0.138
35				0.0169	0.0169	0.0190	0.106
		TOTAL				159.9339	

The computation of UHOs should be continued until the summation of the values in column 7 equals 160 km²-cm within an acceptable truncation error.

implicit in Equations 6.45 to 6.47, the HEC-1 model (USACE, 1987) allows some latitude in the choice of the value of t_r such that,

$$t_r \leq \frac{1}{4}\, t_p \tag{6.49}$$

or

$$t_r \leq \frac{2}{7}\, t_L. \tag{6.50}$$

d) Snyder's Synthetic Unit Hydrograph. This method uses Equation 6.28 for lag time and the following empirical equations for other unit hydrograph parameters.

TABLE 6.7. Ratios for Dimensionless Unit Hydrograph and Mass Curves.

Time ratios (t/T_p)	Discharge ratios (q/q_p)	Mass curve ratios (Qa/Q)
0	.000	.000
.1	.030	.001
.2	.100	.006
.3	.190	.012
.4	.310	.035
.5	.470	.065
.6	.660	.107
.7	.820	.163
.8	.930	.228
.9	.990	.300
1.0	1.000	.375
1.1	.990	.450
1.2	.930	.522
1.3	.860	.589
1.4	.780	.650
1.5	.680	.700
1.6	.560	.751
1.7	.460	.790
1.8	.390	.822
1.9	.330	.849
2.0	.280	.871
2.2	.207	.908
2.4	.147	.934
2.6	.107	.953
2.8	.077	.967
3.0	.055	.977
3.2	.040	.984
3.4	.029	.989
3.6	.021	.993
3.8	.015	.995
4.0	.011	.997
4.5	.005	.999
5.0	.000	1.000

Source: USDA 1972.

$$t_r = \frac{2}{11} t_L \tag{6.51}$$

$$Q_p = \frac{2.76\, C_p\, A}{t_L} \quad (S.I.\ Units) \tag{6.52a}$$

$$Q_p = \frac{640\, C_p\, A}{t_L} \quad (U.S.\ Customary\ Units) \tag{6.52b}$$

$$Q_{PR} = Q_p \frac{t_L}{t_{LR}} \tag{6.53}$$

$$t_{LR} = t_L + \frac{1}{4}(t_R - t_r) \tag{6.54}$$

$$W_{50} = \frac{5.87}{(Q_{PR}/A)^{1.08}} \quad (S.I.\ Units) \tag{6.55a}$$

$$W_{50} = \frac{770}{(Q_{PR}/A)^{1.08}} \quad (U.S.\ Customary\ Units) \tag{6.55b}$$

$$W_{75} = \frac{3.35}{(Q_{PR}\ /\ A)^{1.08}} \quad (S.I.\ Units) \tag{6.56a}$$

$$W_{75} = \frac{440}{(Q_{PR}\ /\ A)^{1.08}} \quad (U.S.\ Customary\ Units) \tag{6.56b}$$

$$T_b = 72 + 3\ t_L \tag{6.57}$$

in which t_r is the duration of rainfall excess (cm or in) in hours, Q_p is the peak of unit hydrograph in m³/s (cfs), A is the drainage area in km² (mi²), C_p is a coefficient that accounts for flood wave storage in the basin (the value of C_p varies from 0.61 in states bordering the eastern Gulf of Mexico, 0.63 in Appalachian Highlands, to 0.94 in Southern California), t_R is the duration of rainfall excess other than the standard value, t_r, in hours; t_{LR} is lag time from the center of t_R to peak of the corresponding unit hydrograph, Q_{PR} is the peak of t_R - hour unit hydrograph in m³/s (cfs), W_{50} is the width of unit hydrograph at 50% of peak discharge in hours, W_{75} is the width of unit hydrograph at 75% of peak discharge in hours, and T_b is the time base of unit hydrograph (including interflow) in hours.

Note that this method provides two points on the base, one point at the peak, and widths (durations) of the unit hydrograph at 50 and 75% of the peak discharge rather than the complete hydrograph. To generate the complete unit hydrograph, the HEC-1 model (USACE, 1987) uses the Clark's method described earlier. Use of this method in the HEC-1 model requires the time–area curve, t_L, Δt (unit duration of the unit hydrograph), and C_p as input. Using this information, the program estimates Clark's parameters, t_c, and K (see subsection (b)) and computes a trial unit hydrograph. Using Q_p and t_p (time to peak) for this unit hydrograph, Snyder's C_{p1} and t_{L1} are estimated as follows:

$$C_{p1} = Q_p \frac{t_p - 0.5\ \Delta t}{2.78\ A} \quad (S.I.\ Units) \tag{6.58a}$$

$$C_{p1} = Q_p \frac{t_p - 0.5\ \Delta t}{640\ A} \quad (U.S.\ Customary\ Units) \tag{6.58b}$$

$$t_{L1} = 1.048\ (t_p - 0.75\ \Delta t). \tag{6.59}$$

The coefficient 1.048 in Equation 6.59 converts the standard unit duration, t_r, of Snyder's unit hydrograph to the specified duration, Δt. If the computed C_{p1} and t_{L1} are not within 1% of the specified values, then t_c and K are adjusted to compensate for the differences and the computations repeated up to 20 iterations.

e) USBR Dimensionless Unit Hydrograph (USBR, 1987). The United States Bureau of Reclamation has developed tables of dimensionless unit hydrographs and S-graphs for the six regions listed in Table 6.3. The ordinates and abscissas of the dimensionless unit hydrographs are:

$$t^a = t/(t_L + 0.5\ \Delta t) \tag{6.60}$$

$$Q^a = 2.54\ Q(t_L + 0.5\ \Delta t)/U \quad (U\ in\ m^3/s\text{-}day\ for\ 1\ cm\ of\ rain.) \tag{6.61a}$$

$$Q^a = Q(t_L + 0.5\ \Delta t)/U \quad (U\ in\ cfs\text{-}day\ for\ 1\ in\ of\ rainfall) \tag{6.61b}$$

in which t^a is dimensionless time, Q^a is dimensionless discharge, Q is UHO at time t in m^3/s (cfs), U is the unit runoff volume due to 1 cm (1 inch) of effective rainfall in a day in m^3/s-day, i.e., $0.11574A$ (cfs-day, i.e., $26.89A$), and A is the watershed area in km^2 (mi^2). Knowing A, Δt, and t_L, the UHOs can be computed from the tabulated values or by interpolations between the tabulated values. Note that t, t_L, and Δt in these equations must be expressed in hours.

The ordinates and abscissas of the dimensionless S-graphs are:

$$t^a = 100 \times t/t_L \tag{6.62}$$

$$Q^a = 254 \times Q/Q_u \quad (Q \text{ and } Q_u\ in\ m^3/s) \tag{6.63a}$$

$$Q^a = 100 \times Q/Q_u \quad (Q \text{ and } Q_u\ in\ cfs) \tag{6.63b}$$

in which Q is the S-graph ordinate at time t in cfs and Q_u is the ultimate discharge defined as the limiting value when the ordinate of the S-graph equals effective rainfall intensity. For 1 mm(im) of effective rainfall in Δt hours,

$$Q_u\ (m^3/s) = 2.78\ A/\Delta t \tag{6.64a}$$

$$Q_u\ (cfs) = 645.3\ A/\Delta t. \tag{6.64b}$$

In this case, the time step of computations should be Δt so that the value of each time level, t, is an integer multiple of Δt. The UHOs for unit duration Δt are obtained by subtracting two consecutive S-graph ordinates, i.e.,

$$U(t) = Q(t) - Q(t - \Delta t) \tag{6.65}$$

in which $U(t)$ is UHO at time t in cfs for unit duration Δt (1 cfs = 35.32 m^3, 1 mi^2 = 2.59 km^2, and 1 in = 2.54 cm).

G. Unit Hydrograph Durations

Some of the models, e.g., the National Weather Service River Forecast System (NWSRFS) model (NWS, 1986), do not include algorithms for the development of unit hydrographs and require computational time steps that may be different from the t_r computed from equations like 6-51 and computer models like the HEC-1 (USACE, 1987). This involves derivation of the unit hydrograph for one unit duration, t_r, and transforming it to one of another unit duration, t_R. Snyder's method described previously can be used to develop unit hydrographs for a nonstandard unit duration, t_R. In addition, there are two commonly used methods to derive unit hydrographs of other durations from one of a given unit duration, t_r (Chow, 1964; Ponce, 1989).

1) Superposition Method. This method is applicable only if the new unit duration, $t_R = nt_r$, in which

n is an integer. If $U_r(t)$ are the given unit hydrograph ordinates (UHOs) arranged at t_r intervals and $U_R(t)$ are the UHOs for unit duration, t_R, then,

$$U_R\ (t) = \sum_{i=1}^{n} U_r\ [t - (i-1)t_r] \div n, \quad t \geq (i-1)\ t_r. \tag{6.66}$$

Note that the computed new UHOs are at time intervals t_r, the base of the new unit hydrograph is $T_b + (n - 1)\ t_r$, in which T_b is the time base of given unit hydrograph, and the first UHO is 0 at t = 0.

2) S-Hydrograph Method. This method is applicable for all positive values of n but involves more computational steps than the Superposition Method. The S-hydrograph is a hydrograph produced by a continuous effective rainfall at a constant rate for an indefinite period of time. It is derived by lagging the unit hydrograph ordinates by t_r and summing them over a duration greater than the time base of the unit hydrograph.

$$S(t) = \sum_{i-1}^{m} U_r\ [t - (i-1)\ t_r], \quad m = \frac{t}{t_r} + 1, \quad t \geq (i-1)t_r \tag{6.67}$$

in which $S(t)$ is the S-hydrograph with unit duration t_r at time t. It has an S-shape, its effective rainfall has a unit duration t_r, and the ultimate ordinate approaches the rate of effective rainfall either as a limit or at the time of equilibrium. Minor errors in UHOs may result in oscillations in the ultimate ordinates of the S-hydrograph. These oscillations should be adjusted by judgment to achieve a continuous smooth curve. Note that $S_i(t) = S(t)/t_r$ represents an S-hydrograph for a unit rate or intensity of effective rainfall.

The computational steps to obtain a unit hydrograph of unit duration t_R are as follows:

a) Tabulate the S-hydrograph ordinates at time intervals equal to the highest common factor (HCF) of t_r and t_R, say, D.

b) Lag the ordinates by t_R and perform the following computations,

$$U_R\ (t) = [S(t) - S(t-t_R)]\ \frac{t_r}{t_R} \tag{6.68}$$

H. Instantaneous Unit Hydrograph

Instantaneous Unit Hydrograph (IUH) is the hydrograph, $U_i(t)$, at the outlet of a watershed resulting from an instantaneous input of effective rain, i.e., $1 \times \delta(t)$, where $\delta(t)$ is the dirac delta function such that $\delta(t = 1$ at $t = 0$ and $\delta(t) = 0$ at $t \neq 0$. The effective rain is assumed to be uniformly distributed over the watershed. Theoretically, the IUH can be used to develop the surface runoff or unit hydrograph; however, its practical applications are relatively less than the unit hydrograph described previously. In fact, the Clark unit hydrograph described previously and included in the HEC-1 model (USACE, 1987) is a form of an instantaneous unit hydrograph.

If $U_i(O,t)$ = IUH ordinate at time t, $I(\tau)$ is the function defining the variation of effective rainfall rate with time (τ), and t_0 is the duration of effective rainfall, then

$$Q(t) = \int_0^{t \le t_0} U_i\ (0, t - \tau) \quad I(\tau)\ d\ \tau \tag{6.69}$$

in which $Q(t)$ is the ordinate of surface runoff hydrograph at time t, $t' = t$ when $t \leq t_0$, and $t' = t_0$ when $t > t_0$.

The watershed is conceptualized as a series of n identical linear reservoirs such that the outflow from one is the inflow to the other. Thus, the outflow from the nth reservoir or from the watershed (simulated by n linear reservoirs) due to an instantaneous unit rainfall excess is given by (Nash, 1957)

$$U_i(t) = O_n(t) = \frac{1}{K(n-1)!}\left(\frac{t}{K}\right)^{n-1} e^{-t/K}. \tag{6.70}$$

This is the gamma function form of IUH. K is the storage coefficient as defined in Equation 6.41.

If the S-hydrograph, $S_i(t)$, represents unit rate or intensity of effective rainfall, then (Wang et al., 1970; Dooge, 1973)

$$U_i(t) = \frac{d\, S_i(t)}{dt} \tag{6.71}$$

and

$$Si(t) = \int_0^t U_i(t)\, dt \tag{6.72}$$

Also, setting $t_r = 1$ and $t_R = \Delta t$ in Equation 6.68,

$$\begin{aligned} U(t) &= \frac{1}{\Delta t}\,[S_i(t) - S_i(t-\Delta t)] \\ &= \frac{1}{\Delta t}\left[\int_0^t U_i(t)\, dt - \int_0^{t-\Delta t} U_i(t)\, dt\right] \\ &= \frac{1}{\Delta t}\left[I_G\left(\frac{t/K}{\sqrt{n}}, n-1\right) - I_G\left(\frac{(t-\Delta t)/K}{\sqrt{n}}, n-1\right)\right] \end{aligned} \tag{6.73}$$

in which Equation 6.70 is used to define IUH and the integral of the gamma function is represented by I_G, the incomplete gamma function of order $(t/K)/\sqrt{n}$ at $n - 1$. Equations 6.70 and 6.73 were used to develop IUH and unit hydrographs for a number of Hawaiian small watersheds (Wang et al., 1970). The parameters n and K for each watershed were estimated by the method of moments (Chow, 1964):

$$M_{SRH1} - M_{ERH1} = nK \tag{6.74}$$

$$M_{SRH2} - M_{ERH2} - 2nK\, M_{ERH1} = n\,(n + 1)\, K^2 \tag{6.75}$$

in which M_{SRH1} is the first moment of surface runoff hydrograph about the origin, M_{SRH2} is the second moment of surface runoff hydrograph about the origin, M_{ERH1} is the first moment of effective rainfall hyetograph about the origin, and M_{ERH2} is the second moment of effective rainfall hyetograph about the origin. The following regression equations were developed using data for 29 watersheds from about 121.4 to 12,141 hectares (300 to 30,000 acres) in size.

$$K = 0.01785\, A^{0.55048} \quad (S.I.\ Units) \tag{6.76a}$$

$$K = 0.01085\, A^{0.55048} \quad (U.S.\ Customary\ Units) \tag{6.76b}$$

$$n = 11.292\, A^{-0.28098} \quad (S.I.\ Units) \tag{6.77a}$$

$$n = 14.56\, A^{-0.28098} \quad (U.S.\ Customary\ Units) \tag{6.77b}$$

in which A is the watershed area in hectares (acres) and K is the storage coefficient in hours.

A similar study of four watersheds in the Godavari River basin and two in the Krishna River basin in India resulted in the range of values shown in Table 6.8 (Vishwakarma et al., 1991). These studies

TABLE 6.8. Range of Values of IUH Parameters for Selected Basins.

Parameter	Range of Values	
	Godavari River Basin	Krishna River Basin
Watershed Area (km^2)	297.984–3315	271.410–2926.698
L^a (km)	30.898–108.145	27.036–115.869
L^a_{ca} (km)	10.620–48.279	12.390–52.624
Basin Slope (%)	2.623–3.540	2.340–4.507
n	2.400–5.300	2.830–3.300
K (hrs)	7.212–14.360	10.960–14.640

demonstrate the application of the IUH concept to develop unit hydrographs for gaged watersheds and their use for hydrologic studies of similar ungaged watersheds.

I. Runoff Hydrograph Development

Development of runoff hydrograph or flood hydrograph for a basin involves hydrograph convolution and addition of baseflow and recession limb to the computed surface runoff hydrograph. Convolution is the process of deriving surface runoff hydrograph from effective rainfall hyetograph and unit hydrograph. The steps of runoff hydrograph computations are as follows: 1) Determine time step of computations (approx. 0.29 t_L), Δt; 2) Obtain rainfall hyetograph for the desired storm with ordinates (bars) representing rainfall intensities, $I(i)$, in successive Δt time durations; 3) Estimate losses and subtract them from the rainfall to obtain effective rainfall hyetograph; 4) Develop unit hydrograph with unit duration equal to the time step of computations; 5) Convolute effective rainfall hyetograph and unit hydrograph to obtain surface runoff hydrograph, i.e.,

$$Q_r(t) = \sum_{i=1}^{n} U[\Delta t, t - (i - 1)\Delta t]\, I(i)\, \Delta t \tag{6.78}$$

in which $Q_r(t)$ is the surface runoff hydrograph ordinate at time t, $U(\Delta t, t)$ = UHO at time t, $I(i)$ is effective rainfall intensity (ordinate of effective rainfall hyetograph) during the time interval, and n is the number of blocks of duration Δt in which effective rainfall hyetograph is divided. The time base, T_r, of the surface runoff hydrograph is given by

$$T_r = T_b + (n - 1)\Delta t \tag{6.79}$$

in which T_b is the time base of the unit hydrograph of unit duration, Δt. 6) Compute the recession limb and baseflow component and add them to the surface runoff hydrograph to obtain the complete runoff or flood hydrograph.

Generally, runoff hydrograph computations are made using computer programs [e.g., HEC-1 (USACE, 1987); TR-20 (USDA, 1982); SEDIMOT-II (University of Kentucky, 1981); etc.]. The computational steps of Equation 6.78 are illustrated in the following example.

Use the 30-minute unit hydrograph of Table 6.6, in the previous example to compute the first 20 ordinates of the surface runoff hydrograph for the same watershed of a 6-hour storm with the following mass curve of rainfall:

Time (hrs)	0.5	1.0	1.5	2.0	2.5	3.0	3.5	4.0	4.5	5.0	5.5	6.0
Cumulative Rainfall (cm)	0.01	0.09	0.79	0.81	0.87	2.37	2.77	2.81	2.83	2.89	2.91	2.92

Assume zero baseflow.

Solution: Here, $\Delta t = 0.5$ hr and $n = 12$. Since cumulative rainfall depths rather than rainfall intensities are given, compute incremental rainfall depths, $I(i) \cdot \Delta t$, by successive subtraction. The resulting values are shown in the second row of Table 6.9. The UHOs of Table 6.6 are reproduced in column 3. Successive computations using Equation 6.78 are shown in columns 4 through 15. The computed runoff hydrograph ordinates (RHOs) are shown in column 16. Column 4 is the product of the first incremental rainfall depth, 0.01 cm, with the UHOs in column 3. Column 4 is the product of the second incremental rainfall depth, 0.08 cm, with the UHOs lagged by one time step, and so on. For $t > 0$ and $[t - (i - 1)\Delta t] > 0$ in Equation 6.79, n is 1 for RHO No. 2, 2 for RHO No. 3, . . . , and 12 for RHO No. 13 and onwards. The computed values in each row represent the respective summation terms of Equation 6.78.

IV. OVERLAND FLOW

A. Sources

Overland flow is surface runoff upstream of recognizable channels. Topologically, overland flow is a stream of zero order. Overland flow ranges from thin laminar sheet flow to small turbulent rivulets. Overland flow persisting for more than 300 ft. is sometimes identified as shallow concentrated flow (USDA, 1986). Overland flow divided by grass or vegetal matter producing essentially uniform depth is known as subdivided flow. Overland flow may concentrate into small braided rivulets conveying as much as two-thirds of the total runoff (Nelson et al., 1986). While flow concentration is generally not incorporated in overland flow computations, an effective increase in both depth and velocity may be required for problems regarding sediment stability.

Saturation overland flow hypothesizes a dynamic area within the watershed in which the soil surface is saturated upward from a restricting boundary (Musgrave and Holton, 1964). Rainfall on this area becomes overland flow. As the areal extent of saturation increases, overland flow likewise increases. As soil drains and superficial infiltration capacity is restored, overland flow decreases.

The saturation runoff concept has been refined into the *variable source area*, or the *dynamic concept*. Hillslope hydrology frequently employs this model (Gupta, 1989). The rising limb of saturation runoff is closely akin to the step function, exponential, or probabilistic spill from surface depressions (Tholin and Keifer, 1959).

The *partial-area* concept combines subsurface interflow with overland flow. This concept has been employed in hillslope hydrology where hydraulic mechanisms change between upper and lower elevations (Ponce, 1989).

B. Use in Runoff Modeling

As overland flow occurs on a relatively small portion of a drainage basin and can achieve equilibrium relatively quickly in a large basin, its impact on the overall basin hydrograph is often small. Where basin discharge is the principal objective of modeling, overland flow may be ignored in such cases.

Overland flow may be a significant mechanism in urban hydrology. To evaluate hillslope processes, sediment yield, or other hydrologic aspects of basins less than a few acres, overland flow should be considered.

C. Steady-State Solutions

The distributed nature of watersheds makes overland flow a complex physical process. Classical analysis relies on steady-state hydraulics of uniform rainfall on an idealized inclined plane.

TABLE 6.9. Runoff Hydrograph Computation.

RHO no. (1)	t (hr) (2)	UHO (m^3/s) (3)	Incremental 30-min Rainfall Depth (cm)												$Q_r(t)$ (m^3/s) (16)
			0.01 (4)	0.08 (5)	0.70 (6)	0.02 (7)	0.06 (8)	1.50 (9)	0.40 (10)	0.04 (11)	0.02 (12)	0.06 (13)	0.02 (14)	0.01 (15)	
			COMPUTED VALUES FOR n = 1 TO 12 (m^3/s)												
1	0	0	0												0
2	0.5	9.877	0.099	0											0.099
3	1.0	37.312	0.373	0.790	0										1.163
4	1.5	78.403	0.784	2.985	6.914	0									10.683
5	2.0	110.363	1.104	6.272	26.118	0.198	0								33.692
6	2.5	115.468	1.155	8.829	54.882	0.746	0.593	0							66.205
7	3.0	109.562	1.096	9.237	77.254	1.568	2.239	14.816	0						106.210
8	3.5	95.091	0.951	8.765	80.828	2.207	4.704	55.968	3.951	0					157.374
9	4.0	73.960	0.740	7.607	76.693	2.309	6.622	117.605	14.925	0.395	0				226.896
10	4.5	57.524	0.575	5.917	66.564	2.191	6.928	165.545	31.361	1.493	0.198	0			280.772
11	5.0	44.741	0.447	4.602	51.772	1.902	6.574	173.202	44.145	3.136	0.746	0.593	0		287.119
12	5.5	34.799	0.348	3.579	40.267	1.479	5.705	164.343	46.187	4.415	1.568	2.239	0.198	0	270.328
13	6.0	27.066	0.271	2.784	31.319	1.150	4.438	142.637	43.825	4.619	2.207	4.704	0.746	0.099	238.799
14	6.5	21.051	0.211	2.165	24.359	0.895	3.451	110.940	38.036	4.383	2.309	6.622	1.568	0.373	195.312
15	7.0	16.373	0.164	1.684	18.946	0.696	2.684	86.286	29.584	3.804	2.191	6.928	2.207	0.784	155.958
16	7.5	12.734	0.127	1.310	14.736	0.541	2.088	67.112	23.010	2.958	1.902	6.574	2.309	1.104	123.771
17	8.0	9.905	0.099	1.019	11.461	0.421	1.624	52.199	17.896	2.301	1.479	5.705	2.191	1.155	97.550
18	8.5	7.704	0.077	0.792	8.914	0.327	1.263	40.599	13.920	1.790	1.150	4.438	1.902	1.096	76.268
19	9.0	5.992	0.060	0.616	6.934	0.255	0.982	31.577	10.826	1.392	0.895	3.451	1.479	0.951	59.418
20	9.5	4.660	0.047	0.479	5.393	0.198	0.764	24.560	8.420	1.083	0.696	2.684	1.150	0.740	46.214

1. ***Hortonian Runoff*** Identification of the overland flow process is credited to Horton (1933). If f is infiltration rate (typically taken as an exponentially decaying function, held constant over a short time step) and i is rainfall intensity, precipitation less infiltration $(i - f)$ becomes overland flow. Equilibrium discharge per unit width is

$$q = (i - f)L \cos \theta \tag{6.80}$$

in which L is the length of the flow plane and θ is the arctan of the plane slope S_o. In most problems, cos θ is 1.0.

Horton visualized overland flow as sheet flow derived from the entire basin. Interception, detention, and such hydraulic complications as interflow and protruding vegetation are neglected or incorporated in the f coefficient. Hortonian runoff appropriately describes overland flow from plane surfaces either impervious or of slight and uniform permeability (Chow et al., 1988).

Example 1: Rainfall intensity is 30 mm/hr on a field 150 m long. Loss rate is 10 mm/hr. What is equilibrium unit discharge per unit width from the field? By using Equation 6.80, where θ is zero,

$$q = (30 - 10)\frac{mm}{hr}\ 150\,m\left(\frac{m}{10^3\,mm}\right)\left(\frac{hr}{3600\ sec}\right) = 0.0008.\ m^3/sec/m$$

2. ***Stage-Discharge Relationships*** Overland flow is laminar at a Reynold's number ($R_e = uy/\nu$, where u is mean velocity, y is depth, and ν is kinematic viscosity) below 300 and becomes fully turbulent when R_e exceeds 770 (Chow, 1959). The limit corresponds to unit discharges of roughly 0.0001 and 0.0002 $m^3/s/m$. The flow concentration mentioned earlier additionally contributes to turbulence. While laminar conditions may exist over short reaches of the flow path, overland flow typically behaves as mixed or turbulent flow.

Storage routing for overland flow is the same as that employed in reservoir routing,

$$q = ay^m \tag{6.81}$$

$$u = ay^{m-1} \tag{6.82}$$

where a and m are constants. In Hortonian runoff, y is the mean depth over the plane.

For mixed or turbulent conditions,

$$a = 1.49\frac{\sqrt{S_o}}{n},\ \text{Manning's Eq., U.S. Customary Units,}\ \frac{\sqrt{S_o}}{\text{n}},\ \text{SI units} \tag{6.83}$$

$$a = C\sqrt{S_o}\ ,\ \text{Chezy Equation} \tag{6.84}$$

where n and C are Manning and Chezy roughness, respectively. Table 6.10 shows typical values of n.

Manning n values for overland flow tend to be greater than those associated with channelized flow over similar surfaces. Values of the exponent m are shown in Table 6.11.

Force balance for laminar conditions indicates that

$$y = \left(\frac{f\,q^2}{8\,g\,S_o}\right)^{1/3} \tag{6.85}$$

TABLE 6.10. Manning n for Overland Flow.

Watershed cover	Manning n
Packed Clay	0.03
Fallow	0.05
Sparse vegetation	0.05–0.13
Cultivated ground	
residue cover < 20%	0.06
residue cover > 20%	0.17
Asphalt or concrete paving	
depths > 6 mm	0.05–0.10
depths < 6 mm	0.10–0.15
Grass	
Range, short prairie	0.10–0.20
Light turf	0.20
Bluegrass sod	0.20–0.50
Dense turf	0.24–0.35
Pasture	0.30–0.40
Bermuda	0.43

Sources: Crawford and Linsley (1966), USDA (1986)

where g is gravitational acceleration and f is a Darcy–Weisbach type roughness estimated as $C_L/(4\ R_e)$. This f exceeds the roughness commonly associated with pipe flow due to topography, vegetation, and raindrop impact (Chow et al., 1988). C_L is a resistance coefficient,

$$C_L = 96 + 29.6\ i^{0.4} \tag{6.86}$$

where i is measured in mm/hr. Algebraic manipulation reduces Equation 6.85 to the form of Equation 6.81 where m is equal to 3 and

$$a = \frac{32\ g\ S_o}{C_L\ v}. \tag{6.87}$$

Example 2: Determine the depth and velocity for 0.001 m³/s/m through light turf sloped at 0.02. Viscosity ν is 1.3×10^{-6} m²/s.

$$R_e = \frac{0.001}{1.3 \times 10^{-6}} = 770, \text{ fully turbulent}$$

$$a = \frac{\sqrt{0.02}}{0.2} = 0.71 \qquad m = 1.67$$

$$y = \left(\frac{0.001}{0.71}\right)^{1/1.67} = 0.019 \text{ m}$$

$$u = 0.71\ (0.19)^{0.67} = 0.05 \text{ m/s}$$

TABLE 6.11. Values of m.

Turbulence	Manning equation	Chezy equation
100%	1.67	1.5
75%	2.0	1.87
50%	2.33	2.25
25%	2.67	2.62
0% (laminar)	3.0	3.0

Source: Ponce (1989)

Example 3: Same as above, but q is 0.0007 m³/s/m.

$$R_e = \frac{0.0007}{1.3 \times 10^{-6}} = 538, \text{ mixed flow. Assume 50 \% turbulence.}$$

$$a = \frac{\sqrt{0.02}}{0.2} = 0.71 \qquad m = 2.33$$

$$y = \left(\frac{0.0007}{0.71}\right)^{1/2.33} = 0.052 \text{ m}$$

$$u = 0.71\,(0.052)^{1.33} = 0.014 \text{ m/s}$$

Example 4: Same as above, but q is at 0.0003 m³/s/m over a smooth surface. Rainfall intensity is 10 mm/hr.

$$R_e = \frac{0.0003}{1.3 \times 10^{-6}} = 230, \text{ laminar flow}$$

$$C_L = 96 + 29.6\,(10)^{0.4} = 170$$

$$a = \frac{32 \times 9.81 \times 0.02}{170 \times 1.3 \times 10^{-6}} = 2640 \qquad m = 3$$

$$y = \left(\frac{0.0003}{2840}\right)^{1/3} = 0.0047 \text{ m}$$

$$u = 2840\,(0.0047)^2 = 0.063 \text{ m/s}$$

While such shallow, high-velocity flow may exist for short reaches, surface impediments will initiate turbulence.

Crawford and Linsley (1966) add surface detention to depth y, modifying Equation 6.81 to

$$q = 2.17\,ay^m \tag{6.88}$$

for fully turbulent conditions. With i measured in in/hr, L in feet, and D_e as unit-width storage in ft³/ft, the Stanford Watershed Model SWM-IV employs

$$D_e = 0.000818\left(\frac{i^{0.6}n^{0.6}L^{1.6}}{S_o^{0.3}}\right). \tag{6.89}$$

D. Unsteady Flow Problems

Mass and force balance provide that

$$\frac{\partial y}{\partial t} + u\frac{\partial y}{\partial x} + y\frac{\partial u}{\partial x} - (i - f) = 0, \text{ and} \tag{6.90}$$

$$\left(\frac{1}{g}\right)\frac{\partial u}{\partial t} + \left(\frac{u}{g}\right)\frac{\partial u}{\partial x} + S_f - S_o + \frac{(i - f)\,u}{g\,y} = 0 \tag{6.91}$$

where t is time, x is downstream distance, and S_f is friction slope. Solutions proposed for Equation 6.91 are discussed below.

1. Time to Equilibrium Time to equilibrium t_e is commonly employed in estimates of time to peak when there is a constant rainfall excess. In many cases, complete solution of nonsteady behavior is not required. Rather, only an estimate is needed of t_e, the duration before equilibrium conditions are established. Alternative estimates of t_e in seconds for laminar flow are

$$t_e = 2\left(\frac{3\,v\,L_1}{g\,S_o\,i^2}\right)^{1/3} \tag{6.92}$$

where L_1 is the laminar length of the flow plane (Ponce, 1989), and

$$t_e = 3(0.0007i + c)\left(\frac{43{,}200^2 L_1}{S_o\,i^2}\right)^{1/3} \tag{6.93}$$

where c is Izzard's friction factor, i is in in/hr, and L_1 is in feet (Singh, 1988).

For turbulent flow, using any consistent system of units,

$$t_e = \frac{5}{3}\,i^{-0.4}\,(L^{0.6} - L_1^{0.6})\left(\frac{1}{a}\right)^{0.6} \tag{6.94}$$

$$t_e = i^{-0.4}\left(\frac{L}{a}\right)^{0.6} \tag{6.95}$$

$$t_e = 2\,i^{1/m-1}\left(\frac{L}{a}\right)^{1/m}. \tag{6.96}$$

SWM-IV employs Equation 6.95. It should be noted that Equations 6.94 (Singh, 1988), 6.95 (Crawford and Linsley, 1966) and 6.96 (Ponce, 1989) for fully turbulent flow ($m = 1.67$) differ only in the initial constants if L_1 is 0. The differences stem from the respective boundary assumptions. TR-55 employs Equation 6.97.

$$t_e = \frac{34.7}{\sqrt{P_2}}\left(\frac{L}{a}\right)^{0.8} \tag{6.97}$$

where P_2 is the 2-year 24-hour rainfall in inches and L is in feet (USDA, 1986). Where there is no compelling reason to employ a single formulation, it is suggested that Equations 6.94 to 6.97 be employed together to bracket a range of estimates.

Example 5: Find time to equilibrium for a 200 m flow path over light turf. Slope is 0.02 and rainfall intensity is 30 mm/hr. Infiltration is 5 mm/hr.

$$q = (30 - 5)\frac{mm}{hr}\ 200\ m\left(\frac{m}{10^3\ mm}\right)\left(\frac{hr}{3600\ sec}\right) = 0.00139\ m^3/s/m$$

$$R_e = \frac{0.00139}{1.3 \times 10^{-6}} = 1069\text{, turbulent flow}$$

$$a = \frac{\sqrt{0.02}}{0.2} = 0.71$$

Choosing Equation 6.95,

$$t_e = \left(\frac{0.025}{3600}\right)^{-0.4}\left(\frac{200}{0.71}\right)^{0.6} = 3410\text{ sec.}$$

E. Other Solutions

Complete solutions of Equations 6.90 and 6.91 are mathematically complex and dependent on boundary conditions that may be difficult to establish in field situations. The following discussion reviews several analytical approaches in summary fashion.

1. *Linear and Nonlinear Reservoirs* A short reach of unit width of a plane can be a taken as a nonlinear reservoir. Flow cascades from one reservoir to the next. Continuity provides that

$$\frac{\partial y}{\partial t} + \frac{q}{L} - (i - f) = 0 \tag{6.98}$$

where q is a function of x (Equation 6.80). Equation 6.98 is Equation 6.90 where q is uy at any L. If m is 1.0, the problem becomes that of a linear reservoir, a classical, if oversimplified, overland flow conceptual model solved by the Muskingum method with the Muskingum x being 0. If m is other than 1, $y(x)$ can be solved by finite differences.

2. *Kinematic Wave* Unlike the Hortonian or reservoir models that relate discharge to planar storage, kinematic theory adjusts Equation 6.91 for the flow depth at a given point,

$$\frac{\partial y}{\partial t} + mu\frac{\partial y}{\partial x} - (i - f) = 0. \tag{6.99}$$

Equation 6.99 forms the kinematic description of overland flow. A kinematic wave is a variation in flow that maintains a unique function relating discharge to stage. Inertial and pressure forces are insignificant, providing explanation of unsteady uniform free surface flow. This wave progresses downstream at celerity $c = mu$. An infinitesimal kinematic wave front theoretically steepens as it progresses downstream, having no diffusion mechanism to disperse it. Practically, surface irregularities introduce sufficient diffusion to mitigate form transformation, leading to a steady-state monoclinal wave known as *kinematic shock.*

According to Woolhiser and Liggett (1967), kinematic flow occurs when

$$\frac{S_o L}{y F_r^2} > 10\ to\ 20. \tag{6.100}$$

It can be shown that the rising and falling limbs of a kinematic wave hydrograph are (Singh, 1988),

$$\frac{q}{q_e} = \left(\frac{t}{t_e}\right)^m, t < t_e, \text{ and} \tag{6.101}$$

$$\frac{t}{t_e} = \frac{1}{m}\left(\frac{q_e}{q} - 1\right)\left(\frac{q}{q_e}\right)^{1/m}, t > \text{storm duration} \tag{6.102}$$

where q_e is equilibrium discharge. Solution for $q(t)$ describes the kinematic wave passage at any position x. HEC-1 employs a kinematic wave option for overland flow (USACE, 1987).

Kinematic waves can be solved analytically for the case of constant rainfall excess by the method of characteristics, tracing the space–time locus of discontinuities of Equation 6.99; however, an analytical solution is difficult for realistic runoff problems with nonstationary parameters. A finite difference scheme is preferred, wherein a Taylor series is used to expand Equation 6.99 and a central difference approximation used to estimate discrete points in the space-time domain (Bedient and Huber, 1988).

3. ***Diffusion Wave*** In diffusion wave theory, the flow depth gradient $\delta y/\delta x$ diffuses the runoff waveform. The governing equation adds a second-order term to Equation 6.99.

$$\frac{\partial y}{\partial t} + mu\frac{\partial y}{\partial x} - \left(\frac{uy}{2S_o}\right)\frac{\partial^2 y}{\partial x^2} - (i - f) = 0 \tag{6.103}$$

As S_o approaches zero, the diffusion term becomes unrealistically large and Equation 6.103 is rendered invalid; however, for slopes greater than 0.0001, a large diffusion component is justified. At very mild slopes, diffusion is the principal mechanism of overland flow. Like the kinematic wave problem, a finite difference scheme can be used for solution (Ponce, 1989).

V. STREAM FLOW ROUTING

Stream flow routing is a procedure by which the outflow hydrograph at a point on the stream is computed using known inflow hydrograph and channel storage data for one or more points upstream. The inflow and outflow hydrographs may represent continuous daily or monthly stream flows for several years in succession or flood flows resulting from relatively shorter duration storm events. Reservoir routing is a procedure to compute the outflow hydrograph from a reservoir using a given inflow hydrograph, elevation-storage data, and rating table for its outlet works (e.g., spillways and low-level outlets). Stream flow routing for natural channels differs from reservoir routing because the channel or valley storage, which is the volume of water in temporary storage in the channel reach under consideration at any instant, is not a unique function of outflow both for rigid bed and mobile bed streams.

Stream flow routing is required 1) for reservoir planning studies, 2) to prepare emergency evacuation plans, 3) to design flood control structures, e.g., levees, flood walls, and bypass channels, 4) to size hydraulic structures in the downstream reaches, e.g., intakes, outfalls, bridges, and culverts, etc., 5) to plan waterfront developments, 6) to estimate water yields for storage or diversion at a specified location, 7) for proper allocation of water to satisfy adjudicated water rights, treaties, or compacts, 8) to ensure adequate dilution of regulated industrial and municipal discharges into the stream, and 9) to demonstrate compliance with specified in-stream flow requirements.

In general, there are two types of methods for stream flow routing. Methods involving solutions of the basic differential equations for the movement of water and sediment in open channels are known as hydraulic methods. Those approximating solutions of the flow equations without their direct use are known as hydrologic methods.

This chapter describes equations governing the movement of water and sediment in open channels, commonly used methods of stream flow routing with specific applications, and selected public domain stream flow routing models.

A. Open Channel Flow Principles

Stream flow routing involves solutions of the equations of gradually varied unsteady flow coupled with a uniform flow formula of the Chezy or Manning type. In case of mobile bed routing, a continuity equation for sediment along with a sediment transport formula has to be included in the system of equations.

Commonly used uniform flow formulas include the following:

$$\text{Chezy Formula: } V = C\sqrt{R\,S_f} \tag{6.104}$$

$$\text{Manning Formula: } V = \frac{1}{n} R^{2/3} S_f^{1/2} \tag{6.105}$$

in which V is the mean velocity in m/sec, R is the hydraulic radius in m, S_f is the energy slope in m/m, C is Chezy factor of flow resistance, and n is the Manning coefficient of channel roughness. Uniform flow is generally considered to be steady. In this flow, the energy slope, S_f, the water surface slope, S_w, and the channel bed slope, S_o, are all equal.

The Chezy factor can be expressed in terms of Manning's n and Darcy–Weisbach friction factor as follows:

$$C = \frac{R^{1/6}}{n} = \sqrt{\frac{8g}{f}}. \tag{6.106}$$

Typical values of Manning n are reproduced in Tables 6.12 (USBR, 1987) and 6.13 (Lee and Essex, 1983). In gradually varied unsteady flow, the curvature of the wave profile is mild, the change in water depth is gradual, the vertical component of acceleration is small compared to the total acceleration, and the effect of channel friction is significant.

The continuity and momentum equations for gradually varied unsteady flow in an open channel, commonly known as the Saint-Venant equations, are (Chow, 1959):

$$y\frac{\partial V}{\partial x} + V\frac{\partial y}{\partial x} + \frac{\partial y}{\partial t} - q_L = 0 \tag{6.107}$$

$$\frac{1}{g}\frac{\partial V}{\partial t} + \frac{\alpha V}{g} \times \frac{\partial V}{\partial x} + \frac{\partial y}{\partial x} = S_o - S_f - \frac{q_L\,V}{g\,y} \tag{6.108}$$

in which S_o is the bed slope, q_L is the lateral inflow per unit length of channel, y is the water depth, α is the energy coefficient, g is gravitational acceleration, x is the space coordinate in the direction of flow, and t is time. Equation 6.108 is known as the general dynamic equation for gradually varied unsteady flow in prismatic channels. Stream flow routing based on simultaneous solutions of Equations 6.107 and 6.108 with appropriate boundary and initial conditions is called dynamic routing.

Equation 6.108 represents a force balance due to local and convective accelerations, pressure gradient, gravity, and friction and momentum transfer due to lateral inflow. A simpler solution of gradually varied unsteady flow problems is obtained by the diffusion wave approximation. In this case, friction slope is

TABLE 6.12. Typical Values of Manning's Roughness Coefficient,[a] n.

Value of n	Channel condition
0.016–0.017	Smoothest natural earth channels, free from growth, with straight alignment.
0.020	Smooth natural earth channels, free from growth, little curvature.
0.0225	Average, well-constructed moderate-sized earth channels in good condition.
0.025	Small earth channels in good condition, or large earth channels with some growth on banks or scattered cobbles in bed.
0.030	Earth channels with considerable growth. Natural streams with good alignment, fairly constant section. Large floodway channels, well-maintained.
0.035	Earth channels considerably covered with small growth. Cleared but not continuously maintained floodways.
0.040–0.050	Mountain streams in clean loose cobbles. Rivers with variable section and some vegetation growing in banks. Earth channels with thick aquatic growths.
0.060–0.075	Rivers with fairly straight alignment and cross section, badly obstructed by small trees, very little underbrush or aquatic growth.
0.100	Rivers with irregular alignment and cross section, moderately obstructed by small trees and underbrush. Rivers with fairly regular alignment and cross section, heavily obstructed by small trees and underbrush.
0.125	Rivers with irregular alignment and cross section, covered with growth of virgin timber and occasional dense patches of bushes and small trees, some logs and dead fallen trees.
0.150–0.200	Rivers with very irregular alignment and cross section, many roots, trees, bushes, large logs, and other drift on bottom, trees continually falling into channel due to bank caving.

[a]USBR, 1987

assumed to be equal to the water surface slope and Equation 6.108 is replaced by a uniform flow formula (Equation 6.104 or 6.105). Thus, (Ponce, 1989):

$$Q = K\left(S_o - \frac{dy}{dx}\right)^{1/2} \tag{6.109}$$

and

$$\frac{1}{T}\frac{\partial A}{\partial x} + M\,Q^2 - S_o = 0 \tag{6.110}$$

TABLE 6.13. Conditions for Selecting n Values.

n	Observable conditions[b]
0.030	Smooth, well maintained earth channel. No vegetation except grass on channel, bank, and overbank flow area. Constant cross section.
0.050	Smooth earth channel. Shrubs or small trees in channel and on bank. Constant cross section.
0.080	Smooth earth channel. Trees and shrubs in channel and on bank. Constant cross section.
0.045	Earth channel with some irregularity in channel giving minor changes in shape and area of cross section over relatively short distance (100 ft [30.5 m]). No vegetation except grass on channel, bank, and overbank flow area.
0.065	Earth channel with some irregularity in channel giving minor changes in shape and area of cross section over relatively short distances (100 ft [30.5 m]). Shrubs or small trees in channel and on bank.
0.900	Earth channel with some irregularity in channel giving minor changes in shape and area of cross section over relatively short distances 100 ft [30.5 ml]). Trees and shrubs in channel and on bank.
0.055	Earth channel with large rocks and boulders on channel bottom. May have logs and other large debris in channel. Somewhat irregular channel alignment. Significant changes in channel cross section over short distances caused by rock outcrops or tree root growth. With essentially a grassy overbank flow area except for vegetation along edge of channel.
0.085	Earth channel with rocks and boulders on channel bottom. May have logs and other large debris in channel. Somewhat irregular channel alignment. Significant changes in channel cross section over short distances caused by rock outcrops or tree root growth. Moderately dense trees and shrubs in channel and on bank.
0.120	Earth channel with rocks and boulders on channel bottom. May have logs and other large debris in channel. Somewhat irregular channel alignment. Significant changes in channel cross section over short distances caused by rock outcrops or tree root growth. Very dense thicket trees and shrubs in channel, on bank, and on overbank flow area.

[b]Lee and Essex, 1983.

in which Q is discharge, $K = 1/\sqrt{m}$, which is the channel conveyance, and T is the top width of flow so that $dA = T\,dy$. Neglecting lateral inflow, Equation 6.107 is rewritten as follows:

$$\frac{\partial Q}{\partial x} + \frac{\partial A}{\partial t} = 0 \tag{6.111}$$

Using perturbation analysis, Equations 6.110 and 6.111 are combined and linearized to yield the following parabolic equation,

$$\frac{\partial Q}{\partial t} = \frac{Q_o}{2TS_f} \times \frac{\partial^2 Q}{\partial x^2} - \frac{\partial Q}{\partial A} \times \frac{\partial Q}{\partial x}. \tag{6.112}$$

The term $Q_o/(2TS_f)$ is hydraulic diffusivity, $\partial Q/\partial A$ is kinematic wave celerity, and Q_o is a reference discharge. Equation 6.112 is the diffusion wave equation for gradually varied unsteady flow. It does not fully account for momentum effects but does provide for peak flow attenuation. An alternative derivation of the diffusion wave equation without the use of perturbation analysis is presented by Miller and Cunge (1975).

A further simplification of Equation 6.108 is the kinematic wave approximation where it is assumed that all the terms on the left-hand side are negligible compared with those on the right-hand side. This means that friction forces are completely balanced by gravity forces only. This results in

$$S_f = \frac{q^2}{C^2 y^3} = S_o \tag{6.113}$$

or

$$q = \alpha_1 \, y^m \tag{6.114}$$

$$Q = \alpha_2 \, A^m \tag{6.115}$$

in which q is discharge per unit width of the channel. Substitution of Equation 6.110 in 6.111 results in

$$\frac{\partial Q}{\partial t} + m\,V\,\frac{\partial Q}{\partial x} = 0. \tag{6.116}$$

Equation 6.116 is the kinematic wave equation for gradually varied unsteady flow. Kinematic waves travel with wave celerity, $\partial Q/\partial A = mV$, and do not attenuate. Wave attenuation can only be described by a second order partial differential equation, e.g., Equation 6.112.

The hydrologic equation of continuity is obtained by setting $q_L = O$ in Equation 6.107 and integrating it over a channel reach. This gives

$$I - O = \frac{dS}{dt} \tag{6.117}$$

in which I is the inflow, O is the outflow, and S is the storage in the reach at a given time.

The hydrologic equation corresponding to the dynamic equation (Equation 6.108) is the uniform flow equation,

$$Q = CA\sqrt{RS_f} = CA\sqrt{R\left(S_o - \frac{\partial y}{\partial x} - \frac{\alpha V}{g} \times \frac{\partial V}{\partial x} - \frac{1}{g}\frac{\partial V}{\partial t}\right)} \tag{6.118}$$

or simplified versions of this relationship.

The continuity equation for sediment movement in a mobile bed stream is (Garde and Ranga Raju, 1978; USACE, 1991; and Simons and Senturk, 1992):

$$\frac{\partial z}{\partial t} + \frac{1}{1-\lambda} \times \frac{\partial q_s}{\partial x} = 0 \tag{6.119}$$

where z is the elevation of mobile bed above a horizontal datum, q_s is the volume rate of bed load transport per unit width, and λ is the porosity of the bed material. Equation 6.119 is known as the Exnen equation. Sediment transport relationships to be used in conjunction with Equation 6.119 include Toffaleti, Modified Laursen, Yang, DuBoy, Colby, Einstein, and other empirical methods suitable for specific site conditions (Garde and Ranga Raju, 1978; USACE, 1991; Simons and Senturk, 1992). The selection of an appropriate formula for bed-material transport depends on sediment concentration, Froude number, and energy slope. Yang and Wan (1991) compared the accuracy of several bed-material transport formulas by comparing the respective predicted values with measured laboratory flume and natural river channel data. The results of their analyses may be used as a guide in selecting suitable formulas for specific applications.

B. Methods of Stream Flow Routing

1. *Rigid Bed Hydraulic Routing* Rigid bed routing assumes that the channel bed and bank do not undergo any geometric change due to erosion, aggradation, or degradation during the passage of the hydrograph being routed. In most streams, such changes are small compared to the corresponding flow areas and water depths. Therefore, most stream flow routing problems are solved assuming rigid bed conditions. The hydraulic methods of stream flow routing involve solutions of the dynamic, diffusion wave, or kinematic wave equations. The hydrologic methods involve solutions of the simplified continuity and storage discharge relationships for different stream reaches.

2. *Dynamic Routing* Methods of dynamic stream flow routing can be divided into three groups: method of characteristics, finite-difference methods, and finite-element methods.

3. *Method of Characteristics* In this method, the two basic partial differential equations (Equations 6.107 and 6.108) are replaced by four ordinary differential equations known as the characteristic equations. These equations with appropriate initial and boundary conditions, are solved in the (x,t) plane that is filled with forward and backward characteristics. The dependent variables are defined at the intersections of these characteristics. For the case with $q_L = O$, the characteristic equations for Equations 6.107 and 6.108 are (Streeter, 1971; Liggett and Cunge, 1975):

$$\frac{dx}{dt} = V \pm c \tag{6.120}$$

$$\left[(V \pm c)\frac{\partial}{\partial x} + \frac{\partial}{\partial t}\right][V \pm w] = g\,(S_0 - S_f) \tag{6.121}$$

where $c = \sqrt{gy}$ and $w = 2c$.

If the values of the dependent variables at points L and R on the characteristics are known, then the values at the intersection point M are computed by the following equations (see Fig. 6.17):

$$x_m = x_L + \int_{t_L}^{t_M} (V + c)\, dt \tag{6.122}$$

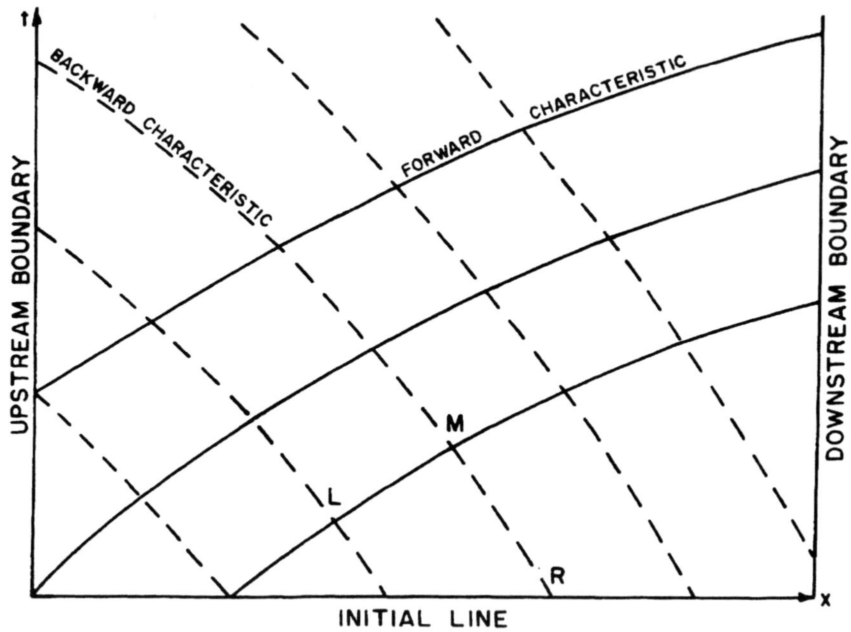

Figure 6.17—Characteristics in the x-t Plane.

$$V_M + W_M = V_L + W_L + \int_{t_L}^{t_M} g(S_o - S_f)\, dt \tag{6.123}$$

$$x_M = x_R + \int_{t_R}^{t_M} (V - c)\, dt \tag{6.124}$$

$$V_M - W_M = V_R - W_R + \int_{t_R}^{t_M} g(S_o - S_f)\, dt. \tag{6.125}$$

The systems of Equations 6.122 to 6.125 for all intersections in the domain of flow ($x - t$ plane) are solved for the unknowns x_M, t_M, V_M, and W_M. The solution advances along a forward characteristic line from the upstream towards the downstream boundary. At any instant of time, conditions along the (k − 1) forward characteristic are known and those along the k characteristic are to be determined. Dependent variables at point (k,1) are determined from upstream boundary conditions. Then, from known values at points $L = (k, m - 1)$ and $R = (k - 1, m)$, those at point $M = (k,m)$, $m = 2,3, \ldots$, are determined (Strelkoff, 1970; Sakkas and Strelkoff, 1973; Liggett and Cunge, 1975). Chen and Armbruster (1980) also used the method of characteristics to solve the Saint-Venant Equations 6.107 and 6.108, and applied the model to simulate the dam-break scenario for Laurel Run dam near Johnstown, Pennsylvania. The shape of the channel cross-section was approximated by an asymmetric trapezoid. The model used a linear explicit finite-difference scheme along with the Newton–Raphson or Secant Method.

a. Finite-Difference Methods There are two groups of finite-difference methods used for the solution of Equations 6.107 and 6.108, namely, the explicit and implicit methods. The solution domain is the (x,t) plane, which is divided into a number of rectangular cells (finite-difference grids) by a series of rows and columns. In the explicit method, the solution proceeds from one point at time level i to the other such that the dependent variables at the point $i + 1,j$ are determined independently of other unknown points in row $i + 1$. The row numbers, i, denote different time levels and the column numbers, j, denote different points in space in the x-direction.

In the implicit method, a group of simultaneous finite-difference equations is formed including the unknown dependent variables at all points in the solution domain.

Assuming a wide channel with no lateral flow and $\alpha = 1$, the simplest explicit finite-difference approximations of Equations 6.107 and 6.108 are (see Fig. 6.18):

$$\frac{y(i+1,j) - y(i,j)}{\Delta t} + V(i,j)\left[\frac{y(i,j+1) - y(i,j-1)}{2\,\Delta x}\right] + y(i,j)\,\frac{V(i,j+1) - V(i,j-1)}{2\,\Delta x} = 0 \qquad (6.126)$$

and,

$$\frac{V(i+1,j) - V(i,j)}{\Delta t} + V(i,j)\left[\frac{V(i,j+1) - V(i,j-1)}{2\,\Delta x}\right] + gy\left[\frac{y(i,j+1) - y(i,j-1)}{2\,\Delta x}\right] = g(S_o - S_f)\,(i,j) \qquad (6.127)$$

Another formulation with limited stability is the Dronkers' Explicit Scheme (Dronkers, 1965; Liggett and Cunge, 1975). The scheme is valid only when variations in flows and channel geometry are slow, both in space and time. Equations 6.107 and 6.108 are recast in the following forms assuming $\alpha = 1$ and $q_L = O$:

$$\frac{\partial(AV)}{\partial x} + B\frac{\partial y}{\partial t} = 0 \qquad (6.128)$$

$$\frac{\partial V}{\partial t} + V\frac{\partial V}{\partial x} + g\frac{\partial y}{\partial x} - g\frac{V/V/A^2}{K^2} = 0 \qquad (6.129)$$

in which y is water stage measured above a fixed datum, B is constant channel width, and K is channel conveyance. The explicit formulations for these equations are (see Fig. 6.19):

$$V(2i+1,2j) = V(2i-1,2j) - \frac{\Delta t}{2\Delta x} \times V(2i+1,2j)\,[V(2i-1,2j+2) - V(2i-1,2j-2)] + g \times \frac{\Delta t}{\Delta x}\,[y(2i,2j+1) - V(2i,2j-1)] + 2g\Delta t\,\frac{|V(2i-1,2j)|}{K^2/A^2\,(2i,2j)}\,[V(2i+1,2j)] \qquad (6.130)$$

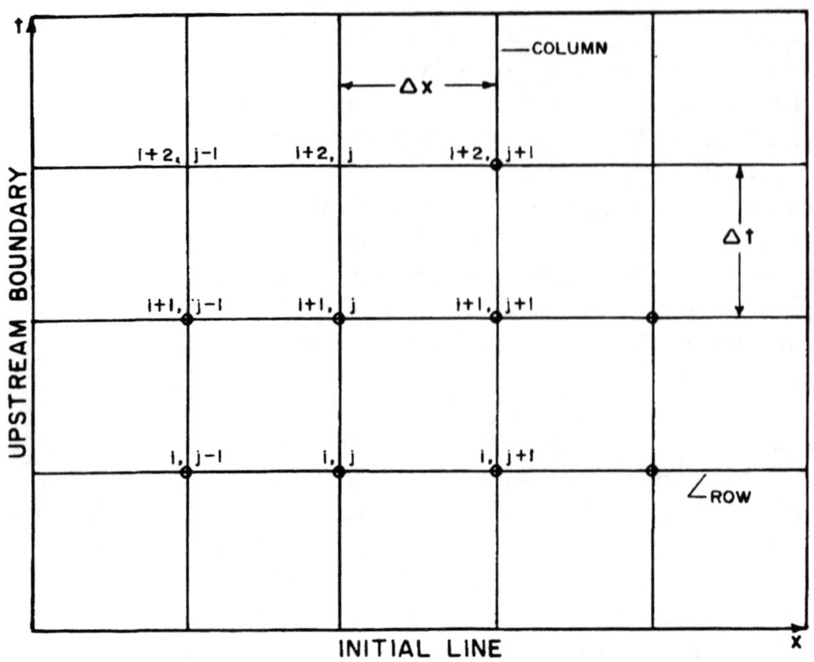

Figure 6.18—Finite-Difference Grid in x-t Plane.

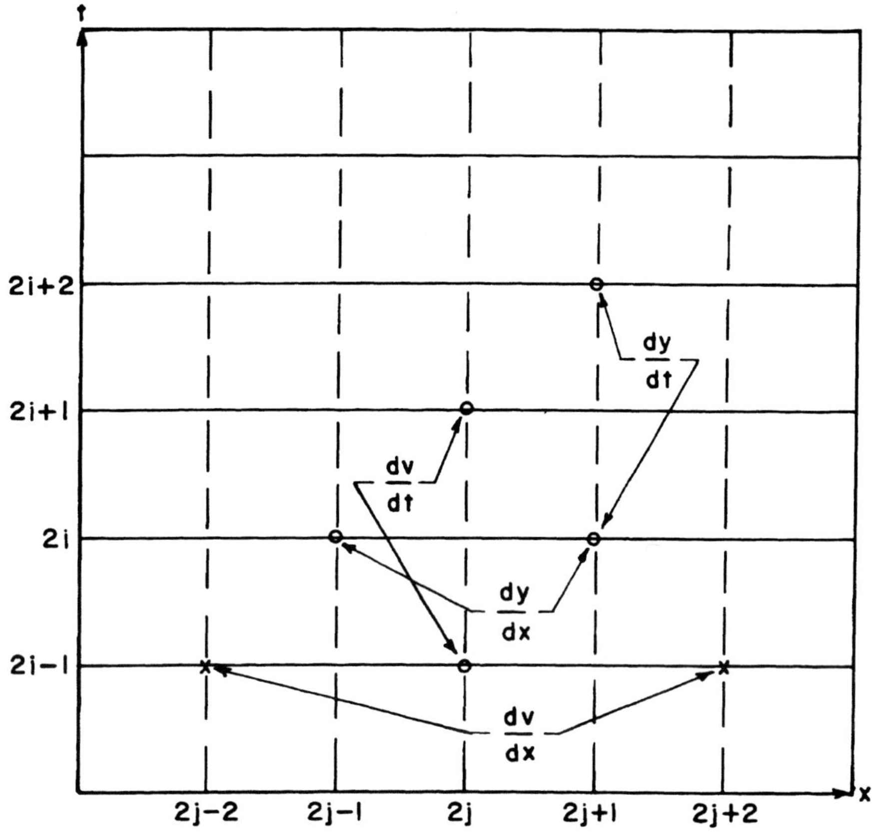

Figure 6.19—Dronkers Finite-Difference Discretization Scheme.

and

$$y(2i + 2,2j + 1) = y(2i,2j + 1) - \frac{\Delta t}{\Delta x} \frac{1}{B(2i,2j + 1)} \times \\ [V(2i + 1,2j + 2) \times A(2i,2j + 2) - V(2i + 1,2j) \times A(2i,2j)]. \tag{6.131}$$

The solution proceeds with the assumption that $V(\Delta t, x) = V(O, x)$. Then, $y(2\Delta t,x)$ is computed from Equation 6.131, after which $V(3\Delta t,x)$ is computed from Equation 6.130. Note that the points where velocities are computed are not the same as those where water levels are computed.

Because of stability problems associated with this formulation, several other finite-difference schemes have been adopted by different investigators (Liggett and Cunge, 1975). The explicit centered difference leap–frog scheme has been successfully used in some applications by the Tennessee Valley Authority (Garrison et al., 1969). The other explicit finite-difference formulation is based on the Lax–Wendroff second order scheme. Kibler and Woolhiser (1970) used the scheme for the solution of the kinematic wave equation. Finite-difference approximations of the Saint–Venant equations using this scheme are presented by Liggett and Cunge (1975).

Finite-difference formulations of the Saint–Venant equations using three commonly known implicit schemes, namely, Preissmann (Sogreah), Vasiliev, and Abbott's schemes, are described by Liggett and Cunge (1975). Of these, Preissmann's weighted four-point implicit scheme has been successfully used by Chaudhry and Contractor (1973) and Fread (1973a; 1973b). In this scheme (Fread 1984), the time derivatives are approximated by (see Fig. 6.18):

$$\frac{\partial G}{\partial t} = \left[\frac{G(i+1,j) + G(i+1,j+1) - G(i,j) - G(i,j+1)}{2\Delta t}\right] \tag{6.132}$$

in which G represents any variable. The spatial derivatives are approximated by a finite-difference quotient positioned between two adjacent time lines according to weighting factors θ and $1-\theta$, i.e.,

$$\frac{\partial G}{\partial x} \cong \theta\left[\frac{G(i+1,j+1) - G(i+1,j)}{\Delta x}\right] + (1-\theta)\left[\frac{G(i,j+1) - G(i,j)}{\Delta x}\right] \tag{6.133}$$

and

$$G = \theta\left[\frac{G(i+1,j) - G(i+1,j+1)}{2}\right] + (1-\theta)\left[\frac{G(i,j) + G(i,j)}{2}\right]. \tag{6.134}$$

The weighting factor θ varies from 0.5 to 1.0. Substitution of the above finite-difference forms results in two nonlinear algebraic equations in terms of two unknowns at the $i+1$ time line. Similar equations are formed for each location, j, in the x-direction and two additional equations are provided by the boundary conditions at the upstream and downstream ends. The resulting system of nonlinear equations is solved by the Newton–Raphson method. In the Operational Dynamic Wave model (Fread, 1982), which solves the Saint-Venant equations using this method, trial values are assigned to the unknown values. Substitution of these values into the nonlinear algebraic equations yields a set of residuals. The Newton–Raphson method seeks to reduce the residuals to an acceptable tolerance level.

b. Finite-Element Methods The commonly used finite-element method for the solution of Saint–Venant equations is based on the Galerkin approach (Cooley and Moin, 1976) and uses linear finite elements bounded by nodal coordinates, x_i. For this purpose, Equations 6.107 and 6.108 are re-written as follows:

$$\frac{\partial Q}{\partial t} + \frac{\partial (VQ)}{\partial x} + gA\frac{\partial y}{\partial x} - (S_0 - S_f) - u_x q_L = 0 \tag{6.135}$$

and

$$T\frac{\partial y}{\partial t} + \frac{\partial Q}{\partial x} - q_L = 0 \tag{6.136}$$

in which $T = \partial A/\partial y$ = top width of the cross-section at depth y, and u_x is the average component of inflow velocity in the x direction. Using the Galerkin procedure, the following elemental equations can be written.

$$\Sigma_e \int_{L_e} N_i^e \left[\frac{\partial Q}{\partial t} + V\frac{\partial Q}{\partial x} + Q\frac{\partial V}{\partial x} + gA\frac{\partial y}{\partial x} - gA(S_o - S_f) - u_x q_L\right] dx = 0 \tag{6.137}$$

and

$$\sum_e \int_{L_e} N_i^e \left(T\frac{\partial y}{\partial t} + \frac{\partial Q}{\partial x} - q_L\right) dx = 0 \tag{6.138}$$

in which L_e is the river reach within element e and N_i^e is a shape factor defined by

$$N_i = \frac{(x_{i+1} - x)}{(x_{i+1} - x_i)} \tag{6.139}$$

where

$$N_{i+1} = \frac{(x - x_i)}{(x_{i+1} - x_i)} \tag{6.140}$$

such that $Q = N_iQ_i + N_{i+1}\, Q_{i+1}$ and i, $i + 1$ refer to adjacent nodal points. The variables y, V, A, T, and S_f are assumed to be linearly variable like Q within each linear element and U_x and q_L are assumed to be constant within each element. $N_i^e = O$ for all elements not bounded by node i. The differential with respect to time is expressed in terms of finite differences, e.g.,

$$\frac{\partial Q_i}{\partial t} = \frac{Q_i^{n+1}Q_i^n}{\Delta t} \tag{6.141}$$

or

$$\overline{Q}_i = Q_i^{n+1} + \theta(Q_i^n - Q_i^{n-1}). \tag{6.142}$$

The superscripts denote time levels n, $n - 1$, etc., the overbar denotes weighted average between two successive time levels, and $0 \leq \theta \leq 1$. The nonlinear elemental equations are assembled for all elements and solved using the Newton–Raphson, predictor-corrector, or other suitable techniques (Hornberger et al., 1970; Zienkiewicz, 1971).

c. Diffusion Wave Methods The diffusion wave equation (Equation 6.112) for a rectangular channel with constant width, B, and lateral inflow, q_L, can be written as follows (Miller and Cunge, 1975)

$$\frac{\partial Q}{\partial t} = \frac{K^2}{2QB} \times \frac{\partial^2 Q}{\partial x^2} - \frac{Q}{BK} \times \frac{\partial K}{\partial y} + \frac{\partial Q}{\partial x} + \frac{q_L}{BK} \times \frac{\partial K}{\partial y} \times Q \tag{6.143a}$$

or

$$\frac{\partial Q}{\partial t} = \frac{Q}{2BS_f} \times \frac{\partial^2 Q}{\partial x^2} - m\, V\, \frac{\partial Q}{\partial x} + q_L\, m\, V \tag{6.143b}$$

Various investigators have developed analytical and numerical solutions for simplified and modified versions of Equation 6.143 (Chow, 1959; Eagleson, 1970; Dooge, 1973; and Miller and Cunge, 1975).

A comparison of the diffusion wave and kinematic wave models indicates that the former is applicable to a wider range of bedslopes and wave periods (Ponce et al., 1978). Akan and Yen (1981) present finite-difference formulation for diffusion-wave flood routing in channel networks with hypothetical examples to demonstrate its validity and efficiency compared to nonlinear dynamic wave and kinematic wave models. Hromadka et al. (1985) develop a finite-difference dam-break model based on two-dimensional diffusion analogy approximation of the dynamic equation for unsteady flow in open channels.

4. *Kinematic Wave Routing* Kinematic wave routing involves solutions of Equations 6.107 and 6.114 using analytical, finite-difference, finite-element, or characteristics methods.

A commonly used finite-difference formulation using Equations 6.107 and 6.115 is reproduced below (USACE, 1987):

$$\begin{aligned} &\frac{A(i,j) - A(i,j-1)}{\Delta t} + \alpha_2\, m \left[\frac{A(i,j-1) + A(i-1,j-1)}{2}\right]^{m-1} \\ &\quad \times \left[\frac{A(i,j-1) - A(i-1,j-1)}{\Delta x}\right] - \overline{q_L} = 0 \end{aligned} \tag{6.144}$$

in which

$$\overline{q_L} = \frac{q_L(i,j) + q_L(i,j-1)}{2} \tag{6.145}$$

$$Q = B\,V\,y \tag{6.146}$$

$$A = B\,y \tag{6.147}$$

and alternatively,

$$\frac{Q(i,j) - Q(i-1,j)}{\Delta x} + \frac{A(i-1,j) - A(i-1,j-1)}{\Delta t} - \overline{q_L} = 0. \tag{6.148}$$

To ensure numerical stability, Equations 6.144 and 6.145 are applicable to the conditions $c \le \Delta x/\Delta t$ and $c > \Delta x/\Delta t$, respectively. The wave celerity, c, is estimated as the average change of flow divided by the average flow area for the routing reach under consideration. The quantities $A(i,j)$ or $Q(i,j)$ at grid point (i,j) are computed from Equation 6.144 or 6.148. The corresponding $Q(i,j)$ or $A(i,j)$ is then computed from Equation 6.115.

Analytical solutions of the kinematic wave equation for rectangular or wide uniform channels for the rising and recession components of a hydrograph can be obtained using the method of characteristics (Streeter, 1971). The characteristic equations corresponding to Equations 6.107 and 6.114 for the rising limb of the hydrograph are:

$$\frac{dx}{dt} = m\,\alpha_1\,y^{m-1} \tag{6.149}$$

$$\frac{dy}{dt} = q_L. \tag{6.150}$$

With the initial conditions, $t = 0$, $x = x_0$, $y = y_0$, the solutions of Equations 6.149 and 6.150 give

$$q = \alpha_1\,(q_L t + y_0)^m \tag{6.151}$$

$$x = x_o + \frac{\alpha_1}{q_L}\left[(q_L\,t + y_o)^m - y_o^m\right]. \tag{6.152}$$

For a given length, L, of lateral inflow, the time to reach steady-state and the steady-state discharge or water depth can be calculated from Equations 6.151 and 6.152.

The recession limb of the hydrograph commences after the discharge has become q_R, i.e., at $x_R = q_r/q_L$. The characteristic Equation 6.150 now becomes

$$\frac{dy}{dt} = 0. \tag{6.153}$$

Reckoning the time from $t = 0$ at $x = x_R$,

$$x = m\,\alpha_1\left(\frac{q}{\alpha_1}\right)^{\frac{m-1}{m}} t + \frac{q_R}{q_L}. \tag{6.154}$$

The recession limb of the hydrograph can be calculated by Equation 6.154.

The kinematic assumption implies that as far as momentum is concerned, the flow can be considered steady and the unsteadiness of flow is represented by the rate-of-rise term in Equation 6.107. Since Equation 6.107 or 6.116 are first order differential equations, they describe convection but not diffusion, which is a second order process (Ponce, 1991). This means that the kinematic wave equation can simulate translation but not attenuation of the flood wave. Commonly used finite-difference approximations assume that the second-order terms in the Taylor series expansions of variables are negligible. This results in numerical diffusion or attenuation in the calculated runoff hydrograph. Thus the finite-difference solution simulates physical diffusion through numerical diffusion.

5. ***Rigid Bed Hydrologic Routing*** Hydrologic stream flow routing methods include the Muskingum (USACE, 1987), Muskingum–Cunge (Ponce, 1989), Modified Puls (Chow, 1959), Tatum (USACE, 1960; Chow, 1964), Coefficient (Chow, 1964), Straddle–Stagger (USACE, 1960), Working Value (Chow, 1964), Convex (USDA, 1982), and Att–Kin (USDA, 1982) routing.

Some commonly used hydrologic stream flow routing methods are described in the following paragraphs.

The assumptions and approximations inherent in these methods include the following:

a) These methods involve no direct solutions of the dynamic equation of unsteady gradually varied flow but heuristic approximations to the solutions of this equation.
b) They neglect the effects of surges, bores, and backwaters.
c) The channel cross-sections are assumed to have constant physical characteristics, e.g., shape and roughness factor, within a selected routing reach.
d) Tributary inflows, ground water contributions, and lateral surface inflows are accounted for by inflows to the respective routing reaches.
e) The routing proceeds from upstream to downstream on a flow-through basis.
f) Velocity changes along the channel are small so that the acceleration terms of the dynamic equation are negligible.

In general, the storage–outflow relationship for a channel reach is in the form of a loop. For unsteady flow, storage in a channel reach depends on the inflow and outflow discharges and on the geometric and hydraulic characteristics of the channel. Assuming that the stage–discharge and stage–storage relationships for the upstream and downstream end sections of the reach are the same (Chow, 1959),

$$I = a\, y_i^m \tag{6.155}$$

$$O = a\, y_o^m \tag{6.156}$$

$$S_i = b\, y_i^n \tag{6.157}$$

$$S_o = b\, y_o^n \tag{6.158}$$

where a and b are coefficients, m and n are exponents of the respective relationships, and the subscripts i and o refer to the upstream and downstream ends of the channel reach, respectively.

Assuming that the storage in the reach can be expressed as a linear function of S_i and S_o,

$$S = X\, S_i + (1 - X)\, S_o \tag{6.159}$$

in which X is a weighting factor. Then eliminating y in Equations 6.155 to 6.158,

$$S = K[XI^r + (1 - X)\, O^r] \tag{6.160}$$

where $K = (b/a^{n/m})$ and $r = n/m$. For uniform channels, $X = 0.5$. For prismatic rectangular channels, $n = 5/3$ if the Manning formula is used and $3/2$ if the Chezy formula is used and $m = 1$, resulting in $r = 0.6$ to 0.67. In natural channels, r is usually larger than 0.67 and is often assumed to be unity.

a. Muskingum Method This method utilizes the concepts of prism and wedge storages in the stream reach under consideration and assumes that these storages can be treated as linear functions of the instantaneous values of outflow and inflow. During the advance of the flood wave, inflow exceeds outflow and produces a wedge of storage. Conversely, during hydrograph recession, outflow exceeds inflow resulting in a negative wedge storage. The wedge storage is represented by $KX(I - O)$. In addition, there exists prism storage in the channel reach that pertains to steady flow conditions and can be represented by a simple linear function of outflow, KO. Herein, I and O are instantaneous rates of inflow and outflow, K is a coefficient (travel time through the reach), and X is a dimensionless weighting factor. The total storage in the reach is given by:

$$S = KO + KX(I - O). \tag{6.161}$$

This corresponds to $r = 1$ in Equation 6.160.

Two methods of solving Equation 6.161 for different time steps are commonly used. The first method uses the following formulation (USACE, 1987):

$$O_2 - Q_1 = C_1(I_1 - O_1) + C_2(I_2 - I_1) \tag{6.162}$$

in which

$$C_1 = \left[\frac{\Delta t}{K(1 - X) + 0.5\ \Delta t}\right] \tag{6.163}$$

$$C_2 = \left[\frac{0.5\ \Delta t - KX}{K(1 - X) + 0.5\ \Delta t}\right]. \tag{6.164}$$

The second method uses the following alternative formulation (USACE, 1960):

$$O_2 = C'_1O_1 + C'_2I_1 + C'_3I_2 \tag{6.165}$$

in which

$$C'_1 = \frac{K - KX - 0.5\ \Delta t}{K - KX + 0.5\ \Delta t} \tag{6.166}$$

$$C'_2 = \frac{KX + 0.5\ \Delta t}{K - KX + 0.5\ \Delta t} \tag{6.167}$$

$$C'_3 = \frac{-\ KX + 0.5\ \Delta t}{K - KX + 0.5\ \Delta t}. \tag{6.168}$$

O_1, O_2, I_1, I_2 represent the values at successive time steps and Δt is the size of the time step. K is nearly the time interval between the centers of mass of the inflow and outflow hydrographs and X is a dimensionless constant representing the wedge storage in the routing reach. Methods to estimate K and X are described in the section titled Routing of Floods Through River Channels (USACE, 1960).

b. Muskingum-Cunge Method In the Muskingum method the routing parameters K and X are determined by matching calculated outflow hydrographs with the observed ones at the same location and for the same hydrologic event. In the Muskingum–Cunge method (Ponce, 1989), the dimensionless routing parameter, X, is treated as a function of reach length Δx, channel bed slope S_o, kinematic wave celerity c, and stream flow per unit width q, i.e.,

$$X = \frac{1}{2}\left(1 - \frac{q}{S_o\ c\ \Delta x}\right) \tag{6.169}$$

and the storage coefficient K is defined as

$$K = \frac{\Delta x}{c}. \tag{6.170}$$

Introducing the Courant number, $C = c\ \Delta t/\Delta x$, and the cell Reynolds number, $D = q/S_o c \Delta x$, Equations 6.166, 6.167, and 6.168 can be rewritten as follows:

$$C'_1 = \frac{1 - C + D}{1 + C + D} \tag{6.171}$$

$$C'_2 = \frac{1 + C - D}{1 + C + D} \tag{6.172}$$

$$C'_3 = \frac{-1 + C + D}{1 + C + D}. \tag{6.173}$$

As defined in the section on open channel flow principles, $c = \partial Q/\partial A = (1/T)(\partial Q/\partial y) = mV$ and can be estimated from stage–discharge relationship and cross-sectional geometry or known values of m and V. For practical problems, values of q and c may be estimated for peak flow or average flow conditions in each computational reach.

Equation 6.169 is obtained by equating the numerical diffusion coefficient for Equation 6.116 with the hydraulic diffusivity $= Q/2TS_o = q/2S_o$. The numerical diffusion coefficient is the error of the first-order-accurate finite-difference formulation of Equation 6.116 as compared to the second-order-accurate formulation, i.e., $c\Delta x(\frac{1}{2} - X)$.

c. Modified Puls Method This method neglects the variable water surface slope occurring during the passage of the flood wave and assumes a unique storage–outflow relationship for the channel reach under consideration. The degree of attenuation of the flood wave depends on the length of the routing reach. Therefore, this method may yield poor results for stream flow routing, if the selected reach length is not calibrated using observed inflow and outflow hydrographs. Stream flow routing is accomplished by the following equation:

$$\frac{1}{2}(I_1 + I_2)\ \Delta t + S_1 - \frac{1}{2}\ O_1 \Delta t = S_2 + \frac{1}{2}\ O_2 \Delta t. \tag{6.174}$$

The quantity $S_2 + \frac{1}{2} O_2 \Delta t$ at the succeeding time step is computed using the known quantities on the left-hand side. Then O_2 is estimated from the known relationship between O and $S + \frac{1}{2} O \Delta t$ (Chow, 1959).

d. The Working-Value Method The Working-Value Method (Chow, 1964) involves the concept of a virtual working discharge that represents a steady flow that would produce a storage equal to that produced by the actual inflow I and outflow O.

The working discharge D is defined as

$$D = X I + (1 - X) O. \tag{6.175}$$

Combining Equation 6.161 with Equation 6.175,

$$S = KD. \tag{6.176}$$

Also,

$$O_1 = D_1 - \frac{X}{1 - X} (I_1 - D_1) \tag{6.177}$$

$$O_2 = D_2 - \frac{X}{1 - X} (I_2 - D_2) \tag{6.178}$$

from which

$$S_1(1 - X) + \frac{1}{2}\Delta t(I_1 + I_2) - \frac{1}{2}\Delta t D_1 = S_2(1 - X) + \frac{1}{2}\Delta t D_2. \tag{6.179}$$

Setting

$$S_1(1 - X) + \frac{1}{2}\Delta t D_1 = R_1 \tag{6.180}$$

$$S_2(1 - X) + \frac{1}{2}\Delta t D_2 = R_2 \tag{6.181}$$

$$R_2 = R_1 + \frac{1}{2}\Delta t(I_1 + I_2) - \Delta t D_1 \tag{6.182}$$

in which R_1 and R_2 are known as working values and represent indices of channel storage.

The values of K and X are determined the same way as in the Muskingum method. The computational steps of this method are as follows: 1) Compute D_1 for given values of I_1, O_1, and X. Using Equation 6.175, 2) compute S_1 from Equation 6.176 using known value of K_1, 3) compute R_1 from Equation 6.180, 4) compute R_2 from Equation 6.182 using known values of R_1, I_1, I_2, and D_1, 5) compute D_2 from Equation 6.176 using known value of K_2. If K_2 is a function of O_2, then estimate D_2 using an assumed value of O_2. 6) Finally, compute O_2 using Equation 6.178. If it does not match the assumed value, then repeat the last two steps using a modified value of O_2.

This method is considered better than the Muskingum method if tributary inflow or controlled discharges through gate outlets, etc., are to be included in the computations.

e. Modified Attenuation–Kinematic (Att–Kin) Method The Att–Kin method (USDA, 1982) is a method developed by the Soil Conversation Service and incorporated in the TR-20 model. The modified Att–Kin method consists of a two-step process. In the first step, attenuation or storage effects are simulated by routing the inflow hydrograph through a reservoir. In the second step, pure kinematic translation is simulated using the kinematic wave method.

6. Movable Bed Routing In the case of loose boundary channels, the friction factors f, Chezy coefficient C, or Manning coefficient n have been found to vary with the types of bed forms (Garde and Ranga Raju, 1978; Simons and Senturk, 1992) and the stage–discharge relationship is modified by continually occurring aggradation and degradation. The types of bed forms encountered in loose boundary channels include plane beds, ripples, bars, dunes, washed-out dunes, anti-dunes and chutes, or pools. Methods for the solution of problems related to resistance to flow and estimation of stage–discharge relationships for loose boundary channels include Einstein's, Shen's, Engelund and Hansen's, Alan and Kennedy's, Bajoruna's, Simons and Richardson's, and Senturk's approaches, etc. (Simons and Senturk, 1992). Stream flow routing for loose boundary channels involves simultaneous solution of the dynamic equations of flow (Equations 6.107 and 6.108) and Exner equation for continuity of sediment material (Equation 6.119) coupled with appropriate resistance and sediment transport equations (Chang, 1988). The stream flow hydrograph is divided into discrete time intervals and hydraulic variables (e.g., water surface profiles and velocity distribution across each cross-section) are estimated using selected flow resistance coefficients and equations corresponding to each discrete discharge and prescribed boundary conditions. The sediment transport capacity for each cross-section and discrete discharge is estimated using the hydraulic data developed previously, selected sediment transport equation, and known gradation of bed material. The continuity equation for sediment material is then solved with known inflowing sediment load for that discrete discharge. In this way, changes in the total sediment load, channel bed elevation, and sediment outflow are computed for each cross-section of the channel. The elevations of channel cross-sections are adjusted to reflect the estimated aggradation or degradation. These hydraulic and sediment transport computations are repeated for all discrete discharges of the given stream flow hydrograph (Chang, 1988; USACE, 1991).

7. Special Applications In addition to the applications mentioned previously, stream flow routing is used for a number of special situations. Some practical examples are listed below:

a) Stream flow routing for tidal streams (see following section). In this case, the dynamic and continuity equations are solved with a downstream boundary defined either by tidally varying outflows or water surface elevations (Feigner and Harris, 1970; Roesch et al., 1979; Ambrose et al., 1988).
b) Flood routing through alluvial fans. Alluvial fans are fan-shaped deposits of cobbles, gravel, and sand formed by ephemeral streams as they descend from desert mountains onto the valley floor. The surface of the fan has a mild radial and very mild transverse slope with poorly defined migrating channels. During floods, shallow unstable channels are formed through the alluvial material in the fan. Methods to distribute flood flows through the width of the alluvial fan are presented by French (1987; 1992). Knowing the rates of flows and storage characteristics of the fan as it expands radially, storage routing computations may be performed using the methods described previously.
c) Dam-break flood routing. This involves dynamic routing through the downstream channel of the flood hydrograph generated by a sudden breach in a dam. Dam-break flood routing is required to develop inundation maps and to perform risk analysis for areas downstream of existing or proposed dams. Generally, the size and shape of breach and time of breach development are estimated by judgment. Computer models pertaining to dam-break analysis are described in the following section. The information required to perform a dam-break analysis include the inflow hydrograph at the time of failure, initial reservoir water surface elevation, reservoir storage-area-elevation table, size, shape, length, and material characteristics for the dam, rating table for the outlet works, size, shape type (e.g., piping, overtopping, etc.), and time of development of the breach, and cross-sections, bed slopes, and Manning roughness coefficients for different reaches of the downstream channel (Fread, 1984).
d) Stream flow routing through compound sections. Stream flow routing through flood plains and compound sections is performed using numerical models based on the Saint–Venant equations (Fread, 1984) or the Muskingum–Cunge method (Garbrecht and Brunner, 1991). In Fread's model, flow transfer between a flood plain compartment and the river is assumed to occur along the interface between the channel and floodplain and is controlled by the broad-crested weir equation

with submergence correction. In the Garbrecht and Brunner model, compound sections are treated by decoupling main and overbank flow and by routing the flows in each channel portion separately. Flow mixing between main channel and overbank flow takes place at the end of each computational space increment.

e) Flow through structures. Stream flow routing through structures like bridges, culverts, siphons, tunnels or conduits, gated structures, trash racks, screens, and lakes or reservoirs can be performed using the methods described previously with outflow rating tables for the respective structures. The basic data required for these computations include the following: (i) the inflow hydrograph, (ii) storage-elevation data for the impoundment or backwater area, and (iii) outflow rating for the respective structures.

   The inflow hydrograph is obtained from rainfall–runoff simulation and stream flow routing for watersheds and stream reaches upstream of the structure. The storage elevation data are developed from a contoured map of the impoundment or backwater areas. The outflow ratings may be developed using the energy equations for flow through different structures (AISI, 1971; USDOT, 1972; USDOT, 1973; USBR, 1987).

C. Stream Flow Routing Models

Commonly used public domain stream flow routing computer models are described in the following paragraphs.

1. *Flood Hydrograph Package, HEC-1* The stream flow routing component of this model (USACE, 1987) performs channel routing computations using the kinematic wave, Muskingum, Working R and D, Modified Puls, Straddle–Stagger (Progressive Average Lag), or Tatum (Successive Average Lag) method. Stream flow routing using the modified Puls or Working R and D method may be accomplished by providing a storage versus outflow relationship derived from water surface profile studies or normal depth storage–outflow characteristics of the channel.

2. *Computer Program for Project Formulation Hydrology—TR-20* In this model (USDA, 1982), stream flow routing is performed using the modified Att–Kin procedure. The attenuation of the hydrograph due to storage within the reach is determined first, using reach-specific relationships between outflow and average channel area and storage and outflow:

$$O = x A^m \tag{6.183}$$

$$S = KO. \tag{6.184}$$

The routing equation is derived from the equation of conservation of mass,

$$I - O = \frac{dS}{dt}. \tag{6.185}$$

Using values at time steps 1 and 2,

$$I_1 - \frac{O_1 + O_2}{2} = \frac{K(O_2 - O_1)}{\Delta t} \tag{6.186}$$

or

$$O_2 = (1 - C_r)O_1 + C_r I_1 \tag{6.187}$$

in which

$$C_r = \frac{2\Delta t}{2K + \Delta t}. \tag{6.188}$$

The time lag of the inflow and outflow hydrographs is then computed using kinematic wave routing.

3. *The Illinois Urban Drainage Area Simulator—ILLUDAS* This model (Terstriep and Stall, 1980), includes two stream flow routing algorithms capable of routing through circular or rectangular conduits or trapezoidal open channels. These include hydrologic routing using an implicit solution of the continuity equation and a simple time shift routing.

In the implicit method, a linear approximation of the continuity equation is used:

$$\frac{\Delta S}{\Delta t} = \frac{I_1 + I_2}{2} - \frac{O_1 + O_2}{2} = \frac{L}{\Delta t}\left[\frac{F(I_2) + F(O_2)}{2} - \frac{F(I_1) + F(O_1)}{2}\right] \tag{6.189}$$

in which I_1 and O_1 are the initial inflow and outflow, I_2 and O_2 are the final inflow and outflow, F is the flow area function in which the kinematic wave assumption, $S_f = S_o$ and the Manning formula are used, S_f is the friction slope and S_o is the reach slope, and L is the reach length. Equation 6.189 can be rearranged to obtain:

$$I_1 + I_2 - O_1 - O_2 = [F(I_2) + F(O_2) - F(I_1) - F(O_1)] \times \frac{L}{\Delta t} \tag{6.190}$$

$$O_2 + \frac{L}{\Delta t} F(O_2) = I_1 + I_2 - O_1 + [F(I_1) + F(O_1) - F(I_2)] \times \frac{L}{\Delta t}. \tag{6.191}$$

Equation 6.191 must be solved for O_2 and thus requires an implicit solution.

The flow area function (F) is determined for each cross-section by a dimensionless flow area versus dimensionless discharge tabular look-up within the program. This nondimensionalized function is computed from the channel geometry. For the case of trapezoidal open channels, a Newton–Raphson implicit solution of the Manning formula must be performed to develop this dimensionless function.

The time shift routing of the hydrograph is performed for both the time shift and the hydrologic routing techniques. For time shift routing, it is used to determine final outflows (O_2) directly. In hydrologic routing the time shift routing is used to obtain the initial estimates of O_2 to be used in the implicit solution. Time shifting of the inflow hydrograph is performed by computing the velocity from the inflow hydrograph peak. The inflow hydrograph is interpolated at time increments equal to the time of travel of the inflow hydrograph peak discharge through the reach.

4. *WASP4—A Hydrodynamic and Water Quality Model* The stream flow routing component of this model (Ambrose et al., 1988) has the acronym DYNHYD 4. It solves the one-dimensional unsteady flow equations describing the propagation of a long wave through a shallow water system while conserving both momentum or energy and volume or mass. The equation of motion, based on the conservation of momentum, predicts water velocities and flows. The equation of continuity, based on the conservation of volume, predicts water heights or heads and volumes. The model can simulate most natural flow conditions in large rivers and estuaries. It is assumed that flow is predominantly one-dimensional, Coriolis and other accelerations normal to the direction of flow are negligible, channels are nearly rectangular, the wave length is significantly greater than depth, and the bottom slopes are moderate. Inflows and outflows can be specified as constant or variable with time. The downstream boundaries can be defined by specifying outflows or surface elevations as a tidal function. The equations of motion and continuity used in this model are as follows:

$$\frac{\partial V}{\partial t} = - V \frac{\partial V}{\partial x} + a_g + a_f + a_w \tag{6.192}$$

$$\frac{\partial A}{\partial t} = - \frac{\partial Q}{\partial x} \tag{6.193}$$

where a_g is gravitational acceleration, a_f is frictional acceleration, and a_w is wind stress acceleration along axis of channel.

5. National Weather Service Operational Dynamic Wave Model DWOPER This model (Fread, 1982) is based on an implicit finite-difference solution of the complete one-dimensional Saint–Venant equations of unsteady flow and is applicable for flow routing through streams of varying physical features, such as irregular geometry, variable roughness parameters, lateral inflows, flow diversions, off-channel storage, local head losses such as bridge contraction-expansions, lock and dam operations, and wind effects. Also, it can simulate dendritic river systems or channel networks consisting of bifurcations with weir-type flow into the bifurcated channel. A weighted four-point nonlinear implicit finite-difference scheme is used to obtain solutions of the Saint–Venant equations via a Newton–Raphson iterative technique.

6. DAMBRK: The NWS Dam-Break Flood Forecasting Model The complete package of the NWS dam-break flood forecasting model includes three models: DAMBRK, BREACH, and SMPDBK. The DAMBRK (Fread, 1984) model consists of three functional parts, namely, description of dam failure mode, i.e., temporal and geometric representation of the breach, computation of outflow hydrograph through the breach, and routing of the outflow hydrograph through the downstream valley. The routing component of the model is based on a weighted, four-point nonlinear finite-difference solution of the one-dimensional equations of unsteady flow. The model can route both subcritical and supercritical flows and can simulate the changes in the outflow hydrograph due to valley storage, frictional resistance, and downstream bridges or dams. The BREACH model estimates the size and shape of the breach using the D_{50}, unit weight, angle of internal friction, cohesive strength, and Manning roughness coefficient for the embankment material, geometric data for the embankment and downstream channel, elevation-area data for the impoundment, initial reservoir water surface elevation, and the inflow hydrograph at the time of failure. The SMPDBK model is a simplified version of DAMBRK. It approximates the downstream channel as a prism and neglects off-channel storage and backwater effects from downstream bridges or dams. It utilizes dimensionless peak flow routing graphs developed using the DAMBRK model and predicts only peak flows, stages, and travel times.

7. Finite Element Solution of Saint–Venant Equations This model (Cooley and Moin, 1976) solves the Saint–Venant equations for unsteady flow through open channels using Galerkin finite element formulation for linearized versions of these equations. The time integration of the resulting equations is performed using the predictor corrector technique. The model has been tested for wide rectangular channels with hypothetical inflow hydrographs and by routing the January–February 1963 flood on the Truckee River in California and Nevada through a stream reach of approximately 36 km (22.75 miles).

8. Scour and Deposition in Rivers and Reservoirs, HEC-6 This model (USACE, 1991)simulates the interaction between the water–sediment mixture, sediment material forming the bed and banks of the stream, and hydraulics of flow as the stream flow hydrograph is routed through a stream reach. It simulates the ability of the stream to transport sediment considering the conditions embodied in Einstein's bedload function plus silt and clay transport and deposition, and armoring and destruction of the armor layer. The model includes options to use the Toffaleti, Laursen, Yang, Duboy, or a user-specified sediment transport relationship. Long-term trends of scour or deposition in a stream channel and the corresponding water depths and water surface profiles resulting from varying frequency and duration of water discharges or stages or floodplain encroachments can be simulated.

VI. RESERVOIR STORAGE—YIELD ANALYSIS

Seldom, if ever, does the need for water synchronize with the available supply. Reservoirs regulate the availability of water. They stabilize the flow of water, either by regulating a varying supply or by satisfying a varying demand (Linsley and Franzini, 1972). Reservoirs may be used for conservation, control, or flow regulation. Conservation uses include irrigation, hydroelectric power, recreation, low-flow augmentation, water supply, navigation, or a combination of these.

Storage for irrigation is needed to collect water during the wet season for use during the growing season. The growing season may last several months, depending on the crop being cultivated and the number of crops planted during the year. Storage for hydroelectric power is used to maintain generating capacity throughout the year and to provide an easily activated source of power generation during periods of peak demand. Releases from reservoirs may also be used to augment flows whenever dry season discharges in a river are unable to maintain water quality for environmental purposes.

Storage for flood control is used in a slightly different way. A dam is built so that excess waters during flood times are temporarily stored until such time they can be released without flooding downstream reaches. For navigation structures, storage is incidental to the navigation dam's purpose. A navigation dam is constructed to raise the river stage and permit the flow of traffic. Provisions are typically included for locking operation. Storage for recreation is used in still another way. It might be needed to maintain a prescribed lake area or shoreline for recreational purposes. It should not be excessive so as to flood piers for launching boats; however, the reservoir should not be drawn down too low such that the shoreline recedes too far out from the designated beaches.

A conservation reservoir is designed to retain surplus water during periods of excess flows for use during periods of drought. In water supply systems, distribution reservoirs are used to meet the varying rate of demand during the day. This enables the water treatment and pumping facilities to operate at reasonably uniform rates.

These multiple and often conflicting purposes make the regulation of storage a complicated task; however, once the objectives have been identified, storage analysis proceeds by performing a mass balance using the continuity equation.

A. Reservoir Yield

The most important physical characteristic of a reservoir is its storage capacity. The storage capacity in turn determines the yield. Yield is defined as the amount of water that can be supplied during a specified period of time. Given a target yield, the selection of the reservoir capacity depends on how much risk the designer is willing to take that the yield will not be realized. This risk must be low for such critical water needs as municipal water supply but may be higher for irrigation water requirements.

The inclusion of the concept of risk in the design of a reservoir recognizes that the supply of water during the projected economic life of the reservoir is uncertain and that this uncertainty must be explicitly taken into account during the design process.

The problem of design may be posed in terms of the question: How much storage capacity must be provided in order to meet a prescribed release or draft pattern with a given reliability? Thus, the design storage capacity depends on the magnitude of the demand, the variability of the supply, and the reliability that the demand will be met.

B. Preliminary and Final Design Procedures

In the typical situation, several candidate sites for the reservoir may be examined. The sites must be investigated not only from the hydrologic viewpoint but from the structural, environmental, construction, and other perspectives as well.

The methods that are used for a quick assessment of the problem may be called preliminary design procedures. In these procedures, simplifying assumptions are made. These may include assuming a constant release pattern, neglecting evaporation losses, neglecting provisions for reservoir sedimentation, postponing reliability considerations, etc. The preliminary procedures are used to eliminate unsuitable sites so that the remaining candidates can be evaluated using final design techniques. The final design procedure, in general, is more detailed and takes into account features that were ignored or neglected during the preliminary design phase (U.S. Army Corps of Engineers, 1975; McMahon and Mein, 1978).

C. Reservoir Capacity Determination—Mass Curve Analysis

The complexity of reservoir capacity design depends on the type of flow regulation. If regulation is of the over-year type, the analysis is based on annual stream flows and a given degree of development. The degree of development is usually expressed as a percentage of the mean inflow. For annual flows, the assumption of independence often suffices.

If the within-year storage fluctuations are considered, the analysis is usually made with monthly, weekly, or daily stream flows. In dealing with these flows, a nonstationary stochastic process must be considered because observed data for these flows exhibit periodicities in their sample statistics. As will be discussed later, this presents some difficulty in modeling the inflows using the simulation approach.

Historically, the procedure for determining the capacity of a reservoir has evolved from mass curve procedures initially proposed by Rippl (1883). The theoretical basis for Rippl's method derives from the following analysis.

Consider a time period of length T. Let $X(t)$ be the inflow to a reservoir and $I(t)$ be the mass curve of $X(t)$.

$$I(t) = \int_0^t X(t)\,dt \tag{6.194}$$

Neglecting evaporation from and precipitation into the reservoir, the total amount of water that is available at the end of the period is $I(T)$. If all this water is utilized (full development), the maximum constant demand rate $Y(t)$ that can be satisfied is

$$Y(t) = \frac{1}{T}\,I(T) = \overline{X}. \tag{6.195}$$

Let $D(t)$ be the mass curve of demand:

$$D(T) = \int_0^t Y(t)\,dt. \tag{6.196}$$

For full development, $D(T) = I(T)$. Given a record of $X(t)$, it can be shown that the storage required, S_R, in order to satisfy a constant demand rate equal to the mean inflow, is

$$S_R = \textit{Range of}\,[I(t) - D(t)] = \max\,[I(t) - D(t)] - \min\,[I(t) - D(t)] \tag{6.197}$$

that is, the range of the accumulated departures from the mean. It can further be shown that the initial storage requirement is equal to the maximum accumulated deficit, i.e., $max\,[D(t) - I(t)]$. This is illustrated in Fig. 6.20. This analysis is a generalization of Rippl's mass curve technique.

For partial development, the most common approach to design is a modification of the above procedure. Data from the historical period of low flows are used as inflows to the reservoir being designed. The steps in this method are as follows.

For the reservoir site, construct the mass curve of historical flows. Assume a constant draft. This is usually taken as a percentage of the mean inflow. Superimpose on the mass curve cumulative draft lines from the reservoir so that they are tangential to each peak of the inflow mass curve. Determine the

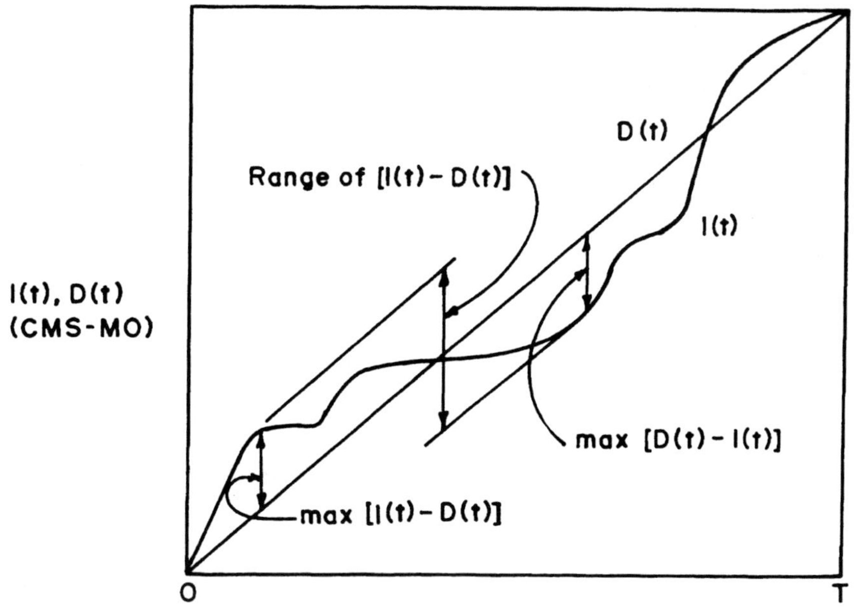

Figure 6.20—Mass Curve Analysis.

maximum deviation between the inflow mass curve and the cumulative draft lines. This is the reservoir design capacity. The procedure is illustrated in Fig. 6.21. The draft is assumed to be 60% of the mean flow. This reservoir, if subjected to the historical inflow sequence, is assumed full at time zero, will be full at A, will be drawn down from A to B, will fill again from B to C, spill from C to D, just empty when it reaches E, and begin refilling again. In this analysis, it is assumed that future inflows will not contain a drier sequence than the historical river flows. The releases in this approach are usually assumed to be constant. Storage capacities estimated using this technique increase with increasing record length. The procedure may be modified to take into account nonconstant draft and to include evaporation losses and precipitation into the reservoir. This method and several other variations suffer from several defects:

1) The analysis is based solely on the historical record that is unlikely to recur in exactly the same sequence in the future.
2) The mass curve approach does not help in establishing or calculating the risk involved, i.e., the probabilities that there will be water shortages.
3) The storage capacity obtained using the Rippl method increases with the record length. Since the length of the historical record is likely to differ from the projected economic life of the reservoir, the storage obtained from this method, while suitable as basis for design, needs to be modified.

Another shortcoming of this approach relates to the initial state of the system. If the mass curve is used to estimate the magnitudes and frequency of flow deficiencies, the initial storage plays a significant role in the results. To address this problem, Thomas and Fiering (1962) proposed the Sequent Peak Algorithm in which a mass curve analysis is done on the historical record concatenated with itself. This approach takes into account the possibility that the depletion of the reservoir may be interspersed by periods of filling. Assuming that the inflows and drafts are repeated cyclically, the Sequent Peak Algorithm determines the minimum storage for no shortage in draft based on the two repeated cycles. The procedure is illustrated in the following example adapted from Fair et al. (1966).

As an example, for the mean monthly runoff shown in column 2, Table 6.14, determine the required storage necessary to satisfy the estimated rates of draft in column 3. Inflow, draft, and storage are in million gallons per square mile (mg/sq mi). Drafts of 27, 30, 33, and 36 mg/sq mile equal 0.89, 1.01, 10.9,

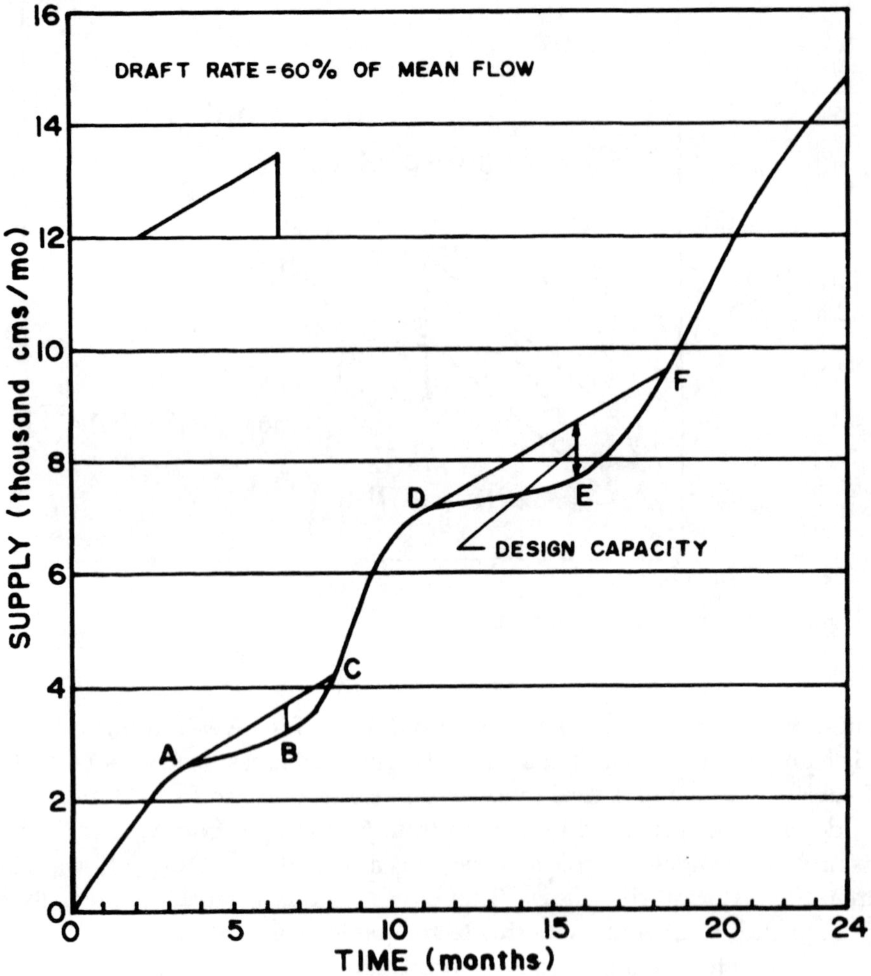

Figure 6.21—Graphical Determination of Reservoir Capacity.

and 1.18 mgd/sq mi, respectively, assuming that each month has 30.4 days (1 mg = 3,785 m^3; 1 mgd/sq mi = 16.9 l/s – sq km).

For a total flow of 500 mg/sq mi over 12 months, the average daily flow is 500/365 = 1.37 mg/sq mi. The total annual draft is 360 mg and the degree of development is 100 × 360/500 = 72%.

Column 4 in Table 6.14 shows the calculation of monthly surpluses and deficits. Column 5 tracks the peaks and troughs of the differential mass curve. P1 is the first peak and T1 is the first trough in the range P1–P2. The required maximum storage is Sm = max (Pj – Tj) = P1 – T1 = 198 – 32 = 166 mg. The fact that P2 – T2 = 338 – 172 = 166 mg also implies that there is seasonal rather than over-year storage.

These improvements on what is basically Rippl's method do not completely resolve the shortcomings enumerated above. To overcome some of these shortcomings, other investigators have suggested further improvements. Some of the currently proposed methods advocate the use of stochastic stream flow models to generate flows or the determination of the distribution of storage using probability matrix methods or from the distribution of the range. These will be discussed in subsequent sections.

D. Reservoir Operation Study

In Rippl's analysis and in many of its variants, the supply is usually assumed to be known either from historical data or obtained from a simulation technique. The release rule is usually assumed to be

TABLE 6.14. Calculation of Required Storage.

Month	Inflow X	Draft Y	X − Y	I − D	Storage S = P − T	Spill	State#
1	96	27	69	69	69	0	R
2	130	27	103	172	166	6	S
3	56	30	26	198 P1*	166	26	S
4	8	30	−22	176	144	0	F
5	7	33	−26	150	118	0	F
6	5	30	−25	125	93	0	F
7	2	27	−25	100	68	0	F
8	5	27	−22	78	46	0	F
9	8	30	−22	56	24	0	F
10	12	36	−24	32 T1	0	0	E
11	80	33	47	79	47	0	R
12	91	30	61	140	108	0	R
13	96	27	69	209	166	11	S
14	130	27	103	312	166	103	S
15	56	30	26	338 P2	166	26	S
16	8	30	−22	316	144	0	F
17	7	33	−26	290	118	0	F
18	5	30	−25	265	93	0	F
19	2	27	−25	240	68	0	F
20	5	27	−22	218	46	0	F
21	8	30	−22	196	24	0	F
22	12	36	−24	172 T2	0	0	E
23	80	33	47	219	47	0	R
24	91	30	61	280	108	0	R

*P = peak; T = trough
#R = rising; F = falling; S = spilling; E = empty

relatively simple so that deficits and surpluses can easily be evaluated. The method is therefore used for preliminary design studies to obtain a rough estimate of the capacity required. At the prefinal or final design stage, a reservoir operation study is usually made to examine in more detail the consequences of alternative release rules. In a reservoir operation study, the changes in storage content of a reservoir are tracked using the following storage form of the conservation equation:

$$S_{t+1} = S_t + I_t - Q_t - E_t - L_t \tag{6.198}$$

where S_{t+1} is storage at the end of the time period t, S_t is storage at the beginning of the time period t, I_t is inflow during the time period t, Q_t is release during the time period t, E_t is net evaporation loss during the time period t, L_t is other losses, and C is capacity subject to $0 < S_t < C$.

The operation study is performed as follows:

1) Arbitrarily choose the reservoir capacity C and assume that it is initially full, i.e., $S_0 = C$.
2) Apply the above equation sequentially for the whole historical record. Q_t may vary seasonally, depend on S_t, or be assumed constant. Estimates must be made of E_t and L_t.
3) Plot S_t against time and estimate the probability of failure by dividing the number of time periods for which the reservoir is empty by the total number of periods.
4) If the probability of failure is unacceptable, choose a new value of the storage C and repeat the process.

This is, therefore, an iterative procedure that permits the choice of the storage capacity corresponding to a particular design probability of failure.

To put this discussion in perspective, it should be noted that the focus of this section is on the hydrology of the reservoir storage–yield relationship. If the scope is expanded to include reservoir operation, then a whole new body of techniques come to the foreground. Indeed, recent work has concentrated on the operational aspects of reservoirs where techniques of operations research are particularly relevant (Loucks, 1968). In the final reckoning, the optimal operation of a reservoir must be tied to the objectives of the project in which the reservoir is but one component. Although minimizing the duration and severity of shortfall is a valid objective in optimization, a more realistic analysis should consider cost and benefits; however, this takes the subject outside the domain of hydrology and is beyond the scope of this section. Yeh (1985) reviewed the then-current state of the art. Mathematical programming techniques both deterministic and stochastic, have been applied in the optimization formulation (Croley, 1974; Kelman et al., 1990). Simulation methods (Johnson et al., 1991) have been shown to be useful in developing operating policies. In many situations, the system may include more than one reservoir whose joint operation needs to be optimized (Marino and Loaiciga, 1985). As engineering systems become more complex, the overall objectives of a project also become more complicated. Multiple objectives, in addition to being noncommensurate at times, often are in conflict with each other. A new set of tools may be brought to bear on such multiple-objective formulations (Haimes and Allee, 1984; Fontaine and Flug, 1991). These topics are beyond the scope of this discussion.

E. Sequential Flow Generation Method

An alternative approach for calculating the probability that the actual storage needed will exceed the adopted design value, is to determine the probability distribution of storages. This technique has led to sequential flow generation methods. The earliest of these techniques date to Hazen (1914) who proposed an operation study using a "standardized" stream flow record. Different storage capacities are tested and the number of water shortage periods are counted. While this modification enables one to approximate the probability of failure, there are many deficiencies in this approach.

Sudler (1927), Barnes (1954) and others employed decks of cards and other sampling devices for generating synthetic stream flows. In Sudler's approach, a representative annual flow was written on each card and by drawing from the deck, it was possible to generate a long artificial record. This long sequence was subdivided into shorter records of duration equal to the economic life of the project. In turn, each of these short records was analyzed using the mass curve approach and a probability distribution of storage capacities was obtained. From this distribution, a design value may be chosen corresponding to a preselected probability of failure.

The essence of this technique is to use a data generation procedure that synthesizes stream flow sequences with the same statistical properties as the historical record. Each sequence is then analyzed using some acceptable method to determine the storage capacity. Since many sequences can be synthesized, a distribution of storage values can be obtained. This is used to give insight on the reliability of the adopted design capacity.

This technique has led to several proposals for generating synthetic stream flow sequences. Although the general objective is to produce stream flow sequences that preserve the statistical characteristics of the historical data, there is no general agreement on how many and which of these characteristics should be preserved (Askew et al., 1971).

A fundamental difficulty with this approach is that the statistical properties are only as good as their estimates from the historical data that is available for model identification and parameter estimation. In identifying the model, assumptions are made regarding the structure of the stream flow record. These include the assumption that the stationarity or nonstationarity of the historical data can be determined. If the data is stationary (statistical properties do not vary with time), then a stationary model for flow generation is developed. If it is varies with time as often happens if daily, weekly, or monthly stream flows are analyzed, then an appropriate time varying (seasonal or periodic) model is used.

After the stationarity issue is resolved, either by assumption or through data preprocessing, the modeler must decide what statistical properties need to be preserved in the model. Early attempts at stochastic stream flow modeling tried to preserve the mean, standard deviations, and serial correlation structure of the historical record (Fiering, 1961; Beard, 1963). In addition to preserving the mean and variance of the data, it is often suggested that the autocorrelation and cross-correlation structure be also preserved. These statistical characteristics must first be estimated from the available record. Later modifications took into account the fact that stream flows do not follow symmetric distributions and therefore, skewness characteristics were included in the data generation algorithm. Still further analysis, especially when within-year storage was being considered, indicated that the synthetic stream flow sequence should include the nonstationary character of the historical data (Quimpo, 1968; Young and Pisano, 1968; Kottegoda, 1980; Salas and Fernandez, 1989). If a time-varying representation is deemed appropriate, there are several ways of modeling time variability. For example, one may represent the process as the sum of harmonics plus a correlated process or one may use a harmonically varying mean and similarly varying variance. There is also no agreement on how "persistence effects" should be incorporated in modeling the structure of the stream flow process. The empirical observation of Hurst (Hurst, 1951; 1956) regarding the behavior of the range has also influenced the evolution of stochastic stream flow modeling (Mandelbrot and Wallis, 1968; Yevjevich, 1972; O'Connell, 1974). Since the true structure is unknown, different investigators can always justify their choice of model as long as they use accepted methods of parameter estimation. This has resulted in the proliferation of stochastic models for generating stream flows. There is no general agreement on which model is best. Depending on the purpose for which the synthesized flow will be used, different models may be appropriate for different situations.

F. Reservoir Design by Simulation

In this method, the procedure for determining the reservoir capacity is the same as described before. The difference is that the inflow (supply) hydrograph is a stream flow series that is synthetically generated using a stochastic stream flow generation model for the river. As discussed above, the generated stream flows are designed to have the same statistical properties as the data.

After a stream flow simulation model is selected, several stream flow sequences are generated. Each sequence is analyzed to determine a storage capacity according to some acceptable method such as the mass curve analysis. These simulated sequences yield a distribution of storage values. This distribution may then be used as the basis for deciding which value to adopt for design.

Since there is a lack of consensus on the choice of generating model (Wallis and Matalas, 1972), this discussion is restricted only to the autoregressive models that have been used for generating annual and monthly stream flows.

The basis for these models is the commonly accepted representation of most geophysical data as composed of a trend, seasonal, and/or cyclical components and a serially correlated random component (Quimpo, 1968). Although these components could interact multiplicatively, in hydrology they are commonly assumed as additive. There are procedures for eliminating the trend component, using differencing for an example. Similarly, the seasonal component may be detected and removed through several methods (Kottegoda, 1980).

In the case of annual stream flows, the earliest application in hydrology assumes that both the trend and cyclical components are either absent or have been removed so that only the serially correlated component remains. A first order autoregressive model, proposed for annual stream flows by Julian (1961), takes the form

$$X_{j+1} = \overline{X} + r(X_j - \overline{X}) + s(1 - r^2)N_j \qquad (6.199)$$

where X_j) is the annual flow for the year j, X_{j+1} is runoff for the year j + 1, $\overline{X}$ is mean annual flow, r is the first autocorrelation coefficient, s is the standard deviation of the annual flows, and N_j is the standard normal random deviate.

From a sequence of randomly generated standard normal numbers, N_j, this model generates an annual stream flow sequence with mean X, variance s^2, and first autocorrelation coefficient r.

One criticism of this algorithm is its use of normally distributed random deviate N_j. Stream flow data usually exhibit a skewed distribution. In order to preserve this nonsymmetric characteristic, a random number with a skewed distribution should be used. The skewness coefficient of the historical data can be preserved if N_j in Equation 6.199 is first transformed to

$$W = \frac{2}{g}\left[1 + \frac{gN_j}{6} - \frac{g^2}{36}\right]^3 - \frac{2}{g} \tag{6.200}$$

where g is the skewness coefficient of the historical data. There are some limitations on the use of the above transform (Thomas and Burden, 1963). Other transformations have also been proposed (Kirby, 1972).

In some storage problems, monthly stream flows must be synthetically generated. Thomas and Fiering (1962) proposed a model of the same form as Equation 6.199 but that also takes into account the seasonal variation of the mean and variance. Their model may be written as

$$X_i = \overline{X}_j + b_j(X_{i-1} - \overline{X}_{j-1}) + W_{i-1}s_j(1 - r_{j-1}^2)^{1/2} \tag{6.201}$$

where X_i is generated flow during the season (month) i, X_{i-1} is generated flow during the season i−1, $\overline{X}_j$ is the mean flow during the season j, b_j is the regression coefficient between the flows for seasons j and j−1 = $r_{j-1}s_j/s_{j-1}$, W_{j-1} is the skewed random deviate, s_j is the standard deviation of flows during the month j, s_{j-1} is the standard deviation of flows during the month j−1, and r_{j-1} is the correlation of flows between months j and j − 1.

For monthly flows, this model requires the estimation of 36 parameters; 12 values each of X_j, s_j and r_j. Problems of parameter estimation and parsimony have led to the suggestion of many other models for generating flows. Lawrance and Kottegoda (1977) present a comprehensive discussion of the models of the 1970's.

G. Probability Matrix Methods

These methods seek to determine the probability distribution of storage. Depending on the complexity of the release rule and distribution of the inflows, several solutions have been proposed. The reservoir storage capacity is divided into states or zones, e.g., empty, half-full, and full. It is usually suggested that the capacity be divided into 8 to 12 states S_i. The probability that storage will be at any state S_k is equal to the sum of the products of (1) the probability that it is initially at state S_i and (2) the probability that it will go from state i to k where the summation is done over all values of i. The probability in (2) depends on the release rule and the distribution of the inflows. A system of equations may be set up in the following form:

$$[P_{t+1}] = [T][P_t] \tag{6.202}$$

where $[P_t]$ is a vector of probabilities corresponding to the states S_i and $[T]$ is a transition matrix of probabilities that depend on the release rule and the distribution of the inflows.

The solution of this equation yields the probability distribution of the states S_i, which represents the distribution of storage contents. Among other things, this distribution also yields the probability that the reservoir will be empty and hence, the probability of failure. Langbein (1958) used this method to analyze storage–yield relationships. The theory was formalized by the work of Moran (1959) and subsequent investigators who considered seasonality and serial correlation of the flows (Lloyd, 1963).

H. Methods Based on the Distribution of the Range

Under full development, the range of the accumulated departures from the mean is equal to the required storage capacity as derived in Equation 6.198. This relationship between the range and the storage requirement has prompted many investigators to examine the statistical properties of the range (Yevjevich, 1965). Theoretically, the quantity defined in Equation 6.198 is called the adjusted range. This differs from the true range R_T, which is defined as

$$R_T = M - m \tag{6.203}$$

where M is the maximum of the cumulative sums of a sequence of random variables with zero mean and m is the minimum of the cumulative sums. Feller (1951) derived the asymptotic mean and variance of the range. Anis and Lloyd (1953) also obtained results that agreed with Feller's. Interest in the range was further heightened by the observation of Hurst who derived the asymptotic expectation of the adjusted range R_a as

$$E(R_a) = s\sqrt{\frac{\pi n}{2}} \tag{6.204}$$

where s is the standard deviation of the data and n is the length of record. Hurst further found that the rescaled mean adjusted range, R_a/s, increased not as the square root of n as Equation 6.204 suggests but with an exponent whose mean value is 0.729. This observation has in turn led to a whole new theory for stream flow synthesis designed to explain the so-called Hurst phenomenon (Mandelbrot and Wallis, 1968; Klemes, 1974).

The preceding results on the expected value of the range are based on the assumption that the partial sums are those of independent variables. When dealing with within year storage, the inflows (stream flow) are generally nonstationary. This introduces two additional difficulties in the analysis: the serial correlation of the stream flows and the effect of the seasons. Although seasonal effects may also be measured by the correlation structure of the flows, it is more convenient to treat seasonality in a different manner. Salas (1972; 1974) derived some relationships for the range of periodic–stochastic processes which he suggests may be used to determine the required storage capacity.

I. Dependability and Risk Analysis

In the classical mass curve approach to storage reservoir design, the derived capacity does not give the hydrologist an indication of the risk. In this context, the risk is equal to the probability that the selected reservoir capacity will be unable to meet the future water demand. Although they are related, this risk is not the same as the probability of failure that was defined earlier. One approach that has been suggested to determine this risk is to carry out N simulation analyses as described previously. These analyses yield N estimates of the required storage size. The N estimates define a frequency histogram that may be used to estimate the frequency distribution of future storage requirements. The selected reservoir storage capacity can therefore be compared with this distribution in order to estimate the probability that the future demand will not be satisfied. Hydrologists have approached the problem of risk and failure probability in different ways.

Insight on the yield of reservoirs may be gained by interpreting the yield in terms of associated risks. A reservoir may be designed to meet a period of low flow or drought associated with a recurrence interval T. Of course, the reservoir is subject to other droughts which may be less or more severe than the design drought. By developing an array of probabilities of occurrence of various droughts for different design horizons, the hydrologist can obtain a better insight into the concept of risks associated with a selected design.

Consider a drought event of magnitude E. Since E is a random variable, let P denote the probability that an equal or more severe event will occur in any year. The recurrence interval (average time between recurrences) can be shown to be equal to the reciprocal of P (Ang and Tang, 1975). If the design life of a project is n years, it can be shown that the Risk R, defined as the probability that at least one event that is more severe than E will occur during the design life of the project, is

$$R = 1 - \left[1 - \frac{1}{T}\right]^n. \tag{6.205}$$

The complement of R is the reliability Q:

$$Q = 1 - R = \left[1 - \frac{1}{T}\right]^n. \tag{6.206}$$

Values of E corresponding to the chosen return period T may be used in Equation 6.205 to develop yield–probability curves corresponding to different design lifetimes (Stall, 1964). This is shown in Fig. 6.22. This figure shows the reliability of the design if a drought that is more severe than the T-year drought occurs.

J. Sequential and Nonsequential droughts

In the Rippl mass curve analysis, the reservoir inflows used to generate the mass curve of supply are derived from the driest historical stream flow record. The natural sequence of inflows is used to determine the storage capacity. The capacity so obtained is thus derived from a sequential drought. Since only a single record is available, as discussed earlier, no probability can be attributed to the reservoir

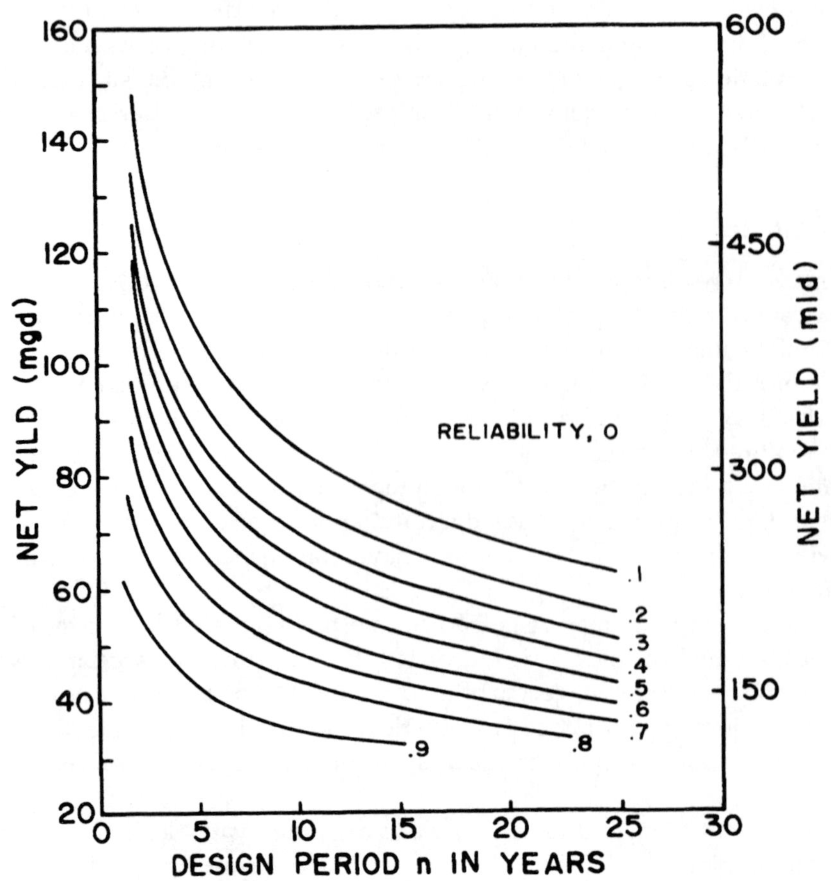

Figure 6.22—Yield–Reliabillity Curve.

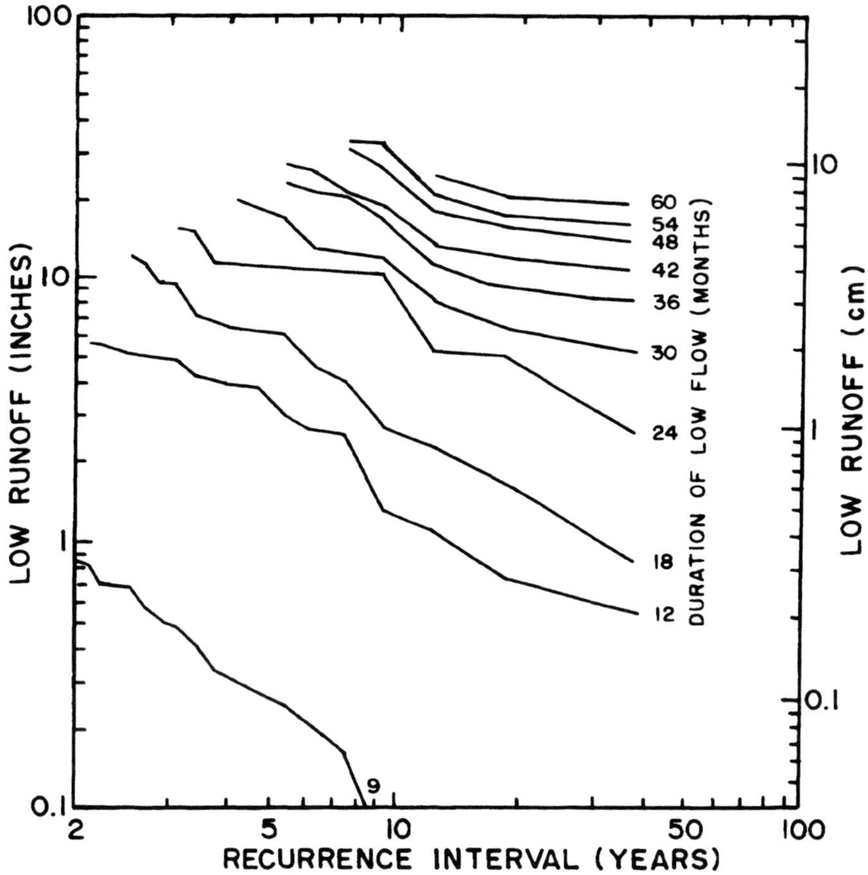

Figure 6.23—Low Flow Frequency Curves.

capacity that is determined using this method. Stall (1962; 1964) proposed a technique that is able to determine the required storage capacity corresponding to different recurrence intervals.

In Stall's method, frequency curves are first developed for several series of low flow indexes, each of different duration, derived from the data. For each duration, the low flow series is identified by scanning the entire historical record of the subject stream. The low flow values are ranked, their plotting positions calculated and then plotted on probability paper. This procedure yields a family of low-flow frequency curves as shown in Fig. 6.23. Then for a selected recurrence interval T, the T-year low-flow values may be read from the curves for each duration. If these values are plotted versus the duration as abscissa, they will define a nonsequential mass curve for the selected recurrence interval. This may be treated as the mass curve of inflows in Rippl's analysis. If the mass curve for a given draft rate is superimposed on this plot, then the required storage capacity may be read off as equal to the maximum deviation between the mass curves of supply and demand (Fig. 6.24). The draft rate is expressed as a percentage of the mean flow. Several draft rates may be used to develop a set of draft-storage-recurrence curves (Fig. 6.25).

To determine the net yield for a given recurrence interval, the mass curve is analyzed during the critical drawdown period. It is assumed that the reservoir is full at the beginning of the critical drawdown period and is empty at the end of the period. The total gross draft minus the reservoir capacity is equal to the reservoir inflow during this drawdown period. The reservoir capacity is corrected for evaporation and other losses. The total net draft is equal to the corrected reservoir capacity plus the reservoir inflow. The net yield is equal to the total net draft divided by the length of the critical drawdown period.

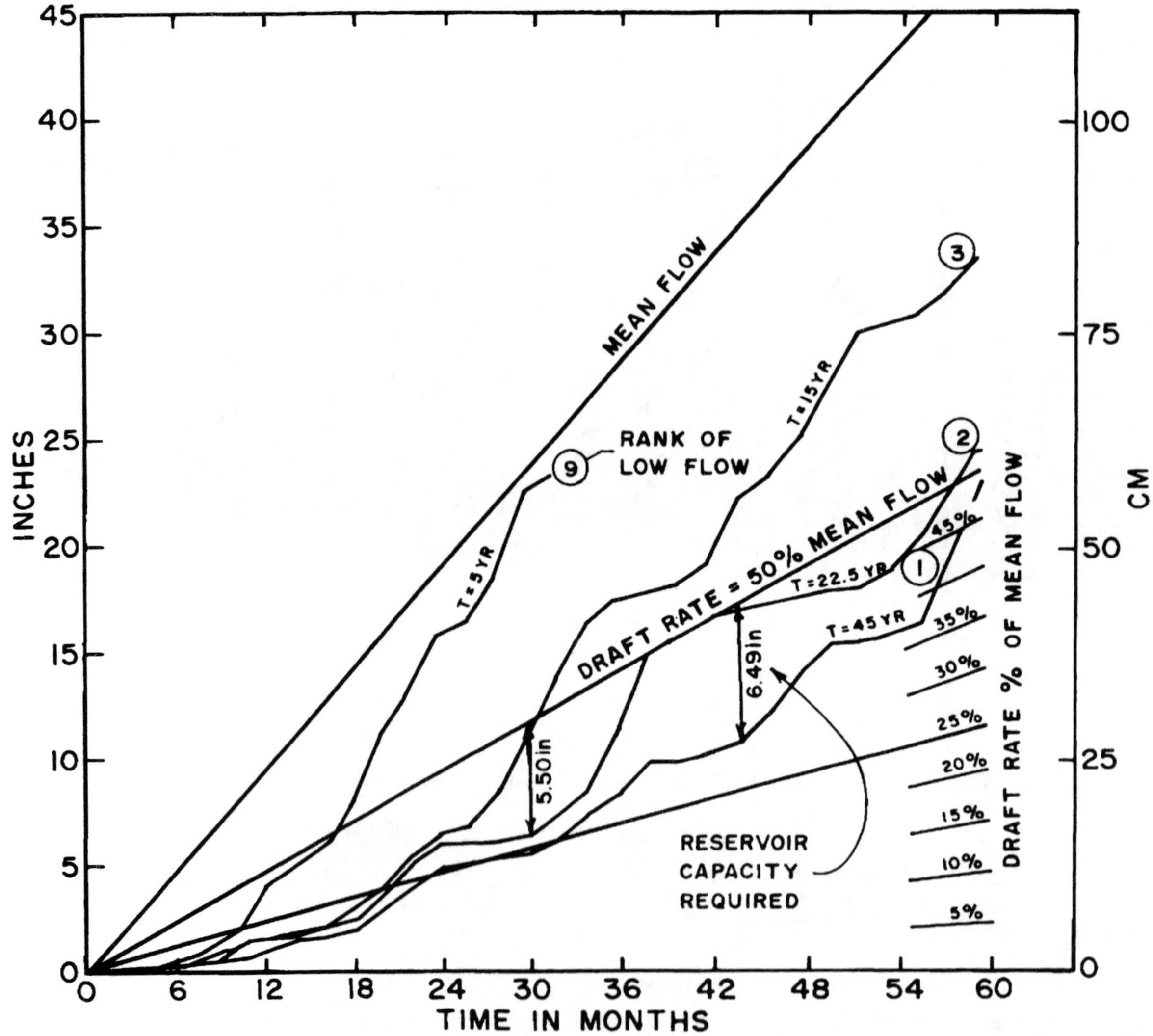

Figure 6.24—Nonsequential Mass Curves.

K. Flow Duration Curves

If the sequence of flows is a secondary consideration to the frequency that certain flows are exceeded, a common method of describing flow variability during a year, or any prescribed period, is through the flow duration curve. This curve is a plot of discharge versus the percent of time that it is equaled or exceeded during the period being examined. Although numerical procedures for developing the flow duration curve have been prescribed (Searcy, 1959), the flow duration curve may be obtained graphically as shown in Fig. 6.26. The flow series is sequenced in decreasing order, with each flow value being assigned an order number. For each flow value, the percent time is defined as the ratio of its order number to the total number of days, expressed in percentage. The flow duration curve is obtained by plotting flow (in the ordinates) vs. percent time (in the abscissas). The flow duration curve is an average curve for the period on which it is based. The flow duration curve for the year of lowest runoff is radically different from that for the year of maximum runoff.

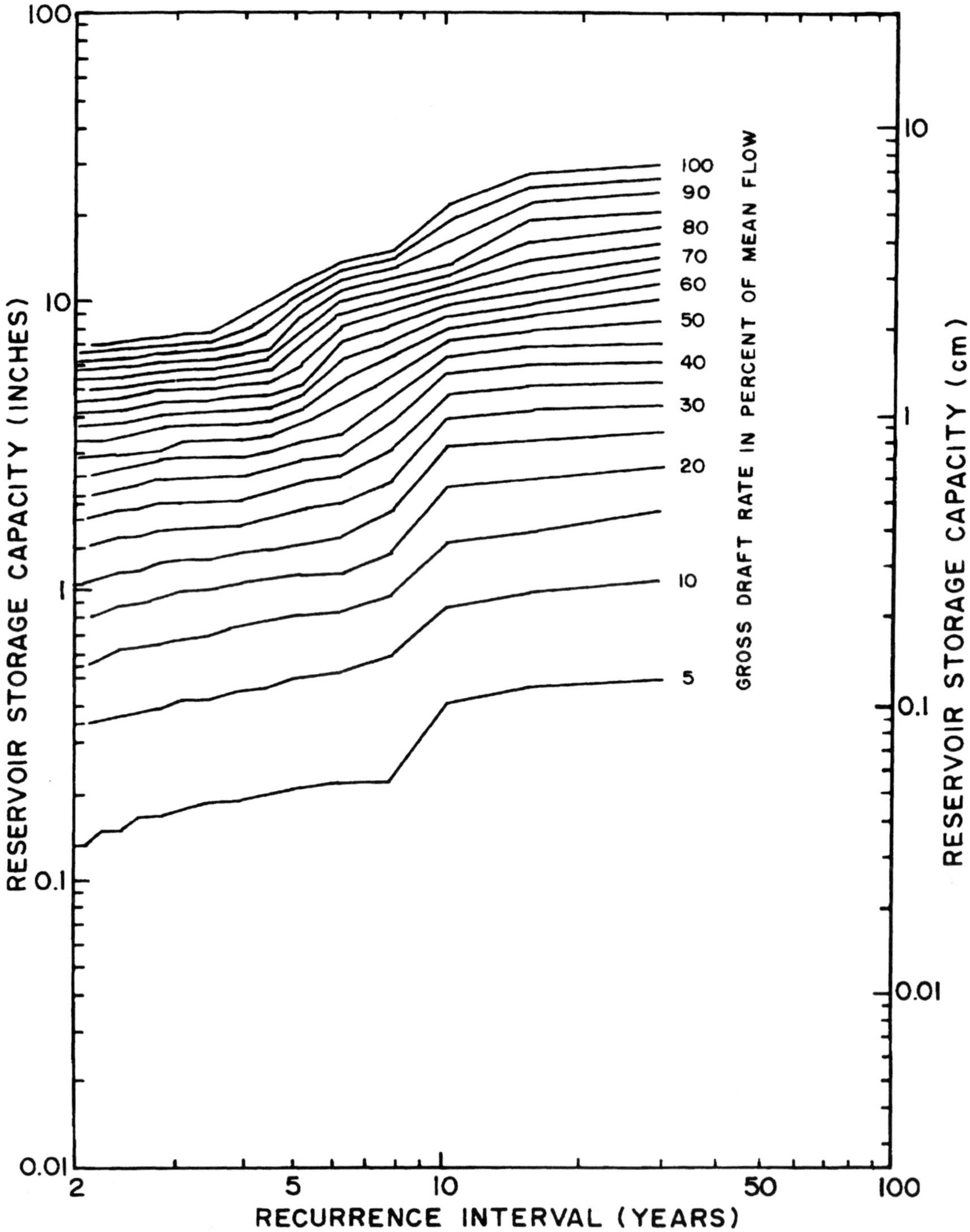

Figure 6.25—Draft-Storage-Recurrence Curves.

VII. RUNOFF QUALITY

A. Overview

This section describes processes that affect runoff quality, sources of surface water contamination, water quality standards, methods of water quality monitoring, commonly known models for surface water quality simulation, and methods to control the quality of surface runoff.

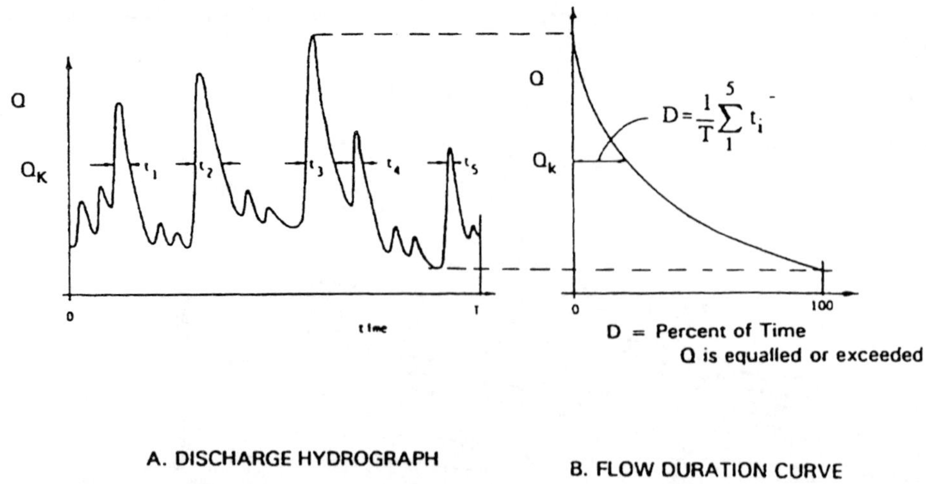

Figure 6.26—Graphical Derivation of Flow Duration Curve.

Surface water bodies, e.g., streams and lakes, are used to supply water for municipal, agricultural, and industrial purposes and serve as receptacles of municipal, agricultural, and industrial wastewater. Also, surface runoff either recharges ground water aquifers or is augmented by ground water depending on whether a particular stream segment or lake bottom is a gaining or losing system. Surface runoff quality studies, analyses, and data are required to assess the suitability of a particular surface water body to serve as a viable source for the previously mentioned uses, the degree of water treatment required for specific uses, and the limitations to be imposed on the quantity and quality of wastewater discharges to avoid significant water quality degradation of the receiving surface water body.

In the United States, an institutionalized national approach toward the control of water pollution began with the passage of the Water Pollution Control Act (PL80–845) in 1948. This act was amended in 1952 and 1955. Subsequently, the Water Quality Act of 1965 and the Clean Waters Restoration Act of 1966 set water quality standards for interstate waters and Federal Water Pollution Control Act Amendments of 1972 created a regulatory mechanism requiring compliance with uniform effluent standards and national permits for all point-source discharges into the waters of the nation. The 1972 amendments to the Federal Water Pollution Control Act (referred to as the Clean Water Act or CWA) prohibited the discharge of any pollutant into navigable waters from a point source unless the discharge is authorized by a National Pollutant Discharge Elimination System (NPDES) permit. This act articulated several goals and policies (Clark et al., 1977). The goals were to attain a water quality that provided for the protection and propagation of fish, shellfish, and wildlife and for recreation in and on the water by July 1, 1983. The act also aimed to eliminate the discharge of pollutants into navigable waters by 1985. The major policies included prohibiting the discharge of pollutants in toxic amounts; providing federal financial assistance for construction of publicly owned treatment works (POTWs); development and implementation of area-wide waste-treatment management planning; mounting a major research and demonstration effort in wastewater-treatment technology; to recognize, preserve, and protect the primary responsibilities and roles of the states to prevent, reduce, and eliminate pollution; to ensure, where possible, that foreign nations act to prevent, reduce, and eliminate pollution in international waters; and to provide for, encourage, and assist public participation in executing the Act.

The NPDES program focused almost entirely on the goal of reducing pollutants from municipal sanitary sewage and industrial process wastewater. The National Safe Drinking Water Act was enacted in 1974. Major provisions of this act included the establishment of primary regulations for the protection of public health regulations to protect underground drinking water sources by the control of surface injection, and secondary regulations that are related to taste, odor, and appearance of drinking water; initiation of research on health, economic, and technological problems related to drinking water supplies and a survey of rural water supplies; and allocation of funds to states in improving their drinking water programs through technical assistance, training of personnel, and grant support.

The Clean Water Act (CWA) aided by Section 405 of the Water Quality Act (WQA) of 1987 required the Environmental Protection Agency (EPA) to establish regulations setting forth NPDES permit application requirements for stormwater discharges associated with industrial activity and discharges from municipal separate storm sewer systems (Federal Register, 1990).

Under the NPDES permit regulations, all effluents are analyzed for the following chemical parameters: 1) Biochemical Oxygen Demand (BOD), 2) Chemical Oxygen Demand (COD), (3) Total Organic Carbon (TOC), 4) ammonia as nitrogen, 5) temperature, and 6) pH.

Other chemicals required to be analyzed if expected to be present include the following:

- Bromide
- Total residual chlorine
- Fecal Chloroform
- Fluoride
- Nitrate–Nitrite
- Total Organic Nitrogen
- Oil and grease
- Total phosphorus
- Radioactivity
- Sulfate
- Sulfide
- Sulfite
- Surfactants
- Total aluminum
- Total barium
- Total boron
- Total cobalt
- Total iron
- Total magnesium
- Total manganese
- Total molybdenum
- Total tin
- Total titanium

In addition, this program includes testing for a number of priority pollutants that include a number of volatile compounds, acid extractable compounds, base/neutral extractable compounds, pesticides, polychlorinated biphenyls (PCBs), metals, cyanide, and phenols.

1. ***Point and Nonpoint Sources of Pollution*** "Point source" means any discernible, confined, and discrete conveyance, including but not limited to, any pipe, ditch, channel, tunnel, conduit, well, discrete fissure, container, rolling stock, concentrated animal feeding operation, vessel, or other floating craft from which pollutants are or may be discharged. This term does not include return flows from irrigated agriculture (USEPA, 1991a). Point sources of surface water pollution include effluent discharges to lakes, streams or wetlands through pipes, open-channels, multiport diffusers, or other types of outfall structures. Point source discharges of industrial effluents and sanitary sewage are regulated by the National Pollutant Discharge Elimination System.

Nonpoint sources of surface water pollution include overland and ground water flows to lakes, streams, wetlands, or other surface water bodies from agricultural, mining, industrial, urban, developed, undeveloped, or construction areas. These effluents enter surface water bodies along relatively large reaches or areas.

2. ***Runoff Quality Data*** Runoff data for a number of stations on various surface water bodies in the United States are available in U.S. Geological Survey (USGS) Water-Data Reports for different states published for each water year (October 1 through September 30). This information can be accessed through the National Water Storage and Retrieval System (WATSTORE) maintained on the central computer facilities of the USGS at its national center in Reston, Virginia. Additional information on the quality of surface water bodies in various states is available in the STORET data system of the U.S. Environmental Protection Agency maintained at its Environmental Research Laboratory in Athens, Georgia. Site-specific runoff quality data have to be collected to evaluate the impacts of existing and proposed activities on the surface water regime in the site vicinity.

3. ***Runoff from Undisturbed Watersheds*** Surface runoff from undisturbed watersheds is usually considered clean; however, depending on the mineralogical character of the rock mass that comes in contact

with these waters, they may contain substantial quantities of dissolved substances. Transport of rock and soil mass with surface runoff from these watersheds due to sheet, rill, or gully erosion or landslides may result in increasing the total suspended and dissolved solids contents of these waters. Also, overland flow from the vegetated portions of these watersheds may carry fair quantities of biomass and litter with it. Another notable source of pollution for surface runoff from undisturbed watersheds may be pollutants washed from the atmosphere by precipitation. Moisture-laden air masses moving across certain areas may pick up chemicals and particulates from the atmosphere, transform into clouds, and precipitate on local and distant watersheds.

In general, runoffs from undisturbed watersheds are of acceptable quality for agricultural usage unless the total dissolved solids (TDS) content exceeds 500 mg/l; however, some plants may tolerate TDS concentrations up to 5,000 mg/l with careful management practices. The use of these waters for municipal and industrial water supply will require testing for some or all of the parameters set forth by EPA in accordance with the Safe Drinking Water Act.

4. *Runoff from Agricultural Watersheds* To increase agricultural production, farmers use substantial quantities of fertilizers, herbicides, pesticides, and limited quantities of fungicides on their fields. A major portion of these chemicals is consumed by the plants. The remaining portion is leached into the subsoil by infiltration and adsorbed or absorbed by surface soils. The portions contained in soils and plant residue are transported by subsurface flow to ground water bodies or surface water streams. The types, quantities, and character of pollutants in surface runoff from agricultural watersheds are highly variable. Commonly occurring pollutants in these waters include sediments, total suspended solids (TSS), animal wastes, wastes from industrial processing of raw agricultural products, plant nutrients, forest and crop residues, inorganic salts and minerals, pesticides, herbicides (see Table 6.15), and fungicides. Farm animals, livestock, and dairy cattle produce large quantities of manure and excrete significant quantities of undigested nutritive material each day. Animal wastes contain nitrogen, phosphorus, and most of the salt intake of the animals (Clark et al., 1977).

5. *Runoff from Forest Land* Surface runoff from forest land may contain significant amounts of suspended sediment, particularly during heavy storm events. The suspended sediment load may include sediments of different mineralogical character, tree twigs, brush, sawdust, and plant residue.

6. *Runoff from Urban and Industrial Areas* The quality of surface runoff from urban and industrial areas varies from location to location depending upon numerous physical, chemical, and biological processes (Boyd and Gardiner, 1990; Lazaro, 1991). The major processes that may influence the quality of runoff from these areas include the following:

a) Atmospheric Scrubbing. Airborne particles (e.g., dust, soot, aerosols) can be scrubbed out of the atmosphere by precipitation.
b) Surface Washoff. Rainfall and overland flow wash pollutants off of bituminous asphaltic pavements, concrete and synthetic surfaces, dirt roads, lawns, parking lots, buildings, storage and equipment areas, landscaped areas, construction sites, residential areas, and open spaces.
c) Erosion and Scour. Depending on storm intensities, topography, soil types, and vegetative cover, surface runoff may erode pollutants, soil, and minerals from exposed land surfaces and scour them from the beds and banks of unlined ditches and creeks.

The common sources of pollution in runoff from urban and industrial areas include the following:

a) Sand and Salt Used for De-icing. In winter, sands, salts, and mixtures of sands and salts are used for de-icing purposes on roads, streets, parking lots, and airports. These materials are washed off with melting snow and result in TSS and salt loading in surface runoff from these areas. The most commonly used de-icers include calcium chloride, sodium chloride (rock salt and NaCl), potassium chloride, urea, and in certain cases, calcium magnesium acetate. Surface runoff from areas treated with de-icers should be tested for the above mentioned constituents, pH, and conductivity.

TABLE 6.15. List of Commonly Known Pollutants in Pesticides and Herbicides. (*To Be Analyzed in Runoff From Agricultural Watersheds*)

• Aldrin	• Beta-Endosulfan (II)
• Alpha-BHC	• Endosulfan-Sulfate
• Beta-BHC	• Endrin
• Gamma-BHC	• Endrin Aldehyde
• Delta-BHC	• Heptachlor
• Chlordane	• Heptachlor Epoxide
• 4,4′-DDT	• Lindane
• 4,4′-DDE	• Methoxychlor
• 4,4′-DDD	• Toxaphene
• Dieldrin	• 2,4-D
• Alpha-Endosulfan (I)	• 2,4,5-TP (Silvex)

b) Atmospheric Deposition. Atmospheric deposition in urban and industrial areas with significant general or fugitive air emissions results in surface water pollution due to wetfall with precipitation and dryfall with air movements.
c) Erosion. Soil erosion from urban and industrial areas may contribute to increased TSS, nutrients, organic matter, and release of adsorbed pollutants in surface runoff.
d) Construction Materials. Commonly used materials in the construction and maintenance of buildings and infrastructure include flashing and shingles, gutters and downspouts, copper and galvanized piping, metal plating, paints, and wood preservatives. Over time, these surfaces corrode, flake, dissolve, decay, and are subjected to leaching and are transported by stormwater. These processes are exacerbated by acid rain.
e) Manufactured Products. Manufactured products commonly occurring in urban and industrial areas include fertilizers, insecticides, algicides, and fungicides; automobile brake linings, clutch facings, and tire compounds; lubricants, hydraulic fluids, combustion products, and automotive coolants; and solvents, detergents, and other cleaning compounds. These products are transported with overland flow and contaminate surface waters.
f) Plants and Animals. Trees and shrubs deposit pollen, fruit, bark, twigs, and leaves that enter into stormwater runoff. During the growing season, nutrients leach from tree leaves and stems and are transported by stormwater runoff. Animal excrements are normally washed into the storm drainage system.
g) Accidental Spills and Unintended Discharges. These discharges into the storm drainage system may include illicit connections to process water systems, illegal dumping of contaminated effluents, leaking underground storage tanks, faulty septic systems, leachate from sanitary landfills, and improperly managed hazardous waste treatment, storage, and disposal sites (Boyd and Gardiner, 1990).

7. ***Runoff from Construction and Mine Disturbed Watersheds*** Disturbed watersheds include those with ongoing construction and mining activities and those subjected to natural disturbances like landslides, seismic events, and volcanic activities. Generally, surface runoff from construction and mine disturbed areas is controlled by federal, state, and local regulations. Federal permits required for the discharge of surface runoff from these areas include the NPDES permit to be issued by the U.S. Environmental Protection Agency and another permit under Section 404 of the Clean Water Act, issued by the U.S. Army Corps of Engineers, which prohibits the discharge of dredged or fill material into the waters of the United States. Surface runoff from all types of disturbed watersheds includes suspended sediments along with the chemicals associated with such sediments. Under unusual circumstances, surface runoff from construction areas may contain traces of cement, bituminous materials, and other chemicals used by the construction industry that may have potential to be dissolved and washed away by water.

Similarly, surface runoff from mining areas may carry wash-offs of fugitive dust formed by fallouts from on-site smelters and soluble chemicals contained in ore storage areas, coal piles, waste rock storage, and tailings disposal areas. Drainage waters from active and abandoned mine sites may carry dissolved toxic minerals such as heavy metals into surface streams. Normally, these operations are regulated by the Office of Surface Mining and some state and local regulations. Surface runoff from areas disturbed by landslides, earthquakes, and volcanic eruptions may contain large quantities of debris, suspended sediment, different types of mineral particles, and chemicals in the disturbed soils, rocks, and lava that can be dissolved and carried away by water.

8. Acid Rain Industrialization has resulted in widespread use of smoke stacks and chimneys emanating flue gases with different quantities and character of particulate matter. These tiny particles contain different types of chemicals and are transported into the atmosphere by prevailing winds. During this process, they are deposited on tree leaves and twigs and hill slopes, or are kept in suspension. When moist air masses pass by them, they form tiny nuclei for the condensation of water vapor. The number of these condensation nuclei is many orders of magnitude larger than the number of water droplets in a cloud. They vary in size from 0.1 to 10 microns. As a vapor-laden air mass rises, it expands because of the reduced ambient pressure and undergoes cooling. When this cooling brings the mass to the saturation point, excess water vapor begins to condense on these condensation nuclei (Eagleson, 1970). Finally, the chemicals contained in these particles and dissolved by the water droplets reach the ground as precipitation. Because of the predominantly acidic character of these droplets, this phenomenon is commonly known as acid rain. Surface runoff from areas affected by acid rain may have a variety of pollutants (Galloway et al., 1978; Eshleman and Hemond, 1985; and Howells, 1990).

B. Water Quality Monitoring

1. Objective Surface runoff reaching lakes and streams is used for municipal, industrial, and irrigation water supply, for swimming, boating, and fishing, for fish hatching and maintenance of aquatic life, and for other miscellaneous purposes, e.g., dust control and pavement cleaning. The water quality requirements for each of these purposes are different. Depending upon the sources of surface runoff and past and present effluent discharges into the surface water body, it may or may not be polluted with organic matter, suspended solids, floating debris, microorganisms such as bacteria and algae, inorganic matter, acids, alkalis, toxic chemicals, radioactive materials, and heated waste water. The objective of water quality monitoring is to determine the suitability of a particular surface water body for the aforementioned water uses with and without economically and technically viable treatment.

2. Water Quality Standards Water to be used for most of the purposes mentioned previously must be free from substances that will settle to form sludge deposits, floating debris, oil, scum, and materials producing color and bad odor. The National Safe Drinking Water Act, (December 16, 1974) provided for the issuance of primary and secondary drinking water regulations. A summary of these regulations is provided in Table 6.16 (USEPA, 1991b).

As water quality regulations continue to undergo review, revision, and expansion, check with the EPA or local health authority for the latest drinking water standards.

3. Water Quality Sampling Water quality sampling is required to ensure that a certain source of surface water is suitable for an assigned or intended use without endangering human health, survival of aquatic biota and livestock, and sustenance of crops and natural vegetation. Water quality sampling should be performed when contaminated waters are discharged into an existing surface water body through point or nonpoint sources and when significant changes in stream flows, water depths, and weather (e.g., temperature, precipitation, wind velocity, net solar radiation) are likely to cause degradation of water quality. Water quality sampling is useful to evaluate the impacts of specific discharges and withdrawals on the affected surface water body. To assess the impacts of an industrial or domestic wastewater discharge, water quality should be monitored at the outer edge of the mixing zone. The size of the mixing zone is determined by regulatory agencies responsible to maintain the quality of the waters of the state. Mixing zone is the volume of water downstream of a discharge structure in which complete mixing of the discharge and receiving waters is expected to occur (Prakash, 1977). To examine the suitability of a certain water source, it is advisable to install water quality monitoring equipment near the location of the intake outside the mixing zone of any outfall. Normally, intake structures should be located outside the mixing zones of existing wastewater discharges in lakes and streams such that the velocity field created by the intake does not influence the flow field within the mixing zone.

The frequency of sampling may be dictated by regulations. To assess the suitability of a water source, water quality samples should be collected during varying climatic (i.e., very cold, mild, and very hot seasons) and stream flow (i.e., flood flow, average flow, and low flow) conditions. Water samples must be collected in standard sampling bottles. Separate samples should be collected for chemical and bacteriological examination since the methods for their collection, preservation, and shipment are different. A 1- to 2-liter sample is adequate for most water quality analyses. A grab sample is a single sample taken from mid-depth in the middle of the water body. Grab samples may be used to estimate the quality of a well-mixed water body. In the case of stratified or poorly mixed water bodies, integrated or composite samples may be necessary. An integrated sample is a mixture of equal volumes of water taken from different depths in the middle of the stream. A composite sample is collected by an automatic sampler or by mixing a number of flow-weighted grab samples, i.e., volume of composite sample,

$$V = \sum_{i=1}^{N} kq_i \tag{6.207}$$

and

$$v_i = kq_i \tag{6.208}$$

where N is the number of grab samples, k is a constant of proportionality, q_i is the flow rate when the grab sample, i, is collected, and v_i is the volume of the grab sample, i. Normally, water quality sampling kits are provided by the laboratory selected to perform the desired analyses. A sampling kit includes bottles, jars, and other containers with appropriate preservatives for each constituent to be analyzed. Commonly used preservatives include nitric acid (HNO_3), sulfuric acid (H_2SO_2), hydrochloric acid (HCL), and sodium sulfate ($NA_2S_2O_3$). The sampling bottles are placed in a cooler and packed with ice or freezer packs and shipped to the laboratory within the specified holding time for each constituent. The holding times vary from 6 hours for coliforms and fecal streptococci to 2 days for BOD, turbidity, and spectrophotometric analysis of certain chemicals and more than a week for many other constituents. The methods for collection and analysis of water quality samples are documented in *Standard Methods for the Examination of Water and Wastewater* (1985). Commonly used water quality tests include the following (Clark el al., 1977):

a) *Physical and Chemical Examination*
Temperature, turbidity, color, taste, and odor;
Residue, solids after evaporation;
Hardness;
Acidity, alkalinity, pH, and carbon dioxide;
Nitrogen and nitrate;
Chloride, fluoride;
Phosphate;
Chlorine and chlorine demand; and
Oxygen and oxygen demands.
b) Biological examination for traces of coliform group bacteria
c) Radiological examination if such contamination is suspected.

TABLE 6.16. Drinking Water Regulations.

Parameter	Maximum permissible contaminant level (as of April 1991)
Inorganics	
Antimony	0.01/0.005 mg/l
Arsenic	0.05 mg/l
Asbestos (fibers/l 10 μm)	7 MFL
Barium	2 mg/l
Beryllium	0.001 mg/l
Cadmium	0.005 mg/l
Chromium	0.1 mg/l
Copper	1.3 mg/l
Cyanide	0.2 mg/l
Fluoride	4 mg/l
Lead	0.005 mg/l
Mercury	0.002 mg/l
Nickel	0.1 mg/l
Nitrate (as nitrogen)	10 mg/l
Nitrite (as nitrogen)	1 mg/l
Nitrate + Nitrite (both as nitrogen)	10 mg/l
Selenium	0.05 mg/l
Sulfate	400/500 mg/l
Thallium	0.002/0.001 mg/l
Radionuclides	
Beta Particle and Photon Activity	4 mrem
Gross Alpha Particle Activity	15 pCi/l
Radium 226/228	5 pCi/l
Organics	
Adipates (Diethylhexyl)	0.5 mg/l
Alachlor	0.002 mg/l
Aldicarb	0.003 mg/l
Aldicarb Sulfone	0.003 mg/l
Aldicarb Sulfoxide	0.003 mg/l
Atrazine	0.003 mg/l
Benz(a)anthracene (PAH)	0.0001 mg/l
Benzene	0.005 mg/l
Benzo(a)pyrene (PAH)	0.0002 mg/l
Benzo(b)fluoranthene (PAH)	0.0002 mg/l
Benzo(k)fluoranthene (PAH)	0.0002 mg/l
Bromodichloromethane (THM)	0.1 mg/l
Bromoform (THM)	0.1 mg/l
Butylbenzyl-phthalate (PAE)	0.01 mg/l
Carbofuran	0.04 mg/l
Carbon tetrachloride	0.005 mg/l
Chlordane	0.002 mg/l
Chlorodibromomethane	0.1 mg/l
Chloroform (THM)	0.1 mg/l
Chrysene (PAH)	0.0002 mg/l
2,4-D	0.07 mg/l
Dalapon	0.2 mg/l
Dibenz(a,h)anthracene (PAH)	0.0003 mg/l
Dibromochloropropane (DBCP)	0.0002 mg/l
Dichlorobenzene P-	0.075 mg/l
Dichlorobenzene O-, M-	0.6 mg/l
Dichloroethane (1,2-)	0.005 mg/l
Dichloroethylene (1,1-)	0.007 mg/l
Dichloroethylene (cis-1,2-)	0.07 mg/l
Dichloroethylene (trans-1,2-)	0.1 mg/l
Dichloromethane	0.005 mg/l
Dichloropropane (1,2-)	0.005 mg/l
Diethylhexylphthalate (PAE)	0.004 mg/l
Dinoseb	0.007 mg/l
Diquat	0.02 mg/l

TABLE 6.16. *(Continued)*

Parameter	Maximum permissible contaminant level (as of April 1991)
Endothall	0.1 mg/l
Endrin	0.002 mg/l
Ethylbenzene	0.7 mg/l
Ethylene dibromide (EDB)	0.00005 mg/l
Glyphosate	0.7 mg/l
Heptachlor	0.0004 mg/l
Heptachlor epoxide	0.0002 mg/l
Hexachlorobenzene	0.001 mg/l
Hexachlorocyclopentadiene	0.05 mg/l
Indeno(1,2,3,-e,d)pyrene (PAH)	0.0004 mg/l
Lindane	0.0002 mg/l
Methoxychlor	0.04 mg/l
Monochlorobenzene	0.1 mg/l
Oxamyl (Vydate)	0.2 mg/l
Pentachlorophenol	0.001 mg/l
Picloram	0.5 mg/l
Polychlorinated biphenyls (PCBs)	0.0005 mg/l
Simazine	0.001 mg/l
Styrene	0.1 mg/l
2,3,7,8-TCDD (Dioxin)	5E-08 mg/l
Tetrachloroethylene	0.005 mg/l
Toluene	1 mg/l
Toxaphene	0.005 mg/l
2,4,5-TP	0.05 mg/l
Trichlorobenzene (1,2,4-)	0.009 mg/l
Trichloroethane (1,1,1-)	0.2 mg/l
Trichloroethane (1,1,2-)	0.005 mg/l
Trichloroethylene	0.005 mg/l
Vinyl Chloride	0.002 mg/l
Xylenes	10 mg/l
Secondary Maximum Permissible Contaminant Levels	
Aluminum	0.05 to 0.2 mg/l
Chloride	250 mg/l
Color	15 color units
Copper	1 mg/l
Corrosivity	non-corrosive
Fluoride	2 mg/l
Foaming Agents	0.5 mg/l
Hexachlorocyclopentadiene	0.008 mg/l
Iron	0.3 mg/l
Manganese	0.05 mg/l
Odor	3 threshold odor numbers
pH	6.5–8.5 mg/l
Silver	0.10 mg/l
Sulfate	250 mg/l
Total Dissolved Solids	500 mg/l
Zinc	5 mg/l

4. Data Interpretation and Analysis In certain instances, large quantities of water quality data are accumulated to assist in the feasibility evaluation of a water supply source. For meaningful interpretation, these data should be analyzed using standard statistical methods.

A simple statistical analysis of water quality data would include the following computations:

a) *Tests for outliers in data* (USEPA, 1989). i. Calculate the mean $\overline{X}$ and the standard deviation S_x using all observations, n. ii. Estimate the statistic, $T_n = (X_n - \overline{X})/S_x$, where X_n is the largest value suspected to be the high outlier. iii. Compare T_n with tabulated values for the sample size n (e.g., Table 8 Appendix B in USEPA, 1989). If T_n exceeds the tabulated value, X_n is considered an outlier.

b) *Estimation of overall and seasonal means, standard deviations, and skew coefficients of water quality parameters of concern.* Sample estimates of these statistical parameters can be calculated by the following equations (Barfield et al., 1981).

$$\overline{X} = \frac{\sum_{i=1}^{n} X_i}{n} \tag{6.209}$$

$$S_x = \sqrt{\frac{(\Sigma X_i^2 - n\,\overline{X}^2)}{(n-1)}} \tag{6.210}$$

$$C_s = \frac{n^2\,\Sigma X_i^3 - 3n\Sigma X_i\,\Sigma X_i^2 + 2(\Sigma X_i)^3}{n(n-1)(n-2)\,S_x^3} \tag{6.211}$$

where X_i is the data value, n is the sample size, $\overline{X}$ is the mean, S_x is the standard deviation, C_s is the skew coefficient, and all summations are from 1 to n.

c) *Trend Analysis.* It is often desirable to know if water quality parameters exhibit rising or falling trends with time, stream flow, or both. This can be accomplished by simple linear or polynomial bivariate regression between the water quality parameter and time or stream flow. If necessary, stepwise linear multivariate regression analysis may be performed with the water quality parameter as the independent variable and time and flow as the dependent variables (Spurr and Bonini, 1973; Crawford et al., 1983).

d) *Test for Adequacy of Sample Size* (Spurr and Bonini, 1973). The computational steps for this test are:
   i. Take a random sample of any convenient size and compute the sample standard deviation, S_x, as an estimate of the population standard deviation, σ.
   ii. Estimate the standard deviation of the mean, $\sigma_{\overline{X}}$, to obtain the desired precision. For instance, if the population mean is to be estimated within an error, e, at p percent confidence level, find the frequency factor, k, corresponding to p for normal distribution, then

$$k\,\sigma_{\overline{x}} = e. \tag{6.212}$$

   iii. Compute required sample size n from the equation

$$\sigma_{\overline{x}} = \frac{\sigma}{\sqrt{n}}. \tag{6.213}$$

   iv. If the sample makes up more than about 5% of the population, then use the following expression with finite population correction:

$$\sigma_{\overline{x}} = \frac{\sigma}{\sqrt{n}}\sqrt{\frac{1-n}{N}} \tag{6.214}$$

   in which N is the population size.

e) *Test for normality of data* (USEPA, 1989). Estimate the coefficient of variation:

$$CV = \frac{S_x}{\overline{X}}. \tag{6.215}$$

If CV exceeds 1.0, then normal distribution does not fit the data. To test the assumption of normality of data, the Chi-squared test may be used (USEPA, 1989).

f) *Test of difference between two population means.* The computational steps for this test are:
   i. Compute the standard errors of the means of the two samples:

$$S_{\overline{x_1}} = \frac{S_{x_1}}{\sqrt{n_1}} \text{ and } S_{\overline{x_2}} = \frac{S_{x_2}}{\sqrt{n_2}} \tag{6.216}$$

   where S_{x_1} and S_{x_2} are standard deviations of samples 1 and 2, respectively; n_1, n_2 are sizes of samples 1 and 2, respectively; and $S_{\bar{x}_1}$ and $S_{\bar{x}_2}$ are standard errors of means of samples 1 and 2, respectively.

   ii. Compute

$$S_{\bar{x}_1 - \bar{x}_2} = \sqrt{S^2_{\bar{x}_1} + S^2_{\bar{x}_2}}. \tag{6.217}$$

   iii. If e_m is the observed difference in the two means, then compute

$$k = \frac{e_m}{S_{(\bar{x}_1 - \bar{x}_2)}} \tag{6.218}$$

   iv. Read the probability level p for a frequency factor slightly less than k from tables of standard normal deviates. The observed sample difference is significant at a probability level of $2p$ (USWRC, 1981).

If the sample sizes from the two populations are small, then the following simpler method can be used, provided the two populations can be assumed to be normally distributed and their standard deviations can be assumed to be nearly equal. The computational steps of this method are (Spurr and Bonini, 1973):

i. Compute S_{po}, the polled estimate for the standard deviation.

$$S_{po} = \sqrt{\frac{(n_1 - 1)S_1^2 + (n_2 - 1)S_2^2}{n_1 + n_2 - 2}} \tag{6.219}$$

ii. Compute the sampling error of the distribution of differences in sample means, $S_{(\bar{x}_1 - \bar{x}_2)}$.

$$S_{\bar{x}_1 - \bar{x}_2} = S_{po}\sqrt{\frac{1}{n_1} + \frac{1}{n_2}} \tag{6.220}$$

iii. Compute the t-statistic

$$t = \frac{\overline{X}_1 - \overline{X}_2}{S_{\bar{x}_1 - \bar{x}_2}} \tag{6.221}$$

where n_1 and n_2 are sample sizes, S_1 and S_2 are sample standard deviations, and $\overline{X}_1$ and $\overline{X}_2$ are sample means from the two populations.

iv. Read the t value for $n_1 + n_2 - 2$ degrees of freedom from tables (e.g., Spurr and Bonini, 1973) for the desired significance (probability) level. If the computed t is greater than the tabulated value, there is significant difference between the two means.

If the standard deviations of the two populations are not nearly equal, then more sophisticated tests may be necessary (Dixon and Massey, 1969).

g) *Test for equality of population variances (Spurr and Bonini, 1973).* The computational steps of this test are:
   i. Compute the F ratio for n_1-1 and n_2-1 degrees of freedom.

$$F(n_1-1, n_2-1) = \frac{S_1^2}{S_2^2} \tag{6.222}$$

   ii. Read the corresponding F value for the desired level of significance (probability) from tables of F values.
   iii. If the computed value is less than the tabulated value, then the variances of the two populations can be assumed to be statistically equal.

C. Modeling of Runoff Quality

Selected models that simulate runoff water quality and are available in public domain (i.e., have no proprietary constraints) are listed in the following paragraphs (USEPA, 1987b; 1988a).

1. Models for Simulation of Watershed Runoff Quality The objective of these models is to estimate the quantity of surface runoff and pollutants generated by different watersheds.

a) Agricultural Runoff Model (ARM-II) (USEPA, 1978). A continuous simulation model that includes degradation, adsorption/desorption, and reaction of chemical constituents. It requires sediment loading parameters as input and is intended to be used for small agricultural areas.
b) NPS Model (USEPA, 1976). A continuous simulation model that simulates the sediment transport process as the detachment and subsequent transport of fines and requires sediment loading parameters and sediment accumulation and removal rates from impervious areas for dry periods as input. It does not simulate decay or degradation of pollutants.
c) Agricultural Chemical Transport Model (ACTMO) (USDA, 1975). A continuous simulation model that can simulate snowmelt and runoff from one or more catchments. It estimates erosion using the Universal Soil Loss Equation (USLE) and can model adsorption, degradation, or mineralization of chemicals. The model is applicable to agricultural chemicals; however, it may be possible to adapt it to other types of contaminants as well.
d) Model for Chemicals, Runoff, and Erosion from Agricultural Management Systems (CREAMS) (USDA, 1980). Estimates catchment runoff using daily or hourly precipitation data, erosion using USLE, sediment transport using transport capacity, and decay and degradation of chemicals using specified coefficients. The model is applicable to agricultural areas.
e) Storm Water Management Model (SWMM) (USEPA, 1982a). A continuous simulation model designed to simulate multiple urban catchments with conventional and conservative pollutant loadings. It simulates erosion using USLE and deposition and sediment delivery using specified deposition ratios and washoff functions.
f) Hydrologic Simulation Program—Fortran (HSPF) (USEPA, 1984b). A very complex continuous simulation model that uses the kinematic wave approximation for channel routing and can perform reservoir routing as well. The sediment transport simulation includes settling, deposition, and scouring using specified coefficients for soil detachments, sediment influx, and sediment washoff. The chemical quality constituents associated with sediment are modelled using specified washoff and scour potency factors. The model simulates adsorption/desorption processes separately for sand, silt, and clay contained in the bed material or in suspension; however, the data requirements are extensive and the calibration and implementation of the model is tedious and labor-intensive. In the absence of adequate calibration, the accuracy of the results may not be better than other relatively simpler models.

g) SEDIMOT II—Hydrology and Sedimentology Model (U. of Kentucky, 1981). An event-based model that performs flow and sediment routing through channels and sedimentation basins. Also, it simulates sediment erosion using USLE or Modified Universal Soil Loss Equation (MUSLE) for specified storms and watershed characteristics. This model can be used to simulate nonpoint source runoff contamination for the short-term (event-based) condition including the estimation of trap efficiency and concentrations of suspended sediments in reservoirs.

2. ***Models for Simulation of River Water Quality*** The function of these models is to route surface runoff and pollutants through stream channels.

a) Scour and Deposition in Rivers and Reservoirs (HEC-6) (USACE 1991). A one-dimensional, transient numerical model that simulates sediment transport in river and reservoir systems with tributary inflows. It can perform continuous simulation of aggradation and degradation in streams and deposition in reservoirs based on particle sizes of the incoming suspended material.
b) MICHRIV (USEPA, 1984a). An analytical steady-state model to simulate advective transport of a pollutant through river reaches without dispersion but with degradation. It predicts pollutant concentrations in dissolved and particulate forms as a function of distance.
c) Water Quality for River–Reservoir Systems (WQRRS) (USACE, 1978). This model has a stream hydraulics, a stream water quality, and a reservoir water quality component. It is a spatially one-dimensional and horizontally averaged continuous simulation model. The stream hydraulics component is similar to the HEC-2 or HEC-6 models of the U.S. Army Corps of Engineers (USACE, 1973; 1991). The model is very complex and difficult to calibrate. The water quality components of this model are similar to the CE-QUAL-R1 and QUAL 2E models described later.
d) Channel Transport Model (CHNTRN) (Yeh, 1982). A one-dimensional, transient model to simulate the transport of organic pollutants through rivers, lakes, estuaries, and coastal waters. The model has to be coupled with a hydrodynamic model for estimation of flow dynamics. It requires extensive input data and setup time for calibration and implementation. The model has not undergone adequate field testing and peer review at this time.
e) Chemical and Stream Quality Model (TOXIWASP) (USEPA, 1983). A three-dimensional transient model for simulating the transport of organic pollutants with comprehensive second-order decay through rivers, estuaries, and lakes. It is a very data-intensive model and requires extensive labor (150-300 hours) for setup.
f) Enhanced Stream Water Quality Model (QUAL 2E) (USEPA, 1987a). This model can simulate steady-state or dynamic transport of up to 15 chemical constituents in well-mixed streams where advection and dispersion are dominant only in the longitudinal direction. Hydraulically, it is limited to the simulation of time periods during which both stream flow and input waste load are constant.
g) Probabilistic Dilution Model (PDM) (USEPA, 1988b). This model estimates the percent of times a given concentration level may be exceeded in a stream based on statistical distribution of daily volume flow and estimates of dilution based on mass balance.
h) Sediment-Contaminant Transport (SERATRA) (Onishi and Wise, 1982). A transient, two-dimensional (longitudinal and vertical) sediment transport model with algorithms to simulate complex sediment transport and decay mechanisms for organic pollutants for rivers and lakes. The model has been tested with field data. It requires extensive input data and labor (about 750 manhours) for setup, assuming all input data are readily available.
i) Transfer One-Dimensional Degradation and Migration Model (TODAM) (Onishi and Wise, 1982). A transient, one-dimensional model to simulate the transport of contaminants in rivers and estuarine systems with second-order decay mechanisms. It requires channel and flow characteristics to be provided as data developed by another hydrodynamic model as well as data for input.
j) Water Quality Analysis Simulation Program (WASP 4) (USEPA, 1988d). This model simulates contaminant transport in streams in one, two, or three dimensions. The contaminants that can be

simulated include biochemical oxygen demand, dissolved oxygen dynamics, nutrients and eutrophication, bacterial contamination, and toxic chemicals. It is similar to QUAL 2E, described previously.

k) Water Quality Assessment Methodology (WQ AM) (Mills et al., 1982). A simplistic, one-dimensional steady-state model to estimate contaminant concentrations in rivers, lakes, and estuaries. The computations can be made by hand calculators.
l) Estuary and Stream Quality Model (WASTOX) (USEPA, 1985). A transient, three-dimensional model to simulate pollutant transport with sophisticated second-order organic decay kinetics in rivers, lakes, and estuaries. It requires flows between model compartments as input. The flows have to be estimated by a different hydrodynamic flow model. It requires extensive input data for calibration and implementation. The setup time is estimated to be 150-300 man-hours.
m) Finite-Element Transport Model (FETRA) (Onishi, 1981). A transient, two-dimensional (longitudinal and lateral) finite-element model to simulate the transport of contaminants through rivers, estuaries, coastal systems, and completely mixed lakes including second-order decay mechanisms for organic pollutants. The model has to be coupled with a hydrodynamic model to generate flow characteristics. The model has been field validated. It requires extensive data for calibration and extensive labor and time for setup and execution.

3. Models for Simulation of Reservoir Water Quality The purpose of these models is to route surface runoff and pollutants through lakes and reservoirs.

a) Simplified Lake/Stream Analyses (SLSA) (Hydro Qual, 1981). An analytical model that can be used to analyze the steady-state, zero-dimensional mixing of pollutants in Standley and Great Western Reservoirs. The model accounts for sediment suspension, exchange between the water column and bed sediments, and degradation but not for dispersion. The model is a good tool for preliminary screening.
b) A Numerical One-Dimensional Model of Reservoir Water Quality (CE-QUAL-R1) (USAWES, 1986a). A spatially one-dimensional and horizontally averaged continuous simulation model. It conceptualizes the reservoir as a vertical sequence of horizontal layers where the contaminants are uniformly distributed in each layer. The model simulates inflows, outflows, entrainment, vertical diffusion, and interactions of a number of water quality constituents. It is a fairly complex model.
c) A Numerical Two-Dimensional, Laterally-Averaged Model of Hydrodynamics and Water Quality (CE-QUAL-W2) (USAWES, 1986b). A two-dimensional, laterally-averaged, continuous simulation, finite-difference model for reservoir water quality simulation. It accounts for vertical and longitudinal diffusion, entrainment, and interaction among various water quality constituents. The model is more complex and sophisticated and much more difficult to use than CE-QUAL-R1, described previously.
d) Time-Dependent, Three-Dimensional, Variable Density Hydrodynamic Model (USEPA, 1988c). This model estimates the movement of pollutants in thermal discharges in harbors, bays, lake basins, entire lakes, estuaries, and marine coastal areas. In particular, it estimates velocities, temperatures, and salinity.
e) Convective-Dispersion in Perennial Streams (Prakash, 1977). This program analyzes the spatial dispersion and dilution of effluents discharged continuously in a stream through vertical line or plane and horizontal line or plane sources. The computational procedure is based on analytical solutions using the method of images to account for the boundaries provided by the bed, banks, and water surface. The channel cross-section is approximated by an equivalent rectangular section and the flow is assumed to be steady and uniform. The flow field has to be specified by the user. Dispersion or diffusion is computed in three dimensions. Radioactive decay can be simulated by specifying an appropriate decay factor. An algorithm to model nonsteady-state dispersion is also

included. The contribution of adsorption by sediment particles can be accounted for by specifying a retardation factor based on a linear equilibrium isotherm.

VIII. REFERENCES

Akan, A.O. and Yen, B.C. (1981). "Diffusion wave flood routing in channel networks." J. Hydr. Div., ASCE, 107(6), New York, NY, 719-732.

Ambrose, R.B., Wool, T.A., Connolly, J.P., and Schanz, R.W. (1988). "WASP4, a hydrodynamic and water quality model—model theory, user's manual, and programmer's guide." US Environmental Protection Agency, Athens, GA EPA/600/3-87-039.

American Iron and Steel Institute (AISI) (1971). *Handbook of steel drainage & highway construction products*, Washington, DC.

Ang, A.H.S. and Tang, W.H. (1975). *Probability concepts in engineering planning and design, Vol. I–Basic principles*, John Wiley and Sons, New York, NY.

Anis, A.A. and Lloyd, E.H. (1953). "On the range of partial sums of a finite number of independent random variables," Biometrika, Vol. 40, 35-42.

Askew, A.J., Yeh, W.W-G. and Hall, W.A. (1971). "Use of Monte Carlo techniques in the design and operation of a multi-purpose reservoir," Water Resources Research, Vol. 7: 819-826.

Bailey, E.G. and Ray, H.A. (1966). "Definition of stage–discharge relation in natural channels by step-backwater analysis." U.S. Geological Survey Water-Supply Paper 1869-A, 24.

Barfield, B.J. Warner, R.C., and Haan, C.T. (1981). *Applied hydrology and sedimentology for disturbed areas*, Oklahoma Technical Press, Stillwater, Oklahoma.

Barnes, F.B. (1954). "Storage required for a city water supply," Journal of the Institution of Engineers, Australia, 26:198.

Barnes, H.H. Jr. (1967). "Roughness characteristics of natural channels," U.S. Geological Survey Water-Supply Paper 1849, 213.

Beard, L.R. (1963). "Estimating long-term storage requirements and firm yield of rivers," IASH, Pub. #63:151-166.

Bedient, P.B. and Huber, W.C. (1988). *Hydrology and floodplain analysis*, Addison-Wesley, Reading, MA.

Benson, M.A. and Dalrymple, T. (1967). "General field and office procedures for indirect discharge measurements," *U.S. Geological Survey Techniques of Water Resources Investigations*, Book 3, Chapter Al, 30.

Bodhaine, G.L. (1968). "Measurement of peak discharge at culverts by indirect methods," *U.S. Geological Survey Techniques of Water Resources Investigations*, Book 3, Chapter A3, 60.

Bos, M.G., Replogle, J.A., and Clemmens, A.J. (1991). *Flow measuring flumes for open channel systems*. American Society of Agricultural Engineers, St. Joseph, MI.

Boyd, G. and Gardiner, N. (1990). "Urban stormwater, an overview for municipalities," Public Works Journal Corp., Ridgewood, New Jersey, Public Works, Vol. 121, No. 13.

Brater, E.F. and King, H.W. (1976). *Handbook of hydraulics*, McGraw-Hill Book Co., New York, NY.

Carter, R.W. and Anderson, I.E. (1963). "Accuracy of current-meter measurements." American Society of Civil Engineers Journal, v. 89, no. HY4, 105-115.

Chang, H.H. (1988). *Fluvial processes in river engineering*, John Wiley & Sons, New York, NY.

Chaudhry, Y.M. and Contractor, D.N. (1973). "Application of the implicit method to surges in open channels." Water Resource Res. 9(6). 1605-1612.

Chen, C.L. and Armbruster, J.T. (1980). "Dam–break wave model: Formulation and verification." J. Hydr. Div., ASCE, 106(5), 747-767.

Chow, Ven Te (1959). *Open channel hydraulics*. McGraw-Hill Book Co., New York, NY. 680.

Chow, Ven Te (1964). *Handbook of applied hydrology*. McGraw-Hill Book Co., New York, NY. 1454.

Chow, V.T., Maidment, D.R., and Mays, L.W. (1988). *Applied hydrology*, McGraw-Hill Book Co., New York, NY.

Clark, J.W., Viessman, W., and Hammer, M.J. (1977). *Water supply and pollution control*, Harper and Row Publishers, New York, NY.

Clemmens, A.J., Replogle, J.A., and Bos, M.G. (1987). *Flume: A computer model for estimating flow through long-throated measuring flumes*. USDA. Agricultural Research Service, ARS-57. 64.

Clemmens, A.J., Replogle, J.A., and Bos, M.G. (1988). "Broad-crested weirs and long-throated flumes for open channel flow measurement." *Civil Engineering Practice, 2/Hydraulics/Mechanics*, Chapter 8. Technomic Publishing Co., Inc. Lancaster, PA. 137-187.

Collins, D.L. (1977). "Computation of records of stream flow at control structures." U.S. Geological Survey Techniques of Water-Resources Investigations 77-8, 57.

Cooley, R.L. and Moin, S.A. (1976). "Finite element solution of Saint–Venant equations." J. Hydr. Div., ASCE, 102(6), 759-774.

Crawford, C.J., Slack, J.R. and Hirsch, R.M. (1983). "Nonparametric tests for trends in water quality data using the statistical analysis system," USGS, Open-File Report 83-550, Reston, VA.

Crawford, N.H. and Linsley, R.K. Jr. (1966). "Digital simulation in hydrology," Stanford Watershed Model IV, Tech. Rept. 39, Dept. of Civil Engineering, Stanford University, Stanford, CA.

Creager, W.P., Justin, J.D., and Hinds, J. (1945). *Engineering for dams*, John Wiley & Sons, Inc., New York, NY.

Croley, T.E., II (1974). "Sequential deterministic optimization in reservoir operation," J. Hydr. Div., ASCE, 100 (H73).

Dalrymple, T. and Benson, M.A. (1967). "Measurement of peak discharge by the slope–area method." *U.S. Geological Survey Techniques of Water-Resources Investigations*, Book 3, Chapter A-2, 12.

Davidian, Jacob (1984). "Computation of water-surface profiles in open channels." *U.S. Geological Survey Techniques of Water Resources Investigations*, Book 3, Chapter A-15, 48.

Dixon, W.J. and Massey, F.J. (1969). *Introduction to statistical analysis*, McGraw-Hill Book Company, New York, NY.

Dooge, J.C.I. (1973). "Linear theory of hydrologic systems," Technical Bulletin No. 1468, Agricultural Research Service, U.S. Department of Agriculture, Washington, DC.

Dronkers, J.J. (1965). "Tidal computations for rivers, coastal areas and seas." J. Hydr. Div., ASCE, 95(1).

Eagleson, P.S. 1970. *Dynamic hydrology*. McGraw-Hill Book Co. Inc., New York, NY.

Eshleman, K.N. and Hemond, H.F. (1985). "The role of organic acids in the acid-base status of surface waters at Bickford Watershed, Massachussets," Water Resources Research 21:1503-1510.

Fair, G.M., Geyer, J.C., and Okun, D.A. (1966). *Water and wastewater engineering*, Vol. 1, Chapter 8, Storage and Runoff control, John Wiley & Sons, New York, NY.

Federal Register (1990). Environmental Protection Agency, 40 CFR Parts 122, 123, and 124, November 16, 1990.

Feller, W. (1951). "The asymptotic distribution of the range of sums of independent random variables." Annals of Mathematical Statistics, 22: 427-432.

Fiering, M.B. (1961). "Queuing theory and simulation in reservoir design." J. Hyd. Div., ASCE, 87 (HY6): 39-69.

Fontane, D.G. and Flug, M. (1991). "Introduction to multi-criterion methods and selected software," *Water Resources Planning and Management and Urban Water Resources* Ed. by J.L. Anderson, ASCE, New York, NY.

Fread, D.L. (1973a). "Effect of time step size in implicit dynamic routing." Water Resour. Bulletin, AWRA, 9(2), 338-351.

Fread, D.L. (1973b). "Technique for implicit dynamic routing in rivers with major tributaries." Water Resour. Res., 9(4) 918-926.

Fread, D.L. (1982). "Operational dynamic wave model—DWOPER." National Weather Service, NOAA, Silver Spring, MD.

Fread, D.L. (1984). "DAMBRK: The NWS Dam–break flood forecasting model." National Weather Service, NOAA, Silver Spring, MD.

French, R.H. (1987). *Hydraulic processes on alluvial fans*. Elsevier Science Publishers, Amsterdam, The Netherlands.

French, R.H. (1992). "Preferred directions of flow on alluvial fans." J. Hyd. Eng., ASCE, 118(7), 1002-1013.

Galloway, J.N., Cowling, E.B., Gorham, E. and McFee, W.W. (1978). "A national program for assessing the problem of atmospheric deposition (acid rain)." A Report to the Council on Environmental Quality, National Atmospheric Deposition Program NC-141.

Garbrecht, J. and Brunner, G. (1991). "Hydrologic channel-flow routing for compound sections." J. Hydr. Engrg., ASCE, 117(5) 629-642.

Garde, R.J. and Ranga Raju, K.G. (1978). *Mechanics of sediment transportation and alluvial stream problems*, Wiley Eastern, Ltd.

Gupta, R.S., (1989). *Hydrogeology and hydraulic systems*, Prentice Hall, Englewood, Cliffs, N.J.

Garrison, J.M., Granju, J.P., and Price, J.T. (1969). "Unsteady flow simulation in rivers and reservoirs." J. Hydr. Div., ASCE, 95(5), 1559-1576.

Haimes, Y.T. and Allee, D., eds. (1984). *Multi-objective analysis in water resources*, ASCE, New York, NY.

Hall, M.J. (1984). *Urban hydrology*, Elsevier Applied Science Publishers, London, England.

Hawkins, R.H., Hjelmfelt, A.T., and Zevenkergen, A.W. (1985). "Runoff probability, storm depth, and curve numbers." Journal of the Irrigation and Drainage Division, ASCE, Vol. 111, No. 4, December 1985.

Hazen, A. (1914). "Storage to be provided in impounding reservoirs for municipal water supply." Transactions, ASCE, 77: 1539.

Herschy, Reginald W. (1985). *Stream flow measurement*. Elsevier Applied Science Publishers, London and New York. 553.

Holton, H.N, Stitner, G.J., Hensen, W.H., and Lopez, N.C. (1975). "Revised model of watershed hydrology," Technical Bulletin No. 1518, USDAHL-74, Agricultural Research Service, U. S. Department of Agriculture, Washington, DC.

Hornberger, G.M., Ebert, J., and Remsen, I. (1970). "Numerical solutions of the Boussinesq equation for stream aquifer interaction." Water Resour. Res., 6(2), 601-608.

Horton, R.E. (1933). "The role of infiltration in the hydrologic cycle," Trans., (14), AGU, 446-460.

Howells, G. (1990). *Acid rain and acid waters*, Prentice-Hall, Inc., Englewood Cliffs, NJ, 215.

Hromadka II, T.V., Berenbrock, C.E., Freckleton, J.R., and Guymon, G.L. (1985). "A two-dimensional diffusion dam-break model." Advances in Water Resources, Vol. 8, 7-14.

Huber, W.C., Heaney, J.P., Nix, S.J., Dickinson, R.E., and Polmann, D.J. (1981). *Storm water management model user's manual*, Version III, 600/2-84-109a, E.P.A., Athens, GA.

Hulsing, Harry (1967). "Measurement of peak discharge at dams by indirect methods." *U.S. Geological Survey Techniques of Water Resources Investigations*, Book 3, Chapter A5, 29.

Hurst, H.E. (1951). "Long term storage capacity of reservoirs," Transactions, ASCE, 116: 770-799.

Hurst, H.E. (1956). "Methods of using long term storage in reservoirs," Proceedings of the Institution of Civil Engineers, Paper 6059, 5: 519.

Hydro Qual, Inc. (1981). "Analysis of fate of chemicals in receiving waters, phase 1." Prepared for Chemical Manufacturers Association, Washington, DC.

Johnson, S.A., Stedinger, J.R., and Stachus, K. (1991). "Heuristic operating policies for reservoir system simulation," Water Resources Research, Vol. 27(5).

Julian, P.R. (1961). "A study of the statistical predictability of stream runoff in the Upper Colorado River Basin." In Part II of *Past and Probable Variations in Streamflow in the Upper Colorado River*, Univ. of Colorado, Boulder, CO.

Kelman, J., Stedinger, J.R., Cooper, E., Hsu, E., and Yuan, S.Q. (1990). "Sampling stochastic dynamic programming applied to reservoir operation," Water Resour. Res., Vol. 26(3).

Kennedy, E.J. (1984). "Discharge ratings at gaging stations." *U.S. Geological Survey Techniques of Water-Resources Investigations*, Book 3, Chapter A-10, 59.

Kibler, D.F. and Woolhiser, D.A. (1970). "The kinematic cascade as a hydrologic model." Hydrol. Paper No. 39, Colorado State University, Fort Collins, CO.

Kilpatrick, F.A. and Cobb, E.D. (1985). "Measurement of discharge using tracers." *U.S. Geological Survey Water-Resources Investigations*, Book 3, Chapter A-16, 52.

Kirby, W. (1972). "Computer oriented Wilson–Hilferty transformation that preserves the first three moments and the lower bound of the Pearson Type 3 distribution, Water Resources Research, Vol. 8(5), 1251

Klemes, V. (1974). "The Hurst phenomenon: A puzzle?" Water Resources Research, 10 (4): 675-688.

Kottegoda, N.T. (1980). *Stochastic water resources technology*, John Wiley & Sons, New York, NY: 384.

Laenen, (1985). "Acoustic velocity meter systems." *U.S. Geological Survey Techniques of Water-Resources Investigations*, Book 3, Chapter A-17, 38.

Langbein, W.B. (1958). "Queuing theory and water storage." Proceedings, J. Hyd. Div., ASCE, 84: Paper 1811.

Lawrance A. and Kottegoda, N. T. (1977). "Stochastic modelling of river flow time series," Journal of the Royal Statistical Society, Ser. A, Vol 140, Part 1, 1-47.

Lazaro, T.R. (1991). *Urban hydrology*, Technomic Publishing Co., Lancaster, PA.

Lee, L.T. and Essex, T.L. (1983). "Urban headwater flooding damage potential." J. Hydr. Div., ASCE, 109(4), 519-535.

Liggett J.A. and Cunge, J.A. (1975). "Numerical methods of solution of the unsteady flow equations." *Unsteady Flow in Open Channels*, K. Mahmood and V. Yevjevich, eds., Water Resources Public actions, Fort Collins, CO. Vol. 1.

Linsley, R.K. and Franzini, J.B. (1972). *Water resources engineering*. McGraw-Hill Book Co., New York, NY.

Linsley, R.K., Kohler, M.A., and Paulhus, J.L.H. (1958). *Hydrology for engineers*, McGraw-Hill Book Co., New York, NY.

Lloyd, E.H. (1963). "A probability theory of reservoirs with serially correlated inputs." Journal of Hydrology, 1 (2): 99-128.

Loucks, D.P. (1968). "Computer models for reservoir regulation." Journal San. Eng. Div., ASCE, 94(SA4): 657-669.

Mandelbrot, B. and Wallis, J. R. (1968). "Noah, Joseph and operational hydrology," Water Resources Research, 4(5): 909-918

Marino, M.A. and Loaiciga, H.A. (1985). "Dynamic model for multi-reservoir operation," Water Resources Research, Vol. 21(10).

Matthai, H.F. (1967). *Measurement of peak discharge at width contractions by indirect methods*. U.S. Geological Survey Techniques of Water-Resources Investigations, Book 3, Chapter A-4, 44.

McCuen, R.M., Wong, S.L., and Rawls, W.J. (1984). "Estimating urban time of concentration," J. Hydr. Engr., ASCE, Vol. 110, No. 7, July, 1984.

McMahon, T.A. and Mein, R.G. (1978). *Reservoir capacity and yield*. Elsevier Scientific Pub. Co., Amsterdam, The Netherlands: 213.

Miller, W.A. and Cunge, J.A. (1975). "Simplified equations of unsteady flow." *Unsteady Flow in Open Channels*, K. Mahmood and V. Yevjevich, eds., Water Resources Publications, Fort Collins, CO, Vol. 1.

Mills, W.B., et al. (1982). *Water quality assessment methodology*, WQAM, ORD Publications, US EPA, Cincinnati, Ohio.

Moran, P.A.P. (1959). *The theory of storage*. Methuen, London.

Musgrave, G.W. and Holton, H.N. (1964). "Infiltration," in *Handbook of Applied Hydrology*, V.T. Chow, ed., McGraw-Hill Book Co., New York, NY.

Nash, J.E. (1957). *The Form of the instantaneous unit hydrograph*, International Association of Scientific Hydrology, Publication 45, Vol. 3.

National Weather Service (NWS) (1986). *National Weather Service river forecast system (NWSRFS) model, User's Manual*, Hydrologic Research Laboratory, NOAA, Silver Spring, Maryland.

Nelson, J.D., Abt, S.R., Volpe, R.L., van Zyl, D., Hinkle, N.E., and Staub, W.P. (1986). *Methodologies for evaluating long-term stabilization designs of uranium mill tailings impoundments*, NUREG/CR-4620, ORNL/TM-10067, U.S. Nuclear Reg. Com., Washington, DC.

O'Connell, P.E. (1974). "A simple stochastic modeling of Hurst Law," Pub. No. 100, International Association of the Hydrologic Sciences, 169-187.

Onishi, Y. (1981). *Finite-element transport model, FETRA*, Battelle Pacific Northwest Laboratories and ICF Northwest, Richland, WA.

Onishi, Y. and Wise, S.E. (1982). *Sediment-contaminant transport*, SERATRA, ORD Publications, Center for Environmental Research Information, US EPA, Cincinnati, OH.

Ponce, V.M. (1989). *Engineering hydrology, principles and practice*, Prentice Hall, Englewood Cliffs, NJ.

Ponce, V.M. (1991). "The kinematic wave controversy." J. Hydr. Eng., ASCE, 117(4), 511-525.

Ponce, V.M., Li, R.M., and Simons, D.B. (1978). "Applicability of kinematic and diffusion models." J. Hydr. Div., ASCE, 104(3), 353-360.

Prakash, A. (1977). "Convective–dispersion in perennial streams," Journal of the Environmental Eng. Division, ASCE, Vol. 103, No. EE2, 321-340.

Prakash, A. (1987). "Current state of hydrologic modeling and its limitations." In *Flood Hydrology*, ed. V.P. Singh, D. Reidel Publishing Company, Dordrecht, Holland.

Quimpo, R.G. (1968). "Stochastic analysis of daily river flows," Journal of the Hydraulics Div., ASCE, 94(HY1): 43-57.

Quimpo, R.G., Alejandrino, A.A., and McNally, T.A. (1983). "Regionalized flow duration for the Philippines," Journal of Water Resources Planning and Management, ASCE, Vol. 109(4):320-330.

Rantz, S.E. (1982). "Measurement and computation of stream flow: v. 1, measurement of stage and discharge; v. 2, computation of discharge," U.S. Geological Survey Water-Supply Paper 2175, 631.

Replogle, John A. (1968). "Target meters for velocity and discharge measurements in open channels." Trans. ASAE 11(6):854-856, 862.

Replogle, J.A., Clemmens, A.J., and Bos, M.G. (1991). "Measuring irrigation water." Chapter 10. In: Hoffman, G., T.A. Howell, and K.H. Solomon (eds.). *Management of Farm Irrigation Systems*. Am Soc of Agric Engrs, Monograph. 1991.

Riggs, H.C. (1972). "Low-flow investigations." *U.S. Geological Survey Techniques of Water-Resources Investigations*, Book 4, Chapter A-1, 18.

Riggs, H.C. (1976). "A simplified slope–area method for estimating flood discharges in natural channels." U.S. Geological Survey Journal of Research, 4(3), 285-291.

Riggs, H.C. (1985). *Stream flow characteristics*. Elsevier Science Publishers, Amsterdam, 243.

Rippl, W. (1983). "Capacity of storage reservoirs for water supply." Minutes of Proceedings, Institution of Civil Engineers, Vol. 71: 270-278.

Roesch, S.E., Clark, L.J., and Bray, M.M. (1979). *User's manual for the dynamic (Potomac) estuary model*. U.S. Environmental Protection Agency, Annapolis, MD. EPA-903/9-79-001.

Sakkas, J.G. and Strelkoff, T. (1973). "Dam-break flood in a prismatic dry channel." J. Hydr. Div., ASCE, 99(12), 2195-2216.

Salas, J.D. (1972). "Range analysis for storage problems of periodic–stochastic processes," Hydrology paper No. 57, Colorado State University, Ft. Collins, CO.

Salas, J.D. (1974). "Range of cumulative sums, II. Application to storage capacity of reservoirs," J. of Hydrology, 23: 329-339.

Salas, J.D. and Fernandez, B. (1989). "Model for data generation of hydrologic time series: univariate techniques," Proceedings, NATO Advanced Study Institute on Stochastic Hydrology and Water Resources Systems, Peniscola, Spain.

Scott, R.W.W. (1982). *Developments in flow measurements—1*, Applied Science Publishers, London and New Jersey, 333.

Searcy, J.K. (1959). "Flow duration curves, manual of hydrology: Part 2. Low-flow techniques," U.S. Geological Survey Water Supply Paper 1542-A, U.S. Govt. Printing Office, Washington, DC.

Simons, D.B. and Senturk, F. (1992). *Sediment transport technology*. Water Resources Publications, Fort Collins, CO.

Singh, V.P. (1988). *Hydrologic systems, rainfall-runoff modeling*, Vol. 1, Prentice Hall, Englewood Cliffs, NJ.

Smoot, G.F. and Novak, C.E. (1969). "Measurement of discharge by the moving boat method." *U.S. Geological Survey Techniques of Water Resources Investigation*, Book 3, Chapter A-11, 22.

Spurr, W.A. and Bonini, C.P. (1973). *Statistical analysis for business decisions*, Richard D. Irwin, Inc., Homewood, IL.

Stall, J.B. (1962). "Reservoir mass analysis by a low-flow series." J. of the Sanitary Engineering Div., ASCE, 88(SA5): 21-40.

Stall, J.B. (1964). "Low flow of Illinois streams for impounding reservoir design." Bull. 51, Illinois State Water Survey, Urbana. IL.

Standard Methods (1985). *Standard methods for the examination of water and wastewater (16th Edition)*, American Public Health Association, Washington, DC.

Streeter, V.L. (1971). *Fluid mechanics*. McGraw-Hill Book Company, New York, NY.

Strelkoff, T. (1970). "Numerical solution of the Saint-Venant equations." J. Hydr. Div., ASCE, 96(1) 223-252.

Sudler, C.H. (1927). "Storage required for regulation of stream flow." Transactions, ASCE, 61:622.

Terstriep, M.L. and Stall, J.B. (1974). "The Illinois urban drainage area simulator—ILLUDAS." Illinois State Water Survey Bulletin 58.

Tholin, A.L. and Keifer, C.J. (1959). "The hydrology of urban runoff," J. Sanitary Eng., (85:SA2), ASCE, 60-106.

Thomas, H.A. and Burden, R.P. (1963). *Operations research in water quality management*, Harvard Water Group, Cambridge, MA.

Thomas, H.A. and Fiering, M.B. (1962). "Mathematical synthesis of stream flow sequences for the analysis of river basins by simulation." In *Design Of Water Resources Systems*, Ed. by Maass, et al., Harvard University Press, Cambridge, 459-493.

U.S. Army Corps of Engineers (USACE) (1960). "Routing of floods through river channels." *Engineering and Design*. EM1110-2-1408. US Govt. Printing Office, Washington, DC.

U.S. Army Corps of Engineers (USACE) (1973). HEC-2, *Water surface profiles*, The Hydrologic Engineering Center, Davis, CA.

U.S. Army Corps of Engineers (USACE) (1975). *Hydrologic engineering methods for water resources development, Vol. 8 Reservoir Yield*, The Hydrologic Engineering Center, Davis, CA.

U.S. Army Corps of Engineers (USACE) (1978). *Water quality for river–reservoir systems*, WQRRS, The Hydrologic Engineering Center, Davis, CA.

U.S. Army Corps of Engineers (USACE) (1987). "Flood hydrograph package." *HEC-1 User's Manual*, The Hydrologic Engineering Center. Davis, CA.

U.S. Army Corps of Engineers (USACE) (1991). "Scour and deposition in rivers and reservoirs." *HEC-6 User's Manual*, The Hydrologic Engineering Center. Davis, CA.

U.S. Army Waterways Experiment Station (USAWES) (1986a). *A numerical one-dimensional model of reservoir water quality*, CE-QUAL-R1, Vicksburg, MS.

U.S. Army Waterways Experiment Station (USAWES) (1986b). *A numerical two-dimensional laterally-averaged model of hydrodynamics and water quality*, CE-QUAL-W2, Vicksburg, MS.

U.S. Bureau of Reclamation (USBR) (1977, 1987). *Design of small dams*, Denver, CO.

U.S. Department of Agriculture (USDA) (1972). *National engineering handbook, Section 4*, Hydrology. Soil Conservation Service, Washington, DC.

U.S. Department of Agriculture (USDA) (1975). "An agricultural chemical transport model," ACTMO, Publication No. ARS-H-3, Agriculture Research Service, Hyattsville, MD.

U.S. Department of Agriculture (USDA) (1980). "A field scale model for chemicals, runoff, and erosion from agricultural management systems, CREAMS," Conservation Report No. 26, Washington, DC.

U.S. Department of Agriculture (USDA). (1982). "Technical Release 20 (TR-20). Computer program for project formulation—hydrology." Draft 1982, Errata 1984.

U.S. Department of Agriculture (USDA) (1985). *National engineering handbook, Section 4*, Hydrology, Washington DC.

U.S. Department of Agriculture, Soil Conservation Service (1986). "Urban hydrology for small watersheds," Technical Release No. 55 (TR-55), Washington, DC.

U.S. Department of Transportation (USDOT) (1972). "Capacity charts for the hydraulic design of highway culverts," Hydraulic Engineering Circular No. 10, Federal Highway Administration, Washington, DC.

U.S. Department of Transportation (USDOT) (1973). "Hydraulics of bridge waterways." Hydraulic Design Series No. 1, Federal Highway Administration, Washington, DC.

U.S. Environmental Protection Agency (USEPA) (1976). "Modeling pesticides and nutrients on agricultural lands," NPS, EPA-600/2-76-043, Environmental Research Laboratory, Athens, GA.

U.S. Environmental Protection Agency (USEPA) (1978). "Agricultural runoff management (ARM) model," Versions I and II, EPA-600/3-78-080, Environmental Research Laboratory, Athens, GA.

U.S. Environmental Protection Agency (USEPA) (1982). *Storm water management user's manual*, Version III, SWMM, EPA-600/2-84-109A, Cincinnati, OH.

U.S. Environmental Protection Agency (USEPA) (1983). *User's manual for the chemical transport and fate model (TOXI-WASP)*, Version 1, EPA-600/3-83-005, Athens, GA.

U.S. Environmental Protection Agency (USEPA) (1984a). *Technical guidance manual for performing work load allocation*, EPA-440/4-84-022, Washington, DC.

U.S. Environmental Protection Agency (USEPA) (1984b). *Hydrologic simulation program—Fortran, HSPF, User's Manual for Release 8.0*, EPA-600/3-84-006, Environmental Research Laboratory, Athens, GA.

U.S. Environmental Protection Agency (USEPA) (1985). *WASTOX, A framework for modeling the fate of toxic chemicals in aquatic environments, Part 1, Exposure Concentration* (1984) and *Part 2, Food Chain* (1985), Connolly, J.P. and R.V. Thomann, Gulf Breeze, FL and Duluth, MN.

U.S. Environmental Protection Agency (USEPA) (1987a). *The enhanced stream water quality models, QUAL2E and QUAL2E - UNCAS, Documentation and User Model*, EPA/600/3-87/007, Athens, GA.

U.S. Environmental Protection Agency (USEPA) (1987b). *Selection criteria for mathematical models in exposure assessments*, EPA/600/8-87/042, Washington, DC.

U.S. Environmental Protection Agency (USEPA) (1988a). *Superfund exposure assessment manual*, EPA/540/1-88/001, Washington, DC.

U.S. Environmental Protection Agency (USEPA) (1988b). "Probabilistic dilution model (PDM)." In *Risk Assessment Guidelines and Information Directory*, Government Institutes, Inc.

U.S. Environmental Protection Agency (USEPA) (1988c). "Time-dependent, three-dimensional, variable density hydrodynamic model." In *Risk Assessment Guidelines and Information Directory*, Government Institutes, Inc.

U.S. Environmental Protection Agency (USEPA) (1988d). *WASP4, A hydrodynamic and water quality model, Model Theory, User's Manual, and Programmer's Guide*, EPA/600/3-86-034, Athens, GA.

U.S. Environmental Protection Agency (USEPA) (1989). *Statistical analysis of ground water monitoring data at RCRA facilities—Interim Final Guidance*, PB89-151047, Washington, DC.

U.S. Environmental Protection Agency (USEPA) (1991a). *Guidance manual for the preparation of NPDES permit applications for stormwater discharges associated with industrial activity*, Office of Water Enforcement and Permits, April 1991, Washington, DC.

U.S. Environmental Protection Agency (USEPA) (1991b). *Drinking water regulations and health advisories*, Office of Water, Safe Drinking Water Hotline, Washington, DC.

U.S. Water Resources Council (USWRC) (1981). "Guidelines for determining flood flow frequency," Bulletin #17B, Washington, DC.

University of Kentucky (1981). "Sedimot II hydrology and sedimentology model," Department of Agricultural Engineering, Lexington, KY.

Vishawakarma, C.S., Reddy, A.S., and Shastry, P.G. (1991). "Studies on Nash hydrograph model with reference to small watersheds in Krishma–Godavari River Basins," Journal of the Institution of Engineers (India), Vol. 71, Pt CV5.

Wang, R.Y., Wu, I.P., and Lau, L.S. (1970). "Instantaneous unit hydrograph analysis of Hawaiian small watersheds," Water Resources Research Center, Technical Report No. 42, August, 1970, University of Hawaii, Honolulu, HI.

Wallis, J.R. and Matalas, N. (1972). "Sensitivity of reservoir design to the generating mechanism of the inflows," Water Resources Research, 8(3): 634-641.

Woolhiser, D.A. and Liggett, J.A. (1967). "Unsteady one dimensional flow over a plane: The rising hydrograph," Water Resources Res., (3:3), 753-771.

Yang, C.T. and Schenggan, Wan (1991). "Comparisons of selected bed–material load formulas," J. Hydr. Engr., ASCE, Vol. 117, No. 8, pp 973-989.

Yeh, G.T. (1982). "Channel transport model, CHNTRN," Environmental Science Division, Oak Ridge National Laboratory, Oak Ridge, TN.

Yeh, W.W-G. (1985). "Reservoir management and operation models: A state-of-the-art review." Water Resources Research, 21:1797-1818.

Yevjevich, V.M. (1972). *Stochastic processes in hydrology*. Water Resources Publications, Fort Collins, CO, 276.

Young, G.K. and Pisano, W.C. (1968). "Operational hydrology using residuals." J. Hydr. Div., ASCE, 94(HY4): 909-923.

Zienkiewicz, O.C. (1971). *The finite element method in engineering science*. McGraw-Hill Publishing Co. Ltd., London, England.

IX. NOTATION

a	a constant
c	Izzard's roughness
C	Chezy roughness
C_L	resistance coefficient
D_e	surface detention
f	Darcy–Weisbach roughness
f	infiltration rate
F_r	Froude number
g	gravitational constant
i	rainfall intensity
k	constant
L	length of flow plane
L_1	laminar length of flow plane
m	a constant
n	Manning's roughness
P_2	2-yr 24-hr rainfall
q	unit discharge
q_e	equilibrium unit discharge
q_i	flow rate when grab sample *i* is taken
R_e	Reynold's number
S_f	friction slope
S_o	slope
t	time
t_e	time to equilibrium
u	mean velocity
V	volume of composite sample

v_i volume of grab sample i

x distance

x Muskingum constant

y depth

ν kinematic viscosity

θ slope

X. GLOSSARY

Antecedent Moisture—The term antecedent moisture refers to a watershed's level of soil moisture prevailing at the start of a storm event of interest. The antecedent moisture, or antecedent moisture condition determines to a large extent the amount of surface runoff. Watersheds with low antecedent moisture are not conducive to high runoff response. Conversely, watersheds with high antecedent moisture are likely to produce large quantities of surface runoff. Therefore, surface runoff is strongly positively correlated with antecedent moisture.

Baseflow—Baseflow is that portion of stream flow originating in indirect runoff, that is, runoff that has reached the stream or river by passing first through the underlying aquifer, rather than by flowing directly on the ground surface. Baseflow effectively drains the neighboring shallow ground water reservoirs, eventually leading to their depletion in the absence of substantial recharge. Baseflow is characteristically a very slow process, with strong runoff diffusion and very little variability. The presence of baseflow throughout the year (i.e., dry-weather stream flow) is an indication of a humid climate and a shallow ground water table with fast recharge potential. Conversely, the absence of baseflow (e.g., in ephemeral streams) indicates an arid or semiarid climate and a deep ground water table with slow recharge potential.

Channel transmission Losses—This refers to the runoff losses, by percolation to deep ground water reservoirs, typical of flood flows in transit in ephemeral channels. In arid or semiarid climates, ephemeral streams and rivers have the capability to lose substantial quantities of surface runoff during flood events. These losses have the effect of reducing flood peaks while in transit in the ephemeral channels, at the same time leading to ground water replenishment.

Direct Runoff—Direct runoff is that portion of runoff that flows on the surface of the earth, first spread thinly overland in response to effective precipitation (i.e., total precipitation minus hydrologic abstractions), and eventually concentrated in rills, gullies, streams, and rivers. Direct runoff has the capability to produce large flow concentrations in a relatively short period of time. Therefore, direct runoff is largely responsible for flood flows.

Effluent Stream—An effluent stream is one that feeds from ground water during low-flow periods, i.e., a stream that drains ground water reservoirs in order to sustain its baseflow. Accordingly, effluent streams are usually of a perennial nature.

Ephemeral Stream—An ephemeral stream is one that remains dry during most of the year, flowing only during a few days or weeks in response to direct runoff (i.e., effective precipitation). In many areas of the semiarid Southwestern United States, ephemeral streams or arroyos have measurable flow only a few days every year, during and immediately following a major storm. Ephemeral streams do not intercept ground water flow and, therefore, have no baseflow.

Exfiltration—Exfiltration refers to the discharge from subsurface water (interflow and/or ground water flow) to surface water, the latter being either watercourses or water bodies, such as streams, rivers, reservoirs, lakes, or oceans. Exfiltration, the opposite of infiltration, takes place typically in effluent streams.

Flow-duration Curve—A flow-duration curve is a practical way to evaluate daily stream flow variability. It is a plot of Q vs T, Q (ordinates) is the flow level exceeded T% (abscissa) of the time. Such a curve is used to establish the permanence of characteristic low-flow levels, e.g., the flow level Q_{90}, to be exceeded 90% of the time.

Flow-mass Curve—A flow-mass curve is a practical way to evaluate seasonal stream flow variability. It is a plot of cumulative flow volume V (in the ordinates, in cubic meters or cubic feet) vs time T (in the abscissas). Applications of the flow-mass curve are in reservoir design and operation, including the determination of reservoir capacity and the establishment of operating rules for storage reservoirs.

Ground Water Flow—Ground water flow takes place below the ground water table in the form of saturated flow through alluvial deposits and other water-bearing formations located beneath the soil mantle. Ground water flow includes the portion of infiltrated volume that has reached the water table slowly, by percolation, from the overlying soils. Ground water flow is a characteristically slow process. The average residence time of subsurface and ground water has been estimated to vary between 2 weeks and 10,000 years. Like surface runoff, ground water flow is a continuous process, with water constantly moving to lower elevations (or to zones of lower potential) and eventually making its way to the sea. Where the ground water table is close to the surface, ground water flow may be intercepted by streams and rivers and discharge into them.

Hydrograph—A hydrograph, or stream flow hydrograph, is a plot of flow or discharge vs time at a cross-section, station, or gaging station of a stream or river. Time is plotted in the abscissa, and discharge in the ordinates (See also Stream Flow Hydrograph).

Indirect Runoff—Indirect runoff is that portion of surface runoff that has flowed first under the ground surface, either as unsaturated flow (interflow) or saturated flow (ground water flow). Unlike direct runoff, indirect runoff is subject to substantial runoff diffusion, and therefore, it generally lacks the capability to produce flood flows.

Infiltration—Infiltration is the process by which precipitation and surface water seeps into the ground below the land surface. Once below the surface, the water moves either laterally, as interflow, or vertically, by percolation, to join the ground water reservoirs. Interflow and ground water flow eventually discharge into streams, rivers, reservoirs, lakes, or oceans.

Influent Stream—An influent stream is one that feeds to ground water, i.e., a stream that replenishes ground water reservoirs by percolation through its porous sand bed during high-flow periods. Since influent streams tend to lose substantial amounts of their water, they are usually of an ephemeral nature.

Interflow—Interflow is shallow subsurface flow, i.e., flow that takes place in the unsaturated or saturated soil layers located immediately beneath the ground surface. Interflow is characteristically a slow process, but eventually interflow volumes discharge into streams and rivers.

Intermittent Stream—Intermittent streams are of mixed characteristics, behaving as perennial streams at certain times of the year and ephemeral at others. Depending on seasonal conditions, intermittent streams may feed to or from ground water. In subhumid and semiarid regions undergoing desertification, perennial streams may turn intermittent in time, and they later may turn ephemeral. Intermittent streams are found in climatic regions ranging from subhumid to semiarid.

Low Flow—Low flow is the dry-season stream flow which, in the absence of rain and/or snowmelt, is sustained by discharge from ground water.

Non point Source—Overland and ground water flow from agricultural, mining, industrial, urban, developed, undeveloped, or construction areas entering a lake, stream, wetland, sea, or other surface water body along a relatively large reach or flow area.

Overland Flow—Overland flow is surface runoff that occurs in the form of sheet flow over the land surface without concentrating in clearly defined channels. This type of flow is usually the first manifestation of surface runoff, which occurs first as overland flow, then as rill and/or gully flow, eventually concentrating into stream flow.

Perennial Stream—A perennial stream is one that always has flow. During low-flow periods, the flow of perennial streams is baseflow. Perennial streams are typical of humid and subhumid climates, where ground water reservoirs are subject to substantial wet-season recharge, discharging during the dry season.

Point Source—Any discernible, confined, and discrete conveyance, including but not limited to, any pipe, ditch, channel, tunnel, conduit, well, discrete fissure, container, rolling stock, concentrated animal feeding operation, vessel, or other floating craft from which pollutants are or may be discharged.

Rating Curve—A rating curve is the relationship between discharge and stage at a given cross-section or gaging station. Usually, discharge Q is plotted in the abscissa and stage y in the ordinates.

Runoff Concentration—Runoff concentration refers to the property of surface runoff to concentrate on a given watershed, i.e., to attain a constant and maximum (i.e., equilibrium) flow at the outlet of a watershed that has been subjected to a temporally and spatially uniform effective rainfall intensity for a period longer than or equal to the time of concentration. The principle of runoff concentration has particular application to small watersheds, for which the assumption of temporally and spatially uniform effective rainfall intensity is usually justified on practical grounds.

Runoff Diffusion—Runoff diffusion refers to the property of surface runoff to spread the flows in time and space, attenuating the peak flows, smoothing the stream flow hydrograph, and eventually causing a net increase in baseflow. The quantity of runoff diffusion is an inverse function of stream slop, i.e., a mild slope may lead to substantial runoff diffusion. Conversely, a steep slope (e.g., that of a mountain stream) may lead to small (or negligible) runoff diffusion.

Stream Flow Hydrograph—A stream flow hydrograph is a plot of flow or discharge vs time at a cross-section, station, or gaging station of a stream or river. Time is plotted in the abscissa, and discharge in the ordinates. A stream flow hydrograph shows the variability of stream flow, either in an event (short-term) or continuous mode (long-term). Depending on the watershed size, the time frame for short-term is measured in minutes, hours, or days; for long-term it is measured in days, months, or years. Short-term hydrographs are used for flood analysis and design; long-term hydrographs are used to study monthly, seasonal, or annual stream flow variability.

Stream Flow Routing—Stream flow routing refers to a mathematical procedure to evaluate the temporal and spatial variations of stream flow as it passes through a stream channel of known or inferred geometric and physical properties. Stream flow routing usually implies unsteady open channel flow calculations in streams and rivers, although there are a few exceptions, e.g., storm-sewer flow, for which mixed open channel–closed conduit flow conditions may prevail.

Subsurface Flow—Subsurface flow takes place below but close to the ground surface, usually in the form of unsaturated flow. Subsurface flow is usually the primary contributor to interflow, although it may also contribute to ground water flow by percolation to the water table.

Surface Flow—Surface flow takes place above the ground surface in the form of overland, rill, gully, stream, or river flow. Surface flow is driven by the force of gravity in its unrelenting movement toward the sea. Surface flow is a relatively fast process, compared to subsurface and ground water flow.

Time of Concentration—For a given watershed, the time of concentration is the period of time that it takes for runoff to concentrate at the watershed outlet, i.e., to attain a peak equilibrium flow, following the application of a (temporally and spatially) uniform effective rainfall intensity for a period longer than that equal to the time of concentration.

A. Special Terms for Reservoir Storage–Yield Analysis

Some of the terms in the Reservoir Storage–Yield Analysis section are defined somewhat differently in other texts. It is important that the contexts in which they are used in this section are clear to the reader.

Active Storage—Water stored above the lowest intake elevation. Water stored below the lowest intake elevation is considered "dead" storage.

Carry-over Storage—If, on the average, a reservoir fills up only once every few years, the water in one year is carried over to the next. This is called carry-over storage.

Critical Period—The period during which a reservoir goes from a full condition to an empty condition without spilling and back to the full condition again.

Critical Drawdown Period—The period from the full condition to emptiness during the critical period.

Full Development—Development wherein all of the reservoir inflow is used. For a constant demand, full development implies a demand rate equal to the mean flow.

Partial Development—Development where the demand rate is a given percentage of the mean flow.

Probability of Failure P/(F)—Several definitions exist in the literature. The most common is to define the probability as the ratio of the length of the period(s) during which the reservoir is empty to the total period used in the analysis.

Release—Is the volume of controlled water released from the reservoir during a given time interval. It is used synonymously with yield, draft, regulation, or outflow. Release is often expressed as a percentage of the mean flow. It has values generally around 50–70% and because of net evaporation, rarely exceeds 90% of the mean flow.

Release Rule—Usually the amount of water released from a reservoir is equal to the demand. When the reservoir is low, it may be necessary that only part of the demand is released. The rule for determining how much water is released is called the operating rule.

Reliability—The complement of the probability of failure, Q = (1 −) (F).

Spill—Uncontrolled flow when the reservoir is full.

With-in Year Storage—If, on the average, a reservoir fills up and spills several times a year it provides water only for short drawdown period. Storage in this case is referred to as within-year storage.

Chapter 7

SNOW AND SNOWMELT

I. INTRODUCTION

This chapter considers snow as a factor in the hydrologic cycle and, in turn, in engineering applications. Involved are such questions as: 1) whether or not precipitation falls in a liquid or solid state; 2) how much snow (more specifically, "snow water equivalent") is present at the beginning of a snowmelt event; and 3) how rapidly snow on the ground will melt when meteorological conditions are favorable. The hydrologic analysis required to address such questions can include consideration of atmospheric energy fluxes, the seasonal water balance of the watershed, and the effect of the snowpack on the storage and delay of liquid water derived from either snowmelt or rainfall. Techniques are introduced in this chapter to cover such considerations in a practical manner, with options to utilize procedures of varying degrees of sophistication. The techniques described herein are meant to be applied to various types of studies and uses, including: 1) design flood determinations for planning and design; 2) day-to-day and seasonal runoff forecasting of river flows and water levels; and 3) hydrologic studies of a general nature for defining the effect of snow on runoff.

The techniques of analyzing snow in the hydrologic cycle and for quantifying snowmelt runoff rates are described generally in terms of computer modeling of some sort, whether it be a single-event analysis of flood runoff or continuous simulation throughout the accumulation and ablation of the snow. In addition, hydrologic regimes subject to snow lend themselves to statistical modeling principally for seasonal runoff volume forecasting, which is also described. The application of snow hydrology generally involves two distinct types of runoff situations: 1) rain-on-snow events typical of the winter floods in the Sierra Nevada and Cascade Mountains of the western United States and the Appalachians in the East; and 2) snowmelt-dominated runoff that is typified by spring/summer floods on large rivers such as the Columbia, Missouri, and Colorado.

The scope of this chapter is intended to provide only relatively brief descriptions of practical applications in snow hydrology for practicing engineers and hydrologists who do not have detailed experience in this subject. Extensive descriptive coverage of the physical processes involved and in-depth explanations of theoretical concepts are not included. Since the proper application of principles contained herein requires considerable background, the use of reference material cited in this chapter is highly recommended.

II. OVERVIEW OF PHYSICAL PROCESSES

A. Precipitation, Snowfall, and Snow Accumulation

In the middle latitudes, precipitation usually occurs as a result of the colloidal instability of a mixed water–ice cloud at temperatures below 0°C. The formation of snow and rain in the atmosphere is a dynamic process. It has been observed that winter precipitation occurs initially in the form of snow crystals in sub-freezing portions of clouds. As the snowflakes fall through the atmosphere, they later melt into raindrops when they fall through warmer, above-freezing air at lower elevations. The melting level air temperature for snowflakes falling through the atmosphere varies from 0–4°C, but it is usually about 1–2°C. Accordingly, on the earth's surface, snowfall occurs at elevations higher than the melting level, while rainfall occurs at elevations lower than the melting level. The most significant determinant of the occurrence of rain or snow is the elevation of the melting level. This is particularly important in mountainous regions. Factors influencing the amount and distribution of precipitation in the form of snow and the snowpack water equivalent may be classified as being meteorologic and topographic. Meteorologic factors include air temperature, wind, precipitable water, atmospheric circulation patterns, frontal activity, lapse rate, and stability of the air mass. Topographic factors include elevation, slope, aspect, exposure, and vegetation cover.

B. Snow Metamorphism

Freshly fallen dry snow exists in a clearly defined crystalline state, with sharply defined edges and abrupt points in each snow crystal. Metamorphism of the snow occurs over time as the individual crystals lose their original distinct form and become rounded and bound together, ultimately into uniform, coarse, large ice crystals. This transformation may take place in a period of time as short as several hours but commonly involves a period of days, weeks, or months in intercontinental areas with large, deep snowpacks. Accompanying the change in crystalline character is a general increase in the density of the snowpack and a change in the temperature of the snowpack. This entire process, commonly called "ripening," is brought about by heat exchanges at the snow surface and ground, the percolation of rain and melt through the snowpack, and snowpack settlement and wind packing. Fig. 7.1 graphically illustrates a typical seasonal representation of snowpack condition as measured at the Central Sierra Snow Laboratory (USACE, 1956). This chart shows depth, density, and temperature of the snow throughout the season, together with positions of the layers of the snowpack as determined by a special settling gage.

Snow density is generally expressed as the ratio (in percent) of the weight of a unit snow volume to the weight of a unit volume of liquid water. The density of the newly fallen snow is typically on the order of 10%, with variations of 6–30% dependent upon the meteorological conditions involved (primarily air temperature and wind). As metamorphism occurs, density increases, reaching values of 45–50% at the end of the season. A snowpack ripe for melt also contains a small amount of free water, on the order of 3–5%. A ripe snowpack is said to be "primed" to produce runoff when it contains all the water it can hold against gravity. The snowpack temperature usually varies seasonally throughout the winter to spring period. In its early stages, the variation throughout the depth may be marked, from near 0°C at the ground to sub-freezing temperatures at shallower depths. As the snow reaches its primed condition, the snowpack is completely isothermal and near 0°C.

The snow condition is an important consideration in snow hydrology since a cold, dry snowpack can act as a temporary storage reservoir for meltwater or rain; however, it should be noted that the snowpack condition can change rapidly at any season of the year, as the result of heavy winter rain, accompanied by unusual winter snowmelt. In such situations, the snowpack may become fully primed in a matter of hours, with the pack isothermal and liquid water requirements fully met. The change in temperature gradient during the period from the fifth to the fifteenth of January in Fig. 7.1 is an illustration of such an occurrence. Under these conditions, subsequent rainfall and/or snowmelt will pass through the pack, regardless of the snowpack density, with little or no time delay.

C. Snowmelt

The process of melting snow involves the transformation of snow/ice from its solid form to liquid water through the application of heat energy from outside sources. While the latent heat of ice is

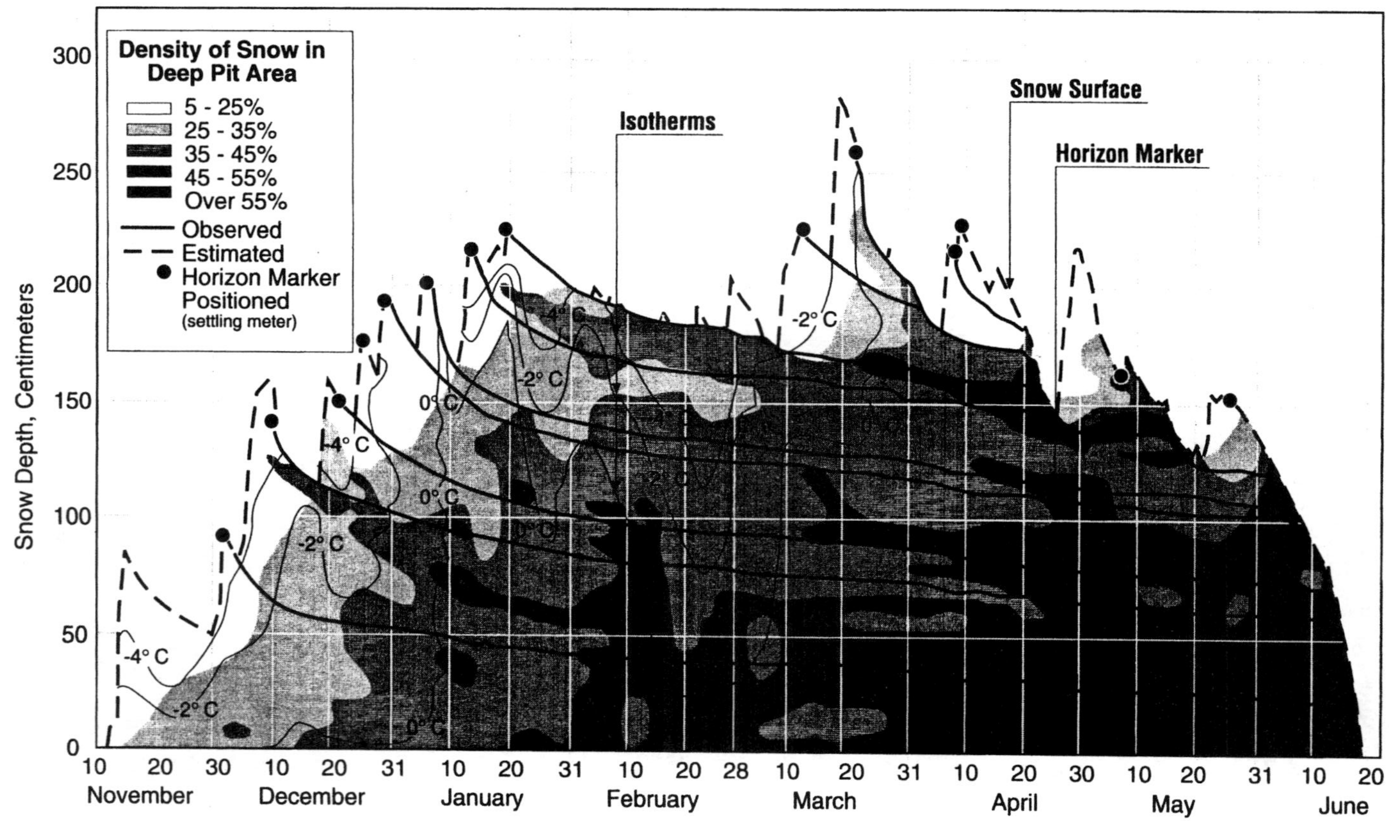

Figure 7.1—Depth, Density, and Temperature Changes in a Snowpack.

established at 80 cal/g (335joules/g), this factor usually must be adjusted to actual snow conditions since the snowpack is not in the form of pure ice at 0°C. The ratio of heat necessary to produce water from snow (and associated free water) to the amount required to melt the same quantity of ice at 0°C is termed the "thermal quality" of the snowpack. For a fully ripe snowpack, the thermal quality can be on the order of .95 to .97.

The rate of snowmelt is dependent upon the many different processes of heat transfer to and from the snowpack, but it is also somewhat dependent upon the snowpack condition. The relative importance of these processes varies seasonally as well as with the day-to-day variation of meteorological factors. The heat transfer processes also vary significantly under various conditions of vegetation, exposure, elevation, and other environmental factors.

The four major natural heat sources in melting snow are: 1) absorbed solar radiation, 2) net long-wave (terrestrial) radiation, 3) convective heat transfer from the air, and 4) latent heat of vaporization by condensation from the air. Two additional minor sources of heat are: 5) conduction of heat from the ground, and 6) heat content of rainwater.

Solar radiation is the prime source of all energy at the earth's surface. The amount of heat transferred to the snowpack by solar radiation varies with latitude, aspect, season, time of day, atmospheric conditions, forest cover, and reflectivity of the snow surface (albedo). The albedo ranges from 40–80%. Long-wave radiation is also an important process of energy exchange to the snowpack. Snow is very nearly a perfect black body with respect to long-wave radiation. Long-wave radiation exchange between the snow surface and the atmosphere is highly variable, depending upon cloud cover conditions, atmospheric water vapor, nighttime cooling, and forest cover. Heat exchange between the snow surface and the atmosphere by convection and by water vapor condensation (items 3 and 4, above) is dependent upon the atmospheric air temperature and vapor pressure gradients, together with the wind gradient in the atmosphere immediately above the snow surface. These processes are particularly important under storm conditions with warm air advection and high relative humidity. In summary, there is no one process of heat exchange with the snowpack that may be universally applied, but the relative importance of each process is dependent on atmospheric, environmental, and geographic conditions for a particular location and a particular time or season. Thorough discussions of the heat exchange processes are given by the USACE (1956), Anderson (1968, 1976), and Male and Gray (1981).

III. DATA REQUIREMENTS, COLLECTION, AND SOURCES

A. Data Requirements

Hydrometeorological data needed for snow hydrology computations include those required for rain-only situations plus the additional variables necessary for the snow accumulation/snowmelt processes involved. These include air temperature data and snow measurements as a minimum but could include wind speed, dewpoint, solar radiation, and other variables if energy budget computations are performed. Since the data requirements for a full energy budget approach are usually unavailable except for hypothetical design conditions and specially instrumented watersheds, analysis and modeling typically use only precipitation and air temperature as input. Table 7.1 summarizes the possible data types that are needed in snow hydrology analysis along with comments on the purpose and application of the data.

Besides the hydrometeorological data requirements stated above, physical data related to the characteristics of the watersheds must be obtained. Area-elevation data, derived from topographic maps or digital elevation models, are of prime importance in mountainous regions. This is particularly necessary for distributed modeling methods that compute hydrometeorological elements on the basis of discretely defined elevation zones or grid cells. Also, data related to the type and density of forest cover, aspect of the watershed elements, exposure to prevailing winds, slopes, etc., are necessary for judging snowmelt parameters for energy budget computations. Overall water balance computations require definition of precipitation variation over the watershed, usually determined from isohyetal maps for the areas involved.

TABLE 7.1. Data Requirements for Snow Analysis.

Data type	Physical process measured or indexed	Analysis application
Streamflow (Q)	a. Continuous discharge b. Runoff volumes	a. Hydrograph analysis, model calibration b. Water supply analysis, forecasting
Precipitation (P)	a. Basin moisture input b. Estimate of SWE	a. Hydrograph analysis, model calibration b. Water supply forecasting
Air Temperature (T_a)	a. Rain/freeze interface b. Index to all energy exchanges c. Factor in energy budget estimates	a. Modeling snow accumulation b. Modeling snowmelt (temp index) c. Modeling snowmelt (energy budget)
Snow Water Equivalent (SWE)	a. Estimate of precipitation b. Index to basin water supply c. Snowpack quantity during ablation	a. Analysis, model calibration b. Water supply forecasting c. Modeling snowmelt
Areal Snow Cover	a. Extent of basin snow cover b. Snow-line elevation	a. Model calibration b. Parameter in forecast models
Snow Fall	a. Estimate of SWE, precipitation b. Accumulation of snow	a. SWE, precip. applications b. Avalanche forecasting
Snow Density	a. Estimate of SWE, precipitation b. Condition of snow	a. SWE, precip. applications b. Avalanche conditions, snow loads
Snow Depth	a. Estimate of SWE, precipitation b. Estimate of weight	a. SWE applications b. Snow load on structures
Snow Albedo	Solar energy absorption	Modeling (energy budget), design floods
Solar Radiation	Solar energy flux	Modeling (energy budget), design floods
Wind Velocity (v)	Factor in estimate of convection/condensation energy flux and sublimation	Modeling (energy budget), design floods
Dewpoint Temperature (T_d)	Factor in estimate of condensation energy flux and sublimation	Modeling (energy budget), design floods

Air temperature data are quite critical in any modeling or analysis effort, since freezing level must be known during the snow accumulation process to determine the form of precipitation in the basin. Further, air temperature often is used as a key index in computing snowmelt. Historic temperature data are normally archived as maximum and minimum for the day. If a computational interval more frequent than 24 hours is used, then intermediate values may have to be estimated by assuming a typical diurnal temperature pattern (Anderson, 1973). An additional parameter that may be needed in modeling is the lapse rate (vertical temperature profile), which must either be assumed to be a fixed value or estimated from observed temperature readings of atmospheric soundings and ground measurements from stations at different elevations.

Snow data are collected in the form of snow water equivalent on a daily or more frequent basis in the case of automated stations using snow pillows, or monthly (sometimes bi-monthly) in the case of manually measured snow courses. Snow water equivalent data as applied to flood runoff analysis are needed as an independent variable for simplified analyses and seasonal runoff forecasting and as data to assist in calibrating and verifying simulation models. Since snow stations may be the only source of high elevation precipitation data, they also can be used to help estimate basin-wide precipitation input to simulation or statistical models. Data on snow characteristics such as depth and density have only secondary application in snow hydrology but may have importance in other engineering applications such as avalanche prediction and control and snow loading on structures.

B. Data Collection and Utilization

1. Precipitation and Temperature The collection of precipitation data in areas subject to snow accumulation presents additional problems in gaging due to gage freezing, "capping" of the gage by snow, and high winds. Equipment and field procedures for such conditions are well-documented (Soil Conservation Service, 1972; Male and Gray, 1981). The hydrologist utilizing archived records should, therefore, exercise particular care in assuring that the data are reasonable under these circumstances. When selecting appropriate precipitation, snow, and temperature gages for analysis in a mountainous environment subject to snow conditions, vertical distribution of precipitation and the vertical temperature profile must be considered. The necessary data depends on the application; for simple indexing applications, for instance, a mid-elevation snow gage may be sufficient; however, for detailed simulation, gages located throughout the elevation range are necessary to define the vertical distribution of precipitation and to give a field reference of snow conditions during critical snowmelt times.

2. Snow Water Equivalent The oldest technique for monitoring snowpack is to take manual measurements of snow depth and water content using a snow-sampling tube (Soil Conservation Service, 1972). In the western United States, this type of snow sampling has been done on an organized basis since the 1930's by the Soil Conservation Service; sampling programs also exist in the eastern United States. At a measurement site (or snow course), several measurements are made along a traverse of 90–300 meters in length, each measurement point being 8–30 meters apart. The snow tube is inserted vertically into the snow to the soil surface, the depth is recorded, and the tube and snow core are weighed. The weight of the snow sample is obtained by subtracting the tare weight of the empty tube, and this weight indicates the quantity of water contained in the snow. This is called the "snow water equivalent" (SWE) and is expressed in units of depth. The density of the snow is calculated as the water equivalent divided by the snow depth. The measurements at the sampling points are averaged to obtain single values of depth, SWE, and density for the snow course.

Snow water equivalent is measured at automated sites by a pressure pillow. The pillow is made of stainless steel or a rubber-like material (hypalon) and filled with an antifreeze fluid. It ranges from 2–11 square meters, with the larger sizes required for higher snow accumulation sites. It is connected by a tube to a pressure transducer, which converts the fluid pressure caused by the weight of the snow atop the pillow to an electric signal. The fluid level in the precipitation gage is also sensed by a pressure transducer. The signals from the pillow, precipitation gage, and thermistor (electronic thermometer) are telemetered to a central computer and are available on a daily or more frequent basis. In the western United States, the Soil Conservation Service operates a network of approximately 600 automated snow-monitoring sites in a program called "SNOTEL" (SNOw TELemetry) (Rallison, 1981).

Another method of measuring snow water equivalent is by overflight along flight lines by aircraft. This is done routinely by the National Remote Sensing Hydrology Program of the National Weather Service (Carroll and Allen, 1988). This technique makes use of the fact that there are natural background emissions of gamma radiation from bare ground that are attenuated by water. Snow water equivalent can be calculated from measurements of the gamma radiation from the snowpack compared with that precalibrated from bare ground. Flight lines are typically 16 km long by 300 m wide, and the values reported are mean snow water equivalent over this area. Flight lines exist throughout the northern and western United States as well as in most of Canada.

3. Areal Snow Cover Areal snow cover has been observed visually from the ground over a number of years as an operational guide in assessing snowmelt runoff potential at several locations in the western United States. Beginning in the early 1950's, specially made snow cover observations during the snowmelt season using light aircraft were instituted by the Corps of Engineers for project operations in the Columbia River, and in certain Sierra Nevada drainages in California. These flights continue to be made to provide an assessment of snowcover in conjunction with operational hydrologic forecasts. Remote sensing of snow-covered areas from satellites has also become feasible, while sensing of other snow attributes is still in the research phase. Snow-covered areas for entire basins have been obtained from LANDSAT imagery in various studies and applications. The National Remote Sensing Hydrology Program of the National Weather Service (Carroll and Allen, 1988) reports snow-covered areas both for entire basins and for each of several elevation zones within selected basins. These data come from the GOES satellite and the NOAA Polar Orbiting Advanced Very High Resolution Radiometer (AVHRR).

4. ***Other Meteorological Variables*** Variables such as albedo, dewpoint temperature, solar radiation, and wind velocity are collected at certain first-order weather stations of the National Weather Service and by research laboratories and other institutions. These data are not widely published, so they may have to be obtained by direct contact with the archiving agency. As these data are observed at only a limited number of sites, it is often necessary in an application to use data from sites somewhat removed from the area of interest. In design flood derivations, one is often interested in the general magnitude of these variables and, in particular, the possibilities for extremes. General information of this type can be found in the technical references already cited and in other technical literature on snow hydrology (see Table 7.2).

C. Data Sources

Table 7.2 summarizes data sources typically used in snow hydrology applications. Virtually all data collection made by agencies with official program responsibilities are available in computer databases. Archive files of precipitation, temperature, and other meteorological variables are maintained by the National Weather Service in Asheville, North Carolina, and by NOAA-administered Regional Climate Centers. The largest repository of snow data in the United States is maintained by the Soil Conservation Service in Portland, Oregon. Other agencies that collect snow data and maintain computer data systems include the U.S. Bureau of Reclamation, U.S. Army Corps of Engineers, California Department of Water Resources, and provincial agencies in Canada. In addition to storing the data, the databases allow for data retrieval, report generation, statistical analysis, model calibration, and input data preparation. The capabilities of the databases vary depending on the requirements of the owner agencies. Access permission also varies, as some databases are more restrictive than others.

The various sources for the entire range of hydrometeorological data needed for energy budget calculations are also given in Table 7.2. These include archive data by the NWS and others, as well as research-related data available in books, technical papers, and research laboratories. Four government research laboratories are listed as possible sources of information. The complete hydrometeorological logs from the Cooperative Snow Investigations snow laboratories, active from 1945–1952, are also available. These three field laboratories, located in western Montana, western Oregon, and central California, provided basic data for the report, Snow Hydrology (USACE, 1956).

IV. SNOW ACCUMULATION AND DISTRIBUTION

A. Overview

From the standpoint of general hydrology, the topic of snow accumulation and distribution is a large one, involving considerations such as the physics of snow formation, the effect of terrain and vegetative cover on precipitation and snow accumulation, the effect of interception, and the determination of areal distribution of snow from point measurements; however, in practical engineering applications, the details of the snow accumulation/distribution process are often relatively unimportant compared with the snowmelt runoff determination involved. In a rain-on-snow design flood derivation, for instance, the magnitude of rain may overshadow the snowmelt quantities involved thereby permitting a relatively simple estimate of snow accumulation prior to the design rainstorm. A water supply analysis or streamflow forecast might employ point measurements of snow or accumulated precipitation as indexes to basin-wide SWE with weighting factors determined by model calibration. Although snow accumulation and distribution is not directly used in these examples, the general principles involved are nevertheless invaluable to the practitioner. The details of snow accumulation do come into play when continuous

TABLE 7.2. Data Sources.

Data type	Sources	Contact
Streamflow, Reservoir Storage	a. USGS Water Supply Papers b. USGS WATSTOR System c. Agency Publications, Databases d. CDROM (private vendors)	a. Local USGS office b. Local USGS office for information c. Local/regional agency offices (see text) d. Earth Info, Inc. Hydrosphere, Inc.
Precipitation, Air Temperature	a. NWS Climatological Bulletins b. NOAA National Archives c. State Climatologists d. NOAA Regional Climate Centers e. CDROM (private vendors)	a. Local NWS offices b. NOAA National Climatic Data Center Federal Building Ashville, NC 28801 c. State offices (not avail. in every state) d. Contact national center for information e. See above
Snow Water Equivalent	a. SCS Snow Summary Bulletins b. SCS SNOTEL database c. State databases	a. State SCS offices b. SCS West National Technical Ctr. 101 SW Main St. Portland, OR 97204 c. State water resources agencies, e.g., California Department of Water Resources 1416 Ninth Street Sacramento, CA 95814
Snow Depth, Snow Fall	a. NOAA Climatological Bulletins b. NOAA National Archives c. NOAA Regional Climate Centers	a. See above b. See above c. See above
Areal Snow Cover	a. NWS Remote Sensing Center b. LANDSAT Photography	a. Nat'l Operational Hydrologic Remote Sensing Ctr. 6301 34th Avenue South Minneapolis, MN 55450 b. National Snow and Ice Data Ctr. Cires Campus Box 449 University of Colorado Boulder, CO 80309
Solar Radiation, Dewpoint, Wind Velocity, Albedo	a. NOAA national/regional archives b. Operational Snow Laboratories c. Textbooks, Handbooks d. (1) Proceedings of the Western Snow Conf. (2) Proceedings of the Eastern Snow Conf. e. Hydrometeorological logs of Cooperative Snow Investigation Laboratories, 1945-1952	a. See above b. (1) USDA Forest Service Central Sierra Snow Lab Soda Springs, CA (2) US Army Corps of Engineers Cold Regions Research & Engineering Lab Hanover, NH (3) USDA Forest Service Rocky Mt. Forest and Range Experiment Sta. Laramie, WY (4) USDA Agricultural Research Service Northwest Watershed Research Center Boise, ID c. (1) US Army Corps of Engineers (1956) (2) Gray and Male (1981) (3) List (1968) (see also, References) d. (1) Western Snow Conference PO Box 2646 Portland, OR 97208-2646 (2) Eastern Snow Conference 161 London Road Newmarket, Ontario L3Y 7A7 Canada e. US Army Corps of Engineers, North Pacific Div. PO Box 2870 Portland, OR 97208-2870

simulation is used for forecasting or analysis. Here, the vertical and spatial distribution of precipitation and snow must be explicitly modeled during periods of snow accumulation and such factors as interception and sublimation must be taken into account. For comprehensive general coverage on the subject of snow accumulation and distribution, the reader is referred to USACE (1956), Schemenauer et al. (1981), and McKay and Gray (1981).

Hydrologic engineering and streamflow forecasting applications basically require an estimate of SWE for the basin being studied as input into a runoff derivation, and this estimate must either directly or indirectly consider the process of snow accumulation and distribution. The choice of methodology to employ depends upon data availability, the amount of effort to be expended and the type of application involved. Alternative approaches range from simple estimates of a single basin-wide average to detailed simulation using a distributed formulation of the basin. Table 7.3 summarizes three possible approaches of varying complexity.

B. Snow Water Equivalent Estimate from Historic Data

For design flood derivations and other applications, the basin's initial SWE at the beginning of snowmelt can be estimated from historic records of precipitation and snowpack. As noted above, the complexity of the analysis used for this process would logically depend upon the application and the relative importance of snowmelt in the runoff hydrograph. The analysis involved is basically the conversion of point historic data to mean basin or sub-basin SWE and the determination of a specified probability or degree of maximization required for a design flood.

In the simplest applications, snow and precipitation gages in and near the basin can form the basis for relatively rough, single-value estimates of SWE. Normal annual precipitation maps can be used to extrapolate to high, ungaged elevations in mountainous regions. Establishment of a SWE/probability relationship might be based upon frequency analysis of a historic SWE record at a snow station, or it might be based upon precipitation records. When using precipitation data—either station records or isohyets of specified frequency—the determination must be made as to how much precipitation falls and accumulates as snow, to be available at the beginning of the design event. In cold climatic regions or at high mountainous elevations, this would be a simple assumption for a winter period.

A further detailed analysis of initial SWE might include a more thorough determination of the snow "wedge"—the characteristic distribution of snow throughout a range of elevation in a mountainous basin. This type of analysis would employ weighting factors for snow/precipitation stations based upon the extent to which a station represents the total drainage, and it might lead to a distribution of SWE by elevation. A detailed SWE determination might also include a water balance analysis, at which estimates

TABLE 7.3. Alternatives for Estimating Snow Water Equivalent (SWE).

Approach	Possible application	Comment
Simple estimate of SWE	1. Single event rain-on-snow computation	1. Simple estimate based upon historical records. May be adequate where rain dominates
	2. Forecasting in rain-dominated areas	2. Estimates of SWE obtained from real-time data reports
	3. Forecasting seasonal runoff volume using regression methods	3. Measurement provide an index to basin snow
Detailed estimate of SWE, considering elevation and spatial distribution	Design flood derivation, snow-dominated basin	More detailed analysis of historic records required
Simulation of snow accumulation through the accumulation season	1. Detailed design flood derivation 2. Long-range streamflow forecasting	Requires a distributed, continuous-simulation model

of basin SWE are developed by balancing all the components of the hydrologic cycle against observed runoff.

In design flood studies, care must be taken in considering the combined probability of the initial SWE in combination with subsequent rainfall and other factors that produce the design flood. A reasonable combination, based upon analysis of historic records, must be used to ensure that the design event is neither under- nor over-maximized. In probable maximum flood determinations involving rain on snow, for instance, the PMP might result from a maximized initial SWE in conjunction with a spring rainfall of moderate magnitude, or the reverse may be true; however, maximizing both would be unrealistic. This question is best addressed by running sensitivity tests and comparing the alternatives.

C. Watershed Definition in Detailed Simulation

Since temperature, and therefore elevation, plays such an important role in defining the conditions of the basin during a precipitation event, the watershed needs to be defined with independent subunits in a detailed snow accumulation analysis. The most common approach is to divide the basin into zones or bands based on elevation. On each band, precipitation, snow, soil moisture, etc., can be simulated independently, as illustrated in Fig. 7.2. Available hydrologic models such as HEC-1 (USACE, 1990) and SSARR (USACE, 1987) provide for the watershed definition to be established relatively easily, employing simplifying assumptions such as defining zone characteristics for the basin through generalized functions. Other models, such as PRMS (Leavesley et al., 1983), divide the watershed into relatively homogeneous, hydrologic response units based on elevation, slope, aspect, and vegetation. (See the section in this chapter regarding computer programs).

D. Simulation of Snow Accumulation

The most thorough approach to estimating SWE is to utilize a continuous simulation model to account for the entire snow accumulation process in relative detail. This requires a distributed basin definition and a frequent (at least daily) calculation time step during which precipitation, temperature, lapse rate, snow line, and runoff are considered. Fig. 7.3 illustrates this process as performed in the SSARR model with a basin subdivided into elevation bands.

V. SNOWMELT ANALYSIS AND SIMULATION

A. Overview of Applications and Approaches

Numerous alternatives present themselves in determining the best approach for computing snowmelt in hydrologic engineering analysis. Ranging from simplified assumptions on discrete storm events to detailed simulation using energy budget principles and distributed definition of the watershed, the choice is dependent upon the application involved, resources available and data availability. Table 7.4 summarizes the options that are possible and how they tend to relate to given types of applications. A typical situation that might be encountered is that of calculating a hypothetical flood from rainfall, either of specified frequency or from a NWS hydrometeorological report. If the meteorological conditions are such that rainfall dominates, and the storm duration is relatively short, it may be quite satisfactory to use a simple approach to estimate snowmelt, e.g., by establishing an antecedent water content with historical analysis, then using an assumed rate of melt or a temperature index applied with a melt–rate factor. The simulation of snow conditioning would not be required since a "ripe" snowpack prior to the storm could be assumed. On the other hand, the derivation of a design flood, or the forecasting of flood runoff in a basin that is snow-dominated, would likely require a more detailed simulation of snow conditioning and snowmelt, perhaps through the use of theoretical or empirical equations as described below. Examples of three applications with differing degrees of complexity appear later in this chapter.

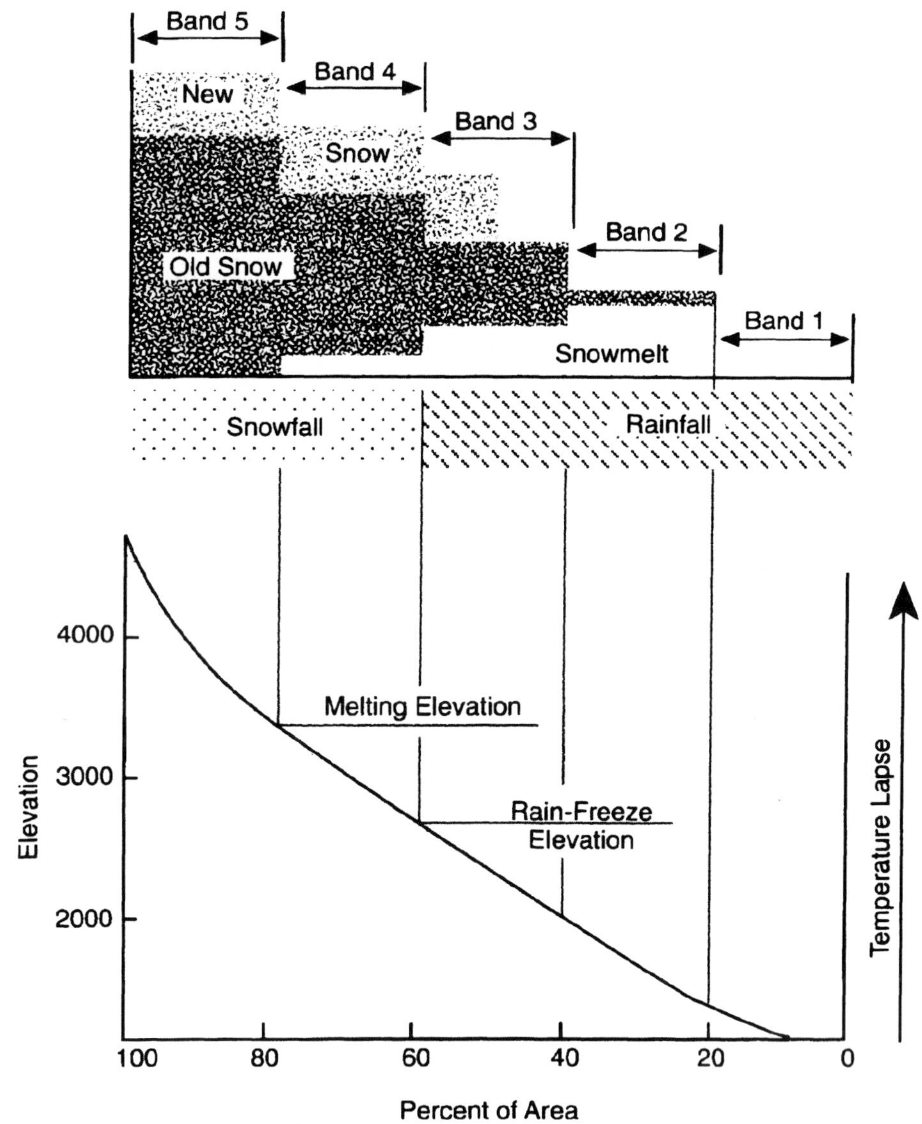

Figure 7.2—Illustration of Distributed Formulation of a Watershed Model using Elevation Bands.

B. Snowmelt

The basic equation defining the amount of energy available for snowmelt can be written as:

$$H_m = H_{sw} + H_{lw} + H_c + H_e + H_r + H_g - d(H_q)/dt \tag{7.1}$$

where H_m is the energy flux available for melt, H_{sw} is the net short-wave radiation flux absorbed by the snow, H_{lw} is the net long-wave radiation flux at snow–air interface, H_c is the convective heat transfer (sensible heat) from the air, H_e is the flux of latent heat released by condensation or sublimation, H_r is the heat flux from rain, H_g is the heat flux from the ground, and H_q is the internal energy content of the snowpack.

While all terms in the above equation are shown as positive, some heat fluxes may in fact be negative, i.e., directed away from the snowpack. For instance, H_{lw} is normally negative in the open and H_e and

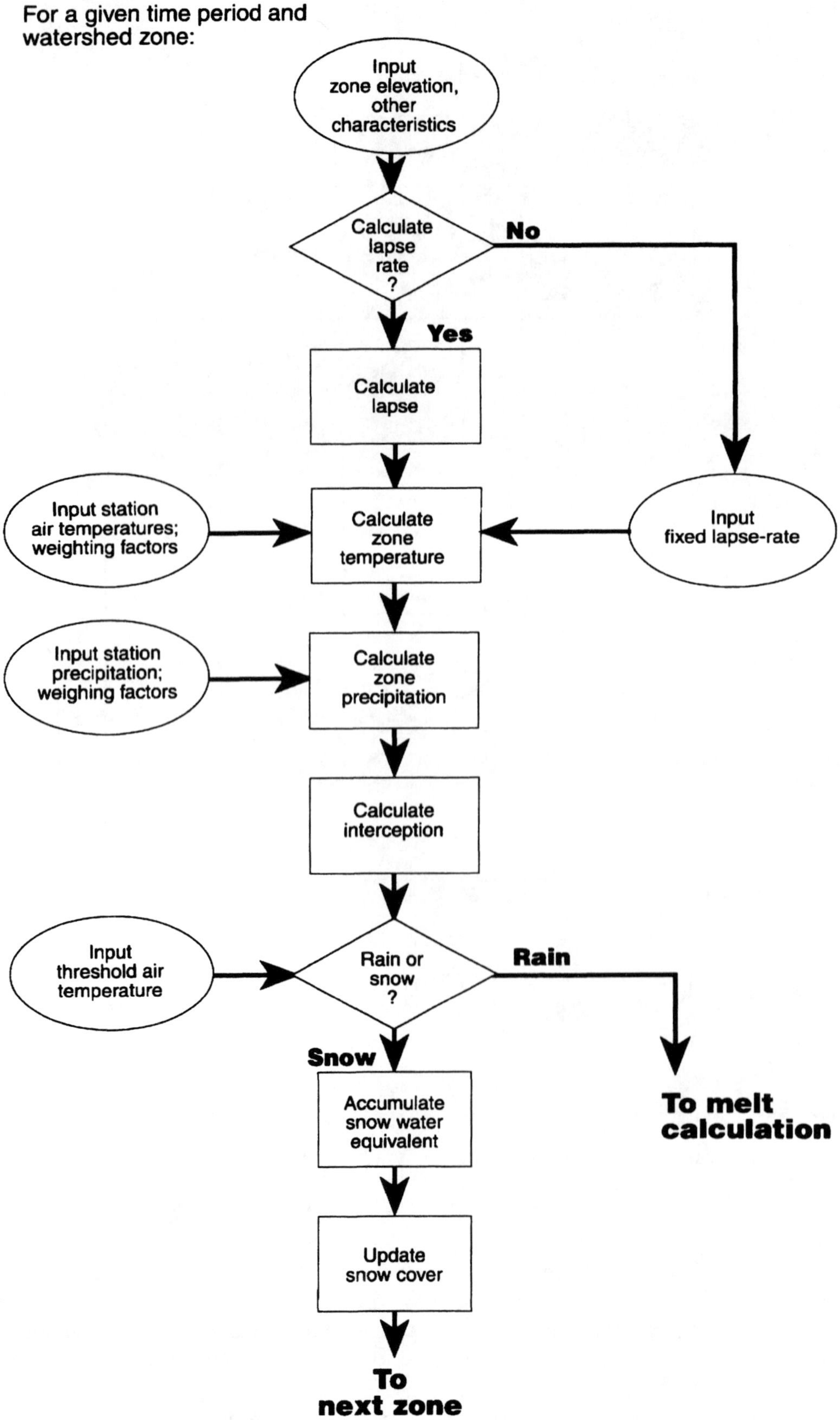

Figure 7.3—Illustration of Snow Accumulation Simulation.

TABLE 7.4. Snowmelt Options.

Application	Example	Basin configuration[1]		Melt calculation			
		Lumped	Distributed	Snow Conditioning	Simplified[2]	Temperature index	Energy Budget
Single-event analysis—Rain-on-snow	Design floods in coastal mountains	Yes	Possibly	Assumed "ripe"	Possibly	Possibly	Possibly
Single-event analysis—Snow (plus rain)	Design floods in interior basins	Yes	Yes	Assumed "ripe"	No	Yes	Yes
Single-event Forecasting Rain-on-snow	Short-term flood forecasting	Yes	Yes	Optional	Possibly[3]	Yes	No
Single-event Forecasting Snow (plus rain)	Short-term flood forecasting	Yes	Yes	Optional	No	Yes	No
Continuous simulation, any environment	Long-term flood & drought forecasting; detailed design analysis	No	Required	Required	No	Yes	Possibly
Detailed simulation in small watersheds	R&D applications; analysis for detailed design; special applications	No	Required	Required	No	No	Yes

[1]Qualitative indicator shown for type of option that might typically be used for application. This is a guideline only. "Yes" or "No" indicates suggested option.
[2]Simplified approach might be to assume a constant or variable moisture input due to snowmelt.
[3]Would be appropriate only in situations where snowmelt is small compared to rain.

dH_q/dt can be either positive or negative. The resulting daily melt produced by a quantity of energy, H_m, is given by:

$$M = H_m/334.9\ B \tag{7.2}$$

where M is the daily snowmelt in mm, H_m is the daily heat flux in kJ/m^2; and B is the thermal quality of the snow. The factor 334.9 is the latent heat of fusion, kJ/kg. As noted previously, the thermal quality of ripe snow is usually between 0.95 and 0.97.

Each of the heat fluxes listed in Equation 7.1 is, in itself, a function of several components, some of which can be difficult to quantify for practical applications. In actual practice then, the theoretical relationships involved are reduced to empirically derived equations that have been found to work satisfactorily in simulation models. Two basic approaches are commonly used: 1) the "energy budget" solution, which employs equations that represent key energy fluxes such as solar radiation, long-wave radiation, wind, heat from condensation of water vapor, etc., and 2) the "temperature index" solution, which uses air temperature as the primary independent variable through the use of a fixed or variable "melt-rate factor." The latter solution is almost exclusively used in practical engineering applications of simulation and analysis for real-time forecasting and reservoir regulation.

1. Energy Budget Equations Numerous researchers have developed equations that represent each of the energy flux components contained in Equation 7.1. These investigations typically include field experiments conducted under a specific physical environment, and often lead to equations that simplify the physical process with empirically derived coefficients. Male and Gray (1981) and Gray and Prowse (1993) contain concise summaries of this work, frequently comparing the findings contained in the literature.

The U.S. Army Corps of Engineers (1956) developed generalized energy budget equations for snowmelt based upon Equations 7.1 and 7.2. These equations employ simplifying assumptions related to the environment in which they are to be used, and make use of variables and coefficients that are feasible in practical engineering applications. The equations were derived primarily for the purpose of developing design floods for major dams, in which prescribed procedures require the use of the combined maximized effect of snow accumulation, snowmelt, storm precipitation, and the effect of the snowpack on runoff. In the case of the maximized snowmelt, there are definite upper limits to the amount of heat exchange by radiative processes or by the advection of heat by the air mass, and these upper limits are determinable only by the use of the energy budget solution. The temperature index solution does not provide the ability to maximize snowmelt rates on a physical basis and, accordingly, it should not be used for design flood determinations for major projects. Two of several generalized equations are presented below with an abbreviated explanation. These and other equations are described in detail in USACE (1956). English unit versions are contained in this reference.

For snowmelt during rain, in which short-wave solar radiation is relatively unimportant and latent heat released by condensation of atmospheric water vapor is relatively high, the following equation applies for open or partly forested areas:

$$M_s = (1.33 + 0.239k_w v + 0.0126P_r)(T_a) + 2.3 \tag{7.3}$$

in which M_s is the total daily snowmelt in mm/day, v is the wind velocity at the 15 m height, in km/hour, k_w is the convection–condensation factor representing the relative exposure of the basin to wind (for unforested areas, $k_w = 1$), T_a is the mean temperature of the saturated air in °C at the 3 m level, and P_r is the daily rainfall in mm.

The constants in the equation are based upon field investigations. The factor 1.33 relates snowmelt due to long-wave radiation to temperature, and the term $0.239k_w v$ represents the effects of convection–condensation melt. The quantity 2.3 accounts for melt from ground heat plus short-wave radiation.

For the case of snowmelt during rain-free periods, direct (short-wave) solar radiation must be accounted for. Several equations were developed by the Corps of Engineers depending upon the degree of forest canopy involved. One, for partly forested areas, is as follows:

$$M_s = k_r(1-F)(2.43I_i)(1-a) + k_w(0.239v)(0.22T'_a + 0.78T'_d) + F(1.33T'_a) \quad (7.4)$$

where M_s is the daily snowmelt in mm/day, T'_a is the difference between the air temperature measured at 3 m and the snow, surface temperature in °C (snow surface temperature is often assumed to be 0°C), T'_d is the difference between the dewpoint temperature measured at 3 m and the snow surface temperature (in °C), I_i is the observed or estimated insolation (solar radiation on horizontal surface) in MJ/m^2 (one langley = 0.04187 MJ/m^2), a is the average snow surface albedo, k_r is the basin short-wave radiation melt factor (it depends upon the average exposure of the open areas to short-wave radiation in comparison with an unshielded horizontal surface, i.e., for a north-facing slope, $k_r < 1.0$, but for a south-facing slope, $k_r > 1.0$), F is an estimated average basin forest canopy cover, effective in shading the area from solar radiation, expressed as a decimal fraction, and k_w is the basin convection–condensation melt factor, as defined above (it depends on the relative exposure of the area to wind). The first term in the above equation is melt due to short-wave radiation. The second term accounts for convection–condensation melt, and the third term represents long-wave radiation melt.

It can be seen that the above energy budget equations have data requirements that may preclude their usage in real-time applications; however, these equations are useful in design flood derivations where variables such as insolation, albedo, etc., can be determined from meteorological tables or upon historic data for the region. These factors would be maximized in the case of a probable maximum flood derivation, for instance. Both of the equations presented above are available in the HEC-1 and SSARR computer programs. The generalized snowmelt equations also provide a useful method of estimating relative magnitudes of melt components. Table 7.5 presents melt quantities calculated from these equations for six hypothetical situations—three with rain, three without.

2. ***Temperature Index Solution*** Because of the practical difficulties of obtaining data needed for the energy budget equations, common practice is to simulate snowmelt by the "temperature index" solution in real-time forecasting and other applications. This approach utilizes the basic equation:

$$M_s = C_m(T_a - T_b) \quad (7.5)$$

where M_s is the snowmelt in mm per period, C_m is a melt–rate coefficient that is often variable, as discussed below, T_a is the air temperature, and T_b is the base temperature, near 0°C.

Given the variables contained in the energy budget equations previously presented, it can be seen that the employment of temperature only as an index to snowmelt results in further approximation of the actual physical processes involved; yet, considering other uncertainties—particularly in forecasting applications—this does not usually preclude its use from real-time applications.

The melt-rate factor, C_m, is an important key in the successful application of the temperature index equation. While temperature is a reasonably good index of energy flux in heavily forested areas, it is less so in open areas where short-wave radiation or wind velocity play a more important role in the melt process. In clear-weather melt situations, the melt-rate factor varies seasonally, typically increasing as the snowmelt season progresses due to factors such as the decrease in albedo, increased short-wave radiation, longer daylight hours, etc. Because of this, a provision is usually made in simulation models to calculate C_m as a variable, perhaps as a function of accumulated runoff or accumulated degree-days of air temperature.

The choice of base temperature depends upon the computation interval involved and the form of the temperature data involved. If maximum daily temperature is the input variable, then this factor would be higher than 0°C, perhaps as high as 4°C. For a more frequent time interval, the factor would be at or near 0°C.

TABLE 7.5. Relative Magnitude of Snowmelt Factors.

(a) Assumed Conditions

Case	Description	Assumed meteorological conditions				
		T_a	T_d	I_i	P_r	v
1.	Clear, hot, summer day. No forest cover. Albedo = 40%	20	7	30	0	5
2.	Same as Case 1, 50% cloud cover	18	10	20	0	5
3.	Same as Case 1, fresh snow. Albedo = 70%	20	7	30	0	5
—		—	—	—	—	—
4.	Heavy wind and rain, warm. No forest cover	10	10	0	80	25
5.	Same as Case 4, but light rain, windy	10	10	0	10	25
6.	Same as Case 5, but light wind	10	10	0	10	5

T_a = Air Temperature, °C
T_d = Dewpoint Temperature, °C
I_i = Solar Insulation, MJ/m^2
P_r = Daily rainfall, mm
v = Mean wind velocity, km/hr

(b) Daily Melt Quantities

Case	Snowmelt components, mm					Total Melt, mm	Rain + Melt mm
	M_{sw}	M_{lw}	M_{ce}	M_r	M_g		
1.	43.7	0.0	11.8	0.0	0.5	56.0	56.0
2.	29.2	0.0	14.1	0.0	0.5	43.7	43.7
3.	21.9	0.0	11.8	0.0	0.5	34.2	34.2
—	—	—	—	—	—	—	—
4.	1.8	13.3	47.8	10.1	0.5	73.5	153.5
5.	1.8	13.3	47.8	1.3	0.5	65.7	74.4
6.	1.8	13.3	9.6	1.3	0.5	26.4	36.4

M_{sw} = Short-wave radiation melt
M_{lw} = Long-wave radiation melt
M_{ce} = Convection/Condensation melt
M_r = Rain melt
M_g = Ground heat melt

The possible range of the melt-rate factor can be illustrated by referring to the hypothetical cases presented in Table 7.5. Using the daily melt quantity calculated by the empirical energy budget equations and the temperatures assumed, the melt-rate coefficients calculated with Equation 7.5 would be as shown in Table 7.6. Table 7.6 generally confirms field experience regarding the range in variation of the temperature index melt-rate factor: for clear-melt conditions, the factor varies between 1.5-3.0 mm/°C and increases as the snowmelt season progresses. For rain-melt conditions, the factor can exhibit wide-ranging variations from 2.5–9.0 mm/°C, depending upon wind velocity and, to a lesser extent, the precipitation quantity. These factors would be higher if the temperature index used is the maximum daily temperature. In forecasting practice, the melt-rate factors are estimated through the process of calibrating

TABLE 7.6. Relative Magnitude of Melt-Rate Factors (Refer to Table 7.5 and Equations 7.3, 7.4 and 7.5).

Case	T_a (°C)	T_b (°C)	Melt (mm)	C_m (mm/°C)	Comment
1.	20	0	56.0	2.80	Clear, low albedo
2.	18	0	43.7	2.43	Case 1, cloud cover
3.	20	0	34.2	1.71	Case 1, fresh snow
4.	10	0	73.5	7.35	Heavy rain, windy
5.	10	0	64.7	6.47	Light rain, windy
6.	10	0	26.4	2.64	Light rain, light wind

a hydrologic model. Once established for known historic conditions, the factor can be modified by judgment to be applied to the design condition or forecast situation under consideration. Equations 7.3 and 7.4 can be useful guides in this process. Additional discussion of the magnitude of the temperature index melt-rate factor can be found in USACE (1956), Anderson (1973), and Male and Gray (1981).

C. Snow Condition

As discussed in the overview of physical processes, the storage effect of a snowpack can be a significant factor in determining the time distribution of snowmelt runoff. The condition of the snow—as indicated by its temperature, crystalline structure, density, etc.—will influence the travel rate of meltwater or rain as it moves through the snowpack. In fresh, subfreezing snow, for instance, an initial quantity of meltwater or rain that enters the snowpack will freeze as it warms the snowpack to 0°C. An additional quantity of water is also required to satisfy a liquid water-holding capacity before the snow will release any water by gravity. A further time delay occurs as the free water makes its way to the ground surface, typically by an indirect path around relatively impermeable ice layers. This entire process results in a delay between the onset of atmospheric conditions that cause melt and the corresponding melt runoff. At the same time, it leads to a metamorphosis or ripening of the snowpack. In contrast to a dry, cold snowpack, one that is isothermal at 0°C and ready to yield water to gravity has a relatively minor storage effect, that is associated with the water movement through the snowpack. Male and Gray (1981) comprehensively cover recent research on water movement in a snowpack, including discussion of the propagation rate of the meltwater flux in both homogeneous and stratified snowpacks.

Fig. 7.4 illustrates the storage and transmission of water in the snowpack for an observed rain-on-snow situation. Here, a snowpack having a water equivalent of 680 mm receives input from a 2-day rainstorm. Since the snowpack was colder than 0°C (the beginning snowpack temperature was apparently unknown), some rain/condensation was taken up in raising the temperature of the snow to 0°C through the process of freezing; this amounted to 8 mm of water. An additional quantity of liquid water, 12 mm, was permanently retained in the snowpack, making the total quantity of stored water 20 mm. Finally, the snowpack also temporarily stored some water input as it progressed through the pack. In this case, given the rate of input involved, a time delay of 12 hours occurred between the beginning of rain/melt and the beginning of runoff.

For practical applications in a rain-on-snow situation, snow condition effects can be thought of as an "initial loss" that is subtracted from input, much in the same way as initial losses in dry soil conditions are simulated in rainfall/runoff situations. For the engineer, the problem is to be able to recognize this potential and to be able to incorporate this time lag in the snow hydrology analysis, where appropriate. Practically speaking, this may not be a major factor in design analysis since the snow can usually be assumed to be fully primed prior to the beginning of significant runoff producing melt; however, in certain forecasting situations, the effect of snow conditioning can be noticeable, and it is definitely a factor that must be considered in continuous simulation models that operate through periods antecedent to active snowmelt periods.

1. Cold Content For practical applications, the concept of cold content is useful in quantifying the condition of the snowpack and determining its impact in delaying snowmelt input to the soil. Cold content defines the amount of energy needed to raise a dry snowpack to 0°C, in terms of the amount of water needed to be produced at the surface to release energy by freezing. This can be calculated by:

$$W_c = 0.5\rho_s d T'_s / 80 \qquad (7.6)$$

where W_c is the cold content (mm), 0.5 is the specific heat of ice (cal/(gm °C)), ρ_s is the average density of the snowpack (g/cc), d is the depth of pack (mm), T'_s is the average temperature deficit of snowpack below 0°C (°C), and 80 is the latent heat of fusion of water (cal/cc).

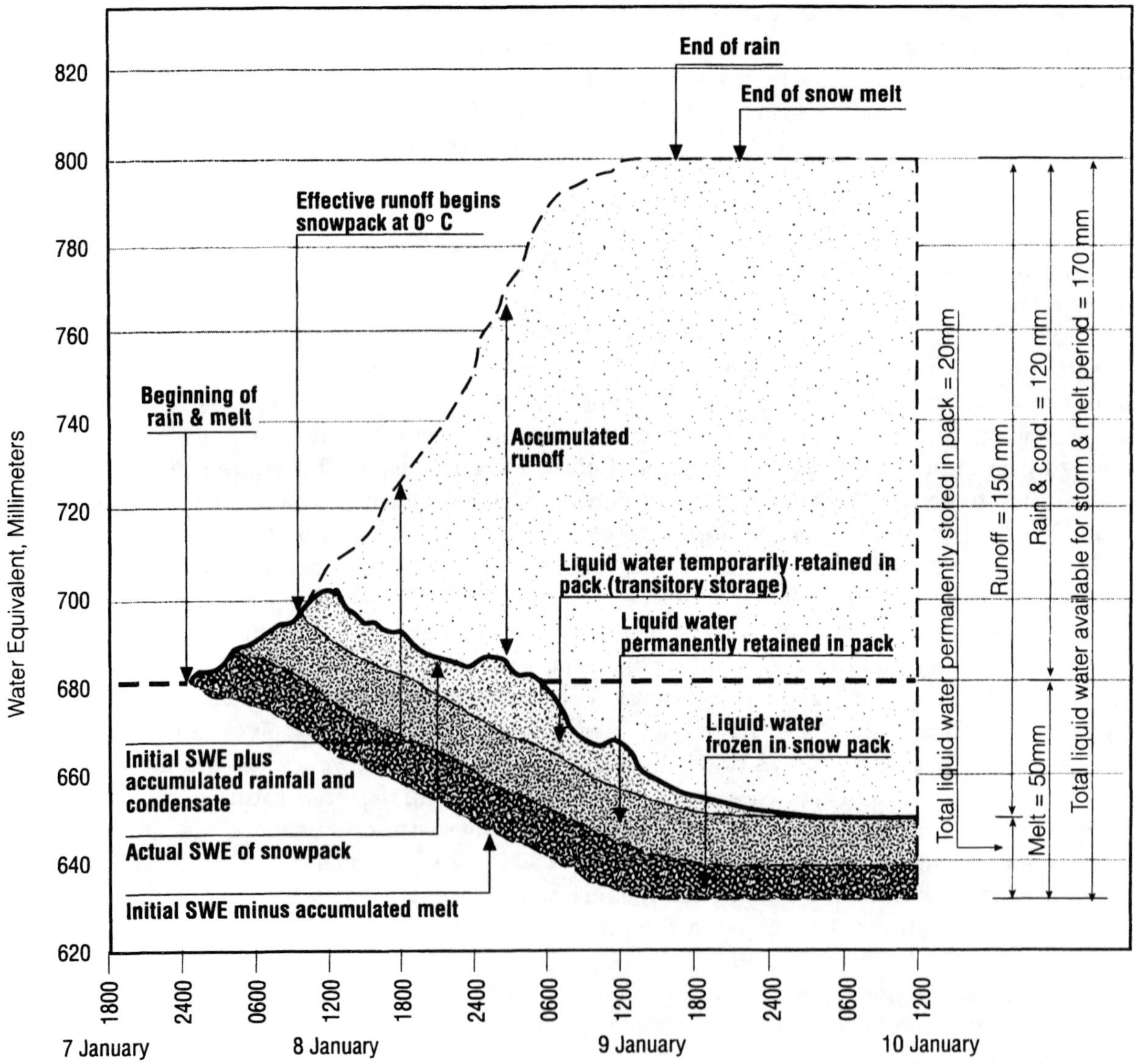

Figure 7.4—Snowpack Water balance during Rain on Snow.

For practical applications, the average snowpack temperature at the time of the application can be estimated based upon the air temperature for 1–3 days prior to the forecast time. Typically, the temperature is close to that of the air at the snowpack surface but approaches 0°C at the ground. For deep snowpacks, a further assumption can be made that only the top 60 cm or so of snow is subject to the influence of air temperature, and that the deeper pack is only 1°C or so below freezing. The density of this layer of snow can also be assumed to be greater than the top layer.

Table 7.7 lists calculated cold content values using Equation 7.6 for various assumed conditions. For the deep snowpack example, the calculation is subdivided into two layers, above and below a 60 cm depth. As can be seen, the cold content is a relatively small factor compared to the potential magnitude of rainfall and snowmelt. It varies from 3% or more of the SWE for "cold" snow to 1% or less of the SWE for deeper snow that is closer to 0°C.

2. ***Liquid Water Holding Capacity*** As shown in Fig. 7.4, the liquid water holding capacity of the snow is a second factor that can be considered an "initial loss" in practical applications of snow hydrology and

TABLE 7.7. Variation in Cold Content.

	Assumed factors			Calculated factors		
Descriptive condition	d (cm)	ρ	T'_s (°C)	SWE (mm)	W_c (mm)	Wc/SWE (%)
Shallow, relatively fresh snowpack. Several days of −8°C temperatures prior to application	40	0.20	6.0	80	3.0	3.8
Same, but warm snowpack	40	0.20	1.0	80	0.5	0.6
Deep snowpack, top 60 cm layer cold	60	0.20	5.0	120	4.0	
	90	0.30	1.0	270	2.0	
	150			390	6.0	1.4
Deep, ripe snowpack. Warmer antecedent conditions	60	0.35	1.0	210	1.0	
	140	0.45	0.5	630	2.0	
	200			840	3.0	0.4

d = snow depth, cm
ρ = snow density
T'_s = average temperature of snow layer, °C below freezing
SWE = beginning snow water equivalent, mm
W_c = cold content from Equation 7.6, mm

forecasting. Unfortunately, very little experimental evidence has led to quantification of this factor; it varies depending upon the depth and density of the snow, the mass of ice layers, and the channelization and honeycombing of the snowpack. It has been observed (USACE, 1956) that at 0°C this factor is approximately 2–5% of the SWE. For most practical applications, a fixed percentage of the SWE is used as an initial loss, in addition to the cold content loss. It should be noted that this magnitude of loss assumes the free drainage of the water. Therefore, in flat areas the snowpack may hold liquid water far in excess of the amount that is associated with mountainous areas.

3. ***Simulating Snow Condition*** Since the computation of cold content hinges on the estimation of the snowpack temperature, the common approach to modeling the snow condition in continuous simulation is to maintain an accounting of the relative temperature of the snowpack below freezing as a function of time. In effect, the snowpack is simulated as an energy reservoir; once the reservoir is full (snowpack isothermal and at 0°C), melt water moves to the ground. This can be done through an index relation such as that proposed by Anderson (1973):

$$T_s(2) = T_s(1) + F_p[T_a(2) - T_s(1)] \tag{7.7}$$

where T_s is the index of the snowpack surface temperature at times (1) and (2) (°C), T_a is the air temperature (°C), and F_p is a factor, varying from 0 to 1, representing the relative penetration of the air temperature into the snowpack. If F_p is close to 1.0, the snow temperature remains close to that of the air; thus, values close to 1.0 would be appropriate for shallow snowpacks. For a deep snowpack, a low value of F_p will result in a slow cooling or warming of the snow. The variable T_s is limited to a maximum of 0°C in the simulation process.

Once a snow surface temperature index is established for a computation period, the cold content can be calculated through an equation such as:

$$W_c(2) = W_c(1) + C_r[T_a(2) - T_s(1)] \tag{7.8}$$

where W_c is the cold content at times (1) and (2) (mm), T_a is the air temperature (°C), T_s is the index of the snowpack surface temperature (°C), and C_r is a conversion factor (mm/degree-day). The value of C_r can be made a variable in simulation models by relating it to calendar periods or to a cumulative temperature index function.

D. Snow Accounting During Snowmelt

As snowmelt progresses, the elevation of the snowline increases, and the areal snow cover of the basin decreases. An accounting of this is necessary to be able to: 1) differentiate between snow-free and snow-covered areas, which have different hydrologic characteristics, and 2) determine the elevation of the snowpack for calculating air temperature for indexing melt. A second computation needed is the accounting of the remaining snow water equivalent of the snowpack.

If, in modeling, the basin has been configured into zones of equal elevation as described in the section titled "Snow Water Equivalent Estimate from Historic Data," the accounting of snow cover and quantity can be done on a zone-by-zone basis. One assumption that can be made is to make the zone homogeneous with respect to elevation and either 100% snow-covered or snow-free. This assumption may require a large number of zones for adequate basin representation. Even with a large number of zones, abrupt changes in snowline can occur as a zone changes from snow-covered to snow-free. Because of this, some models provide an ability to simulate a gradual transition within the zone.

An alternative to the distributed approach in accounting for snow during melt is to employ a "snow cover depletion curve" in conjunction with a "lumped" watershed configuration. A snow cover depletion curve describes the basin's snow-covered area as a function of accumulated snow runoff (percent of seasonal total), as shown in Fig. 7.5. Studies have shown this relationship to be of relatively uniform shape for a basin. Using historic field and remote sensing information, a theoretical curve can be developed for a basin. As shown in Fig. 7.5, if actual observations in real time indicate a departure from the theoretical curve, then they follow a proportionally adjusted curve.

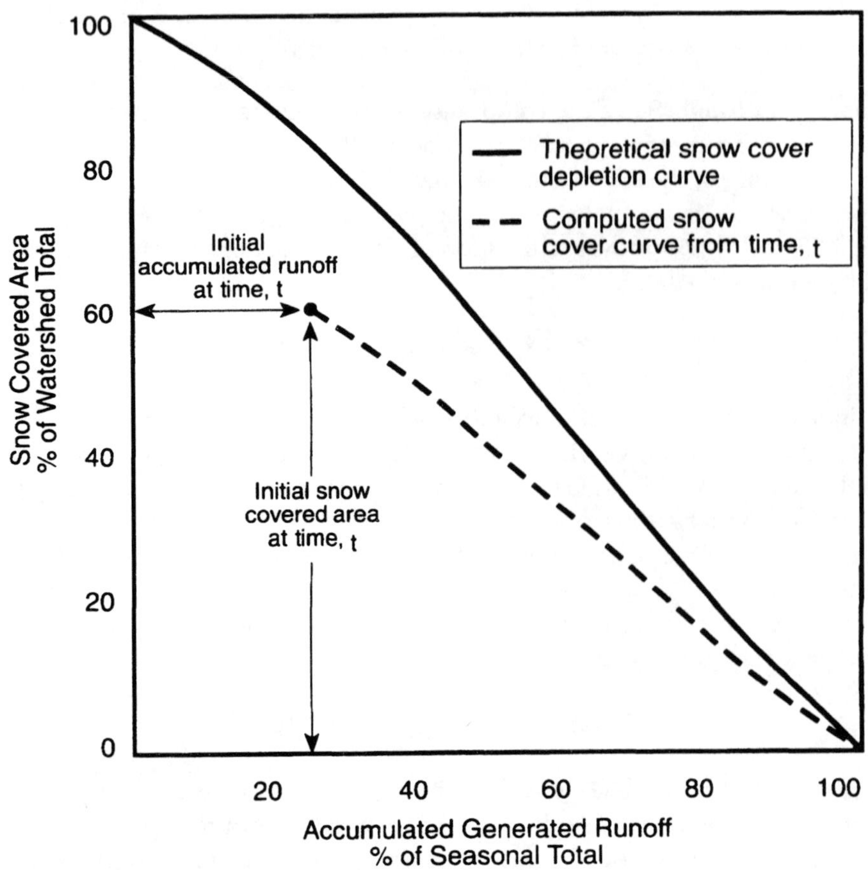

Figure 7.5—Illustration of Snow Cover Depletion Curve.

While the snow cover depletion curve yields an accounting of snow cover, this method still needs to employ an independently-derived estimate of expected total basin snow water equivalent. The typical approach is to use multiple-regression procedures, as described in the section titled "Conceptual Modeling Approach," to determine an initial estimate of the total SWE. The accounting of currently remaining SWE during the snowpack melt is simply a process of subtraction.

During the application of simulation models, use can be made of areal snowcover and ground observations in updating the model. Rango (1988) has effectively employed satellite observations in accounting for snow during snowmelt. In forecasting, adjustments to the estimates of SWE and snowcover can also be based upon model performance in simulating runoff. A technique to accomplish this has been developed by the National Weather Service. Using a geographical information system and optimal interpolation (kriging), mean areal SWE is calculated from snow measurements (snow course and SNOTEL). These values are used to update model-simulated SWE via Kalman filtering (Day et al., 1989; Day, 1990).

E. Snowmelt Simulation

The overall process of snowmelt simulation can be illustrated by the flow chart shown in Fig. 7.6. This represents a typical general algorithm that would be utilized in a hydrologic model in which the watershed has been configured into elevation bands.

VI. WATER SUPPLY FORECASTING

A. Background

Water supply forecasting is the prediction of the volume of water that flows past a given point on a stream during a specific period of time. Such forecasting is usually done for streams in which the primary source of flow is the melting of snow accumulated during the winter. This water storage in the form of snow makes it possible to forecast streamflow, as this provides a natural lead time between the accumulation of the snowpack and its eventual appearance as streamflow several months later. The period, or season, that is forecast typically covers the months when most of the snowmelt and summer runoff occurs, e.g., March–July, April–September. Forecasts are usually prepared shortly after the first of the month from January–May/June, and in some areas, forecasts are also prepared mid-month. In the United States, water supply forecasting is a long-standing practice in the West and, to a lesser extent, in the Northeast. Water supply forecasts are used for operational planning and management of water for irrigation, hydropower, flood control, fisheries, navigation, and recreation. Users of these forecasts range from individual farmers, who use forecasts for crop management, to government agencies operating multiple-purpose reservoir systems.

A number of entities produce water supply forecasts in the United States. Most forecasts for the general public are prepared by the Soil Conservation Service (SCS, U.S. Department of Agriculture) and the National Weather Service (NWS, National Oceanic and Atmospheric Administration, U.S. Department of Commerce). In the West, these two agencies publish forecasts for over 600 points, which appear in *Basin Outlook Reports*, issued by SCS, and in *Water Supply Outlook for the Western United States*, issued jointly by SCS and NWS. In California, forecasts are prepared by the State Department of Water Resources, while in the Northeast, NWS publishes water supply forecasts for the public. Some agencies responsible for the operation of reservoirs use the publicly available forecasts, but often they prepare their own forecasts or use both.

B. Regression Approaches

The traditional approach in developing water supply forecasts is to use multiple-linear regression equations. Input data for these equations include snow water equivalent at one or more snow-monitoring

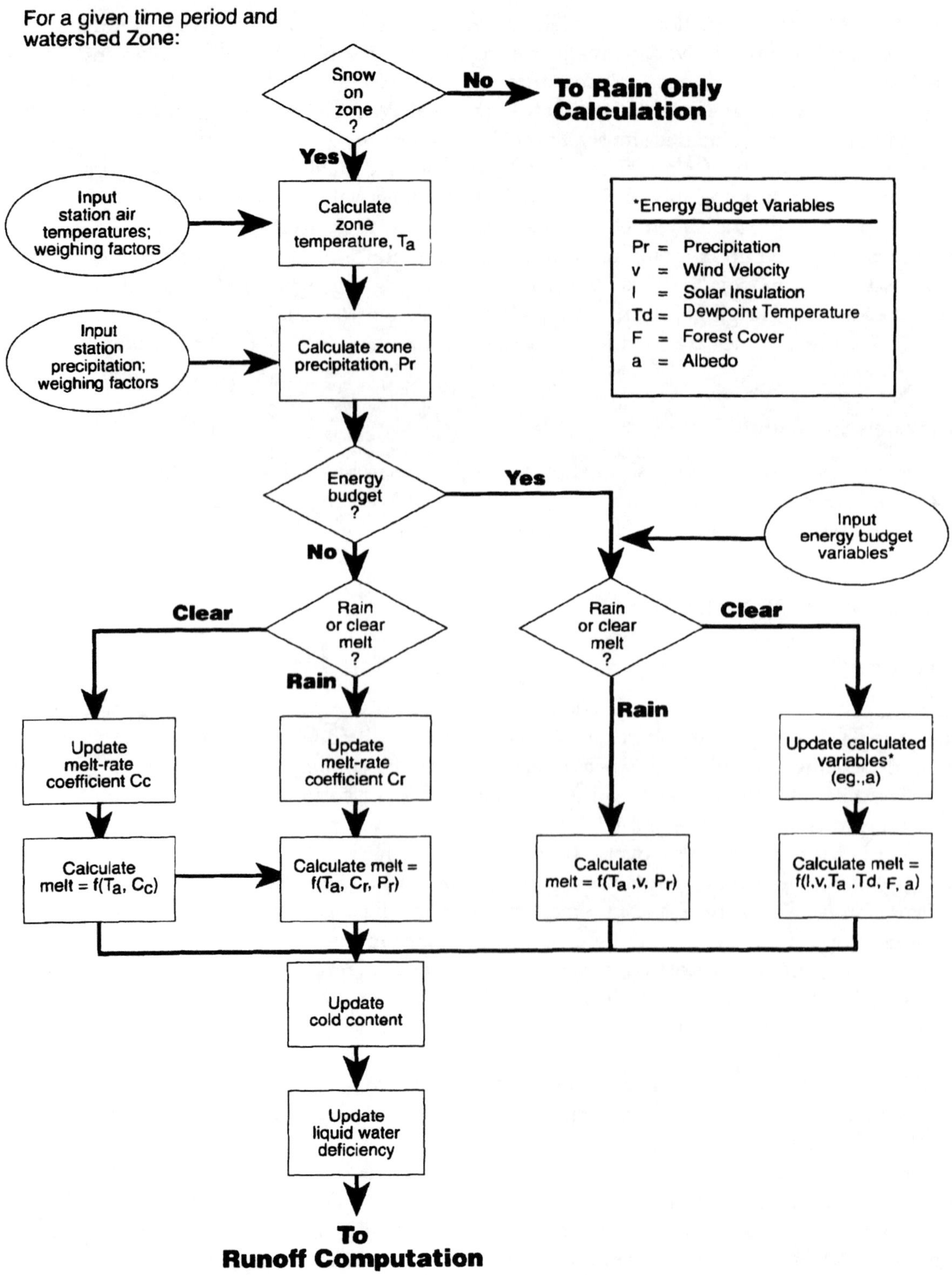

Figure 7.6—Illustration of Snowmelt Simulation.

sites, precipitation at one or more sites, and streamflow for previous months at the forecast point. Other data, such as soil moisture or temperature, are also used occasionally. The standard practice has been to aggregate like data types into weighted sums to develop indices, which become the independent variables in the regression.

A typical equation would be of the form (Soil Conservation Service, 1972):

$$Y = b_0 + b_1 BF + b_2 FP + b_3 S + b_4 SP \tag{7.9}$$

where Y is the seasonal streamflow volume, b_i are the regression coefficients, BF is the base flow index, FP is the fall precipitation index, S is the snow water equivalent index, and SP is the spring precipitation index.

The base flow index is usually the streamflow volume that occurs during the fall or early winter, e.g., October–December, November–January. The fall precipitation index is a sum or weighted sum of monthly precipitation at one or more sites for the fall period, e.g., September–November, October–December. The fall precipitation index and the base flow index are considered to be surrogates for soil moisture. The snow index is a sum or weighted sum of snow water equivalent at several sites for the month usually having the maximum snow accumulation for the season; this is typically April, although it can be March or May. The spring precipitation index is the same as the fall precipitation index except for the spring period, e.g., April–June. Using multiple-linear regression, equations of the above form can be calibrated using observed data. At least 15 years of data are considered necessary for reasonable forecast accuracy.

Recent research and development has shown that the traditional multiple-linear regression approach can be improved by employing several techniques as outlined by Garen (1992). The techniques are briefly described below.

1. Using Different Equations for Each Date of Forecast The traditional approach is to develop a single equation of the form in Equation 7.9 using observed data through the peak of the snow accumulation and on into the spring. When the equation is applied for forecasting, some of these variables are in the future and hence unknown; therefore, long-term averages are substituted. This approach is conceptually appealing, but it results in nonoptimal forecasts because the equation is only optimal when all data are known. A more straightforward and statistically optimal approach is to develop equations that only use data known at forecast time. This results in separate equations for each forecast date. This process eliminates the need to assume anything about future weather, and it allows for a more accurate depiction of forecast accuracy; however, a concern sometimes raised about this practice is that the series of forecasts may show unexplainable jumps up and down due solely to the different variables and coefficients in the equations and not due to changing hydrologic conditions. This does not have to be a problem if care is taken to ensure that the variables included are physically meaningful and that there is similarity in variable usage from equation to equation.

2. Employing Principal Components Regression without Combining Data into Indices The independent variables used in water supply forecasting are usually highly intercorrelated with each other, which causes problems in regression models if they are used as separate variables. The use of indices, where like data types are aggregated, partially circumvents this problem; however, the construction of these indices is partly subjective and, being done outside the regression, is not necessarily optimal. A more satisfactory and statistically rigorous way to deal with intercorrelation is the use of principal components regression. Principal components analysis is a technique that restructures a set of intercorrelated variables into an equal number of uncorrelated variables via a linear transformation. Each new variable (principal component) is a different linear combination of all the original variables. These principal components can themselves be used as independent variables in a regression equation. Only a few of the components are generally retained in the equation, as the bulk of the information contained in the original variables are represented by the first few principal components. After the regression coefficients for the principal components have been computed, the linear transformation can be inverted so that the coefficients are expressed in terms of the original variables.

3. Using Cross-Validation Techniques for Estimating Forecast Error Also called "jackknifing," this procedure is a way of better assessing the ability of a regression equation to perform in an actual forecasting situation than simply looking at the normal standard error. In the cross-validation (jackknife) procedure, one observation is removed from the data set, and the regression coefficients are calculated. These coefficients are used with the data for the withheld observation to make a prediction for the dependent variable of the withheld observation. The withheld observation is returned to the data set, and the next observation is removed. This process is repeated through the entire data set so that when finished, a set of forecasts generated by equations that have not "seen" the observation that they were forecasting is available. A standard error can then be calculated from these "forecasts." By attempting to mimic the actual forecasting situation in this way, one obtains a more realistic measure of the actual forecasting potential of an equation than the standard error where all observations are used to estimate the regression coefficients.

4. Systematic Searching for Optimal or Near-Optimal Combinations of Variables Many independent variables are often available to be potentially included in a regression model. Traditionally, variables for inclusion in indices have been selected before entering the regression using indicators such as correlations, data quality, and subjective judgment concerning how well each variable represents the watershed. Standard variable selection techniques, such as stepwise regression, are inadequate when principal components regression is used. Garen (1992) describes a search procedure that can be used with principal components regression that systematically tests numerous combinations of the available variables to arrive at combinations that have the smallest or nearly the smallest cross-validation (jackknife) standard error. Optimization of the variables used can contribute significantly to improved forecast accuracy.

5. Using Indices of Large-Scale Atmospheric Circulation as Predictor Variables In some areas of the northwestern and southwestern United States, the Southern Oscillation Index (SOI), which is an indicator of the occurrence of the El Niño phenomenon, provides a moderate, but significant, indicator of winter precipitation and snowpack with a lead time of as much as six months (Cayan and Peterson, 1989; Koch et al., 1991; Redmond and Koch, 1991). This can provide an additional increment of forecast accuracy, particularly early in the winter before the majority of snow accumulation has occurred.

These techniques, when used together as a system, can result in significant improvements in forecast accuracy over traditional practice, especially early in the forecast season. These techniques are currently in use by SCS, as well as other agencies, for making water supply forecasts in the western United States.

C. Conceptual Modeling Approach

The need for more detailed streamflow forecasts than seasonal volumes and the limitations of regression have led, in recent years, to the application of conceptual watershed models to water supply forecasting. These models attempt to simulate the various physical processes occurring in the watershed, including snow accumulation and melt, evapotranspiration, infiltration, and the movement of water through the soil to produce streamflow. They typically simulate streamflow using a daily time step, although smaller time steps are sometimes used, especially if the model is also used for flood forecasting.

While regression models provide adequate seasonal volume forecasts in many cases, they have their limitations. As already mentioned, a finer time resolution is desirable. In addition, performance in extreme years (dry or wet) can be questionable, especially in ones beyond the range of the observations used to calibrate the equations. Adding to this problem is the general lack of adequate accounting for soil moisture; measurements are not commonly available and the surrogate variables used may not be good substitutes. By using models that have a greater basis in physical reality than purely statistical techniques, it may be possible to obtain greater seasonal volume forecast accuracy, with the added benefit of greater time resolution of streamflow.

The general process for using conceptual models for water supply forecasting follows the Extended Streamflow Prediction (ESP) procedure of the National Weather Service (Twedt et al., 1977; Day, 1985). To produce a forecast, a calibrated model is first run up to the forecast date using observed input data. Then, for the period between the forecast date and the end of the forecast period, the model is run repeatedly using historical traces of precipitation and temperature (and any other necessary variables) as input data. The model is generally run for all input data traces available in the historical record; this assumes that the occurrence of each trace is equally likely and that the historical record is a representative sample of climate scenarios. Numerous possible future streamflow traces are thereby generated, each beginning at the current hydrologic state of the watershed and each presumably being equally likely. Statistical analyses can be performed on these traces to calculate means and confidence bounds for any of several streamflow quantities, which can include seasonal volume, peak flow, date to reach a specified low flow, duration above or below a specified flow, and average hydrograph.

VII. COMPUTER PROGRAMS AVAILABLE

The following computer programs developed in the United States possess snow simulation routines and have been used in practical hydrology applications. The WMO (1986) also describes other models that participated in a world-wide comparison of snowmelt runoff models.

A. HEC-1, HEC-1F

Contact: U.S. Army Corps of Engineers
Hydrologic Engineering Center
609 Second Street
Davis, California 95616

B. National Weather Service River Forecast System (NWSRFS)

Contact: National Oceanic and Atmospheric Administration
National Weather Service, Office of Hydrology
1325 East–West Highway
Silver Spring, Maryland 20910

C. Precipitation Runoff Modeling System (PRMS)

Contact: U.S. Geological Survey
Branch of Regional Research
P.O. Box 25046
Mail Stop 412
Denver Federal Center
Lakewood, Colorado 80225

D. Snowmelt Runoff Model (SRM)

Contact: U.S. Department of Agriculture
Agricultural Research Service
Hydrology Laboratory
Building 007, Room 104, BARC–West
10300 Baltimore Avenue
Beltsville, Maryland 20705-2350

E. Streamflow Synthesis and Reservoir Regulation (SSARR)

Contact: U.S. Army Corps of Engineers
North Pacific Division
P.O. Box 2870
Portland, Oregon 97208-2870

VIII. SAMPLE APPLICATIONS

A. Example #1—Hypothetical flood, Rain-on-Snow, Temperature Index

1. Setting A reconnaissance study for a potential flood protection project in an ungaged basin requires estimates of the 100-year flood. The project is located in a 75 sq km basin near an urbanized area. The basin is ungaged and is about 25% forest covered. It is estimated to contain about 10% impervious area. The design condition is a mid-winter rainstorm associated with the passage of a warm front. Because the basin frequently is covered with snow, the snowmelt contribution to the flood must be determined. The flood estimate is to be determined by applying a 100-year rainstorm to the basin using the HEC-1 model.

2. *Previously Determined Factors*

a. A synthetic unit hydrograph has been derived. Snyder's factors are: Tp = 2.1 hours; Cp = 0.40.
b. Using NOAA Atlas II, the basin 100-year precipitation has been calculated and disaggregated into 1-hour time increments. Total 24-hour precipitation is 94 mm, and the maximum hourly precipitation is 10 mm.
c. A constant loss rate of 0.90 mm/hour will be used.

3. *Snow-Related Factors*

a. Snow water equivalent at beginning of storm. Weather records at a city near the basin indicate that mid-winter snow depths have reached as much as 70 cm. For purposes of the hypothetical flood derivation, a snow cover nearer the average is to be used. This is estimated to be 30 cm deep in the city and 45 cm averaged over the basin. Assuming a density of 20%, this equates to a snow water equivalent at the beginning of the storm of 9 cm (45 cm × 0.20).

b. Temperature. Winter rainstorms result from a warm front passage that includes a warm, moist air mass. Examination of weather records for a major historic storm indicates air temperatures of 15°C and high winds. For the design assumptions, temperatures lapsed to the mid-elevation of the basin begin at near freezing and increase to a maximum of 12°C during the height of the storm.

c. Melt-Rate Coefficient. For this derivation, it has been decided to use the temperature index approach in computing hourly snowmelt. Since the basin is relatively open and subject to high convection–condensation melt rates, the melt–rate coefficient must be chosen carefully. This is done by using Equation 7.3. Assuming quantities T_a = 12°C, v = 25 km/hr, and P_r = 90 mm, the 24-hour snowmelt would be about 90 mm. This suggests a value for C of 7.0-8.0 mm/°C in Equation 7.5, using a base temperature of 0°C. A coefficient of 7.0 mm/°C will be used initially, and a sensitivity test made to see its relative influence.

d. Snow Condition. The snow condition at the beginning of the storm will be assumed isothermal and ready for melt.

e. HEC-1 Output. A summary of the HEC-1 output is shown in Fig. 7.7, and a plot of the computed hydrograph is shown in Fig. 7.8.

HYDROGRAPH AT STATION 1

DA MON HRMN	ORD	PRECIP	TEMP	SNOMELT	SNOLOSS	SNOEXCS	RAIN	RAINLOS	RAINEXS	SNO+RAIN	LOSS	EXCESS	COMP Q
25 DEC 0000	1	0.00	4.00	0.00	0.00	0.00	0.00	0.00	0.00	0.00	0.00	0.00	1
25 DEC 0100	2	0.00	4.70	1.07	0.90	0.17	0.00	0.00	0.00	1.07	0.90	0.17	1
25 DEC 0200	3	0.00	5.30	1.26	0.90	0.36	0.00	0.00	0.00	1.26	0.90	0.36	2
25 DEC 0300	4	0.00	6.00	1.46	0.90	0.56	0.00	0.00	0.00	1.46	0.90	0.56	3
25 DEC 0400	5	0.00	6.70	1.65	0.90	0.75	0.00	0.00	0.00	1.65	0.90	0.75	5
25 DEC 0500	6	0.00	7.30	1.85	0.90	0.95	0.00	0.00	0.00	1.85	0.90	0.95	8
25 DEC 0600	7	0.00	8.00	2.04	0.90	1.14	0.00	0.00	0.00	2.04	0.90	1.14	11
25 DEC 0700	8	2.00	8.30	2.14	0.47	1.67	2.00	0.43	1.57	4.14	0.90	3.24	16
25 DEC 0800	9	2.20	8.70	2.24	0.45	1.78	2.20	0.45	1.75	4.44	0.90	3.54	25
25 DEC 0900	10	2.40	9.00	2.33	0.44	1.89	2.40	0.46	1.94	4.73	0.90	3.83	36
25 DEC 1000	11	2.50	9.30	2.43	0.44	1.99	2.50	0.46	2.04	4.93	0.90	4.03	46
25 DEC 1100	12	2.60	9.70	2.53	0.44	2.08	2.60	0.46	2.14	5.13	0.90	4.23	54
25 DEC 1200	13	2.80	10.00	2.63	0.44	2.19	2.80	0.46	2.34	5.43	0.90	4.53	62
25 DEC 1300	14	2.90	10.30	2.72	0.44	2.29	2.90	0.46	2.44	5.62	0.90	4.72	69
25 DEC 1400	15	3.20	10.70	2.82	0.42	2.40	3.20	0.48	2.72	6.02	0.90	5.12	76
25 DEC 1500	16	3.60	11.00	2.92	0.40	2.51	3.60	0.50	3.10	6.52	0.90	5.62	83
25 DEC 1600	17	5.50	11.30	3.01	0.32	2.70	5.50	0.58	4.92	8.51	0.90	7.61	92
25 DEC 1700	18	6.30	11.70	3.11	0.30	2.81	6.30	0.60	5.70	9.41	0.90	8.51	106
25 DEC 1800	19	6.50	12.00	3.21	0.30	2.91	6.50	0.60	5.90	9.71	0.90	8.81	121
25 DEC 1900	20	8.00	12.00	3.21	0.26	2.95	8.00	0.64	7.36	11.21	0.90	10.31	137
25 DEC 2000	21	10.00	12.00	3.21	0.22	2.99	10.00	0.68	9.32	13.21	0.90	12.31	155
25 DEC 2100	22	8.00	12.00	3.21	0.26	2.95	8.00	0.64	7.36	11.21	0.90	10.31	173
25 DEC 2200	23	4.00	12.00	3.21	0.40	2.81	4.00	0.50	3.50	7.21	0.90	6.31	180
25 DEC 2300	24	3.50	12.00	3.21	0.43	2.78	3.50	0.47	3.03	6.71	0.90	5.81	173
26 DEC 0000	25	3.00	12.00	3.21	0.47	2.74	3.00	0.43	2.57	6.21	0.90	5.31	162
26 DEC 0100	26	2.80	11.30	3.01	0.47	2.55	2.80	0.43	2.37	5.81	0.90	4.91	150
26 DEC 0200	27	2.70	10.70	2.82	0.46	2.36	2.70	0.44	2.26	5.52	0.90	4.62	140
26 DEC 0300	28	2.50	10.00	2.63	0.46	2.16	2.50	0.44	2.06	5.13	0.90	4.22	130
26 DEC 0400	29	2.40	9.30	2.43	0.45	1.98	2.40	0.45	1.95	4.83	0.90	3.93	121
26 DEC 0500	30	2.20	8.70	2.24	0.45	1.78	2.20	0.45	1.75	4.44	0.90	3.54	112
26 DEC 0600	31	2.00	8.00	2.04	0.45	1.59	2.00	0.45	1.55	4.04	0.90	3.14	104
26 DEC 0700	32	0.00	7.30	1.85	0.90	0.95	0.00	0.00	0.00	1.85	0.90	0.95	93
26 DEC 0800	33	1.00	6.70	1.65	0.56	1.09	1.00	0.34	0.66	2.65	0.90	1.75	80
26 DEC 0900	34	1.00	6.00	1.46	0.53	0.92	1.00	0.37	0.63	2.46	0.90	1.56	69

EXPLANATION OF CODES

DA MON HRMN: DAY, MONTH, HOUR, MINUTE
ORD: ORDINATE NUMBER
PRECIP: PERIOD PRECIPITATION, mm
TEMP: PERIOD TEMPERATURE, , degrees C
SNOMELT: COMPUTED PERIOD SNOWMELT, mm
SNOLOSS: COMPUTED PERIOD SNOWMELT LOSS, mm
SNOEXCS: PERIOD SNOWMELT EXCESS, mm
RAIN: BASIN PERIOD RAINFALL, mm
RAINLOS: COMPUTED PERIOD RAIN LOSS, mm
RAINEXS: COMPUTED PERIOD RAINFALL EXCESS, mm
SNO+RAIN: TOTAL OF SNOWMELT PLUS RAINFALL FOR PERIOD, mm
LOSS: TOTAL OF SNOW LOSS AND RAIN LOSS FOR PERIOD, mm
EXCESS: TOTAL OF SNOW EXCESS AND RAIN EXCESS FOR PERIOD, mm
COMP Q: COMPUTED DISCHARGE FOR PERIOD, cubic meters per sec

Figure 7.7—HEC-1 Output, Example #1.

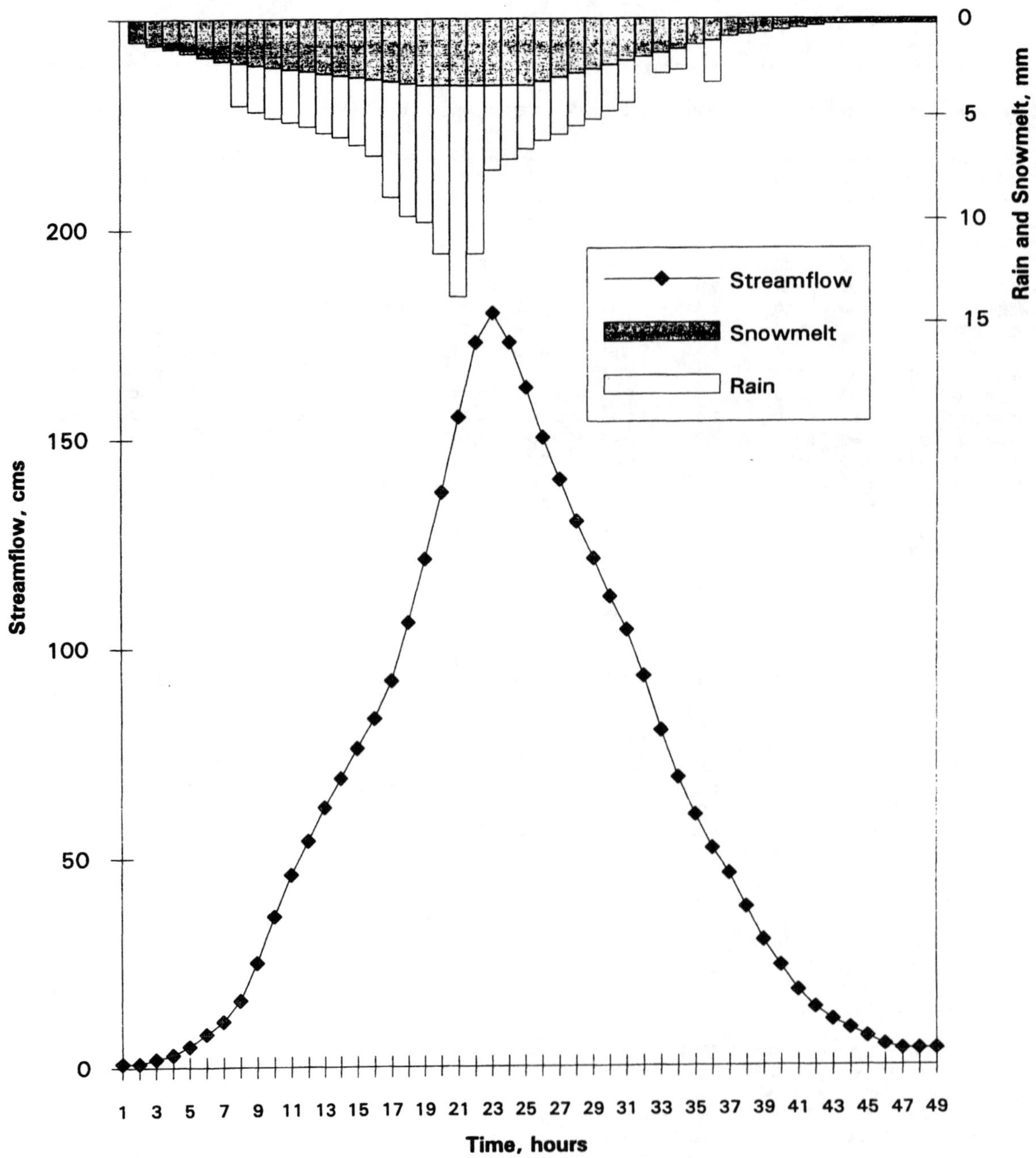

Figure 7.8—Plot of Results, Example #1.

f. Sensitivity Test. Several simulations were made with varying melt-rate factors. The results are shown in Fig. 7.9. It can be seen that an incremental change in C by 1.0 results in about a 5–6% change in the peak of the design flood. Further consideration of the meteorological conditions present led to discarding lower melt factors and accepting a factor on the order of 6–8. The original factor of 7.0 was, therefore, verified.

The assumption of the initial SWE of 90 mm was verified by inspection. Approximately 54 mm of snowmelt occurred prior to the maximum moisture input to the flood, indicating that the SWE could be reduced by 60% and still remain fully contributing to the peak of the runoff. This is greater than the error in estimating the SWE.

C mm/degC	Max Melt mm	PEAK cms
3	1.4	144
4	1.8	153
5	2.3	162
6	2.8	171
7	3.2	180
8	3.7	189

a

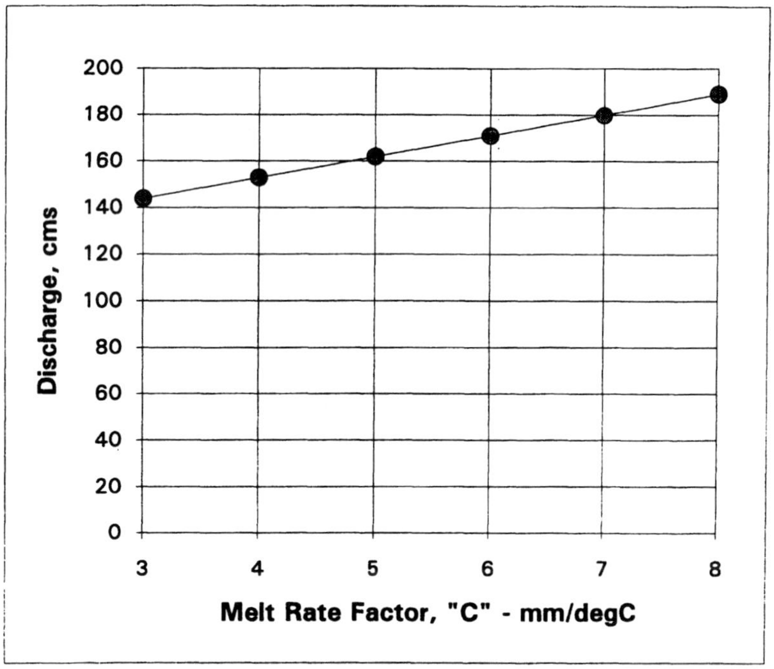

b

Figure 7.9—Sensitivity of Melt–Rate Factor, Example #1.

g. Commentary. The purpose of this example is to illustrate the use of the temperature index method for computing snowmelt. Because the setting is described as relatively unforested, and therefore subject to the effects of wind velocity in producing convection and condensation melt, it would be preferable to use the energy budget approach and include wind velocity as an assumed input. In effect, the energy budget was used to calculate the daily melt (therefore, C), and the temperature index was used to distribute the melt through the 24-hour storm duration.

B. Example #2—Design Flood Derivation for a Partly Forested Basin

1. Setting An inflow design flood of 1/2 Probable Maximum Flood (PMF) magnitude is to be derived for a dam safety analysis. The project is located in a mountainous region of western Montana that is subject to snowmelt runoff during April through July. Snowmelt can be augmented by spring rainstorms. The basin area is 4,000 sq. km, and elevations in the basin range between 900 and 2,500 m. For modeling purposes, the basin is divided into five bands of equal elevation, and snowmelt and rainfall quantities are calculated with a spreadsheet for input into a hydrologic model. For this illustration, computations are made for one elevation zone at 1500–1810 m. It has been decided to use Equations 7.3 and 7.4 as the basis for calculating melt quantities. A 24-hour time step will be used. Based upon analysis of historic records, the design flood should consist of the melting of a large snowpack, combined with the imposition of a rainstorm occurring near the peak of the snowmelt runoff. Snow water equivalent at the beginning of the design event and rainfall quantity appropriate for the design flood magnitude have been previously determined by independent hydrometeorological analysis.

2. *Snowmelt Factors* Meteorological factors for several extreme historic floods have been analyzed and, based upon these factors and other guidance, time series for the variables have been derived as follows. These are judged to reflect reasonably extreme conditions consistent with the 1/2 PMF magnitude flood being derived.

a. Solar Radiation. Fig. 7.10a displays the assumed insolation to be used in Equation 7.4. This is based upon references (USACE, 1956; List, 1968) that give solar radiation as a function of latitude and date. Losses due to atmospheric conditions such as cloud cover are assumed to be relatively low. During the rainstorms, insolation was reduced substantially as shown.

b. Air and Dewpoint Temperatures. Fig. 7.10a shows the assumed temperature and dewpoint pattern for the design event. The air temperature sequence is based upon previous hydrometeorological studies for the region that, in turn, utilized historic temperature records. For dewpoint temperatures, a dewpoint depression of about 8°C is assumed for rain-free periods. The temperatures shown have been determined by factoring using an assumed lapse rate of 0.55°C/100m from a base station having a long period of record. This lapse rate is used to calculate temperatures at other elevations.

c. Wind Velocity. A constant wind velocity of 10 km/hr is assumed during rain-free periods. During the rainstorms, wind speed is increased to 40 km/hr. The wind velocity time series is shown in Fig. 7.10b.

d. Albedo. The assumed pattern of albedo versus date is shown in Fig. 7.10c. It is an idealized shape based upon information from USACE (1956).

e. Forest Cover. This factor (*F*), which affects solar radiation that is effective in providing energy for snowmelt, is assumed to be relatively high for this elevation range; a value of 0.6 is used. Forest cover would be lower for higher elevation bands in the basin.

f. Short-Wave Radiation Melt Factor. This factor (kr) varies with season of the year, slope of the basin, and proportion of basin that is north- versus south-facing. It can vary between 0.8 to 1.2 in the spring season (USACE, 1956). For this example, a value of 1.0 is used.

g. Convection-Condensation Melt Factor. This factor (k_r) is a function of the relative exposure of the basin to wind. For unforested areas, this factor would be 1.0. Given the partial forest cover for this band, a value of 0.5 is used.

3. *Snowpack* An analysis is made of snow station records in and near the basin at the beginning of the snowmelt season, and estimates of SWE probability are made. Given station probability estimates, SWEs for individual elevation bands in the basin model could be made. For this example, a value of 1,600 mm is chosen for the May 1 starting conditions. This represents a relatively extreme condition of approximately 50-year return period. The snowpack is assumed to have been accumulating through part of April, and during the latter half of April it undergoes a ripening process such that it is fully primed for runoff by May 1.

4. *Rainstorms* As shown in the figures, the clear weather conditions are assumed to be interrupted by a major rainstorm in early June, thereby modifying the clear weather melt factors abruptly. Based upon analysis of historic meteorological conditions, the rainstorm is assumed to be 5 days in duration with a total rainfall of 200 mm. Based upon the National Weather Service hydrometeorological reports for PMP of the region, this magnitude represents approximately 1/2 PMP quantity. The rainfall pattern is shown in Fig. 7.10b.

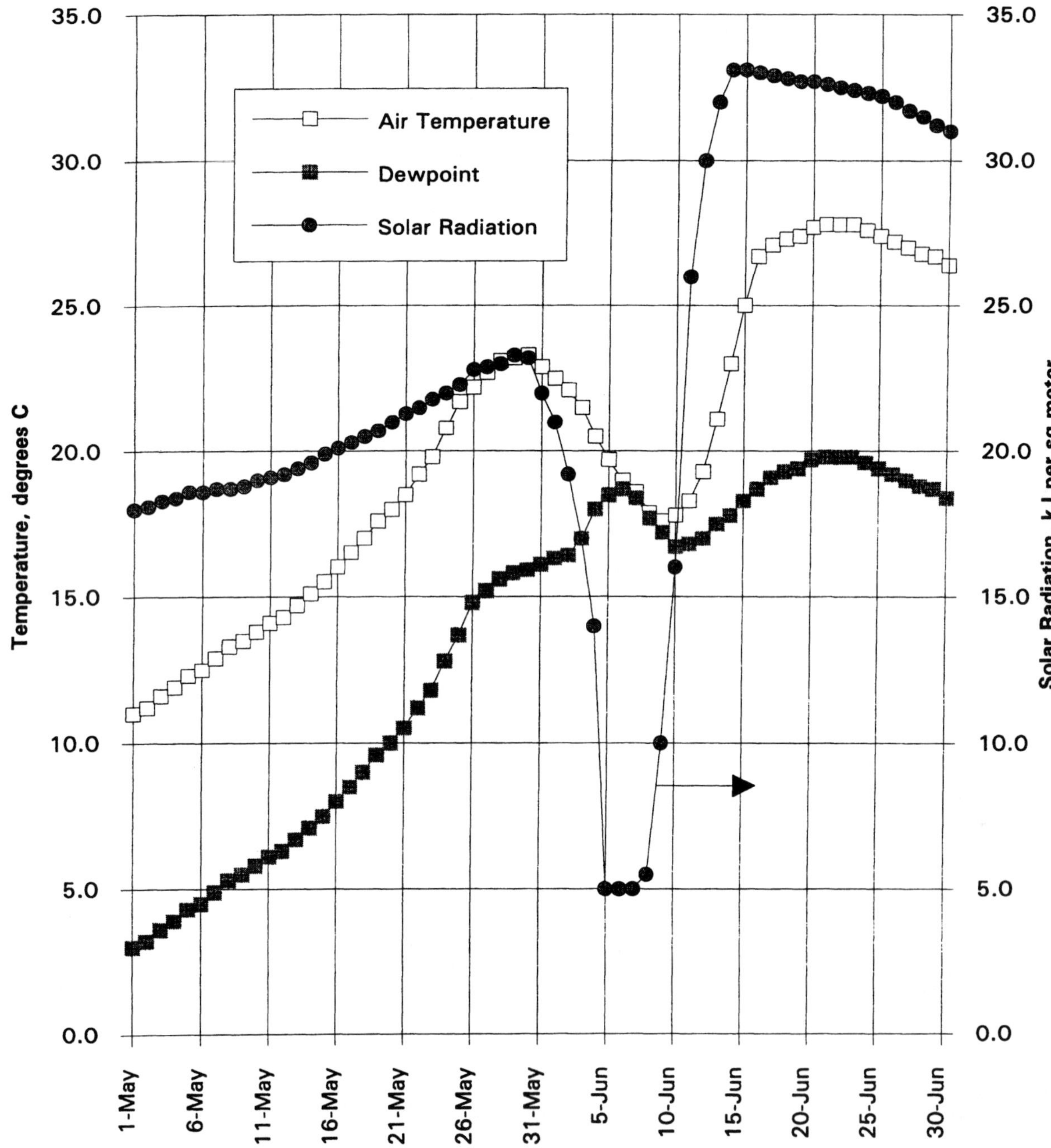

Figure 7.10a—Input and Output Data, Example #2 (Air Temp, Dew Point, Solar Radiation).

5. ***Computations*** The resulting snowmelt and rain is shown in Fig. 7.10d and the spreadsheet calculation for one band is shown in Fig. 7.11. The components of snowmelt are shown separately on the table for illustration of their relative magnitude. Although not shown, Equation 7.3 was used during the 5 days of rainfall, while Equation 7.4 was used at other times. Note that the accumulated melt reached 1,600 mm on June 10th so thereafter snowmelt is zero. This is assumed as being an abrupt transition, although in reality a gradual reduction in melt-rate would occur as the area became partly depleted of snow cover.

6. ***Commentary on Design Flood Development*** The derivation of a hypothetical design flood requires careful attention to the combination of factors that leads to the desired design hydrograph. In the example

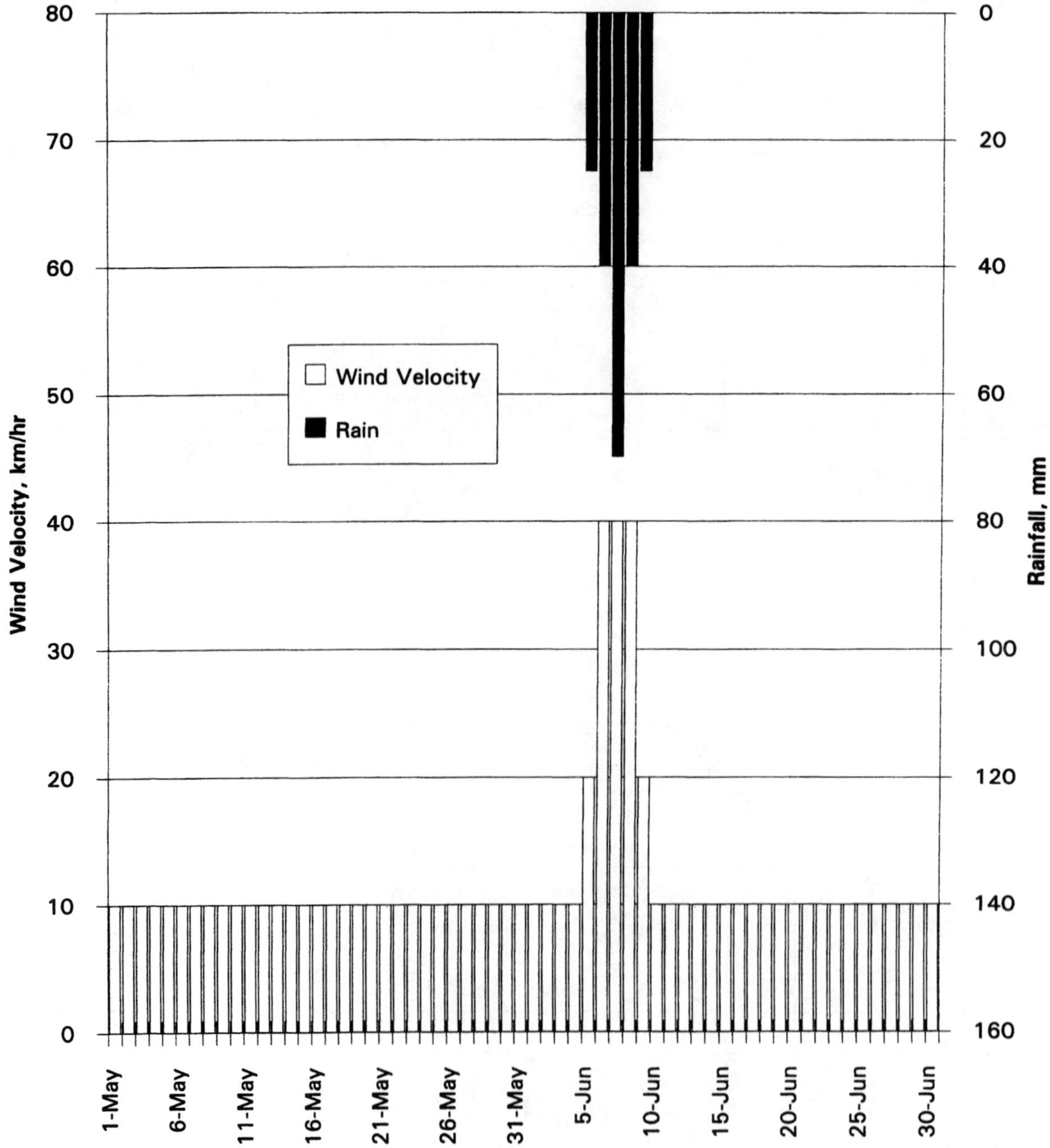

Figure 7.10b—Input and Output Data, Example #2 (Wind Velocity and Rain).

shown, for instance, an important factor is the timing of the rainstorm and the extent of the snowpack depletion prior to the storm. In this case, relatively cool conditions with reduced solar radiation were assumed during the weeks prior to the rainstorm, to assure that there was snow available at the time of the rainstorm so that the peak of the hydrograph is augmented by both components. This assumption must be verified by analysis of meteorological records, considering the relative magnitude of the design event. Another important decision is how to maximize the combination of rain and snow magnitudes. In the example, the snowpack represents about a 50-year event, while the rainstorm is 1/2 PMP magnitude. The alternative might be the reverse—a maximized snowpack with a rainstorm of a specified probability.

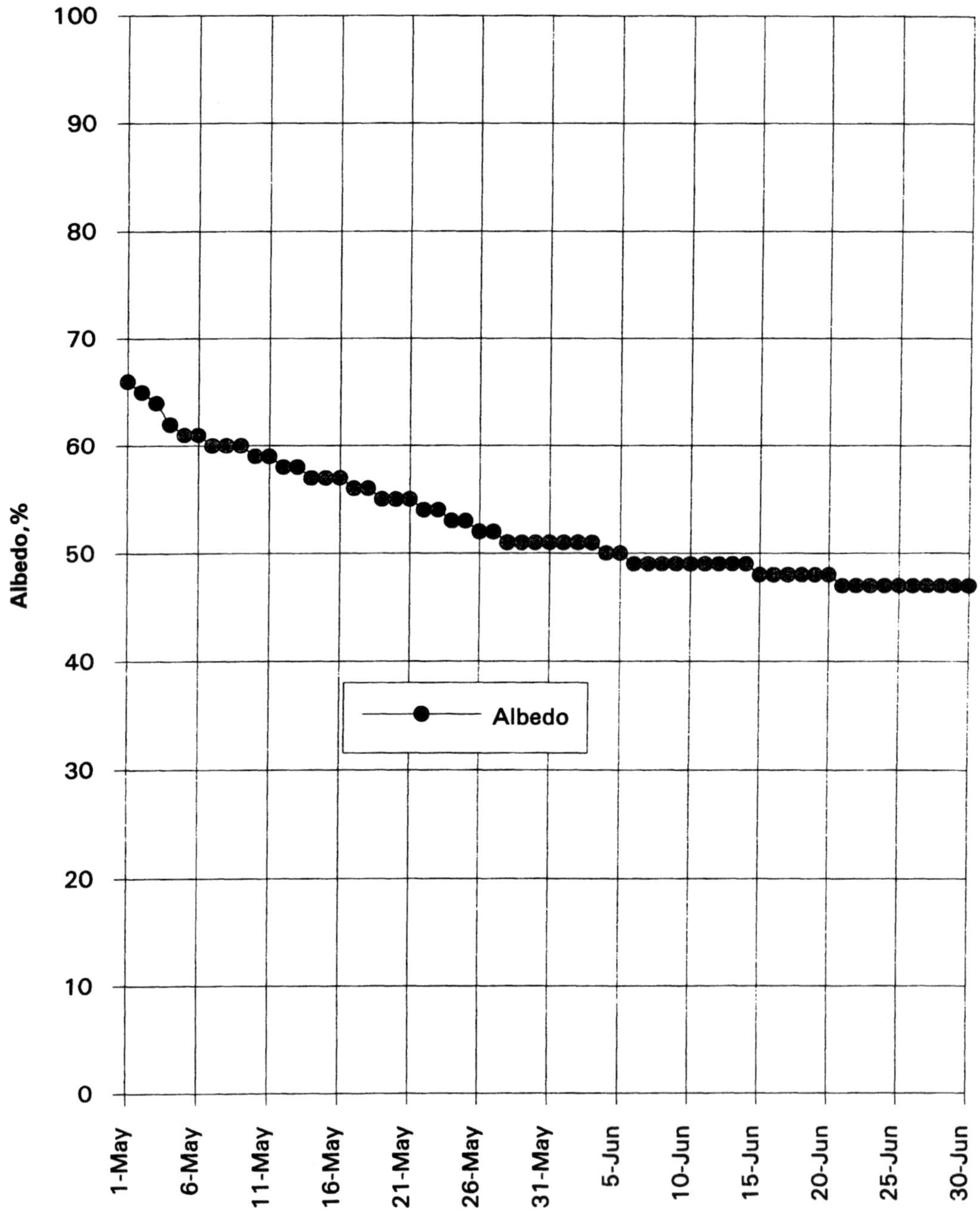

Figure 7.10c—Input and Output Data, Example #2 (Albedo).

This might yield a hydrograph of greater volume but lower peak, which may be more critical where a large storage reservoir is involved to regulate the inflow.

The spreadsheet calculation provides a convenient mechanism for demonstrating the sensitivities of these and other assumptions, to insure that a reasonably sized end-product is achieved. Ultimately, of course, the snowmelt and rainfall needed to input into a hydrologic/reservoir routing model to determine the resulting design flood as it is affects the structure under investigation. The spreadsheet calculation would be unnecessary for those models that incorporate the desired degree of sophistication in computing snowmelt.

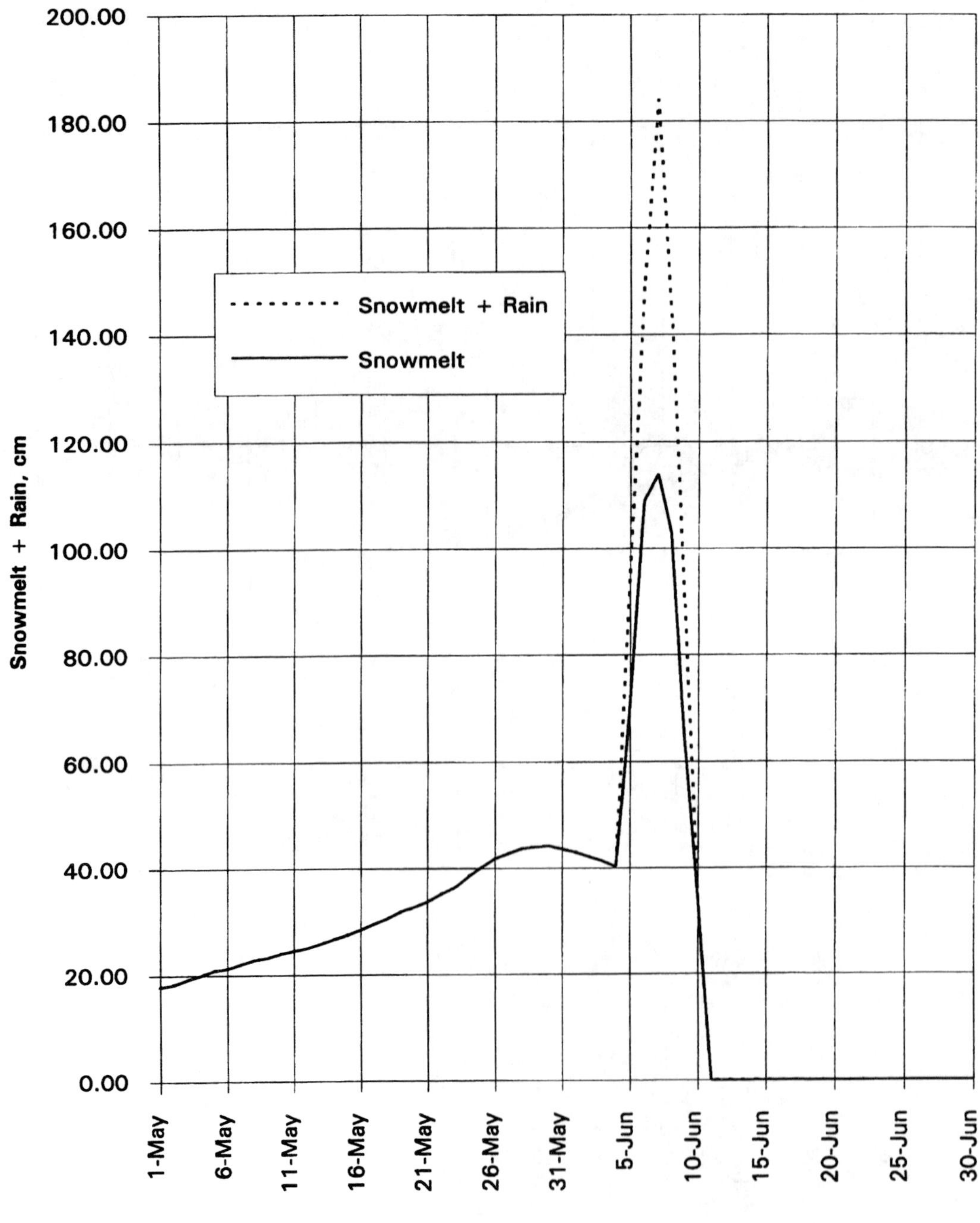

Figure 7.10d—Input and Output Data, Example #2 (Snow and Melt–Rain).

C. Example #3—Model Calibration, Continuous Simulation, Temperature Index

1. Setting A continuous simulation model (SSARR) is used for forecasting on a 1155-sq. km basin that is predominantly snowmelt fed, although rainfall can occur during melt periods. The model is calibrated by simulating 25 continuous years of data. The model has been set up using eight elevation bands upon which independent accounting of precipitation, snow, soil moisture, and runoff are made. This example is a short extraction from model output during the spring rise for 1 year in the record.

DESIGN FLOOD DERIVATION
SOUTH FORK FLATHEAD RIVER
BAND 3, ELEVATION 1500 - 1800 m
Assumed constants: kr = 0.7
F = 0.7
kw = 0.4

EQUATIONS	
M = kr(1-F)(2.43*I)(1-a)	(short-wave)
+kw(2.39V)(0.22Ta+0.78Td)	(conv.-condens.)
+F(1.326Ta)	(long-wave)
M = (1.326+0.239kwV+0.0126)Ta+2.3	(melt during rain)

DATE	INSOL. I kJ/sqm	AIR TMP Ta C	DEW PT Td C	WIND V V km/hr	ALBEDO a %	RAIN Pr mm	...COMPUTED SNOWMELT... Short W. mm	Conv-Cnd mm	Long W. mm	Total mm	RAIN +MELT mm	SWE LEFT mm
1-May	18.0	11.0	3.0	10	66	0	3.12	4.55	10.21	17.88	17.88	1582.12
2-May	18.1	11.2	3.2	10	65	0	3.23	4.74	10.40	18.37	18.37	1563.75
3-May	18.3	11.6	3.6	10	64	0	3.36	5.12	10.77	19.25	19.25	1544.49
4-May	18.4	11.9	3.9	10	62	0	3.57	5.41	11.05	20.02	20.02	1524.47
5-May	18.6	12.3	4.3	10	61	0	3.70	5.79	11.42	20.91	20.91	1503.56
6-May	18.6	12.5	4.5	10	61	0	3.70	5.98	11.60	21.29	21.29	1482.27
7-May	18.7	12.9	4.9	10	60	0	3.82	6.37	11.97	22.16	22.16	1460.11
8-May	18.7	13.3	5.3	10	60	0	3.82	6.75	12.35	22.91	22.91	1437.20
9-May	18.8	13.5	5.5	10	60	0	3.84	6.94	12.53	23.31	23.31	1413.89
10-May	19.0	13.8	5.8	10	59	0	3.98	7.23	12.81	24.01	24.01	1389.88
11-May	19.1	14.1	6.1	10	59	0	4.00	7.51	13.09	24.60	24.60	1365.28
12-May	19.2	14.3	6.3	10	58	0	4.12	7.71	13.27	25.09	25.09	1340.19
13-May	19.4	14.7	6.7	10	58	0	4.16	8.09	13.64	25.89	25.89	1314.30
14-May	19.6	15.1	7.1	10	57	0	4.30	8.47	14.02	26.79	26.79	1287.51
15-May	19.9	15.5	7.5	10	57	0	4.37	8.85	14.39	27.61	27.61	1259.90
16-May	20.1	16.0	8.0	10	57	0	4.41	9.33	14.85	28.59	28.59	1231.31
17-May	20.3	16.5	8.5	10	56	0	4.56	9.81	15.32	29.68	29.68	1201.63
18-May	20.5	17.0	9.0	10	56	0	4.60	10.29	15.78	30.67	30.67	1170.96
19-May	20.7	17.6	9.6	10	55	0	4.75	10.86	16.34	31.95	31.95	1139.01
20-May	21.0	18.0	10.0	10	55	0	4.82	11.24	16.71	32.77	32.77	1106.24
21-May	21.3	18.5	10.5	10	55	0	4.89	11.72	17.17	33.78	33.78	1072.45
22-May	21.5	19.2	11.2	10	54	0	5.05	12.39	17.82	35.26	35.26	1037.20
23-May	21.8	19.8	11.8	10	54	0	5.12	12.96	18.38	36.46	36.46	1000.74
24-May	22.0	20.8	12.8	10	53	0	5.28	13.92	19.31	38.50	38.50	962.23
25-May	22.3	21.7	13.7	10	53	0	5.35	14.78	20.14	40.27	40.27	921.96
26-May	22.8	22.2	14.8	10	52	0	5.58	15.71	20.61	41.90	41.90	880.07
27-May	22.9	22.7	15.2	10	52	0	5.61	16.11	21.07	42.79	42.79	837.28
28-May	23.0	23.1	15.6	10	51	0	5.75	16.49	21.44	43.68	43.68	793.60
29-May	23.3	23.2	15.8	10	51	0	5.83	16.66	21.53	44.02	44.02	749.58
30-May	23.2	23.3	15.9	10	51	0	5.80	16.76	21.63	44.18	44.18	705.39
31-May	22.0	22.9	16.1	10	51	0	5.50	16.82	21.26	43.58	43.58	661.81
1-Jun	21.0	22.5	16.3	10	51	0	5.25	16.89	20.88	43.02	43.02	618.79
2-Jun	19.2	22.1	16.4	10	51	0	4.80	16.88	20.51	42.19	42.19	576.60
3-Jun	17.0	21.5	17.0	10	51	0	4.25	17.20	19.96	41.41	41.41	535.19
4-Jun	14.0	20.5	18.0	10	50	0	3.57	17.73	19.03	40.33	40.33	494.86
5-Jun	5.0	19.7	18.5	20	50	25	0.00	35.88	26.12	70.50	95.50	424.35
6-Jun	5.0	19.0	18.7	40	49	40	0.00	71.76	25.19	108.83	148.83	315.52
7-Jun	5.0	18.6	18.4	40	49	70	0.00	70.53	24.66	113.90	183.90	201.62
8-Jun	5.5	17.9	17.7	40	49	40	0.00	67.85	23.74	102.91	142.91	98.71
9-Jun	10.0	17.6	17.2	20	49	25	0.00	33.05	23.34	64.24	89.24	34.48
10-Jun	16.0	17.8	16.7	10	49	0	4.16	16.20	16.52	34.48	34.48	0.00
11-Jun	26.0	18.3	16.8	10	49	0	6.77	16.38	16.99	0.00	0.00	0.00
12-Jun	30.0	19.3	17.0	10	49	0	7.81	16.74	17.91	0.00	0.00	0.00
13-Jun	32.0	21.1	17.5	10	49	0	8.33	17.49	19.59	0.00	0.00	0.00
14-Jun	33.1	23.0	17.8	10	49	0	8.61	18.11	21.35	0.00	0.00	0.00
15-Jun	33.1	25.0	18.3	10	48	0	8.78	18.90	23.21	0.00	0.00	0.00
16-Jun	33.0	26.7	18.7	10	48	0	8.76	19.56	24.78	0.00	0.00	0.00
17-Jun	32.9	27.1	19.1	10	48	0	8.73	19.94	25.15	0.00	0.00	0.00
18-Jun	32.8	27.3	19.3	10	48	0	8.70	20.13	25.34	0.00	0.00	0.00
19-Jun	32.7	27.4	19.4	10	48	0	8.68	20.23	25.43	0.00	0.00	0.00

Figure 7.11—Spreadsheet Output, Example #2.

2. ***Previously Determined Factors*** Model parameters have been determined by adjustment after inspecting initial runs of the model. The example shown represents the final calibration run.

3. *Snow-Related Factors*

a. Snow Accumulation and Accounting. The basin is subdivided into eight elevation bands. Distribution of precipitation by elevation follows a specified distribution curve, based upon analysis of precipitation gage relationships inside and outside the basin.

b. Snowmelt. The temperature index approach is to be used. Base temperature is 0°C. The melt–rate factor C_m varies as a function of accumulated degree-days above freezing. This permits lower melt rates for fresh snow and a higher value (maximum = 3.3 mm/°C) for late season conditions.

ZONE MM-YY AREA ELEV
5 4 82 173 1916 COMPUTED FLOW, ILLECILLEWAET R., CANADA

MO	DA	HR	PCP	INT	WE	ATI	CC	LIQD	TA	ATI	MR	MELT	MI	PET	ET	SMI	ROP	RO	BII	BFP	SFRO	SSRO	BFRO	LZRO	Z	F
4	27	240			149	0	-0.3			0						3.85			.01							
4	28	240	.30	.30	149	-3	-0.5	4.3	-4	0	.000	.000	.040	.07	.05	3.82	37	.015	.01	89	.000	.001	.009	.004	5	S
4	29	240	.08	.08	149	-4	-0.6	4.3	-5	0	.000	.000	.040	.07	.05	3.79	36	.015	.01	89	.000	.001	.009	.004	5	S
4	30	240	.00	.00	149	-2	-0.4	4.3	-2	0	.000	.000	.040	.07	.05	3.77	36	.015	.01	89	.000	.001	.009	.004	5	S
				TOTALS, CM																						
			.39									.000		.22				.044			.001		.027			
				.39									.120		.16							.004		.012		
5	1	240	.00	.00	149	-2	-0.4	4.3	-2	0	.000	.000	.040	.08	.05	3.74	36	.014	.01	89	.000	.001	.009	.004	5	S
5	2	240	1.19	.41	150	-2	-0.4	4.3	-2	0	.000	.000	.040	.08	.05	3.71	36	.014	.01	89	.000	.001	.009	.004	5	A
5	3	240	.73	.08	150	-4	-0.6	4.3	-5	0	.000	.000	.040	.08	.05	3.69	36	.014	.01	89	.000	.001	.009	.004	5	A
5	4	240	.45	.08	151	-5	-0.6	4.3	-5	0	.000	.000	.040	.08	.05	3.66	36	.014	.01	89	.000	.001	.009	.004	5	A
5	5	240	.00	.00	150	-3	-0.5	4.3	-4	0	.000	.000	.040	.08	.05	3.63	35	.014	.01	89	.000	.001	.009	.004	5	S
5	6	240	1.48	.15	152	-2	-0.5	4.4	-2	0	.000	.000	.040	.08	.05	3.61	35	.014	.01	89	.000	.001	.009	.004	5	A
5	7	240	1.18	.08	153	-1	-0.4	4.4	-2	0	.000	.000	.040	.08	.05	3.58	35	.014	.01	89	.000	.001	.009	.004	5	A
5	8	240	.00	.00	153	-1	-0.4	4.4	-1	0	.000	.000	.040	.08	.05	3.56	35	.014	.01	89	.000	.001	.009	.004	5	S
5	9	240	.00	.00	153	-1	-0.4	4.4	-2	0	.000	.000	.040	.08	.05	3.53	35	.014	.01	89	.000	.001	.009	.004	5	S
5	10	240	.00	.00	153	0	-0.4	4.4	0.5	0	.051	.025	.040	.08	.06	3.5	34	.014	.01	89	.000	.001	.009	.004	5	C
5	11	240	.01	.01	153	0	-0.3	4.4	1.9	2	.055	.106	.040	.11	.07	3.46	34	.014	.01	89	.000	.001	.009	.004	5	C
5	12	240	.74	.40	153	0	0	4.2	1.4	3	.058	.083	.040	.10	.06	3.42	34	.014	.01	89	.000	.001	.008	.004	5	Q
5	13	240	.15	.10	153	0	0	4.1	2	5	.061	.123	.040	.11	.07	3.38	34	.013	.01	89	.000	.001	.008	.004	5	Q
5	14	240	.03	.03	153	0	0	3.8	3.6	9	.069	.249	.040	.14	.09	3.31	33	.013	.01	89	.000	.001	.008	.004	5	Q
5	15	240	.12	.12	153	0	0	3.2	5	14	.113	.559	.040	.19	.12	3.22	33	.013	.01	89	.000	.001	.008	.004	5	Q
5	16	240	.10	.10	153	0	0	2	7	20	.184	1.29	.040	.26	.16	3.09	32	.013	.01	89	.000	.001	.008	.003	5	Q
5	17	240	1.14	.45	153	0	0	0	6.9	27	.253	.000	.500	.26	.15	3.27	31	.157	.08	89	.003	.013	.098	.042	5	D
				TOTALS, CM																						
			7.31									2.438		1.93				0.378			0.008		0.236			
				2.01									1.14		1.26							0.032		0.101		

EXPLANATION OF CODES

MO	Month	PET	Potential evapotranspiration, cm/day
DA	Day	ET	Computed evapotranspiration, cm/day
HR	Hour	SMI	Soil moisture index, cm
PCP	Precipitation, cm	ROP	Runoff percent
INT	Interception, cm	RO	Computed runoff, cm
WE	Snow water equivalent, cm	BII	Baseflow infiltration index, cm
ATI	Antecedent temp. index for cold content	BFP	Percent runoff to baseflow
CC	Cold content of snow, cm	SFRO	Surface runoff, cm
LIQD	Liquid water holding capacity of snow, cm	SSRO	Subsurface runoff, cm
TA	Air temperature at mid elevation of band	BFRO	Baseflow runoff, cm
MR	Computed melt rate, cm/ degree-day	LZRO	Lower zone runoff, cm
MELT	Computed snowmelt, cm	Z	Zone number (1 = lowest)
		F	Flag for composite basin printout

Figure 7.12—Elevation Band Listing from SSARR Model (Band 5), Example #3.

c. Snow Conditioning. Equations 7.7 and 7.8 are used to simulate cold content of the snow. The value of 0.84 is used for the factor F_p and the cold rate factor in Equation 7.8 varies from 0.05 to 0.09 as a function of the antecedent temperature index of accumulated degree-days below freezing.

4. Program Output Examples for a 21-day period are shown. Fig. 7.12 is an example of the optional listings from one of the elevation bands (band 5), showing the details of the snow accumulation and melt for the simulation. Note that snow accumulated during the first half of the run, and that melt occurred during the last 6 days of the run; however, a snow conditioning process is underway on the fresh snow during this period so that melt is limited. Fig. 7.13 is a summary tabulation of output that combines the results from the seven individual snow bands. Note the last column, which is a code indicating the

```
                                        SSARR SNOWBAND MODEL (METRIC) -
        COMPUTED FLOW, ILLECILLEWAET R., CANADA
              AREA        BASE-TEMP         ZONES
APR 1982      1155            0               8                                                     FLAGS BY ZONE
DA  HR  PCPN   INT SNOWL  WE  LR TA     MR    RG    ET SMI ROP BFP   SURF   SUBSF BASEF OWERZ    TOTAL      OBS 1***5***10
27 240                709 124                          4.9       0  5.000  13.800 4.000 4.800   27.600  27.550
28 240  0.29 0.19     709 124 7.3 10 0.242 0.30 0.06 5.0  44  81  5.221  16.041 4.185 4.785   30.232  30.200 DDSSSAAA
29 240  0.08 0.05     709 124 7.3  9 0.303 0.13 0.06 4.9  44  80  3.600  15.230 4.365 4.770   27.965  28.800 DDSSSAAA
30 240     0    0     709 123 7.3 12 0.245 0.27 0.08 5.0  44  81  3.393  13.776 4.540 4.755   26.465  27.400 DDDSSSSS
                           V O L U M E  -  C E N T I M E T E R S
        0.37                              0.71                       0.09           0.1          0.63
             0.25                                0.2                        0.34          0.11             0.65
MAY 1982
 1 240  0.00 0.00     709 123 7.3 12 0.254 0.28 0.08 5.0  44  81  4.115  14.365 4.726 4.741   27.946  28.250 DDDSSSSS
 2 240  1.14 0.30     709 123 7.3 12 0.263 0.40 0.08 5.1  45  80  5.232  16.246 4.934 4.727   31.139  28.150 DDDAAAAA
 3 240  0.70 0.07     709 124 7.3  9 0.311 0.13 0.06 5.1  45  78  4.034  15.994 5.144 4.713   29.885  25.750 DAAAAAAA
 4 240  0.43 0.05     709 124 7.3  9 0.311 0.12 0.06 5.1  45  81  2.360  12.629 5.324 4.699   25.013  24.550 DAAAAAAA
 5 240  0.00 0.00     709 124 7.3 10 0.240 0.18 0.07 5.1  45  83  2.175  10.479 5.487 4.685   22.826  24.900 DDCSSSSS
 6 240  1.41 0.11     709 124 7.3 12 0.247 0.50 0.07 5.2  45  83  4.281  12.318 5.684 4.672   26.956  31.050 DDDAAAAA
 7 240  1.13 0.07     709 125 7.3 12 0.192 0.54 0.08 5.4  46  77  7.222  18.154 5.944 4.659   35.979  36.550 DDDCAAAA
 8 240  0.00 0.00     709 124 7.3 13 0.202 0.39 0.09 5.4  47  74  7.825  22.853 6.226 4.647   41.550  37.000 DDDCSSSS
 9 240  0.00 0.00     709 124 7.3 13 0.211 0.37 0.08 5.4  47  75  7.084  24.008 6.503 4.635   42.230  40.450 DDDCSSSS
10 240  0.00 0.00     709 123 7.3 14 0.180 0.53 0.11 5.5  48  76  8.098  25.551 6.791 4.623   45.063  45.400 DDDQCSSS
11 240  0.01 0.01     709 122 7.3 16 0.164 0.66 0.13 5.6  48  74 11.069  29.932 7.111 4.612   52.723  51.250 DDDQCCAA
12 240  0.70 0.32     709 122 7.3 15 0.167 0.68 0.12 5.6  48  72 13.392  35.360 7.464 4.601   60.816  56.650 DDDQQCAA
13 240  0.14 0.08     709 121 7.3 16 0.176 0.68 0.13 5.7  49  71 14.317  39.818 7.835 4.591   66.562  63.500 LDDQQCAA
14 240  0.03 0.02     709 120 7.3 18 0.175 0.69 0.16 5.7  49  70 14.325  42.876 8.221 4.581   70.004  74.200 LDDQQCCA
15 240  0.11 0.08    1050 119 7.3 19 0.198 0.79 0.19 5.8  50  69 15.741  46.186 8.624 4.572   75.123  82.300 DDDQQCCA
16 240  0.10 0.06    1050 117 7.3 21 0.205 1.07 0.23 6.0  50  69 21.152  52.401 9.069 4.564   87.186  92.250  DDDQQQC
17 240  1.09 0.33    1050 116 7.3 21 0.230 1.39 0.24 6.3  51  66 29.731  63.251 9.594 4.556  107.132 109.000  DDDDQQQ
                           V O L U M E  -  C E N T I M E T E R S
        6.99                              9.39                       1.29           0.86          6.34
             1.50                                1.97                       3.61          0.59             6.37
```

<u>**EXPLANATION OF CODES**</u>

DA	Day	**SURF**	Surface flowrate, cms
HR	Hour	**SUBSF**	Subsurface flowrate, cms
PCPN	Precipitation, cm	**BASF**	Baseflow flowrate, cms
INT	Interception, cm	**LOWERZ**	Lower zone flowrate, cms
SNOWL	Elevation of snowline, meters	**TOTAL**	Total computed discharge, cms
WE	Snow water equivalent, cm	**OBS**	Observed discharge, cms
LR	Lapse rate, degrees C / 1000 m	**FLAGS**	Indicators of snow activity on each elevation band
TA	Air temperature at sea level, degrees C	**D**	Dry weather melt occuring
MR	Melt rate, cm/degrees C-day	**R**	Rain melt occuring
RG	Runoff generated, melt + precip-int-soil loss	**S**	Snow on band, no accumulation nor melt
ET	Evapotranspiration, cm/day	**A**	Snow being accumulated
SMI	Soil moisture index, cm	**L**	Dry melt restricted by band transition
ROP	Computed runoff percent	**Q**	Melt, but no RO because of liquid water deficiency
BFP	Computed baseflow percent	**C**	Melt, but no RO because of cold content

Figure 7.13—Basin Summary Output Listing, Example #3.

activity on each band for the period. This shows the different activity taking place at different elevations in the basin for a given period. On this tabulation, the comparison between observed and computed streamflow can be seen.

5. ***Commentary*** This example is a small illustration of continuous simulation involving snow accumulation and snowmelt. Since the typical calibration run would be 10–30 years of continuous data, it is common to use statistical summaries of the entire calibration run to judge the results. These might include comparisons of computed vs. observed data using monthly average flows, peak flow summaries, flow-duration curves, etc. The detailed output listings shown would only be employed where a more careful analysis had to be made of a specific time series in the total simulation.

IX. REFERENCES

Anderson, E.A. (1968). "Development and testing of snow pack energy balance equations." Water Resources Research, 4(1), 19–37.

Anderson, E.A. (1973). "National Weather Service river forecast system—snow accumulation and ablation model." NOAA Technical Memorandum NWS HYDRO-17, National Weather Service, Silver Spring, MD.

Anderson, E.A. (1976). "A point energy and mass balance model of a snow cover." NOAA Technical Report NWS 19, National Weather Service, Silver Spring, MD.

Carroll, T. and Allen, M. (1988). "Airborne gamma radiation snow water equivalent and soil moisture measurements and satellite areal extent of snow cover measurements—a user's guide." National Remote Sensing Hydrology Program, National Weather Service, Minneapolis, MN.

Cayan, D.R. and Peterson, D.H. (1989). "The influence of North Pacific atmospheric circulation on stream-flow in the West." Geophysical Monograph 55, American Geophysical Union, 375–397.

Day, G.N. (1985). "Extended streamflow forecasting using NWSRFS." J. Water Resources Planning and Management, ASCE, 111(2), 157–170.

Day, G.N., Schaake, J. C., and Ellis, J. H. (1989). "A direction toward improved streamflow forecasting in the western mountains." Proc. Western Snow Conference, 79–89.

Day, G.N. (1990). "A methodology for updating a conceptual snow model with snow measurements." NOAA Technical Report NWS 43, National Weather Service, Silver Spring, MD.

Garen, D.C. (1992). "Improved techniques in regression-based streamflow volume forecasting." J. Water Resources Planning and Management, ASCE, 118(6), 654–670.

Gray, D.M. and Prowse, T.D. (1993). "Snow and floating ice." In D.R. Maidment, ed., *Handbook of Hydrology*, McGraw-Hill Book Co., 7.1–7.58.

Koch, R.W., Buzzard, C.F., and Johnson, D.M. (1991). "Variation of snow water equivalent and streamflow in relation to the El Niño/Southern Oscillation." Proceedings of the Western Snow Conference, 37–48.

Leavesley, G.H., Lichty, R.W., Troutman, B.M. and Saindou, L.G. (1983). "Precipitation–runoff modeling system." Users Manual, U.S. Geological Survey Water Resources Investigations Report B3–4238.

List, R.J. (1968). *Smithsonian meteorological tables*. 6th rev. ed., The Smithsonian Institution, Washington, DC.

Male, D.H. and Gray, D.M. (1981). "Snowcover ablation and runoff." In D.M. Gray and D.H. Male, eds., *Handbook of Snow*. Pergamon Press, 360–436.

McKay, G.A. and Gray, D.M. (1981). "The distribution of snowcover." In D.M. Gray and D.H. Male, eds., *Handbook of Snow*. Pergamon Press, 153–190.

Rallison, R.E. (1981). "Automated system for collecting snow and related hydrological data in mountains of the western United States." Hydrological Sciences Bulletin, 26(1), 83–89.

Rango, A. (1988). "Progress in developing an operational snowmelt–runoff forecast model with remote sensing input." Nordic Hydrol., 19, 65–76.

Redmond, K.T. and Koch, R.W. (1991). "Surface climate and streamflow variability in the western United States and their relationship to large-scale circulation indices." Water Resources Research, 27(9), 2381–2399.

Schemenauer, R.S., Berry, M.O., and Maxwell, J.B. (1981). "Snowfall formation." In D.M. Gray and D.H. Male, eds., *Handbook of Snow*. Pergamon Press, 129–152.

Soil Conservation Service (1972). "Snow survey and water supply forecasting." *National Engineering Handbook*, Section 22.

Twedt, T.M., Schaake, J.C., and Peck, E. L. (1977). "National Weather Service extended streamflow prediction." Proc. Western Snow Conference, 52–57.

U.S. Army Corps of Engineers (USACE) (1956). *Snow hydrology*. North Pacific Division, Portland, OR.

U.S. Army Corps of Engineers (USACE) (1987). *SSARR user's manual*. North Pacific Division, Portland, OR.

U.S. Army Corps of Engineers (USACE) (1990). *HEC-1 flood hydrograph package user's manual*. Hydrologic Engineering Center, Davis, CA.

World Meteorological Organization (WMO) (1986). "Intercomparison of models of snowmelt runoff." Operational Hydrology Report No. 23, World Meteorological Organization, Geneva, Switzerland.

X. NOTATION

The following abbreviations and symbols are used in this chapter.

B = thermal quality of snow

BF = base flow index

C_m = melt-rate coefficient, mm/°C

C_r = conversion factor in cold content equation, mm/degree-day

F = estimated average basin forest canopy cover, effective in shading the area from solar radiation

FP = fall precipitation index

F_p = a factor, varying from 0 to 1, representing the relative penetration of the air temperature into the snowpack

H_c = convective heat transfer (sensible heat) from the air

H_e = flux of latent heat released by condensation, sublimation

H_g = flux of heat from ground

H_{lw} = net long-wave radiation flux at snow–air interface

H_m = energy flux available for melt

H_q = internal energy content of snowpack

H_r = flux of heat from rain

H_{sw} = net short-wave radiation flux absorbed by the snow

I_i = insolation (solar radiation on horizontal surface) in langleys

J = joule; a basic unit of heat quantity. One Joule = 0.239 calorie.

M_s = snowmelt, mm per period

P_r = period rainfall, mm

S = snow water equivalent index

SP = spring precipitation index

SWE = snow water equivalent

T'_a = difference between the air temperature measured at 3 m and the snow surface temperature, in °C

T'_d = difference between the dewpoint temperature measured at 3m and the snow surface temperature, in °C

T_a = air temperature, °C

T_b = base temperature, °C

T'_s = temperature deficit of the snowpack below C

T_s = index of snow surface temperature in snow conditioning equation

W_c = snow cold content, mm

Y = seasonal streamflow volume

a = snow surface albedo

d = depth of snow, cm

k_w = convection–condensation factor representing the relative exposure of basin to wind

k_r = basin short-wave radiation melt factor

v = wind velocity, km/hr

ρ = density of snow or water (g/cc)

Chapter 8
FLOODS

I. INTRODUCTION

This chapter covers approaches and analytical processes commonly employed by practicing engineers and hydrologists in flood evaluations. Many publications cover these subjects in great detail. Consequently, this chapter provides an overview of various aspects of floods, a description of flood problems, and a discussion of flood analyses applied to these problems. A greater emphasis is given to the discussion of design flood for water resources facilities including statistical and deterministic methods for flood analysis and probable maximum floods (PMF). Some discussions are devoted to flood hazards and warnings as well as microcomputer software for flood analyses.

A. Flood and Flood Characteristics

A flood may be defined as an event when waters from a river, lake, ocean, or other surface water feature rise above normal limits. Floods often endanger lives and cause millions of dollars in damage. Floods also have beneficial, but seldom recognized, effects such as replenishing soil moisture, cleaning stream channels, depositing fertile fine sediments on agricultural lands, and restricting upstream movement of salt fronts in estuaries.

In general, the definition of floods stated above is insufficient for engineering and related economic, financial, and social analyses. Engineers and hydrologists commonly utilize an "annual flood series" (AFS) in which only the largest event (in terms of peak flow rate or volume) in a year is recognized. One flood is identified in each year even though the value may be smaller than the second or third largest value in some other years. Also, with the AFS on ephemeral streams, it is possible to have a year with a flood magnitude of zero.

Engineers sometimes use a "partial duration series" (PDS) in the analysis. In the PDS, all independent events exceeding a specified base magnitude are identified as floods regardless of the number of events in any given year. The base magnitude for a PDS is commonly established to identify an average of about three events per year as floods. Thus, the PDS provides a greater number of floods for the engineer to analyze.

B. Causes of Floods and Flooding

Flood magnitude and frequency varies spatially and temporally. To evaluate the variability and to suggest appropriate solutions to flood problems requires a recognition of natural and man-made factors that affect floods. The following paragraphs review major factors that affect river floods and lake and coastal flooding.

1. *River Floods*

a. Watershed Characteristics. Size, shape, slope, and surface topography are watershed characteristics that affect floods. Larger watersheds produce larger floods of longer durations. These floods generally have lesser year-to-year variation in magnitudes than those from smaller watersheds. In general, a round watershed produces larger flood peaks than a long narrow one. A basin with a steep slope also produces larger flood peaks than basins with small slopes. Larger floods occur on basins with rocky or other impervious surfaces than basins with deep and pervious soil cover.

b. Land Use Characteristics. Forested and heavily vegetated drainage basins usually produce floods of smaller peaks and longer durations than comparable bare basins. The effects of forest fire can significantly increase flood runoff. Construction of reservoirs usually reduces flood peaks downstream; however, small uncontrolled reservoirs on tributaries can disrupt natural runoff timing and cause increase in flood peaks on the receiving stream depending on the positions of these reservoirs. Urban and suburban developments can have profound effects on flooding. Placement of impervious surfaces reduces infiltration. Construction of storm sewers and improvement of drainage channels, including streets with curbs and gutters, causes rapid concentration of runoff. The result is usually a much larger flood peak with increased runoff volume during a shortened flow duration.

c. Precipitation. Rainfall usually is the primary source of flood waters. Intense, short-duration thunderstorm rainfalls often cause severe floods on small streams. Larger streams usually experience the largest floods as a result of longer duration and wider spread storms. In addition to the magnitude, duration, and intensity of rainfall, other factors affect the magnitude and frequency of resulting floods. Seasonally, floods respond differently to comparable rainfalls because rainfall is intercepted by vegetation during growing seasons, infiltration is greater during dry seasons and frozen ground reduces infiltration. While rain falling on snow probably increases total runoff, it may either increase or decrease the peak magnitude.

d. Snowpack. In some parts of the United States, melting of seasonal snowpack provides the primary source of flood waters. Snowmelt rates are controlled by the rates of long- and short-wave radiation, wind, and rain. Since the rates of radiation are limited, peak flood flow rates per unit area are often smaller for snowmelt events than for rain storm events. A snowpack that is saturated and isothermal at the freezing temperature is termed "ripe." Rain falling on a ripe snowpack can add to the runoff and produce a large flood peak; however, rain falling on an unripe snowpack may be absorbed, thereby retarding runoff until the pack ripens (see Chapter 7).

e. Ice. Ice can have a significant effect on floods or flooding. During winter, ice can form in the stream channel to reduce conveyance and cause flooding of overbanks. Spring breakup of ice-covered streams often causes ice jams that temporarily dam the stream resulting in backwater flooding of upstream reaches. Spring flood waters carrying ice are more likely to cause structural damage than ice-free flows.

f. Erosion and Sedimentation. Erosion and sedimentation contribute to a large portion of annual damage losses due to floods. Floods carry sediments eroded from the watershed and scoured from the stream channel. As the flood waters recede, the sediment load is often deposited in the stream channel resulting in the reduction of the flow-carrying capacity of the stream. This can cause flooding of overbanks during subsequent storms.

g. Dam Failure. Though infrequently, dams and other water-retaining structures have failed and produced major floods causing severe loss of lives and properties. Although programs exist for inspection and maintenance of dams, a probability of dam failure remains.

2. ***Lake and Coastal Flooding*** Climatic and topographic features control the magnitude and frequency of flooding along the shores of lakes, estuaries, and ocean fronts. A principal component of such flooding is "wind setup" or "storm surge" caused by strong, long-duration, on-shore winds and by reduced atmospheric pressure. Wind generated "wave runup" adds to the wind setup or storm surge, often causing severe destruction of facilities along a waterfront. The magnitude of runup is controlled by the on-shore and near-shore beach slopes and the angle of incidence of the waves upon the beach.

Additional factors affecting lake and coastal flooding may include seiche, tides, tsunamis, and sea-level rise. These subjects are discussed in Chapter 10 of this handbook.

C. Measurement of Flood Magnitude

Hydrologists and engineers use several indices to describe flood magnitude. They commonly identify the magnitude of an event by the instantaneous maximum flow rate or flood "peak" in cubic meters per second (m^3/s) or cubic feet per second (cfs). The "peak stage," or maximum water surface elevation associated with a flood peak, is another common index of magnitude. At some stream gages, the instantaneous peak flow rates and stages are not measured. In such situations, the hydrologist or engineer then uses the "maximum daily" flow rate and/or the associated stage. The maximum daily rate is the average rate over 24 hours and is always less than the peak rate. The difference between the peak and the maximum daily value is greater on small streams than on large ones.

Studies comparing flood magnitude between sites of different size in drainage areas commonly use "unit discharge," which is the ratio of peak flow rate to drainage area in m^3/s per square kilometer (km^2) or cfs per square mile (sq. mi.). Studies comparing flood runoff to causative rainfall often use an index of runoff in mm (or inches), which is computed as the ratio of runoff volume for the event to the drainage area. The volume of flood runoff for various durations of time such as 1-, 3-, 7-, 15-, and 30-days is often used in the design of a storage reservoir.

Flood magnitude is sometimes described on the basis of likelihood of exceedance; such as the flood having a 4% chance of being exceeded in any year. The reciprocal of exceedance probability defines an average length of time, in years, between the occurrences of floods of a specified magnitude or larger. Because of this, it has become common to term the flood having a 4% chance of exceedance as the 25-year flood and the flood having a 1% chance of exceedance as the 100-year flood.

Facilities whose failure due to flood could endanger many lives or result in catastrophic economical losses are normally designed for probable maximum flood (PMF), discussed later in this chapter. The likelihood of experiencing a PMF in any future year cannot be properly determined with the current technique in statistical analysis, although it has been assumed to be as small as one in 1,000,000.

D. Flood Hazards

Space occasionally inundated by floods is often ideal for residential and commercial use during nonflood periods. River flood plains provide relatively flat lands where buildings, roads, and utilities can be easily constructed. Alluvial fans, where flood paths are indeterminate, are also popular locations for land developments because they have flatter slopes than surrounding mountain areas. Coastal and lake shores offer recreational and aesthetic opportunities that encourage development. As a result of these characteristics, much development has occurred in areas subject to high flood hazards.

Flood protection works such as storage reservoirs, levees, diversion channels, or channel improvements can protect developments from floods of a selected magnitude; however, complete protection of most development from floods is seldom practical. Flood plain management programs encourage controlled development in flood hazard areas through flood insurance requirements, open space reservations, and building restrictions.

E. Flood Warnings

The National Weather Service (NWS) has been authorized by Congress to provide flood warning services to prevent loss of life and to reduce potential flood damage. Therefore, the NWS has developed flood forecasting models for various watersheds throughout the United States. These models are designed to provide flood warnings based on observed rainfall and runoff information with updates provided as new rainfall and runoff data become available. Although many of the models may be used to estimate the exceedance probability of runoff, the primary concern is to alert communities and individuals of the immediate flood threat by estimating the likely magnitude and timing of flood discharges and/or stages at specific locations. If flood stages are not forecast, they may often be estimated by comparing the forecast discharge with previous flood discharges and their associated stages. Limited resources prevent the NWS from providing flood forecasts for all concerned communities, especially those with small watersheds that have rapid runoff and involve flash flood warnings. Thus, many communities have provided their own flood warning systems, often with assistance from the NWS, other federal agencies and private companies. Flood warning systems are usually combined with emergency evacuation plans, which provide orderly withdrawal from the flood plain prior to significant flooding.

F. Flood Information

Prior to the implementation of the National Flood Insurance Program (NFIP), federal agencies including the U.S. Army Corps of Engineers (USACE), U.S. Geological Survey (USGS), and Tennessee Valley Authority (TVA) provided communities with flood information maps that are similar to the Flood Insurance Rate Maps (FIRMs) of the NFIP. These maps contain information on floods of various exceedance probabilities. In some cases, floods are identified by the dates of occurrence or other terms without assigning an exceedance probability. The need for such flood information maps diminished as the NFIP expanded and the production of such maps gave way to the use of FIRMs. A new type of flood information map evolved from the National Dam Safety Program conducted by the USACE. Because many dams were found to be unsafe under criteria used in the inspections, information regarding downstream flooding by dam failure is viewed as important to evacuation planning and to alerting downstream inhabitants of potential flooding risk. Thus, many federal and state agencies have required the preparation of dam failure inundation maps.

There are two important types of dam failure flood maps. The first involves a nonstorm condition with the reservoir full of water and a sudden or even a timed structural failure of the dam. This condition is probably the most serious as there would be little, if any, time to warn the downstream inhabitants. The second dam failure scenario involves its occurrence during a storm event large enough to overtop and cause dam failure. Thus, large spillway discharges would generally prevail prior to the failure and would alert downstream inhabitants and dam owners of the potential for dam failure. Estimating the exceedance probability of dam failure hydrographs is highly uncertain because of many unknowns associated with dam failure; however, a range of exceedance probabilities for dam failure can be estimated when a storm event is the basis for potential failure.

II. FLOOD ANALYSIS

The objective of flood analysis is to provide information for beneficial social and economic development in flood prone areas or to determine design floods for various components of a water resources development project. This section discusses basic approaches and general considerations for a flood analysis. A detailed discussion of statistical methods for flood analysis is given in the following section, which is followed by a section on estimating floods from rainfall.

A. Basic Approaches

Flood analyses are generally performed to derive peak flow values or complete flood hydrographs at specific locations on streams. The watershed contributing runoff to the location of interest is determined from topographic maps. The derived peak flows and hydrographs are of limited use unless they are associated with some measure of probability of occurrence or physical meaning of the meteorological conditions associated with the floods. Exceedance probability (such as a 1% chance in any given year) combined with potential damages is the desirable approach for selecting a design flood; however, limitations on available data prevent reliability and accuracy in computing the exceedance probability of extremely rare flood events. Because of this, the concept of hydrometeorology was applied as an alternative approach in an attempt to obtain more consistency and reliability in the design of major structures such as dams. The PMF and standard project flood (SPF) discussed later in this chapter are examples.

B. Design Floods

Design floods are the maximum floods against which the various components of a water resource facility or a system of facilities are protected. In some cases a design flood may consist of a peak flow value, whereas other design floods are defined by a complete runoff hydrograph of one or more storm events.

Some projects may require more than one design flood. For a flood control reservoir project, for example, the safety of the dam may be based on the combined reservoir storage and discharge capability to safely pass the PMF. Another design flood may be the 2% chance exceedance flood for selecting the flood control storage in the reservoir. Still other design floods could be used for determining the minimum elevation of relocated highways and railroads or acquisition of land around the reservoir. Design floods for levees may be of the SPF magnitude or a flood of some exceedance probability with smaller discharge.

Selection of design floods depends on the policies and procedures of the entities involved. Seldom are levees designed for floods greater than the SPF. A series of interior floods or drainage may be considered in sizing the drainage facilities, including gravity pipes, pumping plants, pressure pipes, and ponding volumes. Design floods for channel modifications and/or diversions are generally based on a flood of some specified exceedance probability that would not cause flood damage to the facility or overflow damage along the channel banks. Bridges, culverts, and storm drains are customarily designed to pass a discharge of a specified exceedance probability; however, storage is often provided to reduce the required size of the water passage. In this case, flood routing through the storage is necessary.

As a general rule in selecting design floods, the greater the adverse consequences of exceeding the design flood, the greater the design flood should be. This concept is supported by the notion that a higher degree of flood protection provides reduced tangible flood damage, less threat to loss of life, and other intangible advantages. There has been considerable debate regarding the selection of an inflow design flood for the safety of dams. Many engineering conferences and publications have addressed this subject. General consensus regarding large dams located above highly populated flood plains is that no significant threat of dam failure and loss of life due to floods should exist. If the economics of the project prevent achieving this goal, then construction of the project would be an unwise decision. Less agreement has been reached about the inflow design floods for the safety of dams located above flood plains with little or no development. Use of risk analysis (an economic approach) in the selection of the inflow design flood may appear wise at time of construction; however, future downstream development could seriously alter the dam design concept unless land use regulations are enforced for the areas that would be inundated by the dam's failure.

C. Regulatory Floods

Regulatory floods are design floods designated by regulatory agencies for specific purposes. There are three major activities that involve regulatory floods: the National Flood Insurance Program (NFIP) managed by the Federal Emergency Management Agency (FEMA), the licensing of non Federal hydropower dams managed by the Federal Energy Regulatory Commission (FERC), and the location of power reactor sites, managed by the Nuclear Regulatory Commission. The primary regulatory flood for the NFIP is termed the base flood having 1% exceedance probability in any one year. The flood plain associated with the base flood is identified on the Flood Insurance Rate Map (FIRM), and communities participating in the program must regulate development in the flood plain to reduce future flood damages. Other floods greater and smaller than the flood of 1% exceedance probability are also identified on the FIRM for informational purposes. FERC's regulatory floods are used to assure that construction of a non Federal hydropower dam will not increase the magnitude of natural floods by peaking the flood hydrograph or by dam failure. An individual flood event cannot define all potential flood hazards at nuclear power reactor sites; therefore, all potential sources and magnitudes of floods must be defined to assure a site with essentially no flood risk. This consideration includes artificial floods that could be created by dam failures or other catastrophic events.

III. STATISTICAL ANALYSIS FOR ESTIMATING FLOODS

A. Frequency Analysis

Flood frequency analysis is an important activity in hydrology. The sizing of bridges, culverts, and other facilities, the design capacities of levees, spillways, and other control structures, and the delineation of floodplains depend upon the estimated magnitude of various flood flows that serve as design values. Flood frequency analysis poses a challenge to hydrologists and engineers because available flood records are generally too short to clearly define the frequency of the large flood flows of interest. In such cases, engineers and hydrologists should do their best to use practical knowledge of the processes involved and efficient and robust statistical techniques to develop estimates of flood risk. In the United States, simple flood frequency techniques based primarily on the flood record available at a site are generally restricted to estimating flood flows exceeded with a chance of no less than one in 100.

1. Basic Data and Adjustment for Regulation Streamflow data are collected and stored by many state and federal agencies in the United States. Most cooperate with the U.S. Geological Survey (USGS) in operating a National Water Data Exchange (NAWDEX). A convenient way to access these files is through NAWDEX assistance centers located throughout the United States (contact NAWDEX, U.S. Geological Survey, 421 National Center, Reston, VA, 22092; tel. 703–648–6848). The USGS and its cooperators collect streamflow records at several thousand sites nationwide. Many of these records are published in annual USGS Water Data Reports for each state and are available in major technical libraries and from the National Technical Information Service (NTIS) as well as on CD-ROMs available from private vendors for use with personal computers. Records are stored in the computerized files of the National Water Data Storage and Retrieval System (WATSTORE). The WATSTORE peak-flow records contain annual maximum instantaneous flood-peak discharges and stages, and their dates of occurrence, as well as associated partial duration series. USGS district offices also develop and publish regression relationships for predicting flood quantiles, which are discharges associated with specified exceedance probabilities, for ungaged sites.

If the streamflow record for a site is subject to upstream regulation, the observed flows must be corrected for those activities as well as the effects of watershed changes including urban development, channel modifications, engineering works, and other disturbances.

2. Statistical Terminology The annual peak flood flow is generally thought of as a random variable that can be described by a probability distribution. The parameters of such a distribution are often estimated with the data available for a site. The flood discharge exceeded with probability q is called the 100q% exceedance probability flood, or 100q% chance flood.

The annual exceedance probability of a flood has often been described by its return period T, or the recurrence interval, where

$$T = 1/q. \tag{8.1}$$

The return period concept can be understood in several ways. In particular, the annual 100q% chance flood is exceeded on average once in a T = 1/q-year period, if the distribution of floods does not change over that period. Likewise, the average time between exceedances of the T-year flood is T years; however, this does not mean that one and only one T-year flood will occur every T years. The probability of the T-year flood being exceeded is 1/T in each and every year. To avoid confusion and improve understanding, the reporting of return periods should be avoided. It is clearer to refer, for example, to the 1% exceedance probability flood, the 1% chance flood, or the flood exceeded with a chance of one in 100, rather than to the 100-year flood.

The general character of a distribution can be described by its mean, denoted by the Greek letter μ, its variance, denoted by σ^2, and its coefficient of skewness, denoted by γ. The mean and variance are defined as

$$\text{mean} = \mu = E[X]$$

$$\text{variance} = \sigma^2 = E[(X-\mu)^2] \tag{8.2}$$

where E[] denotes expectation.

Dimensionless measures of variability and skewness are

$$\text{Coefficient of Variation} = CV = \sigma/\mu, \text{ and}$$

$$\text{Coefficient of Skewness} = \gamma = E[(X-\mu)^3]/\sigma^3. \tag{8.3}$$

Fig. 8.1 illustrates the role of the mean, standard deviation, and skew coefficient in defining the center, scale, and shape of a frequency distribution.

*3. **Probability Distributions*** Many families of distributions have been used to describe flood flows in the United States and in other countries. The normal and lognormal distributions, the Gumbel and Generalized Extreme Value distributions, and the Pearson and log–Pearson type III distributions are discussed below because of their widespread use.

a. Normal and Lognormal Distributions. The normal distribution is useful in hydrology for describing well-behaved phenomena, such as the total annual flow. The bell-shaped normal distribution is symmetric and therefore has a skew coefficient equal to zero. Flood distributions are generally positively skewed so that a normal distribution is not a reasonable model. If Q denotes an annual flood, in many cases the logarithms of flood flows log[Q] are adequately described by a normal distribution. If the logarithms of floods are not quite normally distributed, subtraction of a lower-bound parameter, c, before taking logarithms to obtain log[Q-c] can often resolve this problem, yielding a 3-parameter lognormal distribution (Stedinger, 1980; Hoshi et al., 1984). This distribution is often used to describe floods in Japan and monthly flow series in the United States.

b. Gumbel and Generalized Extreme Value (GEV) Distributions. Annual floods correspond to the maximum flows that occurred within a year. This suggests that their distribution is likely to be a member of a general class of Extreme Value (EV) distributions. If individual daily floods have an "exponential" like

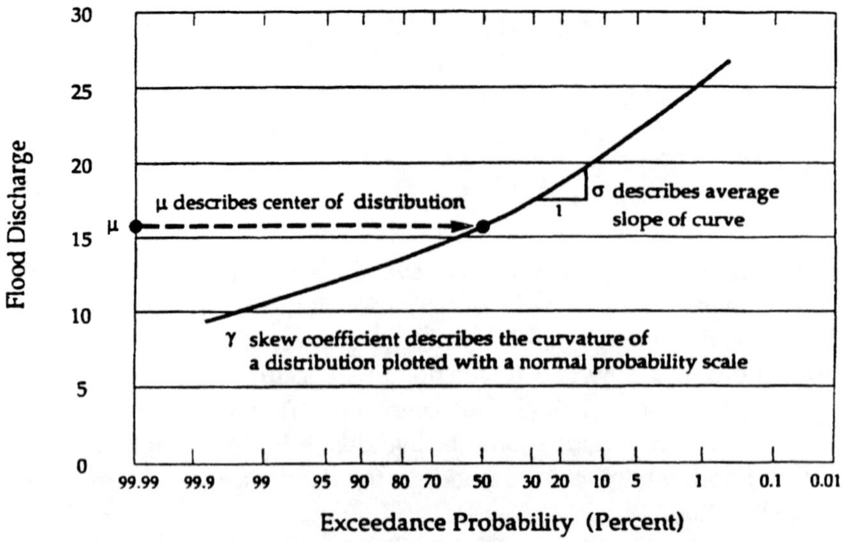

Figure 8.1—The Role of Mean, Standard Deviation, and Coefficient of Skewness in Defining a Flood Frequency Distribution.

upper tail, then the annual maximum is likely to have an Extreme Value type I distribution or Gumbel distribution (Ang and Tang, 1984). Parameter estimation procedures are described by Landwehr et al. (1979) and Phien (1987).

The Gumbel distribution with a fixed skew coefficient of 1.14 is an example of one of many extreme value distributions. The Generalized Extreme Value distribution (GEV) spans three types of extreme value distributions (Gumbel, 1958; Jenkinson, 1969). The GEV distribution is employed in flood studies in the United Kingdom (NERC, 1975; Hosking et al., 1985a,b) and in many other parts of the world (Cunnane, 1988).

c. Pearson Type III and Log–Pearson Type III Distributions. The Pearson type III (P3) distribution is often used to describe hydrologic phenomena. This distribution has a variable skew coefficient, and like the 3-parameter lognormal distribution, has a variable lower bound. For a small coefficient of skewness, it is similar to a normal distribution. When the logarithms of flood flows follow a Pearson type III distribution, the original flows have a log–Pearson type III distribution (LP3). The 2-parameter lognormal distribution is a special case of the LP3 distribution with a log-space skew coefficient of zero. The LP3 distribution is recommended for describing the distribution of floods in the United States in Bulletin 17B by the Interagency Advisory Committee on Water Data (IACWD, 1982) with a regional or weighted estimate of the skewness of the logarithms of the flows. The method for fitting the LP3 distribution is discussed later in this section.

4. Parameter Estimation Fitting a parametric probability distribution to the available flood record provides a compact and smoothed representation of a frequency distribution as well as a systematic procedure for extrapolation of flood risk to discharges beyond the range of the data. Assuming that floods are well described by a particular family of distributions, one needs to estimate the parameters of that distribution so that flood discharges associated with different exceedance probabilities are reasonably approximated by the "fitted" model. The statistical and hydrologic literature contain many methods and philosophies for estimating a distribution's parameters. Perhaps the simplest is the method of moments which computes estimates of the parameters so that the theoretical moments of a distribution match the computed sample moments. The recommended Bulletin 17B procedure for Federal agencies in the United States works with the sample moments of the logarithms of flood flows.

A variation upon the method of moments, which has proved effective in hydrology with the GEV distribution, is the method of probability weighted moments or of L–moments (Hosking et al., 1985a,b).

Estimators of those statistics are linear combinations of the ranked observations and thus do not involve squaring or cubing the observations. As a result they are almost unbiased, whereas the product–moment estimators of the coefficient of variation and coefficient of skewness are both highly-biased and highly-variable (Wallis, 1988; Hosking, 1990).

Another method is the method of maximum likelihood. With this method, parameters are selected that make the fitted distribution most consistent with the available data set in the sense that the selected parameters maximize the probability assigned to the available sample information. Maximum likelihood estimators (MLEs) are known to have good statistical properties in large samples. They are particularly valuable with historical and paleoflood records (discussed later in this section) because of their ability to make efficient use of such data sets (Cohn and Stedinger, 1987). Unfortunately, maximum likelihood estimators are often not available in closed form and thus must be computed using iterative numerical methods. In addition, because MLEs attempt to provide a good fit to all of the observations, they can provide poor estimates of large flood flows when low outliers in a data set are inconsistent with the parametric flood distribution being fit.

Occasionally, modern nonparametric methods are employed to estimate flood-flow frequency relationships (Adamowski, 1985). These methods have the advantage that one does not need to assume that floods are drawn from a particular parametric family of distributions. Modern nonparametric methods have not yet proven themselves in practice and have rarely been used; however, curve fitting procedures that employ plotting positions are simple nonparametric procedures that have a long history of use in hydrology.

B. Selection of a Flood Frequency Distribution

Goodness-of-fit criteria can be used to test whether a selected distribution is inconsistent with a particular data set. Plotting recorded data on the probability paper is an effective way to see what the data looks like and if a fitted frequency curve appears consistent with the available data. Analytical goodness–of–fit tests provides criteria for deciding whether lack of fit is due to sample-to-sample variability or if it reflects a fundamental difference between the fitted distribution and the distribution from which the data were drawn.

Several fundamental issues arise in selecting a distribution for frequency analysis. One could try to determine the true distribution from which the observed flood values were drawn; however, one is unlikely to ever have enough data to answer this question. Moreover, the functional representation of the true distribution of flood discharges would likely have too many parameters to be of practical use. The more practical question is what simple distribution should be used to obtain reasonably accurate estimates of flood flows for exceedance probabilities of interest with the available data sets.

At one time, the distribution that best fit each data set was used for frequency analysis. This approach is not recommended. Employing the best-fitting distribution for each data set provides flow exceedance probability estimates that are too sensitive to sampling variations in the data, and the period of record available. This can result in inconsistencies in flood frequency relationships derived for similar and nearby watersheds. Selection of a family of distributions is best done by a regional analysis using records from many sites. Operational procedures adopted in the United States and by many other countries are based on split-sample/Monte Carlo evaluations of different estimation procedures to find those that yield reliable flood flows and risk estimates for the area in question (Cunnane, 1989, Chapter 6). In the United States, the log–Pearson type III distribution with weighted skew coefficient was adopted based upon a 1974 study by Beard (IACWD, 1982, Appendix 14). In the United Kingdom, the GEV distribution with regional estimators of shape was adopted (NERC, 1975).

1. Plotting Positions and Probability Plots The graphical evaluation of the adequacy of a fitted distribution is often performed by plotting the observations so that they would fall approximately on a straight line if a postulated distribution were the true distribution from which the observations had been drawn. They can also be used to fit a probability distribution to a data set using various curve fitting procedures. Special commercially-available probability papers can be used for some distributions (normal, 2-parameter lognormal, and Gumbel) that have a fixed shape, perhaps after a logarithmic transformation, or with the more general techniques upon which such special papers are based (Loucks et al., 1981; Cunnane, 1978; Stedinger et al., 1993, p. 18). Special printed probability papers are not available for the Pearson and log–Pearson type III distributions. Lognormal probability paper is commonly used to compare observed flood peaks to a fitted log–Pearson type III frequency curve. When the skew coefficient is not zero, the relationship appears as a curved line (Vogel and McMartin, 1991).

A visual comparison of the empirical and fitted distributions is provided by a plot of the ith largest observed flood $x_{(i)}$ versus the estimated exceedance probability, or probability–plotting position q_i assigned to each ranked flood flow value $x_{(i)}$: $x_{(1)} > x_{(2)} > \ldots > x_{(n)}$ using a probability scale. Fig. 8.2 provides an example of a log–Pearson type III distribution fit to a 24-year record for Fishki Creek at Beacon, New York, using lognormal paper (IACWD, 1982, Figure 12-1). The annual exceedance probability of the ith largest flood in a record containing n values can be estimated by any of the formulas in

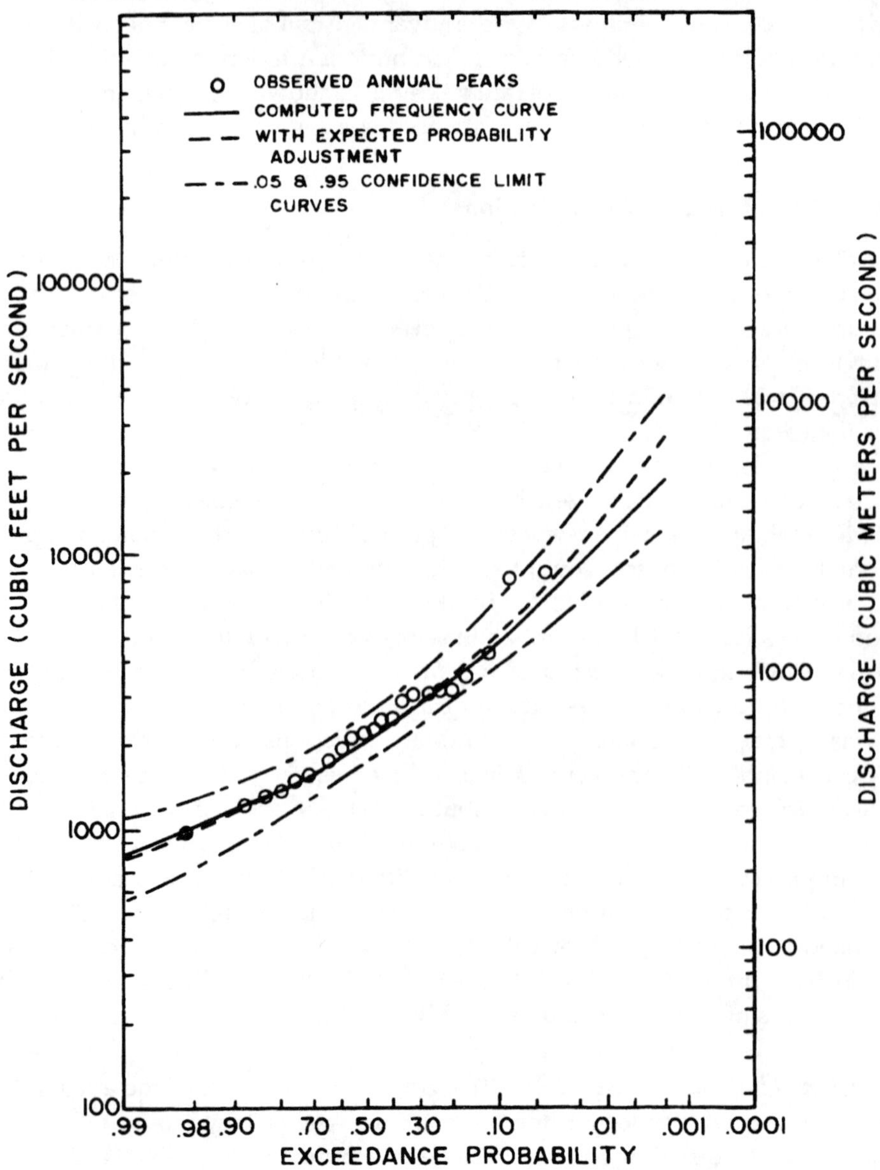

Figure 8.2—An Example of a Flood Frequency Curve. (IACWD, 1982, Figure 12-1.)

TABLE 8.1. Alternative Plotting Positions.

Name	Plotting positions	General comments
Weibull	i/(n + 1)	Employed in Bulletin 17B, Unbiased exceedance probabilities
Median	(i − 0.3175)/(n + 0.365)	Provides median exceedance probability for all distribution
Blom	(i − 0.375)/(n + 0.25)	Unbiased normal quantiles
Cunnane	(i − 0.40)/(n + 0.20)	Approximately quantile unbiased
Hazen	(i − 0.5)/n	A traditional choice

Table 8.1. Hirsch and Stedinger (1987) review plotting positions for combinations of traditional gaged records and historical information.

Different philosophies lead to selection of different plotting position formulas (Cunnane, 1978). The differences between the Hazen, Cunnane, and Weibull formulas in Table 8.1 are modest for floods that are neither among the largest nor among the smallest; however, they can be appreciable for the largest flood, to which the Weibull formula assigns a chance of one in (n + 1), and the Hazen formula assigns a chance of one in 2n. Bulletin 17B (IACWD, 1982, p. 26) makes no explicit recommendation. Weibull plotting positions have been used by many federal agencies, as in Fig. 8.2; however, the Cunnane plotting positions are gaining in popularity as a reasonable intermediate choice. The largest three or four events can be plotted with both Hazen and Weibull plotting positions to illustrate the impact of the choice of plotting position formula.

The analyst should remember that the actual exceedance probability associated with the largest observation in a random sample has a mean of 1/(n + 1) and a standard deviation of nearly 1/(n + 1) (Loucks et al., 1981, p. 109); thus, plotting positions give only crude estimates of the relative range of exceedance probabilities that could be associated with the largest events (Hirsch and Stedinger, 1987). One can use an interval within which the actual exceedance probability falls 90% of the time to represent this uncertainty (Stedinger et al., 1993).

2. ***Analytical Goodness-of-Fit Tests*** Rigorous statistical tests are available to test whether a set of observations is inconsistent with a particular family of distributions. The Probability Plot Correlation test is often a powerful test of whether a sample has been drawn from a normal, lognormal, Gumbel, Pearson type III, or log–Pearson type III distribution (Filliben, 1975; Vogel, 1987; Vogel and Kroll, 1989; Vogel and McMartin, 1991). Recently developed L-moments can also be used to assess if a Gumbel, GEV, normal, and other distributions are consistent with a data set (Hosking, 1990; Chowdhury et al., 1991). These tests can be employed with records from many sites to construct a more definitive test for a region.

3. ***Confidence Intervals*** Confidence intervals are a common vehicle for describing the possible value of a parameter or a flood with a given exceedance probability and the certainty with which the value in question can be estimated. Kite (1988) provides equations for calculating such confidence intervals for a large number of distributions. Bulletin 17B (IACWD, 1982, Appendix 9) discusses special procedures for log–Pearson type III distributions; Fig. 8.2 provides an example.

Stedinger (1983b) developed approximate confidence intervals for Pearson and log–Pearson type III quantiles given a known skew coefficient which appears to perform better than the approximate confidence intervals recommended in Bulletin 17B. However, both procedures ignore the uncertainty in the estimated weighted skew coefficient recommended in Bulletin 17B. Chowdhury and Stedinger (1991) show how this omission can be corrected and how confidence intervals that reflect uncertainty in skew coefficients can be constructed.

C. Bulletin 17B Frequency Analysis Method

Uniform procedures for flood-frequency analyses by United States federal agencies are recommended by the Interagency Advisory Committee on Water Data in Bulletin 17B (IACWD, 1982). These methods evolved over a long period (Thomas, 1985) and were adopted based upon an examination of alternative flood-frequency procedures available in the mid-1970's. The Bulletin 17B procedure estimates two parameters from the data available for a site and a weighted skew coefficient using at-site and regional information. Subsequent research has demonstrated the advantages of making greater use of regional hydrologic information so as to reduce the number of parameters estimated with short at-site records. In particular, use of recently developed index flood procedures for a gaged site with records of modest length is likely to provide more reliable and consistent estimates of flood quantiles (Wallis and Wood, 1985; Landwehr et al., 1987; Potter, 1987, Potter and Lettenmaier, 1990); however, Bulletin 17B procedures are now firmly established in the United States and are not as dependent on the availability of regional analyses.

Bulletin No. 15, "A Uniform Technique for Determining Flood Flow Frequencies," released in December 1967, recommended the log–Pearson type III distribution for use by United States federal agencies. The original Bulletin 17, released in March 1976, extended Bulletin 15 and recommended the log–Pearson type III distribution with a regional estimator of the skew coefficient for the logarithms of the flows. Bulletin 17A followed in 1977, and Bulletin 17B was issued in September 1981 with corrections in March 1982 (IACWD, 1982).

Bulletin 17B prescribes special procedures that address unusual situations, zero-flows or low outliers, historic peaks, regional information, confidence intervals, and expected probabilities. Some features of Bulletin 17B are described here. Those who wish to employ the full Bulletin 17B procedure should refer to that publication or other sources (such as Chow et al., 1988; or Cudworth, 1989).

1. Bulletin 17B Moment Estimators Bulletin 17B describes procedures for computing flood-flow frequency curves using annual flood series with at least 10 years of data. The recommended technique is to fit a Pearson type III distribution to the base 10 common logarithms of the peak discharges. The flood flow estimator Q_q associated with exceedance probability q is then

$$Log\,[Q_q] = \overline{X} + K_q\,S \tag{8.4}$$

where $\overline{X}$ and S are the sample mean and standard deviation, respectively, and K_q is a frequency factor that depends upon the skew coefficient and selected exceedance probability. Values of K_q for a specified skew coefficient can be read from tables in Bulletin 17B.

For $0.01 \le q \le 0.99$ and $|\gamma| < 2$, a good approximation of K_q is:

$$K_q = \left[\frac{2}{Y}\left(1 + \frac{Y}{6} Z_q - \frac{Y^2}{36}\right)^3 - \frac{2}{Y}\right] \tag{8.5}$$

where Z_q is the quantile of a standard zero-mean unit-variance normal distribution exceeded with probability q.

The flood quantile estimator depends on estimators of the mean, variance, and coefficient of skewness of the logarithms of the flood flows. These estimators (denoted $\overline{X}$, S^2 and G) are calculated as:

$$\overline{X} = \sum_{i=1}^{n} X_i/n$$

$$S^2 = \left[\frac{\sum_{i=1}^{n}(X_i - \overline{X})^2}{n-1}\right]$$

$$G = n\,\frac{\sum_{i=1}^{n}(X_i - \overline{X})^3}{(n-1)\,(n-2)\,S^3} \tag{8.6}$$

where X_i is the common logarithm of the ith annual peak discharge and n is the number of years of observation.

2. ***Generalized Skew*** Because of the variability of at-site sample skew coefficients in small samples, Bulletin 17B recommends weighting the station skew with a generalized skew, which is a regional estimate of the skew coefficient for sites in a region. A generalized skew G_g for sites in the United States can be read from Plate I in the Bulletin. Bulletin 17B indicates that the mean square error (MSE) of the weighted skew estimator is minimized by weighting the station and generalized skew inversely proportional to their individual mean-square errors (IACWD, 1982).

3. ***Outliers*** Bulletin 17B defines outliers to be data points that depart significantly from the trend of the remaining data. Procedures for treating outliers require statistical and hydrological judgment. This is an important issue. Because Bulletin 17B works with the logarithms of the observed flood peaks, a single unusual low value, or several such values, can distort the entire frequency curve, particularly its description of the flood-flow frequency relationship for the large events of interest and of importance. A graphical examination of the data and the fitted curve should be performed to check for such problems.

Bulletin 17B suggests that flood peaks identified as low outliers should be discarded from the record and the conditional probability adjustment discussed below be employed. High outliers are retained unless historical information is found to show that such floods are the largest in an extended period. In that case, the Bulletin recommends a historical weighted moments procedure. More recent research has shown that maximum likelihood procedures can make more effective use of historical information, and that it is advantageous to look for such information even when one does not have high outliers (Cohn and Stedinger, 1987).

4. ***Conditional Probability Adjustment*** The Bulletin recommends a conditional probability procedure for frequency analysis at sites whose record of annual peaks is truncated by the omission of peaks below a minimum recording threshold, years with zero flow, or low outliers provided that 75% or more of the record is above the truncation level. Maximum likelihood methods developed for use with historical information are effective when a larger fraction of the observations fall below a detection threshold. A frequency curve can also be fit graphically to the probability plot of the observed annual maximum events above the truncation level using plotting positions based on the total record length.

5. ***Expected Probability*** A fundamental and unresolved issue is whether, when estimating the flood exceeded with probability q, an analyst should: a) calculate a level that will on average (given the uncertainty in estimated parameters) be exceeded with probability q, or b) provide an unbiased estimation of the true unknown flood level exceeded with probability q. Unfortunately, with limited length records the two criteria lead to different estimations because of the effect of the uncertainty in the estimated parameters. Bulletin 17B (IACWD, 1982; Appendix 11-4) provides a formula for the probabilities that unbiased estimations of the 100q% chance flood will be exceeded. The Bulletin suggests that for lognormal or log–Pearson distributions those equations can be used to make an expected probability adjustment so that reported flood-flows will on average actually be exceeded with the specified probabilities. Fig. 8.2 provides an example of such an adjustment. Such corrections are commonly employed by the U.S. Army Corps of Engineers in the formulation of flood control projects, but are not generally used by other Federal agencies. Arnell (1989) and Stedinger (1983a) show that while the expected probability correction can eliminate the bias in the expected exceedance probability, the correction generally increases the bias in estimated average annual damages for existing facilities located in the flood plain. Thus, a policy either to make or not to make the expected probability adjustments should reflect the intended use of the derived flood-flow frequency relationship.

D. Record Augmentation

It is often possible to effectively extend a short record using a longer nearby record with which observations in the short record are highly correlated. In particular, the longer nearby series can be used to improve estimates of the mean and variance of the events that occur at the short-record site. For this purpose it is not necessary to actually construct the extended series; one only needs the improved estimates of the mean and variance (IACWD, 1982, Appendix 7). Generally, one needs cross correlations of at least 0.7 to improve the mean, and 0.85 to improve the variance (Stedinger et al., 1993, p. 18.37).

E. Risk from Coincidental Events

A flood near the mouth of small tributaries to larger rivers or reservoirs can result from flooding on either the tributary, the larger river or reservoir, or both (Dyhouse, 1985). The flood profiles above a given elevation on the tributary should represent the combined probability of flooding from both causes. Assuming that the flood events on the tributary and on the larger river or reservoir are independent but not mutually exclusive, the combined risk q_c is the sum of the annual probability of flooding above that elevation due to a flood generated on the tributary's drainage q_f, plus the annual probability of flooding above that elevation from the large river or reservoir q_r minus the product of the two:

$$q_c = q_r + q_r - q_r q_f. \tag{8.7}$$

Independence can be assumed when flooding on the larger river or reservoir usually results from a storm centered outside the tributary drainage area that does not produce significant flooding on the tributary, and flooding on the tributary usually results from storms centered over the tributary drainage area that do not generally produce significant flooding on the larger river or reservoir. This approach produces exceedance probabilities intermediate between 1) that produced by considering only the probability of backwater flooding or tributary flooding, $\max[q_r, q_f]$, depending upon which is greater and 2) that produced by ignoring the possibility of concurrent events so that q_c equals $q_r + q_f$.

1. Risks on Alluvial Fans Similar calculations can be used in the evaluation of flood discharge probabilities on alluvial fans. The Federal Emergency Management Agency (FEMA) has developed procedures for combining the probabilities of flood peaks at the apex of the fan and probabilities for the flow path channel geometry to derive the probabilities for flooding at different locations on a fan (Dawdy, 1979; FEMA, 1991).

F. Analysis of Mixed Populations

A common problem in hydrology is that annual maximum series are composed of events that may arise from distinctly different processes. For example, precipitation may correspond to different storm types in different seasons (such as summer thunderstorms, snowmelt events, winter frontal storms, and remnants of tropical hurricanes). Floods arising from these different types of precipitation events may have distinctly different distributions (Cudworth, 1989). Equation 8.7 can be used to describe the combined risk from two different sources of floods that are described by separate component distributions.

Some important questions are: When is it advisable to model several different component flood series separately? And when is it just as reasonable to model the composite annual maximum series directly? If several series are modeled, then more parameters must be estimated, but more data are available if the annual maximum series (or the partial duration series) for each type of event is available. Modeling the component series separately is most attractive when the annual maximum series is composed of only two or three component series with distinctly different distributions that are individually described by relatively simple 2-parameter Gumbel or 2-parameter lognormal distributions (Waylen and Woo, 1982).

However, in such cases, the maximum of the different components still defines a valid annual maximum series that can be modeled directly.

G. Regional Analysis

Frequency analysis is a challenge in hydrology because sufficient information is seldom available at a site to adequately determine the frequency of rare floods. Moreover, design floods and estimates of flood risk are commonly required at sites for which there is no nearby gaged flood record. In such instances one may employ regional flood frequency relationships derived using gaged flood records from similar locations.

Regional hydrologic information is also useful at gaged sites because of the imprecision of flood-risk estimators based upon records of the length typically available. For example, Bulletin 17B uses a regional map of skewness coefficients to calculate flood risk. Record augmentation discussed earlier is only useful if one has several years of record at the site of interest and an adequate nearby site.

The National Research Council (1988) put forward three principles for hydrologic and meteorological modelings; 1) substitute space for time, 2) introduce more structure into models, and 3) focus on extremes or tails as opposed to or even to the exclusion of central characteristics. One substitutes space for time by using hydrologic information at different locations to compensate for short records at a single site. This is easier to do for rainfall, whose characteristics in regions without appreciable relief should be fairly uniform over large areas. Regional estimates of rainfall distributions can allow the selection of design rainfall depths which can serve as the base for calculation of design floods. Conversion of design rainfall to runoff is particularly attractive for urban basins and watersheds subject to regulation and significant development. In other cases, one can employ regional relationships for flood characteristics derived directly from available flood records with appropriate methods of determining the effects of basin area, geomorphology, and land cover.

Two successful examples of regionalization are the index flood method and regional regression procedures discussed below. Both use available samples for sites in a region to construct relationships useful for estimating flood risk at a particular site. For such a method to be effective, the available sample must be representative of floods at the site of interest after necessary corrections are made. Problems occur if sites included in a regional analysis have watershed characteristics that are significantly different from the site of interest. Even worse, if significant development has taken place or if flood flows have been regulated at the site of interest, then the regional relationships may not be applicable. Sites with such effects are generally not used in deriving the regional relationships.

Where appropriate, flood-regionalization methods have significant advantages over the derivation of design floods from design rainfalls. Flood-regionalization methods are relatively easy to apply and different individuals obtain essentially the same flood flow estimates once the background regional analyses have been performed. There are no questions about the relationships between rainfall and flood exceedance probabilities, nor are assumptions needed about antecedent watershed conditions. The United States has very rich networks of stream gaging stations on which regional flood analyses can be based.

1. *Index Flood* The index flood procedure is a simple regionalization technique with a long history of use in the United States (Dalrymple, 1960). The basic idea behind the index flood method is that the distribution of floods at every site in a region should be the same except for a scale parameter (the index flood) that reflects the size and runoff characteristics of each basin. At gaged sites, the at-site sample mean can be used to estimate that scale value. At ungaged sites, regional regression methods discussed below are generally used to estimate the mean flood.

The normalized frequency curve, or growth curve, can be estimated in several ways. Regional probability-weighted-moment (PWM) index-flood frequency estimation procedures that employ PWM and L-moments have been studied and often provide more accurate estimates of extreme flood quantiles than would a single-site analysis (Hosking et al., 1985b). The corresponding index-flood procedure based upon L-moments with the GEV distribution used to describe the growth curve should, in practice with reasonably defined regions, be robust and more accurate than procedures that attempt to estimate two or more parameters with the short-records available at many sites (Potter and Lettenmaier, 1990).

The key to the success of the index-flood approach is the identification of reasonable hydrologically similar sets of basins. One may try to group basins geographically, by altitude or by size, recognizing that hydrologic similarity, not geographic proximity, is the important issue. Regional regression can be employed to predict the normalized quantiles using physiographic characteristics of a basin. Modest variability in the normalized quantiles can be tolerated but can degrade the procedure's results (Lettenmaier et al., 1987).

2. ***Regional Regression for Ungaged Sites*** Regression can be used to derive simple models that predict the values of means, standard deviations, discharges exceeded with various probabilities, and normalized index-floods as a function of physiographic river basin characteristics and other parameters. Such models are often used to predict various flood flows at ungaged sites (Jennings and Thomas, 1992). In a nationwide test, these models did as well or better than more complex procedures (Newton and Herrin, 1982).

Consider the traditional log-linear model:

$$\theta_i = a + b_1 \log(\text{Area}_i) + b_2 \log(\text{Slope}_i) + \ldots + e_i \tag{8.8}$$

where drainage area and slope are two of many possible explanatory variables used to estimate θ. A challenge in analyzing such models is that with available flood records, one only obtains sample estimates, denoted as y_i, of the streamflow statistics θ_i. Thus the error e_i is a combination of: a) time-sampling-error $\text{Var}[y_i]$ in sample estimators of y_i and b) underlying model error (lack of fit). The e_i for different sites are cross-correlated if the y_i are based upon concurrent and cross-correlated flow records. Often these problems have been ignored and standard ordinary least squares (OLS) regression has been employed to obtain reasonable estimators of a model's parameters and its accuracy. Such procedures are satisfactory when the flood records are all sufficiently long.

Stedinger and Tasker (1985; 1986) developed the basic theory for Generalized Least Squares (GLS) regression as a regional hydrologic analysis methodology. GLS effectively assigns different weights to the observed streamflow statistics y_i depending upon the estimated time-sampling error in each statistic $\text{Var}[y_i]$ relative to the accuracy of the regression model. Tasker and Stedinger (1986; 1989) and Potter and Faulkner (1987) illustrate GLS use in regionalization studies for the log–space coefficient of skewness and of flow quantiles using physiographic basin characteristics. Advantages of the GLS procedure include more efficient parameter estimates and an unbiased model-error estimator.

Major advantages of regional regression models are that they are relatively easy to employ; an estimate of flood quantile's precision is provided and they embody explanatory variables that reflect the unique physical characteristics of the watershed of interest.

H. Historical Information and Paleofloods

Available at-site systematic records are the traditional and most obvious source of information on the frequency of floods, but they are of limited length. Another source of at-site information is historical flood records. Historical information includes written and other records of large floods left by human observers: newspaper accounts, letters, and flood markers. Paleoflood information describes the many botanical and geophysical sources of information on large floods which are not limited to the locations of past human observations or man-made recording devices. Botanical data can consist of the systematic interpretation of tipped trees, scars, and abnormal tree rings along a water course providing a history documenting the frequency with which different flood stages were exceeded (Hupp, 1987; 1988).

Recent advances in physical paleoflood reconstruction have focused upon the use of slack-water deposits and scour lines as indicators of paleoflood stages, and the absence of large flows that would have left such evidence (Costa, 1987; Stedinger and Baker, 1987; Baker et al., 1988). Such physical evidence of flood stage along a water course with radiocarbon and other dating techniques can provide a relatively accurate and complete catalog of paleofloods in favorable settings with stable channels. A series of studies has documented the potential value of historical information using maximum likelihood procedures (Jin and Stedinger, 1989; Cohn and Stedinger, 1987).

Bulletin 17B contains a less efficient weighted moment procedure that can be used when all floods that exceed a threshold over an extended historical period have been observed. The procedure assumes that the observed gaged-record floods less than a threshold are representative of the unobserved floods which occurred in the historical period outside of the gaged-record period. The weighted moments procedure tends to de-emphasize a large flood, recorded during a short gaged-record, when such a flood is known to be the largest in an extended historical period.

I. Partial Duration Series

There are two general and equally valid approaches for the statistical modeling of flood series. The annual flood series (AFS) approach employs the largest flood in each year and is the most common. The partial duration series (PDS), or peaks-over-threshold approach, considers all independent flood peaks above a truncation or threshold level (Shane and Lynn, 1964; NERC, 1975; Todorovic, 1978). Flood damage studies that consider the possibility of more than one damage-causing flood occurring in a year (or in a season) need to employ partial duration series models; however, such studies should also consider if adequate time passed between damage-causing floods for repairs to occur or another crop to be planted to avoid double-counting. Care must also be exercised to select only the independent peaks for analysis.

An objection to AFS analyses is that they employ only one flood for each year, regardless of whether the second largest flood in a year exceeds the annual flood of other years. In arid regions, there may be no significant flow during an entire water year and thus no significant annual flood to consider. In humid regions, the number of independent and distinct damage-causing flood events may significantly exceed the number of years of record.

Care should be exercised in the frequency interpretation of partial duration series because an n–year series will generally not contain n values. The general frequency implications of a PDS data set can be investigated by plotting the ordered partial-duration peaks $X_{(i)}$ versus an empirical exceedance frequency $f_i = i/n$ for $i = 1, \ldots, m$, when m peaks were observed in an n–year period. Alternatively, $X_{(i)}$ can be plotted against log f_i to obtain a more linear relationship. For m greater than n, the exceedance frequency for some floods exceeds unity because that level is exceeded on average more than once per year. The annual exceedance probability q associated with a flood discharge of exceedance frequency f is essentially (Chow, 1964):

$$q = 1 - \exp[-f]. \tag{8.9}$$

In most design applications, analytical methods are used to describe the frequency of events in the partial duration series. Such PDS flood-frequency models generally describe both the arrival rate of floods larger than a threshold level and the probability distribution for those exceedance events. The natural estimator of the exceedance rate of a threshold is simply the number of times it was exceeded divided by the length of record. Using a Poisson model for the arrival of floods and an exponential distribution to describe the magnitudes of those flood peaks that exceed the threshold leads to a Gumbel distribution for the annual flood series (Shane and Lynn, 1964; NERC, 1975). If a more flexible generalized Pareto distribution is used to model flood peaks, then Equation 8.9 yields a GEV distribution for the annual flood series (Davison, 1984; Hosking and Wallis, 1987; Stedinger et al., 1993).

J. Bayesian Risk Analysis

Bayesian inference provides another philosophical approach to risk and frequency analysis. Advantages of the Bayesian approach are that it provides a theoretically consistent way of thinking about statistical decision making, allows the explicit modeling of uncertainty in parameters, and provides a theoretically consistent framework for integrating information from at-site gaged-flow records with regional hydrologic information and data from other sources.

The Bayesian approach suggests that supplemental regional hydrologic information should be incorporated through a "prior probability distribution" to augment the information provided by the gaged record for a site (Vicens et al., 1975; Kuczera, 1982; Stedinger, 1983a). Unfortunately, the actual updating or integrating step is straightforward for only a few simple distributions, and the use of prior distributions is sometimes controversial.

IV. ESTIMATING FLOOD FROM RAINFALL

When flood records are not available or not long enough at the site of interest, engineers often use selected storm rainfalls to compute a flood for the site. This approach is sometimes termed the "deterministic method" when compared to the "statistical methods" discussed in the previous section. Use of rainfall to estimate floods is particularly valuable in the situations described below:

a) If a site is essentially ungaged so that flood records are not available for a flood frequency analysis, a design flood can be estimated based upon available rain gage data or regional rainfall frequency analysis.
b) If proposed or recent channelization, urbanization, or other changes within a watershed significantly affect the relationship between rainfall and runoff so that existing flood records are a poor description of current or future flood potential, then the use of rainfall frequency information with a rainfall–runoff model may be advantageous.
c) When reservoir and other water resource operating plans require the simulation of runoff hydrographs at one or more sites, concurrent hydrographs can be generated by multi-basin rainfall–runoff models using multi-site rainfall data from historical or design storms.
d) When interest is in very extreme flood events for design of proposed dams or for safety evaluation of existing dams, available flood records are generally not long enough to properly estimate the rare flood of desired probability of occurrence. In such situations, the flood is estimated by transforming the design storm estimated generally by a hydrometeorological analysis.

To estimate a flood from rainfall, the temporal and spatial distributions of the storm rainfall over a particular drainage area are analyzed. The resulting rainfall, whether it is recorded or based on a statistical or hydrometeorological analysis, is then transformed into a flood peak or hydrograph. The technique could be as simple as applying an empirical equation, using precipitation as an input to obtain peak discharge as an output, or as complex as employing a watershed hydrologic model to capture the complete physical process of rainfall transforming into runoff throughout the watershed.

As statistical methods for estimating rainfalls of prescribed exceedance probabilities are similar to those for floods discussed in the previous section, only estimation of rainfall based on hydrometeorological analysis will be discussed in this section. The discussion will start with an introduction to precipitation analyses covering synopsis of major storms, spatial and time distribution of rainfall, snow accumulation and snowmelt. It is followed by a brief discussion of antecedent and subsequent storms, baseflow, and transformation of rainfall excesses to floods. Finally, the movement of the flood through the channel and overbank storage in the watershed's drainage system is discussed. Precipitation losses due to infiltration are discussed in Chapters 3 and 6.

A. Synopsis of Major Historical Rainstorms

In a broad sense, all rainstorms result from moisture (water vapor) in the atmosphere being lifted to a higher elevation. Atmospheric processes do the lifting, aided in some situations by topography. The more water vapor that is in the lifted air, the better the chance of greater rainfall. Further, the warmer the lifted air is, the more water vapor it can hold. As the air is lifted, it cools, becomes saturated, and the water vapor condenses into ice crystals or small water droplets. The crystals or droplets then increase in size until the force of gravity overcomes vertical lifting, causing precipitation.

In mountainous regions, it is often convenient to categorize precipitation as 1) due to atmospheric processes alone, or 2) due to terrain effects or orography. The precipitation due only to atmospheric processes is termed "convergence" precipitation. The increase or decrease due to terrain is termed "orographic."

The term convergence is derived from the general process of moist air converging at lower levels and then lifted. Atmospheric pressure patterns over large areas, with high and low pressure centers well placed, are mostly responsible for converging and lifting. Weather fronts, the junction of two different atmospheric air masses, often are responsible for lifting. For example, an approaching cold air mass will "dig" below a warmer air mass in its path, thereby lifting it.

Interaction between air masses of different characteristics (moisture source and quantity, temperature, humidity, etc.) can cause heavy convergence rainfall over large areas. These have been termed "general" storms. On the other hand, thunderstorms, limited in supply of moisture, can produce extreme convergence rainfall for a relatively small area and short time. These are termed "local" storms.

East of the Continental Divide, the most intense small area–short duration rainfalls occur within large area–long duration storm situations. In mountainous western regions, the most intense small area–short duration storms in some cases have occurred quite independent of a large scale atmospheric process or storm system. These "local" storms usually cover less than 1300 km^2 (500 sq. mi.). The storm types "local" or "general" are adopted to some extent for expediency since nature can produce rains which blend or fall anywhere between the two extremes.

Tropical storms or hurricanes, the latter with wind speeds exceeding 65 knots, are noted for producing large area–long duration rainfalls over eastern states from Texas to Maine. These storms have also produced some of the most intense small area–short duration rainfalls for this region. Table 8.2 shows the greatest known United States rain depths for various durations. Many of the storms affecting the values in this table are due to hurricanes.

Tropical storms have large circulations originating over warm oceans around a low central pressure. The lower the central pressure, the greater the winds converge to the center and cause greater vertical lifting and resultant rainfall.

As tropical storms progress inland from the coast, the winds are reduced because of friction over the land surface; however, should such storms have a strong circulation against mountains such as the Appalachian or Catskill Mountains in the United States the rainfall can be increased due to the orographic effect.

Tropical storms are also indirectly responsible for some of the largest record rainfalls at a considerable distance from the coast. For example, the great Hearne, Texas storm of June 27–July 1, 1899 and the Thrall, Texas storm of September 8–10, 1921 (Table 8.2) were associated with moisture supplied by tropical storms. Similarly, tropical storms have provided moisture, if not a triggering mechanism, for heavy rains west of the Appalachians all the way to the Great Lakes.

Another aspect of tropical storms is that the slower the storm moves, the more the rain falls along the storm path. The Yankeetown, Florida storm of September 3–7, 1950 progressed very slowly by looping. This storm is also responsible for some United States maximum observed depths (Table 8.2).

Tropical storms have also been responsible for some of the greatest rains in the southwest states. Such storms moving from near Baja California, Mexico have provided moisture for heavy rainfall into Utah and western Colorado (National Weather Service Hydrometeorological Report - [HMR] No. 49, 1977).

TABLE 8.2. Maximum known observed depth–area–duration data for the United States (average rainfall in millimetres (and inches).

Area	Duration (h) 6	12	18	24	36	48	72
26 km²	627a	757b	922e	983e	1062e	1095e	1148e
10 mi²	(24.7)	(29.8)	(36.3)	(38.7)	(41.8)	(43.1)	(45.2)
259 km²	498b	668e	826e	894e	963e	988e	1031e
100 mi²	(19.6)	(26.3)	(32.5)	(35.2)	(37.9)	(38.9)	(40.6)
518 km²	455b	650e	798e	869e	932e	958e	996e
200 mi²	(17.9)	(25.6)	(31.4)	(34.2)	(36.7)	(37.7)	(39.2)
1295 km²	391b	625e	754e	831e	889e	914e	947e
500 mi²	(15.4)	(24.6)	(29.7)	(32.7)	(35.0)	(36.0)	(37.3)
2590 km²	340b	574e	696e	767e	836e	856e	886e
1000 mi²	(13.4)	(22.6)	(27.4)	(30.2)	(32.9)	(33.7)	(34.9)
5180 km²	284b	450e	572e	630e	693e	721e	754e
2000 mi²	(11.2)	(17.7)	(22.5)	(24.8)	(27.3)	(28.4)	(29.7)
12950 km²	206bh	282b	358b	394e	475i	526i	620i
5000 mi²	(8.1)	(11.1)	(14.1)	(15.5)	(18.7)	(20.7)	(24.4)
25900 km²	145h	201j	257k	307k	384i	442i	541i
10000 mi²	(5.7)	(7.9)	(10.1)	(12.1)	(15.1)	(17.4)	(21.3)
51800 km²	102h	152i	201k	244k	295i	351i	447i
20000 mi²	(4.0)	(6.0)	(7.9)	(9.6)	(11.6)	(13.8)	(17.6)
129000 km²	64km	107n	135k	160k	201k	251r	335r
50000 mi²	(2.5)	(4.2)	(5.3)	(6.3)	(7.9)	(9.9)	(13.2)
259000 km²	43m	64om	89k	109k	152p	170p	226q
100000 mi²	(1.7)	(2.5)	(3.5)	(4.3)	(6.0)	(6.7)	(8.9)

Storm	Date	Location of Center	
a	17-18 July 1942	Smethport, PA	
b	8-10 September 1921	Thrall, TX	
e	3-7 September 1950	Yankeetown, FL	Hurricane
i	27 June–1 July 1899	Hearne, TX	
k	13-15 March 1929	Elba, AL	
q	5-10 July 1916	Bonifay, FL	Hurricane
n	15-18 April 1900	Eutaw, AL	
m	22-26 May 1908	Chattanooga, OK	
o	19-22 November 1934	Millry, AL	
h	27 June–4 July 1936	Bebe, TX	
j	12-16 April 1927	Jefferson Parish, LA	
r	19-24 September 1967	Cibolo Ck., TX	Hurricane
p	29 September–3 October 1929	Vernon, FL	Hurricane

Source: World Meteorological Organization, 1986.

Extra-tropical storms associated with well-placed high and low pressure centers and air masses with different characteristics (temperature, stability, sources, etc.) have also produced some of the greatest known rainfalls. A wave moving along a front associated with lower pressure creates circumstances that have produced heavy rainfall.

Orographic rain is most notable along the western slopes of the Sierra Nevada in California and the Cascades in Oregon and Washington. Relatively stable air masses in the cool season regularly strike these mountains and the air is lifted, reducing the temperature and causing condensation and precipitation. Examples of unusually severe storms are the January 19–24, 1943 and the December 18–26, 1955 storms, which produced some of the record rains for large regions of the Sierra Nevada (HMR 37, 1962). In many locations, a reduced rain amount is evident due to upwind barriers. It emphasizes that while the air masses tend to show relatively high stability, considerable convergence rainfall is produced. In fact, thunderstorms have been observed in general storm conditions within regions where vertical lifting by the mountain slopes is great.

B. Spatial and Temporal Distribution of Storm Precipitation

The spatial distribution of precipitation varies widely from isolated small area rainfall over a short time period to general rains over large areas continuing for days. There are weather systems ranging from only a few thunderstorms (small area, short duration) to systems producing rains that can last for days over many thousand square kilometers.

Greatest rain depths for various area sizes and durations have been determined in record storms. Computing these depths is an ongoing project of the USACE in cooperation with the NWS (USACE, 1945-). To date, values for over 500 record storms have been published. Similar, unpublished data exist for over 300 additional storms at the U.S. Bureau of Reclamation (USBR) and the NWS. (A summary is available of the greatest values from all analyzed storms within each of 40 sections covering the United States. It is given for the four seasons and several area sizes and durations (Shipe and Riedel, 1976).)

For each storm, rainfall data supplemented by meteorological analysis were plotted on a small-scale topographic map and an isohyetal map was prepared. The temporal distribution was estimated using all rainfall recorder values, observers' comments, and synoptic weather maps. A standardized procedure (World Meteorological Organization, 1969) was then used to compute the greatest rain depths, usually by 6-hour time steps, over area sizes usually from 25.9 km^2 (10 sq. mi.) to total storm area. Results for a storm are shown in Fig. 8.3 to 8.5 as an example. First, the determined isohyetal map is given for a May 1953 storm (Fig. 8.3). Then the computed greatest average rain depths within isohyets are plotted and smooth enveloping curves are constructed (Fig. 8.4) for standard durations. Values taken from this plot are then tabulated (Fig. 8.5) for standard area sizes. This last figure is an example of the published data.

The distribution of incremental rainfall varies widely from storm to storm. For any given storm, hourly rainfall data within or near the drainage area of interest can be obtained from precipitation data, compiled by state, published monthly by the National Climatic Data Center.

C. Snowmelt Contribution

The subject of snow and snowmelt is discussed in more detail in Chapter 7.

At many locations, snowmelt is an important contribution to runoff. Computation of a snowmelt flood hydrograph can be a complicated undertaking, involving numerous variables including air and dew point temperatures and winds prior to, during, and after the rain, snow depth and density, percent of forest cover, and radiation. Each of these factors should be evaluated within elevation bands of the study drainage. Before giving details and data sources, it may be of interest to mention that some organizations, including the Bureau of Reclamation (Cudworth, 1989), have assumed a 100-year snowmelt flood upon which a rainstorm is sometimes superimposed. It is placed timewise to yield the greatest flow.

1. *Snowpack Water Equivalent* The water equivalent measurements of a snowpack, or the depth of water resulting when a sample is melted, are published on a state-by-state basis by the Soil Conservation Service (SCS). The annual summary for Colorado 1987, for example, lists the water equivalent and snow depth for the months of January through June for over 150 snow courses in the state. Similar publications are available for 11 western states. These publications include observations made by agencies such as the National Oceanic and Atmospheric Administration and state offices. For example, a publication for Alaska (USDA, 1951–70) gives snow depth and water equivalent summarized for the period from 1951 to 1970.

When floods are being estimated, the referenced material can be used to determine the mean, most extreme of record, or the value for some particular frequency for application to a drainage area at a certain time of year. Should snow course data not be available, indirect methods can be used to arrive at an estimate based on long-term measurements of monthly or seasonal precipitation and terrain factors.

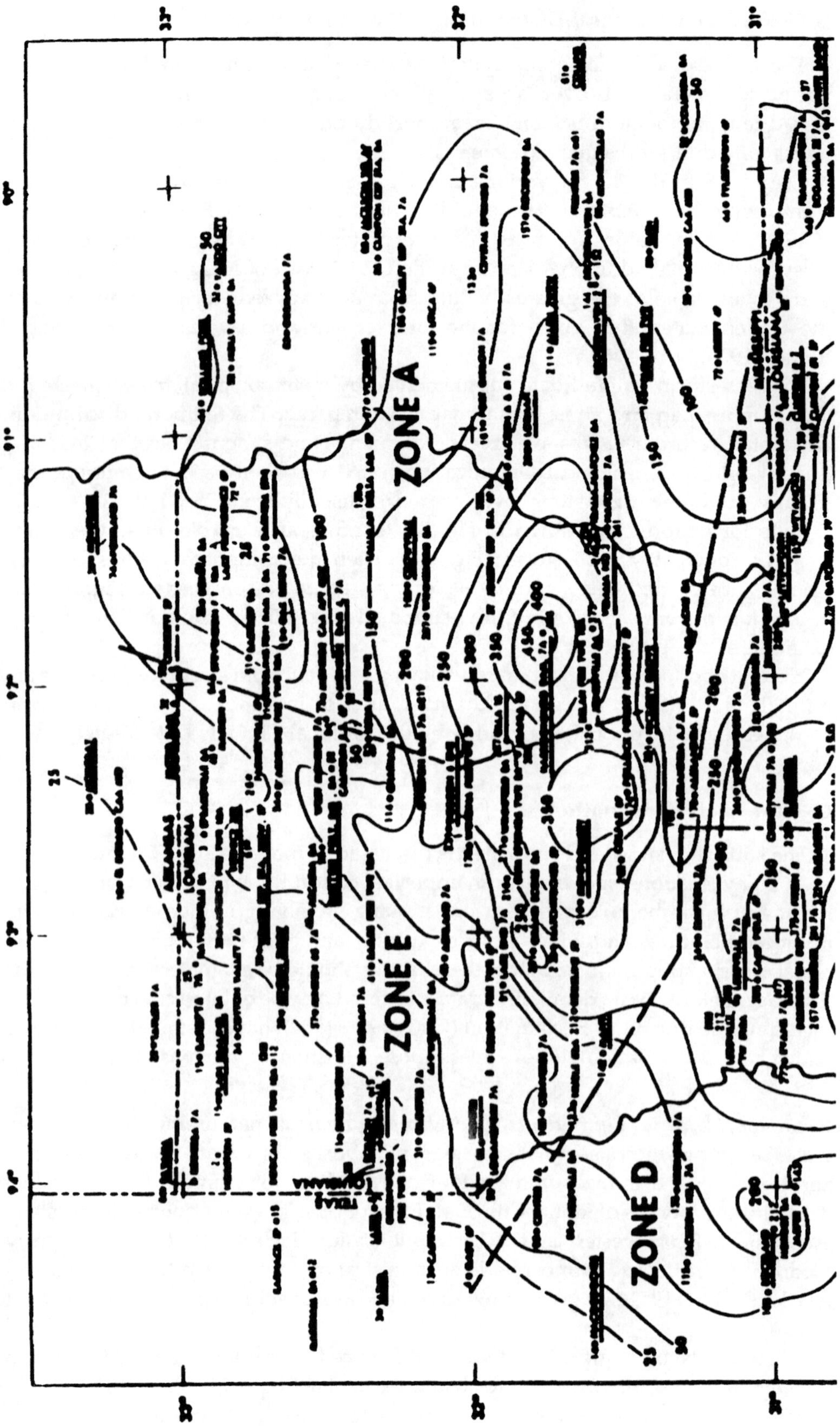
ZONE A
ZONE E
ZONE D
ARKANSAS
LOUISIANA
TEXAS
LOUISIANA

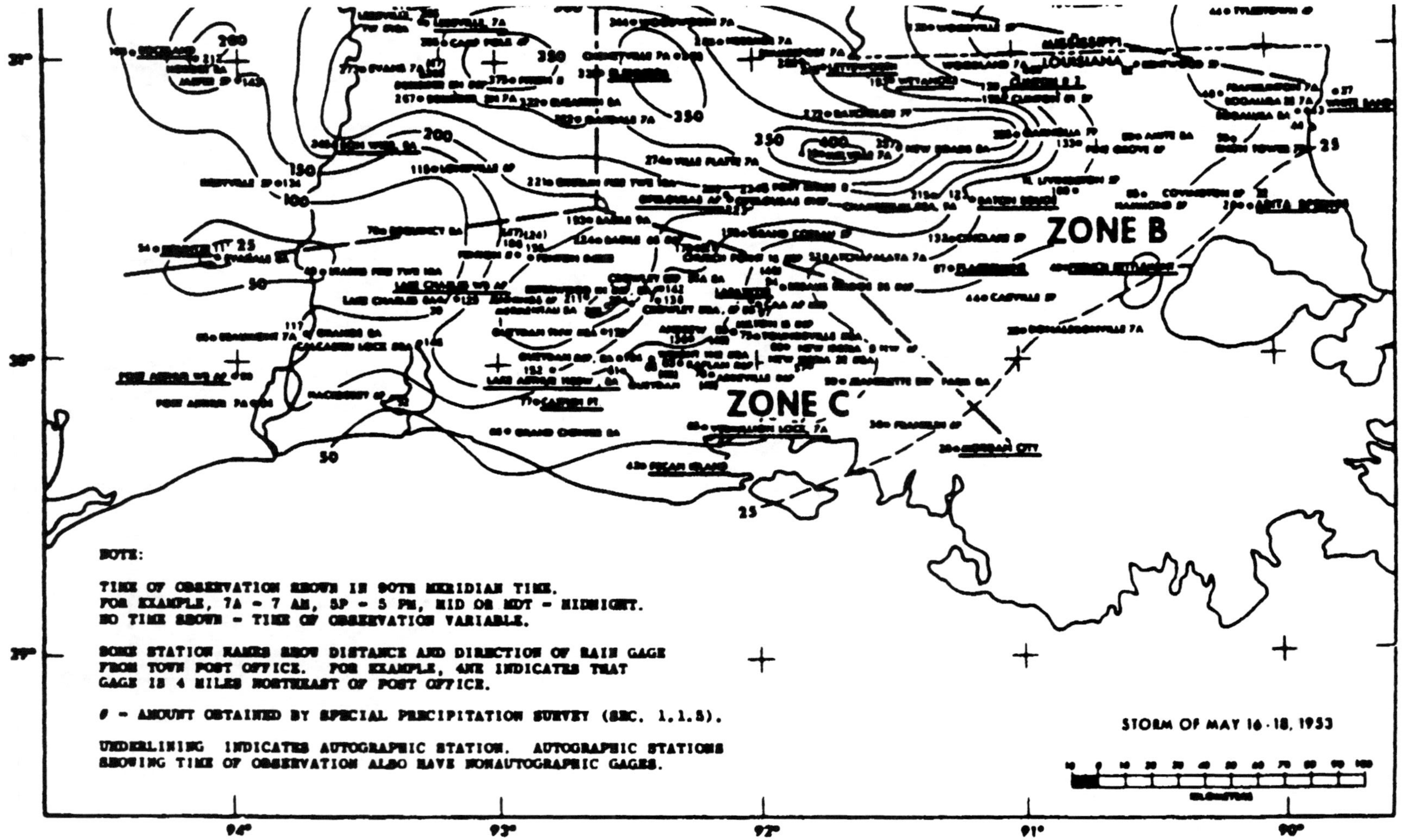

Figure 8.3—Isohyetal Map, Storm of May 16–18, 1953 (WMO, 1969).

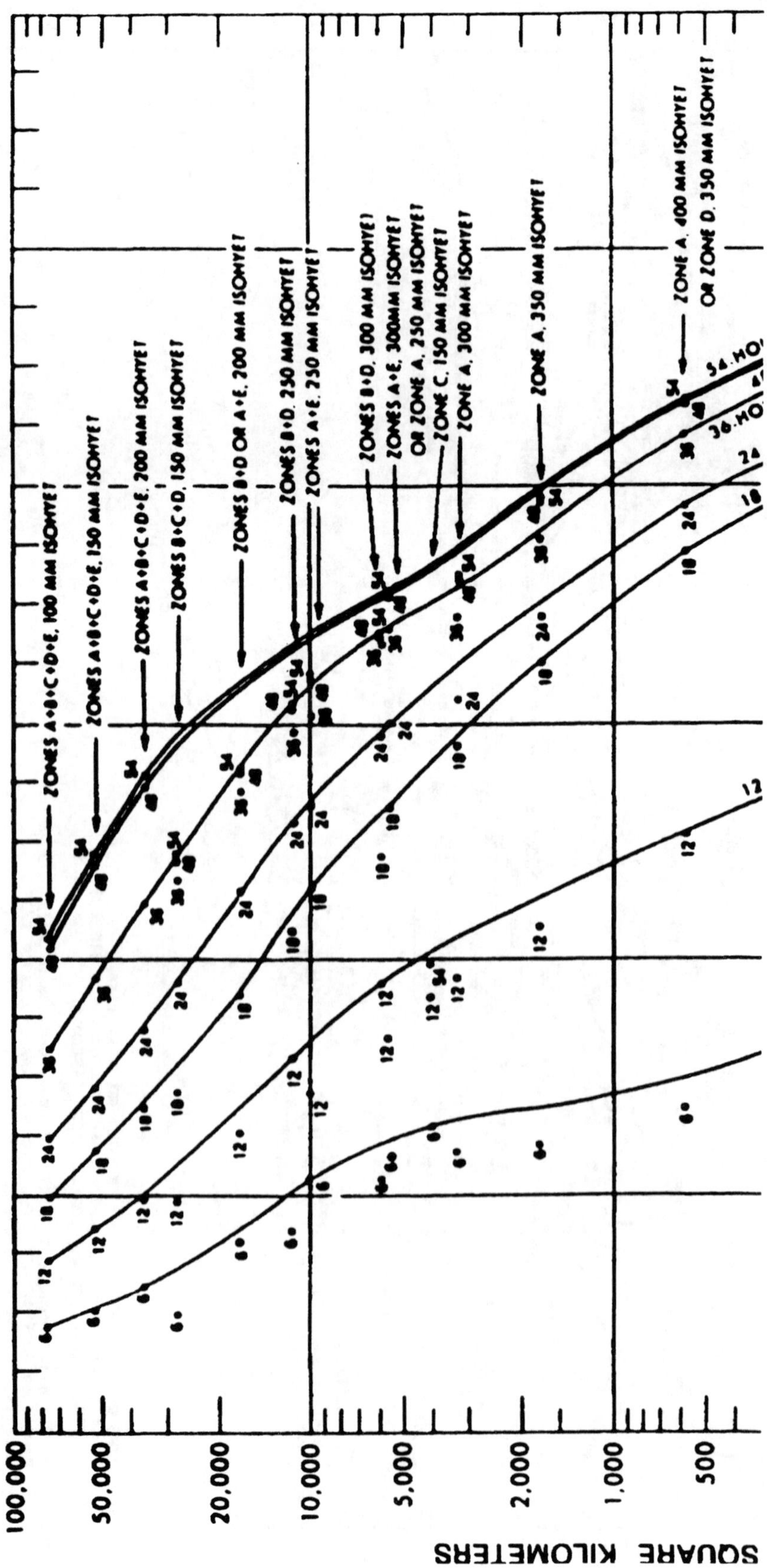
ZONES A+B+C+D+E, 100 MM ISOHYET
ZONES A+B+C+D+E, 150 MM ISOHYET
ZONES A+B+C+D+E, 200 MM ISOHYET
ZONES B+C+D, 150 MM ISOHYET
ZONES B+D OR A+E, 200 MM ISOHYET
ZONES B+D, 250 MM ISOHYET
ZONES A+E, 250 MM ISOHYET
ZONES B+D, 300 MM ISOHYET
ZONES A+E, 300MM ISOHYET
OR ZONE A, 250 MM ISOHYET
ZONE C, 150 MM ISOHYET
ZONE A, 300 MM ISOHYET
ZONE A, 350 MM ISOHYET
ZONE A, 400 MM ISOHYET
OR ZONE D, 350 MM ISOHYET
100,000
50,000
20,000
10,000
5,000
2,000
1,000
500
SQUARE KILOMETERS

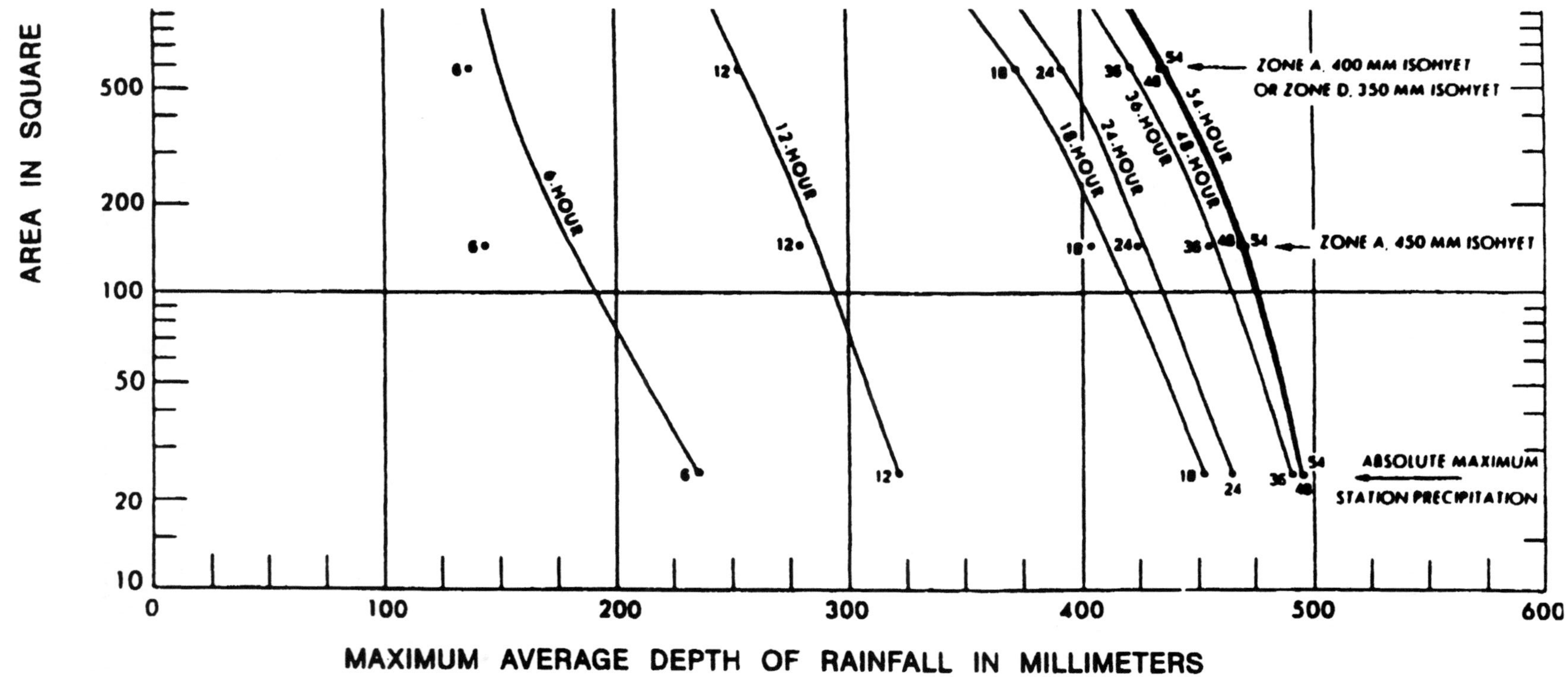

Figure 8.4—Maximum Average Depth of Rainfall for Selected Durations, Storm of May 16–18, 1953. (WMO, 1969).

STORM STUDIES – PERTINENT DATA SHEET

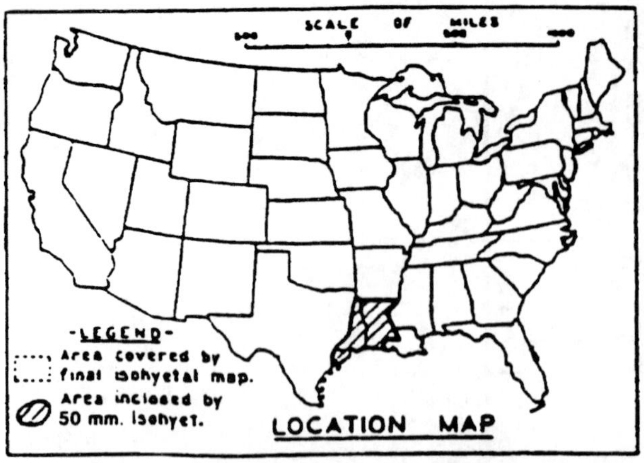

Storm of *May 16–18, 1953*

Assignment____________

Location____________

Study Prepared by:

Part I Reviewed by H. M. Sec. of Weather Bureau,__________

Part II Approved by Office, Chief of Engineers for Distribution of Factual Data,__________

Remarks:

DATA AND COMPUTATIONS COMPILED

PART I

Preliminary isohyetal map, in______sheet, scale____________

Precipitation data and mass curves: (Number of Sheets)

- Form 5001-C (Hourly precip. data)______
- Form 5001-B (24-hour " ")______
- Form 5001-D (" " " ")______
- Miscl. precip. records, meteorological data, etc. *(5001 WB)*______
- Form 5002 (Mass rainfall curves)______

PART II

Final isohyetal maps, in______sheet, scale____________

Data and computation sheets:

- Form S-10 (Data from mass rainfall curves)______
- Form S-11 (Depth-area data from isohyetal map)______
- Form S-12 (Maximum depth-duration data)______
- Maximum duration-depth-area curves______
- Data relating to periods of maximum rainfall______

Area in Km^2	Duration of Rainfall in Millimeters										
	6	12	18	24	30	36	48	54	72	96	126
Max. Station	236	322	453	465		489	496	496			
100	181	286	418	437		471	483	484			
200	164	269	399	422		459	476	478			
500	151	246	367	398		439	457	460			
1,000	146	230	336	369		411	427	429			
2,000	139	213	306	339		383	398	401			
5,000	124	191	261	302		345	361	364			
10,000	108	170	224	264		310	334	337			
20,000	90	135	185	220		275	306	311			
50,000	63	94	130	155		205	246	251			
100,000	37	60	82	102		140	176	181			

Figure 8.5—Pertinent Data Sheet for Storm of May 16–18, 1953 (WMO, 1969).

2. ***Snowmelt Computation*** Chapter 7 presents examples of snowmelt computations.

a. Air Temperature. During an extreme rainstorm, an upper limit to temperature could be based on mapped maximum persisting dew points (USDC, 1968). An alternative is daily temperatures available for numerous stations from the NWS, National Climatic Data Center, Asheville, North Carolina.

During a storm, the dew point should be about the same as the ambient air temperature. For periods before and after the storm, the dew point is usually lower. How much lower depends on the location of the subject drainage area relative to the moisture source.

b. Winds. Investigations of upper–free-air winds during a historical rainstorm in the subject drainage area is the first step in selecting the winds required for snowmelt computation. Besides winds for the rainstorm, winds for a series of days both before and after the rainstorm will be needed. A necessary adjustment (reduction) must then be applied for the winds above the ground or the snow levels at various elevations.

c. Radiation. Radiant energy, as solar (short-wave) radiation reaching the snow surface, and infrared (long-wave) radiation from the air and clouds during a storm are also required for computing snowmelt. These data are available at the NWS, National Climatic Data Center. A publication (USACE, 1960) presents the physical principles involved in snowmelt by radiant energy.

D. Antecedent and Subsequent Storms

"How large a storm should be considered antecedent to the main storm?" is a valid question in many situations. Similarly, the storm following the main storm is sometimes important. An equally important consideration is how much time should elapse between the two events. The U.S. Weather Service has made several studies of this problem for particular regions and storm sizes.

Sequences of recorded major storms for large drainage areas were postulated for the Ohio and Mississippi Rivers. It was found that, in general, a three dry-day interval was required for a reasonable transition from the weather situation at the end of the first storm to the beginning of the second. A similar conclusion was reached for smaller drainage areas in the Tennessee Valley. It was found that the larger the major storm, the smaller the antecedent or subsequent storm.

A study of small basins in Texas also showed that the larger the major storm, the smaller the antecedent and subsequent storms. For example, the greatest 10-day rains surrounding and including maximum 1-day rains were compared. It showed that when the maximum 1-day rain was 381 mm (15 in.), the surrounding 9-day rain was about 114 mm (4.5 in.). When the maximum 1-day rain was 457 mm (18 in.), the surrounding 9-day rain was about 91 mm (3.6 in.). This study was based on selected cases where pairs of storms occurred within a short time period. The outstanding or shortest time between 2 major storms (Mounds, Oklahoma, May 6–12, 1943 and Warner, Oklahoma, May 12–20, 1943) indicated a dry period of 3 days between them. The storm centers were approximately 80 km (50 mi.) apart.

While not covering all regions or storm sizes, it can be gathered that a minimum interval of 3 days between major storms is reasonable. It may also be concluded that the antecedent or subsequent storm should not exceed 30% of the main storm, especially if they are to be centered on the same drainage area.

Some federal agencies have set criteria regarding antecedent conditions. The USBR (Cudworth, 1989) uses a 100-year precipitation or flood prior to the PMP with a 3-day interval for areas east of the Sierra Nevada and Cascade Ranges. West of these ranges, an interval between 2 and 3 days is used. West of the 103rd meridian, no antecedent storm or flood is prescribed, although it is expected that initial soil retention is satisfied by an earlier storm.

In determining the design flood for a reservoir project, the antecedent or subsequent floods may or may not be important, depending on their relative magnitudes compared to the discharge capacity of the reservoir at normal maximum pool. For example, if the discharge capacity at the normal maximum pool is larger than the peak flow due to the antecedent storm, it normally would not be necessary to consider the antecedent storm. Consideration of a subsequent storm also normally would not be necessary if the discharge capacity is large enough to return the high reservoir level, caused by the main flood, to the normal maximum pool before the subsequent storm.

E. Baseflow

The subject of base flow is discussed in Chapter 6. For analysis of floods, base flow is the flow in the channel at the start of the storm runoff (when the hydrograph begins to rise).

The separation of base flow from the balance of the hydrograph is often a matter of judgment or policy. If recorded flood hydrographs are available, the base flow can be obtained from these hydrographs by the methods described in Chapter 6; however, the analysis can be very time-consuming. Thus, the mean of minimum daily flows for the month of occurrence of the flood is often used for simplicity although an overestimate of direct runoff is likely with this method. In the absence of any records, base flow can be transposed from similar, nearby gaged basins, using a drainage area ratio.

The importance of base flow increases as the size of the watershed increases. Base flow is often ignored, or assumed to be zero, for small, urban watersheds. Base flow cannot be ignored for larger streams, with drainage areas in hundreds of square kilometers, or where significant precipitation and runoff has recently occurred. The magnitude of the base flow at the beginning of a flood is dependent on antecedent rainfall and soil moisture conditions. The base flow for a flood should be reasonably higher when recent runoff has occurred, such as during the previous few days. If the time between runoff events is greater, the base flow would decrease. When an antecedent storm is included with the main design storm to simulate the overall design conditions, the base flow should be that prior to the antecedent storm. The intervening flow between the two storms is defined by the recession of the antecedent flood computed from the first storm.

F. Transformation of Rainfall Excess to Flood

The conversion of rainfall excess into a runoff rate requires the determination of an appropriate transformation mechanism. The rainfall to runoff response of a watershed is complex, typically nonlinear and time-variant. However, a nonlinear, time-variant approach has not gained general acceptance for practical problems. Therefore, a linear, time-invariant approach is usually followed to derive a peak discharge or an outflow hydrograph.

Transformation of rainfall and snowmelt can be accomplished by using any of the following approaches: an empirical relationship, unit hydrograph method, kinematic wave method, or a runoff simulation model. The selection of a method depends upon the availability of basic meteorological, hydrologic, and topographic data, desired accuracy of the results, and time and monetary resources available.

1. Empirical Methods These methods are used to determine the peak discharge for small drainage areas and are typically the design basis to analyze culvert, sewer, or bridge openings. The relationship between rainfall and peak discharge has been represented by many empirical and semi-empirical formulas (Chow, 1964). The most widely used formula of this category is the rational method:

$$Q = 0.278C\,I\,A \qquad (8.10)$$

in which Q is the peak discharge (m^3/s), C is the runoff coefficient, reflecting characteristics of the drainage basin and balance of units, I is the rainfall intensity (mm per hour), and A is the drainage area (km^2).

There are many assumptions involved in the rational formula, some appropriate and others known to be in error; however, as long as the area is sufficiently small, estimates based on the rational formula are usually considered satisfactory. The maximum area to which the formula is applicable is a topic for debate, but most practitioners limit the application to an area of less than a few square kilometers.

Although more sophisticated methods are available, the formula is still widely used for airport drainage and storm sewer design. Selection of parameters is discussed in a variety of hydrologic publications, including Chow, et al. (1988, Table 15.1.1). Use of the rational formula is discussed in Chapter 9.

2. *Unit Hydrograph Method* The definition and derivation of a unit hydrograph are discussed in Chapter 6. The assumption of linearity is not strictly correct under natural conditions of rainfall and runoff; however, extensive experience has shown that the limitations of the method are not a major handicap, considering the limited rainfall and snowmelt data that are usually available (USACE, 1973). The method provides acceptable results when the hydrologic data and size of the watershed are carefully selected to approximately meet the assumptions.

The size of the watershed for which a flood hydrograph is required may vary from less than one square kilometer to thousands of square kilometers. Unit hydrographs are best defined for smaller areas. For larger areas, the project basin should be subdivided into sub-basins so that areal variation in rainfall, soil, vegetative cover, and topography within each sub-basin would be as small as practicable.

Ideally, the selected storms for unit hydrograph derivation should be of short durations producing intense and nearly uniform rainfalls. Storms producing at least 15–50 mm (0.5–2.0 inches) of excess are desired; however, due to the scarcity of actual rainfall and runoff data, the engineer often has to use what is available. A number of flood events should be used to derive unit hydrographs and an average unit hydrograph adopted to conform to the shape of the derived graphs. The adopted unit hydrograph must contain a volume equal to unit rainfall excess on the basin.

The selection of a particular method to derive a unit hydrograph depends primarily on the available flood and precipitation data and the desired accuracy of the outflow hydrograph. The two main methods are the direct method and a synthetic method. These methods are discussed in Chapter 6.

3. *Kinematic Wave Method* Because of dissatisfaction with the "black box" type of analysis represented by the unit hydrograph technique, the kinematic wave method, a more physically-based procedure, evolved in the 1950's. Kinematic wave applications have increased significantly since the advent of large high-speed computers in the 1970's. The technique traces the movement of runoff through the watershed to the outlet of the basin as indicated in Fig. 8.6. The main assumption of this technique is that the bed slope and water surface slope are equal and acceleration effects are negligible. Fig. 8.6 also shows conceptually the individual elements used in applying the kinematic wave method.

The kinematic wave equations are an abbreviated form of the full equations of motion, continuity, and momentum. These equations take the following form:

a) Continuity (Conservation of Mass)

$$A \frac{\partial V}{\partial x} + VB\frac{\partial y}{\partial x} + B\frac{\partial y}{\partial t} = q \tag{8.11}$$

b) Conservation of Momentum

$$S_f = S_o - \frac{\partial y}{\partial x} - \frac{V}{g}\frac{\partial V}{\partial x} - \frac{1}{g}\frac{\partial V}{\partial t} \tag{8.12}$$

in which A is the cross sectional flow area (m^2), V is the average velocity of water (m/sec), x is the distance along the channel (m), B is the water surface width (m), y is the depth of water (m), t is time (sec), q is the lateral inflow per unit length of channel (m^3/sec./m), S_f is the friction slope (m/m), S_o is the channel bed slope (m/m), and g is the gravitational constant (9.8 m/sec^2).

The assumption of kinematic flow allows the simplification of the momentum equation to the form of $S_f = S_o$. This equates the frictional forces to the gravitational forces, allowing an assumption of normal

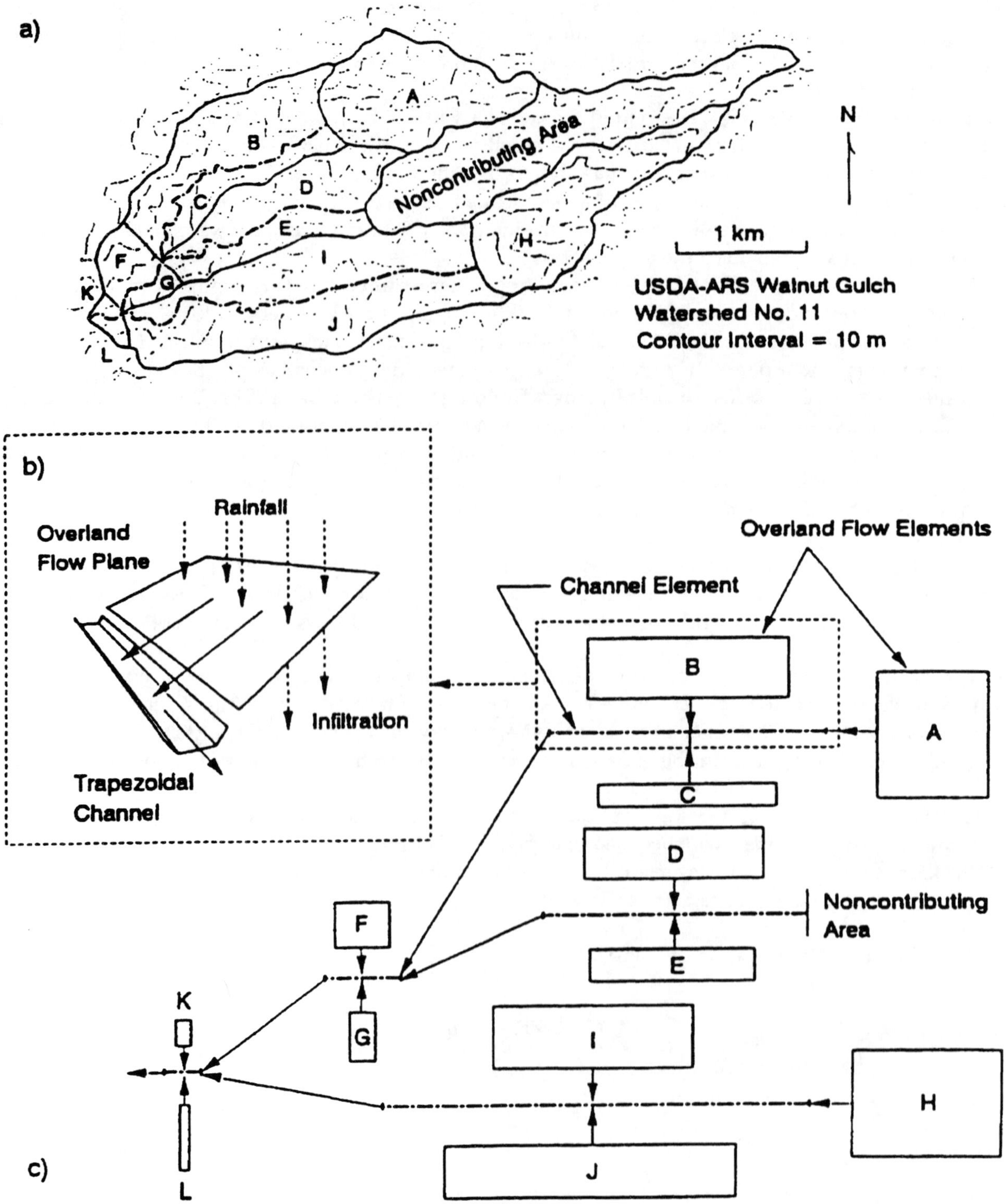

Figure 8.6—Watershed Modeling Using Kinematic Wave Method.
(a) Watershed discretization to define model elements; (b) typical model elements; (c) plan view of model elements forming the watershed representation, where the letters correspond to those in a)

depth and uniform flow. The momentum equation for application in kinematic flow allows Manning's equation to be written as:

$$Q = \alpha A^m \tag{8.13}$$

where α and m are related to the surface roughness and flow geometry. Routing can therefore be described solely by the continuity equation. The kinematic wave method accounts for time translation, but does not attenuate the flood wave. Therefore, the technique is not applicable for all situations. The use of this technique is most common for urban watersheds. These watersheds often have relatively steep slopes and prismatic channels, typically resulting in little attenuation, thus making the kinematic wave technique applicable. The application of this procedure requires considerable judgement in the selection of appropriate parameters for each of the flow strips and to evaluate the discharge hydrograph output for "reasonableness." The main advantage of this technique is that it is physically-based (no black box) and conceptually complete in terms of the physics of runoff. The main disadvantages include the difficulty in determining average strip lengths, slopes and roughnesses, and its applicability in small slope land surfaces and channels.

4. Runoff Simulation Models Estimation of runoff due to rainfall can be accomplished directly with the unit hydrograph or kinematic wave procedures; however, discharge analysis is typically required at many points throughout the basin. Therefore, it is often more efficient to apply a watershed rainfall–runoff model to determine the runoff, with the model using unit hydrograph, kinematic wave, or another method. A number of such models have been developed, primarily since about 1970. Some of the models simulate single event rainfall–runoff and may be designated as event simulation models. Other models provide a continuous hourly or daily simulation of the rainfall–runoff process over long time periods and are designated as continuous models. Generally, the models have been developed by government agencies, universities, or private consultants. Some of the better known models are discussed later in the section on microcomputer software for flood analysis.

For event simulation models, direct runoff is computed from a single storm event using initial retention and an infiltration rate. There is no provision for recovery of initial retention or infiltration during periods of no rainfall. The continuous simulation models have soil moisture accounting procedures. Soil moisture contents, evapotranspiration losses, and recovery of infiltration rate are updated continuously. The channel and reservoir routings, often required in both event and continuous models, are accomplished using either hydrologic (volumetric) routing or hydraulic (dynamic) routing. Some of the most commonly used models are also discussed in the section on microcomputer software for flood analysis.

5. Channel and Reservoir Routing Routing is necessary to trace the hydrograph through a watershed system, typically accounting for changes in shape and timing due to storage in each river reach. A wide variety of both hydrologic and hydraulic routing methods and techniques are available to the analyst, depending on factors relating to desired accuracy, time and funds available, and complexity of the flow situation. A detailed discussion of these models and their application to channel and reservoir routing is given in Chapter 6. In general, hydrologic methods are often employed, as they are considerably simpler than hydraulic methods and give adequate results for most routing situations.

V. PROBABLE MAXIMUM FLOOD

A. Basic Concepts and Definitions

The concept of using meteorological analysis to estimate limiting precipitation values for hydrologic design originated in the 1930's (Paulhus and Gilman, 1953). The Flood Control Act of 1936 assigned to the USACE the responsibility of constructing flood control projects throughout the country. Soon thereafter the USACE sought the assistance of the then Weather Bureau (currently the NWS) in assessing the rainfall potential in prospective project areas. As a result, the Hydrometeorological Research Section (now the Hydrometeorological Branch) was established within the Washington headquarters of the Weather Bureau in 1937.

At the Hydrometeorological Branch, attempts were made, based on meteorological knowledge, to determine physical upper limits to the rainfall that could occur over certain basins during specified time intervals. Such is the conceptual definition of probable maximum precipitation (PMP). It was originally known as maximum possible precipitation (MPP), and this term is found in most reports on estimates of maximum precipitation made prior to about 1950. The reason for the name change to PMP was that MPP carried a stronger implication of physical upper limit of precipitation than does PMP. PMP is preferred because of uncertainties in estimating maximum precipitation.

Based on the definition by the NWS (HMR No. 33, 1956), PMP represents the critical rainfall depth–duration–area relations for a particular area during various seasons of the year. These relations would result if meteorological conditions including precipitable water and wind speed during an actual storm in the region were adjusted to represent the most critical meteorological conditions that are considered probable of occurrence.

In a more recent publication (WMO, 1986), PMP is defined as "the greatest depth of precipitation for a given duration meteorologically possible for a given size storm area at a particular location at a particular time of year, with no allowance made for long-term climatic trends."

In any case, PMP is the estimated precipitation depth for a given duration, drainage area, and time of year for which there is virtually no risk of being exceeded (USNRC, 1977). The word "virtually" is added as recognition that nothing in life is completely and absolutely certain.

Probable maximum flood (PMF) can be defined as the flood resulting from the most severe combination of meteorological and hydrologic conditions that are considered reasonably possible for the drainage basin under study. It is a flood which would be created by the PMP over the basin assuming critical areal and time rainfall distributions as well as other conservative assumptions pertaining to antecedent watershed conditions.

B. Estimation of PMP

1. Critical Duration The magnitude and intensity of PMP varies with duration. The duration of the PMP that causes the most critical flood at the site of interest is the critical duration for that drainage basin. In general, the critical duration is short for a small basin and long for a large basin. The flood regulation capacity of reservoirs, depressions, and flood plains located at or upstream of the site also affects the length of critical duration.

To determine the critical duration, inflow hydrographs resulting from PMP of various durations should be derived and routed through the upstream and project reservoirs, if any. The duration of the PMP that causes maximum flood surcharge and outflow at the project site is the critical duration; however, in practice, the critical duration is often estimated based on the size of the drainage area and the capacity of upstream storage (including reservoirs) to avoid the need for estimating PMP for several durations.

For a small drainage area and a small storage capacity, a short duration of six hours or less may be critical. On the other hand, for a large drainage area and a large upstream storage capacity, a long duration of three days or more may be critical (Wang, 1988). A critical duration of 15 days or more is not uncommon for a very large basin having a drainage area of several hundred thousand km^2 or more.

The general guideline for determining the critical duration is that it should be at least equal to the time of concentration of the basin, and should be longer when a reservoir of substantial storage capacity exists or is being considered upstream. In case of any doubt, a duration sufficiently longer than the time of concentration should be used.

2. *Storm Type* For the purpose of estimating the PMP for a specific basin, rainstorms can be classified as "general-storms" or "local-storms." For a project basin with a critical duration of 6 hours or less, a local-storm PMP generally would produce the PMF. For a critical duration of more than 24 hours, a general-storm PMP normally would be more critical. For a critical duration of 6 to 24 hours, a 6- or 12-hour local storm and a 24-hour general storm should be analyzed to determine whether the local- or general-storm PMP would be more critical.

3. *Generalized Estimates* For most of the United States, generalized PMP has been estimated and published by the NWS in collaboration with other agencies including the USACE, the USBR, the Tennessee Valley Authority (TVA), and the Soil Conservation Service (SCS), (NWS Hydrometeorological Reports No. 33–56, 1956–88). Fig. 8.7 shows the regions covered by these studies.

These publications describe in considerable detail how the generalized PMP estimates were made. The procedures used differ from region to region depending on the physiographic and meteorological characteristics of the region. The resulting Hydrometeorological Reports (HMRs) also differ in the sizes of drainage areas considered as well as the degree of detail given. In the studies for mountainous regions, topographic control sometimes cannot be shown in the detail required for some small drainage areas. For the region east of the 105th meridian, PMP can be estimated for a storm area of any size (HMR 51, 1978) and the spatial as well as temporal distributions can be determined (HMR 52, 1982). Studies for the other regions do not give this much detail.

Some judgment on the level of the generalized PMP estimates can be obtained by comparison with the greatest known rainfalls of record (Riedel and Schreiner, 1980). This study presents maps showing rainfall that is equal to or greater than 50% of the PMP. Fig. 8.8 is an example of the 30 maps, each for a different combination of area sizes (25.9 to 51,800 km^2) and durations (6 to 72 hours). This figure shows that for 25.9 km^2 (10 mi^2) and 24 hours there were 32 separate rainfalls greater than or equal to 50% of PMP east of the Continental Divide and 54 cases for west of the Divide. Some observations from the referenced study are:

a) For a given area and duration, some large regions east of the Continental Divide have no storms greater than or equal to 50% of PMP. However, within the same regions, there often are storms greater than or equal to 50% of PMP in adjacent area sizes and durations.
b) In western states, values greater than or equal to 50% are concentrated along the Sierra Nevada and Cascade Ranges. Storms approaching PMP can be expected more often in these wet regions than in dry regions.
c) The relative lack of storms greater than or equal to 50% of PMP in the east compared to the west is most likely due to a few very extreme storms in the east, which when transposed to compute PMP, far exceed most observed values.

The last observation is supported by the fact that there are 19 storms in the east and 13 storms in the west where the rains are greater than or equal to 70% of PMP for 25.9 km^2 and 24 hours. There are also other factors that enter into the comparisons, such as the number of major storms that have been studied in detail and the number of observation stations and the length of record.

4. *Site-Specific Study* There are basins of unique topographic characteristics where generalized PMP estimates cannot be properly applied. Under such conditions, or in areas where generalized PMP estimates are not available for the size of the drainage area, a site specific study may be needed (Wang, 1986).

The Hydrometeorological Reports are a good source for learning how such a study can be carried out. Further guidance can be obtained from a PMP manual (WMO, 1986). It has been shown that a meteorologist experienced in estimating extreme rainfall is required to properly make a PMP estimate. Major steps to be considered include:

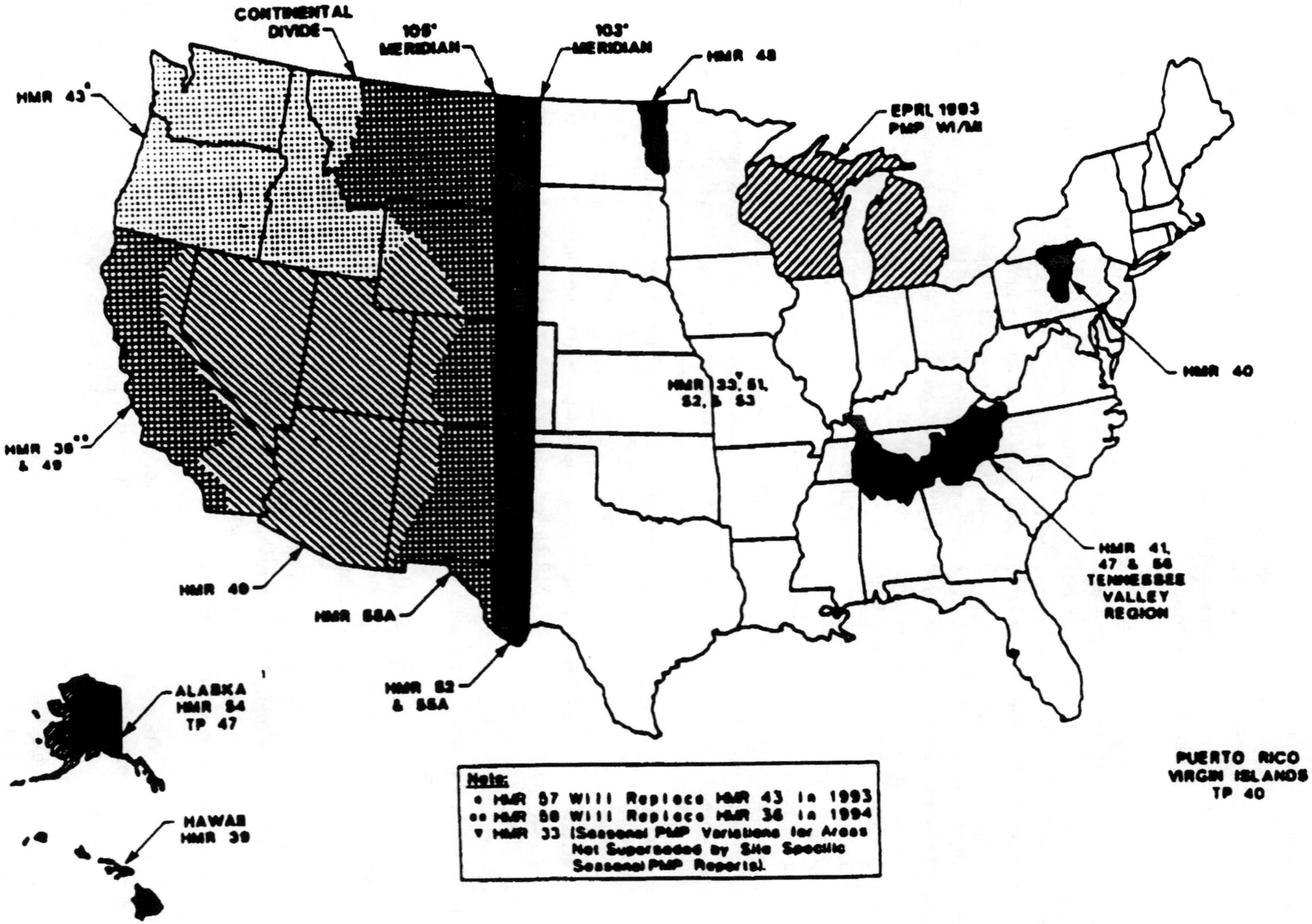

Figure 8.7—Regions Covered by Generalized PMP Studies.

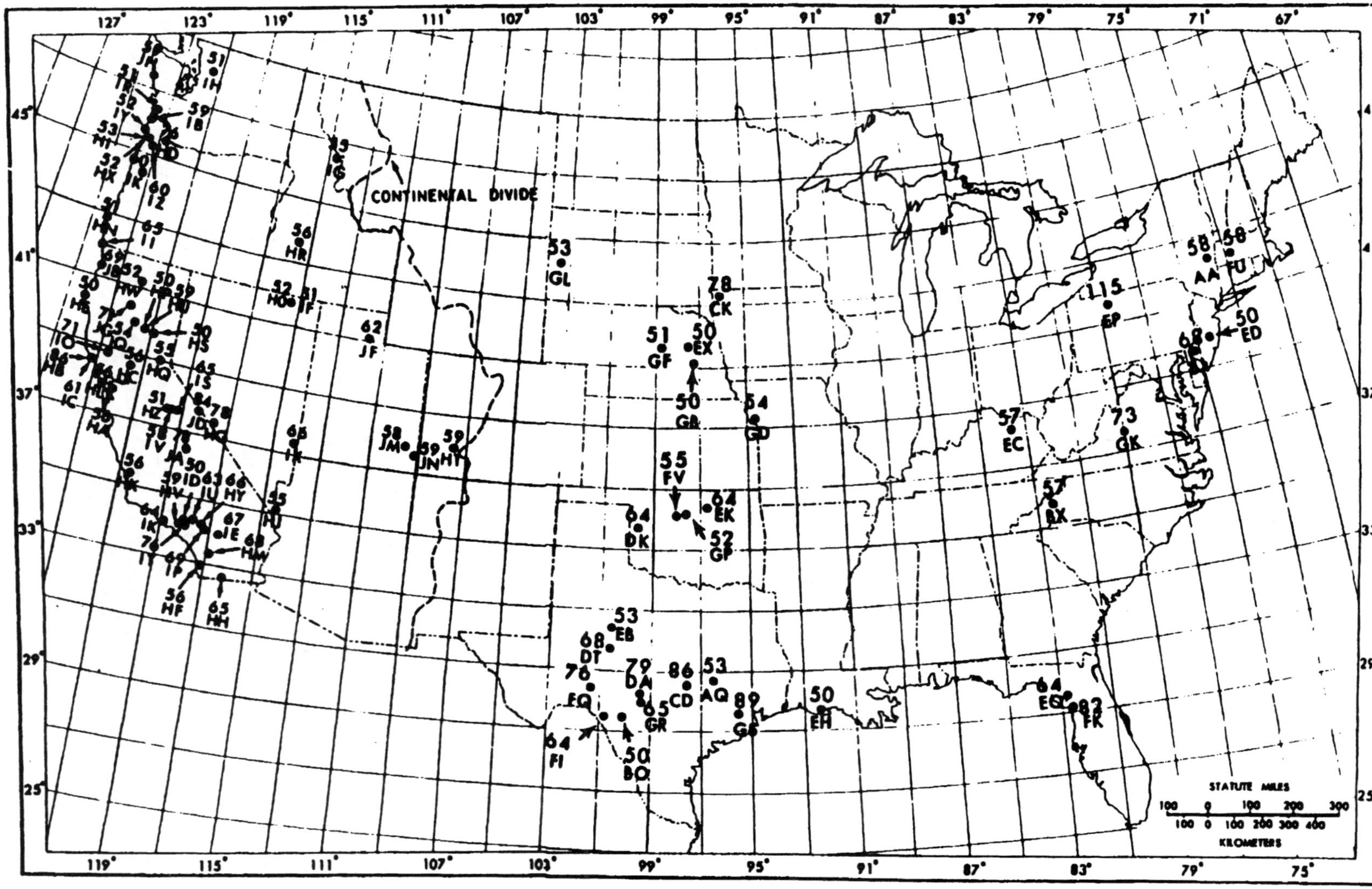

Figure 8.8—Observed 24-hour, 25.9 km^2(10 sq. mi.) Rainfall Quantities Expressed as Percent of All-Season PMP Estimates.

a) determination of critical storm duration for the site,
b) definition of a meteorologically homogeneous region surrounding the site,
c) identification of major storms that have occurred in the region,
d) determination if these storms could occur on the site,
e) preparation of isohyetal maps for the qualifying storms,
f) determination of moisture maximization factors for candidate storms (including elevation adjustments),
g) transposition of candidate storms adjusted for maximum moisture,
h) further adjustment of each storm for differences in topography between the storm location and the site,
i) determination of temporal distribution for critical storms, and
j) envelopment of the transposed and adjusted storm depths to obtain the estimate of PMP.

It should be noted that no two PMP studies have been carried out in exactly the same way. Topographic problems, availability of records on major historic storms and other meteorological data for a "sufficient" length of time are some reasons why each site-specific PMP study differs from others.

C. Transformation of PMP to PMF

Transformation of a PMP to the PMF is generally accomplished by the more conventional unit hydrograph method or the computer-based simulation model discussed earlier in this chapter. In applying either method, a careful reconnaissance of the project basin is needed. It is also important to divide the basin into a number of sub-basins so that the areal variation of rainfall and basin characteristics can be properly incorporated. The basin is divided so that areal variation in rainfall, soil, vegetative cover, and topography within each sub-basin would be as small as practicable. For most cases, five to fifteen sub-basins are required except for very small basins.

The consideration of antecedent and/or subsequent storms, snowmelt, and base flow are important in the transformation of a PMP to the PMF. These subjects are discussed earlier in this chapter.

D. Greatest Rainfalls and Floods of Record

Before a PMF estimate is adopted for design purposes, its reasonableness is often investigated further by comparing the PMP and PMF with greatest storms and floods of record.

Fig. 8.9 shows the world's greatest observed point rainfalls, while Table 8.2 gives the greatest observed rain depths in the United States for selected area sizes and durations. These data can be compared with PMP estimates to check their reasonableness; however, geographical, meteorological, and topographic similarities between the study areas and the record storms used for comparison should be carefully evaluated and taken into consideration. For example, the 2,590 km^2 (1,000 mi^2) 24-hour rainfall of 767 mm (30.2 in.) in the table recorded at Yankeetown, Florida potentially could occur at other locations in the southeastern United States, but not in the Rocky Mountain region of Colorado.

Table 8.3 (Riedel, 1990) is similar to Table 8.2 but also includes maximum values from Taiwan, India, and China, where extreme rainfalls are known to occur. This table also can be used as reference for comparison between PMP estimates and record rainfall.

Figs. 8.10 and 8.11 are plots of the data in Tables 8.2 and 8.3 for easier comparison with PMP determined for a specific area size. Bullard (1986) of the USBR has made a comparison of PMF peaks with historical floods for sites throughout the United States.

E. Conservatism of PMF Estimates

Attempts have been made by many researchers to assign a probability of exceedance to PMF. Many meteorological and hydrologic factors affect the estimate of a PMF. The basic approach is to determine

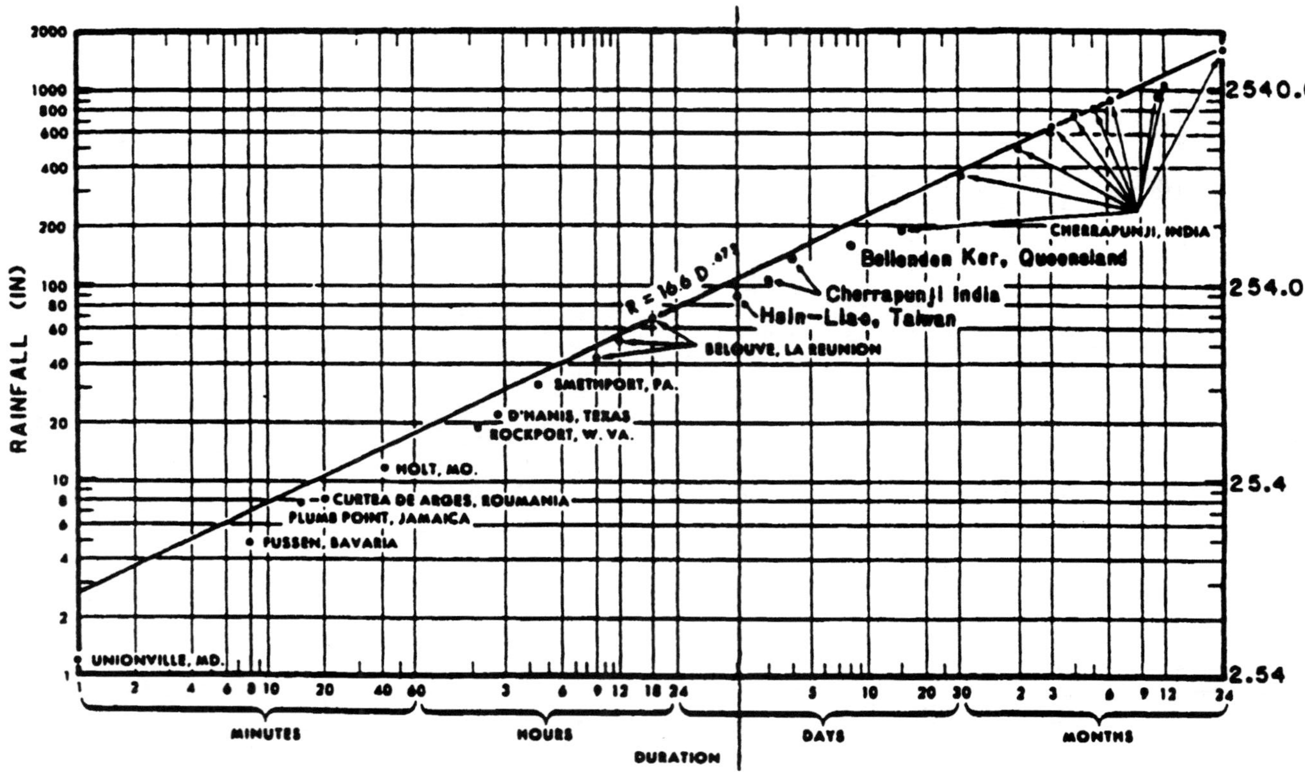

Figure 8.9—World's Greatest Observed Point Rainfalls (WMO, 1969).

TABLE 8.3. Maximum known observed depth–area–duration data for Taiwan, China, India, and the United States (average rainfall in millimetres (and inches).

Area	Duration (h) 6	12	18	24	36	48	72
25.9 km²	627a	757b	1036c	1232c	1488c	1585c	—
10 mi²	(24.7)	(29.8)	(40.8)	(48.5)	(58.6)	(62.4)	—
259 km²	526r	716s	950c	1166c	1405c	1473c	1680c
100 mi²	(20.7)	(28.2)	(37.4)	(45.9)	(55.3)	(58.0)	(66.1)
518 km²	500r	686r	925c	1120c	1354c	1412c	1565c
200 mi²	(19.7)	(27.0)	(36.4)	(44.1)	(53.3)	(55.6)	(61.6)
1295 km²	452d	625c	856c	1036c	1242c	1308c	1380c
500 mi²	(17.8)	(24.6)	(33.7)	(40.8)	(48.9)	(51.5)	(54.3)
2590 km²	381d	594c	744c	881c	1062c	1113c	1240c
1000 mi²	(15.0)	(23.4)	(29.3)	(34.7)	(41.8)	(43.8)	(48.8)
5180 km²	284b	450e	572e	630e	693e	859f	1095c
2000 mi²	(11.2)	(17.7)	(22.5)	(24.8)	(27.3)	(33.8)	(43.1)
12950 km²	206bh	282b	358b	472f	475f	734f	882t
5000 mi²	(8.1)	(11.1)	(14.1)	(18.6)	(18.7)	(28.9)	(34.7)
25900 km²	145h	201j	257k	371f	384i	587f	743t
10000 mi²	(5.7)	(7.9)	(10.1)	(14.6)	(15.1)	(23.1)	(29.2)
51800 km²	102h	152j	201k	264f	295i	424f	595t
20000 mi²	(4.0)	(6.0)	(7.9)	(10.4)	(11.6)	(16.7)	(23.4)
129000 km²	64km	107n	135k	160k	201k	279g	389g
50000 mi²	(2.5)	(4.2)	(5.3)	(6.3)	(7.9)	(11.0)	(15.3)
259000 km²	43m	64mo	89k	109k	152p	170p	226q
100000 mi²	(1.7)	(2.5)	(3.5)	(4.3)	(6.0)	(6.7)	(8.9)

Storm	Date	Location of Center
a	17-18 July 1942	Smethport, PA, US
b	8-10 September 1921	Thrall, TX, US
c	11 September 1963	Peishih, Taiwan
d	7 August 1959	Tahushan, Taiwan
e	3-7 September 1950	Yankeetown, FL, US
f	17-18 September 1880	Uttar Pradish, India
g	26-28 July 1927	Gujarat, India
h	27 June–4 July 1936	Bebe, TX, US
i	27 June–1 July 1899	Hearne, TX, US
j	12-16 April 1927	Jefferson Parish, LA, US
k	13-15 March 1929	Elba, AL, US
m	22-26 May 1900	Chattanooga, OK, US
n	15-18 April 1900	Eutaw, AL, US
o	19-22 November 1934	Millry, AL, US
p	29 September–3 October 1929	Vernon, FL, US
q	5-10 July 1916	Bonifay, FL, US
r	7 August 1975	Linzhuanz, Henan, China
s	1 August 1977	Muduocaidang, NeiMongal, China
t	4 July 1935	Nishi, Hunan, China

Source: World Meteorological Organization, 1986 with addition of Chinese data.

the probability of exceedance of each maximized factor and compute the joint probability of these factors occurring simultaneously; however, this approach is practically impossible because:

1) the recurrence interval of any maximized individual factor that is much longer than the length of historical record is difficult to determine; and
2) the computation of joint probability from the exceedance probabilities of individual factors is difficult when the interactions between the factors are not completely known.

An alternative approach, though less rigorous and somewhat subjective, has been proposed (Wang and Revell, 1983) to measure the relative conservatism of a PMF estimate. By knowing how conservative the

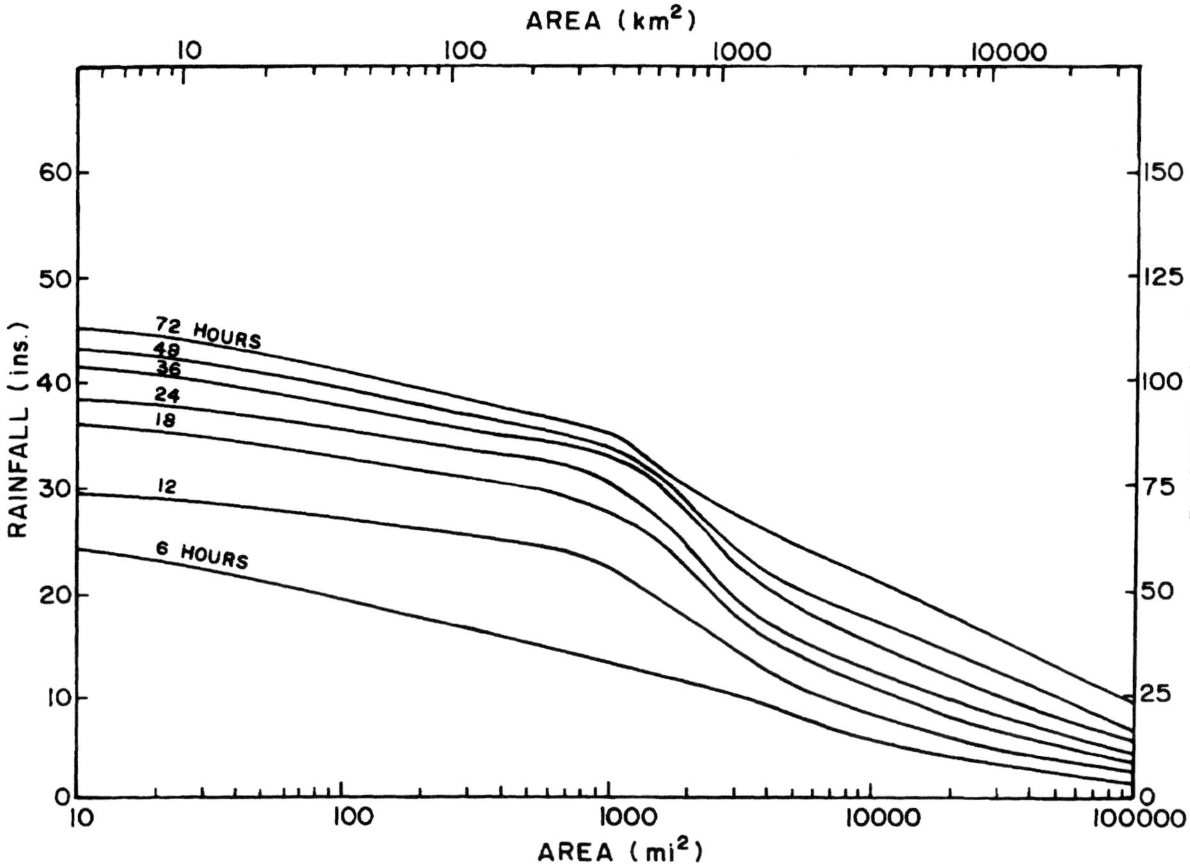

Figure 8.10—Maximum Known Observed Depth–Area–Duration Data for the U.S. (Table 8.2).

estimated PMF is, the engineer will be able to make a better decision as to how the estimate should be used in the design of the project. For example, a smaller or larger freeboard may be considered depending on whether the estimate is conservative or otherwise.

F. Standard Project Flood

The standard project flood (SPF) as defined by the USACE (1952) is a "hydrograph representing runoff from the Standard Project Storm (SPS)." As the PMF, it is a hypothetical flood, estimated by transforming the SPS. The SPS is the most severe storm that is considered "reasonably characteristic" of a region. It often undercuts storms of extraordinary severity and has been assumed to be approximately 40–60% of the PMP.

Estimation of the SPS can be best accomplished by following the procedures prescribed by the USACE (1952). Transformation of an SPS to the SPF can follow the procedure for transforming a PMP to the PMF. The SPF has been used to evaluate certain features of flood control projects and to identify flood plain, primarily by the USACE.

VI. FLOOD HAZARD AND FLOOD WARNING

As defined earlier in this chapter, a flood is an overflow on lands that are not ordinarily covered by water. When the land is in use by humans, the overflow causes damages to crops, property, and infrastructure and, at times, results in loss of human life.

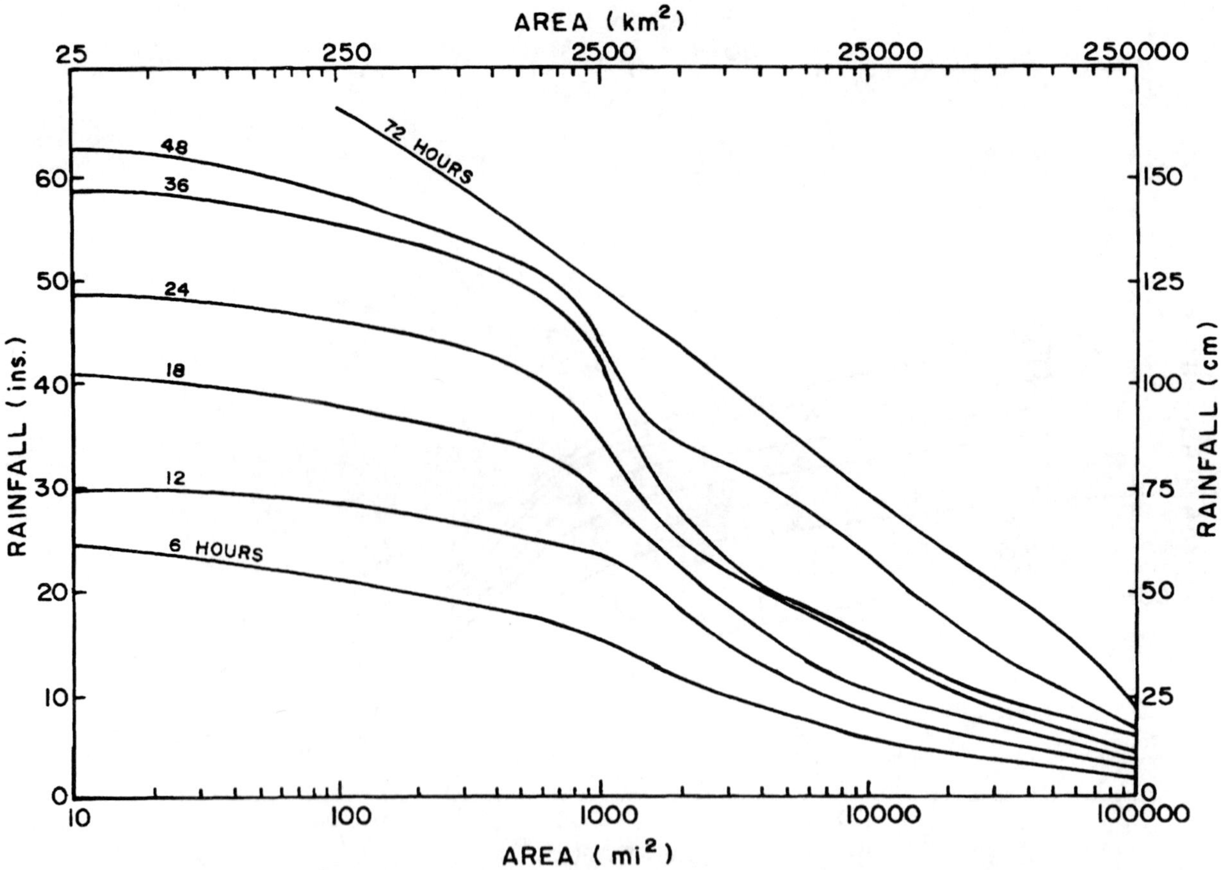

Figure 8.11—Plot of Maximum Known Observed Depth–Area–Duration Data for Taiwan, China, India, and the United States (Table 8.3).

Flood hazards are evaluated by the extent of inundation in the flood plains. The maximum depth and duration of inundation as well the rate of rise in flood level are the major factors that determine the flood severity; however, velocity of flow can be a factor. A slow rising flood may allow sufficient time to relocate the movable properties to safe places, resulting in minimum damages and no loss of life, but a fast rising flood can cause major damages and loss of life.

A series of land-use and flood-severity maps are generally available to assess potential damages due to flood. A real-time flood forecast and warning can predict the timing and extent of inundation prior to the occurrence of a flood and thus help to reduce the potential damages. Emergency action plans are available in many flood-prone communities to evacuate people to predesignated safe locations during catastrophic floods.

A. Evaluation of Potential Hazards

Total flood damage is one important indication of a potential flood hazard. The first step in estimating flood damages is to assess the severity of the flooding produced by a given flood event. The most important measures of flood severity are the areal extent of flooding, the depth of flooding at each selected location (buildings, warehouses, cropped land, etc.), and the duration of flooding. Other factors determining the severity are the time to reach the maximum depth, flow velocity, sediment concentration, and season of occurrence.

Generally, the flood damage potential at critical locations within the flood plain is determined by combining two types of maps (James and Lee, 1971):

1) A series of land-use maps, each indicating details of development including agricultural area (cropping pattern and seasonality), residential and industrial buildings, and other infrastructure. These maps are time-variant and are periodically updated.
2) A series of maps showing lines of equal inundation depth in the flood plain. These maps also are time-variant and should be updated depending upon the changes in the watershed land use and flood plain development. The maps are prepared for major historical floods and floods of various frequencies.

The areal extent and depth of inundation are best determined from historic flood data. These inundation depths and resulting damages can be used to develop depth-damage and depth-frequency data. Generally the historical flood damage data are not available for the full range of desired frequencies. Synthetic inundation boundaries and depths at selected structures are determined for flood peaks of various frequencies. These synthetic data are used to assess potential damages. Flood peaks and/or flood hydrographs are determined using the procedures discussed earlier in this chapter and Chapter 6 of the Handbook.

1. Backwater Computations With flood peak data for various exceedance probabilities available, the inundation boundaries, depths, and average flow velocities can be determined through backwater computations. The objective of these computations is to determine the shape of water surface profiles in the open channel and flood plain. Two widely used computer programs for computing water surface profiles are: WSP-2 (USDA, 1976) and HEC-2 (USACE, 1991).

The programs (described in the following section, Microcomputer Softwares for Flood Analysis) are used in flood plain management and flood insurance studies to delineate flood hazard zones and determine floodways. Fig. 8.12 shows typical results of program application—flood profiles for various frequencies (FEMA, 1983). Estimated water surface elevations at various locations are transferred from the profiles to topographic maps to define inundation boundaries for different levels of flooding. The difference between the ground elevation and the water surface elevation indicate the depth of inundation.

Because the water surface profiles correspond to flood peaks, the duration of inundation at a given structure cannot be determined through backwater computations; however, the duration of inundation at selected locations can be estimated from a flood hydrograph and a stage–discharge relationship. The stage causing inundation is used to estimate the corresponding discharge. The hydrograph is then used to determine the time between the rising and falling limb of the hydrograph that the discharge is greater than the critical value. This time duration would then be available for use with the land use maps to estimate the flood damages. Duration is most important when agricultural damages are to be estimated.

2. Dam Break Analysis Dams are constructed to provide benefits such as water supply, irrigation, hydropower, flood control, and recreation; however, dam failure can cause catastrophic flooding in the downstream valley due to sudden release of impounded water. Depending upon the quantity of the stored water and the height of the dam, the flood peak can greatly exceed all known previous floods. The warning time available can also be much shorter compared to the time for a precipitation-induced flood.

The International Commission on Large Dams (ICOLD, 1973) and the United States Committee on Large Dams, in cooperation with the American Society of Civil Engineers (ASCE/USCOLD, 1975) have reported that about 38% of all dam failures are caused by inadequate spillway capacity with the dam subsequently being breached. About 33% of dam failures are caused by seepage or piping through the dam or along internal conduits, about 23% from foundation problems and about 6% by slope embankment slides, damage, or liquefaction of earthen dams from earthquakes, and landslide-generated waves within the reservoir. All these scenarios can result in dam failure and breaching. Johnson and Illes (1976) and Jansen (1988) have summarized major dam failures.

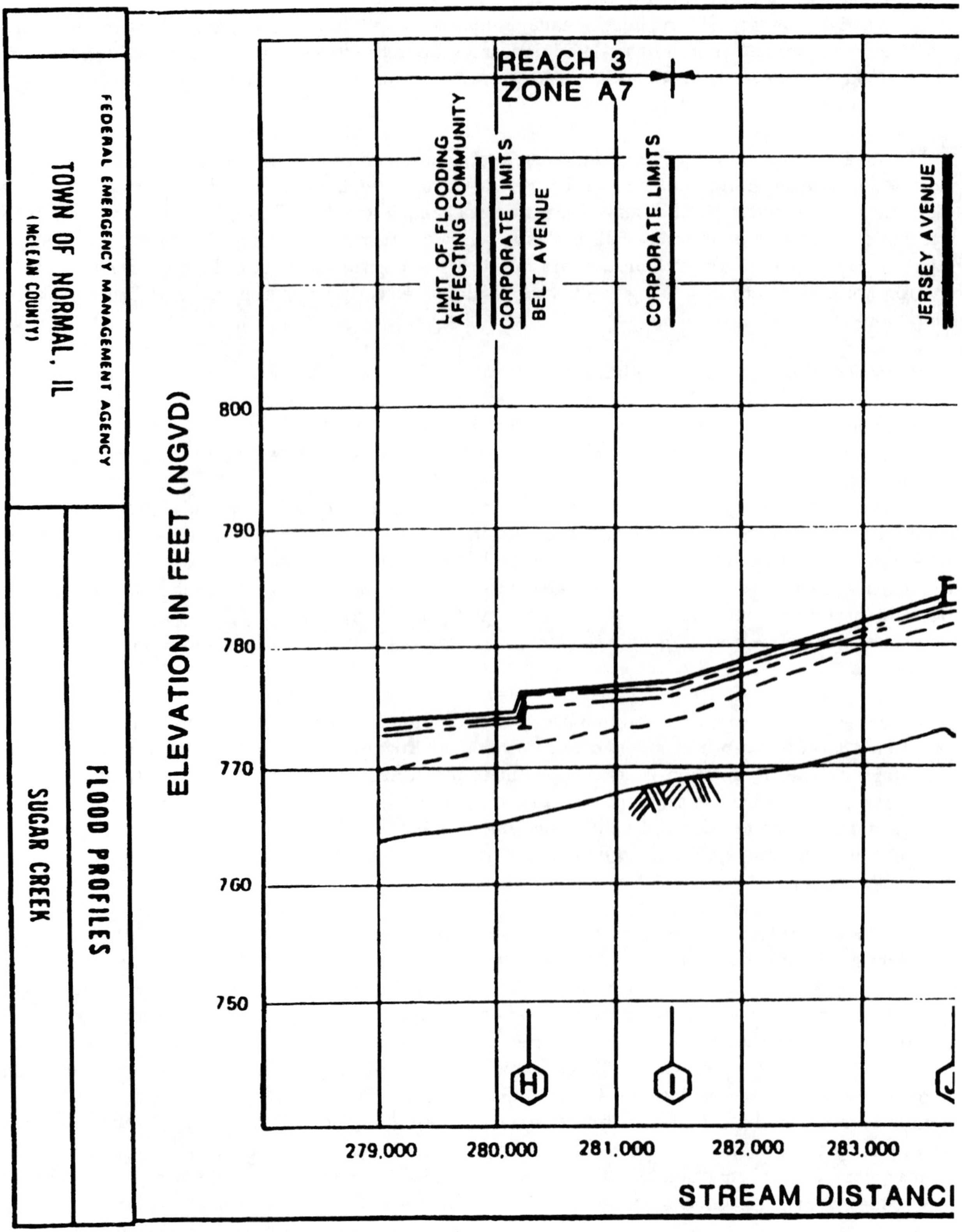

Figure 8.12—Flood Profiles for Various Return Periods.

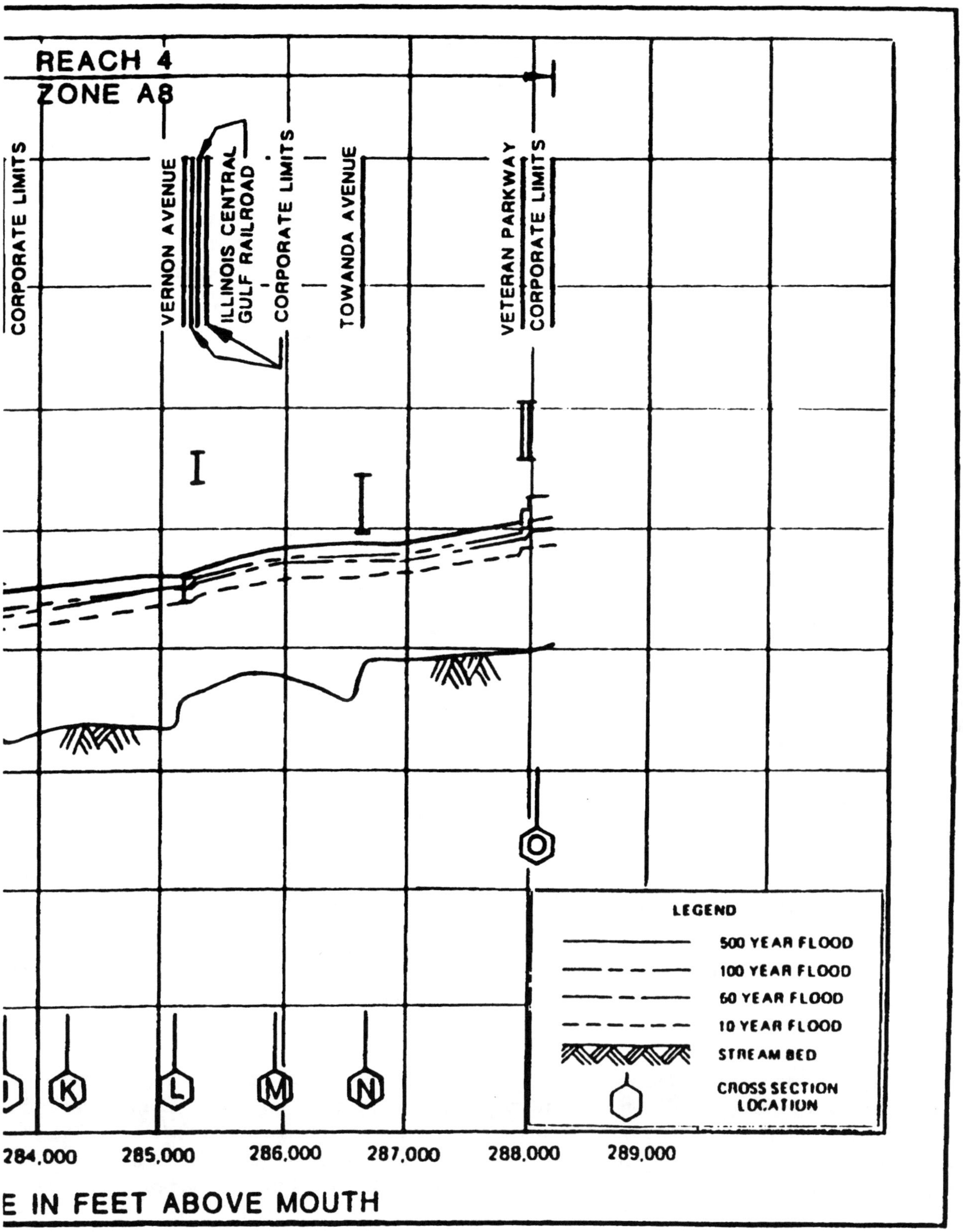
REACH 4
ZONE A8
CORPORATE LIMITS
VERNON AVENUE
ILLINOIS CENTRAL GULF RAILROAD
CORPORATE LIMITS
TOWANDA AVENUE
VETERAN PARKWAY
CORPORATE LIMITS
K
L
M
N
O
LEGEND
500 YEAR FLOOD
100 YEAR FLOOD
60 YEAR FLOOD
10 YEAR FLOOD
STREAM BED
CROSS SECTION LOCATION
284,000
285,000
286,000
287,000
288,000
289,000
E IN FEET ABOVE MOUTH

The NWS has the responsibility to advise the public of downstream flooding in case of a dam failure. The private dam owner is also responsible for informing the appropriate local, state, and federal agencies for three failure conditions: a) failure pending, b) failure imminent, and c) failure in progress. The FERC and state and local laws require that private owners complete hypothetical dam break analyses for the preparation of inundation boundary maps for the downstream valley.

A number of computer models have been developed to simulate the dam break flood hydrograph and its subsequent routing through the downstream valley. The most widely used is probably the DAMBRK model (Fread, 1988) discussed in the following section on Microcomputer Software for Flood Analyses.

3. Inundation Maps Flood hazard information may be provided on maps showing the areal extent of inundation. Flood mapping has been used to document both actual events and hypothetical floods corresponding to a specified probability of flood risk. Mapping has been provided for sites flooded by rivers, dam breaks, lakes, extreme tides, tsunamis, and for alluvial fans where the position of the channel may change during the flood.

a. Historical Development. Large historical floods have been mapped for nearly 100 years, mainly by the USGS. However, wide-scale flood mapping throughout the United States dates primarily from House Document No. 465 of the 89th Congress, Second Session 1966. That report noted inadequacies in the existing federal policy that concentrated on structural measures for flood control and stressed flood plain management and nonstructural solutions for flooding. The act resulted in a need for inundation mapping as a basis for planning and management of flood plains—the impetus for a large national inundation mapping effort.

Congress established the NFIP in 1968, further emphasizing the need for flood mapping. This program provides financial compensation to policy holders for flood damages in communities that regulate development of flood prone lands within the communities' jurisdiction. A basic requirement for effective operation of the NFIP is information on the flood hazard within each community. This includes data on the areal extent and depth of flooding and the velocity of flows for various probabilities for all riverine, lacustrine, and coastal flooding sources. As of 1990, flood information has been provided in flood insurance studies (FIS) for about 19,000 of the Nation's 22,000 communities known to have flood hazards.

The National Dam Safety Act of 1972 also resulted in wide-spread flood hazard mapping. The act required a nationwide inventory of all dams and an assessment of the integrity of non-Federal dams that could have significant downstream hazards. Although hazard mapping was widely discussed, the act produced no unified national program for mapping the hazardous areas that could be inundated downstream from dams of questionable integrity; however, federal agencies used dam break analysis techniques to establish the downstream areas at risk for most federally owned dams, and a few states required similar studies of nonfederally owned structures. These studies commonly resulted in a dam-break water surface profile downstream from the structure. Some studies also defined inundation boundaries.

b. Inundation Mapping Techniques. The report "Flood Insurance Study Guidelines and Specifications for Study Contractors" prepared by the Federal Emergency Management Agency (FEMA, 1991) provides the most comprehensive set of instructions available for producing inundation maps for specific hypothetical floods. Although these guidelines contain information that is unique to flood insurance studies, guidance is also given on standard professional practices for inundation mapping. The guidelines cover techniques for mapping inundation from riverine and coastal sources but not for dam-breaks or for documenting experienced events. The guidelines recommend adopting the most efficient procedures to produce an inundation map of appropriate accuracy. Accuracy needs may range from highly precise data generally applicable to densely developed or developing areas to approximate data appropriate to areas with little development.

Detailed mapping of a riverine flood hazard includes the flood boundary of the 1% annual chance event, with additional hypothetical events as needed. Flood insurance studies include the 10%, 2%, and

0.2% chance floods, along with the 1% chance event. Federal flood damage evaluation studies could include additional flood profiles. When economic damages are to be computed, hypothetical events would be analyzed and possibly mapped to provide the detailed hydraulic information needed for economic analysis.

c. Sources of Inundation Maps. Available information provides a large portion of national needs for the evaluation and warning of the existing flood hazards, but may not be easily accessed. Sources of information are listed as follows:

1) Information about the availability of flood insurance maps is available through the Flood Map Distribution Center, FEMA. This information also may be available by contacting the office of the State Flood Plain Manager, or the community officer responsible for local flood plain land-use regulations.
2) Flood plain information or dam break reports prepared by the USACE may be obtained through the Flood Plain Management Services Unit of the appropriate Corps District or Division Office.
3) Inquiries about flood plain information available from the SCS should be addressed to the State Conservationist, Soil Conservation Service, in the appropriate state office. The SCS also maintains a field office in virtually every county in each state and personnel of these offices may be familiar with local flood study reports.
4) Information on inundation maps defined by the TVA may be obtained from the Floods Protection Section of the authority.
5) Water Resources Division field offices of the USGS exist in every state and those offices can identify all available flood information produced by the USGS for the particular state.
6) In addition to the information produced by federal agencies, a large amount of inundation information has been collected by state and local agencies. A search of the state and local engineering offices such as Flood Control Districts may produce additional data.

4. Flood Damages Major water resource projects, especially those featuring flood control, must show economic justification (benefits exceed costs at an authorized interest rate) for federal funds to be used. Thus, a linkage between depth and frequency of flooding and economic damages is required. Damages due to flooding must be determined and then meshed with depth-frequency to give average annual damages for pre-project and for post-project conditions. The difference between the two is the average annual damages avoided by the project, or project benefits. Fig. 8.13 illustrates this procedure.

Flood damages can be classified into five categories (Breaden, 1973) as direct damages, indirect damages, secondary damages, intangible damages, and uncertainty damages.

Direct damages are losses caused by physical contact with floodwater and are generally evaluated as the cost of replacing, repairing, or rehabilitating the affected property for urban damages and as the net effect on farm income for crop damages.

Indirect damages include the value of lost business and services, including the cost of alleviating hardship, safeguarding health, rerouting traffic, traffic delays, and related phenomena (Breaden, 1973). The USACE (Kates, 1965) has suggested assessing the indirect damages as percentages of direct damages as follows:

Residential	15%	Public Facilities	34%
Commercial	35%	Agriculture	10%
Industrial	45%	Highways	25%
Utilities	10%	Railroads	23%

Secondary damages are defined as the adverse effects on people who depend on output produced by damaged property or on hindered services (Breaden, 1973). These damages are often not evaluated.

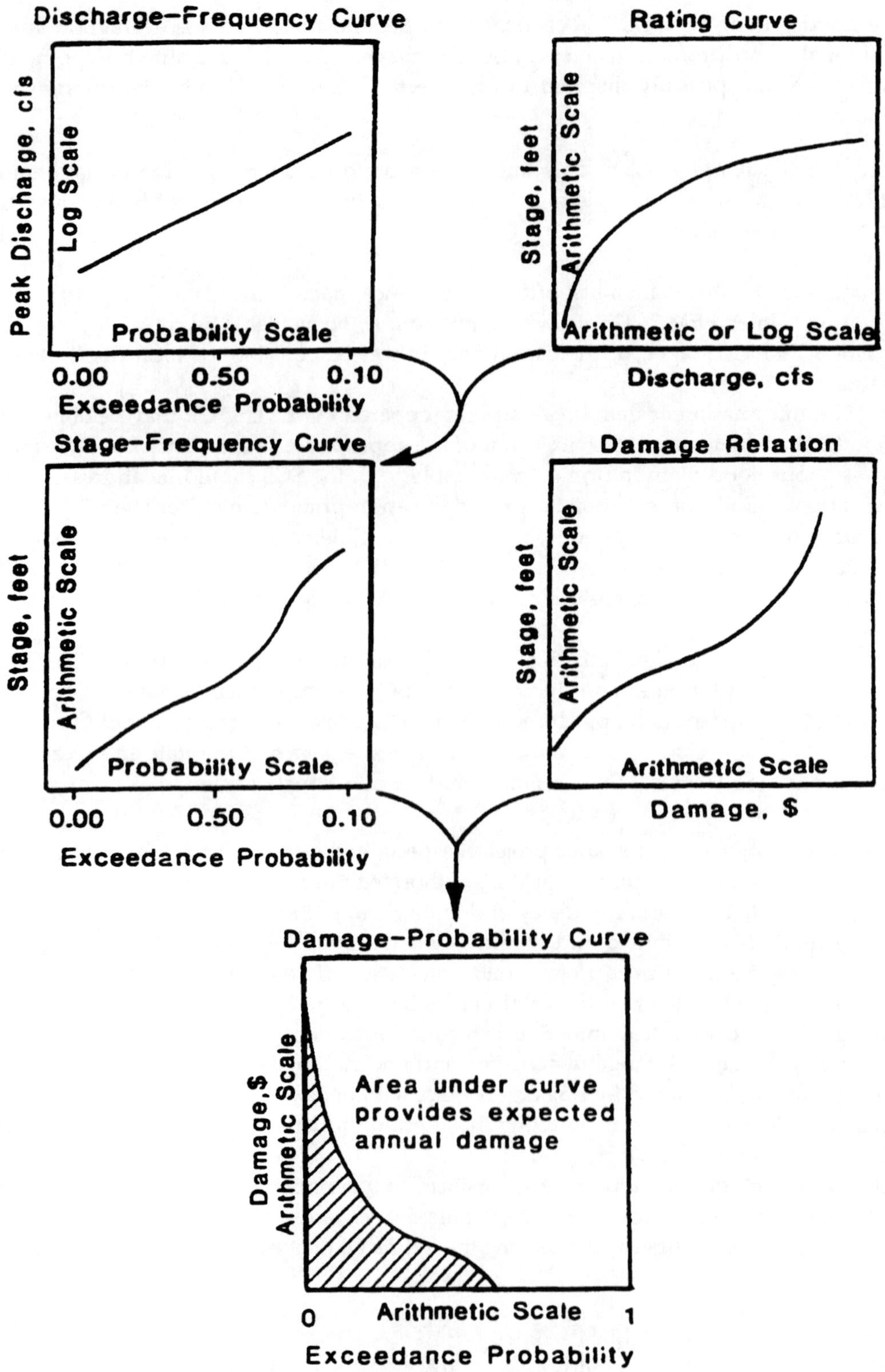

Figure 8.13—Derivation of Damage CDF from Discharge CDF, Rating Function, and Elevation-Damage Function.

Intangible damages may be defined as the impacts on environmental quality, social well-being, and aesthetic values. These may include fear, anxiety, annoyance, distress, insecurity, ill health, and ultimate loss of life (Parker and Penning–Rowsell, 1972). These impacts have drawn considerable attention recently and are part of the analysis for flood mitigation project justification.

The uncertainty damage cost may be calculated as an amount in excess of the expected value of the damages that flood plain occupants are willing to pay to avoid a flood loss (Breaden, 1973). The calculation of uncertainty damages depends upon the flood insurance practices in specific areas.

5. Depth-Damage Relationships Flood damages are a function of several variables, the most important being the land use type and floodwater characteristics. The floodwater characteristics include depth, velocity (including wave action), duration, and sediment load.

Different types of land use obviously have different susceptibility to flooding. An easily measured characteristic is the flood depth. This parameter is often used exclusively to define the depth–damage relationship and is most appropriate for a residential dwelling. Although still applicable, depth–damage is difficult to develop for commercial and industrial establishments without very site-specific damage surveys. Depth–damage must be related to the time of year and the duration of flooding for agricultural damage determination. Recognition of the effect of other variables besides depth (velocity, sediment deposition, etc.) is important, but often cannot be logically applied to the damage relationship.

Two basic approaches have been used to estimate flood damages: a) actual damages approach and b) potential damages approach.

The actual damages approach involves estimating damages resulting from historical floods through field interviews. This technique is commonly used for analysis of agricultural flood damages, especially if a proposed flood control project only provides protection from relatively frequent flood events. The major disadvantage of the method is the lack of sufficient data on large, rare floods to provide accurate damage estimates for extreme floods. The method can be applied to period–of–record analyses using a continuous simulation model, as well as to individual historical events.

The potential damages approach is typically used for the majority of flood damage reduction studies. The approach to determine a damage–frequency relationship involves:

a) Preparation of inundation maps based on hydrologic and hydraulic analyses for floods of various exceedance probabilities.
b) Development of stage–frequency relationships from the inundation maps in hazard areas.
c) Estimation of depth–damage relationships.
d) Linking depth–damage with depth–frequency to obtain a damage–frequency relationship at each desired location.
e) Integration of damage–frequency curves to obtain average annual damages. This process is depicted in Fig. 8.13.

Many government agencies and private organizations have developed computer programs to derive and integrate damage–frequency curves, given the characteristics of the study area. A program frequently used is the Expected Annual Damages Package developed by USACE Hydrologic Engineering Center (1988b).

The depth–damage relationship can be highly variable and difficult to estimate, particularly for indirect and secondary damages. The form of some typical stage–damage curves is shown on Fig. 8.14.

6. Alluvial Fans and Channels Alluvial fans and alluvial channels present difficult problems for an engineer or hydrologist evaluating flood hazards. Flow paths on these alluvial features tend to change shape and position during major flood events. This movement adds to the uncertainty present in any flood evaluation and requires special analytical approaches.

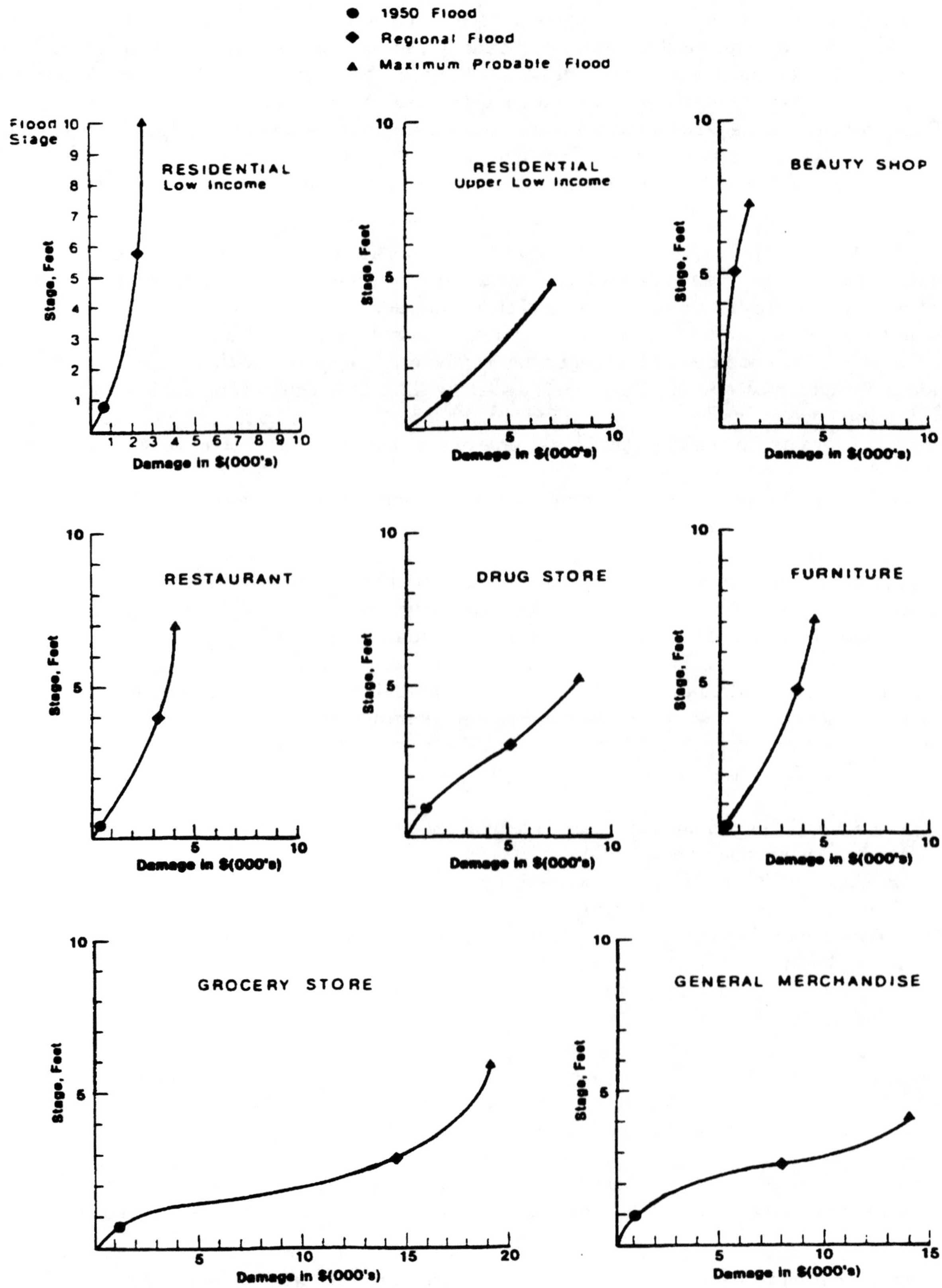

Figure 8.14—Typical Stage–Damage Curves.

a. Alluvial Fan. An alluvial fan commonly forms as sediment deposits where a sediment and debris-laden stream emanates from a steep mountain valley onto a flat land surface such as the flood plain of a large receiving stream. Contours of fan deposits show on a topographic map as arcs radiating from a point at or near the mountain front, and give the appearance of a fan. Fans form most readily in arid and semi-arid mountain regions and are relatively common features in the western United States.

Fans generally form over geologic time when the sediment and debris-laden flows or bank sloughing blocks an existing water course causing overflows that form one or more new channels on the erodible fan surface. Sediments are deposited in the newly formed channels as the flow recedes. As the process is repeated, a relatively smooth fan surface forms. On an active fan, the process of channel forming is random and unpredictable; any point has a roughly equal chance of being the location of a new channel.

The FEMA "User's Guide to Analysis and Mapping of Alluvial Fan Flooding," (in preparation) provides a complete and detailed description on techniques for fan hazard evaluation. The recommended analysis is based upon a procedure proposed by Dawdy (1979) that is currently used with some modifications in preparing Flood Insurance Studies. Because every location on the active fan is subject to flooding, the analysis determines flood hazard by the percent chance that a specified magnitude of depth and velocity will be exceeded in any future year at various locations.

The methodology defines the probability of exceedance, P(%), as:

$$P(\%) = \int_{q}^{\infty} P(H|Q=q)\ f_a(q)\,dq \tag{8.14}$$

where H is the event that any point on a selected contour will be flooded, $P(H|Q = q)$ is the conditional probability of the point being flooded given that a flood of magnitude q has occurred at the apex, and $f_a(q)$ is the probability density function (pdf) of the discharge Q at the apex. Replacing Q with depth D or velocity V, and replacing q with d or v in the equation yields the definition of the P(%) flood depth or velocity, respectively.

Dawdy (1979) employed three assumptions required with the method: 1) the pdf for flood flows has a log–Pearson Type III distribution as recommended in WRC Bulletin 17B, 2) the conditional probability term, $P(H|Q = q)$, on any contour is equal to the width of the flow channel divided by the width of the active fan area measured along the contour, and 3) the width of the flow channel is proportional to the four–tenths power of the flood discharge. This width relationship is based upon field observations in New Mexico, California, and Nevada that floods on alluvial fans flow at critical depth in wide and approximately rectangular channels and that the depth of flow decreases until a further decrease results in a 200–fold increase in width.

Three analytical steps are required in the FEMA fan hazard analysis.

1) Define the flood peak frequency curve at the fan apex. The flood frequency curve is defined by a log–Pearson Type III distribution. Definition of the apex is a somewhat arbitrary judgment that should be made in the field. It is defined as the point below which flow paths are unpredictable. On some fans a channel is deeply incised into the alluvium for a distance from the mountain front to the apex and this deep channel is unlikely to experience the avulsion necessary to produce unpredictable flow paths.
2) Determine the boundaries of the area subject to alluvial fan flooding. One must decide if a fan surface is active and currently subject to flooding or if it is a remnant of previous geologic experience. Fan boundaries commonly are defined from study of maps and aerial photographs and field inspection of such features as channel links, soil types, vegetation, soil color, and desert varnish.
3) Compute and map the flood hazard depth and velocity. A computer program available from FEMA greatly simplifies the computational effort.

Fan analysis is a complex and detailed procedure that is best performed by experienced analysts.

b. Alluvial Channel. The instability of stream channels in erodible alluviums complicates the determination of flood hazards. During passage of a flood wave, heavy sediment loads lead to aggradation and/or degradation that can alter bed slope and result in changing water surface profiles. Lateral migration by bank scouring and calving during the process can also alter the flood hazard area.

Several computer models have been proposed for evaluating flood hazards on alluvial and moveable bed streams. Most of these models evaluate the aggradation/degradation in subreaches before computing the water surface profile. Some models also estimate the magnitude of lateral bed migration. All models are dependent upon a large quantity of detailed input data and upon simplifying assumptions, and therefore have limitations in accuracy.

The National Academy of Sciences (NAS, 1983) studied whether flood stage could be better predicted by fixed bed models than by models that incorporate degradation/aggradation and river bed mobility. The NAS concluded "that the effect of river-bed degradation and aggradation on water–surface elevation during flood passage is much smaller than the effects of the uncertainties of channel roughness or flow friction factor, sediment input, and initial channel geometry. Moreover, the available input data on channel geometry, bed material characteristics, etc., generally are inadequate to permit full utilization of the capabilities of erodible-bed models. Therefore, except in cases of severely disturbed rivers that have experienced extreme local degradation or aggradation through man's intervention, utilization of erodible-bed models instead of fixed-bed models cannot be justified in flood-insurance studies."

The NAS study identified various sources of errors and limitations in use of the moveable-bed models and suggested worthwhile areas for hydrologic research and investigation.

In Flood Insurance Studies, present practice for evaluating flood hazards on alluvial channel streams is to use fixed-bed models to establish the water surface profiles. The effects of lateral channel migration are usually covered in flood plain management plans through "set-backs," which identify areas within a specified distance of the stream bank as hazard areas. These are difficult problems that require that engineers and hydrologists exercise the requisite care to identify reasonable solutions.

B. Real-time Forecast and Warning

Great losses of both property and life have resulted from flooding. Structural flood control measures such as channel enlargement, reservoirs, levees, and bypasses have greatly alleviated these losses. However, these structures can provide a false sense of security and may encourage development in flood-prone areas. The structures are usually designed for a certain level of protection, and occurrence of a flood higher than the protected level can cause great losses. Therefore, nonstructural measures should also be analyzed and included for flood damage mitigation.

The ability to provide sufficient advance warning of flood occurrence is important in reducing the disastrous effects of floods. The warning can provide time for the people to evacuate to safer locations. It should provide advance and reasonably accurate information to federal, state, and local agencies for emergency action coordination.

The NWS is responsible for forecasting the weather and for warning the public about life-threatening weather conditions, including floods. The NWS has flood forecasting centers for major river systems, like the Missouri or Upper Mississippi and their major tributaries. Smaller rivers and streams may receive only a general flood warning or watch, without specific stages or time of peak forecasted. Thus, a local flood forecasting and warning system is often a part of a community's flood damage abatement and public safety strategy. In California, a Joint-Federal State forecasting center is maintained by the California Department of Water.

Essential components of a real-time flood forecasting and warning system may include:

1) An appropriately designed network of precipitation and stream gages, providing a reasonable coverage of the drainage area.

2) Remote data collection or warning devices giving real-time precipitation and/or river stage data.
3) Microwave, telephone, VHF telemetric, meteorburst, satellite, or UHF radio communication networks to transmit real-time data.
4) Base station computers for data storage and processing.
5) A computational procedure for predicting river stages and time to reach the stages at designated locations.
6) Dissemination of information to local, state, and government agencies, the news media, and people living in flood-prone areas.

1. ***Data Acquisition System*** The first phase of the forecasting and warning system requires the collection and transmission of hydrologic data, including precipitation, river stage, and other significant variables in real-time. The system may be simple, relying on manually-read rain gages and stream gages with the information called to a central point for evaluation by office personnel, possibly through simple "look-up" tables. It could also be fully automated with computer modeling of the entire watershed. The system must be designed with the users' capabilities in mind. A town of 500 people with a part-time city staff may not be able to afford or utilize sophisticated electronics or computer systems for a flood warning program. A manual system may be most suitable for a small community. Whether the data acquisition system is manual or automated, the information transfer is similar. Schematic diagrams of an automated system are shown in Figs. 8.15 and 8.16. For a manual system, rain gages would be read by volunteers and the information called in to the base station several times a day, while key river levels would trigger an alarm at one or more locations in the community to alert responsible personnel.

A data acquisition system usually consists of a) precipitation gages in the basin, b) river stage gages on tributaries and/or the main channel, c) a number of "repeater" stations to transmit from the remote sensors to the base station, depending on topography and distance, and d) the base station, where all data are received and processed for the forecast. The number of sensors and repeater stations depend upon the size of the basin and desired accuracy of the forecast. A number of private vendors market data acquisition systems; however, the engineer should not rely entirely on these vendors to design and implement the system, but rather should prepare a preliminary system layout (including number and location of gages, type of forecast model needed for the basin, accuracy and warning time desired, etc.) prior to contacting various vendors. The overall system is often a compromise between forecast accuracy desired and first cost/annual operation and maintenance costs of the system. Annual charges are often appreciable and can be a large portion of a small town's budget. Once implemented, continued maintenance and improvements to the system are necessary, along with maintaining user skills needed to operate the forecasting and warning systems.

2. ***Flood Forecasting Models*** After transmission of the real-time precipitation and river stage data to the base station, the data are processed and input to a watershed hydrologic model to forecast flood levels and time to reach these levels (prediction of flood hydrograph) at key locations. A number of watershed models developed by local, state, and federal agencies, universities, and private consultants are available. Although all have the same basic concept, they are developed to various levels of sophistication by using different mathematical functions to derive watershed responses. An appropriate model should be selected based on the density of data acquisition stations, the desired accuracy of the forecast, the availability and capability of computers, and the ability of the user to operate the model. Some of the models currently in use are described in the following section "Microcomputer Software for Flood Analyses."

3. ***Real-Time Warning*** The final phase in the flood warning and forecasting system is the dissemination of pertinent information and warning. A precise plan of information dissemination must be prepared for flood emergencies and the information given out must be as timely and as accurate as possible. The warning time must be determined for appropriate actions to occur. If warning time is very short, little

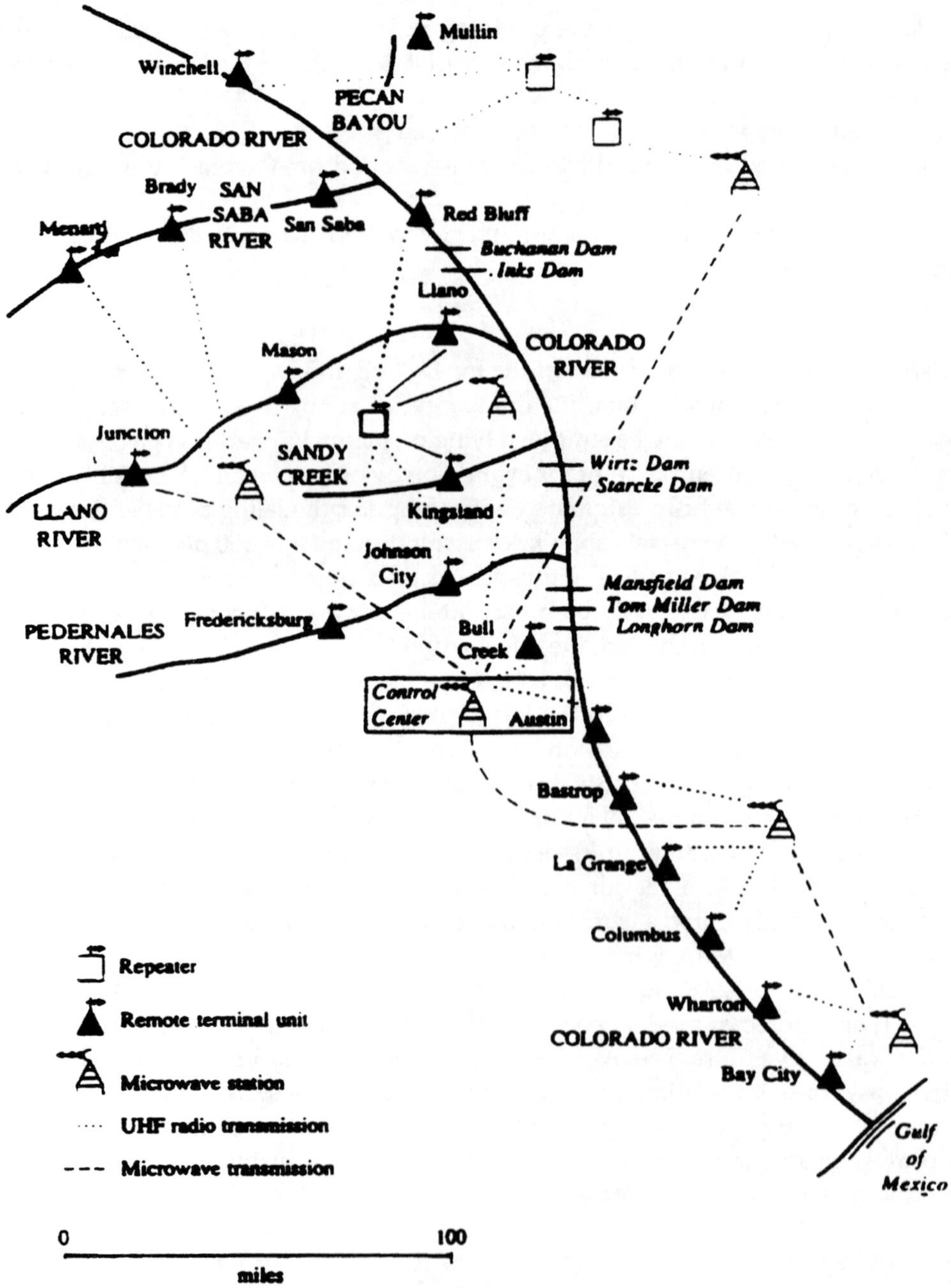

Figure 8.15—Automated Data Transmission System.

action other than life-saving may be possible. If several hours or more are available, additional damage reduction measures are possible. The responsibility as to who issues flood warnings must be clearly established. Those personnel who are authorized to provide input into these decisions also must be clearly identified. The delivery of the warning is critical. A general warning broadcast over the electronic media will often be ignored, unless the individuals receiving the warning have been recently flooded. To have the highest success of a positive response to a flood warning, an individual should receive specific information as to the height of the flood crest, when the crest will occur, and the area that will be inundated (what streets should be evacuated). If possible, personnel issuing the warnings should go door-to-door to insure that the warning is received and that proper action is taken. As floods often occur many years apart, periodic rehearsals of activating the entire system should be conducted. These rehears-

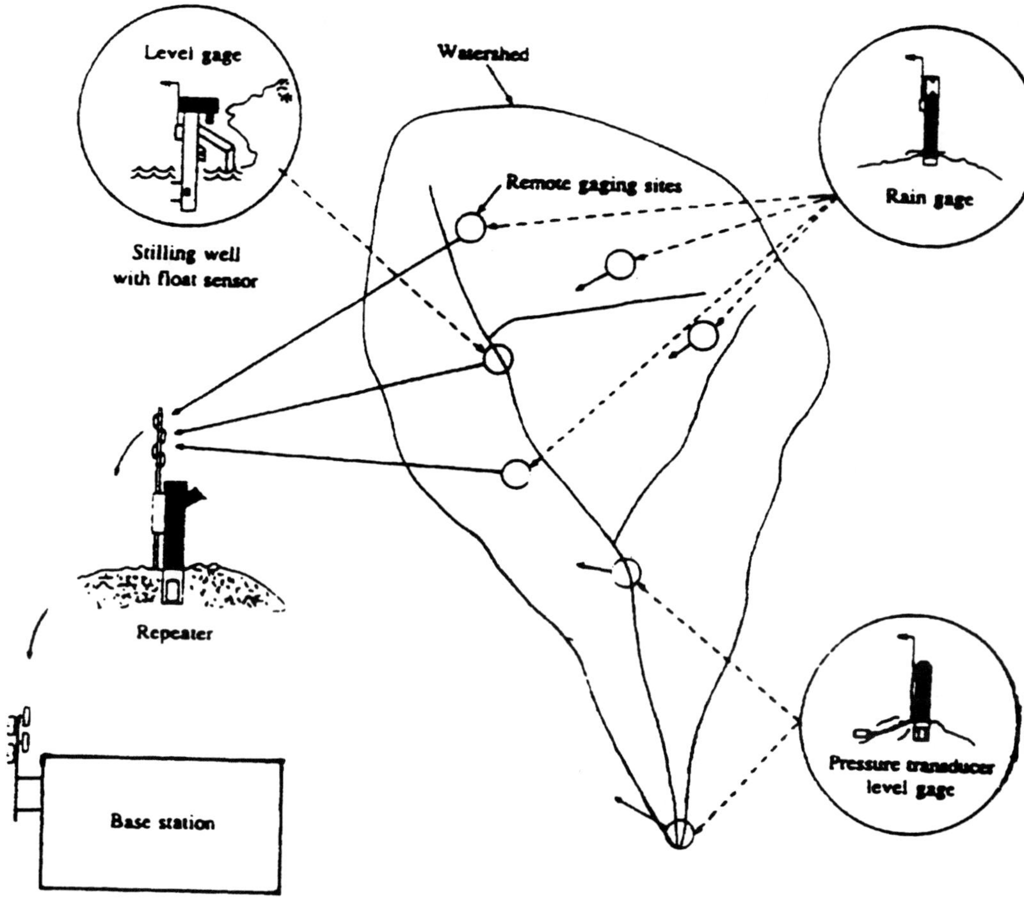

Figure 8.16—Automated Data Transmission System, Urban Watershed.

als should insure that all key personnel are familiar with the system and skilled at the roles they should play during an actual flood emergency.

Guidelines for flood warning and preparedness are available from local, state, or federal agencies (FIAC, 1985; USACE, 1988a). A number of papers on flood forecasting and warning were presented in an ASCE Symposium on Engineering Hydrology (Feldman, 1987).

C. Emergency Action Plan

A flood forecasting and warning system cannot be successfully employed without an emergency action (or flood preparedness) plan in place at all times. Such a plan should set out the community's detailed procedures and actions necessary for response to a flood emergency. Information in the plan should include:

1) warning dissemination procedures, including who issues the warning and the frequency of updates;
2) what areas are inundated by different flood levels;
3) backup means of communication;
4) general evacuation routes and destinations;
5) surveillance procedures at remote locations, during darkness, during adverse weather, etc.;
6) specific evacuation plans for special areas (hospitals, prisons, retirement homes, etc.);
7) river stages that close each bridge through the populated area;

8) direct damage reduction measures, where and when employed;
9) areas that will have access cut off by floods;
10) relocation of critical services prior to the flood;
11) alternate source of potable water;
12) emergency supplies and resources;
13) fire and police protection arrangements;
14) emergency shelters;
15) procedures for re-occupation of the flooded areas, including safety measures; and
16) restoration of services.

A training, rating, and annual review of the flood preparedness plan should be followed. An example of flood preparedness planning is outlined by the USACE (1982).

Emergency action plans for the floods caused by sudden release of water, due to breach of dams, detention ponds, etc., require special provisions. FERC (1988) and FEMA (1985) have mandated guidelines for preparing emergency action plans. All of the above information could be included in the plan, with additional warnings for potential dam failure. Flowcharts must be provided for two conditions: 1) failure is imminent or has occurred or 2) a potentially hazardous situation is developing. A chart shows, in order of priority, who is to be notified and who is responsible for the notification. Names, position, titles, office telephone numbers, alternate contacts, and means of communication are provided.

VII. MICROCOMPUTER SOFTWARE FOR FLOOD ANALYSES

The models covered in this section are mostly in the public domain, developed primarily by U.S. federal agencies. These models are grouped into the following categories based upon the tasks the model accomplishes.

a) Event-oriented precipitation–runoff models.
b) Continuous simulation precipitation–runoff models.
c) Steady-flow flood routing models.
d) Unsteady-flow flood routing models.
e) Reservoir regulation models.
f) Flood frequency analysis models.

Event-oriented precipitation–runoff models simulate rainfall (and snow) over a basin and the resulting flows through the watershed one storm at a time. The storm being modelled can be complex and of long duration, but no provision is made to track the soil moisture changes that occur between storms. Continuous simulation precipitation–runoff models, on the other hand, can provide long-term simulations dealing with multiple-storm events because they include routines that track soil moisture between periods of rain or snow.

Steady-flow flood routing models can calculate the flow depth and velocity at various locations along the channel as the flood moves downstream. The basic assumptions often used include flows being steady, uniform, and one-dimensional. Unsteady-flow flood routing models should be used when steady flow conditions cannot be assumed, such as when a dam breaks, or when large tributary flows enter the main channel. In fact, whole networks of channels can be simulated using unsteady-flow models.

The precipitation–runoff models and the flood routing models can also be referred to as hydrologic and hydraulic models respectively. Hydrologic models simulate the processes within a basin that transform rainfall (and/or snowmelt) into flows in the channels of the system. Routing for hydrologic models deals with the transport of water through the system computing a flow versus time curve for a given location such as the outlet of a sub-basin before proceeding downstream. Hydraulic models use flow data, often the output from a hydrologic model, to compute depths and velocities as they vary along a given channel reach for intervals in time as the flood wave passes through the basin. Routing in this sense tracks the depths along the channel as a function of position in channel length for given intervals in time. Thus, downstream flow conditions are taken into account in the upstream routings.

Reservoir regulation models can provide flood analysis capabilities for reservoirs with gated outlet works. Uncontrolled reservoirs can be modelled by the event-based and continuous simulation hydrologic models. Gated outlet works simulations regulate a reservoir, or series of reservoirs so as to minimize flooding at downstream control points. For these models, flood analysis is only one aspect of their capabilities; the models usually can operate for water supply and hydropower objectives as well.

Flood frequency analysis models provide a statistical analysis of available flood series and other information. Most programs include capabilities to perform statistical analyses such as those based upon the Water Resource Council guidelines outlined in Bulletin 17B.

Table 8.4 lists specific models and their place of development. A short abstract for each model is given below. It should be noted that these models have been and will continue to be updated to take advantage of new technological and theoretical developments.

TABLE 8.4. Flood Analysis Models Presented

I. *EVENT-ORIENTED PRECIPITATION–RUNOFF MODELS*	
Model	*Developing Agency*
HEC–1	USACE
TR–20	SCS
PSRM	Penn State University
ILLUDAS	Illinois State Water Survey
SWMM	EPA
II. *CONTINUOUS PRECIPITATION–RUNOFF MODELS*	
Model	*Developing Agency*
DRRRM	USGS
PRMS	USGS
HSPF	EPA
SSARR	USACE
III. *STEADY-FLOW FLOOD ROUTING MODELS*	
Model	*Developing Agency*
HEC–2	USACE
WSP2	SCS
WSPRO	USGS for FHWA
PSUPRO	Penn State University
IV. *UNSTEADY-FLOW FLOOD ROUTING MODELS*	
Model	*Developing Agency*
UNET	USACE
NETWORK	NWS
DMBRK	NWS
V. *RESERVOIR REGULATING MODELS*	
Model	*Developing Agency*
HEC–5	USACE
VI. *FLOOD FREQUENCY ANALYSIS MODELS*	
Model	*Developing Agency*
HEC–FFA	USACE
MAX	Cornell University
J407	USGS

Developer Acronyms Defined

USACE	United States Army Corps of Engineers
FHWA	Federal Highway Administration
USGS	United States Geological Survey
EPA	Environmental Protection Agency
NWS	National Weather Service
SCS	Soil Conservation Service

A. Event-oriented Precipitation-runoff Models

1. HEC–1 (Flood Hydrograph Package) The HEC–1 model, developed by the USACE Hydrologic Engineering Center, is an event-oriented, deterministic, lumped-parameter model for the determination of flood hydrographs for a given watershed. Originally released in October 1968 and updated frequently since then, the model simulates the watershed as a series of hydraulic and hydrologic components. A variety of options exist for modeling the hydrologic processes of precipitation, infiltration, and watershed runoff, and for routing the flows through the rivers and reservoirs. Special features of HEC–1 include snowpack simulation, dam safety and failure analysis, flood damage analysis, and parameter optimization. The operation of HEC–1 is run through a menu program. Input, and modifications of input for HEC–1 are accomplished through an editor that comes with the program package. The quantity and type of output are specified in the input to the program and are stored in a user-specified output file on disk or in an HEC–DSS file. Report quality graphics and tabulations of outflow hydrographs can be developed through DSPLAY, a routine in the HEC–DSS package. The HEC–DSS package is a data storage system capable of interacting with HEC–1 for both input and output. It also serves to transfer data between HEC–1 and other HEC software packages. HEC–1 is written in FORTRAN and is available for IBM-PC compatible computers.

2. TR–20 (Computer Program for Project Formulation Hydrology) The SCS model TR–20, is a single-event, deterministic, lumped-parameter model to assist in the evaluation of flood events. The model uses sub-basins with homogeneous properties to represent the entire watershed. The SCS curve number is then used to calculate the runoff for an input actual or synthetic cumulative rainfall distribution. Routing through reaches and reservoirs is possible by a modified kinematic method, and storage indication method respectively. The outflow hydrograph is a sum of the hydrographs from the sub-basins. Input to the program can be prepared through any editor using the program's specified format. Output is written to an output file with some components available for use with other SCS programs. TR–20 is written in FORTRAN IV and is available for IBM-PC compatible computers.

3. PSRM (Penn State Runoff Model) The Pennsylvania State University model PSRM is a single-event, deterministic, lumped-parameter model for the simulation of runoff from watersheds. Originally released in 1976, with the most recent update in 1991, the model computes the runoff from each sub-basin in a given basin and routes the cumulative flow to the outlet. Infiltration is accounted for using the SCS method. Overland flow is calculated using a depth-averaging technique for both pervious and impervious surfaces. Routing of flows through the watershed is accomplished by using a time lagging technique, and reservoir routing is done using the modified-Puls method. Input is handled through an interactive program that prompts users for the necessary data. Output consisting of outflow hydrographs, flooded areas, and tabulated results is written to an output file. PSRM is written in GW BASIC and is available for IBM-PC compatible computers.

4. ILLUDAS (Illinois Urban Drainage Area Simulator) The Illinois State Water Survey model ILLUDAS is a single-event, deterministic, lumped-parameter model for urban drainage. Released in 1979, the model can be used to assist in the design of a drainage system, or evaluate an existing system and scenarios for its improvement. The drainage area is divided up into pervious areas, impervious areas directly connected to the drainage system, and impervious areas separated from the drainage system by a pervious area. Infiltration is computed using the Holtan method, and flows are found using the modified RRL method. Routing of flows is accomplished using a time lag without storage attenuation, or an implicit solution of the continuity equation. Input is entered using an interactive program that also allows editing. Output is written to output file tables, and selected hydrographs can be plotted using program IPLOT, which is included in the program package. ILLUDAS is written in BASIC and is available for IBM-PC/XT compatible computers.

5. SWMM (Storm Water Management Model) The Environmental Protection Agency (EPA) model, SWMM, is a single-event, deterministic, distributed-parameter model for the analysis of water quantity and quality for urban runoff, or rural watersheds. Originally released in 1971, with the most recent update in 1989, the model can be used to assess the design of a new drainage system, or evaluate the capabilities of a current system and its proposed improvements. Storm sewers, combined sewers, and natural channels can be simulated with this model. Special features of SWMM include a continuous operation option, a statistics package for use with the continuous simulation routine, a hydraulic design routine, subsurface flow routing, and a storage and treatment routine. Input to the program is done through any editor in the format specified in the users manual. SWMM is written in FORTRAN and is available for use on IBM-PC compatible computers.

B. Continuous Precipitation–Runoff Models

1. NWSRFS (NWS River Forecasting System) The NWS model, NWSRFS, is a continuous simulation, conceptual, lumped-parameter model for streamflow simulations. Originally released in 1972, with the most recent version in 1992, the model is capable of making historical streamflow simulation for various types of design studies. The model contains a variety of hydrologic models and procedures that can be connected in whatever sequence is needed for a particular application. These include a snow accumulation and ablation model using temperature to index snowmelt, two conceptual soil moisture accounting models using a two layer representation of the soil profile, several variations of an empirical rainfall–runoff model using an antecedent precipitation index (API), the unit hydrograph method, four storage routing procedures, and a reservoir simulation model containing various modes of reservoir operations. Automatic parameter optimization methods are available. Input for the program consists of continuous time series input from a disk file system or ASCII flat files and various parametric information. Output includes selected time series, plots, tables, and statistical summaries. NWSRFS is written in FORTRAN and is available for IBM-PC compatible computers.

2. DRRRM (Distributed Routing Rainfall–Runoff Model) The USGS model, DRRRM, is a continuous simulation, deterministic, distributed-parameter model for small watersheds less than 26 km^2 (10 mi^2) in area. Originally released in 1978, with the most recent update in 1982, the model computes runoff for primarily urban watersheds. Limited application has been made to rural watersheds. Three solution schemes are available to choose from for a given simulation. Two types of time steps are used depending on whether or not a precipitation event is occurring. Infiltration is calculated using the Green and Ampt formula, and flows are routed using the kinematic wave approximation. Reservoir routing is accomplished by using either the linear storage or modified Puls method. Soil moisture is tracked using a two-layer system. The upper layer stores water that infiltrates into the ground during precipitation events, and the lower layer receives water from the upper layer both during and between storm events. Evapotranspiration from the soil layers is estimated using pan evaporation data. Input to the program is prepared through any editor in the specified format. Output is written to an output file, with some output available for input into DRRRM–QUAL. DRRRM is written in FORTRAN and is available for use on IBM-PC compatible computers.

3. PRMS (Precipitation–Runoff Modeling System) The USGS model, PRMS, is a continuous simulation, deterministic, distributed-parameter, modular program for use in small forested headwater watersheds. Originally released in 1978, with the most recent version in 1983, the model is used to simulate mean daily flows, and shorter time interval streamflow hydrographs. A finite difference approximation is used for the solution scheme. Infiltration is calculated using the Green and Ampt formula, and channel routing is done only during storm flows using the kinematic wave approximation. Reservoir routing is accomplished using either the linear reservoir or modified Puls method. Soil moisture accounting is based upon a two-layered system. The upper layer handles infiltration from storm events, and the lower layer handles drainage from the upper layer. Three methods are available to compute evapotranspiration, which comes from both soil layers. Special features available in PRMS include: sediment transport calculations and ground water flow calculations. The program is designed to run data from the USGS WATSTORE system. Input files from an editor can also be used. The content of the output file is specified from a list of available options in the input file. Plots of observed and predicted flow hydrographs can be included in the output. PRMS is written in FORTRAN, except for the data management package which is written in PL/I, and is available for IBM-PC compatible computers.

4. HSPF (Hydrological Simulation Program-FORTRAN) The EPA model, HSPF, is a continuous simulation, deterministic, distributed-parameter model for precipitation–runoff, and pollutant transport simulations. With the latest version released in 1991, HSPF provides basin-scale analyses for one-dimensional channels. Moisture is tracked at three levels: surface, subsurface, and base flow. Overland flow can be modelled using a time lag method for planes up to approximately 1,000 m (3,000 ft). Routing of flows through channels is based upon a modified storage method. Special features available in HSPF include: sediment transport and several transport and reaction processes. The USGS has HSPF within an expert system (HSPEXP) that allows editing of input parameters, simulations, output graphics, error statistics, and calibration advice all in an interactive environment. Input for the program includes continuous records for some parameters, while other parameters have default values if no data is available. Output from a simulation can include time histories of flows, sediment, and pollutants. HSPF is written in FORTRAN and is available for IBM-PC compatible computers.

5. SSARR (Streamflow Synthesis and Reservoir Regulation) The USACE/NWS model, SSARR, is a continuous simulation, deterministic, lumped-parameter model for streamflow simulations. Originally released in 1978, with the most recent update in 1986, the model is capable of making year-round simulations for flood forecasting, and for reservoir design and operation. The program contains two complete models for watershed hydrology, and a model for river system and reservoir routing. The difference between the two watershed models is in the treatment of snow accumulation and melt. Soil moisture is accounted for using a soil moisture index. Flows are routed through channels using the time-lag method. Special features available in SSARR include: snow accumulation and melting routine, baseflow infiltration, and interception. Input for the program is accomplished using an editor following the format specified in the user's manual. Output from the program is written to an output file which includes plots of selected parameters. SSARR is written in FORTRAN and is available for IBM-PC compatible computers.

6. SWM (Stanford Watershed Model) There are several versions of this model available in the market, with the best-known possibly being the proprietary Hydrocomp model. A basic philosophy of this model is to recognize the spatial variability of infiltration, interflow prediction, surface runoff, and evapotranspiration. The soil moisture accounting procedure used in the model has 16 parameters. The model can be calibrated manually or using automatic optimization technique. The HSPF version (Johnson, et al., 1980) includes water quality modeling.

C. Steady-Flow Flood Routing Models

1. HEC–2 (Water Surface Profiles) The USACE Hydrologic Engineering Center model, HEC–2, computes water surface profiles for one-dimensional, gradually-varied, steady flows in natural or man-made channels with small slopes. Released originally in 1964, with the most recent version in 1991, HEC–2 can be used to determine the effects of obstructions or structures in the flood plain. Calculations are made using the standard step procedure with iterations made until the error is less than 0.01 meters or smaller as specified. Supercritical as well as subcritical flows can be modelled. Various methods are available to calculate the frictional losses. Expansion and contraction losses are also included. Bridges, culverts, and weirs can be modelled as well as structures such as buildings built in the flood plain. Input to the program, and editing is done through a package editor in a format described in the user's manual. Output from the program is sent to an output file, and graphics are plotted using program PLOT2 which is included in the HEC–2 package. HEC–2 is written in FORTRAN and is available for IBM-PC compatible computers.

2. *WSP2 (Water Surface Profile 2)* The SCS model, WSP2, is a water surface profile program for one-dimensional, gradually-varied, steady flow applications. The most recent version was released in 1982. However, a draft version of a more robust model is available from some of the SCS offices. The 1982 model determines flood elevations and flood areas for certain flow rates. Calculations are made using the standard step method with iterations made until the error is less than 2.5 centimeters (cm). Only subcritical flows can be modelled, and bridge calculations are limited as well. Frictional loss is the only loss calculated in the program. Input to the program is handled with any editor using the format specified in the user's manual. Output from the program includes summary tables and rating curve plots. WSP2 is written in FORTRAN IV and is available for IBM-PC compatible computers.

3. *WSPRO (A Model for Water Surface Profile)* The USGS/Highway Administration model, WSPRO, is a water surface profile program for one-dimensional, gradually-varied, steady flows in open channels. Released originally in 1986, with the latest version in 1988, the model can compute up to 20 profiles per run. Calculations are made using standard USGS techniques described by Davidian (1984). Supercritical as well as subcritical flows can be modelled, although no provision is made for a hydraulic jump. Several types of bridge configurations can be modelled and are described in the user's manual. Other features include backwater calculations, embankment overflow, and multiple-opening stream crossings. Input to the program is handled through any editor in combination fixed–free format. Output from the program is sent to printer compatible output files which can include cross sectional plots. WSPRO is written in FORTRAN and is available for IBM-PC compatible computers.

4. *PSUPRO (Penn State University Encroachment Analysis Program)* The Pennsylvania State University model, PSUPRO, is a water surface profile program for one-dimensional, gradually-varied steady flow. Released in 1991 in conjunction with FEMA and the Philadelphia District of the USACE, the model compares water surface profiles with and without flood plain encroachments. Calculations are made using the standard step method with up to five iterations to resolve the differences between actual and computed energies within 60 cm (2 ft). If the differences are not resolved within the five iterations, the program prompts for more cross sections. Supercritical as well as subcritical flows can be modelled, but no provision is made for hydraulic jumps. Bridges cannot be modelled in this program. Losses are assumed to come from friction, and expansion and contraction of the channel. Input to the program is prepared through any editor using the format described in the user's manual. A routine in the program can convert HEC–2 files into a format compatible with this program. Output from the model includes encroachment widths as well as plots of cross sections and profiles. PSUPRO is written in BASIC and is available for IBM-PC compatible computers.

D. Unsteady-Flow Flood Routing Models

1. *UNET (One Dimensional Unsteady-flow Network)* The model, UNET, developed by Dr. Robert L. Barkaw and used by USACE and others is a one-dimensional unsteady-flow model for a channel network. Released in 1989, the model can simulate flows in a system containing multiple branches and storage areas. The continuity and momentum equations are solved using an implicit finite difference scheme for subcritical flows. For the interior boundary conditions, either continuity of flow or continuity of stage is used. An input discharge hydrograph is used as the upstream boundary condition, and the downstream boundary condition can be one of four options. Hydraulic structures such as levees, spillways and weirs, bridges, culverts, and low water dams can be included in the modeling process. Input to the program can be made using the HEC–DSS data base. HEC–2 input files can be modified to

become input files for UNET. An option exists for the output to be written to an HEC–DSS file so that graphical displays can be generated. UNET is written in F0RTRAN and is available for IBM-PC compatible computers.

2. NETWORK (Dynamic Wave Operational Model) The NWS model, NETWORK, formerly DWOPER, is a one-dimensional, unsteady-flow dynamic wave routing model. Released originally in 1978, with the most recent version in 1989, the model can be used for operational forecasting, or for design and planning purposes. The one-dimensional solution to the full St. Venant equations for unsteady flow can be solved for a full network of channels. Options exist for the input of initial and boundary conditions. Cross sections can be of irregular shape, but must be piecewise linear. Other capabilities of the model include local losses due to expansions and constrictions, wind resistance, lock and dam facilities, diversion facilities, and automatic calibration of the Manning n as a function of discharge. Input to the program is prepared through an editor by setting up two files, one for time dependent data and the other for time independent data. Output from the program is written to an output file for subsequent use or analysis. Graphical representations of stage and discharge hydrographs are available. NETWORK is written in FORTRAN and is available for IBM-PC compatible computers.

3. DAMBRK (Dam–Break Flood Forecasting Model) The NWS model, DAMBRK, is a one-dimensional, unsteady-flow model for the simulation of dam breachings and the routing of the subsequent flood wave or other flood hydrographs. Released originally in 1977, with the most recent version in 1991, the program models the breaching and flood wave routing for a single channel. Dams are categorized as being concrete arch, concrete gravity, or earthen with the breaching assumed to take place in a finite time interval. Downstream constrictions, channel sinuosity, tributary flows, and tidal effects can be simulated. Downstream structures such as bridges, levees, and dams can also be considered. The flow can be pressurized or free-surface. The dynamic wave method, based on a 4-points implicit difference solution of the complete St. Venant equations, is used to route the flood wave. The flow may be supercritical, subcritical, or a spatial/temporal combination of each. Through steep reaches, the Muskingum–Cunge method can be used. Input to the program can be done through an editor using the specified format. Output includes an output file and plots of items such as peak discharge profiles, hydrographs at selected times, and flood peak travel times. DAMBRK is written in FORTRAN and is available for IBM-PC compatible computers.

4. BRANCH (Branch–Network Flow Model) The USGS program, BRANCH, is a one-dimensional, unsteady-flow routing model for single, looped, or dendritic channels. Originally released in 1981, and updated in 1992, the model approximates the solution to the unsteady flow equations using a four–point, implicit, finite-difference scheme. Special features of the model include: wind shear, which is treated as a forcing function, a particle tracking routine, a dimensioning program that allows the user to specify array sizes, and a default initial condition routine. Input to the program is accomplished through an interactive program, and a data base from the Time Dependent Data System (TDDS). Output from the program can be written to a file in TDDS, and graphical representations are possible for stage and discharge hydrographs. BRANCH is written in FORTRAN and is available for IBM-PC compatible computers.

E. Reservoir Regulation Models

1. HEC–5 (Simulation of Flood Control and Conservation Systems) The USACE Hydrologic Engineering Center model, HEC–5, is a reservoir regulation model for the design and planning of a proposed reservoir, and for the management of a series of reservoirs. Released originally in 1973, with the most recent update in 1991, the model can evaluate the impacts of flood flows on a reservoir system as a simulation, or in real-time operation. The program uses a series of reservoirs, control points, and diversions to simulate a system. Each point in the system is connected to the next point by specifying a channel reach. Routing through each channel can be done using one of seven methods. Reservoir routing is accomplished using one of three methods. Time steps within a simulation can vary from one minute to one month. Other capabilities of HEC–5 include conservation, and hydropower. Input to the program can be created through an interactive program, INFIVE, and edited with program MOD5. Input to and output from the program can be read from and written to a file in HEC–DSS for storage or to generate plots of rating curves and hydrographs. HEC–5 is written in FORTRAN 77 and is available for IBM-PC compatible computers.

F. Flood Frequency Analysis Models

1. HEC–FAA (Flood Frequency Analysis) Formerly HECWRC, HEC-FAA is a flood frequency analysis model for the statistical analysis of annual extreme events based upon the Water Resources Council's Bulletin 17B guidelines. Released originally in 1978, with the most recent version in 1992, the model computes the frequency curve ordinates with the option of including the expected probability adjustment. The log-Pearson type III distribution is the only distribution used as specified in the guidelines. The Weibull, median, or Hazen formula can be used for the plotting position computations. Zero flood years are deleted and the record is treated as an incomplete record. Automatic adjustments are made for high and low outliers, and the computed station skew coefficient is weighted with an input general skew coefficient. Historic events can also be included in the analysis. Input into the program is through a package editor in the format outlined in the user's manual. Output includes a report quality frequency curve plot with user specified confidence limits. HEC–FAA is written in FORTRAN and is available for IBM-PC compatible computers.

2. MAX (Flood Frequency Analysis Using Maximum Likelihood Analysis) The Cornell University model, MAX, is a flood frequency program using the Maximum Likelihood Estimates method. Released in 1988, the program integrates any systematic gaged flood record with historical, botanical, and physical paleohydrologic records as well as regional analyses for flow frequency analyses. Seven distributions, among them the log-Pearson Type III, are available to fit the data. Plotting position computations are made based upon a general formula with coefficients specified by the user. Fitted quantiles are calculated and their standard errors are computed. Input to the program is accomplished through the interactive program MAXSETUP, or the required data can be input through any editor. The editor can be used to make any corrections to the input data. Output from the program is sent to a user-specified output file. MAX is written in FORTRAN and is available for use on IBM-PC compatible computers with a math coprocessor.

3. J407 (Annual Flood Frequency Using Bulletin 17B Guidelines) The USGS model, J407, is a flood frequency analysis model that follows the guidelines in Bulletin 17B. Released originally in 1976, with the most recent version in 1981, the model calculates the estimated probability of exceedance curve with confidence limits for a given site. Several versions of the model exist depending on the type of computer used and method of data entry. The log-Pearson type III distribution is the only one available due to its specification in the guidelines. Plotting position computations are made using a general formula with user-specified coefficients. The program uses historical and systematic data making adjustments for high and low outliers. Zero flows and below-gage–base flows are reflected in a conditional probability adjustment. A weighted skew coefficient is possible as well. Input to the program is entered through a direct link to the USGS Peak Flow file, card-imaging format or user-supplied input. Output from the program includes a summary output file that can be used as input to the USGS Streamflow/Basin Characteristics file. J407 is written in FORTRAN and is available for IBM-PC compatible computers.

VIII. REFERENCES

Adamowski, K. (1985). "Nonparametric kernal estimation of flood frequencies." Water Resour. Res., 21(11), 1585–1590.

American Society of Civil Engineers/U.S. Committee on Large Dams (ASCE/USCOLD). (1975). "Lessons from Dam Incidents, USA." New York, NY.

Ang, A.H-S. and Tang, W.H. (1984). *Probability concepts in engineering, planning and design, Vol. I, Basic Principles.* John Wiley & Sons, New York, NY.

Arnell, N. (1989). "Expected annual damages and uncertainties in flood frequency estimation." J. of Water Res. Plan. and Manage., ASCE, 115(1), 94–107.

Baker, V.R., Kochel R.C., and Patton P.C. (eds.) (1988). *Flood geomorphology.* John Wiley & Sons, New York, NY.

Breaden, J.P. (1973). "The generation of flood damage time sequences." University of Kentucky, Water Resources Institute Paper No. 32.

Bullard, K.L. (1986). "Comparison of estimated probable maximum flood peaks with historic floods." USBR Engineering and Research Center, Denver, CO.

Chow, V.T. (ed.) (1964). *Handbook of applied hydrology.* McGraw–Hill Book Co., New York, NY.

Chow, V.T., Maidment, D.R., and Mays, L.W. (1988). *Applied hydrology.* McGraw–Hill Book Co., New York, NY.

Chowdhury, J.U. and Stedinger, J.R. (1991). "Confidence intervals for design floods with estimated skew coefficient." J. of Hydr. Div., ASCE, 117(7), 811–831.

Chowdhury, J.U., Stedinger J.R., and Lu, L-H. (1991). "Goodness–of–fit tests for regional GEV flood distributions." Water Resour. Res., 27(7), 1765–1776.

Cohn, T.A. and Stedinger, J.R. (1987). "Use of historical information in a maximum-likelihood framework." J. of Hydrol., 96(1–4), 215–223.

Costa, J.E. (1987). "A history of paleoflood hydrology in the United States, 1800–1970." EOS, AGU, 67, 425–430, 1987.

Cudworth, A.G. (1989). *Flood hydrology manual.* Water Resources Technical Publication, U.S. Department of the Interior, Bureau of Reclamation, Denver Office, CO.

Cunnane, C. (1978). "Unbiased plotting positions—a review." J. of Hydrol., 37(3/4), 205–222.

Cunnane, C. (1988). "Methods and merits of regional flood frequency analysis." J. of Hydrol., 100, 269–290.

Cunnane, C. (1989). "Statistical distributions for flood frequency analysis." Operational Hydrology Report No. 33, World Meteorological Organization WMO—No. 718, Geneva, Switzerland.

Dalrymple, T. (1960). "Flood frequency analysis," Water Supp. Paper, *1543–A,* USGS, Reston, VA.

Davidian, J. (1984). "Computation of water-surface profiles in open channels." *U.S. Geological Survey techniques of water resources investigations, Book 3,* Chapter 15, 48 p., USGS, Reston, VA.

Davison, A.C. (1984). "Modeling excesses over high thresholds, with an application," *Statistical Extremes and Applications.* Edited by J. Tiago de Oliveira, D. Reidel Publishing Co., Dordrecht, Holland, pp. 461–482.

Dawdy, D.R. (1979). "Flood frequency estimates on alluvial fans," J. of Hydr. Div., ASCE, 105(HY11), 1407–1413, 1979.

Dyhouse, G.R. (1985). "Stage–frequency analysis at a major river junction." J. of Hydr. Div., ASCE, 111(4), 565–583.

Federal Emergency Management Agency (FEMA)(1983). "Flood insurance study, town of Normal, Illinois, McLean County." Community Number 170502. Washington, DC.

Federal Emergency Management Agency (FEMA)(1985). "Emergency action planning guidelines for dams." FEMA 64. Washington, DC.

Federal Emergency Management Agency (FEMA)(1991). "Flood insurance study, guidelines and specifications for study contractors." FEMA #37. Flood Insurance Administration, Washington, DC.

Federal Emergency Management Agency (FEMA) (in preparation). "Users guide to analysis and mapping of alluvial fan flooding." Washington, DC.

Federal Energy Regulatory Commission (FERC)(issued February 22, 1988, Addendum September 9, 1988). "Notice of revised emergency action plan guidelines." Washington, DC.

Federal Interagency Advisory Committee (FIAC) on Water Data (1985). "Guidelines on community local flood warning and response systems." NTIS, Springfield, VA.

Feldman, A.D. (1987). "Engineering Hydrology." *Proceedings of the symposium at Williamsburg, VA*, ASCE Hydraulics Division.

Filliben, J.J. (1975). "The probability plot correlation test for normality." Technometrics, 17(1), 111–117.

Fread, D.L. (1988). "The NWS DAMBRK model: theoretical background/user documentation." National Weather Service, NOAA, Silver Spring, MD.

Gumbel, E.J. (1958). *Statistics of extremes*. Columbia University Press, New York, NY., 375 pp.

Hirsch, R.M. and Stedinger, J.R. (1987). "Plotting positions for historical floods and their precision." Water Resour. Res., 23(4), 715–727.

Hoshi, K., Stedinger, J.R., and Burges, S.J. (1984). "Estimation of log normal quantiles: Monte Carlo results and first-order approximations." J. of Hydrol., 71(1/2), 1–30.

Hosking, J.R.M. (1990). "L–Moments: Analysis and estimation of distributions using linear combinations of order statistics." J. of Royal Statistical Society, B, 52(2), 105–124.

Hosking, J.R.M. and Wallis, J.R. (1987). "Parameter and quantile estimation for the generalized pareto distribution." Technometrics, 29(3), 339–349.

Hosking, J.R.M., Wallis, J.R., and Wood, E.F. (1985a). "Estimation of the generalized extreme–value distribution by the method of probability weighted moments." Technometrics. 27(3), 251–261.

Hosking, J.R.M., Wallis, J.R., and Wood, E.F. (1985b). "An appraisal of the regional flood frequency procedure in the UK flood studies report." Hydrol. Sci. J., 30(1), 85–109.

Hupp, C.R. (1987). "Botanical evidence of floods and paleoflood history." *Regional Flood Frequency Analysis*, edited by V.P. Singh, D. Reidel Publishing Co., Dordrecht, Holland, pp. 355–369.

Hupp, C.R. (1988). "Plant ecological aspects of flood geomorphology and paleoflood history." Chapter 20, of *Flood Geomorphology*, edited by V.R. Baker, R.C. Kachel, and R.C. Patton, John Wiley & Sons, New York, NY.

Interagency Advisory Committee on Water Data (IACWD)(1982). "Guidelines for determining flood flow frequency," Bulletin #17B. U.S. Geological Survey, Office of Water Data Coordination, Reston, VA.

International Commission on Large Dams (ICOLD) (1973). *Lessons from dam incident*. Abridged Edition, U.S. Committee on Large Dams, Boston, MA.

James, L. Douglas and Lee, R.R. (1971). *Economics of water resources planning*. McGraw–Hill Book Co., New York, NY.

Jansen, R.B. (ed.) (1988). *Advanced dam engineering*. Van Nostrand, New York, NY.

Jenkinson, A.F. (1969). "Statistics of extremes." *Estimation of Maximum Floods*. WMO No. 233, TP 126 (Tech. Note No. 98), 183–228.

Jennings, M.E. and Thomas, W.O. Jr. (1992). "National flood frequency program," USGS Water Resources Investigation Report 92.

Jin, M. and Stedinger, J.R. (1989). "Flood frequency analysis with regional and historical information." Water Resour. Res., 25(5), 925–936.

Johnson, F.A. and Illes, P. (1976). "A classification of dam failures." Water Power and Dam Construction, pp. 43–45.

Johnson, F.A., Imhoff J.C., and Davis, H.H. Jr. (1980). "User's manual for hydrologic simulation program FORTRAN." Report EPA 600/9-80-015, by Hydrocomp, Environmental Protection Agency, Athens, GA.

Kates, R.W. (1965). "Industrial flood losses, damage estimation in Lehigh Valley," University of Chicago Research Paper No. 98, University of Chicago Press, Chicago, IL.

Kite, G.W. (1988). *Frequency and risk analysis in hydrology*. Water Resources Publication, Littleton, CO.

Kuczera, G. (1982). "Combining site-specific and regional information: An empirical Bayes approach." Water Resour. Res., 18(2), 306–314.

Landwehr, J.M., Matalas, N.C., and Wallis, J.R. (1979). "Probability weighted moments compared with some traditional techniques in estimating Gumbel parameters and quantiles." Water Resour. Res., 15(5), 1373–1379.

Landwehr, J., Tasker, G.D., and Jarrett, R.D. (1987). Discussion of "Relative accuracy of log–Pearson III procedures," by J.R. Wallis and E.F. Wood, J. of Hydr. Div., ASCE, 111(7), 1206–1210.

Lettenmaier, D.P., Wallis, J.R., and Wood, E.F. (1987). "Effect of regional heterogeneity on flood frequency estimation." Water Resour. Res., 23(2), 313–324.

Loucks, D.P., Stedinger J.R., and Haith, D.A. (1981). *Water resource systems planning and analysis*. Prentice–Hall, Englewood Cliffs, NJ, 560 pp.

National Academy of Sciences (1983). "An evaluation of flood-level prediction using alluvial-river models." Washington, DC.

National Research Council, Committee on Techniques for Estimating Probabilities of Extreme Floods (1988). *Estimating probabilities of extreme floods, methods and recommended research.* National Academy Press, Washington, DC.

National Weather Service (NWS) (1956–1988). *Hydrometeorological reports*, NOAA, Silver Spring, MD:

No. 33 "Seasonal variation of the Probable maximum precipitation east of the 105th meridian," 1956.

No. 36 Interim Report—"Probable maximum precipitation (PMP) in California," 1961.

No. 37 "Meteorology of hydrologically critical storms in California," 1962.

No. 39 "PMP in the Hawaiian Islands," 1963.

No. 42 "Meteorological conditions for the PM flood on the Yukon River above Rampart, Alaska," 1966.

No. 43 "PMP, Northwest states," 1966 (Currently being revised, 1991).

No. 49 "PMP, estimates, Colorado and Great Basin Drainages," 1977.

No. 51 "PMP estimates, U.S. east of the 105th Meridian," 1978.

No. 52 "Application of PMP estimates—U.S. east of the 105th Meridian," 1982.

No. 53 "Seasonal variation of 10 sq. mi. PMP estimates, U.S. east of the 105th Meridian," 1980.

No. 54 "Probable maximum precipitation and snowmelt criteria for S.E. Alaska," 1983.

No. 55A "PMP estimates, U.S. between the Continental Divide and the 103rd Meridian," 1988.

No. 56 "Probable maximum and TVA precipitation estimates with areal distribution for Tennessee River Drainages less than 3,000 mi in area."

Natural Environmental Research Council (NERC) (1975). "British, flood studies report, Vol. I, Hydrological studies." London, England.

Newton D.W. and Herrin, J.C. (1982). "Assessment of commonly used flood frequency methods." Transportation Research Record Series, TRR896, Washington, DC.

Parker, D.J. and Penning–Rowsell, E.C. (1972). "Problems and methods of flood damage assessment." Flood Hazard Research Project, Progress Report No. 3, Middlesex Polytech (Enfield College).

Paulhus, J.L.H. and Gilman C.S. (1953). "Evaluation of probable maximum precipitation." AGU, Vol. 34.

Phien, H.N. (1987). "A method of parameter estimator for the extreme value type I distribution." J. of Hydrology, 90, 251–268.

Potter, K.W. (1987). "Research in flood frequency analysis: 1983–1986." Reviews of Geophysics. 25(2), 113–118.

Potter, K.W. and Faulkner, E.B. (1987). "Catchment response times as a predictor of flood quantiles." Water Resour. Bull. 23(5), 857–861.

Potter, K.W. and Lettenmaier, D.P. (1990). "A comparison of regional flood frequency estimation methods using a resampling method." Water Resour. Res. 26(3), 415–424.

Riedel, J.T. (1990). "Maximum known observed depth–area–duration data for Taiwan, China, India and the United States," unpublished personal file, Charlottesville, VA.

Riedel, J.T. and Schreiner, L.C. (1980). "Comparison of generalized estimates of probable maximum precipitation with greatest observed rainfalls." NOAA Technical Report NWS25, NOAA, NWS, Silver Spring, MD.

Shane, R.M. and Lynn, W.R. (1964). "Mathematical model of flood risk evaluation." J. of Hydr. Div., ASCE, 90(HY6), 1–20.

Shipe, A.P. and Riedel, J.T. (1976). "Greatest known areal storm rainfall depths for the contiguous United States." NOAA Technical Memorandum NWS HYRO-33, NWS, Silver Spring, MD.

Stedinger, J.R. (1980). "Fitting log normal distributions to hydrologic data." Water Resour. Res. 16(3), 481–490.

Stedinger, J.R. (1983a). "Design events with specified flood risk." Water Resour. Res. 19(2), 511–522.

Stedinger, J.R. (1983b). "Confidence intervals for design events." J. of Hydr. Div., ASCE, 109(HY1), 13–27.

Stedinger, J.R. and Tasker, G.D. (1985). "Regional hydrologic analysis, 1. Ordinary, weighted and generalized least squares compared." Water Resour. Res. 21(9), 1421–1432.

Stedinger, J.R. and Tasker, G.D. (1986). Correction to "Regional hydrologic analysis, 1. Ordinary, weighted and generalized least squares compared," Water Resour. Res. 22(5), 844.

Stedinger, J.R. and Baker, V.R. (1987). "Surface water hydrology: historical and paleoflood information." *Reviews of Geophysics*. 25(2), 119–124.

Stedinger, J.R., Vogel R.M., and Foufoula–Georgiou, E. (1993). "Frequency analysis for extreme events," in Chapter 18, *Handbook of Hydrology*. Edited by D. Maidment, McGraw–Hill Book Co., New York, NY.

Tasker, G.D. and Stedinger, J.R. (1986). "Estimating generalized skew with weighted least squares regression." J. of Water Resour. Plan. and Manage. ASCE 112(2), 225–237.

Tasker, G.D. and Stedinger, J.R. (1989). "An operational GLS model for hydrologic regression." J. of Hydrol. 111, 361–375.

Thomas, W.O. (1985). "A uniform technique for flood frequency analysis." J. of Water Resour. Plan. and Manage., ASCE *111*(3), 321–337.

Todorovic, P. (1978). "Stochastic models of floods." Water Resour. Res. 14(2), 345–356.

U.S. Army Corps of Engineers (USACE)(1945-). "Storm rainfall in the United States." Washington, DC.

U.S. Army Corps of Engineers (USACE) (1952). "Standard project flood determinations." Civil Engineer Bulletin No. 52–8, EN 1110-2-1411, Office of the Chief of Engineers, Washington, DC.

U.S. Army Corps of Engineers (USACE) (1960). "Runoff from snowmelt," Manual EM 1110–2–1406. Washington, DC.

U.S. Army Corps of Engineers (USACE)(1973). "Hydrologic engineering methods for water resources development," *Vol. 4, Hydrograph Analysis*. Hydrologic Engineering Center, Davis, CA.

U.S. Army Corps of Engineers (USACE)(1982). "Flood preparedness planning, Metropolitan Phoenix Area." Hydrologic Engineering Center, Davis, CA.

U.S. Army Corps of Engineers (USACE)(1988a). "Comprehensive flood warning/preparedness studies." Hydrologic Engineering Center, Davis, CA.

U.S. Army Corps of Engineers (USACE) (1988b). "Expected annual flood damage computation." Hydrologic Engineering Center, Davis, CA.

U.S. Army Corps of Engineers (USACE)(1991). "HEC–2, water surface profile, users manual." Hydrologic Engineering Center, Davis, CA.

U.S. Department of Agriculture (USDA)(1951-1970). "Summary of snow survey measurements for Alaska." Soil Conservation Service, Washington, DC.

U.S. Department of Agriculture (USDA)(1976). "WSP2 computer program, a water surface profile computer program for determining flood elevation and flood areas for certain flow rates." Soil Conservation Service, Washington, DC.

U.S. Department of Commerce (USDC) (1968). "Weather atlas of the United States." Environmental Science Services Administration (ESSA), Environmental Data Service, Washington, DC.

U.S. Nuclear Regulatory Commission (USNRC) (1977). "Design basis flood for nuclear power plants," Regulatory Guide 1.59, Rev. 2. Washington, DC.

Vicens, G.J., Rodriquez-Iturbe, I., and Schaake, J.C. Jr. (1975). "A Bayesian framework for the use of regional information in hydrology." Water Resour. Res. 11(3), 405–414.

Vogel, R.M. (1986). "The probability plot correlation coefficient test for normal, lognormal, and Gumbel distributional hypotheses." Water Resour. Res. 22(4), 587–590.

Vogel, R.M. (1987). "Correction to the probability plot correlation coefficient test for normal, lognormal, and Gumbel distributional hypotheses." Water Resour. Res. 23(10), 2013.

Vogel, R.M. and Kroll, C.N. (1989). "Low-flow frequency analysis using probability–plot correlation coefficients." J. of Water Resour. Plan. and Manage. ASCE 115(3), 338–357.

Vogel, R.M. and McMartin, D.E. (1991). "Probability plot goodness–of–fit and skewness estimation procedures for the Pearson type 3 distribution." Water Resour. Res. 27(12), 3149–3158.

Wallis, J.R. (1988). "Catastrophes, computing and containment: living in our restless habitat." Speculation in Science and Technology. 11(4), 295–315.

Wallis, J.R. and Wood, E.F. (1985). "Relative accuracy of log–Pearson III procedures." *J. Hydr. Div. ASCE* 111(7), 1043–1056, (with discussion and closure J. Hydr. Div. ASCE, 111(7), 1205–1214, 1987).

Wang, B.H. (1986). "Probable maximum flood and its application." Harza Engineering Company, Chicago, IL.

Wang, B.H. (1988). "Determination of design flood for spillways." *Transaction, Sixteenth Congress on Large Dams, Vol. IV.*, ICOLD, San Francisco, CA.

Wang, B.H. and Revell, R.W. (1983). "Conservatism of probable maximum flood estimates." J. Hydr. Div. ASCE 109(3), 400–408.

Waylen, R. and Woo, M.–K. (1982). "Prediction of annual floods generated by mixed processes." Water Resour. Res. 18(4), 1283–1286.

White, G.G. (1964). "Choice of adjustments to floods," University of Chicago, Department of Geography, Research Paper 93.

World Meteorological Organization (WMO) (1969). "Depth-area-duration analysis of storm precipitation," WMO No. 237, TP129. Geneva, Switzerland.

World Meteorological Organization (WMO)(1986). "Manual for estimation of probable maximum precipitation." Operational Hydrology Report No. 1, Geneva, Switzerland.

IX. GLOSSARY

Alluvial Channel—A stream channel whose boundary is composed of appreciable quantities of sediments transposed by the flow and which generally changes its bed forms as the rate of flow changes.

Annual Flood Series—The largest flood in each year constitutes the annual flood series (AFS) upon which frequency analyses are often based. See also partial duration series.

Avulsion—A sudden change in the location of an alluvial stream channel resulting from flood flows.

Barrier—A mountain range that partially blocks the flow of warm humid air from its oceanic source to a basin under study.

Base Flow—That portion of stream discharge derived from ground water. (See Chapter 6 for a more complete definition.)

Bayesian Statistical Analysis—Another philosophical approach to risk and frequency analysis that employs Bayes theorem and sample data to improve a prior distribution describing the likely values of a model's parameters.

Coefficient of Skewness—Denoted γ, a measure of asymmetry or skewness of a probability distribution.

Coefficient of Variation—The ratio of the standard deviation to the mean of a random variable, δ/μ. A measure of the variability of flow.

Conditional Probability Adjustment—A statistical parameter estimation procedure for use at sites whose record of flows is truncated below a threshold, or includes years with zero flow.

Confidence Intervals—A vehicle for describing precision with which a parameter or a flood quantile can be estimated.

Convective Rain—Rainfall that is caused by the vertical motion of an ascending mass of air that is warmer than the environment.

Convergence—Horizontal shrinking and vertical stretching of a volume of air, accompanied by net horizontal inflow and internal upward motion.

Depth–area Curve—A curve showing, for a given duration, the relation of maximum areal rainfall depth to size of the area within a storm or storms.

Depth–area–duration Curves—Combination of depth–area and depth–duration relationships.

Depth–duration Curve—A curve showing the relationship between maximum point or basin average precipitation and duration.

Design Flood—A flood event that is the largest flood against which the various components of a water resources facility of a system of facilities is protected.

Dew Point—The temperature to which a given parcel of air must be cooled at constant pressure and constant water–vapor content in order for saturation to occur.

Discharge—A volume of flow per unit of time passing a point of interest, usually expressed in m^3/s or cfs.

Drainage Area—The area draining to a lake or a specified point on a stream; it lies upstream of the point and is enclosed by a hydrologic surface drainage divide.

Duration—Length of actual or assumed period of time for rainfall or rainfall excess.

Erosion—The wearing away of the land by detachment and movement of soil and rock fragments through the action of moving water and other geological agents.

Exceedance Probability—The probability that a flood level is exceeded in any year.

Expected Probability—Related to a correction made to estimates of flood quantiles so that the average or expected exceedance probability has a target value when uncertainty in estimated parameters is considered.

Flood—An event that causes a river, lake, ocean, or other water feature to rise above normal nondamaging limits.

Flood Hazard—The adverse consequences of a flood with a specific magnitude occurring at a specific location.

Flood Information Maps—Graphic information depicted on maps and charts which illustrates flooded area and/or water surface profiles for specific floods; may also include other relationships between flood characteristics such as exceedance probability and flood magnitude. An area flooded by a dam break hydrograph on a topographic map is also a flood information map.

Flood Magnitude—The instantaneous rate of flow (runoff) in m^3/s or cfs.

Flood Peak—The highest value of the discharge attained during the passage of a flood wave.

Flood Plain—A strip of relatively smooth land bordering a stream, built of sediment carried by the stream and dropped in the slack water beyond the influence of the swiftest currents.

Flood Warning—A weather advisory issued by the National Weather Service for specific streams, rivers, lakes, or other flood potential locations when meteorological or physical conditions such as a dam break indicates potential flooding.

Flood Wave—A distinct rise in stage culminating in a crest and followed by recession to lower stages.

Flood Zone—The land bordering a stream that is subject to floods of about equal frequency.

Floodway—A part of flood plain, otherwise leveed, reserved for emergency diversion of water during floods.

Frequency Analysis—Develops the frequency or probability that different levels of flow, stage, or precipitation depth will be exceeded.

General Storm—A storm event that is associated with a major synoptic weather feature, produces precipitation over areas in excess of 1,360 km^2 (500 sq. mi.) and has a duration longer than 6 hours.

Generalized Extreme Value Distribution—The Generalized Extreme Value distribution (GEV) spans the three types of theoretical extreme value distributions for maxima.

Generalized Skew—A regional estimate of the log–space skewness coefficient recommended by the U.S. Water Resources Council Bulletin 12B.

Goodness–of–fit Tests—Statistical tests employed for assessing whether a selected distribution is consistent with a particular data set. Probability plots are an example.

Gumbel Distribution—The Extreme Value type 1 distribution is often called the Gumbel distribution, or double exponential distribution. See Generalized Extreme Value distribution.

Historical Information—Includes written and other records of large floods left by human observers: newspaper accounts, church records, letters, and flood markers.

Hydrograph—A graphical representation of stage, discharge, velocity, or other characteristics of water at a given point as function of time.

Hyetograph—A bar graph of rainfall, rainfall excess, or both.

Index Flood Procedure—A simple regionalization technique that assumes the probability distribution of floods or rainfall at every site in a "region" is the same except for a scale parameter.

Initial Loss—The maximum amount of precipitation that can occur under specific conditions without producing runoff.

Isohyet—Lines of equal value of precipitation for a given time interval.

Kinematic Wave—A flow condition where the effects of gravity are counter-balanced by the forces of friction, resulting in approximately uniform flow. Inertial and pressure forces are not important and are neglected for this flow condition.

L-moments—Linear combinations of order statistics that summarize the statistical properties of data sets; includes the mean and surrogates for the standard deviation and skew coefficient.

Local Storm—A storm event that occurs over a small area in a short time period. The precipitation rarely exceeds 6 hours in duration and the area covered by the precipitation is normally less than 1,360 km^2 (500 sq. mi.).

Lognormal Distribution—A random variable whose logarithm has a normal distribution and is lognormally distributed. If a constant is subtracted before taking logarithms, the variable has a 3-parameter lognormal distribution.

Log-Pearson Distribution—A random variable when its logarithm follows a Pearson type III distribution has a log-Pearson type III distribution.

Maximum Likelihood Estimator (MLE)—Maximum likelihood estimation is a statistical philosophy that estimates a distribution's parameters so that available data is most consistent in a statistical sense with the observed sample.

Mean—The expected or average value of a random variable in many samples, denoted by μ.

Meteorburst—A type of random data transmission. Data is forwarded to the base site as soon as a threshold value is exceeded, rather than at a regular interval.

Method of Moment—A common procedure used for estimating the parameters of probability distributions for floods and other phenomena.

Mixed Population—When annual maximum series are composed of events that are from processes with distinctly different probability distributions.

Moisture Maximization—The process of adjusting observed precipitation quantities upward based upon an increase in available moisture to the storm.

Normal Distribution—The classical normal or Gaussian distribution is symmetric with a skewness coefficient equal to zero; useful in many applications.

Outliers—Data points that depart significantly from the trend of the other data.

Paleoflood Information—Describes the many botanical and geophysical sources of information (including damaged and tilted trees, slackwater deposits, stranded debris, and scour lines) pertaining to large floods that are not limited to the locations of past human observations or man-made recording devices.

Partial Duration Series (PDS)—Or the peaks-over-threshold (POT) approach. Constitutes all flood peaks above a truncation or threshold level. See also annual flood series.

Pearson Type III Distribution—A useful family of probability distributions for describing flows with an adjustable skewness coefficient.

Percentile—A flood level exceeded with probability q is the 100qth percentile.

Persisting Dew Point—When the dew point value at a station has been equalled or exceeded throughout a period, commonly a duration of 12 or 24 hours.

Plotting Positions—Estimators of the cumulative or exceedance probability associated with observations in a sample. See Weibull plotting position.

Precipitable Water—The total atmospheric water vapor contained in a vertical column of unit cross-sectional area extending between any two specified levels in the atmosphere, commonly expressed in terms of height to which the liquid water would stand if the vapor were completely condensed and collected in a vessel of the same unit cross section.

Probability Plots—A graphical display of data plotted against assigned "plotting positions" using special probability scales. Depending upon scale chosen, reveals if a particular family of distributions is consistent with the observations.

Probable Maximum Flood—A hypothetical flood hydrograph representing runoff from the Probable Maximum Precipitation.

Probable Maximum Precipitation—Estimated precipitation depth for a given duration, drainage area, and time of the year for which there is virtually no risk of being exceeded. The total depth of precipitation is distributed in a systematic manner throughout the storm duration.

Rainfall Excess—The volume of rainfall available for direct contribution to the stream channel, equal to the total rainfall minus detention losses.

Real-time—The time step for obtaining actual stage and/or discharge data. While implying basically an instantaneous time step, real-time, for flood forecasting and warning purposes, could range from a few minutes to a few hours, depending on how rapidly changes in stage of discharge occur in a particular watershed.

Record Augmentation—A procedure to improve the estimates of the mean and variance of flows at a short-record site by effectively extending the record using a longer record with which observations in the short record are highly correlated.

Recurrence Interval—See return period.

Regional Regression—The use of regression to derive models that predict the values of hydrologic statistics and quantiles as a function of physiographic river basin characteristics and other parameters.

Regional Analysis—The use of hydrologic information from different locations, often to compensate for short records at a particular site.

Regulatory Flood—A flood event that is the largest flood designated by regulatory authorities for specific purposes such as land use or flood insurance.

Return Period—The average time interval between actual occurences of a hydrological event of an equal or greater magnitude.

Saturation—Upper limit of water vapor content in a given space, a function of temperature only.

Sedimentation—The gravitational settling of suspended sediment particles that are heavier than water.

Snowpack—The accumulation of snow at higher elevations usually expressed in terms of water equivalents.

Spatial Distribution—The geographic distribution of precipitation over a drainage according to an actual or idealized storm pattern for the storm area.

Standard Project Flood—A hypothetical flood hydrograph representing runoff from the Standard Project Storm.

Standard Project Storm—The most severe storm that is considered reasonably characteristic of a region; it undercuts storms of extraordinary severity.

Standard Deviation—See Variance.

Storm Surge—The departure from normal water level due to the action of storms.

Synoptic—Showing the distribution of meteorological elements over an area at a given time.

Temporal Distribution—The time order in which incremental rainfall is arranged within the storm.

Unit Hydrograph—The discharge hydrograph representing one unit (mm or inch) of direct runoff resulting from a rainfall of some unit duration and specific areal distribution.

Variance—Describes the variability of a random variable. Equal to the square of the standard deviation.

Wave Runup—The vertical height above the still water level to which water from an incident wave runs up the face of a structure or beach. Runup depends on structure shape and roughness, water depth at structure toe, bottom slope in front of structure, and incident wave characteristics.

Weibull Plotting Position—A commonly used plotting position. The ith largest observation in a sample of size n is assigned probability $i/(n + 1)$.

Wind Setup—The rising water surface on the leeward side of lakes, reservoirs, and smaller bodies of water due to the action of wind (same as storm surge on oceans).

Chapter 9

URBAN HYDROLOGY

I. INTRODUCTION

The process of urbanization and its hydrologic effects are widely known. At an individual development site, trees and natural cover (see Fig. 9.1a) are replaced by buildings and impervious cover (see Fig. 9.1b). These physical changes result in a loss of interception and depression storage, a decrease in the potential for infiltration, and a redirection of principal flowpaths. On a larger scale, natural channels are cleared, straightened and, in some cases, lined or replaced by pipes that are more hydraulically efficient. The hydrologic impact of these alterations to the physical watershed can change the timing, magnitude, volume, and duration of a flood and its corresponding runoff hydrograph. Baseflow rates can decrease, flood peaks and volumes can increase, and the timing of runoff can change, including decreases in the time-to-peak. Additionally, because of the larger volumes of runoff, the duration of out-of-bank flows can increase. In terms of the frequency of flooding, urban development can increase the frequency of out-of-bank or local flood magnitudes. There are other consequences of the changes in flood runoff characteristics. Surface erosion of cleared areas can increase substantially, erosion of previously stable channel banks can begin, and loadings of water borne pollutants can increase.

The effect of urbanization on flood peaks depends on the exceedance probability of a storm. For the larger peak discharges, the watershed is usually saturated whether urbanized or not, and flood peak increases tend to be smaller because urban development does not create more runoff, only a more efficient runoff collection system. For smaller storm events, urbanization produces larger peaks primarily because of the increase in the runoff volume. Thus, the relative change in flood magnitude is greater for more frequent events and smaller for less frequent ones. This is illustrated in Fig. 9.1c for a hypothetical flood frequency curve. Fig. 9.1d illustrates the changes in peak flow.

The particular engineering emphasis in urban hydrology can be listed as follows.

a) Urbanization adjustments to flood frequency curves from gaged records and estimation of flood peaks in small ungaged urban watersheds for floodway delineation or flood insurance studies administered by the Federal Emergency Management Administration.
b) Flood peak estimation and design of stormwater conveyance systems, such as storm sewers and highway culverts.
c) Storage routing and hydraulic design of stormwater detention facilities that require a full hydrograph.
d) Comprehensive stormwater management computer models for watershed-wide analysis of subarea impacts on the flood hydrology of an entire basin.

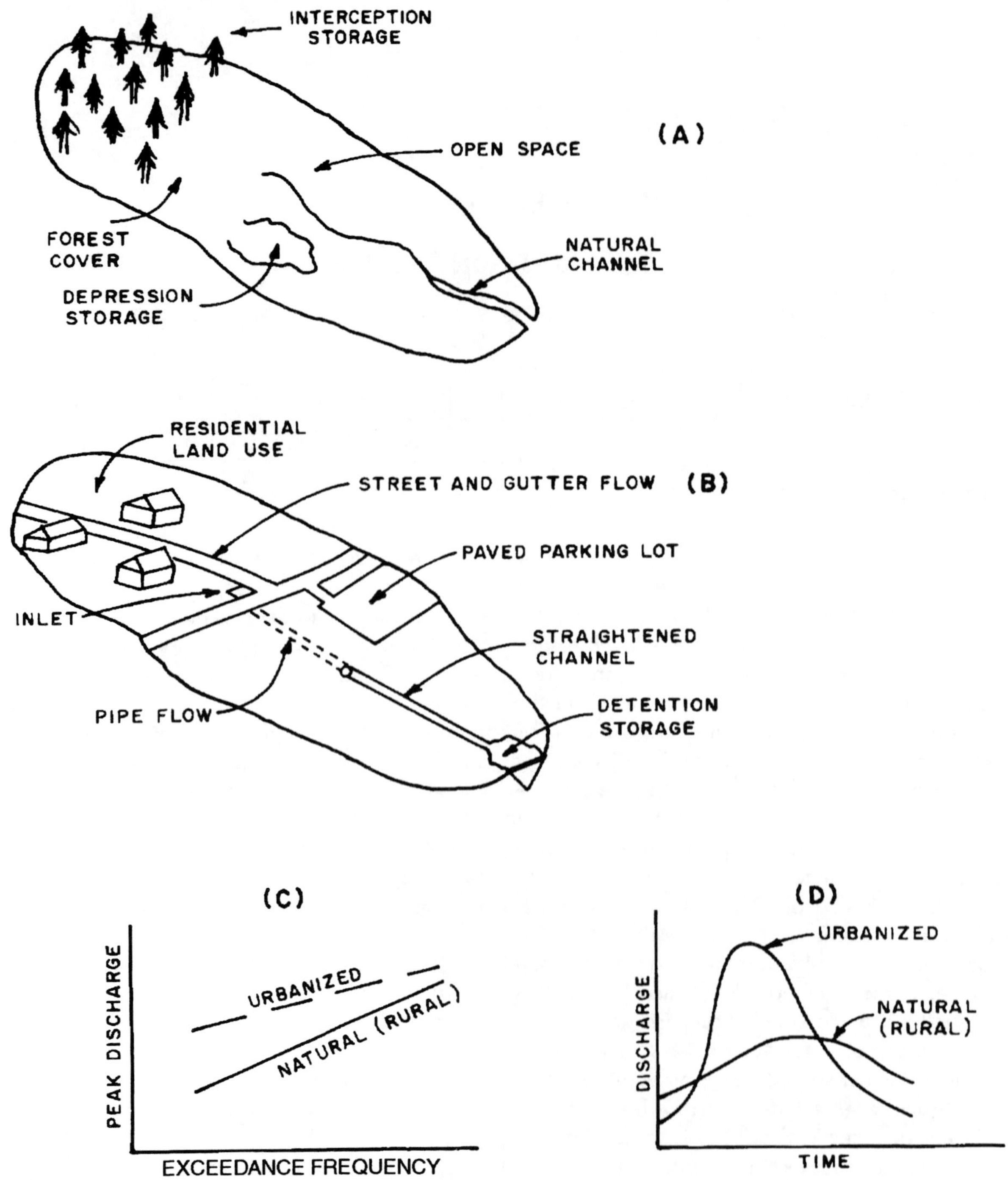

Figure 9.1.—(a) Cover and Storage of a Natural Watershed; (b) Cover and Storage of an Urbanized Watershed; (c) Flood Frequency Curves for Natural and Urbanized Watersheds; (d) Hydrographs for Natural and Urbanized Watersheds.

These engineering and planning applications require that an engineer be knowledgeable regarding rainfall–runoff processes as well as hydraulic principles used in design. Accordingly, this chapter on urban hydrology describes the state-of-the-practice in terms of currently used methods for flood frequency adjustment, precipitation analysis, hydrologic losses, and urban runoff generation. It concludes with two sections on typical urban drainage design calculations and computer model applications. While the chapter addresses current advances in urban hydrology, it is not intended to be a research document. The reader is encouraged to investigate new research directions through appropriate references which are cited in this chapter. A more detailed overview of each of the major sections follows in the next section.

A. Overview of Urban Hydrology Methods and Processes

A statistical flood frequency analysis is based on the assumption of a homogeneous and time-invariant annual flood record in which rainfall and watershed characteristics remain constant over time. Significant changes in land use are a source of nonhomogeneity in flood statistics. A flood frequency analysis based on a nonhomogeneous record results in inaccurate flood estimates for any return period. Therefore, nonhomogeneities must be identified and the flood record adjusted for the nonhomogeneity prior to making a frequency analysis. An adjustment procedure is provided in this section.

Rainfall information used in urban hydrologic analysis and design most often takes one of two forms: 1) a characteristic rainfall depth and intensity from intensity–duration–frequency curves (IDF) and 2) a time distribution of the design storm depth. These are discussed in the next section on precipitation in the urban watershed.

In addition to changes in flood peak frequency and magnitude, urbanization can cause significant changes to the flood hydrograph. While Fig. 9.1d shows an increase in the peak discharge from the urbanized watershed, other changes are evident, including decreases in the time-to-peak and the base-flow rate, and increases in the runoff volume. These changes to the hydrograph reflect the physical changes to the watershed, with decreases in both interception storage and the potential for infiltration. These changes are reflected in rainfall losses. Methods of accounting for losses on urbanized watersheds are discussed in the section entitled Hydrologic Losses in Developing Watersheds.

The hydrologic methods most frequently used in design work are usually separated into two groups: peak-discharge methods and hydrograph methods. The former is commonly used for design problems on small watersheds where storage effects are unimportant; this would include street drainage, inlet design, and storm sewer design. Hydrograph methods are most often applied to larger watersheds where the effect of storage must be taken into account, such as in the design of regional detention facilities. The fundamentals of these two approaches are discussed in the section on urban runoff methods. Examples of urban drainage design, emphasizing urban storm sewers and detention basins, are presented in the section on urban drainage design calculations. Finally, the application of computer models to problems of storm sewer system design and analysis is contained in a separate section on computer model applications.

B. The Effects of Urbanization on Flood Peaks

A number of hydrologic models are used to evaluate the effect of urbanization on peak discharges. With some models, it is difficult to develop a general statement of the effect of urbanization. For example, with the rational method, urban development would affect both the runoff coefficient and the time of concentration (t_c). The value of t_c is used, in turn, to obtain the design rainfall intensity that appears in the peak discharge formula Q = ciA. Thus, because of the many factors involved, it is not possible to make a general statement that a 5% increase in imperviousness will cause an x% increase in the peak discharge for a specific return period.

Other models allow the effect of urbanization to be evaluated directly. For example, a number of regression equations include the percentage of imperviousness as a predictor variable. With such models it is possible to develop a general statement on the effect of urbanization. Sarma et al. (1969) provided one such example:

$$q_p = 2.441A^{0.723}(1 + U)^{1.516} P_E^{1.113} T_R^{-0.403} \tag{9.1}$$

in which A is the drainage area (km^2), U is the imperviousness as a decimal fraction, P_E is the excess rainfall depth (cm), T_R is the rainfall excess duration (hours), and q_P is the peak discharge (m^3/sec). Since the equation has the power model form, the specific effect of urbanization depends on the values of the other predictor variables (i.e., A, P_E, and T_R). However, the relative sensitivity of Equation 9.1 can be used as a measure of the effect of urbanization. The relative sensitivity is given by

$$R_s = \frac{\partial q_p}{\partial U} \cdot \left(\frac{U}{q_p} \right). \tag{9.2}$$

Evaluation of Equation 9.2 yields a relative sensitivity of 1.516 U/(1+U). Thus, at U = 0.1, a 1% change in U (from 10% to 10.1%) causes a change of about 0.1% in the peak discharge. At U = 0.4, a 1% change in U causes a change of about 0.4% in the peak discharge. The importance of U increases as U itself increases. This estimate is an average effect since the actual sensitivity varies with the values of A, P_E, T_R, and U inputs.

Based on the work of Carter (1961) and Anderson (1970), Dunne and Leopold (1978) provided the following equation for estimating the effect of urbanization:

$$f = 1 + 1.5\,U \tag{9.3}$$

in which f is a factor that gives the relative increase in peak discharge for a decimal fraction of imperviousness, U. Thus a 1% increase in U will increase the peak discharge by 1.5%, which is similar to the effect determined from Equations 9.1 and 9.2.

The USGS urban peak discharge equations (Sauer et al., 1981) provide another alternative for assessing the effects of urbanization. The equations are given in Table 9.1. Fig. 9.2 shows the ratio of the urban-to-rural 2-year peak discharge as a function of the percentage of imperviousness and a basin development factor. The peak factors are shown in Fig. 9.3 for the 2-year and 100-year events. For the 2-year event, the ratio ranges from 1 to 4.5, with the latter value indicating complete development. For the 100-year event,

TABLE 9.1. USGS Urban Flood Peak Equations (Sauer et al., 1981).

Regression equations*	R^2	Standard error of regression: Log units	Standard error of regression: Average percent
Seven-parameter equations			
$UQ2 = 2.35A^{0.41}SL^{0.17}(RI2 + 3)^{2.04}(ST + 8)^{-0.65}(13 - BDF)^{-0.32}IA^{0.15}RQ2^{0.47}$	0.93	0.1630	±38
$UQ5 = 2.70A^{0.35}SL^{0.16}(RI2 + 3)^{1.86}(ST + 8)^{-0.59}(13 - BDF)^{-0.31}IA^{0.11}RQ5^{0.54}$	0.93	0.1584	37
$UQ10 = 2.99A^{0.32}SL^{0.15}(RI2 + 3)^{1.75}(ST + 8)^{-0.57}(13 - BDF)^{-0.30}IA^{0.09}RQ10^{0.58}$	0.93	0.1618	38
$UQ25 = 2.78A^{0.31}SL^{0.15}(RI2 + 3)^{1.76}(ST + 8)^{-0.55}(13 - BDF)^{-0.29}IA^{0.07}RQ25^{0.60}$	0.93	0.1705	40
$UQ50 = 2.67A^{0.29}SL^{0.15}(RI2 + 3)^{1.74}(ST + 8)^{-0.53}(13 - BDF)^{-0.28}IA^{0.06}RQ50^{0.62}$	0.92	0.1774	42
$UQ100 = 2.50A^{0.29}SL^{0.15}(RI2 + 3)^{1.76}(ST + 8)^{-0.52}(13 - BDF)^{-0.28}IA^{0.06}RQ100^{0.63}$	0.92	0.1860	44
$UQ500 = 2.27A^{0.29}SL^{0.16}(RI2 + 3)^{1.86}(ST + 8)^{-0.54}(13 - BDF)^{-0.27}IA^{0.05}RQ500^{0.63}$	0.90	0.2071	49
Three-parameter equations			
$UQ2 = 13.2A^{0.21}(13 - BDF)^{-0.43}RQ2^{0.73}$	0.91	0.1797	±43
$UQ5 = 10.6A^{0.17}(13 - BDF)^{-0.39}RQ5^{0.78}$	0.92	0.1705	40
$UQ10 = 9.51A^{0.16}(13 - BDF)^{-0.36}RQ10^{0.79}$	0.92	0.1720	41
$UQ25 = 8.86A^{0.15}(13 - BDF)^{-0.34}RQ25^{0.80}$	0.92	0.1802	43
$UQ50 = 8.04A^{0.15}(13 - BDF)^{-0.32}RQ50^{0.81}$	0.91	0.1865	44
$UQ100 = 7.70A^{0.15}(13 - BDF)^{-0.32}RQ100^{0.82}$	0.91	0.1949	46
$UQ500 = 7.47A^{0.16}(13 - BDF)^{-0.30}RQ500^{0.82}$	0.89	0.2170	52

*A = basin area (0.2 to 100 sq. mi.); SL = channel slope (3 to 70 ft/mile, with any SL > 70 set to 70); RI2 = basin rainfall depth for 2-yr, 2-hr duration (0.2 to 2.8 inches); ST = basin (detention) storage (0 to 11 percent); BDF = basin development factor (0 to 12); IA = percentage of impervious area (3 to 50%); RQ = rural peak discharge for same exceedance probability (ft^3/sec).

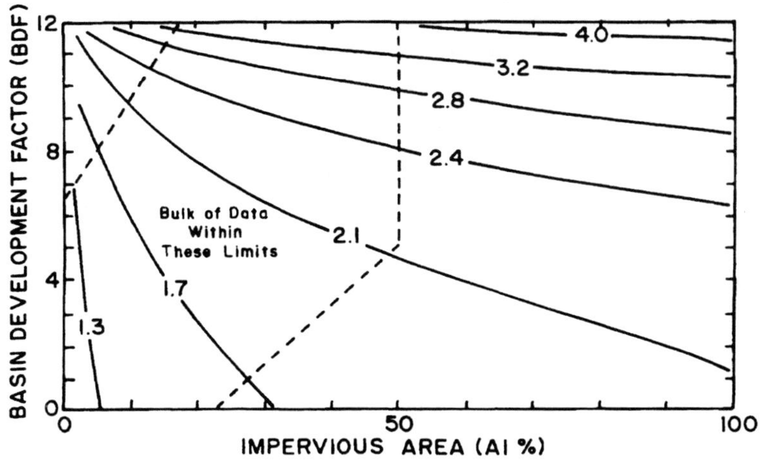

Figure 9.2.—Ratio of the Urban to Rural 2-Year Peak Discharge as a Function of Basin Development Factor (BDF) and Impervious Area (IA) (Sauer et al., 1981).

the ratio has a maximum value of 2.7. For purposes of illustration, assume that basin development occurs in direct proportion to changes in imperviousness. The values of R_s in Table 9.2 will then show the effect of urbanization on peak discharge. The average change in peak discharge due to a 1% change in urbanization is 1.75% and 0.9% for the 2- and 100-year events, respectively. While the methods discussed previously provide an effect of about 1.5%, the USGS equations suggest that the effect is slightly higher for the more frequent storm events and slightly lower for the less frequent storm events. This effect is also evident from the conceptual frequency curve of Fig. 9.1c.

Rantz (1971) provided a method for assessing the effect of urbanization on peak discharges using the simulated data of James (1965) for the San Francisco Bay area. Urbanization is characterized by two variables: 1) the percentage of channels sewered and 2) the percent development of the basin. The percentage of basin developed is approximately twice the percentage of imperviousness. The data in Table 9.2 show the relative sensitivity of the peak discharge to: a) the fraction of imperviousness (U) and b) the combined effect of fraction of channels sewered and basin developed (D_b). For urbanization as measured by the fraction change in imperviousness, the means of the relative sensitivities are 2.6%, 1.7%, and 1.2% for the 2-, 10-, and 100-year events, respectively. When both the fractions of channel sewered

TABLE 9.2. Relative Sensitivity (R_s) of Peak Discharge to Urbanization (U) and Basin Development (D_b) for the USGS Seven-Parameter Equations of Table 9.1 and The Peak Factors of Rantz (1971).

T (yrs)	U	USGS R_s	T (yrs)	U	Rantz R_s	T (yrs)	D_b	Rantz R_s
2	0.20	0.016	2	0.10	0.025	2	0.10	0.040
	0.25	0.016		0.20	0.025		0.20	0.060
	0.30	0.018		0.30	0.026		0.30	0.085
	0.35	0.020		0.40	0.029		0.40	0.100
100	0.20	0.010	10	0.10	0.015	10	0.10	0.022
	0.25	0.008		0.20	0.017		0.20	0.040
	0.30	0.008		0.30	0.018		0.30	0.050
	0.35	0.010		0.40	0.018		0.40	0.092
			100	0.10	0.011	100	0.10	0.010
				0.20	0.012		0.20	0.025
				0.30	0.013		0.30	0.050
				0.40	0.012		0.40	0.055

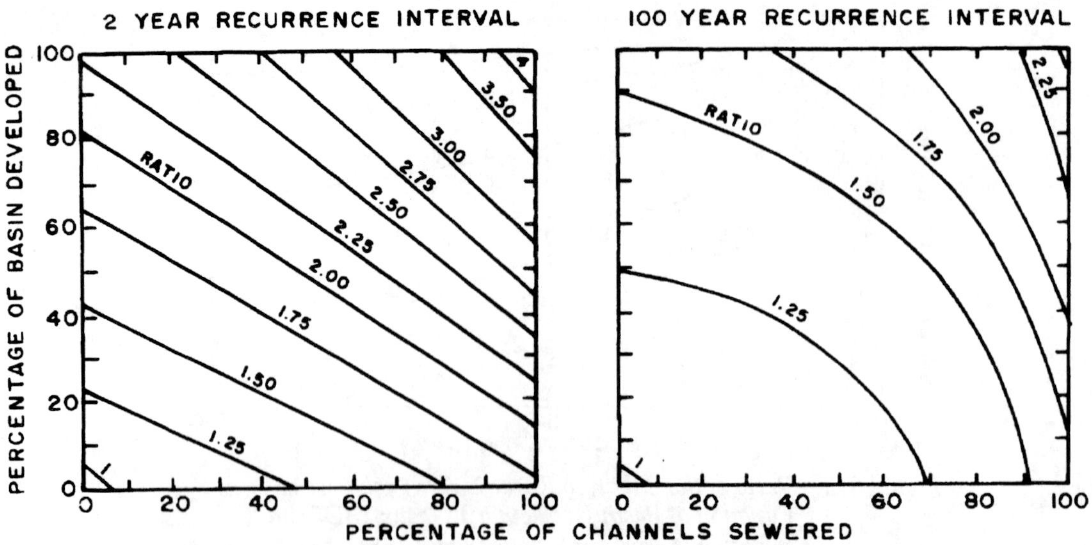

Figure 9.3.—Effect of Percentages of Imperviousness and Channels Sewered on the Ratio of the Urbanized to Undeveloped Peak Discharges (Rantz, 1971).

and basin developed are used as indices of development, the relative sensitivities are considerably higher. The means of the relative sensitivities are 7.1%, 5.1%, and 3.5% for the 2-, 10-, and 100-year events, respectively. These values are larger than the values suggested by the other methods discussed in the preceding paragraphs. The procedures cited above are presented as examples of the techniques available for urban flood frequency analysis. Many other statewide adjustment procedures are available and the interested reader is urged to consult a state office of the U.S. Geological Survey or a State Department of Water or Environmental Resources for the appropriate regional method.

C. A Method for Adjusting a Flood Record

The literature does not identify a single method that is considered the best for adjusting a flood record for the effects of urbanization. Each method depends on the data used to calibrate the prediction model. Unfortunately, the database used to calibrate the models is very sparse. The sensitivities suggest that a 1% increase in urbanization causes an increase in peak discharge of about 1 to 2.5% for the 100-year event and the 2-year events, respectively; however, there is considerable variation at any return period.

Based on the general trends of the available data, the following method is suggested for adjusting a flood record. Fig. 9.4 shows the peak adjustment factor (f_i) as a function of the exceedance probability for percentages of urbanization up to 60%. In Fig. 9.4, urbanization can be defined as the imperviousness or the degree of basin development. The greatest effect is for the more frequent events and the highest percentage of urbanization. Given the exceedance probability or return period of a flood peak for a nonurbanized watershed, one can assess the effect of increasing urbanization by multiplying the discharge by the peak adjustment factor for the corresponding return period and percentage of urbanization. Where it is necessary to adjust a discharge from a partially urbanized watershed to a discharge for another watershed condition, an engineer/hydrologist can divide the discharge by the peak adjustment factor for the existing condition and then multiply the resulting "rural" discharge by the peak adjustment factor for the second watershed condition. The first operation (i.e., division) adjusts the discharge to a magnitude representative of a natural or nonurban condition. The second operation (i.e., multiplication) adjusts the discharge to a newly developed condition. The adjustment method of Fig. 9.4 requires an exceedance probability obtained from a plotting position formula. The following procedure can be used to adjust a flood record for which the individual flood events have occurred on a watershed that is undergoing a continuous change in the level of urbanization.

1) Identify both the percentage of urbanization for each event in the flood record and the percentage of urbanization for which an adjusted flood record is needed.
2) Compute the rank (m) and exceedance probability for each event in the flood record. The Weibull plotting position formula can be used to compute the probability, where the plotting position (pp) for events given in descending order equals the rank (m) divided by 1 plus the sample size (N); i.e., pp = m/(N + 1) = estimated annual exceedance probability of event having rank m.
3) Using the exceedance probability estimate and the actual percentage of urbanization, one then finds from Fig. 9.4 the peak adjustment factor (f_1) to transform the measured peak from the actual level of urbanization to a nonurbanized condition.
4) Using the exceedance probability and the percentage of urbanization for which a flood series is needed, one finds from Fig. 9.4 the peak adjustment factor (f_2) that is necessary to transform the nonurbanized peak to a discharge for the desired level of urbanization.
5) Compute the adjusted discharge (Q_a) by

$$Q_a = \frac{f_2}{f_1} Q \tag{9.4}$$

in which Q is the measured discharge.

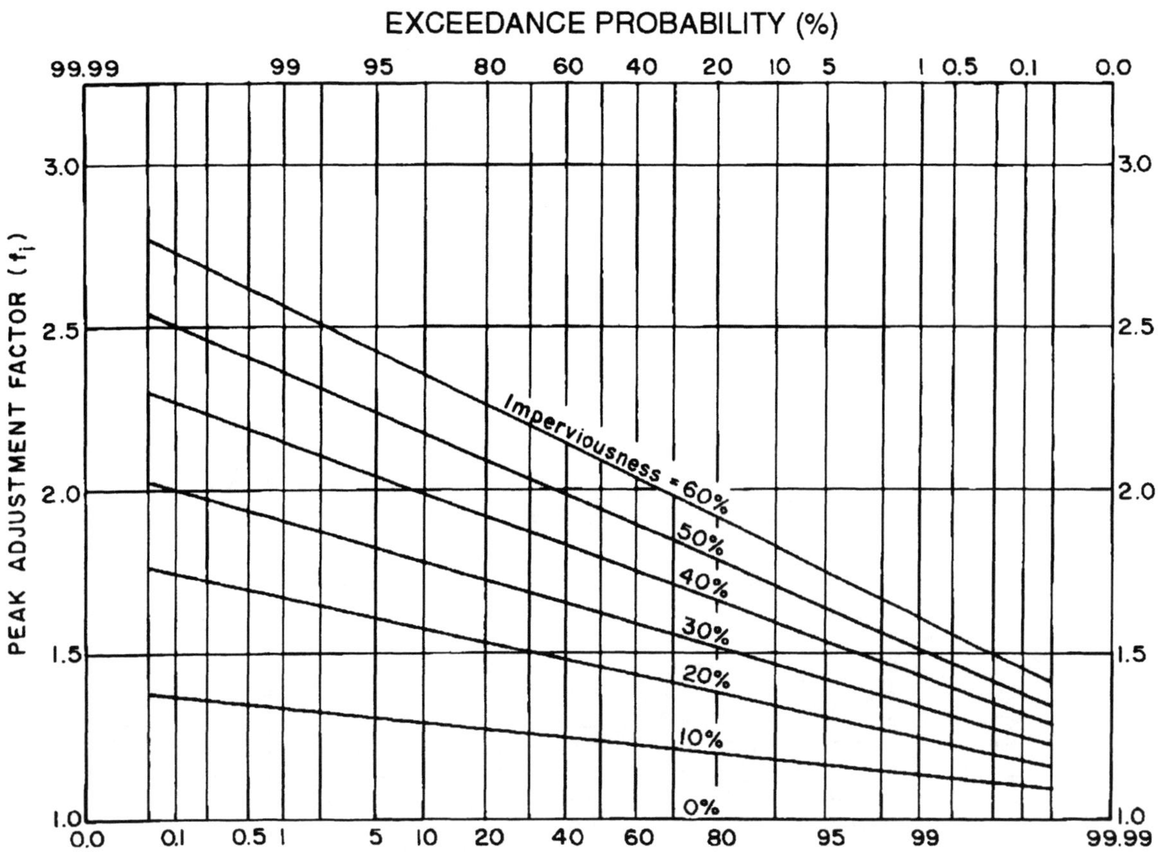

Figure 9.4.—Peak Adjustment Factors for Urbanizing Watersheds (McCuen, 1989).

6) Repeat steps 3, 4, and 5 for each event in the flood record and rank the adjusted series.
7) If there are changes in the ranks of the measured (Q) and adjusted (Q_a) flood series, repeat steps 2 through 6 until the ranks do not change.

The 13-year annual maximum flood record shown in column 3 of Table 9.3 for the Alhambra Wash watershed, Los Angeles, California, is used to illustrate the application of Fig. 9.4. The percentage of imperviousness, used as a measure of urbanization, increased from 26 to 44% over the thirteen-year period (see column 2 of Table 9.3). Ultimately, the percentage increased to 45%. The ranks (column 4) and the Weibull exceedance probability and percentage of imperviousness were used to obtain the peak adjustment factor (f_1 of Equation 9.4) that converts the measured flow to an estimated flow for nonurbanized watershed conditions. The exceedance probability and the ultimate percentage (45%) were used to obtain the peak adjustment watershed condition to the flow for the ultimate development condition. The ratio f_2/f_1 converts the measured flow to a flow for ultimate development. Columns 6 through 8 of Table 9.3 give initial estimates of the adjustment factors and the adjusted discharges. Since the ranks of the adjusted discharges (column 9) are different from the ranks of the measured series (column 4), a second iteration is necessary; this is shown in columns 10 to 14. The final estimates for ultimate conditions are given in column 13. Although the length of the annual series is too short in the example for flood frequency analysis, that would be the next step if the annual series exceeded 20 years of record.

D. Design–Storm Approach

Models such as the USGS regression equations of Table 9.1 are very useful for developing a frequency curve for a watershed. For design problems involving large nonhomogeneous watersheds where watershed subdivision is necessary, the regression-equation approach is inadequate. In such cases, the hydraulic model/design storm approach is the most frequently used method for evaluating the effect of urban/suburban development on peak discharge rates. In this approach, it is necessary for the hydrologic engineer to select: 1) a regional rainfall intensity–duration–frequency (IDF) curve, 2) a design–storm hyetograph, 3) a technique for computing the volume of rainfall excess, 4) a unit hydrograph or kinematic overland flow model, 5) a time-of-concentration procedure, and 6) a channel routing method. The total watershed can then be subdivided so that the effects of local development within individual subwatersheds can be assessed for the entire watershed.

TABLE 9.3. Application of Flood Frequency Adjustment Method: Alhambra Wash Watershed, 1939–1951.

Water year (1)	% Imp. (2)	Annual max Q cfs (3)	Rank, m (4)	$\frac{m}{N+1}$ (5)	f_1 (6)	f_2 (7)	Adj. Q (8)	Adj. rank m (9)	$\frac{m}{N+1}$ (10)	f_1 (11)	f_2 (12)	Final adj Q, cfs (13)	Adj. rank m (14)
1939	26	4224	1	.071	1.39	1.61	4893	1	.071	1.39	1.61	4893	1
1940	26	2048	6	.429	1.55	1.84	2431	5	.357	1.52	1.81	2439	5
1941	27	3375	2	.143	1.42	1.69	4017	2	.143	1.42	1.69	4017	2
1942	29	1125	9	.643	1.65	1.92	1309	9	.643	1.65	1.92	1309	9
1943	30	2816	3	.214	1.52	1.75	3242	3	.214	1.52	1.75	3242	3
1944	32	2675	4	.286	1.59	1.77	2978	4	.286	1.59	1.77	2978	4
1945	33	2125	5	.357	1.63	1.82	2373	6	.429	1.67	1.84	2341	6
1946	34	1664	8	.571	1.72	1.91	1848	8	.571	1.72	1.91	1848	8
1947	36	1792	7	.500	1.73	1.87	1937	7	.500	1.73	1.87	1937	7
1948	39	1024	11	.786	1.90	2.01	1083	11	.786	1.90	2.01	1083	11
1949	42	1100	10	.714	1.93	1.98	1128	10	.714	1.93	1.98	1128	10
1950	43	900	13	.929	2.10	2.11	904	13	.929	2.10	2.11	904	13
1951	44	1024	11	.786	2.00	2.01	1029	12	.857	2.03	2.05	1034	12

There are numerous computer models available for urban stormwater analysis and design, each having its own input data structure. As an example, the SCS TR–20 program uses the 24-hour SCS design hyetograph, the SCS runoff equation, and the SCS curvilinear unit hydrograph (USDA SCS, 1969). An IDF curve is used to obtain the rainfall depth, which is used to dimension the ordinates of the SCS 24-hour dimensionless cumulative hyetograph. The SCS runoff equation is then used to compute the time distribution of rainfall excess. The incremental rainfall excess is then applied to the unit hydrograph in a convolution process to obtain the subarea design hydrograph. If necessary, the discharge hydrograph can be routed through channel reaches and combined with flood hydrographs from other subareas of the watershed. Urbanization is represented in the process by the runoff curve number (used to compute the rainfall excess) and by the time of concentration (used to dimension the unit hydrograph). It is noted that TR–20 is not particularly well-suited for storm sewer flow because of its inability to represent surcharge. While the details of the calculations vary, other computer models are also widely used to analyze urbanization effects on the flood hydrograph. Hydrologic models such as EPA SWMM and HEC–1 use the kinematic wave routing method for overland and channel/pipe flow. Like TR–20, HEC–1 is useful as an analysis tool but has limited capability to represent storm sewer flow. Selected models are described in a later section of this chapter.

II. PRECIPITATION IN THE URBAN WATERSHED

The management of urban storm runoff requires that the engineer/hydrologist conduct a careful study of historical rainfall data before selecting the design storm to be used in sizing the new project or in analyzing an existing one. It is presumed that a project will function properly if it can accommodate the design flow at full capacity. The project will fail to function as intended whenever the design event is exceeded. Some risk of failure must be accepted for economic reasons in selecting the magnitude of the design event. Thus, a storm sewer system designed to convey a 10-year storm runoff will surcharge and probably flood the street during a 25-year flood.

A return period, T, is defined as the average number of years between successive occurrences of a hydrologic event with a certain magnitude. The inverse of this return period is called the exceedance probability or the probability that this magnitude will be exceeded in any one year. Statistical flood frequency methods are used to analyze historical records for the return periods of different magnitude hydrologic events. The risk probability that the T-year event will occur at least once in N consecutive years is: $R = 1 - (1 - 1/T)^N$. The reliability probability is $1 - R$.

Because urban stormwater projects deal with storm runoff, the engineer ideally will choose an observed runoff event as the design event; however, urban runoff data is usually unavailable at most sites in the quantity and quality needed to obtain the peak discharge-return period and the volume-return period relations. Local rainfall data, on the other hand, are generally available. Thus, one approach to stormwater analysis and design is to transform a long-term rainfall record into a record of runoff before a design runoff event is identified. This can be achieved using a rainfall–runoff model, as discussed in a later section.

Another approach is to adopt national and regional rainfall intensity–duration–frequency (IDF) data found in various rainfall atlases. The U.S. Weather Bureau, now the National Weather Service, prepared isohyetal maps of rainfall depth for the entire United States for specified durations and return periods in a publication commonly known as TP 40 (Hershfield, 1961). These maps cover durations from 30 minutes to 24 hours and return periods from 1 to 100 years. TP 40 also includes interpolation procedures to determine rainfall depths for durations and return periods not included in the maps. Miller et al. (1973) updated the data in TP 40 and presented isohyetal maps for the 11 western states for 6- and 24-hour durations and return periods of 2 to 100 years in NOAA Atlas 2. It is noted that frequency analysis of local rainfall data may produce rainfall depths which differ significantly from NOAA Atlas 2. This is attributed to the regional averaging process in NOAA Atlas 2. The U.S. National Weather Service publication known as HYDRO 35 (Frederick et al., 1977) is particularly useful to urban hydrologists because of the short duration rainfalls included. Many urban storm drainage projects deal with small drainage watersheds where storm durations less than 30 minutes need to be considered. HYDRO 35 includes isohyetal maps of rainfall depth for the eastern and central United States for 5-, 15-, and 60-minute durations and return periods of 2 and 100 years. HYDRO 40 provides comparable rainfall data for the southwest United States.

The purpose of this section is to describe how the available rainfall data can be used to identify a design runoff event. Basically, there are two fundamentally different approaches, namely the continuous simulation and the single-event design storm approaches. Both are discussed here. The single-event design–storm method is then elaborated further because of its simplicity and widespread use. Subsequent sections explain how to determine the elements of a design storm. The reader may refer to Chapter 2 for additional information on design storms.

A. Continuous Simulation and Single Events

In the continuous simulation approach, a chronological record of rainfall is used as input to a mathematical rainfall–runoff model of the drainage system being considered. The continuous rainfall record should be several decades long, usually at 1-hour time intervals, although the interval must be shorter for small drainage systems. The output from the mathematical model should be a chronological record of runoff. This record is then analyzed using statistical methods to determine the magnitude of the runoff that has a designated return period or exceedance probability. Since a moisture-accounting routine is incorporated in a continuous simulation, antecedent moisture conditions prior to rainfall events can be determined by the model itself after the initial start-up. Continuous simulation is particularly useful in detention projects in which the sequence of events and the inter-event time are important; however, the results of continuous simulation are meaningful only if an accurate rainfall–runoff model is used. The main disadvantages of continuous simulation are the large effort needed to prepare the input data, the long computer runs involved when hydrodynamic-based rainfall–runoff models are used, and the lack of short-duration rainfalls (15 minutes or less) required for small drainage systems.

In the single-event approach, a historical rainfall record is first analyzed to obtain the rainfall–return period relations. These are usually available in the form of intensity–duration–frequency or IDF curves. Then a single storm with return period equal to the design-life or maximum protection capacity of the project is chosen as the design storm. This event is used as input to a single-event rainfall–runoff model, along with an assumed antecedent condition for the urban watershed prior to the storm. The output is a design runoff hydrograph. An important assumption inherent in this approach is that the return period of the calculated runoff is the same as that of the design storm. This assumption is rarely met because the antecedent condition of the urban watershed, together with changes in the hydraulic behavior of the drainage system, can affect the runoff process significantly from one event to the next. Nevertheless, the assumption of equal rainfall and runoff frequency must be made in the design-storm approach.

A logical compromise between the design-storm and the continuous simulation approach is a pseudo-continuous method using a series of discrete major historical storms. In this approach, as summarized by Walesh (1989), a rainfall hyetograph and associated antecedent conditions are assembled for major storm events using a historical rainfall record of several decades. These hyetographs are input as multiple/discrete events to a rainfall–runoff model and the corresponding runoff hydrographs are produced. From the model output, the engineer/hydrologist then obtains the annual peak discharges and maximum volumes and these are analyzed statistically to develop discharge-probability and volume–probability relationships.

The continuous simulation method is data-intensive and time-consuming. Also, the mathematical representation of an urban watershed in a rainfall–runoff model may not be accurate enough to warrant continuous simulation. The design-storm approach, by comparison, is much simpler and is used extensively despite its limitations.

B. Elements of a Design Storm

A design storm is characterized in terms of a return period (or frequency), storm duration, average intensity (or total depth), spatial distribution, storm movement and direction, and time pattern of rainfall intensity (hyetograph). The selection of the design return period depends on several factors among which are the importance of the structure being designed, the project cost, the level of protection it will provide, and the damages that would result from failure of the structure. Typical design return periods vary from 2 to 25 years for storm sewers and 25 to 100 years for detention basins. Design return periods of 5 to 10, 10 to 25, and 25 to 50 years are common for culverts under highways carrying light, medium, and heavy traffic, respectively. Major highway bridges are designed to pass Q_{50} (the 50-year flood peak). Floodway delineation calculations, as presently administered by the Federal Emergency Management Administration (FEMA), are based upon Q_{100}. Design for erosion control in natural channels is usually based upon Q_2 or Q_5, since this flow range is regarded as dominant in forming stable channels. While street gutter design is commonly based on Q_2 or Q_5, it is obvious these appurtenances must be able to operate consistently with the main storm sewer system. Local drainage manuals and regulations must be used to determine the design return period for specific projects.

The design storm duration depends on the type of the project. For the design of hydraulic structures based on peak discharge, such as a storm sewer or culvert, the design storm duration should produce the largest peak discharge for a given return period. This will normally be the time of concentration for the drainage area above the design point. For flow detention projects, the duration causing the largest detention volume is the most critical. Generally, the critical storm duration can be determined only after trying several different storm durations and investigating the sensitivity of the peak discharge and the runoff volume to variable storm duration. For the design of storm sewers, the engineer/hydrologist should first estimate t_c, the time of concentration, and then obtain the average rainfall intensity from local intensity–duration–frequency relationships for a given return period and storm duration equal to t_c. These IDF relationships are discussed in a later section.

The spatial distribution of rainfall within an urban watershed can be highly nonuniform. The effect of watershed size on average areal rainfall is often accounted for by applying a reduction factor to the point rainfall values. Fig. 9.5 shows the percent reductions recommended by the U.S. Weather Bureau (1957) for the eastern United States. Most urban drainage catchments are sufficiently small that a rainfall reduction factor is not required. When large urban watersheds are investigated, the recommended reductions given in Fig. 9.5 can be employed. The U.S. Army Corps of Engineers has developed an equation for the reduction factors in Fig. 9.5. (USACE, 1990). Alternatively, the design storm can be assumed to occur only over part of the watershed with no areal reduction factor. Storm movement and direction of travel can have a significant impact on runoff peak, depending on the shape of the watershed. An elongated basin will experience higher runoff peaks when rainfall moves downstream from the headwater area rather than in the upstream direction. The location of the design storm center and the storm duration should be selected so that peak runoff rates and volumes will replicate real-world experience for a specified return period. Several locations must be tried before the most critical storm can be identified. The distribution of the total rainfall within the design storm can have a significant effect on the peak runoff rate. An early peaking storm may produce less runoff than a later-peaking one, having the same total rainfall. Several methods available to determine the time distribution of design rainfall are discussed in a later section. The interested reader also is referred to discussions by McPherson (1958), Preul and Papadakis (1973), Marsalek (1978), and Yen and Chow (1980; 1983).

C. Intensity–Duration–Frequency Relations

The relationships between average rainfall intensity, duration, and return period are usually presented as a graph as illustrated for Norfolk, Virginia in Fig. 9.6. These curves are obtained from local rainfall records using statistical frequency analysis techniques. Annual maximum or annual exceedance series of rainfall depths are extracted from historical records compiled by the National Weather Service or other

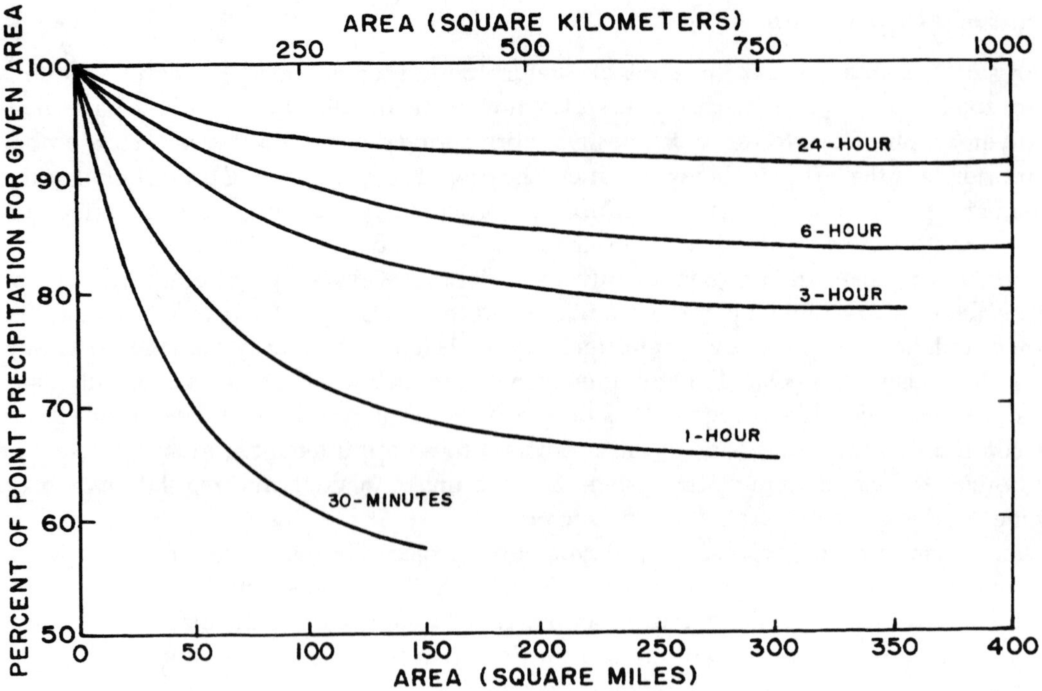

Figure 9.5.—Reductions in Point Rainfall to Obtain Areal Average Values (After Miller et al., 1973).

sources for the durations selected. Storm durations as small as 5 minutes are desirable for the urban drainage system. Then a frequency analysis is performed for each series. The Extreme Value Type I or the Gumbel distribution is commonly used for this purpose. The results are presented in chart form after dividing the depths by the respective durations to obtain the average intensities. These curves are commonly referred to as the intensity–duration–frequency (IDF) relations.

The IDF charts have already been developed for most major cities around the country, and they can be found in local drainage manuals and ordinances. Regional rainfall frequency maps reported in various publications, some of which were summarized earlier, can be used in the absence of local rainfall records or IDF relationships.

IDF relationships can also be expressed in equation form. Two common forms are

$$\bar{i} = \frac{A}{t_d^B + C} \tag{9.5}$$

and

$$\bar{i} = \frac{a}{(t_d + b)^c} \tag{9.6}$$

where $\bar{i}$ is the average intensity (cm/hr), t_d is the storm duration (minutes), and A, B, C, a, b, and c are fitting parameters. Wenzel (1982) presented the constants A, B, C, a, b, and c from a number of cities in the United States for storms of 10-year return periods. Table 9.4 shows these constants. Aron (1989) reported that for any return period, the constant *b* of Equation 9.6 is 15 and *a* and *c* can be determined from,

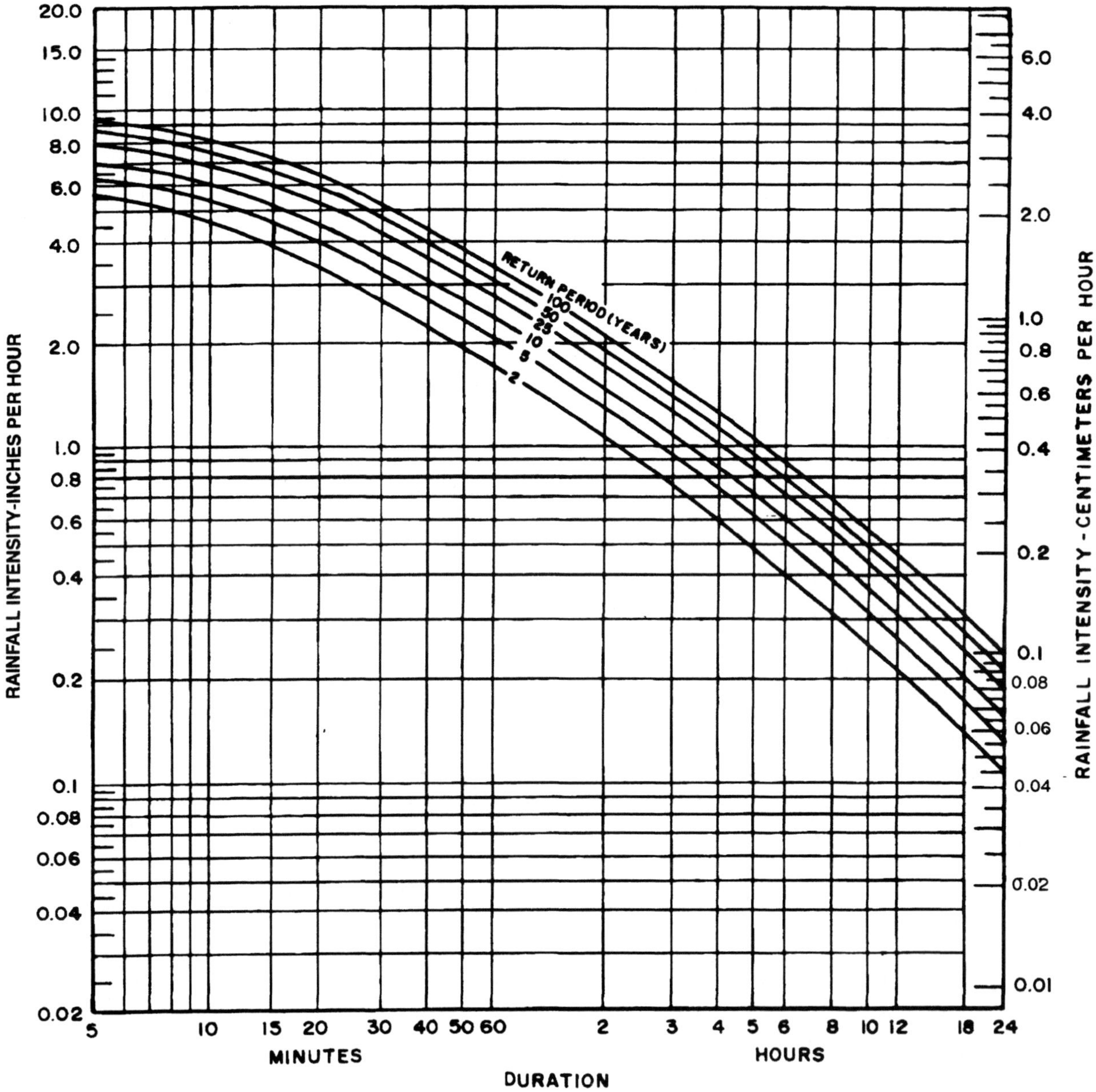

Figure 9.6.—Intensity-Duration Return Period Relationships for Norfolk, Virginia (VDOT, 1980).

$$a = 213.4 + 50.80\,(e^{0.394\,i_{60}}) \tag{9.7}$$

$$c = 0.2316\,ln\left(\frac{a}{i_{60}}\right) \tag{9.8}$$

where i_{60} is the 60 minute rainfall intensity in cm/hr for the return period being used. Aron's expressions were obtained by fitting Equation 9.6 to the generalized intensity–duration curves developed by Yarnell (1935). They provide a good fit to the Yarnell IDF curves for the range 5 min $\leq t_d \leq$ 120 min, but should not be extended beyond 120 minutes.

TABLE 9.4. Constants for IDF Relationships for 10-Year Return Period At Various Locations (After Wenzel, 1982).

Location	A	B	C	a	b	c
Atlanta	247.6	0.83	6.88	162.8	0.76	8.16
Chicago	241.0	0.88	9.04	154.7	0.81	9.56
Cleveland	187.2	0.86	8.25	120.9	0.79	8.86
Denver	245.4	0.97	13.90	129.0	0.84	10.50
Houston	247.4	0.77	4.80	249.7	0.80	9.30
Los Angeles	51.6	0.63	2.06	27.7	0.51	1.15
Miami	315.5	0.81	6.19	202.9	0.73	7.24
New York	198.4	0.82	6.57	130.6	0.75	7.85
St. Louis	265.9	0.89	9.44	154.9	0.78	8.96

Constants correspond to $\bar{i}$ in cm./hr. and t_d in min. in Eqs. 9.5 and 9.6.

D. Temporal Distribution of Rainfall and Design Storms

The time distribution of total rainfall within a storm is usually nonuniform. This distribution can have a significant effect on the computed peak runoff rate corresponding to a given design return period. Therefore, the time pattern of rainfall intensity needs to be specified for a complete description of a design storm. The uniform intensity distribution is the simplest form of the design storms, and it is associated with the Rational Method discussed in later sections. Other standard distributions employed extensively in urban storm drainage projects are nonuniform. Some of these distributions are discussed in the following subsections.

E. Soil Conservation Service Distributions

The U.S. Department of Agriculture, Soil Conservation Service (1986), developed four 24-hour rainfall distributions called Types I, IA, II, and III, respectively. These standard distributions represent different geographical locations within the United States as shown in Fig. 9.7. All design storms of shorter duration are included in these 24-hour rainfall hyetographs, and therefore these distributions are meant for both small and large watersheds. The four distributions are displayed in Fig. 9.8, and the fractions of the total rainfall are tabulated as a function of time in Table 9.5 in which P_T is total rainfall, and P is rainfall occurring between time zero and t.

An example 24-hour Type II hyetograph is developed in Table 9.6 for Norfolk, Virginia. A return period of 10 years is selected, and for this return period the average rainfall intensity for a 24-hour rainfall is found from the local IDF curves in Fig. 9.6, as 0.53 cm/hr (0.21 in./hr). The total rainfall depth, P_T, is 0.53 cm/hr. × 24 hours = 12.72 cm (5.04 in). The peak portion of the calculated hyetograph is displayed in Fig. 9.9.

F. Other Design Storm Hyetographs

Various other rainfall distributions have been proposed in the literature. Hershfield (1962) reported time–distribution relationships based on hourly rainfall data from 50 widely separated stations with different rainfall regimes. Durations of 6, 12, 18, and 24 hours were considered. An average relationship was obtained for these four durations as shown in Fig. 9.10. Hershfield (1962) points out that such an average curve gives no indication of the distribution from an individual storm. Therefore, a rearrangement of the rainfall distribution for a particular event is not unreasonable. An example 10-year, 2-hour storm hyetograph is developed for Norfolk, Virginia in Table 9.7 assuming the average Hershfield distribution can be used for 2-hour durations also. The total rainfall depth of P_T = 7.47 cm (2.99 in) can be obtained from Fig. 9.6 as described in the preceding paragraph. The resulting design storm hyetograph is displayed in Fig. 9.9. Other balanced design storm distributions are described by the U.S. Army Corps of Engineers (1982).

Huff (1967) presented a thorough analysis of large storms that occurred in east central Illinois between 1955 and 1966. The storm durations varied from 3 to 48 hours. The storms were grouped according to the

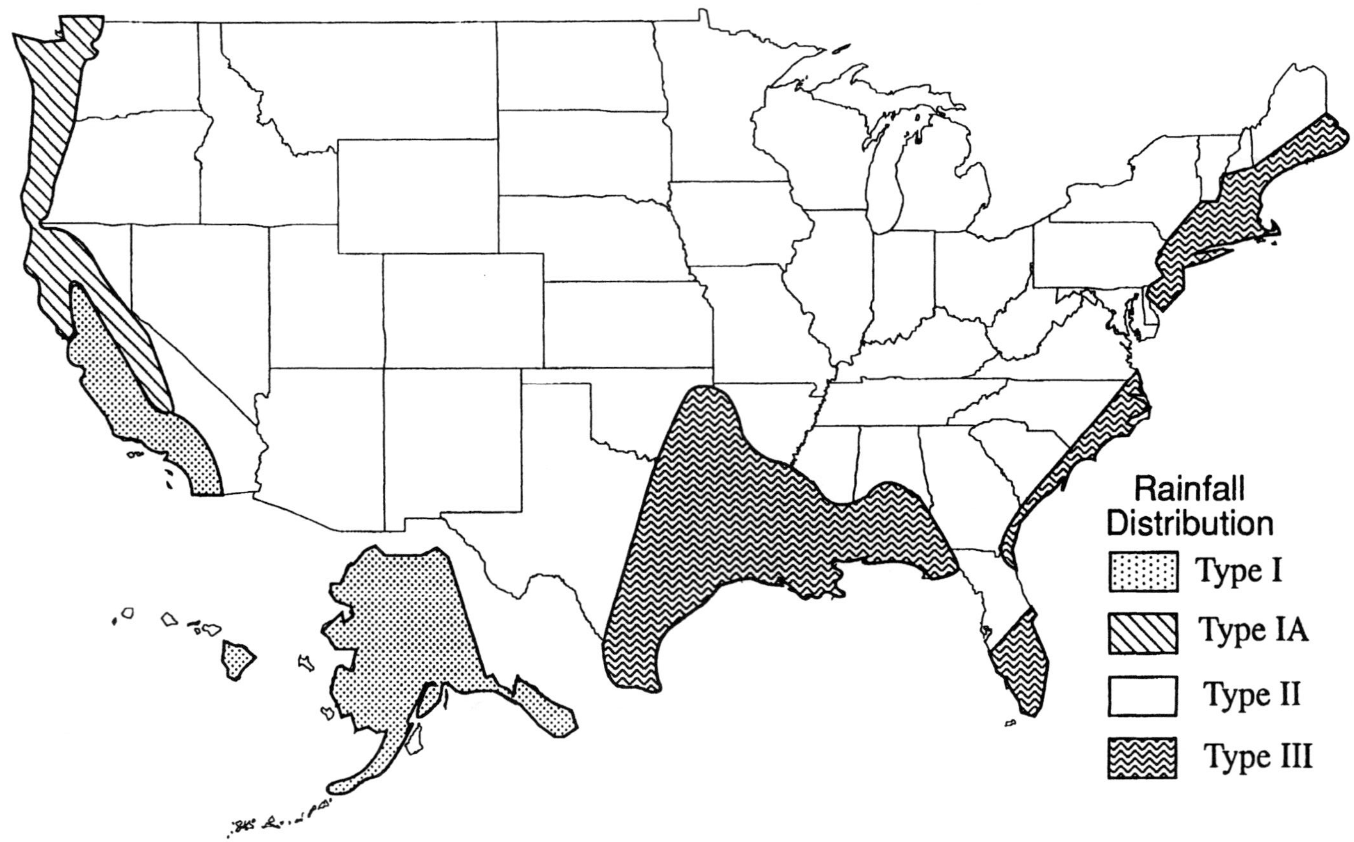

Figure 9.7.—Approximate Geographical Boundaries for SCS 24-hr Rainfall Distributions (After USDA SCS, 1986).

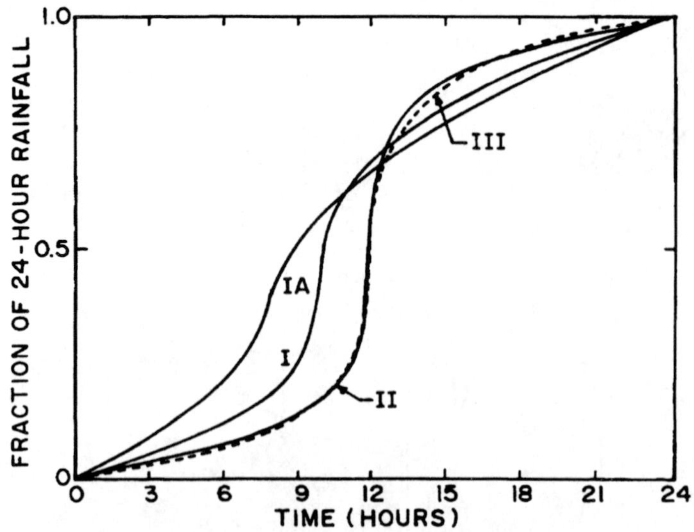

Figure 9.8.—Soil Conservation Service 24-hr Rainfall Distributions (After USDA SCS, 1986).

quartile in which rainfall was the heaviest. Then probability distributions were obtained for the storms in each quartile. For small areas and short durations, first-quartile storms were the most frequent. The median distribution for all the four quartiles is shown in Fig. 9.10. An example second quartile Huff hyetograph is developed in Table 9.8 for Norfolk using a return period of 10 years and a duration of 2 hours. As before, the total rainfall is P_T = 7.47 cm. The resulting hyetograph is plotted in Fig. 9.9.

In Fig. 9.9 it is clear that there are major discrepancies in the three rainfall distributions. Since these are all statistically derived, there is no way to assign a measure of absolute accuracy to any of the distribu-

TABLE 9.5. Soil Conservation Service Rainfall Distributions.

	Fraction of Total Rainfall, P/P_T			
Hour t	Type I	Type IA	Type II	Type III
0	0	0	0	0
2.00	0.035	0.050	0.022	0.020
4.00	0.076	0.116	0.048	0.043
6.00	0.125	0.206	0.080	0.072
7.00	0.156	0.268	0.098	0.089
8.00	0.194	0.425	0.120	0.115
8.50	0.219	0.480	0.133	0.130
9.00	0.254	0.520	0.147	0.148
9.50	0.303	0.550	0.163	0.167
9.75	0.362	0.564	0.172	0.178
10.00	0.515	0.577	0.181	0.189
10.50	0.583	0.601	0.204	0.216
11.00	0.624	0.624	0.235	0.250
11.50	0.654	0.645	0.283	0.298
11.75	0.669	0.655	0.357	0.339
12.00	0.682	0.664	0.663	0.500
12.50	0.706	0.683	0.735	0.702
13.00	0.727	0.701	0.772	0.751
13.50	0.748	0.719	0.799	0.785
14.00	0.767	0.736	0.820	0.811
16.00	0.830	0.800	0.880	0.886
20.00	0.926	0.906	0.952	0.957
24.00	1.000	1.000	1.000	1.000

TABLE 9.6. Example SCS 24-hr Type II Hyetograph for Norfolk, VA.

t (hrs)	P/P_T	P (cm)	Δp (cm)	Δt (hrs)	i (cm/hr)
0	0	0			
			0.611	4.00	0.153
4.00	0.048	0.611			
			0.915	4.00	0.229
8.00	0.120	1.526			
			0.344	1.00	0.344
9.00	0.147	1.870			
			0.203	0.50	0.406
9.50	0.163	2.073			
			0.229	0.50	0.458
10.00	0.181	2.302			
			0.293	0.50	0.586
10.50	0.204	2.595			
			0.394	0.50	0.788
11.00	0.235	2.989			
			0.611	0.50	1.222
11.50	0.283	3.600			
			0.941	0.25	3.764
11.75	0.357	4.541			
			3.892	0.25	15.568
12.00	0.663	8.433			
			0.916	0.50	1.832
12.50	0.735	9.349			
			0.471	0.50	0.942
13.00	0.772	9.820			
			0.343	0.50	0.686
13.50	0.799	10.163			
			0.267	0.50	0.534
14.00	0.820	10.430			
			0.764	2.00	0.382
16.00	0.880	11.194			
			0.915	4.00	0.229
20.00	0.952	12.109			
			0.611	4.00	0.153
24.00	1.00	12.720			

tions. Rather, it is left to the engineer/hydrologist to select a rainfall distribution and duration that will produce a conservative design event for the particular project at hand. As a minimum, the rainfall duration should be twice the time of travel through the drainage area to allow development of a full hydrograph. The use of a 24-hour storm is considered highly conservative in the case of local stormwater detention projects. This may be readily justified depending on the downstream damage area.

III. HYDROLOGIC LOSSES IN DEVELOPING WATERSHEDS

Hydrologic losses refer to processes which act on rain or snow before it turns into rainfall excess, snowmelt, and surface runoff. These processes generally include evaporation of water from the land and vegetative leaf surface, interception of rainfall by vegetation, depression storage on the land surface (paved or unpaved) and infiltration of water into the soil matrix. If the effects of a single storm event of short duration are of primary interest, evaporation losses can be neglected; however, it has been shown by Horton (1933) that during a sequence of low-intensity rainfalls, evaporation can enhance considerably the interception capacity of a dense leaf canopy.

Interception, as part of surface retention, varies depending on the type of vegetation and maturity and extent of the canopy cover. Depression storage, which is the major part of surface retention in urban

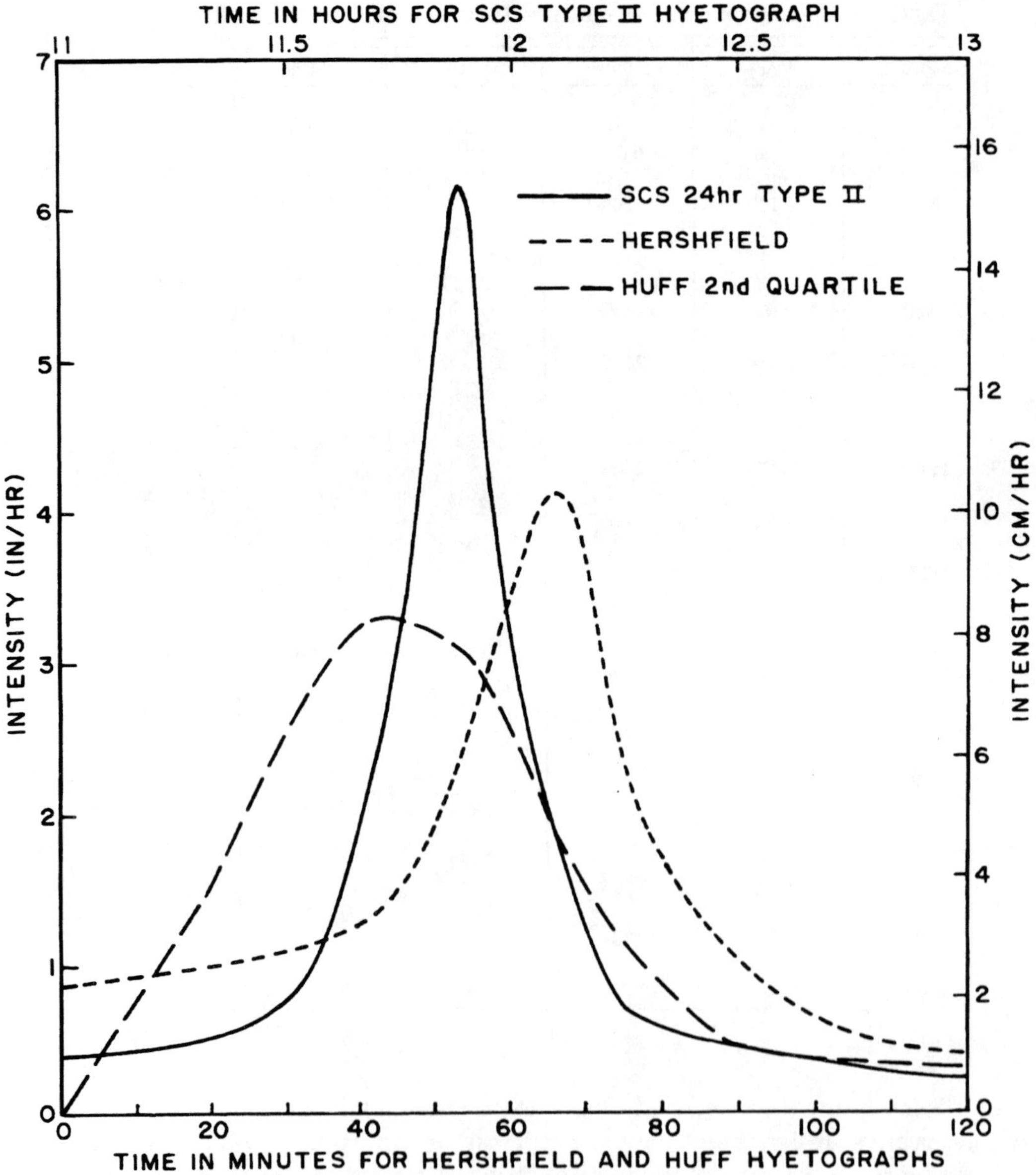

Figure 9.9.—Comparison of SCS Type II, Hershfield, and Huff Second Quartile Hyetographs.

basins, depends on the irregularities in the surface topography. Finally, infiltration comprises the major portion of abstractions from pervious land surfaces. Like interception, infiltration tends to be strongly enhanced by warm temperatures and a dense vegetation.

A. Interception

References to interception can be found in most hydrology textbooks and some professional journal papers, but almost all of these refer to Horton's (1919) observations taken in his laboratory and published in a highly detailed 20-page article. Horton concluded that evaporation can play a major role in the interception process, if the total leaf surface of the projected area is dense and the storm endures at least several hours. For these conditions, Horton proposed the equation:

$$J = S_j + k_j \, E \, D_s \tag{9.9}$$

TABLE 9.7. Example Hershfield Hyetograph for Norfolk, VA.

t (min)	t/t_d	P/P_T	P (cm)	ΔP (cm)	ΔP/Δt (cm/hr)
0	0	0	0		
				0.448	2.24
12	0.1	0.06	0.448		
				0.486	2.43
24	0.2	0.125	0.934		
				0.560	2.80
36	0.3	0.20	1.494		
				0.672	3.36
48	0.4	0.29	2.166		
				1.196	5.98
60	0.5	0.45	3.362		
				2.091	10.46
72	0.6	0.73	5.453		
				0.971	4.86
84	0.7	0.86	6.424		
				0.523	2.62
96	0.8	0.93	6.947		
				0.299	1.50
108	0.9	0.97	7.246		
				0.224	1.12
120	1.0	1.0	7.470		

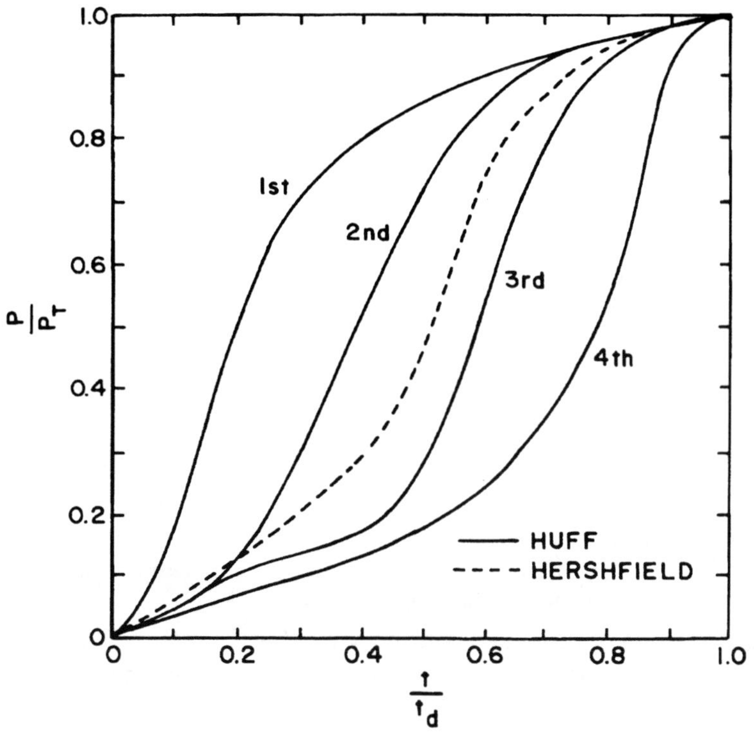

Figure 9.10.—Hershfield and Huff Rainfall Distribution.

TABLE 9.8. Example Second Quartile Huff Hyetograph for Norfolk, VA.

t (min)	$\frac{t}{t_d}$	$\frac{P}{P_t}$	P (cm)	ΔP (cm)	$\frac{\Delta P}{\Delta t}$ (cm/hr)
0	0	0.0	0		
				0.247	1.24
12	0.1	0.033	0.247		
				0.687	3.44
24	0.2	0.125	0.934		
				1.307	6.54
36	0.3	0.300	2.241		
				1.681	8.41
48	0.4	0.525	3.922		
				1.568	7.84
60	0.5	0.735	5.490		
				0.934	4.67
72	0.6	0.860	6.424		
				0.486	2.43
84	0.7	0.925	6.910		
				0.231	1.16
96	0.8	0.956	7.141		
				0.195	0.98
108	0.9	0.982	7.336		
				0.134	0.67
120	1.0	1.000	7.470		

in which J is the interception (cm), S_j is the interception storage capacity (cm), on the projected area, k_j is the ratio of the evaporation surface to the projected area, E is the evaporation rate (cm per hour), and D_s is the storm duration (hours).

A modification of the Horton equation was later proposed, reflecting the notion that not all of the interception capacity is immediately filled. The modified equation reads:

$$J = S_j \left[1 - \exp\left(-\frac{P}{S_j}\right)\right] + k_j\ E\ D_s \tag{9.10}$$

in which P is cumulative precipitation (cm).

For situations in which evaporation is negligible or difficult to evaluate, Horton proposed the alternative equation:

$$J = a + b\ P^n \tag{9.11}$$

in which a, b, and n are parameters to be found experimentally. Horton's parameter values for selected tree types and for short crops are shown in Table 9.9 (for J and P in inches).

Two problems can be found with Equation 9.11 and the parameter values. In the first place, interception can exceed precipitation for small values of P. Secondly, the suggested value of the exponent n is 1.0 for most tree types, which implies that interception continues to increase at an undiminished rate as precipitation increases. Because of these deficiencies, Aron (1992) analyzed a set of data collected from summer storms by Horton (1919). Equation 9.12 was developed by linear regression on the logarithmic values of P and J.

$$J = c\ P^m \tag{9.12}$$

TABLE 9.9. Parameter Values for Interception Equations.

	Equation: $J = a + b P^n$			Equation: $J = c P^m$	
Tree type	a	b	n	c	m
Apple	0.04	0.18	1.0	0.25	0.73
Ash	0.015	0.23	1.0	0.26	0.88
Beech	0.02	0.23	1.0	0.21	0.65
Chestnut	0.04	0.20	1.0	0.30	0.77
Elm	0.0	0.23	0.5	0.15	0.48
Hemlock	0.0	0.20	0.5	0.32	0.74
Maple	0.03	0.23	1.0	0.29	0.77
Oak	0.03	0.22	1.0	0.24	0.66
Pine	0.0	0.20	0.5	0.30	0.70
Willow	0.02	0.40	1.0	0.43	0.85

Note: Parameter values for J and P in inches

Values of c and m are shown in Table 9.9 for J and P in inches. The equation and the parameter values avoid the drawbacks cited for the Horton equation.

Equations 9.11 and 9.12, in association with the suggested parameter values, are valid only for the fraction of the land surface actually covered by the trees or crop. In case of incomplete coverage the computed interception depth should be multiplied by the coverage fraction.

B. Depression Storage

Depression storage consists essentially of water trapped in puddles formed by the irregularity of a surface. The most frequently quoted values for depression storage, D_p, are 0.16 cm and 0.64 cm for paved and unpaved surfaces, as shown in Table 9.10.

For impervious surfaces of various slopes between 0.005 m/m and 0.04 m/m, Kidd (1978) analyzed data from Holland, the United Kingdom, and Sweden and proposed the equation

$$D_p = 0.0076\, S^{-0.5} \tag{9.13}$$

in which D_p is the potential depression storage (cm) and S is the dimensionless land slope. For modeling purposes, Linsley et al. (1975) proposed gradual depression storage buildup, Equation 9.14, in which P_e is excess precipitation or cumulative rainfall minus interception and infiltration. D_p is computed from Equation 9.13.

$$D = D_p \left[1 - \exp\left(\frac{P_e}{D_p}\right)\right] \tag{9.14}$$

TABLE 9.10. Depression Storage Estimates in Urban Areas (Bedient and Huber,© 1988. Reprinted by permission of Addison-Wesley Publishing Co.).

		Depression storage depth		
City	Cover	inches	cm	Reference
Chicago	Pervious	0.25	0.64	Tholin and Keifer (1960)
	Impervious	0.06	0.16	
Los Angeles	Sand	0.20	0.51	Hicks (1944)
	Loam	0.15	0.38	
	Clay	0.10	0.25	

C. Infiltration

Infiltration is a highly complex process. Even though it is generally accepted that a given soil will have its largest infiltration capacity when it is at its driest state, this is not necessarily true. It can frequently be observed that an extremely dry soil sets up electrostatic forces that repel water. Ion concentration in the water can also have a major effect on infiltration rates. And finally, the type and density of vegetation can enhance infiltration rates. All fall into the two basic categories of: 1) watershed loss function and 2) infiltration component model. The former combines all initial and infiltration losses in a single loss function, while the latter provides for separate estimation of initial losses and infiltration.

Most of the currently used infiltration models and equations have been presented in Chapter 3. In this chapter, discussion will be restricted to three models, namely the ϕ-index loss function, and the infiltration equations proposed by Horton (1933) and Green and Ampt (1911). Soil Conservation Service rainfall–runoff relationships, including the runoff curve number and the soil cover complex method, are presented in detail in Chapter 3.

D. The ϕ-Index

The ϕ-index represents a uniform loss rate and is used frequently in the generation of unit hydrographs from rainfall and runoff data. As shown in the example illustrated in Fig. 9.11, volumes of measured rainfall and runoff are compared. In the example, a runoff volume of 5.5 m^3/s-hrs is converted to 33 mm

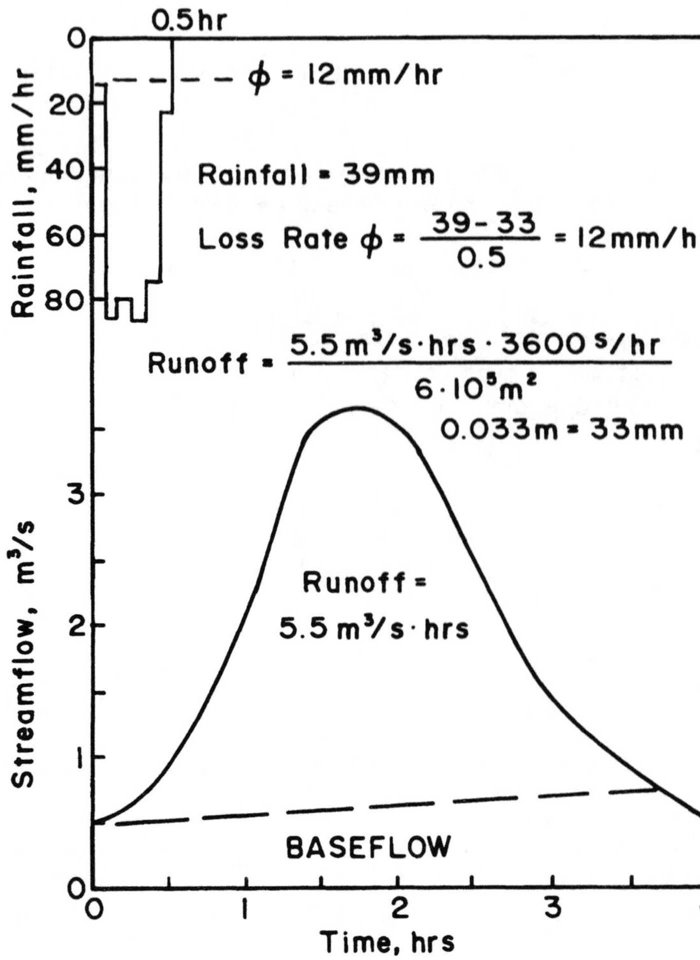

Figure 9.11.—Example Development of a ϕ-Index Loss Rate.

of runoff from a 0.6 km^2 drainage basin. The storm event causing the runoff had a rainfall total of 39 mm, which fell reasonably uniformly over a time span of 30 minutes. Comparing rainfall and runoff, it appears that 6 mm were lost. The total loss is then assumed to be distributed uniformly over the half-hour as a ϕ-index of $6/0.5 = 12$ mm/hr. This ϕ-index, or uniform loss rate, can then be applied to other historic or synthetic storm events to construct a hyetograph of rainfall excess, which in turn can be used to develop a design hydrograph.

E. The Horton Equation

Horton (1933) proposed the exponential decay equation

$$f = f_c + (f_o - f_c)e^{-kt} \tag{9.15}$$

in which f is the potential infiltration rate at any time, t, subject to the availability of ponded water (cm/hr or inches/hr), f_o and f_c are the initial and final infiltration rates, respectively (cm/hr or inches/hr), and k is the exponential decay coefficient (1/hr). Equation 9.15 is also illustrated in Fig. 9.12.

The main problem with empirical equations like Horton's is the general lack of experimentally-determined values of the parameters. Among the most detailed listings of Horton parameter values obtained by calibration is that by Rawls et al. (1976), shown in Table 9.11. Chapter 3 has additional parameter ranges for the Horton infiltration equation.

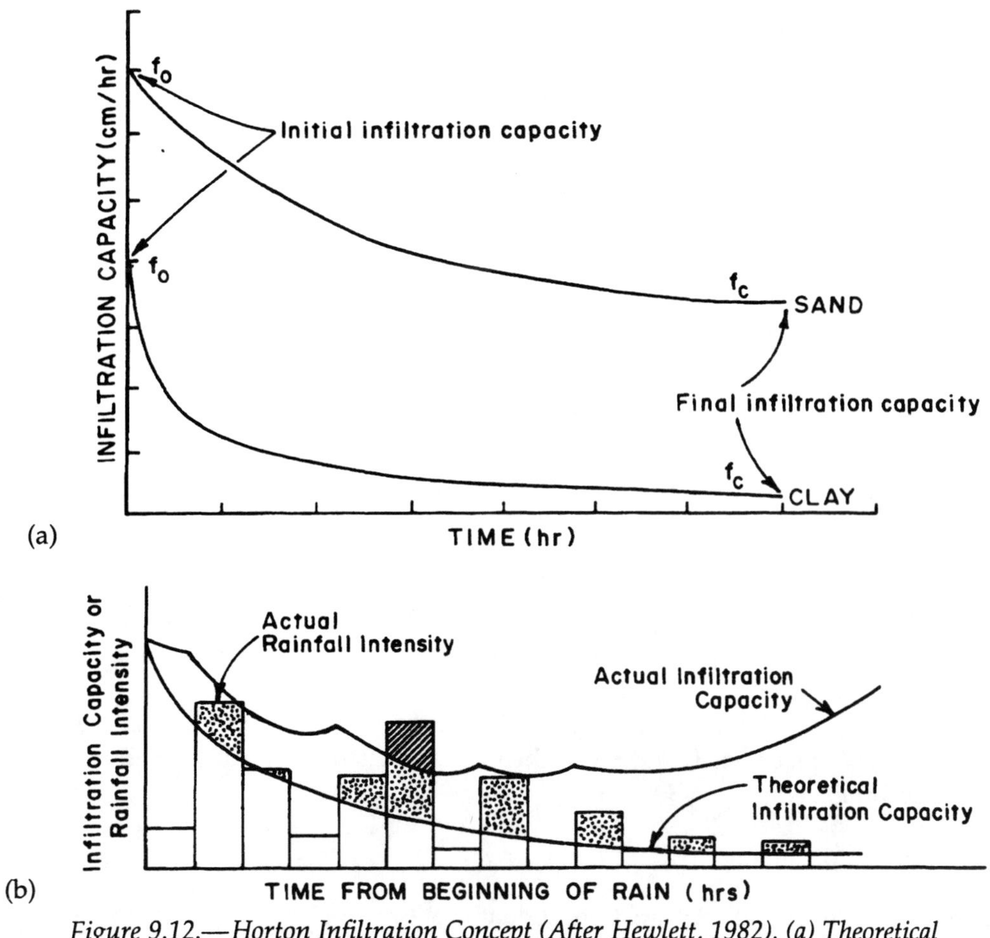

Figure 9.12.—Horton Infiltration Concept (After Hewlett, 1982). (a) Theoretical Capacity Curves. (b) Theoretical Capacity Curves Modified by Actual Rainfall.

TABLE 9.11. Typical Values of the Parameters of f_c, f_o, and k of the Horton Model (after Rawls et al., 1976).

	f_c		f_o		k
Soil type	(in/hr)	(cm/hr)	(in/hr)	(cm/hr)	(hr^{-1})
Alpha loamy sand	1.40	3.56	19.00	48.3	32.29
Carnegie sandy loam	1.77	4.50	14.77	37.5	19.64
Cowarts loamy sand	1.95	4.95	15.28	38.8	10.65
Dothan loamy sand	2.63	6.68	3.47	8.8	1.40
Fuquay pebbly loamy sand	2.42	6.15	6.24	15.8	4.70
Leefield loamy sand	1.73	4.39	11.34	28.8	7.40
Robersdale loamy sand	1.18	3.00	12.41	31.5	21.75
Stilson loamy sand	1.55	3.94	8.11	20.6	6.55
Tooup sand	1.63	4.14	9.67	24.6	7.28

Aside from the difficulty of obtaining reliable values for the three parameters, f_c, f_o, and k, the original Horton equation has two weaknesses: it expresses infiltration capacity as a function of time and it has no provision for the recovery of soil storage capacity during dry periods. Although the infiltration rate decreases with time if the water supply exceeds the infiltration rates during a storm event, time in itself is not a physical process that affects infiltration capacity. Infiltration is dependent on soil moisture, and as water drains down to lower strata there is a recovery of the soil's infiltration capacity. Since f_c is the equilibrium infiltration rate reached eventually after a long rainfall, it can be considered equal to the deep percolation rate. Thus, the soil water storage accumulation rate is expressed in the form:

$$\frac{dS}{dt} = f - f_c = (f_o - f_c)\, e^{-kt} . \tag{9.16}$$

The soil water in storage, S, at any time is

$$S = \frac{f_o - f_c}{k}(1 - e^{-kt}) . \tag{9.17}$$

As t approaches infinity, the soil becomes saturated and the maximum soil storage capacity is

$$S_c = \frac{f_o - f_c}{k} . \tag{9.18}$$

Combining Equations 9.16 and 9.17, the potential infiltration rate can be expressed as

$$f = f_o - kS . \tag{9.19}$$

Equation 9.19 has the advantage over Equation 9.15 in that the infiltration rate is a function of soil water storage instead of time. As long as the water available on the ground surface is sufficient to supply the potential infiltration rate, the two equations yield the same result. When available water depth, y (rainfall and surface ponding), is lower than the potential infiltration, the potential f is reduced to an effective rate $f_e = y/\Delta t$. The quantities f_e and S can then be computed in suitably small time steps, as demonstrated in the example below.

F. Modified Horton Example

The conventional and modified Horton equations, Equations 9.15 and 9.19, respectively, were applied to a storm consisting of two symmetrical hyetographs separated by 20 minutes of zero rainfall. As shown in Table 9.12, the chosen infiltration parameters were f_o = 80 mm/hr, f_c = 20 mm/hr, and k = 0.8 hr^{-1}.

TABLE 9.12. Infiltration Computed by Modified Horton Equation.

Time min. (1)	Horton infiltration mm/h (2)	Rain intensity mm/h (3)	Potential infiltration mm/h (4)	Effective infiltration mm/h (5)	Soil moisture mm (6)	Ponded depth mm (7)
0	80.00	0.00	80.00	0.00	0.00	0.00
10	72.51	50.00	78.80	50.00	1.50	0.00
20	65.96	100.00	72.33	72.33	9.58	1.08
30	60.22	100.00	65.76	65.76	17.80	6.20
40	55.20	50.00	60.01	60.01	24.98	8.18
50	50.81	0.00	54.99	54.99	31.27	2.73
60	46.96	0.00	53.61	0.00	32.99	0.00
70	43.59	0.00	53.61	0.00	32.99	0.00
80	40.65	50.00	52.41	50.00	34.49	0.00
90	38.07	100.00	48.33	48.33	39.58	4.08
100	35.82	100.00	44.78	44.78	44.03	12.96
110	33.84	50.00	41.66	41.66	47.92	18.24
120	32.11	0.00	38.94	38.94	51.32	15.67
130	30.60	0.00	36.56	36.56	54.30	9.36
140	29.28	0.00	34.48	34.48	56.90	3.43
150	28.12	0.00	33.36	0.00	58.30	0.00
160	27.11	0.00	33.36		58.30	0.00

Parameters: f_c = 20.0 mm/h, f_o = 80.0 mm/h, k = 0.8/hr^{-1}

Computations were performed in 1-minute increments and are shown for every 10 minutes. Column 2 in Table 9.12 contains the traditional time-dependent Horton infiltration capacity from Equation 9.15. Column 3 contains the rainfall intensity, while Column 4 contains the potential infiltration rate according to Equation 9.19. Column 5 shows the actual infiltration rate, equal to the available rain and ponded water or to the potential infiltration rate, whichever is less. Column 6 contains the accumulated soil moisture $\Sigma\,(f_e - f_c)\Delta t$. Column 7 contains the water depth, y, left on the ground at any time.

It can be seen from Table 9.12 that the infiltration capacity does not decrease continuously as the traditional Horton equation, but is dependent on the cumulative soil moisture and available rainfall and ponded water. In a real situation, some of the ponded water may run off depending on the slope, width and roughness of the runoff path. The upper soil storage capacity is 75 mm using Equation 9.18.

$$S_c = \frac{f_o - f_c}{K} = \frac{60}{.80} = 75\,\text{mm}$$

In the example, the soil moisture reached a maximum of 58.3 mm, roughly 17 mm short of full saturation. The difference between the original and modified Horton infiltration rates is also shown in Fig. 9.13.

G. Green and Ampt Equation

Green and Ampt (1911) developed the infiltration equation

$$f = K\left(\frac{H_c + L_f}{L_f}\right) \tag{9.20}$$

in which f is the infiltration rate (L/T), K is the hydraulic conductivity of the soil behind the wetting front, in the same units of (L/T) used for f, H_c is the capillary suction head at the wetting front (L), and L_f is the depth from soil surface to wetting front (L).

The infiltration concept envisioned by Green and Ampt is presented in Fig. 9.14. It is based physically on the Darcy equation for flow through porous media, plus the assumptions that: 1) there is a well-de-

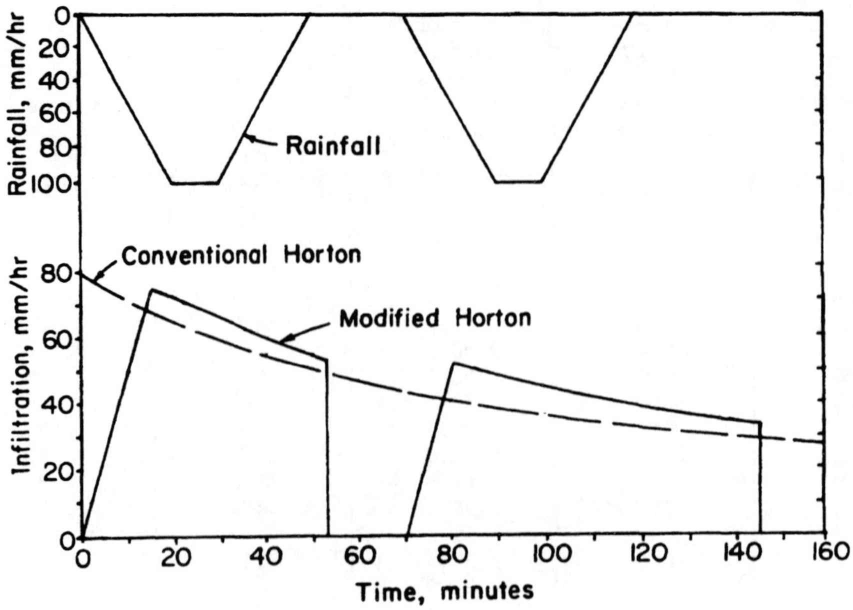

Figure 9.13.—Comparison Between Conventional and Modified Horton Infiltration.

fined wetting front and capillary suction at depth L_f and 2) that the capillary suction head and the hydraulic conductivity behind the wetting front are constant.

The equation has been used successfully at experimental stations, but without detailed field data, it is again virtually impossible to estimate the values of K, H_c, and L_f for any given soil and location. The test data reported by Rawls et al. (1983), reproduced in Table 9.13, are probably the most useful for general application of the Green and Ampt equation.

From Equation 9.20 it can be shown that the depth to the wetting front advances at the rate

$$\frac{dL_f}{dt} = \frac{f}{\theta} = \frac{K\,(H_c + L_f)}{\theta\, L_f} \tag{9.21}$$

in which θ is the effective porosity of the soil. Equation 9.21 can be solved for L_f either by a stepwise computation, as shown in Table 9.14, or by the integration of Equation 9.21.

A minor problem in the stepwise solution of Equation 9.21 is encountered in the initial condition, in which L = 0 and the infiltration rate is momentarily infinite. This problem can be avoided by applying the following integral of Equation 9.21

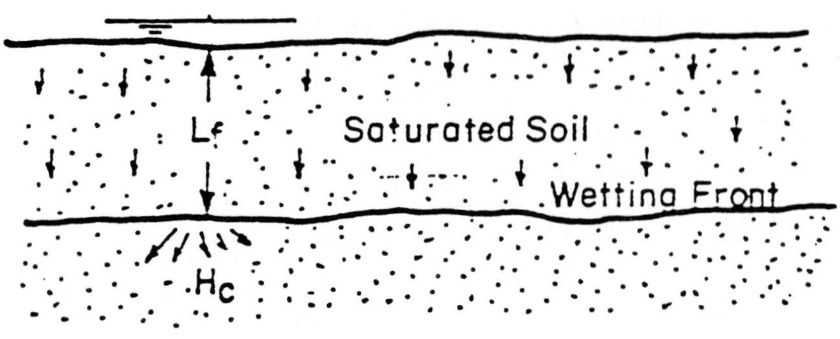

Figure 9.14.—Green and Ampt Infiltration Concept.

TABLE 9.13. Green and Ampt Parameters According to Soil Texture Classes and Horizons (Rawls et al., 1983).

Soil texture class (1)	Horizon (2)	Sample size (3)	Effective porosity, θ, cc/cc (4)	Capillary suction H_c, cm (5)	Hydraulic conductivity K cm/hr (6)
Sand[a]		762	0.417	4.95	11.78
	A	370	0.431	5.34	
	B	185	0.421	6.38	
	C	127	0.408	2.07	
Loamy sand		338	0.401	6.13	2.99
	A	110	0.424	6.01	
	B	49	0.412	4.21	
	C	36	0.385	5.16	
Sandy loam		666	0.412	11.01	1.09
	A	119	0.469	15.24	
	B	219	0.428	8.89	
	C	66	0.389	6.79	
Loam		383	0.434	8.89	0.34
	A	76	0.476	10.01	
	B	67	0.498	6.40	
	C	47	0.382	9.27	
Silt loam		1206	0.486	16.68	0.65
	A	361	0.514	10.91	
	B	267	0.515	7.21	
	C	73	0.460	12.62	
Sandy clay		498	0.330	21.85	0.15
	A	—[b]	—	—	
	B	198	0.330	26.10	
	C	32	0.332	23.90	
Clay loam		366	0.309	20.88	0.10
	A	28	0.430	27.00	
	B	99	0.397	18.52	
	C	55	0.400	15.21	
Silty clay loam		689	0.432	27.30	0.10
	A	65	0.477	13.97	
	B	191	0.441	18.56	
	C	39	0.451	21.54	
Sandy clay		45	0.321	23.90	0.06
	A	—[b]	—	—	
	B	23	0.335	36.74	
	C	—[b]	—	—	
Silty clay		127	0.423	29.22	0.05
	A	—[b]	—	—	
	B	38	0.424	30.66	
	C	21	0.416	45.65	
Clay		291	0.385	31.63	0.03
	A	—[b]	—	—	
	B	70	0.412	27.72	
	C	23	0.419	54.65	

[a]Values for the texture class
[b]Insufficient sample size

$$t = \frac{\theta}{K}\left[L_f - H_c \ln\left(1 - \frac{L_f}{H_c}\right)\right] \tag{9.22}$$

to a chosen small value of L_f and subsequently solving Equation 9.21 in small time steps. This procedure is followed in the example below. For the values of K = 3 cm/hr, H_c = 20 cm, and θ = 0.2, an original wetting front depth of L_f = 3 cm was chosen. The time to reach this depth is

TABLE 9.14. Computed Green and Ampt Infiltration Rates and Depths to the Wetting Front.

Time min.	f cm/hr	L_f cm
0.82	23.00	3.00
2.0	15.57	4.85
3.0	13.05	6.03
4.0	11.59	7.05
6.0	9.88	8.82
8.0	8.84	10.37
10.0	8.13	11.77
12.0	7.61	13.08
16.0	6.89	15.49
20.0	6.40	17.70
30.0	5.66	22.69
40.0	5.21	27.20
50.0	4.92	31.41
60.0	4.70	35.41

Parameters: K = 3.0 cm/hr, H_c = 20 cm, θ = 0.2

$$t = \frac{0.2}{3}\left[3 - 20 \ln\left(1 - \frac{3}{20}\right)\right] = 0.014\ hrs = 0.82 \text{ min.}$$

The computations proceed as shown in Table 9.14, starting with these values.

As in the Horton equation, the validity of Equation 9.21 as well as the computations in Table 9.14 are based on the assumption that sufficient rain is available to satisfy the infiltration capacity f. Another assumption is that the parameters H_c, K, and θ remain constant over the total depth of interest. Frequently the soil will consist of various layers of thickness L with different parameters. If these layers are labeled i = 1, 2, 3 . . . j, with j denoting the deepest layer, the total head in Equation 9.20 can be expressed as

$$\Sigma L_i + H_{cj} = f\, \Sigma \frac{L_i}{K_i} \tag{9.23}$$

and the infiltration capacity is

$$f = \frac{\Sigma L_i + H_{cj}}{\Sigma \frac{L_i}{K_i}} \tag{9.24}$$

As in the previous example, an initial infiltration time can be computed using Equation 9.22, followed by a stepwise computation of f and L using Equation 9.24.

Many other infiltration equations have been discussed in Chapter 3, including a summary of current rainfall–runoff models and their rainfall excess components. The purpose of this section has been to identify the models that are used most frequently in the urban environment and to provide a few simple example calculations.

IV. URBAN RUNOFF ESTIMATING METHODS

The purpose of this section is to describe the conceptual basis for several urban runoff methods currently in widespread use in the United States. The section begins with a brief discussion of the kinematic flow equations for overland and channel/pipe flow, which serve as the physical–mathematical bases for the major urban storm water models in general use for larger urban systems. This is followed by the Rational Formula (and its variations) which is widely used for sub-division level drainage design and small urban storm sewer systems. The Soil Conservation Service dimensionless unit hydrograph is also introduced because it is representative of current United States practice for small and intermediate watersheds which are uncomplicated by extensive sewer/channel networks. This section includes a brief discussion of the time of concentration calculation which is an inherent feature of nearly all simplified runoff procedures, such as the Rational Method and the SCS unit hydrograph. It concludes with a discussion of storage routing through stormwater detention facilities. The next section on typical urban drainage design calculations provides a working example of the basic steps in the Rational Method for a 5.67 hectare (14 acre) sub-division, including the sizing of storm sewers and the design of a stormwater detention facility by storage routing.

A. Overland Flow Routing by Kinematic Wave Technique

The conceptual basis for overland flow routing in the urban watershed is illustrated in Fig. 9.15 where a complex basin geometry and topology in the physical system have been replaced by uniform overland flow planes and simple conduits. The rainfall–runoff process in the physical system has been reduced to a level that can be addressed by a set of one-dimensional equations, known as the kinematic approximation. These equations can be applied to both the overland flow and to the simple channel/pipe elements shown in Fig. 9.15. They are solved easily by computer and form the basis of the major urban stormwater models in common use today. Rovey and Woolhiser (1977) have further applied the kinematic cascade to a 67 hectare (165 acre) urban watershed in the Denver area, with individual planes in the cascade representing roof top, lawn, street, and parking lot surfaces.

The kinematic wave procedure for generating subarea outflows by routing of overland runoff has long been recognized as a valid approach in the urban watershed. It has a firm basis in physical process modeling which is developed from the one-dimensional mass conservation equation and a uniform flow equation. The latter may be solved for either turbulent or laminar flow and has the parametric form:

$$V = \alpha Y^{m-1} \tag{9.25}$$

where V is average flow velocity (m/s), Y is depth of flow (m), and α and m are friction parameters.

The two friction parameters, α and m, depend on the specific friction relationship. In laminar flow, the following expression may be used:

$$\alpha = \frac{gS}{3v} \tag{9.26}$$

where g is a gravitational constant (m/s^2), S is overland slope (m/m), ν is kinematic viscosity (m^2/s), and m = 3. For turbulent flow, which is widely assumed, the Manning formula may be used:

$$\alpha = \frac{1}{n} S^{0.5} \tag{9.27}$$

with n = overland resistance coefficient and m = 1.667.

Typical overland flow resistance coefficients have been reported by USACE (1990) and by Engman (1986). These are listed in Table 9.15. Recent work on overland flow resistance on textured surfaces under simulated rainfall is reported by Kibler and Reed (1992). The working relationships for kinematic overland flow depth are obtained by the characteristic method as:

$$Y_1 = Y_o + i\,(t_1 - t_o) \tag{9.28}$$

$$X_1 - X_o = \frac{\alpha}{i}\{[Y_o + i\,(t_1 - t_o)]^m - Y_o^m\} \tag{9.29}$$

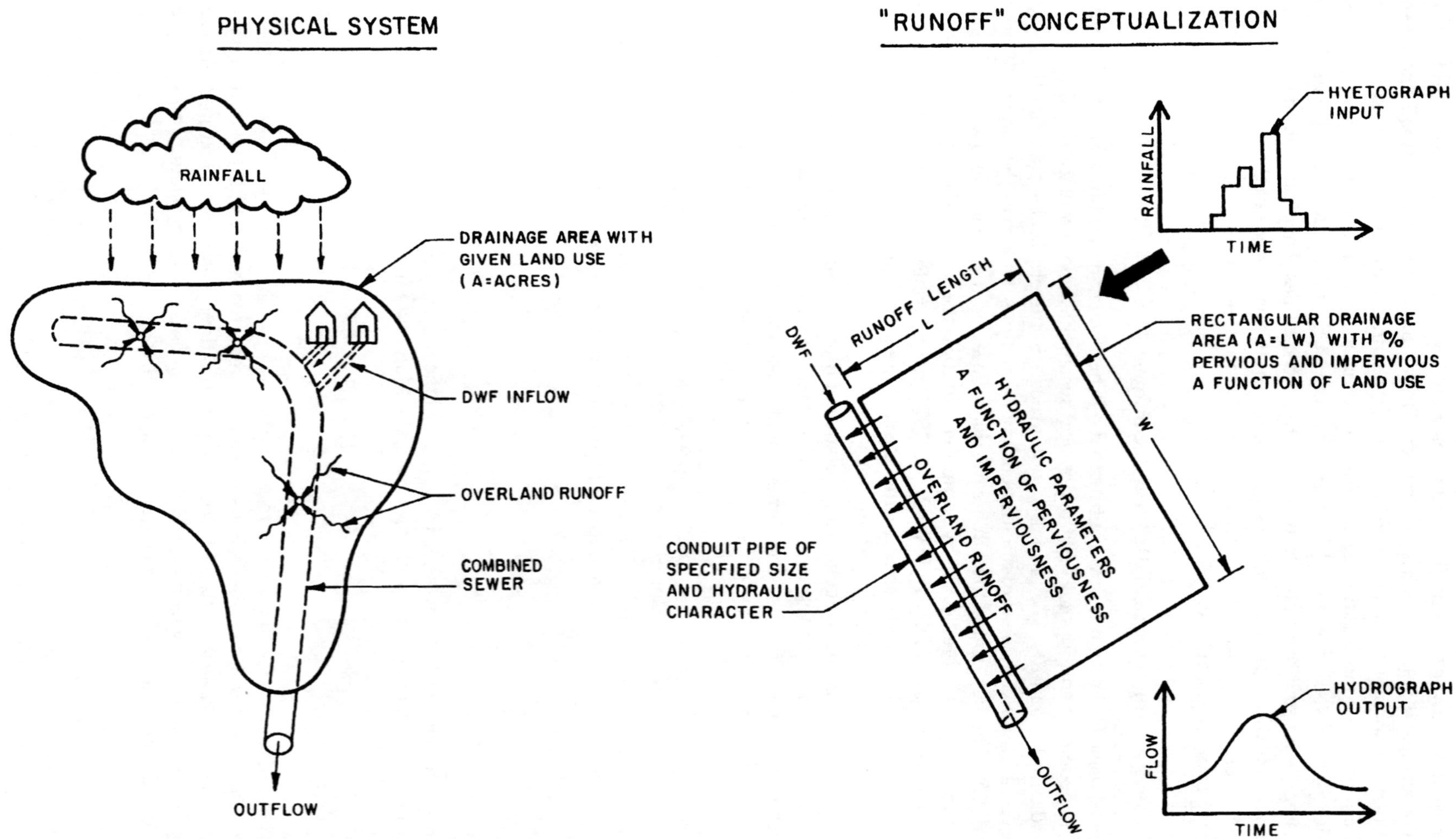

Figure 9.15.—Hydraulic Conceptualization of an Urban Drainage Area (After Roesner, 1982).

TABLE 9.15. Typical Overland Flow Resistance Factors. Compiled by Hydrologic Engineering Center (USACE, 1990).

Surface	N value
Asphalt/Concrete	0.05–0.15
Bare Packed Soil Free of Stone	0.10
Fallow—No Residue	0.008–0.012
Conventional Tillage—No Residue	0.06–0.12
Conventional Tillage—With Residue	0.16–0.22
Chisel Plow—No Residue	0.06–0.12
Chisel Plow—With Residue	0.10–0.16
Fall Disking—With Residue	0.30–0.50
No Till—No Residue	0.04–0.10
No Till (20–40% residue cover)	0.07–0.17
No Till (60–100% residue cover)	0.17–0.47
Sparse Rangeland with Debris:	
0% Cover	0.09–0.34
20% Cover	0.05–0.25
Sparse Vegetation	0.05–0.13
Short Grass Prairie	0.10–0.20
Poor Grass Cover, Moderately Rough	0.30
Bare Surface	
Light Turf	0.20
Average Grass Cover	0.40
Dense Turf	0.17–0.80
Dense Grass	0.17–0.30
Bermuda Grass	0.30–0.48
Dense Shrubbery and Forest Litter	0.40

Note: N is equivalent to n in kinematic overland flow equations.

where Y_1 and Y_o are depths along the flow path defined for a single characteristic at distances X_1 and X_o, respectively; t_1 and t_o are times at the beginning and end of a time step along a characteristic; i is the rainfall excess rate during time $t_1 - t_o$; and m and α were previously defined. Discharge is then computed as $Q = \alpha Y^m$.

Equations 9.27 to 9.29 are derived by substituting Equation 9.25 in the one-dimensional form of the mass conservation equation for overland or channel flow. The fundamental assumption of kinematic flow is that friction and gravity forces are equal and that they dominate the flow itself. Therefore, a uniform flow equation is sufficient to represent the momentum equation. Hence, the overland slope, S, is assumed equal to the energy slope. This means that backwater effects cannot be simulated and all flow disturbances are assumed to move downstream. This assumption has been found to be very reasonable for overland flow. Many cases of channel/pipe flow in urban drainage systems can also be analyzed under the assumption of kinematic flow; however, the presence of backwater will affect the accuracy of the computations.

The great advantage of the kinematic wave method is that it is physically-based while providing a fairly simple solution procedure. The characteristic solution of the kinematic equations in particular provides a near-analytical solution, and therefore is often used as a bench-mark for other solution methods. The success of the kinematic method depends on a number of factors including: rainfall excess determination, overland flow friction relationship (laminar or turbulent), and the numerical solution technique applied to the basic equations. The characteristic method has been used extensively because of its ability to produce a stable solution to the kinematic flow equations.

As a simple example, three numerical schemes for solving the one-dimensional kinematic overland flow equations are compared by application to a basic building block in urban hydrology—the asphalt parking lot. The three methods of solution are: characteristic method, Newton–Raphson as used in EPA SWMM Version 4.05, and a fixed step version of the explicit forward-in-time finite difference scheme

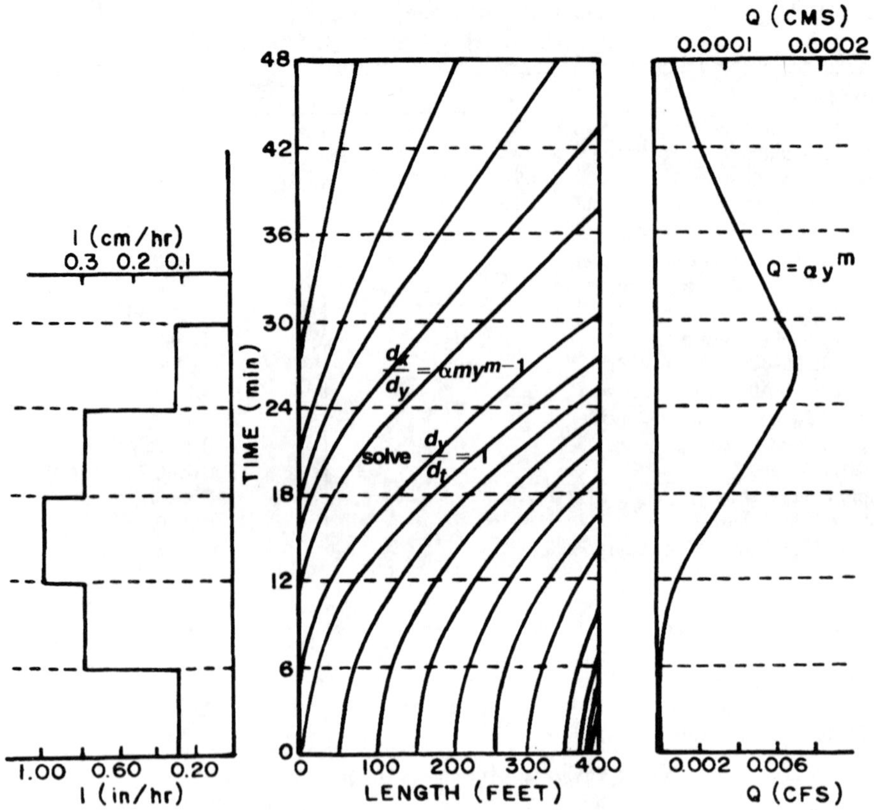

Figure 9.16.—Characteristic Solution for Outflow Hydrograph.

used in HEC–1 (USACE, 1990). In this instance, the asphalt surface is 122 m (400 feet) long, slopes at 0.0025 m/m and has a Manning, n, of 0.025. The surface is assumed to be totally impervious and is a modified version of the problem developed by Overton and Meadows (1976). The rainfall distribution and the characteristic paths are shown in Fig. 9.16.

The outflow hydrograph obtained by tracking characteristics across the 400-foot surface is shown in Fig. 9.17, together with the outflow hydrographs generated by applying the Newton scheme in EPA SWMM and the FD method in HEC–1. There is a significant loss in the peak flow region caused by the very long distance step of 122 m (400 feet) in the latter two methods. This is improved dramatically in the case of the HEC–1 solution by shortening the distance step to 15.2 m (50 feet) and in fact the 1987 and 1990 versions of HEC–1 operate with a variable time step to converge on the characteristic solution.

Clearly, different model assumptions have a major impact on the outflow hydrograph from this very simple flow surface. For this reason, it is important that engineers learn to distinguish the effects of fundamental model assumptions on the one hand from the effects of physical properties, such as impervious fraction, land slope, and friction on the other. Otherwise, the hydrologic significance of land-use parameters is masked by the modeling artifacts, thereby making model calibration extremely difficult.

While an attempt is made here to demonstrate the application of commonly used models, such as HEC–1 and EPA SWMM, no conclusions can be drawn regarding their overall applicability to a given design/analysis problem. These conclusions require the judgment of an experienced modeler who understands the respective hydraulic/hydrologic limitations of each model.

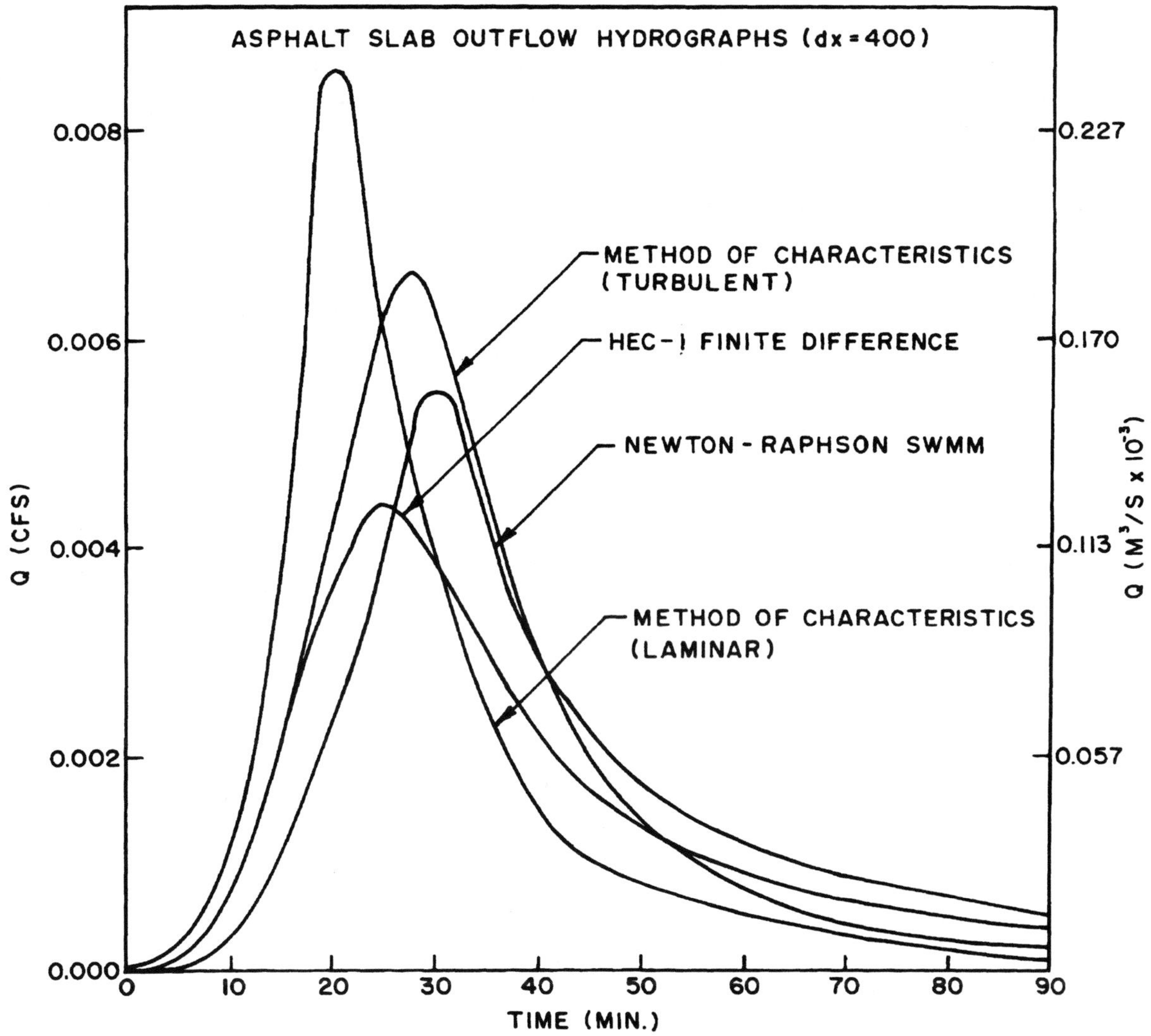

Figure 9.17.—Outflow Hydrographs by Different Methods for Simple Asphalt Surface.

B. Overview of the Rational Formula

The Rational Formula has served as the basis for United States storm drain design practice since the turn of the twentieth century. It is essentially a peak discharge method based on the following formula:

$$Q_T = kCi_TA \tag{9.30}$$

where Q_T is the peak flow rate in m^3/s for return interval T years, C is the runoff coefficient dependent on land use, i_T is the design rainfall intensity in cm per hour for return period of T years and duration equal to the time of concentration for the basin, A is the drainage area in hectares, and k = 0.0278, the number of m^3/s in one hectare-cm/hour.

The underlying principle of the Rational Method is that under steady rainfall intensity, maximum discharge will occur at a basin outlet at a time when the entire area above the outlet is contributing runoff. This is a time commonly known as the time of concentration, t_c, and is defined as the time required for runoff to travel from the most distant point hydraulically in the basin (in time) to the outlet.

Other key assumptions are that: 1) the frequency or return period of the computed peak flow is the same as that for the rainfall intensity, i and 2) the rainfall intensity i is constant over a storm duration and spatially uniform for the area under analysis. It is further assumed that the necessary basin characteristics can be identified and that the runoff coefficient does not vary during a storm. The limits of applicability for the Rational Method traditionally have been kept at small urban basins less than 100 hectares (247 acres). In larger basins, the sewer or channel system is more complex and usually requires a full-hydrograph method involving the analysis of flow routing and channel storage effects, as well as possible backwater effects. Steps in storm sewer design by the Rational Formula are explained by example calculation in the next section, Typical Urban Drainage Design Calculations.

C. Modified Rational Method

An interesting modification of the Rational Method that has particular relevance for the design of detention storage facilities for areas of 8–12 ha (20–30 acres) or less is the trapezoidal hydrograph shown in Fig. 9.18. The trapezoidal hydrographs in Fig. 9.18 were obtained by setting rising and receding limbs equal to t_c and computing the peak discharge by the Rational Method. A trapezoidal hydrograph is then constructed such that the peak flow has a duration equal to the rainfall duration minus t_c. The process is illustrated in Fig. 9.18 for three rainfalls having durations of t_c, $2t_c$, and $3t_c$, respectively. The volume of required detention storage is the area between the trapezoidal inflow hydrograph and the outflow hydrograph. The latter is constructed from some specified maximum release from the undeveloped area, under the assumption that it will be reached on a rising limb that intersects the recession side of the corresponding inflow hydrograph—as in a normal reservoir routing operation. The pond volume can then be computed as:

$$V_p = Q_p \cdot D - Q_o\left[\frac{D + t_c}{2}\right] \tag{9.31}$$

where D is the duration of rainfall, Q_p is the runoff peak for that duration, and Q_o is the maximum allowable discharge from the area.

The major difficulty with pond sizing by the modified rational method is that the critical rainfall duration producing the required storage volume is never known in advance. Rarely does it coincide with the storm of t_c duration. Therefore, an iterative process is required.

As an example, assume a drainage area of 40.5 ha (100 acres), a runoff coefficient, C, of 0.5 for developed conditions, and a t_c of 30 minutes. The maximum allowable release is 1.42 m^3/s (50 cfs) and the 25-year 1-hour rainfall depth is 5.08 cm (2.0 inches). Using the Yarnell intensity–duration–frequency relationships in a previous section, we can determine the Rational Method peaks for a series of rainfalls,

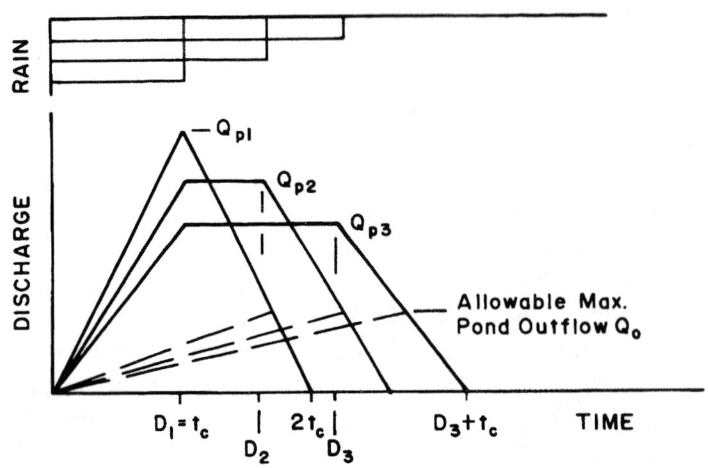

Figure 9.18.—Modified Rational Formula Hydrographs.

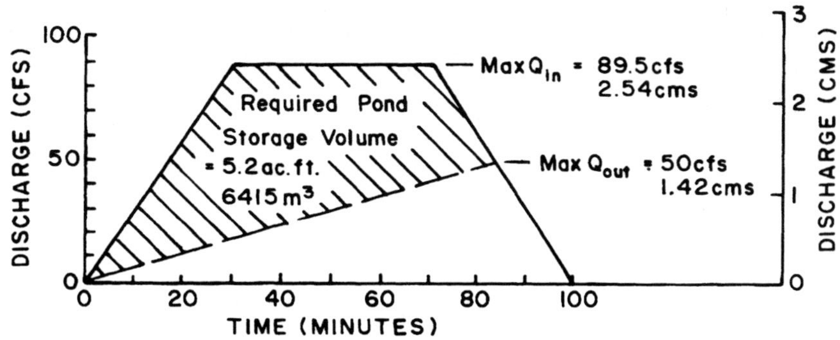

Figure 9.19.—Preliminary Estimate of Required Detention Pond Volume.

beginning with 30 minutes and running through 75 minutes in increments of 5 minutes. The computations of required storage in this incremental duration process are shown in Table 9.16. It can be seen clearly that the storm producing the maximum 25-year discharge peak is not the one that produces the maximum storage requirement. In fact, the 70-minute storm produces the controlling storage volume of 6415 m^3 (5.2 acre-feet), and has an inflow peak of only 2.54 m^3/s (89.5 cfs). The storage volume requirement is shown graphically in Fig. 9.19.

Aron and Kibler (1990) have developed an analytical technique for locating the critical storm duration in the modified rational procedure for sizing small detention structures. This procedure eliminates the incremental method illustrated in Table 9.16. Details on this procedure can be found in the above reference.

As with all simplified detention storage procedures, the authors urge that the modified rational method be considered only as a first approximation. The final pond volume should be established by selecting a trial outlet structure and routing the trapezoidal hydrograph through the detention basin, revising the outlet as needed to meet downstream peak flow targets.

D. Universal Rational Method

Another variation of the rational method that is in widespread use for 4–8 ha (10–20 acre) sites is the so-called "universal" rational method in which a curvilinear hydrograph is fit to the peak computed by the standard rational method. In this version, the peak flow rate occurs at a time equal to three times the

TABLE 9.16. Solution for Modified Rational Method Pond Sizing Problem by Incremental Duration Method (Aron and Kibler, 1990).

Storm duration (min)	Max. runoff rate (m^3/s)	Storm runoff volume (m^3)	Release flow volume (m^3)	Required storage volume (m^3)	Required storage volume (ac-ft)
30	4.42	7964	2549	5416	4.39
35	4.02	8475	2762	5714	4.63
40	3.71	8913	2974	5939	4.81
45	3.43	9293	3186	6107	4.95
50	3.20	9629	3398	6231	5.05
55	3.00	9928	3611	6318	5.12
60	2.83	10,198	3823	6375	5.17
65	2.68	10,442	4036	6407	5.19
70	2.54	10,666	4248	6418	5.20
75	2.42	10,872	4460	6411	5.20

Return period = 25 years; 1-hr rain intensity = 5.08 cm/hr; maximum outflow = 1.42 m^3/s; drainage area = 40.5 ha; time of conc. t_c = 30 min; Rational coef. c = 0.5.

time of concentration, while the base length equals eleven times the time of concentration. Hydrograph ordinates are computed from a design storm hydrograph having the same frequency as the hydrograph peak. The hydrograph ordinates are computed as the product CiA over an interval of rainfall having a duration equal to the time of concentration. The resulting shape approximates the SCS dimensionless unit hydrograph described later. A solution for the problem shown earlier in Table 9.16 is presented in Table 9.17. The Yarnell IDF in Equation 9.6 is applied as before to a 25-year 1-hour storm depth of 5.08 cm (2.0 inches). The required pond volume of 6464 m^3 (5.24 acre-feet) is almost identical to Table 9.16. It is computed as the area above 1.42 m^3/s (50 cfs) in the pond inflow hydrograph in Table 9.17, assuming a rising outflow hydrograph that intersects the inflow hydrograph at 1.42 m^3/s (50 cfs), in accordance with storage routing principles.

The advantage of this method is that it allows sizing of a detention facility using a hydrograph in which the peak corresponds to that used in the design of the upper-lying storm drain system. This is not the case in the modified rational method described previously because of the iterative nature of the method in searching for the rainfall duration producing a maximum storage requirement. The universal hydrograph can of course be routed through detention facilities for final sizing analysis.

E. Concluding Comment on Rational Method

The deficiencies of the Rational Method are well-known and have been described by McPherson (1969) and Rossmiller (1980). The method is intended primarily for small urban basins where rainfall intensities can be assumed spatially uniform throughout the time of concentration. It is highly sensitive to t_c, the time of concentration, and also to the runoff coefficient. Calibration of the Rational Method has been limited to the runoff coefficient for small parking lots instrumented at the Johns Hopkins University (Schaake, 1967). Most C-tables have been developed for the frequent events associated with storm sewer design. Since the C-value should increase with storm magnitude, the use of a C-table based on the 10-year storm may underestimate the discharge peak for the 100-year event. Finally, the Rational Method is a design storm method with all of its inherent limitations. Nonetheless, the Rational Method, properly applied, remains an extremely useful tool for the drainage engineer faced with design of simplified storm drain systems. Further design details are illustrated in later sections of this chapter.

F. Synthetic Unit Hydrographs for Urban Watersheds

Since the early work of Snyder (1938) on synthetic unit hydrographs for ungaged areas, several investigators have developed empirical relationships between UH parameters and physical watershed

TABLE 9.17. Inflow Hydrograph for Universal Rational Method Pond Sizing Problem.

Time, min.	Ave. int. cm/hr	Total rain, cm	Incr. rain cm	Incr. int. cm/hr	Re-arranged incr. int. cm/hr	Pond inflow	
						m^3/s	ft^3/s
0	0.00	—	—	0.00	0.00	0.00	0.0
30	7.92	3.96	3.96	7.92	0.46	0.26	9.3
60	5.08	5.08	1.12	2.24	1.22	0.67	23.6
90	3.78	5.69	0.61	1.22	7.92	4.42	156.2
120	3.05	6.10	0.41	0.82	2.24	1.24	43.8
150	2.56	6.40	0.30	0.60	0.82	0.45	15.8
180	2.21	6.63	0.23	0.46	0.60	0.33	11.8
210	1.96	6.81	0.18	0.37	0.37	0.22	7.7
240	1.75	6.99	0.18	0.36	0.36	0.19	6.6
270	1.57	7.14	0.15	0.30	0.30	0.16	5.8
300	1.45	7.26	0.12	0.25	0.25	0.14	5.1

Return period = 25 years; 1-hour rain intensity = 5.08 cm/hr; ave. intensity obtained from Yarnell equation 9.6, with a and b coefficients from Aron equations; re-arranged incremental intensities placed such that peak occurs at 3 t_c, alternating after that; max pond outflow = 1.42 m^3/s; drainage area = 40.5 ha; t_c = 30 minutes; Rational c = 0.5.

properties. The Colorado urban hydrograph procedure (Wright–McGlaughlin Engineers, 1969) is a good example in which Snyder C_p and C_t coefficients are modified for the urban conditions found in the Denver region.

Because of its geographical coverage, the Espey and Altman (1978) 10-minute unit hydrograph is an important synthetic UH technique for urban basins. It was developed from 41 basins located in Texas (16), Tennessee (1), Mississippi (2), Pennsylvania (1), North Carolina (9), Colorado (2), Kentucky (6), and Indiana (4). These basins range in size from 0.036 to 38.9 km^2 (0.014 to 15.0 mi^2) and have impervious fractions ranging from 2 to 100%. The Espey UH method has been documented in a U.S. Environmental Protection Agency report (Espey and Altman, 1978).

Because of its simplicity, the Soil Conservation Service dimensionless curvilinear UH has been used frequently in urban drainage design studies. It was developed by Victor Mockus of the USDA Soil Conservation Service and first appeared in its National Engineering Handbook in 1964. The corresponding reference is Chapter 16 of Section 4 Hydrology (USDA SCS, 1969; 1985 reprinted). Chapter 6 of this handbook contains discussion of the SCS dimensionless UH.

The SCS dimensionless curvilinear UH has found its way into several well-known hydrograph synthesis packages. The SCS TR–20 watershed modeling program is built around the SCS soil–cover–complex method and includes the dimensionless UH, together with a so-called attenuated kinematic wave channel routing method. The U.S. Army Corps of Engineers hydrologic simulation package HEC–1 has also incorporated the SCS dimensionless UH, in addition to other synthetic UH techniques. While there is no strict upper limit on drainage area size, the SCS UH is best applied to areas smaller than 64.8 km^2 (25 square miles). On the lower end, it is clear the method could be applied to areas that have a t_c less than 1 minute, although it is impractical to do so, given that rainfall is usually not available in less than 5-minute intervals. As with all synthetic UH procedures, the SCS UH is tedious to apply to 24-hour storm events on small watersheds that are frequently encountered in urban drainage facility design. Nevertheless, the availability of software for the SCS UH method makes it an attractive alternative when it is desirable to avoid the approximations of the Rational Method and its modifications.

One of the important uses of the SCS dimensionless UH and TR–20 in particular has been to serve as the parent model for an SCS procedure known as TR–55 (USDA SCS, 1986). This method incorporates runoff generation and channel routing operations in a set of tables and permits the computation of a full hydrograph. The particular method is called the tabular hydrograph method in TR–55. While it is simpler than TR–20, TR–55 nevertheless requires input data for the drainage area that is similar to that for the parent model. With the advent of improved software for hydrograph generation and channel/pipe routing, the future of the TR–55 tabular hydrograph and other highly simplified methods is uncertain. The interested reader is referred to Chapter 16, NEH 4 (USDA SCS, 1969; reprinted 1985) and Technical Release 55 (USDA SCS, 1986) for full details on the dimensionless unit hydrograph and tabular hydrograph, respectively.

G. Time of Concentration in the Urban Watershed

The preceding discussion of SCS and Rational Methods illustrates the sensitivity of most present-day urban runoff procedures to t_c estimation. This timing parameter appears in one form or another in virtually all current urban runoff models and methods. In terms of impact on the outflow hydrograph, it is probably the most critical, after impervious fraction and rainfall distribution effects are considered. Yet t_c remains an elusive hydrologic variable because it is so difficult to measure. At present there are perhaps a dozen different t_c estimating formulas in use. The more important ones in the urban environment are summarized in Table 9.18. Note that they appear here in customary United States units to remain consistent with the original documentation. Each was developed under specific laboratory or field conditions and should only be used in those portions of the urban watershed where it applies. Limited testing of t_c formulas has been reported by Kibler et al. (1981), McCuen et al. (1984), and Papadakis and Kazan (1987). Other t_c formulas in general use are discussed in Chapter 6.

TABLE 9.18. Summary of Time of Concentration Methods.

Method and date	Formula for T_C (minutes)	Remarks
Kirpich (1940)	$T_c = 0.0078\left(\frac{L^2}{S}\right)^{0.385}$ L = length of channel/ditch from headwater to outlet, feet S = average gully slope, ft/ft.	Developed from SCS data for 7 rural basins in Tennessee with well-defined channel and steep slopes (3% to 10%). For overland flow on concrete or asphalt surfaces multiply T_c by 0.4. For concrete channels multiply by 0.2. No adjustment for overland flow on bare soil or flow in roadside ditches. Reference: Civil Engineering, Vol. 10, No. 6, June 1940.
Izzard (1946)	$T_C = \frac{41.025\,(0.0007\,i + c)\,L^{0.33}}{S^{0.333}\,i^{0.667}}$ i = rainfall intensity, in/hr. c = retardance coef. L = length of flow path, feet S = slope of flow path, ft/ft.	Developed in laboratory experiments by Bureau of Public Roads for overland flow on roadway and turf surface. Values of the retardance coef. range from 0.0070 for very smooth pavement; c = 0.012 for concrete pavement; c = 0.06 for dense turf. Solution is extremely tedious and requires iteration. Product i times 1 should be < 500. Reference: Proc. Highway Research Board Vol. 26, pp. 129–146, 1946.
Fed. Aviation Agency (1970)	$T_C = \frac{1.8\,(1.1 - C)L^{0.50}}{S^{0.333}}$ C = rational method runoff coeff. L = length of overland flow, ft. S = average overland slope, %	Developed from air field drainage data assembled by the Corps of Engineers. Method is intended for use on airfield drainage problems, but has been used frequently for overland flow in urban basins. Reference: "Airport Drainage," Federal Aviation Agency, Dept. of Transportation Advisory Circular, A/C 150–5320–5B, Washington, D.C., 1970).
Kinematic Wave Formulas, Morgali and Linsey(1965)	$T_C = \frac{0.94\,L^{0.6}\,n^{0.6}}{i^{0.4}\,S^{0.3}}$ L = length of overland flow, ft. n = Manning roughness coef. i = rainfall intensity in/hr. S = average overland slope, ft./ft.	Overland flow equation developed from kinematic wave analysis of surface runoff from developed surfaces. Method requires iteration since both i (rainfall intensity) and T_c are unknown. References: Morgali and R. K. Linsley, "Computer Simulation of Overland Flow," J. Hydr. Div. ASCE, Vol. 91(81), May 1965.
Old SCS Lag Equation (1975) USDA SCS 1975	$T_C = \frac{L^{0.8}\left[\frac{1000}{CN} - 9\right]^{0.7}}{190\,S^{0.5}}$	

TABLE 9.18. *(Continued)*

Method and date	Formula for T_c (minutes)	Remarks
	L = hydraulic length of watershed (longest flow path), CN = SCS runoff curve number S = average watershed slope, ft/ft	Equation developed by SCS from agricultural watershed data. It has been adapted to small urban basins under 2000 acres. Found generally good where area is completely paved. For mixed areas, it tends to overestimate. Adjustment factors are applied to correct for channel improvement and impervious area. The equation assumes that $T_c = 1.67 \times$ basin Lag. Reference: Soil Conservation Service Tech. Release No. 55, U.S. Dept. of Agriculture, January 1975.
New SCS Sheet Flow Equation USDA SCS 1975 revised 1986	$T_C = \frac{0.42\,(nL)^{0.8}}{P_2^{0.5}\,S^{0.4}}$	
	L = overland flow length, ft. n = Manning n for sheet flow P_2 = 2 yr. - 24 hr. rainfall, inches S = overland slope, ft/ft.	Equation developed by SCS for Revised TR 55, June 1986.

Kibler et al. (1981) and Kibler and Aron (1983) concluded that the segmental approach to t_c estimation was superior, based on dye travel measurements and hydrograph comparisons in a number of gaged urban basins in Pennsylvania. The segmental approach divides the total flow path into overland, concentrated, and pipe/channel phases and computes a separate flow time for each. The sum of the segmental flow times gives an estimate of the total flow time from the most time-distant point in the contributing drainage area and by definition of the t_c.

The segmental approach has been incorporated in SCS TR 55 (USDA SCS, 1986). It is based on a modified kinematic wave formula for sheet flow (as shown in Table 9.18), a Manning uniform flow assumption for concentrated flow in a gutter or swale, and the Manning formula for open channel/pipe flow. The interested reader is referred to Chapter 3 of TR 55 (USDA SCS, 1986).

H. Example Travel Time Computation

The watershed sketch in Fig. 9.20 is useful in illustrating the steps in t_c computation by a number of methods listed in Table 9.18. The physical properties of the flow path are also given in Fig. 9.20. The t_c computations are summarized in Table 9.19, along with the 25-year flood peaks by Rational Formula. Assuming a single storm sewer at the outfall location, the pipe diameters computed by Manning equation under full flow range in size from 90 to 140 cm (36 to 54 inches). These correspond to a design flow ranging from 3.68 to 7.65 m^3/s (130 to 270 cfs), respectively, once again demonstrating the importance of t_c in storm sewer design.

I. Storage Routing Through Stormwater Detention Ponds

Detention storage facilities are designed to receive the stormwater collected by a drainage system for a development and release it at a specified rate. The primary design objective is thus to determine the necessary storage volume to contain the flood with an outlet structure to release it at some specified nondamaging rate. The design process is by storage routing of the design flood through a trial detention basin and outlet. If infiltration and evaporation from the pond are neglected during the runoff period,

Overland Flow Element

Length = 21.3 m (70 ft)
Slope = 0.037 m/m
Cover is meadow with dense grass
P (2-yr, 24-hour) = 7.62 cm (3.0 inches)

Concentrated Flow Element

Length = 381 m (1250 ft)
Slope = 0.042 m/m
Cover is grass swale

Channel Flow Element

Length = 701 m (2300 ft)
Slope = 0.025 m/m
Manning n = 0.040 natural channel
Wetted section area = 0.929 m^2 (10.0 ft^2)
Wetted perimeter = 2.58 m (8.47 ft)

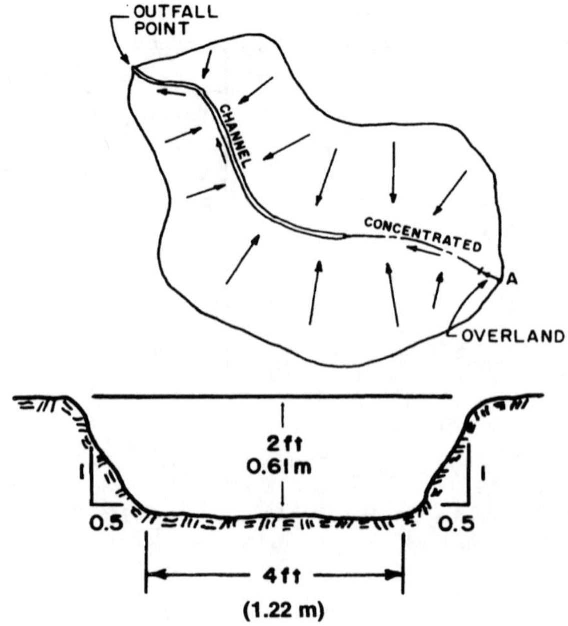

Figure 9.20.—Watershed Flow Paths and Channel Geometry Travel Time (t_c) Example.

the continuity equation for a storage facility may be written for a pond (just as for a major flood control reservoir):

$$I(t) - O(t) = \frac{\Delta S}{\Delta t} \tag{9.32}$$

where I(t) is the inflow to the pond from the sewer network at time t (m^3/s), O(t) is the outflow from the pond into the downstream drainage network at time t (m^3/s), and ΔS is the change in storage (m^3) in the time interval Δt (sec). Equation 9.32 may also be written as:

TABLE 9.19. Summary of Travel Time (t_c) Calculations by Various Methods for Example Watershed (Fig. 9.20).

Method	Overland time, min	Swale time, min	Channel time, min	Total flow time, min	Design rain[7] i_{25}, in/hr	Design rain[7] i_{25}, cm/hr	$Q_{25} = Ci_{25}A$ ft^3/s	$Q_{25} = Ci_{25}A$ m^3/s	Req'd pipe diam.[9] inches	Req'd pipe diam.[9] cm
SCS Seg.[1]	8.7	6.3	5.8	21	4.7	11.94	235	6.66	48	122
FAA[2]	8.3	—	—	20	4.6	11.68	230	6.51	48	122
SCS Lag Eq.[3]	—	—	—	59	2.6	6.60	130	3.68	36	91
SCS Lag Eq.[4]	—	—	—	42	3.3	8.38	165	4.67	42	107
SCS Lag Eq.[5]	2.6	—	—	15	5.4	13.72	270	7.65	54	137
Kirpich[6]	—	—	—	15	5.4	13.72	270	7.65	54	137

[1]Overland time by SCS modified kinematic wave using n = 0.24; concentrated flow time by SCS velocity chart for unpaved surface; channel time by Manning eq. (L/V)
[2]Based on Rational C-factor of 0.25; swale time and channel time by SCS segmental.
[3]SCS Lag eq. assuming one flowpath of 1103 meters (3620 ft.), sloping at 0.040 m/m; average CN = 71.
[4]SCS Lag eq. assuming one flow path of 1103 meters (3620 ft.), sloping at 0.040 m/m; average CN = 82.
[5]SCS Lag eq. applied only to overland path; swale and channel time from SCS segmental.
[6]Kirpich eq. applied to total length of 1103 meters (3620 ft.), sloping at 0.040 m/m.
[7]Design rain for 25-yr. event produces 6.6 cm/hr (2.6 in/hr) over 1-hour in SW VA; use Yarnell eqs. to develop intensities for associated t_c times.
[8]Q_{25} based on area of 80.9 hectares (200 acres) and average C-value of 0.25 for meadow.
[9]Pipe size assumes full flow by Manning eq., with slope of 0.040 and n = 0.015.

$$(I_1 + I_2)\frac{\Delta t}{2} - (O_1 + O_2)\frac{\Delta t}{2} = S_2 - S_1 \tag{9.33}$$

or

$$(I_2 + I_2) + \left(\frac{2S}{\Delta t} - 0\right)_1 = \left(\frac{2S}{\Delta t} + 0\right)_2 \tag{9.34}$$

where subscripts 1 and 2 denote the flows, volumes, and storage at times t_1 and t_2. The inflow hydrograph is determined using runoff procedures outlined earlier. The outflow hydrograph is determined using a storage routing procedure such as the modified Puls or storage-indication method. In order to use this procedure, a trial elevation–discharge–storage relationship is needed.

The storage indication routing operation involves solving for $(2S/\Delta t + 0)_2$ on the right-hand side of Equation 9.34. This term is called the indicated storage, hence the name of the routing method. A second independent relation between $(2S/\Delta t + 0)_2$ and 0_2 is needed to obtain 0_2, the basin outflow at the end of the time step. This relationship is developed from storage–elevation and discharge–elevation data for the trial basin. The elevation–storage relationship is determined by developing preliminary geometry and grading for the storage site, calculating the area at each elevation, then converting the incremental areas to a cumulative storage vs. elevation table.

The elevation–discharge curve is determined by evaluating the hydraulics of a proposed outlet structure. The outlet structure can consist of a weir, orifice, pipe, or a combination of these outlets. Ideally, the composite outlet is designed to reduce the peak rate for the frequent as well as the infrequent design storm. Each of the above outlets and the governing equations are described individually in the following paragraphs.

An orifice is a submerged circular or rectangular opening of a prescribed shape and perimeter through which water flows. The flowrate depends upon the height of water h above the orifice centerline and the type of orifice. The equation used to calculate the flow rate for a free jet is:

$$Q = C_d A\sqrt{2gh} \tag{9.35}$$

where C_d is the discharge coefficient, A is the area of the orifice (m^2), g is the acceleration due to gravity (9.81 m/s^2), and h is the height of water above the orifice center line (m). Fig. 9.21 is a typical circular orifice defining the above variables and showing discharge coefficients for various types of orifices.

A weir is a depression or cutout in a wall or channel through which water flows. Weirs are often used for controlling overflows. Two types of sharp-crested weirs will be discussed: rectangular and triangular. For a typical rectangular weir with dimensions as shown in Fig. 9.22. The equation for the discharge is:

$$Q = C_d \frac{2}{3} \sqrt{2g}\, L_w\, h^{3/2} \tag{9.36}$$

where L_w is the length of the weir (m) and C_d is the discharge coefficient that is approximately 0.6, but should be checked in hydraulic references for special conditions, such as contractions. The head, h, is the height of water (m) measured above the crest at a point upstream of the crest, which is unaffected by draw–down at the weir.

For the triangular weir shown in Fig. 9.23, Equation 9.37 is used to calculate the flowrate.

$$Q = \frac{8}{15} C_d \sqrt{2g} \tan \theta h^{5/2} \tag{9.37}$$

where θ is half the total weir angle, h is the height of water measured above the weir apex (m), and C_d is the discharge coefficient, which is approximately 0.6. Again, standard hydraulic references should be checked for special conditions.

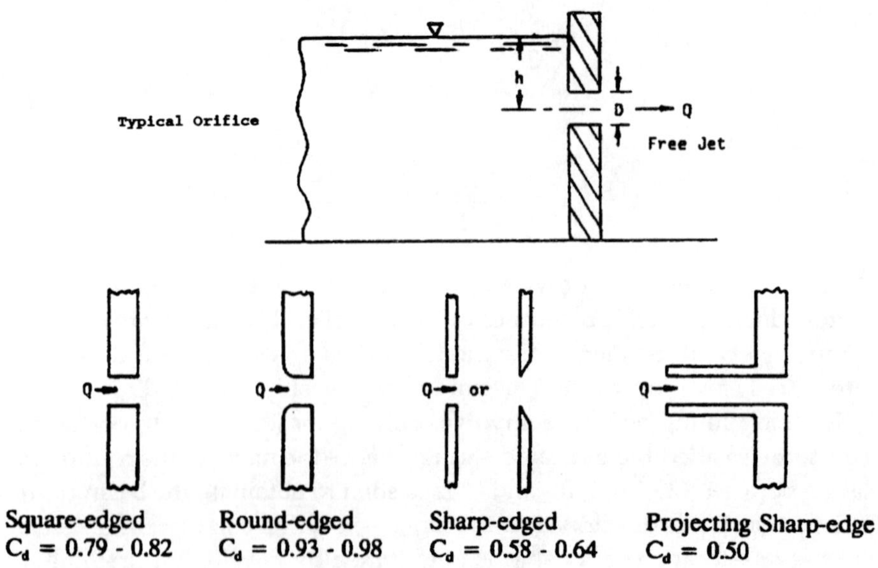

Figure 9.21.—Typical Circular Orifice and Discharge Coefficients C_d for Various Entrance Types (Kuhl, 1977).

For a pipe flowing full but not under pressure, Manning's equation may be used to calculate flowrate. As soon as the depth of water develops above the pipe crown the flowrate must be calculated using a modified form of the Manning equation and must include entrance and exit losses, as well as submergence effects. The reader is referred to U.S. Department of Transportation, Hydraulic Circular No. 5, "Hydraulic Design of Highway Culverts" (FHWA, 1985).

The steps in the routing operation can now be summarized as follows:

1) Select a routing time stop, Δt, such that there is a minimum of six steps on the rising side of the inflow hydrograph, the last step coinciding with the inflow hydrograph peak.
2) Develop a $(2S/\Delta t + 0)$ versus 0 table from storage–elevation and discharge–elevation data assembled for the trial basin and outlet works.
3) Perform routing by solving Equation 9.34 at Δt intervals for the routed outflow, 0_2, at the end of each time step.

The design of a stormwater detention facility is thus one of repetitive analysis, whereby the routed outflow peak is compared against a specified maximum release rate and the outlet structure modified as needed to ensure that the detention basin discharges no more than the specified maximum. A separate $(2S/\Delta t + 0)$ curve and routing table must be developed for each trial basin until the outflow target is

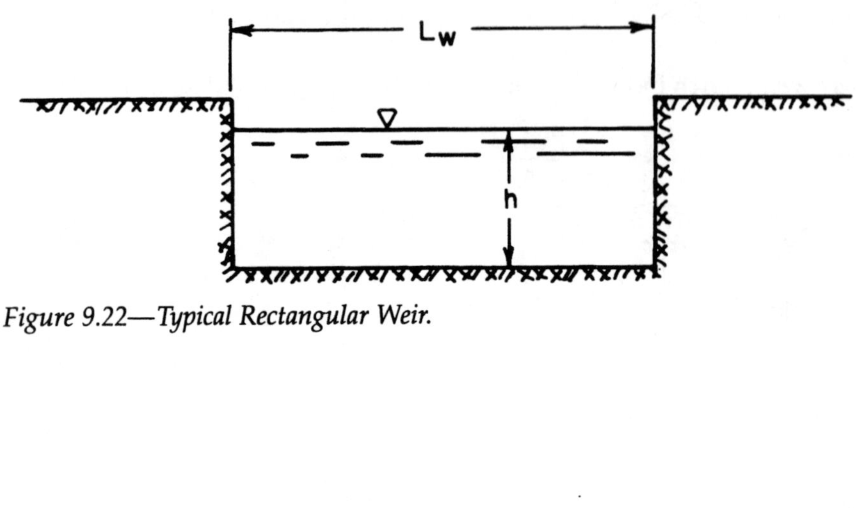

Figure 9.22—Typical Rectangular Weir.

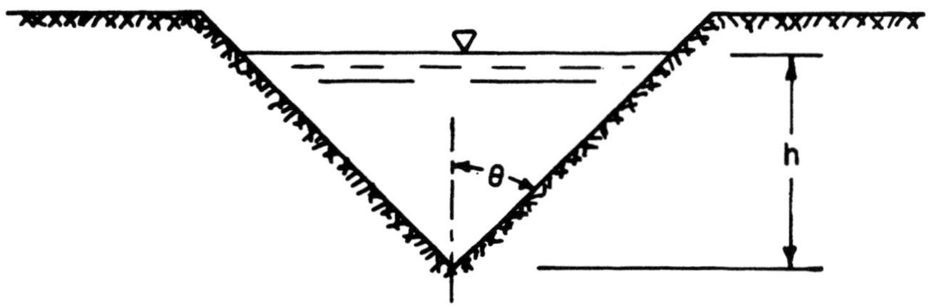

Figure 9.23.—Typical Triangular Weir.

achieved. This is readily accomplished by computer as illustrated in the next section. Details of the storage routing operation by the storage indication method can be found in any hydrology text.

V. TYPICAL URBAN DRAINAGE DESIGN CALCULATIONS

Urban stormwater drainage projects typically consist of a collection and conveyance system designed to carry stormwater runoff so as to minimize property damage and disruption to traffic and a storage facility to reduce downstream flooding. Over the past several decades, many methodologies and computer models have been developed to size these drainage system components. For most subdivision applications and highway drainage projects, the Rational Method remains a commonly accepted procedure. Advanced computer models, on the other hand, are always required for the analysis of complex drainage networks involving multiple sub-areas, backwater, looped sewers, and flow diversion structures.

For the sizing of storage facilities, the SCS procedures discussed earlier are widely used because of their simplicity. They are particularly useful for obtaining initial estimates of detention basin size that can then be analyzed by more exact hydrologic methods. The 1986 revision of TR–55, along with updated procedures for calculating curve numbers and times of concentration, has resulted in a more coherent application of the SCS curve number method for urban hydrology. TR–20 and TR–55 (with a companion routing program) are useful procedures for sizing and evaluating the performance of stormwater storage facilities.

A. Sizing the Collection and Conveyance Systems

The ASCE *Manual of Practice for the Design and Construction of Sanitary and Storm Sewers* (ASCE, 1976) describes some general design considerations for both sanitary and storm sewers. This includes the system layout, material type, allowable depths, acceptable velocities, and a procedure for sizing the pipes. Specific design guidance for storm sewers and detention systems can be found in the ASCE Manual of Engineering Practice No. 77/WEF Manual of Practice FD–20, (ASCE/WEF, 1992). No attempt is made in this handbook to duplicate this information. The reader is referred to these documents for design-related details.

The collection system consists of inlets, catch basins or other appurtenances that allow the entry of stormwater into a conveyance or sewer system. The latter may consist of closed conduits (i.e., storm sewers or tunnels) and open channels (i.e., ditches, swales, or lined channels). The collection and conveyance system may be designed to carry storm runoff from Q_2 to Q_{25}, with Q_{10} being the nominal design basis. In order to size each of these components, a methodology must be used that incorporates hydrologic and hydraulic principles. For a typical urban application, the necessary physical data include: a preliminary roadway layout with low points, a concept grading plan showing the overland flow routes, the pipe material and loading, minimum slopes for grassed and paved areas, and the outlet elevation.

Using this information, a preliminary layout can be prepared that will provide initial estimates of storm sewer invert elevations and reach lengths.

Next, the following data are needed in determining the peak rate of stormwater runoff (regardless of the methodology used): 1) tributary area draining to an inlet or catch basin, 2) percent imperviousness, 3) travel lengths, slopes, and surface characteristics, and 4) rainfall depths and/or intensities for a range of storm durations and frequencies.

For a typical urban drainage system, the size and spacing of inlet grates and curb openings is determined to verify that the runoff can enter the conveyance system without flooding the roadway to an unacceptable depth or encroachment width. The most current references for sizing these components is *Design of Urban Highway Drainage* (FHWA, 1979) and *Design and Construction of Urban Stormwater Management Systems* (ASCE/WEF, 1992).

The Rational Method design calculations proceed from upstream to downstream with a successive determination of pipe slope, pipe crown and invert elevations and hydraulic grade line using Manning's equation. Full pipe flow is usually assumed in the sizing of closed conduits. Once the sizes of the collection and conveyance system are established, the nonsewered part of the system is evaluated using a 50 to 100-year runoff peak to determine the width and slope of channels, swales, and floodways.

B. Rational Method Pre-Design Data

The Rational Method is one of the oldest and most widely used methods employed in the determination of peak discharges from a small developing watershed. The Rational Formula is written simply as Q = CiA, where all terms have been described earlier. The selection of the C-coefficient and travel time are critical to proper use of the Rational Formula.

Values of the C runoff coefficient are given in Table 9.20 for rural areas and Tables 9.21 and 9.22 for urban areas. Table 9.21 presents runoff coefficients for particular types of urban areas and Table 9.22 gives coefficients that can be used to compute a weighted C based on the actual percentage of lawns, streets, roofs, etc.

As mentioned before, the C-coefficient includes many factors such as type of cover, soil types, infiltration, evaporation, evapotranspiration, and any antecedent moisture condition. It has been shown that C actually does not remain constant during the storm duration. Thus the strong dependence on engineering judgment in selecting a runoff coefficient is one of the main weaknesses of the Rational Method (Rossmiller, 1980).

After the C-coefficient, travel time, or t_c, is most important. A basic premise of the Rational formula is that an area will deliver maximum discharge when rainfall duration equals t_c and the entire drainage

TABLE 9.20. Rural Runoff Coefficients (Schwab et al., © 1966. Reprinted by permission of John Wiley & Sons, Inc).

Vegetation and topography	Soil texture: Open sandy loam	Clay and silt loam	Tight clay
WOODLAND			
Flat 0–5% Slope	0.10	0.30	0.40
Rolling 5–10% Slope	0.25	0.35	0.50
Hilly 10–30% Slope	0.30	0.50	0.60
PASTURE			
Flat	0.10	0.30	0.40
Rolling	0.16	0.36	0.55
Hilly	0.22	0.42	0.60
CULTIVATED			
Flat	0.30	0.50	0.60
Rolling	0.40	0.60	0.70
Hilly	0.52	0.72	0.82

TABLE 9.21. Urban Runoff Coefficients for the Rational Method (ASCE WEF, 1992).

Description of area	Runoff coefficients
Business	
Downtown	0.70 to 0.95
Neighborhood	0.50 to 0.70
Residential	
Single-Family	0.30 to 0.50
Multi-Units, Detached	0.40 to 0.60
Multi-Units, Attached	0.60 to 0.75
Residential (Suburban)	0.25 to 0.40
Apartment	0.50 to 0.70
Industrial	
Light	0.50 to 0.80
Heavy	0.60 to 0.90
Parks, Cemeteries	0.10 to 0.25
Playgrounds	0.20 to 0.35
Railroad Yard	0.20 to 0.35
Unimproved	0.10 to 0.30

contributes runoff. Thus, the designer must search for the most time-distant point in the contributing drainage area. This means that the inlet time for each sub-basin must be compared to the travel times from all upstream sub-basins and the longer time selected as the time of concentration. Failure to understand this concept is a common source of error. Inlet times are then added to the flow time in the pipe or channel to determine the travel time at the next point of interest. However, if the local inlet time for this point is greater than the upstream travel time, the local inlet time is used as t_c in the subsequent calculation. The sewer flow time is usually calculated by choosing a pipe or channel configuration and calculating the velocity. The travel time is then found by $t = L/V$, where t is travel time, L is pipe length, and V is the velocity.

C. Steps in Use of Rational Method for Storm Sewer Design

The choice of parameters in the Rational Method is sometimes arbitrary and leaves much to individual interpretation and preference. Consequently, variations in solutions from one designer to another may occur. Another associated problem is the lack of verification. Since rainfall intensity values are derived

TABLE 9.22. Values Used to Determine a Composite Runoff Coefficient for an Urban Area (ASCE/WEF, 1992).

Character of surface	Runoff coefficients
Pavement	
Asphalt and Concrete	0.70 to 0.95
Brick	0.70 to 0.85
Roofs	0.75 to 0.95
Lawns, Sandy Soil	
Flat, 2%	0.05 to 0.10
Average, 2 to 7%	0.10 to 0.15
Steep, 7%	0.15 to 0.20
Lawns, Heavy Soil	
Flat, 2%	0.13 to 0.17
Average, 2 to 7%	0.18 to 0.22
Steep, 7%	0.25 to 0.35
Water Impoundment	1.00

Note: ASCE/WEF (1992) indicates the range of the C values shown above is typical for return periods of 2–10 years. Higher values are appropriate for larger design storms.

from statistical analysis and do not represent any given storm event, it is virtually impossible to observe a storm of a specified design intensity and duration. Regardless of its shortcomings, the Rational Method remains in wide-spread use among engineers; however, because it can be misused, the authors recommend the following procedure for application of the Rational Method.

Step 1. Determine the contributing basin area A by using topographic maps developed from a survey of the area, or plans made specifically for the basin. This area can then be found by use of a planimeter or digitizer.

Step 2. By the use of Table 9.20 for rural areas or Tables 9.21 and 9.22 for urban areas, find the appropriate value of C. If the area has multiple land uses, a composite C value may be determined by:

$$C_{comp} = \frac{C_1A_1 + C_2A_2 + \cdots C_nA_n}{A_t} \tag{9.38}$$

where $C_1, C_2, \ldots C_n$ are the runoff coefficients associated with component areas $A_1, A_2, \ldots A_n$, and $A_t = A_1 + A_2 + \ldots A_n$.

Step 3. Determine the time of concentration as the travel time from the most distant point to design location. The time of concentration for the basin can be determined by using one of the equations listed in Table 9.18. Where possible, t_c should be calculated from segmental flow lengths, slopes, and cover condition.

Step 4. Determine the design rainfall intensity. Enter an intensity–duration–frequency curve with the storm duration equal to the time of concentration and, by choosing a curve with the appropriate return period, read off the intensity value. This is then multiplied by the area and runoff coefficient to determine the peak discharge.

If there is another sub-area downstream (sub-area 2), the first time of concentration is added to the travel time in the channel or pipe to the mid-point of sub-area 2. This is then compared to the local inlet time of the second sub-area and the larger of the two is used as the new time of concentration for the channel/pipe serving sub-area 2. This process is repeated for each lower sub-area.

The step-by-step application of the Rational Method for storm sewer design is illustrated below in the storm sewer design for Fair Oaks Estates.

D. Fair Oaks Estates Storm Sewer Design

Fair Oaks Estates is a subdivision of Carol Stream, Illinois, a suburb west of Chicago. This example involves the design of the storm sewer system for Fair Oaks Estates, using the Rational Method. The soil is classified as Type B SCS hydrologic soil group and the topography is flat. The total area of the subdivision, which includes 30 lots, is 5.42 ha (13.4 acres).. The imperviousness is 28%. The planned drainage system is shown in Fig. 9.24. Inlets are located in the low spots and at the corners of intersections. The drainage sub-basins that contribute to the inlets are also delineated in Fig. 9.24. The storm sewer discharges into a detention pond located in the northeast corner. The design requirements are: 1) the storm sewer must pass a ten-year storm, based on the intensity–duration–frequency curve for Chicago, Illinois, 2) the top of the pipe must have a 1.07 m (3.5 ft) cover, 3) the full-pipe velocity must be greater than 0.61 m/sec (2 ft/sec) and less than 3.66 m/sec (12 ft/sec), 4) the minimum pipe diameter is 30 cm (12 in), 5) the pipe is commercial reinforced concrete pipe with a Manning's n of 0.013, 6) the outfall pipe invert must be at or above 230.43 m (756.00 ft) MSL, 7) when the pipe size changes at a structure, the top of the pipes shall be matched, 8) subarea inlet times shall be assumed to be 20 minutes.

From the plan drawings and topographic map, the ground elevations and pipe lengths were determined. The Rational Method was used to determine the required pipe diameters and slopes. The

BASIN No.	AREA (Ac.)	BASIN No.	AREA (Ac.)
1	0.68	12	0.70
3	0.53	13	0.43
4	1.31	14	0.38
5	0.22	15	0.88
7	1.54	17	0.79
8	0.48	18	1.55
9	2.85	19	0.52
10	0.83		

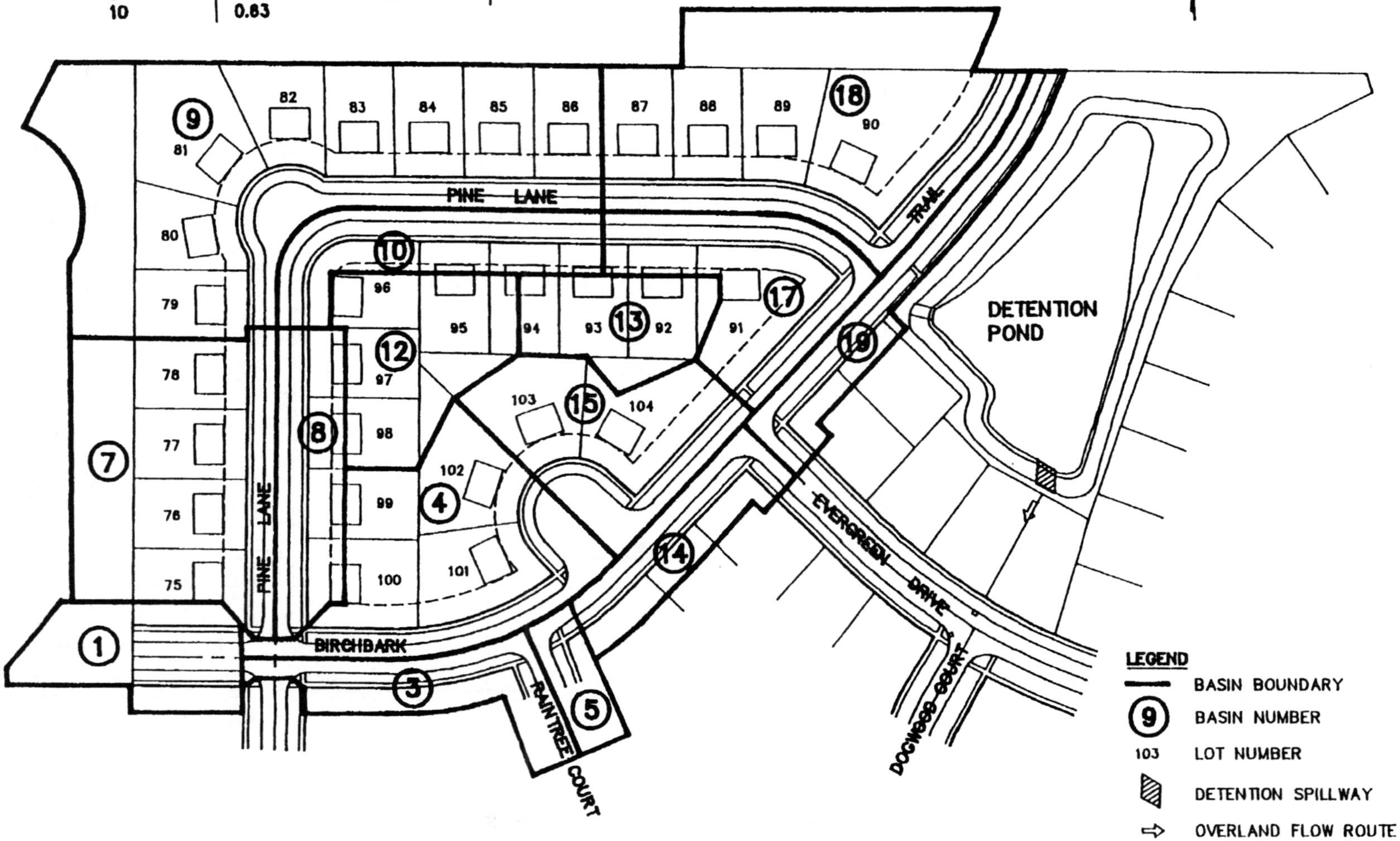

Figure 9.24.—Fair Oaks Estates Hydrologic System.

TABLE 9.23. Storm Sewer Design for Fair Oaks Estates Based on 10-Year Storm, Manning n = 0.013.

Line no.	U/S man hole	D/S man hole	Length (m)	C_j	A_j (ha)	C_jA_j	ΣC_jA_j	t_j (min)	t_{cum} (min)	Intensity (cm/hr)	Peak flow Q (m^3/s)	Peak flow Q (cfs)	Pipe diameter (cm)	Pipe diameter (in)
1	2	3	4	5	6	7	8	9	10	11	12		13	
1	21	20	6.71	0.28	0.25	0.07	0.07	20	20.0	10.16	0.020	0.7	30.5	12
2	20	19	98.76	—	—	—	0.07	—	20.2	10.13	0.020	0.7	30.5	12
3	19	18	23.16	0.44	0.21	0.09	0.16	20	20.9	10.01	0.045	1.6	30.5	12
4	18	17	88.39	—	—	—	0.38	—	21.2	9.96	0.102	3.6	38.1	15
5	17	16	59.44	—	—	—	1.47	—	27.1	8.89	0.365	12.9	53.3	21
6	16	15	9.14	—	—	—	1.97	—	27.7	8.81	0.479	16.9	53.3	21
7	14	13	8.53	0.32	1.15	0.37	0.37	25	25.0	9.30	0.093	3.3	30.5	12
8	13	12	35.05	0.49	0.25	0.12	0.49	20	25.1	9.30	0.125	4.4	38.1	15
9	12	11	7.92	—	—	—	0.81	—	25.5	9.22	0.204	7.2	38.1	15
10	11	10	91.44	0.25	0.28	0.07	0.88	20	25.6	9.19	0.221	7.8	45.7	18
11	10	9	23.16	0.25	0.17	0.04	0.92	20	26.7	8.99	0.227	8.0	45.7	18
12	9	17	15.85	0.38	0.36	0.14	1.12	20	27.0	8.92	0.272	9.6	45.7	18
13	8	7	9.75	0.32	0.53	0.17	0.17	20	20.0	10.16	0.045	1.6	30.5	12
14	7	18	7.01	0.47	0.09	0.04	0.21	20	20.2	10.11	0.057	2.0	30.5	12
15	6	5	13.72	0.45	0.32	0.14	0.14	20	20.0	10.16	0.040	1.4	30.5	12
16	5	4	13.41	0.37	0.63	0.23	0.37	20	20.3	10.08	0.102	3.6	30.5	12
17	4	16	7.62	0.50	0.21	0.11	0.48	20	20.5	10.06	0.130	4.6	30.5	12
18	3	2	8.53	0.33	0.62	0.20	0.20	20	20.0	10.16	0.057	2.0	30.5	12
19	2	12	43.89	0.54	0.19	0.10	0.30	20	20.1	10.11	0.085	3.0	30.5	12
20	1	9	17.07	0.39	0.16	0.06	0.06	20	20.0	10.16	0.017	0.6	30.5	12

Table 9.23. *(Continued)*

Pipe slope (%)	Pipe cap. (m^3/s)	Pipe vel. (m/s)	Travel time (min)	Ground elev U/S (m)	Ground elev D/S (m)	Invert elev U/S (m)	Invert elev D/S (m)	Pipe cover U/S (m)	Pipe cover D/S (m)
14	15	16	17	18	19	20	21	22	23
0.002	0.045	0.619	0.18	235.89	235.90	234.52	234.51	1.07	1.09
0.024	0.157	2.152	0.76	235.90	233.52	234.51	232.14	1.09	1.08
0.010	0.103	1.408	0.27	233.52	233.44	232.14	231.90	1.08	1.23
0.005	0.140	1.225	1.20	233.44	232.96	231.82	213.31	1.23	1.26
0.006	0.368	1.649	0.60	232.96	232.74	231.24	230.76	1.19	1.44
0.011	0.479	2.140	0.07	232.74	OUTFALL	230.76	230.66	1.44	—
0.008	0.094	1.292	0.11	234.17	234.17	232.80	232.72	1.07	1.14
0.006	0.152	1.335	0.44	234.17	233.85	232.65	232.40	1.14	1.07
0.016	0.232	2.036	0.06	233.85	233.73	232.40	232.28	1.07	1.07
0.005	0.221	1.347	1.13	233.73	233.43	232.20	231.70	1.07	1.28
0.014	0.353	1.387	0.28	233.43	232.89	231.70	231.37	1.28	1.06
0.008	0.275	1.673	0.16	232.89	232.96	231.37	231.24	1.06	1.26
0.002	0.048	0.664	0.25	233.32	233.32	231.95	231.93	1.07	1.09
0.003	0.060	0.820	0.14	233.32	233.44	231.93	231.90	1.09	1.23
0.002	0.045	0.619	0.37	232.67	232.71	231.29	231.26	1.07	1.15
0.010	0.103	1.414	0.16	232.71	232.68	231.26	231.12	1.15	1.25
0.017	0.132	1.814	0.07	232.68	232.74	231.12	230.99	1.25	1.44
0.003	0.058	0.796	0.18	234.74	234.74	233.37	233.35	1.07	1.09
0.019	0.142	1.945	0.38	234.74	233.85	233.35	232.48	1.09	1.07
0.002	0.045	0.619	0.46	232.93	232.89	231.56	231.52	1.07	1.07

TABLE 9.24. Explanations of Entries Shown in Fair Oaks Storm Sewer Design Sheet.

Column number	Entry value	Explanation comments	
1	3	Line Number	
2	19	Upstream Manhole	
3	18	Downstream Manhole	
4	23.16	Length Between Structures (from Fig. 9.25)	
5	0.44	C_j	Runoff coefficient for subbasin. This is a weighted composite based on the type of cover in the subbasin.
6	0.21	A_j	Area (ha) of subbasin under consideration.
7	0.09	C_jA_j	Product of Area and Runoff Coefficient for subbasin ($0.44 \times 0.21 = 0.09$)
8	0.16	ΣC_jA_j	Sum of all Area–Runoff Coefficient products contributing to line #3 ($0.07 + 0.09 = 0.16$)
9	20	t_j	Time of Concentration (minutes) of subbasin. (Assumed to be 20 min.)
10	20.9	t_{cum}	The longest Travel Time (minutes) from the most distant basin to contribute ($20 + 0.18 + 0.76 = 20.94$ min.)
11	10.01	i	Rainfall Intensity (cm/hr) for Chicago, Illinois for a ten-year return period and storm duration of 20.94 min. (10.01 cm/hr)
12	0.045 m^3/s 1.6 cfs	Q	Peak Flow (m^3s) (cfs) into line #3, found by $Q = i\Sigma C_jA_j$,
13	30.5 cm 12 in	Pipe Diameter (cm) (in) selected to convey Q.	
14	0.010	Pipe Slope at which the pipe selected above is placed.	
15	0.103	Full Pipe Capacity (m^3/s) of the pipe with the selected diameter and slope determined by Manning's Equation. (This must be greater than or equal to the peak flowrates, Q)	
16	1.408	Velocity of flow (m/s) in pipe ($Q/A = V$)	
17	0.27	Travel Time (minutes) for the water to flow from Structure 19 to Structure 18. ($t = L/60V = 0.27$ min.)	
18	233.52	Ground Elevation Upstream (meters)	
19	233.44	Ground Elevation Downstream (meters)	
20	232.14	Invert Elevation Upstream (meters)	
21	231.90	Invert Elevation Downstream (meters)	
22	1.08	Pipe Cover Upstream (Ground elev.—Top of Pipe elev.), meters	
23	1.23	Pipe Cover Downstream (Ground elev.—Top of Pipe elev.), meters	

calculations are shown in Table 9.23. An explanation of each column and entry value for storm sewer line number 3 is listed in Table 9.24.

The schematic of the drainage system designed with the Rational Method is shown in Fig. 9.25 The peak flow rate is found to be 0.48 m^3/s (16.90 cfs) and the time of concentration for the basin is 27.78 min. The minimum pipe is 0.30 m (12 in) and the maximum diameter is 0.53 m (21 in). The minimum pipe cover is 1.07 m (3.5 ft) and the maximum is 1.44 m (4.74 ft). Velocities are in the range of 0.62 m/sec (2.04 ft/sec) to 2.04 m/sec (6.68 ft/sec).

E. Sizing of Stormwater Storage Facilities

It was shown earlier that the response of an urbanized watershed to a storm is considerably different than that of rural, undeveloped land. Ground that once allowed infiltration of rainwater is covered with impervious parking lots, streets and homes, causing a higher percentage of rain to become surface runoff. Improvements such as streets, curbs, and storm sewers collect and convey this runoff to its point of disposal more rapidly than in the undeveloped watershed, resulting in a higher runoff volume in a shorter period of time.

These increased runoff rates produced by the urbanization of a watershed may overtax lower portions of the larger drainage network and cause flooding problems downstream. To circumvent this difficulty, the drainage engineer should provide storage facilities to receive the runoff from the developed watershed while releasing water to the larger drainage network at a reduced rate. This reduced rate is determined by local ordinance or by calculating the available capacity of the downstream network. Although space does not permit discussion, stormwater storage facilities can also be used for improving water quality (see Urbonas and Roesner, 1986; Urbonas and Stahre, 1993).

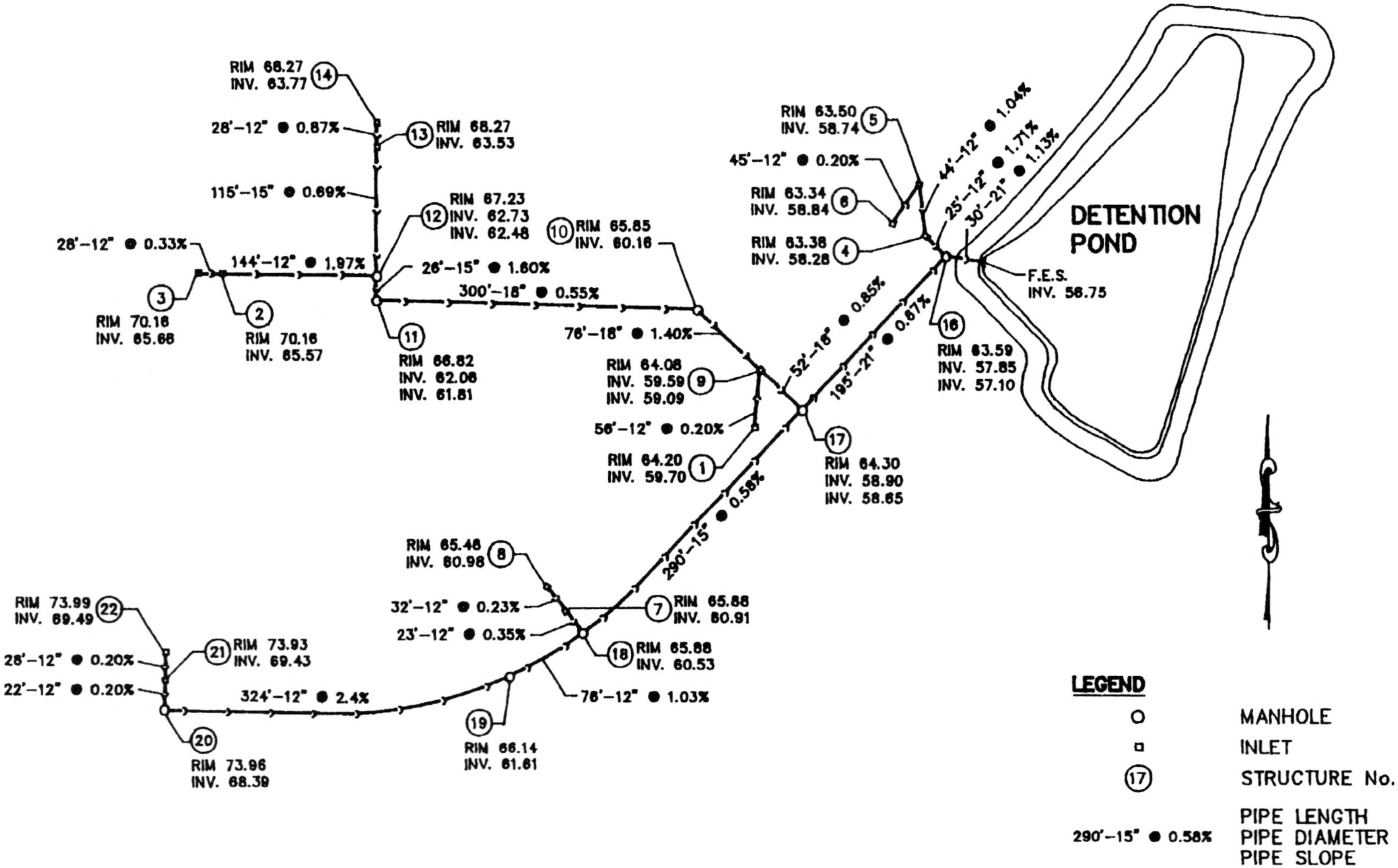

Figure 9.25.—Layout of Storm Sewer System Designed by Rational Method.

This section concludes with a discussion of the various types of storage facilities and the methods of computing required storage volumes for stormwater quantity control. Sizing considerations are outlined in a design example for a stormwater detention pond in the Fair Oaks Estates subdivision.

F. Types of Urban Stormwater Storage Facilities

Storage facilities can be broken down into two general categories: detention and retention. These two basin types are often described as "dry" and "wet" ponds, depending on the presence of a permanent pool.

Detention is the temporary storage of the incoming runoff which is in excess of the permissible release. At some time after the storm, the facility empties and resumes its normal function. Parking lots, rooftops, and parks are common detention facilities which are designed to store runoff temporarily.

Retention facilities on the other hand, retain the runoff for an indefinite period of time. Water level changes with evaporation and infiltration only. Ponds and lakes are examples of retention facilities that may be implemented in a development and in many cases, may enhance the overall project.

Where favorable soil conditions exist, stormwater may be allowed to seep into the soil using infiltration basins or ditches. These facilities recharge the ground water supply, treat the stormwater as it infiltrates through the soil and reduce surface runoff in parking lot and low density highway situations. Porous asphalt is another option which is currently receiving considerable attention. Voids in the asphalt allow the water to seep into the underlying gravel base course where it is detained. See *Design and Construction of Stormwater Management Systems* (ASCE/WEF, 1992) for further discussion of porous pavement and infiltration facilities.

G. Detention Basin Design Considerations

In this section, general guidelines are presented for the design of storage facilities. Commonly used design criteria are first presented for each type or class of facility.

Retention (wet) ponds should always be constructed for both aesthetic and practical use. Serious investigation of potential water quality problems associated with low flows in dry weather, direct pollution from surface runoff, or high nutrient levels is necessary. The latter may cause algae growth and subsequent eutrophication. If any of these problems could arise, a detention pond should be considered instead. The following is a list of some design considerations for retention ponds: 1) When possible, retention ponds should be stocked with fish. To provide a habitat suitable for aquatic life, it is recommended that the pond have a surface area of at least .20 ha (0.5 acre) and a minimum depth of at least 3.05 m (10 ft) over at least 25% of the total area. The average depth of the remaining area should be at least 1.52 m (5 ft). This is needed to inhibit insect breeding and weed growth although periodic maintenance may still be necessary to completely control this growth. A minimum side slope of 3:1 is also recommended. 2) An emergency overflow spillway or weir should be installed to insure that flows exceeding the design capacity of the pond do not damage surrounding property. 3) Fencing may be required to ensure the safety of children. 4) All inlet openings larger than 15 cm (6 in) should have bars or screens on them to protect children and animals.

Detention (dry) ponds should be designed for multiple purposes. In some parts of the country the ponds are used as athletic fields or parks during dry weather. Regardless of their dry weather use, there are certain guidelines that must be followed to insure that the ponds are properly maintained and do not become detriments or hazards. 1) The pond should not have side slopes greater than 3 m horizontal to 1 m vertical. 2) An underdrain system should be constructed to minimize the wetness of the pond bottom. This is to insure that the bottom does not harbor insects or prohibit recreational dry weather use. This underdrain system may discharge directly into the receiving stream or into a wet well where it is then pumped out to the receiving stream. 3) An alternative to the underdrain is the sloping of the pond bottom with at least a 2% grade from inlet to outlet. Another means of eliminating the wetness is to slope the pond to a gutter that flows through the pond. This gutter may also be used to convey dry weather flow. 4) The pond bottom and side slopes should be finished with at least four inches of topsoil and seeded.

This seed or sod should be capable of withstanding periodic flooding conditions. 5) Outlets should be provided with rip-rap protection to prohibit erosion. 6) An overflow spillway or weir should be placed at the high water elevation and a minimum of 8 cm (3 in) of freeboard should be provided. 7) Screens or bars should be placed over all inlets and outlets greater than 15 cm (6 in) to protect children and animals.

A summary of general water quality design considerations for the design of both wet and dry bottom detention facilities is given in Table 9.25. Also see Urbonas and Roesner (1986) and Urbonas and Stahre (1993).

H. Detention Storage Calculations for Fair Oaks Estates Subdivision

A dry detention pond is to be built at the site shown in Fig. 9.24. Local ordinance requires that it be designed such that Q_{100} in post-development condition is attenuated to no more than 0.10 m^3/s per hectare or 0.057 m^3/s (0.15 cfs/acre or 2.01 cfs). The SCS TR 55 tabular hydrograph (USDA SCS, 1986) is used to generate the post-development inflow hydrograph to the detention pond. The design data are given as follows: drainage area = 5.42 hectares (13.4 acres); allowable release rate = 0.057 m^3/s (2.01 cfs); CN = 79; rainfall depth = 14.73 cm (5.8 in) for 100 year, t_c = 0.42 hours; rainfall distribution = SCS Type II; storm duration = 24 Hours.

The volume–elevation data in Table 9.26 were developed by digitizing the area from field surveys of the site at 0.3048 m (1-foot) intervals. The discharge data were computed for a 0.114 m (4½ in) circular orifice with C_d = 0.80, and invert elevation 229.88 m (754.2 ft). The routing time step Δt was set at 0.10 hours for convenience, although 0.20 hours would be sufficiently accurate. The $2S/\Delta t + 0$ values were then computed at each elevation. The basin floor elevation is 229.88 m.

The routing table, showing the 100-year inflow hydrograph by TR55 and the individual steps in the routing operation, is presented in Table 9.27. Because of space limitations the table is truncated at hour 15.4 in the 24-hour period. A summary of the routing operation is: peak inflow = 1.218 m^3/s (43.00 cfs) at hour 12:30; peak outflow = 0.050 m^3/s (1.76 cfs) at hour 17:10; peak WSEL = 231.81 m (760.54 ft) at hour 17:10; and max storage = 3084 m^3 (2.50 acre feet) at hour 17:10.

TABLE 9.25. Summary of Detention Basin Considerations For Water Quality (NIPC, 1986).

Design consideration	Water quality enhancement suggestion
1. Basin Size	Wet—Determine pool volume based on actual solids settling rate and runoff volumes, rates, and frequencies. Typically, between 1.57 and 3.14 cm (0.25 and 0.5 inches) of water volume per hectare (acre) of watershed depending on land cover characteristics and design settling velocity. Storage needed for flood control is then added to this volume. Dry—Again, actual solids settling rates should be determined. Basin volume needed for flood control should be computed and then overflow rates should be checked for frequent runoff events to be sure that overflow rate or detention time does not exceed design rate. If overflow rate is too large the basin area would be increased while holding depth constant until satisfactory overflow rates are obtained.
2. Inlet/Outlet Orientation	Opposite ends of the basin.
3. Basin Shape	Elliptical, rectangular, or triangular with inlet and outlet at opposite longitudinal ends.
4. Length to Width Ratio	3:1 to 5:1 or greater.
5. Velocity Dissipation	Baffles and presedimentation basins located in flow path between inlet and outlet.
6. Outlets	Wet—Surface withdrawal controlled by a weir. Dry—Perforated riser, buried in gravel or underdrain system with no direct release.
7. Depth	Wet—Deep enough to prevent resuspension of settled sediment due to wind or velocity currents. Typically at least 1.22–1.52 meters (4–5 feet) average length.
8. Shorelines	Stabilized with vegetation or rip rap to prevent erosion. Safety ledge should be provided. Slope 5:1 or flatter.
9. Bank Slope	Wet—Safety ledge should be provided. Slope may be important if fish life is to be maintained. Dry—3:1 or flatter for safety and to encourage establishment of vegetative cover.
10. Maintenance	Provision for maintenance equipment access. Provision of a presedimentation basin to localize sediment deposition. Provision of an alternate bottom outlet in wet basins to allow the pond to be drained for maintenance.

TABLE 9.26. Elevation–Volume–Discharge Data for Fair Oaks Estates Detention Site.

Elevation–Volume–Discharge Data							
Elevation		Volume		Discharge		$\frac{2S}{\Delta t} + 0$	
meters	feet	m^3	acres-feet	m^3/s	ft^3/s	m^3/s	ft^3/s
229.88	754.2	0.0	0.000	0.000	0.00	0.00	0.0
230.43	756.0	379	0.307	0.027	0.94	2.13	75.2
230.73	757.0	863	0.700	0.033	1.17	4.83	170.6
231.04	758.0	1433	1.162	0.039	1.36	8.00	282.6
231.34	759.0	2051	1.663	0.043	1.53	11.44	404.0
231.65	760.0	2711	2.198	0.048	1.68	15.11	533.6
231.95	761.0	3411	2.765	0.052	1.82	19.00	671.0

Since the pond is initially dry, the total storage required to reduce Q_{100} of 1.218 m^3/s (43.0 cfs) to 0.050 m^3/s (1.76 cfs) is 3084 m^3 (2.50 acre-feet). This is obtained either by interpolation in the elevation–volume table for the Fair Oaks detention site, or by back-solving for S_{max} at Q_{max} of 0.050 m^3/s (1.76 cfs) in Table 9.27. A slightly larger orifice could be employed, since maximum allowed outflow is 0.057 m^3/s (2.01 cfs). A 6-inch orifice would be attempted next. This requires that the discharge elevation and $2S/\Delta t + 0$ tables be re-computed. Then the routing in Table 9.27 would be performed to determine whether the basin outflow met the target value or exceeded it. This example demonstrates the need for computerized algorithms for design of detention facilities. The use of computers in urban hydrology is discussed in the next section.

VI. COMPUTER MODEL APPLICATIONS

Computer modeling of the rainfall–runoff response of watersheds to land use change has been a topic of intense interest to hydrologists for many years. The first major attempt in the United States to model overland runoff and combined sewer flows in a single hydrologic simulation package for urban catchments was by the EPA in 1972. The result was the EPA Stormwater Management Model, otherwise known as EPA SWMM or SWMM, which is widely used today. The original SWMM program was developed for mainframe applications by a consortium of the University of Florida, Water Resources Engineers, Inc., and Metcalf and Eddy, Inc. The most recent version released by the EPA in 1994 is maintained by the University of Florida and the EPA and now runs efficiently on high-end personal computers.

The rainfall–runoff–sewer flow processes represented in SWMM and many other urban models have been presented in earlier sections of this chapter. It is sufficient to emphasize here that urban stormwater models should be viewed as powerful tools which, in the hands of an experienced professional, can be used to find engineering solutions to problems of street flooding, out–of–bank flows, combined sewer overflows, and non-point source pollutant management. The purpose of this section is to provide a brief overview of the array of urban hydrology computer modeling packages available today and to present the application of three selected models to the Fair Oaks Estates subdivision described earlier.

A. Overview of Urban Hydrology Software

In view of the dramatic increase in urban hydrology software packages in the past ten years, ASCE formed a task committee (TC) to report on the general availability and properties of microcomputer software in urban hydrology. The primary mission of the TC was to compile an inventory of microcomputer software having the capability to: 1) generate runoff peaks and hydrographs, 2) analyze/design storm sewers, and 3) design stormwater detention basins. The TC compiled a list of 40 vendors and developed a summary listing of the operational features of each software package, based on personal knowledge and vendor-supplied information. This was completed in 1990 and published in 1991 (Kibler et al., 1991).

TABLE 9.27. Storage Routing Table for Fair Oaks Detention Basin.

Inflow hydrograph			Routing computations									
TIME	INFLOW		I_1+I_2		2S/DT−O		2S/DT+O		OUTFLOW		ELEVATION	
(hrs)	(cfs)	(m³/s)	(cfs)	(m³/s)	(cfs)	(m³/s)	(cfs)	(m³/s)	(cfs)	(m³/s)	(ft)	(m)
11.00	1	0.028	2	0.057	0.0	0.000	0.0	0.000	0.00	0.000	754.20	229.88
11.10	1	0.028	3	0.085	2.0	0.057	2.0	0.057	0.02	0.001	754.25	229.90
11.20	2	0.057	4	0.113	4.8	0.136	5.0	0.142	0.06	0.002	754.32	229.92
11.30	2	0.057	4	0.113	8.6	0.244	8.8	0.249	0.11	0.003	754.41	229.94
11.40	2	0.057	5	0.142	12.3	0.348	12.6	0.357	0.16	0.005	754.50	229.97
11.50	3	0.085	6	0.170	16.9	0.479	17.3	0.490	0.22	0.006	754.61	230.01
11.60	3	0.085	7	0.198	22.3	0.632	22.9	0.649	0.29	0.008	754.75	230.05
11.70	4	0.113	9	0.255	28.6	0.810	29.3	0.830	0.37	0.010	754.90	230.09
11.80	5	0.142	11	0.312	36.6	1.037	37.6	1.065	0.47	0.013	755.10	230.15
11.90	6	0.170	16	0.453	46.4	1.314	47.6	1.348	0.59	0.017	755.34	230.23
12.00	10	0.283	30	0.850	60.9	1.725	62.4	1.767	0.78	0.022	755.69	230.33
12.10	20	0.566	54	1.529	88.9	2.518	90.9	2.574	0.98	0.028	756.16	230.48
12.20	34	0.963	77	2.181	140.7	3.985	142.9	4.047	1.10	0.031	756.71	230.65
12.30	43	1.218	85	2.407	215.2	6.094	217.7	6.165	1.25	0.035	757.42	230.86
12.40	42	1.189	74	2.096	297.4	8.422	300.2	8.502	1.38	0.039	758.15	231.08
12.50	32	0.906	54	1.529	368.5	10.436	371.4	10.518	1.48	0.042	758.73	231.26
12.60	22	0.623	38	1.076	419.4	11.877	422.5	11.965	1.55	0.044	759.14	231.39
12.70	16	0.453	28	0.793	454.2	12.863	457.4	12.954	1.59	0.045	759.41	231.47
12.80	12	0.340	22	0.623	478.9	13.562	482.2	13.656	1.62	0.046	759.60	231.53
12.90	10	0.283	18	0.510	497.7	14.095	500.9	14.185	1.64	0.046	759.75	231.57
13.00	8	0.227	15	0.425	512.3	14.508	515.7	14.605	1.66	0.047	759.86	231.61
13.10	7	0.198	13	0.368	524.0	14.840	527.3	14.933	1.67	0.047	759.95	231.63
13.20	6	0.170	12	0.340	533.6	15.112	537.0	15.208	1.68	0.048	760.02	231.65
13.30	6	0.170	11	0.312	542.2	15.355	545.6	15.451	1.69	0.048	760.09	231.68
13.40	5	0.142	10	0.283	549.8	15.570	553.2	15.667	1.70	0.048	760.14	231.69
13.50	5	0.142	9	0.255	556.4	15.757	559.8	15.854	1.71	0.048	760.19	231.71
13.60	4	0.113	8	0.227	562.0	15.916	565.4	16.012	1.71	0.048	760.23	231.72
13.70	4	0.113	8	0.227	566.6	16.046	570.0	16.142	1.72	0.049	760.27	231.73
13.80	4	0.113	8	0.227	571.1	16.174	574.6	16.273	1.72	0.049	760.30	231.74
13.90	4	0.113	7	0.198	575.7	16.304	579.1	16.400	1.73	0.049	760.33	231.75
14.00	3	0.085	6	0.170	579.2	16.403	582.7	16.502	1.73	0.049	760.36	231.76
14.10	3	0.085	6	0.170	581.7	16.474	585.2	16.573	1.73	0.049	760.38	231.76
14.20	3	0.085	6	0.170	584.3	16.547	587.7	16.644	1.74	0.049	760.39	231.77
14.30	3	0.085	6	0.170	586.8	16.618	590.3	16.717	1.74	0.049	760.41	231.77
14.40	3	0.085	5	0.142	589.3	16.689	592.8	16.788	1.74	0.049	760.43	231.78
14.50	2	0.057	4	0.113	590.8	16.731	594.3	16.831	1.74	0.049	760.44	231.78
14.60	2	0.057	4	0.113	591.4	16.748	594.8	16.845	1.74	0.049	760.45	231.79
14.70	2	0.057	4	0.113	591.9	16.763	595.4	16.862	1.74	0.049	760.45	213.79
14.80	2	0.057	4	0.113	592.4	16.777	595.9	16.876	1.74	0.049	760.45	231.79
14.90	2	0.057	4	0.113	592.9	16.791	596.4	16.890	1.74	0.049	760.46	231.79
15.00	2	0.057	4	0.113	593.4	16.805	596.9	16.904	1.74	0.049	760.46	231.79
15.10	2	0.057	4	0.113	593.9	16.819	597.4	16.918	1.75	0.050	760.46	231.79
15.20	2	0.057	4	0.113	594.4	16.833	597.9	16.933	1.75	0.050	760.47	231.79
15.30	2	0.057	4	0.113	594.9	16.848	598.4	16.947	1.75	0.050	760.47	231.79
15.40	2	0.057	4	0.113	595.4	16.862	598.9	16.961	1.75	0.050	760.48	231.79

In terms of hydrologic capabilities, the software reviewed by the TC tended to support simple desk-top procedures for a single storm. For example, the SCS curve-number and rational C-coefficient dominated the list of hydrologic loss methods, followed by Green and Ampt and Horton infiltration equations. Hydrograph synthesis tended toward use of the SCS unit hydrograph, followed by the modified rational method, kinematic wave, and other synthetic unit hydrograph techniques. Hydrograph routing was performed primarily by time-of-travel translation, Muskingum routing, or kinematic routing, with occasional use of dynamic routing by the gradually-varied unsteady flow equations. This level of computation is very much in line with the observation (by 72% of the vendors) that private consulting firms are the most frequent users of their software. Consultants are followed by state and federal agencies in terms of user frequency; however, many of these agencies were committed to microcomputer versions of larger packages such as EPA SWMM, USGS DR3M, ILLUDAS, and HSPF. A brief listing of the most commonly used models is presented in Table 9.28. The details on individual software capabilities, together with vendor information, are provided in the article by the ASCE TC (Kibler et al., 1991). A more general review of runoff quantity and quality models is presented in Chapter 6.

The selection of an urban hydrologic model is still a matter of experience in assessing the needs of a particular design project and matching them with appropriate hydrologic and hydraulic procedures. A general premise is that the simplest model that will accomplish the design or analysis objective will drive the model selection process. In many instances, it is necessary to employ two or three different types of models, particularly in the case of small urban watersheds where there are no calibration data.

Clearly, the biggest factors in choice of model are user experience with certain classes of models, knowledge of the system to be simulated, and understanding of the design project objectives. If, for example, the problem at hand requires model output at interior locations of the storm sewer system to reduce flooding, then obviously a model that uses a distributed subarea approach must be selected. The degree of watershed disaggregation and the number of subareas, pipes, and channels then must be decided to properly adapt the model to the prototype system. If the problem involves only storm sewer sizing in a simple system, then there is little need to go beyond a model based on the rational formula and its variations. In the case of larger systems requiring the sizing of detention basins, storm sewers, and outfall structures, then the use of comprehensive models, such as EPA SWMM, is justified. In projects involving planning and management studies for watershed-wide control of stormwater, it is appropriate to use intermediate level models such as ILLUDAS or HEC–1. Specific example applications are provided below.

B. Model Application to Basin-wide Stormwater Management and Master Planning

Basin-wide planning is a particular application that relies heavily on computer models. A detailed watershed simulation provides the hydrologist with a better understanding of individual subbasin response and the interrelation of the total watershed with its various subbasins. With this understanding, the engineer can quantify the impacts of land use changes and evaluate proposed runoff control measures. The simulation can then be repeated with the appropriate input changes reflecting altered land uses and the implementation of selected control measures to verify the qualitative assessment.

An example application is the development of a stormwater management master plan that includes facilities to mitigate the impacts of development within a watershed. To mitigate such impacts, the watershed release rate concept has frequently been applied. This concept requires that the peak rate of runoff from a lower larger watershed not increase as a result of upstream development. The implementation of this watershed-level stormwater management objective is generally accomplished through the use of detention storage, at strategic locations, preceded by computer simulation of the total system. Achieving the target discharge rate now is a function of detention basin release rates and inflow from other, possibly unregulated, subbasins within the watershed. To understand these relationships and to demonstrate compliance with the target watershed release rate, the engineer must use a computer model with the capability to simulate subarea hydrograph timing effects within the total watershed.

The situation is illustrated in Fig. 9.26, where on-site detention is deployed in subwatershed A in an effort to maintain post-development flood peaks at the predevelopment level. The detention facility was designed to release the predevelopment peak at the site; however, because of the extended release from the detention basin and the hydrograph timing from subwatershed B, the existing watershed experiences an increase in peak discharge of approximately 4 m^3/s in the controlled post-development condition. A target release rate lower than the predevelopment peak is required here to maintain predevelopment flow peaks for the entire system. Release rate is defined as the subarea predevelopment discharge at the time of the total watershed peak, divided by the subarea predevelopment hydrograph peak. Thus, in the case of Fig. 9.26, a release rate of 75% (7.5 $m^3/s \div 10.0\ m^3/s$) $\times$ 100 is required to maintain the total

Table 9.28. Common Urban Hydrology Models.

Model name	Agency name	Precip. excess method	Hydrograph synthesis method	Channel or sewer routing method	Detention basin option	Runoff quality
TR 20	USDA Soil Conservation Service, Washington, DC	Curve number and SCS soil-cover complex	SCS dimensionless unit hydrograph (UH)	Attenuated kinematic wave	Storage–indication routing	No
HEC–1	Hydrologic Engrg. Center, US Army Corps. of Engrs. Davis, CA 95616	SCS curve number; Holtan eq.; Variable loss rate	SCS UH, Snyder UH, Clark UH, Kinematic overland flow	Muskingum; Muskingum–Cunge; Kinematic	Storage–indication routing	No
HYMO	USDA/ARS; also Ontario Provincial Government, Ottawa	SCS curve number	SCS UH	Convex routing method	Storage–indication routing	No
ILLUDAS	Illinois State Water Survey, Champaign, IL 61820	Holtan eq.; SCS curve number	Time–area UH	Kinematic by Manning eq.	Storage–indication routing	Yes
SWMM	US EPA, Center for Exposure Assessment, Athens, GA 30613	Horton; modified Horton; SCS curve number	Kinematic overland flow	Kinematic by Manning eq.; also by complete dynamic eqs. in EXTRAN	Storage–indication routing	Yes
PSRM/ QUAL	Penn State U. Dept. Civil Engrg. University Park,PA 16801	Horton; SCS curve number	Kinematic overland flow	Muskingum	Storage–indication routing	Yes
DR3M	US Geological Survey Urban Studies Program, Austin, TX	Phillips eq.	Kinematic overland flow	Kinematic by Manning eq.	Storage–indication routing	Yes
KINEROS	USDA Agricultural Research Service, Washington, DC	Smith and Parlange	Kinematic overland flow	Kinematic by Manning, Chezy or Darcy–Weisbach eq.	Storage–indication routing	Sediment only

predevelopment hydrograph at the confluence of subwatershed A and B. Careful application of a distributed subarea model will identify the combined effects of subarea timing and detention releases so that expensive stormwater facilities can be used effectively to meet watershed-level outflow targets.

Storm water management measures such as detention basins, storm sewers, and channel improvements can then be tested using the model. Computer simulation can be used to determine the cumulative effects of such facilities on achieving the desired watershed release rate at critical control locations. If the target discharge is not met, facility sizes or locations are modified or new facilities are added. Success in meeting the target values can be determined by repeated simulation. If the target values are matched, a system of storm water management components has been identified which utilizes the watershed release rate concept to mitigate the impacts of development within the watershed. The results of this final simulation can also be used to size facilities within the watershed for the control of local flows.

C. Storm Sewer Analysis and Design

Computer models are readily applied in the analysis and design of storm sewer systems. The models perform subarea runoff computations and routes flow through the system. By evaluating the characteristics of the routed flow, the engineer can determine the adequacy of existing or proposed facilities. If the facilities are undersized such that there is frequent storm sewer surcharge and street flooding, the engineer can enlarge the facilities and the model rerun to verify sizing. To illustrate this process, the Fair Oaks subdivision presented in the last section has been analyzed using EPA SWMM and ILLUDAS, the Illinois Urban Drainage Area Simulation developed by the Illinois State Water Survey.

Fig. 9.27 shows a sketch of the Fair Oaks Estates, used in an earlier section as the example subdivision. Included in the figure is a delineation of the subbasin boundaries, a layout of the storm sewer system and identification of manhole/inlet locations, as well as designations used in the various models. This is a

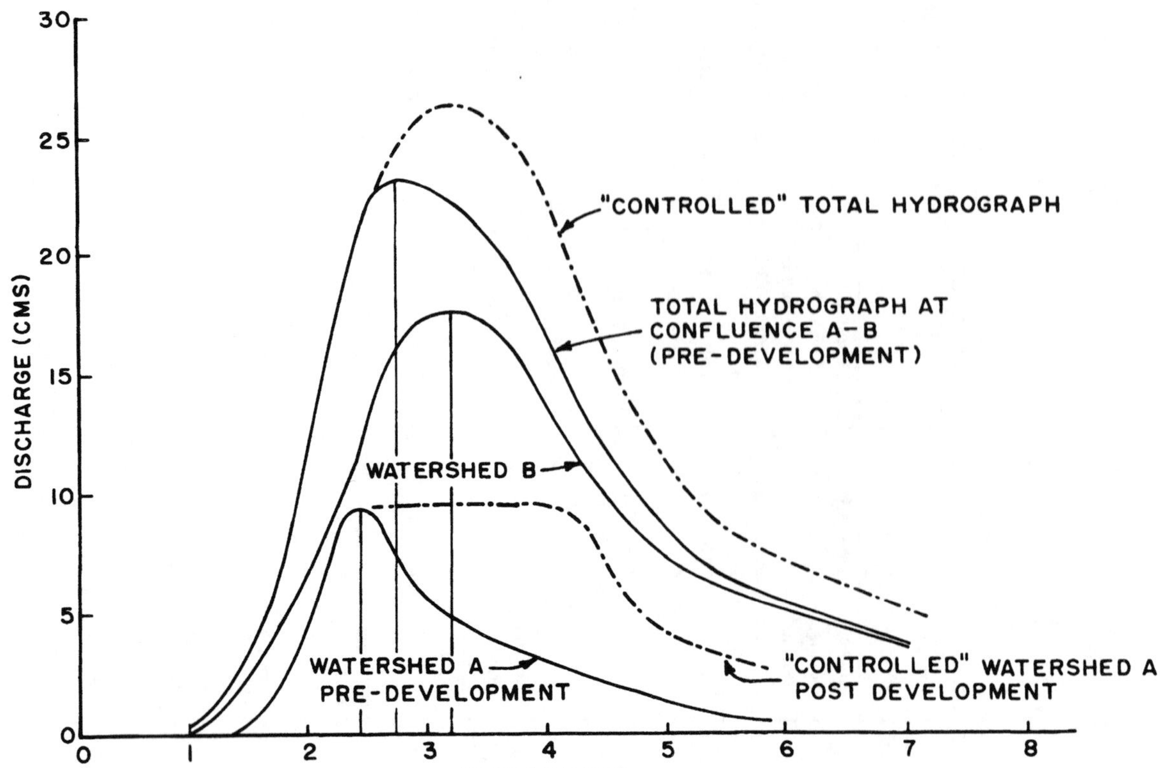

Figure 9.26.—Release Rate Concept Applied to Controlled Post-Development Hydrographs (After Lehigh-Northampton Joint Planning Commission, 1988).

highly simplified drainage system used here only to illustrate the basic computational features of SWMM and ILLUDAS.

D. EPA Stormwater Management Model

Version 4.05 of the EPA Stormwater Management Model (SWMM), used in the Fair Oaks Estates storm sewer example, is maintained by EPA in its Athens, Georgia, Office of Exposure Assessment, by the University of Florida in Gainesville, and by the Oregon State University in Corvallis (USEPA, 1990). SWMM 4.05 consists of a Runoff Block, a Transport Block, and a Storage/Treatment Block. The Runoff Block simulates runoff quantity and quality by means of kinematic overland flow routing within subareas, which are connected to small gutter/pipe elements leading to an outfall location. The Transport Block receives subarea hydrographs and uses the kinematic approximation to route sewer flows and to compute hydraulic head at key junctions in large sewer systems. (Note that the Transport Block is not to be confused with EXTRAN, the Extended Transport model, developed by Camp Dresser and McKee as a model for large complex sewer networks, with backwater and flow-reversal capability). The Storage/Treatment Block (S/T) is used to represent detention basins and combined sewer overflow treatment units. Discussion of EXTRAN and S/T block are beyond the scope of this chapter. Because of the simplicity of the present Fair Oaks Estates system, only the Runoff Block is actually needed; however, both the Runoff and the Transport Block have been used for illustration purposes here.

The Runoff Block requires rainfall input, soil cover and land use information, and subarea geometrics. Routing and combining the sub-area runoff requires a description of basin structure and conveyance geometry. This is normally performed in the Transport Block, although very small sewers can also be included in the Runoff Block.

Table 9.29 presents example input and output files of the SWMM model for Fair Oaks Estates using SWMM 4.05. For purposes of discussion, the input required to produce a SWMM model is subdivided

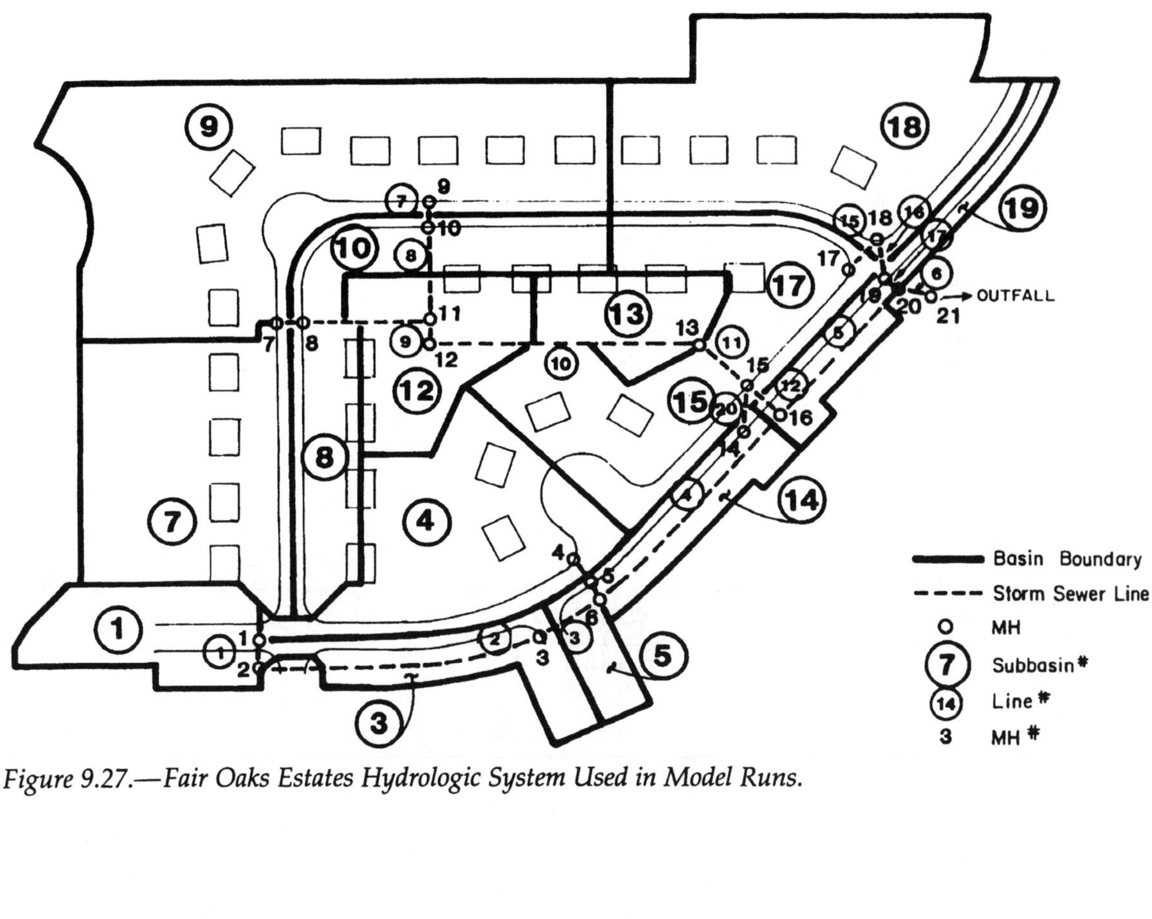

Figure 9.27.—Fair Oaks Estates Hydrologic System Used in Model Runs.

TABLE 9.29. SWMM 4.05 Input/Output for Fair Oaks Estates.

```
                 **************************************************
                 *        Environmental Protection Agency         *
                 *         Storm Water Management Model           *
                 *                Version 4.05                    *
                 **************************************************

                                Developed by

                 **************************************************
                 *             Metcalf & Eddy, Inc.               *
                 *            University of Florida               *
                 *       Water Resources Engineers, Inc.          *
                 *               September 1970                   *
                 **************************************************

                                 Updated by

                 **************************************************
                 *            University of Florida               *
                 *          Camp Dresser & McKee, Inc.            *
                 *                                                *
                 *    March     1975           November  1977     *
                 *    November  1981           December  1990     *
                 **************************************************


* EPA Storm Water Management Model version 4.05 Input Data File
*
* Computations performed by VIPER(c), hydrologic data organization system
* Copyright(c) 1992, Aaron B. Small
*
* Fair Oaks Estates Subdivision containing  15 subareas
*  10-yr storm event
* File = FAIROAKS.SWM  created:  06-21-1993  13:19:58
*
*
*
SW 3 0 10 10 20 20 30
MM 7 1 2 3 4 12 13 14
@ 30 'FAIROAKS.SHY'
*
$RUNOFF
*
*
A1 'Fair Oaks Estates Subdivision containing  15 subareas 10-yr storm event'
A1 'File = FAIROAKS.SWM  created:  06-21-1993  13:20:01'
*
*  Metric Snow   No.  Infilt.Type  Quality    Evap
*   0=no  0=no  Hyet.  0=Horton     0=no      0=no    Hr    Min   Day  Month Yr
B1   0     0      1       0           0         0      0      0    21     6  93
*
*  Print control lines
*   input   graphs   totals
B2   0        0        2
*
*
*             Time steps                 simulation length
*      WET          WETDRY       DRY       UNITS    LENGTH
B3       60           300        600         1         120
*
*
*
*-------------------------------------------------------------------------
*                               HYTCREAT.BAS
*                          Hyetograph file creator
*                                Version 1.0
*                                April 1993
*
*                     EPA SWMM 4.05 Rainfall cards (D1-E3)
*                            from file:  FAIROAKS.SE1
*
```

TABLE 9.29. *(Continued)*

```
*                      0=       # Vary  units     type    # Pts. delta time  initial
*                   suppress   time    0=min   0=inten.         in minutes   time
*    KTYPE  KINC    KPRINT    KTHIS   KTIME    KPREP   NHISTO   THISTO     TZRAIN
E1     2      1        0         0       0        0        61       1         0.0
*
*  Rainfall Hyetograph cards
*
*      Time (min)   Intensity (in/hr)
E3          0.00          4.800
E3          1.00          4.800
E3          2.00          5.340
E3          3.00          5.700
E3          4.00          5.700
E3          5.00          5.880
E3          6.00          5.820
E3          7.00          5.280
E3          8.00          5.160
E3          9.00          4.740
E3         10.00          4.080
E3         11.00          4.080
E3         12.00          3.540
E3         13.00          3.540
E3         14.00          2.880
E3         15.00          2.400
E3         16.00          2.340
E3         17.00          2.040
E3         18.00          2.040
E3         19.00          1.740
E3         20.00          1.680
E3         21.00          1.620
E3         22.00          1.440
E3         23.00          1.440
E3         24.00          1.200
E3         25.00          1.200
E3         26.00          1.200
E3         27.00          1.200
E3         28.00          1.200
E3         29.00          1.200
E3         30.00          0.960
E3         31.00          0.960
E3         32.00          0.960
E3         33.00          0.960
E3         34.00          0.960
E3         35.00          0.960
E3         36.00          0.840
E3         37.00          0.840
E3         38.00          0.840
E3         39.00          0.840
E3         40.00          0.840
E3         41.00          0.840
E3         42.00          0.780
E3         43.00          0.780
E3         44.00          0.780
E3         45.00          0.780
E3         46.00          0.780
E3         47.00          0.780
E3         48.00          0.780
E3         49.00          0.780
E3         50.00          0.780
E3         51.00          0.780
E3         52.00          0.780
E3         53.00          0.780
E3         54.00          0.720
E3         55.00          0.720
E3         56.00          0.720
E3         57.00          0.720
E3         58.00          0.720
E3         59.00          0.720
E3         60.00          0.000
*----------------------------------
*--- End of Rainfall Data Cards ---
*
```

TABLE 9.29. *(Continued)*

```
* ------------------- SWMM Runoff Block Subarea data cards ------------------------------------------
*ID HYT Subarea Inlet     Width         Area  Imperv.  Slope    Imperv     Perv.  Imperv Perv     Max
*No No     no     no       (ft)         (Ac)  percent (ft/ft)          'n'          Deprs. Stor. Horton
*----------------------------------------------------------------------------------------------------


H1   1     1    6001      370.0         0.60  30.00  0.02000 0.020000 0.200000 0.100 0.200    7.5000
H1   1     3    6003      320.0         0.50  40.00  0.02500 0.020000 0.200000 0.100 0.200    7.5000
H1   1     9    6007      450.0         2.80  20.00  0.03600 0.020000 0.200000 0.100 0.200    7.5000
H1   1    10    6008      410.0         0.60  48.00  0.01400 0.020000 0.200000 0.100 0.200    7.5000
H1   1    12    6010      250.0         0.70   9.00  0.01500 0.020000 0.200000 0.100 0.200    7.5000
H1   1    13    6011      200.0         0.40   7.00  0.01500 0.020000 0.200000 0.100 0.200    7.5000
H1   1    15    6012      200.0         0.90  31.00  0.01200 0.020000 0.200000 0.100 0.200    7.5000
H1   1     4    6013      400.0         1.30  20.00  0.02700 0.020000 0.200000 0.100 0.200    7.5000
H1   1     5    6014       60.0         0.20  45.00  0.02000 0.020000 0.200000 0.100 0.200    7.5000
H1   1    17    6015      440.0         0.80  41.00  0.02100 0.020000 0.200000 0.100 0.200    7.5000
H1   1    18    6016      320.0         1.50  28.00  0.01600 0.020000 0.200000 0.100 0.200    7.5000
H1   1    19    6017      450.0         0.50  50.00  0.01200 0.020000 0.200000 0.100 0.200    7.5000
H1   1     7    6018      300.0         1.50  22.00  0.01700 0.020000 0.200000 0.100 0.200    7.5000
H1   1     8    6019      300.0         0.50  56.00  0.01300 0.020000 0.200000 0.100 0.200    7.5000
H1   1    14    6020      220.0         0.40  31.00  0.01500 0.020000 0.200000 0.100 0.200    7.5000
*
*    Inlet Print Control
*     NPRNT     INTERV
M1       0          0
*
*
*****************END OF RUNOFF BLOCK DATA INPUT**************
*
. ---------------------------------------------------------------------------------------------------

$TRANSPORT
*
*
A1 'Fair Oaks Estates Subdivision containing  15 subareas 10-yr storm event'
A1 'File = FAIROAKS.SWM  created:  06-21-1993  13:21:31'
*
B0 1 0 0 1
*
*       NDT    NINPUT  NNYN   NNPE  NOUTS  NPRINT NPOLL NITER IDATEZ  METRIC  INTPRT
B1      120       0      0      0      1      0      0     4  000000     0       0
*
*      DT     EPSIL  DWDAYS  TZERO   GNU      TRIBA
B2      60   .0001     0       0   .000001     0
*
*   NCNTRL    NINFIL   NFILTH   NDESN
B3   0          0        0        0
*
*   NKLASS   KPRINT
C1   0          0
*
```

TABLE 9.29. *(Continued)*

```
*
* ------------------- SWMM Transport Block Element data cards -----------------------
*ID  NO  1stup 2ndup 3rdup Type Length  Geom1   Slope    'n'   Geom2 Barrel Geom3
*-----------------------------------------------------------------------------------
*
*---------------------------------- Manhole cards-----------------------------------
E1  6001     0     0     0  19      0.00    0.00   0.000 0.0000    0.00      0    0.00
E1  6002  1001     0     0  19      0.00    0.00   0.000 0.0000    0.00      0    0.00
E1  6003  1002     0     0  19      0.00    0.00   0.000 0.0000    0.00      0    0.00
E1  6004  1003  1014     0  19      0.00    0.00   0.000 0.0000    0.00      0    0.00
E1  6005  1004  1012     0  19      0.00    0.00   0.000 0.0000    0.00      0    0.00
E1  6006  1017  1005     0  19      0.00    0.00   0.000 0.0000    0.00      0    0.00
E1  6007     0     0     0  19      0.00    0.00   0.000 0.0000    0.00      0    0.00
E1  6008  1007     0     0  19      0.00    0.00   0.000 0.0000    0.00      0    0.00
E1  6009  1008  1019     0  19      0.00    0.00   0.000 0.0000    0.00      0    0.00
E1  6010  1009     0     0  19      0.00    0.00   0.000 0.0000    0.00      0    0.00
E1  6011  1010     0     0  19      0.00    0.00   0.000 0.0000    0.00      0    0.00
E1  6012  1011  1020     0  19      0.00    0.00   0.000 0.0000    0.00      0    0.00
E1  6013     0     0     0  19      0.00    0.00   0.000 0.0000    0.00      0    0.00
E1  6014  1013     0     0  19      0.00    0.00   0.000 0.0000    0.00      0    0.00
E1  6015     0     0     0  19      0.00    0.00   0.000 0.0000    0.00      0    0.00
E1  6016  1015     0     0  19      0.00    0.00   0.000 0.0000    0.00      0    0.00
E1  6017  1016     0     0  19      0.00    0.00   0.000 0.0000    0.00      0    0.00
E1  6018     0     0     0  19      0.00    0.00   0.000 0.0000    0.00      0    0.00
E1  6019  1018     0     0  19      0.00    0.00   0.000 0.0000    0.00      0    0.00
E1  6020     0     0     0  19      0.00    0.00   0.000 0.0000    0.00      0    0.00
*
*---------------------------------- Outlet Manhole----------------------------------
E1  5000  1006     0     0  19      0.00    0.00   0.000 0.0000    0.00      0    0.00
*
*---------------------------------- Element cards-----------------------------------
E1  1001  6001     0     0   1     22.00    1.00   0.002 0.0130    1.00      1    0.00
E1  1002  6002     0     0   1    324.00    1.00   0.024 0.0130    1.00      1    0.00
E1  1003  6003     0     0   1     76.00    1.00   0.010 0.0130    1.00      1    0.00
E1  1004  6004     0     0   2    290.00    1.25   0.005 0.0130    1.30      1    0.00
E1  1005  6005     0     0   2    195.00    1.75   0.006 0.0130    1.80      1    0.00
E1  1006  6006     0     0   2     30.00    1.75   0.011 0.0130    1.80      1    0.00
E1  1007  6007     0     0   2     28.00    1.00   0.008 0.0130    1.00      1    0.00
E1  1008  6008     0     0   2    115.00    1.25   0.006 0.0130    1.30      1    0.00
E1  1009  6009     0     0   2     26.00    1.25   0.016 0.0130    1.30      1    0.00
E1  1010  6010     0     0   2    300.00    1.50   0.005 0.0130    1.50      1    0.00
E1  1011  6011     0     0   1     76.00    1.50   0.014 0.0130    1.50      1    0.00
E1  1012  6012     0     0   2     52.00    1.50   0.008 0.0130    1.50      1    0.00
E1  1013  6013     0     0   2     32.00    1.00   0.002 0.0130    1.00      1    0.00
E1  1014  6014     0     0   2     23.00    1.00   0.003 0.0130    1.00      1    0.00
E1  1015  6015     0     0   2     45.00    1.00   0.002 0.0130    1.00      1    0.00
E1  1016  6016     0     0   2     44.00    1.00   0.010 0.0130    1.00      1    0.00
E1  1017  6017     0     0   2     25.00    1.00   0.017 0.0130    1.00      1    0.00
E1  1018  6018     0     0   2     28.00    1.00   0.003 0.0130    1.00      1    0.00
E1  1019  6019     0     0   1    144.00    1.00   0.019 0.0130    1.00      1    0.00
E1  1020  6020     0     0   1     56.00    1.00   0.002 0.0130    1.00      1    0.00
*-----------------------------------------------------------------------------------
*
*  Outfall hydrograph sent to COMBINE
H1 5000
*-----------------------------------------------------------------------------------
*
*   Create a Hydrograph File Named 'FAIROAKS.SHY' on the current drive
$COMBINE
*
A1 4
B1 'Fair Oaks Estates Subdivision  created:  06-21-1993  13:21:45'
B1 '15 subareas 10-yr storm event'
B2 0 0
C1 0 0
*
$ENDPROGRAM
*
```

TABLE 9.29. *(Continued)*

Outflow Hydrograph from Line No. 6 (node 5000)		
Time, Minutes	Flow, cfs	Flow, m^3/s
1	0.119	0.0033
2	1.260	0.0357
3	4.311	0.1221
4	10.01	0.2835
5	15.53	0.4400
6	18.86	0.5341
7	20.43	0.5786
8	20.74	0.5874
9	20.41	0.5780
10	19.60	0.5550
11	18.31	0.5185
12	17.02	0.4820
13	15.72	0.4452
14	14.59	0.4132
15	13.51	0.3826
16	12.19	0.3452
17	10.92	0.3092
18	9.735	0.2757
19	8.814	0.2496
20	8.112	0.2297
21	7.490	0.2121
22	6.978	0.1976
23	6.502	0.1841
24	6.102	0.1728
25	5.715	0.1618
26	5.329	0.1509
27	5.015	0.1420
28	4.766	0.1350
29	4.609	0.1305
30	4.528	0.1282
31	4.407	0.1248
32	4.230	0.1198
33	4.044	0.1145
34	3.857	0.1092
35	3.719	0.1053
36	3.643	0.1032
37	3.565	0.1010
38	3.471	0.0983
39	3.376	0.0956
40	3.280	0.0929
41	3.205	0.0908
42	3.161	0.0895
43	3.119	0.0883
44	3.071	0.0870
45	3.023	0.0856
46	2.975	0.0842
47	2.936	0.0831
48	2.913	0.0825
49	2.899	0.0821
50	2.891	0.0819
51	2.887	0.0818
52	2.887	0.0818
53	2.883	0.0816
54	2.883	0.0816

into four basic groups: 1) Input/Output Control, 2) Rainfall Information, 3) Basin Characteristics, and 4) Conveyance Characteristics.

Input and output control is generally a function of the specific hardware used to run the model and the specific information desired for evaluation. The SWMM 4.05 users manual provides detailed information for executing the simulation with the specific hardware available and is not repeated here. Specific information desired by the modeler is unique to the particular application and can be requested by referring to the users manual. This information will not be discussed in detail in this handbook.

Rainfall information for the Fair Oaks Estates example is required. The rainfall depth and distribution are established using local recommendations. A one-minute time increment was selected due to the small size and rapid runoff response of the watershed. This version of SWMM requires that rainfall data be entered as a rainfall intensity in units of inches/hour. One value of intensity is entered for each time increment during the rainfall event. In this example, a 2.0-inch 60-minute rainfall representing a 10-year event was applied, with a 1-minute early peaking distribution.

Subarea characteristics used by SWMM to simulate basin response include connectivity, basin geometry and hydrologic abstraction information. Basin connectivity information includes the subarea designation and the channel/pipe element number into which subarea runoff flows. This information is based upon a naming convention for subareas and conveyance elements. Basin geometry includes a subarea width that can be approximated as the area divided by the average overland flow length for runoff to reach the main collector channel. Area, percent imperviousness of the basin, and slope of the overland surface are input. A resistance factor for both pervious and impervious surfaces is also required for the simulation. These resistance factors must represent shallow surface flows, rather than the conventional Manning's roughness for open channels and pipe flow (see Table 9.15).

Rainfall losses are represented by a surface depression depth and, for pervious surfaces, infiltration parameters based on Horton's equation. Impervious surface storage is represented by surface depression depth, which must be satisfied before runoff from impervious surfaces can begin. The storage on pervious surfaces is greater, reflecting the higher degree of surface irregularity. Infiltration rates used in Horton's equation are a function of soil type and antecedent moisture condition and may be determined from tests or soil surveys.

Conveyance input in the Transport Block consists of connectivity information, a description of type of conveyance section, the geometry of the section, and the hydraulic properties. For each conveyance element, a downstream connection must be identified which establishes the connectivity and the structure of the conveyance system. The type of conveyance structure must be identified, as well as other geometric information such as size, shape, slope, length, and the Manning roughness coefficient. We note again that very simple storm sewer systems involving minor pipes and gutters can also be simulated by the Runoff Block.

The sample input and output for SWMM in Table 9.29 is only an abbreviated summary. Other more detailed summaries presenting subbasin hydrographs and flow velocity in the conveyance elements can be requested. The users manual provides the necessary instructions.

E. Illinois Urban Drainage Area Simulation

ILLUDAS, the acronym for the Illinois Urban Drainage Area Simulator, is a hydrologic model developed and supported by the Illinois State Water Survey (Terstriep and Stall, 1974). This program simulates the hydrologic and water quality response of a watershed through the application of a time–area unit hydrograph and a simple channel routing scheme. The program combines individual subarea hydrographs through the use of a defined conveyance system. Backwater effects cannot be simulated by ILLUDAS. Once again, the Fair Oaks Estates subdivision was modeled to demonstrate the application of ILLUDAS. The model requires the definition of rainfall, basin parameters, and conveyance elements.

Table 9.30 presents example input/output files for the ILLUDAS model of the Fair Oaks Estates subdivision described earlier. The menu-driven program generates an input file, executes the model, and

TABLE 9.30. ILLUDAS Input/Output for Fair Oaks Estates.

```
ILLUDAS --- ILLINOIS URBAN DRAINAGE AREA SIMULATOR
            Version 2.1   Updated Sep 1988

(C)Copyright -  Illinois State Water Survey - DENR, 1987

          TIME-SHIFT  ROUTING METHOD ACTIVATED


          FAIR OAK ESTATES
          EXAMPLE HYDROLOGY
          ILLUDAS MODEL


                    RAINFALL PATTERN

0.000 0.080 0.080 0.089 0.095 0.095 0.098 0.097 0.088 0.086
0.079 0.068 0.068 0.059 0.059 0.048 0.040 0.039 0.034 0.034
0.029 0.028 0.027 0.024 0.024 0.020 0.020 0.020 0.020 0.020
0.020 0.016 0.016 0.016 0.016 0.016 0.016 0.014 0.014 0.014
0.014 0.014 0.014 0.013 0.013 0.013 0.013 0.013 0.013 0.013
0.013 0.013 0.013 0.013 0.013 0.012 0.012 0.012 0.012 0.012
0.012

                    USER-DEFINED

          Histogram of  Distributed Rainfall

Time since start:  0      12      24      36      48      60 Mn
Accumulated rain:  0   1.023   1.468   1.684   1.846   1.996 In
Percent of storm:  0   51.25   73.55   84.37   92.48   100.0  %

    RUN NUMBER   BASIN AREA   TIME INCREMENT   SOIL TYPE
                   ACRES         MINUTES       1234=ABCD

        1          13.40           1.0             2

  TOTAL RAIN   FREQUENCY   DURATION   AMC   IMP ABS   PER ABS
    INCHES       YEARS     MINUTES    PCT   INCHES    INCHES

     2.00          0          60      33.    0.10      0.20

                    STORM SEWER LINE NUMBER

---- BEGIN BRANCH   1 - REACH   0 ----         1


- B-R -  1 -  0 -----------------------------------------------------
LOCAL     AREA : CPA =   0.20        SPA =   0.06       CGA =   0.37
UPSTREAM AREA : CPA =   0.20        SPA =   0.06       CGA =   0.37
RAINFALL  FACTOR = 1.000      BASE FLOW =   0.00 CFS     SOIL  TYPE = 2

REACH :    LENGTH =     221 FT       SLOPE = 0.20 %       ROUGH = 0.0130
PAVED FLOW :   LEN =    270 FT       SLP =  1.970 %    ENTRY =  3.27 MIN
GRASS FLOW :   LEN =     20 FT       SLP =  2.000 %    ENTRY = 15.72 MIN

PEAKS(CFS) :   INLET =    1.164   DESIGN =   1.164   OUTLET =   1.161
VOLS :   ROUTED =          1535   GROSS =         1535   %PAVED = 88.95

-------------------  DESIGN   CHARACTERISTICS   ------------------
PIPE :   DIAMETER =  12 INCHES                  ROUGHNESS = 0.0130
DESIGN :   CAP =   1.592 CFS     VEL = 2.027 FPS    TRV =   1.664 MIN
--------------------------------------------------------------------


ROUTED DESIGN HYDROGRAPH
0.000 0.000 0.000 0.074 0.330 0.664 0.959 1.104 1.150 1.159
1.161 1.153 1.113 1.058 1.005 0.950 0.905 0.830 0.753 0.683
0.632 0.599 0.565 0.539 0.511 0.475 0.420 0.357 0.311 0.279
0.258 0.247 0.235 0.220 0.206 0.195 0.192 0.192 0.190 0.182
```

TABLE 9.30. *(Continued)*

```
- B-R -  4 -  1 --------------------------------------------------
LOCAL     AREA : CPA =    0.32        SPA =   0.12       CGA =   1.11
UPSTREAM AREA : CPA =    0.58        SPA =   0.18       CGA =   1.58
RAINFALL FACTOR = 1.000      BASE FLOW =   0.00 CFS     SOIL  TYPE = 2

REACH :   LENGTH =       44 FT       SLOPE = 1.04 %      ROUGH = 0.0130
PAVED FLOW :   LEN =    330 FT       SLP =  2.300 %   ENTRY =  3.43 MIN
GRASS FLOW :   LEN =    160 FT       SLP =  1.300 %   ENTRY = 36.46 MIN

PEAKS(CFS) :   INLET =    1.852   DESIGN =   3.350   OUTLET =   3.340
VOLS :   ROUTED =         4370   GROSS =         4370   %PAVED = 90.59

------------------  DESIGN   CHARACTERISTICS  ------------------
PIPE :   DIAMETER =  12 INCHES                   ROUGHNESS = 0.0130
DESIGN :    CAP =   3.631 CFS     VEL = 4.623 FPS    TRV =   0.140 MIN
-------------------------------------------------------------------

ROUTED DESIGN HYDROGRAPH
0.000  0.000  0.447  1.258  2.198  2.907  3.239  3.340  3.304  3.235
3.132  2.950  2.787  2.586  2.443  2.261  2.038  1.828  1.634  1.513
1.408  1.319  1.245  1.172  1.120  1.052  0.993  0.949  0.925  0.920
0.919  0.890  0.850  0.793  0.744  0.718  0.701  0.675  0.646  0.622
0.610  0.607  0.607  0.600  0.590  0.573  0.543  0.512  0.488  0.471
0.460  0.453  0.452  0.452  0.452  0.445  0.435  0.425  0.419  0.418
0.418  0.328  0.209  0.088  0.016  0.001


---- BEGIN BRANCH   4 - REACH   2 ----   17

 TRAVEL TIME SO SMALL THAT ROUTED APPROXIMATELY = DESIGN HYDROGRAPH


- B-R -  4 -  2 --------------------------------------------------
LOCAL     AREA : CPA =    0.20        SPA =   0.06       CGA =   0.26
UPSTREAM AREA : CPA =    0.78        SPA =   0.24       CGA =   1.84
RAINFALL FACTOR = 1.000      BASE FLOW =   0.00 CFS     SOIL  TYPE = 2

REACH :   LENGTH =       25 FT       SLOPE = 1.71 %      ROUGH = 0.0130
PAVED FLOW :   LEN =    200 FT       SLP =  0.450 %   ENTRY =  3.97 MIN
GRASS FLOW :   LEN =     60 FT       SLP =  2.000 %   ENTRY = 23.29 MIN

PEAKS(CFS) :   INLET =    1.180   DESIGN =   4.515   OUTLET =   4.506
VOLS :   ROUTED =         5930   GROSS =         5930   %PAVED = 89.79

------------------  DESIGN   CHARACTERISTICS  ------------------
PIPE :   DIAMETER =  12 INCHES                   ROUGHNESS = 0.0130
DESIGN :    CAP =   4.656 CFS     VEL = 5.928 FPS    TRV =   0.062 MIN
-------------------------------------------------------------------

ROUTED DESIGN HYDROGRAPH
0.000  0.000  0.589  1.642  2.861  3.866  4.343  4.506  4.485  4.417
4.282  4.033  3.821  3.552  3.354  3.116  2.814  2.541  2.272  2.103
1.963  1.841  1.743  1.641  1.572  1.482  1.401  1.343  1.306  1.299
1.285  1.223  1.146  1.056  0.978  0.936  0.908  0.868  0.829  0.798
0.779  0.775  0.775  0.765  0.753  0.734  0.701  0.670  0.646  0.628
0.616  0.610  0.608  0.608  0.608  0.599  0.586  0.573  0.564  0.562
0.562  0.444  0.290  0.132  0.023  0.002


---- BEGIN BRANCH   1 - REACH   5 ----   6

 TRAVEL TIME SO SMALL THAT ROUTED APPROXIMATELY = DESIGN HYDROGRAPH
```

TABLE 9.30. *(Continued)*

```
- B-R -  1 -  5 ----------------------------------------------------
LOCAL      AREA : CPA =    0.00        SPA =    0.00       CGA =    0.00
UPSTREAM AREA : CPA =    2.83        SPA =    0.98       CGA =    9.63
RAINFALL  FACTOR = 1.000       BASE FLOW =    0.00 CFS     SOIL  TYPE = 2

REACH :     LENGTH =       30 FT        SLOPE = 1.13 %          ROUGH = 0.0130

PEAKS(CFS) :     INLET =    0.000   DESIGN =   15.824    OUTLET =   15.823
VOLS :    ROUTED =          21347   GROSS =         21347    %PAVED = 90.49

------------------     DESIGN   CHARACTERISTICS   ------------------
PIPE :    DIAMETER =  21 INCHES                        ROUGHNESS = 0.0130
DESIGN :     CAP =  16.833 CFS     VEL = 6.998 FPS      TRV =    0.063 MIN
--------------------------------------------------------------------


ROUTED DESIGN HYDROGRAPH
 0.00    0.00    0.63    2.20    4.89    8.21   11.27   13.62   15.08   15.76
15.82   15.44   14.87   14.11   13.36   12.55   11.66   10.73    9.75    8.88
 8.13    7.49    6.98    6.54    6.18    5.84    5.52    5.22    4.95    4.74
 4.56    4.37    4.19    3.96    3.74    3.56    3.43    3.33    3.25    3.17
 3.08    2.99    2.91    2.84    2.76    2.66    2.55    2.45    2.38    2.34
 2.31    2.29    2.27    2.26    2.25    2.23    2.21    2.17    2.14    2.11
 2.08    1.94    1.67    1.26    0.82    0.46    0.22    0.08    0.02


--------------------------------------------------------------------

OUTFALL HYDROGRAPH IN CFS   RUNOFF VOLUME IN CU FT =          21347

 0.00    0.00    0.63    2.20    4.89    8.21   11.27   13.62   15.08   15.76
15.82   15.44   14.87   14.11   13.36   12.55   11.66   10.73    9.75    8.88
 8.13    7.49    6.98    6.54    6.18    5.84    5.52    5.22    4.95    4.74
 4.56    4.37    4.19    3.96    3.74    3.56    3.43    3.33    3.25    3.17
 3.08    2.99    2.91    2.84    2.76    2.66    2.55    2.45    2.38    2.34
 2.31    2.29    2.27    2.26    2.25    2.23    2.21    2.17    2.14    2.11
 2.08    1.94    1.67    1.26    0.82    0.46    0.22    0.08    0.02
```

creates an output file for printing or review. ILLUDAS operates in both the analysis mode and the design mode. In the analysis mode, an existing system can be defined and the hydrologic characteristics of the system determined. In the design mode, ILLUDAS recommends pipe sizes necessary to convey the flows predicted during the hydrologic simulation. For this application, ILLUDAS was used in the design mode.

The input required in the various menus is thoroughly described in the user's manual. In general, the input is divided into three main categories: 1) input/output/evaluation control, 2) rainfall data, and 3) basin and conveyance characteristics.

The program prompts for various file designations and output level and requires user input. It also requests information regarding the design or evaluation mode for the analysis. Once this information has been supplied, the program prompts for specific simulation data. To modify or alter input files, the user returns to the main menu to select various options.

For the Fair Oaks Estates ILLUDAS simulation, the rainfall depth and distribution were manually input as incremental rainfall depths for the 2.0-inch 60-minute storm that was used previously in the SWMM simulation. This information is repeated and printed in the output file.

ILLUDAS defines basins and conveyance lengths through a branch and reach designation. Each stem of the storm sewer is defined by a unique combination of branch and reach identifiers. For the purpose of clarification, the branch and reach identifiers in the output file are appended with the conveyance element designation as shown in Fig. 9.27. A local tributary subarea is defined for each branch and reach. Runoff input data must be supplied for those areas that are considered directly connected paved areas, such as roadways, parking lots and some roof tops and driveways. The areas that are paved but are not directly connected are also identified, as are the contributing grassed areas. Information defining the character of overland flow is also requested and includes the length of paved and grassed overland flow and the respective slopes. A soil type is also needed to verify the appropriate losses from the grassed areas.

In the design mode, no pipe diameters are specified; however, the pipe roughness, length, and slope are input requirements. Using this information, the program generates a subarea hydrograph which is routed through the pipe system. In areas where pipe sizes or slopes change, but where there is no additional tributary area, only pipe sizing parameters are provided and the program routes upstream flows. The program computes a routed design hydrograph for each branch and reach.

The results of the storm sewer computer simulations by Rational Formula, EPA SWMM Runoff Block, EPA SWMM Runoff plus Transport Blocks, and ILLUDAS for the 10-year storm on the Fair Oaks Estates watershed are summarized for four different locations in Table 9.31. The outfall is located at line 6. The models generally produce discharge peaks that are substantially the same even though they are based on conceptually different rainfall–runoff methods. In the absence of gaged runoff records, it is important to make this kind of comparison between dissimilar hydrologic techniques. It is especially interesting that the explicit treatment of storm sewers in the EPA Transport Block does not provide any substantial improvement over the Runoff Block by itself in this simple system.

F. Detention Basin Analysis and Design

Computer models can be used very effectively to assist in the evaluation and design of detention basins. Detention basins have long been recognized as an important component of a storm water management program. In fact, most communities now require through their stormwater ordinances that detention facilities be provided to control future basin discharges and limit releases to levels no greater than historic or predevelopment levels. Most ordinances require that discharges be limited to historic levels for multiple events, such as the 2-year, 10-year, 25-year, and 50-year storms. This requires the design of a multi-stage outlet structure.

The following example presents the sizing of a multiple-stage detention structure that regulates flow from the Fair Oaks Estates development. A simple computer program called SWIRM–ROUTE (Small Watershed Interactive Rational Method) is used to size the required storage volume and the outlet geometry for controlling the 1-year, 10-year and 100-year events. The objective is to limit developed basin flows to predevelopment levels.

The initial step in this process is to compute the predevelopment and developed discharge peaks from Fair Oaks Estates. This has been computed by SWIRM as shown in Table 9.32. SWIRM is a model developed at Penn State University for instructional purposes. It uses the universal Rational Method to generate hydrographs from very small urban sub-basins and then routes them through trial stormwater detention facilities by the storage-indication method. A simple rational formula computation for the aggregated basins is used to determine the historic peak flows. Since volummetric requirements are not imposed by the discharge regulations, no hydrograph generation for the historic condition is necessary. A detailed example of the input data for the 10-year event is presented in Table 9.32. Note that the

TABLE 9.31. Summary of 10-year Discharge Peaks from Fair Oaks Estates by Different Models.

	SWIRM rational method		SWMM runoff + transport		SWMM runoff block		ILLUDAS	
Location (line no.)	m^3/s	cfs	m^3/s	cfs	m^3/s	cfs	m^3/s	cfs
4	0.113	4	0.113	4	0.113	4	0.113	4
10	0.226	8	0.255	9	0.226	8	0.198	7
15	0.028	1	0.056	2	0.056	2	0.028	1
6	0.396	14	0.595	21	0.566	20	0.453	16

TABLE 9.32. SWIRM Input/Output for Fair Oaks Estates.

%%%%%%%%%%%% Predevelopment Flow Computations %%%%%%%%%%%%

Subarea ID	acres	TC min.	Wght'd C	Rain in/hr.	Qp cfs
1	13.4	60.0	0.15	2.15	4.3

%%

%%%%%%%%%%%% Postdevelopment Flow Computations %%%%%%%%%%%%

Subarea ID.	Subarea ac.	Overland Flow L,ft	Overland Slope ft/ft	Mann. n	C	Tc min.	Total Area ac.	Wght. C	Comb. Tc min.	Rain Int. in/hr.	Qp cfs
1	0.6	100	0.050	0.200	0.28	20.0	0.6	0.28	20.0	4.03	0.7
2	0.0	100	0.050	0.200	0.28	8.5	0.6	0.28	20.4	3.99	0.7
3	0.5	100	0.050	0.200	0.44	20.0	1.2	0.35	24.3	3.66	1.5
4	1.3	100	0.050	0.200	0.32	20.0	1.3	0.32	20.0	4.03	1.7
5	0.2	100	0.050	0.200	0.47	20.0	1.5	0.34	20.2	4.01	2.1
6	0.0	100	0.050	0.200	0.32	8.5	2.7	0.35	25.0	3.61	3.4
7	1.5	100	0.050	0.200	0.33	20.0	1.5	0.33	20.0	4.03	2.0
8	0.5	100	0.050	0.200	0.54	20.0	2.0	0.38	20.2	4.01	3.1
9	2.8	100	0.050	0.200	0.32	25.0	2.8	0.32	25.0	3.61	3.3
10	0.6	100	0.050	0.200	0.49	20.0	3.5	0.35	25.2	3.59	4.4
11	0.0	100	0.050	0.200	0.32	8.5	5.5	0.36	25.8	3.55	7.1
12	0.7	100	0.050	0.200	0.25	20.0	6.2	0.35	25.8	3.55	7.7
13	0.4	100	0.020	0.200	0.25	20.0	6.6	0.34	27.4	3.44	7.8
14	0.4	100	0.050	0.200	0.39	20.0	0.4	0.39	20.0	4.03	0.6
15	0.9	100	0.020	0.200	0.38	20.0	7.9	0.35	27.6	3.43	9.4
16	0.0	100	0.050	0.200	0.38	8.5	10.6	0.35	27.9	3.41	12.6
17	0.8	100	0.050	0.200	0.45	20.0	0.8	0.45	20.0	4.03	1.4
18	1.5	100	0.050	0.200	0.37	20.0	2.3	0.40	20.4	3.99	3.7
19	0.5	100	0.020	0.200	0.50	20.0	2.8	0.42	20.7	3.97	4.7
20	0.0	100	0.050	0.200	0.50	8.3	13.4	0.36	28.9	3.34	16.3
21	0.0	100	0.020	0.200	0.50	10.8	13.5	0.36	29.0	3.33	16.3

%%%%%%%%%%%%% Culvert Details %%%%%%%%%%%%

Inlets From	Inlets To	Ground Elev. Upst.	Ground Elev. Dwnst.	Invert Elev. Upst.	Invert Elev. Dwnst.	Culvert L,ft	Culvert D,in	Slope ft/ft	Design Q, cfs
	***** For Ditches and Gutters, D= depth in inches *****								
1	2	773.90	774.00	769.40	769.33	22	12	0.0030	0.7
2	3	774.00	766.10	769.08	761.60	324	12	0.0231	1.1
3	6	766.10	765.90	761.35	761.12	76	12	0.0030	1.5
4	5	765.50	765.50	761.00	760.90	32	12	0.0030	1.8
5	6	765.50	765.90	760.65	760.59	23	15	0.0030	2.1
6	16	765.90	764.30	760.34	759.47	290	15	0.0030	3.4
7	8	770.20	770.20	765.45	765.37	28	15	0.0030	2.5
8	11	770.20	767.20	765.12	762.70	144	12	0.0168	3.1
9	10	768.30	768.30	763.30	763.22	28	18	0.0030	3.7
10	11	768.30	767.20	762.97	762.45	115	15	0.0045	4.4
11	12	767.20	766.80	762.01	761.69	26	15	0.0125	7.3
12	13	766.80	765.80	761.19	760.29	300	21	0.0030	7.8
13	15	765.80	764.10	760.04	758.89	76	15	0.0151	8.0
14	15	764.20	764.10	759.70	759.14	56	12	0.0100	0.9
15	16	764.10	764.30	758.14	757.98	52	24	0.0030	9.4
16	20	764.30	763.60	757.73	757.15	195	27	0.0030	12.6
17	18	763.30	763.50	758.55	758.41	45	15	0.0030	2.4
18	19	763.50	763.40	758.16	758.03	44	18	0.0030	4.1
19	20	763.40	763.60	757.78	757.65	25	15	0.0052	4.7
20	21	763.60	763.20	756.65	756.56	30	27	0.0030	16.3

TABLE 9.32. *(Continued)*

```
%%%%%%       Inout Hydrographs by    Univ.R.M.       %%%%%%

Return Periods:    1    10    100   yrs.

Hydrograph Routing Interval =  7.0 minutes

Ret. Per.     1 yrs.     Pre-dev. Q =  2.2 cfs
Discharge Rates, cfs
  0.0     0.2     0.3     0.5     0.6     0.7     0.8     0.8     0.9     2.5
  4.1     5.7     7.3     6.0     4.7     3.3     2.0     1.8     1.6     1.4
  1.1     1.0     0.9     0.8     0.7     0.7     0.6     0.6     0.6     0.6
  0.6     0.5     0.5     0.5     0.5     0.5     0.5     0.5     0.5     0.4
  0.4     0.0     0.0     0.0     0.0     0.0     0.0     0.0     0.0     0.0

Ret. Per.    10 yrs.     Pre-dev. Q =  3.7 cfs
Discharge Rates, cfs
  0.0     0.4     0.8     1.2     1.5     1.7     1.8     2.0     2.1     5.0
  8.0    10.9    13.8    11.5     9.1     6.8     4.4     4.0     3.6     3.2
  2.7     2.5     2.3     2.0     1.8     1.7     1.6     1.5     1.4     1.4
  1.3     1.3     1.3     1.2     1.2     1.2     1.1     1.1     1.1     1.1
  1.1     0.0     0.0     0.0     0.0     0.0     0.0     0.0     0.0     0.0

Ret. Per.   100 yrs.     Pre-dev. Q =  6.0 cfs
Discharge Rates, cfs
  0.0     0.7     1.4     2.1     2.8     3.1     3.4     3.6     3.9     9.1
 14.2    19.4    24.6    20.7    16.8    12.8     8.9     7.9     7.0     6.0
  5.1     4.6     4.2     3.7     3.3     3.1     2.9     2.7     2.5     2.5
  2.4     2.3     2.3     2.2     2.2     2.2     2.1     2.1     2.0     2.0
  1.9     0.0     0.0     0.0     0.0     0.0     0.0     0.0     0.0     0.0

 Detention Pond Data
  El. =  754.00  Top =  761.00  Length =  300 ft

Interpolated Areas and Volumes
  Elev.         Area        Volume
               acres        ac-ft
  754.0        0.010         0.00
  755.0        0.110         0.05
  756.0        0.160         0.19
  757.0        0.193         0.36
  758.0        0.230         0.57
  759.0        0.269         0.82
  760.0        0.310         1.11
  761.0        0.360         1.45
  762.0        0.414         1.83


Starting Water Surface Elevation in Pond =  754.0
Detention Pond Outlet Structure
Orif.  No.of   Tilt   Culv.  Width  Height   Area    Cd    Invert
 No.   orif.  angle   type    ft     ft      sqft
  1      1     0.0    circ   0.00    0.66    0.34   0.60    754.0
  2      1     0.0    rect   1.00    0.25    0.25   0.60    757.0

Spillway    Side     Cd    Invert
No. L,ft   slope
 1     0     1.0   3.00     759.5
 2    20     0.5   2.50     760.5

Starting W.S. Elevation in Pond =  754.0
```

TABLE 9.32. *(Continued)*

Univ.R.M. Time min.	Hydrogr. for Q-in cfs	2S/dt-Q cfs	1 - year Ret.Per. 2S/dt+Q cfs	Q-out cfs	Storage ac-ft	Elev. ft.
0.0	0.0	0.0		0.0	0.0	754.0
7.00	0.2	0.1	0.2	0.0	0.00	754.03
14.00	0.3	0.4	0.6	0.1	0.00	754.09
21.00	0.5	0.8	1.2	0.2	0.00	754.19
28.00	0.6	1.4	2.0	0.3	0.01	754.31
35.00	0.7	1.9	2.7	0.4	0.01	754.43
42.00	0.8	2.3	3.4	0.5	0.01	754.51
49.00	0.8	2.8	3.9	0.6	0.02	754.54
56.00	0.9	3.2	4.5	0.6	0.02	754.57
63.00	2.5	4.8	6.5	0.9	0.03	754.69
70.00	4.1	8.7	11.3	1.3	0.05	754.97
77.00	5.7	15.2	18.5	1.7	0.08	755.25
84.00	7.3	24.5	28.2	1.9	0.13	755.60
91.00	6.0	33.6	37.8	2.1	0.17	755.91
98.00	4.7	39.9	44.3	2.2	0.20	756.11
105.00	3.3	43.4	47.9	2.2	0.22	756.21
112.00	2.0	43.4	47.9	2.2	0.22	756.21
119.00	1.8	42.7	47.2	2.2	0.22	756.19
126.00	1.6	41.6	46.1	2.2	0.21	756.16
133.00	1.4	40.2	44.5	2.2	0.20	756.11
140.00	1.1	38.3	42.6	2.2	0.20	756.06
147.00	1.0	36.2	40.5	2.1	0.18	756.00
154.00	0.9	34.0	38.2	2.1	0.17	755.92
161.00	0.8	31.7	35.8	2.0	0.16	755.85
168.00	0.7	29.3	33.3	2.0	0.15	755.77
175.00	0.7	26.9	30.8	1.9	0.14	755.69
182.00	0.6	24.5	28.3	1.9	0.13	755.61
189.00	0.6	22.2	25.8	1.8	0.12	755.53
196.00	0.6	19.8	23.4	1.8	0.10	755.44
Maximum Q,cfs,		Stor,ac-ft,	Elev. =	2.2	0.22	756.21
		Pre-Development Qp =		2.2 cfs		

Univ.R.M. Time min.	Hydrogr. for Q-in cfs	2S/dt-Q cfs	10 - year Ret.Per. 2S/dt+Q cfs	Q-out cfs	Storage ac-ft	Elev. ft.
0.0	0.0	0.0		0.0	0.0	754.0
7.00	0.4	0.3	0.4	0.1	0.00	754.06
14.00	0.8	1.0	1.4	0.2	0.01	754.23
21.00	1.2	2.1	2.9	0.4	0.01	754.46
28.00	1.5	3.4	4.8	0.7	0.02	754.59
35.00	1.7	4.8	6.6	0.9	0.03	754.70
42.00	1.8	6.2	8.4	1.1	0.04	754.80
49.00	2.0	7.6	10.0	1.2	0.04	754.89
56.00	2.1	9.1	11.7	1.3	0.05	754.99
63.00	5.0	13.0	16.2	1.6	0.07	755.16
70.00	8.0	22.4	26.0	1.8	0.12	755.53
77.00	10.9	37.0	41.3	2.1	0.19	756.02
84.00	13.8	56.9	61.7	2.4	0.29	756.59
91.00	11.5	76.6	82.1	2.8	0.38	757.10
98.00	9.1	90.3	97.2	3.4	0.45	757.45
105.00	6.8	98.6	106.2	3.8	0.49	757.64
112.00	4.4	102.0	109.8	3.9	0.51	757.72
119.00	4.0	102.6	110.4	3.9	0.51	757.73
126.00	3.6	102.3	110.2	3.9	0.51	757.72
133.00	3.2	101.3	109.1	3.9	0.51	757.70
140.00	2.7	99.5	107.2	3.8	0.50	757.66
147.00	2.5	97.2	104.7	3.8	0.49	757.61

TABLE 9.32. *(Continued)*

161.00	2.0	91.8	98.9	3.5	0.46	757.48
168.00	1.8	88.9	95.6	3.4	0.44	757.41
175.00	1.7	86.0	92.3	3.2	0.43	757.33
182.00	1.6	83.1	89.2	3.0	0.42	757.26

Maximum Q,cfs, Stor,ac-ft, Elev. = 3.9 0.51 757.73
Pre-Development Qp = 3.7 cfs

Univ.R.M. Hydrogr. for 100 - year Ret.Per.

Time min.	Q-in cfs	2S/dt-Q cfs	2S/dt+Q cfs	Q-out cfs	Storage ac-ft	Elev. ft.
0.0	0.0	0.0		0.0	0.0	754.0
7.00	0.7	0.5	0.7	0.1	0.00	754.11
14.00	1.4	1.8	2.6	0.4	0.01	754.41
21.00	2.1	3.9	5.4	0.8	0.02	754.63
28.00	2.8	6.6	8.8	1.1	0.04	754.82
35.00	3.1	9.8	12.6	1.4	0.05	755.02
42.00	3.4	13.1	16.3	1.6	0.07	755.17
49.00	3.6	16.6	20.1	1.7	0.09	755.31
56.00	3.9	20.6	24.1	1.8	0.11	755.47
63.00	9.1	29.5	33.5	2.0	0.15	755.77
70.00	14.2	48.2	52.9	2.3	0.24	756.35
77.00	19.4	76.4	81.9	2.8	0.38	757.10
84.00	24.6	112.1	120.4	4.1	0.56	757.94
91.00	20.7	147.7	157.4	4.8	0.74	758.66
98.00	16.8	174.6	185.1	5.2	0.87	759.16
105.00	12.8	193.2	204.2	5.5	0.96	759.49
112.00	8.9	203.6	215.0	5.7	1.01	759.66
119.00	7.9	208.7	220.4	5.8	1.03	759.74
126.00	7.0	211.8	223.7	5.9	1.05	759.79
133.00	6.0	212.9	224.8	5.9	1.06	759.81
140.00	5.1	212.2	224.0	5.9	1.05	759.80
147.00	4.6	210.1	221.8	5.9	1.04	759.77
154.00	4.2	207.3	218.9	5.8	1.03	759.72
161.00	3.7	203.7	215.1	5.7	1.01	759.66
168.00	3.3	199.5	210.7	5.6	0.99	759.59
175.00	3.1	194.8	205.8	5.5	0.97	759.51
182.00	2.9	189.9	200.8	5.4	0.94	759.43
189.00	2.7	184.7	195.5	5.4	0.92	759.34
196.00	2.5	179.4	190.0	5.3	0.89	759.24
203.00	2.5	173.9	184.4	5.2	0.86	759.15
210.00	2.4	168.5	178.8	5.2	0.84	759.05
217.00	2.3	163.1	173.2	5.1	0.81	758.96
224.00	2.3	157.7	167.7	5.0	0.78	758.85
231.00	2.2	152.4	162.2	4.9	0.76	758.75
238.00	2.2	147.2	156.9	4.8	0.73	758.66
245.00	2.2	142.1	151.5	4.7	0.71	758.56

Maximum Q,cfs, Stor,ac-ft,. Elev. = 5.9 1.06 759.81
Pre-Development Qp = 6.0 cfs

1 Circular Orifice(s) with D = 1.00 ft, Invert at El. 575.00 cd = 0.60
1 Rectang.Orif., W = 1.00 ft, H = 0.25 ft, Invert El. 757.25 cd = 0.60

Outlet Culvert Hydraulics
Circular Culvert, Diam.= 2.00 ft Length = 100 ft Inv. 753.0 to 752.0

Depth	Normal Q	Crit.Q	Manning n = 0.014
0.20	0.4	0.3	

numbering convention is the same as shown in Fig. 9.27. The peak discharges from SWIRM, summarized in Table 9.31, compare reasonably well with the other model results presented.

Sizing the detention reservoir requires the estimation of storage volume necessary to control each event. This can be estimated for each storm by plotting the developed basin hydrograph and superimposing the desired post development hydrographs. Inflow hydrographs and target discharges for this example are summarized in the output file in Table 9.32. The required storage volume is the area between the two hydrographs as depicted in Fig. 9.28.

The required storage estimate for each event can now be used to size the storm water pond, given the site topography. Note that the volume of the 100-year event controls the sizing of the reservoir. A storage volume–elevation relationship must be developed for the specific site and the reservoir geometry. The detention reservoir data is summarized in Table 9.32 and in Fig. 9.29.

Fig. 9.29 shows the storage–elevation relation for the Fair Oaks Estates site. Outlet structures are sized for each of the storage volumes. The control for the 1.01-year event is set at the bottom of the pond or at the normal pool elevation. This outlet is generally an orifice that limits discharge to the 1.01-year historic flow at the elevation of the 1.01-year storage volume. The invert of the second orifice, for the 10-year control, is set at or slightly above the 1.01-year water surface elevation such that, in combination with the lower level orifice, it limits discharge to the 10-year predevelopment discharge. The same process is used for the 100-year event; however, a spillway is frequently used as it can more easily convey flows resulting from extreme events in excess of the 100-year event. In this example, a narrow spillway is used to convey flows from the 100-year event and a larger spillway section is provided for larger emergency flows.

A review of the output in Table 9.32, indicates that the 1.01-year and 100-year target discharges have been met while that of the 10-year target is exceeded by 0.0057 m^3/s (0.2 cfs) but is generally within acceptable limits. The total storage required is 1308 m^3 (1.06 acre-feet) to limit Q_{100} flows to 0.17 m^3/s (6 cfs), which is the predevelopment target. This, unfortunately, cannot be compared with the 3084 m^3 (2.50 acre-feet) storage requirement computed earlier by the TR 55 tabular hydrograph method because in that example the outflow was restricted to 0.05 m^3/s (1.76 cfs) versus 0.17 m^3/s (6.0 cfs) in the current case. Also, a 24-hour storm was applied earlier that produced a Q_{100} inflow peak of 1.24 m^3/s (43.7 cfs) versus only 0.70 m^3/s (24.6 cfs) by SWIRM-ROUTE.

It is currently a matter of great interest that a 24-hour storm always gives a higher detention storage requirement than the modified Rational or universal Rational Method and raises the issue of appropriate design storm duration. This issue has never been satisfactorily resolved from an engineering standpoint,

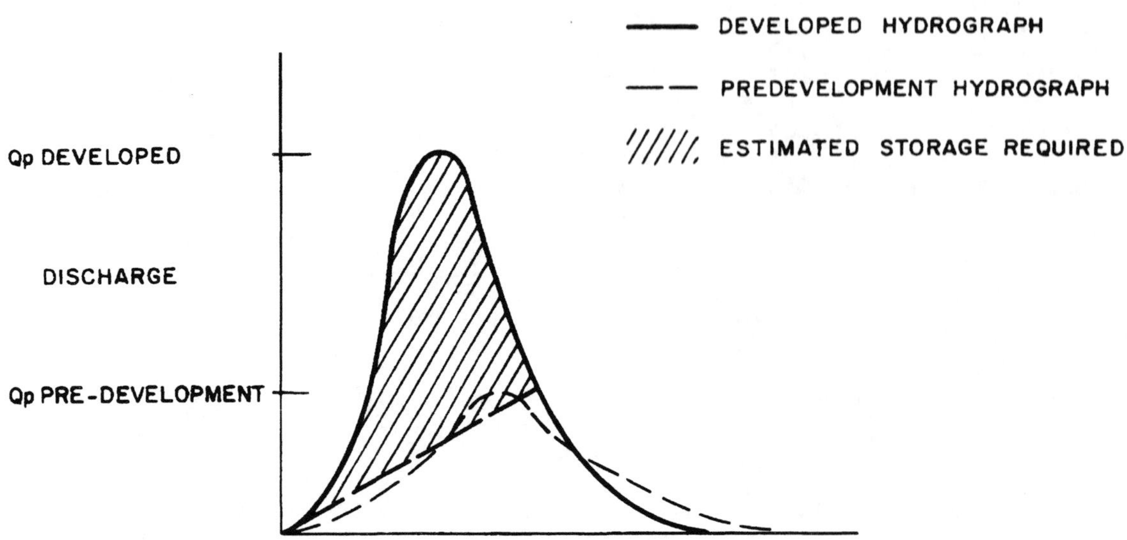

Figure 9.28.—Estimate of Required Storage Volume.

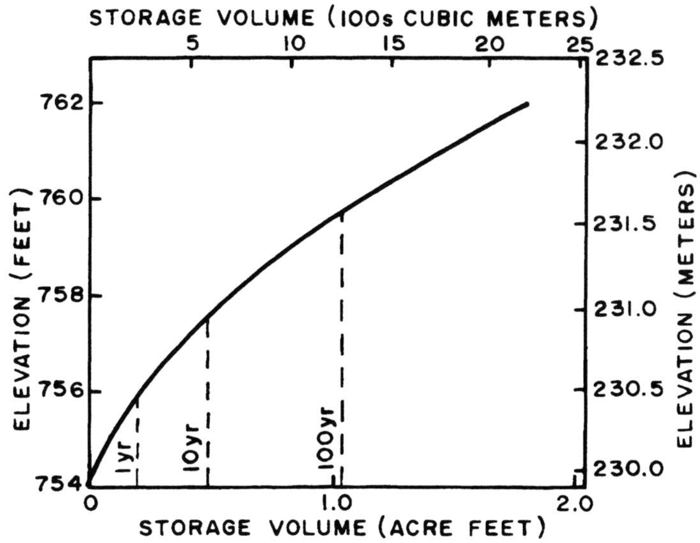

Figure 9.29.—Storage Elevation Relationship.

and consequently the matter generally depends on local regulations and ordinances that usually favor more conservative approaches. Clearly, as detention facility design becomes better adapted to stormwater quantity *and* quality control, the question of design storm size and duration must be addressed. This is an area where continuous simulation, with its inherent ability to account for inter-event draw-down, appears to have a clear advantage over the single-event approach.

VII. REFERENCES

American Society of Civil Engineers, (ASCE) (1976). "Design and construction of sanitary and storm sewers," ASCE Manual of Practice No. 37, New York, NY.

American Society of Civil Engineers, (ASCE/WEF) (1992). "Design and construction of urban stormwater management systems." ASCE Manual of Practice No. 77, WEF Manual of Practice FD0-20, New York, NY.

Anderson, D. G. (1970). "Effects of urban development on floods in northern Virginia," USGS Water Supply Paper 2001–C.

Aron, G. (1989). "Sensitivity of storm runoff hydrographs to hyetograph shape. Channel flow and catchment runoff," Proceedings of the International Conference for the Centennial of Manning's Formula and Kuichlingg's Rational Formula, Charlottesville, VA, May 22–26, pp. 1–9.

Aron, G. (1992). Personal Communication.

Aron, G. and Kibler, D.F. (1990). "Pond sizing for rational formula hydrographs," American Water Resources Association Water Resources Bulletin, Vol. 26, No. 2, pp. 255–258.

Bedient, P.B. and Huber, W.C. (1988). *Hydrology and floodplain analysis*. Addison–Wesley Publishing Co., p. 343.

Carter, R.W. (1961). "Magnitude and frequency of floods in suburban areas," USGS Prof. Paper 424-B, pp. B9–B11.

Dunne, T. and Leopold, L.B. (1978). *Water in environmental planning*, W.H. Freeman and Co., San Francisco, CA.

Espey, W.H. and Altman, D.G. (1978). "Nomographs for ten-minute unit hydrographs for small urban watersheds," Addendum 3 of urban runoff control planning, Rep. EPA-600/9-78-035, Environ. Protection Agency, Washington, DC.

Engman, E.T. (1986). "Roughness coefficients for routing surface runoff." J. Irrig. and Drainage Div., ASCE, 112 (1), 39–53.

Federal Aviation Agency (1970). "Advisory circular on airport drainage," Rep. A/C 050–5320–5B, Department of Transportation, Washington, DC.

Federal Highway Administration (FHWA) (1979). "Design of urban highway drainage: State of the art," Rep. FHWA–TS–79–225, Washington, DC.

Federal Highway Administration (FHWA) (1985). "Hydraulic design of highway culverts", Hydraulic Circular No. 5, Washington, DC.

Frederick, R.H., Myers, V.A., and Auciello, E.P. (1977). "Five to 60-minute precipitation frequency for the eastern and central United States." NOAA Technical Memo. NWS HYDRO–35, National Weather Service, Silver Spring, MD.

Green, W.H. and Ampt, G.A. (1911). "Studies on soil physics, I. The flow of air and water through soils." J. Agriculture Science 4(1): 1–24.

Hershfield, D.M. (1961). "Rainfall atlas of the United States for durations from 30 minutes to 24 hours and return periods from 1 to 100 years." Technical Paper 40, U.S. Dept. of Comm., Weather Bureau, Washington, DC.

Hershfield, D.M. (1962). "Extreme rainfall relationships." J. Hydr. Div., ASCE, Vol. 88, No. HY6, pp. 73–92.

Hewlett, J.D. (1982). *Principles of forest hydrology*. University of Georgia Press, Athens, GA.

Hicks, W.I. (1944). "A method of computing urban runoff." Transactions, ASCE. Vol. 109: 1217–1253.

Horton, R.E. (1919). "Rainfall interception." U.S. Monthly Weather Review, Vol. 47.

Horton, R.E. (1933). "The role of infiltration in hydrologic cycle." Transactions Amer. Geophys. Union 14: 446–460.

Huff, F.A. (1967). "Time distribution of rainfall in heavy storms," Water Resources Research, Vol. 3, No. 4, pp. 1007–1019.

Izzard, C.F. (1946). "Hydraulics of runoff from developed surfaces," Proc. Highway Res. Board, 26, 129–146.

James, L.D. (1965). "Using a computer to estimate the effects of urban development on flood peaks," Water Resources Research, Vol. 1: 223–234.

Keifer, C.J. and Chu, H.H. (1957). "Synthetic storm pattern for drainage design," J. Hydr. Div., ASCE, 83(4), pp. 1–25.

Kibler, D.F. and Aron, G. (1983). "Evaluation of T_c methods for urban watersheds," Proc. ASCE Hydraulics Specialty Conference, MIT.

Kibler, D.F., Aron, G., Riley, K.A., Osei–Kwadwa, G., and White, E.L. (1981). "Recommended hydrologic practices for computing urban runoff from small developing watersheds in Pennsylvania." Technical report, Penn State Univ. Inst. for Res. on Land and Water Resources, Univ. Park, PA.

Kibler, D.F., Jennings, M.E., Lewis, G.L., Tschantz, B.A., and Walesh, S.G. (1991). "Microcomputer software in urban hydrology," HYDATA published by American Water Resources Assoc., 27–32, September.

Kibler, D.F. and Reed, J.R. (1992). *Runoff resistance of a textured surface under simulated rainfall, in catchment runoff and rational formula*, edited by B. C. Yen, Water Resources Publications, Littleton, CO.

Kidd, C.H.R. (1978). "Rainfall–runoff processes over urban surfaces." Proceedings of the International Workshop held at the Institute of Hydrology. Wallingford, Oxon, United Kingdom, April.

Kirpich, Z.P. (1940). "Time of concentration of small agricultural watersheds," Civ. Eng., 10(6), 362, 1940.

Kuhl, R.A. (1977). "On-site stormwater detention basin design," presented at Illinois Society of Professional Engineers Stormwater Detention Seminar, adopted by Metropolitan Sanitary District of Greater Chicago.

Lehigh–Northhampton Counties Joint Planning Commission (1988). "Monocacy creek watershed stormwater management plan," Allentown, PA.

Linsley, R.K., Kohler, M.A., and Paulhus, J.L.H. (1975). *Hydrology for engineers*. McGraw–Hill Publishing Co., p. 260.

Marsalek, J. (1978). "Research on the design storm concept," Technical Memorandum 33, ASCE, Urban Water Resources Research Program, New York, NY.

McCuen, R.H. (1989). *Hydrologic analysis and design*, Prentice Hall, Englewood Cliffs, NJ.

McCuen, R.H., Wong, S.L., and Rawls, W.J. (1984). "Estimating urban time of concentration," ASCE J. Hydraulics Div., Vol. 110, (7): 887–904.

McPherson, M.B. (1958). "Synthetic storm pattern for drainage design." Discussion., J. Hydr. Div., 84(1), pp. 49–57.

McPherson, M.B. (1969). "Some notes on the rational method of storm drainage design," Tech. Memo. 6, 76 pp., ASCE, Urban Water Resour. Res. Program, New York, NY.

Miller, J.F., Frederick, R.H., and Tracey, R.J. (1973). "Precipitation–frequency atlas of the conterminous western United States (by states)," NOAA Atlas 2, 11 Vols., National Weather Service, Silver Spring, MD.

Morgali, J.R. and Linsley, R.K. (1965). "Computer simulation of overland flow," J. Hydr. Div., ASCE, 90(30), 81–100.

Northeastern Illinois Planning Commission (NIPC) (1986). "Stormwater detention for water quality benefits," 25 pp.

Overton, D.E. and Meadows, M.E. (1976). *Stormwater modelling*, Academic Press, New York, NY.

Papadakis, C.N. and Kazan, M.N. (1987). "Time of concentration in small rural watersheds." Proc. ASCE Engineering Hydrology Symposium, 633–638, Aug. 3–7, Williamsburg, VA.

Preul, H.C. and Papadakis, C.N. (1973). "Development of design storm hyetographs for Cincinnati, Ohio." Water Resources Bulletin, Vol. 9, No. 2, pp. 291–300.

Rantz, S.E. (1971). "Suggested criteria for hydrologic design of storm–drainage facilities in the San Francisco Bay Region, California." USGS Open File Report, Menlo Park, CA.

Rawls, W.J., Brakensick, D.L., and Miller, N. (1983). "Green–Ampt infiltration parameters from soils data." ASCE J. Hydr. Div. 109(1).

Rawls, W.J., Yates, J.P., and Rasmussen, L.E. (1976). "Calibration of selected infiltration equations for Georgia Coastal Plain." ARS–S–113, 110 pp.

Roesner, L.A. (1982). Chapter 5 *Urban runoff processes in urban stormwater hydrology*, edited by D. F. Kibler, American Geophysical Union, Water Resources Monograph No. 7, Washington, DC.

Rossmiller, R.L. (1980). "The Rational formula re-visited." Proceedings of the Int'l Symposium on Urban Runoff, July 28–31, Lexington, KY.

Rovey, E.W. and Woolhiser, D.A. (1977). "Urban storm runoff model." J. Hydr. Div., ASCE 103 (HY11):1339–1351.

Sarma, P.G.S., Delleur, J.W., and Rao, A.R. (1969). "A program in urban hydrology, Part II," Tech. Rept. No. 99, Purdue University, Water Resources Research Center, West Lafayette, IN.

Sauer, V.B., Thomas, W.O. Jr., Stricker, V.A., and Wilson, K.V. (1981). "Magnitude and frequency of urban floods in the United States," USGS, Reston, VA.

Schaake, J.C., Jr., (1967). "Experimental examination of the rational method," J. Hydr. Div. ASCE 93 (HY6), 353–370.

Schwab, G.O., Frevert, R.K., Barnes K.K., and Edminister, T.W. (1966). *Elementary soil and water engineering*, John Wiley & Sons, Inc., New York, NY.

Snyder, F.F. (1938). "Synthetic unit hydrographs." Transactions of the American Geophysical Union, vol. 19, part 1, 447–454.

Terstriep, M.L. and Stall, J.B. (1974). "The Illinois urban drainage area simulator, ILLUDAS." Bul. 50, Illinois State Water Survey, Urbana, IL.

Tholin, A.L. and Keifer, C.J. (1960). "Hydrology of urban runoff. Trans.", ASCE. Vol. 125: 1308–1379.

Urbonas, B.R. and Roesner, L.A., eds, (1986). "Urban runoff quality: impact and quality enhancement technology." *Proc. ASCE Engineering Foundation Conference*, New York, NY.

Urbonas, B.R. and Stahre, P. (1993). *Stormwater best management practices and detention for water quality, drainage and CSO management*, Prentice Hall, Englewood Cliffs, NJ.

U.S. Army Corps of Engineers, (USACE) (1982). "Hydrology analysis of ungaged watersheds using HEC–1." Report TD–15, Hydrologic Engineering Center, Davis, CA.

U.S. Army Corps of Engineers, (USACE) (1990). *HEC–1* "Flood hydrograph package, users manual", Hydrologic Engineering Center, Davis, CA.

U.S. Department of Agriculture, Soil Conservation Service (USDA SCS) (1969). "National engineering handbook", Section 4, Hydrology, Washington, DC. Reprinted 1985.

U.S. Department of Agriculture, Soil Conservation Service (USDA SCS) (1986). "Urban hydrology for small watersheds," Tech. Release 55, Washington, DC., 1975, revised 1986.

U.S. Environmental Protection Agency (USEPA) (1990). "Users manual for EPA stormwater management model version 4.05." Authors are W.C. Huber and R.E. Dickinson EPA Office of Environmental Assessment, Athens, GA.

U.S. Weather Bureau (1957). "Rainfall-intensity-frequency regime," Technical Paper 29, Washington, DC.

Virginia Department of Transportation (VDOT) (1980). *Drainage manual*, Locations and Design Division, Hydraulics Section, Richmond, VA.

Walesh, S.G. (1989). *Urban surface water management*, John Wiley & Sons, Inc., New York, NY.

Wenzel, S.G. (1982). "Rainfall for urban stormwater design." Urban Stormwater Hydrology, American Geophysical Union, Water Resources Monograph 7, pp. 35–68.

Wright–McLaughlin Engineers (1969). *Urban storm drainage criteria manual*, Denver Regional Council of Governments, Denver, CO.

Yarnell, D.L. (1935). "Rainfall intensity-frequency data." U.S. Dept. of Agri., Misc. Pub. 204, Washington, DC.

Yen, B.C. and Chow, V.T. (1980). "Design hyetographs for small drainage structures," J. Hydr. Div., ASCE, Vol. 106, No. HY6, pp. 1055–1076.

Yen, B.C. and Chow, V.T. (1983). "Local design storm," Vol. II, Report No. FHWA/RD–82–064, Federal Highway Administration, Washington, DC.

VIII. NOTATION

The following symbols are used in this chapter:

A = basin area $[L^2]$; orifice area $[L^2]$

C = Rational Method runoff coefficient

C_d = discharge coefficient for weir, orifice

CN = SCS curve number

D = duration of rainfall or rainfall excess [T]

D_b = basin development factor

D_p = potential depression storage [L]

D_s = storm duration [T]

E = evaporation rate [L/T]

F = cumulative infiltration [L]

$f(t)$ = infiltration rate at any time t [L/T]

f_c = final (minimum) infiltration rate, as used in the Horton equation [L/T]

f_i = urban flood peak adjustment factor

f_o = the initial (maximum) infiltration rate, as used in the Horton equation [L/T]

$H_{c,f}$ = capillary suction head at the wetting front [L]

h = head of water on weir or orifice [L]

IA = initial abstraction, consisting of interception and depression storage [L]

I = inflow hydrograph values $[L^3/T]$

i = rainfall intensity [L/T]

J = interception [L]

K = hydraulic conductivity of soil [L/T]

k = exponential decay coefficient [1/T]

k_j = ratio of evaporation surface to projected area [L/T]

L = length of flow path [L]

L_f = depth from soil surface to wetting front [L]

L_W = weir length [L]

O = outflow hydrograph value [L^3/T]

P = cumulative precipitation [L]

P_E = depth of excess rainfall [L]

Q = runoff rate [L^3/T]

q_p = peak rate of discharge [L^3/T]

R = reliability probability

R_s = relative sensitivity

S = soil water storage [L]; land slope [L/L]; storage volume [L^3]

S_c = maximum soil storage capacity [L]

S_j = interception storage capacity [L]

T = average return period of a storm, years

T_b = base length of SCS unit hydrograph [T]

T_p = time to peak of SCS unit hydrograph [T]

T_R = duration of rainfall excess, [T]

t_c = time of concentration [T]

U = impervious fraction as a decimal

V = velocity of flow [L/T]

V_p = volume of detention pond [L^3]

X = distance along flow path [L]

Y = depth of flow [L]

ϕ = phi Index loss rate [L/T]

θ = wetted porosity of the soil as decimal fraction; weir angle (degrees)

Chapter 10

WATER WAVES

I. INTRODUCTION

Surface water waves are a mechanism for transferring energy from a source across the water surface to a shoreline or structure. Common sources of energy are the wind, the gravitational pull of the moon and sun, seismic activity, or a moving vessel. These waves range in period from a few seconds for wind-generated waves up to twelve and twenty-four hours for the major components of the tide. They are commonly the most important design factor for shoreline stabilization works, marinas housing moored vessels, and navigation control structures.

The Water Waves chapter consists of eleven related sections. The level of presentation in these sections varies. For example, the sections considered less important to hydrologists and hydraulic engineers are kept brief while others significant to their practice are more comprehensive. The presentation is a first-level treatment in that the state of the art is reviewed and pertinent references are given for detailed analyses. Example problems are included where necessary to illustrate applications of the theories and principles.

The sections in the chapter are separated for an orderly presentation of several complex subjects in water waves. The chapter begins with a review of basic water wave theory and the calculation of wave properties. Wind waves are discussed in the following section, including wave generation, prediction, and statistical properties in deep and shallow waters. Ship-generated waves are briefly described in the next section. The effects of waves on structures is reviewed in the section titled Wave–Structure Interaction. Wave–current interaction is outlined in the following section. The longer period phenomena, including water level fluctuations, astronomical tides, storm generated surges, tsunamis, and basin resonances, are described in the next three sections. Water surface probability analysis follows. The selection of design waves and water levels and water surface elevation probability analysis is reviewed in the last section of the chapter.

It is noted that each of these topics is quite broad in scope, and it is virtually impossible to present in this handbook all aspects of water waves of greatest interest to hydrologists and hydrologic engineers. Appropriate references are included and listed at the end of the chapter to provide more details on the subject of water waves. The Shore Protection Manual (SPM) (USACERC, 1984), is a good overall reference.

II. WAVE THEORY

A. Fundamentals and Classification

Wave motions can be calculated using wave theories, which have been developed since 1845. These theories are simplified versions of nature, often involving such assumptions as horizontal bottoms or

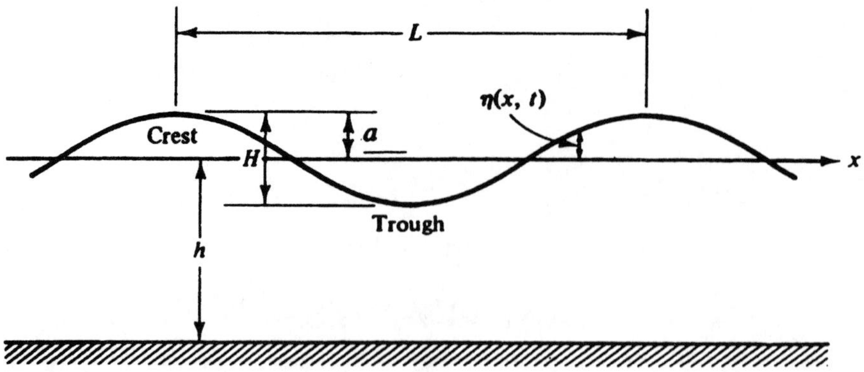

Figure 10.1.—Schematic of a Water Wave.

inviscid flow. There are a variety of these wave theories that are useful for different situations; for example, there are shallow water theories and deep water theories and there are small amplitude wave theories and large amplitude theories. This section will discuss the background of some of these theories and illustrate their uses.

Almost all wave theories have a common theoretical basis, such as water is incompressible and fluid motions are irrotational. Further the bottom is assumed to be flat and impermeable. The equations governing these wave theories are either the Laplace equation or Euler's equations of motion.

Important wave parameters include height, H, which is the vertical distance from the wave crest to the wave trough (as illustrated in Fig. 10.1), wave length, L, which is the distance between two crests, and the period, T, the time for two waves to pass a point. (L and T are later demonstrated to be related.) Another important parameter is water depth, denoted by d hereafter, except in Figs. 10.1, 10.2, and 10.3, where it is denoted as h. Dimensional analysis shows that the following dimensionless ratios are important:

$$H/L, d/L \tag{10.1}$$

The first dimensionless number is the wave steepness, and the second is the relative water depth. Steeper waves tend to be more forceful and also require more elaborate theories to describe them. The relative water depth is used to determine whether the wave is in deep or shallow water, requiring a different theory. Dividing these two dimensionless numbers yields a third number, H/d, which often occurs in wave breaking calculations.

B. Linear (Airy) Wave Theory

The simplest and surprisingly most versatile wave theory is the Airy theory. It is also called linear theory due to the mathematical linearization of the nonlinear wave problem. A full description of the linear wave theory is provided by Dean and Dalrymple (1991).

The water surface displacement of this wave has the following trigonometric form:

$$\eta(x,t) = \frac{H}{2}\cos[2\pi(x/L - t/T)] = \frac{H}{2}\cos\left(\frac{2\pi}{L}(x - Ct)\right) \tag{10.2}$$

at location x and at time t. The presence of the 2π in the argument of the cosine function ensures that the wave repeats itself in a distance equal to a wave length and at a time equal to a wave period. The speed of the wave in the positive x direction is C, which is the distance the wave travels in one wave period divided by the wave period, or $C = L/T$.

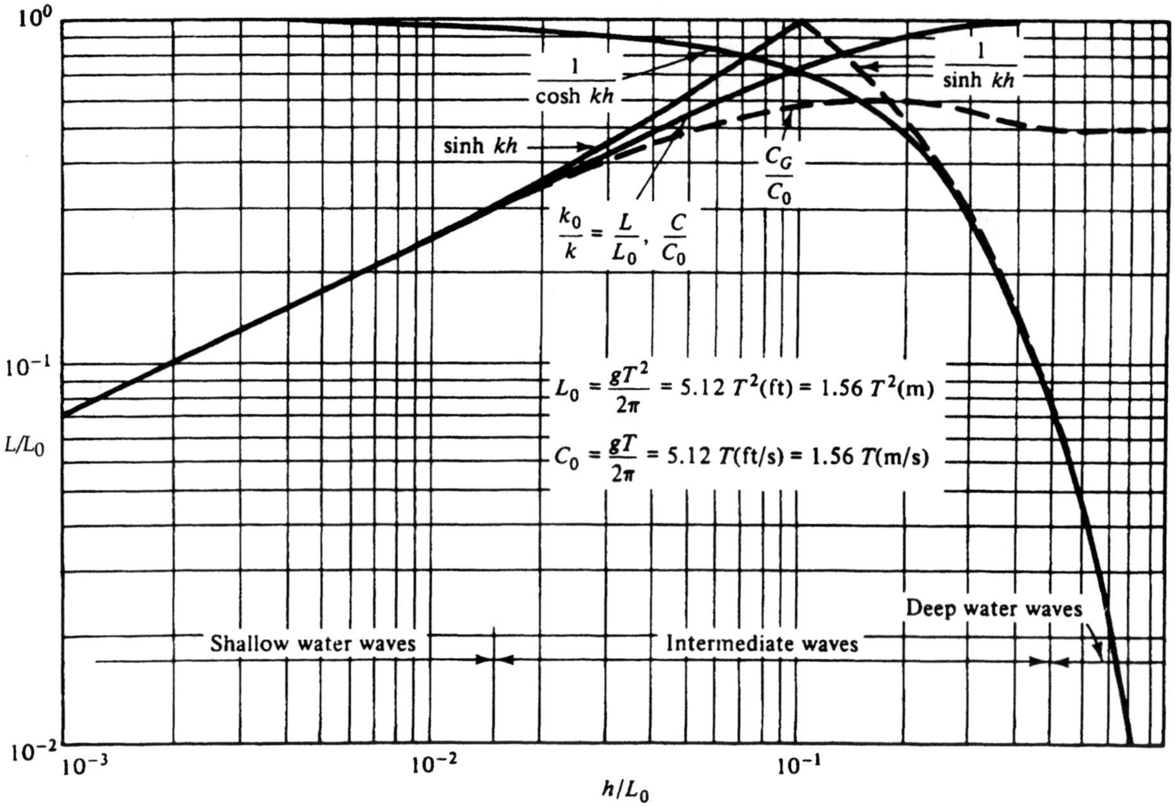

Figure 10.2.—Graphical Means to Calculate Wave Number, Wave Speed, and Group Velocity (Dean and Dalrymple, 1991).

For linear theory, the wave length and wave period are related through the (frequency) dispersion relationship

$$L = \frac{gT^2}{2\pi}\tanh\left(\frac{2\pi d}{L}\right) \tag{10.3}$$

where g is the acceleration of gravity. This equation involves the wave length in the argument of the hyperbolic tangent function as well, and therefore the equation must be solved numerically (often the Newton–Raphson method is used) or graphically. Fig. 10.2 provides a means to determine the wave number.

Note that as the relative water depth decreases, the wave length decreases as well. For deep or shallow water, the equation can be solved directly. For waves in deep water, the relative water depth is large, $d/L > 1/2$. In this case, the hyperbolic tangent has a value approaching unity, and the above equation becomes

$$L_o = \frac{g}{2\pi}T^2. \tag{10.4}$$

Example The wave period is known to be 5 seconds and the water depth is 3.05 m (10 ft.). What is the wave length, L?

First, the deep water wave length is calculated: $L_o = g\,T^2/2\pi = 9.81\,(5)^2/\,(2\times3.1415\ldots) = 39$ m. Therefore $d/L_o = 0.078$. Using Fig. 10.2, $L/L_o = 0.65$. This means that the local wave length is L = 0.65 (39) = 25.4 m, which is very close to the exact value of 25.1 m (about a 1% error in reading the graph).

In shallow water, the argument of the hyperbolic tangent is small, $d/L < 1/20$, such that the function approaches its argument in value, and the dispersion relationship becomes

$$L_s = \sqrt{gd}\,T. \tag{10.5}$$

An interesting result for shallow water is the wave speed, C_s ($= L_s/T$) is simply $\sqrt{gd}$, independent of the wave period.

As another alternative to the graphical method (or as an initial guess for the Newton–Raphson solution of Equation 10.3), an accurate approximation to the dispersion relationship is that of Fenton and McKee (1989), which has a maximum error less than 1.7%, well within most design criteria. Their relationship is

$$L = \frac{g}{2\pi}T^2\left(\tanh\left[(2\pi\sqrt{d/g}T)^{3/2}\right]\right)^{2/3}. \tag{10.6}$$

For the previous example problem, this formula gives L = 25.48 m, which differs only from the exact solution by 1.5%.

The rate at which energy is carried by the waves is different from the speed of the wave form, C. The speed at which energy is transported is the group velocity, C_g, and it is related to the wave speed. In deep water, $C_g = C/2$, while in shallow water, $C_g = C$. The ratio of the two is

$$n = \frac{C_g}{C} = \frac{1}{2}\left(1 + \frac{\frac{4\pi d}{L}}{\sinh\frac{4\pi d}{L}}\right). \tag{10.7}$$

Fig. 10.2 can be used to calculate the group velocity conveniently. For the previous case, where d/L_o = 0.078, the value of $C_g/C_o = 0.56$. Since $C_o = g\,T/(2\pi) = 7.8$ m/s, $C_g = 4.1$ m/s (about a 3.5% error).

The particle velocities under the Airy wave depend on the particle's position relative to the wave crest and the bottom. The horizontal and vertical velocities are

$$u(x,z,t) = \frac{H\pi}{T}\frac{\cosh\frac{2\pi}{L}(d+z)}{\sinh\frac{2\pi}{L}d}\cos\left(\frac{2\pi}{L}(x-Ct)\right) \tag{10.8}$$

$$w(x,z,t) = \frac{H\pi}{T}\frac{\sinh\frac{2\pi}{L}(d+z)}{\sinh\frac{2\pi}{L}d}\sin\left(\frac{2\pi}{L}(x-Ct)\right) \tag{10.9}$$

where z is measured upwards from the water surface. The velocities are maximum at the water surface, u is in-phase with the wave form, and w goes to zero at the bottom ($z = -d$) to satisfy the impermeability condition.

The pressure under the wave is composed on the hydrostatic term due to the presence of the water and a dynamic term due to the presence of the wave

$$p(x,z,t) = -\rho g\left(z - \eta(x,t)\frac{\cosh\frac{2\pi}{L}(d+z)}{\cosh\frac{2\pi}{L}d}\right) \tag{10.10}$$

C. Nonlinear Wave Theories

Mathematically, the linear theory is valid for infinitesimally small steepness, that is, for $(2\pi/L)a << 1$. In practice, most waves have greater steepness and should be treated with a more sophisticated wave theory to correctly model wave characteristics. For example, in linear theory, the wave length does not depend on the wave height, while for nonlinear theories higher waves go faster than smaller waves.

The choice of a correct theory depends on the water depth of interest. For deeper water, the Stokes theories are often used, while for shallow water the Boussinesq theories are most often used.

D. Stokes Theory

The Stokes theories come about through a perturbation expansion in wave steepness. This means the steeper the waves, the more terms must be added to the original linear theory to provide a correct representation of the wave motion.

As an example, the second order Stokes theory provides a water surface displacement of the following form:

$$\eta(x,t) = \frac{H}{2}\cos 2\pi(x/L - t/T) - \frac{H^2\pi \cosh(2\pi d/L)}{8L \sinh^3(2\pi d/L)}(2 + \cosh(4\pi d/L))\cos 4\pi(x/L - t/T). \quad (10.11)$$

The added term oscillates twice as fast as the linear term adding to the linear term at the crest and trough, causing the crests to steepen and the troughs to flatten. There is also a change in the mean water level, referred to as the set-down.

$$\overline{\eta}(x,t) = -\frac{\pi H^2}{4L \sinh(4\pi d/L)} \quad (10.12)$$

The set-down is negligible in deep water and becomes more pronounced in shallow water. The fifth order Stokes theory adds three more contributions to the expression of the wave form. Fenton (1985) provides a description of the fifth order Stokes theory.

Higher order theories are solved at the cost of considerable mathematics. It is far easier today to use a numerical procedure for solving these theories. For example, Dean (1965) developed the Stream Function wave theory, which allows for very high order theory that is valid in both deep and shallow water. Dean (1974) has tabulated the results of 40 different wave cases, spanning the range from deep to shallow water for a variety of wave heights (up to breaking). These tables can be used by interpolation to provide accurate predictions of wave characteristics, velocities, and accelerations. Dalrymple (1974b) developed a version of the Stream Function theory for the case of currents that converged on wave height.

E. Shallow Water Theories

In shallow water, where the water depth to wave length ratio is very small, an alternative theoretical formulation provides a more compact expression of the wave field. This is the so-called Boussinesq theory, which has two well-known formulations, the solitary wave and the cnoidal wave theories.

F. Solitary Wave Theory

The solitary wave consists solely of a wave crest and has an infinite length. Russell (1838) was the first to discuss this type of wave motion, which he observed in canals where horsedrawn canal boats were being used. The free surface of the solitary wave is described by

$$\eta(x,t) = a\ sech^2\left(\sqrt{\frac{3a}{4d}}\frac{x}{d}\right). \tag{10.13}$$

The volume of water contained in the wave crest is

$$V = 4d\sqrt{\frac{ad}{3}}. \tag{10.14}$$

Most of this wave volume (95%) is contained in a finite distance, 2 ℓ, about the wave crest, where

$$\ell = \frac{2.12d}{\sqrt{a/d}}. \tag{10.15}$$

The speed of the solitary wave is approximately

$$C = \sqrt{gd}\left(1 + \frac{a}{2d}\right). \tag{10.16}$$

G. Cnoidal Wave Theory

The cnoidal wave theory provides for a periodic wave train in shallow water. These waves, described by Korteweg and deVries (1895), are characterized by a steepness factor, K. The wave form is

$$\eta = a\ cn^2\left(\sqrt{\frac{3a}{4dK^2}}\frac{x}{d}\right). \tag{10.17}$$

The cn function is a Jacobean elliptic function with modulus, K. Wiegel (1964) shows the variations of the wave form with K. The mathematical complexities of the elliptic functions usually result in the use of numerical models in shallow water design.

H. Numerical Models

The wave-induced motion in shallow water is described by the Boussinesq equations for variable depth (Peregrine, 1967). Solutions of these equations are obtained by a variety of methods—finite differences (Abbot et al., 1978), spectral means (Freilich and Guza, 1984), and parabolic approaches for small angle of wave incidence (Liu et al., 1985). Computational models are now available commercially that describe the motion of shallow water waves over irregular bathymetry, around structures, and in harbors.

I. Shoaling/Refraction

Waves travel slower in shallow water than in deep water due to the decrease in wave length (see Equation 10.3). If a long-crested wave encounters variable-depth water, the portions of the wave crest in deeper water will travel faster than those in shallower water, resulting in changes in the local wave direction, called refraction. For simple bathymetry, the wave direction change can be computed from the wave speed with Snell's Law

$$\frac{\sin\theta_1}{C_1} = \frac{\sin\theta_2}{C_2} \tag{10.18}$$

where the subscripts denote the wave angle and wave speed at one location compared to those on the other side of a change of depth. The wave angle is measured from a line perpendicular to a bottom contour. Following a section of wave crest as the wave propagates towards shore allows the construction of wave rays, which are lines that are always parallel to the wave direction. Graphical techniques exist to compute the wave rays by hand (JSPM) (USACERC, 1984); however, this is easier done by computer.

The change in wave height due to changes in wave length and direction between two points, denoted by 1 and 2, is obtained by

$$H_2 = H_1\sqrt{\frac{C_{g,1}}{C_{g,2}}}\sqrt{\frac{b_1}{b_2}} = K_s K_r \tag{10.19}$$

where K_s is the shoaling coefficient, involving the ratio of the group velocities at the two locations and K_r is the refraction coefficient, which is a function of the ray spacing, b.

For the simplest case of straight and parallel contours in the alongshore direction, Fig. 10.3 provides a means to calculate wave height and direction change at a given depth as a function of the deep water wave height and direction and a dimensionless ratio, $d/(gT^2)$.

Example A wave with a deep water wave height of 1 m and a wave period of 8 seconds encounters a shoreline characterized by straight and parallel contours. The deep water wave angle of incidence (measured from the shoreline perpendicular) is 30°. Find the wave height and direction in 2.5 m water depth.

> For this case, $d/(gT^2) = 2.5/(9.81 \times 8^2) = 0.004$. Entering Fig. 10.3 from the right where $\theta_o = 30°$, move horizontally to the vertical line corresponding to $d/(gT^2) = 0.004$. Interpolating between the solid lines corresponding to θ, the wave direction in shallow water is found to be $\theta = 11°$ and $K_r K_s = 1.10$, which gives H (1.10) = 2.2m.

For more complicated bathymetry, computer programs exist for refraction and shoaling based either on ray tracing algorithms (Noda, 1974) or rectangular grid methods (Dalrymple, 1988).

J. Diffraction

Diffraction is the change in wave direction effected by strong variations in wave height. For example, waves passing a breakwater have very low wave heights behind the breakwater. Diffraction causes a wave direction change to allow energy to flow to the low wave height regions. Penney and Price (1952) showed how the analogy to optics can be used for wave calculations. Graphic examples of waves passing very long breakwater sections or through gaps in breakwaters, all for the case of constant depth, are available (USACERC, 1984). The important variables are gap spacing and the wave length.

For variable depth in the vicinity of the breakwaters, both refraction and diffraction must be computed. The combined effects are conveniently described by the mild-slope equation, developed by Berkhoff (1972). Numerous models of the mild slope have been developed.

Parabolic modelling of the mild-slope equation (Radder, 1979) has led to the development of a class of models that permit the rapid calculation of wave fields where both refraction and diffraction is important (Kirby and Dalrymple, 1983, 1984; Liu and Tsay, 1983).

K. Wave Breaking

In very shallow water, waves break as the wave length decreases monotonically with depth and the height increases until finally the wave is unstable and breaks. The simplest criterion for wave breaking is that the wave height is a significant proportion of the water depth, $H = \kappa d$, where κ is the breaker

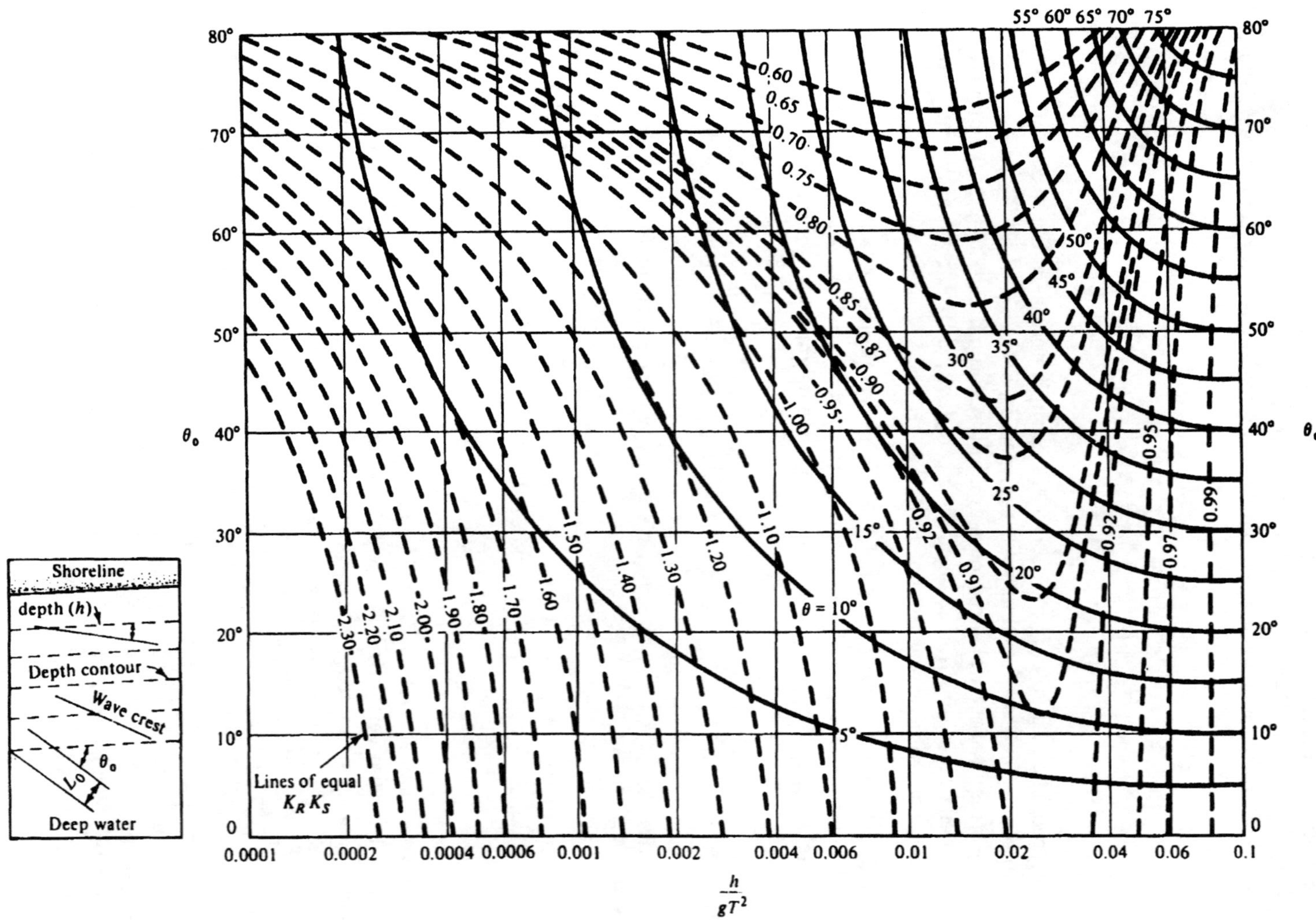

Figure 10.3.—Changes in Wave Direction and Height Caused by Refraction on a Shoreline with Straight and Parallel Bottom Contours (USACERC, 1984).

index that has a range of $0.78 < \kappa < 1.5$. The particular value for κ depends on the slope of the bottom and the wave steepness, H/L. For mild slopes, the smaller values apply. This implies that over a flat bottom, the maximum wave height is 0.78 times the water depth.

Battjes (1974b) introduced the surf similarity parameter, ζ, to describe a number of nearshore phenomena, including wave breaking. This parameter is defined as:

$$\zeta = \frac{m}{\sqrt{H/L_o}} \tag{10.20}$$

where m is the local beach slope. The surf similarity parameter has a typical range of $0.1 < \zeta < 5.0$, ranging from very mild slopes to steep slopes. In Table 10.1, the breaking index is shown as a function of the surf similarity parameter.

Waves break in a variety of ways. On mild slopes spilling breakers occur where the foaming crest of the wave slides down the wave front. Plunging breakers (surfers' delight) occurring on steeper beaches have crests that fall forward of the wave front in a big jet, often trapping air. Surging breakers occur on very steep beaches where breaking is dominated by wave reflection.

For a vertical wall, the waves do not break but are totally reflected, leading to standing waves in front of the wall, with a doubling of wave height.

Example Determine the breaking depth for a 1.5-m high wave on a beach that has a 1:10 slope. The wave period is 6 seconds.

> The beach slope m = 0.1. $L_o = 1.56(6)^2 = 56.16$ m (from Equation 10.4). The surf similarity ζ in Equation 10.20 is $0.10/\sqrt{1.5/56.16} = 0.61$. From Table 10.1, $\kappa \sim 1.0$. Therefore, the breaking depth would be 1.5 m.

More sophisticated wave models take into account the details of wave breaking (Dally et al., 1985) or the spectral nature of waves (Battjes and Janssen, 1978; Thornton and Guza, 1983).

III. WIND WAVES

A. Description of Irregular Waves

Contrary to the assumptions of wave theory presented in the preceding section, wind waves in nature are a statistical process. The height and period of individual waves passing a fixed point in space during a stationary sea state vary considerably and more or less randomly with time. Useful statistical descriptions and parameters of a real sea state have been developed so that wave information can be conveniently used in engineering applications. The descriptions and parameters are discussed in this section. General references on wind waves and more complete and detailed information is given elsewhere (Dean and Dalrymple, 1991; Goda, 1985; IAHR, 1986; Ochi, 1982; Sorensen, 1978; USACE, 1989).

A closely related engineering concern is the long-term description of the ensemble of sea states that can occur at a site. The long-term description includes wave climate and extreme wave analysis. These topics are addressed in the section titled Selection of Design Water Waves and Levels.

TABLE 10.1. Surf Similarity Parameter and Breaker Index and Reflection Coefficient.

$\zeta = \frac{m}{\sqrt{H/L_o}}$	0.1	0.5	1.0	2.0	3.0	4.0
Breaker Type	Spilling		Plunging		Surging	No Breaking
κ	0.8	1.0	1.1	1.2		
R	0.001	0.01	0.1	0.4	0.8	

The concept of a "significant wave height" and a "significant wave period" that can be used to characterize a sea state is appealingly simple. It suggests a direct transition from the experimental results in a laboratory wave tank and regular wave theory to the phenomena that occur in nature. The concept of a significant wave height and period was first introduced when sailors were asked to report the height and period of the larger, well-formed waves, and omit entirely the low and poorly-formed waves as part of the synoptic weather reports from ships. Comparisons of early wave gage records with observations led to the conclusion that the wave height given by observers was approximately equal to the average height of the one-third highest individual waves, $H_{1/3}$. The parameter $H_{1/3}$ is referred to as "significant wave height," and the corresponding period is the "significant wave period," $T_{1/3}$. Thus, formally, significant height is the average height of the one-third highest individual waves in a record, and significant period is the average period of the one-third highest individual waves in a record. In practice, significant wave height and period can be estimated by several approaches, as discussed in the following paragraphs.

B. Wave Measurements

The primary information from any wave observation technique is generally an estimate of significant wave height and period. Many techniques provide additional valuable data. Commonly-used techniques are described in the remainder of this section. Their particular advantages and disadvantages are given in Table 10.2.

Wave staff observations are obtained with a gage that pierces the water surface. In its simplest form, the wave staff is a vertical rod with visible marks on it at measured intervals. It allows an observer to obtain better quantitative estimates of wave characteristics by watching the undulating water motions at the staff; however, most wave staffs operate as electrical sensors. Additional information about wave staffs is available from Parker (1982).

In addition to the advantages and disadvantages of the wave staff listed in Table 10.2, the staff is often inexpensive relative to other instrumental observation techniques. The visually-observed staff is especially economical, and it can be very useful in low-budget projects. In low energy environments, the staff gage can be mounted on a spar buoy as an alternative to a rigid mounting structure.

Pressure cell observations are obtained with a pressure-sensitive gage mounted under the water surface. Pressure cells can be situated anywhere in the water column as long as they are below the elevation of the lowest expected wave trough at the lowest expected tide level. Pressure cells are often placed on a small tripod which rests submerged on the bottom. The pressure cell senses dynamic pressure fluctuations created by passing surface waves. The magnitude of the fluctuations decreases exponentially with distance below the surface. The pressure fluctuations are converted to an electronic

TABLE 10.2. Wave Observation Techniques (USACE, 1989).

Technique	Water Depth[1]	Advantages[2]	Disadvantages[3]
Shore observations	3	1	1
Staff gage	1,2,3	2,5	2,3
Pressure gage (shore-connected by cable or telemetry)	2,3	3,5	3,4,5
Pressure gage (internally recording)	2,3	3,4	3,4,5,6,7
Accelerometer buoy	1,2	4,5	5,8,9

[1]Water Depths: 1 = deep, 2 = intermediate, 3 = shallow

[2]Advantages: 1 = inexpensive; 2 = direct measurement of surface waves; 3 = relatively reliable for unattended use; 4 = relatively simple installation; 5 = adaptable to real time monitoring.

[3]Disadvantages: 1 = low accuracy; 2 = sturdy support structure required for exposed sites; 3 = subject to fouling by marine growth; 4 = divers needed for installation and maintenance; 5 = indirect measurement of desired surface waves; 6 = prone to loss due to mechanical failure (leaking) or damage by bottom trawlers; inaccurate positioning and loss of marker buoys; 7 = frequent maintenance required for power supply and recorder; 8 = prone to loss due to collision or failed mooring; 9 = relatively expensive.

signal which can be recorded at the gage or sent to a shore station by either an armored electrical cable or a surface buoy transmitter. Additional information about pressure cells is available (Forristall, 1982).

Accelerometer buoy observations are obtained with a surface-following buoy which is usually spherical in form. Buoys are routinely moored in water depths of less than 200 m. Buoys sense vertical acceleration, which is usually integrated twice electronically to give a record of surface displacements. The observations may be recorded in the buoy or transmitted to a shore station either directly or via a satellite link. Some buoys are equipped with additional motion sensors so that wave direction information may be obtained.

A variety of other wave recording instruments are available for practical use although those discussed in the preceding subsections are most widely used. The instruments are categorized as in-situ instruments and remote sensing devices. A variety of ground-based and airborne remote sensing devices are available but most are not suited for routine data collection.

In-situ instruments include acoustic, ultrasonic, optical (laser), and radar instruments, all of which transmit a signal toward the water surface from above or below. The reflected signal is then received and interpreted. Another approach to in-situ data recording consists of operating several instruments together at a site and analyzing the records jointly to get additional information, particularly wave direction. Typical combinations are a pressure cell or staff with two orthogonal horizontal current meters, and a spatial array of staffs or pressure cells. More information on these in-situ instruments is available (Parker, 1982; Seymour, 1982).

C. Wave Analysis and Statistics

Wave records are usually collected as a digital time series of surface elevation or subsurface pressure. Typical steps in analyzing a digital wave record are schematized in Fig. 10.4 and briefly discussed in this section. The editing step should include a check for waves at short and long periods outside the range of wind wave periods, which is 1 to 30 seconds in the ocean. Apparent sources of waves at undesired periods are tides, water level oscillations, surf beats, electronic drift, electronic noise, and transmission

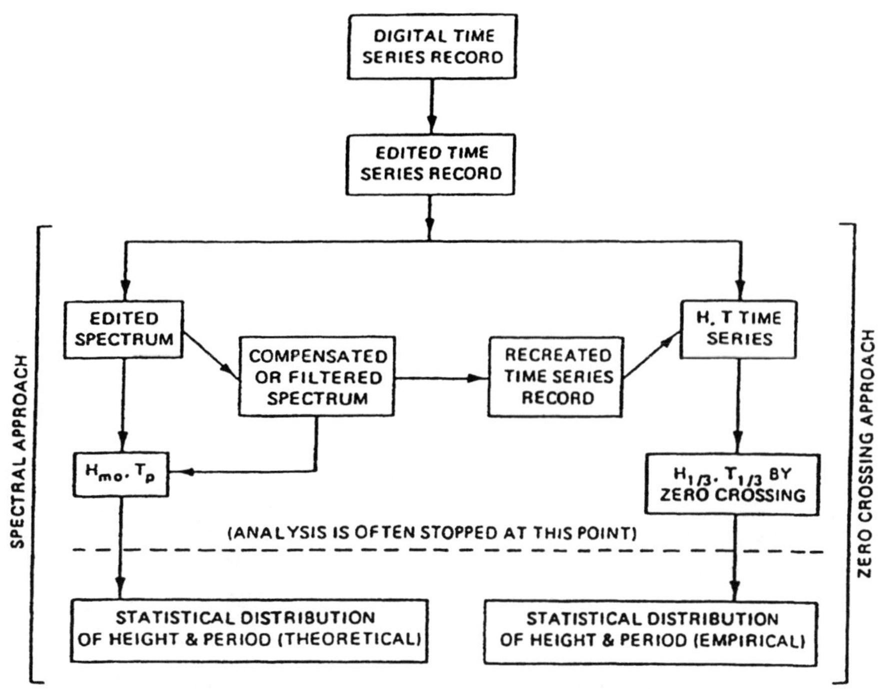

Figure 10.4.—Analysis of a Digital Wave Record (USACE, 1989).

interference. If any of these apparent waves are significant in the record, they may distort estimates of wind wave characteristics. Short- and long-period contamination can be identified visually or by numerical tests. It can be removed by filtering or by considering the spectrum discussed in the following subsections.

A sea state can be analyzed by extracting information from the time series or by transforming the time series into a frequency spectrum and extracting information from the spectrum (spectral analysis). Although spectral analysis is most common, both approaches are widely used and give complementary information.

Significant height and period may be estimated directly from the time series by identifying all individual waves in the record. The procedure that is most widely used and most easily applied on a digital computer is known as the zero-crossing method. Each wave in the record is identified as an event between two successive points at which the wave trace crosses the mean in a downward moving direction (Fig. 10.5). The time between crossings is the wave period, T. Wave height, H, is defined as the elevation difference between the highest point (crest) and lowest point (trough) of each wave. Significant wave height is computed as the average height of the highest one-third zero-crossing waves. Other wave height statistics, such as the root-mean-square wave height, are also easily computed. Significant period is computed as the average period of the one-third highest waves. This procedure is called the zero downcrossing analysis procedure because each wave is defined by two downcrossings of the mean.

By the spectral approach, significant height is estimated from the spectral density integrated over the range of wind wave frequencies (IAHR, 1986). A standardized analysis package for wind waves should be used when the spectral approach is desired. A comprehensive package has the advantages of automatically removing nonwind wave energy from the results and options for compensating pressure or acceleration signals to give estimates of surface waves. The significant height estimate is sometimes referred to as H_{m0} to clearly identify that it was obtained by a spectral approach. Wave period estimated by the spectral approach is the period corresponding to the highest energy density in the spectrum. It is called significant or peak period, T_p. In relating spectral parameters to traditional time series parameters, it is generally assumed that

$$H_{mO} = H_{1/3}. \tag{10.21}$$

This relationship is reasonable as long as the wave shapes are not severely deformed by shallow water depth or high wave steepness. In very shallow water, an approximation may be used (Hughes and Borgman, 1987). The accepted approximate relationship for wave period estimates is

$$T_p = 1.05\, T_{1/3}. \tag{10.22}$$

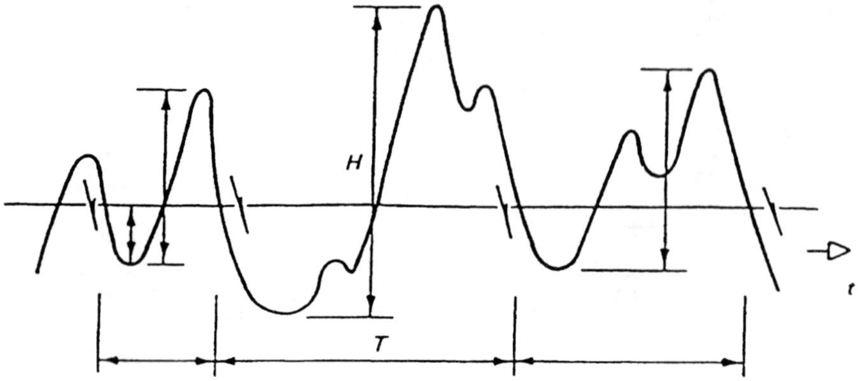

Figure 10.5.—Zero-Downcrossing Waves.

This relationship may not be realistic at exposed ocean locations where several wave trains may arrive simultaneously from different generation areas. Wave direction is an additional parameter that is often of considerable engineering importance. Wave direction can be estimated from measurements only through spectral analysis. The term H_s is commonly used to designate a generalized significant wave height. When it is used, the method for estimating it and whether it represents $H_{1/3}$ or H_{m0} should be clarified.

The distribution of individual wave heights in a sea state is almost always well-approximated by the Rayleigh distribution function. The Rayleigh distribution is given by

$$P(H \geq \hat{H}) = e^{-\left(\frac{\hat{H}}{H_{rms}}\right)^2} \tag{10.23}$$

where $P(H \geq \hat{H})$ = probability that a wave in the distribution will exceed height $\hat{H}$, $\hat{H}$ = given wave height, H_{rms} = root-mean-square wave height, and e = 2.7183. Other statistical wave height parameters can be easily related to H_{rms} through the Rayleigh distribution. The more commonly used parameters are listed in Table 10.3.

An estimate of the maximum wave height in a sea state can also be obtained from the Rayleigh distribution as

$$H_{max} = H_{rms}\sqrt{\ln N} = 0.71\, H_{1/3}\sqrt{\ln N} \tag{10.24}$$

where H_{max} is the maximum individual wave height, ln is the logarithm with respect to the base e, and N is the number of waves occurring. The parameter N can be estimated for a storm as the duration of the intense part of the storm divided by the peak or significant wave period. The distribution of individual wave periods is much more variable than the distribution of individual wave heights. It is also of less engineering importance than the height distribution.

D. Wind Parameters and Fetch

Simplified methods have been developed for estimating wave parameters during wind wave generation. The simplified methods are particularly useful when quick, low-cost estimates are needed. They provide accurate results for areas with relatively simple wind conditions, fetches, and water depths. The methods presented here are taken from the SPM (USACERC, 1984).

Estimation of the wind speed is a critical, and often difficult, first step. Winds over the wave growth area are needed for estimation. Winds can be estimated from measurements or from surface pressure information on synoptic weather charts. Wind speed must be adjusted to be consistent with the wave growth equations. That is, the wind speed should be adjusted for elevation, averaging time, air–water temperature differences, local land effects (if winds are from a measurement on or near land), and nonconstant coefficient of drag. Details of the adjustments are given in the SPM (USACERC, 1984). At a minimum, the elevation and drag coefficient adjustments should be made. The elevation adjustment can be approximated by the following equation if the observation elevation is less than 20 m:

$$U_{10m} = U_z \left(\frac{10}{z}\right)^{1/7} \tag{10.25}$$

TABLE 10.3. Wave Height Relationships Based on the Rayleigh Distribution.

Average of 1% highest waves = 2.36 H_{rms} = 1.67 $H_{1/3}$
Average of 10% highest waves = 1.80 H_{rms} = 1.27 $H_{1/3}$
Average of 1/3 highest waves = 1.42 H_{rms} = 1.00 $H_{1/3}$
Average of all waves = 0.89 H_{rms} = 0.63 $H_{1/3}$

where U_{10m} is the wind speed at 10 m elevation (needed for wave prediction), U_z is the wind speed observed at elevation z, and z is the elevation of the wind observation, in meters. The drag coefficient correction is:

$$U_A = 0.71\, U^{1.23} \tag{10.26}$$

where U_A is the wind-stress factor (adjusted windspeed) and U is the wind speed after other adjustments have been made, such as elevation, etc., in meters per second.

Estimates of the wind duration may also be needed for wave prediction if the fetch is relatively long. The duration usually must be estimated by judgment and interpolation using available information. A fetch is defined as the length of an overwater region in which the wind speed and direction are reasonably constant. A fetch should be defined such that wind direction variations do not exceed 15 degrees and wind speed variations do not exceed 2.5 m/s from the mean. When the point of interest is located downwind from a coastline, the coastline limits the fetch. In inland waters (bays, rivers, lakes, and reservoirs), fetches are limited by landforms surrounding the water body. Shorelines are usually irregular. In such cases, fetch can be defined as a radial average over an arc of 24 degrees centered on the wind direction with radials placed at 3-degree intervals (Fig. 10.6).

E. Deep Water Wave Prediction

Significant wave height and peak period generated by local winds can be estimated from the wind speed (corrected as prescribed in the SPM, USACERC, 1984), duration, and fetch. The estimation is based

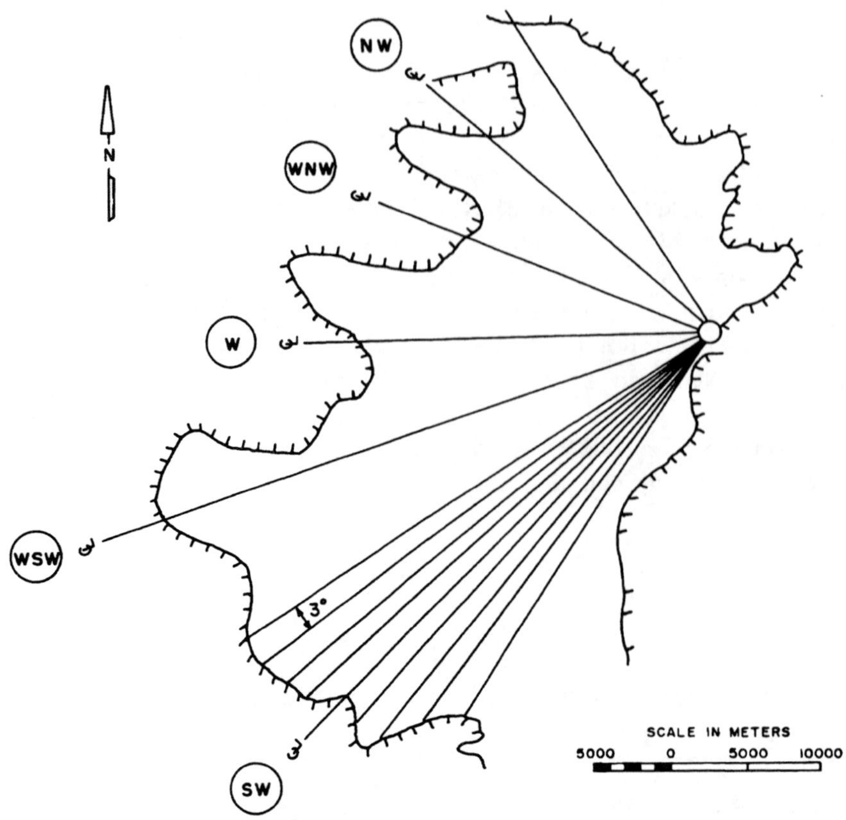

Figure 10.6.—Fetch Estimation on Small Water Bodies with Irregular Shape (USACE, 1989).

on equations developed from the Joint North Sea Wave Project (JONSWAP) experiment. The equations are:

$$H_{mO} = 0.01616\, U_A\, F^{1/2} \tag{10.27}$$

$$T_p = 0.6238\, (U_A\, F)^{1/3} \tag{10.28}$$

where F is the fetch (in km), provided H_{m0} is in meters, T_p in seconds, and U_A in meters per second.

On lakes and reservoirs, wave conditions are almost always limited by the wind speed and fetch. On the ocean, where fetches can be very long, the time duration over which the wind blows often limits wave growth more than the fetch. The minimum duration which will not limit wave growth is approximated as:

$$t = 0.893 \left(\frac{F^2}{U_A}\right)^{1/3} \tag{10.29}$$

where t is duration in hours. Special procedures for use with hurricanes and other tropical storms are available (USACERC, 1984).

F. Shallow Water Wave Growth

If the predominant depth of water over the fetch is less than one-half the deepwater wavelength, wave growth is affected by the bottom. When shallow water wave growth occurs, the known wind speed, duration and fetch can be used to predict significant wave height and period from the following equation:

$$H_{1/3} = 0.283 \left(\frac{U_A^2}{g}\right) \tanh\left[0.530\left(\frac{gd}{U_A^2}\right)^{3/4}\right] \tanh\left[\frac{0.00565\left(\frac{gF}{U_A^2}\right)^{1/2}}{\tanh\left[0.530\left(\frac{gd}{U_A^2}\right)^{3/4}\right]}\right] \tag{10.30}$$

where d is water depth, usually determined as a rough average of the depths in the downstream half of the fetch, g is acceleration due to gravity = 9.80 m/sec^2, and tanh is the hyperbolic tangent function. The minimum duration required to reach fetch-limited wave growth is given by:

$$t = 0.0537 \left(\frac{U_A}{g}\right)\left(\frac{gT}{U_A}\right)^{7/3}. \tag{10.31}$$

Dimensional units in the above equations must be consistent with the units used for gravity.

G. Computer Modeling

Because of the continually increasing capabilities and availability of digital computers, comprehensive numerical wave models are becoming essential tools for practical engineering work. Models vary greatly in complexity. The simplest wave growth model is a computer program to solve the wave growth equations presented above. The most complex models are comprehensive spectral models which operate on a grid and simulate natural wave processes including growth, dissipation, propagation, and nonlinear interactions between different wave frequencies. The models can be used for both wave hindcasting (wind input based on data from past events) and wave forecasting (wind input based on predictions for the future). A microcomputer software package for coastal engineering applications, called the Automated Coastal Engineering System (ACES), is available (Leenknecht et al., 1992).

IV. SHIP-GENERATED WAVES

As water flows back past a moving vessel, a variable pressure distribution develops along the hull. The pressure rises at the bow, then falls to below the free stream pressure over most of the side of the hull and rises again at the stern. The water surface profile along the hull responds to this pressure distribution causing a rapid rise and fall of the water surface at the bow and to a lesser extent (owing to flow separation) at the stern. Inertial effects cause the water surface to "overshoot" its equilibrium position and generate sets of waves that propagate out from the bow and stern of the vessel. The hull pressure distribution and resulting height of generated waves depend on the water velocity relative to the hull (i.e., the ship speed and the clearance between the hull and the channel bottom and sides) and the bow and stern geometry. The periods and propagation directions of the generated waves depend only on the ship speed and the water depth.

For a detailed discussion of the process of ship wave generation and the resulting wave characteristics, see references section (Robb, 1952; Sorensen, 1973a). Only an overview of the pattern of ship waves and their characteristics are provided in this chapter.

A. Ship Wave Patterns

Fig. 10.7 shows the pattern of wave crests generated by the bow of a ship moving over deep water. It consists of symmetrical sets of diverging waves that move obliquely out from the sailing line and a single set of transverse waves that move in the same direction as the ship. The transverse and diverging waves meet to form cusps located along lines that are 19 degrees 28 minutes out from the sailing line independent of the hull design and ship speed. The largest wave amplitudes are found at the cusp points. If

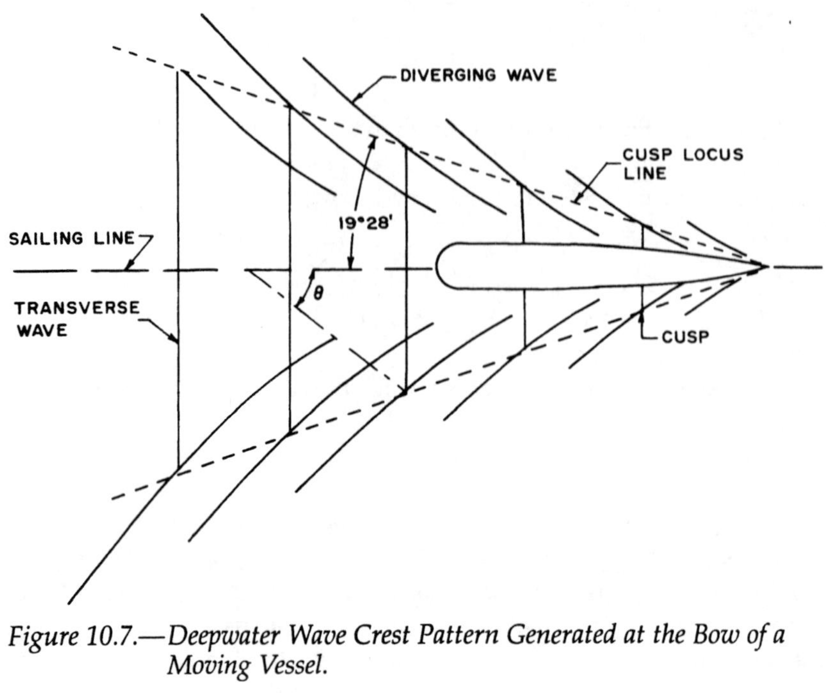

Figure 10.7.—Deepwater Wave Crest Pattern Generated at the Bow of a Moving Vessel.

the ship speed is increased, this pattern retains the same form but expands in size as the individual wave lengths increase. A similar pattern of waves, typically with lower amplitudes, would be generated at the ship's stern.

Since the wave system remains steady with respect to the ship as it moves forward at a constant speed, the waves that form the pattern must have a phase celerity C given by

$$C = V \cos \theta \tag{10.32}$$

where V is the ship speed and θ is the angle between the sailing line and the direction of wave propagation (Fig. 10.7). For diverging waves in deep water, the theoretical value for θ is 35 degrees 16 minutes.

For a given ship speed, at increasing distances from the bow, diffraction causes the wave crest length to increase and, consequently, the wave height to decrease. Havelock (1908) demonstrated analytically that the wave heights at the cusp points should decrease at a rate that is inversely proportional to the cube root of the distance from the bow, while the transverse wave heights along the sailing line decrease at a rate that is inversely proportional to the square root of the distance from the bow. Thus, at greater distances from the ship, the diverging waves become relatively higher than the transverse waves.

Wave amplitudes increase exponentially as the ship speed increases. When ship speed increases to a point where the length of the diverging waves exceeds twice the water depth, the waves "feel" the channel bottom and their characteristics and the wave crest pattern begins to change. This "shallow water" condition occurs for a vessel Froude number, F, in excess of approximately 0.7 where

$$F = V/\sqrt{gd}, \tag{10.33}$$

d is the still water depth, and g is the acceleration of gravity.

In shallow water, as the Froude number increases from 0.7 to 1.0, several changes occur. Transverse wave heights increase at a faster rate than the diverging wave heights, so they become relatively more prominent as the Froude number approaches unity. The cusp locus angle increases from the deep water value of 19 degrees 28 minutes to 90 degrees at a Froude number of unity. At F = 1, with a cusp locus angle of 90 degrees, the diverging and transverse waves have coalesced with their crests oriented perpendicular to the sailing line. Also, most of the energy in the wave system is concentrated in the first wave at the bow. Self-propelled vessels cannot exceed a Froude number of unity and most vessels operate at speeds such that their Froude number is less than 0.9 (Schöfield, 1974).

Very light vessels, moving at increasing speeds, may develop sufficient hydrodynamic lift to cause them to plane. After the onset of planing there is usually no significant increase in generated wave heights for further increases in the vessel speed (Sorensen, 1973b).

B. Ship Wave Characteristics

For the design of bank protection works or marina mooring areas, expected ship wave conditions must be known as a function of vessel draft, speed, and distance from the point of concern. The wave conditions of primary interest are the dominant approach direction, the dominant wave period and a meaningful representative wave height. When designing for wind waves, a representative height commonly used is the significant height. Vessel-generated wave spectra differ from wind wave spectra and have not been studied in sufficient detail to define an appropriate representative height.

The highest ship-generated waves are at the cusp point and have the direction and period of the diverging waves at that point. Based on the theoretical limits of 35 degrees 16 minutes and 0 degrees for deep water and travel at a Froude number of unity along with available empirical data, Weggel and Sorensen (1986) developed the following relationship for the direction of wave propagation θ (in degrees):

$$\theta = 35.27\,(1 - e^{12(F-1)}). \qquad (10.34)$$

Thus, for a given vessel speed and water depth the wave direction can be determined from Equation 10.34 and then the diverging wave celerity can be determined from Equation 10.32. Then, the diverging wave period T can be determined from a modified form of the wave dispersion equation

$$C = \sqrt{\frac{gCT}{2\pi} \tanh \frac{2\pi d}{CT}} \qquad (10.35)$$

by trial and error. For common water depths and ship speeds, the diverging wave period has typical values between one and three seconds.

More than a dozen laboratory and field investigations of ship-generated waves have been conducted during the past three decades (Sorensen and Weggel, 1984; Weggel and Sorensen, 1986). Typically, the water surface time history was measured for each vessel passage to yield a record similar to that in Fig. 10.8. Usually, the maximum wave height (H_m in Fig. 10.8) from each record was reported along with the vessel speed and type, the water depth, and the distance from the ship's sailing line. An occasional author has presented an average or a root mean square wave height from each record, but these data are quite limited. Sorensen (1967; 1973b) reports on measured values of H_m for a variety of vessel sizes and shapes in open water. For ship speeds between 6 and 14 knots, H_m values range from 0.2 to 0.9 meters at a distance of thirty meters from the sailing line and from 0.1 to 0.8 meters at a distance of 150 meters from the sailing line. These values are representative of H_m values found in most common navigation situations.

At least six procedures have been developed for the prediction of H_m. They are all quasi-empirical and most are restricted to a specific range of vessel and/or channel conditions. Sorensen (1989) references these procedures and briefly discusses them. A comparison calculation of H_m for a selected ship speed, water depth, and distance from the sailing line showed significant deviation of results. The best approach appears to be to use the method(s) that most closely compares to design conditions and temper the result with field measurements discussed above.

V. WAVE-STRUCTURE INTERACTION

Water waves and their kinematics are important in hydrology and hydraulic engineering. Waves interact with boundaries of estuaries, causeways, tidal inlets/entrances, river embankments, coastal

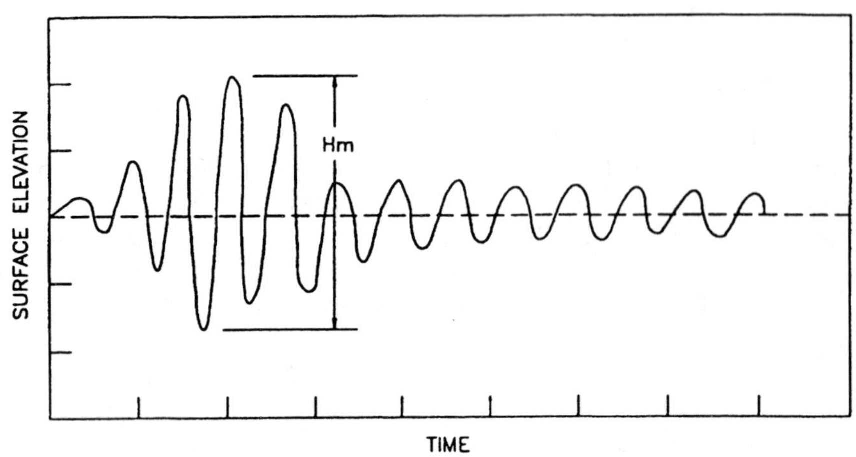

Figure 10.8.—Typical Ship-Generated Water Surface Time History.

beaches, marinas, and small boat harbors. These and other man-made hydraulic structures are built for attenuating wave impact and providing a safer and more navigable environment in the waterways. The ability to quantify the interaction of regular and random waves with surrounding lands and structures is essential in design work. This interaction is often described in terms of parameters such as wave runup or rundown, wave overtopping of structures, wave transmission and reflection coefficients or sometimes in the form of fluid loads acting on structures. The literature includes a large number of publications on each of these topics, and therefore only a brief overview of the subject will be presented.

A. Regular Wave Runup and Rundown

Reflection of waves by structures can cause significant fluctuations in the water level. Similar variations in water level may also occur when waves travel up inlets or entrances to waterways and estuaries, or lakes and reservoirs. Incident waves generated by winds, ships, or sliding lands are modified by structures and land boundaries through wave reflection or transmission and wave breaking. The difference between a wave breaking on a structure and a gentle beach is the steep slope of the former. As waves break on a slope, the wave momentum pushes the wave profile to rush up the slope. The runup, R_u, is the local maximum or peak vertical elevation reached by wave elevation on the face of a structure (Fig. 10.9). This maximum elevation includes both the wave set-up, the superelevation of the mean water level caused by the energy of waves known as the radiation stress, and the swash that represents the fluctuations about that mean. The runup depends both on slopes of the bed and structure, and reflection from these slopes. For wind waves, runup is a function of the incident wave characteristics, including wave height, wave period, wave steepness, and the distribution of wave energy in the wave energy spectrum. The water depth at the toe of a structure and wave breaking also influence the runup. For a fully-reflective and impermeable vertical wall, wave runup is theoretically twice the incident wave height.

For regular waves, an empirical relationship may be derived between runup and wave properties, geometry of a structure, water depth, slope angle, structural surface roughness, porosity, and beach face slope. Wave breaking on slopes reduces the energy of incident waves through reflection and runup. The mechanics of wave breaking and degree of reflected and consumed momentums are controlled by the steepness of waves (H/L) and slope of the beach ($\tan\alpha$). Using the subscript 0 for values of deep water wave height and wavelength, these may be combined into one dimensionless variable called the "surf similarity parameter" or "Iribarren's number" defined comparable to Equation 10.20 as

$$\xi = \tan\alpha/(H_0/L_0)^{1/2}. \tag{10.36}$$

The influences of toe depth and slope are obvious; steep slopes cause greater reflection, resulting in little or no runup. On relatively flat or gentle slopes, the up-rush of waves is retarded by the bottom friction over the distance water travels, limiting the vertical reach of water usually to be less than the length of the sloping surface. On mild slopes, it is generally difficult to define the wave runup precisely due to either the foamy water front after wave breaking, or the presence of a very thin film of water on

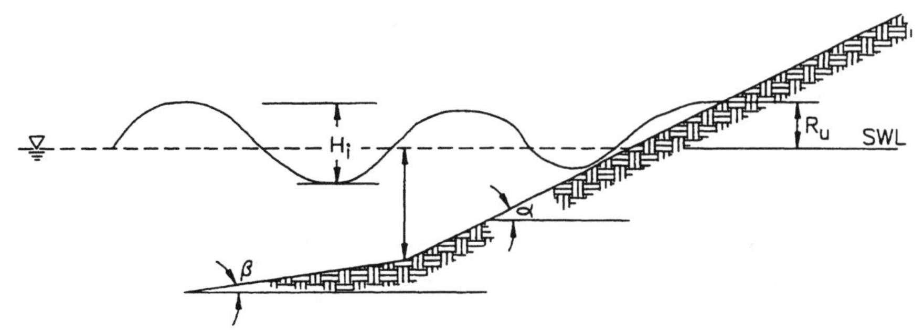

Figure 10.9—Definition Sketch for Wave Runup on a Sloping Structure

the slope. In this case, runup may either be taken as the maximum or some average rush-up water level. This definition gives rise to a large difference between runup measurements from laboratory and field studies. The temporal error in the measured runup at any time depends on how accurate the position of water elevation on a structure may be located. Waves may break before reaching the toe of a structure or while on its sloping surface. Thus wave breaking is the essence of runup and rundown calculations, and will next be briefly discussed.

The height of a wave moving over a sloping beach may increase while its length decreases; waves are deformed by a shoaling process before breaking. Thus, the steepness of waves, defined as the ratio of wave height to wavelength H/L, approaching a beach increases. Waves break when they reach a limiting steepness. In general, wave steepness depends on the relative water depth defined as d/L, and the beach slope, $\tan\alpha$. The theoretical wave breaking limit for constant water depth is (Miche, 1944)

$$(H/L)_{max} = 0.142 \tanh kd. \tag{10.37}$$

This equation may also be used for very mild slopes.

The form of waves at breaking is known as the breaker type, characterized by spilling, plunging, collapsing, and surging types. The surf similarity parameter, based on deepwater wave assumption (i.e., $d/L > 1/2$), can be used to distinguish breaker types (Galvin, 1967; Weggel, 1972; Battjes, 1974a). Wave breaking types as a function of the surf similarity parameter are listed in Table 10.4.

Waves of very low steepness on a slope may not break but instead reflect from the slope to form a standing wave in front of the slope. In general, steep waves on flat beaches or low slopes spill; waves of moderate steepness plunge on steeper slopes, while low steepness waves surge on the steeper slopes. Plunging breakers produce high local impact pressures on sloping surfaces. Winds also influence the breaking of waves; onshore winds tend to trigger breaking in deeper depths whereas offshore winds retard breaking, forcing it to occur in shallower depths.

If waves do not break until reaching a structure, the sloping face of the structure may itself cause waves to break. The slope of a structure increases the breaking wave height and water depth at the point of breaking as (Le Mehaute and Koh, 1967)

$$H_b/H_0 = 0.76(\tan\alpha \cos\alpha_b)^{1/7} (H_0/L_0)^{-1/4} \tag{10.38}$$

where H_b is the breaking height at the toe of structure, H_0 and L_0 are the deepwater unrefracted incident wave height and wavelength, respectively, and α_b is the breaking angle to the shore. The deepwater wavelength may be estimated in units of meter by $L_0 = 1.56\ T^2$. Collins (1970) shows that depth at breaking is roughly $1.3H_b$. For slopes less than 1 on 8, this may be estimated by

$$H_b/d_b = 0.72 + 5.6 \tan\alpha \tag{10.39}$$

Different formulas have subsequently been derived for breaking wave height and water depth (Weggel, 1972; Komar and Gaughan, 1973) from laboratory studies on smooth, plane slopes using linear wave theory (USACERC, 1984). For irregular waves, the breaking root-mean-square (rms) wave height is approximately half the local water depth at the breaking point.

The wave runup elevation on a slope is different for nonbreaking and breaking wave conditions. In

TABLE 10.4. Classifications of Wave Breaking.

Surf Similarity Parameter ξ	Wave Breaking Type
0.5 or less	Spilling
0.5 to 3.3	Plunging
3.3 or larger	Surging and Collapsing

the absence of information about wave breaking, runup may be estimated for both nonbreaking and breaking waves, using the following respective formulas (Hunt, 1959):

$$R_u = 2\, a_c\, A \tag{10.40}$$

$$R_u = (H_0 L_0)^{1/2}\alpha \tag{10.41}$$

in which a_c is the incident wave crest elevation above the still water level, and α is the face slope in radians. The parameter A varying with $K = (kd/\alpha)(L/L_0)$ is the surface amplification factor (Fig. 10.10) that determines if waves break on a slope.

When the toe of a structure with a steep slope is placed in deep water, the nonbreaking wave runup may be approximated by (Miche, 1944)

$$R_u = H_0\,(\pi/2\alpha)^{1/2} \tag{10.42}$$

Wave steepness generally increases the runup elevation on a structure with a sloping face. This increase may be estimated from the standing wave theory (Miche, 1944). The resulting wave runup is then determined from (Takada, 1970; 1977)

$$R_u = H_0\, K_s \{(\pi/2\alpha)^{1/2} + (R_v/H_t - 1)\}(\cot \alpha_c \tan \alpha)^{2/3} \tag{10.43}$$

$$R_v/H_t = 1 + (\pi H_t/L_t)\coth k_t\, d_t \{1 + 0.75\, \mathrm{cosech}^2\, k_t d_t - 0.25\, \mathrm{sech}^2\, k_t d_t\} \tag{10.44}$$

where H_0 is the deep water wave height, R_v is the wave runup elevation on a vertical wall ($\alpha = \pi/2$), K_s is the shoaling coefficient (USACERC, 1984), H_t is the wave height at the toe of the slope, and α_c is the limit angle for waves not to break on a slope. Here k_t, L_t, and d_t are the wave number, wavelength, and water depth at the toe location.

Data on the effect of wave steepness on runup is available (USACERC, 1984) for various ratios of d_s/H_0' where $d_s = d_t$ and $H_0' = H_0$. Samples of the theoretical and experimental runup estimates for different values of structural slopes fronted by a 1 on 10 beach slope are summarized in Figs. 10.11, 10.12, and 10.13.

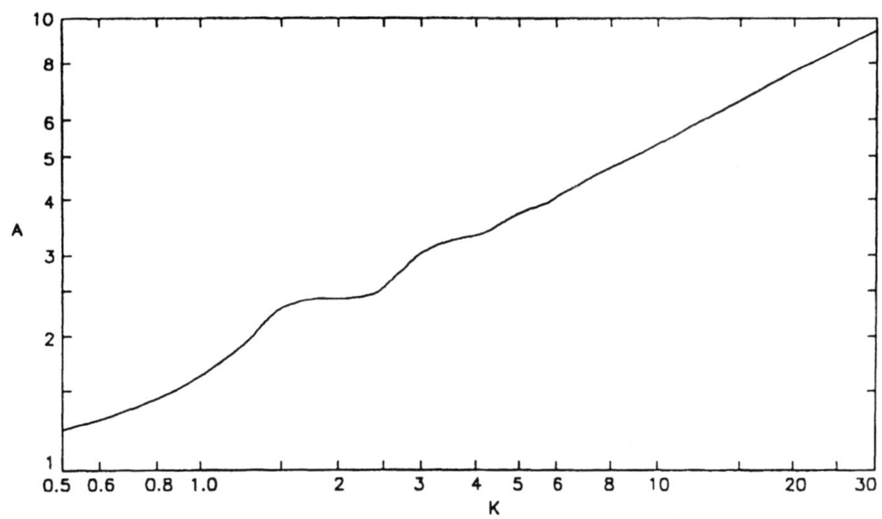

Figure 10.10.—Amplification Factor for Wave Runup Estimates

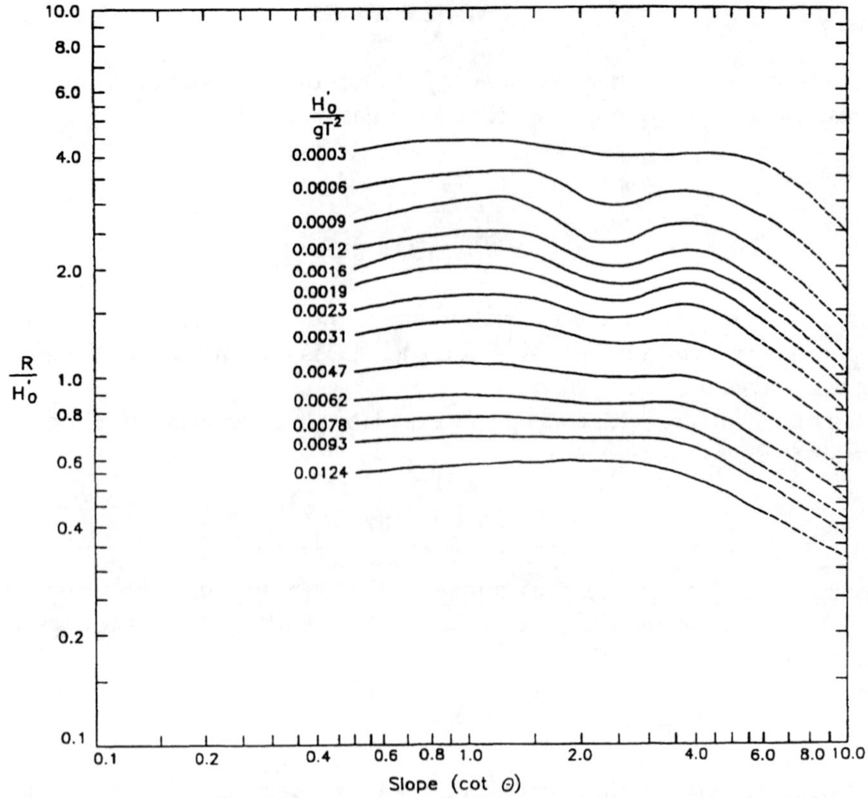

Figure 10.11.—Wave Runup on Smooth, Impermeable Structural Slopes Fronted by a 1:10 Beach Slope for d_s/H_0 = 0.0 (USACERC, 1984).

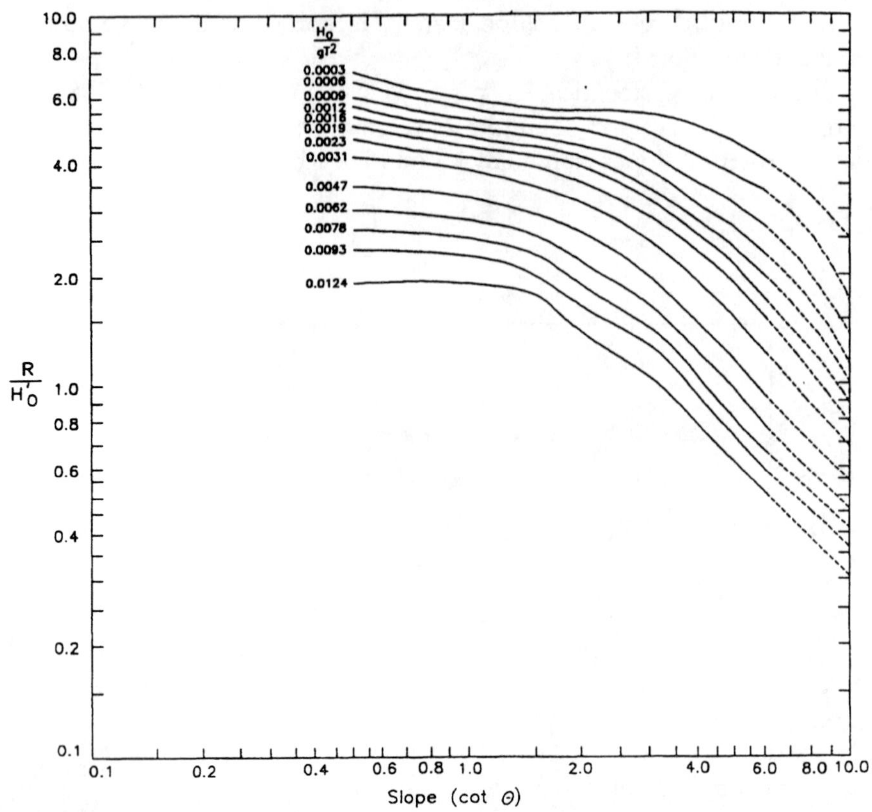

Figure 10.12.—Wave Runup on Smooth, Impermeable Structural Slopes Fronted by a 1:10 Beach Slope for d_s/H_0 = 0.8 (USACERC, 1984).

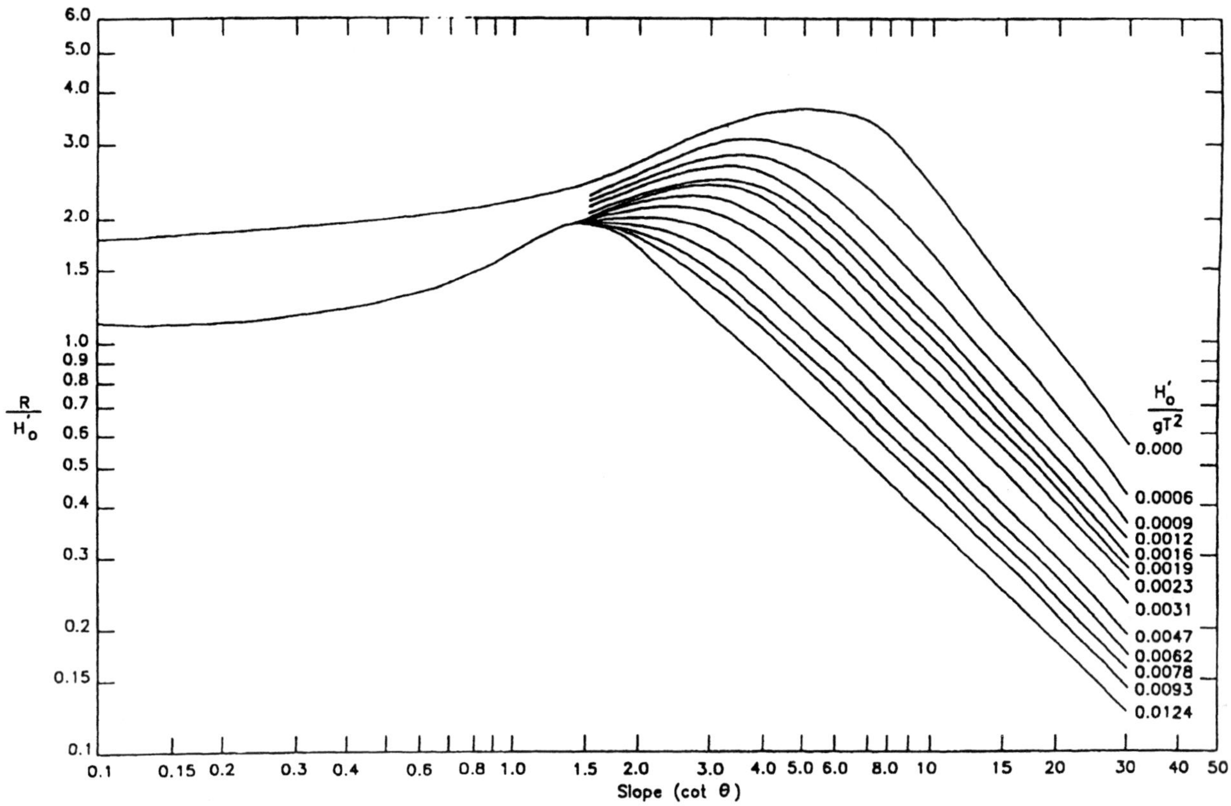

Figure 10.13.—Wave Runup on Smooth, Impermeable Structural Slopes Fronted by a 1:10 Beach Slope for $d_s/H_0 \geq 3.0$ (USACERC, 1984).

When waves break on the slope, the following equation has been derived from experimental results for runup estimates (Hunt, 1959):

$$R_u = 1.01\, H_t \tan \alpha\, (H_t/L_0)^{-1/2} \tag{10.45}$$

To determine whether or not overtopping is a serious problem in the design of mound structures (riprap and rubble mound) requires the prediction of wave runup on these irregular shape structures. Most of these protective structures have rough and highly permeable surfaces which absorb wave energy and reduce wave runup. The fundamental difference between runup on rubble mound structures and riprap revetments is in their structural design. On rubble mound slopes, runup and rundown exhibit characteristics similar to the smooth slope results reduced by a coefficient representing effects of surface roughness and permeability (USACERC, 1984). Comparison of wave runup on smooth slopes with runup on permeable rubble slopes is presented in Fig. 10.14.

A correction factor, r defined as the ratio of runup on a rough permeable slope or other slopes to the runup on a smooth impermeable slope should be used when applying smooth-slope design curves to the roughened and porous structures. This factor is specified in Table 10.5.

The determination of wave rundown for regular waves on sloping surfaces often is based on the assumption that rundown on these surfaces occurs on a wedge profile. The water surface runs down the wedge, varies with time, and attains its maximum velocities before the face slope is dry. Based on experimental data, a formula has been suggested (Roos and Battjes, 1976) for rundown on smooth slopes. This formula does not allow rundown to penetrate below the still water level at low values of ζ.

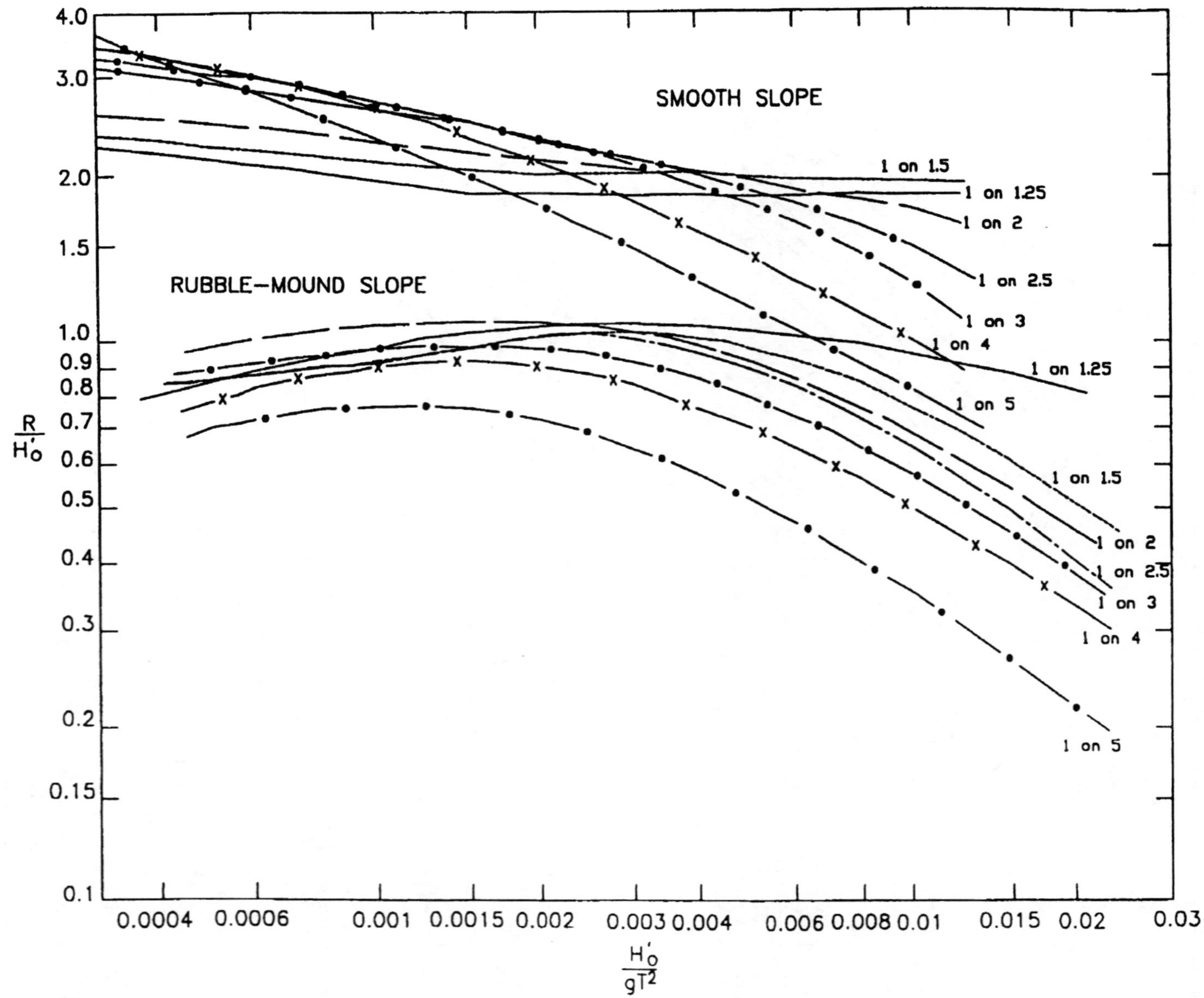

Figure 10.14.—Comparison of Wave Runup on Smooth Slopes with Runup on Rubble Slopes for $d_s/H_0 \geq 3.0$ (USACERC, 1984).

TABLE 10.5. Roughness and Porosity Correction Factor.

Sloping Surface	r
Smooth	1.0
Concrete, Basalt and Gobi Blocks	0.85–0.90
Grass	0.90
Quarrystone (one layer)	0.70–0.80
Quarrystone (rounded)	0.65–0.70
Quarrystone (layers)	0.50–0.60
Concrete Armor Units (50% void ratio)	0.40–0.50

Assuming that rundown is independent of wave height on smooth slopes for low values of ξ, say for $0.3 < \xi < 3.0$, an empirical formula curve-fitting the model data for rundown is

$$R_d/H = (1 - 0.45\xi)\xi \tag{10.46}$$

Physical model results are correctly represented by this equation (Bruun, 1985). Fig. 10.15 presents wave rundown and wave runup information for a graded riprap with slope 1 on 2 over an impermeable base.

B. Irregular Wave Runup and Rundown

The majority of runup and rundown results have been obtained from tests with regular waves. Hydraulic structures in nature are exposed to irregular waves, and therefore design estimates should be based on irregular sea states. In this case, since irregular wave properties vary randomly, it is necessary to treat runup and rundown as stochastic variables which depend on wave characteristics for a given slope, including the wave energy spectrum and its statistical parameters. The preferred approach in practice has been to relate the wave runup and rundown of irregular seas to their counterparts obtained from regular waves. The transfer of results from regular waves to irregular waves uses the so-called "hypothesis of equivalence" (Saville, 1962; Van Oorschot and d'Angremond, 1968; Battjes, 1971; Gunbak, 1977, 1978) that assumes each wave component in a spectrum generates runup/rundown values which would be similar to those by a regular wave of the same height and period.

In hydraulic design studies, a wave spectrum may be approximated by a monochromatic wave of period T_p which is the period associated with peak spectral energy. Consequently, for irregular waves, the significant wave height, $H_{1/3}$ (or H_s) and T_p should be used in the formulas and figures for regular waves. A conservative estimate of the component wave runup resulting from irregular waves is plausible assuming that wave runup may be represented by the Rayleigh distribution. The probabilistic runup estimate then becomes (Van Oorschot and d'Angremond, 1968; Battjes, 1974a, 1974b)

$$R_p/R_s = H_p/H_s = \{\ell n(1/p)/2\}^{1/2} \tag{10.47}$$

in which R_p is the runup for a "p-percent" probability of exceedance (i.e., the elevation exceeded by p-percent of runups), and R_s is the runup for H_s. As an example, $R_{0.01}$, or the 1% runup, would be 1.52 times greater than R_s (Fig. 10.16). The percent runup means the percent exceedance, and it should not be confused with the average value for a given percent probability.

C. Wave Overtopping

When the crest elevation of a structure is lower than the maximum runup, water will spill onto or over it. Overtopping does not refer to some spray or splash over a structure; it represents overrun by clear water up-rushing the structure. For quantifying the consequences of overtopping, it is customary to define overtopping as the volume of flow per unit time and crest length. This may also be expressed as a damage criterion for sea walls, a required armor units weight criterion for mound structures, or a water

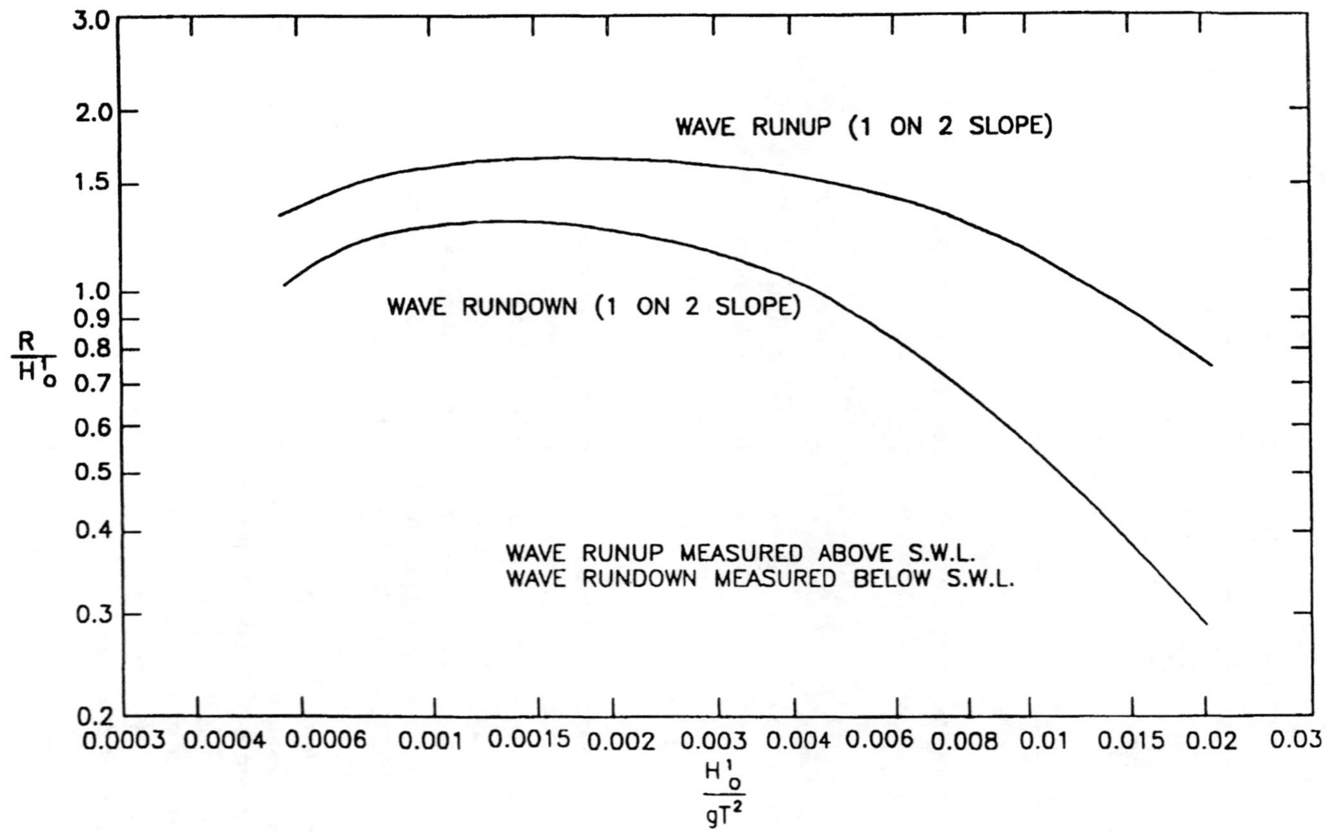

Figure 10.15.—Wave Runup and Rundown on Graded Riprap of Slope 1:2 on an Impermeable Base for $d_s/H_0 \geq 3.0$ (USACERC, 1984).

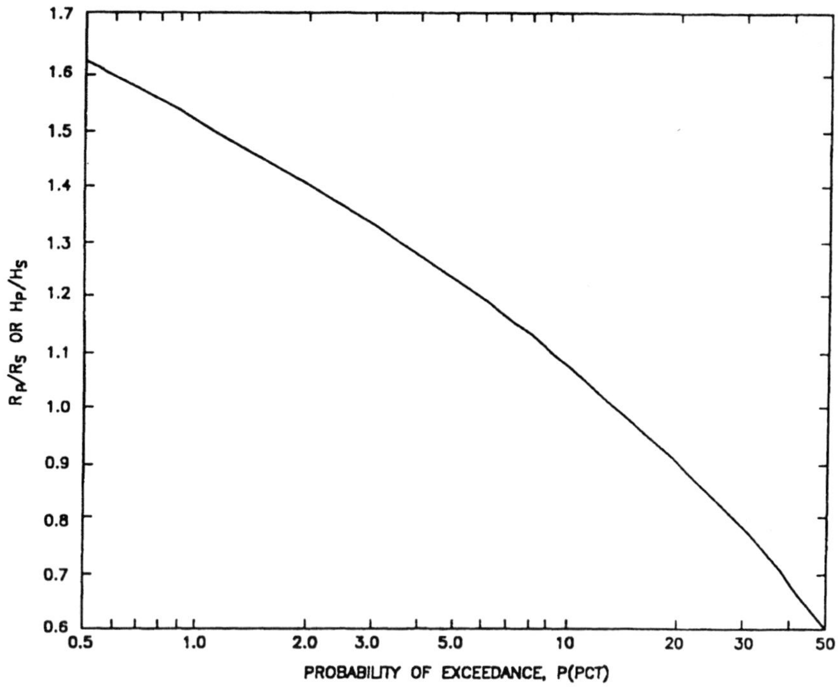

Figure 10.16.—Relative Runup R_p/R_s or Relative Wave Height H_p/H_s as a Function of the Probability of Exceedance p (Battjes, 1974).

removal criterion for sizing drainage systems. In general, it is prohibitively expensive to build hydraulic structures to preclude overtopping from the largest waves, and consequently, a certain quantity of overtopping flow may have to be allowed in the planning and design. The quantity of overtopping can be established beforehand from the extent of a flood area and by the required capacity of pumping facilities for dewatering the downstream side of the overtopped structure.

The prediction of overtopping for regular and irregular waves is described in detail in SPM (USACERC, 1984). The factors that have the greatest influence on overtopping include the face slope and water depth at the toe of a structure and the wind speed. In general, the overtopping rate increases with increasing face slope of a structure, attaining the maximum rate when slope is roughly 1:2 to 1:3. The toe depth determines whether or not the structure is positioned in the standing or the breaking wave zone. The wind supplies more energy into waves, and thus overtopping increases proportionally to the wind speed. Recent research studies (Takada, 1970; 1977; Sato and Goda, 1981) have modified the earlier results (Saville, 1956; 1958; 1960; Savage, 1958) to significantly improve overtopping predictions. The following formulas may be used for estimating overtopping of regular and irregular waves, respectively:

$$Q_m = (gQ_0H_0'^3)^{1/2} \exp\{-u\} \qquad (10.48)$$

where for $0 \le (h-d_s)/R_u \le 1.0$, and

$$u = (0.217/\alpha) \tanh^{-1}(h-d_s)/R_u \qquad (10.49)$$

$$Q_i = [gQ_0(H_0'^3)_s]^{1/2} \exp\{-uR_u/(R_sK_i)\} \qquad (10.50)$$

$$K_i = R_p/R_s = (-\ln p/2)^{1/2} \qquad (10.51)$$

$$Q_r = 1/199\, \Sigma\, Q_i \qquad (10.52)$$

Q_m is the overtopping rate due to regular waves or volume per unit time per unit structure length; Q_i is the overtopping rate due to irregular waves or volume per unit time per unit structure length associated with a particular value of K_i and $(H'_0)_s$, where $(H'_0)_s$ is the equivalent deep water significant wave height; index "i" refers to ranges of probabilities desired to be generated for runup with a Rayleigh distribution; index "p" refers to particular probability of exceedance; R_p is the runup associated with a particular exceedance limit; R_s is the runup corresponding to a deep water significant wave height and period, predicted by conventional regular wave studies; $Q_{1/2}$ is the average overtopping rate for runup with a Rayleigh distribution having a probability of exceedance of $p = 0.005$ or 0.5%; g is gravitational acceleration; H_0' is the equivalent deep water wave height; h is the height of the structure crest above the bottom; d_s is water depth at the structure toe; R_u is runup on the structure if it were high enough to prevent overtopping; $h - d_s/R_u$ is relative freeboard; and α and Q_0 are empirical coefficients (see USACERC, 1984 for values of these parameters).

Note that for a 0.5% exceedance ($p = 0.005$), the corresponding overtopping rate, $Q_{1/2}$, is for a runup of 63% greater than R_s ($K_i = 1.63$). Therefore, it is customary to refer to $Q_{1/2}$ as the *peak overtopping rate.* Fig. 10.17 shows a family of curves for ratios of Q_r/Q_m and $Q_{1/2}/Q_m$ for values of α ranging from 0.04

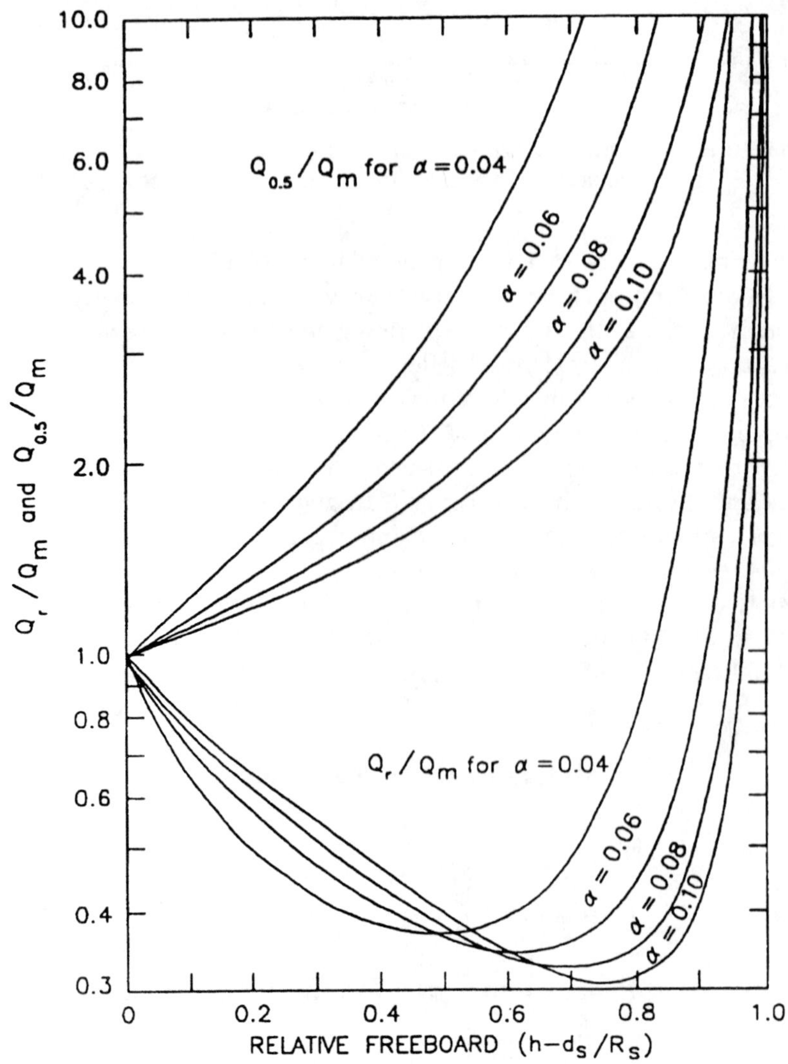

Figure 10.17.—Ratios of the Wave Overtopping Rates (see Eqs. 10.41 through 10.45) as a Function of the Relative Freeboard for Different Values of α (USACERC, 1984).

to 0.1. These are typical values of α in applications, and clearly this parameter can strongly influence overtopping estimates. A different method may be used for prediction of the overtopping rate due to irregular waves using the discharge formula for flow over a weir.

The effect of winds on overtopping can be important, though little information is available to determine this effect. A correction is necessary for including the effect of wind in overtopping estimates. This correction may be estimated by (Takada, 1977)

$$Q(v) = \exp(\lambda v / c_0)\, Q(0) \tag{10.53}$$

where Q(0) is the overtopping rate in a windless condition and v and c_0, respectively, are the wind velocity and deepwater wave celerity. The value of λ for nonbreaking waves is 0.112, and 0.088 in the zone between the breaking point and the toe of a structure. An approximate value of λ in the shallower waters where depth is less than the toe depth of a structure is 0.231.

The prediction of overtopping on gravel islands may be based on data and semi-empirical methods applicable to the design of straight ripraps and dikes. This approach assumes that runup height around gravel islands would be the same as the runup height on an equivalent two-dimensional structure, and therefore the variation of overtopping rates around the periphery of gravel islands is neglected.

D. Wave Transmission

The determination of wave transmission past hydraulic structures may be required for some projects. Important parameters for wave transmission estimates include the incident wave characteristics (height H_i and period T), water depth d, crest elevation z_c, and crest width of a structure. The angle, roughness and permeability of the slope may also be important for gentle slopes when the crest height of a structure is greater than 10 m. For a fully- or partially-submerged structure, the controlling parameter is the relative submergence of the crest height, z_c/d. The effect of this parameter on wave transmission coefficient varies with the wave steepness and a large transmission coefficient is usually associated with longer waves. In addition, when the crest of a structure is relatively low compared to the wave height, the resulting overtopping may create quite large waves on the lee side of the structure.

For short waves past a structure whose crest is near the still water level, a more appropriate parameter for quantifying wave transmission is the ratio z_c/H_i, where H_i is the incident wave height. In order to accurately predict wave transmission coefficients due to short waves, it may be necessary to solve three-dimensional hydrodynamic diffraction/radiation equations of the flow field around structures. In the preliminary engineering calculations, the wave transmission coefficient, T_r, defined as the ratio of the transmitted wave height to the incident wave height ($T_r = H_t/H_i$), may be approximated by Tr $\approx$ 0 for $z_c/d \geq 0.75$. For $z_c/H_i \approx 0$, it may be assumed as $T_r \leq 0.5$, while $T_r \geq 0.75$ if $z_c/H_i < -0.5$. Due to lack of reliable data, it is difficult to give specific design guidance for wave transmission around rubble mound structures. This is mainly due to the presence of high porosity and mixed-size rocks. For these structures, wave transmission may either be numerically estimated using finite element or boundary integral techniques (Garrison, 1978; Mei, 1991) or determined experimentally by physical model studies.

E. Wave Forces on Structures

The knowledge and prediction of wave forces is useful for analysis and maintenance of various types of hydraulic structures, including seawalls, bulkheads, and revetments that protect lands from waves, storm surges, and currents. These structures have critical functional roles. For example, groins provide protection against land erosion resulting from waves and currents. Jetties prevent shoaling within navigational channels and control both depth and direction of the flow in river mouths, estuaries, and ports. Breakwaters offer protection to shores, develop beaches, reduce wave energy, and provide safe berthing in marinas, ports, and harbors. The calculation of wave forces on these structures is not a trivial task, and it may be performed by three different theoretical methods.

The first method is the Morison formula, (Demirbilek, 1987) (Dean and Dalrymple, 1991) suitable for estimating the fluid loading on slender pile type structures where the net force on a pile is represented by sum of the inertia and drag components using two experimentally determined semi-empirical coefficients. The method is applicable to the slender structures that cause least disturbance to waves since their size is small compared to the wavelength. The drag loads resulting from vortical flows are important in the design of piles.

For a small and non-slender structure, the drag force is usually negligible compared to the inertia (or mass) force. In this case, a second method called the Froude–Krylov (F–K) hypothesis (Garrison, 1978; Mei, 1991) is applicable for determination of wave forces. The F–K technique is a pressure–area method accounting concurrently both for the effects of added mass and diffraction in the computation of wave loads. Experimentally determined force coefficients are also required for the F–K method, but for a thin, elongated structure, the strip theory may be used to obtain this coefficient. For other situations, a third method is necessary for determining wave forces on hydraulic structures: the diffraction theory. The diffraction theory (Garrison, 1978; Mei, 1991) is a numerical technique that uses boundary integral or finite element or a combination of both methods (hybrid). The third method is more accurate than the first two methods, but it requires large and powerful computers in its numerical implementation. Thus, the first two methods continue to be used in practice. The primary difficulty in applying the first two methods is the selection of appropriate force transfer coefficients for typical hydraulic structures of complex geometry and configuration. These coefficients are available only for simple structures, for example, spheres and cylindrical bodies, and consequently, all three methods are needed for engineering design estimates.

The range of applicability of the three wave force methods for a pile structure is illustrated in Fig. 10.18 where dominant regions of the drag and inertia forces are defined in terms of the Keulegan–Carpenter number ($KC = u_0T/D$) and diffraction parameter ($KA = \pi D/L$). For this purpose, deepwater conditions have been assumed. Here D is a characteristic length of the structure (pile diameter), T is the wave period, u_0 is the maximum horizontal water particle velocity, and L is wavelength. Regions I–III are inertia dominated with region II representing the diffraction theory, while the F–K theory may be used in all three regions if force coefficients are known. Morison's formula should be used in regions IV–VI. Key features of each method will next be described for completeness.

F. Morison Equation

Using experimentally obtained drag and inertia force coefficients and theoretical values of the water particle kinematics, Morison et al. (1950) developed the following equation to calculate wave forces on slender piles:

$$F = F_i + F_d = K_m \, du/dt + K_d \, |u| \, u \qquad (10.54)$$

where $K_m = \rho C_m (\pi/4D^2)$, $K_d = 1/2\rho C_d D$, F_i and F_d are the inertia and drag components of the total fluid force per unit length of a pile, ρ is the mass density of water, and u and du/dt are horizontal water particle velocity and acceleration, respectively. Parameters C_m and C_d are experimentally obtained inertia and drag coefficients, and D is the pile diameter. The wave kinematics, u and du/dt, may be approximated by the linear wave theory (Airy theory) as

$$u = (\pi H/T) \{\cosh k(z + d)/\sinh kd\} \cos (kx - \omega t) \qquad (10.55)$$

$$du/dt = (2\pi^2 H/T^2) \{\cosh k(z + d)/\sinh kd\} \sin (kx - \omega t) \qquad (10.56)$$

where H and T are wave height and period, $\omega = 2\pi/T$, d is water depth, k is the wave number $= 2\pi/L$, and L is the wavelength. Deepwater condition corresponds to $d/L \geq 1/2$ where wavelength is given by $L = L_0$

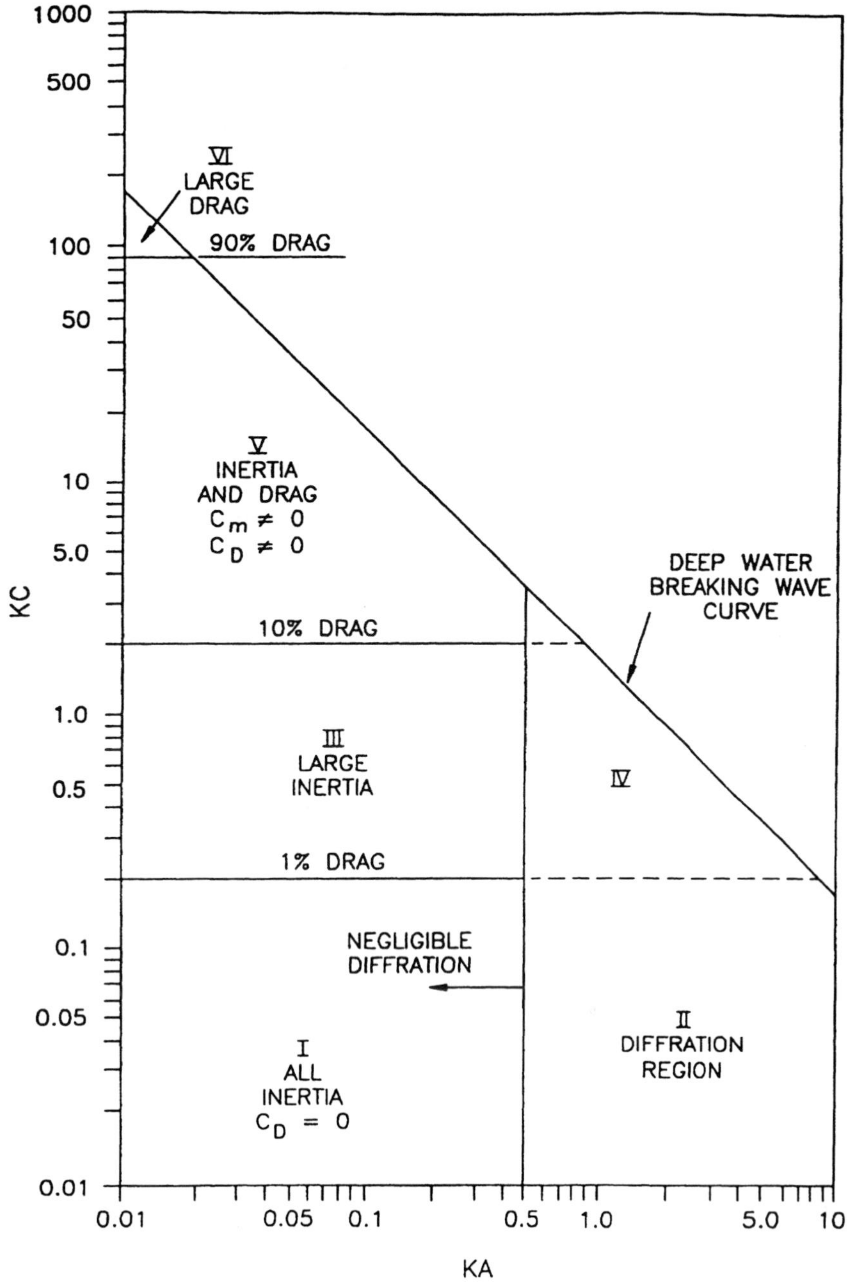

Figure 10.18.—Regions of Application of the Wave Force Formulas (Adapted from Garrison 1978).

$= (gT^2/2\pi)$. For intermediate water depth (i.e., $1/2 > d/L > 1/20$), the wavelength is found from $L = L_0 \{\tanh 2\pi d/L_0\}^{1/2}$, while $L = T\,(gd)^{1/2}$ in shallow water depth if $d/L \leq 1/20$. The two hydrodynamic force coefficients, C_m and C_d may either be determined from laboratory or field measurements made with smooth and rough cylinders. There can be a large difference between these coefficients from similar tests performed at different facilities (Demirbilek, 1987; Chakrabarti, 1987). Values of these coefficients for piles may range from 1.5 to 2.5 for C_m and 0.6 to 1.3 for C_d, respectively. Extreme care is needed when selecting these coefficients for other type structures. Note that Morison's equation can also be used for estimation of wave forces on inclined piles, provided normal velocity and acceleration components are used.

G. Froude–Krylov Theory

Hydraulic structures whose sizes are large compared to the wavelength would in general have negligible drag loads, and these are said to be inertially loaded structures. These structures cause a significant disturbance to the waves around them. Fluid forces on such structures may be estimated by the Froude–Krylov method (MacCamy and Fuchs, 1954) using the incident wave pressure information in a surface integral. The total fluid force on the submerged surface of an arbitrarily shaped structure may be computed by

$$F_{x,y} = C_{x,y} \int\int p dA \tag{10.57}$$

where $F_{x,y}$ are horizontal and vertical force components on the submerged surface of the structure in x- and y-directions, respectively. The integrated force results from the action of wave pressure p on the elemental surface area dA. The coefficients $C_{x,y}$ are available only for some elementary geometries, which are listed in Table 10.6.

For large structures, the diffraction parameter ka ($= \pi D/L$), characteristic length D and wavelength L may exceed the approximate values specified here. These are indications that diffraction effects are important and that wave forces must be calculated using the diffraction theory.

H. Diffraction Theory

A large structure can greatly disturb the incident wave field, scatter waves and may even radiate waves if it is free to move. These three types of waves combine to create a complex flow pattern around the structure. The solution of this problem is now well-established, and there are a number of computer programs for prediction of wave forces on structures of different shape and size. For axisymmetric structures (cylindrical and spherical), wave forces may also be estimated by simpler methods. For example, a two-dimensional strip theory method used routinely in naval hydrodynamics and offshore design applications may also be used for horizontal, elongated, and long hydraulic structures such as sea walls and barriers. For breakwaters and rubble mound structures of different shape and orientation, three-dimensional boundary integral or finite element methods should be used to predict fluid loads and reflection and transmission coefficients (Wehausen and Laitone, 1960; Garrison, 1978; Faltinsen and Demirbilek, 1989; Mei, 1991). The details of the diffraction theory are beyond the scope of this handbook.

I. Wave Forces on Vertical Walls

An engineering estimate of the wave forces on vertical walls with an underlying rubble mound foundation can be made using the static and wave dynamic pressure data. This simplification assumes that a standing wave pattern, called clapotis, forms in front of a wall where incident and reflected waves combine. Sainflou (1928) used the trochoidal wave theory to calculate wave pressure on walls in order to consider variability in the pressure from trough to crest of a standing wave (Fig. 10.19). The comparison of Sainflou's method with other experiments (Nagai, 1969) has shown that this method can greatly underestimate wave forces when $d/L \leq 1/10$. For determining wave forces on a vertical wall, three

TABLE 10.6. Inertia Coefficients for Froude–Krylov Method.

Structure/Geometry	ka	Horizontal Force Coefficient	Vertical Force Coefficient
Cylinder (horizontal)	0.0–1.0	2.0	2.0
Half-cylinder	0.0–1.0	2.0	1.1
Sphere	0.0–2.0	1.5	1.0
Hemisphere	0.0–0.8	1.5	1.1
Block (rectangular)	0.0–0.5	1.5	6.0

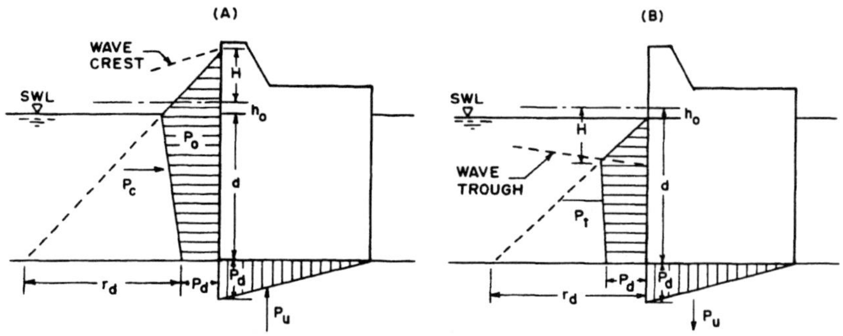

Figure 10.19.—Standing Wave Pressures for Sainflou's Method for (a) Wave Crest, or (b) Wave Trough at the Wall (Nagai, 1969).

subregions varying with the parameter d/L have been defined by Nagai. This modification and other formulas based on higher-order wave theories for nonbreaking wave loads on a vertical wall compare reasonably better with laboratory measurements than the Sainflou method.

The maximum force on a wall occurs at the instant a wave breaks in front of the wall due to combined effects of the sea bed slope and incident wave characteristics. Over a short time interval, wave breaking can give rise to very high wave pressure on the wall. Bagnold (1946) and Minikin (1950) developed approximate formulas for estimating shock pressures on walls assuming that the reflection coefficient is known. These formulas cannot easily be applied in practice due to scarcity of data and large scatter in the values of reflection coefficient obtained from laboratory studies.

Li et al. (1983) have recently developed and tested a different model that includes the effects both of a rubble mound and the sea bed slope, designated by i in Fig. 10.20, on the pressure distribution in front of the sea wall. Using extensive laboratory model test results, Li et al. (1983) identified two breaking wave force regions. In their model, the effect of rubble mound can be neglected if the height of rubble mound above the sea bed is small (i.e., $d_1/d \approx 1$ in Fig. 10.20). The breaking point is located at the antinode $x = L/2$ from the wall. A wave pressure diagram is then constructed by specifying pressures at four elevations depending on values of height of the rubble mound and bed slope. These and other variables are defined in Fig. 10.20 where b denotes the width of the wall.

The first defined region of the force corresponds to these range of parameters: $d_1/d > 2/3$, $d < 2H$, $\tan \alpha \le 0.1$, and $14 \le L/H \le 30$. In the first region, Li et al. (1983) define pressures as:

$P_0 = K_1K_2\gamma H$	wave pressure at the SWL	(10.58)
$K_1 = 1 + 3.2\,(i)^{0.55}$	factor for bed slope	(10.59)
$10K_2 = (-1 + L/H) - 0.015\,(L/H)^2$	factor for the ratio of L/H	(10.60)
$P_d = 0.6P_0$ for $d/H \le 1.7$	wave pressure at sea bed, d	
$0.5P_0$ for $d/H > 1.7$		(10.61)
$P_{h/2} = 0.7P_0$ at $H/2$	wave pressure at H/2 below the SWL	(10.62)
$P_u = 0.7\,(Pd_1b)/2$	lifting force at the bottom of the wall	(10.63)

In the second region, Li et al. (1983) account for the effects of mound, and its permeability on waves in front of the wall are considered. Laboratory tests in this case showed that waves break at a distance less

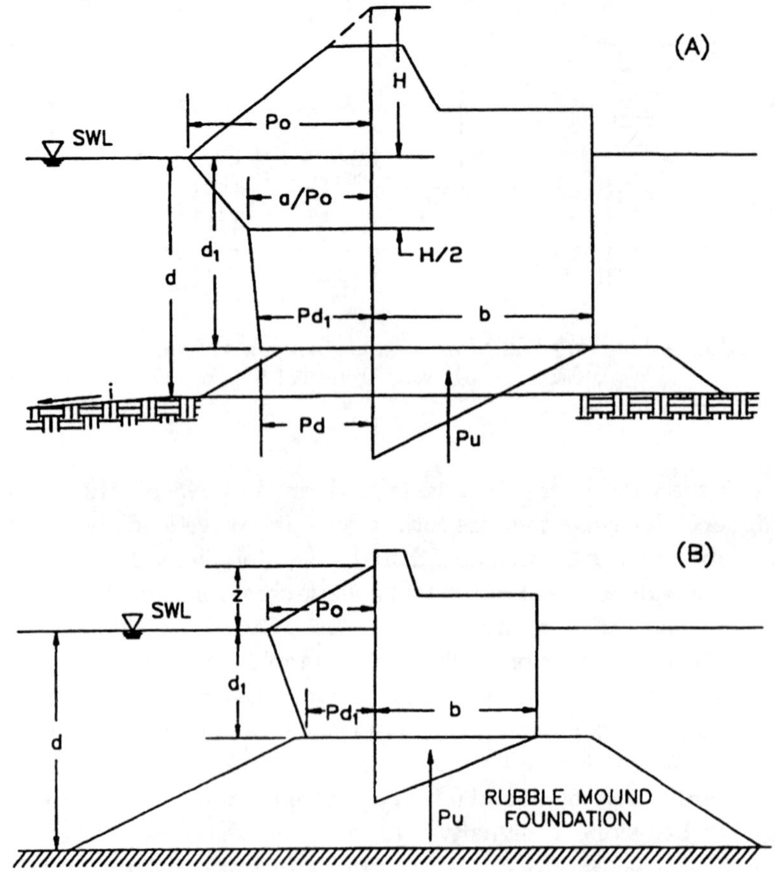

Figure 10.20.—Pressure Distribution Due to Breaking Waves on a Vertical Wall with (A) Low, and (B) High Rubble Mound Foundation (Li et al., 1983).

than L/2 from the wall as the height of the rubble mound increases (Li et al., 1983). The pressures in the second region where $1/4 < d_1/d \leq 2/3$ and $0.6H \leq d_1 < 1.8H$ are defined as

$$P_0 = 1.25\gamma H\,(1.8H/d_1 - 0.16)\,(1.0 - 0.13H/d_1) \qquad (10.64)$$

$$Pd_l = 0.6P_0 \qquad \text{wave pressure at the bottom of the wall} \qquad (10.65)$$

$$P_u = 0.7\,(Pd_l\, b)/2 \qquad (10.66)$$

Results from an alternative method developed by Goda (1974) also based on experimental data are in agreement with the estimates obtained from the above equations (Li et al., 1983).

J. Prediction of Irregular Wave Forces

A difficulty arises when the wave force formulas developed for regular waves are used with irregular waves which consist of several wave heights and periods. For irregular seas, the structural design should consider the behavior of structures under various combinations of wave conditions, including standing and breaking waves. The expected sliding distance concept (Ito et al., 1966) may be used for estimating irregular wave pressure on a wall (Fig. 10.21) by assuming that wave pressure is continuous in the region between standing and breaking waves. The method is similar to Li et al. (1983) and Sainflou (1928) where various pressure terms have been defined as

$$P_1 = 2\eta^*\gamma(\beta_1 + \beta_2 \cos^2 \alpha)/3 \tag{10.67}$$

$$P_2 = P_1 \text{ sech } kh \tag{10.68}$$

$$P_3 = \beta_3 P_1 \tag{10.69}$$

$$\beta_1 = 0.5\,[1.2 + (2kh)^2 \text{ cosech}^2\, 2kh] \tag{10.70}$$

$$\beta^2 = \min\{(h_0 - d)/3h_0\,(H_m/d)^2, 2d/H_m\} \tag{10.71}$$

$$\beta_3 = 1 - h'/h + h'/h \text{ sech } kh \tag{10.72}$$

where d is the water depth above the crest of a rubble mound, γ is the specific weight of water, h' is the draft of the caisson from SWL, h_0 is the water depth at a distance of $5H_{1/3}$ offshore side of the wall, $H_{1/3}$ is the significant wave height, α is the incident wave angle, and $\eta^* = 0.75(1 + \cos\alpha)H_m$ (see Fig. 10.21). The maximum wave height and wave period may be approximated by $H_m = 1.8H_{1/3}$, and $T_m = T_{1/3}$ where $H_{1/3}$ and $T_{1/3}$ are the significant wave height and period. In the case where both breaking and standing irregular waves occur in front of a wall, a correction to Ito et al.'s formula has been proposed as (Tanimoto et al., 1976)

$$\eta^* = 0.75\lambda H_m\,(1 + \cos\alpha) \tag{10.73}$$

$$P_1 = 0.5\beta_1\gamma H_m\,(1 + \cos\alpha) \tag{10.74}$$

where P_2 is still given by Equation 10.68. The parameter λ is a wave pressure reduction factor, typically assumed to have a value of 0.8.

The wave–structure interaction is a complex subject, and due to space limitations for this handbook, only a summary of the topics discussed could be presented here. To ensure proper applications of the principles and formulas outlined in this section, it would be necessary to supplement the information with pertinent references.

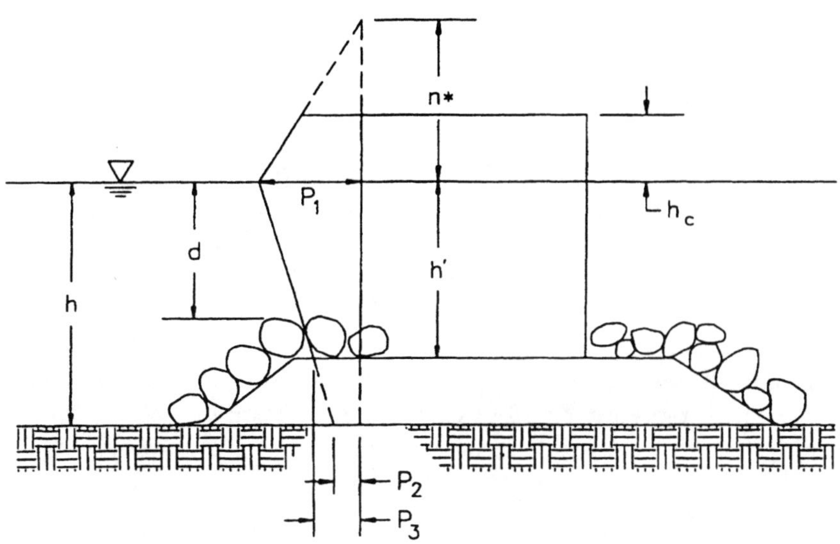

Figure 10.21.—Pressure Distribution of Breaking Waves on a Vertical Wall with Rubble Mound Foundation (Ito, 1966).

VI. WAVES AND CURRENTS

A. Nearshore Currents

Nearshore currents in the vicinity of hydraulic structures are composed of many scales, driven by several processes. These currents are generated by the action of winds, waves, and the gradient of temperature, density, and salinity of different water layers. In the context treated herein, the term "nearshore currents" refers only to the wave-induced currents, such as longshore currents, shoreward-directed currents, and rip currents. Types and characteristics of nearshore currents have been extensively treated in detail elsewhere (e.g., Johnson, 1919; Shepard et al., 1941; Putnam et al., 1949; Shepard and Inman, 1950; Longuet–Higgins and Stewart, 1960; 1961; 1962; 1964; Homma et al., 1962; Galvin and Eagleson, 1965; Galvin, 1967; Bowen, 1969a and 1969b; Thornton, 1970; Horikawa and Sasaki, 1972; Liu and Mei, 1974; Komar, 1976; Dalrymple, 1974; 1977; 1978; 1980; Phillips, 1977; Mei, 1991). A strong variation in the magnitude and direction of nearshore currents occurs over space and time due to the complex interplay of several factors: winds, waves, tides, topography, air–sea interaction, and sediments. Most of these factors continually vary with time in nature and therefore several simplifications and assumptions have to be made in the analytical and experimental methods for describing currents.

Other meteorological events besides winds and waves also influence nearshore currents and water levels. These include the seasonal variation of water temperatures, salinity content of coastal water, precipitation, and fresh water from river flows. Consequently, nearshore waters are inhomogeneous, stratified, and characterized by complicated flow patterns all of which are simply called currents. The currents control to a large extent the nearshore sediment movement. The wave-induced longshore currents, undertow and rip currents are well-known by fishermen, sailors, and others living near the coast. The existence of a strong and narrow seaward current in the offshore direction that returns the sea water transported shoreward by wave action is sometimes observed in the nearshore zone. Shepard et al. (1941) named these "rip currents," concentrated within the surface layer with strong velocities and carrying away fine sediments from the surf zone to deeper water regions. Rip currents and their generation mechanisms have been extensively studied.

Fig. 10.22 shows typical schematic representations for the concept of a nearshore current pattern based on extensive field observations (Shepard and Inman, 1950; Shadrin, 1961; Harris, 1969). In addition to structures, the current pattern is also controlled by the beach topography that molds itself to the currents in the area. There are three common nearshore current patterns depending on the wave approach angle to a beach. These include an alongshore system for large oblique wave approach angle, a system consisting of two symmetric cells resulting from shore-normal wave approach and an unsymmetric cellular system associated with small wave approach angle. The conditions leading to the development of longshore and rip currents are depicted in Fig. 10.22.

High values of longshore currents up to 1 m/sec are possible in storms. These currents ordinarily are 0.3 m/sec or less, remain fairly constant as the water depth changes and flow parallel to the shoreline. The magnitude of longshore currents is controlled by the wave height and the angle between the wave crest and the shoreline. Unlike the longshore currents, the cross-shore currents are not constant over the water depth, and their magnitudes are generally less than 10% of the wave speed. Rip currents are the most visible of all, manifesting strong and narrow patterns to form circulatory cells near beaches; thus, rip currents often pose hazard to swimmers. The high speeds of rip currents can also cause severe scouring problems to the foundation of hydraulic structures. Various mechanisms leading to the generation of the three current systems are described elsewhere (Dalrymple, 1974a; 1974b; Phillips, 1977; Mei, 1991).

B. Mathematical Modeling

Nearshore currents may be estimated from momentum equations (Phillips, 1977; Dean and Dalrymple, 1991), provided that forcing terms are included in these equations for winds, pressure gradients due to

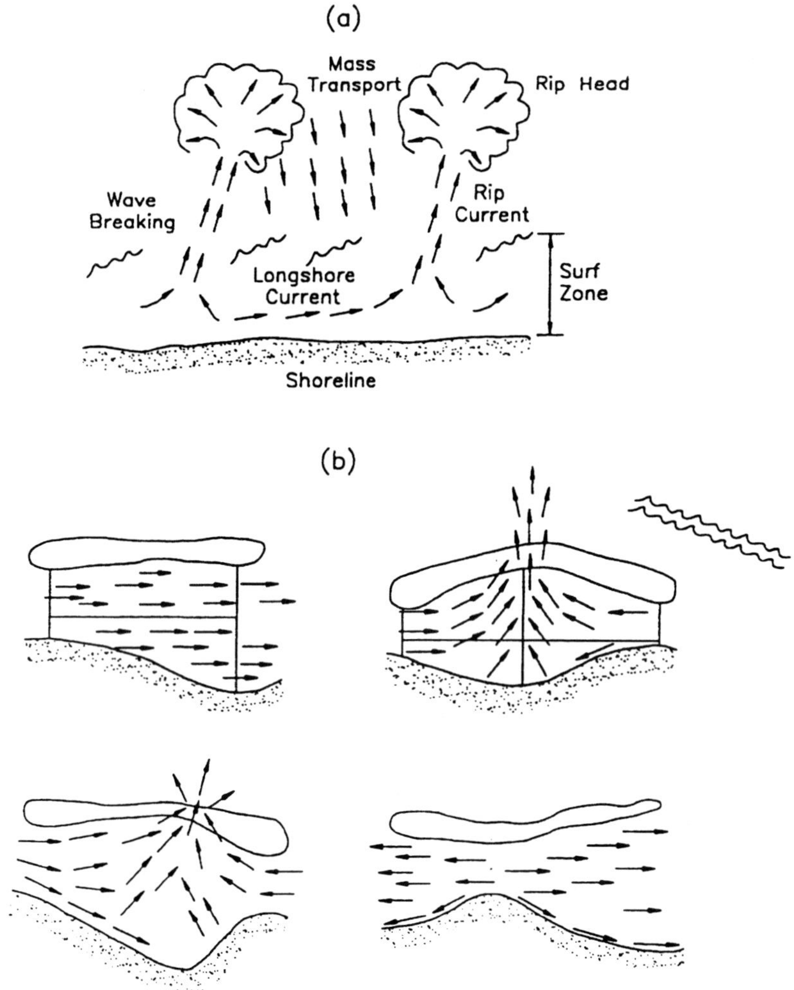

Figure 10.22.—Schematic Representation for (a) Nearshore Current System, (b) Development of a Longshore Current (Shadrin, 1961).

mean water level fluctuations, bottom friction due to waves and currents, and lateral mixing of the currents. The primary driving forces are usually the gradient in the momentum flux due to decay or breaking of obliquely incident waves and the longshore component of the wind. This gradient is called the radiation stress (Longuet–Higgins and Stewart, 1962; 1964; Bowen, 1969a; 1969b). The radiation stress, proportional to the wave height squared, is the cause both of longshore and cross-shore currents. The currents resulting from the radiation stress are the greatest in the regions of steep wave height decay. The lateral mixing is caused by the turbulent eddies, and it tends to diffuse the effect of radiation stress beyond these regions through the exchange of momentum. Other forcing terms play comparatively a lesser role than the radiation stress. For example, the bottom friction, which is the resisting force to the currents, may be significant only when the bottom roughness or the speeds of wave and currents are high.

It is necessary to know horizontal and vertical longshore current velocity distribution in the nearshore region to establish a relationship between the longshore current and coastal sediment transport rate. The distance from the shoreline to the offshore bar combined with amplitude and direction of incident waves are the controlling parameters for both alongshore and offshore components of the nearshore current system. In the last three decades, much research has focused on understanding the horizontal velocity, but the vertical longshore current velocity distribution is an unsolved problem.

The presence of nearshore currents influences waves arriving at the coasts, giving rise to a complex interaction between the two. The concept of radiation stress, introduced by Longuet–Higgins and Stewart (1960; 1961; 1962; 1964) for analysis of wave–current and wave–wave interactions, is now extended to such phenomena as wave set-down, wave set-up, longshore currents, and rip currents. The radiation stress corresponds to the excess momentum flux induced by the wave motion. A physical interpretation of the radiation stress for plane, sinusoidal waves propagating over water of uniform depth, is that it represents the difference between the mean momentum fluxes with and without wave-induced motion. The radiation stress describes the wave-induced hydrodynamic force upon a structure such as a breakwater that causes a change in the direction and magnitude of the wave momentum.

Numerous analytic solutions and various formulas have been proposed before and after the introduction of the radiation stress concept to evaluate the longshore current mean velocity (Horikawa, 1978). A detailed derivation of the depth-integrated and time-averaged mass, momentum, and energy equations for nearshore currents is available (Phillips, 1977; Mei, 1991). Some recent numerical models for nearshore wave transformation include the effects of currents, breaking, dissipation, shoaling, refraction, and diffraction for the propagation of waves in nearshore waters (e.g., Tsay and Liu, 1982; 1989; Kirby and Dalrymple, 1993).

C. Wave–Current Interaction

The influence of currents on waves has been known for centuries, though Unna (1942) was the first to consider the change of phase velocity as waves propagated with currents. Arthur (1950) studied the kinematics of shallow-water waves encountering rip currents, and Johnson (1947) tried to quantify the change of deep water wave height due to currents. These early studies were followed by some pioneering research (Longuet–Higgins and Stewart, 1960; 1961; 1962; 1964; Bretherton and Garrett, 1969) wherein the concepts of radiation stress and wave action were introduced for analysis of wave–current interaction. Later studies (Christofferson and Jonsson, 1980; Booij, 1981; Tsay and Liu, 1982, Kirby, 1984) incorporated the effect of variable depth, breaking and dissipation mechanisms in the governing equations.

It may suffice for engineering purposes to assume that when waves propagate onto a faster or slower current, wave frequency will remain constant, but both wave height and wavelength will either increase or decrease. However, if waves and currents are oblique during interaction, waves will be refracted and an energy transfer between current and waves occurs. In such cases, the effect of current on waves can be quantified as a Doppler shift, that is a change in the wave period results in a change in the wavelength. When waves are intercepted by currents, wave height is also modified by the currents. It is necessary to differentiate between wave parameters obtained with and without currents in the engineering design estimates. In this chapter the distinction will be made by representing wave parameters with the subscript c to denote their values in the presence of currents. Note that such parameters are measured relative to a fixed reference frame. The nonsubscripted variables pertain to the absence of currents.

The simplest case to consider in a reference frame where there is no current is a wave of frequency ω interacting with a uniform current of speed U. The wave frequency is relative to the current present. This same wave has an absolute or intrinsic frequency $\omega_c = \omega + kU$ in the frame of reference where current is present. The wavelength L and wavenumber k $(=2\pi/L)$ are unaltered by a change in reference frame so that $T_c^{-1} = T^{-1} + (2\pi)^{-1}kU$, where $T^{-1} = (2\pi)^{-1}\,(gk \tanh kd)^{1/2}$ for small amplitude waves at water depth d. The solution for wave number k will determine the wavelength L in the presence of the current. The nomogram in Fig. 10.23 provides the ratio of modified wavelength or celerity to those in still water condition (i.e., no current condition) using linear wave theory.

The change in the height of monochromatic waves due to a uniform current can also be estimated using the conservation of wave action (Longuet–Higgins and Stewart, 1961; Bretherton and Garrett, 1969). The maximum wave height estimated is shown in Fig. 10.24. Numerical models on computers are also available (Tsay and Liu, 1982; 1989; Kirby and Dalrymple, 1993) for estimating effect of currents on waves.

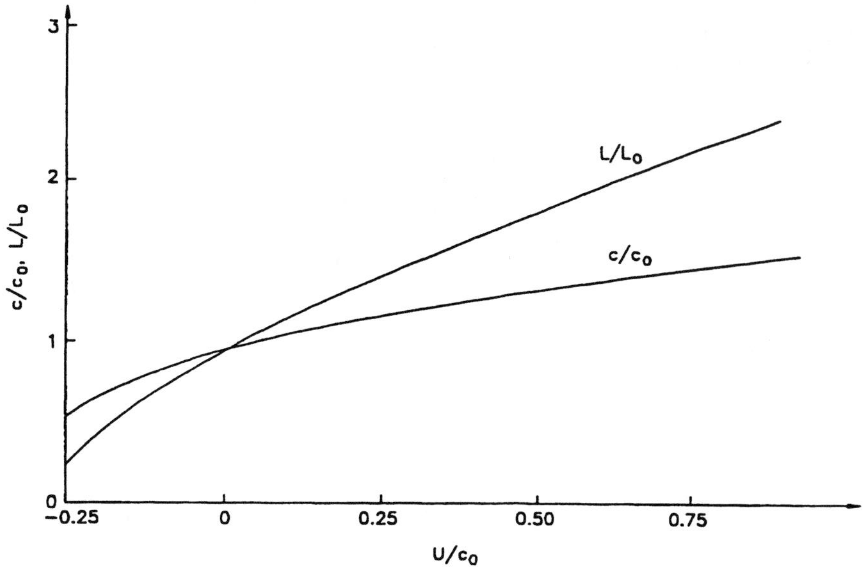

Figure 10.23.—Ratio of the Wavelength and Celerity of Linear Waves with Collinear and Opposing Currents to Those Without Currents (Longuet–Higgins and Stewart, 1961).

In the case of interaction between spectral waves and currents, it is also necessary to calculate wave parameters in the relative frequency domain. The relative frequency of the i-th spectral component may be determined by

$$(\omega_r)_i = (\omega_s)_i - k_i U. \tag{10.75}$$

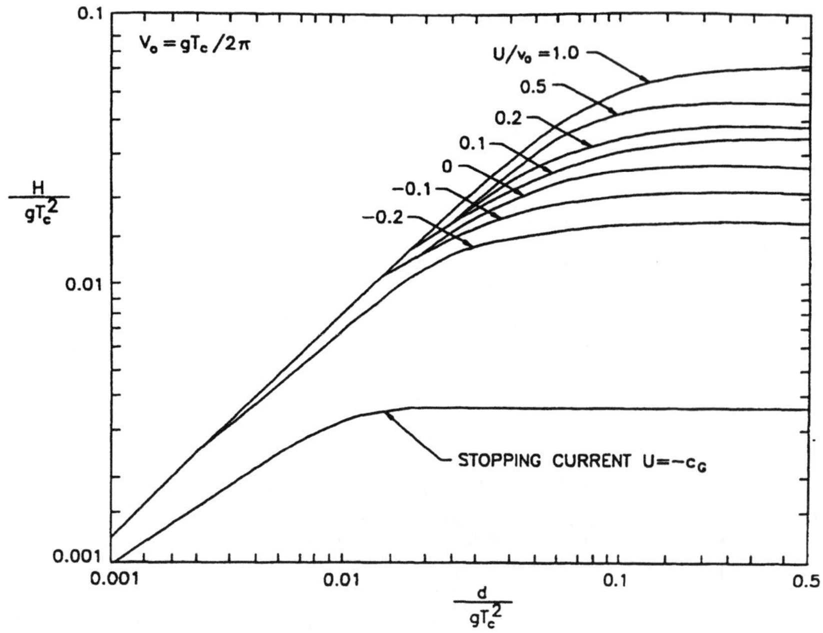

Figure 10.24.—Maximum Wave Height in the Presence of a Uniform Current (Dalrymple and Dean, 1975).

The modified wave spectral density in a current is then related to the current-free spectra by a transfer function, Y. The following expression is then obtained.

$$S\eta\eta(\omega_r)_i = Y_\eta(\omega_r)_i^2\, S\eta\eta(\omega_s)_i. \qquad (10.76)$$

The transfer function Y is different for deep and shallow water depth (Tung and Huang, 1973; Lambrakos, 1981). The velocity and acceleration spectra of waves are also different from the spectrum of the surface elevation as shown in Fig. 10.25. Comparison of spectral density curves of the surface elevation,

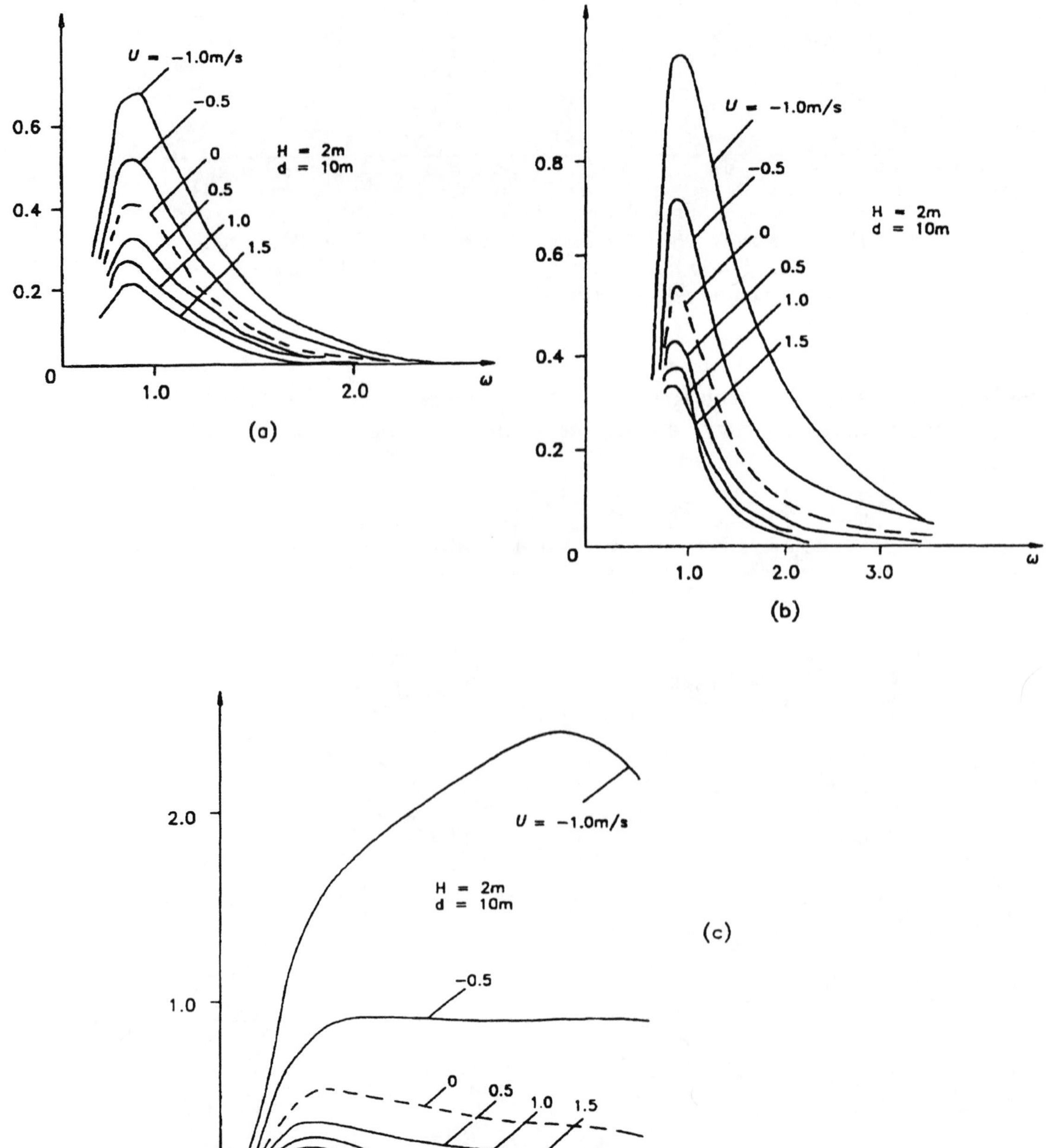

Figure 10.25.—Wave Energy Spectral Densities in the Presence of Currents for (a) Elevation, (b) Particle Speed, and (c) Particle Acceleration at the Surface (Tung and Huang, 1973).

velocity, and acceleration indicates that peak frequencies may be shifted, and that the amount of shift is determined by the current speed.

A complete and comprehensive treatment of the wave–current interaction and nearshore current systems is beyond the scope of this handbook. The overview presented here includes additional references necessary for engineering applications.

VII. TIDES AND TIDAL DATUMS

This and the following sections address water surface fluctuations with periods substantially longer than those of wind waves. The fluctuations have periods ranging from several minutes to many years. Because of their long periods, water level fluctuations may have important consequences in hydrology, including elevating coastal water levels long enough to cause flooding and other damage, and providing an increased nearshore water level that allows large wind waves to penetrate further shoreward before and during breaking.

Types of water level fluctuations important to engineering practice include astronomical tides, storm effects, seiches, tsunamis, and long-term variations. A brief overview of these will be presented in this section because more detailed information is available elsewhere (e.g., USACERC, 1984; USACE, 1989).

A. Astronomical Tides

The astronomical tide is a periodic rising and falling of sea level caused by gravitational attraction of the moon, sun, and, to a much lesser extent, other astronomical bodies acting on the rotating Earth. Tides most closely follow the moon, which exerts the most powerful force. Typically, the dominant tidal cycle repeats every 12 hours and 25 minutes. Thus, nearly two complete tidal cycles can be expected in a 24-hour day. The high and low waters occur about 50 minutes later each day. Tide-generating forces are well understood and can be predicted many years in advance. For completeness, some of the most common terms used for description of tides are identified in Fig. 10.26 and defined in the Glossary. A more complete set of terms has been published by Hicks (1989).

Water level change in response to the tidal forces is modified by depth variations and boundaries in coastal regions. Tides can vary substantially along a coast. Local response to the tide-generating forces can be determined from tide gage records. Astronomical tides are important in predicting water levels along ocean coasts and in bays, estuaries, and seaward parts of rivers and channels. They are not important in enclosed lakes or in rivers and channels above the fall line or sufficiently distant from open water. General references on tides include Godin (1972); Schureman (1976); Harris (1981); Pugh (1987); and Reid (1990). The United States government agency responsible for tide information is the National Ocean Service (NOS), National Oceanic and Atmospheric Administration (NOAA), Department of Commerce.

Several important features of the astronomical tide are shown by the record in Fig. 10.26. The record is a semidiurnal tide, the most common type along the United States Atlantic coast. It appears as a series of nearly sinusoidal oscillations with an average period near 12 hours and 25 minutes. The amplitude of these oscillations varies from one to the next, and the average range of any two successive tide waves rises and falls with a cycle of about 14.5 days. A cycle with a period near 29 days (one lunar month) is also present. The largest amplitudes (spring tides) coincide approximately with the time of a new moon and a full moon. The smallest (neap tides) occur when the moon is in its first or third quarter.

Tidal curves for several representative United States coastal locations are shown in Figs. 10.27 and 10.28. Fig. 10.28A is a typical example of a semidiurnal tide. Two high tides and two low tides occur each tidal day. Fig. 10.28E is a typical example of a diurnal tide. Only one high and one low are clearly discernible for each tidal day. Figs. 10.28B through 10.28D illustrate intermediate types of tides. Two highs and two lows during a tidal day are evident in Figs. 10.28B and 10.28C. The amplitudes of

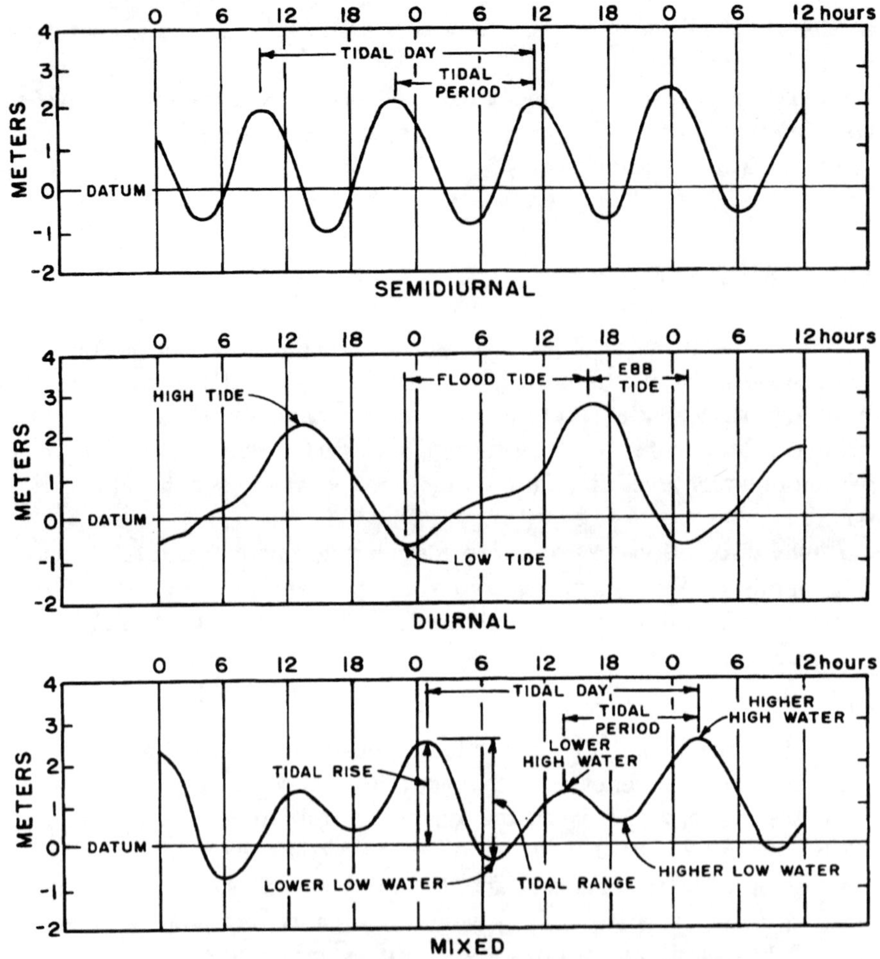

Figure 10.26.—Types of Tides (Wiegel, Oceanographical Engineering, 1964, Fig. 12.4, Reprinted by permission of Prentice-Hall, Inc., Englewood Cliffs, N.J.).

consecutive tide waves for these locations are very unequal, except for short periods near January 15 and 30 in 10.28B. Tidal curves of this type are called mixed tides. Fig. 10.28D is intermediate between the mixed and diurnal types. Two distinct, unequal lows and highs are recognized on most days, but there are several days when one tide wave vanishes. The tide appears to stand for a time at an intermediate value between the daily high and low, although only one distinct high and low can be identified. This phenomenon is often called a vanishing tide. The tidal range is generally low where these tides occur.

The tide-generating force can be expressed as the sum of a number of periodic terms determined from the astronomical parameters pertaining to the orbit of the moon around Earth and the orbits of Earth and the moon around the sun. The periods of major importance in predicting tides at most locations are less than 30 days. However non-negligible contributions are often present at longer periods. The tide is predicted as a linear summation of each of the important frequencies (including frequencies that arise due to nonlinear interactions between primary forcing frequencies). The tide prediction equation is expressed as:

$$h_{ys}(t) = h_o + \sum_{n=1}^{N} f_{ny} A_{ns} \cos(\sigma_n t - v_{ny} - \kappa_{ns}) \tag{10.77}$$

where $h_{ys}(t)$ is the tide at station s during year y at time t, h_o is the height of the local MSL datum above the reference datum, N is the number of constituents, f_{ny} is the node factor, A_{ns} is the amplitude, σ_n is the

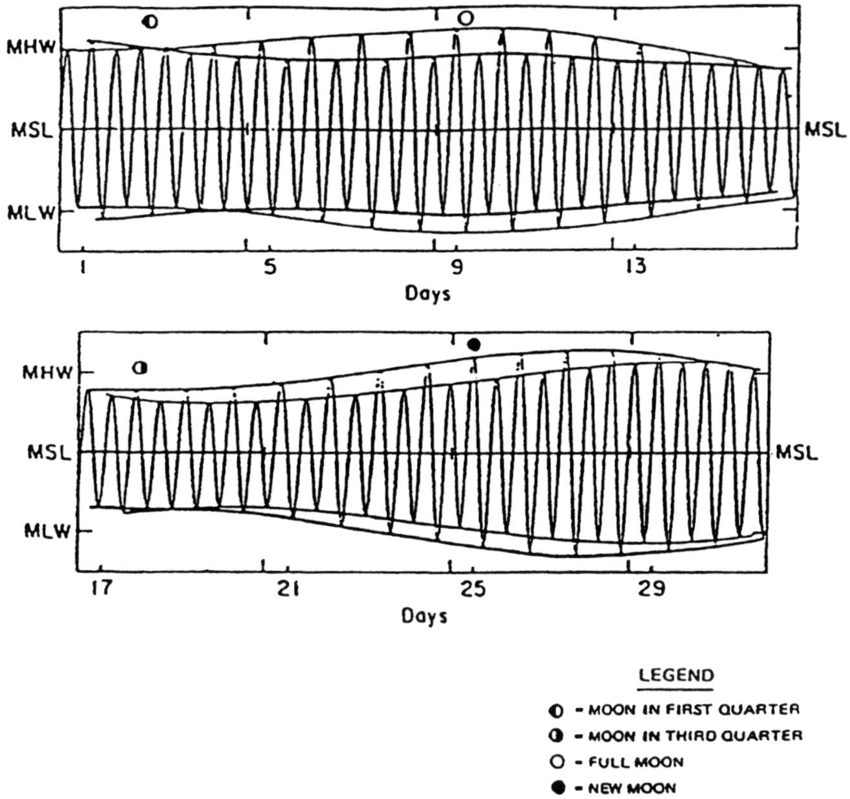

Figure 10.27.—Tide Predictions for Boston, Mass., January 1963 (Harris, 1981).

frequency, or angular speed, t is time reckoned from some initial epoch (phase), ν_{ny} is the equilibrium argument, and κ_{ns} is the phase lag, or epoch. The subscript y indicates the parameter may change yearly but is independent of location. The subscript s indicates the parameter depends on location of the tide station.

The parameters σ_n, f_{ny}, and ν_{ny} are determined from astronomical theory. The parameters f_{ny} and ν_{ny} represent the effect of periodicities longer than 1 year on the amplitude and phase, respectively. Yearly values of these parameters have been calculated and tabulated through the year 2000 (Schureman, 1976). The parameters A_{ns} and κ_{ns} are standardized for a particular location based on past measurements at the site. A minimum record length of 29 days is needed, although a 369-day record is preferred. The parameters A_{ns} and κ_{ns} as well as the MSL datum adjustment, h_o, are available from the NOS.

The number of trigonometric terms needed to describe the tide varies with location. More terms are needed at locations where the tide must travel a great distance through shallow water than when the tide station is near the open sea. Additional terms may be needed to obtain an adequate representation when the tidal range is large. In the United States, 37 standard constituents are adequate for most tide stations. The four constituents that are typically most important are listed in Table 10.7. Symbols listed are standard conventions. Subscripts in the constituent symbols indicate the approximate number of cycles in the tidal day.

Because of the strong modification of tides in shallow water, tide predictions (and measurements) at one site should be transferred to other sites with great care. For nearby sites along the open coast, tide information may usually be transferred directly. When the site for which tide information is available is inland from the open coast, the information should not be directly transferred to other sheltered areas or the open coast. For example, tide measurements inside a sheltered bay should not be directly used as an estimate of instantaneous water levels for a dredge operating several miles offshore.

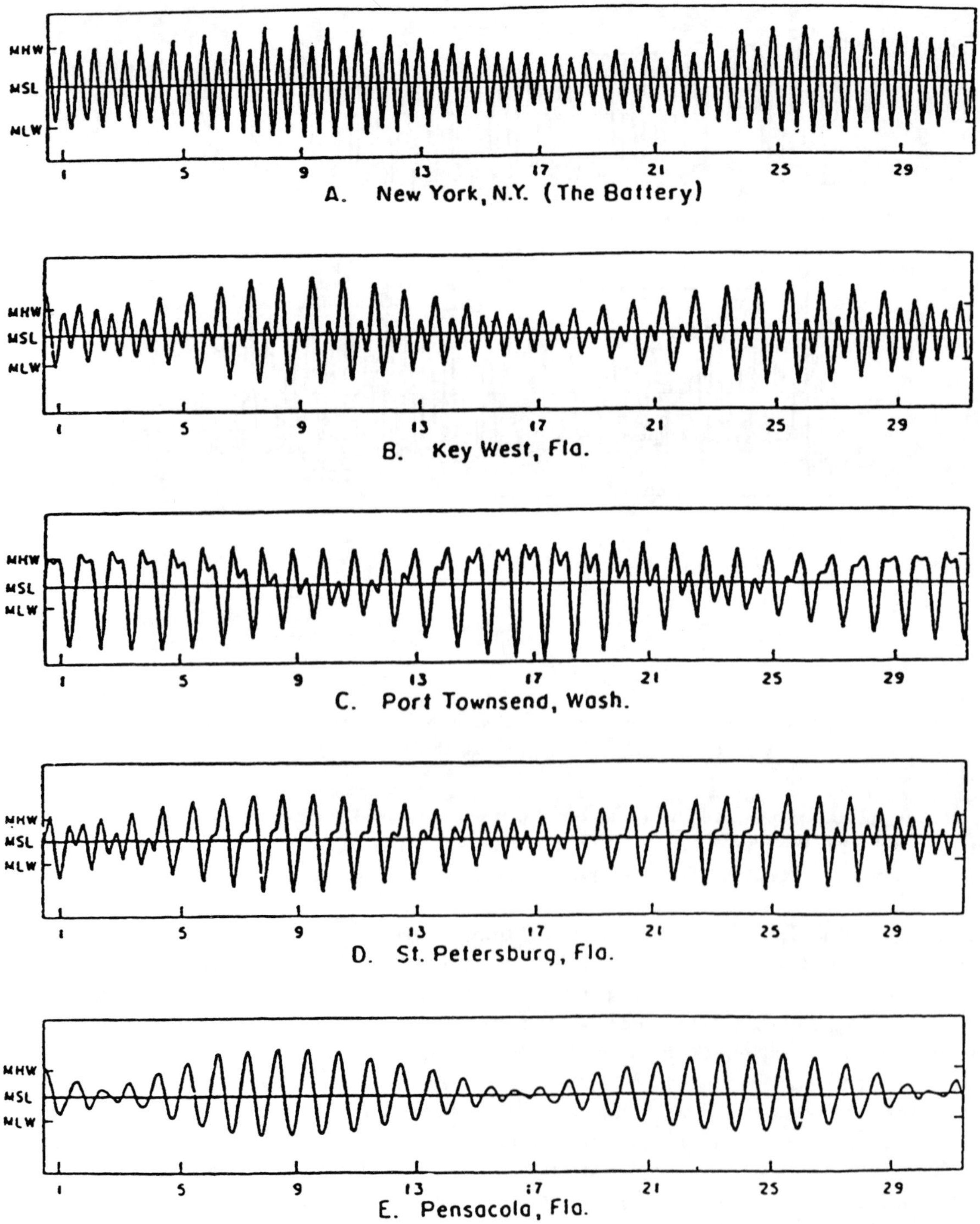

Figure 10.28.—Tidal Curves for Several Representative U.S. Coastal Locations for January 1963 (Harris, 1981)

Tide tables and tidal current tables are published annually by the National Ocean Service (NOS). The Atlantic and Gulf coasts of the United States are together in one volume and the Pacific coast is in another. The tide tables provide predicted time and elevation for each high and low water at reference stations around the United States for the entire year. Information for adjusting the predictions so they can be applied to nearby secondary stations is also provided.

TABLE 10.7. Some Important Tidal Constituents.

Tide	Symbol	Period (hr)
Semidiurnal	M_2	12.42
	S_2	12.00
Diurnal	K_1	23.93
	O_1	25.82

B. Tidal Datums

A critical element of water level analysis is the establishment of a clear reference level, or datum. This task is more challenging than one might expect. Confusion about datums often arises in practical studies. Datums that are often encountered are defined in the Glossary. They include MSL, MHW, MLW, MTL, MHHW, MLLW, and NGVD. These datums are illustrated in Fig. 10.29.

For activities related to coastal land ownership and development, the most important datums are MHW and MHHW. These datums usually define the limit of private property along the coast (Fig. 10.30). The low water datums are primarily used in relation to navigation. The MLW and MLLW are used on NOS charts, although MLLW is being adopted for all coastal areas as NOS charts are revised.

The NGVD datum is usually preferred as the primary datum for engineering design. The NGVD shown in Fig. 10.29 is the only datum that is fixed, regardless of location. The other tidal datums (e.g., MSL, MLLW) apply only locally. They also change slowly with time. A special datum plane, referred to as the International Great Lakes Datum (IGLD), is used in the Great Lakes. It is based on a level surface established over each lake and referenced to a single point in the Great Lakes system.

The NOS collects, analyzes, and maintains tide and datum information pertaining to United States coasts. In addition to producing tide tables and other published products, the NOS maintains an archive of tidal constituents, water level time series, and datum information. Datum information includes index maps of tidal datum benchmarks and lists of the established relationships between NGVD and other datums. Requests for information can be addressed to: NOAA National Ocean Service, Office of Ocean and Earth Sciences, Ocean and Lake Levels Division, Datums Section N/OES 231, 6001 Executive Blvd., Rockville, MD 20852.

VIII. STORM SURGES

Storm surges are increases in normal water levels along a shoreline brought about by meteorological events such as hurricanes, storms associated with weather fronts, and "northeasters" associated with

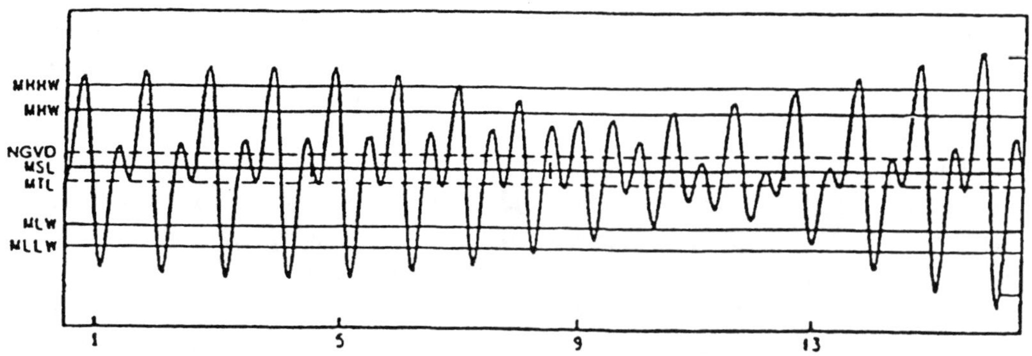

Figure 10.29.—Illustration of Tidal Datums, Los Angeles, California, January 1973 (Harris, 1981).

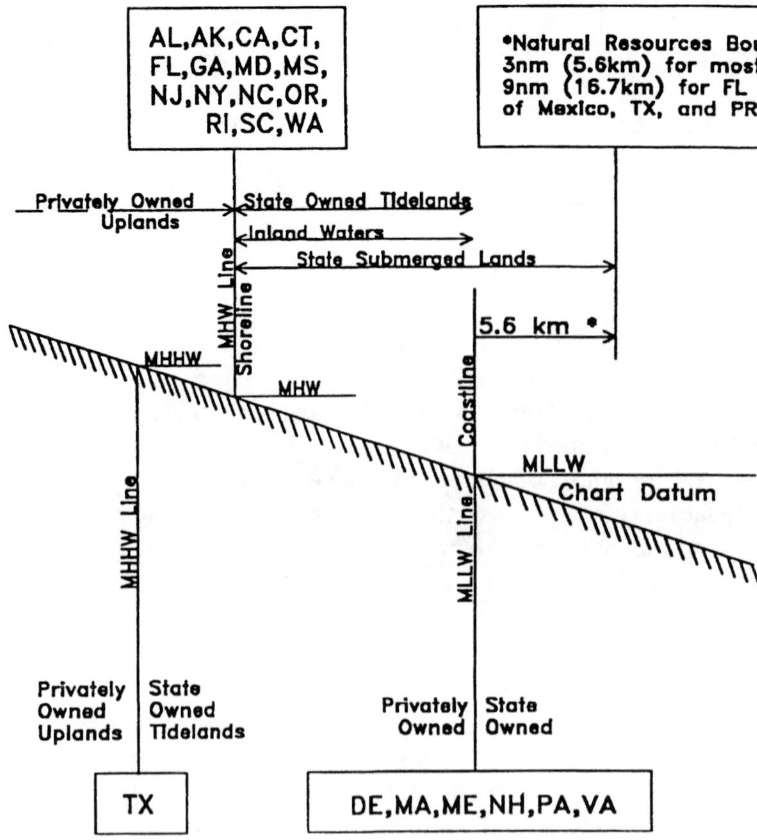

Figure 10.30.—Coastal Boundaries (Hicks, 1988).

extratropical low pressure systems. Storm surge is defined in SPM (USACERC, 1984) as "a rise above normal water level on the open coast due to the action of wind stress on the water surface. Storm surge resulting from a hurricane also includes that rise in level due to atmospheric pressure reduction as well as that due to wind stress." The glossary to this chapter includes a number of definitions related to hurricanes and storm surges. In enclosed or semi-enclosed water bodies such as natural lakes and man-made reservoirs, there is also a reduction in water level associated with storm surges. Lower water levels prevail at the windward end of the lake while higher levels prevail at the leeward end. Knowledge of storm surges and an ability to estimate the surge associated with a given set of meteorological conditions is important because surges can result in coastal flooding. The increased water levels due to surges also allow waves to attack structures located at higher elevations. Structures that might normally be safe from wave attack may be vulnerable during storm surges. In reservoirs, strong winds blowing toward a dam can increase the water level and, if sufficient freeboard is not provided, there is a probability that the waves associated with strong winds might overtop the dam.

On the open coast along the East Coast of the United States, record storm surges are most often associated with hurricanes; however, any extended period of strong onshore winds can result in elevated water levels. Fortunately, storm surges associated with hurricane landfall are relatively local, extending only over tens of miles. Surges associated with storms other than hurricanes are usually lower in magnitude, but may extend over larger stretches of coast and for longer durations.

For example, the maximum storm surge associated with hurricane "Camille" in August 1969 was about 24.2 feet at Pass Christian, Mississippi, but the surge was only 5 feet high 15 miles west and 60 miles east of the location of the peak surge. Thus, the stretch of coast that experienced a surge of more than 5 feet was only about 75 miles long. In contrast, the surge associated with the "Ash Wednesday" storm of 1962 along the northeast coast of the United States was only about 5 to 6 feet but it extended from Long Island, New York to Hampton Roads, Virginia and lasted for 2 1/2 days. The storm's effects were felt from Maine to northern Florida. With five unusually high astronomical tides associated with it, the "Ash Wednesday" storm caused extensive damage along the coast.

A. Charactristics of Storm Surges

A typical hurricane surge hydrograph is shown in Fig. 10.31 along with the various components that make up the hydrograph. There are several components of storm surges occurring on the open coast. They include the wind setup, pressure setup, wave setup, and Coriolis effects. In addition, there is a little-understood component, termed the "forerunner," which has been observed to arrive several hours in advance of the main storm surge.

Wind setup is generally the largest component of storm surge. It results from shear stresses caused by wind on the water surface and by bottom shear stresses that resist the offshore return flow of the water piled up on shore. In shallow coastal areas, onshore winds result in a flow toward shore and a build-up of water there. Elevated water levels at the shore cause a seaward return flow along the bottom toward deeper water. The return flow induces bottom shear stresses that resist the flow of water away from shore. Thus, for onshore winds, both the wind stresses and the bottom stresses act toward shore to keep the water there. The magnitude of the setup depends on the magnitude of the wind and bottom stresses.

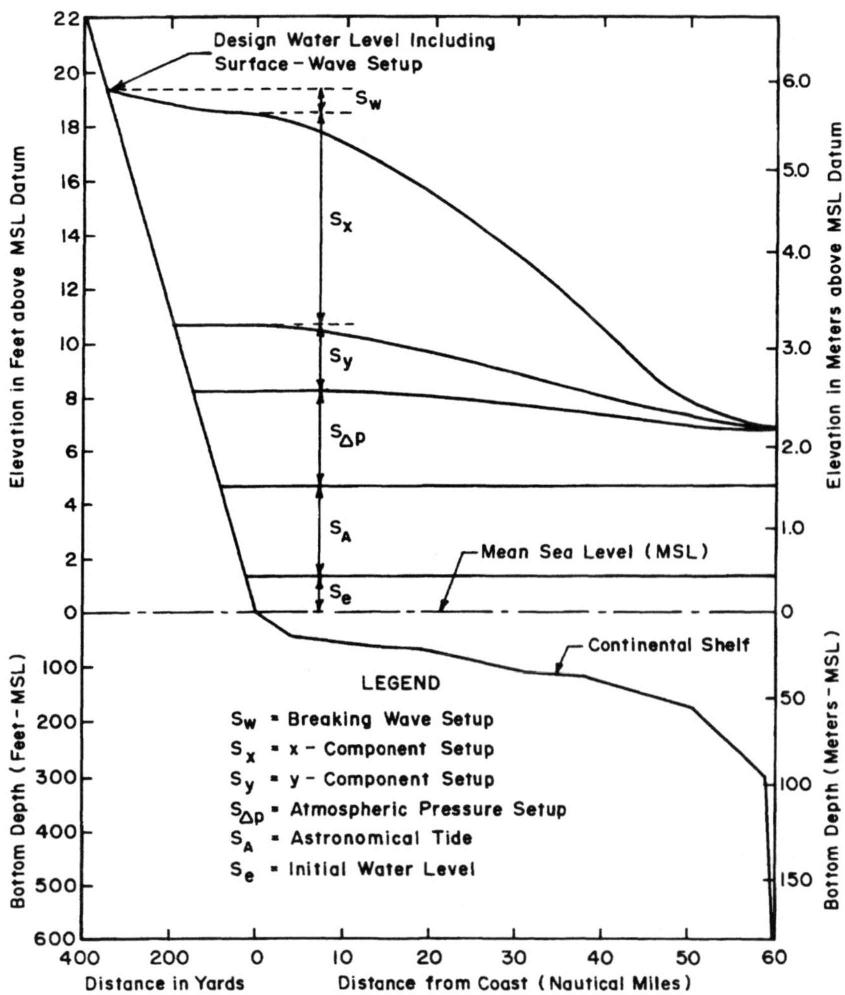

Figure 10.31.—Setup Components Over the Continental Shelf.

A one-dimensional static analysis is depicted in Fig. 10.32. This is a somewhat simplified picture since the phenomenon is three-dimensional, dynamic, and there are a number of other components that complicate the picture.

For the simplified picture of Fig. 10.32, horizontal forces are in equilibrium so that the difference in hydrostatic pressure across the free-body of water is balanced by the wind shear stress on the surface and the bottom shear stress due to the return flow. The summation of horizontal forces gives:

$$\frac{\gamma(d + \Delta S_w)^2}{2} - \frac{\gamma d^2}{2} = (\tau_s + \tau_b)\,\Delta x \tag{10.78}$$

in which ΔS_w is the increase in water depth across Δx, the length of the free body, γ is the unit weight of the water, d is the water depth, τ_s is the wind shear stress, and τ_b is the bottom shear stress. The surface and bottom stresses can be expressed by,

$$(\tau_s + \tau_b) = K\rho U^2 \tag{10.79}$$

in which K is a calibration factor involving both wind stress and bottom stress. Solving Equation 10.78 for ΔS_w gives

$$\Delta S_w = d\left[\sqrt{\frac{2KU^2\Delta x}{gd^2} + 1} - 1\right]. \tag{10.80}$$

Or, if the wind blows at an angle θ to the x direction, Equation 10.80 can be written as:

$$\Delta S_w = d\left[\sqrt{\frac{2(\tau_s + \tau_b)\cos\theta}{\gamma d^2} + 1} - 1\right]. \tag{10.81}$$

or,

$$\Delta S_w = d\left[\sqrt{\frac{2KUU_x\Delta x}{gd^2}} - 1\right]. \tag{10.82}$$

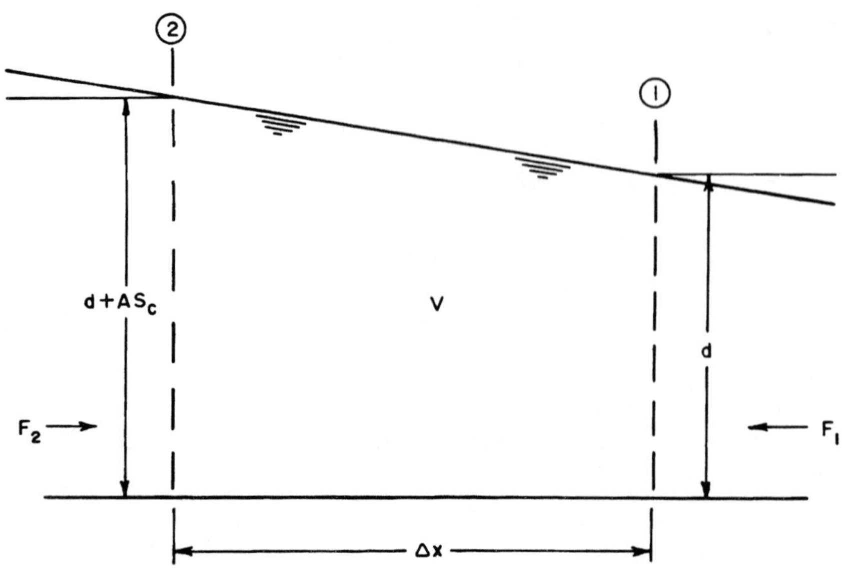

Figure 10.32.—Simplified Free-Body Diagram for Wind Induced Setup.

in which U_x is the component of the wind velocity in the x direction. Note that ΔS_w depends on the water depth in the reach. The amount of wind setup is thus related to the stage of the astronomical tide at the time the surge-producing winds occur. Equation 10.82 can be solved in a stepwise manner by positioning a storm wind field over a bathymetric chart of a coastal area, determining the wind speeds and directions over a reach of water, Δx, and computing ΔS at the landward end of the reach. This allows the water depth at the seaward end of the next Δx reach to be determined. Computations proceed shoreward in Δx steps from the seaward end of the storm.

Pressure setup is generally associated only with the intense low pressure centers of hurricanes. It is often termed the "inverse barometer" effect since the water level rise is calculated as the difference in pressure, expressed in terms of the height of a column of water, between the central pressure of the hurricane and the ambient barometric pressure in the atmosphere surrounding the hurricane. As the atmospheric pressure increases with radial distance from the eye of the hurricane, the pressure setup decreases proportionately. The pressure setup is given by

$$S_p = \frac{p_a - p_r}{\gamma} \tag{10.83}$$

in which S_p is the pressure setup, p_a is the ambient atmospheric pressure outside of the hurricane eye, p_r is the central pressure index of the hurricane (the atmospheric pressure in the eye is P_0; see Fig. 10.35), and γ is the specific weight of the water. Equation 10.83 can be used with any consistent system of units. Note that the inverted barometer effect does not come into play for relatively small bodies of water where the atmospheric pressure gradient cannot be developed over the body of water. (There must be an area of relatively high atmospheric pressure over the water body to drive the flow toward the low pressure area.)

For large bodies of water such as the oceans and large lakes, Coriolis effects also come into play. Coriolis effects arise because any reference frame fixed with respect to the earth's surface is actually accelerating due to the earth's rotation. Water and air in apparent rectilinear motion with respect to an earth-fixed reference frame are actually experiencing acceleration. Conversely, what is really rectilinear motion with respect to a reference frame fixed in space appears to be curvilinear in an earth-fixed reference frame. To account for the observed curvilinear motion in the earth-fixed reference frame, an apparent acceleration, proportional to the velocity and perpendicular to it, is introduced. Coriolis accelerations cause flows to deflect to the right in the northern hemisphere and to the left in the southern hemisphere.

The observed counter-clockwise flow around meteorological atmospheric low pressure systems in the northern latitudes is a manifestation of the Coriolis effect. Winds blowing parallel to the shoreline cause a shore-parallel flow of water which is deflected to the right. If the shoreline is to the left for an observer facing into the wind, the Coriolis effect causes an increase in water level along the shore. If the shoreline is to the right, a decrease in water level occurs. For the counter-clockwise winds around hurricanes approaching the East Coast of the United States, Coriolis effects cause an increase in surge levels as the hurricane approaches landfall (front of the hurricane) and a decrease after the eye of the hurricane has crossed the shoreline. Fig. 10.33 depicts the effect of the Coriolis forces.

The equation for the force balance in Fig. 10.33 is,

$$-\frac{\rho g d^2}{2} + \frac{\rho g (d + \Delta S_w)^2}{2} - [2\omega V \sin \Phi]\rho d \Delta x = 0 \tag{10.84}$$

in which ω is the angular rotation of the earth (7.28×10^{-5} radians/sec), Φ is the angle of latitude, and V is the magnitude of the shore-parallel current which is perpendicular to the free body in Fig. 10.33. Solving for ΔS_w,

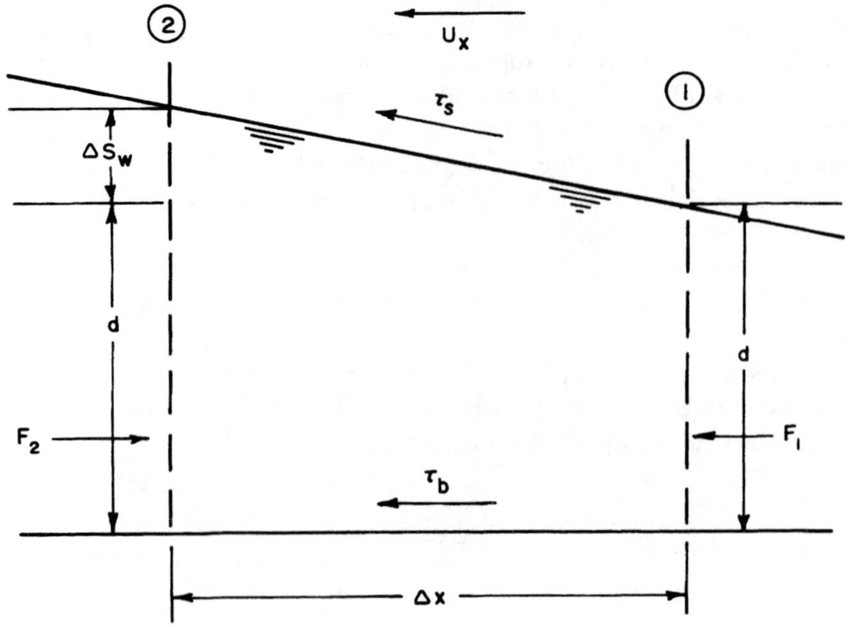

Figure 10.33.—Simplified Free-Body Diagram for Coriolis Setup.

$$\Delta S_w = \frac{2\omega}{g} V \sin \Phi \Delta x. \tag{10.85}$$

The wind-generated shore-parallel current, V, can be estimated by,

$$V = Ud^{1/6} \sqrt{\frac{k}{14.6\ n} \sin \theta} \tag{10.86}$$

in which n is Mannings roughness coefficient ($\approx$ .035), θ is the angle between the wind and a perpendicular to shore, and k is the wind stress coefficient.

As waves propagate toward shore they first depress and then elevate the mean water level. As the water depth decreases the water level is depressed to below the still water level (SWL) seaward of the depth where the waves break. This is termed wave setdown. Landward of the point where the waves break, where wave energy dissipation occurs, the mean water level is elevated. This is termed wave setup. Wave setup contributes to the increase in water levels experienced along a shoreline during a storm. The amount of wave setup depends on the height of the incoming waves, the wave period, and the local beach slope. Larger waves and flatter beach slopes result in higher setup. Since high waves are associated with both hurricanes and other storms, setup can be a significant contribution to the water levels experienced during storms. The wave setup can be estimated by the equation,

$$S_{wave} = 0.15 d_b - \frac{g^{1/2} H_o^2 T}{64 \pi d_b^{3\backslash 2}} \tag{10.87}$$

in which S_{wave} is the wave setup, H_o is the unrefracted deep water wave height, g is the acceleration of gravity, d_b is the water depth in which the waves break, and T is the wave period. The slope enters the setup expression only indirectly by determining the water depth in which the incident waves break.

Fig. 10.34 gives the wave setup in dimensionless terms as a function of breaker steepness, H_b/gT^2, and beach slope where H_b = the breaking wave height.

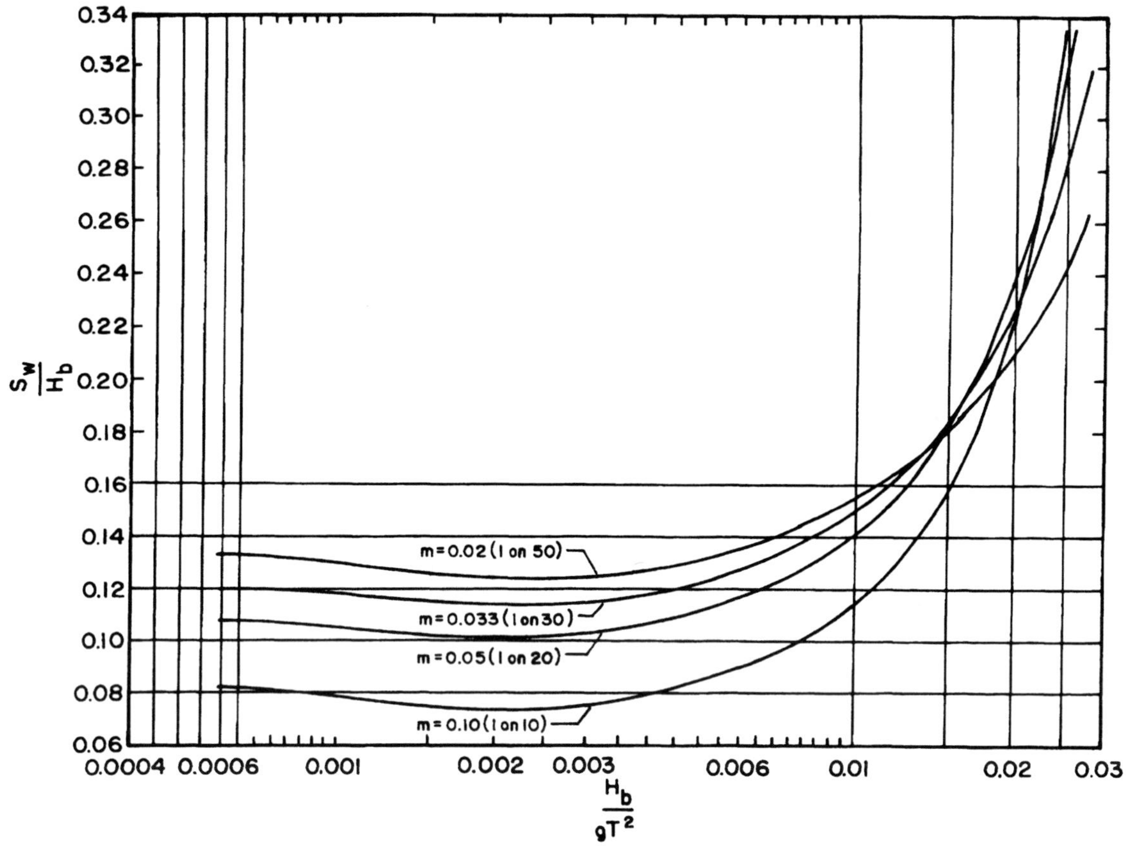

Figure 10.34.—Wave Setup as a Function of Breaker Steepness and Beach Slope (USACERC, 1984).

Observations of hurricane storm surges have shown that above-normal water levels precede, by several hours, the arrival of the hurricane itself. The cause of this surge "forerunner" is not known and is the topic of current research. This component of the surge can be on the order of 2 feet or more. In the absence of criteria for estimating the magnitude of the forerunner, an initial surge of between 1 and 2 feet is assumed; the other components of surge are subsequently computed and added to this forerunner. Other forerunner depths, more or less conservative, can be assumed depending on the type of facility being designed and the consequences of its failure. Note that this phenomenon is associated only with hurricanes.

Storm surges occur in addition to the astronomical tides that are independent of meteorological conditions. Consequently, the arrival of maximum storm surge levels at the same time as high astronomical tides results in the extreme water levels critical for some design problems. Hence, the time of arrival of a storm surge at the same time of high astronomical tide is critical in determining the water levels to which an area will be subjected. The two phenomena are not truly independent because water depths due to astronomical tides have some influence on the magnitude of wind and wave setup. Consequently, terms that account for depths resulting from astronomical tides are included in the equations used to model hurricane surges. However, storm surges and astronomical tides can often be assumed independent and the water levels added.

B. Storm Surge Generation and Prediction

The storms creating the most severe surges are generally hurricanes; however, any storm with strong, persistent onshore winds can cause damaging surges. Hurricanes are storms that orginate over the tropical oceans and are characterized by strong counter-clockwise rotating wind patterns (Fig. 10.35). If the maximum winds exceed 33.5 meters per second (65 knots or 75 miles per hour) the storm is classified

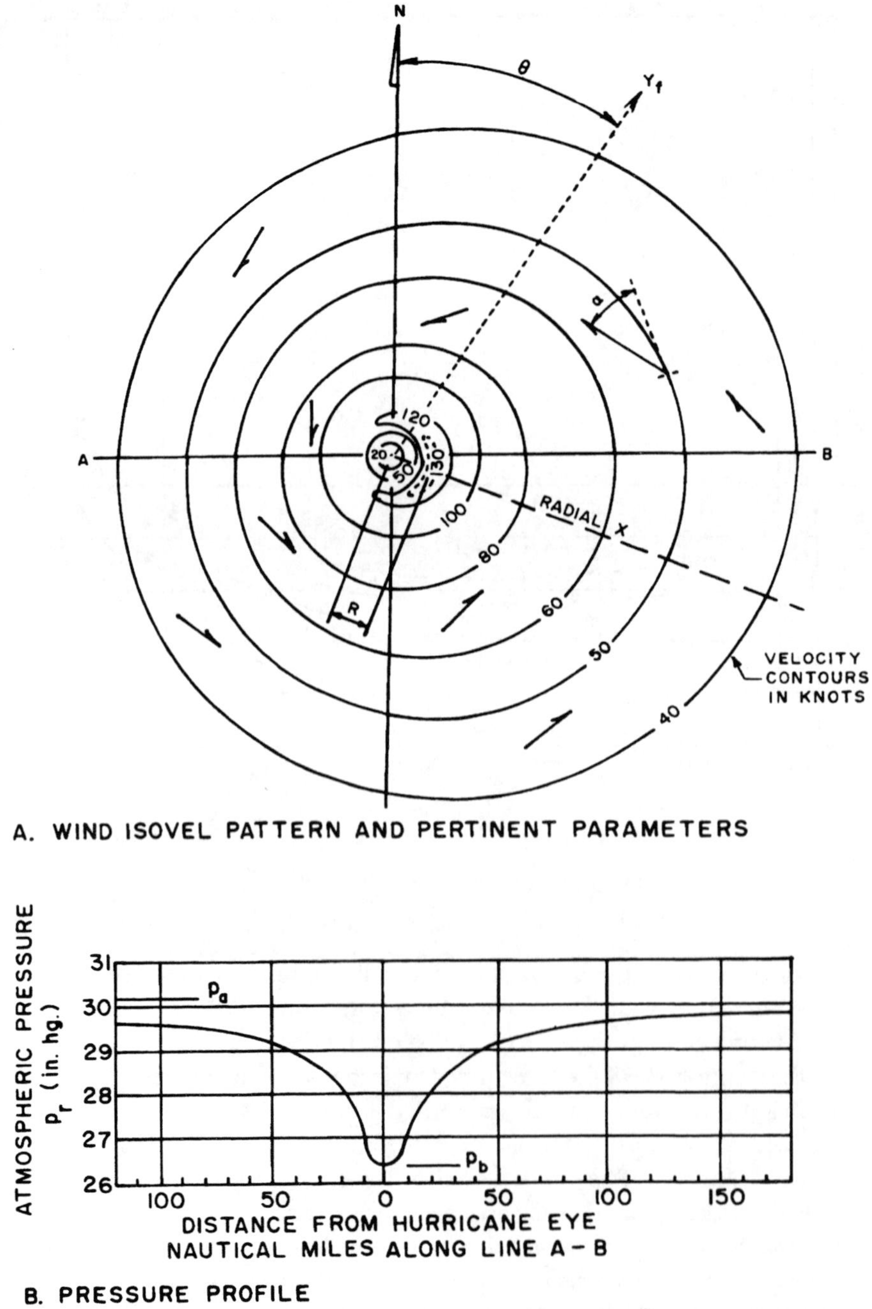

Figure 10.35.—Typical Wind Patterns and Atmospheric Pressure Distribution within a Hurricane.

as a hurricane. If the wind speed is less than 33.3 meters per second the storm is classified as a tropical storm. Hurricanes have a forward motion that causes the maximum winds to be to the right of the hurricane eye when facing in the direction of the forward motion. Consequently, maximum surges are experienced to the right of the eye when the hurricane makes landfall.

Extratropical low pressure systems can also cause damaging storm surges. "Northeasters," for example, are stationary low pressure systems in the North Atlantic that result in sustained northeast winds with resulting storm surges along the East Coast of the United States.

Storms associated with passing weather fronts are generally less important in producing surges since they do not produce long periods of sustained winds. Fronts usually produce only short periods of high winds. Passing frontal systems can produce seiching (standing wave oscillations) in enclosed bodies of water, however. For example, frontal systems moving across Lakes Erie and Ontario can cause oscillations that produce alternating elevated and lowered water levels at the ends of the lakes.

Two approaches to storm surge prediction are available. They are: a) the statistical approach in which historical data of water levels at a site are used to estimate the frequency of occurrence of given water levels, and b) the numerical modelling approach in which the storm surge caused by a hypothetical design hurricane is calculated. The first approach uses frequency analysis procedures similar to those outlined elsewhere in this chapter. The latter approach uses the hydrodynamic equations of fluid motion and appropriate hurricane characteristics to calculate the storm surge. The first approach is usually limited by how much water level data is available for a site. Coastal water level records are often not available for many sites.

In numerically modelling storm surges, two forms of the hydrodynamic equations are typically used. Some investigators prefer to use the vertically integrated average flow velocities while others prefer to use the fluid transport rates. Both methods are equivalent. The momentum equations in the x and y directions, respectively, are given by

$$\frac{\partial U}{\partial t} + \frac{U}{D}\frac{\partial U}{\partial x} + \frac{V}{D}\frac{\partial U}{\partial y} = fV - gD\frac{\partial}{\partial x}(\eta - \xi - \zeta) + \frac{1}{\rho}(\tau_{sx} - \tau_{bx}) + W_x R + \epsilon_{xx}\frac{\partial^2 U}{\partial y^2} + \epsilon_{xy}\frac{\partial^2 U}{\partial y^2} \qquad (10.88)$$

$$\frac{\partial V}{\partial t} + \frac{U}{D}\frac{\partial V}{\partial x} + \frac{V}{D}\frac{\partial V}{\partial y} = -fU - gD\frac{\partial}{\partial y}(\eta - \xi - \zeta) + \frac{1}{\rho}(\tau_{sy} - \tau_{by}) + W_y R + \epsilon_{xy}\frac{\partial^2 V}{\partial x^2} + \epsilon_{yy}\frac{\partial^2 V}{\partial y^2} \qquad (10.89)$$

in which U is the discharge per unit width in the x direction, V is the discharge per unit width in the y direction, D is the water depth, g is the acceleration of gravity, f is the Coriolis parameter ($f = 2\Omega \sin \Phi$, where Ω is the angular velocity of the earth's rotation and Φ is the geographical latitude), η is the elevation of the water surface above the mean level, ξ is the atmospheric pressure deficit, ζ is the astronomical tide potential expressed as an equivalent head of water, τ_{sx} is the wind shear stress in the x direction, τ_{sy} is the wind shear stress in the y direction, τ_{bx} is the bottom shear stress in the x direction, τ_{by} is the bottom shear stress in the y direction, W_x and W_y is the wind velocity in the x and y directions, respectively, and ϵ_{xx}, ϵ_{yy}, and ϵ_{xy} are the eddy shear stresses, ρ is the fluid density, and R is the rate at which water is added to or removed from the system, e.g., by precipitation or evaporation.

The equation of continuity is given by

$$\frac{\partial \eta}{\partial t} + \frac{\partial U}{\partial x} + \frac{\partial V}{\partial y} = R \qquad (10.90)$$

in which the variables are as defined above. The wind shear stresses are given by

$$\tau_{sx} = kW^2 \cos \theta$$

$$\tau_{sy} = kW^2 \sin \theta \qquad (10.91)$$

in which W is wind speed, k is a nondimensional wind stress coefficient usually taken to be constant or a function of wind speed, and Θ is the angle between the wind velocity vector and the x axis. The bottom shear stresses are given by

$$\tau_{bx} = KU(U^2 + V^2)^{\frac{1}{2}} D^{-2}$$

$$\tau_{by} = KV(U^2 + V^2)^{\frac{1}{2}} D^{-2} \qquad (10.92)$$

in which K is a bottom shear stress coefficient.

A number of storm surge models based on variations of the above equations are available for both open coasts and inland flooding. They include: (1) **Open Coast Models**; SPLASH, SSURGE, TSURGE, WIFM, and TWO-D-SURGE and (2) **Inland Flooding Models**; SURGE I, SURGE II, WIFM, TWO-D-SURGE, and SLOSH.

SPLASH was developed by the National Weather Service (NWS) for real-time prediction of hurricane surges. It employs a curvilinear grid with variable mesh spacing. SLOSH is a more recent model developed by the NWS to deal with inland flooding problems. SSURGE, developed for the Corps of Engineers, is an explicit finite difference model that also employs a curvilinear grid. TSURGE, developed for the Federal Emergency Management Agency, is also an explicit finite difference scheme using a uniform rectangular grid. SURGE II and TWO-D-SURGE are extensions of the SURGE I model. SURGE II includes subgrid channels with the possible inclusion of levees. TWO-D-SURGE is a modification of SURGE I similar to SURGE II but with a different numerical scheme to evaluate channel flows. WIFM, developed by the Corps of Engineers, employs an implicit numerical scheme. In addition, there are a number of proprietary numerical models developed by private consulting firms.

Types of basic storm surge data available for a given area may include: statistical data on high water levels including the frequency of occurrence of given water levels, surge levels associated with a specific storm, meteorological information on historical hurricanes or other surge-producing storms, and historical hurricane tracks and landfall data.

Historical data on storm surge levels can often be obtained from the Corps of Engineers District Office that services a given area. Data on the characteristics of specific hurricanes can be obtained from the U.S. Department of Commerce, National Oceanic and Atmospheric Agency, National Weather Service, and National Hurricane Office. Other meteorological data can be obtained from the Weather Service's National Climatic Data Center, Asheville, North Carolina.

In addition to basic meteorological and water level data, the Corps of Engineers, the Federal Emergency Management Agency and the National Weather Service have conducted several emergency management and engineering studies of potential storm surges at various locations around the United States. The results of these studies are available in reports published by the Corps of Engineers and Weather Service (i.e., Delaware, 1990). For example, as a part of the "Delaware and New Jersey Hurricane Evacuation Studies" the Storm Surge Group at the National Hurricane Center (1990) published a "Storm Surge Atlas for Delaware Bay," which presents the results of numerous computer runs of the SLOSH model for hypothetical hurricanes.

IX. BASIN OSCILLATIONS AND TSUNAMIS

A. Basin Oscillations

A body of water such as a lake or a harbor that is open to the sea will have a fundamental resonant frequency and related higher harmonic frequencies at which it will oscillate, if excited. This oscillation is simply the reflection of a wave back and forth in the water body. Typically this wave has a relatively long period and low amplitude. The fundamental and harmonic frequencies of basin oscillation are defined by the basin geometry that controls the speed and pattern of wave propagation. Which frequency or frequencies that are excited depends on the nature of the forcing function.

A closed body of water can be excited by storm activity disturbing the water surface through the horizontal pressure gradient and/or the surface wind stress developed by the storm. These forces pile the water up at one end of the basin and, upon release of the water as the storm moves on, an oscillation of the water surface commences. Local seismic activity may also generate closed basin oscillations. If the basin has an opening to a larger water body, the appropriate frequencies of the spectrum of wave energy in that larger body will set the basin in oscillation. These frequencies may have a seismic, tidal, or storm generated origin.

When basin oscillations occur, the resulting water motion is cyclic, typically having a small vertical range of motion but large horizontal motions at certain locations. This, for example, can cause large horizontal oscillations of moored vessels, particularly if the basin oscillation period is close to the resonant period of the vessel and mooring system. Damaged mooring lines and fender systems can result. If the horizontal water motions occur in constricted areas significant current velocities can be generated, which may or may not be harmful.

B. System Resonance

Preliminary insight into some of the basic characteristics of basin oscillations can best be achieved by considering the behavior of a single degree of freedom oscillating system (such as a cantilevered elastic rod). This behavior is depicted in Fig. 10.36. The amplification, which is the system displacement divided by the displacement of the exciting force is plotted against the natural or resonant period of the system divided by the period of the exciting force (T_n/T). The different curves represent increased damping of the system. For a basin of water, the resonant period could be established by tilting the water surface and measuring the frequency of oscillation that develops as the surface is released. (The resonant period is usually determined by calculations that will be discussed below.) Damping of the system would result owing to dissipation of wave energy by bottom friction or incomplete reflection of the wave by the basin boundaries. If the water surface is tilted and then released, it will oscillate at T_n with exponentially decreasing amplitude until all of the energy in the reflecting wave is dissipated.

If the system is excited at the resonant period by a continuing cyclic force ($T_n/T = 1$) the oscillations will grow to a level of amplification established by a balance of the rate of incoming energy and energy dissipation. The greater the level of system damping the lower this equilibrium amplification will be as shown in Fig. 10.36. If the system is excited at other than the resonant period by a continuing cyclic force the resulting amplification will be less than at the resonant period as shown. And, if the system is excited by a spectrum of periods the system will selectively amplify those frequencies around the resonant period. Fig. 10.36 depicts the response of a system that has a single mode of oscillation. Water bodies commonly have several oscillation modes, each having a different resonant period and pattern of behavior as depicted in the figure.

The primary concerns when analyzing basin oscillations are to determine the fundamental and harmonic periods of resonant oscillation, and the patterns of surface motion and water flow for each of these resonant modes. This is done by first considering two-dimensional open and closed basins which give a

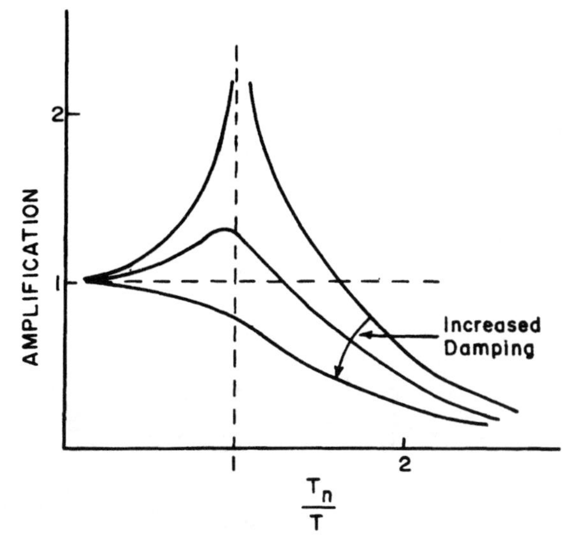

Figure 10.36.—Behavior of an Oscillating System.

good insight into the basic phenomena and closely approximate some natural systems. Then the more complex three-dimensional basins are considered. More detailed coverage of the material discussed in this section is given by Raichlen (1966) and Wilson (1972).

C. Two-Dimensional Basins

Fig. 10.37 shows profile views of idealized two-dimensional open and closed basins. They are defined by their depth d and length L_b. For most basins the ratio of the water depth to the length of the reflecting wave is sufficiently small that the wave celerity is given by $C = \sqrt{gd}$. The resonant period $T_n = L/C$ where L is the length of the reflecting wave. For the fundamental mode or first harmonic, $L = 2L_b$ for a closed basin and $L = 4L_b$ for an open basin (Fig. 10.37). Consequently the following equations apply:

$$\text{Closed basin} \quad T_n = \frac{2\,L_b}{n\sqrt{gd}} \tag{10.93}$$

$$\text{Open basin} \quad T_n = \frac{4\,L_b}{(2n+1)\sqrt{gd}} \tag{10.94}$$

where n is the harmonic mode (1 for the fundamental, 2 the next harmonic, etc.).

Fig. 10.37 shows the envelopes of surface position for the first three harmonics; the water surface position is shown at time 0, $T_n/2$, and T_n, which completes one cycle of oscillation. Note that the larger body of water to which the open basin opens causes a node at the opening and the vertical walls cause anti-nodes. At a node the water surface is fixed and flow is always horizontal and at an anti-node the

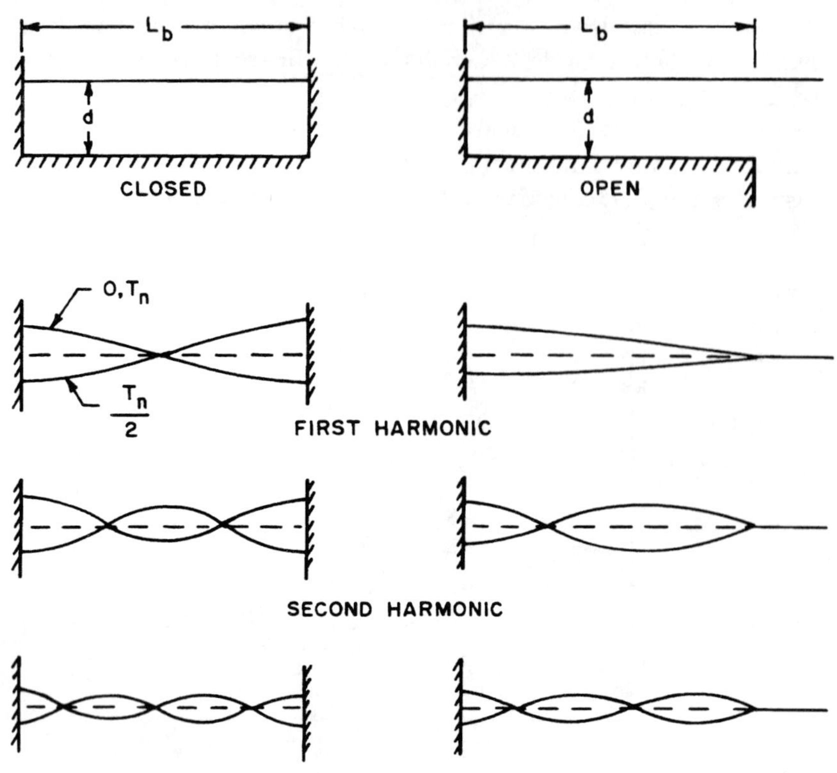

Figure 10.37.—Ideal Two-Dimensional Basin Oscillations.

water surface moves up and down but horizontal flow is prevented. The full oscillatory flow pattern for a closed basin is shown in Fig. 10.38.

It is instructive to develop and apply simple equations for the water particle velocity and extent of horizontal motion where it is largest, i.e., under the node. Consider Fig. 10.38 and assume that the wave surface has a sinusoidal profile. The volume of water that passes through a vertical line (and unit width out of the page) under the node is

$$Vol = 2\int_{L_4}^{L_2} \frac{H}{2} \sin\left(\frac{2\pi x}{L}\right) dx = \frac{HL}{2\pi} \tag{10.95}$$

where H is the wave height and x is the horizontal distance measured along the still water line from one side of the basin. From continuity, the average horizontal flow velocity under the node is

$$V_{avg} = \frac{(Vol/T_n/2)}{d(1)} = \frac{HL}{\pi d t_n}. \tag{10.96}$$

And, for sinusoidal motion the maximum velocity is

$$V_{max} = \frac{\pi V_{avg}}{2} = \frac{HL}{2dT_n} = \frac{HC}{2d} = \frac{H}{2}\sqrt{\frac{g}{d}}. \tag{10.97}$$

The maximum horizontal excursion of a water particle under the node is

$$X = \frac{V_{avg}\,T_n}{2} = \frac{HL}{2\pi d} = \frac{HT_n}{2\pi}\sqrt{\frac{g}{d}}. \tag{10.98}$$

For example, consider a closed basin with H = 0.5 feet, L_b = 3000 feet, and d = 20 feet. Equations 10.93, 10.97, and 10.98 respectively yield $T_n = T_1 = 3.94$ min., $V_{max} = 0.32$ ft/sec, and X = 23.8 feet. This demonstrates the small maximum flow velocity that develops but the relatively large horizontal excursion of the water, which can be quite troublesome.

Some real irregular water bodies that do not have the idealized shape depicted in Fig. 10.37, are still long and narrow so that oscillations along the long axis are the dominant resonant modes. The resonant periods for these basins can be determined by approximating the basin geometry by a series of constant depth sections as shown in Fig. 10.39. The depth d_i at any section of length x_i is the average depth over that section. Then the respective resonant periods, analogous to Equations 10.93 and 10.94, are

$$\text{Closed} \quad T_n = \frac{2}{n} \Sigma \frac{x_i}{\sqrt{g d_i}} \tag{10.99}$$

$$\text{Open} \quad T_n = \frac{4}{2n+1} \Sigma \frac{x_i}{\sqrt{g d_i}}. \tag{10.100}$$

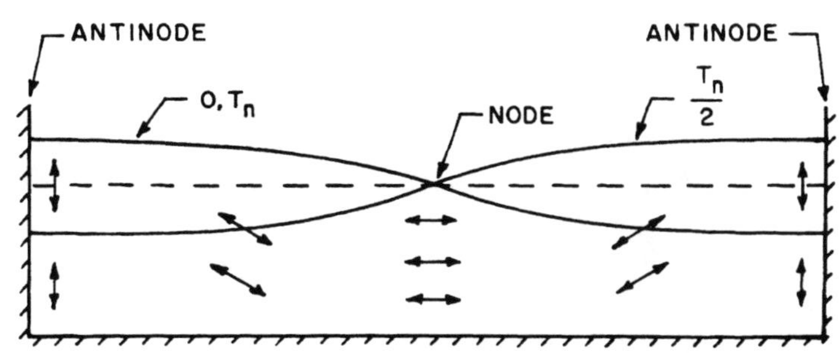

Figure 10.38.—Particle Motions in an Oscillating Closed Basin.

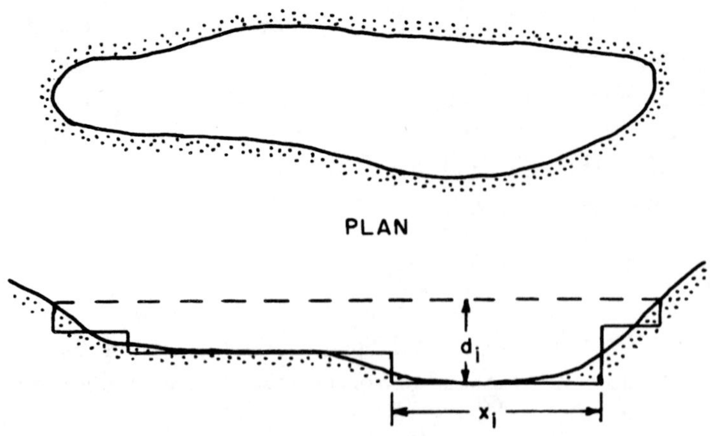

Figure 10.39.—Two-dimensional Irregular Basin.

D. Three-Dimensional Basins

Most open or closed basins have significant dimensions in both width and length. Consequently, resonant oscillations can be established across both the width and length as well as in directions that are oblique to the main width and length axes. Some analytical solutions are available for basins with a regular form such as closed and open rectangular basins having a constant depth and closed circular basins of constant depth (Raichlen, 1966). To provide some insight into the nature of three-dimensional basin resonance, the solution for a closed rectangular basin of constant depth will be presented.

Consider the basin shown in Fig. 10.40 that has horizontal dimensions A by B and a depth d. There might be a small opening in one side, but this opening should be sufficiently small that the side essentially acts as an anti-node. The x, y, and z coordinates are as shown.

The solution yields a resonant period, $T_{n,m}$ where the values n and m are the number of nodes in the x and y directions respectively. This is

$$T_{n,m} = \frac{2}{\sqrt{gd}}\left[\left(\frac{n}{A}\right)^2 + \left(\frac{m}{B}\right)^2\right]^{-1/2}. \quad (10.101)$$

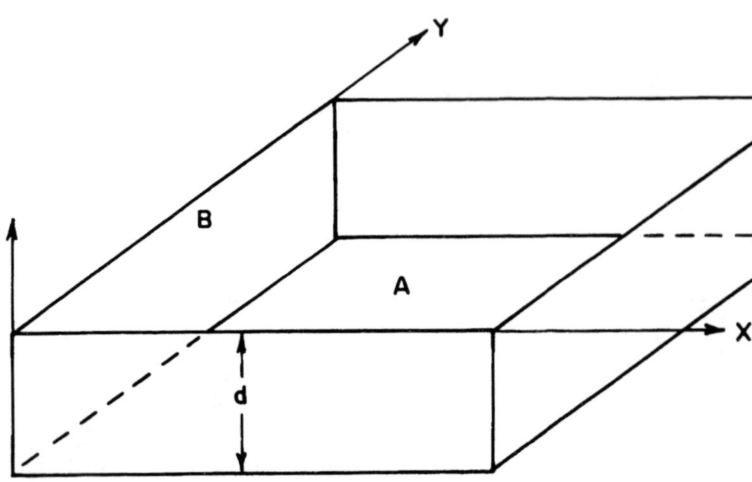

Figure 10.40.—Rectangular Basin.

Note that for n = 1 and m = 0, Equation 10.101 reduces to Equation 10.93 where $A = L_b$. The surface elevation, measured with respect to the still water position, is given by

$$\eta = \frac{H}{2}\cos\frac{n\pi x}{A}\cos\frac{m\pi y}{B}\cos\frac{2\pi t}{T_{n,m}} \tag{10.102}$$

where t is the time and H/2 is the wave amplitude. Again note that for n = 1 and m = 0 this defines the surface oscillation for the two-dimensional basin depicted in Fig. 10.37.

Equation 10.102 can be used to define the nodal line pattern for a basin. At a nodal line the surface elevation is zero at all times. This can only be if

$$\cos\frac{n\pi x}{A},\ \cos\frac{m\pi y}{B} = 0 \tag{10.103}$$

or

$$\frac{n\pi x}{A},\ \frac{m\pi y}{B} = \frac{\pi}{2},\ \frac{3\pi}{2}\ldots \tag{10.104}$$

Thus, the nodal line coordinates are given by

$$x = \frac{A}{2n},\ \frac{3A}{2n}\ldots \qquad y = \frac{B}{2m},\ \frac{3B}{2m}\ldots \tag{10.105}$$

where the termination of the series is dictated by the extent of the basin dimensions.

As a simple example, consider a square basin (A = B = L) and the resonant mode n = m = 1. The resonant period would be

$$T_{1,1} = \sqrt{\frac{2}{gd}}\,L \tag{10.106}$$

and the surface profile would be defined by

$$\eta = \frac{H}{2}\cos\frac{\pi x}{L}\cos\frac{\pi y}{L}\cos\frac{2\pi t}{T_{1,1}} \tag{10.107}$$

where nodal lines are at x = L/2 and y = L/2. The surface pattern for the instant t = T/2 is shown in Fig. 10.41.

The solid lines are positive surface contours and the dashed lines are negative contours. The surface elevation would be H/2 at the corners x = L, y = 0 and y = L, x = 0. It would be −H/2 at the other two corners. At t = 0 and T the pattern would be the same but with the positive and negative contours reversed. At t = T/4 and 3T/4 the water surface would be flat. The arrows indicate the reversing flow that develops across the nodal lines.

For basins having more complex and irregular forms, either a numerical or physical model is required to determine the nodal and resulting water motion patterns. A number of basin resonance physical model studies have been conducted, primarily for harbors open to the sea. See Weggel and Sorensen (1980) for a study concerning the reconstruction of a small boat basin, and Outlaw et al. (1977) for a study of proposed changes to the extensive Los Angeles–Long Beach Harbor complex. For a thorough discussion of the design and operation of physical models of basin resonance consult Hudson (1979).

The basin and necessary adjacent coastal area is constructed to a reduced undistorted or, if necessary, distorted scale. The model scale may need to be distorted to eliminate scale effects, have the model fit in a limited space, and allow for accurate water surface fluctuation measurements. Froude similitude is

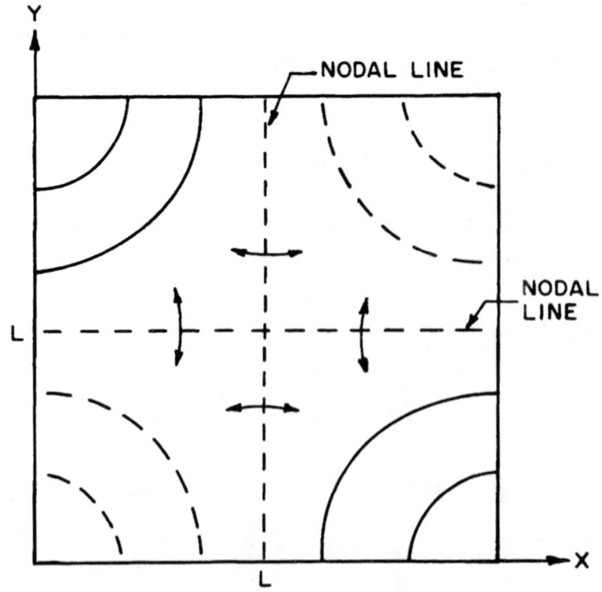

Figure 10.41.—Resonant Pattern of a Square Basin for the First Natural Period.

used to relate model/prototype conditions. If the model is undistorted the resonant period ratio T_r is related to the length ratio L_r by

$$T_r = \sqrt{L_r}. \tag{10.108}$$

If the model is distorted with a horizontal scale L_{rh} and a vertical scale L_{rv}, the resonant period ratio would be

$$T_r = \frac{L_{rh}}{\sqrt{L_{rv}}}. \tag{10.109}$$

Waves having the range of model periods of interest are generated at the harbor, and the harbor response for each period is measured. It is important that the intervals between the sequential periods investigated be small so that no resonant modes are missed. Typically, a wave gage is placed offshore to measure the incident wave condition, and gages are placed at several positions inside the harbor to measure the resonant response. The ratio of harbor to incident wave amplitudes for each incident wave period shows which periods are amplified by the harbor at each point of interest. Time exposures of small floats on the water surface demonstrate the patterns of horizontal water motion and the resulting node/anti-node patterns.

Usually, the amplitudes of the incident long wave spectrum are not adequately known for the prototype, and frictional dissipation is not quantified in the prototype or simulated in the model. As a consequence the actual response amplitude cannot be scaled to prototype conditions, but only the periods at which resonant amplification occurs are defined. And, the nodal/anti-nodal line patterns are defined but not the actual scale of prototype horizontal motion. From these results, troublesome areas can be identified and the effect of proposed basin modifications can be qualitatively determined.

Another limitation on physical models is that the portion of the wave basin representing the sea is of relatively small extent compared to the prototype and has reflecting boundaries. Long low waves have a high reflectivity and thus are slow to dissipate. So extreme care is required when model studies of harbors open to the sea are conducted. See Raichlen (1966) for a more thorough discussion of this subject.

Numerical models may also be employed to investigate resonance in irregular three-dimensional open or closed basins. This involves a solution of the continuity equation and the equations of motion for the two perpendicular horizontal directions. For basin oscillations the vertical acceleration is negligible reducing the vertical equation of motion to the hydrostatic equation. See Dean and Dalrymple (1991) for a development of these equations. Typically the Coriolis and bottom stress terms are neglected in the horizontal equations of motion, leaving a balance between the surface slope and the convective and local acceleration terms.

These equations can be written in finite difference form and solved for a grid overlying the basin. Appropriate boundary conditions must be applied at the grid segments lying on the boundaries. The opening to a larger body of water can be particularly troublesome (Raichlen and Naheer, 1976; Demirel and Wang, 1986) as it is often not a pure nodal line. The conditions in the basin and larger water body must be matched at the entrance to account for incident and reflected energy as well as energy radiated from the basin.

The numerical model calculations yield the time dependent flow velocities at the grid element boundaries and water surface fluctuations at each grid element. As with the physical model investigation, the basin response is investigated for a range of incident wave periods and amplitudes and the surface response at points of interest in the basin is evaluated. The horizontal flow patterns also define nodal and anti-nodal patterns.

E. Helmholtz Resonance

For basins open to the sea through a channel there is an additional possible mode of oscillation, known as the Helmholtz mode, which is analogous to that of a Helmholtz resonator in acoustics. For this mode of oscillation, the water surface in the basin uniformly rises and falls and the water in the entrance to the basin oscillates in and out. The resonant period T_H is (Carrier et al., 1971)

$$T_H = \sqrt{\frac{(L_c + L_c')A_b}{g\ A_c}}. \tag{10.110}$$

where A_b is the basin surface area, A_c is the channel cross section area, L_c is the channel length, and L_c' is an additional length to account for the mass of water outside of each end of the channel that is involved in the resonant motion (Miles, 1948; Sorensen, 1978).

T_H is greater than T_n for an open basin. The Helmholtz mode of oscillation appears to be important for harbors responding to tsunami excitation (Miles, 1974) and for harbors on the Great Lakes responding to storm-generated resonant oscillations in the lakes (Sorensen and Seelig, 1976).

F. Tsunamis

The term "tsunami" is used to denote relatively long period waves generated by coastal and undersea seismic disturbances (earthquakes) and related landslides, bottom slumping, and volcanic eruptions. These waves have caused several major catastrophes including extensive loss of life and property in coastal areas exposed to the effects of tsunami events. The areas in United States where tsunamis are most likely to be a concern are the Pacific coasts of California, Oregon, Washington, Alaska, and Hawaii. General references on tsunamis include Camfield (1980; 1990).

Most tsunamis are caused by earthquakes, but not all submarine earthquakes generate tsunamis. The earth disturbance must be in sufficiently shallow water and the vertical displacement large enough to generate a large wave. A tsunami typically consists of a group of waves having periods of 5–60 minutes and irregular amplitudes. In the deeper ocean areas, a wavelength of 200 kilometers and an amplitude of 1 meter is reasonable.

Nearshore tsunamis are strongly modified by coastal shape and bathymetry. It is not unusual for a nearshore tsunami to cause an increase in water level of 10 meters or more. In extreme cases the water level increase at the coast can exceed 30 meters. Thus, a tsunami that is imperceptible to a vessel in the mid-ocean can devastate a coastal town! The drawdown of water when a tsunami trough arrives at the shore can also cause damage such as seaward collapse of seawalls, expose seawater intake pipelines, and damage to ships in harbors. Because of their long periods, tsunamis may cause damaging harbor resonance.

Tsunamis generated by landslides are an unusual but potentially very destructive type of wave. Wiegel (1964) cites a classic example from Miller (1960) of landslide-generated waves in Lituya Bay, Alaska. The slide occurred along one side of the head of a T-shaped bay. The resulting surge at the opposite side reached a maximum elevation of 365 meters above water level. A wave extending about 60 meters above water level moved down the bay and out to sea. The wave easily stripped large trees from the terrain in its path.

Tsunami damage to structures may be caused by the direct force of the waves. Damage may also be caused by strong currents forming around a structure that interferes with incoming waves and by hydrostatic pressure against the backside of a flooded structure during the extreme drawdown of water level between wave crests. Additional causes of damage include wave-induced erosion around a structure base and debris carried by the tsunami.

Tsunamis are rare events. However, tsunami estimates are needed for coastal flood zone planning and for operation of the tsunami warning system to evacuate people from potentially dangerous areas. Systematic tsunami flooding estimates obtained by numerical modeling have been published for much of the United States Pacific coast, including Alaska and Hawaii (Crawford, 1987; Houston et al., 1977; Houston and Garcia, 1974; 1978). Typically, estimates of the 100- and 500-year water level elevation due to the combined effect of astronomical tides and earthquake-generated tsunamis are given at numerous points along the coast.

X. WATER SURFACE PROBABILITY ANALYSIS

Knowledge of the frequency of occurrence of various water surface elevations is important because of coastal flooding, wave damage, and shoreline erosion. High water levels, in addition to flooding coastal areas, allow waves to act higher up on the beach profile causing shore erosion and damage to structures at higher elevations. Evaluation of the economic impact of flooding, erosion, and wave action as well as the potential benefits of providing protection require that the frequency of occurrence of these events be quantified.

A. Open Coast Water Levels

Along the world's ocean coastlines, water level changes result from astronomical tides, storm surges, wave setup, wave runup, tsunamis and long-term changes in sea level. Astronomical tides are the rise and fall in water levels caused by the moon and the sun. As briefly described in an earlier section, tides arise due to the earth's motion about the center of mass of the earth–moon system (the barycenter) and the gravitational attraction of the moon. The rotation about the barycenter and gravitational attraction result in two tidal bulges on the surface of the earth, one facing the moon and one on the side opposite the moon. A similar relationship exists for the movement of the earth about the earth–sun system. Thus, there are four tidal bulges, two due to the moon and two due to the sun. While the sun is by far the larger of the two tide producing bodies, the tides due to the sun are only about 46% of the lunar tides because of the great distance between the earth and sun. Depending on the relative position of the earth, moon, and sun the lunar and solar bulges may reinforce or cancel each other. They reinforce each other when the earth, moon, and sun are generally in alignment during a full or new moon. Tides occurring at this time are termed "spring" tides.

When the earth, sun, and moon are in quadrature, during the first and last quarter phases, the lunar and solar tidal bulges tend to cancel each other and the tides are termed "neap" tides. Consequently, there are two periods of spring tides and two periods of neap tides in a lunar month of approximately 29 1/2 days. Other variations in tidal heights occur when the moon's orbit takes it far from earth during apogee and when it's orbit brings it close to earth during lunar perigee. Tides during this period are termed perigean high tides. Coupled with spring tides, lunar perigee results in higher than usual tides. The earth's rotation on its axis results in the movement of the tidal bulge over the earth's surface and results in the diurnal (one high and one low tide per day) or semidiurnal (two high and two low tides per day) tides. Since there are two tidal bulges on the earth, most areas experience two high and two low tides per day (actually 24 hours and 50 minutes because the earth–moon system moves around the sun and must move a little farther each day for the same point to be under the tidal bulge). The tidal day is thus about 24 hours and 50 minutes long.

Based on astronomical observations of the moon and sun, the relative phases of the tidal components are known quite accurately and the times of occurrence of daily high and low tides and the corresponding tidal elevations can be predicted at any given site based on observations taken over a period of time. Ideally, predictions are based on observations taken over a 19-year period; however, reasonably accurate predictions can be based on shorter periods of observation, e.g., one year. Tidal predictions for numerous sites around the world are made by the National Ocean Survey (NOS, 1992) and published annually for the coming year as "Tide Tables." Some tidal observation sites, termed reference stations, have hourly predictions made for them although only the time and height of high and low tides are published. Other sites have their tides referred to conditions at the reference stations. Corrections for the time and height of high and low tides at the station are given in relation to a reference station. A typical page from the 1992 "Tide Tables" for Sandy Hook, New Jersey is given in Fig. 10.42. A typical page showing the corrections for other sites along the New Jersey coast is given in Fig. 10.43. The features of tides are discussed in some detail in other sections of this chapter.

Statistics of tidal water levels at NOS Reference Stations in the United States have been presented by Harris (1981). A typical entry (Sandy Hook, New Jersey) is shown in Fig. 10.44. While the times of high and low tides and the corresponding water levels are deterministic, other components contributing to local water levels are not; consequently, the probability that the astronomical tide at a given site will be above a given level is important to estimate the joint probability of high astronomical tides that might coincide with other events producing high water levels such as tsunamis and storm surges. Similarly, the statistics of low water levels are important for navigation.

Storm surges are meteorologically-caused increases or decreases in water levels. They are discussed in some detail earlier in this chapter. While the water levels resulting from a given set of meteorological conditions can be estimated using numerical models, in general, storm surge levels cannot be predicted far in advance. The National Weather Service does some real-time surge level computations when an actual hurricane approaches the coastline; however, the characteristics of hurricanes that will occur at some given time in the future cannot be predicted so that the surge they will produce also cannot be predicted. However, a statistical analysis can be made of the historical water levels recorded at a site to obtain an estimate of the return period of a given storm surge level.

Wave setup is an increase in the mean water level at the shoreline that occurs because of the presence of waves. Wave setup causes higher water levels on which breaking waves travel and from which they can run up on a beach or coastal structure. Runup is the height above the still water level that water will run up on a beach or structure. Since runup is measured from the still water level, measurements of wave runup generally include wave setup as a component.

Local changes in relative sea level must also be considered in design. Sea level changes can occur because of either an increase in the global sea level or by local land subsidence. Subsidence is a regional phenomenon and subsidence data are site specific. For example, Galveston, Texas and coastal Louisiana are some of the regions where land subsidence has caused a significant change in relative sea level. Many developed barrier islands, where significant groundwater withdrawal has taken place have also experienced subsidence. Eustatic sea level rise has been predicted as a result of global warming. If global warming predictions are correct, sea level is expected to rise because of melting glaciers and other

January

Day	Time (h m)	Height (ft)	Height (cm)
1 W	0500	4.9	149
	1120	0.1	3
	1723	3.8	116
	2326	0.1	3
2 Th	0550	5.0	152
	1209	0.0	0
	1811	3.9	119
3 F	0013	0.1	3
	0632	5.1	155
	1255	-0.1	-3
	1853	4.0	122
4 Sa ●	0058	0.1	3
	0711	5.1	155
	1338	-0.2	-6
	1932	4.1	125
5 Su	0141	0.1	3
	0749	5.1	155
	1418	-0.2	-6
	2009	4.1	125
6 M	0223	0.1	3
	0825	5.0	152
	1456	-0.2	-6
	2048	4.1	125
7 Tu	0300	0.2	6
	0902	4.8	146
	1531	-0.1	-3
	2123	4.0	122
8 W	0334	0.3	9
	0938	4.6	140
	1603	0.0	0
	2202	4.0	122
9 Th	0408	0.4	12
	1015	4.4	134
	1637	0.1	3
	2240	4.0	122
10 F	0442	0.5	15
	1053	4.3	131
	1706	0.2	6
	2322	4.1	125
11 Sa	0525	0.7	21
	1133	4.1	125
	1746	0.3	9
12 Su ◐	0006	4.2	128
	0620	0.8	24
	1223	3.9	119
	1835	0.3	9
13 M	0057	4.3	131
	0729	0.8	24
	1319	3.8	116
	1938	0.3	9
14 Tu	0156	4.4	134
	0843	0.6	18
	1424	3.7	113
	2042	0.2	6
15 W	0301	4.6	140
	0945	0.3	9
	1535	3.8	116
	2145	0.0	0
16 Th	0409	4.9	149
	1043	0.0	0
	1643	4.0	122
	2245	-0.2	-6
17 F	0512	5.3	162
	1142	-0.4	-12
	1744	4.4	134
	2344	-0.5	-15
18 Sa	0606	5.7	174
	1237	-0.7	-21
	1839	4.7	143
19 Su ○	0042	-0.7	-21
	0658	5.9	180
	1330	-1.0	-30
	1930	5.0	152
20 M	0138	-0.9	-27
	0748	6.0	183
	1420	-1.2	-37
	2019	5.2	158
21 Tu	0230	-1.0	-30
	0837	5.9	180
	1509	-1.3	-40
	2110	5.3	162
22 W	0322	-1.0	-30
	0928	5.7	174
	1555	-1.2	-37
	2202	5.3	162
23 Th	0411	-0.8	-24
	1019	5.3	162
	1640	-0.9	-27
	2254	5.2	158
24 F	0503	-0.5	-15
	1111	4.9	149
	1729	-0.6	-18
	2346	5.0	152
25 Sa	0558	-0.1	-3
	1201	4.5	137
	1820	-0.2	-6
26 Su ◑	0039	4.8	146
	0657	0.2	6
	1257	4.1	125
	1919	0.1	3
27 M	0134	4.6	140
	0802	0.4	12
	1352	3.8	116
	2018	0.3	9
28 Tu	0230	4.5	137
	0905	0.5	15
	1453	3.5	107
	2116	0.4	12
29 W	0332	4.4	134
	1001	0.4	12
	1558	3.5	107
	2210	0.4	12
30 Th	0430	4.5	137
	1053	0.3	9
	1658	3.6	110
	2302	0.3	9
31 F	0524	4.6	140
	1142	0.2	6
	1748	3.8	116
	2350	0.3	9

February

Day	Time (h m)	Height (ft)	Height (cm)
1 Sa	0609	4.8	146
	1229	0.0	0
	1832	4.0	122
2 Su	0039	0.2	6
	0649	4.9	149
	1312	-0.1	-3
	1911	4.2	128
3 M ●	0121	0.1	3
	0727	5.0	152
	1352	-0.2	-6
	1947	4.3	131
4 Tu	0203	0.0	0
	0803	4.9	149
	1429	-0.2	-6
	2022	4.4	134
5 W	0240	0.0	0
	0838	4.8	146
	1502	-0.2	-6
	2056	4.4	134
6 Th	0314	0.1	3
	0912	4.7	143
	1534	-0.1	-3
	2131	4.4	134
7 F	0347	0.2	6
	0949	4.5	137
	1603	0.0	0
	2205	4.4	134
8 Sa	0419	0.3	9
	1023	4.3	131
	1632	0.1	3
	2245	4.5	137
9 Su	0458	0.4	12
	1104	4.1	125
	1704	0.2	6
	2329	4.5	137
10 M	0546	0.5	15
	1154	3.9	119
	1751	0.3	9
11 Tu ◐	0022	4.5	137
	0655	0.6	18
	1247	3.8	116
	1855	0.4	12
12 W	0121	4.6	140
	0813	0.6	18
	1355	3.7	113
	2014	0.4	12
13 Th	0231	4.7	143
	0921	0.4	12
	1513	3.8	116
	2124	0.2	6
14 F	0343	4.9	149
	1023	0.0	0
	1625	4.1	125
	2230	-0.1	-3
15 Sa	0452	5.2	158
	1123	-0.3	-9
	1728	4.5	137
	2329	-0.4	-12
16 Su	0551	5.5	168
	1216	-0.7	-21
	1821	5.0	152
17 M	0028	-0.7	21
	0642	5.8	177
	1309	-1.0	30
	1912	5.4	165
18 Tu ○	0123	-1.0	30
	0731	5.9	180
	1357	-1.2	-37
	2000	5.6	171
19 W	0215	-1.1	-34
	0819	5.8	177
	1444	-1.2	-37
	2047	5.7	174
20 Th	0304	-1.1	-34
	0907	5.6	171
	1528	-1.1	-34
	2136	5.6	171
21 F	0352	-0.9	-27
	0955	5.2	158
	1611	-0.8	-24
	2224	5.4	165
22 Sa	0439	-0.5	-15
	1044	4.8	146
	1656	-0.4	-12
	2313	5.2	158
23 Su	0529	-0.1	-3
	1132	4.4	134
	1743	0.0	0
24 M	0004	4.9	149
	0623	0.3	9
	1223	4.0	122
	1836	0.4	12
25 Tu ◑	0055	4.6	140
	0723	0.6	18
	1318	3.7	113
	1940	0.7	21
26 W	0151	4.4	134
	0828	0.7	21
	1418	3.5	107
	2042	0.8	24
27 Th	0251	4.2	128
	0928	0.7	21
	1523	3.5	107
	2142	0.8	24
28 F	0356	4.3	131
	1022	0.6	18
	1627	3.6	110
	2236	0.6	18
29 Sa	0451	4.4	134
	1110	0.4	12
	1721	3.9	119
	2326	0.5	15

March

Day	Time (h m)	Height (ft)	Height (cm)
1 Su	0540	4.6	140
	1155	0.2	6
	1805	4.2	128
2 M	0011	0.3	9
	0622	4.8	146
	1239	0.0	0
	1845	4.5	137
3 Tu	0056	0.1	3
	0701	4.9	149
	1319	-0.1	-3
	1920	4.7	143
4 W ●	0138	0.0	0
	0738	5.0	152
	1357	-0.2	-6
	1953	4.8	146
5 Th	0216	-0.1	-3
	0812	4.9	149
	1431	-0.1	-3
	2027	4.9	149
6 F	0253	-0.1	-3
	0847	4.8	146
	1502	-0.1	-3
	2101	4.9	149
7 Sa	0328	0.0	0
	0922	4.6	140
	1533	0.0	0
	2136	4.9	149
8 Su	0403	0.1	3
	1001	4.4	134
	1603	0.1	3
	2216	4.9	149
9 M	0440	0.3	9
	1045	4.2	128
	1640	0.3	9
	2304	4.9	149
10 Tu	0530	0.4	12
	1135	4.0	122
	1725	0.4	12
	2358	4.8	146
11 W ◐	0634	0.6	18
	1234	3.9	119
	1833	0.6	18
12 Th	0100	4.7	143
	0753	0.6	18
	1343	3.9	119
	1957	0.6	18
13 F	0211	4.7	143
	0903	0.4	12
	1457	4.0	122
	2111	0.4	12
14 Sa	0324	4.9	149
	1004	0.1	3
	1609	4.4	134
	2217	0.1	3
15 Su	0433	5.1	155
	1059	-0.3	-9
	1710	4.9	149
	2316	-0.3	-9
16 M	0533	5.4	165
	1152	-0.6	-18
	1805	5.4	165
17 Tu	0013	-0.6	-18
	0624	5.6	171
	1243	-0.8	-24
	1853	5.8	177
18 W ○	0106	-0.8	-24
	0713	5.7	174
	1331	-0.9	-27
	1938	6.0	183
19 Th	0157	-0.9	-27
	0758	5.6	171
	1417	-0.9	-27
	2024	6.0	183
20 F	0244	-0.9	-27
	0843	5.4	165
	1501	-0.7	-21
	2107	5.9	180
21 Sa	0330	-0.7	-21
	0929	5.0	152
	1543	-0.4	-12
	2152	5.6	171
22 Su	0415	-0.4	-12
	1016	4.7	143
	1624	0.0	0
	2239	5.3	162
23 M	0500	0.0	0
	1102	4.3	131
	1709	0.4	12
	2330	4.9	149
24 Tu	0549	0.4	12
	1153	4.0	122
	1759	0.8	24
25 W ◑	0018	4.6	140
	0647	0.7	21
	1246	3.8	116
	1900	1.0	30
26 Th	0111	4.4	134
	0748	0.9	27
	1342	3.6	110
	2007	1.2	37
27 F	0209	4.2	128
	0849	0.8	24
	1444	3.7	113
	2111	1.1	34
28 Sa	0311	4.2	128
	0942	0.7	21
	1548	3.8	116
	2204	0.9	27
29 Su	0412	4.3	131
	1030	0.5	15
	1643	4.1	125
	2254	0.7	21
30 M	0505	4.5	137
	1115	0.4	12
	1731	4.5	137
	2341	0.4	12
31 Tu	0550	4.7	143
	1200	0.2	6
	1811	4.8	146

Time meridian 75° W. 0000 is midnight. 1200 is noon.
Heights are referred to mean lower low water which is the chart datum of soundings.

Figure 10.42.—Typical Page from Tide Tables *for Sandy Hook, NJ (NOS, 1992).*

land-supported ice, and thermal expansion of the ocean's waters. Predictions are that sea level will rise by anywhere from 0.6 to 2.0 meters in the next century. Historical rates have been about 0.3 meters per century. A typical sea level record showing the rising trend is given in Fig. 10.45 (Lyles et al., 1987).

Tsunamis, seismically generated waves, are discussed at some length in a previous section of this handbook. Tsunamis are long period waves and can provide a mechanism that causes changes in water level with durations of from 5 to 60 minutes—the usual range of tsunami wave periods.

NO.	PLACE	POSITION Lat.	Long.	DIFFERENCES Time High water	Time Low water	Height High water	Height Low water	RANGES Mean	Spring	Mean Tide Level
		° ′ N	° ′ W	h. m.	h. m.	ft	ft	ft	ft	ft
	NEW JERSEY Sandy Hook Bay Time meridian, 75°W			on SANDY HOOK, p.64						
	Shrewsbury River									
1681	Highlands	40 24	73 59	+0 34	+0 57	*0.82	*0.82	3.8	4.6	2.1
1683	Red Bank, Navesink River	40 21	74 04	+1 47	+2 25	*0.64	*0.64	3.0	3.6	1.6
1685	Normandie	40 23	73 59	+1 08	+1 47	*0.62	*0.62	2.9	3.5	1.5
1687	Sea Bright	40 21	73 59	+2 09	+2 40	*0.36	*0.36	1.7	2.1	0.9
1689	Branchport, Pleasure Bay	40 19	74 00	+2 59	+3 28	*0.36	*0.36	1.7	2.1	0.9
	Outer Coast									
1691	Sea Bright	40 22	73 58	-0 35	-0 43	*0.94	*0.94	4.4	5.3	2.4
1693	Long Branch (fishing pier)	40 18	73 59	-0 28	-0 36	*0.95	*0.95	4.4	5.3	2.4
1695	Asbury Park	40 13	74 00	-0 35	-0 43	*0.92	*0.92	4.3	5.2	2.3
1697	Shark River Inlet (entrance)	40 11	74 01	-0 19	-0 34	*0.86	*0.86	4.0	4.8	2.2
1699	Municipal Boat Basin, Shark River	40 11	74 02	+0 26	+0 38	*0.79	*0.79	3.7	4.4	2.0
1701	Sea Girt	40 08	74 02	-0 35	-0 43	*0.92	*0.92	4.3	5.2	2.3
1703	Manasquan Inlet	40 06	74 02	-0 13	-0 34	*0.86	*0.86	4.0	4.8	2.2
	Manasquan River									
1705	Railroad bridge	40 06	74 03	+0 19	+0 07	*0.75	*0.75	3.5	4.2	1.9
1707	Riviera Beach	40 06	74 05	+0 50	+1 27	*0.67	*0.67	3.1	3.8	1.6
1709	Seaside Park (ocean)	39 55	74 05	-0 34	-0 42	*0.90	*0.90	4.2	5.1	2.3
	Barnegat Bay									
1711	Mantoloking	40 02	74 03	+5 33	+5 36	*0.11	*0.11	0.5	0.6	0.2
1713	Coates Point, highway bridge	39 57	74 07	+4 18	+4 30	*0.11	*0.11	0.5	0.6	0.2
1715	Toms River (town)	39 57	74 12	+4 36	+4 49	*0.13	*0.13	0.6	0.7	0.3
1717	Waretown	39 47	74 11	+2 32	+2 51	*0.13	*0.13	0.6	0.7	0.3
1719	Oyster Cr. Chan. (off Sedge Island)	39 47	74 08	+2 15	+2 19	*0.13	*0.13	0.6	0.7	0.3
1721	Barnegat Inlet	39 46	74 06	-0 21	-0 19	*0.67	*0.67	3.1	3.8	1.6
1723	Harvey Cedars	39 42	74 08	+3 14	+4 04	*0.17	*0.17	0.8	1.0	0.4
	Little Egg Harbor									
1725	Manahawkin Bridge	39 39	74 11	+2 32	+3 22	*0.32	*0.32	1.5	1.8	0.8
1727	Long Point	39 36	74 16	+1 47	+1 58	*0.47	*0.47	2.2	2.7	1.2
1729	Parker Run	39 37	74 19	+2 00	+2 35	*0.46	*0.46	2.1	2.6	1.1
1731	Tuckerton Creek entrance	39 35	74 20	+1 39	+1 56	*0.52	*0.52	2.4	2.9	1.3
1733	Tuckerton, Tuckerton Creek	39 36	74 20	+1 41	+2 11	*0.47	*0.47	2.2	2.6	1.2
1735	Beach Haven	39 34	74 15	+1 46	+2 03	*0.47	*0.47	2.2	2.7	1.2
1737	Beach Haven Coast Guard Station	39 33	74 15	+1 12	+1 17	*0.48	*0.48	2.2	2.7	1.2
1739	Holgate	39 32	74 16	+1 10	+1 09	*0.56	*0.56	2.6	3.1	1.4
	Great Bay									
1741	Little Egg Inlet	39 30	74 18	-0 02	-0 01	*0.79	*0.79	3.7	4.5	2.0
1743	Little Sheepshead Creek	39 31	74 19	+0 30	+0 40	*0.68	*0.68	3.2	3.8	1.7
1745	Seven Islands	39 31	74 20	+0 11	+0 18	*0.73	*0.73	3.4	4.1	1.8
1747	Graveling Point	39 32	74 23	+0 38	+1 11	*0.69	*0.69	3.2	3.9	1.7
1749	Cramers Boatyard, Mullica River	39 33	74 28	+1 23	+1 57	*0.65	*0.65	3.0	3.6	1.7
1751	New Gretna	39 36	74 26	+1 47	+2 02	*0.68	*0.68	3.1	3.8	1.7
1753	Green Bank, Mullica River	39 37	74 35	+2 59	+3 16	*0.66	*0.66	3.1	3.7	1.7
1755	Main Marsh Thorofare	39 29	74 23	+1 03	+1 32	*0.71	*0.71	3.3	4.0	1.7
1757	Brigantine Channel	39 27	74 21	0 00	+0 05	*0.75	*0.75	3.5	4.2	1.9
1759	Grassy Bay	39 26	74 24	+1 07	+1 13	*0.73	*0.73	3.4	4.1	1.8
1761	Absecon Creek entrance, Absecon Bay	39 25	74 29	+1 03	+1 19	*0.77	*0.77	3.6	4.4	2.0
1763	Broad Creek, Middle Thorofare	39 24	74 26	+0 54	+0 35	*0.73	*0.73	3.4	4.1	1.8
1765	Absecon Channel	39 23	74 26	+0 11	-0 09	*0.84	*0.84	3.9	4.7	2.1
1767	Beach Thorofare (railroad bridges)	39 22	74 27	+0 51	+0 42	*0.82	*0.82	3.8	4.6	2.1
1769	Atlantic City	39 21	74 25	-0 27	-0 35	*0.88	*0.88	4.1	5.0	2.2
1771	Chelsea (highway bridge)	39 21	74 28	+0 48	+0 47	*0.86	*0.86	4.0	4.8	2.2
1773	Beach Thorofare (Shelter Island)	39 21	74 30	+0 38	+0 34	*0.84	*0.84	3.9	4.7	2.1
1775	Dock Thorofare, Risley Channel	39 21	74 32	+0 26	+0 24	*0.86	*0.86	4.0	4.8	2.2
1777	Longport (inside)	39 18	74 32	+0 04	+0 01	*0.79	*0.79	3.9	4.7	2.2
1779	Great Egg Harbor Inlet	39 18	74 34	+0 11	-0 03	*0.82	*0.82	3.8	4.6	2.1
1781	Ocean City (9th Street bridge)	39 17	74 35	+0 23	+0 21	*0.79	*0.79	3.7	4.5	2.0
1783	Great Egg Harbor Bay	39 18	74 38	+0 43	+0 59	*0.77	*0.77	3.6	4.4	2.0
	Great Egg Harbor River									
1785	Scull Landing	39 22	74 43	+1 42	+1 56	*0.79	*0.79	3.7	4.5	2.0
1787	Mays Landing	39 27	74 44	+2 25	+2 38	*0.87	*0.85	3.9	4.7	2.1
1789	Cedar Swamp Creek	39 15	74 43	+2 50	+2 28	*0.68	*0.68	3.1	3.7	1.8
1791	Peck Bay (34th Street bridge)	39 15	74 38	+0 50	+1 04	*0.79	*0.79	3.7	4.5	2.0
1793	Devils Island, Crook Horn Creek	39 14	74 39	+0 36	+0 24	*0.77	*0.77	3.6	4.4	2.0
1795	Corson Inlet (bridges)	39 13	74 39	+0 08	+0 06	*0.84	*0.84	3.9	4.7	2.1
1797	Ben Hands Thorofare	39 12	74 40	+0 47	+0 34	*0.79	*0.79	3.7	4.5	2.0
1799	Sea Isle City (Ludlam Thoro. bridge)	39 09	74 42	+0 44	+0 51	*0.82	*0.82	3.8	4.6	2.1
1801	Sea Isle City (beach)	39 09	74 41	-0 20	-0 17	*0.88	*0.88	4.1	5.0	2.2
1803	Townsend Sound	39 09	74 45	+0 40	+1 03	*0.79	*0.45	3.8	4.5	2.0
1805	Townsends Inlet	39 07	74 43	+0 05	+0 06	*0.82	*0.82	3.8	4.6	2.1
1807	Long Reach	39 06	74 45	+0 52	+0 55	*0.82	*0.82	3.8	4.6	2.1
1809	Great Sound (ent. to Cresse Thoro.)	39 05	74 47	+1 02	+1 07	*0.88	*0.88	4.1	5.0	2.2
1811	Stone Harbor (Great Chan. bridge)	39 03	74 46	+0 41	+0 28	*0.88	*0.88	4.1	5.0	2.2
1813	Hereford Inlet (North Wildwood)	39 01	74 48	+0 01	+0 04	*0.88	*0.88	4.1	5.0	2.2
1815	Wildwood (beach)	38 59	74 48	-0 16	-0 17	*0.88	*0.88	4.1	5.0	2.2
1817	Grassy Sound Channel (hwy. bridge)	39 02	74 49	+0 39	+0 30	*0.88	*0.88	4.1	5.0	2.2
1819	West Wildwood (Grassy Sound bridge)	39 00	74 50	+0 44	+0 31	*0.92	*0.92	4.3	5.2	2.3
1821	Swain Channel	38 59	74 52	+0 53	+0 29	*0.94	*0.94	4.4	5.3	2.4
1823	Cape May Harbor	38 57	74 53	-0 03	-0 14	*0.94	*0.94	4.4	5.3	2.4
1825	Cape May, Municipal Pier	38 56	74 55	+0 01	-0 15	*0.92	*0.92	4.3	5.2	2.3

Endnotes can be found at the end of table 2.

Figure 10.43.—Typical Page from Tide Tables *Showing Corrections for Stations Referred to Sandy Hook, NJ Tide Predictions (NOS, 1992)*

Distribution Function, Sandy Hook, NJ, 1963–81. Lower limit of class interval shown, all heights are normalized with respect to half the mean range, 2,333 feet.

	Extreme mo. high water			Extreme mo. low water		
Class no.	Lower limit	Freq.	Cum. freq.	Lower limit	Freq.	Cum. freq.
101	1.7519	.0044	.0044	−1.1242	.0044	.0044
100	1.7464	.0000	.0044	−1.1297	.0000	.0044
99	1.7410	.0044	.0088	−1.1353	.0000	.0044
98	1.7355	.0000	.0088	−1.1409	.0000	.0044
97	1.7300	.0000	.0088	−1.1465	.0000	.0044
96	1.7245	.0088	.0175	−1.1521	.0044	.0088
95	1.7191	.0175	.0351	−1.1577	.0000	.0088
94	1.7136	.0044	.0395	−1.1633	.0000	.0088
93	1.7081	.0044	.0439	−1.1689	.0000	.0088
92	1.7026	.0088	.0526	−1.1745	.0000	.0088
91	1.6972	.0088	.0614	−1.1801	.0000	.0088
90	1.6917	.0219	.0833	−1.1857	.0044	.0132
89	1.6862	.0088	.0921	−1.1913	.0000	.0132
88	1.6807	.0263	.1184	−1.1969	.0044	.0175
87	1.6752	.0132	.1316	−1.2025	.0132	.0307
86	1.6698	.0219	.1535	−1.2081	.0000	.0307
85	1.6643	.0132	.1667	−1.2137	.0044	.0351
84	1.6588	.0219	.1886	−1.2193	.0044	.0395
83	1.6533	.0263	.2149	−1.2248	.0044	.0439
82	1.6479	.0175	.2325	−1.2304	.0088	.0526
81	1.6424	.0219	.2544	−1.2360	.0088	.0614
80	1.6369	.0132	.2675	−1.2416	.0132	.0746
79	1.6314	.0132	.2807	−1.2472	.0000	.0746
78	1.6260	.0307	.3114	−1.2528	.0044	.0789
77	1.6205	.0088	.3202	−1.2584	.0088	.0877
76	1.6150	.0088	.3289	−1.2640	.0132	.1009
75	1.6095	.0088	.3377	−1.2696	.0044	.1053
74	1.6041	.0263	.3640	−1.2752	.0088	.1140
73	1.5986	.0219	.3860	−1.2808	.0088	.1228
72	1.5931	.0175	.4035	−1.2864	.0088	.1316
71	1.5876	.0088	.4123	−1.2920	.0175	.1491
70	1.5822	.0044	.4167	−1.2976	.0088	.1579
69	1.5767	.0132	.4298	−1.3032	.0263	.1842
68	1.5712	.0132	.4430	−1.3088	.0088	.1930
67	1.5657	.0219	.4649	−1.3144	.0044	.1974
66	1.5602	.0000	.4649	−1.3199	.0175	.2149
65	1.5548	.0219	.4868	−1.3255	.0219	.2368
64	1.5493	.0175	.5044	−1.3311	.0088	.2456
63	1.5438	.0044	.5088	−1.3367	.0175	.2632
62	1.5383	.0219	.5307	−1.3423	.0132	.2763
61	1.5329	.0044	.5351	−1.3479	.0132	.2895
60	1.5274	.0219	.5570	−1.3535	.0175	.3070
59	1.5219	.0263	.5833	−1.3591	.0132	.3202
58	1.5164	.0088	.5921	−1.3647	.0088	.3289
57	1.5110	.0132	.6053	−1.3703	.0175	.3465
56	1.5055	.0175	.6228	−1.3759	.0307	.3772
55	1.5000	.0132	.6360	−1.3815	.0132	.3904
54	1.4945	.0175	.6535	−1.3871	.0263	.4167
53	1.4891	.0132	.6667	−1.3927	.0088	.4254
52	1.4836	.0307	.6974	−1.3983	.0307	.4561

Figure 10.44.—Statistics of NOS Tidal Reference Station at Sandy Hook, NJ (Harris, 1981).

Class no.	Extreme mo. high water			Extreme mo. low water		
	Lower limit	Freq.	Cum. freq.	Lower limit	Freq.	Cum. freq.
51	1.4781	.0044	.7018	−1.4039	.0263	.4825
50	1.4726	.0263	.7281	−1.4095	.0351	.5175
49	1.4671	.0219	.7500	−1.4151	.0219	.5395
48	1.4617	.0088	.7588	−1.4206	.0132	.5526
47	1.4562	.0175	.7763	−1.4262	.0219	.5746
46	1.4507	.0175	.7939	−1.4318	.0307	.6053
45	1.4452	.0088	.8026	−1.4374	.0175	.6228
44	1.4398	.0219	.8246	−1.4430	.0132	.6360
43	1.4343	.0000	.8246	−1.4486	.0132	.6491
42	1.4288	.0000	.8246	−1.4542	.0044	.6535
41	1.4233	.0088	.8333	−1.4598	.0219	.6754
40	1.4179	.0044	.8377	−1.4654	.0219	.6974
39	1.4124	.0088	.8465	−1.4710	.0088	.7061
38	1.4069	.0132	.8596	−1.4766	.0307	.7368
37	1.4014	.0088	.8684	−1.4822	.0088	.7456
36	1.3960	.0088	.8772	−1.4878	.0175	.7632
35	1.3905	.0044	.8816	−1.4934	.0132	.7763
34	1.3850	.0088	.8904	−1.4990	.0175	.7939
33	1.3795	.0044	.8947	−1.5046	.0044	.7982
32	1.3740	.0000	.8947	−1.5102	.0132	.8114
31	1.3686	.0088	.9035	−1.5157	.0088	.8202
30	1.3631	.0088	.9123	−1.5213	.0088	.8289
29	1.3576	.0132	.9254	−1.5269	.0175	.8465
28	1.3521	.0000	.9254	−1.5325	.0044	.8509
27	1.3467	.0000	.9254	−1.5381	.0175	.8684
26	1.3412	.0044	.9298	−1.5437	.0088	.8772
25	1.3357	.0088	.9386	−1.5493	.0044	.8816
24	1.3302	.0044	.9430	−1.5549	.0219	.9035
23	1.3248	.0088	.9518	−1.5605	.0088	.9123
22	1.3193	.0044	.9561	−1.5661	.0044	.9167
21	1.3138	.0000	.9561	−1.5717	.0088	.9254
20	1.3083	.0000	.9561	−1.5773	.0000	.9254
19	1.3029	.0000	.9561	−1.5829	.0000	.9254
18	1.2974	.0044	.9605	−1.5885	.0088	.9342
17	1.2919	.0000	.9605	−1.5941	.0088	.9430
16	1.2864	.0088	.9693	−1.5997	.0088	.9518
15	1.2809	.0044	.9737	−1.6053	.0000	.9518
14	1.2755	.0044	.9781	−1.6108	.0044	.9561
13	1.2700	.0000	.9781	−1.6164	.0044	.9605
12	1.2645	.0088	.9868	−1.6220	.0088	.9693
11	1.2590	.0044	.9912	−1.6276	.0132	.9825
10	1.2536	.0044	.9956	−1.6332	.0000	.9825
9	1.2481	.0000	.9956	−1.6388	.0044	.9868
8	1.2426	.0000	.9956	−1.6444	.0044	.9912
7	1.2371	.0000	.9956	−1.6500	.0000	.9912
6	1.2317	.0000	.9956	−1.6556	.0000	.9912
5	1.2262	.0000	.9956	−1.6612	.0044	.9956
4	1.2207	.0000	.9956	−1.6668	.0000	.9956
3	1.2152	.0000	.9956	−1.6724	.0000	.9956
2	1.2098	.0000	.9956	−1.6780	.0000	.9956
1	1.2043	.0044	1.0000	−1.6836	.0044	1.0000

Figure 10.44.—(Continued)

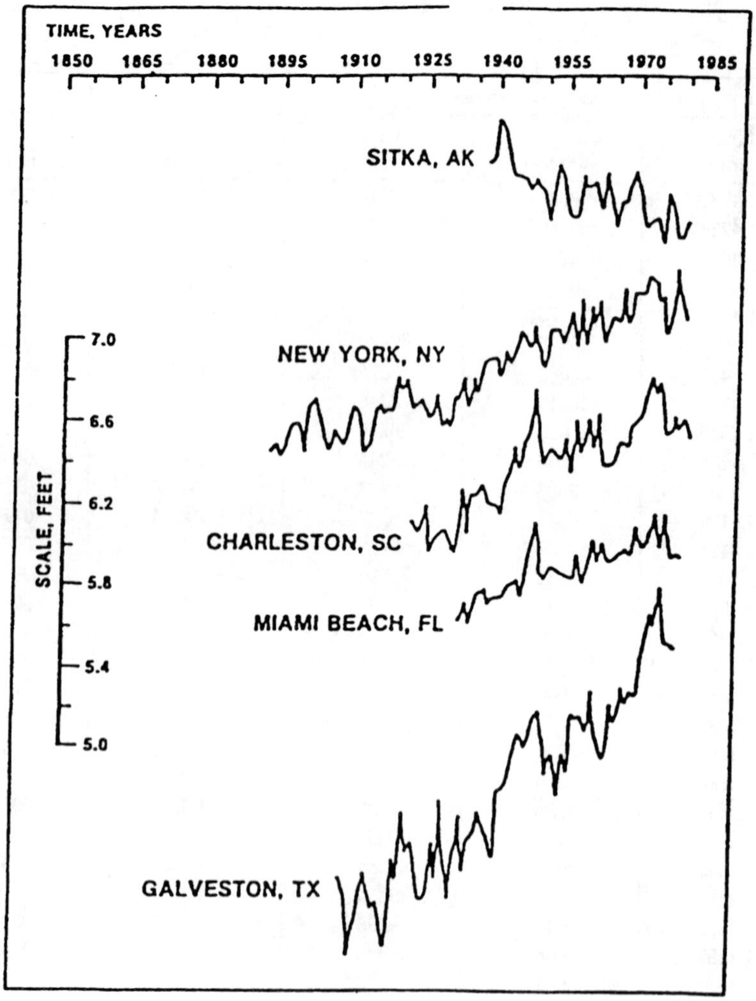

Figure 10.45.—Typical Sea Level Records Showing Rising Trend (Lyles et al., 1987).

B. Lakes and Inland Waters

Water level changes in lakes and inland waters are due primarily to meteorological events. Astronomical tides are generally negligible on inland waters. On the Great Lakes, astronomical tides are less than 0.05 meter in amplitude. In smaller bodies of water they are much smaller. Storm surges due to winds blowing along the axis of a lake are by far the most important short term water level changes in inland waters. Wind-induced storm surges in lakes, unlike storm surges in the ocean, also result in a set down at the windward end of the lake. This setdown can be important from the standpoint of navigation.

In addition, hydrologic factors such as seasonal and long-term variations in precipitation, runoff, and evaporation can lead to seasonal and long-term changes in water levels in inland water bodies. For example, the Great Lakes experience seasonal water level variations with higher lake levels in late spring and early summer. In addition, the Great Lakes experience persistent periods of high levels lasting for several years followed by periods of normal or low levels. Statistics of Great Lakes water levels are compiled by the U.S. Army Corps of Engineers, Detroit District (USACE, 1987).

C. Statistics

Wave statistics and water level statistics are described elsewhere in this chapter. Generally, at any instant in time, the distribution of wave heights can be described by the Rayleigh probability distribution. The Rayleigh distribution is defined by a single parameter, usually the root mean squared wave height or the significant wave height (the average height of the highest 1/3 of the waves). Wave climatologies are constructed by compiling the statistics of the characteristic wave height used to define the Rayleigh probability distribution.

Wave climatologies for some areas often involve two distinct populations corresponding with two different wave generating mechanisms. For example, along the East Coast of the United States waves might be generated by usual wind conditions and also by the unusual winds associated with hurricanes. These two populations are often handled separately in defining wave conditions for the site. Similarly, some wave hindcasts for the Pacific Coast have considered the large waves generated by storms to be from a different population than the usual day-to-day wave conditions. Water level statistics are usually analysed using standard frequency analysis techniques. There is no theoretical preference for any particular probability distribution. Selection of a distribution is based on goodness of fit to the observed data. The data used in the analysis are the storm surge levels, observed water levels with the predicted astronomical tide level removed (Ebersole, 1982). Fig. 10.46 shows the yearly storm surge water level statistics for Atlantic City, New Jersey plotted on Gumbel paper.

D. Flood Insurance Considerations

The 100-year water levels in coastal areas are determined by the Federal Emergency Management Agency's Flood Insurance Administration. These levels are usually established by calculating coastal storm surge levels resulting from storm conditions with a return period of 100 years. Flood Insurance Rate Maps (FIRMs) are prepared for coastal communities showing the boundaries of the 100-year flood

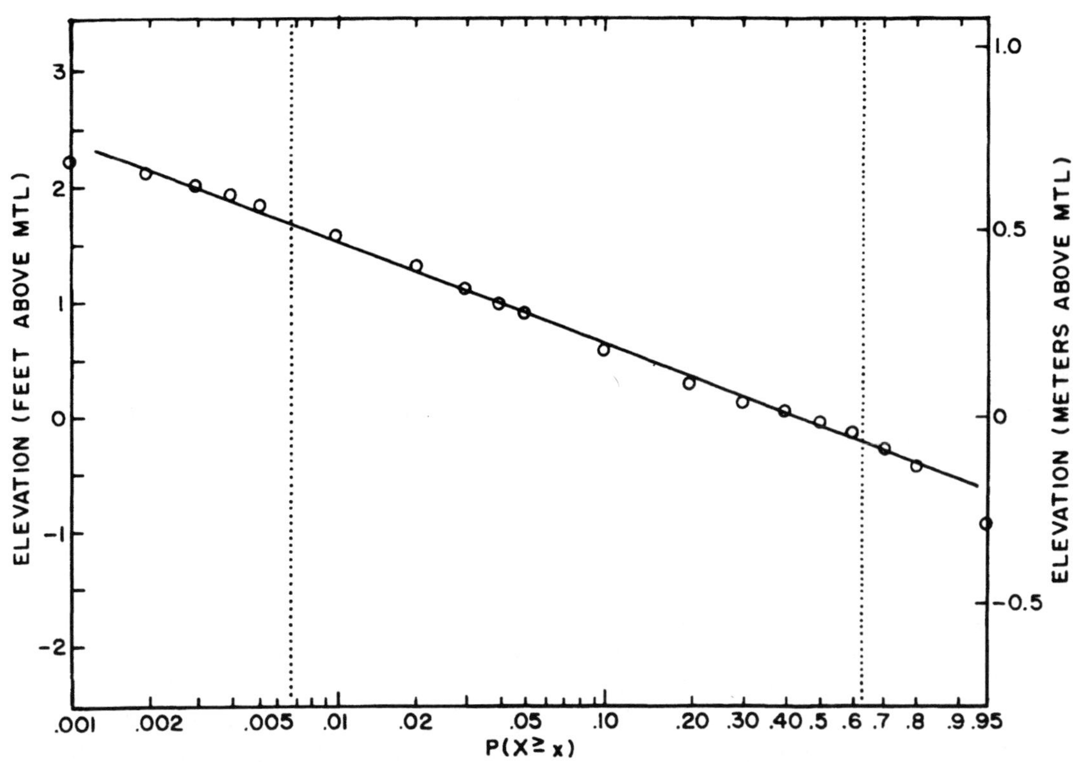

Figure 10.46.—Yearly Storm Surge Water Level Statistics at Atlantic City, NJ (Ebersole, 1982).

plain. In coastal areas subjected to waves, the 100-year water level is supplemented by considering the maximum wave height that the given water level will support. Generally, the crest of the maximum wave is 1.6 times the calculated storm surge depth since in the shallow water near shore the maximum wave height is 0.78 times the water depth while approximately 78% of the wave profile is above the still water level.

XI. SELECTION OF DESIGN WAVES AND WATER LEVELS

This section deals with the selection of waves and water levels for the design of water resources/coastal projects. Like all projects designed to function within and to withstand the natural environment, there will always be a small probability that the conditions for which the project is designed will be exceeded sometime during its lifetime. The usual response to this situation is to adopt higher and higher levels of design in order to minimize the risk of damage and failure. However, as higher and higher levels of design are adopted, the cost of a project increases until its implementation eventually becomes prohibitive. Consequently, there is usually some intermediate level of design which maximizes the economic net benefits of a project. Thus, the level of design adopted for a project is basically an economic decision, with overriding considerations imposed by the public's health, welfare, and safety. The U.S. Army Corps of Engineers (USACE, 1989) discusses the selection of wave and water level conditions for design.

A. Design Philosophy and Design Criteria

There are several factors that influence the level of design and thus the wave heights and water levels used in a design. They include the expected lifetime of the project, the type of facilities/structures involved in the project and their response to conditions that exceed design conditions, and, most importantly, the consequences of the failure of various elements of a project. For example, some structures such as rip-rap revetments or rubble mound groins and breakwaters are able to withstand wave conditions that exceed design conditions with little damage, while other types of structures might experience significant damage or even collapse if design conditions are exceeded. The former type of structure is termed a "flexible" structure while the latter type is termed a "rigid" structure.

The process of failure is also a factor. Some failures are progressive so that the structure continues to function even though damaged, e.g., rubble mound breakwaters. Other failures are catastrophic with complete loss of function. The consequences of a project's failure or the failure of an element of the project is an important consideration in selecting the level of design. The potential for loss of life in the event of failure dictates a higher level of design with a correspondingly small risk of failure although the risk of damage and/or failure can never be reduced to zero. There are also social and political considerations that may override economic considerations. For example, the perception of what constitutes failure by the public may be more important than what actually constitutes a loss of function. This is often a problem in the design of beach nourishment projects where economically successful projects are perceived by the public as failures because nourishment sand may be lost over a relatively short period of time.

Because the physical environment to which a project or project element will be subjected cannot be predicted with certainty, a statistical description of the specific environment is necessary. The selection of a design wave and/or design water level is therefore the process of selecting a frequency of occurrence or return period of design conditions. The process is further complicated in the case of waves. At any given instant in time, the wave conditions prevailing at a site are characterized by a distribution of heights and periods and not by the single height and period which is required in many design procedures. The height distribution is thus often described by a single, characteristic wave. This wave is often the "significant wave" which is defined as the average height of the highest 1/3 of the waves in the distribution. Fortunately, wave heights are almost always described by a Rayleigh distribution which is specified by a single parameter, a characteristic wave height. The distribution is usually specified in terms of the root–mean–squared wave height (Equation 10.23) although the significant wave height can also be used. The relationship of significant height and other commonly used wave height parameters to H_{rms} is given in Table 10.3.

Further, the wave environment at a site can be described by the variation of the single characteristic wave height (either the significant height or rms height) that defines the height distribution with time. This defines the wave climatology for a site. For a wave climatology, the significant height or rms height is given at specified intervals in time such as daily values or values every 3 hours. Often these values are summarized to provide a climatology in which the number of hours per given time period (usually a year) that the characteristic wave heights are within a given interval is tabulated, e.g., the number of hours per year that the significant wave height is between 1 and 2 meters.

Wave climatologies for sites can be developed based on wave measurements; however, wave measurements are expensive to obtain and rarely are records sufficiently long enough to obtain a reliable climatology. No more than a few years of record are usually available. An alternative method of developing the wave climatology at a site is "hindcasting" wave conditions using historical meteorological data. Historical weather maps showing barometric pressures and wind fields are used to create the wind field and a numerical wave generation model is used to determine resulting sea conditions described earlier in the Wind Waves section of this chapter. The waves thus generated are propagated from each generating area (area where the winds are active in generating waves), and their energies are summed to determine the wave conditions at a given site at the time of the historical weather conditions. In addition to the generating mechanism, the processes of wave refraction and shoaling are considered in transforming the wave conditions from offshore sites to nearshore coastal sites. The Corps of Engineers has developed wave climatologies for the East and West Coasts of the United States as well as for the Great Lakes based on twenty years of hindcasts (Jensen, 1983; Hubertz et al., 1993). Typical output for one coastal site, the WIS station #68 (39.25° N, 74.25° W) at 18 meters water depth near Atlantic City, New Jersey is given in Fig. 10.47.

The statistical analysis of climatological wave data is similar to flood frequency analyses. Analysis of annual maximum significant wave heights from twenty years of hindcast data for the East Coast of the United States (Jensen, 1983) suggests that they fit a Gumbel distribution (Fig. 10.48).

Fig. 10.48 shows annual maxima at Atlantic City, New Jersey plotted on Gumbel paper—graph paper with its probability scale distorted in such a way that data following a Gumbel distribution will lie on a straight line. The Gumbel (Extremal Type I) distribution is given by

$$P(h \leq H) = e^{-e^{-\left(\frac{a+x}{c}\right)}} \tag{10.111}$$

and the exceedance probability by

$$P(h \geq H) = 1 - e^{-e^{-\left(\frac{a+x}{c}\right)}}. \tag{10.112}$$

The Weibull plotting position formula is used to assign probabilities to the annual maximum significant wave heights. The Weibull formula is given by

$$P(H \geq h) = \frac{m}{N+1} \tag{10.113}$$

in which $P(H \geq h)$ is the probability that the maximum annual significant wave height, H, will equal or exceed a given value, h, m is the rank of the given value of significant wave height with $m = 1$ for the largest, $m = 2$ for the second largest, etc., and N is the total number of years of record. The constants a and c in Equation 10.112 are given by

WIS ATLANTIC REVISION 1956 - 1975
LAT: 39.25 N, LONG: 74.25 W, DEPTH: 18 M
SUMMARY OF WAVE INFORMATION BY MONTH

STATION: 68

OCCURRENCES OF WAVE HEIGHT BY MONTH FOR ALL YEARS

Hmo(m)	JAN	FEB	MAR	APR	MAY	JUN	JUL	AUG	SEP	OCT	NOV	DEC	TOTAL
0.00 - 0.49	348	369	357	449	501	432	684	752	486	406	430	377	5591
0.50 - 0.99	1591	1335	1661	1925	2578	3161	3374	3377	2815	2220	1629	1511	27177
1.00 - 1.49	1352	1214	1508	1341	1245	941	711	684	913	1345	1268	1259	13781
1.50 - 1.99	805	791	772	627	464	202	161	116	324	534	798	868	6462
2.00 - 2.49	401	429	372	280	127	36	15	18	141	262	388	508	2977
2.50 - 2.99	241	231	153	96	40	23	4	13	71	111	158	225	1366
3.00 - 3.49	101	75	63	61	5	2	2	.	31	43	67	142	592
3.50 - 3.99	52	39	30	10	.	1	4	.	12	18	42	46	254
4.00 - 4.49	46	19	19	9	.	1	3	.	5	16	14	11	143
4.50 - 4.99	13	13	10	2	.	1	2	.	2	5	5	7	60
5.00 - 5.49	8	5	6	.	.	.	.	.	.	.	1	4	24
5.50 - 5.99	2	.	2	.	.	.	.	.	.	.	.	2	6
6.00 - 6.49	.	.	2	.	.	.	.	.	.	.	.	.	2
6.50 - 6.99	.	.	5	.	.	.	.	.	.	.	.	.	5
7.00 - 7.49	.	.	.	.	.	.	.	.	.	.	.	.	0
7.50 - 7.99	.	.	.	.	.	.	.	.	.	.	.	.	0
8.00 - 8.49	.	.	.	.	.	.	.	.	.	.	.	.	0
8.50 - 8.99	.	.	.	.	.	.	.	.	.	.	.	.	0
9.00 - 9.49	.	.	.	.	.	.	.	.	.	.	.	.	0
9.50 - 9.99	.	.	.	.	.	.	.	.	.	.	.	.	0
10.00 - GREATER	.	.	.	.	.	.	.	.	.	.	.	.	0
TOTAL	4960	4520	4960	4800	4960	4800	4960	4960	4800	4960	4800	4960	58440

STATION: 68

OCCURRENCES OF PEAK PERIOD BY MONTH FOR ALL YEARS

Tp(sec)	JAN	FEB	MAR	APR	MAY	JUN	JUL	AUG	SEP	OCT	NOV	DEC	TOTAL
3.0 - 3.9	375	407	420	494	385	350	450	461	320	366	382	323	4733
4.0 - 4.9	1051	883	1032	900	719	636	627	662	690	931	927	859	9917
5.0 - 5.9	846	807	850	624	722	787	603	677	527	674	821	847	8785
6.0 - 6.9	417	436	474	510	732	941	727	842	547	462	480	424	6992
7.0 - 7.9	334	343	428	463	667	981	1171	1069	883	682	433	376	7830
8.0 - 8.9	378	377	459	490	772	676	773	663	780	765	419	350	6902
9.0 - 9.9	548	387	477	539	644	282	416	358	518	509	457	508	5643
10.0 - 10.9	379	316	352	435	207	92	138	111	294	285	372	610	3591
11.0 - 11.9	357	255	244	201	72	41	31	65	168	143	260	363	2200
12.0 - 12.9	160	185	128	93	24	13	6	32	60	86	155	195	1137
13.0 - 13.9	66	101	60	31	9	1	7	15	10	46	52	65	463
14.0 - 14.9	30	20	18	16	7	.	10	2	3	11	15	26	158
15.0 - 15.9	18	3	10	4	.	.	1	1	.	.	17	12	66
16.0 - 16.9	.	.	6	.	.	.	.	2	.	.	9	2	19
17.0 - 17.9	1	.	2	.	.	.	.	.	.	.	1	.	4
18.0 - 18.9	.	.	.	.	.	.	.	.	.	.	.	.	0
19.0 - 19.9	.	.	.	.	.	.	.	.	.	.	.	.	0
20.0 - LONGER	.	.	.	.	.	.	.	.	.	.	.	.	0
TOTAL	4960	4520	4960	4800	4960	4800	4960	4960	4800	4960	4800	4960	58440

STATION: 68

OCCURRENCES OF PEAK DIRECTION BY MONTH FOR ALL YEARS

Dp(deg) DIRECTION BAND & CENTER	JAN	FEB	MAR	APR	MAY	JUN	JUL	AUG	SEP	OCT	NOV	DEC	TOTAL
348.75 - 11.24 (0.0)	132	109	124	117	62	41	52	76	105	91	115	90	1114
11.25 - 33.74 (22.5)	63	75	66	50	39	32	16	66	73	83	53	53	669
33.75 - 56.24 (45.0)	127	201	102	69	86	66	31	117	173	191	124	99	1386
56.25 - 78.74 (67.5)	114	164	133	81	92	77	34	83	162	211	159	101	1411
78.75 - 101.24 (90.0)	473	507	524	406	332	194	172	239	626	630	682	460	5245
101.25 - 123.74 (112.5)	603	418	452	562	853	487	667	770	1361	1052	664	865	8754
123.75 - 146.24 (135.0)	566	403	502	474	884	1171	1198	1168	968	750	490	454	9028
146.25 - 168.74 (157.5)	370	644	790	726	774	726	636	719	321	417	381	570	7074
168.75 - 191.24 (180.0)	612	447	620	890	768	976	978	749	271	292	484	586	7673
191.25 - 213.74 (202.5)	175	159	215	268	283	288	304	222	149	135	268	180	2646
213.75 - 236.24 (225.0)	130	100	121	184	226	241	269	248	146	135	186	174	2160
236.25 - 258.74 (247.5)	113	79	98	100	133	148	199	151	76	121	93	105	1416
258.75 - 281.24 (270.0)	262	220	187	274	201	201	217	153	107	251	255	227	2555
281.25 - 303.74 (292.5)	352	398	354	238	90	92	90	89	92	217	344	324	2680
303.75 - 326.24 (315.0)	467	372	413	206	63	37	57	58	74	229	304	399	2679
326.25 - 348.74 (337.5)	401	224	259	155	74	23	40	52	96	155	198	273	1950
TOTAL	4960	4520	4960	4800	4960	4800	4960	4960	4800	4960	4800	4960	58440

Figure 10.47.—Typical Wave Hindcast Data for Atlantic City, NJ (Hubertz et al., 1993).

WIS ATLANTIC REVISION 1956 - 1975
LAT: 39.25 N, LONG: 74.25 W, DEPTH: 18 M
OCCURRENCES OF WAVE HEIGHT AND PEAK PERIOD FOR 45-DEG DIRECTION BANDS

STATION: 68 (337.50 - 22.49) 0.0 DEG

Hmo(m)	Tp(sec) 3.0-4.9	5.0-6.9	7.0-8.9	9.0-10.9	11.0-12.9	13.0-14.9	15.0-16.9	17.0-18.9	19.0-20.9	21.0-LONGER	TOTAL
0.00 - 0.99	1327	.	.	.	.	.	.	.	.	.	1327
1.00 - 1.99	587	195	.	.	.	.	.	.	.	.	782
2.00 - 2.99	.	50	.	.	.	.	.	.	.	.	50
3.00 - 3.99	.	1	.	.	.	.	.	.	.	.	1
4.00 - 4.99	.	.	.	.	.	.	.	.	.	.	0
5.00 - 5.99	.	.	.	.	.	.	.	.	.	.	0
6.00 - 6.99	.	.	.	.	.	.	.	.	.	.	0
7.00 - 7.99	.	.	.	.	.	.	.	.	.	.	0
8.00 - 8.99	.	.	.	.	.	.	.	.	.	.	0
9.00 - GREATER	.	.	.	.	.	.	.	.	.	.	0
TOTAL	1914	246	0	0	0	0	0	0	0	0	2160

STATION: 68 (22.50 - 67.49) 45.0 DEG

Hmo(m)	Tp(sec) 3.0-4.9	5.0-6.9	7.0-8.9	9.0-10.9	11.0-12.9	13.0-14.9	15.0-16.9	17.0-18.9	19.0-20.9	21.0-LONGER	TOTAL
0.00 - 0.99	603	338	3	.	.	.	.	.	.	.	944
1.00 - 1.99	127	767	160	.	.	.	.	.	.	.	1054
2.00 - 2.99	.	74	244	10	.	.	.	.	.	.	328
3.00 - 3.99	.	.	67	17	.	.	.	.	.	.	84
4.00 - 4.99	.	.	4	39	.	.	.	.	.	.	43
5.00 - 5.99	.	.	.	4	1	.	.	.	.	.	5
6.00 - 6.99	.	.	.	.	1	.	.	.	.	.	1
7.00 - 7.99	.	.	.	.	.	.	.	.	.	.	0
8.00 - 8.99	.	.	.	.	.	.	.	.	.	.	0
9.00 - GREATER	.	.	.	.	.	.	.	.	.	.	0
TOTAL	730	1179	478	70	2	0	0	0	0	0	2459

STATION: 68 (67.50 - 112.49) 90.0 DEG

Hmo(m)	Tp(sec) 3.0-4.9	5.0-6.9	7.0-8.9	9.0-10.9	11.0-12.9	13.0-14.9	15.0-16.9	17.0-18.9	19.0-20.9	21.0-LONGER	TOTAL
0.00 - 0.99	656	728	1075	1564	987	176	52	3	.	.	5241
1.00 - 1.99	73	1074	617	646	482	149	9	.	.	.	3050
2.00 - 2.99	.	77	357	204	109	18	11	.	.	.	776
3.00 - 3.99	.	.	80	109	45	18	2	.	.	.	254
4.00 - 4.99	.	.	.	58	34	8	.	.	.	.	100
5.00 - 5.99	.	.	.	2	7	4	.	.	.	.	13
6.00 - 6.99	.	.	.	.	.	6	.	.	.	.	6
7.00 - 7.99	.	.	.	.	.	.	.	.	.	.	0
8.00 - 8.99	.	.	.	.	.	.	.	.	.	.	0
9.00 - GREATER	.	.	.	.	.	.	.	.	.	.	0
TOTAL	729	1879	2129	2583	1664	379	74	3	0	0	9440

STATION: 68 (112.50 - 157.49) 135.0 DEG

Hmo(m)	Tp(sec) 3.0-4.9	5.0-6.9	7.0-8.9	9.0-10.9	11.0-12.9	13.0-14.9	15.0-16.9	17.0-18.9	19.0-20.9	21.0-LONGER	TOTAL
0.00 - 0.99	791	1482	5703	3044	669	64	5	.	.	.	11758
1.00 - 1.99	113	1234	1244	1054	436	102	5	.	.	.	4188
2.00 - 2.99	.	61	381	401	151	41	1	1	.	.	1037
3.00 - 3.99	.	.	28	148	63	15	.	.	.	.	254
4.00 - 4.99	.	.	.	10	40	8	.	.	.	.	58
5.00 - 5.99	.	.	.	.	9	3	.	.	.	.	12
6.00 - 6.99	.	.	.	.	.	.	.	.	.	.	0
7.00 - 7.99	.	.	.	.	.	.	.	.	.	.	0
8.00 - 8.99	.	.	.	.	.	.	.	.	.	.	0
9.00 - GREATER	.	.	.	.	.	.	.	.	.	.	0
TOTAL	904	2777	7356	4657	1368	233	11	1	0	0	17307

STATION: 68 (157.50 - 202.49) 180.0 DEG

Hmo(m)	Tp(sec) 3.0-4.9	5.0-6.9	7.0-8.9	9.0-10.9	11.0-12.9	13.0-14.9	15.0-16.9	17.0-18.9	19.0-20.9	21.0-LONGER	TOTAL
0.00 - 0.99	1335	2712	2285	644	78	3	.	.	.	.	7057
1.00 - 1.99	167	2137	1447	876	142	4	.	.	.	.	4773
2.00 - 2.99	.	133	651	324	71	2	.	.	.	.	1181
3.00 - 3.99	.	.	85	78	12	.	.	.	.	.	175
4.00 - 4.99	.	.	.	2	.	.	.	.	.	.	2
5.00 - 5.99	.	.	.	.	.	.	.	.	.	.	0
6.00 - 6.99	.	.	.	.	.	.	.	.	.	.	0
7.00 - 7.99	.	.	.	.	.	.	.	.	.	.	0
8.00 - 8.99	.	.	.	.	.	.	.	.	.	.	0
9.00 - GREATER	.	.	.	.	.	.	.	.	.	.	0
TOTAL	1502	4982	4468	1924	303	9	0	0	0	0	13188

Figure 10.47.—(Continued)

WIS ATLANTIC REVISION 1956 - 1975
LAT: 39.25 N, LONG: 74.25 W, DEPTH: 18 M
OCCURRENCES OF WAVE HEIGHT AND PEAK PERIOD FOR 45-DEG DIRECTION BANDS

STATION: 68 (202.50 - 247.49) 225.0 DEG

Hmo(m)	Tp(sec) 3.0-4.9	5.0-6.9	7.0-8.9	9.0-10.9	11.0-12.9	13.0-14.9	15.0-16.9	17.0-18.9	19.0-20.9	21.0-LONGER	TOTAL
0.00 - 0.99	1752	292	.	.	.	.	.	.	.	.	2044
1.00 - 1.99	150	1639	25	.	.	.	.	.	.	.	1814
2.00 - 2.99	.	87	233	.	.	.	.	.	.	.	320
3.00 - 3.99	.	.	35	.	.	.	.	.	.	.	35
4.00 - 4.99	.	.	.	.	.	.	.	.	.	.	0
5.00 - 5.99	.	.	.	.	.	.	.	.	.	.	0
6.00 - 6.99	.	.	.	.	.	.	.	.	.	.	0
7.00 - 7.99	.	.	.	.	.	.	.	.	.	.	0
8.00 - 8.99	.	.	.	.	.	.	.	.	.	.	0
9.00 - GREATER	.	.	.	.	.	.	.	.	.	.	0
TOTAL	1902	2018	293	0	0	0	0	0	0	0	4213

STATION: 68 (247.50 - 292.49) 270.0 DEG

Hmo(m)	Tp(sec) 3.0-4.9	5.0-6.9	7.0-8.9	9.0-10.9	11.0-12.9	13.0-14.9	15.0-16.9	17.0-18.9	19.0-20.9	21.0-LONGER	TOTAL
0.00 - 0.99	2869	.	.	.	.	.	.	.	.	.	2869
1.00 - 1.99	1059	345	.	.	.	.	.	.	.	.	1404
2.00 - 2.99	.	118	.	.	.	.	.	.	.	.	118
3.00 - 3.99	.	3	.	.	.	.	.	.	.	.	3
4.00 - 4.99	.	.	.	.	.	.	.	.	.	.	0
5.00 - 5.99	.	.	.	.	.	.	.	.	.	.	0
6.00 - 6.99	.	.	.	.	.	.	.	.	.	.	0
7.00 - 7.99	.	.	.	.	.	.	.	.	.	.	0
8.00 - 8.99	.	.	.	.	.	.	.	.	.	.	0
9.00 - GREATER	.	.	.	.	.	.	.	.	.	.	0
TOTAL	3928	466	0	0	0	0	0	0	0	0	4394

STATION: 68 (292.50 - 337.49) 315.0 DEG

Hmo(m)	Tp(sec) 3.0-4.9	5.0-6.9	7.0-8.9	9.0-10.9	11.0-12.9	13.0-14.9	15.0-16.9	17.0-18.9	19.0-20.9	21.0-LONGER	TOTAL
0.00 - 0.99	1528	.	.	.	.	.	.	.	.	.	1528
1.00 - 1.99	1513	1665	.	.	.	.	.	.	.	.	3178
2.00 - 2.99	.	533	.	.	.	.	.	.	.	.	533
3.00 - 3.99	.	32	8	.	.	.	.	.	.	.	40
4.00 - 4.99	.	.	.	.	.	.	.	.	.	.	0
5.00 - 5.99	.	.	.	.	.	.	.	.	.	.	0
6.00 - 6.99	.	.	.	.	.	.	.	.	.	.	0
7.00 - 7.99	.	.	.	.	.	.	.	.	.	.	0
8.00 - 8.99	.	.	.	.	.	.	.	.	.	.	0
9.00 - GREATER	.	.	.	.	.	.	.	.	.	.	0
TOTAL	3041	2230	8	0	0	0	0	0	0	0	5279

STATION: 68 ALL DIRECTIONS

Hmo(m)	Tp(sec) 3.0-4.9	5.0-6.9	7.0-8.9	9.0-10.9	11.0-12.9	13.0-14.9	15.0-16.9	17.0-18.9	19.0-20.9	21.0-LONGER	TOTAL
0.00 - 0.99	10861	5552	9066	5252	1734	243	57	3	.	.	32768
1.00 - 1.99	3789	9056	3493	2576	1060	255	14	.	.	.	20243
2.00 - 2.99	.	1133	1866	939	331	61	12	1	.	.	4343
3.00 - 3.99	.	36	303	352	120	33	2	.	.	.	846
4.00 - 4.99	.	.	4	109	74	16	.	.	.	.	203
5.00 - 5.99	.	.	.	6	17	7	.	.	.	.	30
6.00 - 6.99	.	.	.	.	1	6	.	.	.	.	7
7.00 - 7.99	.	.	.	.	.	.	.	.	.	.	0
8.00 - 8.99	.	.	.	.	.	.	.	.	.	.	0
9.00 - GREATER	.	.	.	.	.	.	.	.	.	.	0
TOTAL	14650	15777	14732	9234	3337	621	85	4	0	0	58440

STATION: 68

OCCURRENCES OF WIND SPEED BY MONTH FOR ALL YEARS

WS(m/sec)	JAN	FEB	MAR	APR	MAY	JUN	JUL	AUG	SEP	OCT	NOV	DEC	TOTAL
0.00 - 2.49	108	93	108	158	252	264	327	331	236	188	167	112	2344
2.50 - 4.99	1083	1057	1225	1574	2259	2667	2986	3038	2413	1791	1237	1235	22565
5.00 - 7.49	1139	945	1237	1348	1548	1379	1208	1205	1262	1387	1177	1116	14951
7.50 - 9.99	1191	1179	1346	1184	795	453	419	360	684	1047	1146	1122	10926
10.00 - 12.49	609	588	565	330	85	32	10	19	156	358	604	607	3963
12.50 - 14.99	528	485	375	147	20	2	8	7	43	158	346	495	2614
15.00 - 17.49	127	93	51	24	1	2	2	.	6	24	69	145	544
17.50 - 19.99	139	62	32	30	.	1	.	.	.	7	48	95	414
20.00 - GREATER	36	18	21	5	.	.	.	.	.	.	6	33	119
TOTAL	4960	4520	4960	4800	4960	4800	4960	4960	4800	4960	4800	4960	58440

Figure 10.47.—(Continued)

WIS ATLANTIC REVISION 1956 - 1975
LAT: 39.25 N, LONG: 74.25 W, DEPTH: 18 M

STATION: 68

OCCURRENCES OF WIND DIRECTION BY MONTH FOR ALL YEARS

WD(deg) DIRECTION BAND & CENTER	JAN	FEB	MAR	APR	MAY	JUN	JUL	AUG	SEP	OCT	NOV	DEC	TOTAL
337.50 - 22.49 (0.0)	620	571	592	503	455	340	360	527	644	600	520	559	6291
22.50 - 67.49 (45.0)	318	426	361	248	418	299	214	394	640	569	350	353	4590
67.50 - 112.49 (90.0)	206	335	390	261	306	290	161	212	405	517	277	317	3677
112.50 - 157.49 (135.0)	284	327	441	365	454	356	338	348	415	426	350	361	4465
157.50 - 202.49 (180.0)	412	326	458	669	673	774	770	628	504	396	473	478	6561
202.50 - 247.49 (225.0)	568	466	563	816	1106	1334	1522	1282	806	595	736	606	10400
247.50 - 292.49 (270.0)	1033	925	926	1048	956	992	1169	988	722	968	960	1028	11715
292.50 - 337.49 (315.0)	1519	1144	1229	890	592	415	426	581	664	889	1134	1258	10741
TOTAL	4960	4520	4960	4800	4960	4800	4960	4960	4800	4960	4800	4960	58440

STATION: 68

SUMMARY OF MEAN Hmo(m) BY MONTH AND YEAR

YEAR	JAN	FEB	MAR	APR	MAY	JUN	JUL	AUG	SEP	OCT	NOV	DEC	MEAN
1956	1.75	1.07	1.12	1.12	0.80	0.88	0.81	0.78	0.96	1.35	1.15	1.03	1.07
1957	1.02	1.06	1.00	1.13	0.75	0.89	0.69	0.71	0.81	1.11	1.25	1.47	0.99
1958	1.40	1.41	1.35	1.42	1.03	0.91	0.74	0.84	0.75	1.44	1.11	1.21	1.13
1959	1.44	1.14	1.20	1.07	0.68	0.70	0.86	0.77	0.68	1.09	1.29	1.13	1.00
1960	1.13	1.65	1.30	0.99	0.95	0.89	0.74	0.70	0.90	0.98	0.95	1.18	1.03
1961	1.23	1.22	1.06	1.09	1.07	0.92	0.65	0.73	1.07	1.17	1.13	1.12	1.04
1962	1.30	1.10	1.58	1.16	0.75	0.73	0.77	0.86	0.86	0.94	1.75	1.60	1.12
1963	1.15	1.31	1.20	0.95	0.96	0.77	0.68	0.68	1.01	1.03	1.35	1.12	1.02
1964	1.70	1.42	1.32	1.04	0.99	0.74	0.88	0.75	1.14	0.93	1.00	1.38	1.11
1965	1.46	1.29	0.96	0.92	0.61	0.83	0.72	0.76	0.84	1.14	1.13	0.98	0.97
1966	1.55	1.07	1.00	0.87	0.91	0.94	0.73	0.71	1.04	0.91	1.29	1.36	1.03
1967	1.18	1.45	1.11	1.10	1.03	0.72	0.75	0.75	1.23	0.95	1.10	1.40	1.06
1968	1.36	1.28	1.30	0.97	0.90	0.71	0.51	0.51	0.64	0.99	1.28	1.44	0.99
1969	1.50	1.79	1.45	1.10	1.14	0.75	0.73	0.81	0.88	1.10	1.39	1.64	1.19
1970	1.17	1.51	1.06	1.31	0.93	0.83	0.79	0.79	0.77	1.28	1.27	1.37	1.09
1971	1.27	1.36	1.33	1.07	1.02	0.66	0.75	0.81	0.85	1.23	1.08	1.26	1.06
1972	1.16	1.29	1.25	0.94	1.34	1.03	0.66	0.67	1.03	1.04	1.14	1.31	1.07
1973	1.37	1.37	1.32	1.19	1.03	0.84	0.81	0.64	0.80	0.97	1.29	1.63	1.10
1974	0.92	1.29	1.21	1.19	0.88	0.83	0.57	0.61	0.80	0.83	1.16	1.46	0.98
1975	1.31	1.12	1.46	1.20	0.74	0.77	0.91	0.49	0.95	1.06	1.18	1.46	1.05
MEAN	1.32	1.31	1.23	1.09	0.92	0.82	0.74	0.72	0.90	1.08	1.22	1.33	

STATION: 68

MAX Hmo(m)*10 WITH ASSOCIATED Tp(sec) AND Dp(deg/10) BY MONTH AND YEAR

YEAR	JAN	FEB	MAR	APR	MAY	JUN	JUL	AUG	SEP	OCT	NOV	DEC	MAX
1956	5210 7	38 821	31 813	41 9 7	21 622	18 715	21 718	17 7 8	4711 9	32 8 7	29 721	30 631	5210 7
1957	29 7 7	34 8 7	37 9 7	30 916	18 621	241113	18 623	15 514	18 720	4610 9	34 8 6	37 8 3	4610 9
1958	31 811	30 917	4511 9	38 9 9	22 916	181112	17 623	20 619	221017	4510 9	28 720	451112	451112
1959	361016	27 721	41 917	28 8 9	14 518	16 530	21 716	18 620	16 714	301013	31 817	32 819	41 917
1960	441010	451110	35 8 8	25 721	20 814	16 8 8	29 722	14 6 6	271016	41 9 5	25 720	4510 9	4510 9
1961	4310 8	511213	26 812	35 811	311114	18 623	14 623	14 618	401010	4610 5	25 917	35 819	511213
1962	38 9 8	25 8 5	6914 9	29 721	17 618	15 621	18 621	20 917	36 9 8	25 722	44 9 7	4610 8	6914 9
1963	31 630	31 916	29 918	22 816	23 916	26 915	24 718	26 722	26 8 6	27 8 7	381016	28 722	381016
1964	561310	4510 8	36 918	32 819	20 622	15 621	22 7 4	23 716	351211	26 915	34 818	33 917	561310
1965	4911 9	421015	25 719	22 718	12 521	29 8 7	17 619	17 619	19 717	29 714	29 722	26 818	4911 9
1966	4810 9	301014	27 917	24 710	24 722	28 913	17 6 5	18 619	31 910	23 817	381113	351010	4810 9
1967	28 818	4610 7	26 9 8	33 634	29 8 4	14 519	14 522	15 619	30 914	33 819	21 530	401114	4610 7
1968	461110	5110 5	38 912	32 819	271012	13 512	9 517	13 5 3	18 812	24 718	38 9 9	391117	5110 5
1969	451312	45 9 3	4911 9	28 816	30 918	15 618	17 616	20 721	191210	27 724	481214	361017	4911 9
1970	34 819	511415	28 717	401017	18 916	16 621	19 618	29 8 5	27 721	34 9 7	33 8 8	37 912	511415
1971	32 818	341016	47 9 3	4710 9	27 818	13 814	17 814	29 916	231013	28 810	37 913	27 719	4710 9
1972	29 818	30 811	29 819	28 7 3	30 8 7	301016	17 619	20 619	35 8 3	24 811	28 8 9	26 914	35 8 3
1973	32 819	43 9 4	37 821	29 7 4	29 816	18 624	19 620	16 717	24 814	31 912	35 819	351015	43 9 4
1974	27 719	33 819	3310 9	35 819	22 614	19 7 8	15 624	17 620	21 718	22 719	3010 8	581212	581212
1975	42 9 7	38 822	531215	4510 9	15 621	4510 9	46 9 6	12 430	28 721	26 7 8	511111	4910 4	531215
MAX	561310	511415	6914 9	4710 9	311114	4510 9	46 9 6	29 916	4711 9	4610 5	511111	581212	

MAX Hmo(m): 6.9 MAX Tp(sec): 14. MAX Dp(deg): 86. DATE(gmt): 62030712

MAX WIND SPEED(m/sec): 27. MAX WIND DIRECTION(deg): 20. DATE(gmt): 62030712

Figure 10.47.—(Continued)

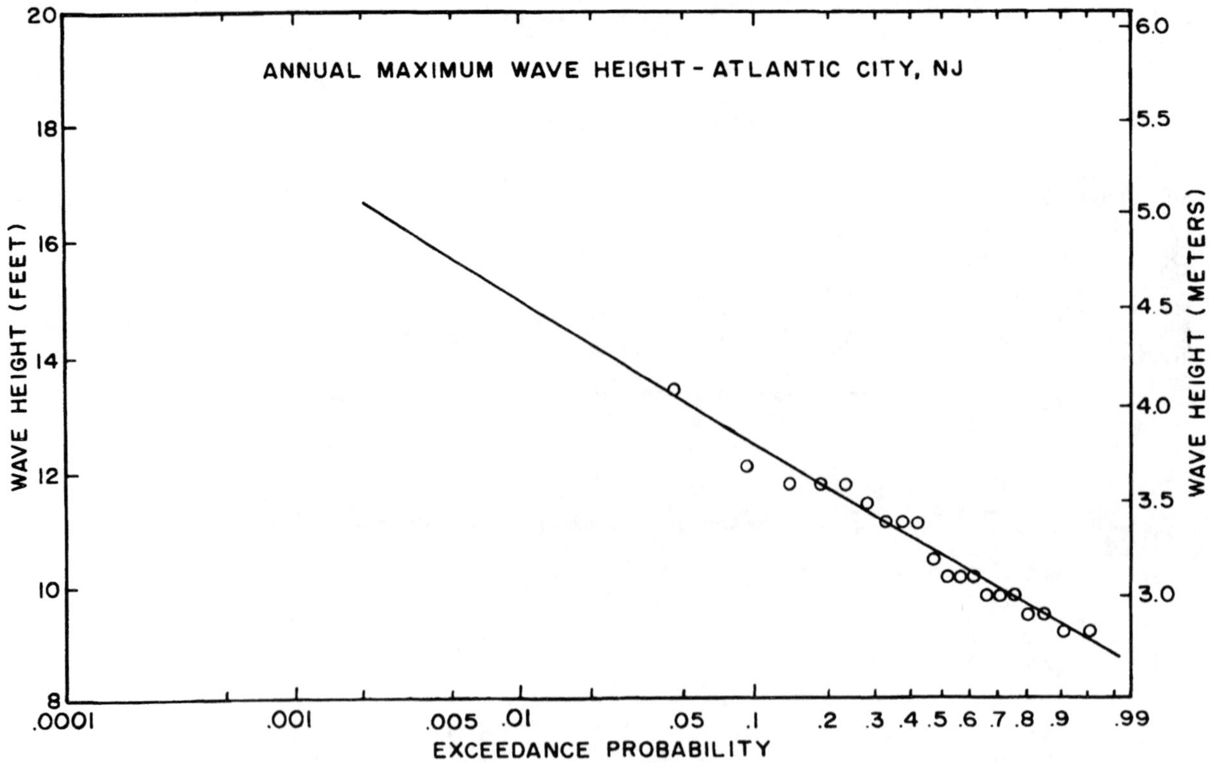

Figure 10.48.—Annual Maximum Wave Data at Atlantic City, NJ Based on Hindcast Data (Jensen, 1983).

$$a = \gamma c - \mu; c = \frac{\sqrt{6}\sigma}{\pi} \tag{10.114}$$

where $\gamma = 0.5772$ (Euler constant), μ is the mean of the distribution, and σ is the standard deviation. In terms of the return period of the given wave height, Equation 10.112 may be written as

$$\frac{1}{T_m} = 1 - e^{-e^{-\left(\frac{a+x}{c}\right)}} \tag{10.115}$$

in which T_m is the return period in years. Equations 10.112 and 10.115 plot as a straight line on Gumbel paper (Fig. 10.48).

B. Design Wave Conditions

Design wave conditions include tsunamis, shallow-water waves nearshore, and deepwater wave conditions. The selection of design values for these waves is briefly discussed in this section.

Tsunamis are impulsively generated waves usually resulting from earthquakes in coastal or offshore areas. They are more common along the Pacific Coast of the United States where earthquakes in remote coastal areas can produce tsunamis. These waves then propagate across the ocean to fairly distant locations. For example, the tsunami generated by the 1964 Alaskan earthquake resulted in damages to coastal areas in northern California. While in deep water, tsunamis are barely noticeable because of their small height and long periods. When they enter shallow water, however, their height increases dramatically and they become very destructive. Fortunately, not all coastal or offshore earthquakes result in tsunamis. Even facilities on relatively high ground can be inundated or damaged by tsunamis because of tsunami runup on the shore. In tsunami prone areas runup studies have been conducted to determine the water levels that will occur for a tsunami with a given return period, usually the 100-years interval. These studies are limited to the states of Alaska, Washington, Oregon, California, and Hawaii where tsunamis have historically occurred in the United States (Houston and Garcia, 1978). It is not possible to recommend a general tsunamis design criterion valid for all sites since there is a large variability in the characteristics of tsunamis. For engineering guidance concerning tsunamis in the Pacific Ocean, appropriate publications should be consulted (Camfield, 1980; 1990).

In shallow water nearshore, wave heights are limited by water depth. The maximum wave height is limited to approximately 0.78 times the water depth. The actual value of this factor depends on the local beach slope and the wave height to wave length ratio, termed the wave steepness. For steeper beach slopes the factor is greater than 0.78. As waves propagate toward shore they undergo transformations including diffraction, refraction, and shoaling. Diffraction is the propagation of waves around barriers such as breakwaters, islands, etc. Refraction involves the bending of the wave crests in response to variations in wave speed along the crest. Waves generally travel faster in deeper water. Thus, different portions of the same wave in different water depths travel at different speeds. Refraction can either increase or decrease the wave height depending on whether the wave energy is concentrated (focused) or spread. Shoaling, however, generally results in an increase in wave height as waves move from deep into shallow water. As the wave height increases and the water depth decreases, the ratio of height to depth eventually exceeds 0.78 and the wave breaks. The actual value of this ratio as a function of wave steepness and beach slope is shown in Fig. 10.49. Local changes in water level thus establish the location of breaking waves on the beach profile. For example, during high tides or during storm surges, waves can travel further across the profile and attack facilities that would not normally be subjected to waves. Additionally, deeper water supports higher waves thus larger waves attack a structure during periods of high water levels.

Consequently, in shallow water, design wave heights are often based on water levels. The return period of the maximum breaking wave height is thus related to the return period of a given water level. The maximum breaking wave height a structure can experience depends only on the maximum water depth at the seaward toe of the structure. High water levels and large waves are both usually associated with strong local winds so that they are correlated; however, astronomical tide levels are independent of the occurrence of large waves and storm surges. An inherent assumption is that high waves are always associated with high water levels. In the case of water levels resulting from storm surges this may be true; however, in the case of astronomical tides it is not true. Some idea of the probability of occurrence of the extreme waves determined using maximum water levels can be had by transforming the waves from the nearshore site back to deep water and comparing the resulting deepwater wave heights with a wave climatology for the area.

In deep water offshore, design waves conditions are usually selected as those wave conditions with a given return period. The return period selected for design depends on the type of structure, the consequences of its failure, the cost of failure, and the acceptable risk of failure. For example, a structure might simply be designed for wave conditions with a 100-year return period with the full knowledge that these conditions might be exceeded during the lifetime of the structure. For example, if the structure is expected to have a lifetime of 50 years, the probability of exceeding the design conditions at least once during that 50 years is 39%. The risk of exceeding design conditions at least once during an N year period is given by

$$R = 1 - \left[1 - \frac{1}{T_m}\right]^N \tag{10.116}$$

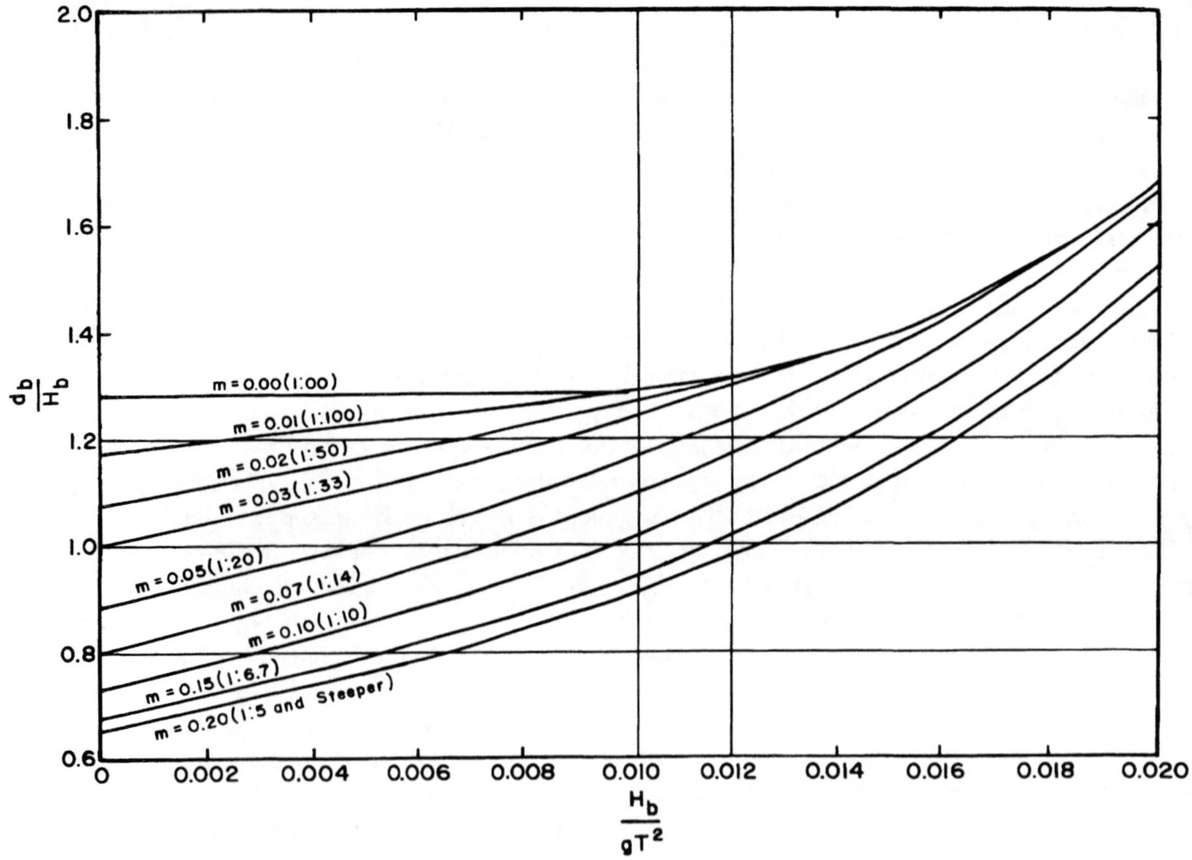

Figure 10.49.—Values of the Ratio d_b/H_b as a Function of Wave Steepness, H_b/gT^2, and Beach Slope (USACERC, 1984).

in which N is the number of years, T_m is the return period of the design wave conditions, and R is the risk that the design conditions will be exceeded at least once during the N years.

Because of the dearth of actual wave records available for design, the 100-year wave conditions (or wave conditions with some other desired return period) would most likely be developed from wave hindcasts. Generally, it is the significant wave height that is predicted by wave hindcasting procedures. Alternatively, a structure might be designed for a relatively high wave in the wave height distribution. For example, the average height of the highest 10% of the waves (H_{10}) or the average height of the highest 1% of the waves (H_1) might be selected. The relationships between H_{10}, H_1, and H_s are given by

$$H_s = 1.1416H_{rms}$$

$$H_{10} = 1.27H_s$$

$$H_1 = 1.67H_s. \tag{10.117}$$

For rigid structures where a single large wave in a wave train can damage a structure, the expected value of the maximum wave height might be used for design in conjunction with a large factor of safety. The expected value of the highest wave, H_{max}, is given by

$$\left(\frac{H_{\max}}{H_s}\right)_{mean} = 0.706\left[\sqrt{\ln N_o} + \frac{\gamma}{\sqrt{2\ln N_o}}\right] \tag{10.118}$$

where H_s is the significant wave height, N_o is the number of waves, and γ is the Euler's constant (γ = 0.5772).

The number of waves can be estimated from the average wave period determined from a hindcast (or measured) and the storm duration. There is still a probability that H_{max} will be exceeded during a given storm since it represents only the mean height of the highest waves taken from a population of storms. If the highest wave with a given probability of exceedance is needed, it can be obtained from

$$\left(\frac{H_{\max}}{H_s}\right)_p = 0.706\sqrt{\ln\left[\frac{N_o}{\ln 1/(1-p)}\right]} \tag{10.119}$$

in which p is the probability of exceedance. For example, if the H_{max} that is exceeded by only 1% of the maximum heights from a population of storms is desired, p = 0.01. Obviously, smaller values of p yield higher estimates for H_{max}.

C. Design Water Levels

Predicting water levels near shore is important for estimating the frequency of flooding, and also in determining maximum wave heights for design. In nearshore areas maximum water levels also produce conditions that allow large waves to attack dunes and coastal facilities. Statistical information on water levels is available for some coastal areas in the form of tables giving the probability that a given water level will be equalled or exceeded at a given site (Ebersole, 1982). These tables are available for each month of the year as well as for the total year (Figs. 10.50 and 10.51). The tables include water levels due to astronomical tides as well as storm surges. Similar information for astronomical tides alone is also available (Harris, 1981). A sample is provided in Fig. 10.52.

Water levels on the Great Lakes vary seasonally with higher levels generally occurring in June and July and lower levels in December, January, February, and March. However, there are also other variations due to hydrologic cycles that result in long periods of high lake levels and long periods of low levels. High lake levels may persist for a period of several years followed by several years of low levels. The Corps of Engineers, Detroit District maintains records of lake levels dating back to about the turn of the twentieth century. They also make predictions six months in advance of monthly average lake levels.

Fig. 10.53 shows water levels of Lake Erie for the month of July 1991 (USACE, 1991). Similar predictions are included in this publication for each of the five Great Lakes and Lake St. Clair. Predicted water levels for the months of July 1991 through December 1991 are shown as a dashed line with an estimate of the prediction error indicated by the shaded area. Also shown on the figure is the average lake level for each month of the 90-year period from 1900 to 1990. The maximum and minimum water levels of record for each month are also shown along with the year when the record high or low level was observed. The design of coastal facilities on the Great Lakes must take into account the observed range of water levels. The most recent record high levels were experienced during the mid-1980s on all lakes except Lake Ontario.

ATLANTIC CITY N.J.

YEARLY STATISTICS

ASTRONOMICAL TIDE — MEAN 0.01 STND DEV 1.53

I	X	P(X)	F(X)
1	-4.00	0.	0.
2	-3.90	0.	0.
3	-3.80	0.	0.
4	-3.70	0.	0.
5	-3.60	0.	0.
6	-3.50	0.0000	0.0000
7	-3.40	0.0001	0.0002
8	-3.30	0.0004	0.0005
9	-3.20	0.0007	0.0013
10	-3.10	0.0014	0.0027
11	-3.00	0.0023	0.0049
12	-2.90	0.0035	0.0084
13	-2.80	0.0047	0.0131
14	-2.70	0.0056	0.0187
15	-2.60	0.0070	0.0257
16	-2.50	0.0081	0.0339
17	-2.40	0.0111	0.0450
18	-2.30	0.0127	0.0576
19	-2.20	0.0151	0.0728
20	-2.10	0.0176	0.0903
21	-2.00	0.0194	0.1097
22	-1.90	0.0218	0.1315
23	-1.80	0.0223	0.1538
24	-1.70	0.0233	0.1771
25	-1.60	0.0240	0.2011
26	-1.50	0.0234	0.2244
27	-1.40	0.0239	0.2483
28	-1.30	0.0222	0.2705
29	-1.20	0.0222	0.2927
30	-1.10	0.0209	0.3135
31	-1.00	0.0195	0.3330
32	-0.90	0.0195	0.3525
33	-0.80	0.0193	0.3718
34	-0.70	0.0174	0.3891
35	-0.60	0.0174	0.4065
36	-0.50	0.0172	0.4236
37	-0.40	0.0165	0.4402
38	-0.30	0.0164	0.4566
39	-0.20	0.0168	0.4734
40	-0.10	0.0161	0.4895
41	0.	0.0159	0.5053
42	0.10	0.0170	0.5224
43	0.20	0.0172	0.5396
44	0.30	0.0171	0.5566
45	0.40	0.0175	0.5742
46	0.50	0.0175	0.5917
47	0.60	0.0192	0.6109
48	0.70	0.0182	0.6291
49	0.80	0.0200	0.6490
50	0.90	0.0199	0.6689
51	1.00	0.0224	0.6914
52	1.10	0.0232	0.7146
53	1.20	0.0236	0.7382
54	1.30	0.0244	0.7625
55	1.40	0.0224	0.7849
56	1.50	0.0223	0.8071
57	1.60	0.0214	0.8286
58	1.70	0.0200	0.8486
59	1.80	0.0199	0.8685
60	1.90	0.0185	0.8870
61	2.00	0.0160	0.9030
62	2.10	0.0157	0.9187
63	2.20	0.0135	0.9322
64	2.30	0.0130	0.9452
65	2.40	0.0106	0.9558
66	2.50	0.0093	0.9651
67	2.60	0.0076	0.9727
68	2.70	0.0061	0.9788
69	2.80	0.0057	0.9845
70	2.90	0.0041	0.9856
71	3.00	0.0031	0.9917
72	3.10	0.0028	0.9945
73	3.20	0.0021	0.9966
74	3.30	0.0015	0.9981
75	3.40	0.0009	0.9990
76	3.50	0.0007	0.9996
77	3.60	0.0003	0.9999
78	3.70	0.0001	1.0000
79	3.80	0.	1.0000
80	3.90	0.	1.0000
81	4.00	0.	1.0000

STORM SURGE — MEAN 0.00 STND DEV 0.56

I	X	P(X)	F(X)
1	-4.00	0.	0.
2	-3.90	0.	0.
3	-3.80	0.	0.
4	-3.70	0.	0.
5	-3.60	0.	0.
6	-3.50	0.	0.
7	-3.40	0.	0.
8	-3.30	0.	0.
9	-3.20	0.0000	0.0000
10	-3.10	0.	0.0000
11	-3.00	0.0000	0.0001
12	-2.90	0.0001	0.0001
13	-2.80	0.0001	0.0002
14	-2.70	0.0002	0.0003
15	-2.60	0.0003	0.0006
16	-2.50	0.0003	0.0009
17	-2.40	0.0003	0.0011
18	-2.30	0.0004	0.0015
19	-2.20	0.0004	0.0020
20	-2.10	0.0007	0.0026
21	-2.00	0.0008	0.0034
22	-1.90	0.0009	0.0043
23	-1.80	0.0011	0.0053
24	-1.70	0.0015	0.0068
25	-1.60	0.0018	0.0086
26	-1.50	0.0021	0.0107
27	-1.40	0.0028	0.0135
28	-1.30	0.0034	0.0169
29	-1.20	0.0045	0.0213
30	-1.10	0.0053	0.0267
31	-1.00	0.0078	0.0345
32	-0.90	0.0117	0.0462
33	-0.80	0.0159	0.0621
34	-0.70	0.0226	0.0847
35	-0.60	0.0335	0.1182
36	-0.50	0.0445	0.1627
37	-0.40	0.0604	0.2231
38	-0.30	0.0753	0.2984
39	-0.20	0.0904	0.3888
40	-0.10	0.0945	0.4832
41	0.	0.0915	0.5747
42	0.10	0.0820	0.6567
43	0.20	0.0689	0.7256
44	0.30	0.0559	0.7815
45	0.40	0.0476	0.8291
46	0.50	0.0368	0.8660
47	0.60	0.0286	0.8946
48	0.70	0.0224	0.9169
49	0.80	0.0181	0.9350
50	0.90	0.0139	0.9488
51	1.00	0.0113	0.9601
52	1.10	0.0089	0.9690
53	1.20	0.0071	0.9761
54	1.30	0.0050	0.9811
55	1.40	0.0042	0.9853
56	1.50	0.0030	0.9883
57	1.60	0.0023	0.9906
58	1.70	0.0021	0.9927
59	1.80	0.0017	0.9944
60	1.90	0.0014	0.9959
61	2.00	0.0012	0.9970
62	2.10	0.0007	0.9977
63	2.20	0.0006	0.9983
64	2.30	0.0004	0.9988
65	2.40	0.0004	0.9991
66	2.50	0.0002	0.9993
67	2.60	0.0001	0.9994
68	2.70	0.0001	0.9995
69	2.80	0.0001	0.9997
70	2.90	0.0001	0.9997
71	3.00	0.0001	0.9998
72	3.10	0.0000	0.9998
73	3.20	0.0000	0.9999
74	3.30	0.0001	0.9999
75	3.40	0.0000	0.9999
76	3.50	0.0000	1.0000
77	3.60	0.	1.0000
78	3.70	0.0000	1.0000
79	3.80	0.0000	1.0000
80	3.90	0.	1.0000
81	4.00	0.	1.0000

TOTAL WATER LEVEL — MEAN 0.01 STND DEV 1.58

I	X	P(X)	F(X)
1	-5.00	0.0001	0.0001
2	-4.80	0.0001	0.0002
3	-4.60	0.0001	0.0003
4	-4.40	0.0001	0.0004
5	-4.20	0.0004	0.0009
6	-4.00	0.0007	0.0015
7	-3.80	0.0010	0.0025
8	-3.60	0.0015	0.0040
9	-3.40	0.0029	0.0069
10	-3.20	0.0048	0.0117
11	-3.00	0.0067	0.0184
12	-2.80	0.0107	0.0290
13	-2.60	0.0151	0.0441
14	-2.40	0.0211	0.0652
15	-2.20	0.0257	0.0909
16	-2.00	0.0338	0.1247
17	-1.80	0.0373	0.1619
18	-1.60	0.0427	0.2046
19	-1.40	0.0438	0.2484
20	-1.20	0.0432	0.2916
21	-1.00	0.0416	0.3332
22	-0.80	0.0385	0.3717
23	-0.60	0.0375	0.4092
24	-0.40	0.0357	0.4449
25	-0.20	0.0366	0.4815
26	0.	0.0354	0.5169
27	0.20	0.0350	0.5519
28	0.40	0.0383	0.5902
29	0.60	0.0383	0.6285
30	0.80	0.0405	0.6690
31	1.00	0.0421	0.7112
32	1.20	0.0417	0.7529
33	1.40	0.0424	0.7953
34	1.60	0.0397	0.8351
35	1.80	0.0354	0.8704
36	2.00	0.0313	0.9017
37	2.20	0.0267	0.9284
38	2.40	0.0214	0.9497
39	2.60	0.0169	0.9666
40	2.80	0.0123	0.9789
41	3.00	0.0079	0.9868
42	3.20	0.0052	0.9920
43	3.40	0.0033	0.9953
44	3.60	0.0020	0.9972
45	3.80	0.0014	0.9986
46	4.00	0.0006	0.9992
47	4.20	0.0003	0.9995
48	4.40	0.0002	0.9997
49	4.60	0.0001	0.9998
50	4.80	0.0001	0.9999
51	5.00	0.0001	1.0000
52	5.20	0.0000	1.0000
53	5.40	0.0000	1.0000
54	5.60	0.0000	1.0000
55	5.80	0.	1.0000
56	6.00	0.	1.0000
57	6.20	0.	1.0000
58	6.40	0.	1.0000
59	6.60	0.	1.0000
60	6.80	0.	1.0000
61	7.00	0.	1.0000

I - INTERVAL NUMBER X - INTERVAL CENTER VALUE

P(X) - PROBABILITY MASS FUNCTION F(X) - CUMULATIVE DISTRIBUTION FUNCTION

Figure 10.50.—Probability that a Given Water Level at Atlantic City, NJ Will Be Equalled or Exceeded in a Given Year (Ebersole, 1982).

ATLANTIC CITY N.J.

STATISTICS FOR JANUARY

ASTRONOMICAL TIDE

MEAN-0.26 STND DEV 1.53

I	X	P(X)	F(X)
1	-4.00	0.	0.
2	-3.90	0.	0.
3	-3.80	0.	0.
4	-3.70	0.	0.
5	-3.60	0.	0.
6	-3.50	0.0001	0.0001
7	-3.40	0.0006	0.0007
8	-3.30	0.0008	0.0015
9	-3.20	0.0020	0.0035
10	-3.10	0.0039	0.0074
11	-3.00	0.0049	0.0123
12	-2.90	0.0059	0.0183
13	-2.80	0.0073	0.0255
14	-2.70	0.0087	0.0343
15	-2.60	0.0112	0.0455
16	-2.50	0.0140	0.0595
17	-2.40	0.0198	0.0793
18	-2.30	0.0197	0.0990
19	-2.20	0.0225	0.1215
20	-2.10	0.0242	0.1457
21	-2.00	0.0254	0.1712
22	-1.90	0.0249	0.1960
23	-1.80	0.0245	0.2205
24	-1.70	0.0253	0.2459
25	-1.60	0.0227	0.2686
26	-1.50	0.0227	0.2913
27	-1.40	0.0214	0.3127
28	-1.30	0.0192	0.3319
29	-1.20	0.0196	0.3515
30	-1.10	0.0162	0.3677
31	-1.00	0.0163	0.3841
32	-0.90	0.0181	0.4022
33	-0.80	0.0161	0.4183
34	-0.70	0.0185	0.4368
35	-0.60	0.0166	0.4534
36	-0.50	0.0160	0.4694
37	-0.40	0.0183	0.4877
38	-0.30	0.0153	0.5030
39	-0.20	0.0142	0.5173
40	-0.10	0.0153	0.5326
41	0.	0.0168	0.5494
42	0.10	0.0163	0.5658
43	0.20	0.0198	0.5856
44	0.30	0.0203	0.6059
45	0.40	0.0160	0.6219
46	0.50	0.0203	0.6421
47	0.60	0.0214	0.6635
48	0.70	0.0211	0.6846
49	0.80	0.0242	0.7088
50	0.90	0.0227	0.7315
51	1.00	0.0213	0.7528
52	1.10	0.0242	0.7770
53	1.20	0.0227	0.7997
54	1.30	0.0256	0.8254
55	1.40	0.0205	0.8459
56	1.50	0.0170	0.8629
57	1.60	0.0175	0.8804
58	1.70	0.0161	0.8965
59	1.80	0.0166	0.9131
60	1.90	0.0139	0.9270
61	2.00	0.0109	0.9378
62	2.10	0.0100	0.9478
63	2.20	0.0091	0.9569
64	2.30	0.0090	0.9658
65	2.40	0.0067	0.9726
66	2.50	0.0057	0.9783
67	2.60	0.0040	0.9823
68	2.70	0.0051	0.9875
69	2.80	0.0035	0.9909
70	2.90	0.0028	0.9937
71	3.00	0.0021	0.9959
72	3.10	0.0022	0.9981
73	3.20	0.0011	0.9992
74	3.30	0.0008	1.0000
75	3.40	0.	1.0000
76	3.50	0.	1.0000
77	3.60	0.	1.0000
78	3.70	0.	1.0000
79	3.80	0.	1.0000
80	3.90	0.	1.0000
81	4.00	0.	1.0000

STORM SURGE

MEAN 0.15 STND DEV 0.70

I	X	P(X)	F(X)
1	-4.00	0.	0.
2	-3.90	0.	0.
3	-3.80	0.	0.
4	-3.70	0.	0.
5	-3.60	0.	0.
6	-3.50	0.	0.
7	-3.40	0.	0.
8	-3.30	0.	0.
9	-3.20	0.	0.
10	-3.10	0.	0.
11	-3.00	0.	0.
12	-2.90	0.	0.
13	-2.80	0.	0.
14	-2.70	0.	0.
15	-2.60	0.0003	0.0003
16	-2.50	0.0001	0.0004
17	-2.40	0.0001	0.0005
18	-2.30	0.0005	0.0010
19	-2.20	0.0001	0.0011
20	-2.10	0.0009	0.0020
21	-2.00	0.0009	0.0028
22	-1.90	0.0004	0.0032
23	-1.80	0.0005	0.0037
24	-1.70	0.0009	0.0046
25	-1.60	0.0015	0.0060
26	-1.50	0.0031	0.0091
27	-1.40	0.0038	0.0129
28	-1.30	0.0058	0.0187
29	-1.20	0.0080	0.0267
30	-1.10	0.0103	0.0370
31	-1.00	0.0139	0.0508
32	-0.90	0.0183	0.0691
33	-0.80	0.0220	0.0911
34	-0.70	0.0252	0.1163
35	-0.60	0.0311	0.1474
36	-0.50	0.0354	0.1828
37	-0.40	0.0434	0.2262
38	-0.30	0.0471	0.2732
39	-0.20	0.0562	0.3294
40	-0.10	0.0557	0.3851
41	0.	0.0602	0.4453
42	0.10	0.0611	0.5064
43	0.20	0.0584	0.5648
44	0.30	0.0578	0.6225
45	0.40	0.0665	0.6890
46	0.50	0.0488	0.7378
47	0.60	0.0467	0.7845
48	0.70	0.0388	0.8234
49	0.80	0.0366	0.8600
50	0.90	0.0320	0.8920
51	1.00	0.0233	0.9153
52	1.10	0.0181	0.9334
53	1.20	0.0135	0.9469
54	1.30	0.0092	0.9561
55	1.40	0.0082	0.9644
56	1.50	0.0057	0.9700
57	1.60	0.0057	0.9757
58	1.70	0.0044	0.9801
59	1.80	0.0028	0.9829
60	1.90	0.0037	0.9866
61	2.00	0.0039	0.9905
62	2.10	0.0017	0.9923
63	2.20	0.0020	0.9942
64	2.30	0.0016	0.9958
65	2.40	0.0012	0.9971
66	2.50	0.0005	0.9975
67	2.60	0.0007	0.9983
68	2.70	0.	0.9983
69	2.80	0.0005	0.9988
70	2.90	0.0003	0.9990
71	3.00	0.0006	0.9996
72	3.10	0.	0.9996
73	3.20	0.	0.9996
74	3.30	0.0001	0.9998
75	3.40	0.0001	0.9999
76	3.50	0.0001	1.0000
77	3.60	0.	1.0000
78	3.70	0.	1.0000
79	3.80	0.	1.0000
80	3.90	0.	1.0000
81	4.00	0.	1.0000

TOTAL WATER LEVEL

MEAN-0.10 STND DEV 1.63

I	X	P(X)	F(X)
1	-5.00	0.	0.
2	-4.80	0.	0.
3	-4.60	0.	0.
4	-4.40	0.0001	0.0001
5	-4.20	0.0006	0.0007
6	-4.00	0.0004	0.0011
7	-3.80	0.0016	0.0027
8	-3.60	0.0028	0.0055
9	-3.40	0.0052	0.0107
10	-3.20	0.0091	0.0198
11	-3.00	0.0098	0.0296
12	-2.80	0.0162	0.0458
13	-2.60	0.0202	0.0660
14	-2.40	0.0263	0.0923
15	-2.20	0.0294	0.1217
16	-2.00	0.0334	0.1551
17	-1.80	0.0379	0.1930
18	-1.60	0.0385	0.2314
19	-1.40	0.0365	0.2679
20	-1.20	0.0408	0.3087
21	-1.00	0.0375	0.3462
22	-0.80	0.0423	0.3885
23	-0.60	0.0364	0.4249
24	-0.40	0.0361	0.4610
25	-0.20	0.0382	0.4993
26	0.	0.0385	0.5377
27	0.20	0.0360	0.5737
28	0.40	0.0392	0.6129
29	0.60	0.0422	0.6551
30	0.80	0.0450	0.7001
31	1.00	0.0407	0.7408
32	1.20	0.0393	0.7801
33	1.40	0.0363	0.8164
34	1.60	0.0356	0.8520
35	1.80	0.0307	0.8827
36	2.00	0.0259	0.9087
37	2.20	0.0226	0.9313
38	2.40	0.0173	0.9486
39	2.60	0.0141	0.9628
40	2.80	0.0116	0.9743
41	3.00	0.0077	0.9821
42	3.20	0.0052	0.9872
43	3.40	0.0050	0.9923
44	3.60	0.0034	0.9957
45	3.80	0.0018	0.9975
46	4.00	0.0005	0.9980
47	4.20	0.0009	0.9989
48	4.40	0.0006	0.9995
49	4.60	0.0004	0.9999
50	4.80	0.	0.9999
51	5.00	0.0001	1.0000
52	5.20	0.	1.0000
53	5.40	0.	1.0000
54	5.60	0.	1.0000
55	5.80	0.	1.0000
56	6.00	0.	1.0000
57	6.20	0.	1.0000
58	6.40	0.	1.0000
59	6.60	0.	1.0000
60	6.80	0.	1.0000
61	7.00	0.	1.0000

I - INTERVAL NUMBER X - INTERVAL CENTER VALUE

P(X) - PROBABILITY MASS FUNCTION F(X) - CUMULATIVE DISTRIBUTION FUNCTION

Figure 10.51.—Probability that a Given Water Level at Atlantic City, NJ in January Will Be Equalled or Exceeded (Ebersole, 1982).

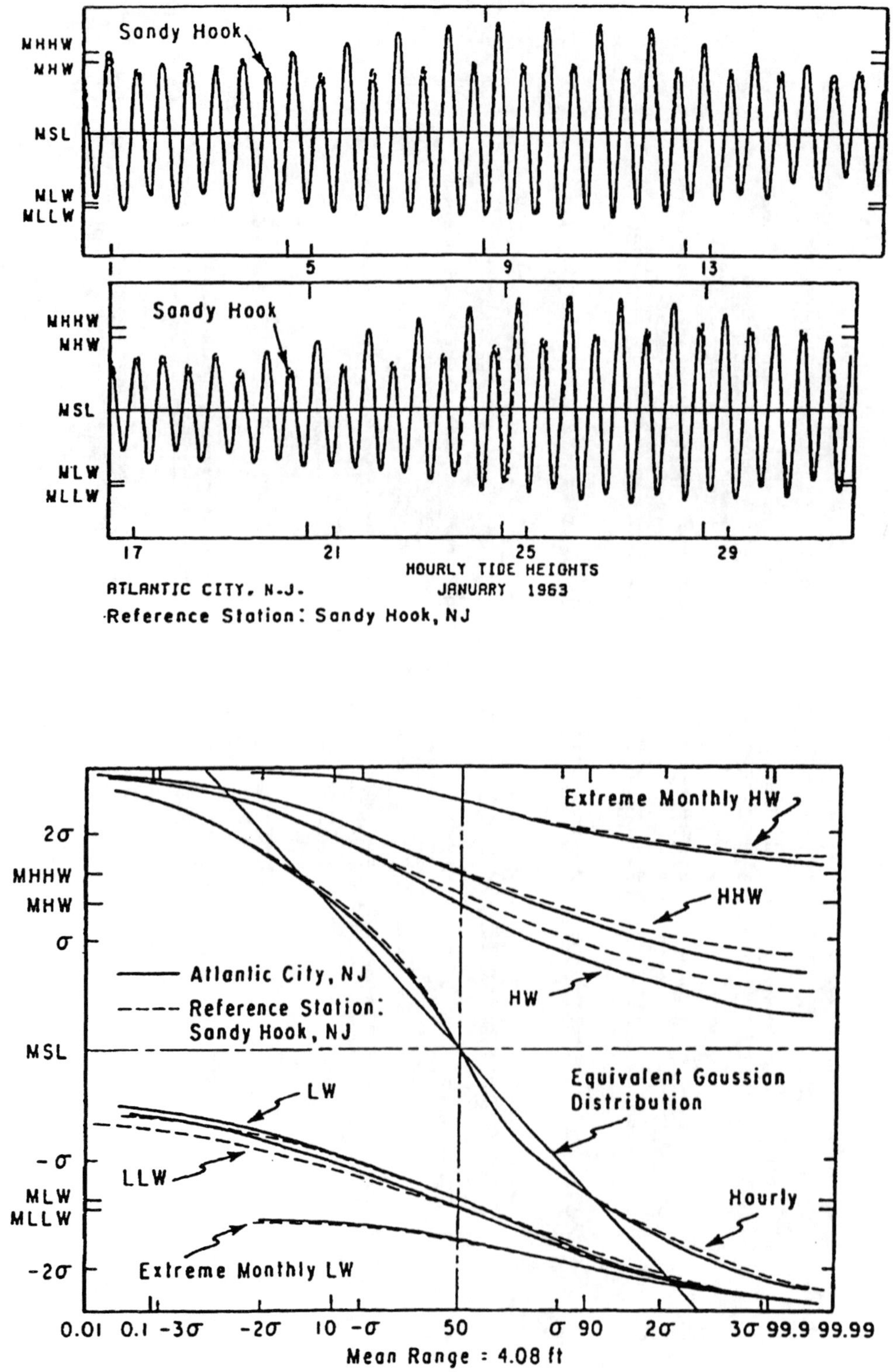

Figure 10.52.—Astronomical Tide Water Level Probabilities at Atlantic City, NJ (Harris, 1981).

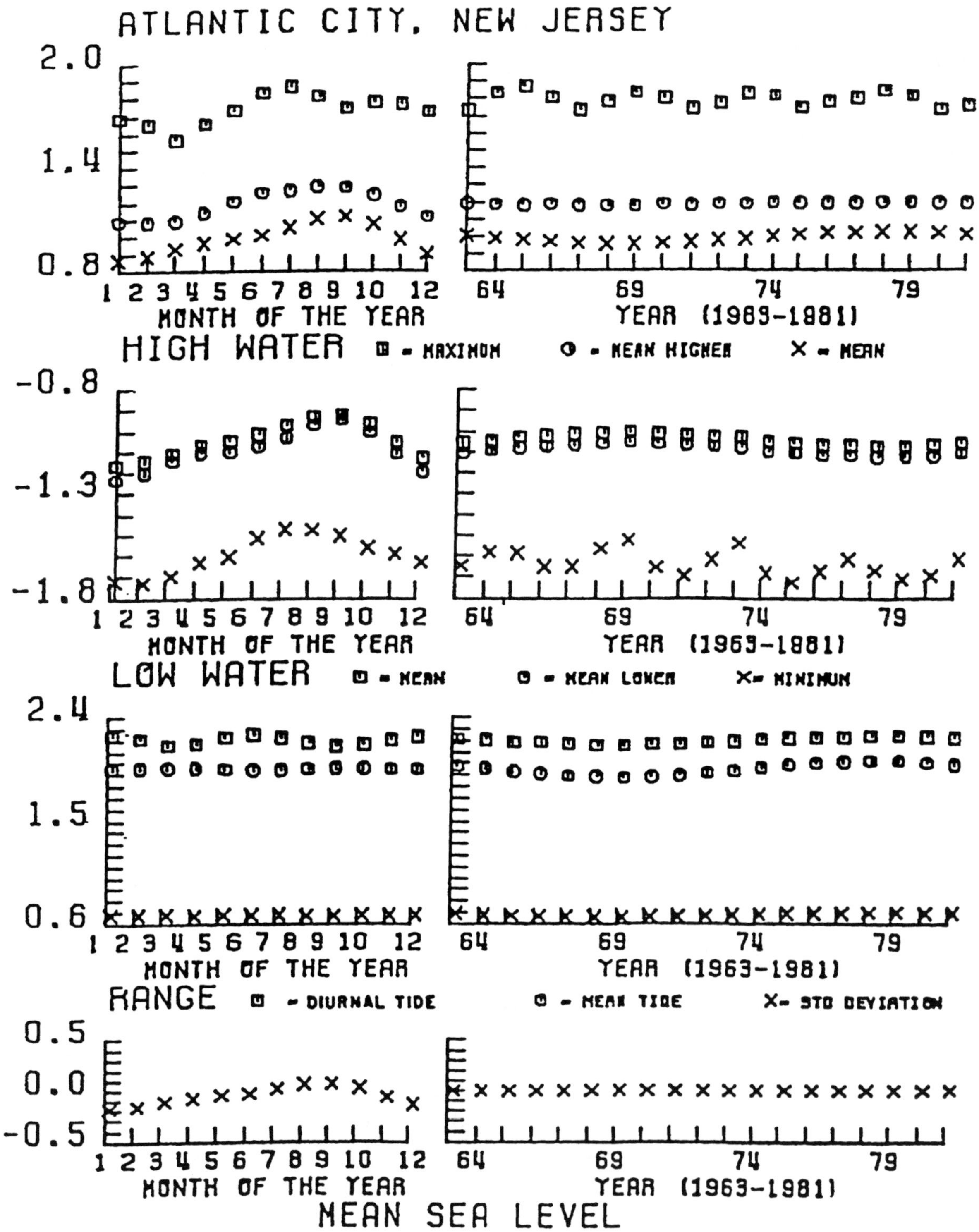

Figure 10.52.—(Continued)

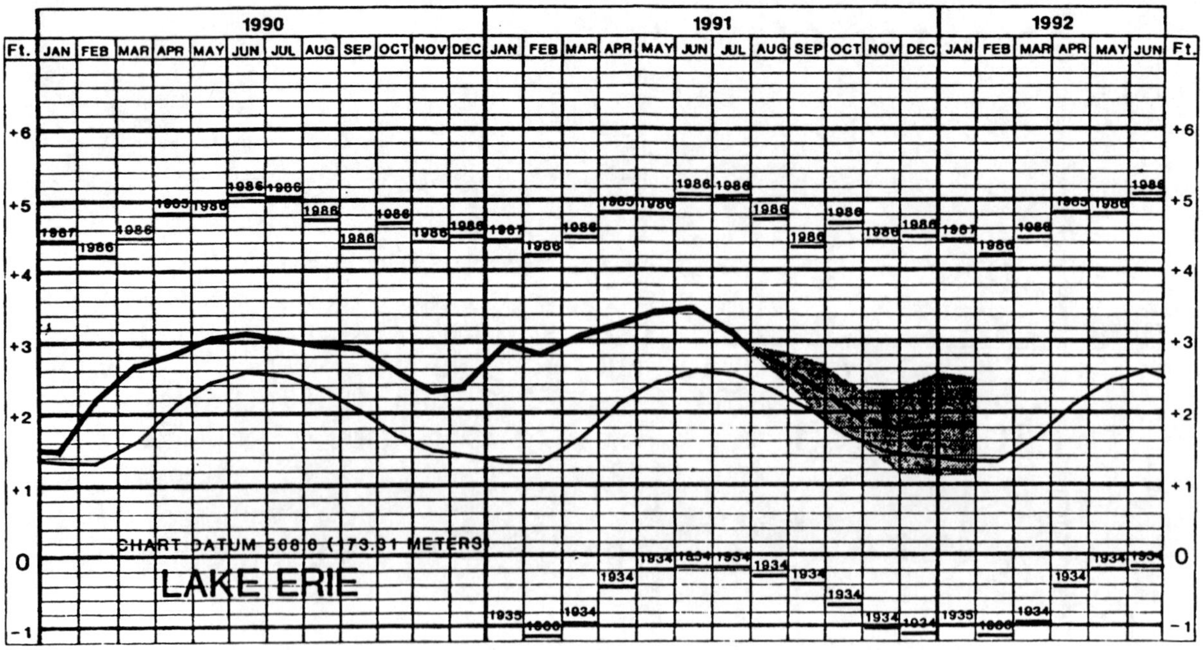

JULY MEAN LAKE LEVELS

	Superior	Mich.-Huron	St. Clair	Erie	Ontario
*1991	600.67	578.85	574.64	571.74	245.68
1990	600.40	578.38	574.35	571.60	245.58
††MAX. YEAR	601.91 1986	581.27 1986	576.56 1986	573.66 1986	247.74 1947
†† MIN. YEAR	598.99 1926	575.96 1964	571.88 1934	568.46 1934	242.75 1934
1900-1990 AVG.	600.92	578.84	574.06	571.15	245.52

* provisional

†† Maxima and Minima for period 1900-1990

Figure 10.53.—Historical and Predicted Lake Levels for Lake Erie, July 1991 (USACE, 1991).

XII. REFERENCES

Abbott, M.B., Petersen, H.M., and Skovgaard, O. (1978). "On the numerical modelling of short waves in shallow water," J. Hydraulic Research, 16, 173–204.

Arthur, R. S. (1950). "Refraction of shallow water waves: the combined effect of currents and underwater topography," Trans. Am. Geophys. Union, Vol. 31, 549–552.

Bagnold, R.A. (1946). "Motion of waves in shallow water: interaction between waves and bottoms," *Proc. Royal Soc. of London, Series A*, Vol. 187, 1–15.

Battjes, J.A. (1971). "Run up distribution of waves breaking on slopes," J. Waterways and Harbors Division, ASCE, Vol. 97, No. WW1.

Battjes, J.A. (1974a). "Surf similarity," *Proc. 14th Intl. Coastal Engrg. Conf.*, ASCE, 466–480.

Battjes, J.A. (1974b). "Computation of setup, longshore currents, runup and overtopping due to wind generated waves," PhD Thesis, Delft Tech. Univ., Holland.

Battjes, J.A. and Janssen, J.P.F.M. (1978). "Energy loss and set-up due to breaking of random waves," *Proc. 16th Intl. Coastal Engrg. Conf.*, ASCE.

Berkhoff, J.C.W. (1972). "Computation of combined refraction–diffraction," *Proc. 13th Intl. Coastal Engrg. Conf.*, ASCE, Vancouver, 471–490.

Booij, N. (1981). "Gravity waves on water with non-uniform depth and current," Comm. Hydraul. Rep. No. 81-1, Delft, Holland.

Bowen, A. J. (1969a). "Rip currents, 1: theoretical investigations," J. Geophys. Res. Vol. 74, 5467–5478.

Bowen, A. J. (1969b). "The generation of longshore currents on a plane beach," J. Mar. Res., Vol. 37, 206–215.

Bretherton, F. P. and Garrett, C. J. R. (1969). "Wave trains in homogeneous moving media," *Proc. Roy. Soc. London, Series A*, Vol. 302, 529–554.

Bruun, P. (1985). *Developments in geotechnical engineering*, Vol. 37 (ed. Bruun, P.), Elsevier, New York, NY.

Camfield, F.E. (1980). "Tsunami engineering," CERC Special Report No. 6, U.S. Army Engineer Waterways Experiment Station, Vicksburg, MS.

Camfield, F.E. (1990). "Tsunami," Chap. 10, *Handbook of coastal and ocean engineering, Vol. 1*, edited by J.B. Herbich, Gulf Publishing Co., Houston, TX.

Carrier, G.F., Shaw, R.P., and Miyata, M. (1971). "Channel effects in harbor resonance," J. Engr. Mech. Div., A S C E, 1703–1716.

Chakrabarti, S.K. (1987). *Hydrodynamics of offshore structures*, Springer-Verlag, London.

Christofferson, J. B. and Jonsson, I. G. (1980). "A note on wave action conservation in a dissipative current wave motion," Appl. Ocean Res. Vol. 2, 179–182.

Collins, I.A. (1970). "Probabilities of breaking wave characteristics," *Proc. 12th Coastal Engrg. Conf.* ASCE, Vol. 1, 399–414.

Crawford, P.L. (1987). "Tsunami predictions for the coast of Alaska, Kodiak Island to Ketchikan," Tech. Rept. CERC-87-7, U.S. Army Engineer Waterways Experiment Station, Vicksburg, MS.

Dally, W. R., Dean, R.G., and Dalrymple, R.A. (1985). "Wave height variation across beaches of arbitrary profile," J. Geophys. Research, 90, C6, 11917–11927.

Dalrymple, R.A. (1974a). "A finite amplitude wave on linear shear current," J. Geophys. Research, 79, 4498–4504.

Dalrymple, R. A. (1974b). "Water waves on a bilinear shear current," *Proc. of the 14th Coastal Engrg. Conf.*, Vol. 1, ASCE 626–641.

Dalrymple, R. A. (1977). "A numerical model for periodic finite amplitude waves on a rotational fluid," J. Comput. Phys., Vol. 24, 29–42. 140.

Dalrymple, R. A. (1978). "Rip currents and their genesis," *Summaries 16th Coastal Engr. Conf.*, Paper No. 140, ASCE, New York, NY.

Dalrymple, R. A. (1980). "Longshore currents with wave–current interaction," J. Waterways, Port, Coastal and Ocean Div., ASCE, Vol. 106, 414–420.

Dalrymple, R.A. (1988). "Model for refraction of water waves," J. Waterways, Port, Coastal, and Ocean Div., ASCE, 114, 4, 423–435.

Dalrymple, R.A. and Dean, R.G. (1975). "Waves of maximum height on uniform currents," J. Waterways, Harbors, and Coastal Eng. Div., ASCE, Vol. 101, 259–268.

Dean, R.G. (1965). "Stream function representation of water waves," J. Geophys. Res., 70, 18, 4561–4572.

Dean, R.G. (1974). "Evaluation and development of water wave theories for engineering application," Spec. Rpt. 1, U.S. Army Coastal Engineering Research Center.

Dean, R.G. and Dalrymple, R.A. (1991). *Water wave mechanics for engineers and scientists*, 2nd Printing, World Scientific, Teaneck, NJ.

Delaware. (1990). "Delaware hurricane evacuation study," Technical Data Report, Delaware Division of Emergency Planning and Operations, Federal Emergency Management Agency, National Weather Service and U.S. Army Corps of Engineers, Philadelphia District.

Demirbilek, Z. (1987). "Morison's formula: relative velocity versus independent flow fields formulation for a case representing fluid damping," *Intl. Conf. on Offshore Mechanics and Arctic Engineering*, Vol. II, 25–32.

Demirel, V. and Wang, S. (1986). "A numerical solution to transient wave induced harbor oscillations using boundary element technique," *Proc., 20th Intl. Conf. on Coastal Engrg.*, ASCE, Taipei, 124–135.

Ebersole, B.A. (1982). "Atlantic coast water level climate," Wave Information Study (WIS) Report 7, U.S. Army Corps of Engineers Waterways Experiment Station, Vicksburg, MS, April 1982.

Faltinsen, O. M. and Demirbilek, Z. (1989). "Hydrodynamic analysis of tension leg platforms, TLPs," in *Tension Leg Platform: A State of the Art Review* (ed. Demirbilek, Z.), Chapter 4, ASCE Monograph, New York, p 335.

Fenton, J.D. (1985). "A fifth order stokes theory for steady waves," J. Waterways, Port, Coastal, and Ocean Div., ASCE, 111, 216–234.

Fenton, J.D. and McKee, W.D. (1989). Unpublished. See reference by Fenton "Nonlinear Wave Theories" in *The Sea*, Vol. 9, Le Mehaute, B. and D.M. Hanes, eds., Wiley-Interscience, New York, NY 1990.

Forristall, G.Z. (1982). "Subsurface wave-measuring systems," *Proc. Symposium and Workshop on Wave-Measurement Technology*, Marine Board, U.S. National Research Council.

Freilich, M.H. and Guza, R.T. (1984). "Nonlinear effects on shoaling surface gravity waves," Phil. Trans. Roy. Soc. London, A, 31, 1–41.

Galvin, C. J. (1967). "Longshore current velocity: a review of theory and data," Revs. in Geophys., Vol. 5, No. 3, 287–304.

Galvin, C. J. and Eagelson, P.S. (1965). "Experimental study of longshore currents on a plane beach," U.S. Army Coastal Engr. Res. Center Tech. Memo. No. 10, 80.

Garrison, C.J. (1978). "Hydrodynamic loading of large offshore structures, three-dimensional source distribution methods," *Numerical Methods in Offshore Engineering*, John Wiley and Sons, London, 87–140.

Goda, Y. (1974). "New pressure formula for composite breakwaters," *Proc. 14th Coastal Engr. Conf.*, ASCE Vol. 3, 1702–1720.

Goda, Y. (1985). *Random seas and design of maritime structures*, University of Tokyo Press, Tokyo, Japan.

Godin, G. (1972). *The analysis of tides*, University of Toronto Press, Toronto, Canada.

Gunbak, A.R. (1977). "Rubble mound breakwaters," Tech. Univ. of Norway, Port and Ocean Engr. Dept., Rep. No. 12/77.

Gunbak, A.R. (1978). "Wave mechanics principles on the design of rubble mound breakwaters," *Proc. of the POAC'79 Conf.*, Trondheim, Norway.

Harris, D.L. (1981). "Tides and tidal datums in the United States," Special Report No. 7, U.S. Army Engineer Waterways Experiment Station, Vicksburg, MS.

Harris, T. F. W. (1969). "Field and model studies of the nearshore circulations," PhD Thesis, Dept. of Physics, Univ. of Natal, 188.

Havelock, T.H. (1908). "The propagation of groups of waves in dispersive media, with application to waves on water produced by a traveling disturbance," *Proc., Royal Society of London, Series A.*

Hicks, S.D. (1988). Unpublished review of tidal datums, National Ocean Service, U.S. Dept. of Commerce.

Hicks, S.D. (1989). "Tide and current glossary," National Ocean Service, U.S. Dept. of Commerce.

Homma, M. K., Horikawa, K., and Sonu, C. (1962). "Field investigation at Tokai, Japan, conducted by combined procedure of macroscopic and microscopic approaches," Coastal Eng. in Japan, Vol. 5, 93–110.

Horikawa, K. (1978). *Coastal engineering*, University of Tokyo Press, Tokyo, Japan, 402.

Horikawa, K. and Sasaki, T. (1972). "Field observations of nearshore current system," *Proc. 13th Coastal Engr. Conf.*, ASCE, 635–652.

Houston, J.R., Carver, R.D., and Markle, D.G. (1977). "Tsunami-wave elevation frequency of occurrence for the Hawaiian Islands," Tech. Rept. H–77–16, U.S. Army Engineer Waterways Experiment Station, Vicksburg, MS.

Houston, J.R. and Garcia, A.W. (1974). "Type 16 flood insurance study: tsunami predictions for Pacific Coastal Communities," Tech. Rept. H–74–3, U.S. Army Engineer Waterways Experiment Station, Vicksburg, MS.

Houston, J.R. and Garcia, A.W. (1978). "Type 16 flood insurance study: tsunami predictions for the west coast of the continental United States," Tech. Rept. H–78–26, U.S. Army Engineer Waterways Experiment Station, Vicksburg, MS.

Hubertz, J.M., Brooks, R.M., Brandon, W.A., and Tracy, B.A. (1993). "Hindcast wave information for the US Atlantic Coast," Wave Information Study (WIS) Report 30, Coastal Engineering Research Center, U.S. Army Waterways Experiment Station, Vicksburg, MS.

Hudson, R.Y. (1979). "Coastal hydraulic models," Special Report No. 5, U.S. Army Coastal Engineering Research Center, Ft. Belvoir, VA, May.

Hughes, S.A. and Borgman, L.E. (1987). "Beta–Rayleigh distribution for shallow water wave heights," *Proc. Coastal Hydrodynamics*, ASCE, 17–31.

Hunt, I.A. (1959). "Design of seawalls and breakwaters," J. Waterways and Harbors Division, ASCE, Vol. 85, No. WW3, 123–152.

IAHR. (1986). "List of sea state parameters," *Suppl. to PIANC Bull. 52*, pub. jointly by IAHR and PIANC, Brussels, Belgium. Also *J. Waterway, Port, Coast., and Ocean Div.*, ASCE, 115(6), 793–808.

Ito, Y., Fujishima, M., and Kitatani, T. (1966). "Study on the stability of breakwaters," Rep. Port and Harbor Inst., Vol. 5, Part 4.

Jensen, R.E. (1983). "Atlantic coast shallow-water significant wave information," Wave Information Study (WIS) Report 9, Hydraulics Laboratory, U.S. Army Waterways Experiment Station, Vicksburg, MS.

Johnson, D. W. (1919). *Shore processes and shoreline development*, John Wiley & Sons, New York, NY, 584.

Johnson, J. W. (1947). "The refraction of surface waves by currents," Trans. Am. Geophys. Union, Vol. 28, 867–874.

Kirby, J. T. (1984). "A note on linear surface wave–current interaction over slowly varying topography," J. Geophys. Res., Vol. 89C, 745–747.

Kirby, J.T. and Dalrymple, R.A. (1983). "A parabolic equation for the combined refraction–diffraction of Stokes waves by mildly varying topography," J. Fluid Mechanics, 136, 453–466.

Kirby, J.T. and Dalrymple, R.A. (1984). "Verification of a parabolic equation for propagation of weakly-nonlinear waves," Coastal Engrg., 9, 219–232.

Kirby, J. T. and Dalrymple, R. A. (1993). "Combined refraction/diffraction model—REFDIF1: Documentation Manual," Center for Applied Coastal Res. Report No. 92-4, University of Delaware, Newark, pp 88.

Komar, P. D. (1976). *Beach processes and sedimentation*, Prentice-Hall, Englewood Cliffs, NJ, 429.

Komar, P.D. and Gaughan, M. K. (1973). "Airy wave theory and breaker height prediction," *Proc. 13th Coastal Engr. Conf.*, ASCE, pp 405–418.

Korteweg, D.J. and de Vries, G. (1895). "On the change of form of long waves advancing in a rectangular channel and on a new type of long stationary waves," Philosophical Mag., 39, 422–443.

Lambrakos, K. F. (1981). "Wave–current interaction effects on water velocity and wave spectra," J. Geophys. Res., Vol. 86.

Leenknecht, D.A., Szuwalski, A., and Sherlock, A.R. (1992). "Automated coastal engineering system: User's Guide and Technical Reference," U.S. Army Waterways Experiment Station, Vicksburg, MS.

Le Mehaute, B. and Koh, R.C.Y. (1967). "On the breaking of waves arriving at an angle to the shore," J. Hydr. Res., Vol. 5, 67–80.

Li, Y.C., Yu, Y.X., and Xoy, M.T. (1983). "Investigation of wave pressure on vertical wall," *Proc. Intl. Conf. on Coastal and Port Engr. in Developing Countries*, Vol. 1, 755–776.

Liu, P.-L. F. and Mei, C. C. (1974). "Effects of a breakwater on nearshore currents due to breaking waves," MIT Ralph M. Parsons Lab. Rep. No. 192, 265.

Liu, P.L.-F. and Tsay, T.-K. (1983). "On weak reflection of water waves," J. Fluid Mechanics, 131, 59–71.

Liu, P.L.-F., Yoon, S.B., and Kirby, J.T. (1985). "Nonlinear refraction-diffraction of waves in shallow water," J. Fluid Mechanics, 153, 184–201.

Longuet–Higgins, M. S. and Stewart, R. W. (1960). "Changes in the form of short gravity waves on long waves and tidal currents," J. Fluid Mech., Vol. 8, 565–583.

Longuet–Higgins, M. S. and Stewart, R. W. (1961). "The changes in amplitude of short gravity waves on steady non-uniform current," J. Fluid Mech., Vol. 10, 529–549.

Longuet–Higgins, M. S. and Stewart, R. W. (1962). "Radiation stress and mass transport in gravity waves, with applications to surf beats," J. Fluid Mech., Vol. 13, 481–504.

Longuet–Higgins, M. S. and Stewart, R. W. (1964). "Radiation stress in water waves: a physical discussion with applications," Deep-Sea Res., Vol. 11, 529–562.

Lyles, S.D, Hickman, L., and Debaugh, H. (1987). "Sea level variations in the United States," National Ocean Service, Rockville, MD.

MacCamy, R.C. and Fuchs, R.A. (1954). "Wave forces on piles: a diffraction theory," U.S. Army Corps of Engineers, Beach Erosion Board, Tech. Memo. No. 69, 1–17.

Mei, C. C. (1991). *The applied dynamics of ocean surface waves*, World Sci. Pub., Teaneck, NJ.

Miche, R. (1944). "Mouvements ondulatores de la mer in profondeur constante ou decriossante," Annals des Ponts et Chaussees, Paris, Vol. 114, 369–406.

Miles, J.W. (1948). "Coupling of a cylindrical tube to a half space," J. Acoustic Society of America, 652–664.

Miles, J.W. (1974). "Harbor seiching," Annual Review of Fluid Mechanics, Vol. 6, 17–35.

Miller, D.J. (1960). "Giant waves in Lituya Bay, Alaska," U.S. Geological Survey Professional Paper 354–C.

Minikin, R.R. (1950). *Winds, waves and maritime structures*, Chas. Griffin & Co., London.

Morison, J.R., O'Brien, M.P., Johnson, J.W., and Schaaf, S.A. (1950). "The force exerted by surface waves on piles," Petroleum Transactions, AIME, Vol. 189, 149–157.

Nagai, S. (1969). "Pressures of standing waves on vertical walls," *Proc. ASCE*, Vol. 95, No. WW1, 53–76.

National Hurricane Center. (1990). "A storm surge atlas for Delaware Bay, Delaware and New Jersey hurricane evacuation studies," Storm Surge Group, National Hurricane Center, Coral Gables, FL 33146.

Noda, E.K. (1974). "Wave-induced nearshore circulation," J. Geophys. Res., 75, 27, 4097–4106.

NOS. (1992). "Tide tables for the East Coast of North and South America," National Ocean Survey, NOAA, Washington, DC.

Ochi, M.K. (1982). "Stochastic analysis and probabilistic prediction of random seas," in *Advances in Hydroscience*, Chow, V.T., Ed., V. 13, Academic Press, New York, NY, 218–376.

Outlaw, D.G., Durham, D.L., Chatham, C.E., and Whalin, R.W. (1977). "Los Angeles and Long Beach Harbors model study; model design," Tech. Report H–75–4, U.S. Army Waterways Experiment Station, Vicksburg, MS.

Parker, A.G. (1982). "Wave measurements using surface-mounted instruments," *Proc. Symposium and Workshop on Wave-Measurement Technology*, Marine Board, U.S. National Research Council.

Penney, W.G. and Price, A.T. (1952). "The diffraction theory of sea waves and the shelter afforded by breakwaters," Philos. Trans. Roy. Soc., A, 244(882), 236–253.

Peregrine, D.H. (1967). "Long waves on a beach," J. Fluid Mechanics, 27, 815–827.

Phillips, O. M. (1977). *The Dynamics of the Upper Ocean*, Camb. Univ. Press, London, 336.

Pugh, D.T. (1987). *Tides, surges, and mean sea-level: A Handbook for Engineers and Scientists*, John Wiley & Sons, New York, NY.

Putnam, J. A., Munk, W.A., and Traylor, M.A. (1949). "The prediction of longshore currents," Trans. Ame. Geophys. Union, Vol 30, 337–345.

Radder, A.C. (1979). "On the parabolic equation method for water-wave propagation," J. Fluid Mechanics, 95, 159–176.

Raichlen, F. (1966). "Harbor resonance," Chapter 7, *Estuary and Coastline Hydrodynamics*, A.T. Ippen, Editor, McGraw-Hill, New York, NY.

Raichlen, F. and Naheer, E. (1976). "Wave induced oscillations of harbors with variable depth," *Proc. 15th Intl. Conf. on Coastal Engr.*, ASCE, Honolulu, 3536–3556.

Reid, R.O. (1990). "Tides and storm surges," Chap. 9 in *Handbook of Coastal and Ocean Engineering*, Vol. 1, edited by J.B. Herbich, Gulf Publishing Co., Houston, TX.

Robb, A.M. (1952). *Theory of naval architecture*, Charles Griffen and Co., London, 58–62.

Roos, B. and Battjes, J.A. (1976). "Characteristic of flow in runup periodic waves," *Proc. 15th Coastal Engr. Conf.* ASCE.

Russell, J.S. (1838). "Report on waves," Report to the 14th British Association Advancement of Science, 417–496.

Sainflou, G. (1928). "Essai sur le digues martimes verticales," Ann. Ponts et Chauss, Vol. 98, No. 4, Paris.

Sato, S. and Goda, Y. (1981). *Coasts, ports and harbors*, 2nd ed., Shokokusha, Japan.

Savage, R.P. (1958). "Laboratory data on wave runup on rough and permeable slopes," J. of Waterways and Harbors Div., ASCE, Vol. 84, No. WW3, 852–870.

Saville, T. (1956). "Wave runup on shore structures," J. of Waterways and Harbors Div., ASCE, Vol. 82, No. WW2.

Saville, T. (1958). "Wave runup on composite slopes," *Proc. of the 6th Coastal Engr. Conf.*, ASCE.

Saville, T. (1960). "Discussion: laboratory investigation of rubble mound breakwater by R.Y. Hudson," J. of Waterways and Harbors Div., ASCE, Vol. 86, No. WW2.

Saville, T. (1962). "An approximation to wave runup frequency distribution," *Proc. 8th Coastal Engr. Conf.*, ASCE.

Schofield, R.B. (1974). "Speed of ships in restricted navigation channels," J. Waterways, Harbors and Coastal Eng. Div., ASCE, May.

Schureman, P. (1976). "Manual of harmonic analysis and prediction of tides," Special Publication No. 98, Coast and Geodetic Survey, U.S. Dept. of Commerce.

Seymour, R.J. (1982). "Review of wave measurement using in situ systems," *Proc. Symposium and Workshop on Wave-Measurement Technology*, Marine Board, U.S. National Research Council.

Shadrin, I.F. (1961). "The possibility of predicting longshore currents in tideless seas," Dyna. and Morpho. of Sea Coasts, 350–364.

Shepard, F. P. and Inman, D. L. (1950). "Nearshore circulation related to bottom topography and wave refraction," Trans. Am. Geophys. Union, Vol. 31, 555–565.

Shepard, F. P., Enery, O.K., and LaFond, E.C. (1941). "Rip currents: a process of geological importance," J. Geol., Vol. 49, No. 4, 337–369.

Sorensen, R.M. (1967). "Investigation of ship-generated waves," J. Waterways and Harbors Div., ASCE, Feb.

Sorensen, R.M. (1973a). "Ship-generated waves," *Advances in Hydroscience*, Academic Press, New York, NY, Vol. 9, 49–83.

Sorensen, R.M. (1973b). "Water waves produced by ships," J. Waterways, Harbors and Coastal Eng. Div., ASCE, May.

Sorensen, R.M. (1978). *Basic coastal engineering*, John Wiley & Sons, New York, NY.

Sorensen, R.M. (1989). "Port and channel bank protection from ship waves," *Proc. Ports '89*, ASCE, Boston.

Sorensen, R.M. and Seelig, N.W. (1976). "Hydraulics of Great Lakes inlet-harbor systems," *Proc. 15th Conf. on Coastal Engr.*, ASCE, Honolulu, HI, 1646–1665.

Sorensen, R.M. and Weggel, J.R. (1984). "Development of ship wave design information," *Proc. 19th Intl. Conf. on Coastal Engr.*, ASCE, Houston, TX.

Takada, A. (1970). "Wave runup, overtopping and reflection," *Proc. Japan Soc. of Civil Engrs.*, Vol. 182, 19–30.

Takada, A. (1977). "Wave runup, and overtopping," *Lecture Notes on Hydraulic Engineering*, Vol. B, Japan Soc. of Civil Engrs.

Tanimoto, K., Moto, K., Ishizuka, S., and Goda, Y. (1976). "Investigation on design formula of wave force on a breakwater," *23rd Japanese Conf. on Coastal Engr.*, Japan Soc. of Civil Engrs., 11–16.

Thornton, E. B. (1970). "Variation of longshore current across the surf zone," *Proc. 12th Conf. on Coastal Engr.*, ASCE, 291–308.

Thornton, E.B. and Guza, R.T. (1983)."Transformation of wave height distribution," Jour. Geophys. Res., Vol. 88, 5925–5938.

Tsay, T. K. and Liu, P. L.-F. (1982). " Numerical solution of water–wave refraction and diffraction problems in the parabolic approximation," J. Geophys. Res., Vol. 87, 7932–7940.

Tsay, T. K. and Liu, P. L.-F. (1989). " Numerical model computing wave propagation in an open coast," US Army Waterways Experiment Station, Coastal Engineering Research Center Report No. CERC–89–14, pp 244.

Tung, C. C. and Huang, N. E. (1973). "Combined effects of currents and waves on fluid force," Intl. J. Ocean Engr., Vol. 2.

U.S. Army Coastal Engineering Research Center (USACERC). (1984). *Shore protection manual*, 4th Edition, Volumes I & II, Waterways Experiment Station, Vicksburg, MS; available from the Superintendent of Documents, U.S. Government Printing Office.

US Army Corps of Engineers (USACE). (1987). "Great Lakes water level facts," U.S. Army Corps of Engineers, Detroit District, Detroit MI, January 1987.

U.S. Army Corps of Engineers, (USACE). (1989). "Water levels and wave heights for coastal engineering design," Engineer Manual No. 1110-2-1414, Washington, DC.

U.S. Army Corps of Engineers (USACE). (1991). "Lake levels for the Great Lakes," Bulletin published monthly by U.S. Army Corps of Engineers, Detroit District, P.O. Box 1027, Detroit, MI.

Unna, P.J.H. (1942). "Waves and tidal streams," Nature, Vol. 149, 219–220.

Van Oorschot, J.H. and d'Angremond, K. (1968). "The effect of wave energy spectra on wave runup," *Proc. 11th Coastal Engr. Conf.*, ASCE.

Weggel, J. R. (1972). "Maximum breaker height," J. Waterways, Port, Coastal and Ocean Engr. Div., Vol. 98, No. WW4, pp 529–548.

Weggel, J.R. and Sorensen, R.M. (1980). "Surging in the Shark River boat basin," *Proc. Ports '80*, ASCE, May, 640–660.

Weggel, J.R. and Sorensen, R.M. (1986). "Ship wave prediction for port and channel design," *Proc. Ports '86*, ASCE, Oakland.

Wehausen, J.V. and Laitone, E.V. (1960). *Surface waves*, Encyclopedia of Physics, Vol. 9, Springer-Verlag, Berlin, 522–553.

Wiegel, R.L. (1964). *Oceanographical engineering*, Prentice-Hall, Englewood Cliffs, NJ.

Wilson, B.W. (1972). "Seiches," *Advances in Hydroscience*, V.T. Chow, Editor, Volume 8, Academic Press, New York, NY.

XIII. NOTATION

A,B	basin horizontal dimensions
A	parameter related to the wave runup
A_b	basin surface area
A_c	channel cross section area
a	wave amplitude (= half the wave height)
a_c	wave crest elevation or incident wave amplitude
C	wave celerity
d	water depth
d_i	average section depth

F	Froude number
g	acceleration of gravity
H	wave height
H_b	breaking wave height
H_m	maximum wave height in a ship wave record
H_s	significant wave height
$H_{1/3}$	average wave height for the one third highest waves
K	elliptic function for enoidal waves
K_d	drag parameter in the Morison equation
K_s	shoaling coefficient
k	wave number
k_t	wave number at the toe depth of a structure
L	wave length
L_b	basin length
L_c	channel length
L_c''	additional channel length to account for end effects
L_r	length ratio
L_{rh}	horizontal length ratio
L_{rv}	vertical length ratio
m,n	harmonic oscillation modes
Q_o	empirical coefficient for overtopping estimate
Q_m	overtopping for regular waves
Q_i	overtopping for irregular waves
$Q_{1/2}$	average overtopping rate
$Q_{(v)}$	wind effect on overtopping
R_p	runup for p-percent probability of exceedance
R_s	runup associated with H_s
R_u	runup on a slopping beach or structure
T	wave period
T_H	Helmholtz period
$T_n, T_{n,m}$	resonant periods
T_p	wave period for the peak spectrum energy value
T_r	resonant period ratio or wave transmission coefficient
t	time
U_A	wind speed adjustment based on frictional drag effect
V	ship speed
V_{avg}	average flow velocity under a node
V_{max}	maximum flow velocity under a node
X	horizontal water particle excursion
x,y	horizontal coordinates

x_i section length

θ direction of wave propagation from the sailing line

XIV. GLOSSARY

A. Terms Pertaining to Ship-Generated Waves

Cusp locus angle—the horizontal angle between the sailing line of a ship and the cusp locus line.

Cusp locus line—a line that connects the points of intersection of transverse and diverging waves generated by a moving ship.

Diverging waves—waves generated by a moving ship that propagate out at an oblique angle to the ship sailing line.

Transverse waves—waves generated by a moving ship that propagate in the same direction that the ship is moving.

Vessel planing—when a moving ship is lifted by hydrodynamic forces to just skim the water surface and generate lower waves.

B. Terms Pertaining to Tides

Diurnal tide—A tide with one high water and one low water in a tidal day (Fig. 10.26).

Ebb tide—The period of tide between high water and the succeeding low water; a falling tide (Fig. 10.26).

Flood tide—The period of tide between low water and the succeeding high water; a rising tide (Fig. 10.26).

High tide—The maximum elevation reached by each rising tide.

Low tide—The minimum elevation reached by each falling tide.

Mixed tide—A type of tide intermediate to those predominantly semidiurnal and those predominantly diurnal (Fig. 10.26).

Neap tide—A tide occurring approximately midway between the time of new and full moon. The neap tidal range is usually 10 to 30% less than the mean tidal range.

Semidiurnal tide—A tide with two high and two low waters in a tidal day with comparatively little diurnal inequality (Fig. 10.26).

Spring tide—A tide that occurs at or near the time of new or full moon and which rises highest and falls lowest from the mean sea level.

Tidal day—The time of the rotation of the Earth with respect to the Moon, approximately 24 hours and 50 minutes (Fig. 10.26).

Tidal range—The difference in height between consecutive high and low (or higher high and lower low) waters (Fig. 10.26).

C. Terms Pertaining to Water Levels

Datum—The vertical reference to which soundings, ground elevations, or water surface elevations are referred.

Mean High Water (MHW)—The average height of the high waters over a 19-year period.

Mean Higher High Water (MHHW)—The average height of the higher high waters over a 19-year period.

Mean Low Water (MLW)—The average height of the low waters over a 19-year period.

Mean Lower Low Water (MLLW)—The average height of the lower low waters over a 19-year period.

Mean Sea Level (MSL)—The average height of the sea surface for all stages of the tide over a 19-year period, usually determined from hourly height readings.

Mean Tide Level (MTL)—A plane midway between Mean High Water and Mean Low Water. Not necessarily equal to Mean Sea Level.

National Geodetic Vertical Datum (NGVD)—A fixed reference adopted as a standard geodetic datum for elevations determined by leveling. Established in 1929. Also referred to as National Geodetic Vertical Datum of 1929 and Sea Level Datum of 1929.

D. Terms Pertaining to Hurricanes and Storm Surges

A limited number of definitions relative to hurricanes and storm surges are listed here. A more comprehensive list is also available (USACERC, 1984).

Central Pressure Index (CPI)—The estimated minimum barometric pressure in the eye (approximate center) of a particular hurricane. The CPI is considered the most stable index to intensity of hurricane wind velocities in the periphery of the storm; the highest wind speeds are associated with storms having the lowest CPI.

Design Hurricane—A representation of a hurricane with specified characteristics that would produce hurricane surge hydrographs and coincident wave effects at various key locations along a proposed project alignment. It governs the project design after economics and other factors have been duly considered. The design hurricane may be more or less severe than the SPH, depending on economics, risk, and local considerations.

Eye—In meteorology, usually the "eye of the storm" (hurricane); the roughly circular area of comparatively light winds and fair weather found at the center of a severe tropical cyclone.

Hurricane—An intense tropical cyclone in which winds tend to spiral inward toward a core of low pressure, with maximum surface wind velocities that equal or exceed 33.5 meters per second (75 mph or 65 knots) for several minutes or longer at some points. Tropical storm is the term applied if maximum winds are less than 33.5 meters per second.

Hurricane Path or Track—Line of movement (propagation) of the eye through an area.

Hurricane Stage Hydrograph—A continuous graph representing water level stages that would be recorded in a gage well located at a specified point of interest during the passage of a particular hurricane, assuming that effects of relatively short-period waves are eliminated from the record by damping features of the gage well. Unless specifically excluded and separately accounted for, hurricane stage hydrographs are assumed to include effects of astronomical tides, barometric pressure differences and all other factors that influence water level stages within a properly designed gage well located at a specified point.

Hurricane Surge Hydrograph—A continuous graph representing the difference between the hurricane stage hydrograph and the water stage hydrograph that would have prevailed at the same point and time if the hurricane had not occurred. (That portion of the hurricane stage hydrograph caused by the effects of the hurricane itself exclusive of the effects caused by the astronomical tides.)

Hurricane Wind Pattern (Isovel Pattern)—An actual or graphical representation of near-surface wind velocities covering the entire area of a hurricane at a particular instant. Isovels are lines connecting points of simultaneous equal wind velocities, usually referenced 9 meters (30 feet) above the surface, in meters per second, knots, or meters per hour; wind directions at various points are indicated by arrows or deflection angles on the isovel charts. Isovel charts are usually prepared at each hour during a hurricane, but for each half hour during critical periods.

Hypothetical Hurricane Based on Generalized Parameters—Hypothetical hurricane estimates based on various logical combinations of hurricane characteristics used in estimating hurricane surge magnitudes corresponding to a range of probabilities and potentialities. The Standard Project Hurricane is most commonly used for this purpose, but estimates corresponding to more severe or less severe assumptions are important in some project investigations.

Probable Maximum Hurricane—A hypothetical hurricane that might result from the most severe combination of hurricane parameters that is considered reasonably possible in the region involved, if the hurricane should approach the point under study along a critical path and optimum rate of movement. This estimate is substantially more severe than the SPH criteria.

Probable Maximum Water Level—A hypothetical water level (exclusive of wave runup from normal wind-generated waves) that might result from the most severe combination of hydro-meteorological, geoseismic, and other geophysical factors and that is considered reasonably possible in the region involved, with each of these factors considered as affecting the locality in a maximum manner. This level represents the physical response of a body of water to maximum applied phenomena such as hurricanes, moving squall lines, other cyclonic meteorological events, tsunamis, and astronomical tide combined with maximum probable ambient hydrological conditions such as wave setup, rainfall, runoff, and river flow. It is a water level with virtually no risk of being exceeded.

Radius of Maximum Winds—Distance from the eye of a hurricane, where surface and wind velocity are zero, to the place where surface wind speeds are maximum.

Standard Project Hurricane (SPH)—A hypothetical hurricane intended to represent the most severe combination of hurricane parameters that is reasonably characteristic of a specified region, excluding extremely rare combinations. It is further assumed that the SPH would approach a given project site from such direction, and at such rate of movement, to produce the highest hurricane surge hydrograph, considering pertinent hydraulic characteristics of the area. Based on this concept, and on extensive meteorological studies and probability analyses, a tabulation of "Standard Project Hurricane Index Characteristics" mutually agreed upon by representatives of the U.S. Weather Service and the Corps of Engineers is available.

Storm Surge—A rise above normal water level on the open coast due to the action of wind stress on the water surface. Storm surge resulting from a hurricane also includes that rise in level due to atmospheric pressure reduction as well as that due to wind stress.

Synoptic Chart—A chart showing the distribution of meteorological conditions over a given area at a given time. Popularly called a weather map.

Tidal Wave—In popular usage, any unusually high and destructive water level along a shore. It usually refers to storm surge or tsunami.

Transposed Hurricane—A hypothetical hurricane based on the storm transposition principle, assumed to have wind patterns and other characteristics basically comparable to a specified hurricane of record, but transposed to follow a new path to serve as a basis for computing a hurricane surge hydrograph that would be expected at a selected point. Moderate adjustments in timing or rate of forward movement may also be made, if these are compatible with meteorological considerations and study objectives.

Tropical Storm—A tropical cyclone with maximum winds less than 33.5 meters per second (75 miles per hour or 65 knots).

Typhoon—The term used to describe a tropical cyclone of hurricane strength in the Western Pacific.

Wave Setup—Superelevation of the water surface over normal surge elevation due to onshore mass transport of the water by wave action alone.

Wind Setup—On reservoirs and smaller bodies of water (1) the vertical rise in the still water level on the leeward side of a body of water caused by the wind stresses on the surface of the water; (2) the difference in still water levels on the windward and leeward sides of a body of water caused by wind stresses on the surface of the water.

E. Terms Pertaining to Basin Oscillations

Anti-node—The location in an oscillating basin of water where water particle motion is vertical and surface oscillation amplitudes are maximum.

Helmholtz Resonance—A mode of oscillation in a basin connected to a larger water body in which flow oscillates horizontally in the entrance and the basin water surface oscillates vertically while remaining essentially horizontal.

Node—The location in an oscillating basin of water where particle motion is horizontal and the surface oscillation amplitude is minimum.

Resonant Period—The period of free oscillation of a given mode of oscillation in a basin of water

Chapter 11

HYDROLOGIC STUDY FORMULATION AND ASSESSMENT

I. INTRODUCTION

Hydrologic investigations are usually complex, requiring the collection, management, and analysis of large amounts of information. The reliability of study results is always a major concern. Time and financial resources are often limited, but study objectives must be achieved within these constraints. This chapter focuses on four practical considerations in conducting hydrologic investigations: a) formulating the study, b) managing data, c) calibrating and verifying hydrologic models, and d) assessing the accuracy and reliability of study results.

The first topic is concerned with establishing the direction, scope, and resource requirements for a hydrologic study. Clear definition of these parameters for a study will help to ensure that objectives are properly formulated and met and that resources are utilized in an optimal manner.

The second topic deals with the best ways to organize and utilize data. Efficient data management is important because the acquisition and processing of hydrologic data often consume a substantial portion of available study resources. The nature and role of Geographical Information Systems (GIS) are a major focus of this section.

The third topic deals with the problem of identifying and validating values for model parameters. Parameter estimation is a requirement for many types of analysis. For example, aquifer transmissivities are needed for a well yield study, ground water flow rates must be estimated for a contamination study, and soil types and other parameters must be identified to generate surface runoff for a flood analysis.

Finally, this chapter addresses how to evaluate the accuracy and reliability of study results. Concepts and information provided in this chapter are applicable across the range of technical areas covered elsewhere in this handbook.

II. STUDY FORMULATION

Several key elements should be addressed when formulating a hydrologic study, including: a) study purpose and scope, b) level of detail, and c) methods and tools. Each of these elements is discussed in the following sections. Fig. 11.1 is a schematic diagram showing how these elements fit into the formulation process, which culminates in the preparation of a technical study work plan, and leads into performance of a successful hydrologic study.

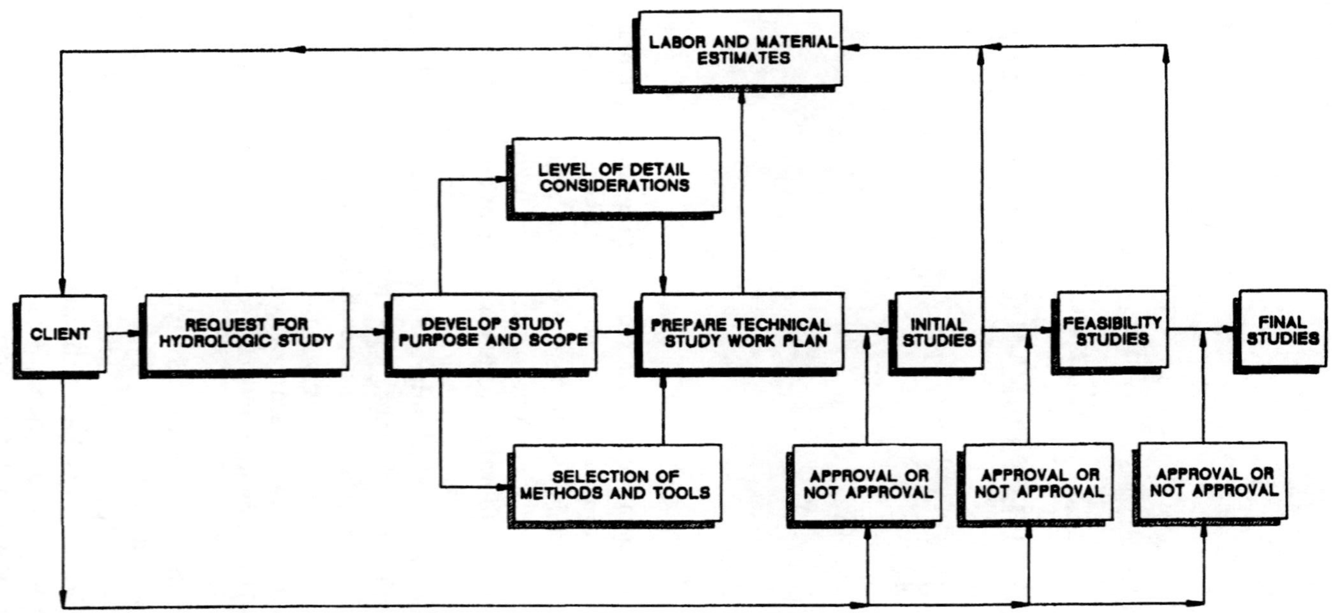

Figure 11.1.—Hydrologic Study Formulation—Process and Stages.

A. Study Purpose and Scope

The typical purpose of a hydrologic study is to provide the technical information that enables decisions to be made on reducing or eliminating existing or potential water-related problems. Examples of problems are:

1) Recurring flood damages in urban and rural areas.
2) Urbanization increases runoff volumes and rates to downstream development.
3) Population growth increases needs for water supply and/or hydroelectric power.
4) Pumping ground water during drought periods depresses ground water levels.
5) Increases in watershed erosion and accumulated sediments result in a loss of reservoir storage or channel carrying capacity.
6) Existing or projected water supply demands exceed reservoir carryover storage.

Hydrologic studies are also conducted to define potential flood hazard areas for regulatory and/or planning purposes.

Generally, the purpose of the hydrologic study will be established prior to the involvement of the hydrologist; however, a hydrologist should be involved in the development of the scope of work, as it must be designed to meet the purpose of the study. The scope of work defines activities required to satisfy the purpose of the study within constraints imposed by schedule, funding, personnel, and/or technical considerations. The scope forms a key part of the Technical Study Work Plan. A tentative scope may be provided by a client, for example, in a request for proposal. Part of the hydrologist's work may involve refining the client's scope to develop a complete scope of work.

There will likely be several specific constraints that must be considered during the development of a scope of work. Examples of constraints are:

1) Time—Planning, design, and construction schedules require hydrology studies to be completed within a specific time period. Institutional concerns and the need for environmental documentation can also affect the time available.
2) Funding—Rarely is a hydrologic study completed without working within funding constraints. Typically, hydrologic studies are funded as part of overall planning studies. These planning funds are generally predetermined based on work previously completed, availability of mapping and data, funds required for similar type and size projects, and estimated construction costs for proposed alternatives to satisfy the study purpose.
3) Personnel—The availability of sufficient qualified personnel who can complete the work within the time schedule may impact the scope of work. If required personnel are not available, it may be necessary to subcontract to other agencies or consultants. In this case, there must be adequate time and funds to advertise, negotiate, and administer outside work activities. It may be necessary to reduce the scope of work to maintain the schedule or negotiate for more time within the original planning schedule.
4) Technical aspects—The methodologies to be used and level of detail of the hydrologic study are very important technical aspects. Both must be appropriate for the time, funding and personnel available, while providing sufficient information for rational decisions. These aspects are discussed in more detail in subsequent sections.

After considering the original purpose of the study and potential constraints, the hydrologist develops a scope that presents the tasks needed to obtain data and apply the selected methodologies at the appropriate level of detail. The scope should include data needs and sources, analyses to be conducted, interactions among different disciplines, documentation to be prepared, and client-coordination points.

B. Level of Detail

Probably the most significant action in formulating a hydrologic study is establishing the level of detail required to satisfy the purpose of each phase of the study. Because the level of detail directly affects the schedule, budget and results of a study, it is vital that an experienced hydrologist be involved at this stage.

The ultimate application of a hydrologic study is a key factor in establishing the appropriate level of detail. Therefore, it is prudent to consider all of the anticipated studies and in some cases the required operation and maintenance activities during the study formulation process. An experienced hydrologist should determine if the hydrologic analyses are needed for a reconnaissance (initial) type of study, for a more detailed feasibility study, or for final design of a project. Obviously, the level of detail required for a reconnaissance type of study is not expected to be as extensive as for a feasibility study or final design.

Typically, during reconnaissance-type studies, the data are inadequate or the funds are insufficient to determine specific design values (e.g., for peak discharges, sediment yields, or runoff volumes). Since the purpose of the reconnaissance-type study is normally to determine whether the project is close to being economically feasible and worth further studies and refinement, solutions with upper- and lower-bound estimates of key parameters may provide a sufficient level of detail. An example is sizing drainage facilities based on estimated peak discharges from similar basins in the region (see Fig. 11.2). A relatively simple analysis (that could be refined during later studies) would be to use upper- and lower-bound peak discharge estimates corresponding to particular drainage areas. A range of sizes for project components would then be determined. If the full range of sizes appears feasible or would not be feasible, then additional refinement is not required because the feasibility question has been answered. Even if the upper- and lower-bound analysis is not definitive, it still provides a better understanding of the sensitivity of a key parameter (peak discharge in this example) to sizing of project facilities.

Even though specific design values would not be generated in a reconnaissance-type study, the level of detail is important. Results from a reconnaissance study are critical, as they may indicate either that a project should be eliminated or that studies should be expanded. Therefore, while expenditures are relatively small for reconnaissance-type studies, their ultimate value can be significant. For example, eliminating a potentially feasible flood control project could have significant and continuing impacts on a community in need of flood protection. Conversely, large expenditures to continue studies on a project that should have been dropped reduces available funds that could have been used on more cost-effective projects.

When selecting the level of detail for each level of study, the use of data and methodology that can be expanded upon or improved as the study progresses should be considered. For example, it may be

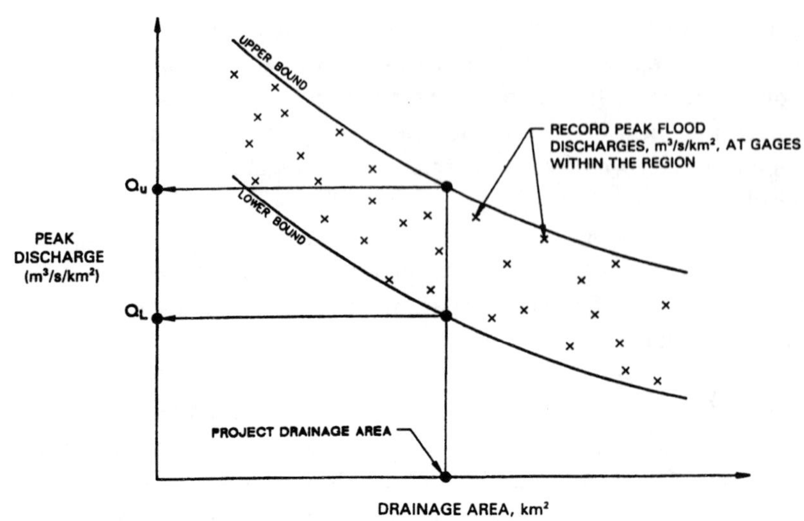

Figure 11.2.—Envelope Curves—Peak Discharge vs. Drainage Area.

reasonable to include rather detailed items such as aerial photography, topographic mapping, field surveys, field samples, and gage readings for a reconnaissance-type study that will later be expanded, because such items can be used at all levels of a study.

The selection of appropriate levels of detail for the various study phases not only facilitates development of the Technical Study Work Plan, but also helps to ensure that only those analyses essential to satisfy the purpose of the study are performed.

C. Selection of Methods and Tools

The selection of methods and tools is particularly important in developing a Technical Study Work Plan. Over the last two decades, the development of computer programs to assist in performing hydrologic analyses has been dramatic. The routine use of desk top computers has brought this technology to essentially all practicing hydrologists, where in past decades only those with access to major computer facilities could utilize available computer programs. The proliferation of computer software provides more alternatives to assist in performing the required studies; however, the wide variety of software makes the selection of the appropriate technology a more difficult task.

Each study may have a number of project-specific items to consider before selecting the methods and tools that best fit the particular study. The following is a list of items that may serve as a starting point for the selection process:

1) project goals and objectives;
2) scope and magnitude of the project;
3) availability of data;
4) difficulty of obtaining essential field data;
5) accuracy requirements;
6) usability of methodology during subsequent studies;
7) staff technical capabilities and training requirements;
8) time constraints; and
9) applicability of computer programs to a specific study (including software documentation and support, and compatibility with other study software packages).

Once the methodology is selected, its application must be clearly defined in the scope of work and incorporated into the Technical Study Work Plan.

D. Preparation of a Technical Study Work Plan

The Technical Study Work Plan is a key document for guiding a hydrologic study. It incorporates the scope of work, which, as discussed previously, defines major work activities and special studies. The work plan should also include a critical path schedule, estimates of required labor and materials, and at least a preliminary content outline of the report(s). Deliverables and milestones should be identified. Key personnel should be shown in a project organization chart, which includes client and other important agency contacts, sub-consultants, and in-house management and technical support staff.

Developing a preliminary outline of the anticipated report(s) can be useful in making sure that all of the major activities have been included in the work plan. Relatively simple flow diagrams and/or bar charts are also useful in preparing and presenting a work plan, as they show the interrelationship of activities and schedule and also serve as a basis for estimating staffing requirements.

Activities that would assist the hydrologist during the preparation of a Technical Study Work Plan include:

1) reviewing available data, mapping, and reports;
2) making site visits, if appropriate;

3) reviewing available methods and tools;
4) coordinating with interested private groups and governmental agencies;
5) evaluating schedule constraints;
6) considering staff constraints;
7) considering funding constraints; and
8) meeting with interdisciplinary team members.

Fig. 11.3 is a schematic diagram that links the activities used to prepare a Technical Study Work Plan to subsequent studies. The intent is to show that the work plan should consider reconnaissance through final studies when developing the total project work plan. Note that the original work plan for studies following the reconnaissance study should be reviewed and refined as more information and reconnaissance study results become available.

The Technical Study Work Plan serves as a guide to the people involved in the performance of the work. With the work plan, the manager responsible for the study has a "blueprint" to work from that is extremely useful in keeping the study focused on critical-path items and on the appropriate level of detail. The work plan is also helpful to the staff working on specific elements of the study by indicating the expected level of effort required, the projected schedule, and where each element fits into the overall study effort.

III. DATA MANAGEMENT

Hydrologic analyses tend to be data intensive. The physical systems being analyzed are often large and complex, so that substantial quantities of data are required for their representation. Also, the analyses themselves are complex, with a variety of interdependent computational elements. The need to transfer information between computational elements can be significant. Because the acquisition, processing, and management of data and information can require a substantial portion of the resources allotted for a study, performance of these tasks in an efficient and reliable manner is necessary.

Hydrologic analyses are performed in both the time and space domains. Data variables routinely treated as time series include precipitation and other meteorologic variables, water discharge of various types (e.g., streamflow, ground water flow, etc.), water stage (or elevation) and storage, water quality parameters, and sediment loads. Spatial variables include topographic and geologic characteristics of basins, soil type, land use, vegetation, and hydraulic characteristics of rivers and flood plains. The following sections deal first with data management considerations in general and then with geographic information systems in particular. The latter are especially powerful for storing and manipulating spatial data.

A. Data Management Concepts

Fig. 11.4 illustrates aspects of data management for a water resource study. Elements of the system include 1) a data-loading module for entering data from various sources into a database, 2) applications software that can retrieve data from, and store results in, the database, and 3) utility programs that perform functions such as data editing, display of data in graphical and tabular form, and mathematical transformation and statistical analyses of data. With such a system, basic data such as precipitation and streamflow can be loaded into data storage and reviewed and perhaps edited with utility programs. Interdependent applications programs can be used to perform the analyses, using the database to pass information between computational elements. Utility programs can be used to view and summarize results in various forms, including report-quality tables and graphs.

There are a number of types of data management systems, each having properties designed to meet specific needs. Because hydrologic investigations often require manipulation of large quantities of time-

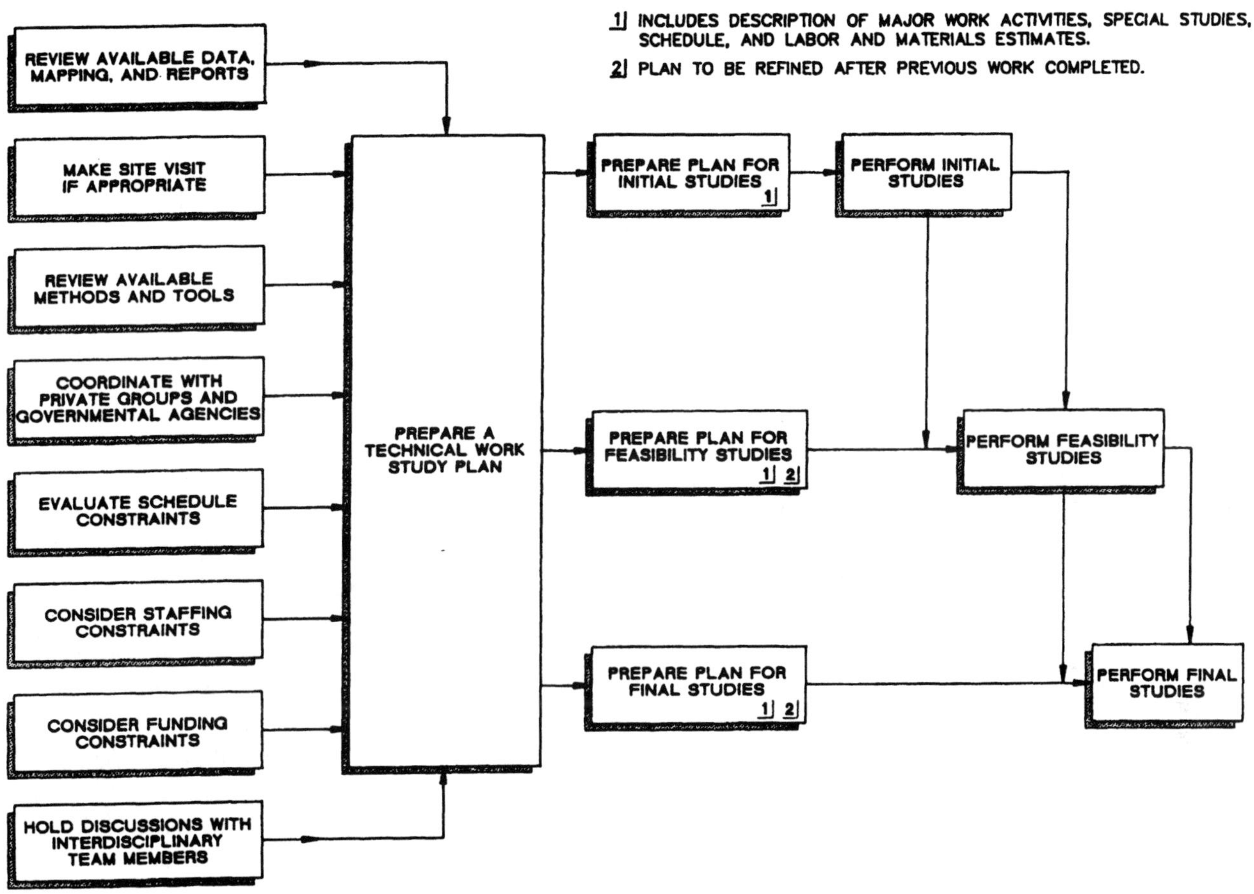

Figure 11.3.—Technical Study Work Plan.

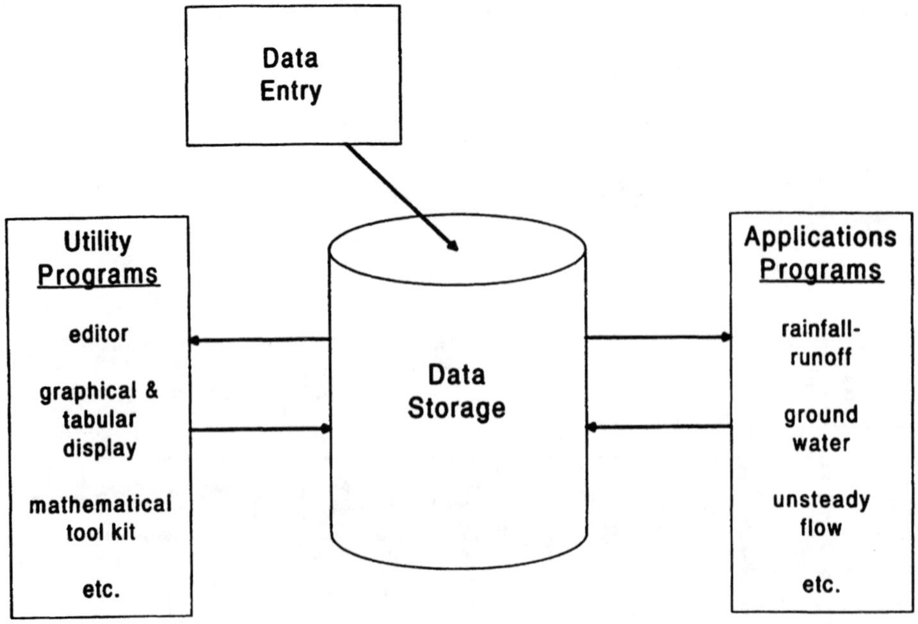

Figure 11.4.—Data Management for Hydrologic Studies.

series data, a data management system that deals with blocks of data can provide significant benefits. For example, individual blocks (records) of data might consist of daily-flow values for a year. By storing data in this manner, an entire year of daily flows can be handled as a single element and accessed with one "search" of the database. A reference by the U.S. Army Corps of Engineers (USACE, 1990) describes a nonproprietary data management system designed for efficient handling of time-series data for which a number of utility programs have been developed.

Data collection and processing are generally essential elements of a hydrologic investigation. Fortunately, much data are available on electronic media (e.g., compact disk), so that data entry can be performed relatively efficiently. For example, software can be used to retrieve data that is in the United States Geological Survey's WATSTORE format, in formats used by the National Weather Service, or in the Standard Hydrological Exchange Format used by various federal agencies (NWS, 1983). Once data have been entered into data storage, utility programs can be used to plot, tabulate, or develop statistical summaries of the data. Such initial analysis can greatly facilitate subsequent use of the data.

Fig. 11.4 shows applications programs communicating with the "data storage." This communication requires that such programs be developed or adapted with specific capabilities for this purpose (USACE, 1991). Alternatively, pre- and post-processors can be developed to enable such communication.

B. Geographic Information Systems (GIS)

A geographic information system (GIS) has been defined in various ways, usually related to its ability to process spatial data through capture, storage, retrieval, analysis, and display. A GIS, apart from computer mapping systems, does not simply display input-mapped data but creates new information through updating, query, and modeling capabilities. The most significant attribute of a GIS is its ability to relate pieces of spatial information with a relational database. This facilitates efficient handling of two generic classes of spatial data: cartographic data, describing the location and topography of map features, and attribute data, describing the characteristics of these features.

Hydrology is concerned with spatial data, namely the movement of water through space and time. Since a GIS is designed to manage and manipulate spatial data, such a system provides the tools to better assess and model the spatial relationships inherent in the hydrologic process. For example, a GIS may be used to evaluate how land use changes may affect the movement of water through a watershed. A GIS enables the user to manipulate data and then view the results graphically and statistically all from one database system.

1. Emergence of GIS Computer mapping and analysis systems have been in existence since the early 1960s when grid-based software packages were developed. The 1970s brought larger and more complex systems such as the Dual Independent Map Encoded (DIME) file, developed by the U.S. Census Bureau, and SYMAP, developed at Harvard University. As more complex software packages were developed, the hardware necessary to run the statistical calculations and store the large data files could not keep pace with increasing processing requirements. However, within the last five to ten years, computer technology has improved and what once required a large main frame computer can now be stored in much smaller and affordable desk top workstations, personal computers, and disk or tape drives. With this capability came the ability to run sophisticated GIS packages such as ARC/INFO (Environmental System Research Institute (ERSI), Redlands, California) on relatively inexpensive hardware platforms. Smaller public agencies and private companies have taken advantage of the technology because affordability, technological limitations and disk space are no longer limiting factors.

2. Definition and Basic Components of a GIS There is a wide variety of computer software products on the market referred to as GIS, but a true GIS must contain all of the following major components (Marble, 1987):

a) A data input subsystem that collects and/or processes spatial data derived from existing maps, remote sensing, or other digital data files.
b) A data storage, retrieval, and update subsystem that organizes spatial data in a form permitting quick retrieval, as well as the ability to perform updates and modifications to the spatial data.
c) A data manipulation and analysis subsystem that performs a variety of tasks to create new data by changing the form of the data through user-defined aggregation, query, or overlay parameters.
d) A data reporting subsystem that is capable of displaying all or part of the original database as well as manipulating data and the output in tabular and map form.

What sets a GIS apart from computer mapping systems, such as Computer Aided Design (CAD) systems, is the ability to both automatically synthesize existing layers of geographic data and to update a database of spatial entities. This ability stems from the way information is stored in a GIS. A GIS is composed of two basic pieces of information: locational data represented by points, lines, polygons, or grid cells, and thematic attributes associated with these features. These two pieces of information are automatically linked in the GIS, and when either the mapped data or the tabular data are manipulated, both sets of data are updated and adjusted to maintain the relationship between them.

For example, a map of a hydrologic stream network is input into the GIS. Physical attributes used to describe and characterize a stream segment are assigned to the individual graphic element used to display the locational features of the stream segment. Certain attribute information, such as length, is automatically calculated while other attribute information, such as stream type, stream name and roughness coefficient, can be input as needed. The attribute data are stored in a file associated with or linked to the graphic (locational) information.

a. Data Structure Types. A GIS stores spatial data in one of two basic storage structures: raster and vector. Raster-based systems store data as a series of grid cells or in a picture element (pixel) format. A grid is essentially placed over the mapped features, and the dominant data type is assigned to an individual cell. Raster systems are more efficient in data storage and operation of analytical data manipulation; however, the resolution and accuracy of the data are limited to the size of the grid cell. The data resolution can be improved by reducing the cell size, but data volume subsequently increases. In addition, graphic output may be unacceptable since the grid cells produce a stair–step effect, making it difficult to represent linear features. Since digital remote sensing data is in raster format, raster-based systems are used to process satellite data.

Vector-system structures store spatial data as a series of points, lines, and polygons that define point, linear, or aerial map features. Vector systems can portray any size of information and are limited by the resolution of the input data. The output displays from a vector-based system look like a standard map product.

The distinction between raster and vector systems is becoming increasingly fuzzy. Vector-based systems are integrating and adding raster-based capabilities, and vice–versa. Both raster and vector systems have distinct capabilities and the trend in software development is to integrate both approaches to take advantage of the capabilities inherent in both data structure types.

b. Data Layering. A key concept with GIS is data layering. Information is separated into "layers" based on a common thematic characteristic, such as soils, parcels, roads, well points, etc. Data layers are defined and differentiated depending upon the types of applications and the resolution of the data.

c. Coordinate Referencing. Data input to a GIS is internally referenced and externally registered to an earth-based coordinate system such as State Plane or Universal Transverse Mercator (UTM). Tic marks with known coordinate values are established, and all graphic features are related to these tic marks. This process provides a common reference between data layers, allowing the GIS to display separate data layers on top of each other. The accuracy of a data base can be highly dependent on how well map features are located in relation to a coordinate reference system. As a result, an important consideration with GIS database development is the quality of base maps and coordinate referencing.

d. Data Input. GIS data input and database development can be accomplished in a variety of ways including manual digitizing, scanning, data transfer, and data purchase. The most common form of data input is manual digitizing. Digitizing involves the use of a digitizing table and a hand-held cursor. The user traces the map features with the cursor, and the computer stores the information as coordinates. Digitized input can be labor-intensive, and the quality of the data is dependent on the digitizer's skill.

The use of scanning has increased with improved technology in recent years. The immediate product from a scanner is a pixel image that can be used as a graphic or picture background if needed. If data from the map are to be assigned attribute information and be used for GIS analysis, the scanned image must be vectorized (i.e., defined as points and lines in separate data layers). Although the capability of the computer to distinguish information from scanned images has greatly improved, a human operator is still required to help the computer make decisions.

Existing digital information can also be transferred into a GIS database from other computerized mapping systems (such as CAD) and between GIS databases. Data transfers save data input time but can be costly to acquire as well as time-consuming to convert into a compatible format. Commonly used data transfer formats include DXF, DLG, and ASCII. Because there are often fundamental differences between computerized mapping software programs, data transfers can result in the loss of certain types of data.

Digital data can also be purchased from a number of data sources. The United States Geological Survey (USGS) sells planimetric data (roads, streams, jurisdictional boundaries) in DLG format and elevation data as Digital Elevation Models (DEM). The U.S. Census TIGER files include road and address information as well as demographic and economic information. The U.S. Soil Conservation Service is in the process of preparing soils data in digital format. In addition, local state, county, and city agencies may have digital data, such as land use or land ownership, available for purchase. Issues related to using existing digital files include proprietary issues, cost, compatibility, and quality of data.

3. ***GIS Capabilities*** A GIS includes many map manipulation and analysis capabilities that can be used in hydrologic applications. The following list displays GIS capabilities.

a) Spatial query involves selecting particular features using Boolean queries and/or algebraic manipulations. For instance, soils with a high erosion potential can be highlighted by selecting soil categories with a soil erosion value greater than a certain criterion.
b) Polygon–overlay analysis involves overlaying two or more data layers to create a new layer of information. For example, this process may be used to create a coverage representing constraints to development of a pipeline by combining and rating information on slope, soils, and geology.
c) Buffering involves creating a new polygon around a map feature or features based upon a specified distance value. An engineer or planner may use the GIS buffering capabilities (along with overlay techniques) to investigate the type of land uses that occur within 500 feet of a stream channel, for example.
d) Surface analysis involves the capabilities to generate slope and aspect information from elevation data. A GIS uses an algorithm such as triangulation and user-specified parameters to generate slope or aspect coverages that can then be overlain on other data layers such as geological units to identify suitable locations for buried pipes.
e) A GIS includes advanced computer graphic capabilities to allow the display of data in a variety of formats and quality. Results can be displayed graphically on the computer screen or on hard copy maps generated by pen or electrostatic plotters. Information can be displayed at any user-specified size or scale limited only by the graphic hardware device.

4. GIS Applications in Hydrology GIS technology provides a rich set of tools for processing data for hydrology-related applications. The ability of the GIS to manage, manipulate, and display spatial data provides the GIS user with new and enhanced ways to investigate a problem that are not necessarily available or practical with CAD or traditional manual methods. Because a significant percentage of the data collected and evaluated in relation to water resource issues have a spatial component, the potential for applying GIS to hydrologic modeling, ground water monitoring, watershed management, water quality assessment, and other hydrology applications is considerable.

The following describes some general categories of GIS applications to hydrology issues. A more detailed discussion of the application of GIS with hydrologic modeling (specifically the integration of GIS with the SCS curve number model for runoff volume estimation) is provided since this is a straightforward and common application of GIS technology to hydrology issues. Undoubtedly, the number and types of applications will continue to grow in the future.

a. Hydrologic Modeling. Many hydrologic models have parameters defined in terms of land use, land cover, soils, slope, and other factors that generally vary across space. The SCS runoff curve number model (described in Chapter 3) is a widely used and accepted model that considers the combined effects of soil type, cover type, treatment, antecedent moisture conditions, and other hydrologic conditions to predict runoff volume. Several recent studies have used GIS to a variable degree to help store and manipulate soils, land cover, and other hydrologic input data (Stuebe and Johnston, 1990; Warwick and Haness, 1994). In some instances, the GIS was used simply for storing data on physical land characteristics and rainfall data, and an associated software program was used to retrieve information from the GIS to compute several runoff parameters. In other studies, many of the calculations and computations were performed using the GIS.

To illustrate a practical application of GIS, the following outline describes typical ways that a GIS can be used to generate and store data associated with the SCS curve number model.

1) Input soils polygons from SCS soil survey maps by scanning, manually digitizing, or purchasing in digital form. Each soils polygon is labeled with its SCS soils classification identification. Using the tabular data management capabilities of the GIS, each polygon is reclassified into one of the four hydrologic soils groups used with the SCS method. Boundaries between like hydrologic soils groups are eliminated to create a hydrologic soils data layer.

2) Input land cover information from existing land use maps or through interpretation of satellite data. Remote sensing has been used successfully to derive land cover estimates for the SCS curve number model in past studies (Ragan and Jackson, 1980). Land cover data are reclassified and dissolved, if necessary, into the cover classification used for the SCS method.
3) Overlay soils and land cover data layers to identify areas with unique combinations of hydrologic soil groups and land cover attribute data (hydrologic response units). Based upon the attribute data, runoff curve numbers are assigned to each hydrologic response unit polygon. The GIS is then used to efficiently and accurately determine the total acreage of each hydrologic response unit per watershed. Each individual hydrologic response unit is divided by the area of the total watershed to determine a spatially-weighted runoff curve number. The curve numbers are input into the runoff equation to estimate runoff. The computations can be programmed into the GIS to be calculated automatically.

The advantages of using the GIS to generate watershed data include gains in precision and speed of calculation over manual procedures (Berry and Sailor, 1987). The benefits of using a computerized system are most evident with repeated iterations of calculations. Repeated simulations of a variety of proposed levels of development can be made rapidly, and the effects on runoff can be assessed and better understood. The "what if" questions a GIS can effectively answer provides valuable insights into the spatial ramifications of planned activities.

b. Other Water Resource Applications. The spatial processes related to water quality are similar to hydrologic (runoff) processes. Typically, the same input variables used for hydrologic modeling (land use, land cover, soils, topography) are also used as input variables for water quality and soil loss assessment and modeling. Several studies have used a GIS to implement the Universal Soil Loss Equation (USLE) or similar model types to calculate potential soil loss (Gilliland and Baxter–Potter, 1987; Evans and Miller, 1988; DelRegno and Atkinson, 1988). GIS has useful applications with respect to water quality modeling and nutrient loading assessment by helping to incorporate spatial factors such as distance to streams into the modeling process (Levine and Jones, 1990).

For less technical applications, GIS has been successfully applied to water resource management and planning programs that often deal with the review and manipulation of mapped resource data for large geographic areas (Hamlett et al., 1991; Baker and Panciera, 1990). The data storage and manipulation capabilities of the GIS allow users to view an entire study area at one time or to focus in on a particular area of interest. The ability to view and combine separate data layers to create new information is particularly effective for planning applications.

Remote sensing can provide information for large geographic areas and is directly compatible with raster-based GIS. The combined capabilities of remote sensing and GIS have been utilized to develop watershed-level information on evapotranspiration, canopy cover, land use, and vegetation.

Most of the applications described above deal with polygon- (or area-) related information. GIS is also an effective tool for handling line or network information sets such as water distribution pipelines or stormwater drainage networks. GIS technology has been used to prioritize, schedule, and coordinate infrastructure maintenance programs. Urban runoff pollution studies have used GIS to track the locations and characteristics of storm drain outfalls.

C. Conclusion

Existing data management technology has the potential to dramatically alter the way hydrologic studies are performed and to contribute to new levels of understanding of hydrologic processes (Lanfear, 1989). As the technology continues to improve, the ability to efficiently manage the tremendous quantities of spatial data associated with hydrologic modeling and analysis will likely allow for a more complete consideration of the spatial relationships inherent in hydrologic processes.

IV. CALIBRATION AND VERIFICATION OF HYDROLOGIC MODELS

A. What is Calibration?

Earlier chapters of this handbook describe various models for analysis of hydrologic systems. All of these models "interrelate, in some time reference, an input, cause or stimulus, of matter, energy, or information to an output, effect, or response, of information, energy, or matter . . ." (Dooge, 1973). Some of the models described are causal—the interrelationship of input to output is based on functional representation of the laws of thermodynamics or conservation of mass, momentum, or energy. The unsteady streamflow routing models of Chapter 6 are causal. They solve equations that describe motion of fluid in an open channel. Other models are empirical—they represent only the functional relationship of observed output to observed input. The direct correlation model described in Chapter 8 for estimating runoff volume is an example of such an empirical model. The Chapter 8 model does not consider the process by which rainfall is converted to runoff. Instead, it uses historical data to establish a relationship with which an engineer can predict runoff, given rainfall estimates.

To use a selected model of either type, the analyst must estimate parameters of the functions that relate output to input. A parameter represents a property or characteristic of the system under specified conditions (McCuen and Snyder, 1986). For example, in the ground water models presented in Chapter 5, hydraulic conductivity is a parameter that describes a property of the aquifer. Model parameters may be classified as physically-based parameters or as calibration parameters. Physically-based parameters are those that can be observed or estimated directly from measurements of the system characteristics. Calibration parameters, on the other hand, are lumped single-valued parameters that cannot be quantified from direct measurements. They represent integrated, spatially- and temporally-averaged conceptual approximations to system components.

How each calibration parameter is estimated depends on the data available. If input and simultaneously-observed output data are available, the calibration parameters may be found by comparing results with alternative parameter values to find those that optimize the fit of predicted output to observed output. For example, the parameters of a unit hydrograph model may be found by computing runoff hydrographs with alternative calibration parameter values. These computed hydrographs are compared with the observed hydrograph. The "best" parameters are those that yield the "best" fit. This optimization procedure is described in further detail in the following subsection.

Unfortunately, as Loague and Freeze (1985) point out, "when it comes to models and data sets, there is a surprisingly small intersecting set." That is, the input and output data necessary to search for the calibration parameters often are not available. This is referred to herein as an ungaged case. Data may be missing, sparse, or unreliable. Furthermore, for many studies, estimates of hydrologic system output are required for the forecasted future and for with-project conditions. Data are never available to permit parameter estimation via optimization for these conditions. When data are available at other sites in the homogeneous region where the estimate is required, a model can be fitted for use at ungaged sites in the region. The section titled Estimating Model Parameters in the Ungaged Case presents alternatives for parameter estimation for such cases.

B. Calibrating a Model with Process Input and Output Data Available

If observed system input and simultaneously-observed output data are available, the calibration parameters can be estimated via model parameter optimization. To accomplish this, the analyst must select 1) the criterion for measuring fit and 2) the procedure to identify the best (optimal) parameter estimates, given the criterion selected.

1. Criteria for Measuring Fit The assessment of goodness of fit may be either subjective, such as is possible with a graphical comparison, or objective, such as is possible with statistical indices. The purpose of such assessment is to evaluate the accuracy of a model. Factors that cause inaccuracy in model predictions may introduce either systematic or nonsystematic (i.e., random) errors. Bias is the term used to indicate systematic errors; precision is the term used to indicate random errors. Accuracy is the cumulative effect of bias and precision.

A visual comparison of observed values to values predicted with alternative parameters provides a quick assessment of model fit. Certain errors, such as those in peak flow rates or volumes, are evident immediately from such a comparison. Further, graphical comparison often provides the analyst with a better understanding of model capabilities and sensitivity to parameters. Figs. 11.5a to d show four plots that illustrate combinations of the effects of bias and imprecision, with the 45° line indicating equality of predicted and measured magnitudes. Both Fig. 11.5b and 11.5d show a tendency to overpredict; thus, they show a positive bias. Neither Fig. 11.5a nor 11.5c show a significant overprediction or underprediction. Thus, they are unbiased since there is no systematic error evident in the plots.

While Fig. 11.5c may indicate that the prediction model is not biased, there still is sufficient scatter that suggests the model is not accurate, i.e., the model does not yield precise estimates. Thus, the scatter of points in Fig. 11.5c indicates a model that provides unbiased but imprecise estimates. The scatter of Fig. 11.5a indicates a prediction model that provides precise, unbiased estimates; thus, it is an accurate model. The scatter of Fig. 11.5b suggests that the model provides precise but biased estimates because there is minimal scatter about the central tendency (dashed line). The model for Fig. 11.5d provides biased, imprecise estimates and is thus highly inaccurate.

In addition to the overall biases evident in Figs. 11.5b and 11.5d, models may provide local biases. That is, a model may provide accurate estimates for some range of predicted values but inaccurate values in other ranges. For example, in Fig. 11.5e, the model provides reasonable estimates for low values but shows an increasing bias for the large values. In Fig. 11.5f, the bias is negative for high values of the magnitude and positive for low values. The model of Fig. 11.5d always overpredicts because the predicted value is greater than the measured value.

Other types of data may require other forms of graphical analysis; however, in these cases, bias, precision, and accuracy should still be assessed. James and Burges (1982) suggest four types of plots for comparing model results with recorded data:

a) A continuous time-series plot of predicted flow versus observed flow. Fig. 11.6 is an example of such a plot. The model provides biased estimates of the flood peaks; all of the measured peaks are underpredicted by the model. However, the model gives unbiased, precise estimates of the recession rate of the hydrograph. The baseflow also appears to be unbiased. Based on these subjective assessments of the goodness of fit, model coefficients related to peak discharges need to be adjusted to eliminate the bias. Recession and baseflow coefficients appear to be accurate and do not require adjustment.
b) A continuous time-series plot of the difference between observed and predicted flow. This is similar to the first plot, but the significance of differences is more obvious. With this plot, local biases can be detected. This can lead to improved parameter estimation.
c) A plot of cumulative sum of departures from the observed mean for the predicted series versus cumulative sum of departures of the observed series. This will show the cumulative impact of errors and may lead to improved parameter estimates. For example, if the infiltration parameter of a rainfall–runoff model is incorrectly estimated, the error in the predicted total volume will accumulate. This can be seen with a plot of cumulative departures.
d) A scattergram of predicted versus measured values, which was discussed previously.

These plots can be prepared to compare results from models of other hydrologic systems. Although assessments made with graphical comparisons are subjective, they do provide information regarding goodness–of–fit that may not otherwise be apparent with statistical measures.

A variety of numerical measures have been proposed to objectively quantify the fit of predicted to measured values. These indices are typically used for one of two functions: single-valued variables or

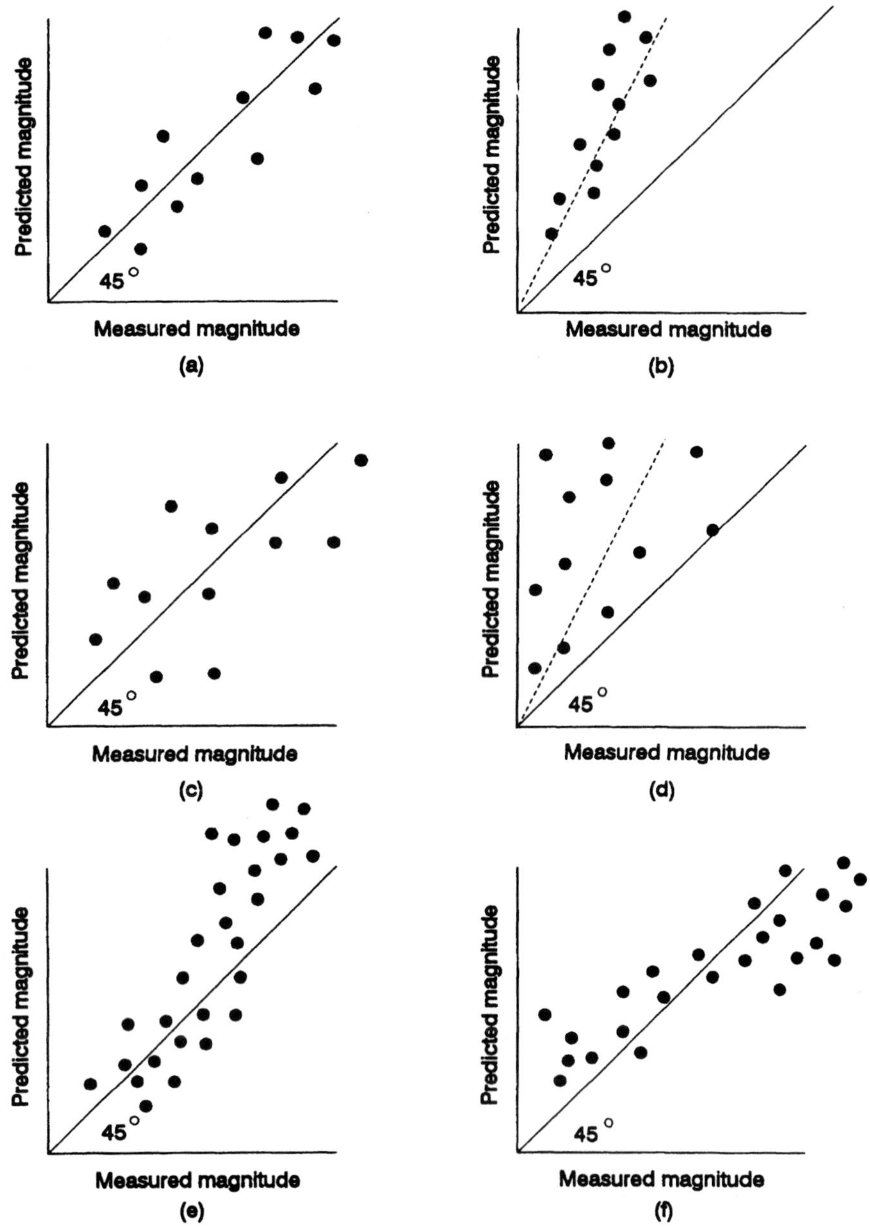

Figure 11.5.—Scattergrams of Measured vs. Predicted Magnitudes.

multi-ordinate functions. The error in the peak discharge of a runoff hydrograph is an example of the single-valued variable. The sum of squares of the differences between the predicted and measured hydrograph ordinates is an example of the multi-ordinate function. In addition to the type of function, the indices can be presented in either dimensioned or dimensionless form. For example, the error in the peak discharge relative to the measured peak discharge would be dimensionless. When comparing different model outputs, such as peak discharges and volumes of runoff, especially for cases of different storm magnitudes, dimensionless values make for more rational assessments. Dimensioned values are preferred when assessing the engineering effects of an error. For example, if the design engineer knows the expected error in a predicted peak discharge is X m^3/s, then the magnitude of the error may influence the design.

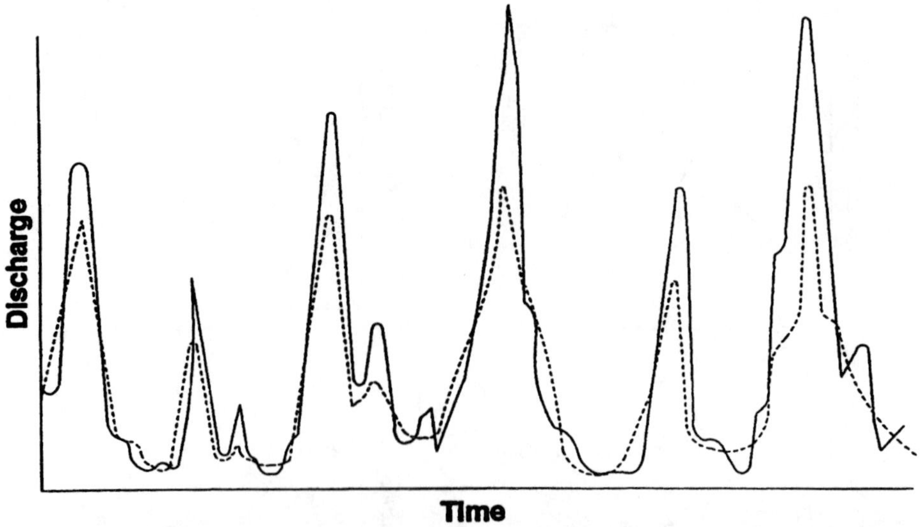

Figure 11.6.—Graphical Comparison of Observed and Predicted Time Series.

The goodness-of-fit indices may be used to measure either bias or accuracy. General forms of indices are given in Table 11.1; specific examples are given in Table 11.2 (Green and Stephenson, 1986. Reprinted with permission from IAHS Press).

With frequency or probability models, the criteria for comparison are slightly different, as direct comparison of time series is not applicable. Instead, statistics or probability estimates of the observed series are compared with those predicted with a chosen model. For example, the method of moments identifies the optimal parameters of the normal probability model by seeking parameters that will yield predicted values of the mean and standard deviation that are equal to the mean and standard deviation of the observed data. Similarly, the method of maximum likelihood seeks parameters that will maximize the probability of predicting the observed values, given the form of the model. However, objective criteria for assessing the goodness-of-fit of data to a computed frequency curve are rarely used; generally, a visual, subjective assessment of goodness-of-fit is the method of choice.

2. Procedures for Identifying Optimal Parameter Estimates The procedure employed to identify the best optimal parameter estimates, given the criterion selected, depends on the mathematical characteristics of the model and of the criterion. Himmelblau (1972) identifies the following general classes of optimization methods appropriate for model-parameter estimation:

a) *Graphical methods* in which a plot of the criterion as a function of the parameters is prepared, and the optimal parameters are identified by inspection of the plot. These methods are limited to models with one or two variables. Helsel and Hirsh (1992) provide a comprehensive discussion of graphical analysis.
b) *Analytical methods* that make use of classical calculus techniques. The method of ordinary least squares used to estimate parameters of a linear model with the minimum sum of squares criterion is an example. In that case, the optimization method finds parameters that cause the derivative of the criterion function with respect to the parameters to vanish. Haan (1977) provides background information and applications on analytical methods.
c) *Case study methods* in which the criterion is evaluated for a number of alternative parameters. From this set, the "best" is selected.
d) *Numerical methods* that use past information to generate better solutions to the optimization problem by means of iterative procedures. Numerical methods can be used to identify optimal parameters in cases where other methods are unusable. Numerical optimization methods are described in

TABLE 11.1. Goodness-of-Fit Assessment Criteria—General Forms.

Type of function (1)	Form of function (2)	Function (3)	Measure of (4)
Single-valued	Dimensioned	$e = (\hat{X} - X)$	Bias
	Dimensionless	$e = \dfrac{(\hat{X} - X)}{X}$	Relative bias
Multi-ordinate	Dimensioned	$\bar{e} = \dfrac{1}{n}\sum_{i=1}^{n}(\hat{X}_i - X_i)$	Bias
	Dimensionless	$\bar{e}_r = \dfrac{1}{n\bar{X}}\sum_{i=1}^{n}(\hat{X}_i - X_i)$	Relative bias
	Dimensioned	$S_e = \left[\dfrac{1}{n-p}\sum_{i=1}^{n}(\hat{X}_i - X_i)^2\right]^{0.5}$	Accuracy
	Dimensionless	$\dfrac{S_e}{S_x} = \dfrac{\left[\sum_{i=1}^{n}(\hat{X}_i - X_i)^2\right]^{0.5}}{\left[\sum_{i=1}^{n}(X_i - \bar{X})^2\right]^{0.5}}$	Relative accuracy

n number of pairs of ordinates compared in a single event
p number of predictor variables
X observed value
$\hat{X}$ estimated value
$\bar{X}$ mean of the observed values

a variety of operations research texts. McCuen (1992) applies a numerical algorithm for the solution of hydrologic analyses.

Table 11.3 and Fig. 11.7 illustrate the search for the optimal parameter value for a simple rainfall runoff model of a 295 km^2 watershed. The model used in this example is a tank model. The tank (reservoir) represents the lumped effects of all storage in the watershed. Tank inflow is the product of areal–average rainfall (column 2) and area, and simulated runoff, $q_s(t)$, is the tank outflow. The simulated outflow is a linear function of average storage, $S(t)$, as defined by $q_s(t) = kS(t)$. In this model, k is a calibration parameter; it must be found by fitting the model to available data. To do so, a trial value, $k = 0.00$/h, was selected, and the corresponding runoff hydrograph was computed with the model. That hydrograph is tabulated in column 4. The goodness-of-fit with this parameter estimate was quantified by computing squared residuals (column 5) and summing these to find G (see Table 11.2). This process was repeated for other values of k, and G was plotted vs. k in Fig. 11.7. When G is minimized, the fit is "optimal," and the corresponding value of the parameter k could be used with the model to predict output from other input. By inspecting the plot, one could conclude that $k = 0.03$/hr is the optimal value; additional trials would "home in" on $k = 0.033$/hr.

C. Estimating Model Parameters in the Ungaged Case

To estimate hydrologic system output in cases in which observed input and output data are not available, the analyst can do the following: 1) use a model that includes only parameters that can be observed or inferred from measurements, or 2) extrapolate calibration parameters from those found for gaged systems. In practice, some combination of these solutions typically is employed, because most models include both physically-based and calibration parameters.

TABLE 11.2. Goodness-of-Fit Assessment Criteria—Examples (Green and Stephenson, 1986, Reprinted with permission from IAHS Press).

No. (1)	Criterion (2)	Equation (3)	Remarks (4)	Reference (5)
1	Sum of squared residuals	$G = \sum_{i=1}^{n} \left[q_o(t) - q_s(t)\right]_i^2$	Most common	Diskin & Simon (1977)
2	Sum of m-powered residuals	$G = \sum_{i=1}^{n} \left[\left(q_o(t)\right)^b - \left(q_s(t)\right)^b\right]_i^c$	Changes vertical scaling	Johnston & Pilgrim (1976)
3	Sum of absolute errors	$G = \sum_{i=1}^{n} \left\lvert q_o(t) - q_s(t)\right\rvert_i$	Suited to optimization	Stephenson (1979)
4	Model efficiency	$R^2 = \frac{F_o^2 - F^2}{F_o^2}$ with $F^2 = \sum_{i=1}^{n} \left[q_o(t) - q_s(t)\right]_i^2$ and $F_o^2 = \sum_{i=1}^{n} \left[q_o(t) - \bar{q}\right]_i^2$	Dimensionless	Nash & Sutcliffe (1970)
5	Normalized objective function	$P = \frac{1}{\bar{q}}\left(\frac{F^2}{n}\right)^{1/2}$	Coefficient of variance	Ibbitt & O'Donnell (1971)
6	Root mean square error	$RMSE = \left(\frac{1}{n}\sum_{i=1}^{n}\left(q_o(t) - q_s(t)\right)_i^2\right)^{1/2}$	$P = \frac{RMSE}{\bar{q}}$	Patry & Marino (1983)
7	Automatic parameter optimization	$G = \sum_{i=1}^{n}\left[\frac{q_o(t) - q_s(t)}{\left(q_o(t)\right)^b}\right]^c$	b, c are control parameters	Wood (1974)
8	Reduced error estimate	$REE = \left[\frac{\sum_{i=1}^{n}\left(q_o(t) - q_s(t)\right)_i^2}{\sum_{i=1}^{n}\left(q_o(t) - \bar{q}\right)_i^2}\right]^{1/2}$	Biased to larger flows	Manley (1978)
9	Proportional error of estimate	$PEE = \left[\sum_{i=1}^{n}\left(\frac{q_o(t) - q_s(t)}{\left(q_o(t)\right)}\right)_i^2\right]^{1/2}$	Equal weight to equal proportional errors	Manley (1978)

10	Standard error of estimate	$SEE = \left(\sum_{i=1}^{n} \frac{\left(q_o(t) - q_s(t)\right)_i^2}{(n-2)} \right)^{1/2}$	Dimensional and independent of number of points	Jewell et al. (1978)
11	Coefficient of persistence	$CP = \sum_{i=1}^{n} \left(\frac{a^2}{F^2} \right)_k$	Dimensional	Wallis & Todini (1975)
12	Percent error in peak	$PEP = \frac{q_{ps} - q_{po}}{q_{po}} \text{ X } 100$		
13	Percent error in volume	$PEV = \frac{V_s - V_o}{V_o}$		
14	Percent error in mean	$PEM = \frac{\bar{q}_s - \bar{q}_o}{q_o}$		
15	Total overall sum of squared residuals	$TSSR = \sum_{j=1}^{n} \left\{ \sum_{i=1}^{n} \left[q_o(t) - q_s(t) \right]_i^2 \right\}_j$		
16	Total overall sum of absolute residuals	$TSAR = \sum_{j=1}^{m} \left\{ \sum_{i=1}^{n} \left\| q_o(t) - q_s(t) \right\|_i \right\}_j$		
17	Sum of absolute areas of divergence	$A = \sum_{i=1}^{n} \left\| \frac{(residual\ )_i + (residual)_{i+1}}{2} \Delta t \right\|_i$		
18	Total sum of absolute areas of divergence	$TSAA = \sum_{j=1}^{m} A_j$		
19	Variance	$S^2 = \frac{1}{n} \sum_{i=1}^{n} \left[q_o(t) - q_s(t) \right]_i^2$		
20	Mean deviation	$MD = \frac{1}{n} \sum_{i=1}^{n} \left[q_o(t) - q_s(t) \right]_i$		
21	Absolute area/ordinate divergence	$\frac{A}{n}$		

a area of an individual segment of deviation
k number of runs of successive over- or under-predictions
m number of events
n number of pairs of ordinates compared in a single event
q_o (t) observed flow rate at time t
q_s (t) simulated flow rate at time t
$\bar{q}$, $\bar{q}_o$ mean of observed flow rate
$\bar{q}_s$ mean of simulated flow rate
q_{pc} observed peak flow rate
q_{ps} simulated peak flow rate
V_o observed volume
V_s simulated volume

TABLE 11.3. Rainfall and Discharge Data For Calibration Example.

Time (h) (1)	Avg. rainfall for period (mm) (2)	$q_o(t)$ (m^3/s) (3)	$q_o(t)$ with k = 0.0 (m^3/s) (4)	Squared residual (5)
0	1.57	2.22	0	4.93
3	1.92	5.8	0	33.64
6	2.39	11.9	0	141.61
9	0	15.2	0	231.04
12	0	15.9	0	252.81
15	0	12.7	0	161.29
18	0.14	10.5	0	110.25
21	1.64	10.2	0	104.04
24	0	10.9	0	118.81
27	0	12.3	0	151.29
30	0	11.2	0	125.44
33	0	9.6	0	92.16
36	0	9.0	0	81.00
39	0	8.4	0	70.56
42	0	8.0	0	64.00
45	0	7.4	0	54.76
48	0	6.7	0	44.89
51	0	6.1	0	37.21
54	0	5.5	0	30.25
57	0	5.1	0	26.01
60	0	4.6	0	21.16
63	0	4.1	0	16.81
66	0	3.6	0	12.96
69	0	3.1	0	9.61
72	0	2.6	0	6.76
75	0	2.1	0	4.41
78	0	1.7	0	2.89
81	0	1.3	0	1.69
84	0	0.9	0	0.81
87	0	0.6	0	0.36
90	0	0.3	0	0.09
			Sum, G	2013.54

With the first solution, all model parameters can be estimated directly from measurable characteristics. For example, the parameters of the Muskingum–Cunge streamflow routing model (Cunge, 1969; Dooge et al., 1982; Ponce, 1989) are channel geometry, reach length, roughness coefficient, and slope. These parameters may be estimated with topographic maps, field surveys, photographs, and site visits. Therefore, this model may be used for analysis of an ungaged system.

With the second solution, parameters are estimated indirectly through extrapolation of gaged-system results. This extrapolation is accomplished by developing equations that predict the calibration parameters for the gaged systems as a function of measurable system characteristics. The assumption is that the resulting predictive equations apply for systems other than those from which data are drawn for development of the equations but have the same characteristics as the gaged sites. For example, calibration-parameter predictive equations for a rainfall–runoff model of an ungaged watershed might be developed as follows:

1) Collect rainfall and discharge data for gaged watersheds in the region. The watersheds selected should have hydrological characteristics similar to the ungaged watershed of interest. For example, the gaged and ungaged watersheds should have similar geomorphological and topographical characteristics. They should have similar land use, vegetative cover, and agricultural practices. The

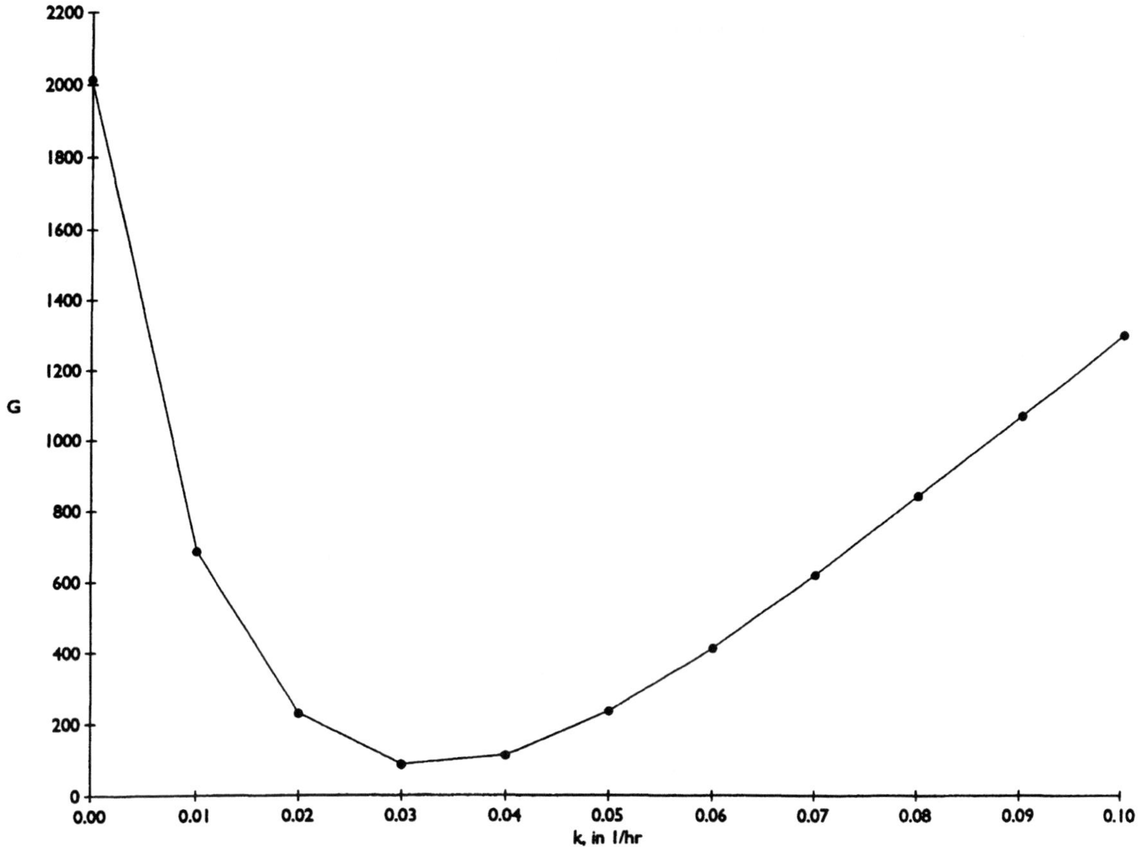

Figure 11.7.—Sum of Squared Residuals vs. Tank-Model Parameter.

watersheds should be of similar size. Rainfall distribution and magnitude and factors affecting rainfall losses should also be similar. If possible, data should be collected for several flood events. These rainfall and discharge data should represent, if possible, events consistent with the intended use of the model of the ungaged watershed. If the rainfall–runoff model will be used to predict runoff from large design storms, data from large historical storms should be used to estimate the calibration parameters.

2) For each gaged watershed individually, use the data to estimate the calibration parameters for the selected rainfall–runoff model. Procedures for doing so are described earlier.
3) Select and measure or estimate physiographic characteristics of the gaged watershed to which the rainfall–runoff model parameters may be related. Table 11.4 lists candidate watershed characteristics. Some of these characteristics, such as the watershed area, are directly measured. Others, such as the Horton ratios, are computed from measured characteristics.
4) Develop predictive equations that relate the calibration parameters found in step 2 (the dependent variables) with characteristics measured or estimated in step 3 (the independent variables). In a simple case, the results of steps 2 and 3 may be plotted with a rainfall–runoff model parameter on the ordinate and a watershed characteristic selected in step 3 on the abscissa. Each point of the plot represents the value of the parameter and the selected characteristic for one gaged watershed. With such a plot, a relationship can be visually fitted and sketched on the plot. Regression analysis is an alternative to the subjective graphical approach to defining a predictive relationship. Regression procedures determine analytically the optimal predictive equation. Details of regression analysis are presented in hydrologic statistics texts, including those by Haan (1977) and McCuen and Snyder (1986). Table 11.5 shows example equations derived for unit–hydrograph parameters.

TABLE 11.4. Watershed Characteristics for Calibration-Parameter Predictive Equations.

Total watershed area
Area below lowest detention storage
Stream length
Stream length to watershed centroid
Watershed shape (e.g., circularity ratio)
Average watershed slope
Average conveyance slope
Conveyance slope measured at 10% and 85% of stream length (from mouth)
Height differential
Elevation of watershed centroid
Average of elevation of points at 10% and 85% of stream length
Permeability of soil profile
Soil–moisture capacity average over soil profile
Hydrologic soil group
Population density
Street density
Impervious area
Directly-connected impervious area
Area drained by storm sewer system
Land use
Detention storage
Rainfall depth for specified frequency, duration
Rainfall intensity for specified frequency, duration
Horton's ratios (Horton, 1945)
Drainage density (Smart, 1972)
Length of overland flow (Smart, 1972)

To apply a parameter-predictive equation for an ungaged system, the independent variables are measured or estimated for the ungaged system. Solution of the equation with these values yields the desired calibration parameter. This parameter is used then to predict output from the ungaged system.

In some hydrologic engineering studies, the goal is to define frequency relationships. Chapter 2 describes procedures for precipitation frequency studies, and Chapter 8 summarizes procedures and describes statistical models for discharge frequency analysis. All of the models described are empirical.

TABLE 11.5. Example Calibration-Parameter Prediction Equations for Unit Hydrograph Model.

Equation	Reference
$C_t = 7.81/I^{0.78}$	WME (1969)
$C_p = 0.89\ C_t^{0.46}$	WME (1969)
$R = c\ T_c$	Russell et al. (1979)
$T_c = 8.29\ (1.00 + I)^{-1.28}\ (A/S)^{0.28}$	USACE (1982)

Note: In the above equations, C_t = calibration coefficient for Snyder's unit hydrograph (UH); C_p = calibration coefficient for Snyder's UH; T_c = time of concentration (h); R = Clark's instantaneous UH storage coefficient (h); I = impervious area (%); L = length of channel/ditch from headwater to outlet (mi); S = average watershed slope (ft/ft); c = coefficient, for forested watershed = 8–12, for rural watersheds = 1.5–2.8, and for developed watershed = 1.1–2.1; A = watershed area (mi^2).

Observed data are necessary for calibration. Consequently, these statistical models cannot be applied directly to an ungaged watershed. Options available to the analyst requiring frequency estimates for an ungaged stream include the following: 1) develop frequency-distribution parameter-predictive equations, or 2) develop distribution-quantile-predictive equations.

Procedures used by the Corps of Engineers illustrate the first option. The Corps and other federal agencies use the log–Pearson type III statistical distribution (model) for annual maximum discharge-frequency studies. This model has three parameters. Following federal guidelines (USWRC, 1982), these parameters are estimated from the mean, standard deviation, and skew coefficient of the logarithms of observed peak discharges. In the absence of flow data, regional frequency analysis procedures may be applied to develop parameter-predictive equations. As with the equations for rainfall–runoff model parameters, these equations relate model parameters to watershed characteristics. For example, for the Shellpot Creek watershed, Delaware, the following predictive equation was developed (USACE, 1982):

$$S = 0.311 - 0.05 \log A \tag{11.1}$$

in which S is the standard deviation of logarithms and A is the drainage area, in sq. mi. With similar equations, other parameters can be estimated. To apply a distribution parameter-predictive equation for an ungaged watershed, the independent variables in the equation are measured or estimated for the ungaged watershed. Solution of the equation with these values yields the desired statistical-distribution parameter. The frequency curve is computed with the model.

With the second option, the frequency distribution quantiles for an ungaged watershed are defined directly with predictive equations. Such a predictive equation is developed by defining the frequency distributions for streams with gaged data, identifying from the distributions specified quantiles, and using regression analysis procedures to derive a predictive equation. For example, for the Red Lion Creek watershed, Delaware, the following quantile predictive equation was developed (USACE, 1982):

$$Q_{100} = 1040\, A^{0.91} \tag{11.2}$$

in which Q_{100} is 100-year (0.01 exceedance probability) discharge and A is drainage area in sq. mi. To use this or a similar predictive equation for an ungaged watershed, the independent variable(s) in the equation are measured or estimated for the ungaged watershed. Solution of the equation yields the desired quantile. The USGS publishes regional regression equations for virtually the entire United States.

D. Validating Estimated Model Parameters

The final stage in preparing for model application is validation of estimated model parameters. If input and output data are available, the predictive ability of the model should be evaluated with data other than those used in calibration-parameter estimation. Criteria for goodness-of-fit in calibration can be employed for validation. If the model predicts well with these test data, the model and parameters can be accepted conditionally and used to predict hydrological system responses to design inputs. However, the validation must be an on-going activity, as the model and parameters must be reassessed as system conditions change and as additional data are available for calibration. For example, if the land use changes, the analyst must recalibrate the parameters (or perhaps, the model functional relationships) to reflect the modified conditions. The continuing process of model assessment with new data indicates if such modification is necessary.

Validation is more difficult in the ungaged case, as no data are available for comparison. Consequently, the analyst must be more creative in searching for confirmation of output estimates. One possible method of validation in the case of a rainfall–runoff model is illustrated in Fig. 11.8. This procedure was used in El Dorado County, California, to validate a rainfall–runoff model intended to estimate design flows of

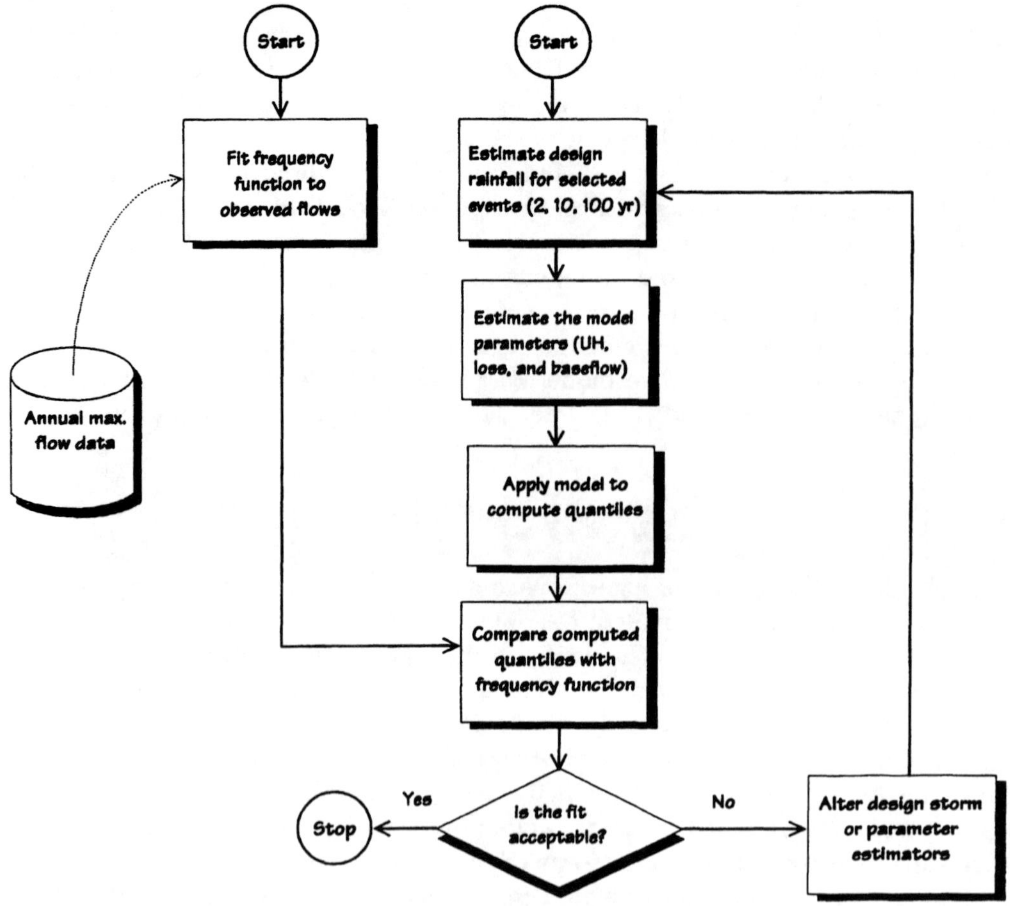

Figure 11.8.—Possible Validation Procedure for Ungaged Watershed Model.

specified frequency from rainfall of the same frequency, a common application and assumption. Short-interval records were not available to permit calibration of the rainfall–runoff model, so parameter-predictive relationships were developed following procedures similar to those suggested herein. Annual maximum flow data were available, so a frequency function was fitted to these annual maximum flows, and selected quantiles were determined. Then the rainfall–runoff model was used with the design storm and parameter estimates to predict exactly the same quantiles. On comparison, the rainfall–runoff model was found to yield quantile estimates within the 90% confidence limits of the fitted frequency functions, so it was accepted for design. Otherwise, the model, its parameters, or even the design storm could be adjusted.

V. ASSESSING ACCURACY AND RELIABILITY OF STUDY RESULTS

Factors that influence the accuracy and reliability of results of hydrologic investigations include the following:

a) inherent complexity and variability of physical systems and hydrologic processes;
b) analysis framework used as a basis for investigation (i.e., analytical treatment of essential characteristics of physical systems and hydrologic processes);
c) assumptions regarding the future nature and condition of physical systems;

d) size and quality of data samples used to represent characteristics of physical systems and hydrologic processes;
e) errors in processing and interpreting data;
f) limitations of analytical procedures, models, etc.; and
g) errors in applying and interpreting results of analytical procedures.

Numerous assumptions and judgments must be made in performing virtually any hydrologic investigation. They may be explicit in the initial formulation of study procedures, and also implicit in the analytical procedures used. The analyst should be cognizant of sources of uncertainty and should, if possible, evaluate their magnitude and significance.

An initial step in reviewing study results is to determine whether they are logical and consistent. The following questions might be raised.

a) Are results physically possible and reasonable? For example, are basic principles of continuity satisfied (in those situations where they apply)?
b) Are values for model parameters physically reasonable and internally consistent? For example, are adopted infiltration rates appropriate for the soil types and land uses represented? Are values for time of concentration consistent with watershed properties? Are values for hydraulic roughness reasonable for the surfaces they represent?
c) Are results for alternative sets of boundary conditions consistent in relation to each other? For example, do increasing rates of pumping produce logical and consistent effects on water levels?

Questions such as these can serve as an initial screening mechanism for validating study results. Although consideration of such fundamental questions may seem obvious, it is surprising how often they are overlooked, sometimes with considerable cost in terms of validity of study results.

Use of independent estimation methods can be effective for validating study results. Such an approach might be used, for example, when determining peak and minimum discharges, flood runoff volumes, surface and ground water yields, flow-duration curves, discharge-frequency curves, etc. Information that might be used for estimation includes regional relationships and information derived from watersheds similar to the study watershed (e.g., Fig. 11.2).

Sensitivity analysis is a particularly valuable tool for assessing model reliability. The remainder of this chapter is devoted to a description of quantitative measures of reliability and to sensitivity analysis.

A. Quantitative Measures of Reliability

Before using a model to make predictions, the model should be examined for its potential reliability, i.e., goodness-of-fit. It is important to assess how well the model can be expected to predict. Since it is rare in hydrologic analysis to have sufficient data to make a test that is independent of the model calibration, it is usually necessary to use the calibration data as the basis for assessing the goodness-of-fit. That is, the reliability is evaluated with statistics derived from the calibration results. Just as the model coefficients are used as sample estimates of the population coefficients, the sample goodness-of-fit statistics are used as estimates of the reliability of future predictions. Where large databases are available, split-sample testing is sometimes used, which uses part of the data for calibration and the remainder for an independent test on reliability. The data should be divided using random-sampling procedures.

There are several criteria that can be used to assess the reliability of a fitted model, including the following: 1) the rationality of the coefficients and, for a multiple-predictor variable model (i.e., an equation with two or more independent variables), the relative importance of the predictor variables, both of which can be assessed using the standardized partial regression coefficients; 2) the degree to which the underlying assumptions of the model are met; 3) the standard error of estimate (S_e); 4) the correlation coefficient, R, or the coefficient of determination, R^2; and 5) the model bias, $\bar{e}$, and relative bias, $\bar{e}/\bar{Y}$. All of these criteria may not apply to every model, but they are important criteria commonly computed in model development. Each of these terms is discussed in the following sections.

1. ***Model Rationality*** Rationality is probably the single most important criterion in assessing a model. Two methods are commonly used for assessing rationality. First, a model should provide rational predictions over all possible values of the predictor variable(s). Second, when a model coefficient is directly related to a predictor variable, such as in a linear multiple regression model, the coefficient should accurately reflect the effect of X on Y. A model that either provides irrational predictions or indicates an irrational relationship between the criterion (i.e., dependent) variable and a predictor (i.e., independent) variable should be used very cautiously, if at all.

A linear multiple regression model provides a good example. The intercept coefficient of a linear model (b_o of Equation 11.3, below) should be rational. As the values of the predictor variables go toward specific target values, the intercept coefficient should approach a physically rational value. For example, if the project cost, Y, is related to the number of inlets in the drainage system, X_1, the total length of pipe, X_2, and the total area being developed, X_3, the following model form could be used:

$$\hat{Y} = b_o + b_1X_1 + b_2X_2 + b_3X_3. \tag{11.3}$$

As the values of X_1, X_2, and X_3 approach zero, the estimated project cost should approach zero. The fitted value of the intercept coefficient for these conditions should be nearly zero so that the estimated value $\hat{Y}$ will be near zero when the three predictor variables equal zero. If the resulting predictions at these target conditions are not rational, the intercept coefficient will be irrational.

The slope coefficients (b_1, b_2, and b_3 of Equation 11.3) should also be rational. Since the slope coefficients represent the change in Y for a given change in the value of the predictor variables (X_1, X_2, and X_3), it is easy to check for rationality in sign. In Equation 11.3, b_1, b_2, and b_3 should be positive. While it is usually easy to assess the rationality of the sign of a coefficient, it is more difficult to check the rationality of the magnitude of a slope coefficient. Since the slope coefficients are a function of the units of both Y and the corresponding predictor variable, it may be easier to interpret the rationality of the coefficient by converting a slope coefficient b_i to the corresponding value of the standardized partial regression coefficient, t_i:

$$t_i = \frac{b_i S_i}{S_y} \tag{11.4}$$

in which S_i is the standard deviation of predictor variable X_i and S_y is the standard deviation of the criterion variable. The t_i values are dimensionless and independent of the units of Y and X_i; thus, they are easier to interpret than the b_i values. A t_i value should be between 0 and 1 in absolute value, with the sign indicating the direction of the relationship. A value near 1 indicates an important predictor; a value near 0 indicates that the corresponding predictor variable is not important. Thus, the larger the value, the more important a predictor variable is considered. Where the intercorrelations between the predictor variables are significant, the absolute value of a t_i may exceed 1. When the t_i value is irrational, the corresponding b_i value (and the model) should be considered irrational. A model that has an irrational coefficient should be used with caution, if at all; this is especially true when the model is used outside the range of the data used for calibrating the model. If the b_i value is not rational, it implies that the model does not give a true indication of the rate of change of Y with respect to change in X_i. Models that include irrational coefficients may still yield accurate predictions within the range of the data that were used to fit the model. They may be good prediction equations but poor models of the underlying process.

2. ***Bias in Estimation*** While models are sometimes developed to make a single prediction, a model will most often be used to make several predictions, either in time or in space. For example, a snowmelt

runoff model developed for a watershed can be used to make runoff projections for several successive years before the model is recalibrated with additional data. Peak discharge models can be developed for a region and used to make design discharge estimates at many locations in the region. Whether the multiple estimates are made in time or space, and recognizing that models cannot be perfectly accurate on every estimate, we would hope that, on the average, the model does not overestimate or underestimate the true value. A model that would consistently overestimate or underestimate is said to be biased. The bias is the difference between the long-term predicted value and the true value. Unfortunately, the true value is not known, but the average of the differences between the predicted values, $\hat{Y}_i$, and the measured values, Y_i, in a sample is used as an estimate of the bias:

$$bias = \bar{e} = \frac{1}{n}\sum_{i=1}^{n}\left(\hat{Y}_i - Y_i\right) = \frac{1}{n}\sum_{i=1}^{n} e_i. \tag{11.5}$$

A model that consistently overestimates the measured value is said to have a positive bias, while consistent underestimation is a negative bias. Ideally, a model should be unbiased, but bias is only one criterion for assessing a model. Some bias may be acceptable in cases where the use of an unbiased model would lead to poorer reliability with respect to other criteria.

The linear model of Equation 11.3 always provides models where the sample bias (Equation 11.5) equals zero. Other model structures do not necessarily yield unbiased estimates. For example, the following power-model form, which is widely used in hydrology, often yields biased estimates:

$$\hat{Y} = b_o X_1^{b_1} X_2^{b_2} \cdots X_p^{b_p} \tag{11.6}$$

in which p is the number of predictor variables. If Equation 11.6 is fit to data using a logarithmic transform, then the estimated values of the logarithms will be unbiased (i.e., sum to zero), but, the estimated values of Y will not sum to zero (i.e., the predicted values will be biased).

The concept of bias is best illustrated with an example. Table 11.6 contains data from 18 drainage projects. The criterion variable is the total project cost (1963 dollars). The three predictor variables are: X_1 = the number of inlets, X_2 = the total length of pipe (feet), and X_3 = the total area developed (acres). Linear multiple regression analysis yields the following model:

$$\hat{Y} = -12670 + 3747\, X_1 + 5.471\, X_2 - 43.35\, X_3. \tag{11.7}$$

The predicted values of Y are given in Table 11.6. Since the sum of the residuals (i.e., $\hat{Y} - Y$) equals zero, the model is said to be unbiased. The same data were used to fit a power model, with the following results:

$$\hat{Y} = 792.2\, X_1^{0.9265}\, X_2^{0.2420}\, X_2^{0.2420}\, X_3^{-0.06146}. \tag{11.8}$$

The predicted values of Y and the errors are also given in Table 11.6. Since Equation 11.8 was fit with logarithms, the sum of the residuals does not equal zero. On the average, the model underpredicts the total project cost by \$5736 (i.e., the bias is −5736). It is of interest to note that the intercept of Equation 11.7 is a large negative value, which is not rational. It did not lead to negative estimates of Y (see column 5 of Table 11.6) so the model should provide reasonable estimates within the range of the data.

3. ***The Standard Error of Estimate*** Given measurements of a single random variable Y, the mean of the sample is the most accurate estimate when a prediction is needed. The mean is an unbiased estimate in that the deviations from the mean sum to zero. In comparison to any other estimate made with the sample, the sum of the squares of the deviations from the mean will be smaller than the sum of squares for any other value. Thus, the standard deviation, S_y, of the values is a minimum (because S_y is just the

TABLE 11.6. Error Analysis of Linear and Power Models of Drainage Project Cost Data.

				Linear Model		Power Model	
No. of inlets (1)	Length of pipe (ft) (2)	Area developed (ac) (3)	Actual cost ($) (4)	Predicted cost ($) (5)	Error ($) (6)	Predicted cost ($) (7)	Error ($) (8)
32	7100	164	191311	138985	−52325	122769	−68541
29	3400	89	151499	110750	−40748	97368	−54130
19	2600	26	96291	71629	−24661	66516	−29774
18	2200	35	90341	65303	−25037	59660	−30680
17	1900	133	84576	55666	−28909	50308	−34267
30	2225	181	63980	104080	40100	86806	22826
15	1285	38	56644	48925	−7718	44017	−12626
24	2029	144	55540	82127	26587	70013	14473
12	1992	31	49492	41854	−7637	40303	−9188
9	1295	18	47740	27362	−20377	28763	−18976
18	1965	50	47029	63367	16338	56792	9763
15	2607	76	39049	54511	15462	50056	11007
12	2115	44	35822	41964	6142	40020	4198
18	1653	28	34462	62614	28152	56441	21979
23	6400	56	34067	106110	72043	94185	60118
10	882	73	23778	26465	2687	26515	2737
4	1054	4	14653	7913	−6739	14159	−493
8	640	19	13348	19987	6639	21674	8326
				Sum = 0		Sum = −103249	
				Bias = 0		Bias = −5736	

square root of the sum of squares divided by the degrees of freedom). In summary, for a single random variable the mean is the best estimate in that it is unbiased and has a minimum error variance, and the standard deviation of the values is a measure of the accuracy of the mean.

One way to reduce the error variance of the estimated value is to relate Y to one or more predictor variables. Then the objective is to provide a relationship between Y and the X variables such that the estimated values, $\hat{Y}$, are unbiased and have a sum of squares of the deviations that are a minimum. Dividing the sum of squares of the errors (i.e., $\Sigma e^2 = \Sigma(\hat{Y} - Y)^2$) by the degrees of freedom yields the error variance, S_e^2, the square root of which is the standard error of estimate, S_e. If the equation between Y and X is worthwhile, then S_e will be significantly smaller than S_y. That is, the equation reduces the error deviation from S_y to S_e. Thus, the ratio S_e/S_y is a measure of the improvement in the accuracy of prediction due to the prediction equation. When S_e/S_y is near 0.0, the model significantly improves the accuracy of prediction over predictions made with the mean. When S_e/S_y is near 1.0, the model provides little improvement in prediction accuracy when compared to just using the mean as the estimate of Y.

If there are two prediction models fitted using the same data but with different S_e/S_y ratios, then it is necessary to decide if the smaller ratio is significantly lower than the larger ratio. This is usually easiest to determine by examining the competing values of S_e. If the smaller S_e is physically better, then the corresponding model is preferred. This concept is best illustrated using an example. The project cost data of Table 11.6 has a standard deviation of \$46,456. This means that, if the mean is used as the estimated value and assuming the errors are normally distributed about the mean, 68% of the projects will have an error that is less than \$46,456. Using the data of Table 11.6, the linear and power models produced standard errors of estimate of \$33,743 ($S_e/S_y = 0.726$) and \$34,198 ($S_e/S_y = 0.736$), respectively. These correspond to a savings of about \$12,000 per project, which is considered important. However, the difference in the standard errors of the two models (i.e., \$455) may not be meaningful, so the models are considered to be equally accurate. Now assume that a third model is developed that uses only one predictor variable from the data matrix:

$$\hat{Y} = 9096\, X_1^{0.4412}. \tag{11.9}$$

This model yields a S_e of \$42,241, which gives $S_e/S_y = 0.909$. Compared to S_y the reduction in S_e of \$4,000 is not highly meaningful, so Equation 11.9 provides only a slight improvement in accuracy over the mean. Other criteria, such as the ease of data collection, could be the deciding factor in selecting a model.

Mathematically, the standard error of estimate (S_e) is:

$$S_e = \left(\sum e^2/\nu\right)^{0.5} = \left(\sum\left(\hat{Y} - Y\right)^2/\nu\right)^{0.5} \tag{11.10}$$

in which $\hat{Y}$ is the predicted value, Y is the measured value of the random variable, and ν is the degrees of freedom, which depends on the model and sample size. As a general statement, the degrees of freedom equals the sample size minus the number of unknowns fitted with the data. If all of the predicted values equal the corresponding measured values, then the S_e is zero, which means that the model provides perfect predictions. If all of the predicted values equal the mean, then, depending on the degrees of freedom, the S_e may equal the standard deviation of Y. The model of Equation 11.9, for example, does not significantly improve the accuracy of predictions.

It is of interest to note that S_e can actually be greater than S_y because S_y and S_e are based on $n-1$ and $n-p-1$ degrees of freedom, respectively, where p is the number of predictor variables. If the model has fewer than $n-1$ degrees of freedom and the Σe^2 for the model is nearly equal to the Σe^2 for the mean, then S_e is greater than S_y. This implies that the mean provides better predictions of Y than the model does.

4. The Correlation Coefficient The correlation coefficient, R, is an index of the degree of linear association between two random variables. The magnitude of R indicates whether or not the model will provide accurate predictions of the criterion variable. Thus, R is often computed before the regression analysis is performed in order to determine whether it is worth the effort to perform the regression; however, R is always computed after the regression analysis because it is an index of the goodness-of-fit. The correlation coefficient measures the degree to which the measured and predicted values agree and is used as a measure of the accuracy of future predictions. It must be recognized that, if the measured data are not representative of the population (i.e., data that will be observed in the future), the correlation coefficient will not be indicative of the accuracy of future predictions.

The square of the correlation coefficient, R^2, equals the percentage of the variance in the criterion variable that is explained by the predictor variable. Because of this physical interpretation, R^2 is considered a more meaningful index than R.

Strictly speaking, the correlation coefficient applies only to the bivariate and multiple linear models with an intercept. It also applies to curvilinear models, such as the polynomial model, that are based on a linear separation of variation. It does not apply to the power model ($\hat{Y} = b_o X^{b_1}$) except when the power model is expressed in the log–transform structure; but then the numerical value applies only to the prediction of the logarithms and not the predicted values of Y for which estimates are required.

The standard error of estimate is a much better measure of goodness-of-fit than the correlation coefficient. The standard error has the advantages of a) having the same units as Y, b) properly accounting for the degrees of freedom, and c) being valid for nonlinear models as well as linear models. The correlation coefficient is a dimensionless index, which makes it easy to compare the goodness-of-fit of different models, but the value of R may be high only because there is very little total variation such that the explained variation seems large.

A major problem with the correlation coefficient is that the degrees of freedom equal the sample size, n. This is acceptable only for large samples and models with few coefficients that need fitting. In modeling, one degree of freedom should be subtracted from the sample size for each coefficient that is quantified with the data. For small samples and models with a large number of fitted coefficients, the number of degrees of freedom may be small; in such cases, the correlation coefficient may be large, but the reliability of the model is still poor. This is best illustrated with an example. Assume that we have six pairs of Y and X measurements, with a systematic trend not evident from a graph of Y versus X. If we fit a linear model, the fit would be poor, with a near-zero correlation coefficient resulting. If we successively added terms of a polynomial, the equation would provide a better fit but the structure would not show a rational trend. We know that a fifth-order polynomial with six coefficients would fit the data points exactly, and so the explained variance would equal the total variance and the correlation coefficient would equal 1. But the polynomial does not reflect the true relationship between Y and X, and thus, the correlation coefficient is an inaccurate measure of the reliability of prediction. In the case of a fifth-order polynomial, there would be no degrees of freedom, so the standard error is infinitely large, i.e., undefined; in this case, the S_e is a more rational measure of reliability than the correlation coefficient of 1.

Another problem with the correlation coefficient is that it assumes the model is unbiased. For a biased model, a correlation coefficient computed with the common computational method may be erroneous. It is better to compute the standard error of estimate and use it as a measure of model reliability. A better index of correlation can be computed using the standard error of estimate and the following equation:

$$R = \left[1 - \left(\frac{n - \nu}{n - 1}\right)\left(\frac{S_e}{S_y}\right)^2\right]^{0.5} \tag{11.11}$$

in which ν is the number of fitted coefficients in the model. This equation will not yield the same numerical value as the correlation coefficient computed with the standard formula, but it is a more correct measure of goodness-of-fit because it is based on the correct number of degrees of freedom. It is sometimes referred to as the adjusted correlation coefficient.

The correlation coefficient was computed for the models fitted with the data of Table 11.6. For the linear model of Equation 11.7, R equals 0.752 (i.e., $R^2 = 0.566$); this indicates that the model explains a little over 50% of the variation. However, this is misleading because the sample size is small (i.e., 18) and there are four coefficients to fit; thus, there are only 14 degrees of freedom. The standard error ratio (S_e/S_y) of 0.726 indicates only a fair level of accuracy and contradicts the more favorable indication given by R. The standard error ratio is the better indicator of prediction accuracy.

The correlation coefficients for the power models of Equations 11.8 and 11.9 are 0.744 and 0.471, respectively. These were computed using Equation 11.11 because the power model is being used. The squares of these coefficients yield: $R^2 = 0.554$ and $R^2 = 0.222$. Thus, the three-predictor model of Equation 11.8 explains 33.2% more variation than the one-predictor model of Equation 11.9.

5. ***Analysis of the Residuals*** The principle of least squares assumes that the errors, that is, the differences between the predicted and measured values of the criterion variable, a) are independent of each other, b) have zero mean, c) have a constant variance across all values of the predictor variables, and d) are normally distributed. If any of these assumptions are violated, the model structure can be assumed to be incorrect. Violations of these assumptions are easily identified using statistical analyses of the residuals.

When the sum of the residuals does not equal zero, it reflects a bias in the model. The regression approach, when applied analytically, requires that the sum of the residuals for a linear model equal zero; however, if an inadequate number of significant digits are computed for the regression coefficients, the sum of the residuals may not equal zero. A model may be biased even when the sum of the residuals equals zero. For example, when the trend of the data is noticeably nonlinear, but the linear regression model is used, local biases may be evident. If the errors e_i are computed and plotted versus the corresponding values of the predictor variable, a noticeable trend appears. The errors may be positive for both low and high values of X, while the errors show a negative bias of prediction for intermediate values of X. The trends in the residuals suggest a model that has local biases; such a model should be replaced by a model structure that does not give local biases.

The residuals for the models of Equations 11.7 and 11.8 are given in Table 11.6. In both cases, there are local biases. For the linear model, the overall bias is zero, but eight of the ten residuals for the ten largest project costs are negative. For the eight smallest projects, seven of the residuals are positive. Thus, the model underpredicts for large projects and overpredicts for small projects. This suggests that the model is not rational, and possibly another model structure may be more appropriate.

Although an exact statistical test does not exist for testing a data set for a nonconstant error variance, it is not unreasonable to separate the errors into two or more sets based on the value of X and use a hypothesis test for comparing variances to detect a significant difference. A systematic method for selecting either the number of groups or the points at which the separation will be made does not exist; therefore, one should attempt to have large subsamples and a rational explanation for choosing the points of separation.

The linear regression procedure also assumes that the errors are normally distributed across the range of the values of the predictor variables. If the errors are not normally distributed, the model structure should be examined. An exact procedure for testing for normality does not exist; however, an analysis or a hypothesis test, such as a rank-order frequency analysis, the chi-square goodness-of-fit test, or the Kolmogorov–Smirnov test (Siegel, 1956), could be performed to study the normality question.

The fourth assumption deals with the independence of the observations. This is not usually a problem except when time or space is the predictor variable. In such cases, the measurement of Y for one time period may not be completely independent of the measurement for an adjacent time period. The run test and/or the serial correlation coefficient may be used to check for statistical independence. Statistical independence should not be used in place of an examination of the physical independence of the data.

The same concepts about the residuals that underlie the linear bivariate model apply to other model structures. A model defines the effect of the predictor variables on the criterion variable; therefore, the residuals represent the variation in the criterion variable that is not explained by the predictor variables. There is sometimes more information contained in the residuals than there is explained by the model. Therefore, the residuals should be examined very closely. The residuals will often be helpful in identifying biases of the model. For example, if the residuals are correlated with any of the predictor variables, the linear model does not adequately describe the relationship between the criterion and predictor variables. The residuals should be plotted against each of the predictor variables and the criterion variable because nonlinear relationships in the plots of the residuals will not be detected by computing the correlation coefficients between the residuals and the $(p + 1)$ variables.

In addition to detecting bias in a model structure, the residuals can be used to identify variables that were not, but should have been, in the data set. This requires that the user have a complete understanding of the physical system being modeled.

6. Application: Monthly Temperature Time Series To illustrate the usefulness of the concepts and methods discussed, a sinusoid model will be fitted using monthly temperature data. Then other data will be used to make independent tests of the model. Temperature data are required for numerous types of hydrologic analyses. Models for estimating evaporation and evapotranspiration require temperature data. Many water quality modeling problems include temperature as an input variable. Watershed models use daily or monthly temperature values for continuous simulation. Snowmelt runoff models generally include temperature as a variable. Temperature is, of course, a central factor in global warming modeling.

A 13-year record (October 1954 to September 1966) of average monthly temperature at Lawrenceville, Georgia was used to fit the following model:

$$\hat{T} = C_1 + C_2(\sin 2\pi t + C_3) \tag{11.12}$$

in which $\hat{T}$ is the estimated average monthly temperature, t is the month of the water year (0 for October, . . . , 11 for September), f is the frequency ($f = 1/12$), and C_1, C_2, and C_3 are coefficients that are fitted with the data. The coefficient C_1 is the average of the temperature sinusoid, C_2 is the amplitude of the sinusoid, and C_3 is the phase-shift angle. A numerical optimization scheme (McCuen, 1992) was used to fit the values of the coefficients with the following result:

$$\hat{T} = 61.3 - 18.55\left(\sin 2\pi\left(\frac{1}{12}\right) t - 0.048\right). \tag{11.13}$$

The fitted model of Equation 11.13 is shown in Fig. 11.9 with the 13-year record of measurements.

The goodness-of-fit statistics indicate that Equation 11.13 provides a good fit to the data. The model is unbiased since the sum of the errors equals zero. The standard error of estimate is 3.16°F, which is 23.4% of the standard deviation of 13.53°F; i.e., the standard error ratio equals 0.234. The correlation coefficient equals 0.973; thus, the coefficient of determination equals 0.946, which indicates that the model explains 94.6% of the variation in the measured temperature values.

While the overall model bias is zero, there are noticeable local biases. In some years, especially at the beginning of the record, the model underpredicts the measured temperatures. In the first four water years, the errors ($\hat{T} - T$) are negative for 33 of the 48 months. In the last four water years of the record, the model underpredicts 19 of the 48 months. There is no noticeable bias in the errors for any given month or period of the year; for example, the model does not tend to overpredict or underpredict during the summer or during the winter. Thus, the local biases in the various years either represent random variation or the presence of a trend towards increasing temperature.

The model of Equation 11.13 was tested using five years of record at the same site (October 1966 to September 1971). The predicted and measured values are shown in Fig. 11.10. While the model follows the general shape of the measured data, it seems to overpredict throughout the five-year period. Only 23 of the 60 errors are negative. The average error is 0.82°F, which indicates a bias towards overprediction. In spite of the bias, the overall accuracy is reasonably good, with a standard error of estimate of 2.93°F, which yields a standard error ratio of 0.216. These are comparable to the values obtained in fitting the model, which suggests that the model is reasonably accurate for predicting average monthly temperature in Lawrenceville, Georgia. The correlation coefficient equals 0.977 and R^2 equals 0.955.

The model was also tested on five years of average monthly temperature at Clayton, Georgia, which is approximately 75 miles northeast of Lawrenceville. Clayton is also at a higher elevation than Lawrenceville, so the temperatures would be expected to be lower at Clayton. The temperatures predicted with Equation 11.13 are shown in Fig. 11.11 versus the measured values. Ninety percent of the predicted values exceeded the measured values, which indicates that the model is biased. The average of the errors was 4.43°F. The standard error of estimate of 5.6°F yields a standard error ratio of 44%, which is larger than that for the fitted model. This suggests that the model does not provide accurate estimates when it is used at a location sufficiently removed from the site where the model was calibrated. The correlation coefficient equals 0.902 ($R^2 = 0.814$). In spite of the large bias and moderate standard error ratio, the correlation coefficient suggests a reasonably accurate fit. In general, the standard error is a more reliable indicator of model accuracy than the correlation coefficient.

B. Sensitivity Analysis

Sensitivity is defined as the rate of change of one factor with respect to change in another factor. Thus, it can be used to measure either the effect of one factor on another or the importance of one element with respect to some criterion. In terms of modeling, the factors could be model variables, coefficients, components, or criteria used to measure model reliability. Additionally, the factor of interest might be a measurement. Examples that illustrate the factors that might be considered in performing sensitivity analyses are given in Table 11.7.

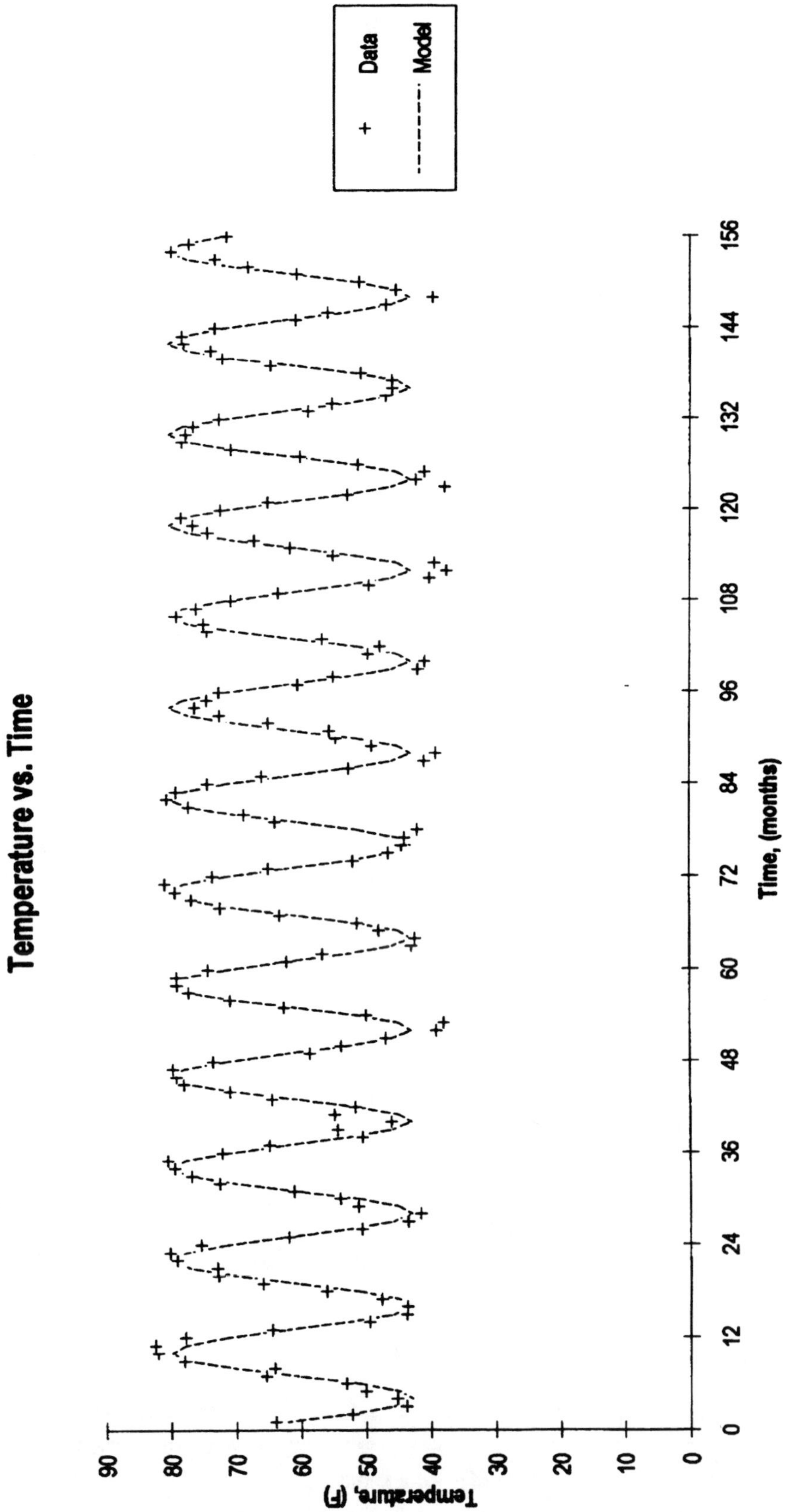

Figure 11.9.—Average Monthly Temperature Data and Fitted Model for 13-years (October 1954 to September 1966), Lawrenceville, GA.

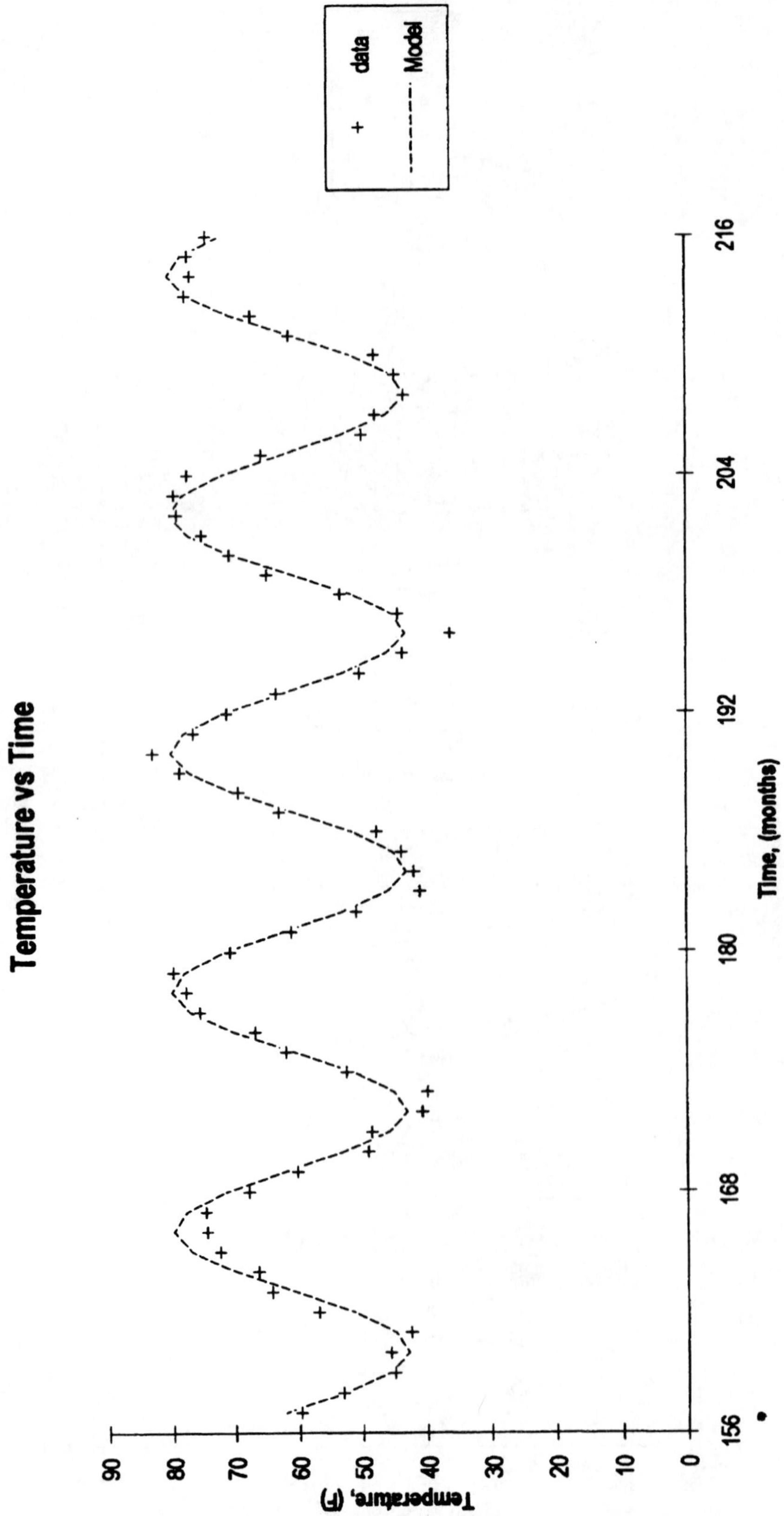

Figure 11.10.—Average Monthly Temperature Data and Model Results for Five-Year Test Period (October 1966 to September 1971), Lawrenceville, GA.

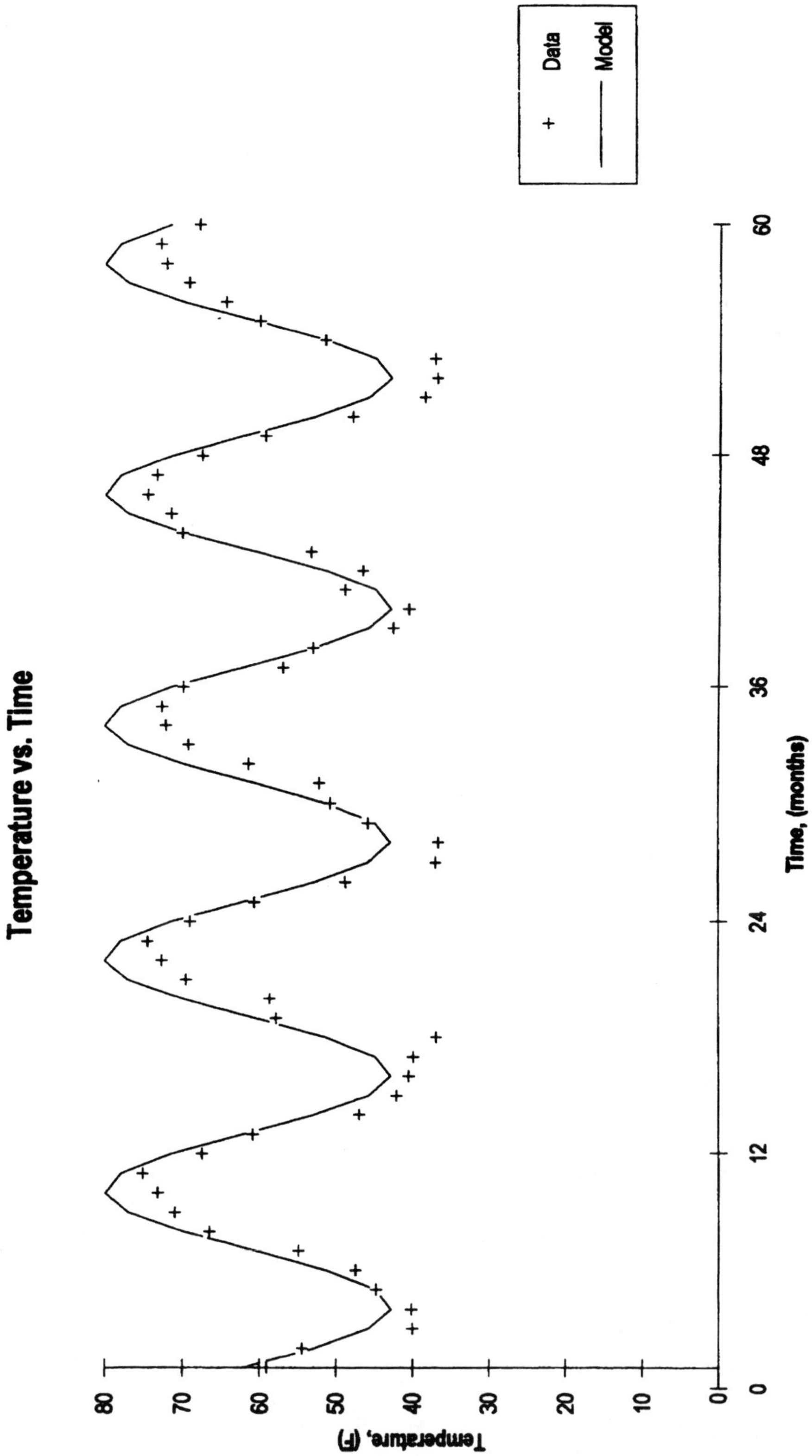

Figure 11.11.—Average Monthly Temperature Data and Model Results for Five-Year Test Period at Clayton, GA.

1. Forms of Model Sensitivity Based on a linearization of a one-dimensional Taylor series expansion, the change in one variable Y with respect to a second variable X is:

$$\Delta Y = \frac{dY}{dX}\Delta X \tag{11.14}$$

or

$$\frac{\Delta Y}{\Delta X} = \frac{dY}{dX}. \tag{11.15}$$

Equation 11.15 suggests that the sensitivity can be computed either analytically (dY/dX) or numerically (ΔY/ΔX). The numerical approach is used by necessity for complex models, while the analytical approach is easier to use with simple models.

Equation 11.15 gives the absolute sensitivity. It has units equal to the ratio of the units of Y to units of X. For example, the sensitivity of the time of concentration (t_c) model of Table 11.7 to precipitation (P) would have units of hours/inches; for the slope (S) the units of the absolute sensitivity would be hours. Thus, it is not meaningful to compare the absolute sensitivity values for different variables. The absolute sensitivity values indicate the effect of the different X on Y, but their magnitudes are not comparable.

Unfortunately, a single method of comparing the sensitivity of different model variables or inputs is not available. Two methods, relative and deviation sensitivity, will be introduced in this section. The relative sensitivity, R_S, is the ratio of the relative change in Y (dY/Y) to a relative change in X (dX/X):

$$R_s = \frac{dY/Y}{dX/X} = \left(\frac{dY}{dX}\right)\left(\frac{X}{Y}\right) \sim \left(\frac{\Delta Y}{\Delta X}\right)\left(\frac{X}{Y}\right) \tag{11.16}$$

Since the numerator (dY/Y) and denominator (dX/X) of Equation 11.16 are dimensionless, then R_S is dimensionless. Thus, in a multi-factor system, the relative sensitivity can be used to compare the relative effects of different factors X. To use Equation 11.16, it is necessary to select values of X and Y. If the relationship between Y and X is nonlinear, then the relative sensitivity will vary with the magnitudes of Y and X. Thus, the importance of the various X depends on their values, and the selection of X values should be based on a rational assessment of conditions for which the model will be used. Mean values are commonly used.

Deviation sensitivity is a second form of sensitivity that can be used to compare the importance of model elements. Deviation sensitivity is the deviation in the criterion variable that results from a standardized measure of variation in a model element, such as a predictor variable or fitted coefficient. Thus, Equation 11.14, rather than Equation 11.15, is used as the basis for deviation sensitivity estimates.

Relative sensitivity has the disadvantage that each predictor variable is incremented by a constant percentage. It may be necessary to use a constant percentage when more precise knowledge of the variables is not available. Where more detailed knowledge is available, it may not be appropriate to assume that the incremental change is constant for all variables. Where different variations are appropriate for measuring importance, deviation sensitivity should be used instead of relative sensitivity. To compute the deviation sensitivity, the following procedure is used:

a) Select values of all model elements, including fitted coefficients and input values of predictor variables; mean values of the latter are commonly used. These values of the model elements are referred to as base point values.
b) Compute the value of the criterion variable (i.e., model output) using the base point values. This value is referred to as the base point value of the criterion variable, Y_o.

TABLE 11.7. Examples: Sensitivity Analyses of Selected Model Elements (Factor Y to Factor X).

Type of factor			Example		
Y (1)	X (2)	Example model (3)	Y (4)	X (5)	Sensitivity analysis (6)
Output variable	Variation in model input	Horton's equation $f = f_c + (f_o - f_c)e^{-Kt}$	f	K	show variation in f for range of K values
Output variable	Error in model input	Time of concentration $t_c = 0.007\ (nL)^{0.8}p^{-0.5}S^{-0.4}$	t_c	n	show error in computed t_c that results from inaccuracy of n
Output variable	Measured value	Flood frequency analysis of peak discharge (q_p)	100-yr q_p	X_i	show how q_{100} would change if the outlier X_i is censored
Output variable	Model coefficient	Schoklitsch bedload $q_b = 25\ S^{1.5}\ d^{-0.5}\ (q-q_c)$	q_b	1.5	show how q_b would change for variation in the exponent of S
Output variable	Model component	SCS TR-20 program	q_p	channel routing method	show how the downstream q_p would vary if an alternative routing method were used (Convex vs. Mod. AttKin)
Intermediate model variable number	Model input	SCS TR-55 graphical peak discharge method	Q	CN	show variation in runoff (Q) due to variation in curve number (CN)
Reliability criterion	Model coefficient	Peak discharge regression $q_p = b_o X_1^{b1}\ X_2^{b2}$	$\bar{e}$	b_o	show how the model bias ($\bar{e}$) varies as b_o is changed
Reliability criterion	Elimination of predictor variable	Evaporation model $D = b_o\ T^{b1}\ H^{b2}\ W^{b3}$	R	W	show how the correlation coefficient (R) varies if the wind speed (W) is deleted from the model

c) Select measures of variation for each model element to be compared. This might be the standard error of a model coefficient or the standard deviation of measured data for predictor variables.
d) Successively compute the value of the criterion variable, denoted as Y_1, using all base point values of step a except for one model element; for this element, increment its base point value by the measure of variation selected in step c.
e) The difference $\Delta Y = Y_1 - Y_o$ is the deviation sensitivity for that model element.

For complex models where the model output is a vector of values rather than a single point estimator, a single-valued index would have to be selected to represent the model output.

The power model of Equation 11.8 can be used to demonstrate both the relative and deviation sensitivity. For a power model, the relative sensitivity for predictor variable X_i is b_i. The deviation sensitivity of the power model is:

$$\Delta\ Y_i = (b_i\ \hat{Y}/X_i)\ \Delta X_i. \tag{11.17}$$

For demonstration purposes, assume a design project that has the mean characteristics of the data in Table 11.6 (i.e., $\bar{X}_1 = 17$, $\bar{X}_2 = 2408$ feet, $\bar{X}_3 = 67.2$ acres). If planning estimates for a proposed development are assumed to be accurate to 3 inlets, 200 feet of pipe, and 0.5 acres in the developed area, then Equation 11.17 yields deviation sensitivities of:

$$\Delta Y_1 = 0.9265\ (\$55{,}577)\ (3)/17 = \$9{,}087$$

$$\Delta Y_2 = 0.2420\ (\$55{,}577)\ (200\ \text{ft})/2408\ \text{ft} = \$1{,}117$$

$$\Delta Y_3 = -0.06146\ (\$55{,}577)\ (0.5\ \text{ac})/67.2\ \text{ac} = -\$25.$$

Thus, planning should concentrate more on developing accurate estimates of the number of inlets than either the length of pipe or the area being developed. The relative sensitivities are 0.93, 0.24, and −0.06 for X_1, X_2, and X_3, respectively. While both forms of sensitivity yield the same order of importance, the deviation sensitivity has the advantage of providing values that have physical meaning (in this case, economic meaning).

2. ***Sensitivity in Model Formulation*** While most sensitivity analyses are made as part of model verification and model testing studies, it is important to give consideration to model sensitivity in the formulation (development) phase of modeling. At the least, the following four criteria should be recognized: a) a model structure that yields an appropriate, rational sensitivity function, b) model inputs that accurately reflect physical processes involved in the modeling problem, c) recognition that insensitive model coefficients may not approach their true values, thus producing irrational estimates of importance of model effects, and d) recognition that the sensitivity functions can be used to indicate the type of watershed or hydrometeorological characteristics that will provide the highest accuracy in relationships between the model coefficients and the characteristics.

The structure of a model determines the structure of the sensitivity functions of the model. Where alternative model structures can be selected, it is important to select the structure that yields sensitivity functions that have desirable characteristics. For example, a linear model between peak discharge q_p and drainage area A ($q_p = a_0 + a_1A$) yields a sensitivity function that is both constant and not a function of drainage area: $dq_p/dA = a_1$. A power model $q_p = b_oA^{b1}$ yields a nonconstant sensitivity function $dq_p/dA = b_oB_1\,A^{b1-1}$. Since b_1 is usually less than 1, the sensitivity function of the power model indicates that the rate of change of q_p with A decreases as A increases. This agrees with measured data; for example, Crippen and Bue (1977) showed that measured maximum flood flows showed a decreasing relationship with drainage area. Thus, the linear model structure, which may result in prediction accuracy comparable to that from the power model, would be a poor choice because it produces an irrational sensitivity function. In this case, the sensitivity function may be a more important criterion for selecting a model structure than the goodness-of-fit statistics for the alternative model structures.

Insensitive model coefficients are very difficult to calibrate. Whether using an analytical (e.g., ordinary least squares) or numerical (e.g., nonlinear least squares) fitting technique, the value of the objective function is most sensitive to the important coefficients. Even a large change in the value of an insensitive coefficient produces an insignificant change in the value of the objective function. For example, when calibrating a multiple linear regression model with hydrologic data, the values of the coefficients for unimportant predictor variables may be irrational because of the effects of intercorrelations between the predictor variables. The coefficients corresponding to the important predictor variables will be accurate estimators of the population coefficients, but intercorrelation between predictor variables inhibits the ordinary least squares estimators of unimportant coefficients from approaching their true values. The same concept applies when a numerical algorithm such as nonlinear least squares is used to fit the coefficients of a complex model. For an m-coefficient model, the (m + 1)-dimensional response surface will have a shallow-sloped valley in the direction of any insensitive coefficient. Thus, the convergence criterion will limit the ability of the algorithm to approach the true value of the coefficient. This was illustrated by Dawdy and O'Donnell (1965) in the fitting of a watershed model.

Regionalization of hydrologic model coefficients is a common calibration requirement in hydrologic analysis. This involves relating the fitted model coefficients to watershed or hydrometeorological characteristics so that the model can be used for prediction on ungaged watersheds. Consideration of the sensitivity functions of the model can lead to more accurate regionalization. An attempt to regionalize the log–Pearson type III skew coefficient produced poor results (McCuen, 1979) because the physical processes that affect the skew were not reflected in the predictor variables used when fitting the regional model. McCuen and Hromadka (1988) produced a more accurate model for estimating skew by relating the skew values of urban watersheds to a watershed storage characteristic that reflected the cause of skew. This shows the importance of selecting predictor variables that have a physical relationship to the criterion (or dependent) variable of the regionalized model.

As further illustration of the point, regional unit hydrographs are commonly used in hydrologic design work. The coefficients that control the shape of the unit hydrograph are regionalized by relating them to watershed characteristics. As an example, the gamma distribution is commonly used as a model for a unit hydrograph:

$$U(t) = \frac{t^{n-1}e^{-t/K}}{K^n\,\Gamma(n)} \tag{11.18}$$

in which n and K are coefficients that define the shape and scale of the distribution, respectively, t is time, and Γ(n) is the gamma function with argument n. This is sometimes called the Nash unit hydrograph model. The sensitivity functions for the gamma unit hydrograph U(t) are shown in Fig. 11.12. These functions indicate that the model is most sensitive along the rising limb of the unit hydrograph, not at the peak or along the recession. Thus, because of the magnitude of the sensitivity functions, the coefficients n and K would be expected to correlate most accurately with variables that reflect the physical processes governing the rising limb of the unit hydrograph. Sarma et al. (1969) derived the values of n and K for a small region in Indiana and found that the coefficient K was a function of the drainage area, the fraction of urbanization, and the depth and duration of rainfall excess. While drainage area is considered a size factor, the other three characteristics are usually associated with the rising limb of the

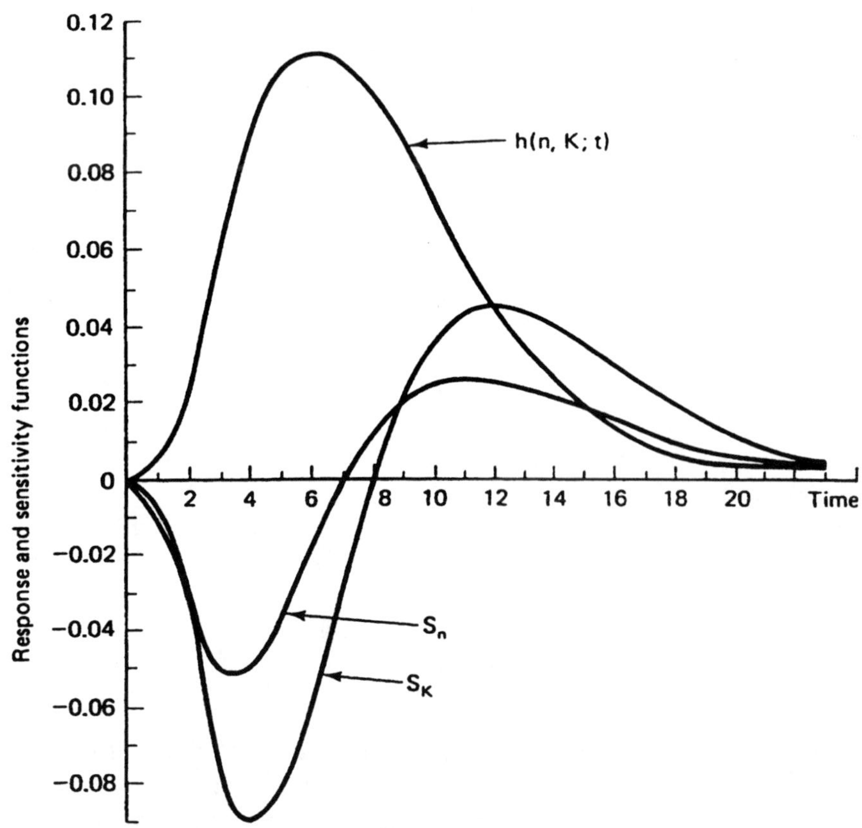

Figure 11.12.—Nash Model Sensitivity: n = 4.0; K = 2.0.

unit hydrograph. Imperviousness causes rapid runoff, which would be reflected in a rapid rise of the unit hydrograph. Rainfall characteristics also affect the shape of the rising limb of the storm hydrograph, but they have less effect on the recession. The inclusion of characteristics that affected the rising limb of the gamma unit hydrograph possibly contributed to the high correlation that resulted (R = 0.852). The inclusion of variables that reflect physical processes associated with the sensitivity functions should produce regionalized models that have the greatest possible accuracy; this was reflected in the high correlation of the equations reported by Sarma et al. (1969).

3. ***Sensitivity in Model Calibration*** It is not widely recognized that sensitivity is the basis for model-calibration algorithms. Whether an analytical method like ordinary least squares or a systematic but subjective optimization ("turning the dials") is used to fit a model to data, the fitting is made with sensitivity functions. In ordinary and numerical least squares, the optimum coefficients are those that minimize the derivatives of the objective function F with respect to the unknown coefficients b_i. The derivatives dF/db_i reflect the sensitivity of the fitting criterion to unknowns b_i. If the response surface (F versus all b_i) is considered, the derivatives dF/db_i are the slopes of the surface in the ith direction. The steepness of the slope is a reflection of the importance of the coefficient to the fitting.

When the optimum solution is found, one- and two-dimensional response surfaces can be computed. The one-dimensional surfaces can be used to assess the relative importance and accuracy of the fitted coefficients. Two-dimensional surfaces can be used to assess the degree to which the coefficients interact (i.e., correlate). A one-dimensional response surface is computed with either Equation 11.14 or a variation of it. Typically, the response surface is for the objective function F and the unknowns b_i. Thus, Equation 11.14 is written for a response surface as $\Delta F_i = (dF/db_i)\,\Delta b_i$. For example, the one-dimensional response surfaces for a multiple regression model based on least squares could be presented as the sum of squares of the errors as a function of a set fraction change in the value of a regression coefficient b_i. Alternatively, the change in the standard error of estimate (S_e) or relative standard error (S_e/S_y) could be used. Typically, fractions of ±5% and ±10% of b_i are used to compute F. Thus, the values of F could be plotted as a function of b_i to show the sensitivity of the coefficient. For two-dimensional response surfaces, two coefficients are incremented simultaneously.

One-dimensional S_e and $\bar{e}$ response surfaces were computed for the power model of Equation 11.8. Using an increment of 2.5%, the values of S_e and $\bar{e}$ were computed for each coefficient; the computed values are given in Table 11.8 and shown in Fig. 11.13. The one-dimensional response surfaces show three important considerations. First, b_1 is the most sensitive coefficient, with the 10% change causing the greatest variation in both S_e and $\bar{e}$. The coefficient b_2 is next in importance followed by b_o and b_3. It appears that X_3 is an unimportant variable because the corresponding response surface of b_3 is so flat. Second, the fitted coefficients do not actually provide the optimum solution. The S_e is smaller for a b_1 of approximately 0.97, with the other coefficients held constant at the values shown in Equation 11.8, or for a b_o value of 871, with the other coefficients held constant. Third, an unbiased model can be achieved by changing the coefficients, such as setting the value of b_1 to about 0.96; however, this will not yield a model with a smaller standard error.

Fig. 11.14 shows the two-dimensional response surface that results from fitting two coefficients of a unit hydrograph model. One coefficient represents the time of concentration (t_c) and the other the peak rate factor (K) of the model:

$$q_p = \frac{KAQ}{t_c}. \tag{11.19}$$

As suggested by the form of the equation, K and t_c are highly correlated, which is reflected by the narrow valley of the response surface. Numerous combinations of K and t_c that give almost identical values of the objective function could be selected. The two coefficients are both important but it might be possible to simplify the model using a single variable with little loss in accuracy.

TABLE 11.8. One-dimensional Response Surfaces of Standard Error of Estimate (S_e) and the Bias ($\bar{e}$) for the Power Model Coefficients (b_i, i = 0, 1, 2, 3) of Equation 11.8.

b_0 (1)	S_e (\$) (2)	$\bar{e}$ (\$) (3)	b_1 (4)	S_e (\$) (5)	$\bar{e}$ (\$) (6)	b_2 (7)	S_e (\$) (8)	$\bar{e}$ (\$) (9)	b_3 (10)	S_e (\$) (11)	$\bar{e}$ (\$) (12)
871	32997	−15	1.019	35783	12400	0.266	33403	6235	−0.0676	34804	−7159
851	33153	−1441	0.996	33625	7382	0.260	32899	3029	−0.0661	34643	−6803
832	33406	−2867	0.973	32767	2705	0.254	32912	−28	−0.0645	34487	−6444
812	33753	−4293	0.950	33038	−1655	0.248	33369	−2941	−0.0630	34337	−6082
792	34198	−5736	0.926	34198	−5736	0.242	34198	−5736	−0.0615	34198	−5736
772	34719	−7145	0.903	35968	−9508	0.236	35302	−8367	−0.0599	34051	−5352
753	35330	−8571	0.880	38142	−13040	0.230	36628	−10891	−0.0584	33918	−4984
733	36020	−9997	0.857	40540	−16334	0.224	38108	−13297	−0.0569	33789	−4613
713	36786	−11423	0.834	43041	−19405	0.218	39693	−15591	−0.0553	33666	−4240

4. *Sensitivity in Model Assessment* Model formulation and calibration are the first two phases of modeling; model assessment is the third phase. Just as sensitivity analysis is important in the first two phases, it is also important in model assessment. It is specifically used a) to make error and stability analyses for assessing model uncertainty and b) to measure the relative importance of both model coefficients and input variables.

a. Error and Stability Analyses. Model uncertainties are important in decision making. Errors in inputs lead to the inaccuracy of model outputs, which could lead to inadequate designs or inaccurate assessments. Knowledge of the sensitivity of designs to model uncertainties can lead to safer designs.

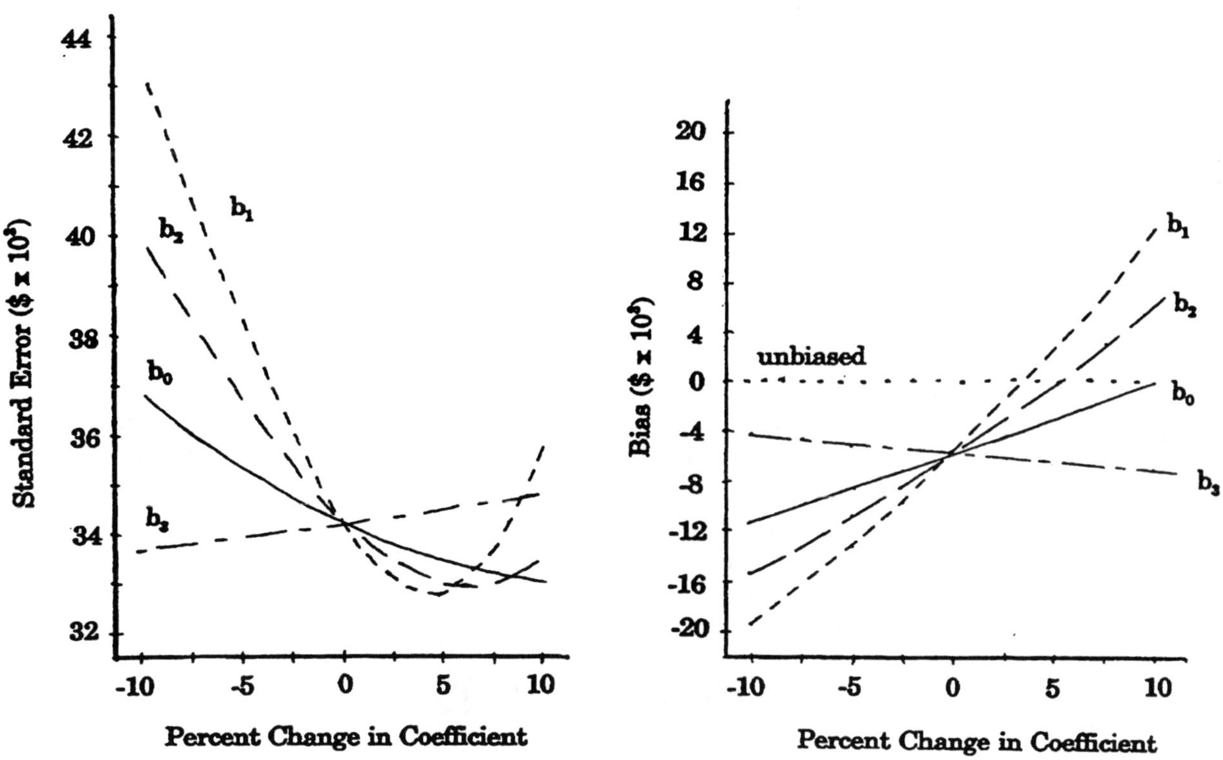

Figure 11.13.—One-Dimensional Response Surfaces of the Standard Error of Estimate and the Bias [illegible] the Power Model Coefficients (b_i, i = 0,1,2,3) of Equation 11.8.

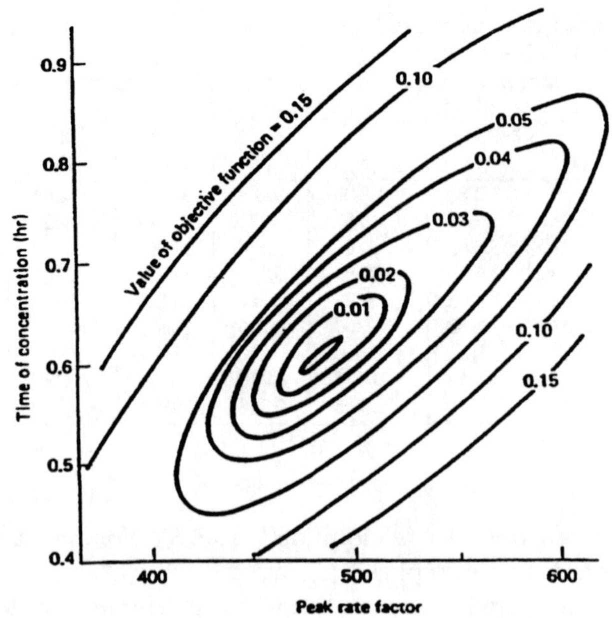

Figure 11.14.—Peak Rate Factor Response Surface for May 31, 1962 Storm on Powells Creek Watershed (McCuen and Snyder, 1986).

When using planning models to compare alternatives, model uncertainties may lead to the rejection of feasible alternatives. A primary use of sensitivity analysis in model assessment is to enable modelers to evaluate the end-result effects of model uncertainties.

If the error ΔY in output Y is defined as the difference between values of Y predicted for inputs X + ΔX and X, then the error can be estimated using a truncated Taylor series:

$$\Delta Y = Y(X + \Delta X) - Y(X) \sim \frac{\partial Y}{\partial X} \cdot \Delta X \tag{11.20}$$

where ΔX is the error in model input X. Thus, the error in Y is equal to the product of the absolute sensitivity and the error in X. The absolute sensitivity of Equation 11.20 can be computed analytically or numerically. The error could also be expressed in relative form, $\Delta Y/Y$. The error ΔY of Equation 11.20 is essentially the deviation sensitivity with ΔX being the error.

Consider Manning's equation as an example. The relative sensitivities (Equation 11.16) of the velocity V to the roughness coefficient n and to the slope S are −1 and 0.5, respectively. But the estimation of n is more error prone than that of S. An error of 20% in n can be expected, while the slope can be estimated with an accuracy of about 10%. Thus, the error in a computed velocity due to error in n is four times greater than the error due to error in the slope. It is necessary to estimate both the sensitivity and the input error when evaluating the effect of input error on the accuracy of the output.

Consider the following two bedload equations, which were developed by Meyer–Peter and Schoklitsch (McCuen, 1989), respectively:

$$q_{bl} = \left[39.25\, S_c\, q^{2/3} - 9.95\, d\right]^{1.5} \tag{11.21}$$

and

$$q_{b2} = 25\, S_c^{1.5}\, d^{-0.5}\left(q - 0.0638\, d\, S_c^{-1.33}\right) \tag{11.22}$$

TABLE 11.9. Error Analysis of Meyer–Peter and Schoklitsch Bedload Equations for $S_c = 0.01$ ft/ft, $q = 0.15$ cfs/ft width, $d = 0.0015$ ft.

Variable (1)	Absolute Sensitivity: Meyer–Peter (2)	Absolute Sensitivity: Schoklitsch (3)	Relative Sensitivity: Meyer–Peter (4)	Relative Sensitivity: Schoklitsch (5)	Errors in Variables (6)	Relative Error (7)	Bedload Error (lb/sec/ft): Meyer–Peter (8)	Bedload Error (lb/sec/ft): Schoklitsch (9)	Relative Error*: Meyer–Peter (10)	Relative Error*: Schoklitsch (11)
q	0.229	0.646	1.16	1.41	0.045 ft³/ft	0.30	0.010	0.029	0.347	0.423
S_c	5.147	14.046	1.73	2.05	0.001 ft/ft	0.10	0.005	0.014	0.173	0.205
d	−4.621	−41.830	−0.23	−0.91	0.0004 ft	0.27	0.002	0.017	0.062	0.244

*$q_{b1} = 0.0297$ lb/sec/ft, $q_{b2} = 0.0686$ lb/sec/ft

where S_c is the channel slope (ft/ft), q is the water discharge (cfs/ft width), d is the mean particle diameter (ft), and q_b is the bedload (pound/sec/ft. width). Using the conditions shown in Table 11.9 the absolute sensitivities were computed. Using errors typical for each of the inputs, which are shown in Table 11.9, the errors in the estimated bedloads for both Equations 11.21 and 11.22 were computed using Equation 11.20; the errors relative to q_{b1} and q_{b2} are also shown in Table 11.9. The errors in bedload can be compared both within and between models; it is probably best to make the comparisons using the relative errors in bedloads. The sensitivities of the input variables were computed using the derivatives of Equations 11.21 and 11.22 with Equations 11.15 and 11.16.

For the Meyer–Peter equation, slope has the highest relative sensitivity; discharge is second in relative sensitivity, while the mean particle diameter has a comparatively low relative sensitivity. The relative error in bedload due to error in discharge is much greater than for the slope because more accurate estimates of slope can be obtained. Even though the relative error in d is almost equal to that of q, the relative error in bedload is much smaller because q_{b1} is insensitive to d. These comparisons show that, when making an error analysis of model inputs, both the error in the input variable and the sensitivity of the output to the input must be considered.

For the Schoklitsch equation, the order of the relative sensitivities are the same as for the Meyer–Peter equation, but the magnitudes are larger, especially for d. Because of the greater sensitivities, the relative errors in bedload are larger than for the Meyer–Peter equation. Also, it is important to note that, while slope has the highest relative sensitivity, it causes the smallest relative error in bedload.

As another illustration of the general concept, the sensitivity of the computed output from a time-dependent watershed model may be of interest, especially where the effect of error in assumed initial conditions is of interest. A plot of the sensitivity versus time or the error in the output due to assumed initial conditions versus time would be useful in identifying the time period required to eliminate the effect of error in the assumed initial conditions before the simulation period commences. Horton's infiltration capacity equation can be used as a simple example: $f = f_c + (f_o - f_c)e^{-kt}$ where f_o is the initial capacity, f_c is the ultimate capacity, and K is the capacity decay coefficient. The value of f_o is the most difficult of the three coefficients to estimate. The absolute sensitivity df/df_o equals e^{-kt}, so the sensitivity is time dependent. Fig. 11.15 shows the absolute sensitivity, e^{-k}. The error in f that results from an error of Δf_o is $\Delta f_o e^{-k}$.

b. Sensitivity and Relative Importance. The absolute sensitivities of different variables or model components cannot be used to compare the relative importance because the sensitivities would have different dimensions. Relative sensitivity values can be used for comparing the importance of predictor variables when equal relative changes in the predictor variables are expected; however, this is not usually the case. In such situations, it is better to use the deviation sensitivity to show the variation in the criterion variable. However, to use deviation sensitivity it is necessary to have reasonably accurate estimates of the causative factors (i.e., predictor variables or model components).

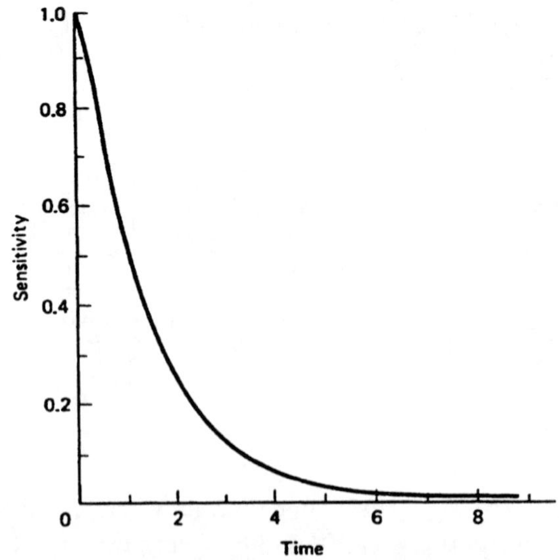

Figure 11.15.—Time Variation of Sensitivity of Initial Estimate (McCuen and Snyder, 1986).

The multiple linear regression model is used frequently in hydrology. It can be shown that the deviation sensitivity (S_{di}) for predictor variable i equals the product of the standardized partial regression coefficient t_i and the standard deviation of the criterion variable (S_y): $S_{di} = t_i S_y$. Since S_y is a constant, the t values are used directly as a measure of the relative importance of the predictor variables of a multiple linear regression model.

For nonlinear models, such as the multiple-variable power model, the deviation sensitivity is nonlinear so the t values computed for the logarithmic transform may not accurately reflect the importance of the predictor variables. In such cases, the deviation sensitivity should be computed and plotted over a range.

To illustrate the use of deviation sensitivity for measuring the relative importance of predictor variables in a nonlinear model, a peak discharge model based on 58 watersheds in Indiana will be used:

$$Q_p = 5.436\, A^{0.6794}\, S^{0.5003}\, L^{0.1686}\, P^{0.7011}\, R^{0.1495}\, D^{0.1652}\, C^{1.2608} \tag{11.23}$$

in which A is the drainage area (mi²), S is the channel slope (ft/mi), L is the channel length (mi), P is a precipitation index (in.), R is the watershed relief (ft), D is the drainage density (mi/mi²), C is the soil runoff coefficient, and $\bar{Q}_p$ is the 10-year peak discharge (cfs). The t values and deviation sensitivity values for $\bar{X} - S$, $\bar{X} - 0.5S$, $\bar{X} + 0.5S$, and $\bar{X} + S$ are given in Table 11.10; the relative

TABLE 11.10. Relative Importance of Predictor Variables for a Multiple-Variable Power Model.

		Deviation Sensitivity (ft³/s) for				Relative Sensitivity for	
Variable (1)	t (2)	$\bar{X} - S$ (3)	$\bar{X} - 0.5S$ (4)	$\bar{X} + 0.5S$ (5)	$\bar{X} + S$ (6)	$\bar{X} - S$ (7)	$\bar{X} + S$ (8)
A, area	0.425	−2177	−1022	935	1807	0.37	0.31
S, channel slope	0.310	−2262	−979	823	1549	0.39	0.27
L, channel length	0.072	−496	−218	181	336	0.09	0.06
P, precipitation index	0.203	−1131	−551	524	1035	0.19	0.18
R, relief index	0.087	−647	−265	206	375	0.11	0.06
D, drainage density	0.079	−548	−236	192	356	0.09	0.06
C, soil runoff coefficient	0.393	−1865	−954	991	2016	0.32	0.35

deviation sensitivities are also given in Table 11.10 for $\bar{X} - S$ and $\bar{X} + S$. The values indicate three groups of importance: 1) A, S, and C, 2) P, and 3) L, R, and D. The first group includes the most important predictor variables, with relative deviation sensitivities of about 35%. The variable in the second group has a relative deviation sensitivity of slightly less than 20%. The three least important predictors have values of about 10%. In all cases, the variation is greater for the smaller values ($\bar{X} - S$) of the predictor variables. It is also important to note that the ranks for the positive deviation sensitivities are not the same as that for the negative values. For example, the slope has the largest magnitude (−2262) for the negative deviation but the third largest magnitude (1549) for the positive deviation. This illustrates that the sensitivity of a nonlinear model such as Equation 11.23 is not constant and varies with the values of the predictor variables.

VI. REFERENCES

Baker, C.P. and Panciera E.C. (1990). "A geographic information system for ground water protection planning," Journal of Soil and Water Conservation, March-April, 246–248.

Berry, J.K. and Sailor, J.K. (1987). "Use of a geographic information system for storm runoff prediction from small urban watersheds," Environmental Management, 11 (1), 21–27.

Crippen, J.R. and Bue, C.D. (1977). "Maximum floodflows in the conterminous United States." U.S.G.S. Water-Supply Paper 1887, USGPO, 52 pp.

Cunge, J.A. (1969). "On the subject of a flood propagation computation method (Muskingum method)," Journal of Hydraulic Research, 7 (2), 205–230.

Dawdy, D.R. and O'Donnell, T. (1965). "Mathematical models of catchment behavior," J. Hydr Div, ASCE, 91 (HY4), 123–137.

Diskin, M.H. and Simon, E. (1977). "A procedure for the selection of objective functions for hydrologic simulation models," J. Hydrol., 34, 129–149.

DelRegno, K.J. and Atkinson, S.F. (1988). "Nonpoint pollution and watershed management: a remote sensing and geographic information system (GIS) approach," Lake and Reservoir Management, 4 (2), 17–25.

Dooge, J.C.I. (1973). "Linear theory of hydrologic systems," Technical Bulletin 1468, USDA Agriculture Research Service, Washington, DC.

Dooge, J.C.I., Strupczewski, W.B., and Napiorkowski, J.J. (1982). "Hydrodynamic derivation of storage parameters of the Muskingum model," Journal of Hydrology, 54, 371–387.

Evans, B.M. and Miller, D.A. (1988). "Modeling nonpoint pollution at the watershed level with the aid of a geographic information system," *Nonpoint Pollution: 1988—Policy, Economy, Management, and Appropriate Technology*, American Water Resources Association.

Gilliland, M.W. and Baxter–Potter, W. (1987). "A geographic information system to predict nonpoint source pollution potential," Water Resources Bulletin. 23(2), 281–291.

Green, I.R.A. and Stephenson, D. (1986). "Criteria for comparison of single event models," Hydrological Sciences Journal, 31(3), 395–411.

Haan, C.T. (1977). *Statistical methods in hydrology*, The Iowa State University Press, Ames, IA.

Hamlett, J.M., Peterson, G.W., Baumer, G.M., Miller, D.A., and Day, R.L. (1991). "GIS-based watershed rankings for nonpoint pollution in Pennsylvania," *Proc., GIS/LIS Annual Conference.*

Helsel, D.R. and Hirsch, R.M. (1992). *Statistical methods in water resources*, Elsevier Science Publishers, Amsterdam.

Himmelblau, D.M. (1972). *Applied nonlinear programming*, McGraw–Hill Book Co., New York, NY.

Horton, R.E. (1945). "Erosional development of streams and their drainage basins: hydrophysical approach to quantitative morphology." Bull. Geol. Soc. of America, Vol. 56, 275–370.

Ibbitt, R.P. and O'Donnell, T. (1971). "Fitting methods for conceptual catchment models," J. Hydr. Div., ASCE, 1331–1342.

James, L.D. and Burges, S.J. (1982). "Selection, calibration, and testing of hydrologic models," *Hydrologic Modeling of Small Watersheds*, C.T. Haan, et al, eds. American Society of Agricultural Engineers, St. Joseph, MO.

Jewell, T.K., Nunno, T.J., and Adrian, D.D. (1978). "Methodology for calibrating stormwater models," J. Environ. Eng. Div., ASCE, 485–501.

Johnston, P.R. and Pilgrim, D.H. (1976). "Parameter optimization for watershed models," Water Resources Research, 12(3), 477–486.

Lanfear, K.J. (1989). "Geographic information systems and water resources applications," (editorial), Water Resources Bulletin, 25 (3), r5–r6.

Levine, A. D. and Jones, W.W. (1990). "Modeling phosphorus loading to three Indiana reservoirs: a geographic information system approach," Lake and Reservoir Management, 6 (1), 81–91.

Loague, K.M. and Freeze, R.A. (1985). "A comparison of rainfall–runoff modeling techniques on small upland catchments," Water Resour. Res., AGU, 21(2), 229–248.

Manley, R.E. (1978). "Calibration of hydrological model using optimization technique," J. Hydr. Div., ASCE, 189–202.

Marble, D. (1987). "Geographic information systems: an overview," *GIS: For Resource Management: A Compendium*, American Society of Photogrammetry and Remote Sensing and American Congress on Surveying and Mapping.

McCuen, R.H. (1979). "Map Skew ???," J. Water Resources Planning and Management, ASCE, 105 (WR2), 269–277.

McCuen, R.H. (1989). *Hydrologic analysis and design*, Prentice–Hall, Inc., Englewood Cliffs, NJ.

McCuen, R.H. (1992). *Microcomputer applications in statistical hydrology*, Prentice–Hall, Inc., Englewood Cliffs, NJ.

McCuen, R.H. and Hromadka, T.V. (1988). "Flood skew in hydrologic design on ungaged watersheds," J. Irrigation and Drainage, ASCE, 114(2), 301–310.

McCuen, R.H. and Snyder, W.M. (1986). *Hydrologic modeling: statistical methods and applications*. Prentice–Hall, Inc., Englewood Cliffs, NJ.

Nash, J.E. and Sutcliffe, J.V. (1970). "River flow forecasting through conceptual models. Part I—A discussion of principles," J. Hydrol., 10, 282–290.

National Weather Service (NWS) (1983). "Standard hydrometeorological exchange format (SHEF) version I," Technical Memorandum NWS WR-180, Salt Lake City, UT.

Patry, G.G. and Marino, M.A. (1983). "Nonlinear runoff modelling: parameter identification," J. Hydr. Div., ASCE, 865–880.

Ponce, V.M. (1989). *Engineering hydrology: principles and practices*, Prentice–Hall, Inc., Englewood Cliffs, NJ.

Ragan, R.M. and Jackson, T.J. (1980). "Runoff synthesis using LANDSAT and SCS model," J. Hydr Div., ASCE, 106, 667–678.

Russell, S.O., Kenning, B.F.I., and Sunnell, G.J. (1979). "Estimating design flows for urban drainage," J. Hydr. Div., ASCE, 105(HY1), 43–52.

Sarma, P.B.S., Delleur, J.W., and Rao, A.R. (1969). *A program in urban hydrology, Part II, an evaluation of rainfall–runoff models for small urbanized watersheds and the effect of urbanization on runoff*, Water Resources Research Center, Purdue University.

Siegel, S. (1956). *Nonparametric statistics for the behavioral sciences*, McGraw–Hill Book Co., New York, NY.

Smart, J.S. (1972). "Channel networks," *Advances in hydroscience*, Academic Press, Orlando, FL, Vol. 8, 305–346.

Stephenson, D. (1979). "Direct optimization of Muskingum routing coefficients," J. Hydrol., 41, 161–165.

Stuebe, M. and Johnston, D.M. (1990). "Runoff volume estimation using GIS techniques," Water Resources Bulletin, 26 (4), 611–620.

U.S. Army Corps of Engineers (USACE) (1982). *Hydrologic analysis of ungaged watersheds using HEC–1*, Training Document No. 15, Hydrologic Engineering Center, Davis, CA.

U.S. Army Corps of Engineers (USACE) (1990). *HECDSS—User's guide and utility program manuals*. Hydrologic Engineering Center, Davis, CA.

U.S. Army Corps of Engineers (USACE) (1991). *HECLIB–HECDSS Subroutines—programmer's manual*. Hydrologic Engineering Center, Davis, CA.

U.S. Water Resources Council (USWRC) (1982). "Guidelines for determining flood flow frequency," Bulletin #17B,, Washington, DC.

Wallis, J.R. and Todini, E. (1975). "Comment upon the residual mass curve coefficient," J. Hydrol., 24, 201–205.

Warwick, J.J. and Haness, S.J. (1994). "Efficacy of ARC/INFO GIS application to hydrologic modeling," J. Water Res. Planning and Management, ASCE, 120 (3), 366–381.

Wood, S.R. (1974). "A catchment simulation model developed for urbanizing catchments with particular reference to the use of automatic optimization techniques," *Proc., IFPF Conf. on Computer Simulation of Water Resources Systems* (Ghent, Belgium), 209–217.

Wright-McLaughlin Engineers (WME) (1969). *Urban storm drainage criteria manual.* Denver Regional Council of Governments, Denver, CO.

VII. GLOSSARY

Boolean—A type of expression based on a true or false condition or used to refer to logical expressions such as "flow greater than 5 cfs." A Boolean operator is a key word that specifies how to combine simple logical expressions into complex expressions. For example, the use of AND in "depth > 100 AND flow > 5 cfs."

Calibration parameters—Parameters that cannot be quantified from direct measurements of system characteristics. Instead, calibration parameters must be estimated with measured input and output or by extrapolation of parameters from other systems, input or output.

Causal model—Model in which interrelationship of input to output is based on functional representation of laws of thermodynamics or conservation of mass, momentum, or energy.

Criterion function—Formal, mathematical representation of the goal of optimization. In the case of model calibration, it is typically a goodness-of-fit measure.

Empirical model—Model that represents only the functional relationship of output to input without regard to the physical processes.

Goodness-of-fit measure—Statistical measure of how well a given model and parameters predict reality.

Input—Driving force or stimulus of system behavior. Input may be matter, energy, or information.

Instantaneous unit hydrograph—Hydrograph of runoff due to one unit of excess precipitation distributed instantaneously over a watershed.

Method of maximum likelihood—Procedure for estimating statistical distribution parameters, based on principle of selecting parameter values that maximize the probability of drawing a given sample from the parent population.

Method of moments—Procedure for estimating statistical distribution parameters, based on assumption that moments of a sample are estimates of moments of the parent population.

Model—Relates known input with output from system of interest. Hydrological models are either *physical*, such as a scaled-down facsimile of a river channel, *analog*, such as an electrical resistance–capacitance representation of an aquifer, or *mathematical*. Mathematical models are now most common; in these models, the behavior of the system is represented with a set of equations that numerically simulate the system behavior or response to the stimulus, with the goal of predicting what will happen.

Output—Result, effect, or response of system to input.

Parameter—Property or characteristics of a system under specified conditions.

Parameter-predictive equation—Equation or relationship derived to permit selection of model parameters based on measurable system characteristics. The SCS curve number table is a well-known example; it predicts the parameter of an infiltration model from soil and land-use characteristics.

Physically-based parameter—Parameter that can be observed or estimated directly from measurements of system characteristics.

Picture element (pixel)—The smallest unit of information in an image or raster map. Referred to as a cell in an image or grid.

Quantile—Magnitude of specified frequency or probability. In the case of flood-frequency analysis, a quantile is a flow of specified frequency.

Rainfall–runoff model—Model that predicts peak flow, volume, or hydrograph of watershed runoff, given observed or hypothetical rainfall.

Raster—A cellular data structure composed of rows and columns, with the value of each cell representing a feature value. Groups of cells are used to represent a feature (for example, a lake).

Scattergram—Plot of pairs of values.

Skew coefficient—Statistical-distribution shape parameter.

Thematic attributes—Characteristic of certain map features. For example, attributes of a stream might include length, name, width, and flow rate for a particular flood.

Ungaged watershed—Watershed for which input and output measurements necessary for calibration are not available.

INDEX